T0140582

Advances in Intelligent Systems and Computing

Volume 1058

The series "Advances in Intelligent Systems and Computing" contains publications on theory, applications, and design methods of Intelligent Systems and Intelligent Computing. Virtually all disciplines such as engineering, natural sciences, computer and information science, ICT, economics, business, e-commerce, environment, healthcare, life science are covered. The list of topics spans all the areas of modern intelligent systems and computing such as: computational intelligence, soft computing including neural networks, fuzzy systems, evolutionary computing and the fusion of these paradigms, social intelligence, ambient intelligence, computational neuroscience, artificial life, virtual worlds and society, cognitive science and systems, Perception and Vision, DNA and immune based systems, self-organizing and adaptive systems, e-Learning and teaching, human-centered and human-centric computing, recommender systems, intelligent control, robotics and mechatronics including human-machine teaming, knowledge-based paradigms, learning paradigms, machine ethics, intelligent data analysis, knowledge management, intelligent agents, intelligent decision making and support, intelligent network security, trust management, interactive entertainment, Web intelligence and multimedia.

The publications within "Advances in Intelligent Systems and Computing" are primarily proceedings of important conferences, symposia and congresses. They cover significant recent developments in the field, both of a foundational and applicable character. An important characteristic feature of the series is the short publication time and world-wide distribution. This permits a rapid and broad dissemination of research results.

** Indexing: The books of this series are submitted to ISI Proceedings, EI-Compendex, DBLP, SCOPUS, Google Scholar and Springerlink **

More information about this series at http://www.springer.com/series/11156

Aboul Ella Hassanien · Khaled Shaalan ·
Mohamed Fahmy Tolba
Editors

Proceedings of the International Conference on Advanced Intelligent Systems and Informatics 2019

 Springer

Editors
Aboul Ella Hassanien
Cairo University
Giza, Egypt

Khaled Shaalan
The British University in Dubai
Dubai, United Arab Emirates

Mohamed Fahmy Tolba
Ain Shams University
Cairo, Egypt

ISSN 2194-5357   ISSN 2194-5365   (electronic)
Advances in Intelligent Systems and Computing
ISBN 978-3-030-31128-5   ISBN 978-3-030-31129-2   (eBook)
https://doi.org/10.1007/978-3-030-31129-2

This Springer imprint is published by the registered company Springer Nature Switzerland AG
The registered company address is: Gewerbestrasse 11, 6330 Cham, Switzerland

Preface

This volume constitutes the refereed proceedings of the 5th International Conference on Advanced Intelligent Systems and Informatics (AISI 2019), which took place in Cairo, Egypt, during October 26–28, 2019, and is an international interdisciplinary conference covering research and development in the field of informatics and intelligent systems. In response to the call for papers for AISI 2019, 133 papers were submitted for the main conference, 70 for two workshops, and 32 for five special sessions, so the total is 235 papers submitted for presentation and inclusion in the proceedings of the conference. After a careful blind refereeing process, 103 papers were selected for inclusion in the conference proceedings. The papers were evaluated and ranked on the basis of their significance, novelty, and technical quality by at least two reviewers per paper. After a careful blind refereeing process, 104 papers were selected for inclusion in the conference proceedings. The papers cover current research in robot modeling and control systems, deep learning, smart grid, sustainable energy, sentiment analysis and Arabic text mining, cloud computing, data mining, visualization and E-learning, and intelligence swarms and optimization.

In addition to these papers, the program included one keynote talk by Professor Farid Meziane from the Salford University, UK, on ultrasound report standardization using rhetorical structure theory and domain ontology. We express our sincere thanks to the plenary speakers, workshop chairs, and International Program Committee members for helping us to formulate a rich technical program. We would like to extend our sincere appreciation for the outstanding work contributed over many months by the Organizing Committee: local organization chair and publicity chair. We also wish to express our appreciation to the SRGE members for their assistance. We would like to emphasize that the success of AISI 2019 would not have been possible without the support of many committed volunteers who generously contributed their time, expertise, and resources toward making the conference an unqualified success. Finally, thanks to Springer team for their support in all stages of the production of the proceedings. We hope that you will enjoy the conference program.

Organization

Honorary Chairs

Fahmy Tolba   Egypt

General Chair

Khaled Shaalan   British University in Dubai

Program Chairs

Aboul Ella Hassanien   Egypt
Ahmad Taher Azar   Egypt

International Advisory Board

Swagatam Das   India
Seyedali Mirjalili   Australia
Fatos Xhafa   Spain
Nadia Hegazy   Egypt
Nagwa Badr   Egypt
Vaclav Snasel   Czech Republic
Janusz Kacprzyk   Poland
Tai-hoon Kim   Korea

Publicity Chairs

Saurav Karmakar   India
Nour Mahmoud   Egypt
Mohamed Hamed   Egypt

Technical Program Committee

Howaida Shedeed	Egypt
Khaled Hossny	Egypt
Yudong Zhang	UK
Alok Kole	India
Thamer Ba Dhafari	UK
Eman Nashnush	UK
Tooska Dargahi	UK
Sana Belguith	UK
Santosh More	UK
Julian Bass	UK
Ibrahim A. Hameed	Norway
Siddhartha Bhattacharyya	India
Subarna Shakya	Nepal
Fatos Xhafa	Spain
Kazumi Nakamatsu	Japan
P. K. Mahanti	Canada
Xiaohui Yuan	USA
Kumkum Garg	India
Ahmed Sharaf Eldin	Egypt
Thomas Loruenser	Austria
Feihu Xu, Cambridge	UK
Vaclav Snasel	Czech Republic
Janusz Kacprzyk	Poland
Tai-hoon Kim	Korea
M. K. Ghose	India
Ahmed Abdel Rehiem	Egypt
Sebastian Tiscordio	Czech Republic
Natalia Spyropoulou	Hellenic Open University, Greece
Dimitris Sedaris	Hellenic Open University, Greece
Vassiliki Pliogou	Metropolitan College, Greece
Pilios Stavrou	Metropolitan College, Greece
Eleni Seralidou	University of Piraeus, Greece
Stelios Kavalaris	Metropolitan College, Greece
Litsa Charitaki	University of Athens, Greece
Elena Amaricai	University of Timisoara, Greece
Qing Tan	Athabasca University, Canada
Pascal Roubides	Broward College, Greece
Alaa Tharwat	Germany
Amira S. Ashour	KSA
Pavel Kromer	Czech Republic
Jan Platos	Czech Republic
Ivan Zelinka	Czech Republic
Sebastian Tiscordio	Czech Republic

Adelkrim Haqiq	Hassan 1st University, Morocco
A. V. Senthil Kumar	Hindusthan College of Arts and Science, India
Benjamin Apraku Gyampoh	Ghana
R. S. Ajin	India
Brian Galli	Long Island University, USA
Camelia Pintea	TU Cluj-Napoca, Romania
Chakib Bennjima	University of Sousse, Tunisia
Christos Volos	Aristotle University of Thessaloniki, Greece
Faisal Talib	Aligarh Muslim University, India
Hajar Mousannif	Cadi Ayyad University, Morocco
Irene Mavrommati	Hellenic Open University, Greece
Jaouad Boumhidi	Sidi Mohammed Ben Abdellah University (USMBA), Morocco
Jesus Manuel Munoz-Pacheco	Autonomous University of Puebla, Mexico
Jihene Malek	Higher Institute of Applied Sciences and Technology, Sousse, Tunisia
Jin Xu	Behavior Matrix LL, USA
Kusuma Mohanchandra	Dayananda Sagar College of Engineering, India
Laura Romero	University of Seville, Spain
Mariem Ben Abdallah	University of Monastir, Tunisia
Marius Balas	Aurel Vlaicu University of Arad, Romania
Mario Pavone	University of Catania, Italy
Mohamed Khalgui	University of Carthage, Tunisia
Muaz A. Niazi	COMSATS Institute of Information Technology, Pakistan
Nickolas S. Sapidis	University of Western Macedonia, Greece
Nilanjan Dey	Techno India College of Technology, India
Nizar Banu P. K.	B.S. Abdur Rahman University, India
Nizar Rokbani	University of Sousse, Tunisia
Peter Gczy	National Institute of Advanced Industrial Science and Technology (AIST), Japan
Philip Moore	University College Falmouth, UK
Valentina Balas	Aurel Vlaicu University of Arad, Romania
Viet-Thanh Pham	Hanoi University of Science and Technology, Vietnam
Salvador Hinojosa	Universidad Complutense de Madrid, Spain
Alberto Ochoa	Universidad Autonoma de Ciudad Juarez, Mexico
Jorge Ruiz-Vanoye	Universidad Autonoma del Estado de Hidalgo, Mexico
Marco Perez-Cisneros	Universidad de Guadalajara, Mexico
Erik Cuevas	Universidad de Guadalajara, Mexico
Daniel Zaldivar	Universidad de Guadalajara, Mexico
Valentin Osuna-Enciso	Universidad de Guadalajara, Mexico

Eman Nashnush   University of Salford, Greater Manchester, UK
Tooska Dargahi   University of Salford, Greater Manchester, UK
Sana Belguith   University of Salford, Greater Manchester, UK
Santosh More   University of Salford, Greater Manchester, UK
Julian Bass   University of Salford, Greater Manchester, UK
Omar Alani   University of Salford, Greater Manchester, UK
Kaja Mohideen   PSG College of Technology, Anna University,
                                     India
Fernando Fausto   Universidad de Guadalajara, Mexico
Adrián González   Universidad de Guadalajara, Mexico

Local Arrangement Chairs

Mohamed Abd Elfattah   Egypt
  (Chair)
Mourad Rafat   Egypt
Hebe Aboul Ella Hassanien   Egypt

Keynote Speaker

Farid Meziane obtained a PhD in computer science from the University of Salford on his work on producing formal specification from natural language requirements. He is currently holding Chair in data and knowledge engineering and is Director of the Informatics Research Centre at the University of Salford, UK. He has authored over 100 scientific papers and participated in many national and international research projects. He is Co-chair of the International Conference on Application of Natural Language of Information Systems and in the programme committee of over ten international conferences and in the editorial board of three international journals. He was awarded the Highly Commended Award from the Literati Club, 2001, for his paper on Intelligent Systems in Manufacturing: Current Development and Future Prospects. His research expertises include natural language processing, semantic computing, data mining and big data and knowledge engineering.

Contents

Robotic and Control Systems

Sentiment Analysis, E-learning and Social media Education

Machine and Deep Learning Algorithms

Recognition and Image Processing

Mobile Computing and Networking

Micro-Grid and Power Systems

Machine Learning and Applications

Optimizing Self-Organizing Maps Parameters Using Genetic Algorithm: A Simple Case Study

Reham Fathy M. Ahmed[1]([⊠]), Cherif Salama[1,2]([⊠]),
and Hani Mahdi[1]([⊠])

[1] Computer and Systems Engineering Department, Ain Shams University,
Cairo, Egypt
eng_rehamfathy@hotmail.com,
{cherif.salama,hani.mahdi}@eng.asu.edu.eg
[2] Computer Science and Engineering Department,
The American University in Cairo, Cairo, Egypt

Abstract. A Self-Organizing Map (SOM) is a powerful tool for data analysis, clustering, and dimensionality reduction. It is an unsupervised artificial neural network that maps a set of n-dimensional vectors to a two-dimensional topographic map. Being unsupervised, SOMs need little input to be successfully deployed. The only inputs needed by a SOM are its own parameters such as its size, number of iterations, and its initial learning rate. The quality and accuracy of the solution offered by a SOM depend on choosing the right values for such parameters. Different attempts have been made to use the genetic algorithm to optimize these parameters for random inputs or for specific applications such as the traveling salesman problem. To the best knowledge of the authors, no roadmaps for selecting these parameters were presented in the literature. In this paper, we present the first results of a proposed roadmap for optimizing these parameters using the genetic algorithm and we show its effectiveness by applying it on the classical color clustering problem as a case study.

Keywords: Gray color clustering · Self-Organizing maps · Genetic algorithm

1 Introduction

Objects' clustering in machine learning can be carried out according to many factors known as attributes or features. The higher the number of these features the harder to visualize the training set and then work on it. Additionally, when the dimensionality increases, we need much more samples to avoid overfitting and much more computational power to train the target machine learning model. Luckily, in some cases, many of the features are correlated and hence redundant. In such cases, dimensionality reduction algorithms can be used to ignore irrelevant features. Many dimensionality reduction algorithms have been developed to fulfill these tasks such as Principal Components Analysis (PCA), Generative Topographic Mapping (GTM), and Self Organizing Maps (SOMs).

In this article, we present a case study where SOM is used to cluster similar gray colors. The SOM learning algorithm is presented with a set of gray colors to cluster.

A. E. Hassanien et al. (Eds.): AISI 2019, AISC 1058, pp. 3–12, 2020.
https://doi.org/10.1007/978-3-030-31129-2_1

Another set will be reserved to test the clustering obtained after the SOM is trained. The success of the SOM formation is critically dependent on the selection of its parameter values. In many cases, trial-and-error is used to tune these parameters. However, trial-and-error is time consuming and is not guaranteed to find optimal or even near optimal parameter values. Many methods have been made to use the genetic algorithm to optimize these parameters for random inputs or for specific applications such as the traveling salesman problem [11, 12]. In this paper we, suggest the usage of the genetic algorithm as the SOM parameters optimization algorithm. The rest of the paper is organized into 4 Sections. Section 2 is used to briefly explain the basic SOM and the basic genetic algorithms. Section 3 presents the gray color clustering problem and the proposed solution. The implementation and results are discussed in Sect. 4. Finally, Sect. 5 concludes the paper.

2   Overview of Used Algorithms

2.1   The Self-Organizing Maps (SOM) Algorithm

Many clustering algorithms were proposed for data clustering. One of these algorithms is the Self-Organizing Map (SOM) algorithm. It is one of the most popular Artificial Neural Network (ANN) models that are trained using unsupervised learning where no human intervention is needed during the learning process. SOM, also known as the Kohonen network, was first proposed by T. Kohonen Professor on 1980s [5]. SOM is a powerful clustering tool which is used for data analysis [2]. It is widely used in various application domains including but not limited to text mining, image clustering, economy, and geography. SOM offers a topology that maps a set of n-dimensional input vectors to a much lower-often one, two, or three- dimensional space called a map. In particular, a two-dimensional space is frequently used as the SOM output.

The SOM algorithm computation is divided into two stages: Training and Testing [1]. The training of the SOM does not differ from the training of other types of artificial neural networks. If the training examples' size is huge, the data is divided into small batches. An epoch consists of a number of batches that covers the whole training data. This number of batches is referred to as the number of iterations. If the epoch consists of one batch, as in our simple case study, the number of iterations will be one. However, hereafter we use the number of iterations to refer to the total number of iterations across the epochs or equivalently to the number of epochs.

In the first iteration of the training stage, the weights of the nodes to the inputs are initialized randomly. Then randomly-selected samples are presented as inputs to the network. For each input pattern there will be a winning node in the output layer. This winning node is called the Best Matching Unit (BMU) which has the closest value to this input pattern. The weights of this BMU are updated to the input value and their topological neighbors are updated according to the closeness to this BMU. In the following iterations weights will continue to be updated but at a lower rate as explained later.

After a large number of iterations, all the output layer weights will be updated and each node in the output layer will be sensitive to a specific cluster. The nodes of the clusters that are close in their characteristics will be close in position to each other.

The relationship between output layer nodes and each input pattern can be determined after training the SOM network. Only then the network can map all the input patterns onto the nodes in the output layer [1]. Figure 1 shows a SOM matrix of 3×3 nodes connected to the input layer shown in red and representing a two-dimensional vector. As shown, each neuron in the output layer is fully connected to the input layer. Each node has a specific position (x and y coordinates) and its weights vector dimension is equal to the dimensionality of input vectors. So if the training data consists of m-dimensional vectors V, i.e. $V = \{v_1, v_2, \ldots, v_m\}$, then the corresponding weight vector for each node of m-dimensions will be $W = \{w_1, w_2, \ldots, w_m\}$.

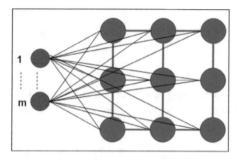

Fig. 1. A simple self-organizing map network

Basic SOM Model Training Steps [3, 4, 10]

1. Randomly initialize the weight vectors for each node. To converge quickly, it is a good practice to initialize them with values from the input data.
2. Choose a random input from the training data and present it to the SOM matrix.
3. Calculate the distance between each node's weight vector W_i and the selected input vector V using Euclidean distance. Equation 1 is used to find the BMU which is the node with the minimum distance.

$$BMU = \operatorname*{argmin}_{i} \|V - W_i\| \qquad (1)$$

4. Find the nodes in the neighborhood of the BMU based on the neighborhood radius. The green arrow in Fig. 2 below represents the neighborhood radius. The radius diminishes each time-step according to Eq. 2 known as the exponential decay function, where k is the current time-step (iteration of the loop), T is the time constant presented in Eq. 3, K number of iterations and R_0 is the initial value of the radius of the neighborhood.

$$R_k = R_0 * e^{\left(\frac{-k}{T}\right)} \qquad (2)$$

$$T = \frac{K}{\log(R_0)} \qquad (3)$$

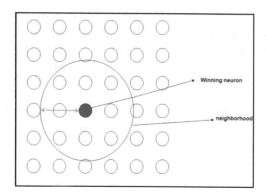

Fig. 2. The BMU's neighborhood

5. Adjust weights of each neighboring node according to Eq. 4 to make them more like the input vector. The closer a node is to the BMU the more its weights get altered.

$$\mathbf{W}_j(k+1) = \mathbf{W}_j(k) + L(k)\big[\mathbf{x}(k) - \mathbf{W}_j(k)\big] \qquad (4)$$

where

$\mathbf{W}_j(k)$: The weight vector before the neuron is updated.
$\mathbf{W}_j(k+1)$: The weight vector after the neuron is updated.
$\mathbf{x}(k)$: The training vector from input data set.
$L(k)$: The decay of the learning rate.

6. Go to Step 2 looping for K iterations to get the final weight matrix for the output layer.

2.2 The Genetic Algorithm

The Genetic Algorithm (GA) is a heuristic search method based on the principles of genetics and natural selection [6]. The GA is used in artificial intelligence and computing. It is inspired by Charles Darwin's theory of natural evolution and is used to solve optimization problems. GAs are excellent for searching through large and complex data sets. They are considered a very good fit for finding sensible answers for complex issues as they are highly capable of solving unconstrained and constrained optimization issues. A GA represents the problem's search space as a population of individuals [7]. Each individual represents a solution for the problem and is evaluated by a fitness function. Individuals are called chromosomes and a population is a group of chromosomes. A GA keeps on iterating and the last generation will be the one that has the fittest solutions. The GA applies the following major steps [8, 9] which are also represented in Fig. 3.

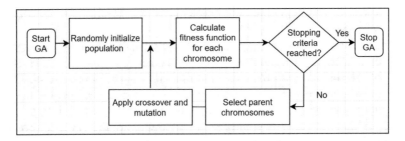

Fig. 3. A flowchart of a genetic algorithm system

GA Major Steps

(1) *Initialization:* Each individual represents the set of variables in the problem. Each variable is known as a gene. Genes are joined to form a chromosome as shown in Fig. 4. A set of parameters have to be selected for the GA itself. These parameters are the number of chromosomes per population (N), the crossover probability (P_c), and the mutation probability (P_m). The initialization step is completed by generating a random initial population of N chromosomes.

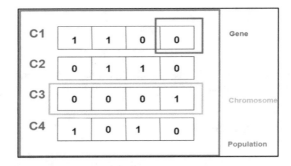

Fig. 4. The basic terms of the GA: population, chromosome, gene

(2) *Evaluation:* Evaluate the fitness function f(x) of each chromosome x in the population. The fitness function is the problem function which determines how fit an individual is. The fitness value of each individual measures its suitability as a solution. A higher fitness value for a chromosome means a higher probability of being selected in the next generation as explained in the next step.

(3) *New Population Generation:* A new empty pool for the new generation is created then the following steps are repeated until the size of the new population becomes equal to the size of the initial population:

   a. *Selection:* According to the fitness function results, two chromosomes (called parents) are selected from a population. The main idea of this step is to make use of the fittest individuals in the next generation. There are many methods to select the best chromosomes such as rank selection and roulette wheel selection.

    b. ***Crossover:*** This is the most important phase in a GA. According to a crossover probability (P_c) we can decide whether the new offspring will be exact copies of the selected parents or the result of applying the crossover operation on them. The crossover operation involves generating a random number to determine the crossover point within the gene as illustrated in Fig. 5(a).

    c. ***Addition:*** Place the new offspring chromosomes in the new population pool.

(4) ***Mutation:*** The mutation operator produces small random changes to newly generated population by randomly choosing a single gene then changing its value. Typically, mutation has a low probability (P_m). A mutation example is illustrated in Fig. 5(b).

(5) ***Replacement:*** Replace the old population with the new population in the new pool and empty the pool of the new population.

(6) ***Stopping Criteria Check:*** After having the new generation, check if the stopping criteria have been met or if the maximum number of iterations was reached, if not go to Step 2.

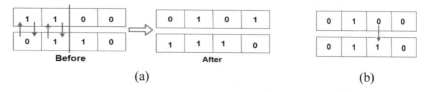

              (a)   (b)

Fig. 5. (a) A pair of chromosomes before and after applying crossover and (b) A chromosome before and after applying mutation.

3   The Proposed Approach

The success of the SOM clustering strongly depends on choosing the right values for its main parameters. As mentioned earlier, trial-and-error methods are usually used to select such values; however, trial-and-error will consume a lot of time without any guarantees to find good values. The proposed approach is to use the GA to tune the SOM parameters. We try this proposal on clustering gray colors as a simple case study.

    The training process of the Self-Organizing Map (SOM) network has some main parameters that will affect its final results. These are the initial radius, the decay ratio of the radius, the SOM matrix dimensions (width and height), the learning rate, the decay ratio of the learning rate, and the number of iterations. Inspecting all these parameters, it is easy to see the main hyper independent parameters are the SOM matrix dimensions (width and height), the initial learning rate, and the number of iterations. For example, the initial radius mainly depends on SOM matrix size and its decaying rate can be computer from Eq. 2.

The learning rate decides the amount of changes to SOM weight matrix at the end of every iteration. If it is too high, we will keep making drastic changes to the SOM and might never settle on a solution. If it is too low, we will only make very small changes and the SOM training will be extremely slow to converge. In practice, it is best to start with a larger learning rate and reduce it slowly over time. The learning rate decay equation shows that the decay rate mainly depends on initial learning rate as shown in the below equation

$$\text{decayed learning rate} = initial\ learning\ rate\ *\ e^{\frac{-k}{K}} \tag{5}$$

where "k" is the current iteration number and K is the number of iterations. The number of iterations can be set to a large value to ensure that enough training is done and that the SOM has reached convergence but sometimes this huge number of iterations is not needed and smaller one is enough.

We can conclude from the above discussion that the main parameters that can affect SOM quality and that need to be tuned either manually or using the suggested GA are the initial learning rate, number of iterations, and the SOM matrix dimensions. In the next section we elaborate on the results of using the GA to optimize these parameters.

The Proposed Algorithm Steps:

1. Generate an initial GA population. Genes values will be randomly generated within the ranges listed in Table 1.
2. Generate 16 random gray colors for the training input data set and apply the GA for the generated data set.
   a. Calculate the fitness for each chromosome by simulating SOM using the parameter values in that chromosome. After reaching the maximum number of iterations use the final weight matrix to calculate average for absolute distances between each of the 256 test data inputs and its BMU value which will be the fitness value.
   b. After getting fitness values for each chromosome the rest of the GA steps illustrated in Sect. 2 will be executed until we find the optimal SOM parameter values for the generated data set.

Figure 6 shows a block diagram of the proposed algorithm.

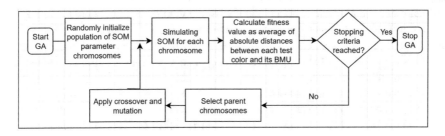

Fig. 6. The block diagram of the proposed algorithm steps

4   Experimental Results

The proposed method was implemented using Python and was tested on an entry-level laptop equipped with a Core i5 processor and 5 GB RAM. The laptop runs a 64-bit Windows 7 operating system.

As discussed in the previous section, we used the GA to optimize the values of the initial learning rate, number of iterations, and SOM matrix dimensions (including width and height). Therefore, the GA chromosome size will be 4. The value ranges for these parameters are shown in Table 1 while the manually chosen parameters used to run the GA itself are shown in Table 2. The proposed algorithm randomly generates 16 gray colors for SOM training data and for testing we use all gray colors from 0 (black) to 255 (white).

Table 1. SOM parameters ranges

Parameter	Value range	Step
Number of iterations	100 to 300	50
Initial learning rate	0.01 to 1	0.01
SOM matrix length	3 to 10	1
SOM matrix width	3 to 10	1

Table 2. GA parameters

Parameter	Value
Population size	16
Chromosomes size	4
Crossover probability	0.8
Mutation probability	0.01

Using the GA as described above lead to the selection of the following SOM parameter values: 0.69 for the initial learning rate, 9 for width, 3 for length, and 100 for the number of iterations. Table 3 shows the result of running the GA-optimized SOM algorithm on some sample test colors. The table lists the SOM-selected BMU for each test color and the absolute distance between that test color and its corresponding BMU.

From Table 3, we can see that for most test colors, the selected BMUs and the corresponding distances are small. These small distance values indicate that the genetically-optimized SOM algorithm managed to map each of the test inputs data items to its proper cluster.

We repeated the same experiment five times using different randomly generated training colors and we recorded the average and standard deviation of absolute distance from BMU for all 256 test cases for each run in Table 4. Table 4 indicates that the average and standard deviation of absolute distance for all 5 runs is reasonably low. On the laptop we used, this experiment took about 3 h to run. For more complex inputs and runs, higher computational capabilities will be needed.

Table 3. Samples of absolute distances between test data and BMU

Test Data	BMU value	Distance
57	67	10
79	80	1
74	77	3
80	80	0
82	80	2
94	96	2
95	96	1
96	96	0
107	112	5
108	112	4
109	112	3
110	112	2
132	133	1
133	133	0
134	133	1
249	228	21
254	228	26

Table 4. Five run results

Run number	Mean of the absolute distance from BMU	Standard deviation of the absolute distance from BMU
First run	12.394	17.263
Second run	3.467	6.693
Third run	8.547	10.79
Fourth run	3.402	4.799
Fifth run	8.816	15.16

5 Conclusion and Future Work

In this paper, we proposed a roadmap for optimizing the SOM parameters using the genetic algorithm. We also applied the proposed roadmap on the grayscale color clustering problem. Experimental results confirm the effectiveness of the genetically-optimized SOM in solving the color clustering problem. In fact, the results expectedly suggest that using the genetic algorithm to optimize the SOM parameters, we are effectively optimizing the clustering algorithm itself. The proposed roadmap is not considering random inputs or a specific type of problem like the traveling salesman problem. Instead, it uses the grayscale range to cover the whole spectrum of values. As a future work this algorithm can be extended for more complex inputs.

References

1. Liu, Y.-C., Liu, M., Wang, X.-L.: Application of self-organizing maps in text clustering: a review. InTechOpen (2012)
2. Zhang, X.-Y., Chen, J.-S., Dong, J.-K.: Color clustering using self-organizing maps. In: International Conference on Wavelet Analysis and Pattern Recognition, Beijing, pp. 986–989 (2007)
3. Maia, J., Barreto, G., Coelho, A.: On Self-Organizing Feature Map (SOFM) formation by direct optimization through a genetic algorithm. In: Proceedings - 8th International Conference on Hybrid Intelligent Systems (2008)
4. Yuan, L.: Implementation of self-organizing maps with Python. Open Access Master's Theses (2018)
5. Natita, W., Wiboonsak, W., Dusadee, S.: Appropriate learning rate and neighborhood function of Self-Organizing Map (SOM) for specific humidity pattern classification over Southern Thailand. Int. J. Model. Optim. 6(1), 61 (2016)
6. Kaya, Y., Uyar, M., Tekin, R.: A novel crossover operator for genetic algorithms: ring crossover. CoRR. abs/1105.0355 (2011)
7. Lingaraj, H.: A study on genetic algorithm and its applications. Int. J. Comput. Sci. Eng. **4**, 139–143 (2016)
8. Negnevitsky, M.: Artificial Intelligence A Guide to Intelligent Systems, 2nd edn, pp. 222–256. Pearson Education, London (2005)
9. Sastry, K., Goldberg, D., Kendall, G.: Genetic algorithms. In: Burke, E.K., Kendall, G. (eds.) Search Methodologies. Springer, Boston (2005)
10. Su, M.-C., Liu, T.-K., Chang, H.-T.: Improving the self-organizing feature map algorithm using an efficient initialization scheme. Tamkang J. Sci. Eng. 5(1), 35–48 (2002)
11. Polani, D.: On the optimization of self-organizing maps by genetic algorithms (1999)
12. Jin, H.-D., Leung, K.-S., Wong, M.-L., Xu, Z.-B.: An efficient self-organizing map designed by genetic algorithms for the traveling salesman problem. IEEE Trans. Syst. Man Cybern. Part B: Cybern. **33**(6), 877–888 (2003)

Rough Sets Based on Possibly Indiscernible Classes in Incomplete Information Tables with Continuous Values

Michinori Nakata[1(✉)], Hiroshi Sakai[2], and Keitarou Hara[3]

[1] Faculty of Management and Information Science, Josai International University,
1 Gumyo, Togane, Chiba 283-8555, Japan
nakatam@ieee.org
[2] Department of Mathematics and Computer Aided Sciences, Faculty of Engineering,
Kyushu Institute of Technology, Tobata, Kitakyushu 804-8550, Japan
sakai@mns.kyutech.ac.jp
[3] Department of Informatics, Tokyo University of Information Sciences,
4-1 Onaridai, Wakaba-ku, Chiba 265-8501, Japan
hara@rsch.tuis.ac.jp

Abstract. Rough sets under incomplete information with continuous domains are examined based on possible world semantics. We focus on possible indiscernibility relations, although the traditional approaches are done under possible tables. This is because we only obtain a finite number of possible indiscernibility relations even if infinite number of possible tables are derived from an incomplete information table. A possibly indiscernible class for an object is derived from a possible indiscernibility relation. The family of possibly indiscernible classes for the object is a lattice for inclusion. Lower and upper approximations are derived from using the minimal and the maximal elements in the lattice. Therefore, there is no computational complexity for the number of objects with incomplete information. Furthermore, the approach based on possible world semantics gives the same approximations as ones obtained from our extended approach, which is proposed in the previous work using indiscernible classes. Therefore, the approach developed in this paper justifies our extended approach.

Keywords: Neighborhood rough sets · Possible world semantics · Incomplete information · Possibly indiscernible classes · Lower and upper approximations · Continuous values

1 Introduction

Big data consists of various types of data. Focusing on data composed of characters, data is roughly divided into discrete data and continuous data.

© Springer Nature Switzerland AG 2020
A. E. Hassanien et al. (Eds.): AISI 2019, AISC 1058, pp. 13–23, 2020.
https://doi.org/10.1007/978-3-030-31129-2_2

Rough sets, constructed by Pawlak [15], are used as an effective method for data mining. The framework is usually applied to complete information tables with nominal attributes and creates fruitful results in various fields. However, attributes taking continuous values frequently appear, when we describe properties of an object in the real world. Furthermore, incomplete information ubiquitously occurs in the real world. We cannot sufficiently utilize information obtained from the real world unless we deal with continuous and incomplete information. Therefore, extended versions of rough sets are proposed to deal with incomplete information in continuous domains [5, 12, 13, 18–20].

An approach, which is most frequently used [5, 18–20], is to handle in the way that Kryszkiewicz applied to nominal attributes [6]. The approach fixes indiscernibility between an object with incomplete information and another object. However, it is natural that an object characterized by incomplete information has two possibilities; namely, the object is indiscernible with another object and not so. To fix the indiscernibility corresponds to taking into consideration only one of the two possibilities. Therefore, the approach creates poor results and information loss occurs [9, 17]. Furthermore, the fixing is not compatible with the approach by Lipski in the field of incomplete databases, because Lipski handles all possibilities of objects with incomplete information [7].

Another is to directly use indiscernibility relations that are extended to deal with incomplete information [12]. Yet another is to use possible classes obtained from the indiscernibility relation [13]. These approaches have the same order of computational complexity as the one in complete information. However, no justification is shown, although it is known in the case of discrete data that these give the same results as the approach based on possible world semantics [10]. To give these approaches a correctness criterion, it is required to develop an approach based on possible world semantics. So far the approaches under possible world semantics use possible tables derived from an incomplete information table. Unfortunately, infinite number of possible tables can be derived from an incomplete information table with continuous values. Possible world semantics is unavailable as long as we use possible tables.

Rough sets are based on the indiscernibility relation on a set of attributes. The number of possible indiscernibility relations is finite, even if the number of possible tables is infinite, because the number of objects is finite. Therefore, we focus on possible indiscernibility relations, not possible tables. A possibly indiscernible class for an object is derived from a possible indiscernibility relation, while another possibly indiscernible class from another possible indiscernibility relation.

In this work, we develop an approach based on possible world semantics by using the family of possibly indiscernible classes derived from possible indiscernibility relations in an incomplete information table with continuous values.

The paper is organized as follows. In Sect. 2, an approach using indiscernible classes from the indiscernibility relation on an attribute is briefly addressed in a complete information table. In Sect. 3, we develop an approach in an incomplete information table under possible world semantics. In Sect. 4, conclusions are addressed.

2 Rough Sets by Using Indiscernible Classes in Complete Information Systems with Continuous Values

A data set is represented as a two-dimensional table, called an information table. In the information table, each row and each column represent an object and an attribute, respectively. A mathematical model of an information table with complete information is called a complete information system. The complete information system is expressed by triplet $(U, AT, \{D(a_i) \mid a_i \in AT\})$. U is a non-empty finite set of objects, which is called the universe. AT is a non-empty finite set of attributes such that $a_i : U \rightarrow D(a_i)$ for every $a_i \in AT$ where $D(a_i)$ is the continuous domain of attribute a_i. We have two approaches for dealing with attributes taking continuous values. One approach is to discretize a continuous domain into disjunctive intervals in which objects are regarded as indiscernible [3]. The discretization has a heavy influence over results. The other approach is to use neighborhood [8]. The indiscernibility of two objects is derived from the distance between them. When the distance of the two objects is within a given threshold, two objects are regarded as indiscernible. Results gradually change as the threshold changes. Thus, we adopt the latter approach.

Binary relation R_{a_i}[1] expressing indiscernibility of objects on attribute $a_i \in AT$ is called the indiscernibility relation for a_i:

$$R_{a_i} = \{(o, o') \in U \times U \mid |a_i(o) - a_i(o')| \leq \delta_{a_i}\}, \tag{1}$$

where $a_i(o)$ is the value for attribute a_i of object o and δ_{a_i}[2] is a threshold that denotes a range in which $a_i(o)$ is indiscernible with $a_i(o')$.

Proposition 1
If $\delta 1 \leq \delta 2$, then $R_{a_i}^{\delta 1} \subseteq R_{a_i}^{\delta 2}$, where $R_{a_i}^{\delta 1}$ and $R_{a_i}^{\delta 2}$ is indiscernibility relations with thresholds $\delta 1$ and $\delta 2$, respectively.

From the indiscernibility relation, indiscernible class $[o]_{a_i}$[3] for object o is obtained:

$$[o]_{a_i} = \{o' \mid (o, o') \in R_{a_i}\}. \tag{2}$$

Proposition 2
If $\delta 1 \leq \delta 2$ on a_i, then $[o]_{a_i}^{\delta 1} \subseteq [o]_{a_i}^{\delta 2}$, where $[o]_{a_i}^{\delta 1}$ and $[o]_{a_i}^{\delta 2}$ is indiscernible classes under thresholds $\delta 1$ and $\delta 2$, respectively.

Using indiscernible class $[o]_{a_i}$, lower approximation $\underline{apr}_{a_i}(\mathcal{O})$ and upper approximation $\overline{apr}_{a_i}(\mathcal{O})$ for a_i of set $\mathcal{O}$ of objects are:

$$\underline{apr}_{a_i}(\mathcal{O}) = \{o \mid [o]_{a_i} \subseteq \mathcal{O}\}, \tag{3}$$

$$\overline{apr}_{a_i}(\mathcal{O}) = \{o \mid [o]_{a_i} \cap \mathcal{O} \neq \emptyset\}. \tag{4}$$

---

[1] When threshold δ requires to denote, $R_{a_i}^{\delta}$ is used.
[2] δ_{a_i} is expressed by δ omitting a_i for simplicity if no confusion.
[3] When threshold δ requires to denote, $[o]_{a_i}^{\delta}$ is used.

Proposition 3 [12]

If $\delta 1 \leq \delta 2$ on a_i, then $\underline{apr}^{\delta 1}_{a_i}(\mathcal{O}) \supseteq \underline{apr}^{\delta 2}_{a_i}(\mathcal{O})$ and $\overline{apr}^{\delta 1}_{a_i}(\mathcal{O}) \subseteq \overline{apr}^{\delta 2}_{a_i}(\mathcal{O})$, where $\underline{apr}^{\delta 1}_{a_i}(\mathcal{O})$ and $\overline{apr}^{\delta 1}_{a_i}(\mathcal{O})$ are lower and upper approximations under threshold $\delta 1$ and $\underline{apr}^{\delta 2}_{a_i}(\mathcal{O})$ and $\overline{apr}^{\delta 2}_{a_i}(\mathcal{O})$ are lower and upper approximations under threshold $\delta 2$.

3   Rough Sets by Possibly Indiscernible Classes in Incomplete Information Systems with Continuous Domains

An information table with incomplete information is called an incomplete information system. In incomplete information systems, $a_i : U \rightarrow s_{a_i}$ for every $a_i \in AT$ where s_{a_i} is the union of the set of values over domain $D(a_i)$ of attribute a_i and the set of intervals on $D(a_i)$. Single value v with $v \in a_i(o)$ or $v \subseteq a_i(o)$ is a possible value that may be the actual one as the value of attribute a_i in object o. The possible value is the actual one if $a_i(o)$ is a single value.

We have two kinds of indiscernibility relations from an incomplete information table[4]. One is the certain indiscernibility relation. The others are possible indiscernibility relations. Certain indiscernibility relation $C(R_{a_i})$ is:

$$C(R_{a_i}) = \{(o, o') \in U \times U \mid (o = o') \vee (\forall u \in a_i(o) \forall v \in a_i(o') | u - v| \leq \delta_{a_i})\}. \quad (5)$$

In this binary relation that is unique, two objects o and o' of $(o, o') \in C(R_{a_i})$ is certainly indiscernible with each other on a_i. Such a pair is called a certain pair. On the other hand, we have lots of possible indiscernibility relations. The number of possible indiscernibility relations grows exponentially as the number of values with incomplete information increases. Family $P(R_{a_i})$ of possible indiscernibility relations is:

$$P(R_{a_i}) = \{pr \mid pr = C(R_{a_i}) \cup e \wedge e \in \mathcal{P}(MPP(R_{a_i}))\}, \quad (6)$$

where each element is a possible indiscernibility relation and $\mathcal{P}(MPP(R_{a_i}))$ is the power set of $MPP(R_{a_i})$ and $MPP(R_{a_i})$ is:

$$MPP(R_{a_i}) = \{\{(o', o), (o, o')\} \mid (o', o) \in MP(R_{a_i})\}, \quad (7)$$

$$MP(R_{a_i}) = \{(o, o') \in U \times U \mid \exists u \in a_i(o) \exists v \in a_i(o') | u - v| \leq \delta_{a_i})\} \backslash C(R_{a_i}). \quad (8)$$

A pair of objects that is included in $MP(R_{a_i})$ is called a possible pair. $P(R_{a_i})$ is a lattice for set inclusion. $C(R_{a_i})$ is the minimal possible indiscernibility relation in $P(R_{a_i})$, which is the minimal element, whereas $C(R_{a_i}) \cup MP(R_{a_i})$ is the maximal possible indiscernibility relation, which is the maximal element. One of

---

[4] For the sake of simplicity and space limitation, We describe the case of an attribute, although our approach can be easily extended to the case of more than one attribute.

possible indiscernibility relations is the actual indiscernibility relation, although we cannot know it without additional information.

Example 1

	$T0$	
U	a_1	a_2
o_1	0.71	$\{1.25, 1.31\}$
o_2	$[0.74, 0.78]$	$[1.47, 1.53]$
o_3	0.81	1.51
o_4	$[0.84, 0.94]$	1.56
o_5	$\{0.66, 0.68\}$	$\{1.32, 1.39\}$

In incomplete information table $T0$, let threshold δ_{a_1} be 0.05 on attribute a_1. The set of certain pairs of indiscernible objects on a_1 is:

$$\{(o_1, o_1), (o_1, o_5), (o_2, o_2), (o_3, o_3), (o_4, o_4), (o_5, o_5), (o_5, o_1)\}.$$

The set of possible pairs of indiscernible objects is:

$$\{(o_1, o_2), (o_2, o_1), (o_2, o_3), (o_3, o_2), (o_3, o_4), (o_4, o_3)\}.$$

Using formulae (5)–(7), the family of possible indiscernibility relations and each possible indiscernibility relation are:

$$P(R_{a_1}) = \{pr_1, \cdots, pr_8\},$$

$pr_1 = \{(o_1, o_1), (o_1, o_5), (o_2, o_2), (o_3, o_3), (o_4, o_4), (o_5, o_5), (o_5, o_1)\},$

$pr_2 = \{(o_1, o_1), (o_1, o_5), (o_2, o_2), (o_3, o_3), (o_4, o_4), (o_5, o_5), (o_5, o_1),$
$\qquad (o_1, o_2), (o_2, o_1)\},$

$pr_3 = \{(o_1, o_1), (o_1, o_5), (o_2, o_2), (o_3, o_3), (o_4, o_4), (o_5, o_5), (o_5, o_1),$
$\qquad (o_2, o_3), (o_3, o_2)\},$

$pr_4 = \{(o_1, o_1), (o_1, o_5), (o_2, o_2), (o_3, o_3), (o_4, o_4), (o_5, o_5), (o_5, o_1),$
$\qquad (o_3, o_4), (o_4, o_3)\},$

$pr_5 = \{(o_1, o_1), (o_1, o_5), (o_2, o_2), (o_3, o_3), (o_4, o_4), (o_5, o_5), (o_5, o_1),$
$\qquad (o_1, o_2), (o_2, o_1), (o_2, o_3), (o_3, o_2)\},$

$pr_6 = \{(o_1, o_1), (o_1, o_5), (o_2, o_2), (o_3, o_3), (o_4, o_4), (o_5, o_5), (o_5, o_1),$
$\qquad (o_1, o_2), (o_2, o_1), (o_3, o_4), (o_4, o_3)\},$

$pr_7 = \{(o_1, o_1), (o_1, o_5), (o_2, o_2), (o_3, o_3), (o_4, o_4), (o_5, o_5), (o_5, o_1),$
$\qquad (o_2, o_3), (o_3, o_2), (o_3, o_4), (o_4, o_3)\},$

$pr_8 = \{(o_1, o_1), (o_1, o_5), (o_2, o_2), (o_3, o_3), (o_4, o_4), (o_5, o_5), (o_5, o_1),$
$\qquad (o_1, o_2), (o_2, o_1), (o_2, o_3), (o_3, o_2), (o_3, o_4), (o_4, o_3)\}.$

These possible indiscernibility relations have the following lattice structure for inclusion:

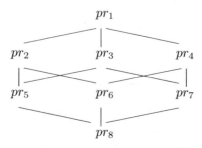

pr_1 is the minimal element, whereas pr_8 is the maximal element.

Possibly indiscernible class $[o]_{a_i}^{pr}$ on attribute a_i for object o is derived from a possible indiscernibility relation pr by using formula (2):

$$[o]_{a_i}^{pr} = \{o' \mid (o, o') \in pr \wedge pr \in P(R_{a_i})\}. \tag{9}$$

Proposition 3
If $pr_k \subseteq pr_l$, then $[o]_{a_i}^{pr_k} \subseteq [o]_{a_i}^{pr_l}$.

From this proposition, the family of possibly indiscernible classes for an object is also a lattice for inclusion.

Example 2
Using formula (2), possibly indiscernible classes of objects are obtained in each indiscernibility relation pr_i with $i = 1, \ldots, 8$. For example, in pr_1 $[o]_1 = \{o_1, o_5\}$, $[o]_2 = \{o_2\}, [o]_3 = \{o_3\}, [o]_4 = \{o_4\}, [o]_5 = \{o_1, o_5\}$ and in pr_8 $[o]_1 = \{o_1, o_2, o_5\}$, $[o]_2 = \{o_1, o_2, o_3\}, [o]_3 = \{o_2, o_3, o_4\}, [o]_4 = \{o_3, o_4\}, [o]_5 = \{o_1, o_5\}$.

By using $[o]_{a_i}^{pr}$ in formulae (3) and (4), lower and upper approximations of set $\mathcal{O}$ of objects in possible indiscernibility relation pr are:

$$\underline{apr}_{a_i}(\mathcal{O})^{pr} = \{o \mid [o]_{a_i}^{pr} \subseteq \mathcal{O} \wedge pr \in P(R_{a_i})\}, \tag{10}$$

$$\overline{apr}_{a_i}(\mathcal{O})^{pr} = \{o \mid [o]_{a_i}^{pr} \cap \mathcal{O} \neq \emptyset \wedge pr \in P(R_{a_i})\}. \tag{11}$$

Proposition 4
If $pr_k \subseteq pr_l$ for possible indiscernibility relations $pr_k, pr_l \in P(R_{a_i})$, then $\underline{apr}_{a_i}(\mathcal{O})^{pr_k} \supseteq \underline{apr}_{a_i}(\mathcal{O})^{pr_l}$ and $\overline{apr}_{a_i}(\mathcal{O})^{pr_k} \subseteq \overline{apr}_{a_i}(\mathcal{O})^{pr_l}$.

This proposition shows that the families of lower and upper approximations under possible indiscernibility relations are also lattices for set inclusion.

We aggregate the lower and upper approximations under possible indiscernibility relations. Certain lower approximation $\underline{Capr}_{a_i}(\mathcal{O})$ of set $\mathcal{O}$ of objects is:

$$\underline{Capr}_{a_i}(\mathcal{O}) = \{o \mid \forall pr \in P(R_{a_i}) \ o \in \underline{apr}_{a_i}(\mathcal{O})^{pr}\}. \tag{12}$$

Possible lower approximation $\underline{Papr}_{a_i}(\mathcal{O})$ is:

$$\underline{Papr}_{a_i}(\mathcal{O}) = \{o \mid \exists pr \in P(R_{a_i})\ o \in \underline{apr}_{a_i}(\mathcal{O})^{pr}\}. \tag{13}$$

Certain upper approximation $\overline{Capr}_{a_i}(\mathcal{O})$ is:

$$\overline{Capr}_{a_i}(\mathcal{O}) = \{o \mid \forall pr \in P(R_{a_i})\ o \in \overline{apr}_{a_i}(\mathcal{O})^{pr}\}. \tag{14}$$

Possible upper approximation $\overline{Papr}_{a_i}(\mathcal{O})$ is:

$$\overline{Papr}_{a_i}(\mathcal{O}) = \{o \mid \exists pr \in P(R_{a_i})\ o \in \overline{apr}_{a_i}(\mathcal{O})^{pr}\}. \tag{15}$$

From Proposition 4, these approximations are transformed into the following formulae:

$$\underline{Capr}_{a_i}(\mathcal{O}) = \underline{apr}_{a_i}(\mathcal{O})^{pr_{max}}, \tag{16}$$

$$\overline{Capr}_{a_i}(\mathcal{O}) = \overline{apr}_{a_i}(\mathcal{O})^{pr_{min}}, \tag{17}$$

$$\underline{Papr}_{a_i}(\mathcal{O}) = \underline{apr}_{a_i}(\mathcal{O})^{pr_{min}}, \tag{18}$$

$$\overline{Papr}_{a_i}(\mathcal{O}) = \overline{apr}_{a_i}(\mathcal{O})^{pr_{max}}, \tag{19}$$

where pr_{min} and pr_{max} are the minimal and the maximal possible indiscernibility relations. These formulae show that we can obtain the four approximations without computational complexity, no matter how many possible indiscernibility relations.

Example 3

We go back to Example 1. Let set $\mathcal{O}$ of objects be $\{o_2, o_3, o_4\}$. Using formulae (10) and (11), lower and upper approximations are obtained in each possible indiscernibility relation. For example, in pr_1 $\underline{apr}_{a_1}(\mathcal{O})^{pr_1} = \{o_2, o_3, o_4\}$, $\overline{apr}_{a_1}(\mathcal{O})^{pr_1} = \{o_2, o_3, o_4\}$, and in pr_8 $\underline{apr}_{a_1}(\mathcal{O})^{pr_8} = \{o_3, o_4\}$, $\overline{apr}_{a_1}(\mathcal{O})^{pr_8} = \{o_1, o_2, o_3, o_4\}$. By using formulae (16)–(19), $\underline{Capr}_{a_1}(\mathcal{O}) = \{o_3, o_4\}$, $\overline{Capr}_{a_1}(\mathcal{O}) = \{o_2, o_3, o_4\}$, $\underline{Papr}_{a_1}(\mathcal{O}) = \{o_2, o_3, o_4\}$, $\overline{Papr}_{a_1}(\mathcal{O}) = \{o_1, o_2, o_3, o_4\}$.

As with the case of nominal attributes [10], the following proposition holds.

Proposition 5

$\underline{Capr}_{a_i}(\mathcal{O}) \subseteq \underline{Papr}_{a_i}(\mathcal{O}) \subseteq \mathcal{O} \subseteq \overline{Capr}_{a_i}(\mathcal{O}) \subseteq \overline{Papr}_{a_i}(\mathcal{O})$.

Using four approximations denoted by formulae (16)–(19), lower and upper approximations are expressed in interval sets, as is described in [11]:

$$\underline{apr}^{\bullet}_{a_i}(\mathcal{O}) = [\underline{Capr}_{a_i}(\mathcal{O}), \underline{Papr}_{a_i}(\mathcal{O})], \tag{20}$$

$$\overline{apr}^{\bullet}_{a_i}(\mathcal{O}) = [\overline{Capr}_{a_i}(\mathcal{O}), \overline{Papr}_{a_i}(\mathcal{O})]. \tag{21}$$

Certain and possible approximations are the lower and upper bounds of the actual approximation. The two approximations $\underline{apr^\bullet_{a_i}}(\mathcal{O})$ and $\overline{apr^\bullet_{a_i}}(\mathcal{O})$ depend on each other; namely, the complementarity property $\underline{apr^\bullet_{a_i}}(\mathcal{O}) = U - \overline{apr^\bullet_{a_i}}(U - \mathcal{O})$ linked with them holds, as is so in complete information systems.

Furthermore, the following proposition is valid from formulae (16)–(19).

Proposition 6

$$Capr_{a_i}(\mathcal{O}) = \{o \mid [o]^{max}_{a_i} \subseteq \mathcal{O}\},$$
$$C\overline{apr}_{a_i}(\mathcal{O}) = \{o \mid [o]^{min}_{a_i} \cap \mathcal{O} \neq \emptyset\},$$
$$Papr_{a_i}(\mathcal{O}) = \{o \mid [o]^{min}_{a_i} \subseteq \mathcal{O}\},$$
$$P\overline{apr}_{a_i}(\mathcal{O}) = \{o \mid [o]^{max}_{a_i} \cap \mathcal{O} \neq \emptyset\},$$

where $[o]^{min}_{a_i}$ and $[o]^{max}_{a_i}$ are the minimal and the maximal possibly indiscernible classes of object o on a_i which are derived from formula (2) in possible indiscernibility relations pr_{min} and pr_{max}, respectively.

This proposition shows that our extended approach using possibly indiscernible classes [13] is justified. Namely, results from the extended approach using possibly indiscernible classes are the same as the ones from possible world semantics. A criterion for justification is formally represented as

$$q(R_{a_i}) = \bigodot q'(P(R_{a_i})),$$

where q' is the approach for complete information, which is described in Sect. 2, and q is an extended approach of q', which directly deals with incomplete information, $\bigodot$ is an aggregate operator, and $P(R_{a_i})$ is the family of possible indiscernibility relations from the original indiscernibility relation R_{a_i} under possible world semantics. This is schematized in Fig. 1.

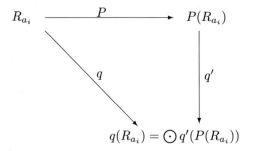

Fig. 1. Correctness criterion of extended method q

This type of correctness criterion is usually used in the field of databases dealing with incomplete information [1, 2, 4, 14, 21].

When we describe the case where $o \in \mathcal{O}$ is specified by numerical attribute a_j with incomplete information. Set $\mathcal{O}$ is specified by an interval where precise values of a_j are used.

$$\underline{Capr}_{a_i}(\mathcal{O}) = \underline{apr}_{a_i}(CO_{[a_j(o_m),a_j(o_n)]})^{pr_{max}}, \tag{22}$$

$$\overline{Capr}_{a_i}(\mathcal{O}) = \overline{apr}_{a_i}(CO_{[a_j(o_m),a_j(o_n)]})^{pr_{min}}, \tag{23}$$

$$\underline{Papr}_{a_i}(\mathcal{O}) = \underline{apr}_{a_i}(PO_{[a_j(o_m),a_j(o_n)]})^{pr_{min}}, \tag{24}$$

$$\overline{Papr}_{a_i}(\mathcal{O}) = \overline{apr}_{a_i}(PO_{[a_j(o_m),a_j(o_n)]})^{pr_{max}}, \tag{25}$$

where

$$CO_{[a_j(o_m),a_j(o_n)]} = \{o \in \mathcal{O} \mid a_j(o) \subseteq [a_j(o_m),a_j(o_n)]\}, \tag{26}$$

$$PO_{[a_j(o_m),a_j(o_n)]} = \{o \in \mathcal{O} \mid a_j(o) \cap [a_j(o_m),a_j(o_n)] \neq \emptyset\}, \tag{27}$$

where $a_j(o_m)$ and $a_j(o_n)$ are precise and $a_j(o_m) \leq a_j(o_n)$.

4   Conclusions

We have described rough sets based on possible world semantics in incomplete information tables with continuous domains. We focus on that the number of possible indiscernibility relations is finite, although the number of possible tables is infinite. The family of possibly indiscernible classes for an object is expressed by a lattice having the minimal and the maximal elements. The families of lower and upper approximations that are derived from possible indiscernibility relations are also a lattice for inclusion. The number of possible indiscernibility relations increases exponentially as the number of attribute values with incomplete information grows. However, approximations are obtained by using the minimal and the maximal possibly indiscernible classes. Therefore, we have no difficulty of computational complexity. By using the minimal and the maximal possibly indiscernible classes, four types approximations: certain lower, certain upper, possible lower, and possible upper approximations are obtained, as is so in incomplete information tables with nominal attributes. These approximations are the same as those obtained from an extended approach using possibly indiscernible classes. Therefore, the approach based on possible world semantics gives justification of the extended approach.

References

1. Abiteboul, S., Hull, R., Vianu, V.: Foundations of Databases. Addison-Wesley Publishing Company, Boston (1995)
2. Grahne, G.: The Problem of Incomplete Information in Relational Databases. Lecture Notes in Computer Science, vol. 554. Springer, Heidelberg (1991)
3. Grzymala-Busse, J.W.: Mining numerical data a rough set approach. In: Peters, J.F., Skowron, A. (eds.) Transactions on Rough Sets XI. LNCS, vol. 5946, pp. 1–13. Springer, Heidelberg (2010). https://doi.org/10.1007/978-3-642-11479-3_1

4. Imielinski, T., Lipski, W.: Incomplete information in relational databases. J. ACM **31**, 761–791 (1984)
5. Jing, S., She, K., Ali, S.: A universal neighborhood rough sets model for knowledge discovering from incomplete heterogeneous data. Expert Syst. **30**(1), 89–96 (2013). https://doi.org/10.1111/j.1468-0394.2012.00633_x
6. Kryszkiewicz, M.: Rules in incomplete information systems. Inf. Sci. **113**, 271–292 (1999)
7. Lipski, W.: On semantics issues connected with incomplete information databases. ACM Trans. Database Syst. **4**, 262–296 (1979)
8. Lin, T.Y.: Neighborhood systems: a qualitative theory for fuzzy and rough sets. In: Wang, P. (ed.) Advances in Machine Intelligence and Soft Computing, vol. IV, pp. 132–155. Duke University, Durham (1997)
9. Nakata, M., Sakai, H.: Applying rough sets to information tables containing missing values. In: Proceedings of 39th International Symposium on Multiple-Valued Logic, pp. 286–291. IEEE Press (2009). https://doi.org/10.1109/ISMVL.2009.1
10. Nakata, M., Sakai, H.: Twofold rough approximations under incomplete information. Int. J. Gen. Syst. **42**, 546–571 (2013). https://doi.org/10.1080/17451000.2013.798898
11. Nakata, M., Sakai, H.: Describing rough approximations by indiscernibility relations in information tables with incomplete information. In: Carvalho, J.P., Lesot, M.-J., Kaymak, U., Vieira, S., Bouchon-Meunier, B., Yager, R.R. (eds.) IPMU 2016, Part II. CCIS, vol. 611, pp. 355–366. Springer, Cham (2016). https://doi.org/10.1007/978-3-319-40581-0_29
12. Nakata, M., Sakai, H., Hara, K.: Rules induced from rough sets in information tables with continuous values. In: Medina, J., Ojeda-Aciego, M., Verdegay, J.L., Pelta, D.A., Cabrera, I.P., Bouchon-Meunier, B., Yager, R.R. (eds.) IPMU 2018, Part II. CCIS, vol. 854, pp. 490–502. Springer, Cham (2018). https://doi.org/10.1007/978-3-319-91476-3_41
13. Nakata, M., Sakai, H., Hara, K.: Rule induction based on indiscernible classes from rough sets in information tables with continuous values. In: Nguyen, H.S., et al. (eds.) IJCRS 2018. LNAI, vol. 11103, pp. 323–336. Springer, Heidelberg (2018). https://doi.org/10.1007/978-3-319-99368-3_25
14. Paredaens, J., De Bra, P., Gyssens, M., Van Gucht, D.: The Structure of the Relational Database Model. Springer, Heidelberg (1989)
15. Pawlak, Z.: Rough Sets: Theoretical Aspects of Reasoning about Data. Kluwer Academic Publishers, Dordrecht (1991). https://doi.org/10.1007/978-94-011-3534-4
16. Skowron, A., Stepaniuk, J.: Tolerance approximation spaces. Fundam. Inform. **27**, 245–253 (1996)
17. Stefanowski, J., Tsoukiàs, A.: Incomplete information tables and rough classification. Comput. Intell. **17**, 545–566 (2001)
18. Yang, X., Zhang, M., Dou, H., Yang, Y.: Neighborhood systems-based rough sets in incomplete information system. Inf. Sci. **24**, 858–867 (2011). https://doi.org/10.1016/j.knosys.2011.03.007
19. Zenga, A., Lia, T., Liuc, D., Zhanga, J., Chena, H.: A fuzzy rough set approach for incremental feature selection on hybrid information systems. Fuzzy Sets Syst. **258**, 39–60 (2015). https://doi.org/10.1016/j.fss.2014.08.014
20. Zhao, B., Chen, X., Zeng, Q.: Incomplete hybrid attributes reduction based on neighborhood granulation and approximation. In: 2009 International Conference on Mechatronics and Automation, pp. 2066–2071. IEEE Press (2009)

21. Zimányi, E., Pirotte, A.: Imperfect information in relational databases. In: Motro, A., Smets, P. (eds.) Uncertainty Management in Information Systems: From Needs to Solutions, pp. 35–87. Kluwer Academic Publishers, Dordrecht (1997)

Intelligent Watermarking System Based on Soft Computing

Maha F. Hany[1($\boxtimes$)], Bayumy A. B. Youssef[2], Saad M. Darwish[2], and Osama Hosam[1]

[1] City for Scientific Research and Technology Applications, Alexandria, Egypt
mhany@srtacity.sci.eg
[2] Institute of Graduate Studies and Research, University of Alexandria, Alexandria, Egypt
saad.darwish@alex-igsr.edu.eg

Abstract. Digital watermarking for various multimedia contents is an effective solution for copyright protection. The techniques of 3D Watermarking carefully insert some secret bits of data in 3D object. These techniques contain contradictory parameters as watermarked object quality (invisibility), capacity and watermark robustness. Thus these parameters should be expressed in fitness functions then trying to estimate their value via utilizing an optimization technique. Thus this paper suggests an optimized watermarking algorithm dependent on non-dominated sorting genetic algorithm II (NSGA-II) optimizer. In this framework, watermark is hided in low frequency coefficients in spectrum domain utilizing an inserting technique dependent on Dither Modulation (DM) technique. The aim of optimization is to look for optimal quantization step, so as to enhance both quality and robustness. As well as, performance of proposed algorithm is analyzed in expressions of signal to noise ratio and correlation coefficient. The investigational results indicate that proposed algorithm can accomplish a good robustness for most of included studied attacks in this research.

Keywords: Watermarking · Copyright protection · Soft computing · NSGA2

1 Introduction

The rapid evolution of internet that aids us to duplicate, move, modify, store and distribute digital data as electronic documents, images, sounds, videos and 3D models [1]. There is need for developing techniques for protection of various multimedia contents copyright to prevent unauthorized use or illegal apportionment of data. Watermarking is an effective solution for protection of copyright. The techniques of 3D Watermarking carefully insert some secret bits of data in 3D object. This data can be utilized for ownership and copyright [2, 3]. Comparing to cryptography, it protects media after transmission and allowed access [4, 5].

An optimal 3D watermarking schema utilizing a multi-objective genetic algorithm is suggested in this paper. The suggested multi-objective technique directly treats problem of optimization under non-dominated meaning (solutions), consequently it avoids the difficulty of estimating optimally weighted factor for single-objective watermarking systems effectively. These solutions can balance effectively robustness and imperceptibility, therefore, provides flexibility in selecting the most suitable parameters for designing watermarking system relying on practical watermarking requirements [6].

The rest of paper is arranged as: Concisely reviews to ideas about suggested system are in Sect. 2. In Sect. 3, the suggested system is discussed. Investigational results containing comparisons with existing techniques are in Sect. 4. Conclusions are in Sect. 5.

2 Literature Review and Related Work

Techniques for watermarking are arranged to robust and fragile techniques relying upon the purposes of application. A robust technique is utilized for the protection to copyright. These techniques are ordered into spatial and spectral techniques. A robust and blind system utilizing Levenberg–Marquardt for optimization strategy is introduced [7]. Surface error is minimized in inserting stage of watermark. This strategy is strong while reducing the surface error in watermarking. Yet the calculation of the technique is very complicated and dominates the algorithm cost. Authors in [8] offered another blind and strong framework. This strategy is fragile to re-meshing and simplification. Another blind technique was introduced [9]. This scheme is robust to similarity transformation, subdivision, cropping and simplification, but it is fragile to noise addition and smoothing.

Authors in [10] introduced a framework relying upon vertex combination and mapping purpose to advance robustness. Another framework is presented [11]. A statistical technique utilized to hide signature by altering the vertex mean. The scheme is strong to cropping attacks. Authors in [12] introduced framework relying upon protective object. Using vertices norm distribution and solving error problem for inserting. The treated objects have a smooth surface and robustness for various geometric attacks. A blind framework relying upon classification and arranging of faces is introduced [13]. The framework has robustness for mesh smoothing, quantization and noise attacks. A semi-blind framework utilizing law of Weber is shown [14]. The data is generated from original object. The framework is robust fort geometric combination of attacks. Distortion error is minimized utilizing genetic algorithm [15]. The primitive is selected by clustering of k-means. Yet genetic algorithm loss the diversity with time.

3 The Proposed System

The proposed non-blind algorithm inserts watermark by modulating spectral coefficients amplitude by Dither Modulation (DM) (low complexity of quantization index modulation) [16–18]. The idea in the proposed technique is calculating Eigen-decomposition to object Laplacian matrix, then eliminate high frequency eigenvectors and get low frequency eigenvectors for hiding watermark to obtain watermarked model. The diagram of suggested algorithm is displayed in Fig. 1, embedding and extraction processes.

3.1 Problem Formulation

Watermarking is viewed as a multi-objective optimization problem as the goal is to maximize robustness and minimize perceptible distortion. There are two strategies: one which consists of aggregating many fitness functions through one utilizing weighted sum, however, this method usually favors one fitness function in detriment of others [19]. The other treats many opposing objectives simultaneously. Multi-objective optimization techniques as NSGA-II depends on Pareto dominance and is employed to mitigate problem of favoring one fitness function.

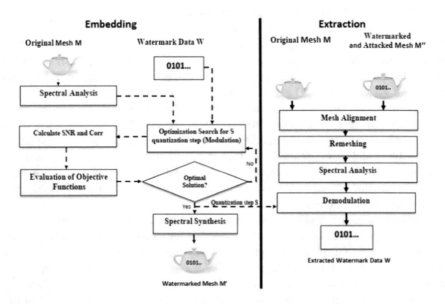

Fig. 1. Overview of the proposed system

3.2 Watermark Embedding Stage

Considering where and how to hide data satisfying various requirements of cover meshes. It can take watermark w and original mesh M as input to output watermarked mesh M'.

Mesh Normalization: 3D objects normalization as pre-processing before hiding of data makes the watermarked media robust for modifications due to affine and scaling transformations. The mass center is relocated to the origin and the mesh is scaled to fit in a unit cube.

Mesh Partitioning: The calculation of eigenvalues and eigenvectors of a huge $N \times N$ Laplacian matrix is complicated $O(N^3)$. To overcome this drawback, the partition of huge mesh into smaller sub-meshes is utilized. The embedding and extraction algorithms are then calculated for each sub-mesh.

Spectral Domain Watermarking: The mesh is a group of polygonal faces aiming to represent suitable approximation of real object. It has three components: vertices, edges, and facets. Spectral coefficients are obtained from Eigen analysis of mesh Laplacian matrix $\mathbf{L}$. Matrix of Laplacian is presented by Eq. (1):

$$L = D - A \tag{1}$$

Where D is diagonal matrix, its diagonal element $D_{ii} = d_i$ is degree of vertex and A is adjacency matrix, its components a_{ij} are defined by Eq. (2):

$$a_{ij} = \begin{cases} 1 \text{ if } \text{ vertices } i \text{ and } j \text{ are adjacent} \\ 0 \qquad\qquad\quad \text{Otherwise} \end{cases} \tag{2}$$

A group of eigenvalues-eigenvector couples is obtained from eigenvalues analysis to L, a sub-group of eigenvalue eigenvector couples in low frequency band is utilized for watermarking, because it represents good approximation for coarse shape [9]. The Eigen- analysis to Laplacian matrix L is as in Eq. (3):

$$L = B \Lambda B^T \tag{3}$$

where $B = (b_1 b_2 \ldots b_N)$ is orthogonal matrix, its columns b_i are eigenvectors which refer to as Laplacian base functions, and $\Lambda = \text{diag} \{\lambda_i : 1 \leq i \leq N\}$ is diagonal matrix of eigenvalues, arranged ascending of magnitude. So mesh vertices is expressed by Eq. (4):

$$V^T = C^T B^T \tag{4}$$

where $C = (c_1 c_2 \ldots c_N)^T$ is an $N \times 3$ coefficients matrix, (projection of vertices onto the base vectors). The coefficients are estimated using Eq. (5) and coefficient amplitudes *amp* are estimated by Eq. (6).

$$C = B^T V \tag{5}$$

$$amp = \sqrt{c_x^2 + c_y^2 + c_z^2} \tag{6}$$

Optimization Search for S: Data to be hidden is an n dimensional bit vector $w = (w_1, w_2, \ldots, w_n)$, where $w \in \{1, 0\}$. In experiments, watermark length n is 64. Each data bit is inserted 3times (watermarking capacity is 64×3 bits). Quantization step S of DM controls balance among visibility and robustness of watermark. A greater S offers better robustness, but watermark becomes more visible, and vice versa. Manual adjustment of S in testing is impractical. Hence, NSGA-II is utilized to alleviate the limitation by automatically searching and adjusting the parameters to enhance both robustness and quality in treated object.

Algorithm 1: Proposed Algorithm

Input: watermark *w*; mesh *M* with group of vertices *V* and faces *F*
Output: watermarked mesh M_w
L= Calculate Laplacian matrix (F,V) // Calculate Laplacian matrix *L*
[E_{vec} , E_{val}] = eig(L) //Calculate eigenvector and eigenvalue of (basis functions) *L*
$C = E_{vec}{}^T * V$ // Project vertices on base functions for coefficients matrix *C*

$amp = \sqrt{c_x^2 + c_y^2 + c_z^2}$ // Calculate coefficient amplitudes

amp_w = Watermark Embedding(W, s, amp)
        // *s optimized quantization step obtained from NSGA-II*
C_w=(amp_w/amp)*C // calculate modified coefficients matrix *C_w*
V_m=(E_{vec}*C_w); // calculate watermarked vertices matrix *V_m*
F_1 = *SNR* // estimate first objective function, (Transparency).
For □ *i* ∈ *[1, m]* // m no. of attacks
     Perform the attack*i* utilizing benchmark[22]
End
W' = Extract watermark() // *Extract the watermark*
F_2 = *Corr* // estimate the second objective function, the robustness.

NSGA-II is computationally effective multi-objective algorithm. It uses a selection method relaying on classes of dominance of individuals. It utilizes fast non-dominated sorting procedure and crowding distance to select and rank solution fronts. Finally, the best in diversity and non-dominance are chosen as solutions (Pareto front). The advantages of NSGA-II are: decreases the computational difficulty of non-dominated sorting, introduces elitism, and replaces sharing with crowded-comparison to reduce computations and need for user defined sharing parameter. Because of space limitations a brief details of NSGA-II, see [20, 21]. The proposed algorithm is given in Algorithm 1.

To insert one bit value w_i, amplitude is shifted so *Q* is an even value for bit value 0, or an odd value for 1. The integer quotient *Q* and remainder R are as in Eqs. (7) and (8). To make Q_i *Mod*2 $= w_i$ always hold, amplitude is modulated as in Eq. (9).

$$Q = \frac{amp}{S} \tag{7}$$

$$R = amp\ Mod\ S \tag{8}$$

where *Mod* is Modular Arithmetic, amp is coefficients amplitude and *S* is quantization step.

$$amp' = \begin{cases} Q * S + \frac{S}{2} & if\ Q\ Mod\ 2 = w_i \\ Q * S - \frac{S}{2} & if\ Q\ Mod\ 2 = 1 - w_i\ and\ R < \frac{S}{2} \\ Q * S + \frac{3S}{2} & if\ Q\ Mod\ 2 = 1 - w_i\ and\ R \geq \frac{S}{2} \end{cases} \tag{9}$$

Where *amp'* is watermarked amplitude.

To qualify watermarked meshes, amount of distortion is determined by Signal-to-Noise Ratio (SNR) that quantifies the visual differences among original and watermarked media as in Eq. (10).

$$SNR = \frac{\sum_{i=1}^{N}\left[x_i^2 + y_i^2 + z_i^2\right]}{\sum_{i=1}^{N}\left[(x_i - x_i')^2 + (y_i - y_i')^2 + (z_i - z_i')^2\right]} \tag{10}$$

where N is number of vertices, x_i, y_i and z_i are Cartesian coordinates of vertex v_i before inserting of data and x_i', y_i', z_i' are vertex coordinates after the inserting of data.

Also, the distortion among watermarked item and original one is estimated by maximum root mean square error (MRMS) or by mesh structural distortion measure (MSDM) [23]. Robustness is estimated by correlation coefficient (*Corr*) among extracted watermark w_n' and originally inserted one w_n as in Eq. (11).

$$Corr = \frac{\sum_{i=1}^{n-1}\left[(w_i' - w^{-'})(w_i - w^{-})\right]}{\sqrt{\sum_{i=1}^{n-1}\left[(w_i - w^{-'})^2\right]\sum_{i=1}^{n-1}\left[(w_i - w^{-'})^2\right]}} \tag{11}$$

Where $w^{-'}$ and w^- indicate respectively averages of watermark w_n' and w_n.

Optimization of Quantization Step (S) Using NSGA-II: Using NSGA-II, suitable *S* is found easily for various applications relying on which one is more important, transparency or robustness. For NSGA-II, the subsequent parameters are used: size of population = 50; Generations = 20; crossover possibility (Pc) = 0.9; mutation possibility (Pm) = 0.1. Objective functions are F_1 and F_2 as in Eqs. (12) and (13). F_1 for robustness and F_2 is for transparency. Target is to acquire maximization of both functions simultaneously.

$$F_1 = Corr \tag{12}$$

$$F_2 = SNR \tag{13}$$

3.3 Watermark Extraction Stage

It is done in a reverse manner to inserting procedure as in Fig. 1. It's a non-blind process (normalized original object and normalized watermarked one are required for extraction) which consist of subsequent steps:

Mesh Registration: Iterative closest point (ICP) process [14] is utilized to get two objects surface alignment. Occasionally it is necessary to get original alignment, particularly for cropping attack.

Re-meshing: A re-meshing process is adopted [14] by finding a ray through every vertex of original item in same direction of vertex normal vector. If meeting point is not established, generate a vertex of equal coordinate as its reference in original object.

Demodulation Using DM: Spectral analysis is done as that for inserting procedure and bits w_i' are extracted as in Eqs. (14) and (15). Extraction is so simple and direct.

$$Q = \lfloor amp''/s \rfloor \tag{14}$$

$$w_i' = Q\%2 \tag{15}$$

4 Experimental Results

Several simulations are performed on 3D objects to estimate proposed approach performance for robustness and imperceptibility. Table 1 displays some of tested models with number of vertices and faces. Horse and Venus are from benchmark [22]. The testes were carried out with MATLAB on FUJITSU workstation with sixteen 2.53 GHz CPU processors (Intel(R) Xeon(R) CPU E5630).

Table 1. The 3D models

Model	No. of vertices	No. of faces
Cow	2904	5804
Horse	112642	225280
Venus	100759	201514
Bunny	34835	69666
Mushroom	226	448

4.1 Watermark Invisibility

Data inserted in every model is invisible and does not produce perceptible distortion. This remark is also confirmed by objective metrics. Table 2 indicates visibility results for watermarked items as MRMS and MSDM.

Table 2. Watermark invisibility as MRMS and MSDM

Model	MRMS (10^{-3})	MSDM
Horse	0.24	0.077
Venus	0.35	0.021
Bunny	0.15	0.078

4.2 Watermark Robustness

Robustness is evaluated under various kinds of attacks [22]. Percentage of correlation among recovered watermark and original is 100% for unattacked watermarked model.

Noise Attack: Watermarked item is attacked with pseudo random noises with different amplitude (0.05%, 0.1%, or 0.3%) [22]. Noise attack doesn't disturb vertices connectivity, so, watermark is extracted although the presence of noise attack. Table 3 presents robustness for noise attack.

Table 3. Robustness and quality results for noise attack

Model	Amplitude (%)	MRMS (10^{-3})	Corr
Cow	0.05	0.15	0.95
	0.1	0.24	0.88
	0.3	0.77	0.74
Mushroom	0.05	0.13	0.79
	0.1	0.26	0.66
	0.3	0.66	0.51

Smoothing Attack: Smoothing attack doesn't disturb vertices connectivity, so data is extracted, thus improving robustness for smoothing attack.

Simplification Attack: Watermarked item is attacked by simplification attack [22] (fewer vertices and faces), vertices connectivity is affected so data cannot be extracted as well. Table 4 displays robustness for simplification attack.

Table 4. Robustness and quality results for simplification attack

Model	Reduction ratio	MRMS (10^{-3})	Corr
Cow	30	0.048	0.51
	50	0.061	0.49
	70	0.082	0.44
Mushroom	30	0.052	0.61
	50	0.059	0.59
	70	0.081	0.55

Cropping Attack: Cropping is strong attack [22] where portion of item is delete and lost, that affects the inserted data. However, recovery of watermark data rate is acceptable because of redundancy of data inserting, further, it is inserted throughout item instead of being concentrated in one region only. The cropping experiments results are displayed in Table 5.

Table 5. Robustness results for cropping attack

Model	Cropping (%)	Corr
Cow	10	0.78
	30	0.66
	50	0.54
Mushroom	10	0.85
	30	0.72
	50	0.45

Similarity Transformation: As establishing invariant space, the relative positions of vertices are unchanged, and therefore, a rotation, scaling, or translation attack does not weaken robustness.

4.3   Robustness Comparison with Other Methods

Tables 6 and 7 show quality and robustness comparison by scheme in [23] for noise and smoothing attacks measured in Corr and MRMS. Obtained results verify the great imperceptibility and great robustness of proposed system comparing to [23]. This because of using NSGA-II (Pareto front) in determining S which controls balance among robustness and visibility.

Table 6. Robustness and visibility compared with scheme in [23] for noise attack

Model	Amplitude (%)	MRMS (10^{-3})	Corr
Cow	0.05	**0.15**/0.17	**0.95**/0.95
	0.1	**0.24**/0.28	**0.88**/0.82
	0.3	**0.77**/0.80	**0.74**/0.71
Bunny	0.05	**0.13**/0.11	**0.87**/1
	0.1	**0.24**/0.22	**0.81**/0.73
	0.3	**0.66**/0.63	**0.61**/0.58

Table 7. Robustness and visibility compared with scheme in [23] to smoothing attack.

Model	Iteration	MRMS (10^{-3})	Corr
Cow	5	**0.17**/0.19	**0.77**/0.73
	10	**0.32**/0.35	**0.67**/0.65
	30	**0.69**/0.84	**0.61**/0.56
Bunny	5	**0.26**/0.37	**0.76**/0.72
	10	**0.49**/0.67	**0.71**/0.55
	30	**0.91**/1.62	**0.69**/0.52

5   Conclusion and Future Work

A robust and non-blind mesh watermarking approach is proposed; watermark hiding is attained by modulation of spectral coefficients using DM. The proposed algorithm is relying on NSGA-II for improving imperceptibility of watermarked mesh and robustness for watermark through defining optimal quantization step. The proposed algorithm has the advantage of utilizing NSGA-II which rely on Pareto dominance and is employed to mitigate problem of favoring one goal in detriment of others. Using Pareto-front, proper S is found easily for various applications relying on which one is more important, transparency or robustness. However, subdivision (add novel vertices and edges) and simplification (fewer vertices and faces) attacks disturb connectivity of vertices, so causing dangerous interference in watermark detection. Thus, performance of proposed system is limited in subdivision and simplification attacks. Further improvements associated to these attacks will be considered in future work.

References

1. Tamana, S., Deshmukh, R., Jadhavpatil, V.: Optimization of blind 3D model watermarking using wavelets and DCT. In: Proceedings of IEEE International Conference on Intelligent Systems, Modelling and Simulation, Thailand, pp. 270–275 (2013)
2. Tamane, S., Deshmukh, R.: Watermarking 3D surface models into 3D surface models based on ANFIS. Adv. Comput. **2**(3), 29–34 (2012)
3. Liu, J., Wang, Y., Li, Y., Liu, R., Chen, J.: A robust and blind 3D watermarking algorithm using multiresolution adaptive parameterization of surface. Neurocomputing **237**(3), 304–315 (2017)
4. Wang, J., Feng, J., Miao, Y.: A robust confirmable watermarking algorithm For 3D mesh based on manifold harmonics analysis. Vis. Comput. **28**(11), 1049–1062 (2012)
5. Elzein, O., Elbakrawy, L., Ghali, N.: A robust 3D mesh watermarking algorithm utilizing fuzzy C-means clustering. Future Comput. Inform. J. **2**(2), 148–156 (2017)
6. Ciro, G., Dugardin, F., Yalaoui, F., Kelly, R.: A NSGA-II and NSGA-III comparison for solving an open shop scheduling problem with resource constraints. In: Proceedings of the Conference on Manufacturing Modelling, Management and Control, France, vol. 49, no. 12, pp. 1272–1277 (2016).
7. Bors, A., Luo, M.: Optimized 3D watermarking for minimal surface distortion. IEEE Trans. Image Process. **22**(5), 1822–1835 (2013)
8. Feng, X., Zhang, W., Liu, Y.: Double watermarks of 3D mesh model based on feature segmentation and redundancy information. Multimed. Tools Appl. **68**(3), 497–515 (2014)
9. Mouhamed, M., Solima, M., Darwish, A., Hassanien, A.: Robust and blind watermark to protect 3D mesh models against connectivity attacks. In: International Proceedings on Intelligent Computing and Information Systems, Egypt, pp. 23–29 (2017)
10. Li, S., Ni, R., Zhao, Y.: A 3D mesh watermarking based on improved vertex grouping and piecewise mapping function. J. Inform. Hiding Multimed. Signal Process. **8**(1), 97–108 (2017)
11. Medimegh, N., Belaid, S., Atri, M., Werghi, N.: A novel robust statistical watermarking of 3D meshes. In: International Workshop on Representations, Analysis and Recognition of Shape and Motion from Imaging Data, Cham, pp. 27–38 (2016)
12. Son, J., Kim, D., Choi, H., Jang, H., Choi, S.: Perceptual 3D watermarking using mesh saliency. In: International Proceedings on Information Science and Applications, Singapore, pp. 315–322 (2017)
13. Molaei, A., Ebrahimnezhad, H., Sedaaghi, M.: Robust and blind 3D mesh watermarking in spatial domain based on faces categorization and sorting. 3D Res. **7**(2), 1–18 (2016)
14. Anbarasi, J., Narendra, M.: Robust watermarking scheme using weber law for 3D mesh models. Imaging Sci. J. **65**(7), 409–417 (2017)
15. Soliman, M., Hassanien, A.E, Onsi, H.: A robust 3D mesh watermarking approach using genetic algorithms, vol. 323, no. 2, pp. 731–741 (2015)
16. Hosam, O.: Side-informed image watermarking scheme based on dither modulation in the frequency domain. Open Signal Process. J. **5**(1), 1–6 (2013)
17. Hossam, O., Alraddadi, A.: A novel image watermarking scheme based on dither modulation. J. Innov. Eng. **1**(1), 1–10 (2013)

18. Ranade, S., Dhawan, D.: Multi-bit watermark embedding using dither modulation. Int. J. Comput. Sci. Technol. **8491**(2), 248–253 (2015)
19. Vellasques, E., Granger, E., Sabourin, R.: Intelligent watermarking systems: a survey. In: Pattern Recognition Computer Vision, pp. 687–724 (2010)
20. Deb, K., Pratap, A., Agarwal, S., Meyarivan, T.: A fast and elitist multiobjective genetic algorithm: NSGA-II. IEEE Trans. Evol. Comput. **6**(2), 182–194 (2002)
21. Emmerich, M., Deutz, A.: A tutorial on multiobjective optimization: fundamentals and evolutionary methods. Nat. Comput. **17**(3), 585–609 (2018)
22. Wang, K., Lavoué, G., Denis, F., Baskurt, A., He, X.: A Benchmark for 3D mesh watermarking. In: Proceedings of International Conference on Shape Modeling, France, vol. 10, pp. 231–235 (2010)
23. Amar, Y., Trabelsi, I., Dey, N., Bouhlel, M.: Euclidean distance distortion based robust and blind mesh watermarking. Int. J. Interact. Multimed. Artif. Intell. **4**(2), 46–51 (2016)

Predicting Student Retention Among a Homogeneous Population Using Data Mining

Ghazala Bilquise[1]([email]) , Sherief Abdallah[2] , and Thaeer Kobbaey[1]

[1] Higher Colleges of Technology, Dubai, UAE
{gbilquise,thaeer.kobbaey}@hct.ac.ae
[2] British University in Dubai, Dubai, UAE
sherief.abdallah@buid.ac.ae

Abstract. Student retention is one of the biggest challenges facing academic institutions worldwide as it does not only affects the student negatively but also hinders institutional quality and reputation. In this paper, we use classification techniques to predict retention at an academic institution based in the Middle East. Our study relies solely on pre-college and college performance data available in the institutional database to predict dropouts at an early stage. We built a predictive model to study retention until graduation and compare the performance of five standard algorithms and five ensemble algorithms in effectively predicting dropouts as early as possible. The results showed that ensemble predictors outperform standard classification algorithms by effectively predicting dropouts using enrollment data with an Area Under the Curve (AUC) of 88.4%.

Keywords: Data mining · Retention · Dropout · Attrition · Higher education

1 Introduction

In today's knowledge society, education is the key to creativity and innovation, which are essential elements of progress. Despite the enormous socio-economic benefits of earning a college degree, nearly 30% of students leave college without earning any credential [1]. This situation is prevalent worldwide [4] with the Middle East being no exception [8].

Previous works have shown that the decision to drop out mostly occurs in the first year of college as students struggle to cope with the challenges of an academic environment and transition from high school [9]. Intervention strategies within the first year of studies can improve retention rates by up to 50% [11]. Therefore, an early identification of students at risk of premature departure enables the institution to target their resources to benefit students who need it the most.

© Springer Nature Switzerland AG 2020
A. E. Hassanien et al. (Eds.): AISI 2019, AISC 1058, pp. 35–46, 2020.
https://doi.org/10.1007/978-3-030-31129-2_4

A promising avenue to address the challenge of early dropouts is the use of data mining and machine-learning techniques. While some studies have used Educational Data Mining (EDM) techniques to study retention, this subject lacks critical investigation in the Middle East.

Our study goes beyond previous retention works using EDM in several ways. First, while other studies have differentiated students based on culture, race, ethnicity and more [14], our study is based on a homogeneous population with similar social and cultural background.

Second, our study utilizes enrollment data and longitudinal data of two consecutive terms to predict retention until graduation, unlike other papers that focus on persistence by utilizing student data in one term only [9].

Third, we use a larger dataset than most studies and focus on student retention across multiple disciplines rather than a single discipline [7,13]. We also apply balancing techniques to get more reliable and unbiased results.

2   Related Work

Several studies have undertaken the task of investigating student success in undergraduate programs by predicting student performance [3,10,12] or ability to progress from one term to another [9,15]. On the other hand, only a few studies have focused on retention until graduation in higher education [7,13,14]. Moreover, to the best of our knowledge, no such study has been conducted till date in the Middle East.

Student demographic data and prior performance scores have been used frequently to predict their success. Weekly performance grades and demographic data were used to predict early failures in a first year programming course at a Brazilian University [6]. Potential dropouts were identified in an introductory engineering course using demographic, academic and engagement data [2]. Similarly, the work in [10] evaluated the performance in a high-impact engineering course to predict dropouts using cumulative GPA, pre-requisite course grades and course work assessment grades.

An EDM approach was used to predict the successful graduation of first-year students in an Electrical Engineering program in a Netherlands-based University [7]. The study utilized 648 student records collected over a ten year period. The data consisted of pre-university and university performance scores and achieved a maximum accuracy of 80%.

Student retention was investigated in a System Engineering undergraduate program using 802 student records collected from the year 2004 to 2010 [13]. The dataset of the study comprised of attributes such as admission data, course grades, and financial aid per term. The study achieved an AUC score of 94%. Although the reported performance is excellent, the study is based on a single discipline and the dataset size is too small for the results to be generalized.

Factors of retention leading to graduation were investigated using Logistic Regression and Neural Networks in a US-based University using 7,293 records [14]. Data was collected for a ten year period and consisted of pre-college and first-term college data.

3   Methodology

Our study aims to answer the following three research questions:

Research Question 1: Can machine-learning algorithms effectively predict retention/dropouts in a homogeneous group of students using performance data?

To answer this research question, we generate predictive models using five standard and five ensemble classification algorithms. Due to the stochastic nature of machine-learning algorithms, predictive models cannot perform with 100% accuracy. Therefore, an effective model is one with a reasonable performance in a given domain. The reviewed literature on retention showed that predictive models have achieved accuracies ranging from 72% to 77% [14]. Therefore, we evaluate the predictions of our models against a minimum threshold requirement of 75% AUC to be labeled as effective.

Research Question 2: How early can we predict potential dropouts using machine learning?

To answer this research question, we use three datasets with pre-college, college term 1 and college term 2 features. The performance of the algorithms on each dataset is compared using a pair-wise t-test to determine if the difference is significant or merely due to chance.

Research Question 3: Which attributes are the top predictors of retention?

To answer this research question, we analyze the relevance of each attribute to the class label. We begin by calculating the feature weight of each attribute using information gain, gain ratio, gini index, correlation and chi-squared statistic. We then examine the top predictors of retention identified by the Decision Tree model. We also explore the weights assigned to each attribute by the predictive models of the Support Vector Machine (SVM), Logistic Regression, Random Forest and Gradient Boosted Trees algorithms. We summarize and report the topmost predictive features for each dataset by the frequency of its appearance.

3.1   The Dataset

The enrollment dataset consists of 22,000 enrollment records for the academic year 2011–2013, inclusive, for each term (excluding the summer term since new enrollments do not take place in summer). The academic year 2013 is chosen as a cut off period to allow the nominal degree completion time which is a period of 6 years.

Each enrollment record is described by 84 features, which includes student demographic data, personal data, High School performance scores, IELTS performance scores, English and Math placement scores and the current term GPA. From this dataset we filter out the newly enrolled students in each term, which results in 4,056 newly enrolled student records and perform the initial preparation tasks in MS Excel and MS Access as shown in Fig. 1.

In addition to the enrollment records we also extract the dataset of graduation records of all students who have graduated till date. We then integrate

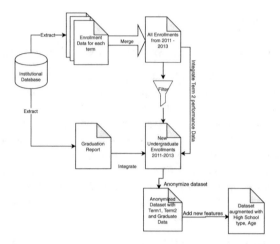

Fig. 1. Dataset extraction and preparation tasks performed in MS Excel and MS Access

the graduation records with the enrollment records and generate a class label to identify new enrollments that graduated or dropped out in the program of study that had enrolled in.

Next we extract the term 2 GPA by joining enrollment records of subsequent terms. Nearly 30 attributes pertaining to student personal data are removed to anonymize the dataset. These include details such as Student ID, name, contact numbers, parent data, sponsor data, advisor details, and more.

We also impute two new features called High School Type (public or private) and Age at enrollment. High school type is generated by crosslisting the high school name attribute against a list of all high schools in the city downloaded from a public website and age is computed using the date of birth feature.

3.2   Data Preprocessing

The quality of the dataset determines the performance of the data mining algorithms. Therefore, pre-processing is a crucial step that is needed to ensure the success of the algorithms as well as the validity of the results. Three main preprocessing tasks are performed in Rapid Miner, which include handling missing values, balancing the dataset, and feature selection. Figure 2 shows the tasks performed with Rapid Miner to pre-process the data, generate, apply and evaluate the predictive models.

Missing Values: First, any feature that has more than 20% of missing values is excluded from the study. This results in removal of around 20 attributes. Second, all other attributes are imputed using the K Nearest Neighbor (k-NN) algorithm with the value of k set to 3.

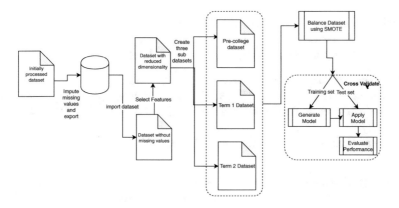

Fig. 2. Pre-processing tasks performed in Rapid Miner

Feature Selection: We further reduce the features by removing irrelevant attributes such as enrollment status, grading system, campus codes and more and select only those features that are related to student demographic data, pre-college performance and college performance. Moreover, any feature that is used to generate new attributes, such as date of birth and high school name, is excluded from the study. The resultant dataset consists of the 20 features shown in Table 1.

Table 1. Dataset features

Type	Attribute	Data type
Demographic data	Age	Numeric
	Gender	Binary
	HS Type	Binary
	HS Stream	Nominal
Pre-college performance	HS Avg	Numeric
	HS Math	Numeric
	HS English	Numeric
	HS Arabic	Numeric
	IELTS Band	Numeric
	IELTS Listening	Numeric
	IELTS Writing	Numeric
	IELTS Speaking	Numeric
	IELTS Reading	Numeric
	English Placement	Numeric
	Math Placement	Numeric
College	Term	Numeric
	Program of Study	Nominal
	Term 1 GPA	Numeric
	Term 2 GPA	Numeric
Class label	Graduated	Binary

Creating Three Sub-datasets: The pre-processed dataset is divided into three distinct datasets to determine at what stage we can effectively predict students who are likely to drop out. The three datasets are described below:

1. **Pre-college Dataset:** Consists of demographic data, pre-college performance data (High School, IELTS, English and Math placement test scores)
2. **College Term 1 Dataset:** Pre-college dataset + Term 1 GPA + program of study
3. **College Term 2 Dataset:** College Dataset 1 + Term 2 GPA

Balancing Datasets: Balancing the dataset is essential to reduce the bias caused by the majority class and improve classification performance [16]. While there are many techniques of balancing a dataset, specifically, in this study we have chosen the Synthetic Minority Oversampling Technique (SMOTE) proposed by [5]. The SMOTE algorithm is a popular technique used in several studies [6].

Training and Validation: We employ the ten-fold cross-validation with stratified sampling to split the original dataset into ten subsets while preserving the ratio of the minority and majority samples. The machine-learning algorithms are trained using nine subsets, while testing is performed using the remaining subset. This process is repeated ten times by holding out another subset and training with the remaining nine. The final performance is reported as an average of all the iterations.

3.3   Predictive Modelling

Our study uses five standard classification algorithms, namely, Decision Tree, Naïve Bayes, Logistic Regression, SVM and Deep Learning. In addition, we also employ five ensemble algorithms – Random Forest, Voting, Bagging, AdaBoost and Gradient Boosted Trees to achieve a robust prediction. The algorithms classify each instance of the testing data in into two classes Y (graduated) or N (did not graduate).

To ensure that the algorithms perform effectively for our requirement, we use five evaluation metrics to measure and compare the performance of our predictive models, namely – Accuracy, True Negative Rate (TNR), True Positive Rate (TPR), AUC of the Receiver Operator Characteristic (ROC) curve and F-Measure. The TNR measures the percentage of correctly classified dropouts, where as the TPR measures the percentage of accurately classified graduates. The AUC of the ROC curves focuses on correctly identifying the graduates (TPR) while lowering the misclassification of the dropouts (TNR).

4   Results and Discussion

In this section, we discuss our experimental findings and answer each research question that was presented in Sect. 3.

4.1   Research Questions

Question 1 - Can Machine-Learning Algorithms Effectively Predict Retention/Dropouts in a Homogeneous Group of Students Using Performance Data? To answer this research question, we generate predictive models on three datasets using five standard classification algorithms as well as five ensemble algorithms. The performance of each model is evaluated to determine if the algorithm performs effectively, with an AUC of at least 75%, as described in Sect. 3.

Standard Classifiers. Table 2 shows the result of the standard classifiers across all three datasets.

None of the standard classifiers are able to meet the threshold requirement of AUC 75% when the pre-college dataset is used. However, when the dataset is enhanced with college term 1 performance, all the standard algorithms perform effectively with an AUC score above the threshold requirement ranging from 77% to 84.8%. The Logistic Regression and SVM algorithms perform the best, achieving accuracies of up to 77.3%. Both algorithms predict graduates and dropouts equally well with TPR of 79.2% and TNR of 75.4%. The use of College Term 2 data further enhances the performance of the algorithms with an AUC score of above 80%. Again, the Logistic Regression and SVM algorithms perform the best with an AUC score of 86%, and overall accuracy of 77.9%.

Ensemble Classifiers. Table 3 shows the result of the ensemble classifiers across all three datasets.

Overall, the ensemble algorithms perform better than the standard algorithms when the pre-college dataset is used. Except for Voting and AdaBoost all ensemble algorithms are able to meet the threshold requirement of AUC 75%.

The Gradient Boosted Trees classifier consistently performs the best across all the three datasets. It achieves an AUC score of upto 92.2% with an accuracy of 84.7%. All other classifiers have also improved in their performance with the lowest accuracy of 78.9% achieved by the Voting algorithm. AdaBoost and Bagging algorithms are the best at predicting dropouts at 88.4%.

Question 2 - How Early Can We Predict Potential Dropouts Using Machine Learning? We answer this research question by examining the performance of each classifier across all the datasets. Figures 3 and 4 show the performance of the Standard and Ensemble Algorithms respectively.

The predictive capabilities of all the standard algorithms increases when the College Term 1 and College Term 2 datasets are used. The results indicate that although pre-college data can provide a good initial prediction of students likely to dropout, the Term 1 and Term 2 performance data can produce more effective and accurate predictions.

Table 2. Standard classifier results

Model	Accuracy	TNR	TPR	AUC	F-Measure
Pre-college dataset					
Decision Tree	**71.59%**	67.17%	76.02%	73.80%	**72.80%**
Naïve Bayes	67.22%	**77.10%**	57.34%	73.90%	63.60%
Logistic Regression	67.92%	70.22%	65.63%	**74.40%**	67.16%
SVM	67.09%	72.64%	61.53%	73.30%	65.13%
Deep learning	62.80%	46.40%	**79.19%**	70.16%	68.02%
College Term 1 Dataset					
Decision Tree	75.77%	70.54%	**81.00%**	79.40%	76.96%
Naïve Bayes	73.97%	**77.49%**	70.45%	81.80%	73.01%
Logistic Regression	77.20%	75.82%	78.57%	**84.80%**	77.51%
SVM	**77.31%**	75.46%	79.23%	84.70%	**77.77%**
Deep Learning	69.68%	61.14%	78.21%	77.50%	72.07%
College Term 2 Dataset					
Decision Tree	**80.41%**	**88.40%**	72.41%	81.70%	**78.70%**
Naïve Bayes	75.23%	76.93%	73.53%	83.33%	74.81%
Logistic Regression	77.42%	76.51%	78.34%	**86.00%**	77.62%
SVM	77.95%	76.38%	79.52%	85.80%	78.27%
Deep Learning	75.20%	67.01%	**83.39%**	82.90%	77.08%

Table 3. Ensembler classifier results

Model	Accuracy	TNR	TPR	AUC	F-Measure
Pre-college Dataset					
Random Forest	70.35%	73.75%	66.94%	77.10%	69.21%
Gradient Boosted Trees	**79.31%**	72.35%	**86.27%**	**88.40%**	**80.66%**
Voting	70.54%	**75.00%**	66.09%	71.90%	69.15%
AdaBoost	71.23%	66.74%	75.72%	71.20%	72.47%
Bagging	73.02%	69.10%	76.87%	79.50%	74.02%
College Term 1 Dataset					
Random Forest	77.79%	73.00%	82.57%	85.00%	78.84%
Gradient Boosted Trees	**82.10%**	**79.59%**	**86.04%**	**90.10%**	**83.35%**
Voting	77.26%	74.93%	79.59%	77.70%	77.77%
AdaBoost	75.62%	70.48%	80.77%	78.60%	76.80%
Bagging	76.69%	71.63%	81.75%	83.40%	77.79%
College Term 2 Dataset					
Random Forest	81.16%	82.67%	79.65%	88.80%	80.88%
Gradient Boosted Trees	**84.75%**	83.32%	**86.17%**	**92.20%**	**84.97%**
Voting	78.96%	80.83%	77.10%	81.60%	78.56%
AdaBoost	80.21%	88.40%	72.02%	83.90%	78.44%
Bagging	82.01%	**88.47%**	75.56%	89.10%	80.76%

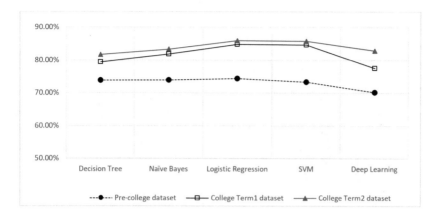

Fig. 3. Standard classifier AUC performance

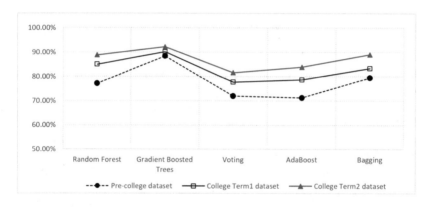

Fig. 4. Ensemble classifier AUC performance

The Logistic Regression and SVM classifiers have shown the most increase in performance among the standard algorithms (up to 11% AUC), while the Decision Trees' performance has improved by 5.6% only. However, among the ensemble algorithms the increase is not very high ranging from 2% to 7%. The Gradient Boosted Trees classifier has shown very little increase proving to be a robust and reliable predictor for all the three datasets.

A pairwise t-test of significance to test the difference in performance between the Pre-college and College Term 1 dataset reveals that the increase in performance is indeed significant ($\alpha < 0.001$, for each algorithm). Although the difference in performance between the College Term 1 and College Term 2 data set is not very large, yet a pair-wise t-test shows that this increase is also significant.

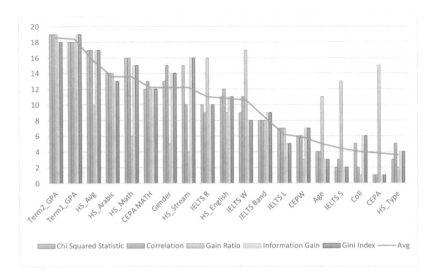

Fig. 5. Feature rankings

Research Question 3: Which Attributes Are the Top Predictors of Retention? To answer this research question, we first analyze the attributes by feature weights to study its relevance with respect to the class label. We begin by calculating the feature weight of each attribute using five feature weight algorithms, namely, information gain, gain ratio, gini index, correlation and chi-squared statistic. We rank the attributes by lowest weight to highest.

Figure 5 shows the feature ranking of all features using the five feature weight algorithms.

Term1 GPA and Term2 GPA are consistently picked as the most significant predictors by all feature weight algorithms, followed by High School average and then High School Math. Some of the least relevant attributes are Age, IELTS Speaking score, College, Placement and High School Type which are ranked the lowest by most of the feature weight algorithms.

We also use the Decision Tree algorithm for its interpretability and ability to identify the top predictor of retention. The root node of the decision tree model shows the most influential attribute in classifying dropouts. The Decision Tree model reveals that, the High School Average, High School Stream and IELTS band are the top predictors of retention when the pre-college data set is used.

We also identify the top predictive attributes using the feature weights assigned by the SVM, Logistic regression, Gradient Boosted Trees and Random Forest algorithm across each dataset. A high weight indicates a higher relevance of the attribute to the prediction. All algorithms also assign highest weight to High School Average and High School Stream and in addition also identify age as the top predictor of retention.

Interestingly, when the College Term 1 Dataset is used, the pre-enrollment features do not play an important role in predicting dropouts. It is the Term

1 GPA and Program of Study that are the top predictors of graduation. When College Term 2 dataset is used, then Term 2 GPA as well as Term 1 GPA are the top predictors of graduation.

5 Conclusion

In this paper, we evaluated ten machine learning algorithms for classifying a student as a successful graduate or a dropout. The ten algorithms were applied to three datasets of student performance at various stages of their academic journey form enrollment to end of the second term.

Overall, the ensemble algorithms have performed better than standard algorithms across all the three datasets, thus proving to be more reliable and better at handling misclassifications. Among these, the Gradient Boosted Trees is the most effective algorithm performing equally well on all datasets, proving to be a robust and reliable predictor. It achieves an AUC score of 88.4% for the pre-college dataset, 90.1% for College Term1 dataset and 92.2% for College Term 2 dataset.

Our research predicts dropouts at a very early stage using pre-enrollment data with an accuracy of 79.31% and AUC of 88.4% using the Gradient Boosted Trees algorithm. Our results will enable the academic institution to start remedial support at an early stage from the first term onwards by directing resources to where they are required the most.

In addition, our study also identified the top predictors of retention at each stage starting from enrollment to end of term 1. High School Average and IELTS Band were found to be the top predictors of retention when a student joins the college. This result indicates that students who did well in High School and possess a good level of English have a better chance of meeting the academic demands of college.

Interestingly, when the College Term 1 dataset is used, the pre-enrollment features do not play an important role in predicting dropouts. It is the Term 1 GPA and Program of Study that are the top predictors of graduation.

The results also show that the program of study is an important predictive factor in determining dropouts. Hence it can be said that if students do not choose their program of study wisely, it is likely they will eventually discontinue their studies. Students often choose their discipline of study based on their interest or prospective career choices without aligning it with their academic capabilities. Therefore, this often leads to academic struggle and abandonment of the studies. It is, therefore, essential for the academic institution to advise students in wisely choosing their programs of study to ensure success.

References

1. NSCRC - National Student Clearinghouse Research Center. https:// nscresearchcenter.org/snapshotreport33-first-year-persistence-and-retention/. Accessed 15 Feb 2019

2. Aguiar, E., Chawla, N.V., Brockman, J., Ambrose, G.A., Goodrich, V.: Engagement vs performance: using electronic portfolios to predict first semester engineering student retention. In: Proceedings of the Fourth International Conference on Learning Analytics And Knowledge, pp. 103–112. ACM (2014)
3. Asif, R., Merceron, A., Ali, S.A., Haider, N.G.: Analyzing undergraduate students' performance using educational data mining. Comput. Educ. **113**, 177–194 (2017)
4. Chalaris, M., Gritzalis, S., Maragoudakis, M., Sgouropoulou, C., Lykeridou, K.: Examining students graduation issues using data mining techniques-the case of TEI of athens. In: AIP Conference Proceedings, vol. 1644, pp. 255–262. AIP (2015)
5. Chawla, N.V., Bowyer, K.W., Hall, L.O., Kegelmeyer, W.P.: SMOTE: synthetic minority over-sampling technique. J. Artif. Intell. Res. **16**, 321–357 (2002)
6. Costa, E.B., Fonseca, B., Santana, M.A., de Araújo, F.F., Rego, J.: Evaluating the effectiveness of educational data mining techniques for early prediction of students' academic failure in introductory programming courses. Comput. Hum. Behav. **73**, 247–256 (2017)
7. Dekker, G.W., Pechenizkiy, M., Vleeshouwers, J.M.: Predicting students drop out: a case study. In: International Working Group on Educational Data Mining (2009)
8. GulfNews. https://www.khaleejtimes.com/nation/new-ratings-system-for-uae-universities-education-quality. Accessed 5 Feb 2019
9. Hoffait, A.-S., Schyns, M.: Early detection of university students with potential difficulties. Decis. Support Syst. **101**, 1–11 (2017)
10. Huang, S., Fang, N.: Predicting student academic performance in an engineering dynamics course: a comparison of four types of predictive mathematical models. Comput. Educ. **61**, 133–145 (2013)
11. Levitz, R.S., Noel, L., Richter, B.J.: Strategic moves for retention success. New Direct. High. Educ. **1999**(108), 31–49 (1999)
12. Miguéis, V.L., Freitas, A., Garcia, P.J., Silva, A.: Early segmentation of students according to their academic performance: a predictive modelling approach. Decis. Support Syst. **115**, 36–51 (2018)
13. Perez, B., Castellanos, C., Correal, D.: Applying data mining techniques to predict student dropout: a case study. In: 2018 IEEE 1st Colombian Conference on Applications in Computational Intelligence (ColCACI), pp. 1–6. IEEE (2018)
14. Raju, D., Schumacker, R.: Exploring student characteristics of retention that lead to graduation in higher education using data mining models. J. Coll. Stud. Retent.: Res. Theory Pract. **16**(4), 563–591 (2015)
15. Rubiano, S.M.M., Garcia, J.A.D.: Formulation of a predictive model for academic performance based on students' academic and demographic data. In: 2015 IEEE Frontiers in Education Conference (FIE), pp. 1–7. IEEE (2015)
16. Thammasiri, D., Delen, D., Meesad, P., Kasap, N.: A critical assessment of imbalanced class distribution problem: the case of predicting freshmen student attrition. Expert Syst. Appl. **41**(2), 321–330 (2014)

The Significance of Artificial Intelligence in Arabian Horses Identification System

Aya Salama$^{(\boxtimes)}$, Aboul Ellah Hassanien, and Aly Fahmy

Faculty of Computers and Artificial Intelligence, Cairo University, Giza, Egypt
Aya.salama@aucegypt.edu, aboitcairo@gmail.com,
aly.fahmy@gmail.com,
http://www.egyptscience.net

Abstract. Over the past decade, increasing attention has been drawn to the application of artificial intelligence (AI) in agriculture market and veterinary-sciences. This paper introduces a new approach for Arabian-horses identification system using their iris significant features and artificial intelligence technology. These features are the retina cogina and the pupil shape. Arabian horses are identified here by using the segmented retina cogina with the pupil region out of iris images using color based k-means clustering. The identification system has been done using different AI algorithms including Artificial Neural Networks (ANN), and Support Vector Machines (SVM) with Scale Invariant Feature Transform (SIFT) and Speeded Up Robust Features (SURF) were tested in the experiments. Moreover, deep learning approach of deep convolutional neural network (CNN) was also implemented. However, segmented features images were given as input to deep convolutional neural network. The experiments were conducted on collected data set of horses' eyes images of 145 Arabian horses. The results of the conducted experiments shows that SVM with SIFT features achieved an accuracy of 97.6%.

Keywords: SURF · SIFT · SVM · ANN · Deep learning · CNN · Artificial intelligence

1 Introduction

Artificial intelligence technology is a multidisciplinary subject, which covers a wide range of contents and intersections with many scientific fields. As it developed, many of the most innovative theories and techniques emerged. At present, the research and application trends of artificial intelligence technology are cognitive intelligence, behavioral intelligence, intelligence thinking, and intelligence learning [14].

Arabian horse recognition is vital in large groups as it is required for management of their breeding programs, sale, and disease detection. Traditional ways of ear tags, tattoos and ear branding cannot be used for Arabian horses as they are precious animals. One Arabian horse can cost millions of dollars for their exceptional beauty. Accordingly,

A. Salama and A. E. Hassanien—Scientific Research Group in Egypt (SRGE).

© Springer Nature Switzerland AG 2020
A. E. Hassanien et al. (Eds.): AISI 2019, AISC 1058, pp. 47–57, 2020.
https://doi.org/10.1007/978-3-030-31129-2_5

traditional methods are actually considered as scars that devaluates the beauty of them. These traditional methods are also vulnerable to loss and manipulation [1, 2].

For Horse identification using iris biometrics, Suzaki et al. used Gabor filters for iris segmentation and hamming distance to classify images [3]. It was a complicated approach that needed manual labelling and special tools. Video tapes were used in their proposed approach. However, the goal of this paper is to have an automatic identification system that can be used as a mobile application by just capturing one image for horse iris. As a result, it can be used in races to rapidly identify the ownership of horses. The iris biometric has already proved its efficiency in automatic identification systems [4]. For this reason, iris features were used in this work for Arabian-horses identification because horses' iris has stable features that won't change over time. Then, it is found that iris is not circular and has a unique component called corpora nigra which is above the pupil [5]. It is the intersection between pupil and iris. Corpora nigra is enough to differentiate an Arabian horse from others. The segmentation approach in [5] segmented the circular region of pupil as it contains the retina cogina. Furthermore, the shape and size of pupil are unique. Color based k-means clustering was used to extract these significant features. This segmentation approach achieved high Jaccard coefficient of value 85%. These segmented images were used in this paper for Arabian horse's recognition. The images of these segmented features were used instead of whole images for classification. For best accuracy results achievement, multiple approaches were tested. CNN was tested as it already proved its efficiency in similar approaches [6]. SURF features were extracted to be used in many animal identification researches [7]. Thus, SVM and ANN with SURF and SIFT features were also tested on both segmented and unsegmented images.

The rest of the paper is organized as follows. Section 2 describes the data set on which the experiments were conducted. Then, Sect. 2 highlights the theoretical background of important methodologies adopted in this work. The proposed approach is then explained in details in Sect. 3. In Sect. 4, experimental results and evaluation are discussed. Finally, conclusions and future work are discussed in Sect. 5.

2   Materials and Methods

2.1   Data Set Collection and Camera Setting

The collected data/images were collected under different transformations: illumination, rotation, quality levels, and image partiality. In 2017, we were able to collect images for horses from Zahraa farm and Obour farm. We took pictures for 145 Arabian horses, 70 females and 75 males. For each horse, 3 images were taken for the left eye and 3 images for the right eye and one front image for the horse. Therefore, total number of images was 780. When we started collecting the images, we noticed that iris of horses eyes is different from other eyes as shown in Fig. 1. It is not circular and it contains what is called corpora nigra which is unique for each horse and not repeated for any other. Corpora nigra was noticed just above the pupil that is considered to be the most significant feature for the iris. The purpose of the corpora nigra is to shade the pupil from glare.

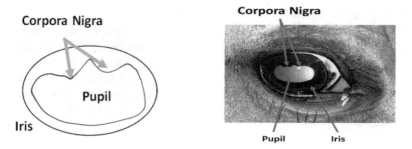

Fig. 1. Corpora nigra

Fig. 2. Images taken by professional camera with no flash

Fig. 3. Images taken by mobile camera with flash

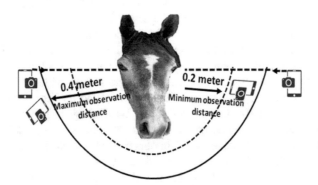

Fig. 4. Settings of the mobile camera

Images were first taken by the canon camera; iris was obvious but the corpora nigra wasn't clear as shown in Fig. 2. However, using high intensity light supported in Iphone 7 enabled us to capture important features as shown in Fig. 3. The flash light showed clearly pupil's shapes and sizes with all its significant unique features. Using flash light also enables classification at night. Moreover, the images were taken at different distances as illustrated in the different settings of the camera shown in Fig. 4.

2.2   SIFT Features

A SIFT feature is simply the circular region in any given image with specific orientation [8]. SIFT detectors extract features and SIFT descriptors compute their appearance. These descriptors are 3-D spatial histograms so that each pixel is represented by three-dimensional gradient vector. SIFT algorithm usually consists of basically four stages; Scale-space Extrema Detection, Key point Localization, Orientation Assignment, and Key Point Descriptors definition. In Scale-space Extrema Detection, Difference of Gaussians is used to detect features at different scales. Difference of Gaussians is an approximation of Laplacian of Gaussian. Gaussian pyramid is obtained by Gaussian difference between blurred images. As a result, local extrema is found by comparing images through different scales and spaces in the Gaussian pyramid. Then these points are refined by Taylor series expansion in the next step of key point localization for accurate detection of edges. Then in the orientation assignment stage, gradient orientation is calculated to all key points. At last, key points descriptors are defined as 128 neighbouring boxes with orientation descriptors and nearest neighbours are matched.

2.3   SURF Features

SIFT and SURF approaches are considered as similar methodologies of features extraction [8]. However, SURF approximates the hessian metric for detection. In DOG, difference of successive Gaussian-blurred images is used so that features are scale invariant. SURF relies on integral images to reduce processing time. Integral images allow usage of box type convolution filters to rapid the computation. SURF also applies the orientation assignment in both the horizontal and vertical directions. The orientation is calculated by adding all the responses under the sliding window to get the most dominant one. Moreover, processing in SURF is speeded up because rotation invariance is unnecessary. SURF also provides only 64 descriptors using integral images which speeds up the computation time and results in more unique features. Descriptors are produced from 20 * 20 boxes of neighbours surrounding each of the extracted key points which are divided by turn in to 4 * 4 sub blocks. SURF descriptors are also represented as a vector $v = \left(\sum d_x, \sum d_y, \sum |d_x|, \sum |d_y| \right)$. Although, lower number of dimensions speeds up the processing; However, there is an extended version of 128 descriptors for more accurate features extraction.

2.4   Artificial Neural Network (ANN)

Neural Network simulates architecture of the human brain as it consists of network of neurons that receives weighted input to compute output. ANN is basically one input layer, one output layer and number of hidden layers in between. The input layer is the first layer in the ANN which receives the given input to be processed by the hidden layers. Each neuron in the hidden layers is fully connected to all the neurons in the previous layer. Then the hidden layers process the weighted inputs using activation function. The hidden layers propagate the output to be transferred to the output layer. The weights represent the relevant importance of any given input and they keep updated to reduce the classification error. Training set of sufficient amount of labelled data is usually needed for ANN to learn the given classification problem. For non-linearly separable problems, multilayer perceptron can be used. Backpropagation is usually used to train multilayer perceptron neural networks. Back propagation consists of entering the training pattern to the input layer. Then, the error is back propagated to the input layer to adjust the difference between the classified output and the actual output. This process keeps on till minimum error is reached.

2.5   Convolutional Neural Network (CNN)

CNN is one form of artificial neural networks that are mainly used to classify images. CNN takes the entire images as input for features extraction [9–11]. Moreover, CNN consists of one or more of the following layers: Convolutional Layer, Pooling Layer, RELU layer and Fully-Connected Layer with an input layer just as the regular ANN. Convolutional Layer is the core layer of CNN as it contains filters to convolve images for features extraction. The convolutional layers contain the different filters that does that heavy job of learning on both forward pass and backward pass. It also contains pooling layer for down sampling by extracting distinctive features to prevent overfitting. Pooling layers also reduce the required computational time. In addition, RELU layer is responsible for running the activation function. Increasing the number of convolutional layers associated with their pooling layer helps to detect more features but with high computational cost. Fully connected layers are connected to all neurons in previous layers to solve non-linear problems.

2.6   Support Vector Machines (SVMs)

Support vector machine is also similar to neural networks because SVM can be represented by two layered AANs with sigmoid activation function. However, it aims to solve quadratic programming problems unlike ANN as it solves non-convex minimization problems. Quadratic Programming is considered as an optimization algorithm that aims to maximize a quadratic function. SVM constructs hyper plane to maximize the separation of classes in higher dimensional space to solve nonlinear problems and give optimum results without being stuck in local optima [12]. Higher dimensional space facilitates classification in linear space. Therefore, the classification problem has to be mapped on the n-dimensional space. Kernel function is used to implement the needed mapping. Support vectors are used to maximize the separation between classes and they are usually closer to decision boundaries.

3   The Proposed Identification System

The proposed system is summarized in Fig. 5.

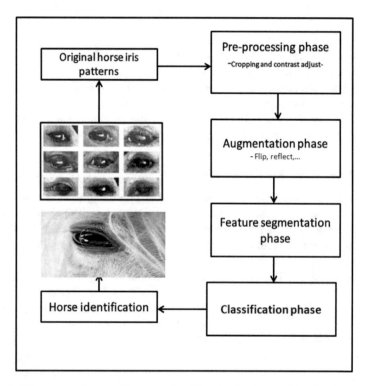

Fig. 5. The proposed general framework of the Arabian-horses identification system

3.1   Pre-processing Phase

Images taken for horses' eyes had different sizes. As a result, all the images were resized to 37 × 37. In addition, eye images may have some additional features that may affect the classification decision such as eye lashes and flyes. For this reason, eye images were cropped to remove these additional features as they are considered to be noise. Furthermore, images were converted to grey-scale images. Examples of horses 's eyes' images and then the preprocessed images are illustrated in Figs. 6 and 7.

3.2   Augmentation Phase

Classifiers usually require more training images for achieving higher classification accuracy. Convolutional neural networks in specific needs more images to achieve satisfying results [13]. As a result, blurring, scaling, rotation, shifting and brightness improvement were applied to already existing images. Scaling and rotation have been applied using random values. After augmentation, each horse had 100 images of eyes. Therefore, after augmentation the total of eyes images were 14,500 images.

Fig. 6. Image samples before pre-processing.

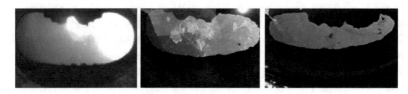

Fig. 7. Image samples after pre-processing.

Fig. 8. Image samples of segmented features.

3.3   Iris Features Segmentation Phase

As mentioned before, this work uses our previous approach of segmentation for the most significant features in iris [5]. It was found that all significant features in horses' eyes are around the pupil in images that were taken with high intensity light [5]. These features with pupil's shape and size are unique. As a result, the circular region that is containing both iris and pupil was extracted using color based k-means clustering since these features had different color than the iris background. Morphological analysis with blob detection was also applied to reduce noise that might occur form flash light and might appear in the eye. Therefore, after segmentation of area of interest, SIFT features were extracted from all the training images. Then these features should be given to a classifier to be trained. However, the number of the segmented features is variant and the number of the inputs to the classifier should be of a fixed size. Therefore, training images were divided into 4 equal parts and only 10 features were extracted from each part.

3.4   Classification Phase

Extracted SIFT features were given as input to SVM classifier to be trained. SIFT created 128 descriptors for each of the ten features extracted from the four quarters of the image. The dataset of 145,00 augmented images were split into training set and validation set. 60% of the images were used for training, while the rest 40% were used for validation. Moreover, radial Bias Function was used as a kernel.

4   Experimental Results and Discussion

In this work, different classification methodologies were tested to find the optimal approach for Arabian horse identification. ANN and SVM were tested on SIFT features and SURF features. CNN was also tested on the whole segmented image of the iris features. In this paper, AlexNet architecture was implemented for CNN. This architecture basically consists of eight layers; five convolutional layers and three fully connected layers with pooling layers in between for activation. For the ANN, only one hidden layer, one input layer and one output layer were used. The middle layer consisted of 7 neurons.

As mentioned before, special features were discovered in horse's eyes during the process of data collection. These features are basically the shape and the size of iris. For this reason, segmentation of this area was a necessary step before the classification process. Segmentation performance of the previously proposed approach in [4] was high as it achieved 85% Jaccard coefficient. Samples of the output are shown in Fig. 8. Then descriptors had to be extracted out of these images to represent the different shapes. In order to measure Jaccard coefficient, segmentation was done manually to images to be able to compare the results of segmentation implemented using the proposed approach.

As illustrated in Table 1, SVM with SIFT features achieved higher accuracy than all the other examined approaches. The proposed approach of SVM with SIFT features achieved 97.6% accuracy. The accuracy metric was used for evaluating the classification. Accuracy predicts how much the model goes right as given in Eq. 1.

$$\text{Classification accuracy} = (CP/TP) * 100 \qquad (1)$$

Where, CP is the number of correct predictions, while TP is the total predictions.

Table 1. Accuracy achieved by all the examined approaches

SVM with SIFT	SVM with SURF	ANN with SIFT	ANN with SURF	CNN
97.6%	70.24%	58.19%	44.52%	76.1%

As shown in Table 1, SVM achieved higher accuracy than ANN using SIFT OR SURF features. Moreover, SIFT features gave higher accuracy when used with SVM more than SURF features. Furthermore, SVM proved to be more efficient than in horse identification than CNN as CNN only achieved 76.1%. The reason behind this is that deep learning approaches usually need huge number of training samples. Therefore, using only 100 images for each class didn't help to achieve accurate results as CNN usually requires at least 1000 images for each category. Moreover, no optimization technique was used to define the architecture of the CNN. However, CNN architecture was decided after several experiments until we used the same number of layers and the same learning rates of AlexNet. The classification methodologies adopted in this work was also tested on the eyes' images itself without iris segmentation. Therefore, segmentation step was skipped to run such experiments before SIFT or SURF extraction. The comparison between the classification on segmented iris and on actual images in Table 2. CNN was also tested on the eyes' image without iris segmentation, so that the whole eye image was given as an input to the CNN for Arabian horses' identification.

Table 2. Accuracy achieved by all the examined approaches on horses' eyes' images without applying the segmentation step.

SVM with SIFT	SVM with SURF	ANN with SIFT	ANN with SURF	CNN
92.15%	60%	47.38%	38.1%	71.1%

As obvious in Table 2, there was a drop in the accuracies of the examined classification techniques on the unsegmented images. However, accuracies of SVM with SIFT and CNN did not drop dramatically as the rest of methodologies as clarified in the graph in Fig. 9. As mentioned before, segmentation of the unique shapes in iris boosts up the performance of classification. The reason behind this is that segmentation based on colours and then biggest blob detection remove the noisy features that leads to misclassification. Therefore, our proposed approach proved its efficiency in focusing on important features for accurate classification results.

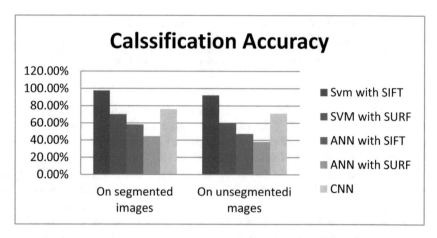

Fig. 9. Graph representation of accuracies of classification techniques in segmented and unsegmented images

5   Conclusion and Future Work

In this paper, we presented a successful approach for the identification of Arabian horses that can also be used also for other types of horses. This approach depends on segmenting an area inside the iris and around the pupil because it was found that all the significant features lie in this area. Therefore, the proposed approach extracted these significant features and used the segmented images for horse identification. Through different experiments and using several Artificial intelligence techniques including SVM, NN, and Deep Learning. We concluded that the SVM with SIFT features achieved an accuracy of 97.6%. For future work, new images will be collected for all horses not only the Arabian breed because we found that all horses have these features. Moreover, this work will be implemented as a mobile application to be used in real time for horse's identification during international races.

References

1. Koik, B.T., Ibrahim, H.: A literature survey on animal detection methods in digital images. Int. J. Future Comput. Commun. **1**, 1–24 (2012)
2. Awad, A.I.: From classical methods to animal biometrics: a review on cattle identification and tracking. Comput. Electron. Agric. **123**, 423–435 (2016)
3. Suzaki, M., Yamakita, O., Horikawa, S., Kuno, Y., Aida, H., Sasaki, N., Kusunose, R.: A horse identification system using biometrics. Syst. Comput. Jpn. **32**, 12–23 (2001)
4. Olatinwo, S.O., Shoewu, O., Omitola, O.: Iris recognition technology: implementation, application, and security consideration. Pac. J. Sci. Technol. **14**, 228–232 (2013)
5. Salama, A., Hassanien, A., Fahmy, A.: Iris features segmentation for arabian horses identification. In: Proceedings of the 1st International Conference on Internet of Things and Machine Learning - IML (2017)

6. Zhao, Z., Kumar, A.: Accurate periocular recognition under less constrained environment using semantics-assisted convolutional neural network. IEEE Trans. Inform. Forensics Secur. **12**, 1017–1030 (2016)
7. Ahmed, Sh., Gaber, T., Tharwat, A., Hassanien, A.E., Snasel, V.: Muzzle-based cattle identification using speed up robust feature approach. In: IEEE International Conference on Intelligent Networking and Collaborative Systems, pp. 99–104 (2015)
8. Panchal, P.M., Panchal, S.R., Shah, K.H.: A Comparison of SIFT and SURF. Int. J. Innov. Res. Comput. Commun. Eng. **1**, 2320–9798 (2013)
9. Krizhevsky, A., Sutskever, I., Hinton, G.F.: ImageNet classification with deep convolutional neural networks. Commun. ACM **60**(6), 84–90 (2017)
10. Chen, L., Guo, X., Geng, C.: Human face recognition based on adaptive deep convolution neural network. In: 2016 35th Chinese Control Conference (CCC), Chengdu, pp. 6967–6970 (2016)
11. Rriya, S.J.: Survey on face recognition using convolution neural network. Int. J. Softw. Hardw. Res. Eng. **5**(9), 6–9 (2017). ISSN-2347-4890
12. Burges, C.J.: A tutorial on support vector machines for pattern recognition. Data Min. Knowl. Discov. **2**, 121–167 (1998)
13. Wang, J., Perez, L.: The effectiveness of data augmentation in image classification using deep learning. Technical report, Stanford University (2017)
14. Brożek, B., Janika, B.: Can artificial intelligences be moral agents? New Ideas Psychol. **54**, 101–106 (2019)

A Neuro-Fuzzy Based Approach for Energy Consumption and Profit Operation Forecasting

Mohamed A. Wahby Shalaby[1,2(✉)], Nicolas Ramirez Ortiz[3],
and Hossam Hassan Ammar[3]

[1] Faculty of Computers and Artificial Intelligence, Cairo University, Giza, Egypt
[2] Smart Engineering Systems Research Center (SESC), Nile University, Giza, Egypt
mwahby@nu.edu.eg
[3] School of Engineering and Applied Science, Nile University, Giza, Egypt
nicolas.ramirez@eu4m.eu, hhassan@nu.edu.eg

Abstract. In recent years, the massive growth in the scale of data is being a key factor in the needed data processing approaches. The efficiency of the algorithms of knowledge extraction depends significantly on the quality of the raw data, which can be improved by employing preprocessing techniques. In the field of energy consumption, the forecasting of power cost needed plays a vital role in determining the expected profit. To achieve a forecasting with higher accuracy, it is needed to deal with the large amount of data associated with power plants. It is shown in the literature that the use of artificial neural networks for the forecast electric power consumption and show short term profit operation is capable of achieving forecasting decisions with higher accuracy. In this research work, a neuro-fuzzy based approach for energy consumption and profit operation forecasting is proposed. First, the main influential variables in the consumption of electrical energy are determined. Then, the raw data is pre-processed using the proposed fuzzy-based technique. Finally, an artificial neural network is employed for the forecasting phase. A comparative study is conducted to compare between the proposed approach and the traditional neural networks. It is shown that the achieved forecasting accuracy of the proposed technique is better than what achieved by employing only the neural network.

Keywords: Energy consumption · Neural networks · Fuzzy sets · Adaptive neuro-fuzzy

1 Introduction

Huge amounts of information surround us today. Technologies like the Internet generate data at an exponential rate thanks to the cheapening and great development of storage and resources net. The current volume of data has exceeded the processing capabilities of the systems classic data mining issues [1]. We have

© Springer Nature Switzerland AG 2020
A. E. Hassanien et al. (Eds.): AISI 2019, AISC 1058, pp. 58–69, 2020.
https://doi.org/10.1007/978-3-030-31129-2_6

entered the era of Big Data or data massive [2], which is defined by the presence of great volume, speed and variety in the data, three characteristics that were Translated by Laney in the year 2001 [3], with the requirement of new systems of high performance processing, new scalable algorithms, etc. Two other important aspects that characterize the massive data are the veracity of data and the intrinsic value of knowledge extracted. Figure 1 shows these five characteristics. The quality of knowledge extracted depends to a large extent on the quality of data.

Fig. 1. Characteristics that define big data [4]

Unfortunately, these data are affected by negative factors such as: noise, missing values, inconsistencies, data superfluous and/or too large a size of any dimension (number of attributes and instances). It is shown that poor data quality leads in most of the cases to a low quality of extracted knowledge. Recently, the term has emerged "Smart Data" that revolves around two characteristics, the veracity and the value of the data, and whose objective is to filter noise and keep valuable data, that can be used to make smart decisions. Three characteristics are associated with this new data paradigm: accurate, processable and agile (accurate, actionable and agile).

The availability of energy is strongly linked to the level of wellbeing, health and lifespan of human beings, currently in the world 80% of energy production is produced through fossil fuels such as coal, natural gas and oil. The availability of electricity, one of the clearest indicators of development, is used as a sign of economic growth. The world demand for electricity is growing rapidly in recent times at a disruptive pace [5]. The possibility of forecasting the future consumption of electricity is a premise for the proper functioning of management systems [6]. This can only be achieved making use of the existing big amounts of Data. The preprocessing of data is an essential stage to the discovery process of Information KDD (Knowledge Discovery in Databases) [6,7]. This stage is responsible for data cleansing, its integration, transformation and reduction for next phase of data mining [4]. A preprocessing phase of raw data is usually needed to enhance the extracted information.

In this research paper, a neuro-fuzzy technique is proposed for the prediction of electricity consumption using Big Data. In this approach, the main influential variables in the consumption of electrical energy are determined. Then, the raw data is pre-processed using the proposed fuzzy-based technique. Finally, an artificial neural network is employed for the forecasting phase. the rest of this paper contains the following. First, a literature review (Sect. 2) about the energy consumption forecast techniques, using data that has been processed, is presented. Then in Sect. 3, the main variables to make prediction of consumption are determined. The proposed approach is then explained in Sect. 4. The experimental results and discussions are presented in Sect. 5. Finally, the conclusions and future work are discussed in Sect. 6.

2   Energy Consumption Forecast Using Data That Has Been Processed

The current rate of data generation is challenging the processing capabilities of the current systems in companies and public bodies. Social networks, the Internet of Things and industry 4.0, and big scale factories are some of the new scenarios with presence of massive amounts of data. The need to process and extract knowledge value of such immensity of data it has become a considerable challenge for data scientists and experts in the matter. The value of extracted knowledge is one of the essential aspects of big data. The importance of the forecast of electricity consumption lies not only in its relation to the availability of primary fuels, but also in the following factors of a technical economic nature:

- Know in advance the future financial expenses for the purchase of fuels for the generation of electric power.
- predict future changes in power grids, substations, transmission lines.
- Predict the increase of the park of electric power plants.
- Need to apply new measures for saving.

The most used methods in the forecast of energy consumption and show profit operation are:

1. **Simple historical trend method** The historical trend method for calculating the future consumption of electricity was used until the end of the 1970s. In many countries, an increase of 7.5% per year in consumption and its doubling per decade was considered. Initially this method was applied to the analysis of finances [6].
2. **End-user method** The end-user method is based on the classical equations that express the consumption of electric power for any location, administrative area, work center, etc., where there are electrical consumers, depending on their number, power and time of use [5].

3. **Parametric methods** The parametric methods are concerned to obtain, from the statistical and mathematical analysis of the real values of the variables, the values that would have the parameters of the models in which these variables act, as well as to check the degree of validity of those models, and see to what extent these models can be used for the prediction of the future change of the variables defined as independent [7].

4. **Methods with application of artificial intelligence** The Artificial Neural Networks (ANN) constitute a powerful instrument for the approximation of nonlinear functions. Its use is especially useful in the modeling of complex phenomena where the presence of non-linear relationships between variables is common. One of the applications of ANN is the prediction of electrical energy consumption. Due to the foregoing, it is advisable to use this research [8]. Condition monitoring systems that collect and analyze data in real time have become a smart and cost-effective way to forecast machinery life in the industry. Big companies such as SIEMENS and start-ups like TWAVE offer real time on-line data monitoring that analyses big amounts of information to predict failures and program pre-failure maintenance [9]. Some neural networks based systems were proposed in [10] and [11]. Both showed that the trained ANNs are capable of forecasting energy consumption considering various factors [12].

3   Variables That Influence the Consumption and Profit Operation of Electrical Energy Production

To make the prediction of the consumption of electrical energy, it is necessary to define the explanatory variables (independent) that will be used, which, directly or indirectly, should have an influence on the consumption of electricity.

List of primary data sources that are helpful for power system modeling of Europe are open source. The upcoming demand for power plants during specific days as well as the day prices are available in [13]. Specific sets of big data and all variables that can be use to find extra knowledge have been put at the disposition of educators and researchers to investigate them. For this work a combine cycle power plant located in Europe has been selected. This power plant is formed by 2 gas turbines and 1 steam turbine with different maximum and minimum power outputs.

The capacity generation limits for each unit are available in Table 1 and are considered constant for the entire days. This data will be used as an example for forecasting consumption and profit operation optimization.

The fuel cost curves of the different operation modes are available in Table 2. Taking into account the following assumptions:

- The fuel cost of the operational modes only depends on the amount of power produced.
- The operational model GT1+ST is equivalent to GT2+ST.
- All the units are working every morning and is not necessary to consider shutdown/start-up times.

Table 1. Generation limits of the exampled power plant

Unit	Min. capacity (MW)	Max. capacity (MW)
GT	80	300
GT	50	230
ST	80	160

Table 2. Fuel cost curves

Unit	$FC_i = f(P, P^2)(E)$	$a(E)$	$b(E/MWh)$	$c(E/MWh^2)$	$d(E/MWh^3)$
GT1	$a + bP + cP^2$	300	2.32	0.002978	0
GT2	$a + bP + cP^2$	250	2.5	0.003378	0
GT+ST	$a + bP + cP^2$	101.3	4.732	-0.0151	$2 * 10^{-5}$
GT1+GT2+ST1	$a + bP + cP^2 + dP^3$	-1633	16.471	-0.036	$3 * 10^{-5}$

The power plant operator is usually provided with the demand forecast by the distribution system operator but only with the average electricity price for the following weeks. For this case is necessary to assume that the data will be used to estimate the hourly electricity price for the following weeks in order to take the right decisions. In this order the variables that will be considered for the study are those showed in Tables 1 and 2, some extra spinning reserve price per hour and the upcoming demand for the power plant during following days day as well as the day ahead prices.

4   The Proposed Approach

Artificial Neural Networks is as a mathematical modelling of mental and brain activities. These systems exploit local processing in parallel and the properties of distributed representation, aspects that apparently exist in the brain. In daily communication, imprecise information is usually employed, which could be processed successfully by the human brain [14]. The neural network itself is not an algorithm, but rather a framework for many different machine learning algorithms to work together and process complex data inputs. ANNs have been used widely to perform complex functions in various fields of application, such as behaviour analysis and human emotion recognition [15], trend recognition [16], optimization of industrial processes [17] and control systems [18]. In matters related to energy processes, ANN can be used to control the operation and efficiency of a cogeneration system, to evaluate the performance of distributed generation units [19].

An artificial Neural network is an interconnected group of nodes, represented like the network of neurons in a brain. Each circular node represents an artificial neuron and an arrow represents a connection from the output of one artificial neuron to the input of another. In order to forecast the values, we designed two different ANNs. The first ANN will be trained using the raw data. In this case, there will be only 2 inputs and one output. Then, the second ANN will be trained with pre-processed data using the proposed fuzzy sets.

4.1 Neural Network (ANN) Raw Data

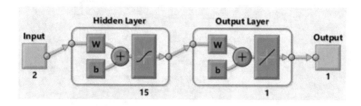

Fig. 2. Artificial neural network configuration and setting up

The system is composed of an input and output layer, with one hidden layer of 15 neurons, which was implemented using MATLAB tool (Fig. 2). As the number of variables are low, one is day from Monday to Sunday, second one is the hour of the day, the process is faster and the performance and results are evident. Besides that, there is just one target goal which is the electricity price (see Fig. 3) where the output is just 1. From the process and to get the result 18 iterations were necessary. To train the neurons the Levenberg-Marquardt algorithm, also known as the damped least-squares (DLS) method was used in [20]. The LMA is used in many software applications for solving generic curve fitting problems. However, as with many fitting algorithms, the LMA finds only a local minimum, which is not necessarily the global minimum. The LMA interpolates between the Gauss Newton algorithm (GNA) and the method of gradient descent.

4.2 The Proposed Neuro-Fuzzy Technique

Fuzzy logic is an extension of traditional Boolean logic that uses concepts of membership sets more similar to the human way of thinking. Exact subsets use Boolean logic with exact values for example: the binary logic that uses values 1 or 0 for your operations. Over the past few decades, fuzzy based approaches have been used successfully in different research areas as noted in [21] and [22].

Fuzzy logic does not use exact values like 1 or 0 but use values between 1 and 0 (inclusive) that can indicate intermediate values. An example of this can be $(0, 0.1, 0.2, ..., 0.9, 1.0, 1.1, ...etc)$. In this case if it also includes values 0 and 1 then it can be considered as a superset or extension of the exact logic [9].

A clear example of a fuzzy set can be given by the complexity of linguistic concepts. For instance small, medium and large characteristics to define states of a variable if speaking about Energy Consumption as seen in Fig. 5. Using the data reduction method know as fuzzy set the intention is to increases the number of inputs and reduce the range of data. Hence, each value of original raw data

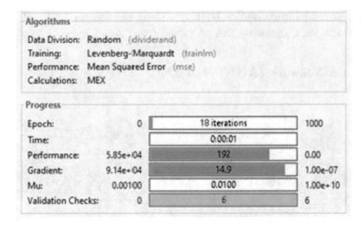

Fig. 3. Result of the training

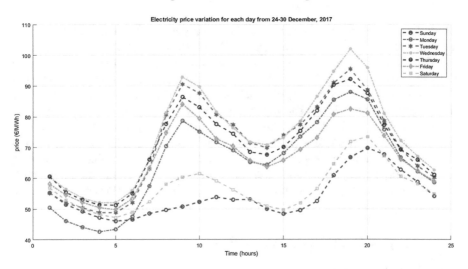

Fig. 4. Forecasted Price values with ANN with 2 inputs

will belong to one or more of fuzzy sets. In that case of this study the second input (the hours of a day) will be replaced by Morning, Noon, Afternoon, Night, Late night categories. The new ANN will have 3 inputs: days, fuzzy set category, membership degree of the day hour within this fuzzy set.

For this specific case the fuzzy sets created were considering the rules that divide the day into overlapped periods. Using the fuzzy toolbox of Matlab these fuzzy sets have been developed. From the raw data: the days, hours and power consumption have been used with five membership functions to develop the fuzzy sets. Therefore, the proposed neuro-fuzzy technique has 125 rules and the significant rules after optimization became 75 rules. In Fig. 3 the previous neural

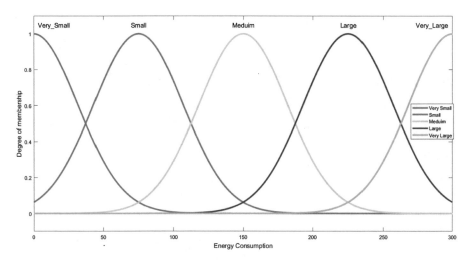

Fig. 5. Energy consmuption in the range [0, 300 MW] conceived as a fuzzy variable [9]

network had two inputs. For the proposed neuro-fuzzy technique, the new ANN will have 3 input with the same number of layers as the previous one. The progress of the training turned to be faster in terms of training time and the Epoch only need it 55 iterations in comparison with the previous test.

5   Results and Discussion

In order to test the proposed neuro-fuzzy technique, we should evaluate with 1×2 arrays, one value for the hour and another for the day. Figures 4 and 6, shows the obtained values for the traditional NN and the proposed neuro-fuzzy technique.

It is clearly seen from both figures that the highest values are between Tuesday and Thursday. also the lowest electricity price was between Saturday and Sunday (weekends). Its been used and applied a mean value approach of each day between October 2017 until January 2017 from [13] isolating each day. First It was necessary to create an input matrix for the input ANNs layer, day and hour, the output layer is composed by the targets values (electricity price). The main limitations of all forecast methods are the probability of appearance of a typical data in the data set (in the developing part, treating data or source data) or on the other hand, the a typical data in the real world at the moment that we want to forecast. However, the results look similar but differences are evident. For instance, the price forecast values are smaller in the case of the proposed neuro-fuzzy technique. In addition, the proposed technique has smoother curves in comparison to the regular neural networks. For example the prediction of change of tendency seems to be anticipated smoother in comparison the use of raw data. This also shows how important is preprocessing of raw data

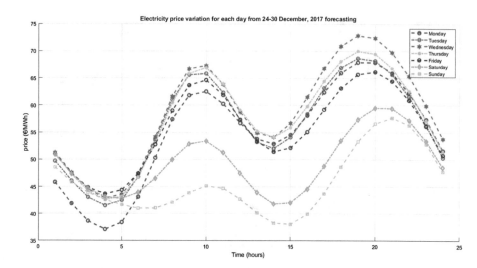

Fig. 6. Forecasted Price values with the proposed neuro-fuzzy technique

before analyzing it. It is of high importance to transform processed data to find patterns and acquire tangible knowledge.

5.1 Maximizing Profit Using the Proposed Neuro-Fuzzy Technique

To maximize the profit of the power plant between morning hours. The first approach was to decide or determine which is the most suitable combination to produce energy, (MHh) knowing that is know that the maximum output of 690 MW according to 1:

1. The best combination depends on the required output power of the turbines, in this case, Fig. 7, shows the response of the plant with a different combination. Its been created a function where the input is the required power, and with the four equations provided in the Table 2, the output will be the fuel cost in euros. Figure 7 plots the combinations as follows:
   a. Red, Combination 1: $GT1 + ST - GT2$, $GT2$ working separately,
   b. Blue: Combination 2: $GT2 + ST - GT1$, $GT1$ working separately,
   c. Green: Full: $GT1 + GT2 + ST1$, all turbines working together.
2. The combination b (blue) provide lest power than the other one, nevertheless the first combination, it is much better than the second one. In the other hand, the green combination, or full mix, provide a cheaper performance if the power demanded is less than 552 MW, so, if we need more power, the first combination is the best one.
3. This solution, have been made, creating matrix with the minimum and maximum output power in all the combination, and then, making a regular scanning with a population of 200 units, so, the combination is made of the possible cases, or two combinations, for that reason the number of iterations is equal to $200^{200} = 40.000$.

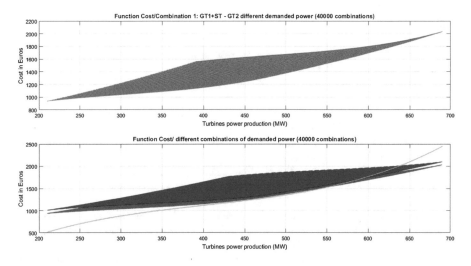

Fig. 7. Cost Function plots of the 3 different combinations

Finally, it is clearly seen that the proposed neuro-fuzzy technique is capable of forecasting the power consumption. Understanding electricity demand is a critical factor that is required for ensuring future stability and security in the present and in the future. For this reason to reduce and optimize the use of electricity is so important.

6   Conclusion and Future Work

The energy consumption is considered a key factor for all any nation economics. Therefore, the forecasting of needed power per a period of time is quite important for determining the expected profit. In this paper, a neuro-fuzzy based approach for energy consumption and profit operation forecasting has been developed. First, the main influential variables in the consumption of electrical energy have been determined. Since there is usually large amount of data associated with power plants, it has been proposed to employ fuzzy-based technique for pre-processing the raw data. The raw data has been categorized using proposed fuzzy sets to reduce the range of values of raw data. Then, the data reduction phase facilitates the use of an artificial neural network for the forecasting phase. A comparative study is conducted to compare between the proposed approach and the traditional neural networks. It has been shown that the achieved forecasting accuracy of the proposed technique is better than what achieved by employing only the neural network.

Finally, the proposed technique could be upgraded with other information in the future, in order to forecast the needed values. Such information could include money currency of a specific country, and environmental issues.

References

1. Wu, X., Zhu, X., Wu, G.-Q., Ding, W.: Data mining with big data. IEEE Trans. Knowl. Data Eng. **26**(1), 97–107 (2014)
2. Kitchin, R.: The real-time city? Big data and smart urbanism. GeoJournal **79**(1), 1–14 (2014)
3. Laney, D.: 3D data management: Controlling data volume, velocity and variety. META Group Res. Note **6**(70), 1 (2001)
4. Gill, D.A.Q.: V5 big data lens (2012). http://aqgill.blogspot.com/2012/06/what-does-bigdata-mean-to-business.html
5. Velázquez, A.S., González, J.O.N., Peña, D.R., García, D.J., et al.: Pronostico de consumo de energía eléctrica usando redes neuronales artificiales. Tlatemoani (16) (2014)
6. Zavala, V.M., Constantinescu, E.M., Krause, T., Anitescu, M.: On-line economic optimization of energy systems using weather forecast information. J. Process Control **19**(10), 1725–1736 (2009)
7. Yu, D., Tao, S.: The method of classification for financial distress prediction indexes of Sinopec Corp. and its subsidiaries based on self-organizing map neural network. In: 2012 Fourth International Conference on Computational and Information Sciences (ICCIS), pp. 590–593. IEEE (2012)
8. R. P. C. 1129409: Industrial compute module (2019). https://www.raspberrypi.org/products/
9. Klir, G., Yuan, B.: Fuzzy Sets and Fuzzy Logic, vol. 4. Prentice hall, New Jersey (1995)
10. Keras Development Team: Keras prioritizes developer experience(2019). https://keras.io/why-use-keras/
11. Loshing, C.T., Thompson, R.J.: System for monitoring, transmitting and conditioning of information gathered at selected locations. US Patent 4,476,535, 9 October 1984
12. Yao, X., Liu, Y.: A new evolutionary system for evolving artificial neural networks. IEEE Trans. Neural Netw. **8**(3), 694–713 (1997)
13. Data source. http://open-power-system-data.org/. Accessed 05 July 2019
14. Neto, M.C.A., Calvalcanti, G.D., Ren, T.I.: Financial time series prediction using exogenous series and combined neural networks. In: International Joint Conference on Neural Networks, IJCNN 2009, pp. 149–156. IEEE (2009)
15. Salama, E.S., El-Khoribi, R.A., Shoman, M.E., Shalaby, M.A.W.: EEG-based emotion recognition using 3D convolutional neural networks. Int. J. Adv. Comput. Sci. Appl. **9**(8), 329–337 (2018)
16. Muyeen, S., Hasanien, H.M., Tamura, J.: Reduction of frequency fluctuation for wind farm connected power systems by an adaptive artificial neural network controlled energy capacitor system. IET Renew. Power Gener. **6**(4), 226–235 (2012)
17. Aiordachioaie, D., Ceanga, E., Mihalcea, R.-I., Roman, N.: Pre-processing of acoustic signals by neural networks for fault detection and diagnosis of rolling mill (1997)
18. Ozgonenel, O., Yalcin, T.: Principal component analysis (PCA) based neural network for motor protection (2010)
19. Nikpey, H., Assadi, M., Breuhaus, P.: Development of an optimized artificial neural network model for combined heat and power micro gas turbines. Appl. Energy **108**, 137–148 (2013)
20. Moré, J.J.: The Levenberg-Marquardt algorithm: implementation and theory. In: Numerical Analysis. Springer, pp. 105–116 (1978)

21. Shalaby, M.A.W.: Fingerprint recognition: a histogram analysis based fuzzy c-means multilevel structural approach. Ph.D. dissertation, Concordia University (2012)
22. Khaled, K., Shalaby, M.A.W., El Sayed, K.M.: Automatic fuzzy-based hybrid approach for segmentation and centerline extraction of main coronary arteries. Int. J. Adv. Comput. Sci. Appl. **8**(6), 258–264 (2017)

Analysis the Consumption Behavior Based on Weekly Load Correlation and K-means Clustering Algorithm

Bo Zhao$^{(\boxtimes)}$ and Bin Shao

Marketing Department of State Grid, Liaoning Electric Power Co., Ltd.,
Shenyang 110006, China
creative0228@163.com

Abstract. There are many factors affecting the user's electricity consumption behavior in China, and the electricity usage behavior of power users is both random and periodic. Therefore, how to more accurately grasp the user's power consumption behavior has always been an important topic for power researchers. This paper proposes to analyze the correlation between users' weekly workdays and weekly rest days on the weekly time scale, with working days and weekly holidays. The correlation coefficient of the daily load data constructs the feature value, combines the variance with the K-means algorithm, determines the number of clusters by the cluster validity index, rapidly clusters the weekly load, and summarizes the weekly power consumption of the user based on the clustering result. The behavior category analyzes the detail the changes in the behavior of users using electricity during the week. The simulation is carried out by using MATLAB software, and the users are divided into four categories. By combining the characteristic value curve of each type of user and the typical user load curve, the characteristics of each user's electricity consumption behavior are analyzed in detail.

Keywords: Electrical consumption behavior · Weekly load · Cluster analysis

1 Introduction

In recent years, with the rapid development of China's electricity information collection system, the power company's access to user electricity data has become a massive trend, which brings new opportunities and challenges for China's power load characteristics analysis and forecasting work. Therefore, the cluster analysis of user electricity usage based on massive power consumption data has received increasing attention from power researchers in recent years. The electricity usage behavior of power users has both periodicity and randomness and has complex correlations with economic factors, meteorological factors, time factors, and political factors. At present, there are many types of research on the influence of temperature on user behavior. However, the temperature factor only has a significant impact on the user's electricity consumption behavior in summer and has certain seasonal restrictions [1, 2]. Economic factors have a significant impact on the electricity consumption behavior of industrial users, but the cycle is longer. Therefore, the impact on the annual power load

© Springer Nature Switzerland AG 2020
A. E. Hassanien et al. (Eds.): AISI 2019, AISC 1058, pp. 70–81, 2020.
https://doi.org/10.1007/978-3-030-31129-2_7

characteristics is generally analyzed. For the time factor, the power load data has significant periodic laws, including annual cycle, monthly cycle, weekly cycle, and daily cycle. Since the production, life, and study of the people are mostly arranged on a weekly basis, the cycle characteristics of the basic units are most significant, and the degree of change in the use of electricity by different types of users on weekdays is also significantly different. At this stage, most scholars often ignore the difference in user's power usage behavior between workdays and weekly vacations, or simply make simple assumptions about the number [3].

The existing research work on user load characteristic analysis mainly takes the user's load characteristic curve as the research object, including the use of data clustering technology to extract the typical characteristic curve of a single user and the user classification of a large number of users. By extracting the typical load curve that best represents the user load characteristics, the users are classified according to different power behavior characteristics, and then provide decision assistance for the demand side management work [4–6]. With the increasing availability of user power data, how to realize the processing and analysis of massive power consumption data is the research direction of user power behavior analysis in the future. At the same time, the important purpose of user clustering behavior analysis is to extract the typical power usage behavior of various users [18, 19]. However, at present, most of the research is based on the daily time scale, which leads to the user's typical power usage behavior cannot accurately distinguish between working days and weekly holidays. Conducting user behavior analysis at weekly time scales will be an important area of focus for power researchers [7].

In summary, the user's power usage behavior has certain rules, but at the same time, it is affected by many factors and may cause large fluctuations at any time.

2 Theory of K-means Clustering Algorithm

Clustering analysis refers to the process of dividing a collection of physical or abstract objects into different groups according to their closeness to each other. The goal of cluster analysis is to classify data with similarities, so that objects in the same category have great similarities, and objects between different categories have great dissimilarity. Therefore, it is better to use cluster analysis to analyze the user's electricity behavior [8].

The K-means clustering algorithm is a hard clustering algorithm based on partitioning. First, we need to select the initial cluster center, then classify all the data points, and finally calculate the average value of each cluster to adjust the cluster center, and continuously iterate the loop, and finally make the object similarity in the class the largest, and the similarity between the classes is the smallest [9–11]. The input of K-means algorithm is the number of clusters K and the database containing N objects. When the cluster center is no longer changed, K cluster clusters are output. The K-means specific steps are described as follows.

- **Step 1.** Initializes. K samples were randomly selected as the initial class center.
- **Step 2.** Sample classification. The distance square of all samples to their cluster center and the value of J(C) are calculated. Each sample is divided into the nearest cluster center, and the same cluster center is a class.

$$J(C) = \min \sum_{m=1}^{k} \sum_{x \in c_m} u_{mn} d^2(c_m, x_n) \tag{1}$$

In the formula: u_{mn} is binary variable, $u_{mn} = 1$ denotes that the nth sample belongs to m class, $u_{mn} = 0$ means not belong to this class, $d^2(c_m, x_n)$ is the distance between the sample and its cluster center, c_m is the sample cluster center. x_n is the other sample data in the class.

- **Step 3.** Class Center updates. According to the partition result of step 2, the least square method and Lagrange principle, the center cm. of K classes is updated.

$$c_m = \frac{\sum\limits_{n=1}^{M} u_{mn} x_n}{\sum\limits_{n=1}^{M} u_{mn}} \tag{2}$$

- **Step 4.** determines whether the convergence condition is satisfied or not, returns step 3 if not, and outputs the result if satisfied [12]. The determination of the best clustering number K depends on the clustering index, and the index can be used as the error square and sum of squared errors (SSE) index, which is a monotone decreasing function. When the decrease is relatively smooth, the optimal clustering number can be reached [20, 21].

3   Application of K-means Clustering Algorithm to Power Consumption Analysis

Data Acquisition: The daily load data of users are obtained from the power company database, and the sample set of user load data is set as X.

$$X = [x_1, x_2, x_3, \ldots, x_k, \ldots, x_m] \tag{3}$$

The x_k is the power data of any one of the M sample data at 96 sampling points (15 min/) per day, as shown in Formula (4).

$$X_k = [x_{k1}, x_{k2}, x_{k3}, \ldots, x_{k96}] \tag{4}$$

Data Preprocessing: Because the similarity of load curve shape is the main consideration of load clustering, in order to eliminate the effect of user load data base value on clustering effect, the maximum load value in one day of each user is considered as the reference value. Normalize the load data is calculated using Eq. (5).

$$x_{k,t}^* = \frac{x_{k,t}}{x_{k,\max}} \tag{5}$$

Where $x_{k,t}^*$ is the normalized load at t time; $x_{k,t}$ and $x_{k,\max}$ are respectively the load size of the k-th user at t time and the maximum load data of 96 sampling points in one day.

In addition, at some time, because of the fault of sampling equipment, it may lead to the sudden rise and drop of load curve, which will affect the effect of load clustering. Smoothing formula is used to correct the abnormal data and given in Eq. (6).

$$x_{k,t}^* = \frac{\sum_{a=1}^{a_1} x_{k,t-a}^* + \sum_{b=1}^{b_1} x_{k,t+b}^*}{a_1 + b_1} \tag{6}$$

Where a, and b represent the forward and backward points.

Data Clustering: Finally, the K-means clustering algorithm is used to cluster the load data. Based on the distance squared and clustering effectiveness evaluation indicators, the K-type user groups are obtained. Since each type of user group has similar load characteristics, the load users belonging to the same class can be classified into the same type of users, and the cluster center is used as the daily equivalent load curve of the users [13].

4   Clustering Analysis of Users' Power Consumption Behavior Based on Weekly Load Correlation

Data Sources and Preprocessing: The daily electricity load data of 600 users in a certain area is selected and recorded every hour. Since the actual data collected has a certain error rate, the data source needs to be preprocessed first [14]. This paper mainly deals with the outliers in the data source, and uses the following three-point smooth data preprocessing method. Daily load data of 600 users in a certain area are recorded once an hour. Because there is a certain error rate in the actual data collected, it is necessary to preprocess the data source first. This paper mainly deals with the outliers in the data source, and adopts the following three smoothing data preprocessing methods.

$$x_i = (x_{i-1} + 2x_i + x_{i+1})/4 \tag{7}$$

The two ends of the data source are:

$$x_1 = (3x_1 + x_2)/4 \tag{8}$$

$$x_{24} = (x_{23} + 3x_{24})/4 \tag{9}$$

X_i is the power consumption of the user during the day. After using the above method to preprocess the data, it not only increases the weight of the data at that time, but also avoids the excessive fluctuation of the data, and removes the error data in the original data to a certain extent. Thus, the reliability and correctness of the final clustering results are effectively guaranteed, which makes the data more suitable for the analysis of electrical behavior [15].

Weekly Load Correlation Feature Extraction: Based on the processed data source, the Pearson correlation coefficient is used as the calculation means to analyze the correlation between the five working days and the weekly rest days of the user's weekly load data, and to represent the user's working days and weekdays according to the degree of correlation [16]. If the user's power load correlation coefficient is similar to the five working days and the weekly rest day, the user's working day and weekly rest have a relatively stable power consumption behavior: if the correlation coefficient is high, the user is There is no significant change in electricity usage during the week; if the correlation coefficient is very low, it means that the user has different electricity usage behaviors on weekdays and weekly vacation days; if the correlation coefficient is at a medium level, then the user week There has been a non-negligible change in the use of electricity on the day off. On the other hand, if the user's power load correlation coefficient is greater than the five working days and the weekly rest day, the user's power usage behavior changes greatly within one week. Therefore, calculating the correlation coefficient between the user's working day and the weekly holiday can specifically show the user's behavior change during the week's rest, which is suitable for the analysis of the electricity usage behavior based on the weekly time scale load studied in this paper, and the correlation clustering feature extraction. The specific process is as follows: Since the weekly holiday is divided into two days on Saturday and Sunday, firstly, the daily electricity load on Saturday and Sunday is used as the reference, and the daily electricity load of 5 working days from Monday to Friday is calculated separately. The Pearson correlation coefficient is then taken from the average on Saturday and Sunday to extract the eigenvalues. First, the definition of the Pearson correlation coefficient is given.

$$r = \frac{\sum_{i=1}^{n} (x_i - x)(y_i - y)}{\sqrt{\sum_{i=1}^{n} (x_i - \bar{x})^2 \sum_{i=1}^{n} (y_i - \bar{y})^2}} \tag{10}$$

In Formula (10) x_i represents the amount of electricity consumed by a person on a working day at the time of the day i. y_i represents electricity consumption on a Saturday or Sunday day i. $\bar{x}\bar{y}$ represents the daily average load for a day takes 24, representing 24 h of the day. The correlation coefficient vectors $R_{sat} = (r_{11}, r_{12}, r_{13}, r_{14}, r_{15})$ and $R_{sun} = (r_{21}, r_{22}, r_{23}, r_{24}, r_{25})$ are calculated for the workday and Saturday and Sunday, respectively.

$$r_m = (r_{1i} + r_{2j})/2 \tag{11}$$

$m, i, j = 1, 2, \ldots, 5$

The calculation results of the eigenvalues of partial user correlation coefficients are given in Table 1.

Table 1. Calculation results of eigenvalues of partial user correlation coefficients

User	r1	r2	r3	r4	r5
1	0.7561	0.7738	0.7488	0.7518	0.7667
2	0.3032	0.3475	−0.0010	0.2491	0.3193
3	0.8860	0.8782	0.8856	0.8710	0.9180
4	0.9280	0.9455	0.9664	0.9370	0.9232
5	−0.1081	−0.0091	−0.1089	0.0423	0.0635
6	0.0598	0.0069	−0.0275	−0.0296	−0.0153
7	0.5579	0.2912	0.0148	0.2022	0.4819
8	0.8991	0.9019	0.9266	0.9235	0.8985
9	0.3169	0.4240	0.2802	−0.2665	0.2160
10	−0.0727	0.3924	0.2257	0.2088	−0.1496

The extracted five eigenvalues can not only indicate the correlation between the working day and the weekly holiday, but also determine the correlation between the five working days according to whether they are similar to each other. Moreover, by using the correlation coefficient as the eigenvalue, it is no longer necessary to normalize, avoiding the error rate generated by normalization, and simplifying the calculation, which is suitable for the research in this paper [17].

Realization of Clustering Analysis Based on Variance and K-means. In order to ensure the reliability of the clustering results, the variance is first used to separate out the user categories with strong randomness of electricity consumption behavior in practice, and then the K-means algorithm is selected to cluster the weekly time scale load. The algorithm flow of applying K-means to the analysis of users' weekly power consumption behavior is as follows:

From the data set $\{R_j\}_{j=1}^{N-N_1}$ of separated irregular users, $K - 1$ is randomly assigned to the initial cluster center $v_1, v_2, \ldots, v_{k-1}, v$, where K represents the number of clustering, N is the total number of users, and is the number of irregular users.

The Euclidean distance between the first sample point of the dataset and the v_j of each cluster center is calculated, and the v_{min} of the nearest cluster between the R_j user and the $K - 1$ cluster is obtained and classified into the cluster v_{min}. Recalculating the $K - 1$ cluster centers:

$$v_{min} = \frac{1}{N'} \sum_{R_j \in v_{min}} R_j \tag{12}$$

Repeat steps, until the criterion function converges. The criterion of convergence is based on the square error criterion, as shown in formula (12).

$$E = \sum_{i=1}^{k-1} \sum_{v_i} \left| R - \overline{R_i} \right|^2 \tag{13}$$

Where E is the sum of square errors for all objects in the database; R is the average of V.

Calculation Method of Optimal Clustering Number Based on Clustering Validity Index: When selecting k-means as clustering means, the number of clusters should be determined first. In this paper, four clustering validity indexes of $DB^{(-)}, Sil^{(+)}$, $CS^{(-)}, SF^{(-)}$ are selected to determine the optimal clustering number, which are respectively recorded as $\alpha_{DB^{(-)}}, \alpha_{Sil^{(+)}}, \alpha_{CS^{(-)}}, \alpha_{SF^{(-)}}$. (+) indicates that the index is a very large index, and the clustering effect of the corresponding cluster number is the best when the value is larger. (−) as an indicator of achievement, the cluster number corresponding to the smaller the value is the best. $DB^{(-)}$ was proposed by Davies et al., defined as the following Formula (14).

$$\alpha_{DB^{(-)}} = \frac{1}{K} \sum_{i=1}^{K} \overset{\max}{jj \neq i} \left(\frac{e_i + e_j}{D_{ij}} \right) \tag{14}$$

Where D_{ij} represents the distance between class i and class j, and e_i, e_j represents the average error of class i and class asdf, respectively. $Sil^{(+)}$ is proposed by Rousseeuw and is defined as the following:

$$\alpha_{Sil^{(+)}} = \frac{1}{N'} \sum_{R_j \in v_{min}} R_j \tag{15}$$

Where a_j denotes the average distance between sample R_j and class i, $a_j = (1/N_i)$ $\sum_{i=1,2,\dots,n.i \neq j} d(R_j, R_i)$, N_i represents the number of samples of class i, and b_j represents the minimum average clustering of all classes C of the sample, $b_j = h = 1, 2, \dots, k; h \neq i^{\min}(1/N_h) \sum_{x_i \in c_h} d(R_j, R_i)$.

Through MATLAB simulation, the values of the above four validity indexes under different clustering numbers are calculated, and the optimal clustering number suitable for this algorithm is finally determined.

5 Experimental Simulation and Analysis

The simulation data is derived from the daily load data of 600 users in a certain area.

Optimal Clustering Number Calculation: For the convenience of calculation and observation, the maximum index $Sil^{(+)}, SF^{(-)}$ of the four indexes $DB^{(-)}, Sil^{(+)}, CS^{(-)}$, $SF^{(-)}$ is transformed into a very small one by simple reciprocal change, and the four indexes are standardized to be taken between [0, 1]. At this point, the number of clusters corresponding to the four indexes 0 is the best clustering number. The cluster number is set to 2, 3, 4, 5, 6 respectively. The simulation results are shown in Figure 1. The experimental results are simulated by MATLAB.

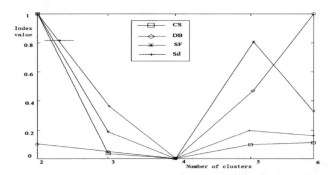

Fig. 1. Clustering effective index evaluation result chart

From Fig. 1, it is obvious that when the cluster number is 4, the four indexes of $DB^{(-)}, Sil^{(+)}, CS^{(-)}, SF^{(-)}$ are all 0. Therefore, the clustering effect is the best when the user is divided into four categories according to the change of the user's electricity consumption on the weekly holiday day. The following MATLAB simulation will be divided into four categories of users, and do specific analysis.

– **Clustering Analysis of Electricity Consumption Behavior Based on Weekly Load Correlation:** According to the above, the number of clusters will be located in 4, and the users are finally divided into four categories. Combine the characteristic value curves of various users (as shown in Fig. 2) and the typical user's weekly power load curve (Fig. 3) to analyze the correlation of the weekly electricity usage behavior of each type of users and the typical electricity consumption behavior.

As shown in Fig. 2(a), the five characteristic values of the first type of users are between 0.8 and 0.9, and the volatility is small, indicating that the user's power consumption behavior is stable within one week. Select a typical user of this class, and the correlation feature value vector is (0.8924, 0.8927, 0.9419, 0.9141, 0.8871). The correlation coefficient between the five working days and the weekly rest day of the user is 0.85 or more, indicating that the daily service load on the weekday and the working day has a high correlation. The user's power load curve for one week is shown in Fig. 3(a). As can be seen from the figure, the daily load curve of the user is consistent in one week, and the peak-to-valley value and peak-to-valley time of the weekday and the working day are basically the same. The behavior of such users on weekdays does not change. It is likely to be an elderly family or a vacant house with stable living schedules. The weekdays have little effect on their electricity use.

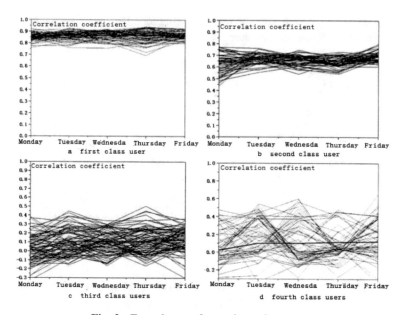

Fig. 2. Four classes of user eigenvalue curves

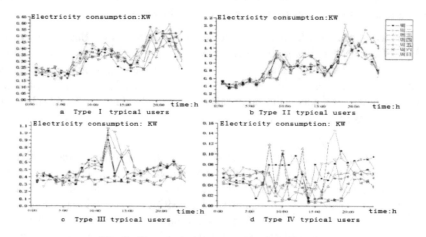

Fig. 3. Four typical users one week load curve

As shown in Fig. 2(b), the characteristic values of the second type of users are about 0.6–0.7. The volatility is low, which indicates that the consumer's electricity consumption behavior still has a strong correlation within a week. The typical users of this class of users are selected and the correlation eigenvalue vector is (0.6965, 0.7188, 0.6957, 0.6226, 0.6882). The one-week load curve is shown in Fig. 3(b). It can be seen from the figure that the user's working day and the weekly rest day are regular, but there are obvious differences: compared with the weekday, the peak and valley value of the daily load curve of the weekly holiday day is out at the same time as the peak and valley value. It is different now. The peak value of power consumption appears at 8 am and 7 pm, and the peak value of weekly holiday is delayed at 12:00 and 9 pm, and the peak value becomes smaller. This kind of consumer's daily power consumption behavior on a weekly break has a regular change, and the change basically occurs at the peak and valley time and the peak and valley value. That is to say, the consumer's weekly daily power consumption behavior is partly changed, which is likely to be a small commercial user affected by the weekly holiday day. An old man with a working family.

As shown in Fig. 2(c), the characteristic values of the third type of users are concentrated between −0.2 and 0.2, and the correlation is small, but the fluctuation is small. Select a typical user in this class and the correlation feature value vector is (0.0483, −0.1067, 0.1755, −0.0707, and 0.0655). Although the correlation coefficient of the user's power load curve for 5 working days and weekly rest days is very low, the volatility is small; indicating that the user has a large difference in power usage behavior between weekdays and weekdays, but each has a strong regularity. The user's weekly power load curve is shown in Fig. 3(c). The working day load curve starts to rise from around 8:00 am, and the highest power peak appears at around 12 noon, and then starts to fall back. On weekdays, the electricity consumption for the whole day is kept at a low level of around 0.3 kWh, and the daily load curve of the weekday and the working day is completely different. The behavior of such users on Sundays and Sundays has completely changed, which may be the office workers who have changed their electricity usage on weekdays.

As shown in Fig. 2(d), most of the characteristic values of the fourth type of users are around 0.2 to 0.4, which is obviously volatility. Select a typical user in this class, and the correlation feature value vector is (0.0483, −0.1067, 0.1755, −0.0707, 0.0655). The user's weekly power load curve is shown in Fig. 3(d). The daily load curve of the user starts to change at around 8:00 in the morning, but some power consumption increases, and some power consumption decreases. The changes are different. The user's power usage behavior has strong randomness, that is, the user's power usage behavior is irregular. The daily power consumption behavior of this type of users varies greatly within a week, so the eigenvalues are very low, which is similar to the range of user eigenvalues shown in Fig. 3(c), but the volatility is significant. The variances of the five eigenvalues of the typical users of Fig. 3(c) and (d) are 0.0056 and 0.0646, respectively, which are significantly different by an order of magnitude. It is likely that the holiday and other factors have a considerable impact on user behavior, resulting in the user's daily electricity usage behavior does not show obvious regularity.

6   Conclusion

The research in this paper is based on the weekly time scale and combined with the clustering algorithm to analyze the user's electricity usage behavior. Firstly, the power consumption area is selected and processed as the data source. Next, the correlation coefficient between the user's working day and the weekly rest day power load is calculated to obtain the eigenvalue. Finally, the cluster validity index analysis is used to determine the most. The number of clusters is excellent, and the users are divided into four categories: the user's weekly power consumption behavior does not change, partial change, complete change, and no obvious rules, and analyzes various users, and the conclusion can be obtained by distinguishing the user's working day and The theoretical basis for the typical electricity use behavior on weekdays and the arrangement of the orderly power-off decision-making measures provide a new idea for the analysis and prediction of power load in China.

References

1. Wu, G., Lin, H., Fu, E., et al.: An improved K-means algorithm for document clustering. In: International Conference on Computer Science and Mechanical Automation (2016)
2. Ahmmed, R., Hossain, M.F.: Tumor detection in brain MRI image using template based K-means and fuzzy C-means clustering algorithm. In: International Conference on Computer Communication and Informatics (2016)
3. Elsheikh, T.M., Kirkpatrick, J.L., Fischer, D., et al.: Does the time of day or weekday affect screening accuracy? A pilot correlation study with cytotechnologist workload and abnormal rate detection using the ThinPrep imaging system. Cancer Cytopathol. 118(1), 41–46 (2010)
4. Novais, R.N., Rocha, L.M., Eloi, R.J., et al.: Burnout syndrome prevalence of on-call surgeons in a trauma reference hospital and its correlation with weekly workload: cross-sectional study. Rev. Colegio Brasileiro Cirurg. 43(5), 314 (2016)
5. Pan, S., Wang, X., Wei, Y., et al.: Cluster analysis for occupant-behavior based electricity load patterns in buildings: a case study in Shanghai residences. Build. Simul. 10(12), 1–10 (2017)
6. Vijay, V., Vp, R., Singh, A., et al.: Variance based moving K-means algorithm. In: IEEE International Advance Computing Conference (2017)
7. Cai, Y., Liang, Y., Fan, J., et al.: Optimizing initial cluster centroids by weighted local variance in K-means algorithm. J. Front. Comput. Sci. Technol. (2016)
8. Hong, H., Tan, Y., Fujimoto, K.: Estimation of optimal cluster number for fuzzy clustering with combined fuzzy entropy index. In: IEEE International Conference on Fuzzy Systems (2016)
9. Vávra, J., Hromada, M.: Determination of optimal cluster number in connection to SCADA. In: Computer Science On-line Conference (2017)
10. Jian, L., Zhao, J., Yan, C., et al.: Analysis of customers' electricity consumption behavior based on massive data. In: International Conference on Natural Computation (2016)
11. Diao, L., Sun, Y., Chen, Z., et al.: Modeling energy consumption in residential buildings: a bottom-up analysis based on occupant behavior pattern clustering and stochastic simulation. Energy Build. 147, 47–66 (2017)
12. Manekar, A.S., Pradeepini, G.: Cloud based big data analytic: a review. Int. J. Cloud-Comput. Super-Comput. 3(1), 7–12 (2016)
13. Hari Krishna, T.: Role of kernel in operating system survey. Int. J. Private Cloud Comput. Environ. Manage. 3(1), 17–20 (2016)
14. Kothapalli, A.: Extraction of patterns on datasets using clustering techniques. Int. J. Private Cloud Comput. Environ. Manage. 3(2), 21–28 (2016)
15. Deepika, C.L.N.: Data access control for multiauthority storage system. Int. J. Private Cloud Comput. Environ. Manage. 4(1), 1–8 (2017)
16. Kavitha Lakshmi, K.: Implementation of different Patterns for human Activities. Int. J. Urban Design Ubiquit. Comput. 4(2), 27–40 (2016)

17. Yang, M., Kim, J., Kim, Y.: A basic study on the planning of the age-integrated facilities as an inter-city exchange space in response to elderly society. Int. J. Urban Design Ubiquit. Comput. **6**(1), 1–6 (2018)
18. Yang, L., Yi, S.J., Mao, X., Li, Y.F., Jiang, S., Xue, L.J.X.: Corn germ the working principle of the directional ordering research. Int. J. Internet Things Big Data. **1**(1), 45–55 (2016)
19. Byun, S.: Design of efficient index management for column-based big databases. Int. J. Internet Things Big Data. **2**(1), 23–28 (2017)
20. Kim, J.B.: An empirical study of effective ways for improving big data project. Int. J. Adv. Res. Big Data Manage. Syst. **1**(2), 23–28 (2017)
21. Sivamani, S., Venkatesan, S.K., Shin, C., Park, J., Cho, Y.Y.: Intelligent food control in a livestock environment. Int. J. Internet Things Appl. **2**(1), 1–6 (2018)

Feature Extraction Using Semantic Similarity

Eman M. Aboelela$^{(\boxtimes)}$, Walaa Gad, and Rasha Ismail

Faculty of Computer and Information Sciences, Ain Shams University,
Cairo, Egypt
{eman_aboelela, walaagad, rashaismail}@cis.asu.edu.eg

Abstract. Recently, opinion mining has become a focus in the field of natural language processing and web mining. Due to the massive amount of users' reviews on the web about some entities or services, opinion mining is appeared to track users' emotions and feelings. Sentiment analysis is a synonym to opinion mining. Feature extraction is an important task in the sentiment analysis process. So, in this paper, a novel model is proposed to extract the most related features to a product from customer reviews using semantic similarity. Wordnet taxonomy and Stanford Part of Speech (POS) tagger are used in the feature extraction process. The extracted features are very important to generate a meaningful feature based product reviews summery which helps the customers to make a decision. The experiments are performed on three different datasets. The proposed model achieves promising results in terms of Precision, Recall and F-measure performance measures.

Keywords: Opinion mining · WorldNet ·
Semantic similarity and stanford POS tagger

1 Introduction

Social networks, such as Facebook, Twitter, Instagram, etc., have many users around the world. The product producers use them to market products and services. Moreover, consumers express opinions and share their experience with the service or product by commenting [1]. Thus, the social network has more and more online comments or reviews about products or services. There are two main review formats on the web which are structured and unstructured reviews [2]. The structured review is a questionnaire about the product or its features. It is written in a form of pros and cons format. The product reviewer writes what is positive and what is negative in pros and cons sections. Moreover, a reviewer can add a detailed review besides the pros and cons [3]. The second format is the unstructured review. The reviewer writes a free text in natural language without pros and cons. The structured reviews are easier than the unstructured reviews in processing and analysis. This work focuses on the unstructured review format. Because of the unstructured reviews are written in natural language, they could have noise or spam words. Therefore, reviews should be analyzed to help the consumers to make a decision whether to buy the product or not and help the producers to know the consumer feedback about their products and services. Such analysis is difficult to be done manually, therefore the Opinion Mining or Sentiment Analysis is proposed. The main objective of Sentiment

© Springer Nature Switzerland AG 2020
A. E. Hassanien et al. (Eds.): AISI 2019, AISC 1058, pp. 82–91, 2020.
https://doi.org/10.1007/978-3-030-31129-2_8

Analysis is to detect the polarity or orientation of the review. The Opinion Mining process can be performed at three different levels. The first one is document (review) level. The whole review is classified as positive or negative. The second level is the sentence level. The sentence is classified as positive or negative. The third one is the aspect (feature) level. The review is classified based on the aspects appeared in the sentences. Aspects of an entity 'e' are the components and attributes of 'e'. The aspect level is more accurate than the other two levels because both document and sentence levels do not identify what exactly users like or dislike. The aspect level solves this problem by classifying review based on each feature or aspect. There are two types of aspects appeared in reviews. The first type is implicit aspects. They are deduced from the review context or meaning. The second one is explicit aspects, which are written explicitly in reviews. This work focuses on the second type 'explicit features'. Two main tasks are involved in the opinion mining process which are features extraction and reviews classification. As the accuracy of classification depends on the result of feature extraction, this paper presents a novel model for the feature extraction process based on the semantic similarity to find the accurate features of a product. The rest of the paper is organized as follows. Section 2 presents related work. The proposed model is introduced in Sect. 3. In Sect. 4, experiments and results are shown. Conclusion and future work are discussed in Sect. 5.

2 Related Work

An integrated hybrid framework is presented in [4] for aspect-based opinion mining. There are four main modules: aspect-sentiment extraction, aspect grouping, aspect-sentiment classification, and aspect-based summary generation. Lexicon based approach with the help of a revised corpus is to improve the accuracy of classification. The feature extraction module is an enhancement of the work proposed by Khan et al. [2]. An extended set of heuristic patterns are used to extract aspects from user reviews. The proposed approach considered nouns, noun phrases and verbs are candidate terms for the aspects. An example of the proposed patterns is, if the first word is adjective 'JJ', the second word is 'TO' and the third word is verb 'VB' such as the sentence "easy to use". Then the word 'use' is considered to be a verb aspect.

A syntactic pattern approach based on features observation is presented in [5] to extract aspects from unstructured reviews. Stanford POS tagger is used to get the noun phrase aspects that match the proposed patterns. It works on POS tag of the current word, POS tag of (current−1) and POS tag of (current+1). It is applied on explicit features only.

In [2], semantic relation and syntactic sequence are used to extract product features from customer reviews. For semantic relation, polarity adjectives are used, and for the syntactic sequence, linguistic patterns are used. The aspect classification task was not performed in this work. The proposed approach achieved more accurate results than the comparing methods.

A pattern knowledge approach is presented in [3] to extract noun/noun phrase features and opinionated phrases from reviews. The patterns consist of two or three words such as (adjective, singular noun/plural noun) and (adverb/comparative adverb/superlative adverb, adjective/adverb/comparative adverb/superlative adverb, singular noun/plural noun). Stanford POS tagger is used to get the POS tag information of each word. The extracted features are compared to the features in the training corpus to calculate the feature extraction Recall, Precision, and F-measures.

Nouns and noun phrases are considered in [6] as product features and adjectives as opinion words that express features. Association mining rules and pruning approach are used to find the most frequent item sets/features. Wordnet is used to get the adjective synonym set and antonym set in order to detect its sentiment orientation.

Unsupervised recommendation based approach is presented in [7]. The proposed approach aims at enhancing the syntactic rule-based method for feature extraction. The recommendation is based on two main tasks. The first one is a semantic similarity and the second one is aspect associations. Recommendation based on semantic similarity aims at using the aspect synonyms. The synonym of an aspect is recommended to be a feature based on the semantic similarity between the two words (aspect and its synonym). Recommendation based on aspect associations aims at recommending words that co-occur with the aspect across domains by using association rules from data mining. The proposed approach aims at improving the double propagation method (DP) through aspect recommendation. Eight review datasets are used to show the impact of the proposed approach.

Most of the previous work assumed that all extracted features have the same degree of importance and relevance. They did not take into considerations the spam words exist in the generated features set which affect the overall performance of the opinion mining process. For this reason, a novel model is proposed to decrease the number of generated spam words in the features set and find the accurate features of a product. The model proposed calculates a score denoting how a feature related to a target product by using the semantic similarity.

3   Proposed Model for Feature Extraction Using Semantic Similarity

The main steps of the proposed model for feature extraction are shown in Fig. 1. Firstly, noun and noun phrase words are extracted from reviews using Stanford Part of Speech (POS) tagger. Secondly, the frequencies of the extracted words are calculated and the most frequent features are identified. Thirdly, a semantic similarity score is measured for the extracted frequent words using Wordnet. Finally, the words/features that have semantic similarity scores exceed the predefined threshold, have been chosen.

3.1   Feature Extraction

In this paper, aspects are defined as the most important product features that most reviewers talk. Nouns and noun phrases are considered to be explicit product features. Example of noun features, "great screen and great sound". The two words "screen" and "sound" have POS = 'NN' and they are noun features of the mobile phone product. Example of phrase features, "the battery life is amazing", "battery life" is a noun phrase

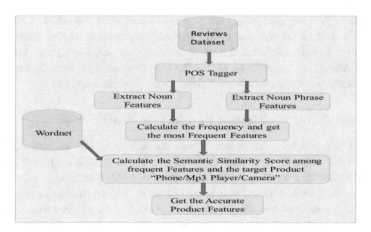

Fig. 1. Proposed model for extracting product features using semantic similarity

feature of the product camera and has POS = 'NP'. Noun phrase features may be a sequence of nouns such as "sound quality" or a combination of adjective and noun such as "signal strength".

Algorithm 1: Feature Extraction Algorithm
Input: Reviews Dataset →RD
Output: list of frequent words →L
1.   **for each** Review R in RD **do**
2.   Get a list of Review Words →RW
3.   Get a POS tag (RW) → Tagged_Rev
4.   Parse (Tagged_Rev) → Parsed_Rev
5.   Define list for Noun Words →NL
6.   Define list for Noun Phrase Words →PL
7.   **for each** Word W in Parsed_Rev **do:**
8.   **if** POS(W)=='NN/NNS'
9.   NL.append(W)
10.   **end if**
11.   **else if** POS(W)=='NP:{<NN><NN> <JJ><NN>}'
12.   PL.append(W)
13.   **end if**
14.   **end for**
15.   **end for**
16.   **for each** Noun Word NW in NL **do**
17.   calculate frequency(NW) → Nf
18.   **if** Nf > threshold
19.   L.append(NW)
20.   **end if**
21.   **end for**
22.   **for each** Phrase Word PW in PL **do**
23.   calculate frequency(PW) →NPf
24.   **if** NPf > threshold
25.   L.append(PW)
26.   **end if**
27.   **end for**
28.   **return** L

Therefore, the proposed model extracts nouns and noun phrases from reviews by using Stanford POS tagger. Then, the frequency of each extracted noun and noun phrase feature is calculated and the most frequent ones are identifying by passing a predefined threshold. Algorithm 1 shows the feature extraction process. The extracted noun and noun phrase words are candidates to be product features.

3.2   Feature Extraction Based on Semantic Similarity

After extracting the frequent words which are significant from products' datasets (mobile phone, camera, mp3 player), there some words which not important (not features of products) are observed. For example, the word 'Samsung' is extracted to be a feature of the mobile phone but it is not a feature. Moreover, the phrase 'extended warranty' in the mp3 player dataset and the phrase 'canon g3' in camera dataset are assumed to be product features but, they are not. These noisy words affect the overall performance of the opinion mining process.

For this reason, the using of semantic similarity is proposed to decrease the number of generated noisy words and get the closest features of each product. Therefore, to get the accurate features of mobile phone product, find the semantic similarity between the word "phone" and each extracted frequent word "noun or noun phrase". The words that exceed a predefined threshold are considered to be the phone features. The same algorithm is performed on the other two products (mp3 player, camera) but, each with a different threshold. For the mp3 player product, the semantic similarity is measured between the phrase "mp3 player" and its extracted frequent features. The same algorithm is used for camera product but the semantic similarity will be calculated between the word "camera" and its extracted frequent features.

The WU-Palmer semantic similarity measure in WorldNet taxonomy is used. The WU-Palmer defines the semantic similarity between two-word senses based on their depth in the taxonomy and their Least Common Subsumer as described in Eq. 1. Algorithm 2 describes how to apply the WU-Palmer semantic similarity on the extracted frequent features.

$$\text{WUP} - \text{Sim} = \frac{2 * \text{depth}(\text{lcs}(c_1, c_2))}{\text{depth}(c_1) + \text{depth}(c_2)} \tag{1}$$

where depth (c1, c2), is the depth of the two senses in the taxonomy. Lcs(c1, c2) is the least common subsume of two senses c1 and c2.

Algorithm 2: Feature Extraction based on semantic similarity

Input: list of frequent words →L
Output: list of related features to the product →FL

1.   **for each** Word W in L **do**
2.   Get product synsets ("phone/camera/mp3 player") →P_Syns
3.   Get word synsets (W) →W_Syns
4.   Get cross product (P_Syns , W_Syns) →CP
5.   Define variable →max= -1
6.   Define variable →score= 0
7.   **for each** items I, J in CP **do**
8.   score= I.wup_similarity(J)
9.   **if** score> max
10.   max=score
11.   **end if**
12.   **end for**
13.   **if** max > predefined threshold
14.   FL.append(W)
15.   **end if**
16. **end for**
17. **return** FL

4   Experiments and Results

4.1   Dataset Description

The proposed model is evaluated using three different product datasets: a mobile phone, an mp3 player and a camera. They are collected from the Amazon website (www.amazon.). Each dataset has approximately 260 reviews written by 325 customers. Figure 2 shows a sample review from mobile phone dataset.

```
1 [t]glad to own .
2 ##i have had this phone for about 5 months .
3 battery[+2]##i treat the battery well and it has lasted .
4 ##at my heaviest usage , i must recharge the battery after 3 days .
5 ##the battery lasts about 5 days otherwise and has lasted up to 10 when i was making very few calls .
6 ##signal strength will affect the battery life .
7 ##frequent signal searches eat up battery power .
8 ##i took great care of the screen , till i realized that the part that might get scratched is part of the cover .
```

Fig. 2.   A sample review from mobile phone dataset

4.2   Performance Measures

Three performance measures are used to assess the proposed model. Precision, Recall, and F-measure defined as follows:

Precision is also called positive predictive value. It is the ratio of correctly predicted positive observations to the total predicted positive observations. High precision value means that the true positive cases are more than false positive cases. Precision can be computed as shown in Eq. 2.

$$\text{Precision} = \frac{T_p}{F_p + F_n} \tag{2}$$

A recall is known as sensitivity. It is the ratio of correctly predicted positive observations to all observations in actual class. Recall formula is shown in Eq. 3.

$$\text{Recall} = \frac{T_p}{T_p + F_n} \tag{3}$$

F-measure is the weighted average of Precision and Recall. Therefore, this score takes both false positives and false negatives into account. F-measure can be computed by Eq. 4.

$$\text{F} - \text{measure} = 2 \ * \ \frac{\text{Recall} \ * \ \text{Precision}}{\text{Recall} \ + \ \text{Precision}} \tag{4}$$

Where TP: is a true positive instance, TN: is a true negative instance, FP: is a false positive instance and FN: is a false negative instance. The performance measures are calculated for each aspect in the feature lexicon.

4.3   Analysis and Results

The feature extraction process is executed on three different products which namely the mobile phone, mp3 player, and camera to extract their frequent words or features which pass a predefined threshold. The type and the number of extracted frequent features before and after applying the proposed model are presented in Table 1.

Table 1. The number of extracted noun and noun phrase words.

Datasets	Word types			
	Before applying the propose model		After applying the proposed model	
	Noun	Noun phrase	Noun	Noun phrase
Camera	13	11	8	7
Mobile phone	27	10	14	3
Mp3 player	12	22	5	10

Moreover, the feature extraction process is evaluated based on three experiments to show the efficiency of the proposed model. The first experiment is extracting the most frequent noun words only from the product reviews. The second experiment is complementary to the first one by adding the most frequent noun phrase words. After applying the first and second experiments, some spam words are found which are not related directly to the product. Therefore, the last experiment is proposed to find the actual aspects/features of products by applying the Wu-Palmer semantic similarity. The experiments are executed on the previously mentioned products (mobile phone, mp3 player, and camera) and the results of each product dataset are shown in Table 2. Three performance measures are used for evaluation which called precision, recall and, f-measure.

Table 2. Performance measures of three products' datasets

Experiment	Datasets								
	Camera			Mobile phone			Mp3 player		
	Precision (%)	Recall (%)	F-measure (%)	Precision (%)	Recall (%)	F-measure (%)	Precision (%)	Recall (%)	F-measure (%)
Extracting frequent noun words only	2.5	21.6	4.6	81.8	100	90	1.2	9.3	2.1
Extracting frequent noun and noun phrase words	1.8	23.7	3.4	4.7	59	8.7	1.2	16.8	2.2
The proposed model	100	92.9	96.3	76.5	72.2	74.3	93.8	83.3	88.2

Figure 3 and Table 3 show the average precision, recall, and f-measure of the three products' datasets after applying the three experiments. As presented in Fig. 3 and Table 3 the average precision, recall, and f-measure achieved '28.5%', '43.6%', and '32.2%' respectively after extracting the frequent noun words from reviews. However, the measures decreased and achieved average precision, recall, and f-measure equal to '2.6%', '33.2%', and '4.8%' respectively after adding the frequent noun phrase words to the frequent noun words. This happened because the extracted noisy words 'which are not features of the product' are increased. The performance of the feature extraction process is improved after applying the Wu-Palmer semantic similarity among the extracted frequent noun and noun phrase words and the target product and achieved average precision, recall, and f-measure equal to '90.1%', '82.8%', and '86.3%' respectively.

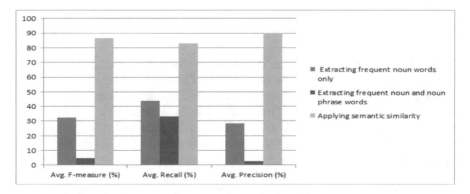

Fig. 3. Performance of feature extraction process using different experiments

Table 3. Performance of feature extraction process using different experiments

Experiment	Measures		
	Avg. Precision (%)	Avg. Recall (%)	Avg. F-Measure (%)
Extracting frequent noun words only	28.5	43.6	32.2
Extracting frequent noun and noun phrase words	2.6	33.2	4.8
The proposed model	90.1	82.8	86.3

The average precision, recall, and f-measure of the presented model are compared to the methods in [2–6] for feature extraction. The methods presented in [2–5] concerned on extracting the opinionated product features based on predefined patterns and in [6] focused on finding the features related to the product by using association mining rules. But, the proposed model aims at eliminating the generated spam words and extracting the closest features to a product by calculating the semantic similarity between the product and its candidate features. The model proposed achieved improved results in terms of precision, recall, and f-measure as given in Table 4.

Table 4. Comparison between the proposed model and other approaches

Method	Measures		
	Avg. precision (%)	Avg. recall (%)	Avg. F-measure (%)
Khan et al. [2]	78.98	71.77	75.19
Haty et al. [3]	73.3	85.7	79
Asghar [4]	83.46	71.0	77.16
Maharani et al. [5]	62.6	72.8	67.2
Hu and Liu. [6]	72	80	75.8
The proposed model	90.1	82.8	86.3

5   Conclusion and Future Work

This work concerned on the feature extraction task in sentiment analysis. The paper aims at finding the closest features to a product. These features help in generating a meaningful feature based reviews summery. Stanford POS tagger is used to extract the noun and noun phrase features from unstructured customer reviews. WU-Palmer semantic similarity measure in Wordnet taxonomy is calculated between the target product and its candidate features to get the most accurate product features. Three different datasets (mobile phone, mp3 player and camera) are used to show the effectiveness of the proposed model. The experiments showed improved results in terms of Recall, Precision, and F-measure performance measures. The next step is to extract the opinion words that express the product features from customer reviews and generate the feature based reviews summery.

References

1. Arunkarthi, A., Gandhi, M.: Aspect-based opinion mining from online reviews. Res. J. Pharm. Biol. Chem. Sci. **7**(3), 494–500 (2016)
2. Khan, K., Baharudin, B., Khan, A.: Identifying product features from customer reviews using hybrid patterns. Int. Arab J. Inform. Technol. **11**(3), 281–286 (2014)
3. Htay, S.S., Lynn, K.T.: Extracting product features and opinion words using pattern knowledge in customer reviews. Sci. World J. **2013**(3), 1–5 (2013)
4. Asghar, M.Z., Khan, A., Zahra, S.R., Ahmad, S., Kundi, F.M.: Aspect-based opinion mining framework using heuristic patterns. Cluster Comput. **20**, 1–19 (2017)
5. Maharani, W., Widyantoro, D.H., Khodra, M.L.: Aspect extraction in customer reviews using syntactic pattern. Proc. Comput. Sci. **59**(Iccsci), 244–253 (2015)
6. Hu, M., Liu, B.: Mining and summarizing customer reviews. In: Proceedings of the Tenth ACM SIGKDD International Conference on Knowledge Discovery and Data Mining, pp. 168–177. ACM (2004)
7. Liu, Q., Liu, B., Zhang, Y., Kim, D.S., Gao, Z.: Improving opinion aspect extraction using semantic similarity and aspect associations. In: Proceedings of the Thirtieth AAAI Conference on Artificial Intelligence, AAAI-16, pp. 2986–2992 (2016)
8. https://wordnet.princeton.edu/
9. http://www.nltk.org/howto/wordnet.html
10. http://nlp.stanford.edu/software/tagger.shtml/
11. Asgarian, E.: Subjective data mining on the web. Ferdowsi University of Mashhad (2014)
12. Vivekanandan, K., Aravindan, J.S.: Aspect-based opinion mining: a survey. Int. J. Comput. Appl. **106**(3), 0975–8887 (2014)
13. Liu, Q., Gao, Z., Liu, B., Zhang, Y.: A logic programming approach to aspect extraction in opinion mining. In: IEEE/WIC/ACM International Conferences on Web Intelligence (WI) and Intelligent Agent Technology, (IAT) (2013)
14. Gupta, N., Kumar, P., Gupta, R.: Automated extraction of product attributes from reviews. CS 224N Final Project (2009)
15. Mishra, N., Jha, C.K.: Classification of opinion mining techniques. Int. J. Comput. Appl. **56**(13), 0975–8887 (2012)
16. Sharma, R., Nigam, S., Jain, R.: Mining of product reviews at aspect level. Int. J. Found. Comput. Sci. Technol. **4**(3), 87–95 (2014)

Big Data Analytics Concepts, Technologies Challenges, and Opportunities

Noha Shehab[(⊠)], Mahmoud Badawy[(⊠)], and Hesham Arafat[(⊠)]

Computers Engineering and Control Systems Department,
Faculty of Engineering, Mansoura University, Mansoura, Egypt
noha.a.shehab@students.mans.edu.eg,
{engbadawy, h_arafat_ali}@mans.edu.eg

Abstract. The rapid observed increase in using the Internet led to the presence of huge amounts of data. Traditional data technologies, techniques, and even applications cannot cope with the new data's volume, structure, and types of styles. Big data concepts come to assimilate this non-stop flooding. Big data analysis process used to jewel the useful data and exclude the other one which provides better results with minimum resource utilization, time, and cost. Feature selection principle is a traditional data dimension reduction technique, and big data analytics provided modern technologies and frameworks that feature selection can be integrated with them to provide better performance for the principle itself and help in preprocessing of big data on the other hand. The main objective of this paper is to survey the most recent research challenges for big data analysis and preprocessing processes. The analysis is carried out via acquiring data from resources, storing them, then filtered to pick up the useful ones and dismissing the unwanted ones then extracting information. Before analyzing data, it needs preparation to remove noise, fix incomplete data and put it in a suitable pattern. This is done in the preprocessing step by various models like data reduction, cleaning, normalization, preparation, integration, and transformation.

Keywords: Big data analytics · Big data preprocessing · Feature selection · Flink

1 Introduction

Internet applications usage is rapidly increasing in most life aspects which boosted life simplicity and resulted in overwhelming quantities of data that needed to be processed and shared between people, applications, and companies. Resulted data poses a great challenge regarding the means of managing and efficiently using it in different applications [1] which led to the "Big data" term [2]. Big data refers to any set of data that require large capabilities in terms of storage space and time to be analyzed [3, 4]. It is the big volume of the data sets in large numbers which may be in-consistence or have incorrect values or both [5]. And the process of gathering, management, analysis of data to generate knowledge and reveal hidden patterns [6].

© Springer Nature Switzerland AG 2020
A. E. Hassanien et al. (Eds.): AISI 2019, AISC 1058, pp. 92–101, 2020.
https://doi.org/10.1007/978-3-030-31129-2_9

Addressing big data is a challenging and time-demanding task that requires a huge computational infrastructure to guarantee effective, successful data processing and analysis [7]. To be addressed, its characteristics are presented in big data multi V Model. Collected big data datasets consists of structured, unstructured, and semi-structured ones that need to be processed [8]. Processing the huge amounts is a trail of madness because of two main reasons the processing itself is not an easy task and data may not be important as it may include noise, uncompleted or incorrect messages that consume processing power and time without any advantage. The improvement of data management, data mining systems, data preprocessing and the reduction was promoted and become essential techniques in current knowledge discovery scenarios, dominated by increasingly large datasets [9].

Big data technologies facilitated and fastened data processing that increased decision-making efficiency and improved companies' overall performance. All Data that collected around of us is considered a big data source [10]. Managing these quantities of data cannot be efficiently processed without understanding the characteristics of big data which can be summarized in the Vs. model (Variety, Velocity, Volume, Veracity, Value) which measures of the correctness and accuracy of information retrieved [11]. As a trending phenomenon, big data poses challenges that can be divided into three main categories data, processing, and management challenges. Figure 1 lists the most common big data challenges [12–14]. The rest of this paper is organized as follows, Sect. 2 discusses big data analytics meaning, architecture, technologies, and challenges. Section 3 discusses data preprocessing related concepts, latest technologies used, and challenges. Section 4 concludes the paper.

2 Big Data Analytics

Big data consists of the large collection of heterogeneous datasets with a very high degree of dimensionality and sparseness that may cause conflicts. Big data analytics is the process of examining big data to uncover hidden patterns, unknown correlations, and other useful information that can be used to make better decisions [15]. The analysis is done using data analytics frameworks which have been built to scale out parallel executions in the purpose of constructing valuable information from big data. As a result, it affords the potential for shifting the collective behavior toward alternative techniques, such as a focus on precise parameter estimation. Big data analytics is a channel of acquisition, extraction, cleaning, integration, aggregation, and visualization, analysis and modeling, and interpretation [16]. The analysis process involves five main steps data integration, data management, data preprocessing, data mining, and knowledge presentation that are independent of the application field analytics used in [5]. Big data analytic concepts can be classified into three main stages descriptive analytics, predictive analytics, and prescriptive analytics [17].

Descriptive analytics helps in understanding what already happened to the data. Predictive analytics helps in anticipating what will happen to it, prescriptive analytics helps in responding now what and so what, and advanced analytics combines predictive and prescriptive analytics to provide a forward-looking perspective that makes use of techniques encompassing a wide range of disciplines including simulation,

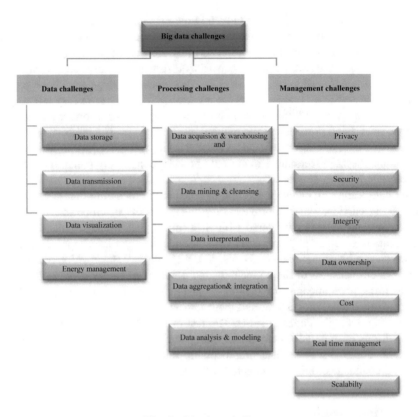

Fig. 1. Big data challenges

learning, statistics, machine, and optimization [18]. Big data analytical methods can be classified into classical methods and modern ones. Classical methods are text analytics, audio analytics, and video analytics [19]. Modern data analytic methods such as random forests, artificial neural networks, and support vector machines are ideally suited for identifying both linear and nonlinear patterns in the data [20].

2.1 Big Data Analytics Architecture

Big data analytics architecture is rooted in the concept of data lifecycle framework that starts with data capture, proceeds via data transformation, and culminates with data consumption. Best practice big data analytics architecture that is loosely comprised of five major architectural layers (1) data, (2) data aggregation, (3) analytics, (4) information exploration, and (5) data governance [20]. Data layer contains all the data sources necessary to provide the insights required to support daily operations and solve business problems [21]. Data is classified into structured data, semi-structured data, and unstructured ones. Data aggregation layer is responsible for handling data from the various data sources. In this layer, data is intelligently digested by performing some steps, data acquisition to read data provided from various communication channels, frequencies, sizes, and formats. Then transformation, cleaning, splitting, translating, merging, sorting, validating the data, After that the storage, data loaded into the target databases such as Hadoop distributed file systems (HDFS) or in a Hadoop cloud for further processing and analysis [22].

Table 1. Big data analytics research issues

Paper	Application	Algorithm/proposed methods	Results
Big data analytics in supply chain management between 2010 and 2016: insights to industries	Handling huge data and facilitates the real-time monitoring that enhances the speed and flexibility of supply chain decision	Machine learning algorithms	They recommend implementing big data analytics in specific industrial applications as it supports supply chain principles
A proposed contextual model for big data analysis using advanced analytics	Proposed a contextual model for resume analytics to determine the match between a job and candidate(s)	Advanced analytics in the form of predictive and prescriptive analytics	The system gives a summary of each resume. Based on it resumes are reordered which facilitate the checking process of them
Mobile agent-based MapReduce framework for big data processing	Mobile agent-based MapReduce	Uses one mobile agent for each split that moves from one location to another based on the scheduler. Performance score of the agent sent to the scheduler to detect straggler	The proposed algorithm shows the best performance based on time with high straggler percentage rather than Hadoop based one
Robust fuzzy neuro system for big data analytics	Enhance big data analytic process results due to data heterogeneity and inconsistency	Combine fuzzy with neural networks and apply Map Reduce and Hadoop expert knowledge and fuzzy systems to represent big data	The proposed system enhanced the power of fault tolerance and explanation competence
Systematically dealing practical issues associated to healthcare data analytics	Implementation of healthcare data analytic process is not straightforward and to effectively implement	Proposed new health insurance portability and accountability act	The system is useful to unlock the insights from patient data, that, improved outcomes and reducing costs

Analytics layer is responsible for processing all kinds of data and performing appropriate analyses. Information exploration layer generates outputs such as various visualization reports, real-time information monitoring, and meaningful business insights derived from the analytics layer to users in the organization. Data governance layer is comprised of master data management (MDM), data life-cycle management, and data security and privacy management. Target data collected from their sources then stored in distrusted database or clouds, so all are ready to start analyzing this done by different analyzing tools to get useful information and data filtering come at the end of the process to extract useful data according to the target requirement [7]. Researchers pay great attention to the analysis process and its role with scientific and life issues. Table 1 lists the most recent big data analytics open issues by giving a brief about the problem, analytic methods they developed or enhanced, and the results [18, 23, 24, 26, 27].

2.2 Big Data Analysis Frameworks and Platforms

Many platforms for large-scale processing data were introduced to support big data analysis. Platforms can be divided into (processing, storage, analytic). Analytic platforms should be scalable to adapt to the increased amount of data from border perspective like Hadoop, Map Reduce, Hive, PIG, WibiData, PLATFORA, and Rapidminer [27]. Analytics tools are scalable in two directions horizontal scaling and vertical scaling. Vertical scaling helps in installing more processors, more memory and faster hardware, typically, within a single server and involves only a single instance of an operating system. Horizontal scaling distributes the workload across many servers. Typically, multiple instances of the operating system are running on separate machines [28]. Map Reduce, Hadoop, Spark, and Apache Flink are examples of big data analytic horizontal scaling platforms [29]. Table 2 summarizes the difference between Hadoop, Spark and Apache Flink [29, 30, 31].

3 Data Preprocessing

Data preprocessing is one of the major phases within the knowledge discovery process in the analytics process which involves more effort and time within the entire data analysis process and considered as one of the data mining Techniques [9]. The traditional data preprocessing methods (e.g., compression, sampling, feature selection, and so on) are not expected to operate effectively in the big data age [32]. Data preprocessing for data mining is a set of techniques used prior to the application of a data mining method. Also, it is the conversion of the original useless data into a new and cleaned data prepared for further mining analysis [3]. Preprocessing concept involves data preparation, integration, cleaning, normalization and transformation of data, and data reduction tasks [33]. Preprocessing stage can be summarized into two main parts data preparation and data reduction. Data Preparation is considered as a mandatory step that includes techniques such as integration, normalization, cleaning, and transformation. Data reduction performs simplification by selecting and deleting redundant and noisy features and/or instances or by discretizing complex continuous feature spaces

[34]. Preprocessing complexity depends on the data source use [3]. Figure 2 summarizes the most common preprocessing stages and the techniques used in each stage according to the data state.

Table 2. Comparison between Hadoop, Spark, and Flink

Technology	Hadoop	Spark	Flink
Purpose	Provides shared storage and analysis infrastructure	Run on the top of Hadoop	Solve problems derived from micro-batch models like a spark
Programming model	Map Reduce	Directed Acyclic Graph Parallel processing	Dataflow
Support real time	It stores data and performs parallel computation with the help of HDFS and MR	Used for real-time stream data processing	It supports real-time a large amount of data with low data latency and high fault tolerance
How it works	Processes data by using parallel and distributed MapReduce programming paradigm	Supports interactive query	It supports stream and batch data processing using datastream and dataset APIs
Batch or stream	Good for large sequential batch processing	It Is a resilient distributed datasets (RDD) works with a static stream of data	Native streaming processing framework works on both batch data and streams
Data utilization	Doesn't support data re-utilization	Support data re-utilization and iterations	
Speed	It is slower than spark	It is 100 times faster than MapReduce in execution	Slower than Spark
Support real time	Not suitable	Can be used to modify the data in real time	
Cost	Low cost and faster in batch processing	Requires more cost, not suitable for batch processing	

Table 3 summarizes the effect of each preprocessing step on data [35, 36]. Dealing with continues incoming data stream is not a straightforward operation that can be done easily due to many aspects [9]. Analysis frameworks make benefit of the programming model and are built on its principles, Hadoop based on the MapReduce programming model. Spark based on Directed Acyclic Graph Parallel processing, [37, 38]. Data preprocessing techniques face many challenges like scaling the techniques to accommodate the current huge amount of data. Many issues still need more research to be solved in data reduction like that all data should be in a balanced state. Dealing with incomplete data streams still an urgent issue as the removed data may have important information.

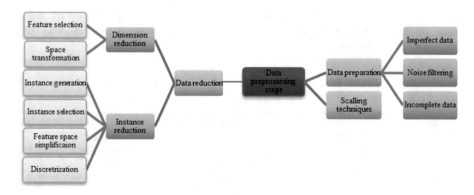

Fig. 2. Data preprocessing stages

Table 3. Preprocessing stages summary

Technique	Data reduction	Data cleaning	Data integration	Data transformation
How it works	Can be achieved via dimensional reduction using a decreasing number of variables in large spaces	Pulls out the information from the underlying sources and express them in a suitable form	Combines data at different autonomous sources and provides the user with a unified view of the data	Scales data into a certain range to be processed more easily
Technologies/techniques	Principle component analysis, singular value decomposition analysis and independent component analysis (ICA) factor analysis	Data mining algorithms	Ontology-based Data access mediator environment for multiple information sources	RDB, XML, JSON and semantic web
Benefits	Extract core data which are simpler and smaller to be stored and processed	Used for data warehousing, mining, and management	Provide reconciled, integrated, virtual view of the data which reduce processing cost	Bring data to sustainable mode to be processed

Big Data Preprocessing Technologies
MLlib, discretization and normalization, feature extraction, feature selection, and feature indexers and encoders are common preprocessing technologies used nowadays. MLlib is a powerful machine learning library that enables the use of Spark in the data analytics field. This library is formed by two packages Mllib and machine learning [39]. Discretization and normalization, Discretization transforms continuous variables using discrete intervals, whereas normalization performs an adjustment of distributions [3]. Feature extraction used to combine the original set of features to obtain a new set of less-redundant variables. Feature selection tries to select relevant subsets of relevant features without incurring much loss of information. Feature indexers and encoders convert features from one type to another using indexing or encoding techniques [40]. Preprocessing technologies performance and usefulness differ from each other from distinct perspectives like effectiveness, time and memory performance and reduction rate [9].

4   Conclusion and Future Directions

Real-world technological revolution and the amount of data generated forced big data to be a challenging compelling task. Big data challenges can be divided into three main categories the data itself, processing, and management. Data analytic is considered as a processing challenge and a mandatory step within the knowledge discovery process. Modern technologies and frameworks like Hadoop, Spark, and Flink were constructed to improve the performance of the analytics process. To achieve the optimal results from the analytics process data preprocessing concept should be applied to remove noise, put data in a suitable form. Computing, quality of results, security and privacy in the analytics process and are challenging issues that needs more efforts.

References

1. Ishikiriyama, C.S., Gomes, C.F.S.: Big data: a global overview. In: Emrouznejad, A., Charles, V. (eds.) Big Data for the Greater Good, pp. 35–50. Springer International Publishing, Cham (2019)
2. Kaisler, S., Armour, F., Espinosa, J.A., Money, W.: Big data: issues and challenges moving forward. In: Proceedings Annual Hawaii International Conference on System Sciences, pp. 995–1004 (2013)
3. García, S., Ramírez-Gallego, S., Luengo, J., Benítez, J.M., Herrera, F.: Big data preprocessing: methods and prospects. Big Data Anal. 1(1), 1–22 (2016)
4. Oussous, A., Benjelloun, F.-Z., Lahcen, A.A., Belfkih, S.: Big data technologies: a survey. J. King Saud Univ. Comput. Inf. Sci. 30(4), 431–448 (2018)
5. Burmester, G., Ma, H., Steinmetz, D., Hartmannn, S.: Big data and data analytics in aviation. In: Durak, U., Becker, J., Hartmann, S., Voros, N. (eds.) Advances in Aeronautical Informatics. Springer International Publishing, Cham (2018)
6. Amini, S., Gerostathopoulos, I., Prehofer, C.: Big data analytics architecture for real-time traffic control. In: 5th IEEE International Conference on Models and Technologies for Intelligent Transportation Systems MT-ITS 2017, pp. 710–715 (2017). Tum Llcm

7. Hashem, I.A.T., Yaqoob, I., Anuar, N.B., Mokhtar, S., Gani, A., Ullah Khan, S.: The rise of big data on cloud computing: review and open research issues. Inf. Syst. **47**, 98–115 (2015)
8. Raghupathi, W., Raghupathi, V.: Big data analytics in healthcare: promise and potential. Heal. Inf. Sci. Syst. **2**, 3 (2014)
9. Ramírez-Gallego, S., Krawczyk, B., García, S., Woźniak, M., Herrera, F.: A survey on data preprocessing for data stream mining: current status and future directions. Neurocomputing **239**, 39–57 (2017)
10. Chen, M., Mao, S., Zhang, Y., Leung, V.C.M
11. Addo-Tenkorang, R., Helo, P.T.: Big data applications in operations/supply-chain management: a literature review. Comput. Ind. Eng. **101**, 528–543 (2016)
12. Miller, K.W., Michael, K.: Big data: new opportunities and new challenges [guest editors' introduction]. Computer **46**(6), 22–24 (2013)
13. Muthulakshmi, P., Udhayapriya, S.: A survey on big data issues and challenges. Int. J. Comput. Sci. Eng. **6**(6), 1238–1244 (2018)
14. Huda, M., et al.: Big data emerging technology: insights into innovative environment for online learning resources. Int. J. Emerg. Technol. Learn. **13**(1), 23–36 (2018)
15. Aggarwal, V.B., Bhatnagar, V., Mishra, D.K.: Big Data Analytics. Advances in Intelligent Systems and Computing, vol. 654. Springer, Cham (2015)
16. Maxwell, S.E., Kelley, K., Rausch, J.R.: Sample size planning for statistical power and accuracy in parameter estimation. Annu. Rev. Psychol. **59**, 537–563 (2008)
17. Sivarajah, U., Kamal, M.M., Irani, Z., Weerakkody, V.: Critical analysis of big data challenges and analytical methods. J. Bus. Res. **70**, 263–286 (2017)
18. Ramannavar, M., Sidnal, N.S.: A proposed contextual model for big data analysis using advanced analytics. Adv. Intell. Syst. Comput. **654**, 329–339 (2018)
19. Vashisht, P., Gupta, V.: Big data analytics techniques: a survey. In: Proceedings 2015 International Conference Green Computing and Internet Things, ICGCIoT 2015, pp. 264–269 (2016)
20. Gandomi, A., Haider, M.: Beyond the hype: big data concepts, methods, and analytics. Int. J. Inf. Manage. **35**(2), 137–144 (2015)
21. Wang, Y., Kung, L.A., Byrd, T.A.: Big data analytics: understanding its capabilities and potential benefits for healthcare organizations. Technol. Forecast. Soc. Change **126**, 3–13 (2018)
22. Dumka, A., Sah, A.: Smart ambulance system using concept of big data and internet of things. In: Dey, N., Ashour, A.S., Bhatt, C., Fong, S.J. (eds.) Healthcare Data Analytics and Management. Elsevier Inc., Amsterdam (2018)
23. Tiwari, S., Wee, H.M., Daryanto, Y.: Big data analytics in supply chain management between 2010 and 2016: insights to industries. Comput. Ind. Eng. **115**, 319–330 (2018)
24. Kumar, U., Gambhir, S.: Mobile agent based mapreduce framework for big data processing. Adv. Intell. Syst. Comput. **654**, 391–402 (2018)
25. Taneja, R., Gaur, D.: Robust fuzzy Neuro system for big data analytics. Adv. Intell. Syst. Comput. **654**, 543–552 (2018)
26. Ahmed, Z., Liang, B.T.: Systematically dealing practical issues associated to healthcare data analytics, vol. 70, pp. 599–613. Springer International Publishing (2020)
27. Praveena, A., Bharathi, B.: A survey paper on big data analytics. In: 2017 International Conference on Information Communication and Embedded Systems ICICES 2017 (2017)
28. Singh, D., Reddy, C.K.: A survey on platforms for big data analytics. J. Big Data **2**(1), 1–20 (2015)
29. Fu, C., Wang, X., Zhang, L., Qiao, L.: Mining algorithm for association rules in big data based on Hadoop. In: AIP Conference Proceedings, vol. 1955 (2018)

30. Abdel-Hamid, N.B., ElGhamrawy, S., El Desouky, A., Arafat, H.: A dynamic spark-based classification framework for imbalanced big data. J. Grid Comput. **16**(4), 607–626 (2018)
31. Alcalde-Barros, A., García-Gil, D., García, S., Herrera, F.: DPASF: a flink library for streaming data preprocessing (2018)
32. Furht, B., Villanustre, F.: Big Data Technologies and Applications, vol. 2, no. 21. Springer, Cham (2016)
33. García, S., Luengo, J., Herrera, F.: Data preparation basic models. In: Data Preprocessing in Data Mining. Intelligent Systems Reference Library, vol. 72. Springer, Cham (2015)
34. Russom, P.: Big data analytics - TDWI best practices report. Introduction to Big Data Analytics. TDWI Research, vol. 1, pp. 3–5 (2011)
35. Di Martino, B., Aversa, R., Cretella, G., Esposito, A., Kołodziej, J.: Big data (lost) in the cloud. Int. J. Big Data Intell. **1**(1/2), 3 (2014)
36. ur Rehman, M.H., Liew, C.S., Abbas, A., Jayaraman, P.P., Wah, T.Y., Khan, S.U.: Big data reduction methods: a survey. Data Sci. Eng. **1**(4), 265–284 (2016)
37. Zhang, W., He, B., Chen, Y., Zhang, Q.: GMR: graph-compatible mapreduce programming model. Multimed. Tools Appl. **78**(1), 457–475 (2019)
38. Ramírez-Gallego, S., Fernández, A., García, S., Chen, M., Herrera, F.: Big data: tutorial and guidelines on information and process fusion for analytics algorithms with MapReduce. Inf. Fusion **42**, 51–61 (2018)
39. Chang, Y.S., Lin, K.M., Tsai, Y.T., Zeng, Y.R., Hung, C.X.: Big data platform for air quality analysis and prediction. In: 2018 27th Wireless Optical Communication Conference WOCC 2018, pp. 1–3 (2018)
40. Zhao, L., Chen, Z., Hu, Y., Min, G., Jiang, Z.: Distributed feature selection for efficient economic big data analysis. IEEE Trans. Big Data **4**(2), 164–176 (2016)

Plantar Fascia Ultrasound Images Characterization and Classification Using Support Vector Machine

Abdelhafid Boussouar[1], Farid Meziane[1(✉)], and Lucy Anne Walton[2]

[1] School of Computing, Science and Engineering,
University of Salford, Salford M5 4WT, UK
A.Boussouar1@edu.salford.ac.uk,
F.Meziane@salford.ac.uk
[2] School of Health and Society, Directorate of Radiology,
University of Salford, Salford M5 4WT, UK
L.A.Walton@salford.ac.uk

Abstract. The examination of plantar fascia (PF) ultrasound (US) images is subjective and based on the visual perceptions and manual biometric measurements carried out by medical experts. US images feature extraction, characterization and classification have been widely introduced for improving the accuracy of medical assessment, reducing its subjective nature and the time required by medical experts for PF pathology diagnosis. In this paper, we develop an automated supervised classification approach using the Support Vector Machine (Linear and Kernel) to distinguishes between symptomatic and asymptomatic PF cases. Such an approach will facilitate the characterization and the classification of the PF area for the identification of patients with inferior heel pain at risk of plantar fasciitis. Six feature sets were extracted from the segmented PF region. Additionally, features normalization, features ranking and selection analysis using an unsupervised infinity selection method were introduced for the characterization and the classification of symptomatic and asymptomatic PF subjects.

The performance of the classifiers was assessed using confusion matrix attributes and some derived performance measures including recall, specificity, balanced accuracy, precision, F-score and Matthew's correlation coefficient. Using the best selected features sets, Linear SVM and Kernel SVM achieved an F-Score of 97.06 and 98.05 respectively.

Keywords: Plantar fascia ultrasound images ·
Features selection and characterization · SVM · K-folded cross validation ·
Matthew's correlation coefficient

1 Introduction

Ultrasound (US) imaging offers a significant potential in the diagnosis of plantar fascia (PF) injuries and monitoring treatments. It offers a real-time effective imaging technique that can reliably confirm structural changes, such as thickening, rupture and

© Springer Nature Switzerland AG 2020
A. E. Hassanien et al. (Eds.): AISI 2019, AISC 1058, pp. 102–111, 2020.
https://doi.org/10.1007/978-3-030-31129-2_10

identify changes in the internal echo structure associated with diseased or damaged tissues. PF US images are usually examined and analysed by health specialists based on visual perceptions and some manual biometric measurements, such as the thickness estimation of the PF region, to identify the presence of any kind of lesions and abnormalities such as plantar fasciitis (inflammation of the plantar fascia).

As reported in the literature, thickening, bi-convexity, rough surface, heterogeneous texture, decreased echogenicity, loss edge sharpness and hypoechoic deformities of the PF are considered as part of the diagnostic criteria and characteristic features of symptomatic PF; whereas, surface smoothness, texture homogeneity and uniform hyperechogenicity are characteristics of asymptomatic PF subjects [8, 15, 19, 24] and this is shown in Fig. 1.

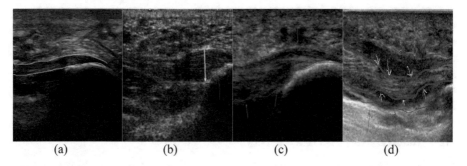

(a)   (b)   (c)   (d)

Fig. 1. Asymptomatic and symptomatic PF region comparison: (a) Asymptomatic PF region (green contours), (b–d) Symptomatic PF region: (b) and (c) a thickened PF sections (red arrows) compared to a normal PF in (a) due to planar fasciitis disorder, (d) a huge partial tear of the PF region: the outer red contour clearly shows a surrounding inflammation (plantar fasciitis), while the inner contour (bold red) shows the irregular outline and disrupted PF region fibres.

Despite the advantages of US imaging, the acquired images interpretation and analysis are time consuming. This is mainly due to the large number of patients, the large medical data history accumulated in the DICOM systems and the large number of physicians required for the analysis and interpretation. The exploration of such massive medical data requires highly efficient and sophisticated techniques capable of finding the class separation between asymptomatic and symptomatic ultrasound images of the plantar fascia. These techniques are highly required to classify different PF US images into normal and abnormal subjects and to prune the huge accumulated data and take into consideration only the symptomatic data with the possibility of plantar fasciitis or other disorders. Therefore, it is a requirement to devise an automated system to characterize and classify PF US images that allows better abnormalities detection and easier interpretation during medical image analysis.

This paper proposes an automated supervised classification approach using linear-SVM and Kernel-SVM to facilitate the detection and the characterization of the plantar fascia region for the classification of PF US images dataset into symptomatic PF subjects and asymptomatic subjects; and the possibility of the identification of patients with normal plantar fascia but at risk of plantar fasciitis disorder.

2   The Methodology and Proposed Approach

The proposed PF classification model consists of the following modules as shown in Fig. 2. (i) preprocessing phase employing speckle noise reduction filtering and image enhancement operations to reduce the effects of undesirable speckle noise phenomenon and improve the contrast of the PF US images using dual tree complex wavelet transform with soft thresholding (DT-CWT_S) and contrast-limited adaptive histogram equalization filter (CLAHE), respectively; (ii) artificial neural networks supervised segmentation phase applying different features measures, a features ranking module and trained radial basic function neural network (RBF-NN) classifier as discussed in [3], to automatically segment the PF region and calculate its thickness; (iii) texture features extraction and analysis introducing 6 sets of feature extraction measures (for extracting a total of 40 features), features ranking and selection operation using an unsupervised infinity feature selection method [17] to select and analyse the most discriminating and suitable features for the classification process; (iv) the classifier module using Linear-SVM and Kernel-SVM to distinguish between asymptomatic and symptomatic plantar fascia subjects; and (v) classification performance analysis using different performance measures such as recall, specificity, balanced accuracy, precision, F-score and MCC.

2.1   Materials and PF Data Collection

Following ethical approval from the University of Salford Research's Ethics Panel (ST1617-48), written informed consent was collected from all patients' participants. Various PF US images, acquired from a patient's footprint area in the prone position were used in the classification approach; more specifically, a total of 284 (252 normal and 32 abnormal taken from diabetic patients with plantar fasciitis) PF US images, were acquired from a patient's footprint area in the prone position were used. The images were obtained from 45 patients for different PF anatomical structures including rear-foot, mid-foot and fore-foot sections with 256 gray levels, a size dimension of 512×512 pixels and a resolution of 28.35 pixels/cm. These images were obtained from the Health Sciences Department, University of Salford, acquired by two expert clinicians according to a precise protocol using a Venue 40 musculoskeletal US system (GE Healthcare, UK) with a 5–13 MHz wideband linear array probe 12.7 mm $\times$ 47.1 mm. All the methods used in the proposed approach were implemented using Matlab R2017b (The Math-Works Inc., Natwick, USA).

2.2   Preprocessing

The preprocessing phase aims to prepare the PF US images for further processing including segmentation and classification and improve their accuracy, efficiency, and scalability. This is achieved by (i) minimizing the effects of the multiplicative speckle noise without losing any valuable information (such as tiny lines, edges) using a selected dual tree complex wavelet-based despeckling filter (DT-CWT_S) [13, 16]. DT-CWT_S filter integrates homomorphic transformation (using log compression and exponent decompression to transform the multiplicative noise to an additive one) and multi-scale DT-CWT decomposition and composition employing the BayesShrink subband thresholding using soft thresholding to reduce or suppress the speckle noise (noisy coefficients) in PF US images. DT-CWT_S has demonstrated a superior edge preserving behaviour and a good visual appearance [4]; (ii) enhancing the PF region contrast and visually improving the global appearance of the PF US images using CLAHE method after the despeckling operation to adjust the intensity of the PF region and to avoid noise amplification in PF US images using different implemented steps as reported in [26].

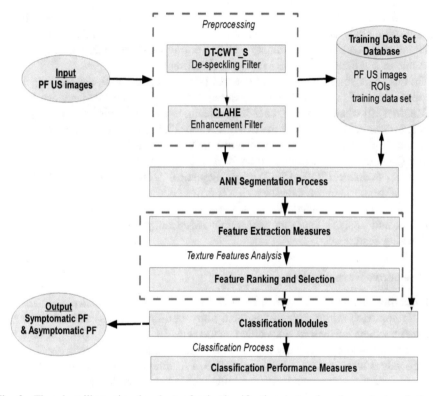

Fig. 2. Flowchart illustrating the plantar fascia classification system based on a texture features analysis and different classifiers modules

2.3 Segmentation

Automated segmentation is one of the most important tasks in medical image processing and analysis, including, pattern recognition, supervised or unsupervised subjects classification and novelty detection; it is mainly used to locate the desired region of interest objects in the input images dataset. As reported in [3], an automated ANNs supervised segmentation approach was introduced in this study to segment different PF regions. The proposed segmentation approach uses the radial basic function neural network (RBF-NN) classifier [9] to automatically segment the PF region and estimate its thickness. Two different quantitative evaluation metrics namely: the region-based metrics and distance based metrics were used to evaluate the segmentation method. For the region-based metrics, the highest precision obtained was 97.70 ± 1.20, whereas for the distance-based metrics it was 0.10 ± 0.07. The full description of the segmentation process and steps and analysis are reported in detail in [3].

2.4   Feature Extraction

In most classification tasks, feature extraction is an important step to extract the relevant information (reduced input dataset representation) from the input dataset in order to perform the remaining tasks. Thus, the main goal of feature extraction in this classification study is to extract a set of textual features from the PF segments (using different measures) that discriminate between one input pattern and another, and then feed this into different classifiers for a classification task.

In this stage six different sets of features (40 features in total) were extracted from the segmented PF region including: (i) Haralick spatial gray level dependence matrices (SGLDM) [10] where a total of 12 SGLDM features were computed and averaged for a selected distance d = 1 (3 × 3 matrices) and four different orientation angles 0°, 45°, 90°, and 135°; (ii) Region based features; (iii) Neighbourhood gray tone difference matrix (NGTDM) for a kernel window of 3 × 3 [2]; (iv) Histogram based features or first-order statistics (FOS) [5, 21]; (v) Statistical feature matrix (SFM) [5, 25]; (vi) Laws' texture energy measures [5, 11, 25].

All the features extracted may have some redundancy, thus we introduced a feature selection and analysis stage to reduce this redundancy and to select the most discriminating feature sets.

2.5   Feature Normalization

All features in this study were normalized using the mean variance normalization (MVN) approach which helps in reducing any non-linear distortion and scaling all features so they fall within a specified range (e.g. [0 1] or [−1 1]) to prevent high measurement values (especially region based measurements) from outweighing other feature values with smaller values (e.g. SGLDM features) [7]. The normalized features $NX_{j,n}$ are computed by calculating the difference between the features and their mean values, and then dividing them by their standard deviation values as given by Eq. 1 [7].

$$NX_{i,n} = \frac{X_{j,n} - \mu_{j,n}}{\sigma_{j,n}} \tag{1}$$

where $\mu_{j,n}$ is the mean value of the feature vector x_i and $\sigma_{j,n}$ its standard deviation.

2.6   Feature Ranking and Selection

A common deficiency in most pattern recognition and classification tasks is the high dimension of the extracted feature space compared to the number of the input samples (40 features × 284 observations). This will lead to some common problems such as:

over-fitting, poor generalization and high computation cost. In order to minimize the aforementioned problems, a combination feature ranking and feature selection unsupervised infinity techniques [17, 18] were introduced to reduce the correlated measurements and to select the most discriminating features. Different selected feature sets were analysed to choose the best discriminating features for SVM classification modules based on high F-score values.

2.7   Classification

Following feature ranking and selection analysis, the feature classification approach was implemented using Linear-SVM and Kernel-SVM. To obtain a good classification result, three main conditions were taken into consideration during the classification process: (1) careful selection of features; (2) a good classifier; and (3) suitable training samples [22]. All the analysed feature sets described in the previous sections, were treated as input vectors to the selected classifier modules and their results were evaluated using different classification measures. To overcome the over-fitting problem and to validate the robustness of the classifiers, a cross-validation task was also introduced using the k-folded (k = 10 folds) approach to randomly select the training and testing instance classes.

3   Support Vector Machines

Support vector machines (SVM) [23] are widely used in bioinformatics and medical studies for pattern recognition related problems [12]. The main concept of SVM is that, firstly, it differentiates between two class samples according to the optimal maximum margin (distance between each set) hyperplane (or decision boundary) search result [21]; secondly, if the hyperplane fails to split the previous linear class samples, the SVM makes use of different kernel functions such as polynomial kernel, Gaussian-RBF and sigmoids-NN instead of linear SVM [6, 14, 23]. This aims to achieve high dimensional feature space when translating original data samples [20]. In this study, both Linear-SVM and Kernel-SVM classifiers were tested and the Gaussian-RBF kernel function is used in the Kernel-SVM main function. For the PF US 2D training dataset T_S with N_L labelled instances (X_j, Y_j), where X_j denotes the feature instances and Y_j is the class label with 1 for normal and -1 for abnormal PF class, and N is the total number of samples (252 normal and 32 abnormal samples with 40 extracted features).

4   Experimental Results and Discussion

For the classification experimental results, a total of 284 (252 symptomatic and 32 asymptomatic) US images of the PF regions (rear-foot, mid-foot and fore-foot sites) were analysed. Six different sets of features representing a total of 40 features were computed both from symptomatic and asymptomatic US images of the PF segments. For all extracted features, feature selection approach was introduced, and their means, weights and ranking orders were computed and analysed for normal and abnormal PF US images.

4.1 Feature Extraction and Selection Analysis

The main reason for feature selection analysis is to eliminate similar or highly co-dependent features and to find the best discriminatory features that predict the best classification results. Feature selection analysis results of the top ranked features calculated from the 284 US images of the segmented PF region are represented in Fig. 3. For each feature, the weight predictor was calculated and its rank order was assigned accordingly. The best features (with the highest weight and ranked predictor) were found to be LS, Contrast, Variance, LE, Energy, SumSquare, AngSecMoment, LL, EE, DiffVariance, Strength, ES, Complexity, Correlation, DiffEntropy, SS, SumAverage, MajAxLength, Periodicity, Business, Mean, Skewness, Kurtosis, Orientation, Roughness, Convex Area, Extent, EquivDiameter and Area.

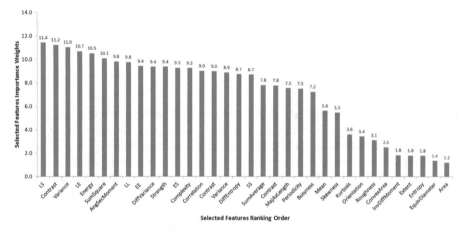

Fig. 3. Graph representation of 34 ranked predictors (features importance) based on their importance weights

Furthermore, to determine the best features for each classifier, the F-score measures were computed for the Linear-SVM and Kernel-SVM classifiers using different selected feature sets (from 1 to 40, starting with the highest ranked features). Six selected feature sets were defined using the highest F-score measure. The F-score result obtained by Kernel-SVM and Linear SVM were 98.05% and 97.06% respectively when using 34-features. It is also clearly evident from this analysis, that there are differences in feature weights values between asymptomatic and symptomatic PF subjects. From this interpretation, symptomatic PF texture tends to be darker with high contrast, high variance, high shape measures (high thickness) (due to the accumulation of the inflammation fluid), more extent, high convex area (due to irregularity of the PF surface and outline disruption), high complexity (more heterogeneous), low strength, less periodicity, more roughness and low grey intensity. While on the other side, asymptomatic PF texture are brighter with low contrast, low variance, less shape measures, less extent, less convex area, low complexity (more homogeneous), high strength, more periodicity, more smoothness and high grey intensity.

4.2 Classification Analysis

For the classification task, Linear-SVM and Kernel-SVM classifies were trained and tested using the same training and testing datasets. To overcome the over-fitting problem during the training stage and to assess the performance of various classification modules, 10-fold cross-validation was introduced. The main concept of the cross-validation approach is that each sample is added in both training and testing samples. For 10-fold cross validation approach, datasets (252 asymptomatic PF subjects and 32 symptomatic subjects) were randomly partitioned into 10 different equal splits (folds) (i.e. $10 - 1 = 9$ folds were used for training task and the remaining fold is used for testing, with an iteration of 10 times dropping one-fold out for testing each time). Six different classification performance measures were computed and analysed for the two classifiers. For each classification the mean value of the 10-cross validations was computed. The results of the classifications using the best selected features are summarized in Table 1.

Table 1. Linear-SVM and Kernel-SVM classification performance measures using the best selected feature sets

Classifier	Recall	Specificity	B-accuracy	Precision	F-Score	MCC[a]
Linear-SVM	95.75	84	89.88	98.41	97.06	71.46
Kernel-SVM	96.18	100	98.09	100	98.05	81.32

[a]The Matthew's correlation coefficient

In all the six classification performances, Kernel-SVM performed better than Linear SVM. For specificity and Precision, Kernel-SVN achieved 100%.

The classification results are very high for this first experiment. However, we experimented with a further two classifiers to have an idea on the performance of the Linear and Kernel SVM algorithms The radial basic function neural network (RBF-NN) classifier and the linear discriminant analysis (LDA) were used with the best selected features set. The results are summarised in Table 2.

Table 2. RBF-NN and LDA Classification performance measures using the best selected feature sets

Classifier	Recall	Specificity	B-accuracy	Precision	F-Score	MCC[a]
RBF-NN	98.82	96.67	97.74	99.60	99.21	92.82
LDA	97.62	81.25	89.43	97.62	97.63	78.87

[a]The Matthew's correlation coefficient

5   Conclusions

In this study we developed a new automatic supervised classification system for discriminating different ultrasound plantar fascia images using Linear-SVM and Kernel-SVM classifiers. This will help medical experts to improve the efficiency of the PF pathology diagnosis and minimize the time required for the diagnosis. Six different feature set measures were used to extract and analyse the texture features. Additionally, the infinity selection method was successfully adopted to rank and characterize asymptomatic and symptomatic features, based on their weights importance. The results of the feature selection stage revealed that the top selected features can represent the characteristics of asymptomatic and symptomatic PF subjects ultrasound images well. The Inf selection method to select the best features is quite effective. In order to define and compare the best features, the F-score measure was independently computed for both classifiers (Linear-SVM and Kernel-SVM) using different selected feature sets (1–40). The best selected feature set for every classifier were fed to the related classifier as the input vector for the classification task. In the experiments, different performance evaluation measures were used to assess the classification capability of the two classifiers using their best selected features. The results have shown that Kernel-SVM outperformed Linear-SVM in all the performance evaluation measures. For specificity and precision, Kernel-SVM achieved 100%. The results show that for precision, specificity and B-Accuracy, Kernel-SVM still outperforms all other classifiers. RBF-NN obtained better results in Recall, F-Score and MCC. LDA in general performed worst then Kernel-SVM and RBF-NN. We are aiming in the next phase of the development of the system to use more classifiers.

References

1. Abe, S.: Support Vector Machines for Pattern Classification. Advances in Computer Vision and Pattern Recognition. Springer, London (2010)
2. Amadasun, M., King, R.: Textural features corresponding to textural properties. IEEE Trans. Syst. Man Cybern. **19**(5), 1264–1274 (1989)
3. Boussouar, A., Meziane, F., Crofts, G.: Plantar fascia segmentation and thickness estimation in ultrasound images. Comput. Med. Imaging Graph. **56**, 60–73 (2017)
4. Boussouar, A.: Thickness estimation, automated classification and novelty detection in ultrasound images of the plantar fascia tissues. Ph.D. thesis, School of Computing, Science and Engineering, University of Salford, UK (2019)
5. Christodoulou, C.I., Pattichis, C.S., Pantziaris, M., Nicolaides, A.: Texture-based classification of atherosclerotic carotid plaques. IEEE Trans. Med. Imaging **22**(7), 902–912 (2003)
6. Cortes, C., Vapnik, V.: Support-vector networks. Mach. Learn. **20**(3), 273–297 (1995)
7. Dougherty, G.: Pattern Recognition and Classification: An Introduction. Springer, New York (2012)
8. Fabrikant, J.M., Park, T.S.: Plantar fasciitis (fasciosis) treatment outcome study: plantar fascia thickness measured by ultrasound and correlated with patient self-reported improvement. Foot **21**(2), 79–83 (2011)
9. Ham, F.M., Kostanic, I.: Principles of Neurocomputing for Science and Engineering. McGraw-Hill Higher Education, New York (2000)

10. Haralick, R.M., Shanmugam, K., Dinstein, I.H.: Textural features for image classification. IEEE Trans. Syst. Man Cybern. **3**(6), 610–621 (1973)
11. Laws, K.I.: Rapid texture identification. In: Proceedings SPIE 0238, Image Processing for Missile Guidance, pp. 238–244 (1980)
12. Martínez-Trinidad, J., Ochoa, J., Kittler, J.: Progress in pattern recognition, image analysis and applications. In: 11th Iberoamerican Congress on Pattern Recognition, CIARP 2006, Proceedings, Cancún, Mexico, 14–17 November. Springer (2006)
13. Michailovich, O., Tannenbaum, A.: Despeckling of medical ultrasound images. IEEE Trans. Ultrason. Ferroelectr. Freq. Control **53**(1), 64–78 (2006)
14. Osuna, E., Freund, R., Girosi, F.: Support vector machines: training and applications (1997)
15. Park, J.W., Yoon, K., Chun, K.S., Lee, J.Y., Park, H.J., Lee, S.Y., Lee, Y.T.: Long-term outcome of low-energy extracorporeal shock wave therapy for plantar fasciitis: comparative analysis according to ultrasonographic findings. Ann. Rehabil. Med. **38**(4), 534–540 (2014)
16. Rabbani, H., Vafadust, M., Abolmaesumi, P., Gazor, S.: Speckle noise reduction of medical ultrasound images in complex wavelet domain using mixture priors. IEEE Trans. Biomed. Eng. **55**(9), 2152–2160 (2008)
17. Roffo, G., Melzi, S., Cristani, M.: Infinite feature selection. In: IEEE International Conference on Computer Vision (ICCV), pp. 4202–4210 (2015)
18. Roffo, G., Melzi, S., Castellani, U., Vinciarelli, A.: Infinite latent feature selection: a probabilistic latent graph-based ranking approach. In: IEEE International Conference on Computer Vision (ICCV) (2017)
19. Saber, N., Diab, H., Nassar, W., Razaak, H.A.: Ultrasound guided local steroid injection versus extracorporeal shockwave therapy in the treatment of plantar fasciitis. Alexandria J. Med. **48**(1), 35–42 (2012)
20. Shi, X., Cheng, H., Hu, L., Ju, W., Tian, J.: Detection and classification of masses in breast ultrasound images. Digit. Signal Process. **20**(3), 824–836 (2010)
21. Umbaugh, S.: Computer Imaging: Digital Image Analysis and Processing. CRC Press book, Taylor & Francis, Boca Raton, Milton Park (2005)
22. Unger, H., Meesad, P., Boonkrong, S.: Recent Advances in Information and Communication Technology. Advances in Intelligent Systems and Computing. Springer International Publishing, Cham (2015)
23. Vapnik, V.: The Nature of Statistical Learning Theory. Springer Science & Business Media, Heidelberg (2013)
24. Wearing, S.C., Smeathers, J.E., Sullivan, P.M., Yates, B., Urry, S.R., Dubois, P.: Plantar fasciitis: are pain and fascial thickness associated with arch shape and loading? Phys. Ther. **87**(8), 1002–1008 (2007)
25. Wu, C.M., Chen, Y.C., Hsieh, K.S.: Texture features for classification of ultrasonic liver images. IEEE Trans. Med. Imaging **11**(2), 141–152 (1992)
26. Zuiderveld, K.: Contrast limited adaptive histogram equalization, chapter VIII.5. In: Heckbert, P.S. (ed.) Graphics Gems IV, pp. 474–485. Academic Press, Cambridge (1994)

Swarm Optimization and Applications

A Novel EEG Classification Technique Based on Particle Swarm Optimization for Hand and Finger Movements

Nourhan Wafeek$^{(\boxtimes)}$, Roaa I. Mubarak, and Mohamed E. Elbably

Helwan University, Cairo, Egypt
Nourhanwafeek@gmail.com

Abstract. Electroencephalogram (EEG) has gained much attention from researchers recently. EEG classification has many applications such as: classifying brain disorders, helping paralyzed people to control a machine by their own imagery mental tasks and controlling a robot or a remote system with both imagery and actually mental tasks. This paper aims to classify arm and finger movements acquired through EEG signals. The EEG signals have been transformed to frequency domain using discrete wavelet transform (DWT) as a feature extractor. These extracted features are then feed into a novel particle swarm classifier to classify the different movements of arm and fingers. The experimental results showed that this new algorithm gives accuracy of 95% with minimum time delay, which is an essential requirement for all biomedical applications. This research is considered the first step towards implementing an automatic system (surgical robot) that can be used in Telesurgery.

Keywords: EEG · BCI · BMI · Wavelet transform · Particle swarm algorithm

1 Introduction

For the past two decades, the Electroencephalogram (EEG) has gained much attention from researchers and engineers, mostly in 3 main applications: Classifying brain disorders, helping disabled and paralyzed people to control a machine by their own imagery mental tasks (Brain Computer Interface BCI) and controlling a robot or a remote system with both imagery and actually mental tasks (Brain Machine Interface BMI). Many surveys and reviews were published to compare and analyze all applied algorithms of basic steps of EEG system: Data acquisition, feature extraction and finally EEG classification [1–3]. According to previous reviews the combination of mental tasks (cognitive, movement, imagery) with feature extraction techniques-or a hybrid one of them-followed by a powerful classifier to perform a high accuracy control system based on EEG signal is infinite.

Most researches in movement classification used datasets of hand and Leg moves, for both sides (right and left), but rarely a dataset of moving individual fingers or grabbing positions as published in [7] and [8]. Unlike any movement that can be sensed from active muscles using EMG signal, fingers movement can't be sensed as they don't have muscles to do so, however the EEG signals for their movement are significant, so

© Springer Nature Switzerland AG 2020
A. E. Hassanien et al. (Eds.): AISI 2019, AISC 1058, pp. 115–124, 2020.
https://doi.org/10.1007/978-3-030-31129-2_11

in this proposed research, a certain movements of both right and left arms and fingers is captured, processed and classified with farther aim to use this system as a remote controller to a robotic arm, or a surgical robot.

2 Methods

2.1 Data Acquisition

The basic steps of EEG system are EEG acquisition, feature extraction and finally EEG classification, are represented in this section in same order. The first step of the flow is EEG acquisition, but before explaining the experimental protocol a brief background of EEG is needed. EEG offers a safe (non-invasively), cheap, portable yet accurate way to record brain waves, it sense electrical fluctuations through electrodes distributed along the scalp according to the 10–20 international system which places the electrodes in either 10% or 20% of the front-to-back or right-to-left total distance of the skull as shown in Fig. 1:

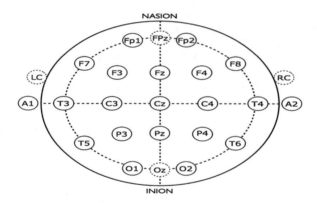

Fig. 1. Position of electrodes in the 10–20 international system

Each group of electrodes sense waves of different lobes of human. Fluctuations corresponding to motor activity sensed by C3 and C4 would be enough to represent a real movement. After EEG is recorded in time domain, a transformation to the frequency domain is applied to extract the 4 main bands (Delta, Theta, Alpha and Beta). Each band contains the information of a known activity as shown in Table 1 [4, 5].

Table 1. Electrical brain bands with corresponding activity

Band name	Frequency range in HZ	Activity
Delta	0.5–3	Deep sleep
Theta	4–7	Imagery, drowsiness
Alpha	8–12	Relaxation, motor activity
Beta	12–30	Active concentration, motor idling
Delta	0.5–3	Deep sleep

In order to fully extract a moving action, alpha and beta bands are used together in feature vector, which are the suitable indicators of a movement. This choice was made in different ways in some literatures either because they contain most of the signal power or even to have the less error ratio [6].

EEG data used in this research was acquired non-invasively from a 26 years old female, using a 32-electrode EEG recording cap and ProFusion EEG4 software as shown in Fig. 2. Each task was repeated on both right and left hands for 5 trials, each trial was performed for 6 s with sampling frequency 256 Hz.

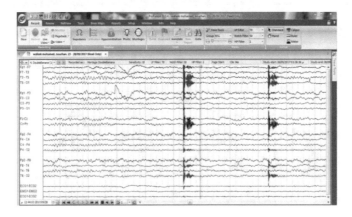

Fig. 2. EEG waves in time domain by ProFusion EEG4 software.

The experiment protocol was done as follows:

Task0: Relaxing with closed eyes.
Task1: Moving up right arm for 90°.
Task2: Moving right thumb & index in grabbing position.
Task3: Moving right thumb & middle in grabbing position.
Task4: Moving up left arm for 90°.
Task5: Moving left thumb & index in grabbing position.
Task6: Moving left thumb & middle in grabbing position.

2.2 Discrete Wavelet Transform

The second step of EEG classification flow is feature extraction, where wavelet transform is known as the most popular yet effective way in EEG systems, because of its highly satisfaction results for non-stationary signals like EEG, so it was used in the proposed system due to its power in multidimensional platforms.

Compared with Fourier transform, it gives accurate time information at high frequency and accurate frequency information at slow time (Multidimensional Analysis). In Wavelet transform the frequency and scale are 2 variables not only frequency as shown in Eq. (1). This key made it very effective in most of the biomedical applications because of the non-stationary nature of these signals [9].

$$X_{WT}(\zeta, s) = \frac{1}{\sqrt{\lfloor s \rfloor}} \int_{-\infty}^{\infty} x(t)\psi\left(\frac{t-\zeta}{s}\right) dt \qquad (1)$$

Where X_{WT} is the transformed signal, which is a function of the translation parameter ζ and the scale parameter s. Wavelet analysis is able to reveal signal aspects that other analysis techniques miss, like: trends, breakdown points and discontinuities, but the best advantage is the possibility to perform a multi-resolution analysis. For discrete data like EEG discrete wavelet transform (DWT) is implemented by using low pass and high pass filters. Low pass filters give an approximate coefficient of the input data, while the high pass filters give the detailed coefficients as illustrated in Fig. 3. Those coefficients are the feature vector of EEG that used as an input to classifier.

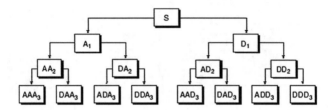

Fig. 3. Discrete Wavelet Transform decomposition.

2.3   Particle Swarm Algorithm

The third and final step of EEG system flow is classification. Here particle swarm Algorithm is transformed from optimizer to classifier. PSO is a nature-inspired algorithm invented by Russell Eberhart and James Kennedy in 1995. Nature-inspired algorithms are a recent trend for optimization that simulates the behavior of nature processes which provide easy, rapid and successful ways of living and productivity. Genetic, Ant Colony and PSO are the most used nature-inspired optimization algorithms. Social animals and particles work [10, 17]. In PSO, the way a swarm of birds or fishes finds a promising food location can be modeled with mathematical equations that can be applied on certain complex problems to find any optimum solution (Fitness Function). Each particle searches the multidimensional space for a solution for the given problem, that is called the local position (Plocal), then all particles share their local solutions and the nearest one to the optimum is defined as the global position (Pglobal) of all swarm. For the next trial, each particle updates its velocity $v(t+1)$ and position $x(t+1)$ as well, according to following equations:

For velocity update:

$$v(t+1) = w * v(t) + c1 * rand * (Plocal - x(t)) + c2 * rand(Pglobal - x(t)) \qquad (2)$$

For position update:

$$x(t+1) = x(t) + v(t+1) \tag{3}$$

Where w is an inertia factor controlling the momentum of the particle by weighting the contributions of its previous velocity, $c1$ is the cognitive parameter and $c2$ is the social parameter. Like any other collaboration algorithms, adequate parameter tuning is important for efficient PSO performance and much work has been done to select the optimum values. In [11] some general directives to choose the good combination where $c1$ in [0,1], with a preference for 0.7 and $c2$ around 1.5 with a preference for 1.43. These steps are repeated for finite number of trials or until a predefined stopping criterion.

3 Proposed System

In this proposed system, wavelet transform is used with Daubechies mother wavelet of 4 decomposition degree to give best results with EEG signals. Wavelet Matlab toolbox was helpful to test EEG data and extract the known alpha and beta bands and use their Coefficients as the feature vector for proposed classifier. In [12] particle swarm is improved to perform as a classifier not only an optimizer. This magnificent study proved that among the many applications of PS as classifier; of many kinds of datasets, but not used with EEG signals. In [13] it was used with EEG in automatic seizures detection with datasets of normal and disordered subjects and yelled high accuracy results. Here the Particle swarm was proved to work as an effective classifier of EEG movement tasks according to the following steps:

For a swarm of M particles in N-dimensional search space, any individual in the swarm is represented by a real-valued vector composed of its position and velocity. The fitness at time t of the ith individual in the swarm with respect to the centroid of a class C is defined as the average Euclidean distance given by Eq. (4):

$$\psi_i^c(t) = \frac{1}{\text{Dtrain}} \sum_{k=1}^{\text{Dtrain}} d(y_k^c x_i^c(t)) \tag{4}$$

Where y_k^c is the kth exemplar of class c in the training set of size Dtrain and $x_i^c(t)$ is the current position of the centroid of class c as determined by the ith particle, both vectors being N-dimensional. So, finding the class centroids is a minimization problem. Improvements of the original particle swarm algorithm are proposed as following and illustrated in Fig. 4, to increase the accuracy and decrease the eclipsed time.

First: The initial position of the particles is set deterministically in a uniform distribution, covering all search space, not randomly as in the original PSO.

Second: Inertia factor is varied during execution, starting ith wmax and linearly decremented according to Eq. (5):

$$w(t) = \text{wmax} - (\text{wmax} - \text{wmin}) * t/\text{Tmax} \tag{5}$$

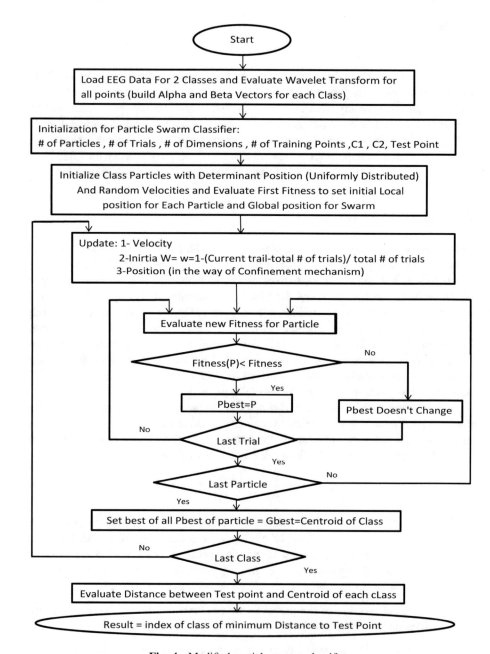

Fig. 4. Modified particle swarm classifier.

Third: Confinement mechanism is used in updating particle's positions [12], it forces the position to change within an infinite interval as described in Eq. (6), without exceeding the search space limits or setting a maximum value for velocity:

$$x(i, k)(t+1) = Min(Max(x(i, k)(t) + v(i, k)(t+1), xmin), xmax) \qquad (6)$$

When the PSC is applied on the acquired EEG it gives very promising results which are differing 2 movement tasks with 95% accuracy. The main concern of PSC was its limitation when the number of classes increases. So, the 95% accuracy of 2 classes is used to differentiate between 3 classes, using a very simple proposed scenario represented in a finite state machine as shown in Fig. 5:

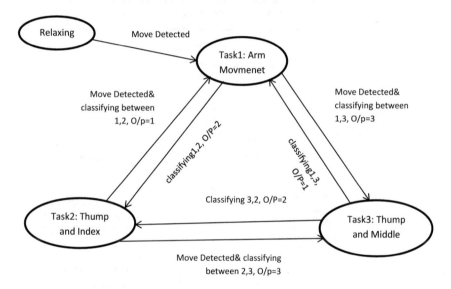

Fig. 5. Simple FSM of 4 states shows transitions between each 2 tasks.

To simplify the system more to guarantee minimum delay, the left or right side of arm and hands is selected from the beginning of training and testing, also with using the PSC as an independent and generic unit, the system is described in the following flow chart as shown in Fig. 6:

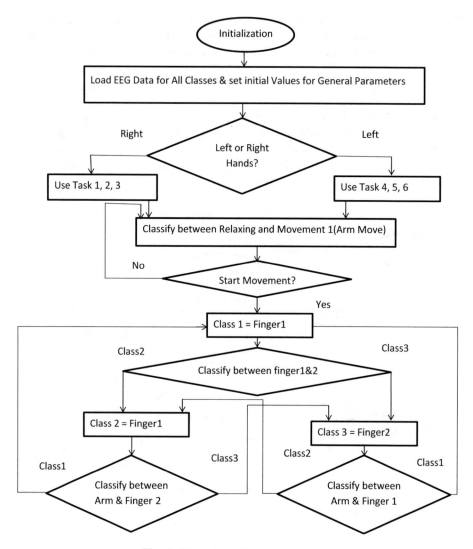

Fig. 6. Flow chart of proposed system.

4 Experimental Results

The recorded trials of each task were fully used in training and testing of PSC according to two scenarios. In the first scenario, the training is done by using 4 trails and use the remaining trail in testing. This is repeated for the second class. The second scenario is done by using only 3 trails in the training and the other 2 trials in testing in both 2 classes. Accuracy is calculated by its original definition; the ratio of right classified trials proportional to all number of trials. The final accuracy is the mean value of accuracy of all testing trials, as found in Table 2.

Table 2. PSC accuracy for each 2 classifications by using 1 and 2 testing points

Accuracy/tasks	Relaxing vs. arm move	Arm move vs. finger 1	Arm move vs. finger 2	Finger 1 vs. finger 2	Average accuracy
With 4 training points, 1 test point for right hand	95%	95%	95%	90%	93.75%
With 4 training points, 1 test point for left hand	95%	90%	90%	90%	91.25%
With 3 training points, 2 test point for right hand	90%	85%	85%	80%	85%
With 3 training points, 2 test point for left hand	87.5%	82.5%	85%	82.5%	84.375%

In terms of Accuracy, the best results is found by using 4 samples in training of the right hand and gets lower with the 3 samples in training especially with the left-hand classification and In terms of elapsed time for all tasks was around 30 s on PC platform which is expected to be less when the PSC runs in high speed platform like FPGA or ASIC.

As mentioned before that most researches in movement classification used datasets of hand and Leg moves, for both sides (right and left), but rarely a dataset of moving individual fingers or grabbing positions so a comparative study on other finger movement in Table 3 regardless of number of classes used against the proposed 3 classes system.

Table 3. Comparison of the PSO on finger classification with methods in the literature

Study	Method	Accuracy
Bera, Sikdar and Mahadevappa [14]	Extra trees classifier (ensemble learning)	74%
Salyers, Dong and Gai [15]	Linear discriminant analysis LDA and slow cortical potentials SCP	50%
Kim, Yoshimura and Koike [16]	Support vector machine	85%
This study	Particle swarm optimization	93.75%

5   Conclusion and Future Work

This paper proposed that particle swarm classifier works effectively in classifying EEG movement tasks of arm and fingers when it used with wavelet transform as feature extractor. This proposed technique gives accuracy tends to 95%. For Future work, Swarm intelligence can be proved to be better alternative for traditional and slow known classifiers according to speed and simplicity when implemented in FPGA or ASIC using simple modules and smart design of Data Path and Controller, which can make good improvements in all BCI and BMI systems in term of delay, layout area and consumed power.

References

1. Aparana, K., Chanadana Priya, R.: A survey on electroencephalography (EEG) – based brain computer interface. IJERCSE, **4** (2017)
2. Bi, L.Z., Fan, X.-A., Liu, Y.: EEG–baised brain controlled mobile robots: a survey. IEEE Trans. Hum.-Mach. **43**(2), 161–176 (2013)
3. Kewate, P., Suryawanshi, P.: Brain machine interface automation system: a review. Int. J. Sci. Technol. **3**(3), 64–67 (2014)
4. Jeannerod, M.J.: Mental imagery in the motorcontext. Neuropsychologia **33**(11), 1419–1432 (1995)
5. Neuper, C., Pfurtscheller, G.: Motor imagery and ERD. In: Lopes da Silva, F.H. (ed.) Event Related Desynchronization. Handbook of Electroencephalography and Clinical Neurophysiology, vol. 6, Revised edn. Elsevier, Amsterdam (1999)
6. Shedeed, H.A., Issa, M.F., Elsayed, S.M.: Brain EEG signal processing for controlling a robotic arm. In: ICCES. IEEE (2013)
7. Kumari, R.S.S., Induja, P.: Wavelet based classification for finger movements using EEG signals. IJCSN **4**(6), 903–910 (2015)
8. Furman, D., Reichart, R., Pratt, H.: Finger flexion imagery: EEG classification through physiologically-inspired feature extraction and hierarchical voting. In: IEEE, 4th International Winter Conference on Brain-Computer Interface (2016)
9. Merry, J.E.: Wavelet Theory and Applications. A literature study. Eindhoven University of Technology, Department Mechanical Engineering, Control Systems Technology Group, Amsterdam, pp. 303–325 (2005)
10. Engebertecht, A.P.: Computational Intelligence: An Introduction, 2nd edn. Wiley, Hoboken (2007)
11. Clerc, M.: L'optimisation par Essaim Particulaire: Versions Parame´triques et Adaptatives. Hermes Science Publications, Paris (2005)
12. Nouaouria, N., Boukadou, M., Proulx, R.: Particle swarm classification: a survey and positioning. Pattern Recogn. **46**, 2028–2044 (2013)
13. Ba-Karait1, N., Shamsuddin1, S., Sudirman, R.: Classification of electroencephalogram signals using wavelet transform and particle swarm optimization. In: ICSI 2014, Part II. LNCS, vol. 8795, pp. 352–362. Springer International Publishing, Switzerland (2014)
14. Sikdar, D., Mahadevappa, M., Roy, R., Bera, S.: An ensemble learning based classification of individual finger movement from EEG. [eess.SP] arXiv:1903.10154v1 (2019)
15. Salyers, J.B., Dong, Y., Gai, Y.: Continuous wavelet transform for decoding finger movements from singe-channel EEG. IEEE Trans. Biomed. Eng. **66**(6), 1588–1597 (2018). https://doi.org/10.1109/TBME.2018.2876068
16. Kim, H., Yoshimura, N., Koike, Y.: Classification of movement intention using independent components of premovement EEG. Front. Hum. Neurosci. **13**, 63 (2019). https://doi.org/10.3389/fnhum.2019.00063
17. El-Shorbagy, M.A., Hassanien, A.E.: Particle swarm optimization from theory to applications. Int. J. Rough Sets Data Anal. (IJRSDA) **5**(2), 1–24 (2018)

PSO-Based Adaptive Perturb and Observe MPPT Technique for Photovoltaic Systems

Nashwa Ahmad Kamal[1], Ahmad Taher Azar[2,3($\boxtimes$)], Ghada Said Elbasuony[1], Khaled Mohamad Almustafa[4], and Dhafer Almakhles[2]

[1] Electrical Power and Machine Department, Faculty of Engineering, Cairo University, Giza, Egypt
nashwa.ahmad.kamal@gmail.com, ghadaelbasuony@gmail.com
[2] College of Engineering, Prince Sultan University, Riyadh, Kingdom of Saudi Arabia
{aazar,dalmakhles}@psu.edu.sa
[3] Faculty of Computers and Artificial Intelligence, Benha University, Benha, Egypt
ahmad.azar@fci.bu.edu.eg
[4] College of Computer Science and Information Systems (CCIS), Prince Sultan University, Riyadh, Kingdom of Saudi Arabia
kalmustafa@psu.edu.sa

Abstract. The classical method of perturb and observe (P&O) is mostly utilized because of it is simple technique, ease of implementation and low cost. Although, it has high oscillations around maximum power point (MPP) at steady state due to perturbation and challenge between step size and the convergence time. To avoid the drawbacks of classical P&O methods, this paper presents adaptive P&O using optimized proportional integral (PI) controller by particle swarm optimization (PSO) algorithm. To evaluate the proposed method, a PV system model is designed with different scenarios under various weather conditions. For each scenario, simulations are carried out and the results are compared with the other classical P&O MPPT methods. The results revealed that the proposed PSO-PI-P&O MPPT method improved the tracking performance, response to the fast changing weather conditions and also has less oscillation around MPP as compared to the classical P&O methods.

Keywords: Photovoltaic systems · Renewable energy control · Maximum power point tracking · Particle Swarm Optimization

1 Introduction

Increasing the awareness of global worming and pollution impact leads to more reliance on renewable clean energy resources like solar, wind, tidal etc. These renewable resources gradually replace the conventional fossil fuel resources which run out by time [8, 21–23, 29]. Also, the rising energy worldwide forces researches to get the maximum benefits from these clean resources [25, 32]. Photovoltaic

© Springer Nature Switzerland AG 2020
A. E. Hassanien et al. (Eds.): AISI 2019, AISC 1058, pp. 125–135, 2020.
https://doi.org/10.1007/978-3-030-31129-2_12

(PV) energy is one of the utmost spread renewable energy resources (RES) as a result of its availability, easy to implement and less maintenance required. PV energy is expected to participate with great share (500 kW) in global energy market by year 2020 [10]. The recent researches in PV focus on obtaining more power by using materials with higher efficiency and extracting more power from the PV. MPPT strategy has an important role in all types of PV systems: stand alone, on-grid and solar pump systems. MPPT is capable to track the maximum power under instantaneous variation in irradiance and temperature, taking into consideration that the PV output power has non-linear variation with these environmental factors [16]. Also partial shading is another obstacle that MPPT must successfully deal with [2]. The MPPT methods can be mainly divided into two types off-line methods (indirect methods) includes short-circuit and open circuit methods. These methods are based on the data of PV modules and don't respond to all the environmental conditions such as partial shading [9,26]. The second type is on-line method (direct method) which relay on instantaneous measurements so they can handle any sudden change in irradiance or partial shading. The most common direct methods are: Perturbation & Observation (P&O) also Incremental Conductance (IncCond). More complex MPPT techniques are introduced in [13] like artificial neural network (ANN) [20], fuzzy logic [27,33], genetic algorithm (GA) [5,19] and differential evolution (DE) [11]. Also in [28], MPPT new methods such as Particle Swarm Optimization (PSO) and Cuckoo Search algorithm (CS), which give high accuracy but with more complexity. P&O method measures the output voltage (or current) and compares it with the its prior value and due to the output, the voltage (or current) is perturbed toward the maximum power operating point. P&O conventional method with fixed perturbation is simple but has disadvantage of low efficiency due to oscillation around the MPP. A modified P&O with fixed perturb, classical P&O with adaptive perturb and modified P&O with adaptive perturb methods are introduced in [1,3,18] and give better performance with partial shading and sudden environmental changes. This research introduces a MPPT method that merge simplicity of P&O technique with the accuracy of PSO method. In the proposed method, particle swarm optimization (PSO) is used to tune the proportional and integral gains of PI controller to enhance the performance of adaptive P&O-PI MPPT technique. The paper is arranged as follows: Modeling of Photovoltaic System is discussed in the second Section. Then in the third Section, MPPT Perturb and Observe with PI control algorithms are described. The forth Section presents the proposed PSO-PI Perturb and Observe MPPT method. The fifth section presents simulation results and discussion. Lastly, the sixth section contains the conclusion of the research work and future directions.

2   Modeling of Photovoltaic System

The principle of the PN semiconductor junction is a basic for photo-voltaic (PV) cell. The researchers utilized model similar to the circuit model of photo-voltaic cell. The PV cell is introduced by a source for current where it is identical to irradiance, also is linearly depend on the temperature of photo voltaic cell [31]. The diode which is parallel with the present source, is representing the phenomena of non-linearity in the photo-voltaic cell. At thus attitude the ideal PV cell is similar the source of current in parallel with a diode [30]. This model is refined by, two resistors, which symbolize the misfortunes, are regularly used: one of them is in series and the second is parallel. A solitary diode show gives enough accuracy less segments. It is embraced to ponder the nonlinear attributes of the PV hotspot for the entire power precision (see Fig. 1). Note that there are likewise models in light of a few diodes for getting better models, along these lines more precise, for which the applications concentrate on PV marvels examines [12].

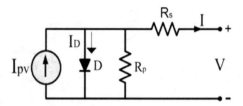

Fig. 1. Single diode RP model. From [14] with permission

With regard to PV control prospectation, the single-diode demonstrates the sufficiently basic. Figure 1 displays the identical circuit model of a PV cell. At Fig. 1, the output current of PV cell could be written as:

$$I = I_{ph} - I_D - I_{sh} \tag{1}$$

where I_{ph} is the light produced the current of pv, I_D the diode current and I_{sh} the Short-circuit current through the parallel resistance R_p.

So, the arithmetical relation between cell current and voltage is expressed as:

$$I = I_{ph} - I_0 \left[\exp\left(\frac{V + I \times R_s}{a \times V_{th}} \right) - 1 \right] - \frac{V + I \times R_s}{R_p} \tag{2}$$

where:
I and V are the output current and output voltage of the PV cell, respectively. I_0 is the reverse saturation current (A), a is the diode ideality constant, R_s is the cell series resistance (Ω), R_p is the cell parallel resistance (Ω) and V_{th} is the cell thermal voltage (V).

3   P&O with PI MPPT Control Algorithm

In the classical P&O method, a fixed perturb value which can be voltage or current is used [4,7]. This value produces reference signal in external control loop. For small perturb values, the oscillation of power/voltage is minimum and the tracking is small and vice versa. The algorithm provides a perturbation (ΔV) in Voltage based on the power change according to the following equations:

$$V_{new} = V_{old} + \Delta V \times slope(if P > P_{old}) \qquad (3)$$
$$V_{new} = V_{old} - \Delta V \times slope(if P < P_{old}) \qquad (4)$$

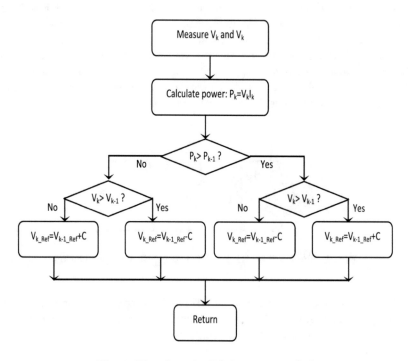

Fig. 2. Flowchart for P&O algorithm [24]

The flowchart of the P&O MPPT technique is displayed in Fig. 2. The step size relys on the slope of powervoltage curve also acquired as [7]:

$$\frac{dP}{dV_{pv}}(n) = \frac{P(n) - P(n-1)}{V_{pv}(n) - V_{pv}(n-1)} \qquad (5)$$

where $dP/dV_{pv}(n)$ is a real derivative of power and voltage of PV, $P(n)$ is real power, $P(n-1)$ is the previous value of power, $V_{pv}(n)$ is real voltage and $V_{pv}(n-1)$ is the previous value of voltage. Diversely if $V < V_{mpp}$, the working point moves towards left and when $V > V_{mpp}$, the working point moves towards right of the curve. For fast convergence, large ΔV is required but this will cause large oscillations in P and vice versa. That is why that improvements of the classical P&O should be introduced like adaptive P&O technique [1]. The P&O-PI technique is an adaptive method which based on utilization of the error between two sequential measured power $P_{pv}(n)$ and $P_{pv}(n-1)$ to calculate the perturb size (PS). The required reference voltage, V_{ref}, can be estimated by using the PI controller as follow [1]:

$$V_{ref} = K_p e(n) + K_I \sum e(n) \qquad (6)$$
$$e(n) \triangleq P_{pv}(n) - P_{pv}(n-1) \qquad (7)$$

Where $P_{pv}(n)$, $P_{pv}(n-1)$, K_p, K_I refer to present power, previous power, proportional gain and integral gain respectively. The flow chart of P&O-PI technique is demonstrated in Fig. 3.

4   Proposed PSO-PI Perturb and Observe MPPT Method

PSO is well-regarded stochastic, population-based algorithms in the research of heuristics and metaheuristics [17]. The PSO technique is depended on the movement and intelligence of swarms. Every particle continues following of its directions in the solution space which are related to the best solution (fitness) that has achieved so far by that particle. This value is called personal best, p_{best}. the other best value tracked by PSO is the best value achieved by any particle in the surroundings of that particle. This value is called g_{best}. The basic concept of PSO lies in accelerating each particle toward its p_{best} and the g_{best} locations. At each iteration i, the positions and velocities of all particles are updated in accordance with the following equations [6, 15].

$$x_i^{k+1} = x_i^k + v_i^{k+1} \tag{8}$$

where v_i is the velocity component and can be estimated as follow

$$v_i^{k+1} = \omega v_i^k + c_1 r_1 \left(P_{besti} - x_i^k \right) + c_2 r_2 \left(G_{best} - x_i^k \right) \tag{9}$$

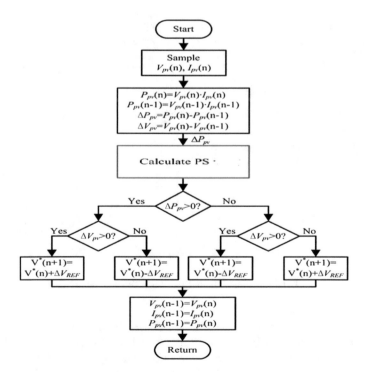

Fig. 3. Flowchart for PI-P&O mechanism. From [1]

Where ω is the inertia weight; c_1 and c_2 are the acceleration constants, while P_{besti} and G_{best} are the individual and global best positions, separately. The block diagram of the suggested PSO-PI-P&O model is presented in Fig. 4. The selection of PI controller gains, K_p and K_I can be done using trial and error methods if the plant under consideration is linear. However, for nonlinear systems like PV system, specific tuning methods are required such as PSO or other metaheuristics optimization techniques. In this proposed PSO-PI-P&O MPPT method, PSO is used to tune K_p and K_I of PI controller to approach the MPP. Under this condition, the PV system can track MPP.

5   Simulation Results and Discussion

To test the validation of the suggested technique, simulations are performed with two various operating conditions: (1) regular irradiance, (2) variation of irradiance, i.e., partial shading condition (PSC). Under uniform condition; the irradiance and the temperature are kept unchanged; while for PSC; the PV irradiance levels are changed keeping temperature constant. MATLAB simulation code is used for PSO-PI-P&O MPPT method. The proposed methods are proved using the model architecture shown in Fig. 5. Because PSO convergence depends on careful parameter's selection to obtain the best solution, the values of PSO parameters are listed in Table 1.

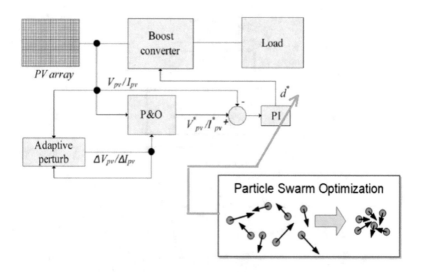

Fig. 4. Block diagram of the suggested PSO-PI-P&O.

The validation of effectiveness of the suggested methods require, ten different shading patterns as sown in Fig. 6 are tested with PI-P&O and PSO-PI-P&O MPPT techniques and the results are summarized in Table 2. Under uniform irradiation, both MPPT methods, PI-P&O and PSO-PI-P&O, reached maximum

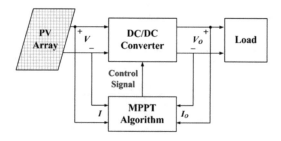

Fig. 5. System components of the suggested methodology

Table 1. Values of PSO parameter

Parameter	PSO-PI-P&O
W_1	0.9
C_1	1.4
C_2	1.8

power of 844.3172 W and 848.8 W, respectively because of single power peak. As noted, PSO-PI-P&O MPPT technique is capable of following the MP value with minimum iterative steps number. Moreover, power oscillation problem of classical P&O MPPT technique is discarded using the suggested PSO-PI-P&O method.

To simulate the partial shading conditions, PS irradiation change is introduced and the results for PI-P&O and PSO-PI-P&O MPPT methods are recorded as displayed Table 2. For classical P&O MPPT mechanism, the duty

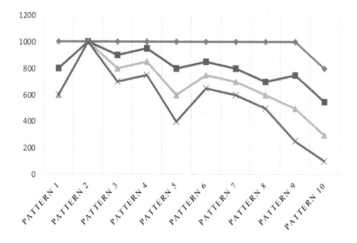

Fig. 6. Various shading patterns for testing the proposed methods

Table 2. Comparing between output power versus PI-P&O and PSO-PI-P&O MPPT technique.

Shading patterns	Solar PV array				PI-P&O (P_{mp})	PSO-PI-P&O (P_{mp})
	PV module 1	PV module 2	PV module 3	PV module 4		
1	1000	800	600	600	636.58	638.6
2	1000	1000	1000	1000	844.31	848.8
3	1000	900	800	700	721.12	723.7
4	1000	950	850	750	752.73	755.2
5	1000	800	600	400	593.26	595
6	1000	850	750	650	689.59	692
7	1000	800	700	600	658.35	660.1
8	1000	700	600	500	593.88	595.7
9	1000	750	500	250	527.93	529.62
10	800	550	300	100	367.35	368.31

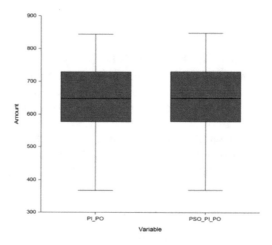

Fig. 7. Comparison of output power of PV array between PI-P&O and PSO-PI-P&O MPPT methods

cycle and PV module output power are oscillating around maximum power point. Under irradiation change while using conventional MPPT methods like P&O or incremental conductance, the PV load line shifts and a new operating point of maximum power could be obtained by increment or decrements of step sizes of duty cycle. Once reaching the MPP, the PV module output power starts oscillating around that point. The results showed that when using PSO in combination with PI-P&O MPPT method, the transient oscillations are avoided as well as better dynamic and steady state performance are obtained.

Statistical analysis was performed to compare the output power means of PV array between PI-P&O and PSO-PI-P&O MPPT methods using paired-sample T-test. The statistical analysis demonstrated that there was a significant improvement in mean of power from PI-P&O to PSO-PI-P&O MPPT method (Power mean: 638.51 ± 131.24 versus 640.70 ± 132.06, respectively; P $= 0.00004$). Figure 7 illustrates the comparison of output power of PV array between PI-P&O and PSO-PI-P&O MPPT methods.

6   Conclusion

In this paper, PSO is used to tune the parameters (proportional and integral) gains of PI controller for improving the effectiveness of adaptive P&O-PI MPPT technique. The Proposed method were tested at uniform and partial shading conditions. The simulation results demonstrated that PSO-PI-P&O MPPT method has good tracking efficiency under uniform and partial shading conditions and without steady state oscillations. The statistical analysis showed that there is a significant improvement in output power mean of PV array when using PSO-PI-P&O MPPT method (P $= 0.00004$). For future direction, other metaheuristic optimization techniques may be used to be compared with the results of the proposed technique. Tuning the metaheuristic parameters is also another future direction when combining it with MPPT methods to guarantee reaching the MPP with fast tracking speed.

Acknowledgement. The authors would like to thank Prince Sultan University, Riyadh, KSA for supporting this work.

References

1. Abdelsalam, A.K., Massoud, A.M., Ahmed, S., Enjeti, P.N.: High-performance adaptive perturb and observe MPPT technique for photovoltaic-based microgrids. IEEE Trans. Power Electron. **26**(4), 1010–1021 (2011)
2. Ahmad, R., Murtaza, A.F., Sher, H.A.: Power tracking techniques for efficient operation of photovoltaic array in solar applications - a review. Renew. Sustain. Energy Rev. **101**, 82–102 (2019)
3. Ahmed, J., Salam, Z.: An enhanced adaptive P&O MPPT for fast and efficient tracking under varying environmental conditions. IEEE Trans. Sustain. Energ. **9**(3), 1487–1496 (2018)
4. Carannante, G., Fraddanno, C., Pagano, M., Piegari, L.: Experimental performance of MPPT algorithm for photovoltaic sources subject to inhomogeneous insolation. IEEE Trans. Industr. Electron. **56**(11), 4374–4380 (2009)
5. Daraban, S., Petreus, D., Morel, C.: A novel MPPT (maximum power point tracking) algorithm based on a modified genetic algorithm specialized on tracking the global maximum power point in photovoltaic systems affected by partial shading. Energy **74**, 374–388 (2014)
6. Eberhart, R., Shi, Y., Kennedy, J.: Swarm Intelligence. Elsevier, Amsterdam (2001)
7. Esram, T., Chapman, P.L.: Comparison of photovoltaic array maximum power point tracking techniques. IEEE Trans. Energy Convers. **22**(2), 439–449 (2007)

8. Ghoudelbourk, S., Dib, D., Omeiri, A., Azar, A.T.: MPPT control in wind energy conversion systems and the application of fractional control (piα) in pitch wind turbine. Int. J. Model. Ident. Control **26**(2), 140–151 (2016)

9. Hart, G., Branz, H., Cox, C.: Experimental tests of open-loop maximum-power-point tracking techniques for photovoltaic arrays. Sol. Cells **13**(2), 185–195 (1984)

10. (IRENA) IREA: Renewable energy statistics 2018. Technical report (2018). https://www.irena.org/publications/2018/Jul/Renewable-Energy-Statistics-2018

11. Ishaque, K., Salam, Z.: An improved modeling method to determine the model parameters of photovoltaic (PV) modules using differential evolution (DE). Sol. Energy **85**(9), 2349–2359 (2011)

12. Ishaque, K., Salam, Z., Syafaruddin: A comprehensive MATLAB simulink PV system simulator with partial shading capability based on two-diode model. Sol. Energy **85**(9), 2217–2227 (2011)

13. Jiang, L.L., Srivatsan, R., Maskell, D.L.: Computational intelligence techniques for maximum power point tracking in PV systems: a review. Renew. Sustain. Energy Rev. **85**, 14–45 (2018)

14. Jordehi, A.R.: Parameter estimation of solar photovoltaic (PV) cells: a review. Renew. Sustain. Energy Rev. **61**, 354–371 (2016)

15. Jordehi, A.R., Jasni, J.: Parameter selection in particle swarm optimisation: a survey. J. Exp. Theor. Artif. Intell. **25**(4), 527–542 (2013). https://doi.org/10.1080/0952813X.2013.782348

16. Kamal, N.A., Ibrahim, A.M.: Conventional, intelligent, and fractional-order control method for maximum power point tracking of a photovoltaic system: a review. In: Azar, A.T., Radwan, A.G., Vaidyanathan, S. (eds.) Fractional Order Systems, Advances in Nonlinear Dynamics and Chaos (ANDC), pp. 603–671. Academic Press, Cambridge (2018). Chap 20

17. Kennedy, J., Eberhart, R.: Particle swarm optimization. In: Proceedings of ICNN 1995 - International Conference on Neural Networks, vol. 4, pp. 1942–1948 (1995)

18. Kollimalla, S.K., Mishra, M.K.: A novel adaptive P&O MPPT algorithm considering sudden changes in the irradiance. IEEE Trans. Energy Convers. **29**(3), 602–610 (2014)

19. Kulaksız, A.A., Akkaya, R.: A genetic algorithm optimized ANN-based MPPT algorithm for a stand-alone PV system with induction motor drive. Sol. Energy **86**(9), 2366–2375 (2012)

20. Lin, W.M., Hong, C.M., Chen, C.H.: Neural-network-based MPPT control of a stand-alone hybrid power generation system. IEEE Trans. Power Electron. **26**(12), 3571–3581 (2011)

21. Meghni, B., Dib, D., Azar, A.T.: A second-order sliding mode and fuzzy logic control to optimal energy management in wind turbine with battery storage. Neural Comput. Appl. **28**(6), 1417–1434 (2017a)

22. Meghni, B., Dib, D., Azar, A.T., Ghoudelbourk, S., Saadoun, A.: Robust adaptive supervisory fractional order controller for optimal energy management in wind turbine with battery storage. In: Azar, A.T., Vaidyanathan, S., Ouannas, A. (eds.) Fractional Order Control and Synchronization of Chaotic Systems, pp. 165–202. Springer International Publishing, Cham (2017b)

23. Meghni, B., Dib, D., Azar, A.T., Saadoun, A.: Effective supervisory controller to extend optimal energy management in hybrid wind turbine under energy and reliability constraints. Int. J. Dyn. Control **6**(1), 369–383 (2018)

24. Mellit, A., Rezzouk, H., Messai, A., Medjahed, B.: FPGA-based real time implementation of MPPT-controller for photovoltaic systems. Renewable Energy **36**(5), 1652–1661 (2011)

25. Moriarty, P., Honnery, D.: 6 - global renewable energy resources and use in 2050. In: Letcher, T.M. (ed.) Managing Global Warming, pp. 221–235. Academic Press, Cambridge (2019)
26. Noguchi, T., Togashi, S., Nakamoto, R.: Short-current pulse-based maximum-power-point tracking method for multiple photovoltaic-and-converter module system. IEEE Trans. Industr. Electron. **49**(1), 217–223 (2002)
27. Ozdemir, S., Altin, N., Sefa, I.: Fuzzy logic based MPPT controller for high conversion ratio quadratic boost converter. Int. J. Hydrogen Energy **42**(28), 17748–17759 (2017). Special Issue on the 4th European Conference on Renewable Energy Systems (ECRES 2016), Istanbul, Turkey, 28–31 August 2016.
28. Rezk, H., Fathy, A., Abdelaziz, A.Y.: A comparison of different global MPPT techniques based on meta-heuristic algorithms for photovoltaic system subjected to partial shading conditions. Renew. Sustain. Energy Rev. **74**, 377–386 (2017)
29. Smida, M.B., Sakly, A., Vaidyanathan, S., Azar, A.T.: Control-based maximum power point tracking for a grid-connected hybrid renewable energy system optimized by particle swarm optimization. In: Azar, A.T., Vaidyanathan, S. (eds.) Advances in System Dynamics and Control, pp. 58–89. IGI Global, Hershey (2018)
30. Tan, Y.T., Kirschen, D.S., Jenkins, N.: A model of PV generation suitable for stability analysis. IEEE Trans. Energy Convers. **19**(4), 748–755 (2004)
31. Villalva, M.G., Gazoli, J.R., Filho, E.R.: Comprehensive approach to modeling and simulation of photovoltaic arrays. IEEE Trans. Power Electron. **24**(5), 1198–1208 (2009)
32. Whiting, K., Carmona, L.G., Sousa, T.: A review of the use of exergy to evaluate the sustainability of fossil fuels and non-fuel mineral depletion. Renew. Sustain. Energy Rev. **76**, 202–211 (2017)
33. Yetayew, T.T., Jyothsna, T.R.: Evaluation of fuzzy logic based maximum power point tracking for photovoltaic power system. In: 2015 IEEE Power, Communication and Information Technology Conference (PCITC), pp. 217–222 (2015)

Particle Swarm Optimization and Grey Wolf Optimizer to Solve Continuous *p*-Median Location Problems

Hassan Mohamed Rabie[(✉)]

Institute of National Planning, Cairo, Egypt
Hassan.rabie@inp.edu.eg

Abstract. The continuous *p*-median location problem is to locate *p* facilities in the Euclidean plane in such a way that the sum of distances between each demand point and its nearest median/facility is minimized. In this paper, the continuous p-median problem is studied, and a proposed Grey Wolf Optimizer (GWO) algorithm, which has not previously been applied to this problem, is presented and compared to a proposed Particle Swarm Optimization (PSO) algorithm. As an experimental evidence for the NFL theorem, the experimental results showed that the no algorithm can outperformed the other in all cases, however the proposed PSO has better performance in most of the cases. The experimental results show that the two proposed algorithms have better performance than other PSO methods in the literature.

Keywords: *p*-median · Particle Swarm · Grey Wolf · Location problem · NFL theorem

1 Introduction

The goal of the continuous *p*-median location problem is to locate *p* medians/facilities within a set of *n* demand points with ($n > p$) in such a way that the sum of the distances between each demand point and its nearest median/facility is minimized [1]. These kinds of location problems include the establishment of the public services including schools, hospitals, firefighting, Ambulance, and etc. The objective function in the median problems is of the minisum type [1]. The demand points and the medians are located in some subset of the *d*-dimensional real space R^d. In many cases, the distance between demand and service points is Euclidean [2]. Therefore, we wish to locate *p* points $X_j = (x_j, y_j)$, $j = 1,..., p$ in the plane R^2 in order to service a set of *n* points at known locations (a_i, b_i), $i = 1,..., n$. The basic version of continuous *p*-median problem with Euclidean distance may be written in the following equivalent form [2]:

$$\min_{X_j} \operatorname{sum}_i \; w_j \min_j \left[\left(a_i - x_j\right)^2 + \left(b_i - y_j\right)^2 \right]^{1/2}, \; j = 1,\ldots,p. \qquad (1)$$

where: *n*: the number of demand points, *p*: the number of facilities, (a_i, b_i): the location of demand points *i* ($i = 1,..., n$), and $wi > 0$: the weight of demand point *i* ($i = 1,..., n$).

© Springer Nature Switzerland AG 2020
A. E. Hassanien et al. (Eds.): AISI 2019, AISC 1058, pp. 136–146, 2020.
https://doi.org/10.1007/978-3-030-31129-2_13

In this paper $w_i = 1$. $X = (X_1, ..., X_p)$: the decision variables related to these p facilities with $X_j = (x_j, y_j)$ representing the location of the new facility j with $X_j \in R^2$; $j = 1, ..., p$. The p-median location problem is NP-hard problem [3], therefore, it is too hard to obtain an optimal solution through polynomial time-bounded algorithms [4].

The p-median location problem is one of the most widely and commonly applied models [5, 6]. Since Weber's book in 1909 on location theory [6], there have been many researchers, such as Hakimi [7, 8] and Cooper [9, 10], who studied location problems. Many extensive studies of location problems were developed after Cooper's research [6]. According to [11], the first well-known heuristic algorithm was developed by Cooper [10], and according to [12] the early attempts to solve this problem by exact methods was through proposing a branch and bound algorithm proposed by Kuenne and Soland in [13] and Ostresh in [14]. We should refer to survey papers in this area such as Melo et al. [15] and ReVelle et al. [16]. A survey on exact and heuristic algorithms can be found in Farahani [1], who reviewed different solution techniques for p-median problems and Mladenovic et al. provided a review on the p-median problem focusing on meta-heuristic methods [17]. Despite the advancement in the methods solving p-median problems, large scale problems found in the literature are still unsolvable by exact methods [12], and, aggregation methods may introduce errors in the data as well as the model output, thus resulting in less accurate results [12]. Optimal solutions may not be found for relatively large problems, and hence meta-heuristic algorithms are usually considered to be the best way for solving such problems [18]. Furthermore, the newer algorithms tend to be highly sensitive to the starting solution, and so, require state-of-the-art heuristics to obtain the best initial solution possible. Thus, advances in heuristic approaches are continually sought [12].

Because PSO has good performance in solving large scale continuous problems [6], in this paper, PSO algorithm is proposed to solve the p-median location problem. However, and due to the No Free Lunch (NFL) theorem which logically proved that there is no meta-heuristic best suited for solving all optimization problems, i.e. a meta-heuristic algorithm may show promising results on a problem, but the same algorithm may show poor performance on another one [19]. This motivates us to compare the performance of PSO versus one of the recent meta-heuristic named GWO.

The paper is organized as follows. In section two, the proposed GWO algorithm and PSO algorithms are described. Section three presents the computational results of the two proposed algorithms, including a comparison with some PSO algorithms in the literature. The last section, the findings are summarized.

2   WGO & PSO for Solving p-Median Location Problem

Meta-heuristic intelligent approaches are high level strategies for exploring search spaces by using different strategies like Simulated Annealing, Tabu Search, Genetic Algorithms and Evolutionary Algorithms, PSO and GWO. Meta-heuristic considered to be more efficient search approaches for solving hard problems [19]. Meta-heuristic algorithms have become remarkably common, that's mainly because of its simplicity, and flexibility [19].

2.1   Grey Wolf Optimizer

Grey Wolf Optimizer (GWO) was introduced in 2014 by Mirjalili et al. [19] as a new meta-heuristic optimization algorithm; by 2019, GWO was found to be the most cited advances in engineering software articles [20]. GWO mimics the leadership hierarchy of wolves. They have very strict social dominant hierarchy. Social hierarchy considered the main feature of the wolves' pack. When hunting a prey, the pack can be categorized into four types [19]: (1) the leader wolf, called alpha (α). Alpha is responsible for making decisions about everything their decisions are dictated to the pack. (2) called beta (β). Beta helps the alpha in decision-making. Beta reinforces the alpha's commands throughout the pack and gives feedback to the alpha. (3) Omega (ω) acts as a scapegoat, they always have to submit to all the other dominant wolves. (4) Deltas (δ) have to submit to alphas and betas, but they dominate the omega.

In order to mathematically model this **Social Hierarchy** of wolf pack, Mirjalili et al. [19] represented the hunting technique of the wolves such that; the best/fittest solution is considered as alpha (α), while the second best solution is considered as beta (β), and the third best solutions is called delta (δ). The other remaining are named omega (ω). **Encircling Prey**: Grey wolves encircling the prey during the hunting. The encircling strategy can be mathematically modelled as following [19]:

$$\vec{D} = \left| \vec{C}.\overrightarrow{X_p}(t) - \vec{X}(t) \right| \tag{2}$$

$$\vec{X}(t+1) = \overrightarrow{X_p}(t) - \vec{H}.\vec{D} \tag{3}$$

$$\vec{H} = 2\vec{h}.\overrightarrow{r_1} - \vec{h} \tag{4}$$

$$\vec{C} = 2.\overrightarrow{r_2} \tag{5}$$

where t indicates the current iteration number, $\vec{H}$ and $\vec{C}$ are two coefficient vectors, $\overrightarrow{X_p}$ represents the position vector of the prey, and $\vec{X}$ represents the position vector of a wolf. r_1, r_2 are two uniformly random vectors in the range [0, 1], and $\vec{h}$ component is linearly decreased from 2 to 0 over the course of iterations.

Hunting: After the wolves recognize the location of a prey and encircle them. The hunt is guided usually by alpha (α), Beta (β) and Delta (δ). Hunting of the grey wolves can be mathematically modelled [19].

$$\overrightarrow{D_\alpha} = \left| \overrightarrow{C_1}.\overrightarrow{X_\alpha} - \vec{X} \right|, \overrightarrow{D_\beta} = \left| \overrightarrow{C_2}.\overrightarrow{X_\beta} - \vec{X} \right|, \overrightarrow{D_\delta} = \left| \overrightarrow{C_3}.\overrightarrow{X_\delta} - \vec{X} \right| \tag{6}$$

$$\vec{X}_1 = \overrightarrow{X_\alpha} - \overrightarrow{H_1}.\overrightarrow{D_\alpha}, \vec{X}_2 = \overrightarrow{X_\beta} - \overrightarrow{H_2}.\overrightarrow{D_\beta}, \vec{X}_3 = \overrightarrow{X_\delta} - \overrightarrow{H_3}.\overrightarrow{D_\delta} \tag{7}$$

$$\vec{X}(t+1) = \frac{\vec{X}_1 + \vec{X}_2 + \vec{X}_3}{3} \tag{8}$$

where X_α, X_β, X_δ are the positions of α, β and δ solutions with the help of Eq. (2).

Attacking Prey: When the prey stops moving, the grey wolves finish the hunting; this can mathematically model by decreasing the value of $\vec{h}$.

For the p-median each candidate solution consists from p (x,y) pairs, which representing the sites of facilities to be located, and p is the number of facilities. The candidate solution can be represented as: $[(x_1,y_1), (x_2,y_2), (x_3,y_3),\ldots, (x_j,y_j),\ldots, (x_p,y_p)]$, where the coordinate (x_j,y_j) denotes the location of the jth facility, $j = 1,\ldots, p$. The initial population is usually constructed by choosing the candidate solution randomly between the lower and upper limit of the coordinates of demand points. In practices, demand points are clustered, therefore, generating the median points randomly without any knowledge about the demand points clusters may result in poor initial locations; since initial solutions could be located in large empty regions or far away from most of demand points [11]. To overcome this problem, k-means cluster algorithm [21] was used to find demand point clusters where $(k = p)$, and then the initial solutions generated within each cluster limits for each p. Figure 1 shows the proposed WGO algorithm.

Step 1. Read the set of demand points (A,B).
Step 2. Let p (number-of-medians), ss (population-size), *iterations* (number-of-iterations).
Step 3. Generate the candidate solutions $X_{ss \times p}$ (wolves) randomly by using k-Means.
Step 4. Evaluate the fitness/objective function, using Equation (1).
Step 5. Update the position of the candidate solutions (wolves).
      for z <= iterations,
– $h=2- z*(2/iterations)$
      for t <= ss;
–   Identify; the fittest/best search wolf (X_α), the second best (X_β), and the third best (X_δ).
–   Generate r_1 and r_2; two random numbers in range [0,1]
    Calculate $H_1=2*h*r_1-h$, **and** $C_1=2*r_2$ and Calculate $D_\alpha = (C_1*X_\alpha - X_t)$ **and** $X_1= (X_\alpha - H_1*D_\alpha)$
–   Generate r_1 and r_2; two random numbers in range [0,1]
    Calculate $H_2=2*h*r_1-h$, **and** $C_2=2*r_2$ and Calculate $D_\beta = (C_2*X_\beta - X_t)$ **and** $X_2= (X_\beta - H_2*D_\beta)$
–   Generate r_1 and r_2; two random numbers in range [0,1]
    Calculate $H_3=2*h*r_1-h$, **and** $C_3=2*r_2$ and Calculate $D_\delta = (C_3*X_\delta - X_t)$ **and** $X_3= (X_\delta - H_3*D_\delta)$
– $X_{t+1}=(X_1+X_2+X_3)/3$
–   if $f(X_{t+1}) \leq f(X_\alpha)$; $f(X_\alpha)= f(X_{t+1})$; $X_\alpha = X_{t+1}$; endif
–   if $f(X_{t+1}) > f(X_\alpha)$ **and** $f(X_{t+1}) < f(X_\beta)$; $f(X_\beta)=f(X_{t+1})$; $X_\beta = X_{t+1}$; endif
–   if $f(X_{t+1}) > f(X_\alpha)$ **and** $f(X_{t+1}) > f(X_\beta)$ **and** $f(X_{t+1}) < f(X_\delta)$;$f(X_\delta)=f(X_{t+1})$; $X_\delta=X_{t+1}$; endif
    end for
    end for
Step 6.   Report X_α and $f(X_\alpha)$

Fig. 1. The basic steps of the proposed WGO for the p-median location problem

2.2   Particle Swarm Optimization

PSO was developed by Kennedy and Eberhart in 1995 [22] as an evolutionary approach, and has become one of the most widely used swarm-intelligence-based algorithms due to its simplicity and flexibility [6]. PSO was inspired from the social behavior of birds flocking. The algorithm employs multiple particles that chase the position of the best particle and their own best positions obtained so far [23]. PSO was originally used to solve continuous optimization problems [24]. The swarm consists from a number of particles that frequently move in the search space [6]. An objective function f must be defined to compare candidate solutions fitness within the search space. Each particle in the population has two state variables: (1) current position X_t, and (2) current velocity V_t. The PSO algorithm starts with initiating the particles of the swarm randomly. The particles are uniformly distributed in decision space. The iterative process optimization starts, such that, each particle's position and velocity is updated according to the following equations:

$$Vel_t^{z+1} = wt_z Vel_t^z + c_1 r_1 \left(Pbest_t^z - X_t^z \right) + c_2 r_2 \left(Gbest - X_t^z \right) \tag{9}$$

$$X_t^{z+1} = X_t^z + Vel_t^{z+1} \tag{10}$$

where, the inertia weight wt, determines the influence of the previous velocity V_t and controls the particle's ability to explore the search space [6]. The variable wt_z is updated as in Eq. (11), where wt_{max} and wt_{min} are the maximum and minimum of wt respectively; k_{max} is the number of iterations [25].

r_1 and r_2 are two random numbers between 0 and 1. c_1 and c_2 are two acceleration constants. These parameters determine whether a particle moves toward its previous *Pbest* or *Gbest* [26]. Vel_t^z: represents the velocity of particle t at iteration z. $Pbest_t^z$, the best previous position of particle t. *Gbest*, the best among all particles. X_t^z represents the position of particle t at iteration z.

$$wt_z = \left(wt_{max} - wt_{min} \right) * \frac{\left(k_{max} - z \right)}{k_{max}} + wt_{min}, \tag{11}$$

The basic steps of the proposed PSO for the continuous p-median location problem is presented in Fig. 2.

Step 1. Read the set of demand points (A,B).

Step 2. Define p (number-of-medians), ss (population-size), $iterations$ (number-of-iterations), wt_{max} (the maximum value of inertia weight), wt_{min} (the minimum value of inertia weight), c_1 (cognitive parameter), and c_2 (social parameter).

Step 3. Generate the candidate solutions $X_{ss \times p}$ (wolves) randomly by using k-Means clustering algorithm. Set $Vel_{ss \times p} = 0$ to initiate the velocity of the swarm.

Step 4. Evaluate the objective function for $X_{ss \times p}$ using Equation (1)

Step 5. Let $Pbest = X$, with $f(Pbest) = f(X)$.

Step 6. Update the swarm fitness/position and velocity

    for z <= iterations,

    for t <= ss,

–   Let $Gbest = Pbest$ (arg(min($f(Pbest)$)))

–   Generate r_1 and r_2; two random numbers in range [0,1]

–   Calculate inertia weight (wt_t) using Equation (11)

–   Calculate the velocity (Vel_t^{z+1}) using Equation (9)

–   Calculate the new position (X_t^{z+1}) using Equation (10)

–   Evaluate $f(X_t^{z+1})$ of the new position using Equation (1)

•   if $f(X_t^{z+1}) < f(Pbest_t^z)$; then $Pbest_t^z = X_t^{z+1}$; $f(Pbest_t^z) = f(X_t^{z+1})$; endif

    end for

    end for

Step 7.   Report $Gbest$ and $f(Gbest)$

Fig. 2. The basic steps of the proposed PSO for the p-median location problem

3 Computational Results

In this section, the computational results for the proposed WGO and PSO algorithms are reported. To investigate the performance of the two proposed algorithms to find good feasible solutions, we considered the following well-known three continuous p-median location problems: the 50-customer problem, the 654-customer problem and the 1060-customer problem, they are listed in the TSP library [27]. According to [28], The first problem has solved optimally in [28]. The second and the third have not been solved by deterministic methods, and only the best known solutions exist in the literature in [28, 29]. The two proposed algorithms coded in MATLAB, 2018a and compiled on Windows 10 running on PC, Intel Core i7@2.2 GHz with 8 GB RAM. Due to the random nature of WGO and PSO; extensive experiments have been conducted by performing 30 runs for each case (p), with total of 1920 experiments. For the two proposed algorithms; the population size is 100 and the number of iterations is 100. The PSO parameters have been chosen after several experimenting by using trial and error approach. $wt_{min} = 0.2$, $wt_{max} = 1.2$, $c_1 = c_2 = 1.3$.

    The numerical results of the two proposed algorithms experiments are summarized in Table 1 for the first problem, Table 2 for the second problem and Table 3 for the third problem. The first column of the tables shows the number of medians/facilities (p) to be located in each problem; the second column lists the optimal or the best-known solution. Other columns are expressed as a percent **deviation** from the optimal/best-known solutions; so that **Best** is the best of the objective function and **Avg.** is the average of the objective function found from 30 runs of the algorithm for each case. The deviation is computed as follows: **deviation** $= (F_{best} - F^*)/F^* 100\%$.

where F_{best} is the value of the objective function found by the two proposed algorithms methods and F^* refers to the optimal or the best-known value of the objective function found in the literature [11]. General conclusions could be inferred from the numerical results in Tables 1, 2 and 3: **(1)** The proposed PSO algorithm found the optimal/best-known solution for all $p \leq 15$ (the number of medians less than 15) for the three problems. **(2)** PSO algorithm can be effectively used to obtain good solutions. The results obtained by the proposed PSO are more effective than the proposed WGO algorithm. **(3)** The comparison between the two proposed algorithms suggests that no one algorithm is the best for all cases. This could be considered as an experimental evidence for the NFL theorem. **(4)** By comparing the average results in Tables 1, 2 and 3, it can be seen that, PSO has better performance than WGO for the first and second problem. As for the average of the best results, the proposed PSO outperform the proposed WGO.

The performance of the two proposed algorithms were compared to five different PSO and local search methods developed by Birto et al. [30]. These methods are explained in details in [30] and they named: (1) StPSO: Standard PSO (2) NPSO: New proposed PSO (3) StPSO + LS: Standard PSO with local search (4) NPSO + LS: New PSO with local search. (5) MS + LS: Multi-start algorithm with local Search. These methods have been used to solve 654-customer problem. Table 4 shows the deviation of the five methods from the best-known solution versus the deviation of the two proposed algorithms. The numerical results of the experiments shows that the two proposed algorithms have better performance than the five methods proposed in [30].

Table 1. Summary results over 30 runs for the 50-customer problem

P	Optimal	Proposed WGO		Proposed PSO	
		Best	Avg.	Best	Avg.
2	135.52	0.00%	0.04%	0.00%	0.00%
3	105.21	0.02%	0.28%	0.00%	0.00%
4	84.15	0.00%	0.09%	0.00%	0.01%
5	72.24	0.00%	0.73%	0.00%	0.01%
6	60.97	0.01%	1.85%	0.00%	0.00%
7	54.50	0.00%	1.28%	0.00%	0.00%
8	49.94	0.00%	2.12%	0.00%	0.66%
9	45.69	0.01%	1.89%	0.00%	1.90%
10	41.69	0.02%	2.96%	0.00%	3.82%
11	38.02	0.01%	3.43%	0.00%	4.86%
12	35.06	0.35%	2.84%	0.00%	0.62%
13	32.31	0.02%	3.48%	0.00%	6.46%
14	29.66	1.54%	4.92%	0.00%	2.68%
15	27.63	0.25%	4.61%	0.00%	5.55%
20	19.36	1.81%	6.84%	4.24%	4.24%
25	13.30	0.02%	9.15%	5.67%	9.97%
Average		0.25%	2.91%	0.62%	2.55%

Table 2. Summary results over 30 runs for the 654-customer problem

P	Best-known	Proposed WGO		Proposed PSO	
		Best	Avg.	Best	Avg.
2	815,313.30	0.00%	0.01%	0.00%	0.00%
3	551,062.88	0.00%	0.13%	0.00%	0.20%
4	288,190.99	0.00%	0.09%	0.00%	0.00%
5	209,068.79	0.00%	0.23%	0.00%	0.00%
6	180,488.21	0.00%	1.61%	0.00%	0.50%
7	163,704.17	0.00%	0.89%	0.00%	0.08%
8	147,050.79	0.00%	0.99%	0.00%	0.73%
9	130,936.12	0.00%	1.66%	0.00%	0.43%
10	115,339.03	0.06%	3.52%	0.00%	0.73%
Average		0.01%	1.01%	0.00%	0.30%

Table 3. Summary results over 30 runs for the 1060-customer problem

P	Best-known	Proposed WGO		Proposed PSO	
		Best	Avg.	Best	Avg.
5	1,851,879.90	0.00%	0.02%	0.00%	0.02%
10	1,249,564.80	0.02%	0.21%	0.00%	0.24%
15	980,132.10	0.06%	1.04%	0.00%	0.68%
20	828,802.00	0.30%	1.24%	0.01%	1.36%
50	453,164.00	1.78%	3.29%	1.04%	2.22%
75	340,242.00	2.55%	3.61%	1.42%	2.51%
150	212,926.00	4.03%	4.30%	2.48%	3.34%
Average		1.25%	1.96%	0.71%	1.48%

Table 4. PSO and local search methods versus the two proposed methods

P	StPSO	NPSO	StPSO+LS	NPSO+LS	MS+LS	WGO	PSO
2	0.09%	0.04%	0.00%	0.00%	0.00%	0.00%	0.00%
3	3.79%	1.18%	0.00%	0.00%	0.00%	0.00%	0.00%
4	31.91%	20.19%	0.00%	0.00%	0.00%	0.00%	0.00%
5	50.53%	36.39%	0.01%	0.00%	0.01%	0.00%	0.00%
6	69.19%	57.50%	0.07%	0.04%	0.03%	0.00%	0.00%
7	64.89%	43.11%	0.24%	0.15%	0.18%	0.00%	0.00%
8	63.04%	48.43%	0.85%	0.40%	0.74%	0.00%	0.00%
9	79.86%	81.38%	0.60%	0.83%	1.31%	0.00%	0.00%
10	104.09%	99.24%	2.09%	1.48%	1.25%	0.06%	0.00%
Avg	51.93%	43.05%	0.43%	0.32%	0.39%	0.01%	0.00%

Figure 3 shows the convergence performance of the two proposed algorithms for the value for $p = 5$ as an illustrative example, the figure contains three charts for the three problems. Each chart shows the convergence curve-Average of the runs- to the optimal/Best-known solution; where the horizontal axis represents the number of iterations and the vertical axis represents the objective function value. Figure 3 shows that the proposed PSO converges much faster than the proposed WGO for the first and second problem, however the proposed WGO converges faster in the third problem.

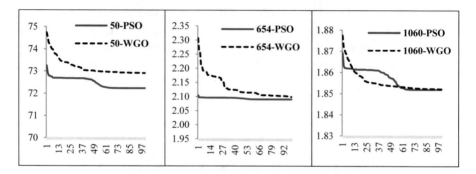

Fig. 3. The proposed PSO&WGO convergence curves for $p = 5$

4   Conclusions

In this paper, the continuous p-median location problem was studied. There are many algorithms have been proposed for solving this problem. However, Grey Wolf Optimizer has not been implemented yet. A proposed WGO algorithm was presented and compared with the proposed PSO algorithm. As an experimental evidence for the NFL theorem; the experimental results showed that the no algorithm can outperformed the other in all cases, however the proposed PSO has better performance in most of the cases. Therefore, for future studies, it will be helpful to investigate hybrid algorithms to solve p-median location problem.

References

1. Farahani, R.Z., Hekmatfar, M.: Facility Location Concepts, Models, Algorithms and Case Studies. Springer-Verlag, Heidelberg (2009)
2. Love, R.F., Morris, J.G., Wesolowsky, G.O.: Facilities Location: Models & Methods. North-Holland, Amsterdam (1988)
3. Megiddo, N., Supowit, K.J.: On the complexity of some common geometric location problems. SIAM J. Comput. **13**(1), 182–196 (1984)
4. Daskin, M.S.: Network and Discrete Location: Models, Algorithms, 2nd edn. Wiley, New York (2013)
5. Laporte, G., Nickel, S., da Gama, F.S.: Location Science. Springer, Cham (2015)

6. Ghaderi, A., Jabalameli, M.S., Barzinpour, F., Rahmaniani, R.J.N.: An efficient hybrid particle swarm optimization algorithm for solving the uncapacitated continuous location-allocation problem. Netw. Spat. Econ. **12**(3), 421–439 (2012)
7. Hakimi, S.L.: Optimum locations of switching centers and the absolute centers and medians of a graph. Oper. Res. **12**(3), 450–459 (1964)
8. Hakimi, S.L.: Optimum distribution of switching centers in a communication network and some related graph theoretic problems. Oper. Res. **13**(3), 462–475 (1965)
9. Cooper, L.: Location-allocation problems. Oper. Res. **11**(3), 331–343 (1963)
10. Cooper, L.: Heuristic methods for location-allocation problems. SIAM Rev. **6**(1), 37–53 (1964)
11. Salhi, S., Gamal, M.: A genetic algorithm based approach for the uncapacitated continuous location–allocation problem. Ann. Oper. Res. **123**(1–4), 203–222 (2003)
12. Brimberg, J., Drezner, Z., Mladenovic, N., Salhi, S.: A new local search for continuous location problems. Eur. J. Oper. Res. **232**, 256–265 (2014)
13. Kuenne, R., Soland, R.: Exact and approximate solutions to the multisource weber problem. Math. Program. **3**, 193–209 (1972)
14. Ostresh, J.L.M.: Multi-exact solutions to the M-center location-allocation problem. In: Rushton, G., Goodchild, M.F., Ostresh Jr, L.M. (eds.) Computer Programs for Location-Allocation Problems. Monograph No. 6. University of Iowa, IA (1973)
15. Melo, M.T., Nickel, S., Saldanha-Da-Gama, F.: Facility location and supply chain management–a review. Eur. J. Oper. Res. **196**(2), 401–412 (2009)
16. ReVelle, C.S., Eiselt, H.A., Daskin, M.S.: A bibliography for some fundamental problem categories in discrete location science. Eur. J. Oper. Res. **184**, 817–848 (2008)
17. Mladenovic, N., Brimberg, J., Hansen, P., Moreno-Perez, J.A.: The p-median problem: a survey of metaheuristic approaches. Eur. J. Oper. Res. **179**, 927–939 (2007)
18. Irawan, C.A., Salhi, S., Scaparra, M.P.: An adaptive multiphase approach for large unconditional and conditional p-median problems. Eur. J. Oper. Res. **237**, 590–605 (2014)
19. Mirjalili, S., Mirjalili, S.M., Lewis, A.: Grey wolf optimizer. Adv. Eng. Softw. **69**, 46–61 (2014)
20. Elsevier: Most Cited Advances in Engineering Software Articles, 16 February 2019. https://www.journals.elsevier.com/advances-in-engineering-software/most-cited-articles
21. Martinez, W.L., Martinez, A.R., Solka, J.: Exploratory Data Analysis with MATLAB. Chapman and Hall/CRC, Boca Raton (2017)
22. Kennedy, J., Eberhart, R.: Particle swarm optimization. In: Proceedings of IEEE International Conference on Neutral Networks, pp. 1942–1948 (1995)
23. Ahmed, H., Glasgow, J.: Swarm intelligence: concepts, models and applications, vol. K7L3N6. Technical report 2012-585, School of Computing, Queen's University, Kingston, Ontario, Canada (2012)
24. Yang, X.-S.: Nature-Inspired Optimization Algorithms. Elsevier, Amsterdam (2014)
25. Shelokar, P.S., Siarry, P., Jayaraman, V.K., Kulkarni, B.D.: Particle swarm and ant colony algorithms hybridized for improved continuous optimization. Appl. Math. Comput. **188**, 129–142 (2007)
26. Sumathi, S., Paneerselvam, S.: Computational Intelligence Paradigms: Theory & Applications Using MATLAB. Taylor and Francis Group, Milton Park (2010)
27. Reinelt, G.: TSLIB—a traveling salesman library. ORSA J. Comput. **3**, 376–384 (1991)
28. Brimberg, J., Hansen, P., Mladenovi, N., Taillard, E.D.: Improvements and comparison of heuristics for solving the uncapacitated multisource weber problem. Oper. Res. **48**(3), 444–460 (2000)

29. Drezner, Z., Brimberg, J., Mladenović, N., Salhi, S.: New heuristic algorithms for solving the planar p-median problem. Comput. Oper. Res. **62**, 296–304 (2014)
30. Brito, J., Martínez, F.J., Moreno, J.A.: Particle swarm optimization for the continuous p-median problem. In: 6th WSEAS International Conference on Computational Intelligence, Man-Machine Systems and Cybernetics, CIMMACS, pp. 14–16 (2007)

Optimization of UPS Output Waveform Based on Single-Phase Bridge Inverter

Guoli Xuan[1], Wenrui Li[2]([⊠]), Dawei Li[1], Jing Xu[1], Xiangluan Dong[2], and Bo Sun[1]

[1] State Grid Chaoyang Electric Power Supply Company, State Grid Liaoning Electric Power Supply Co., Ltd., Chaoyang, China
[2] Graduate Department, Shenyang Institute of Engineering, Shenyang, China
18842333082@163.com

Abstract. Uninterruptible Power Supply (UPS) can provide stable, continuous and uninterruptible power supply, and with the increasing application of UPS, there are higher requirements for the quality of UPS power supply. The inverter is an important part of the UPS system, and its performance directly affects the performance of the whole UPS system. In this paper, the common unidirectional full-bridge inverter is improved. The step wave synthesizer is used to transform the output square wave into step wave, and the number of step wave steps is added to make the output voltage waveform closer to a Sine wave, and finally improve the UPS output voltage waveform to guarantee the power supply quality.

Keywords: UPS · Step wave synthesizer · Transfer term overlay

1 Introduction

In order to ensure the continuity and reliability of power supply in important departments, to eliminate the influence of power network interference on power equipment, and to avoid the load on power network interference, the equipment of Uninterruptible Power Supply (UPS) appears. It is a kind of constant voltage and constant frequency power equipment with energy storage device and inverter as the main part. Its main function is to provide reliable and uninterrupted power supply to computer network system or other power and electronic equipment through UPS system. UPS power supply can solve the damage caused by power supply quality difference to front-end equipment, such as power supply power failure [1–3], voltage sag, power surge, amplitude reduction oscillation, power supply interference, power surge, power surge, power supply fluctuation, switching transient and harmonic distortion.

The inverter unit in UPS transforms the dc or dc power obtained from the municipal power rectification and filtering into the alternating current that is very stable in frequency, the output voltage is less affected by the load, and the waveform distortion factor meets the load requirement [4–6]. As a core part of UPS, its performance is directly related to UPS. Now, the requirements for UPS not only require stable output voltage and accurate output frequency, but also require fast dynamic response of its power supply. In addition, it must be environmentally friendly and pollution-free.

© Springer Nature Switzerland AG 2020
A. E. Hassanien et al. (Eds.): AISI 2019, AISC 1058, pp. 147–155, 2020.
https://doi.org/10.1007/978-3-030-31129-2_14

Therefore, higher requirements should be put on the design of UPS inverter so that it can meet the required output voltage waveform.

At present, the commonly used inverter circuit types include single-phase half-bridge inverter, single-phase full-bridge inverter and three-phase inverter, while single-phase inverter has been widely used due to its simple circuit structure, easy operation and implementation, more stable output waveform and other characteristics [7–10].

In this paper, the common single-phase inverter circuit in UPS is improved to make the output voltage waveform of the improved single-phase inverter circuit more stable and closer to the sine wave, so as to provide stable, continuous and uninterrupted power.

2   Preliminary Knowledge

The DC-AC inverter is an alternating current device that converts the direct current energy into alternating current energy, which can be used for AC load or connected to the AC grid for power generation. The channel through which dc to ac power conversion is completed in the inverter is called the inverter main circuit. It is mainly composed of power switch devices, transformers and electrolytic capacitors. Through controlling the regular on-off of power switch devices, the current is circulated in the predicted way and the dc to ac conversion is realized. With the development of power electronics technology, the inverter usually adopts the device with full control line. The inverter is divided into many kinds according to the working mode of main circuit. In this paper, common inverter circuits are introduced.

2.1   Unidirectional Half-Bridge Inverter Circuit

The schematic diagram of single-phase half-bridge inverter circuit is as follows. It is composed of dc power supply, full-control switch VT1, VT2, and anti-parallel diode VD1 and VD2. In the dc side connection of the circuit diagram, there are two sufficiently large partial voltage capacitors, C01 and C02, and C01 = C02, to ensure that the capacitor voltage is in Ud/2 when the switching device is on or off [11–13]. The control signals of switch devices ug1 and ug2 are complementary to control the on-off of the circuit. Its circuit structure is shown in the Fig. 1.

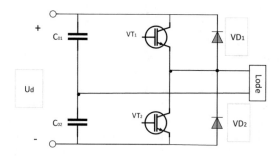

Fig. 1. Unidirectional half-bridge inverter circuit.

During the period from 0 to T/2, $ug_1 > 0$, ug2 < 0, VT_1 on and VT_2 off. During this period: C_{01} discharge, its path is $C_{01}+ \rightarrow VT_1 \rightarrow$ load $\rightarrow C_{01}-$. The path of C_{02} charging is $U \rightarrow VT_1 \rightarrow$ load $\rightarrow C_{02} \rightarrow E$. The Uo = $+U_d/2$.

During T/2 time, $ug_1 < 0$, $ug_2 > 0$, VT_1 turned off and VT_2 switched on. During this time period: C_{02} discharge, its path is $C_{02}+ \rightarrow VT_2 \rightarrow$ load $\rightarrow C_{02}-$. The path of C_{01} charging is $U- \rightarrow VT_2 \rightarrow$ load $\rightarrow C_{01} \rightarrow U$. The Uo = $+U_d/2$.

To sum up, if appropriate control signals are added to the circuit, the circuit will obtain repeated waveforms, so that the circuit can achieve DC conversion to AC purposes.

The output voltage waveform is shown in Fig. 2.

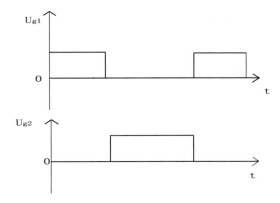

Fig. 2. The output voltage waveform.

2.2 Unidirectional Full Bridge Inverter Circuit

The single-term full-bridge inverter is composed of dc power supply U [14–16], four power switching devices and four diodes. Bridge arms 1 and 4 as a pair, 2 and 3 as another pair. Pairs of two bridge arms conductor close at the same time. Switch devises VT1 and VT4 are complementary with VT2 and VT3. The schematic diagram of the Unidirectional Full Bridge Inverter Circuit is shown in Fig. 3.

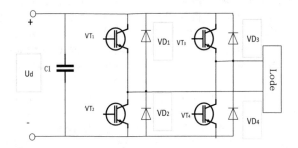

Fig. 3. Unidirectional full bridge inverter circuit.

During the period from 0 to t0/2, VT_1 and VT_4 were switched on, and VT_2 and VT_3 were switched off. At this time, u0 = +Ud

During the period from t0/2 to T_0, VT_2 and VT_3 were switched on, VT_1 and VT_4 were switched off, and U0 = −Ud

Therefore, to realize the change of the DC to AC output voltage, and voltage waveform is 180° wide square wave. Output voltage U_0 developed into Fourier series is defined as follow.

$$U_0 = \sum_{n=1,3,5...}^{\infty} \frac{4U_d}{n\pi} \sin nwt \tag{1}$$

The output voltage waveform is shown in Fig. 4.

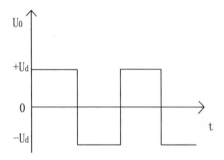

Fig. 4. The output voltage waveform.

2.3 Unidirectional Full Bridge Inverter Circuit

Is composed of three single-phase inverter circuit can three-phase inverter circuit, each single phase inverter circuits can be any form, as long as the size of the three single-phase inverter circuit output voltage is equal, the same frequency and phase difference of 120° to each other. In this paper, the most commonly used three-phase bridge inverter circuit is analyzed. Figure 5 shows the schematic diagram of the Inverter circuit is three-phase.

If voltage type three-phase bridge type inverter circuit works by 180° conduction mode (120° conduction mode) [17–19], where each bridge arm of the conduction Angle is 180°, the same phase fluctuation two alternating bridge arm conduction, began conducting each phase Angle difference in turn 120°, the control signal is shown in figure. So at any time there will be three arms leading at the same time, and the order of conducting is 1, 2, 3 → 2, 3, 4 → 3, 4, 5 → 4, 5, 6 → 5, 6, 1 → 6, 1, 2. It could be the upper bridge arm and the lower bridge arm, or it could be the upper bridge arm and the lower bridge arm.

According to the law can get U'AO, U'BO, U'CO waveform, they are for Ud/2 square wave amplitude, but the difference between phase in turn 120°.

The output voltage waveform is shown in Fig. 6.

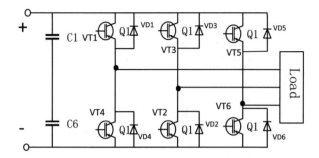

Fig. 5. Three-phase inverter circuit.

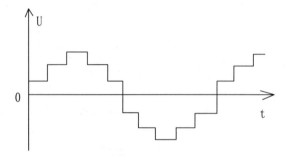

Fig. 6. The output voltage waveform.

Because of the rapid development of power electronic inverter circuit structure has changed a lot, the circuit for the direction of multilevel, multilevel inverter circuit with a switching device voltage stress is small, the output voltage harmonic content is small, switching devices in the process of switching voltage current rate of small advantages, so especially suitable for high power occasions, especially in the aspect of reducing power grid harmonic content has broad prospects. Common have series type and parallel type multiple inverter circuit.

3   Improved Circuit for Single-Phase Full Bridge Inverter

Because the three-phase bridge circuit is similar to the single - phase bridge circuit, this paper discusses the common single-phase you - transform circuit. From the previous analysis, it can be found that the voltage THD output of single-phase full-bridge inverter is too high [20–22], which does not meet the requirements of equipment with high electricity demand. Therefore, to obtain the sinusoidal output voltage waveform, it is necessary to add heavy LC filtering at its output end or to transform the output square wave into a higher-order waveform. Let's talk about two scenarios.

3.1   Add LC Filter to the Output of Inverter

The harmonic content of the output voltage of single-phase bridge circuit is much higher than the ideal value, so low pass filter is added at the output end. Its structure is shown in Fig. 7.

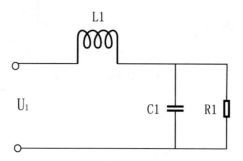

Fig. 7. Circuit structure.

Where the filter circuit parameters are selected based on Eqs. (2) and (3).

$$\sqrt{L_1 C_1} = \frac{1}{2pf_0} \tag{2}$$

$$\sqrt{\frac{L_1}{C_1}} = (0.5 - 0.8)R_1 \sqrt{L_1 C_1} = \frac{1}{2pf_0} \tag{3}$$

Where, R_1 is the load resistance and f0 is the cutoff frequency of the filter circuit.

Because of the heavy size and large loss of the filter circuit, it is not easy to carry and the energy saving is poor.

3.2   The Inverter Circuit of Phase-Shifting Superposition Step Wave

Due to the defect of the LC filter, the method of transforming square wave into a stepped wave is now discussed. The step wave synthesis inverter can realize the transformation from square wave to step wave.

The output waveform of a step wave synthesizer is a step wave. There are many methods of step wave synthesis. For high-power inverters, the common method is to combine the square (or quasi-rectangular) waves of N successively shifted phase shift/N, which is called a phase shift superposition method. As shown in the Fig. 8, the power circuit commonly used in the step wave synthesis inverter. The circuit consists of an oscillator, a phase separation circuit, and a trigger circuit. The power circuit is formed by the bridge circuit in series.

In single phase full bridge inverter, for example, a cycle has 2N number of ladder, need to have an N a single-phase inverter, driver signal is given by split-phase circuit will square wave phase shift in turn PI/N signal, after the single-phase inverter to form

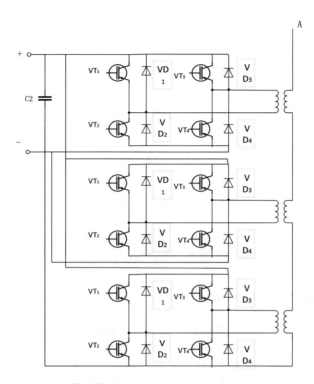

Fig. 8. Improved step wave circuit.

a certain phase Angle difference of rectangular wave, staircase, which is formed by various wave superposition and ladder wave shape is close to Sine wave.

4 Verify the Improved Circuit

Because the output waveform of single-phase inverter circuit is square wave, it has certain error, and the output waveform of equipment with higher requirements cannot meet the strict requirements, so the step-wave synthesis inverter can realize the transformation from square wave to step-wave, and then the sinusoidal wave can be achieved by changing the number of steps. In order to meet the set standard, the improved circuit needs to be verified.

This work takes 12 steps as an example to test the circuit of the improved step-wave synthesis inverter, and the resultant output waveform is shown in the Fig. 9.

According to the figure above, the output waveform of the improved circuit is close to the sine wave, and its waveform changes according to the number of steps. The harmonic content of step wave synthesis inverter is low, and the higher the number of steps, the lower the harmonic content, so the output waveform quality is good. The inverter is composed of several inverter Bridges, each inverter can divide the power equally, which reduces the power requirement of a single inverter bridge and makes it

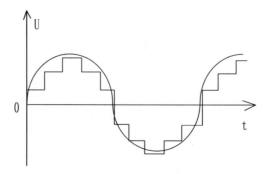

Fig. 9. The output voltage waveform.

easy to realize large power capacity. At the same time, the inverter power tube switching frequency is low, the converter efficiency is high, the reliability is high.

5   Conclusions and Future Work

With the wide application of UPS equipment, on the one hand, people have more and more requirements on the output voltage waveform of UPS inverter. On the other hand, the problem of output waveform distortion of UPS inverter becomes more and more serious due to the characteristics of power electronic circuits and the appearance of a large number of nonlinear loads. In this paper, the requirements for the output voltage waveform of the UPS inverter and the diversity of load are raised, and a reasonably feasible method is proposed to realize the waveform correction of the output voltage of the UPS inverter. In this paper, the original single-phase bridge inverter circuit was improved, because the original bridge circuit output voltage waveform for square wave, the improvement on the circuit design of the ladder wave synthesizer, make the output waveform into a staircase, through the control of the trigger signal phase to change the trigger pulse change the order of the staircase, the resulting waveform ladder more is close to sine wave. The resulting output voltage has less harmonic content.

As more devices are used in the improved circuit design, the consumption and volume of the circuit are increased. Therefore, the control technology of the circuit can be changed in the future design to reduce the volume of the circuit and optimize the performance.

References

1. Anyang, Z.: Optimization design and synchronous control of inverter. Technical report, Zhejiang University (2013). (in Chinese)
2. Escobar, G., Valdez, A., Leyva-Ramos, J., Mattavelli, P.: Repetitive-based controller for a UPS inverter to compensate unbalance and harmonic distortion. IEEE Trans. Ind. Electron. **54**, 504–510 (2007)

3. Zhang, K., Kang, Y., Xiong, J., Chen, J.: Direct repetitive control of SPWM inverter for UPS purpose. IEEE Trans. Power Electron. **18**, 784–792 (2003)
4. Kolar, J., Drofenik, U., Biela, J., Heldwein, M., Ertl, H., Friedli, T., Round, S.: PWM converter power density barriers. In: Power Conversion Conference-Nagoya, 2007, PCC 2007, pp. P-9–P-29 (2007)
5. Kown, B., Choi, J., Kim, T.: Improved single-phase line-interactive UPS. IEEE Trans. Ind. Electron. **4**, 504–811 (2001)
6. Pengfei, S.: Study on single-phase UPS digital waveform control technology. Huazhong University of Science and Technology, Wuhan (2006)
7. Zhaoan, W., Jun, H., et al.: Power Electronics Technology, Version 4, pp. 132–138. Mechanical Industry Press, Beijing (2011)
8. Xixiong, L., Jiaoer, C.: Pulse width modulation technology. Huazhong University of Science and Technology Press, Wuhan (1996)
9. Liangliang, C., Lan, X., Wenbin, H., Yangon, Y.: Causes of the output voltage DC component of the inverter and the suppression method of power electronics technology, vol. 37, pp. 27–29 (2003)
10. Weixun, L.: Modern Electric Power Electronic Circuit, pp. 157–163. Zhejiang University Press, Hangzhou (2002)
11. Jian, C.: Power electronics – power electronic transformation and control technology, pp. 137–142. Advanced Education Press, Beijing (2002)
12. Jingtao, C., Hua, L., Jun, Z., et al.: Will be a load sharing control technique for paralleled inverters. In: IEEE PESC 2003, pp. 1432–1437 (2003)
13. Guleria, M., Rathi, A., Sajwan, D.S., Negi, J.S., Chauhan, N.: DTMF based irrigation water pump control system. Int. J. Smart Device Appl. **5**(1), 19–24 (2017)
14. Han, Y.: Prediction of patient-specific sensitive drug targets using somatic mutation-propagated network modules. Int. J. Digital Contents Appl. Smart Devices **3**(2), 7–14 (2016)
15. Moparthy, N.R.: Advanced m-privacy for collaborative data publishing. Int. J. Secur. Technol. Smart Device **3**(1), 1–8 (2016)
16. Sneha, A., Kavya, P., Raj, S.: A survey on effective analysis and management of a data breach. Int. J. Secur. Technol. Smart Device **5**(1), 1–8 (2018)
17. Ranjan, P., Dhaka, N., Pant, I., Pranav, A.: Design and analysis of two stage op-amp for bio-medical application. Int. J. Wearable Device **3**(1), 9–16 (2016)
18. Murthy, K., Sivalakshmi, B.: Attack as defect against anonymized social networks. Int. J. Commun. Technol. Soc. Netw. Serv. **4**(2), 7–12 (2016)
19. Wang, L., Park, J.: Workload of using digital products while driving. Int. J. Hum. Smart Device Interact. **4**(1), 21–26 (2017)
20. Jang, H., Kang, M., Chung, Y.: An opportunistic routing protocol based on contact ratio. Int. J. Power Devices Compon. Smart Device **3**(1), 21–26 (2016)
21. Bagwari, A., Kanti, J., Tomar, S.: New cooperative spectrum detection technique in cognitive radio networks. Int. J. Wirel. Mobile Commun. Ind. Syst. **3**(1), 43–58 (2016)
22. Park, J., Kim, H.: Analysis of research trends in papers on science education for young children: with a focus on dissertations published between 2008 and 2017. Int. J. Web Sci. Eng. Smart Devices **4**(2), 1–6 (2017)

E-Health Parkinson Disease Diagnosis in Smart Home Based on Hybrid Intelligence Optimization Model

Ahmed M. Anter[1,2] and Zhiguo Zhang[1(✉)]

[1] School of Biomedical Engineering, Health Science Center, Shenzhen University,
Shenzhen 518060, China
Ahmed_Anter@fcis.bsu.edu.eg, zgzhang@szu.edu.cn
[2] Faculty of Computers and Artificial Intelligence, Beni-Suef University,
Benisuef 62511, Egypt

Abstract. The use of internet of things (IoT) in smart home with medical devices within a connected health environment promotes the quick flow of information, the patient's vital parameters are transmitted by medical devices onto secure cloud based platforms where they are stored, aggregated and analyzed. IoT helps to store data for millions of patients and perform analysis and diagnosis in real-time, promoting an evidence-based medicine system. Different intelligence optimization models can be integrated with IoT to improve the patient healthcare. In this paper, an intelligent optimization model is proposed for monitoring patients with Parkinson's disease (PD) based on UPDRS assessment (Unified Parkinson's Disease Rating Scale) from voice records in smart home. Ant lion optimization algorithm (ALO) and adaptive extreme learning machine (ELM) based on differential evaluation (DE) algorithm is proposed; namely (ALO-DEELM), for PD diagnosis. Using this model, home residents will get feedback and keep track on their PD situation. ALO-DEELM model is compared with different machine learning (ML) prediction algorithms and showed the superiority based on different measures. Moreover, the experimental results showed that the proposed model is effective and can significantly reduce the prediction computational time of UPDRS scores. The proposed ALO-DEELM has the potential to be implemented as an intelligent system for PD prediction in healthcare.

Keywords: Regression · Internet of things (IoT) ·
Parkinson's disease (PD) · Differential evaluation (DE) ·
ALO-DEELM · UPDRS

1 Introduction

Research in medical fields is very relevant to clinical advances. In this context, Internet of Things (IoT) or to be more specific, Internet of Medical Things (IoMT), is revolutionizing the healthcare industry. The use of IoMT in smart

© Springer Nature Switzerland AG 2020
A. E. Hassanien et al. (Eds.): AISI 2019, AISC 1058, pp. 156–165, 2020.
https://doi.org/10.1007/978-3-030-31129-2_15

home with medical devices within a connected health environment promotes the quick flow of information, the patient's vital parameters are transmitted by medical devices onto secure cloud based platforms where they are stored, aggregated and analyzed. Within this scenario, machine learning, pattern recognition, and artificial intelligence play a very crucial role. These methodologies are growing rapidly, and have successful applications in the e-Health domain.

Parkinson's disease (PD) is a progressive degenerative nervous system disease which affects the control of movement. PD is the second most common neurodegenerative disorder after Alzheimer's [1]. Schrag et al. estimated 20% of humans with Parkinson's are never diagnosed [2].

Different assessment measures have been developed for PD. One of the gold standard evaluation tool for the characterization of motor impairments in PD is the UPDRS assessment which was introduced in 1987 [3]. The UPDRS is a useful way to keep track of patient function and evaluate disability. Motor-UPDRS refers to the full range of the metric and Total-UPDRS used in the diagnosis of PD [4].

PD is a very complex disorder in which individual motor features vary in their presence and severity over-time. Early PD diagnosis can slow patients' progression of the disease and improve the quality of life [5]. Monitoring PD progression over-time requires repeated clinic visits by the patient. So, an effective screening process, particularly one that doesn't require a clinic visit, would be beneficial. Since PD patients exhibit characteristic vocal features, voice recordings are a useful and non-invasive tool for monitoring and diagnosis. Therefore, an expert and intelligent systems is very important to assist physicians in diagnosing and predicting PD diseases.

This paper will focus on pioneering and innovating in monitoring and diagnosing the patients with PD through smart home based on IoMT. The intelligent optimization model for PD prediction using UPDRS assessment based on ant lion optimization algorithm [6], extreme learning machine [7], and differential evaluation algorithm [8] is proposed. ALO is used to obtain a robust subset of features from the whole PD dataset. DEELM is used to predict the total- and motor-UPDRS scores of PD. The ELM algorithm has been proved effective in dealing with large data set, provide smaller training errors, better performance in fewer iterations and helps to reduce the computational prediction time. DE is used to guide ELM for optimal weights and biases for the regression process. This ALO-DEELM model utilizes the strong ability of the global optimization, and avoids the sensitivity of local optima. In order to minimize error, maximize prediction accuracy, and reduce erroneous clinical diagnostic. The proposed model is evaluated using a real-world PD diagnosis using several measurements and the results compared with the other Ml techniques. The results indicate that the ALO-DEELM is more robust and can achieve high accuracy with fast convergence to optimal solution and lower computational time.

The rest of this paper is organized as follows: Sect. 2 presents the ALO-DEELM model with its phases description in details. Dataset description, experimental results and discussions are presented in Sect. 3. Finally, Sect. 4 presents conclusions and future work.

2 The Proposed ALO-DEELM Model

The aim of this paper is to develop a new hybrid intelligence regression model to predict the diagnosis of PD based on motor and total- UPDRS scores using vocal recordings in an intelligent home based IoT. The main goals of developing this model are to improve the predictability of PD diagnosis, reduce the computational time, comforting the patients and reduce the suffering. The non-linear ant lion optimization algorithm is used for the feature subset selection (ALO). DEELM model is used to predict the Total and Motor- UPDRS scores. The integration between these techniques is able to improve the prediction accuracy of classical methods for PD diagnosis. The proposed ALO-DEELM model for PD diagnosis has three main phases as shown in Fig. 1.

1. **Pre-processing phase:** Firstly, k-Nearest Neighbors (k-NN) algorithm is used to predict the missing values which degraded the prediction accuracy [15]. k-NN substitutes the missing data with suitable values that are as possible as close to the true values. Then, principal component analysis (PCA) method is used to decrease the correlation of this non-linear data.
2. **Feature selection phase:** The proposed binary ALO has high exploitation and high convergence rates, because it has high performance of convergence, decreases the chance of trapping in local minima and guarantees the goodness of the selected features. The stochastic initialization of population of ants and ant lions are presented in the search space and the fitnesses of all ants and ant lions are initialized using cost function based on standard derivation. The fitness values are assigned to each ant and ant lion in the search space, then evaluated in each iteration. The best ant lion (elite) with high fitness is selected as the best solution which can build larger holes and catch ant by ant lion. The best solution represents the best subset of features which indicate the best position and fitness (elite). Then the best solution is normalized using min-max method between [0, 1] for the purpose of increasing the efficiency by smoothing the data points.
3. **Regression and prediction phase:** ELM regression algorithm is used for the prediction of total- and motor-UPDRS scores. As proved in the experiments, ELM needs less training time, better generalization performance, and gives high prediction performance. Moreover, ELM is relatively faster and more computationally efficient than traditional algorithms. The main draw back in ELM is the randomized behavior of neural networks (RNNs). So, random nature gives an unstable forecast, which results in a large number of hidden neurons, its performance must be guaranteed. To address this problem, an evolutionary algorithm based on differential evaluation (DE) is used for improving the ELM algorithm. DE is used to optimize hyper-parameters (weights and biases) in ELM hidden layer.

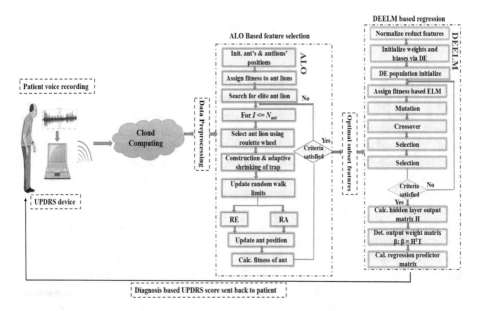

Fig. 1. Architecture of the proposed ALO-DEELM model based IoT.

3   Experimental Results and Discussion

3.1   Dataset Description

Several experiments with a real-world PD data set are conducted to evaluate the proposed model. The dataset used is the PD telemonitoring created by Little et al. of the University of Oxford [9]. The PD data set includes a range of biomedical voice measurements from 42 people with Parkinson's disease. The records were automatically recorded from the home of the patient using IoT service. The dataset contains 5,875 records (approximately 200 records per patient) by 28 men and 14 women. The Total-UPDRS range is 0–176 (0 indicates healthy and 176 total disability) and the Motor-UPDRS range is 0–108 (0 indicates healthy and 108 severe motor impairment).

3.2   Model Performance Measures

To evaluate the proposed regression model, different measures are used such as Mean Absolute Error (MAE), Mean Square Error (MSE), Root Mean Square Error ($RMSE$), Correlation coefficient (R), Willmott's index (WI), Nash-Sutcliffe efficiency (N_{SE}) and Legates McCabe's index (LM) for prediction stage and Mean fitness ($Mean$), Standard deviation (STD), Best fitness (BF), Worst fitness (WF), and Average attribute selection (AAS) for feature selection stage [10–14].

3.3   Numerical Results and Discussion

3.3.1   Feature Selection Phase

The ALO algorithm has high exploitation and high convergence rates, because it has high performance of convergence, decreases the chance of trapping in local minima and guarantees the goodness of the selected features. The main reason for the high speed of exploitation and convergence in the search space is the adaptive shrinking boundaries of the traps of ant lions and elitism. On the other hand, the random selection of ant lions using roulette wheel selection mechanisms and random walks of ants around them guarantees the exploration of the search space. Figure 2 shows the best solution that achieves in less than 12 iterations with stability of fitness value at (0.053979) and the best binary position obtained is (1011000101111101). ALO has consider to be high exploitation and high-speed convergence rates due to the adaptive boundary shrinking and elitism mechanism, as proved with evidence in Fig. 2. The optimal results are achieved in less computational time with very few parameters need to adjust approximately ($\approx$0.156 s) (CPU average time) and the best and worst fitnesses are ($\approx$0.05398, and $\approx$0.08041), respectively. From the results, ALO has excellent results in terms of improved exploration, exploitation, local optima avoidance and high convergence speed. Thus, ALO algorithm-based feature selection has a strong robustness and has the faster convergence speed to find the optimal features.

The computed spearman correlation coefficients between the selected subset of features (Jitter (Var1), RAP (Var2), PPQ5 (Var3), APQ3 (Var4), APQ11 (Var5), DDA (Var6), NHR (Var7), HNR (Var8), RPDE (Var9), and PPE (Var10)) is shown in Fig. 3, which indicates there were positive and negative correlations between various descriptors. From these plots we can see that, the selected descriptors are more efficient and appropriate to the diagnosis process.

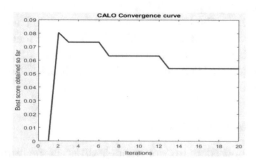

Fig. 2. BALO algorithm convergence curve for PD dataset.

3.3.2   PD Prediction Phase

The prediction model is constructed using adaptive ELM algorithm based on DE algorithm to adapt the hyper-parameters in the ELM hidden layer. In the training step, the initial selection of the weights and biases are done using the DE algorithm which is a global search algorithm, and thereafter they were updated automatically using ELM algorithm. In the experimental analysis, PD data set is divided into 70% and 30% for the training and testing sets for the DEELM algorithm.

In this phase, we have two prediction models are constructed for motor and total- UPDRS assessment based on ALO-DEELM. Figure 4 shows a scatter-plots of the goodness of fit of PD data for total and motor- UPDRS. The correlation coefficient R^2 shows the extent of agreement between motor and total UPDRS Parkinson's disease predicted and observed. The best accuracy confirmed by attaining the larger R^2-value.

Table 1 presents the accuracy results of the proposed ALO-DEELM model for the two UPDRS prediction using different measurements ($MSE, R, WI,$ $N_{SE}, RMSE, MAE,$ $LM,$ and $Time/s$ (CPU average)). The proposed model achieves high prediction accuracy obtained for motor and total- UPDRS scores using different measures with very small training error. Beside the improvement in the prediction accuracy, ALO-DEELM proves to have a good performance in reducing the whole model computational time. From the results ALO-DEELM model is clearly better, more accurate, and has the potential to be used in voice-based PD monitoring for developing a powerful prediction model.

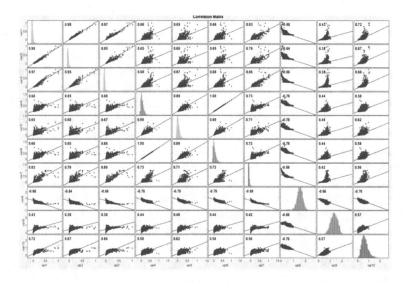

Fig. 3. Spearman correlation coefficient matrix.

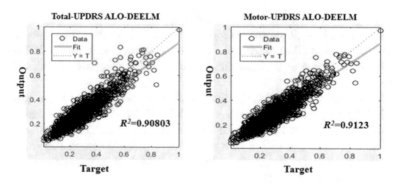

Fig. 4. Scatter-plot and *RMSE* results of the predicted motor-UPDRS PD of the proposed ALO-DEELM model.

Table 1. ALO-DEELM UPDRS matrix predictor results using different measurements.

Measures	MSE	R^2	WI	N_{SE}	$RMSE$	MAE	LM	$Time/s$
Motor	0.003	0.908	0.870	0.825	0.053	0.040	0.587	0.016
Total	0.003	0.912	0.876	0.832	0.052	0.0396	0.592	0.062

3.3.3   Comparison with the Existing Machine Learning Prediction Algorithms and the Previous Work

The comparison results between the proposed ALO-DEELM model and different prediction ML algorithms are provided in this section. The Back Propagation Neural Network (BPNN), Multiple Linear Regression (MLR), Support Vector Regression (SVR), and Adaptive Neuro-Fuzzy Inference Systems (ANFIS) are used for the comparison purpose with the same data partitioning into 70% and 30% for training and testing, respectively. The comparisons are made on the prediction accuracy of the Total and Motor- UPDRS as shown in Fig. 5. It can be seen that the more accurate results are achieved using the proposed algorithm with high prediction accuracy for the two standard Parkinson disease assessment, followed by the results achieved by SVR. The proposed ALO-DEELM model is clearly outperforms the ML algorithms in accuracy, time consuming and fast convergence to the optimal solution in small number of iterations.

In comparison with the previous work, our proposed model achieves higher accuracy in term of R^2. Nilashi et al. in 2016 [16] proposed algorithm based on PCA-SVR with accuracy achieved for(motor: 0.825 and total: 0.831), then proposed HSLSSVR approach in 2018 [4] to increase the accuracy (Motor: 0.885 and Total: 0.868). Eskidere et al. [17] proposed LS-SVM approach for PD prediction with accuracy achieved for (motor: 0.63 and total: 0.65).

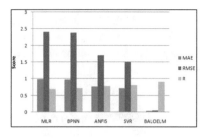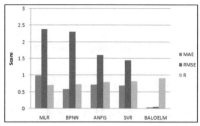

Fig. 5. Comparison between different prediction algorithms and the proposed ALO-DEELM for motor-UPDR (a) and total-UPDR (b) prediction score.

3.3.4   ALO-DEELM Limitations

The main limitation of the methods proposed in this paper is the stagnation or the premature convergence to a local optima. This stagnation occurred because the stochastic and random behaviour of the ALO algorithm in the search space. This limitation, can be controlled carefully by tune the ALO parameter. The main parameter that controls the exploration/exploitation rate in ALO is I factor which control the random walk of the ant lions and ants. So, this parameter should be handled carefully and may be differ from dataset to another. In addition, the proposed model works on the standard derivation kernel function. So, the running time is very fast. But, the rune time will increase if this kernel converted to wrapper-based feature selection approach based regression using KNN, SVR, or MLR classifiers. Therefore, switching to a different classifier should be carefully handled, particularly if the algorithm is adopted in real-time PD diagnosis.

4   Conclusion and Future Work

The topic of e-Health based IoT is still an open field in the machine learning and artificial intelligence community. So, this paper proposed an intelligent optimization regression model for monitoring patients with Parkinson's disease based on IoMT using UPDRS assessment. A new intelligent optimization model was designed for the prediction of Parkinson's disease progression using ALO algorithm and DEELM for the total and motor- UPDRS prediction. The main finding of this study is that the prediction algorithm is integrated with two bio-inspired techniques ALO and DE to improve the prediction accuracy of the PD and reduce the computational time. Other advantages of the proposed model include that, it is effective and efficient in memory requirements, it can significantly reduce the prediction computation time, and it can be effectively implemented for large datasets. Moreover, the proposed model outperforms the ML algorithms in accuracy, time consuming and the fast convergence to the optimal solution in small number of iterations. The superiority of the ALO-DEELM model makes it a potential algorithm to be implemented as an intelligent system for PD prediction in healthcare. In future work, we plan to evaluate the proposed model on the various data sets, in particular on the large data sets for PD diagnosis, to demonstrate the accuracy and effectiveness of the proposed model.

References

1. Tsanas, A., Little, M.A., McSharry, P.E., Ramig, L.O.: Accurate telemonitoring of Parkinson's disease progression by non-invasive speech tests. IEEE Trans. Biomed. Eng. **57**, 884–893 (2010)
2. Schrag, A., Ben-Shlomo, Y., Quinn, N.: How valid is the clinical diagnosis of Parkinson's disease in the community? J. Neurol. Neurosurg. Psychiatry **73**(5), 529–534 (2002)
3. Sakar, B.E., Serbes, G., Sakar, C.O.: Analyzing the effectiveness of vocal features in early telediagnosis of Parkinson's disease. PLoS ONE **12**(8), e0182428 (2017)
4. Nilashi, M., Ibrahim, O., Ahmadi, H., Shahmoradi, L., Farahmand, M.: A hybrid intelligent system for the prediction of Parkinson's Disease progression using machine learning techniques. Biocybern. Biomed. Eng. **38**(1), 1–15 (2018)
5. Farnikova, K., Krobot, A., Kanovsky, P.: Musculoskeletal problems as an initial manifestation of Parkinson's disease: a retrospective study. J. Neurol. Sci. **319**(1–2), 102–104 (2012)
6. Roy, K., Mandal, K.K., Mandal, A.C.: Ant-Lion optimizer algorithm and recurrent neural network for energy management of micro grid connected system. Energy **167**, 402–416 (2019)
7. Sattar, A.M., Ertuğrul, Ö.F., Gharabaghi, B., McBean, E.A., Cao, J.: Extreme learning machine model for water network management. Neural Comput. Appl. **31**(1), 157–169 (2019)
8. Ali, M., Deo, R.C., Downs, N.J., Maraseni, T.: Multi-stage committee based extreme learning machine model incorporating the influence of climate parameters and seasonality on drought forecasting. Comput. Electron. Agric. **152**, 149–165 (2018)
9. Little, M.A., McSharry, P.E., Hunter, E.J., Spielman, J., Ramig, L.O.: Suitability of dysphonia measurements for telemonitoring of Parkinson's disease. IEEE Trans. Biomed. Eng. **56**(4), 1015–1022 (2009)
10. Anter, A.M., Ali, M.: Feature selection strategy based on hybrid crow search optimization algorithm integrated with chaos theory and fuzzy c-means algorithm for medical diagnosis problems. Soft Comput. 1–20 (2019). https://doi.org/10.1007/s00500-019-03988-3
11. Anter, A.M., Hassenian, A.E., Oliva, D.: An improved fast fuzzy c-means using crow search optimization algorithm for crop identification in agricultural. Expert Syst. Appl. **118**, 340–354 (2019)
12. Anter, A.M., Hassanien, A.E., ElSoud, M.A., Kim, T.H.: Feature selection approach based on social spider algorithm: case study on abdominal CT liver tumor. In: 2015 Seventh International Conference on Advanced Communication and Networking (ACN), pp. 89–94. IEEE, July 2015
13. Gupta, H.V., Kling, H.: On typical range, sensitivity, and normalization of mean squared error and nash-sutcliffe efficiency type metrics. Water Resour. Res. **47**(10), 1–3 (2011). https://doi.org/10.1029/2011WR010962
14. Anter, A.M., Hassenian, A.E.: Computational intelligence optimization approach based on particle swarm optimizer and neutrosophic set for abdominal CT liver tumor segmentation. J. Comput. Sci. **25**, 376–387 (2018)

15. Pan, R., Yang, T., Cao, J., Lu, K., Zhang, Z.: Missing data imputation by K nearest neighbours based on grey relational structure and mutual information. Appl. Intell. **43**(3), 614–632 (2015)
16. Nilashi, M., Ibrahim, O., Ahani, A.: Accuracy improvement for predicting Parkinson's disease progression. Sci. Rep. **6**, 34181 (2016)
17. Eskidere, Ö., Ertaş, F., Hanilçi, C.: A comparison of regression methods for remote tracking of Parkinson's disease progression. Expert Syst. Appl. **39**(5), 5523–5528 (2012)

Multi-time Source Selection Optimization Algorithm for Time Synchronization System

Meijie Liu[1(✉)], Guangfu Wang[2], Kai Wei[1], Hong Gang[1], and Hao Wang[2]

[1] State Grid Jinzhou Electric Power Supply Company,
State Grid Liaoning Electric Power Supply Co., Ltd., Jinzhou 121000, China
1441594202@qq.com
[2] Graduate Department, Shenyang Institute of Engineering,
Shenyang 110136, China

Abstract. The optimal time source can provide accurate, stable and continuous time signal output for the time synchronization system, and ensure the power system to operate under a reliable time reference. Aiming at the problem that single time source in substation cannot meet the reliability requirement of system operation on time synchronization, this paper proposes an optimization algorithm of time source selection under the condition of multiple time signal sources input. Firstly, the validity of the time source signal is tested to determine whether the time signal is valid. The priority test is carried out for the effective time source from the aspects of time signal persistence and stability. Finally, the optimal time source is determined by the weighted synthesis method.

Keywords: Time source · Time synchronization device · Selective optimization algorithm · Time synchronization system

1 Introduction

The stable operation of the power system based on a unified time benchmark, with the constant improvement of the power system automation level, the time synchronization accuracy and stability of the demand is higher and higher. At present, the time synchronization system of substations mostly relies on high precision timing signals provided by satellite systems such as BeiDou and GPS [1]. There is no connection between substations at all levels and between substations and dispatching stations, so the isolated time synchronization system is difficult to form a unified time synchronization network [2]. After receiving the satellite timing signal, the time synchronization device conducts the timing signal to the subordinate equipment through analytic coding, but the single timing signal is unstable. When the timing signal fails or the accuracy of timing signal fluctuates greatly due to various reasons, it cannot meet the requirements of safe and stable operation of the power system [3]. An optimal selection algorithm of multiple time sources is designed, and the time signal of each dispatching master station is taken as one of the time sources of time synchronization device [4]. Firstly, the main dispatching station and the time synchronization system of the substation can be connected to form a time synchronization network, so as to avoid the failure of the time synchronization system when the satellite timing signal fails. Secondly, the external time source can be guaranteed to be in the optimal time source state, and the accuracy and stability of timing of the time synchronization device at the end of the substation can be improved [5].

Nowadays, as the satellite timing system is still the mainstream time source, the research on multi-time source selection of time synchronization system mainly focuses on the selection between satellite time sources and the redundant configuration of time sources [6]. The selection of multiple time sources such as satellite time system and ground time source selection optimization algorithm is less research. Aiming at this problem and combining with the future development trend of time synchronization system, this paper studies the multi-time source selection optimization algorithm.

2 Time Source and Time Synchronization Device

2.1 Time Source

The time reference signal of the time synchronization system comes from the time source. The time synchronization device receives the time reference signal from the time source, enters the locking following state, and gradually adjusts the local clock to synchronize with the time source. After compensating the delay caused by various factors, output the reference time signal and frequency signal to realize the timing of the power system. Time source can be divided into wired time source and wireless time source. The wireless time source mainly includes short-wave time system, long wave time system, low-frequency time code time system and satellite time system. GPS satellite timing system and BeiDou satellite timing system are commonly used. Cable time source mainly by Internet or special line as a carrier to pass the time reference signal, common timing signals such as IRIG-B, inhibits PTP.

2.2 Time Synchronization Device Structure

Time synchronization device is an important component of time synchronization system, which is mainly composed of time source receiving unit, internal time source unit and time output unit [7]. Its constituent structure is shown in Fig. 1.

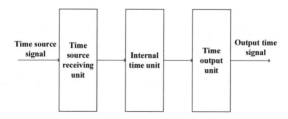

Fig. 1. Time synchronization device structure.

The main function of the time source receiving unit is to receive time and frequency signals from the time source and analyze the received time and frequency signals to obtain time and frequency information. And the parsed information is sent to the internal clock unit of the time synchronization device [8].

The internal clock unit consists of a frequency standard source and a counter. The main function of the internal clock unit is to use the time and frequency information from the time source receiving unit for the calibration of the local clock. According to the frequency information, the frequency standard source is synchronized, and the frequency error of the internal clock is gradually reduced so that the frequency of the frequency standard source is consistent with the frequency information. The purpose of the counter is to record the frequency of vibration of the frequency standard source to master the time information and generate the time information of the internal clock.

The main function of the time output unit is to convert the time information generated by the internal clock unit into time signals and send time signals to devices that need timing. According to the different time signal interfaces of different devices, the time output unit needs to output a variety of time signals to meet the needs of timing [9]. The main time signals include IRIG-B code signal, NTP signal, second pulse signal, serial port message signal, PTP signal, etc.

Since a single time source cannot provide continuous, stable and accurate time signals, multiple time sources are often needed to meet the requirements of time synchronization of power systems. In view of the coexistence of multiple time sources, the time synchronization device needs to complete the comparative analysis of multiple time sources and select the best time source. Therefore, under the conditions of multiple time sources, a time source selection unit is added to the time synchronization device, and multiple time source optimization selection algorithms is used in the time source signal selection unit to improve the timing stability and accuracy of the time synchronization device [9]. The composition and structure of the time synchronization device under the condition of multiple time sources is shown in Fig. 2.

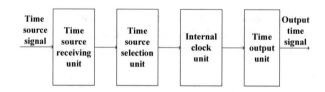

Fig. 2. Multi - time source time synchronization device structure

3   Multi-time Source Selection Optimization Algorithm

Time synchronization system is the source of unified time reference for power plants, substations and main dispatching stations at all levels. When the time synchronization system works normally, it receives and locks the time reference signal of the time source and provides time service to the outside world. In order to ensure the safe and stable operation of power system, at least two more time sources should be set [10]. Under the condition of multiple time sources, a time source selection optimization algorithm is required, which can select the optimal time source in different states and achieve the smooth switching of time sources. The basic principles of time source selection are as follows:

- The satellite timing system is preferred as a reference source;
- The time source with high stability is selected to avoid frequent switching of time source;
- The combination of manual and automatic source selection.

The specific process of time source selection and optimization is shown in Fig. 3.

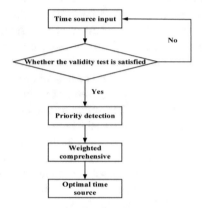

Fig. 3. Multi - time source time synchronization device structure.

3.1 A Validity Test

The purpose of receiving multiple time source signals is to select the best time source and improve the reliability and accuracy of timing. The internal time source of the time synchronization device can only choose an external time source as the time reference. Validity testing is divided into basic validity testing and selective validity testing [11–14].

When multiple time source signals enter the time source receiving unit, the basic validity of the time source signals is firstly detected to determine whether the time source signals are invalid. Basic validity detection is to detect the second pulse and the basic characteristics of time information of the time source signal. The basic characteristics of the second pulse include periodicity and stability, and the basic characteristics of time signal include time format and time continuous increment. When the basic validity test is satisfied, the time source is considered to be basically valid and can continue to participate in the selection of time source. When the basic validity test is not met, the time source is considered invalid and cannot continue to participate in the selection of time source.

Effectiveness to meet the basic testing time source, choose the validity test, choose effective detection is mainly external time between the source and the consistency with the internal clock source, test content includes: the second pulse seconds delay each other difference is smaller than a certain threshold τ, temporal information (year, month, day, hours, minutes, seconds). When these two validity tests are satisfied, the selected time sources are prioritized, and the time source with the highest priority will serve as the external time source benchmark.

3.2 Priority Detection

Priority detection is to compare and evaluate the time source and select the best time source as the time benchmark of an internal clock. Priority detection mainly includes three aspects: type quantity, stability degree, and system quantity.

Type quantity refers to the evaluation of time source according to the classification of the time source. Time source is divided into satellite time source and ground time source. Generally speaking, the timing accuracy of the satellite time source is higher than that of ground time source, so satellite time source is preferred in principle.

Stability is mainly to evaluate whether the time signal of the time source is stable and to choose a time source with high stability can avoid frequent switching of the time source. Stability evaluation is mainly composed of two indicators, TIE (interval error) and MTIE (maximum interval error). TIE refers to the relative delay change between the actual time signal and the ideal time signal over a period of time, while MTIE refers to the maximum phase change of a given sliding time window within a measurement period. TIE and MTIE are important indicators of stability evaluation [15].

System quantity refers to the system default reference source priority order, there are two cases. Under the condition of many different time sources, the default preferred sequence of time sources is GPS satellite, BeiDou satellite, and ground wired time source. The order of priority is defined by the distance of the physical location under the condition of multiple same time sources.

3.3 Weighted Synthesis Algorithm

According to the different importance degrees of each indicator in the priority detection algorithm, the weight of each indicator is comprehensively considered, and the indicators affecting the selected source are converted into mathematical formulas to calculate the optimal time source [16–18]. The weighted synthesis algorithm is shown in Fig. 4.

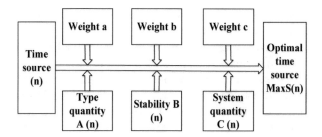

Fig. 4. Weighted synthesis algorithm.

According to the weighted synthesis algorithm, the optimal time source calculation are Eqs. (1) and (2).

$$S(n) = a * A(n) + b * B(n) + c * C(n) \tag{1}$$

$$Best(n) = \underset{1 \leq n \leq N}{Max}[S(n)] \tag{2}$$

Where, n is the reference time sequence, N is the total number of reference time source, $S(n)$ is the time reference source n comprehensive weighted evaluation value, A (n) is a reference time source n type quantity evaluation value, a is the type weight of reference time source n, $B(n)$ is the time reference source n to evaluate the stability of the numerical, c is the time reference source n weight, $C(n)$ is the system quantity evaluation value of reference time source n, c is the system weight of reference time source n. The purpose of selecting the best time source can be realized by controlling the value of weight reasonably manually.

3.4 Time Source Switching

When determining the optimal external time source reference of the internal clock, time source switching is carried out according to the state of the time synchronization device. There are four working states of the time synchronization device, namely

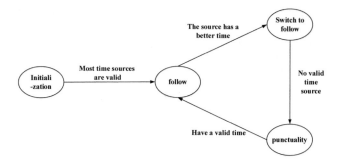

Fig. 5. The working state of time synchronization device

initialization, follow, replacement follow and punctuality. The working state of the time synchronization device is shown in Fig. 5.

The time synchronization device starts for the first time and enters the initialization state. Firstly, the external time source is searched and the validity of the searched time source is tested. At this time, the time output unit is locked and there is no time output. When more than half of the time source satisfy the validity test, shows that the time of the source information is consistent, second pulse seconds delay difference less than τ, each time can be used as an external reference. Select the time source with the highest priority as the reference, establish the internal clock, and the time synchronization device enters the following state [19].

In the following state, the internal clock gradually tamed the frequency standard source with the external time source as the reference. At this time, the time output unit of the device can be unlocked and the standard time signal can be output. When a better time source appears, the device switches smoothly to follow the target and enters the switching and the following state. The internal clock follows the time source in a step mode and outputs time signal to the outside. When no effective time source appears, the device enters the punctuality state and outputs the time signal with the frequency standard source of the internal clock as the reference. When a valid time source appears, the device returns to the following state.

4   Experimental Simulation

4.1   Simulation Design

The design experiment simulation is as follows. The simulated time synchronization device has four time sources, namely BeiDou satellite, GPS satellite, PTP signal and IRIG-B signal. Multiple time source signals are connected to the time synchronization device. The multi-time source system test structure is shown in Fig. 6.

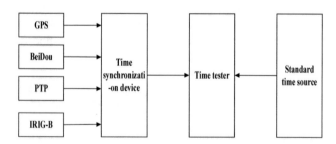

Fig. 6. Multiple time source test system

Time sources were arranged in order of priority of GPS, BeiDou satellite, PTP and IRIG-B to simulate the time source switching process, and turn off the input of GPS satellite antenna, BeiDou satellite antenna and PTP signal successively to observe the data collected by time tester during the time source switching process. The waveform of the TIE curve output by the time tester during the time source switching process is shown in Fig. 7.

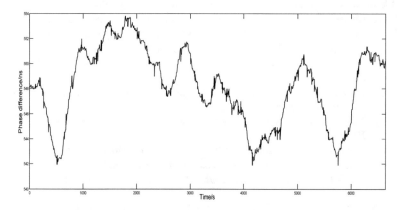

Fig. 7. Time source toggle TIE curve

4.2   Results Analysis

It can be seen from Fig. 7 that the maximum phase difference of each reference source is 12 ns during the switching process and the curve changes little, which meets the time requirement of the power system. Meanwhile, it indicates that the time source has stable and high-precision output effect during the switching process.

5   Conclusion

Ensure the accuracy and stability of time synchronization is the basic requirement of power system development, this paper much time source input conditions, on the analysis of the structure of time synchronization device, time source selection optimization algorithm is proposed. After validity detection, priority detection, and weighted synthesis algorithm, the optimal time source is selected as the time reference of the internal clock. Based on the analysis of simulation results, the multi-time source selection optimization algorithm in this study can ensure the output of time signals with high accuracy and stability. However, the effect of time source switching on the time synchronization accuracy of downstream time synchronization equipment remains to be further studied.

References

1. Shkarbalyuk, M.E., Pil'gave, S.V., Larchenko, A.V.: An emulator of the GPS receiver as an exact-time source. Instrum. Exp. Tech. **54**, 249–253 (2011)
2. Von, A.: EOF: the universal internet time source. Linux J. **138**, 13 (2005)
3. Zhao, W., Ren, X.: Time synchronization with IRIG-B code in intelligent electronic devices. Autom. Electric Power Syst. **34**, 113–115 (2010)
4. Littlestone, N., Warmuth, M.K.: The weighted majority algorithm. In: Proceedings of the Second Annual Workshop on Computational Learning Theory, vol. 108(2), p. 388 (1989)
5. Ma, H.J., Huang, F.J., Zhang, B.: The time synchronization analytical study of the dynamical multi-agent. Appl. Mech. Mater. **198**, 1417–1421 (2012)
6. Johannessen, S.: Time synchronization in a local area network. IEEE Control Syst. Mag. **24**, 61–69 (2004)
7. Tian, Y.P.: Time synchronization in WSNs with random bounded communication delays. IEEE Trans. Autom. Control **99**, 1–7 (2017)
8. Huang, X., Jiang, D.Z.: A high accuracy time keeping scheme based on GPS. Autom. Electric Power Syst. (2010)
9. Ferrari, P., Flammini, A., Rinaldi, S.: Evaluation of time gateways for synchronization of substation automation systems. IEEE Trans. Instrum. Meas. **61**(10), 2612–2621 (2012)
10. Khmou, Y., Safi, S.: A study of 1D quadratic map. Int. J. Adv. Sci. Technol. **121**, 21–30 (2018)
11. Zhu, X.S., Ma, W.L.: A design of aircraft grid harmonic detection device using DSP and ARM. Int. J. Control Autom. **8**(3), 97–98 (2015)
12. Pusuluri, R., Aggarwal, R., Sivachandar.: Simulation study of feedback based adaptive TCP protocol for improving the performance of TCP and high speed data transmission based on congestion window size. Int. J.Grid Distrib. Comput. **11**(1), 1–12 (2018)
13. Khan, S.A., Khan, F.H., Qazi, F., Agha, D., Das, B.: Secure identity-based cryptographic approach for vehicular ah-hoc networks. Int. J. Secur. Appl. **12**(1), 59–68 (2018)
14. Alkhoder, A., Assimi, A., Alhariri, M.: Adaptive retransmission protocol based on mutual information. Int. J. Future Gener. Commun. Netw. **11**(2), 49–70 (2018)
15. Koteswara Rao Devana, V.N.: A novel UWB monopole antenna with defected ground structure. Int. J. Sig. Process.: Image Process. Pattern Recogn. **10**(1), 89–98 (2017)
16. Azam, A., Shafique, M.: An overview of fruits and vegetables trade of China. Int. J. u - and e – Serv. Sci. Technol. **11**(1), 33–44 (2018)
17. Kim, H.G.: SQL-to-MapReduce translation for efficient OLAP query processing with MapReduce. Int. J. Database Theor. Appl. **10**(6), 61–70 (2017)
18. Le, T., Le, C., Jeong, H.D., Jahren, C.: Visual exploration of large transportation asset data using ontology-based heat tree. Int. J. Transp. **6**(1), 47–58 (2018)
19. Guha, S., Dey, A.: Study on morphological changes in and around nayachara tail using remote sensing techniques. Int. J. Disaster Recovery Bus. Continuity **7**, 13–26 (2016)

Optimal Design of PID Controller for 2-DOF Drawing Robot Using Bat-Inspired Algorithm

Ahmad Taher Azar[1,2](✉), Hossam Hassan Ammar[3], Mayra Yucely Beb[3],
Santiago Ramos Garces[3], and Abdoulaye Boubakari[3]

[1] College of Engineering, Prince Sultan University, Riyadh, Kingdom of Saudi Arabia
aazar@psu.edu.sa
[2] Faculty of Computers and Artificial Intelligence, Benha University, Benha, Egypt
ahmad.azar@fci.bu.edu.eg
[3] School of Engineering and Applied Sciences, Nile University, 6th of October City,
Giza, Egypt
HHassan@nu.edu.eg, mayra.y.b@ieee.org
{santiago.ramos,abdoulaye.boubakari}@eu4m.eu

Abstract. Tuning process which is used to find the optimum values of the proportional integral derivative (PID) parameters, can be performed automatically using meta-heuristics algorithms such as BA (Bat Algorithm), PSO (Particle Swarm Optimization) and ABC (Artificial Bee Colony). This paper presented a theoretical and practical implementation of a drawing robot using BA to tune the PID controller governing the robotic arm which is a non linear system difficult to be controlled using classical control. In line with the above and in order to achieve this aim and meet high performance feedback and robust dynamic stability of the system, the PID controller is designed considering the realistic constraints. For faster tuning of the controller parameters, ten individuals and five iterations have been selected. BA, ABC and PSO have been compared and it's noticed that BA is the best choice to achieve good performance control. In the proposed design, MATLAB was used for trajectory reckoning. Afterwards, the value of coordinate position of the shape to be drawn is translated into a joint angle by applying the inverse kinematics to control the two DC motors through the ATMEGA 2560 microcontroller. The suggested technique reveals via simulations and hardware implementation the high efficiency of the applied algorithm. The PID controller approach presents an impressive stability and robustness results. The achieved results demonstrated that high performance can be obtained by tuning a PID controller using nature inspired based algorithms.

Keywords: PID controller · Bat Algorithm · Drawing robot · Inverse kinematics · Metaheuristics

1 Introduction

PID Control is one of the most worldwide used control techniques due to the facility related with its working principle, therefore it is widely used in industrial

© Springer Nature Switzerland AG 2020
A. E. Hassanien et al. (Eds.): AISI 2019, AISC 1058, pp. 175–186, 2020.
https://doi.org/10.1007/978-3-030-31129-2_17

applications [2,3,5–8,13,19,20,23,24]. This type of controller consist of three gains (proportional, integral and derivative) that are tuned with the aim to achieve good control performance. In other words, the modern control problem tries to ensure zero steady state error, small overshoot (depending on the application), small settling time and the minimum possible control effort considering that the performance is related to the specific application [1,4,10,18,20].

The most challenging task in the PID Controller design is tuning the proper gains to reach the desired performance. Many classical techniques are used with this purpose, one of the most known is Ziegler–Nichols which is based on the step response of the system but it is not suitable for all type of systems specially for high-order systems [27]. Manual methods are non-analytical and consist of choosing the gains by trial and error which give poor results and are time consuming. Root locus and frequency methods are also used but, even though these methods are effective, they require some assumptions when the gains are chosen. Based on that, in this paper, the use of metaheuristic algorithm is proposed in order to tune the gains of the PID controller efficiently and avoid suppositions that can affect its performance. Metaheuristic algorithm are becoming popular due to their facility to solve some optimization problems with efficiency and short time. This type of algorithms are nature-inspired i.e inspired by animal behavior and have application in many fields of engineering [9]. Some of the most used search algorithm are genetic algorithm (GA) [25], simulated annealing (SA) [15], and Particle Swarm Optimization (PSO) [14,22]. Nevertheless, new types of algorithms such as Ant Colony Optimization (ACO), Artificial Bee Colony (ABC) and Cuckoo Search Algorithm (CSA) [12] have been growing and are used to solve optimization problems.

In this paper, a two degree of freedom (2-DOF) planar drawing robot that has the ability to draw on an A4 size paper sheets is designed and implemented in order to test the performance of the designed PID controller. This robot is challenging to be controlled due to the high non-linearity between coupling links and high-order from linearization making it hard to control with the conventional methods. Therefore, PSO, ACO and BA nature inspired metaheuristic algorithms are tested to tune the PID gains for motion control of planar Drawing Robot in a proper way. Even though, ACO and PSO are powerful and efficient algorithm, it is shown that BAT algorithm is more suitable for position control of the drawing robot due to the fact that this algorithm converges faster. This characteristic is attributed to the capacity to search in the proximity of the global best particle. The main aim of the Robot is to draw a specific shape with accuracy. In the literature, some drawing robots can be found, but the vast majority of them use stepper motors in open loop and the accuracy depends on the minimum step angle [21]. More sophisticated system implements humanoid robot for artistic tasks, but these works focused on vision and AI algorithm. For example, Betty, the 12-DOF humanoid artist robot [16]. In [17], Computer Numerical control (CNC) code is used to draw specific shapes using servo motors but the motors works also in open loop and there is some control limitation. Considering that the related works does not focus on the control system, the proposed work aims to implement PID controller tuning with the required performance of drawing task using DC motors as actuator.

The paper is organized as follows. Section 2 presents the methodology of the current study. In Sect. 3, PID controller design is discussed. In Sect. 4, results and discussions are addressed. Finally, conclusions is presented in Sect. 5.

2   Methodology

The Hardware of the Drawing robot consists of one Arduino Mega 2560 board. It is a chip based on microcontroller Atmega 2560 that works on a frequency of 16 MHz. Also provide 14 PWM outputs, 16 possible pins for analog inputs, 4 UART pins, that makes this board a suitable solution to implement the control algorithm. The actuators are two DC motor with gearbox (1:75), rated voltage 12 V, no load speed 100 RPM, stall torque 6.5 Kg-cm. The motor is equipped with quadrature encoder with resolution of 825 pulses per revolution. In order to give the necessary power to the motors, one $L298N$ drive was used. For the movement of the pen, a micro servo motor was added to the end of the second link. The mechanical components of the robot body (links and gears) have been designed using SolidWorks software and manufactured with a 3D printer. For the implementation, first the shape is simulated in MATLAB to generate the required coordinates for each point of the system shown in Fig. 1. Then the inverse kinematics helps to get the angle for each joint by taking into account the velocities and acceleration of each joint, the necessary torque can be computed using the mathematical model of the system. After that, the robot is designed with a CAD software and simulated as shown in Fig. 1. Afterwards the whole system is modeled in Simscape and Simulink in order to simulate the behavior and the efficiency of the PID controller tuned with the metaheuristic algorithms. Finally, the control algorithm and the Metaheuristic algorithm are implemented and programmed in the real hardware (Fig. 2).

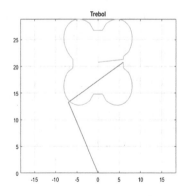

Fig. 1. Generated shape by Trebol

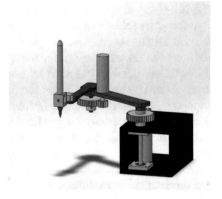

Fig. 2. CAD design of the system

2.1   Dynamic Model of the Manipulator

In order to deduce the dynamic equations, the classical approach deriving the Lagrangian function based on the kinetic and potential energy is carried out. Therefore,

$$
\begin{aligned}
\tau_1 =& [m_1 L_{c_1}^2 + I_1 + m_2(L_1^2 + 2L_1 L_{C_2} Cos(q_2) + L_{C_2}^2) + I_2]\ddot{\theta}_1 \\
&+ [m_2(L_{c_2} + L_1 L_{C_2} Cos(q_2)) + I_2]\ddot{\theta}_2 - m_2 L_1 L_{C_2} Sin(q_2)(2\dot{\theta}_1\dot{\theta}_2 + \dot{\theta}_2^2) \quad (1) \\
&+ m_1 g L_{C1} Cos(q_1) + m_2 g(L_1 Cos(q_1) + L_{C_2} Cos(q_1 + q_2))
\end{aligned}
$$

$$
\begin{aligned}
\tau_2 =& [m_2(L_{C_2}^2 + L_1 L_{C_2} Cos(q_2)) + I_2]\ddot{\theta}_1 + (m_2 L_{C_2}^2 + I_2)\ddot{\theta}_2 \\
&+ m_2 L_1 L_{C_2} Sin(q_2)\dot{\theta}_1^2 + m_2 g L_{C_2} Cos(q_1 + q_2)
\end{aligned} \quad (2)
$$

Where, θ i_{th} is the joint variable; $m i_{th}$ is the link mass; $\overline{I}$ i_{th} is the link inertia, about an axis through the CoM and parallel to z; a_{ith} is the link length; a_{Ci} is the distance between joint i and the CoM of the i_{th} link; τ is the torque on joint i; g is gravity force along y and P_i, K_i is the potential and kinetic energy of the i_{th} link.

On the other hand, the inverse kinematic is a useful tool to transform the position and orientation of the end effector from the cartesian to the joint space, allowing to obtain the angular position of each joint. To solve this problem, geometrical approach is considered. Once the system dynamic is solved, the robot is simulated in MATLAB drawing a desired shape as shown in Fig. 1 in order to compute the required torque, velocities and acceleration. In addition, considering the design specification related to the torque requirements and the geometry (sheet size) to be generated by the robot, CAD design and rigid dynamic simulation has been performed taking into account gravity force and the weight of components, in order to get more accurate results for the torque requirements.

2.2   Manipulator Modeling and Working Principle

The accuracy of the model is crucial in order to achieve good position control, therefore, the robot was simulated using Simscape Multibody library taking into account all the inertia and weight from the CAD software for more realistic simulation. The relevant values for mass of the two links are 100 g and 36 g, moment of inertia are 6.5×10^{-5} kg.m^2 and 6.65×10^{-5} kg.m^2, and length of 7.37 cm and 10 cm from the center of mass for first and second link respectively. The other important component of the system is the DC motor, in order to model this element, it is necessary to get information about the electrical and mechanical features of the real DC motor. Using Parameter estimation in MATLAB, the software compares the real step response velocity curve of the motor with the simulated model, and an optimization method is used to estimate the electrical and mechanical parameters.

On the other hand, the working procedure of the drawing robot consists of 4 fundamental steps. Initially, the desired shape is drawn using MATLAB code, generating the reference values for the trajectory generator. In order to change the shape to be drawn, it is necessary to give the equation of the figure in polar coordinates. After the array for desired position of each angles is transformed to joint space, the control algorithm takes the data and the trajectory generator build a linear trajectory between points in order to achieve soft movement. For the propose of simplifying computational assessment, linear segment between points is chosen. Consequently, the reference angular position for each joint is taken by the control algorithm and compared with the angular position read from the encoders to generate the control signal using the PID control approach tuned by the metaheuristic algorithms. Finally, the control signal is sending to the motor through the drive in a range of 0–5 DCV PWM signal and this device gives the necessary current to the motors in order to move in the required direction.

3   PID Controller Design

The first consideration for the controller is related with the explanation of the BA algorithm. This algorithm is based on the echolocation behavior of bats. It was proposed by Yang in 2010 [26]. The algorithm imitates the sound pulses that bat emit and reflected from prey or obstacles in order to know their locations and reach them as target or obstacles. The pulse rate varies from 0 to 1, where 0 means no emission (close to prey) and 1 means maximum emission. The movement of the bats is related to the frequency of the pulses and is updated about the best actual position. The key of this algorithm is to update randomly the position of the bat close to the bat with best position of finding a prey, making the convergence faster than other algorithms. The pseudo-code of this algorithm is summarized in Algorithm 1.

In the context of the Controller, the position of each bat is a vector with the three gains of the PID. The idea is to find the suitable position (PID gains) to minimize a specific cost function. For the cost function selection, the main performance index was taken into account for trajectory tracking. The idea is to minimize the error for the trajectory tracking, therefore Mean Root Square Error

---

Algorithm 1. BAT Algorithm Pseudo Code [11]

1: **procedure** INITIALIZE RANDOM SOLUTIONS
2:     **While** $t <$ (*Max number of iterations*) Generate new solutions by adjusting frequency, and updating velocities and positions solutions
3:         **if** $(rand > ri)$ Select a solution among the optimum results. Generate a local solution around the selected best solution.
4:             **end if** Generate a new solution by flying randomly.
5:         **if** $(rand < Ai$ & $f(xi) < f(x*))$ Take the new solutions. Increase ri and decrease Ai.
6:             **end if** categorize the bats and find the current best $x*$
7:         **end while** Post procedure results and visualization.
8: **end procedure**

(MRSE) was chosen as a cost function. Once the algorithm finds the suitable gains, they are updated to the control algorithm. However, in the real hardware, the algorithm was adapted to compute the proper discrete PID controller, therefore the process of updating the gains was done in the real hardware showing the flexibility of the algorithm.

4   Results and Discussion

This section shows the results obtained when the PID parameters are tuned via the metaheuristic algorithms and tested on both simulation and real hardware.

4.1   Simulation Results

The first simulated experiment establishes a set-point of 1 rad for the angular position of both joints and compare the response of PSO, ABC and BA algorithms as shown in Figs. 3 and 4. Control Toolbox of MATLAB was used to get the limits of the gains in order to get proper controller. The comparison results PSO, ABC and BA algorithms are summarized in Tables 1 and 2.

Table 1. Comparison results of BA, PSO and ABC algorithms

Algorithms	Joints	kp	ki	kd	Cost
ABC	q1	150	0.1	0.028	0.045
	q2	110.74	0.1	0.0295	**0.0427**
BA	q1	119.97	0.1	0.1	0.0845
	q2	53.2676	0.1	0.0729	0.0698
PSO	q1	111.11	0.1	0.0338	**0.0448**
	q2	123.95	0.1	0.0259	0.0445

It can be seen that BA got largest settling time and cost, but in this type of application the overshoot is critical in achieving drawing accuracy, but BA achieved better result in single position control. On the other hand, the shape to be drawn requires trajectory tracking. The second experiment requires to follow the desired trajectory to draw the trebol shape and testing the algorithms. Even though all the algorithm achieved good trajectory tracking, PSO showed larger overshoot while BA showed good tracking and achieved good performance, therefore BA is most suitable because of its low cost function.

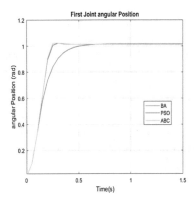

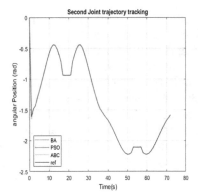

Fig. 3. The first joint angular position   **Fig. 4.** The second joint trajectory tracking

Table 2. Algorithms used to achieve minimum tracking error

Algorithms	Joints	kp	ki	kd	Cost
ABC	q1	200	6.4579	0	0.0002262
	q2	200	9.8039	0	0.00015606
BA	q1	185.4715	7.0574	0	0.0002274
	q2	179.176	2.1483	0	0.0001702
PSO	q1	192.78	6.696	0	0.0002266
	q2	200	8.546	0	0.0001561

4.2 Experimental Results

After achieving good simulation results, the system is tested in real hardware. The sample time for the control algorithm is 25 ms. The first experiment is the same in the simulation but in this case the set-point is set to 351 and 540 degrees for first and second joint. Due to the fact that there is imprecision in the model mostly related with the DC motors, when the controller was discretized the obtained gains were large making the controller unsuitable. To fix this problem, the algorithm calculated the gains in real time getting excellent result which proved the usefulness and flexibility of the algorithms as shown in Table 3.

Table 3. Gains of the controller

Algorithms	Joints	kp	ki	kd
ABC	q1	0.012	0.0001925	0.000684
	q2	0.01323	0.000404	0.001
BA	q1	0.01289	0	0.000143
	q2	0.0137	0.000478	0.0010
PSO	q1	0.01223	0.000314	0.000672
	q2	0.01372	0.0002823	0.0000421

Though, the three algorithms achieved good results, the PSO appeared to be faster than the two others. However, BA has lower cost and the difference in settling time is not considerable compared to PSO, and has very small overshoot. Therefore, it can be concluded that the BA generates the best response for the position control similar to the simulation. Finally, the second experiment required to follow the trajectory of each joint to draw the final shape. Taking into account that the rise time of the system is around 0.3 s i.e 3.33 Hz, the sampling time should be at least 33 HZ i.e 30 ms. Therefore, in the real hardware implemented in Arduino, the control algorithm is performed with an interrupt each 25 ms. The first joint real time angular position is shown in Fig. 5.

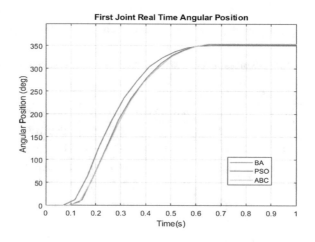

Fig. 5. First joint real time angular position

In Figs. 6 and 7, it can be seen that the off-tracking phenomenon occurs, that is due to the dead zone of the motor which is considered to be a nonlinear system so that the controller can't response for small difference on angle. Other interesting point is related with the integral gain that has to be increased for trajectory tracking in order to follow the curve. The drawback is that it causes instability of the system and can be seen in the response fluctuation of the mentioned figures. However, the system performs good result trajectory tracking for each joint but due to the problems stated before, the accuracy of the real hardware is impacted as shown in Fig. 8.

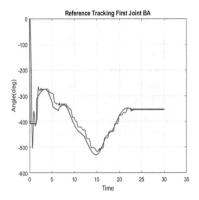

Fig. 6. First joint real-time trajectory tracking BA

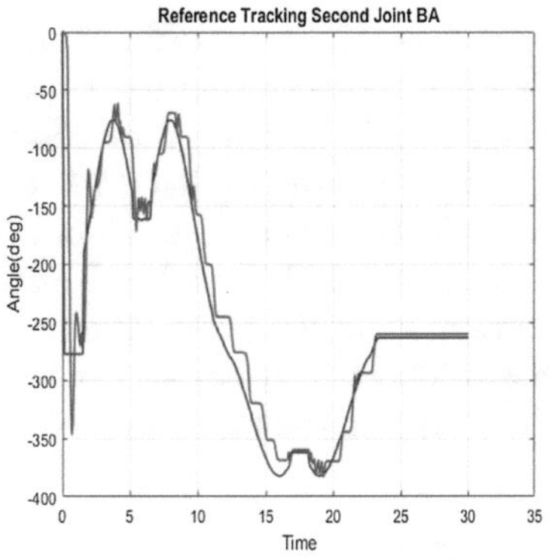

Fig. 7. Second joint real-time trajectory tracking BA

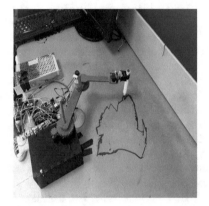

Fig. 8. Shape drawn using bat algorithm

5   Conclusion

The strive to find phase margins and gain of the PID controller are the reasons behind the tuning of its parameters. Even though it is possible to determine these optimal values both manually and automatically, auto-tuning is used for its effectiveness by automatically tuning the parameters and controlling the procedure to its best parameters. This is where the meta-heuristic algorithms comes up for automatic tuning of PID parameters. Even though tuning properly a PID controller is time consuming and has some challenges like achieving good performances analytically, it was demonstrated how useful is the metaheuristic algorithms in solving any optimization problem. It was proven that meta-heuristic is a very helpful tool for solving this type of engineering problems with high effectiveness and achieve the optimal performance. As per the experimental results, it can be seen that BA generated a very good results of tracking angular position of the DC motor controlling the movement of the robot. One of the main reasons is that, BA algorithm apply a random change of bats position close to the global best, accelerating the convergence of this algorithm and hence increases the probability to find a good solution. Despite the effectiveness of metaheuristic algorithm observed in this work, the results of simulation were different from the real implementation, due to some errors in the model and physical constraints like dead zone of DC motor. Nonetheless, the algorithm help to overcome these constraints, tackling this issue while the controller is converted to the discrete domain. This task would have required a lot of efforts if the conventional methods or the analytic solutions method have been used.

References

1. Abdelmalek, S., Azar, A.T., Dib, D.: A novel actuator fault-tolerant control strategy of DFIG-based wind turbines using takagi-sugeno multiple models. Int. J. Control Autom. Syst. **16**(3), 1415–1424 (2018)
2. Ammar, H.H., Azar, A.T.: Robust path tracking of mobile robot using fractional order PID controller. In: Hassanien, A.E., Azar, A.T., Gaber, T., Bhatnagar, R., Tolba, F.M. (eds.) The International Conference on Advanced Machine Learning Technologies and Applications (AMLTA2019), pp. 370–381. Springer, Cham (2020)
3. Ammar, H.H., Azar, A.T., Tembi, T.D., Tony, K., Sosa, A.: Design and implementation of fuzzy pid controller into multi agent smart library system prototype. In: Hassanien, A.E., Tolba, M.F., Elhoseny, M., Mostafa, M. (eds.) The International Conference on Advanced Machine Learning Technologies and Applications (AMLTA2018), pp. 127–137. Springer, Cham (2018)
4. Azar, A.T., Serrano, F.E.: Adaptive decentralised sliding mode controller and observer for asynchronous nonlinear large-scale systems with backlash. Int. J. Model. Ident. Control **30**(1), 61–71 (2018)
5. Azar, A.T., Serrano, F.E.: Fractional order two degree of freedom PID controller for a robotic manipulator with a fuzzy type-2 compensator. In: Hassanien, A.E., Tolba, M.F., Shaalan, K., Azar, A.T. (eds.) Proceedings of the International Conference on Advanced Intelligent Systems and Informatics 2018, pp. 77–88. Springer, Cham (2019)
6. Azar, A.T., Ammar, H.H., Barakat, M.H., Saleh, M.A., Abdelwahed, M.A.: Self-balancing robot modeling and control using two degree of freedom PID controller. In: Hassanien, A.E., Tolba, M.F., Shaalan, K., Azar, A.T. (eds.) Proceedings of the International Conference on Advanced Intelligent Systems and Informatics 2018, pp. 64–76. Springer, Cham (2019)
7. Azar, A.T., Hassan, H., Razali, M.S.A.B., de Brito Silva, G., Ali, H.R.: Two-degree of freedom proportional integral derivative (2-DOF PID) controller for robotic infusion stand. In: Hassanien, A.E., Tolba, M.F., Shaalan, K., Azar, A.T. (eds.) Proceedings of the International Conference on Advanced Intelligent Systems and Informatics 2018, pp. 13–25. Springer, Cham (2019)
8. Azar, A.T., Ammar, H.H., de Brito Silva, G., Razali, M.S.A.B.: Optimal proportional integral derivative (PID) controller design for smart irrigation mobile robot with soil moisture sensor. In: Hassanien, A.E., Azar, A.T., Gaber, T., Bhatnagar, R., Tolba, F.M. (eds.) The International Conference on Advanced Machine Learning Technologies and Applications (AMLTA2019), pp. 349–359. Springer, Cham (2020)

9. Chopard, B., Tomassini, M.: An Introduction to Metaheuristics for Optimization. Natural Computing Series. Springer, Heidelberg (2018)
10. Fekik, A., Denoun, H., Azar, A.T., Hamida, M.L., Zaouia, M., Benyahia, N.: Comparative study of two level and three level PWM-rectifier with voltage oriented control. In: Hassanien, A.E., Tolba, M.F., Shaalan, K., Azar, A.T. (eds.) Proceedings of the International Conference on Advanced Intelligent Systems and Informatics 2018, pp. 40–51. Springer, Cham (2019)
11. Fister, D., Fister, I., Fister, I., Afari, R.: Parameter tuning of PID controller with reactive nature-inspired algorithms. Robot. Auton. Syst. **84**, 64–75 (2016)
12. Gai, W., Qu, C., Liu, J., Zhang, J.: A novel hybrid meta-heuristic algorithm for optimization problems. Syst. Sci. Control Eng. **6**(3), 64–73 (2018)
13. Gorripotu, T.S., Samalla, H., Jagan Mohana Rao, C., Azar, A.T., Pelusi, D.: TLBO algorithm optimized fractional-order PID controller for AGC of interconnected power system. In: Nayak, J., Abraham, A., Krishna, B.M., Chandra Sekhar, G.T., Das, A.K. (eds.) Soft Computing in Data Analytics, pp. 847–855. Springer, Singapore (2019)
14. Kennedy, J., Eberhart, R.: Particle swarm optimization. In: Proceedings of ICNN 1995 - International Conference on Neural Networks, vol. 4, pp. 1942–1948 (1995)
15. Kirkpatrick, S., Gelatt, C.D., Vecchi, M.P.: Optimization by simulated annealing. Science **220**(4598), 671–680 (1983)
16. Lau, M.C., Baltes, J., Anderson, J., Durocher, S.: A portrait drawing robot using a geometric graph approach: Furthest neighbour theta-graphs. In: 2012 IEEE/ASME International Conference on Advanced Intelligent Mechatronics (AIM), pp. 75–79 (2012)
17. Megalingam, R.K., Raagul, S., Dileep, S., Sathi, S.R., Pula, B.T., Vishnu, S., Sasikumar, V., Gupta, U.S.C.: Design implementation and analysis of a low cost drawing bot for educational purpose. Int. J. Pure Appl. Math. **118**(6), 213–230 (2018)
18. Meghni, B., Dib, D., Azar, A.T.: A second-order sliding mode and fuzzy logic control to optimal energy management in wind turbine with battery storage. Neural Comput. Appl. **28**(6), 1417–1434 (2017)
19. Meghni, B., Dib, D., Azar, A.T., Saadoun, A.: Effective supervisory controller to extend optimal energy management in hybrid wind turbine under energy and reliability constraints. Int. J. Dyn. Control **6**(1), 369–383 (2018)
20. Mekki, H., Boukhetala, D., Azar, A.T.: Sliding modes for fault tolerant control. In: Azar, A.T., Zhu, Q. (eds.) Advances and Applications in Sliding Mode Control systems, pp. 407–433. Springer, Cham (2015)
21. Munna, M.S., Tarafder, B.K., Robbani, M.G., Mallick, T.C.: Design and implementation of a drawbot using matlab and arduino mega. In: 2017 International Conference on Electrical, Computer and Communication Engineering (ECCE), pp. 769–773 (2017). https://doi.org/10.1109/ECACE.2017.7913006
22. Sabir, M.M., Ali, T.: Optimal pid controller design through swarm intelligence algorithms for sun tracking system. Appl. Math. Comput. **274**, 690–699 (2016)
23. Smida, M.B., Sakly, A., Vaidyanathan, S., Azar, A.T.: Control-based maximum power point tracking for a grid-connected hybrid renewable energy system optimized by particle swarm optimization. In: Azar, A.T., Vaidyanathan, S. (ed.) Advances in System Dynamics and Control, IGI Global, pp. 58–89 (2018)

24. Soliman, M., Azar, A.T., Saleh, M.A., Ammar, H.H.: Path planning control for 3-omni fighting robot using PID and fuzzy logic controller. In: Hassanien, A.E., Azar, A.T., Gaber, T., Bhatnagar, R., Tolba, F.M. (eds.) The International Conference on Advanced Machine Learning Technologies and Applications (AMLTA2019), pp. 442–452. Springer, Cham (2020)
25. Srinivas, N., Deb, K.: Muiltiobjective optimization using nondominated sorting in genetic algorithms. Evol. Comput. **2**(3), 221–248 (1994). https://doi.org/10.1162/evco.1994.2.3.221
26. Yang, X.S.: A New Metaheuristic Bat-Inspired Algorithm, pp. 65–74. Springer, Heidelberg (2010)
27. Ziegler, J., Nichols, N.B.: Optimum settings for automatic controllers. Trans. ASME **64**(8), 759–768 (1942)

Secondary Virtual Circuit Test Scheme Based on Intelligent Substation SCD File

Yupeng Cai[1,2(✉)], Tongwei Yu[1,2], Hai Qian[2], Yan Lu[1,2],
and Shengyang Lu[1,2]

[1] State Grid Electric Power Research Institute, Shenyang 110003, China
cai8872@126.com
[2] State Grid Liaoning Electric Power Supply Co. Ltd., Shenyang 110004, China

Abstract. The paper proposes a scheme for secondary virtual circuit test based on the characteristics of intelligent substation, and the test scheme is divided into two types including dynamic test and static test, the static test is based on the Substation Configuration Description (SCD) file configured for the whole intelligent substation, firstly virtual terminal connections are exported from SCD by the order of separations, the information associated with each separation is generated as a table type file, checking whether SCD file is consistent with the secondary design drawings. The dynamic test is based on the configure information in SCD as well, then communication condition, message analysis and receiving and dispatching simulation are tested for the network communication, verifying whether the device's communication actions are satisfied to IEC 61850 and the requirement of engineering design. The scheme fully considers the characteristics of intelligent substation network and the communication details that IEC 61850 sets, and strictly checks the important information in the virtual circuit, making sure that virtual circuit is satisfied to the requirement of design. The test scheme provides not only help for the debug and acceptance of the project but also convenient for quick problems searching in operation.

Keywords: Intelligent substation · Secondary virtual circuit · SCD

1 Introduction

The intelligent substation is based on the digital substation, using advanced and reliable intelligent devices and adopting the system architecture of three layers and two networks. One of its important characteristics is shared network information, making multiplexed information compounded, and in the secondary circuit of process layer a few optical-fiber channels will replace a large number of cables, then the conventional secondary hard circuit will never exist [1]. To this end, the industry proposes a concept called secondary virtual circuit, changing the circuit structure constituted by the traditional cables which transmit single signals into the optical-fiber channels which can transmit multiplexed signals, and a lot of research about it is done, among these the problems about secondary virtual circuit test need to be solved quickly [2, 3].

© Springer Nature Switzerland AG 2020
A. E. Hassanien et al. (Eds.): AISI 2019, AISC 1058, pp. 187–193, 2020.
https://doi.org/10.1007/978-3-030-31129-2_18

Now, most studies mainly focus on how to define, design or configure virtual circuits [4]. But when operating and debugging at engineering site, some practical problems such as how to ensure whether substation SCD file is consistent with engineering drawings aren't researched and analyzed deeply [5]. This paper proposes a scheme for secondary virtual circuit test in the intelligent substation, and in practice, it can not only help the debugging staffs test the secondary virtual circuit of intelligent substation quickly and accurately but also reduce the labor intensity of the staffs [6].

2 Intelligent Substation SCD File

The full name of SCD is substation configuration description, and it is an important data source of substation, recording much important information such as device configuration, communication configuration and so on. SCD file mainly consists of five parts, and they are 'Header', 'Substation', 'Communication', 'IED' and 'Data Type Template'. 'Header' is used to describe the basic information of file, and it mainly describe the historical version and revision information of file; 'Substation' is used to describe some functions of substation, including voltage level, interval, topology point and some other information; 'Communication' is used to describe the communication parameters of intelligent substation, such as MMS, GOOSE, SV and so on; 'IED' is used to describe the intelligent electronic devices in substation, including some node elements such as 'services', 'name', 'type', 'access point' and so on, and 'name' is unique in SCD file; 'Data Type Template' describes the data type template that SCD file uses. The SCD file structure in intelligent substation is showed in Fig. 1 [7, 8].

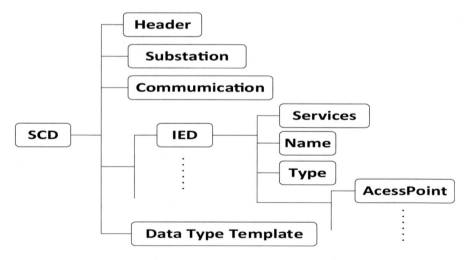

Fig. 1. SCD file structure in intelligent substation.

3 Secondary Virtual Circuit and Virtual Terminal

The secondary virtual circuits in intelligent substation mainly reflect in topological connection relations among the virtual terminals of secondary devices [11]. Irtual terminal is used to describe the input and output signal logical connection points of GOOSE and SV, and it can mark the correspondence between the traditional cubicle terminals and the secondary circuit signals like GOOSE, SV which transmit on the network, making the connection relation of substation devices clearer and producing the necessary text information which GOOSE and SV need. The connection relation of virtual terminals is partly described by 'In-puts' in 'LLN0', and every virtual terminal connection includes the information of input virtual terminal signal in its own logic device and output virtual terminal signal in the opposite device (Refer to Fig. 1). The moment monitored object next poll is defined as in Eq. (1).

$$t_n = t_l + T_i \cdot \frac{\left| TH - \sum\limits_{i=1}^{n} v_{i,tl} \right|}{\sum\limits_{i=1}^{n} \lambda_i} \tag{1}$$

Where t_n is the maximum polling interval period of the monitored object i; $v_{i,tl}$ is the value of the monitored object i at the acquisition time t_l; TH is the fault alarm global threshold; t_l is the current time of the monitored object i polling; t_n is monitored object i next polling moment.

$$\Delta v_i = \left| v_{i,t} - v_{i,t'} \right| \tag{2}$$

Where t' is the acquisition time, and t is the previous information collection time.

The GOOSE link in intelligent substation is equal to the secondary DC cables in conventional substation, and the SV link is equal to the secondary AC cables in conventional substation. And the correspondence between secondary virtual circuit in intelligent substation and traditional secondary circuit is showed in Fig. 2 [12, 13].

4 Scheme for Secondary Virtual Circuit Test in Intelligent Substation

The test scheme is divided into two types including dynamic test and static test, and we will study them deeply in the following section [14, 15].

4.1 Static Test Scheme

The essence of the static test is to check SCD files and design drawings, and we usually adopt the way of artificial contrast, due to the lack of unified conversion specification [16].

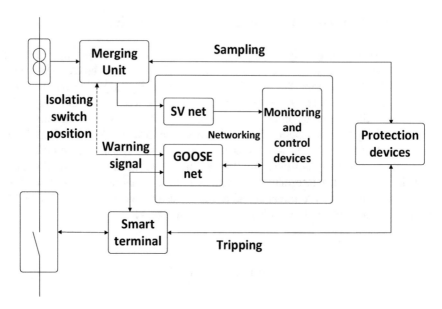

Fig. 2. The correspondence between secondary virtual circuit and traditional secondary circuit

The test scheme will adopt the way of converting SCD file to file form which is easily understood, making the static test more convenient. Firstly, SCD file need to import into the test software, then in the device model which needs to be tested, we should export its 'IEDNAME' model and fill in the table as the input of virtual terminals; then we will retrieve the route of outside input signals that 'IntAddr' quotes based on 'Extref' element in 'LLNO' logic nodes, and export some related communication configuration information, the names of dataset and the names of IEDs assigned by the route, filling in the table as the output of virtual terminals. Finally, the test software will build an information table of the associated virtual terminals for every device in the SCD file [17, 18]. The process of static test is shown in Fig. 3.

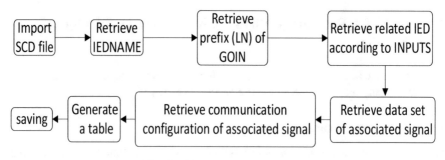

Fig. 3. The process of static test

4.2   Dynamic Test Scheme

Dynamic test scheme needs to test GOOSE virtual circuit and SV virtual circuit respectively based on the difference of message contents, and at the same time it must support network adaptive function [19].

4.2.1   GOOSE Virtual Circuit Test

GOOSE virtual circuit test is based on the SCD file, adopting the way of simulating conventional secondary circuit test. As a subscriber in open circuit, test software will subscribe and open dataset, then we should monitor whether the information such as multicast address of GOOSE message, quantity, and type of dataset is consistent with the SCD file, at the same time we need to monitor whether the published sequence and change of data is consistent with the expected. Then as a publisher in the closed circuit, test software will serve the information which needs published in the SCD file as a publishing source and transmit it by the way of simulating GOOSE message, then we can monitor the subscription behavior of subscription device and judge whether the device is running correctly [20]. The process of GOOSE virtual circuit test as shown in Fig. 4.

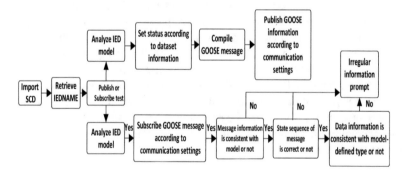

Fig. 4. The process of GOOSE virtual circuit test

4.2.2   SV Virtual Circuit Test

SV virtual circuit test is based on the SCD file as well. Contrary to the SV output from merging unit, we can analyze whether some information such as a multicast address, APPID, the sampling rate is consistent with the SCD file by the way of subscribing SV message, and then judge whether the counter and state flag in the message are normal [21].

Contrary to the devices which subscribe samples, we can use test software to simulate SV message output, then monitor communication recovery warning signals of the devices and judge whether the devices can receive normally. When simulating SV message publishing, we can simulate some irregular states such as loss of synchronization and irregular sampling rate, and analyze whether the tested device is satisfied to the operational requirement. The process of SV virtual circuit test as shown in Fig. 5.

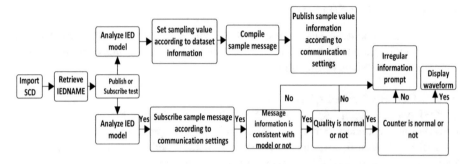

Fig. 5. The process of SV virtual circuit test

5   Engineering Application

In the construction of a 220 kV intelligent substation, the debugging staffs use the test software which adopts the scheme above and test the secondary virtual circuit, when debugging substation system, the debugging staffs find that there is a communication error between smart terminals of bus and bus differential protection. The staffs import the SCD file into the test software, and simulate GOOSE message that smart terminals of protection device publish, then they find that the test software can receive message correctly and display the changes of signals normally, this shows that the message which smart terminals publish is correct; but when simulating smart terminals publishing GOOSE message to protection devices, the protection devices can never prompt communication recovery. Then the staff's change the switch signal type in a published message from two-point data type to single point data type, the protection devices can receive correctly, thus it is sure that the receiving virtual terminals of protection devices can only recognize single point data type, after changing the configuration, communication recovers to normal.

6   Conclusions

The scheme for secondary virtual circuit test in the intelligent substation is based on SCD file. It can reduce the workload of testing intelligent substation secondary virtual circuit obviously. The scheme fully considers the characteristics of the network in intelligent substation and the communication details that IEC 61850 sets, and strictly checks the information in the virtual circuit which needs distinguished, making sure that virtual circuit is satisfied to the requirement of designs after substation is tested. With the quick spread and development of intelligent substation project, the test scheme will provide not only help for the debugging and acceptance of the project but also convenient for quick problems searching in operation.

References

1. Zhang, K., Chen, L.: A smart substation field secondary device testing technique based on recurrence principle. J. Power Energy Eng. **2**(4), 244–251 (2014). (in Chinese)
2. Pan, Y.: Research on IED configurator based on IEC 61850. In: International Conference on Automation and Systems Engineering 04, pp. 1–4 (2011)
3. Zhu, L., Shi, D., Wang, P.: IEC 61850-based information model and configuration description of communication network in substation automation. IEEE Trans. Power Delivery **29**(01), 12–17 (2014)
4. Ke, J.: Intelligent substation relay protection debugging and application. China New Technol. New Prod. **24**, 13–14 (2017). (in Chinese)
5. Yu, L., Wu, H., Huang, J.: Development and application on virtual secondary circuit automatic testing Yu system of smart substation. Inner Mongolia Electric. Power **33**(05), 53–57 (2015). (in Chinese)
6. Liu, B., Lin, J.: Development and application of intelligent software for virtual circuit test in digital substation. Guangxi Electric. Power **34**(02), 5–7 (2011). (in Chinese)
7. Tao, Y., Zhang, M.: Real-time analysis of intelligent substation communication network based on IEC 61850. Electr. Autom. **40**(03), 37–39 (2018). (in Chinese)
8. Hou, L., Li, H., Zhou, J.: Intelligent substation operation and maintenance management problems and solutions. Shandong Ind. Technol. **20**, 170 (2018). (in Chinese)
9. Yoon, J., Lin, M., Kim, B.: MBD-based electric vehicle core part modeling and dynamic property analysis. Int. J. Eng. Technol. Automob. Secur. **1**(2), 1–6 (2017)
10. Lee, S.: Real-time hand tracking using the RGB-D camera. Int. J. Comput. Graph. **8**(2), 13–20 (2017)
11. Wei, W., Suo, L., Peng, X.: Research on transformer fault diagnosis and multi variable parameter decision model. Asia-Pacific J. Adv. Res. Electr. Electron. Eng. **1**(1), 47–54 (2017)
12. Kim, Y.: The hardware accelerated physics engine with operating parameters controlling the numerical error tolerance. Int. J. Adv. Sci. Technol. **119**, 145–152 (2018)
13. Alnefaie, A., Fiaidhi, J., Mohammed, S.: Ontology-based food recommendation system for seniors: a peer to peer networking approach. Int. J. Adv. Sci. Technol. **123**, 41–46 (2019)
14. Hui, L., Rui-li, W.: Experiments and numerical simulation of flow field inner the chain plate fiber dryer. Int. J. Control Autom. **8**(3), 115–126 (2015)
15. Suresh, M., Edward, J.B.: Optimal placement of DG units for loss reduction in distribution systems using one rank cuckoo search algorithm. Int. J. Grid Distrib. Comput. **11**(1), 37–44 (2018)
16. Fardad, N., Soleymani, S., Faghihi, F.: Cyber defense budget assessment in smart grid based on reliability evaluation considering practical data of HV substation. Int. J. Secur. Appl. **12**(1), 1–20 (2018)
17. Mushtaq, S., Mir, A.: Image copy move forgery detection: a review. Int. J. Future Gener. Commun. Networking **11**(2), 11–22 (2018)
18. Dong, X., Han, X., Wang, J., Niu, Q., Shi, W.: Pulse design method for 60 GHz impulse wireless communication system. Int. J. Signal Process. Image Process. Pattern Recogn. **10**(1), 59–70 (2017). (in Chinese)
19. Ali, R., Peng, Y., Gupta, S.: Fusion of total variation filter and weighted bilateral filter in image denoising. Int. J. Serv. Sci. Technol. **11**(1), 17–32 (2018)
20. Petrone, A., Franz, M.: Probe vehicle based trajectory data visualization and applications. Int. J. Transp. **6**(1), 59–74 (2018)
21. Kaur, S., Singh, S.: Modified circular layer caching in wireless sensor networks. Int. J. Sens. Appl. Control Syst. **3**(1), 7–14 (2015)

An Efficient Organizations' Bitcoin Wallet Signature Scheme

Shereen M. Mahgoub$^{(\boxtimes)}$, Fatty M. Salem$^{(\boxtimes)}$, and I. I. Ibrahim$^{(\boxtimes)}$

Department of ECCE, Faculty of Engineering, Helwan University,
Helwan, Egypt
shry_m84@hotmail.com, faty_ahmed@h-eng.helwan.edu.eg,
iiibrahim1953@gmail.com

Abstract. Bitcoin is a totally decentralized digital cryptocurrency in which all the transactions of currencies are recorded and stored in a ledger. The ledger is a publicly available database containing Blockchain. Bitcoins are stored in a wallet and the wallet can be opened only by its secret key. The existing wallet signature schemes are improper for the organization hierarchy since the organization hierarchy requires each employee in the organization to sign independently and the upper management has to prove employee's signature before starting his signing which makes the scheme more realistic. This paper presents a practical signature scheme to organize the access of organization's bitcoin wallet. In the proposed scheme, organizations can create its own rules and manage wallet access without affecting transaction size. In addition, the proposed scheme can keep the structure of organization and manage employee hiring and termination. Moreover, the security analysis of the proposed scheme is provided.

Keywords: Bitcoin · Wallet · Organizations · Digital signature · Access management

1 Introduction

Bitcoin is a software-based online payment system introduced by Satoshi Nakamoto [1] and announced as open-source software in 2009. Bitcoin is a form of digital cryptocurrency, engendered and held electronically by utilizing internet connection peer to peer. It is based on blockchain as a key innovation for transactions integrity as each block contains the brevios block hash. Wallet is an accumulation of addresses with the same security policy; this policy guarantees the access-control list, and the conditions under which bitcoins in the wallet may be spent.

To make bitcoin transaction, you need to get your own wallet address, a private key for signing process [2], other player address, and the amount of bitcoins you will send.

Each transaction has a new random generated address and the output of a transaction is the input of the next one. Miners (verifier nodes) must validate the transaction first before adding it to the block. Proof of work competition is a mathematical problem used to manage a new block addition and prevent double spending attack [3, 4].

There are a lot of methods for securing bitcoin wallet including authentication and various types of digital signature. The two-factor authentication schemes [5] not only

© Springer Nature Switzerland AG 2020
A. E. Hassanien et al. (Eds.): AISI 2019, AISC 1058, pp. 194–203, 2020.
https://doi.org/10.1007/978-3-030-31129-2_19

use password, but also provide at least one more authentication step as email, smart phone authentication application or using a messaging service on cell phone. While the three-factor authentication schemes [6, 7] use the help of methods that include biometrics but anonymity is compromised.

Digital signature schemes have brought more attention in bitcoin wallets. Threshold signature scheme [8] is based on threshold secret sharing where any threshold shared keys or more can reconstruct the secret [9], but smaller subset cannot yields any information about the secret. In the proposed threshold wallet signature scheme, all players get equal shares and signing with equal weight. However, the authors in [10] have introduced a weighted threshold signature scheme where players are given more than one share of the secret key due to user's weight/priority.

Bitcoin multi-signature schemes [11, 12] can provide more than one signer to validate transaction based on wallet policy. The authors in [13] have described a new Schnorr-based multi-signature scheme that makes a group of signers to produce a short, joint signature on a common message called Multi-Sig. The authors in [14] have constructed a new multi-signature scheme to reduce bitcoin blockchain size. Using threshold signature gains some features such as dual control of wallet, detection and delegation of signatures, lost key recovery, and decreasing signature size in wallet signing script. On the other side, using multi-signature [15] transactions that have one clear benefit over using threshold signatures that wallet can be signed independently by each participant in a non-interactive manner, whereas threshold signature [16] requires multiple rounds of interaction.

The existing wallet signature schemes are improper for organizations as the organization's hierarchy requires each employee in the organization to sign independently, and the upper management has to prove employees' signature before starting his signing to make the scheme more realistic. In multi-signature, the number of signers is up to 15 because of script length limit and it can be signed independently. Using threshold signature is improper as organization signature is hierarchal and can't be executed in interaction manner. Moreover; the manager's sign is critical, hence; using the weighted threshold signature is not proper also. In this paper, we propose a practical signature for organizations based upon applying policy for organization structure to different employees without multi-transaction.

The rest of the paper is organized as follows. Proposed organization model and security requirements are discussed in Sect. 2. The proposed organizational wallet signature scheme is presented in Sect. 3. The security of the proposed scheme is analyzed in Sect. 4. Finally, the paper is concluded in Sect. 5.

2 Organization Model and Security Requirements

In this section, our proposed organization model is described and the organization security requirements will be given.

2.1   System Model

All departments of an organization when requesting money, they have to send request to the financial department to verify budget and approve money requests. Hence, in our model, we will divide organizations as shown in Fig. 1 into two branches for accessing wallet: operation branches and financial branch. Operation branch is any department in the organization except the financial one; the operation department team sends the signed money request to the financial department team for approval after verifying the operation team members signatures. The team budget is based on the organization structure and policy. Financial department team is responsible for reviewing any money requests and will be a part of all wallet signature process.

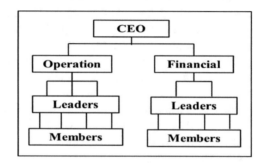

Fig. 1. The proposed organization structure

2.2   Organization Wallet Access Requirements

This paper presents a signature scheme to organize the access of organization's bitcoin wallet. The existing wallet signature schemes are not suitable for the organization hierarchy since the organization hierarchy requires each employee in the organization to sign independently and the upper management must prove employee's signature before his signing which makes the scheme more realistic. The secure bitcoin wallet of organizations must provide the following requirements:

- **Access Management**: Wallet contains money; hence, it must be accessed carefully with right policy (rules).
- **Unforgeability**: Only true signers can create a valid bitcoin wallet signature.
- **Unreliability**: Once true signer can create a valid bitcoin and organization cannot deny the transaction.
- **Transferability**: The transfer of positions in the organization from one person to another with the ability to get related signature to each person and meet organization policy.
- **Detection**: Bookkeeping is a must for detecting fraudulent transactions.
- **Anonymity**: The organization transactions must be kept anonymous in the network.
- **Verifiability**: Any verifier must be able to ensure that the signature is valid.

- **Wallet Recovery**: Ensure recovery of wallet in case of private key loss. In bitcoin, transactions are irreversible which means that once a valid transaction is stored in the Blockchain, it cannot be reversed.
- **Functional Separation of Duties**: Using our signature we can make a separated signature for each organization branch and each employee based on his position and organization hierarchy.

3 The Proposed Organizations' Bitcoin Wallet Signature Scheme

In this section, we present the proposed organizational wallet signature scheme using ECDSA [17] as ECDSA provides the same level of security as RSA but with smaller keys size. The proposed scheme is consisted of four phases including key generation, encapsulated operation signature, encapsulated financial signature, and organizational wallet signature.

3.1 Key Generation

The organization depends on a certificate authority to create the public and private keys for all employees in the organization. Organization must keep a database and a revocation list containing the existing and removed employees' keys. This database will be used by employees to check the keys availability in the signature verification round. In the key generation phase, the certificate authority:

- Selects an elliptic curve E defined over Z_p in which the number of points in $E(Z_p)$ should be divisible by a large prime n, and a point $G \in E(Z_p)$ of order n.
- Chooses $x_{pz} \in [1, n-1]$ for each operation employee where p refers to operation branches and z represents the employee's number in his branch from the lower priority employee to the higher priority top manager employee (from 1 to l). $x_{pz} \in [x_{p1}, \ldots, x_{pl}]$ is an operation department employee' private key where l is the number of employees in the operation branch.
- Computes the operation employee's public key y_{pz} as:

$$y_{pz} = Gx_{pz} \tag{1}$$

- Chooses $x_{ft} \in [1, n-1]$ for each financial team employee where f represents financial employee and j represents the employee's number in his team from lower priority employee to higher top manager employee $[1, j]$. $x_{ft} \in [x_{f1}, \ldots \ldots, x_{fj}]$ is a financial department employee' private key where there are j employees in the financial branch.
- Computes the financial employee's public key $y_{ft} = Gx_{ft}$.
- At the end of this phase, the certificate authority publicly sends the public keys and parameters $(E, G, n, y_{pz}, y_{ft})$, and securely sends the private keys (x_{pz}, x_{ft}) to the operation and financial departments, respectively.

3.2   Encapsulated Operation Signature

In the proposed scheme, the signature is encapsulated by team member's signature and approved by his leader's signature, and then signed by the manager of leader, and so on till last signer of the team from lower to higher priority. Each signer approves the lower employee's signature and encapsulates this signature in his signature round. The encapsulation of operation signature is shown in Fig. 2.

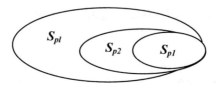

Fig. 2. Operation encapsulated signature

Operation Team Member Part: The first signer who is a member of any operation team requests funds, this member signs the message first by his personal private key x_{p1} to request the money. This member:

- Chooses a random $0 < k_{p1} < n$, to compute the first signature part r_{p1}.
- Calculates $Gk_{p1} = (x, y)$ and uses the x coordinate to compute:

$$r_{p1} = x_{p1} \bmod n \tag{2}$$

- If $r_{p1} = 0$, then go to previous step to re-choose k_{p1}.
- Calculates $k_{p1}^{-1} \bmod n$ to compute the second signature part s_{p1} as:

$$s_{p1} = k_{p1}^{-1}\left(h(m) + x_{p1}r_{p1}\right) \bmod n \tag{3}$$

This member sends his signature (m, s_{p1}, r_{p1}) to the upper manager (team leader) to verify and resign the message m.

Team Leader Signature Round: The team leader first verifies the employee signature by computing:

$$w_{p1} = s_{p1}^{-1} \bmod n \tag{4}$$

$$u_{p1} = h(m)w_{p1} \bmod n \text{ and } q_{p1} = r_{p1}w_{p1} \bmod n \tag{5}$$

$$u_{p1}G + q_{p1}y_{p1} = (x, y), \text{ and } v_{p1} = x_{p1} \bmod n \tag{6}$$

The team leader accepts the signature if $v_{p1} = r_{p1}$, else rejects.

If the signature is accepted, the leader uses his private key x_{p2} to compute (r_{p2}, s_{p2}). To do this, the leader chooses $k_{p2} \in [1, n-1]$ to compute r_{p2}, and then he calculates $k_{p2}^{-1} \bmod n$ to compute:

$$s_{p2} = k_{p2}^{-1} \left(h(m\|s_{p1}) + x_{p2} r_{p2} \right) \bmod n \tag{7}$$

Finally, the team leader sends $(r_{p2}, s_{p2}, s_{p1}, m)$ as his signature to upper manager for verification and resigning.

Top Manager Signature Round: The manager verifies leader's signature by computing $w_{p2}, h(m\|s_{p1}), u_{p2}$, and q_{p2}, and calculating v_{p2}. He accepts the signature if $v_{p2} = r_{p2}$, else rejects.

After verifying the leader's signature and confirming money request, the manager resigns the message using his private key x_{pl}. To resign the message, the manager chooses $k_{pl} \in [1, n-1]$ to compute r_{pl}, and calculates $k_{pl}^{-1} \bmod n$ to compute:

$$s_{pl} = k_{pl}^{-1} \left(h(m\|s_{p1}\|s_{p2}) + x_{pl} r_{pl} \right) \bmod n \tag{8}$$

$(r_{pl}, s_{p1}, s_{p2}, s_{pl}, m)$ is the final operation team's signature with the high priority manager's signature that is sent to the financial team to verify operation team's signature and ensure budget availability.

3.3 Encapsulated Financial Signature

Financial department receives operation team's signature request to be confirmed, and the financial team member who receives the request, reviews the operation team's signatures and ensures budget availability, then member starts to sign the first financial signature.

First Member Signature: The financial team member receives the signed message from operation department manager who requests money approval. This financial member starts by verifying operation department signatures and team budget, then financial team signature round will be started as follows:

When the financial team member receives the request, member starts verifying signature by computing $w_{pl}, h(m\|s_{p1}\|s_{p2}), u_{pl}$, and q_{pl}, and calculating:

$$u_{pl}G + q_{pl}y_{pl} = (x, y), and \ v_{pl} = x_{pl} \bmod n \tag{9}$$

Financial member accepts the signature if $v_{pl} = r_{pl}$, else rejects.

After confirming the operation team signatures, financial member start to sign the message using his personal private key x_{f1}. To do this, the member chooses $k_{f1} \in [1, n-1]$ to compute r_{f1}, and calculates $k_{f1}^{-1} \bmod n$ to compute:

$$s_{f1} = k_{f1}^{-1} \left(h(m) + x_{f1} r_{f1} \right) \bmod n \tag{10}$$

The team member sends (m, s_{f1}, r_{f1}) and $(r_{pl}, s_{p1}, s_{p2}, s_{pl})$ to team leader (reviewer).

Signature Reviewer: The accounting department must have reviewers to review budget and operation signature again for error detection before last confirmation to access. To verify the operation signature, the accounting department reviewer computes w_{f1}, u_{f1}, and q_{f1}, and calculates v_{f1}. The reviewer accepts the signature if $v_{f1} = r_{f1}$, else rejects.

After reviewing and validating budget, the reviewer member uses his personal private key x_{f2} to resign the message. To do this, the reviewer member chooses $k_{f2} \in [1, n-1]$ to compute r_{f2}, and calculates $k_{f2}^{-1} mod\ n$ to compute:

$$s_{f2} = k_{f2}^{-1}\left(h(m||s_{f1}) + x_{f2}r_{f2}\right)mod\ n \qquad (11)$$

Finally, the reviewer sends $(r_{f2}, s_{f2}, s_{f1}, m)$ to the manager as his signature part and $(r_{pl}, s_{p1}, s_{p2}, s_{pl})$ to ensure operation team signature.

Financial Department Manager Signature: The last organization signature was the high financial management team, where manager first verifies reviewer signature by computing $w_{f2}, h(m||s_{f1}), u_{f2}$, and q_{f2}, and calculating v_{f2}. The manager accepts the signature if $v_{f2} = r_{f2}$, else rejects.

Finally, the manager signs the last financial signature using his private x_{fj}. To sign the message, the manager chooses $k_{fj} \in [1, n-1]$, calculates r_{fj} and $k_{fj}^{-1} mod\ n$ to compute:

$$s_{fj} = k_{fj}^{-1}\left(h(m||s_{f1}||s_{f2}) + x_{fj}r_{fj}\right)mod\ n \qquad (12)$$

The financial manager signature is (r_{fj}, s_{fj}, m) will be used to sign wallet script in the first and second scenario illustrated in Sect. 3.4. Also, s_{fj} will be kept in book-keeping server for the organization to be recalled and verified if needed.

Verification: The verifier computes $w_{fj}, h(m||s_{f1}||s_{f2}), u_{fj}, q_{fj}$, and calculates v_{fj}. The manager accepts the signature if $v_{fj} = r_{fj}$, else rejects.

3.4   Organizational Wallet Signature

The proposed signature scheme is compatible with the existing wallet signatures. We introduce three different scenarios to access and sign wallet described as follows:

The First Scenario: The wallet can be accessed using the financial top management key x_{fj} and $s_o = s_{fj}$, where s_o is the organizational wallet signature for wallet transaction. The wallet recovery can be gained only by backup server. This scenario will be danger as the wallet access will depend on one key and it represents a single point of failure.

The Second Scenario: It can be compatible with multi-signing [15, 18] by using one of operation manager's key x_{pl} and financial managers signing keys x_{fl}; hence, the final wallet signature is the concatenated signature 2 out of n (where $s_o = s_{pl}||s_{fj}$). In each signing process, financial signature is a must while different operation manager signature depends on the organization's need. This verifies dual control of wallet, provides easy recover, and reduces the danger of malicious employees.

The Third Scenario: This scenario is compatible with threshold signature [8, 19] as after operation department signature round is completed by computing $s_{pl} = k_{pl}(h(m||s_{p1}||s_{p2}) + x_{pl}r_{pl})mod\ n$ and sending the signature request to the financial department for review and approval, the first team member of financial department signs the message $s_{f1} = k_{f1}^{-1}(h(m) + x_{f1}r_{f1})mod\ n$ and sends signature to the team leaders and managers for reviewing. Then, the remaining members of financial team (reviewers and managers) review operation department signature and the first financial member signature, after verification they will interact to sign the message using threshold signature scheme.

4 Security Analysis

In this section, we analyze the security of our proposed signature scheme predicated on ECDLP as follows:

- *Access Management:* In our proposed scheme, bitcoins' request depends on the organization hierarchy and rules with less interaction round. We made money request more flexible with different scenarios. Each team knows organization process and has a database with all keys required for validation and signature. Each signer will prove lower team member signature in hierarchal way. In our proposed scheme, wallet access has more than one scenario based on organization needs and policy.
- *Unforgeability:* Only true signers who hold private key can create a valid signature. No one except operation team member who hold x_{p1} can create the signature s_{p1}, no one except the leader who holds x_{p2} can create the signature s_{p2}, and no one except the top manager can create the signature $s_{pl} = k_{pl}(h(m||s_{p1}||s_{p2}) + x_{pl}r_{pl})mod\ n$. Hence, no one can forge another member signature if he hasn't the required private keys.
- *Undeniability:* Neither the organization nor the employee can deny the wallet signature where operation and financial members' private keys x_{pz} and x_{ft} generated by the certificate authority; we ensure each member and department signature since the last signature $s_{pl} = k_{pl}(h(m||s_{p1}||s_{p2}) + x_{pl}r_{pl})mod\ n$ encapsulates all lower priority signatures. The operation manager can't deny his signing s_{pl} as he has used his private key x_{pl}, and also the team leader and manager can't deny their signature $s_{p1}||s_{p2}$ as it encapsulated in signing process using their private keys x_{p1} and x_{p2}.
- *Transferability:* As the employee' private key (x_{pz}, x_{ft}) is a part of the operation signature $s_{pz} = k_{pz}^{-1}(h(m) + x_{pz}r_{pz})mod\ n$ and financial signature $s_{ft} = k_{ft}^{-1}(h(m) + x_{ft}r_{ft})mod\ n$; hence, if the person leaves his position, the new signature will depend on the new person private and firing employees' keys must be kept in a revocation list.

- **Detection:** Organization needs to keep assets records for all money transaction. Every transaction marked by employees' identity or private key. A revocation list [20] must be updated by fraudulent or terminated employees' keys. Furthermore, organization must have a database containing all true public keys used for verification. In our proposed scheme, bookkeeping is real; for example, if terminated operation manager tries to make a signature, he first needs to have available s_{p1}. After he signs $s_{p2} = k_{p2}^{-1}\left(h(m\|s_{p1}) + x_{p2}r_{p2}\right) mod\ n$, the operation manager has to review the revocation list before signing.
- **Anonymity:** The organization transactions must be kept anonymous in the network. To ensure anonymity, organization can use mixing methods in [21, 22] as the first and third wallet access scenarios (in Subsect. 3.3) are compatible with the existing mixers.
- **Verifiability:** Any verifier who obtains the related public parameters can check whether the signature is valid. As we have shown above, we have provided a smooth hierarchal verifiability. If $s_o = s_{fj}$ in first scenario, the verifier only needs to know $\left(r_{fj}, s_{fj}, m\|s_{f1}\|s_{f2}, y_{fj}, G, n\right)$ to compute $w_{fj} = s_{fj}^{-1}mod\ n$ and $h(m\|s_{f1}\|s_{f2})$ then computes $u_{fj} = h(m\|s_{f1}\|s_{f2})w_{fj}mod\ n$ and $q_{fj} = r_{fj}w_{fj}mod\ n$. finally, computes $u_{fj}G + q_{fj}y_{fj} = (x,y)$ and $v_{fj} = xmod\ n$. The signature is accepted if $v_{fj} = r_{fj}$.
- **Wallet Recovery:** Ensure recovery of wallet in case of private key loss gained in second scenario where wallet signature $s_o = s_{pl}\|s_{fj}$ is 2 out of n signature. In the third scenario where wallet signature depends on t out of n signers to recover the key. However, in the first scenario in which $s_o = s_{fj}$, the financial manager must be trusted.
- **Functional Separation of Duties:** We can separate the employees' rules by encapsulating employees' signature, ensuring managers signing, and separating branches duty. We enforce operation branch employees' to sign from lower priority member to higher priority manager signature $s_{pl} = k_{pl}(h(m\|s_{p1}\|s_{p2}) + x_{pl}r_{pl})\ mod\ n$ and financial department from lower priority member to higher priority manager signature as $s_{fj} = k_{fj}^{-1}(h(m\|s_{f1}\|s_{f2}) + x_{fj}r_{fj})\ mod\ n$ We provide a smooth way for organization to manage coins requests based on employee's priority and weight with the least possible interaction, effort, and more review steps for verification and without affecting transaction size.

5 Conclusion

This paper has introduced a signature scheme for securing organization's bitcoin wallet access. In the proposed scheme, organization can create its own rules and manage wallet access without affecting transaction size. In addition, the proposed scheme can keep the structure of organization and manage employee hiring and termination. Moreover, security analysis of the proposed scheme has been provided. The proposed scheme is compatible with existing bitcoin wallet signature and can be used with different scenarios based on organization needs. We provided a smooth way for organization to manage bitcoins' request based on employee's priority with the least possible interaction, effort, and more review steps for verification and error detection without affecting transaction size.

References

1. Satoshi, N.: Bitcoin: A Peer-to-Peer Electronic Cash System (2008). https://bitcoin.org/bitcoin. Accessed 10 Jan 2019
2. Kravitz, W.D.: Digital signature algorithm. US Patent 5, 231, 668, 27 Jul 1993
3. Ghassan, O.K., Elli, A., Marc, R., Arthur, G.: Misbehavior in bitcoin: a study of double-spending and accountability. ACM Trans. Inf. Syst. Secur. **18**(1), 1–32 (2015)
4. Iresha, D.R., Kasun, D:. Transaction verification model over double spending for peer-to-peer digital currency transactions based on blockchain architecture. Int. J. Comput. Appl. **163**(5), 24–31 (2017)
5. Transactions. https://en.bitcoin.it/wiki/Transactions. Accessed 10 Jan 2019
6. Lakshmi, S., Annapurna, N.S., Latha, T.S.: Security analysis of three factor authentication schemes for banking. J. Eng. Appl. Sci. **10**(8), 3271–3274 (2015). ISSN 1819-6608
7. Huang, X., Xiang, Y., Chonka, A., Zhou, J., Deng, R.: A generic framework for three-factor authentication: preserving security and privacy in distributed systems. IEEE Trans. Parallel Distrib. Syst. **22**(8), 1390–1397 (2010)
8. Goldfeder, S., Gennaro, R., Kalodner, H., Bonneau, J., Kroll, J.A., Felten, W.E., Narayanan, A.: Securing Bitcoin Wallets via a New DSA/ECDSA Threshold Signature Scheme (2014)
9. Shamir, A.: How to share a secret. Commun. ACM **22**(11), 612–613 (1979)
10. Parajeeta, D., Singh, K.: Weighted threshold ECDSA for securing bitcoin wallet. ACCENTS Trans. Inf. Secur. (TIS) **2**(6), 43–51 (2017). https://doi.org/10.19101/tis.2017.26003
11. Ohta, K., Okamoto, T.: Multi signature scheme secure against active insider attacks. IEICE Trans. Fundam. Electron. Commun. Comput. Sci. **82**(1), 21–31 (1999)
12. Christian, G.A.: Bitcoin Improvement Proposal 11: M-of-N Standard Transactions (2011) https://github.com/bitcoin/bips/blob/master/bip-0011
13. Maxwell, G., Poelstra, A., Seurin, Y., Wuille, P.: Simple Schnorr Multi-signatures with Applications to Bitcoin. Cryptology ePrint Archive, Report 2018/068 (2018)
14. Ohta, K., Okamoto, T.: A Digital Multi Signature Scheme Based on the Fiat-Shamir Scheme. Lecture Notes in Computer Science, vol. 739, pp. 139–148. Springer, Heidelberg, 11–14 November 1993
15. Bitgo Solution Brief: Securing Digital Currency with BitGo Multi-Signature and SafeNet HSM (2014)
16. Maged, H.I., Ali, A.I., Ibrahim, I.I., El-sawi, A.H.: A robust threshold elliptic curve digital signature providing a new verifiable secret sharing scheme, pp. 276–280. IEEE (2003)
17. Liao, H.-Z., Shen, Y.: On the elliptic curve digital signature algorithm. Tunghai Sci. **8**, 109–126 (2006)
18. Pamela, G., Law, E.: Bitcoin's Multi-signature for Corporate. International License, empowered law, 27 June 2014
19. Gennaro, R., Goldfeder, S., Narayanan, A.: Threshold optimal DSA/ECDSA signatures and an application to Bitcoin wallet security. In: International Conference on Applied Cryptography and Network Security, pp. 156–174. Springer, Heidelberg (2016)
20. Solo, D., Housley, R., Ford, W.: Internet X.509 public key infrastructure certificate and certificate revocation list (CRL) profile, April 2002
21. Franc, M., Florian, M.: Anonymous CoinJoin transactions with arbitrary values. In: Conference: IEEE Trustcom/BigDataSE/ICESS (2017)
22. Annette, L.V., Anderson, B.: Blinded, accountable mixes for bitcoin. In: International Conference on Financial Cryptography and Data Security, pp. 112–126 (2015)

Robotic and Control Systems

Chaotic Control in Fractional-Order Discrete-Time Systems

Adel Ouannas[1], Giuseppe Grassi[2], Ahmad Taher Azar[3,4(✉)],
Amina Aicha Khennaouia[5], and Viet-Thanh Pham[6]

[1] LAMIS Laboratory, Department of Mathematics, University of Larbi Tebessi,
12002 Tebessa, Algeria
ouannas.a@yahoo.com, a.ounnes@mail.univ-tebessa.dz
[2] Dipartimento Ingegneria Innovazione, Università del Salento, 73100 Lecce, Italy
giuseppe.grassi@unisalento.it
[3] College of Engineering, Prince Sultan University, Riyadh,
Kingdom of Saudi Arabia
aazar@psu.edu.sa
[4] Faculty of computers and Artificial Intelligence, Benha University, Benha, Egypt
ahmad.azar@fci.bu.edu.eg
[5] Departement of Mathematics and Computer Sciences,
University of Larbi Ben M'hidi, Oum El Bouaghi, Algeria
kaminaaicha@yahoo.fr
[6] Faculty of Electrical and Electronics Engineering, Ton Duc Thang University,
Ho Chi Minh City, Vietnam
phamvietthanh@tdtu.edu.vn

Abstract. In recent years, fractional discrete-time calculus has become somewhat of a hot topic. A few researchers have attempted to develop a framework for the subject and investigate the stability and application of fractional discrete-time chaotic system. In this study, a general method to control fractional discrete-time chaotic systems is proposed. Based on Lyapunov stability theory of fractional-order discrete-time systems, a robust scheme of control is introduced. Numerical results are presented to confirm the findings of the study.

Keywords: Chaos · Discrete fractional systems ·
Lyapunov stability analysis · Chaotic control

1 Introduction

Chaotic systems are dynamical systems that possess some special features, such as being extremely sensitive to small variations of initial conditions, having bounded trajectories in the phase space, and so on. The chaos phenomenon was first observed in weather models by Lorenz in 1963 [17]. This was followed by a discovery of a large number of chaos phenomena and chaos behaviour in physical, medical, social, hydrology, economical, biological, communication, and electrical

© Springer Nature Switzerland AG 2020
A. E. Hassanien et al. (Eds.): AISI 2019, AISC 1058, pp. 207–217, 2020.
https://doi.org/10.1007/978-3-030-31129-2_20

systems [3,4,6,8,9,16,31,33,34]. Discrete chaos have been the subject of study for many researchers in a variety of fields within science and engineering [7]. In recent years, different control schemes and applications of discrete-time chaotic and hyperchaotic systems have been widely investigated [21,22,24,26,27,32,37].

Recently, researchers have diverted their attention to the discrete-time fractional calculus and attempted to put together a complete theoretical framework for the subject [10]. Fractional-order discrete-time systems have a major advantage over their conventional counterparts due to the infinite memory they feature, which allows for more flexibility in modeling and leads to a higher degree of chaotic behavior [1]. More recently, many fractional discrete-time chaotic systems have been proposed [18,36]. These maps quickly found applications in encryption and secure communications once their synchronization became possible [12,15,25,28–30]. The problem of controlling a chaotic system was introduced by Ott et al. in 1990 [19]. The control of chaotic systems is basically to design state feedback control laws that stabilizes the chaotic systems around the unstable equilibrium points to force the system states to zero in sufficient time [13,14,20,23,35]. To the best of authors' knowledge, the study of control for the fractional discrete-time chaotic systems remains a new and mostly unexplored area. This has motivated the authors of this work to examine the phenomenon and develop suitable control laws for stabilization of this type of systems.

In this study, the objective is to propose a general scheme to control chaos in arbitrary fractional-order discrete-time system. Using Lyapunov stability approach, the stability of the zero solution and consequently the convergence of the states is established.

The rest of this paper is organized as follows. Some of the necessary notations and stability theory are introduced in Sect. 2. Section 3 proposes a new control law for the stabilization of general fractional discrete-time chaotic system. Section 4 introduces the control of 2D and 3D fractional-order discrete-time systems, respectively. Finally, Sect. 5 summarizes the main findings of the study.

2   Preliminaries

Throughout this work, we will denote by $^{C}\Delta_a^\nu X(t)$, the ν–Caputo type delta difference of a function $X(t) : \mathbb{N}_a \to \mathbb{R}$ with $\mathbb{N}_a = \{a, a+1, a+2, ...\}$ [1], which is of the form:

$$^{C}\Delta_a^\nu X(t) = \Delta_a^{-(n-\nu)} \Delta^n X(t) = \frac{1}{\Gamma(n-\nu)} \sum_{s=a}^{t-(n-\nu)} (t-s-1)^{(n-\nu-1)} \Delta_s^n X(s),$$

(1)

for $\nu \notin \mathbb{N}$ is the fractional order, $t \in \mathbb{N}_{a+n-\nu}$, and $n = \lceil \nu \rceil + 1$.

The ν–th fractional sum of $\Delta_s^n X(t)$ is defined as [2]:

$$\Delta_a^{-\nu} X(t) = \frac{1}{\Gamma(\nu)} \sum_{s=a}^{t-\nu} (t-s-1)^{(\nu-1)} X(s),$$

(2)

with $t \in \mathbb{N}_{a+\nu}$, $\nu > 0$. The term $t^{(\nu)}$ denotes the falling function defined in terms of the Gamma function Γ as:

$$t^{(\nu)} = \frac{\Gamma(t+1)}{\Gamma(t-\nu+1)}. \tag{3}$$

3 Main Results

By considering the following general fractional discrete-time system

$$^{C}\Delta_{a}^{\nu}X(t) = F\left(X(t-1+\nu)\right), \tag{4}$$

where $t \in \mathbb{N}_{a+1-\nu}$, $X(t) \in \mathbb{R}^{n}$, $0 < \nu \le 1$, $F : \mathbb{R}^{n} \to \mathbb{R}^{n}$.

Assume that F can be divided into two parts:

$$F\left(X(t)\right) = AX(t-1+\nu) + f\left(X(t-1+\nu)\right), \tag{5}$$

where A is a constant matrix, $n \times n$, and $f : \mathbb{R}^{n} \to \mathbb{R}^{n}$ is a nonlinear function. Now, the fractional discrete-time system (4) can be written in the following form:

$$^{C}\Delta_{a}^{\nu}X(t) = AX(t-1+\nu) + f\left(X(t-1+\nu)\right), \tag{6}$$

The aim of controlling the fractional discrete-time system (4) is to derive an adaptive control law such that all of the system's states are stabilized to 0 asymptotically. The controlled fractional discrete-time system (6) is given by:

$$^{C}\Delta_{a}^{\nu}X(t) = AX(t-1+\nu) + f\left(X(t-1+\nu)\right) + U, \tag{7}$$

with $U(t)$ being the adaptive control weight. In order to establish the convergence of the system states to zero, Lyapunov stability theory is applied to fractional discrete-time systems, which can be summarized in the following theorem and lemma.

Theorem 1. *[5] If there exists a positive definite Lyapunov function V such that:*

$$^{C}\Delta_{a}^{\nu}V\left(X\left(t\right)\right) < 0 \text{ for all } t \in \mathbb{N}_{a+1-\nu}, \tag{8}$$

then the trivial solution of the system (7) is asymptotically stable.

Lemma 1. *[5] $\forall t \in \mathbb{N}_{a+1-\nu}$:*

$$^{C}\Delta_{a}^{\nu}X^{T}(t)X(t) \le 2X^{T}(t+\nu-1)^{C}\Delta_{a}^{\nu}X(t). \tag{9}$$

Now, the following theorem presents the result.

Theorem 2. *There exists a feedback gain matrix $C \in \mathbb{R}^{n \times n}$ such that the fractional discrete-time system (7) can be controlled under the following control law:*

$$U(t) = -CX(t) - f\left(X(t)\right). \tag{10}$$

Proof. Substituting the proposed control law (10) into (7) yields the simplified dynamics

$$^C\Delta_a^\nu X(t) = (A - C)\,X(t - 1 + \nu). \tag{11}$$

Lyapunov stability theory as stated in Theorem 1 requires the construction of a positive definite function $V\,(X\,(t))$ that has a strictly negative fractional difference. By considering the function:

$$V\,(X\,(t)) = \frac{1}{2}X(t)X^T(t), \tag{12}$$

which is clearly positive definite. Using Lemma 1, leads to:

$$\begin{aligned}
^C\Delta_a^\nu V\,(X\,(t)) &= \ ^C\Delta_a^\nu\left(\tfrac{1}{2}X^T\,(t)\,X\,(t)\right) \\
&\leq \ X^T(t + \nu - 1)^C\Delta_a^\nu X\,(t) \\
&= \ X^T(t + \nu - 1)\,(B - C)\,X(t).
\end{aligned}$$

If the feedback gain matrix C is selected such that $B - C$ is a negative definite, we obtain:

$$^C\Delta_a^\nu V\,(X\,(t)) < 0. \tag{13}$$

Then, it's noted directly from Theorem 1 that the zero solution of the system (11) is globally asymptotically stable and therefore, the system is controlled.

4 Applications

In this section, two control schemes are presented for a 2D and 3D fractional-order discrete-time systems, respectively.

4.1 Control of the 2D Fractional Discrete Double Scroll

Consider the following 2D fractional-order discrete-time system:

$$\begin{cases}
^C\Delta_a^\nu x_1(t) = -\alpha h(x_2(t - 1 + \nu)), \\
^C\Delta_a^\nu x_2(t) = \beta x_1(t - 1 + \nu) - x_2(t - 1 + \nu),
\end{cases} \tag{14}$$

where α and β are bifurcation parameters and h is the characteristic function defined as:

$$h(x_2) = \frac{1}{2}\left(2m_1 x_2 + (m_0 - m_1)(|x_2 + 1| - |x_2 - 1|)\right), \quad t \in N_{a+1-\nu}. \tag{15}$$

This system is called the fractional discrete double scroll. The fractional discrete double scroll (14) exhibits chaos, for instance, when $\alpha = 3.36$, $\beta = 1.04$, $m_0 = -0.43$, $m_1 = 0.41$ and $\nu = 0.98$, as demonstrated by the phase portraits shown in Fig. 1. It is always helpful to examine the bifurcation diagram and Lyapunov exponents in order to gain a comprehensive understanding of the dynamics of a chaotic map, see Fig. 2.

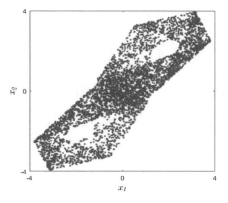

Fig. 1. Phase space plot for fractional discrete double scroll (14).

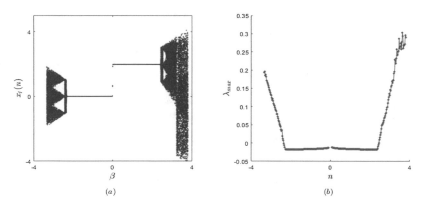

(a)   (b)

Fig. 2. (a) Bifurcation diagram of the fractional map (14); (b) largest Lyapunov exponent

Now, it is easy to see that the linear part of the fractional discrete-time chaotic system (14) is given by:

$$A = \begin{pmatrix} 0 & 0 \\ \beta & -1 \end{pmatrix}, \tag{16}$$

and the nonlinear part is:

$$f = \begin{pmatrix} -\alpha h(x_2) \\ 0 \end{pmatrix}. \tag{17}$$

According to Theorem 2, there exists a control matrix C, which can be chosen as:

$$C = \begin{pmatrix} 1 & 0 \\ 2\beta & 0 \end{pmatrix}. \tag{18}$$

Simply, it's noted that all $A - C$ is a negative definite matrix. Because the matrices C and function f are known, it is rather easy to construct the control law according to (10). The resulting controller is in the form:

$$U(t) = \begin{pmatrix} -x_1(t) + \alpha h(x_2(t)) \\ -2\beta x_1(t) \end{pmatrix}. \tag{19}$$

Numerical simulations are carried out to illustrate the results of Theorem 2. Initial conditions are selected as $x(0) = y(0) = 0.1$. Assuming that $a = 0$, time evolution of states and phase space of the controlled fractional discrete double scroll are depicted in Figs. 3 and 4, respectively.

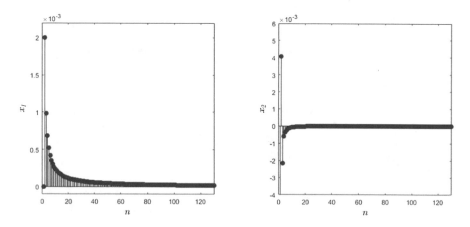

Fig. 3. The controlled states of the discrete double scroll (14).

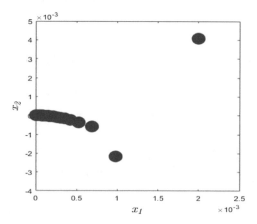

Fig. 4. Attractor of the fractional discrete double scroll (14) after control.

4.2   Control of 3D Fractional-Order Generalized Hénon map

Recently, a new 3D fractional-order discrete-time system has been introduced
[11]. The so called fractional generalized Hénon map is of the form:

$$\begin{cases} {}^{C}\Delta_a^{\nu}x_1(t) = \mathbf{a} - x_2^2(t-1+\nu) + \mathbf{b}x_3(t-1+\nu) - x_1(t-1+\nu), \\ {}^{C}\Delta_a^{\nu}x_2(t) = x_1(t-1+\nu) - x_2(t-1+\nu), \\ {}^{C}\Delta_a^{\nu}x_3(t) = x_2(t-1+\nu) - x_3(t-1+\nu), \end{cases} \tag{20}$$

where $t \in N_{a+1-\nu}$ and $(\mathbf{a}, \mathbf{b}) = (0.7281, 0.5)$. The authors showed that this
system has a chaotic attractor, for instance, when $(x(0), y(0), z(0)) = (1, 0, 0)$,
and fractional order $\nu = 0.98$, the resulting attractors are depicted in Fig. 5.
The bifurcation diagram and Lyapunov exponents of the fractional generalized
Hénon map are shown in Fig. 6.

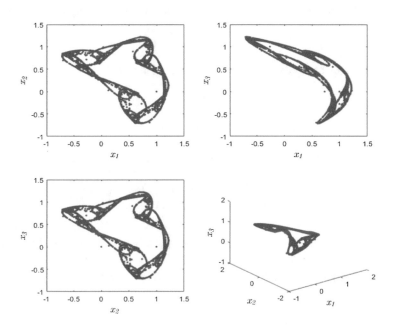

Fig. 5. Phase space plot for fractional generalized Hénon map (20).

The linear and nonlinear parts of (20) are given by:

$$A = \begin{pmatrix} -1 & 0 & \mathbf{b} \\ 1 & -1 & 0 \\ 0 & 1 & -1 \end{pmatrix}, \text{ and } f = \begin{pmatrix} \mathbf{a} - y^2 \\ 0 \\ 0 \end{pmatrix}.$$

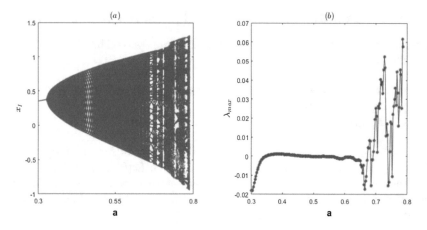

Fig. 6. (a) Bifurcation diagram of the fractional map (20); (b) largest Lyapunov exponent

In this case, the control matrix C is selected as:

$$C = \begin{pmatrix} 0 & 0 & \mathbf{b} \\ 1 & 0 & 0 \\ 0 & 1 & 0 \end{pmatrix} \tag{21}$$

which clearly satisfies condition of Theorem 2. Therefore, the control law can be designed as follow:

$$U(t) = \begin{pmatrix} -\mathbf{b}x_3(t) + x_2^2(t) - \mathbf{a} \\ -x_1(t) \\ -x_3(t) \end{pmatrix} \tag{22}$$

Figures 7 and 8 depict the time evolution of states as well as the phase portraits of the controlled fractional generalized Hénon map. Clearly, the proposed control laws successfully stabilize the states as they converge towards zero asymptotically.

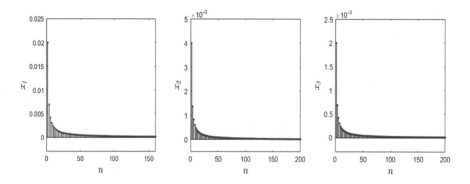

Fig. 7. The controlled states of the fractional generalized Hénon map (20).

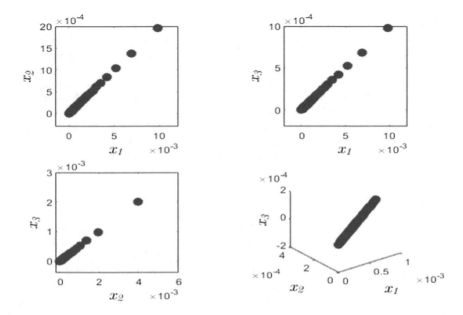

Fig. 8. Attractor of the fractional generalized Hénon map (20) after control.

5 Conclusion

In this study, it has been shown that general control scheme can stabilize arbitrary fractional-order discrete-time chaotic system. Nonlinear control schemes is presented whereby the 2D fractional discrete double scroll and the 3D fractional generalized Hénon map can be stabilized, respectively. The stability of the zero solution and consequently the convergence of the states is established by means of the Lyapunov stability theory. Throughout the paper, numerical solutions are presented to confirm the findings and verify the feasibility of the proposed laws.

References

1. Abdeljawad, T., Baleanu, D., Jarad, F., Agarwal, R.P.: Fractional sums and differences with binomial coefficients. Discrete Dyn. Nat. Soc. **2013**, 1–6 (2013). Article ID 104173
2. Anastassiou, G.A.: Principles of delta fractional calculus on time scales and inequalities. Math. Comput. Model. **52**(3), 556–566 (2010)
3. Azar, A.T., Vaidyanathan, S.: Advances in Chaos Theory and Intelligent Control, vol. 337. Springer, Heidelberg (2016)
4. Azar, A.T., Vaidyanathan, S., Ouannas, A.: Fractional Order Control and Synchronization of Chaotic Systems. Studies in Computational Intelligence, vol. 688. Springer, Heidelberg (2017)
5. Baleanu, D., Wu, G., Bai, Y., Chen, F.: Stability analysis of caputo–like discrete fractional systems. Commun. Nonlinear Sci. Numer. Simul. **48**, 520–530 (2017)

6. Bozoki, Z.: Chaos theory and power spectrum analysis in computerized cardiotocography. Eur. J. Obstet. Gynecol. Reprod. Biol. **71**(2), 163–168 (1997)
7. Elaydi, S.N.: Discrete Chaos: With Applications in Science and Engineering. Chapman and Hall/CRC, Boca Raton (2007)
8. Frey, D.R.: Chaotic digital encoding: an approach to secure communication. IEEE Trans. Circuits Syst. II Analog Digital Signal Process. **40**(10), 660–666 (1993)
9. Garfinkel, A.: Controlling cardiac chaos. Science (1992)
10. Goodrich, C., Peterson, A.C.: Discrete Fractional Calculus. Springer, Cham (2015)
11. Jouini, L., Ouannas, A., Khennaoui, A.A., Wang, X., Grassi, G., Pham, V.T.: The fractional form of a new three-dimensional generalized hénon map. Adv. Differ. Equ. **1**, 122 (2019)
12. Kassim, S., Hamiche, H., Djennoune, S., Bettayeb, M.: A novel secure image transmission scheme based on synchronization of fractional-order discrete-time hyperchaotic systems. Nonlinear Dyn. **88**(4), 2473–2489 (2017)
13. Khan, A., Singh, S., Azar, A.T.: Combination-combination anti-synchronization of four fractional order identical hyperchaotic systems. In: Hassanien, A.E., Azar, A.T., Gaber, T., Bhatnagar, R., Tolba, F.M. (eds.) The International Conference on Advanced Machine Learning Technologies and Applications (AMLTA 2019), pp. 406–414. Springer, Cham (2020)
14. Khan, A., Singh, S., Azar, A.T.: Synchronization between a novel integer-order hyperchaotic system and a fractional-order hyperchaotic system using tracking control. In: Hassanien, A.E., Azar, A.T., Gaber, T., Bhatnagar, R., Tolba, F.M. (eds.) The International Conference on Advanced Machine Learning Technologies and Applications (AMLTA 2019), pp. 382–391. Springer, Cham (2020)
15. Khennaoui, A.A., Ouannas, A., Bendoukha, S., Grassi, G., Wang, X., Pham, V.T.: Generalized and inverse generalized synchronization of fractional-order discrete-time chaotic systems with non-identical dimensions. Adv. Differ. Equ. **1**, 303 (2018)
16. Lau, F., Tse, C.K.: Chaos-Based Digital Communication Systems. Springer, Heidelberg (2003)
17. Lorenz, E.N.: Deterministic nonperiodic flow. J. Atmos. Sci. **20**(2), 130–141 (1963)
18. Megherbi, O., Hamiche, H., Djennoune, S., Bettayeb, M.: A new contribution for the impulsive synchronization of fractional-order discrete-time chaotic systems. Nonlinear Dyn. **90**(3), 1519–1533 (2017)
19. Ott, E., Grebogi, C., Yorke, J.A.: Controlling chaos. Phys. Rev. Lett. **64**, 1196–1199 (1990)
20. Ouannas, A., Azar, A.T., Vaidyanathan, S.: New hybrid synchronization schemes based on coexistence of various types of synchronization between master-slave hyperchaotic systems. Int. J. Comput. Appl. Technol. **55**(2), 112–120 (2017)
21. Ouannas, A., Grassi, G., Azar, A.T., Radwan, A.G., Volos, C., Pham, V.T., Ziar, T., Kyprianidis, I.M., Stouboulos, I.N.: Dead-beat synchronization control in discrete-time chaotic systems. In: 6th International Conference on Modern Circuits and Systems Technologies (MOCAST), pp. 1–4 (2017)
22. Ouannas, A., Odibat, Z., Shawagfeh, N., Alsaedi, A., Ahmad, B.: Universal chaos synchronization control laws for general quadratic discrete systems. Appl. Math. Model. **45**, 636–641 (2017)
23. Ouannas, A., Azar, A.T., Ziar, T.: Control of continuous-time chaotic (hyperchaotic) systems: F-M synchronisation. Int. J. Automa. Control (2018)
24. Ouannas, A., Grassi, G., Karouma, A., Ziar, T., Wang, X., Pham, V.: New type of chaos synchronization in discrete-time systems: the F-M synchronization. Open Phys. **16**, 174–182 (2018)

25. Ouannas, A., Khennaoui, A.A., Grassi, G., Bendoukha, S.: On the Q-S chaos synchronization of fractional-order discrete-time systems: general method and examples. Discrete Dyn. Nat. Soc. **2018**, 1–8 (2018). Article ID 2950357

26. Ouannas, A., Grassi, G., Azar, A.T., Gasri, A.: A new control scheme for hybrid chaos synchronization. In: Hassanien, A.E., Tolba, M.F., Shaalan, K., Azar, A.T. (eds.) Proceedings of the International Conference on Advanced Intelligent Systems and Informatics 2018, pp. 108–116. Springer, Cham (2019)

27. Ouannas, A., Grassi, G., Azar, A.T., Singh, S.: New control schemes for fractional chaos synchronization. In: Hassanien, A.E., Tolba, M.F., Shaalan, K., Azar, A.T. (eds.) Proceedings of the International Conference on Advanced Intelligent Systems and Informatics 2018, pp. 52–63. Springer, Cham (2019)

28. Ouannas, A., Khennaoui, A.A., Zehrour, O., Bendoukha, S., Grassi, G., Pham, V.T.: Synchronisation of integer-order and fractional-order discrete-time chaotic systems. Pramana **92**(4), 52 (2019)

29. Ouannas, A., Grassi, G., Azar, A.T.: Fractional-order control scheme for Q-S chaos synchronization. In: Hassanien, A.E., Azar, A.T., Gaber, T., Bhatnagar, R., Tolba, F.M. (eds.) The International Conference on Advanced Machine Learning Technologies and Applications (AMLTA 2019), pp. 434–441. Springer, Cham (2020)

30. Ouannas, A., Grassi, G., Azar, A.T.: A new generalized synchronization scheme to control fractional chaotic systems with non-identical dimensions and different orders. In: Hassanien, A.E., Azar, A.T., Gaber, T., Bhatnagar, R., Tolba, F.M. (eds.) The International Conference on Advanced Machine Learning Technologies and Applications (AMLTA 2019), pp. 415–424. Springer, Cham (2020)

31. Schiff, S.J., Jerger, K., Duong, D.H., Chang, T., Spano, M.L., Ditto, W.L., et al.: Controlling chaos in the brain. Nature **370**(6491), 615–620 (1994)

32. Shukla, M.K., Sharma, B.B.: Stabilization of Fractional Order Discrete Chaotic Systems, pp. 431–445. Springer, Cham (2017)

33. Sivakumar, B.: Chaos theory in hydrology: important issues and interpretations. J. Hydrol. **227**(14), 1–20 (2000)

34. Strogatz, S.H.: Nonlinear Dynamics and Chaos: With Applications to Physics, Biology, Chemistry, and Engineering. Westview Press, Boulder (2001)

35. Vaidyanathan, S., Jafari, S., Pham, V.T., Azar, A.T., Alsaadi, F.E.: A 4-D chaotic hyperjerk system with a hidden attractor, adaptive backstepping control and circuit design. Arch. Control Sci. **28**(2), 239–254 (2018)

36. Wu, G.C., Baleanu, D.: Discrete chaos in fractional delayed logistic maps. Nonlinear Dyn. **80**(4), 1697–1703 (2015)

37. Xiao-hui, Z., Ke, S.: The control action of the periodic perturbation on a hyperchaotic system. Acta Physica Sinica (Overseas Ed.) **8**(9), 651 (1999)

Synchronization of Fractional-Order Discrete-Time Chaotic Systems

Adel Ouannas[1]([⊠]), Giuseppe Grassi[2], Ahmad Taher Azar[3,4],
Amina–Aicha Khennaouia[5], and Viet-Thanh Pham[6]

[1] LAMIS Laboratory, Department of Mathematics, University of Larbi Tebessi,
12002 Tebessa, Algeria
ouannas.a@yahoo.com,a.ounnes@mail.univ-tebessa.dz
[2] Dipartimento Ingegneria Innovazione, Università del Salento, 73100 Lecce, Italy
giuseppe.grassi@unisalento.it
[3] College of Engineering, Prince Sultan University,
Riyadh, Kingdom of Saudi Arabia
aazar@psu.edu.sa
[4] Faculty of Computers and Artificial Intelligence, Benha University, Benha, Egypt
ahmad.azar@fci.bu.edu.eg
[5] Departement of Mathematics and Computer Sciences,
University of Larbi Ben M'hidi, Oum El Bouaghi, Algeria
kaminaaicha@yahoo.fr
[6] Faculty of Electrical and Electronics Engineering, Ton Duc Thang University,
Ho Chi Minh City, Vietnam
phamvietthanh@tdtu.edu.vn

Abstract. Recently, synchronization in discrete-time chaotic systems attract more and more attentions and has been extensively studied, due to its potential applications in secure communication. This work is concerned with the synchronization of fractional-order discrete-time chaotic systems with different dimensions. In particular, through appropriate nonlinear control, matrix projective synchronization (MPS) can be achieved between different dimensional fractional-order map. Numerical examples and computer simulations are used to show the effectiveness and the feasibility of the proposed synchronization schemes.

Keywords: Chaos · Discrete-time fractional systems ·
Matrix projective synchronization · Different dimensions ·
Lyapunov stability analysis · Chaotic synchronization

1 Introduction

Chaos synchronization is an important issue of chaotic systems, presented by Pecora and Carroll in 1990 [38]. Many types of synchronization have been extensively studied for chaotic systems, such as complete synchronization [38,41], anti-synchronization [13,18], phase synchronization [21,22], projective

© Springer Nature Switzerland AG 2020
A. E. Hassanien et al. (Eds.): AISI 2019, AISC 1058, pp. 218–228, 2020.
https://doi.org/10.1007/978-3-030-31129-2_21

synchronization [4,32], generalized synchronization [31,37], hybrid synchronization [12,28,30,40], compound synchronization [17,39], and so on.

Discrete-time chaotic dynamical systems play a more important role than their continuous parts [42]. In fact, many mathematical models of physical process, biological process, and chemical process were defined using chaotic dynamical systems in discrete time [10,14]. Therefore, it is important to consider chaos (hyperchaos) synchronization in discrete time dynamical systems. Given two systems in the drive-response configuration, the objective in chaos synchronization is to make the response system variables synchronized in time with the corresponding drive system variables [5,29,34,35,43–45]. Recently, more and more attention has been paid to the synchronization of chaos (hyperchaos) in discrete time dynamical systems [26,27].

For a long time, the study and application of fractional calculus were limited to continuous time. Recently, researchers have diverted their attention to the discrete fractional calculus and attempted to put together a complete theoretical framework for the subject [1,3,6,11]. Many fractional order discrete-time chaotic systems have been proposed recently in the literature [7,16,19,20,23,33,36,46]. From what has been reported, fractional chaotic maps are sensitive to variations in the fractional order in addition to their natural sensitivity to variations in the initial conditions and parameters and can exhibit much more dynamics compared to their integer counterparts [9]. This makes them more suitable to applications requiring a higher entropy such as the encryption of data and secure communications [16]. To date, very few of the conventional discrete-time systems listed above have been extended to the fractional difference case. This has motivated the authors to examine the phenomenon and develop suitable control laws for the synchronization of this kind of systems.

In this work, control laws have been proposed for generalized type of synchronization relating to fractional-order discrete-time systems, namely matrix projective synchronization (MPS). MPS is one of the most widely studied synchronization types [24,25]. It refers to the existence of a matrix relationship between the drive states and the response states. Instead of the conventional definition of synchronization, which stipulates that the difference between the drive and response trajectories tends to zero as $t \to \infty$, MPS forces the difference between the response states and a combination of the drive states to zero. The objective of this paper is to establish the convergence of the proposed MPS scheme for a pair of fractional-order discrete systems by means of asymptotic stability results reported recently in the literature.

The rest of this paper is organized as follows. The second section highlights some of the necessary notation and theory related to discrete-time fractional calculus and stability. Section 3 proposes the control law for the matrix projective synchronization of two fractional-order discrete-time systems with non-identical dimensions. Section 4 presents the numerical results related to a particular example. Finally, Sect. 5 provides a general summary of the main findings of this study.

2   Basic Concepts

Some necessary definitions from discrete fractional calculus and some results on the stability of fractional-order discrete-time systems from the literature which are relevant to this work are introduced.

Definition 1 *[1]. Let $\nu > 0$ and $\sigma(s) = s + 1$. The ν-th fractional sum of u is defined by:*

$$\Delta^{-\nu} u(t) = \frac{1}{\Gamma(\nu)} \sum_{s=a}^{t-\nu} (t - \sigma(s))^{(\nu-1)} u(t). \tag{1}$$

Note that u is defined for $s = a \bmod(1)$ and $\Delta^{-\nu} u(t)$ is defined for $t = a + \nu \bmod(1)$, in particular, $\Delta^{-\nu}$ maps functions defined as: $\mathbb{N}_a$ to functions defined as: $\mathbb{N}_{a+\nu}$, where, $\mathbb{N}_t = \{t, t+1; t+2; t+3, ...\}$. In addition $t^{(\nu)}$ is the falling function defined as:

$$t^{(\nu)} = \frac{\Gamma(t+1)}{\Gamma(t+1-\nu)} \tag{2}$$

Definition 2 *[2]. Let $\nu > 0$. The ν-th order Caputo left fractional of a function u defined as: $\mathbb{N}_a$, is defined by:*

$$^{C}\Delta_a^{\nu} u(t) = \Delta_a^{-(n-\nu)} \Delta^n u(t) =$$

$$\frac{1}{\Gamma(n-\nu)} \sum_{s=a}^{t-(n-\nu)} (t - \sigma(s))^{(n-\nu-1)} \Delta^n u(s), \quad t \in \mathbb{N}_{a+n-\nu} \tag{3}$$

$n = [\nu] + 1$, where v is the difference order.

Theorem 1 *[2]. For the fractional difference equation,*

$$\begin{cases} ^{C}\Delta_a^{\nu} u(t) = f(t+\nu-1, u(t+\nu-1)), \\ \Delta^k u(a) = u_k, \ n = [\nu] + 1, \ k = 0, 1, ..., n-1, \end{cases} \tag{4}$$

the equivalent discrete integral equation can be obtained as:

$$u(t) = u_0(t) + \frac{1}{\Gamma(v)} \sum_{s=a+n-\nu}^{t-\nu} (t - \sigma(s))^{(\nu-1)} f(s+\nu-1, u(s+\nu-1)), \quad t \in \mathbb{N}_{a+n}, \tag{5}$$

where

$$u_0(t) = \sum_{k=0}^{n-1} \frac{(t-a)^{(k)}}{\Gamma(k+1)} \Delta^k u(a). \tag{6}$$

Theorem 2 *[8]. The zero equilibrium of the linear fractional-order discrete–time system $^{C}\Delta_a^{\nu} X(t) = \mathbf{M} X(t+\nu-1)$, where $X(t) = (x_1(t), ..., x_n(t))^T$, $0 < \nu \leq 1$, $\mathbf{M} \in R^{n \times n}$ and $\forall t \in \mathbb{N}_{a+1-\nu}$, is asymptotically stable if*

$$\lambda \in \left\{ z \in \mathbb{C}: \ |z| < \left(2\cos \frac{|\arg z| - \pi}{2 - \nu} \right)^{\nu} \ \text{and} \ |\arg z| > \frac{\nu \pi}{2} \right\}, \tag{7}$$

for all the eigenvalues λ of $\mathbf{M}$.

3   Theoretical Results

Consider the following drive system described by:

$$^{C}\Delta^{\nu}X(t) = f(X(t+\nu-1)), \quad t \in \mathbb{N}_{a+1-\nu}, \tag{8}$$

where $X(t) = (x_1(t), ..., x_n(t))^T$ is the state vector, $\mathbf{F} : \mathbb{R}^n \rightarrow \mathbb{R}^n$.

As the response system, consider the following system:

$$^{C}\Delta^{\nu}Y(t) = BY(t+\nu-1) + g(Y(t+\nu-1)) + U, \quad t \in \mathbb{N}_{a+1-\nu}, \tag{9}$$

where $Y(t) = (y_1(k), ..., y_m(k))^T$, $B \in \mathbb{R}^{m \times m}$, $g : \mathbb{R}^m \rightarrow \mathbb{R}^m$ is a nonlinear function and $U = (u_i)_{1 \leq i \leq m}$ is the vector controller to be determined.

Before stating the proposed control law, the proposed synchronization scheme can be defined as:

Definition 3. *The drive system (8) and the response system (9) are said to be matrix projective synchronized (MPS), if there exists a controller $U = (u_i)_{1 \leq i \leq m}$ and the scaling constant matrix $\mathbf{M} \in \mathbb{R}^{m \times n}$ such that the synchronization error*

$$e(t) = Y(t) - \mathbf{M} \times X(t), \tag{10}$$

satisfy the asymptotic rule

$$\lim_{t \longrightarrow +\infty} \|e(t)\| = 0. \tag{11}$$

Thus, the main results can be presented as:

Theorem 3. *Matrix projective synchronized (MPS) is achieved for the pair (8)–(9) subject to*

$$U = -CY(t) - (B-C)\mathbf{M} \times X(t) - g(Y(t)) + \mathbf{M} \times f(X(t)), \tag{12}$$

where $C \in \mathbb{R}^{m \times m}$ is a constant matrix chosen such that all the eigenvalues λ_i of $B-C$ satisfy

$$-2^{\nu} < \lambda_i < 0, \quad i = 1, ..., m. \tag{13}$$

Proof. In order to achieve MPS, the following fractional difference equation of the error vector (10) can be considered.

$$^{C}\Delta_a^{\nu}e(t) = BY(t+\nu-1) + g(Y(t+\nu-1)) + U - \mathbf{M} \times f(X(t)). \tag{14}$$

Substituting the control law given in Eq. (12) into (14), the error system dynamics becomes:

$$^{C}\Delta_a^{\nu}e(t) = (B-C)\,e(t-1+\nu). \tag{15}$$

Next, it will be shown that the zero equilibrium of (15) is globally asymptotically stable. It is easy to note that the eigenvalues of $B-C$, satisfy

$$|\arg \lambda_i| = \pi > \frac{\nu\pi}{2} \text{ and } |\lambda_i| < \left(2\cos\frac{|\arg \lambda_i| - \pi}{2 - \nu}\right)^{\nu}, \quad i = 1, 2, ..., m. \tag{16}$$

This fulfills the criterion of Theorem 3, which means that the zero equilibrium of (15) is globally asymptotically stable. Hence, the drive system (8) and the response system (9) are globally matrix projective synchronized.

4   Numerical and Simulation Example

4.1   Drive-Response Systems Description

The 2D fractional-order discrete-time Lorenz system is considered as a drive system and the 3D fractional generalized Hénon map as a response system. The drive system can be described as follow:

$$\begin{cases} {}^C\Delta_a^\nu x_1(t) = \alpha\beta x_1(t+\nu-1) - \beta x_1(t+\nu-1)x_2(t+\nu-1), \\ {}^C\Delta_a^\nu x_2(t) = \beta\left(x_1^2(t+\nu-1) - x_2(t+\nu-1)\right), \end{cases} \quad (17)$$

where $t \in \mathbb{N}_{a+1-\nu}$ and α, β are bifurcation parameters. The fractional-order discrete-time Lorenz system (17) was proposed in [20]. The authors showed that this map has a chaotic attractor, for instance, when $(\alpha, \beta) = (1.25, 0.75)$ and $\nu = 0.98$. The chaotic behavior of the drive map is depicted in Fig. 1.

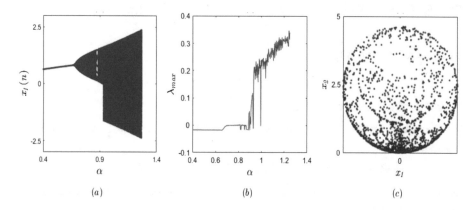

(a)   (b)   (c)

Fig. 1. Chaotic behavior of the fractional-order discrete-time Lorenz system (17).

The response system can be defined as follow:

$$\begin{cases} {}^C\Delta_a^\nu y_1(t) = \mathbf{a} - y_2^2(t-1+\nu) + \mathbf{b}y_3(t-1+\nu) - y_1(t-1+\nu) + u_1, \\ {}^C\Delta_a^\nu y_2(t) = y_1(t-1+\nu) - y_2(t-1+\nu) + u_2, \\ {}^C\Delta_a^\nu y_3(t) = y_2(t-1+\nu) - y_3(t-1+\nu) + u_3, \end{cases} \quad (18)$$

where $u_1(t), u_2(t), u_3(t)$ are controllers. This system exhibits a chaotic behavior subject to certain conditions [15]. The parameters can be chosen for example as $(\mathbf{a}, \mathbf{b}) = (0.7281, 0.5)$ and the fractional order $\nu = 0.98$, which has chaotic trajectories as shown in Fig. 2.
System (18) can be rewritten as:

$$ {}^C\Delta_a^\nu Y(t) = BY(t+\nu-1) + g(Y(t+\nu-1)) + U, \quad (19)$$

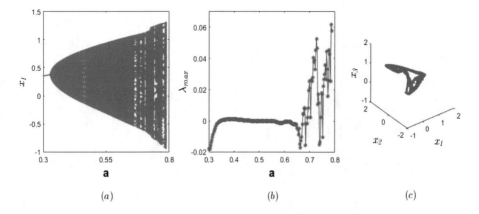

Fig. 2. Chaotic behavior of the fractional generalized Hénon map (18).

where

$$B = \begin{pmatrix} -1 & 0 & \mathbf{b} \\ 1 & -1 & 0 \\ 0 & 1 & -1 \end{pmatrix},$$ (20)

$$g = \begin{pmatrix} \mathbf{a} - y_2^2(t-1+\nu) \\ 0 \\ 0 \end{pmatrix},$$ (21)

and $U = (u_1, u_2, u_3)^T$.

4.2 Synchronization of Fractional Discrete-Time Lorenz System And Fractional Generalized Hénon Map

Based on the approach stated in Sect. 3, the error system between the drive system (17) and the response system (18) can be defined as:

$$(e_1(t), e_2(t), e_3(t))^T = (y_1(t), y_2(t), y_3(t))^T - \mathbf{M} \times (x_1(t), x_2(t))^T,$$ (22)

where the scaling constant matrix $\mathbf{M}$ is given by:

$$\mathbf{M} = \begin{pmatrix} 1 & 0 \\ 0 & -1 \\ 2 & 3 \end{pmatrix}.$$ (23)

According to the proposed approach described in Sect. 3, there exists a control matrix C such that $B - C$ satisfy the condition of Theorem 2. We may simply choose the following:

$$C = \begin{pmatrix} 0 & 0 & \mathbf{b} \\ 2 & 0 & 0 \\ 0 & 2 & 0 \end{pmatrix}.$$ (24)

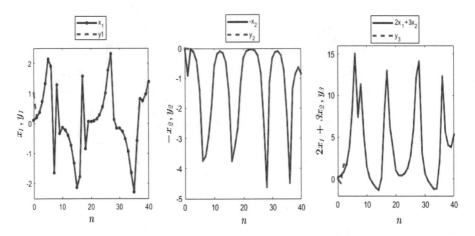

Fig. 3. The time evolution of states of the drive and the response systems (17)–(18) after control.

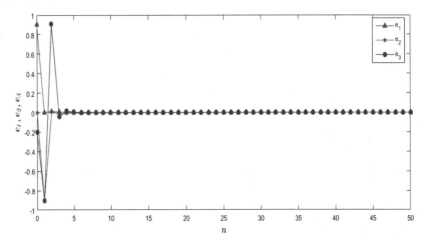

Fig. 4. The errors of synchronization between drive and the response systems (17)–(18)

Now, by using the matrices $\mathbf{M}$ and C, it is rather easy to construct the control law according to (12) as follow:

$$
\begin{cases}
u_1 = -\mathbf{a} + y_2^2 - \mathbf{b}y_3 + (\alpha\beta + 1)\,x_1 - \beta x_1 x_2 \\
u_2 = -2y_1 - \beta\left(x_1^2 - x_2\right) - x_2 + x_1 \\
u_3 = -2y_2 - y_1 + (2\alpha\beta + 3)\,x_1 - 2\beta x_1 x_2 + 3\beta x_1^2 + (-3\beta + 2)\,x_2
\end{cases}
\tag{25}
$$

and the resulting error system is given by:

$$
\begin{cases}
{}^{C}\Delta_a^{\nu}e_1(t) = -e_1(t + \nu - 1), \\
{}^{C}\Delta_a^{\nu}e_2(t) = -e_1(t + \nu - 1) - e_2(t + \nu - 1) - e_3(t + \nu - 1), \\
{}^{C}\Delta_a^{\nu}e_3(t) = -e_1(t + \nu - 1) - e_2(t + \nu - 1) - e_3(t + \nu - 1).
\end{cases}
\tag{26}
$$

The time evolution of the errors is depicted in Fig. 3 for the initial values:

$$\begin{cases} e_1(0) = 0.1 \\ e_2(0) = 0.4 \\ e_3(0) = 1.4 \end{cases} \tag{27}$$

Figure 3 depicts the time evolution of states of the drive and the response systems (17)–(18) after control. The convergence of the errors to zero is depicted in Fig. 4.

5 Conclusion

This work has studied the matrix projective synchronization (MPS) of fractional-order discrete-time chaotic systems of different dimensions. The MPS scheme aims for the general definition of the error towards zero in finite time, thereby covering a range of different synchronization types. A new control law has been derived and the asymptotic stability of their zero solution investigated through the linearization method. Computer simulation results have been presented, whereby a 3-dimensional fractional-order generalized Hénon map response synchronized the 2-dimensional fractional-order discrete-time Lorenz system. The proposed scheme was utilized to achieve 3-dimensional synchronization. The errors have been shown to converge towards zero in sufficient time.

References

1. Abdeljawad, T., Baleanu, D., Jarad, F., Agarwal, R.P.: Fractional sums and differences with binomial coefficients. Discrete Dyn. Nature Soc. **2013**, 1–6 (2013). (Article ID 104173)
2. Anastassiou, G.A.: Principles of delta fractional calculus on time scales and inequalities. Math. Comput. Modell. **52**(3), 556–566 (2010)
3. Azar, A.T., Vaidyanathan, S., Ouannas, A.: Fractional Order Control and Synchronization of Chaotic Systems. Studies in Computational Intelligence, vol. 688. Springer, Germany (2017)
4. Azar, A.T., Adele, N.M., Alain, K.S.T., Kengne, R., Bertrand, F.H.: Multistability analysis and function projective synchronization in relay coupled oscillators. Complexity **2018**, 1–12 (2018). (Article ID 3286070)
5. Azar, A.T., Ouannas, A., Singh, S.: Control of New Type of Fractional Chaos Synchronization, pp. 47–56. Springer, Cham (2018b)
6. Baleanu, D., Wu, G., Bai, Y., Chen, F.: Stability analysis of caputo-like discrete fractional systems. Commun. Nonlinear Sci. Numer. Simul. **48**, 520–530 (2017)
7. Bendoukha, S., Ouannas, A., Wang, X., Khennaoui, A.A., Pham, V.T., Grassi, G., Huynh, V.V.: The co-existence of different synchronization types in fractional-order discrete-time chaotic systems with non-identical dimensions and orders. Entropy **20**(9), 710 (2018)
8. Cermak, J., Gyori, I., Nechvatal, L.: On explicit stability conditions for a linear fractional difference system. Fract. Calc. Appl. Anal. **18**(3), 651–672 (2015)

9. Edelman, M.: On stability of fixed points and chaos in fractional systems. Chaos Interdiscip. J. Nonlinear Sci. **28**(2), 023, 112 (2018). https://doi.org/10.1063/1.5016437

10. Elaydi, S.N.: Discrete Chaos: With Applications in Science and Engineering. Chapman and Hall/CRC, Boca Raton (2007)

11. Goodrich, C., Peterson, A.C.: Discrete Fractional Calculus. Springer, Cham (2015)

12. Hu, M., Xu, Z., Zhang, R.: Full state hybrid projective synchronization of a general class of chaotic maps. Commun. Nonlinear Sci. Numer. Simul. **13**(4), 782–789 (2008)

13. Huang, C., Cao, J.: Active control strategy for synchronization and anti-synchronization of a fractional chaotic financial system. Stat. Mech. Appl. Phys. A **473**(C), 262–275 (2017)

14. Huynh, V.V., Ouannas, A., Wang, X., Pham, V.T., Nguyen, X.Q., Alsaadi, F.E.: Chaotic map with no fixed points: entropy, implementation and control. Entropy **21**(3), 279 (2019)

15. Jouini, L., Ouannas, A., Khennaoui, A.A., Wang, X., Grassi, G., Pham, V.T.: The fractional form of a new three-dimensional generalized hénon map. Adv. Diff. Equ. **1**, 122 (2019)

16. Kassim, S., Hamiche, H., Djennoune, S., Bettayeb, M.: A novel secure image transmission scheme based on synchronization of fractional-order discrete-time hyperchaotic systems. Nonlinear Dyn. **88**(4), 2473–2489 (2017)

17. Khan, A., Budhraja, M., Ibraheem, A.: Multi-switching dual compound synchronization of chaotic systems. Chin. J. Phys. **56**(1), 171–179 (2018)

18. Khan, A., Singh, S., Azar, A.T.: Combination-combination anti-synchronization of four fractional order identical hyperchaotic systems. In: Hassanien, A.E., Azar, A.T., Gaber, T., Bhatnagar, R., F Tolba, M. (eds.) The International Conference on Advanced Machine Learning Technologies and Applications (AMLTA 2019), pp. 406–414. Springer, Cham (2020)

19. Khennaoui, A.A., Ouannas, A., Bendoukha, S., Grassi, G., Wang, X., Pham, V.T.: Generalized and inverse generalized synchronization of fractional-order discrete-time chaotic systems with non-identical dimensions. Adv. Differ. Equ. **1**, 303 (2018)

20. Khennaoui, A.A., Ouannas, A., Bendoukha, S., Grassi, G., Lozi, R.P., Pham, V.T.: On fractional-order discrete-time systems: Chaos, stabilization and synchronization. Chaos, Solitons & Fractals **119**, 150–162 (2019)

21. Luo, Z., Su, M., Sun, Y., Wang, H., Yuan, W.: Stability analysis and concept extension of harmonic decoupling network for the three-phase grid synchronization systems. Int. J. Electr. Power Energy Syst. **89**, 1–10 (2017)

22. Ma, S., Yao, Z., Zhang, Y., Ma, J.: Phase synchronization and lock between memristive circuits under field coupling. AEU - Int. J. Electron. Commun. **105**, 177–185 (2019)

23. Megherbi, O., Hamiche, H., Djennoune, S., Bettayeb, M.: A new contribution for the impulsive synchronization of fractional-order discrete-time chaotic systems. Nonlinear Dyn. **90**(3), 1519–1533 (2017)

24. Ouannas, A., Abu-Saris, R.: On matrix projective synchronization and inverse matrix projective synchronization for different and identical dimensional discrete-time chaotic systems. J. Chaos **2016**, 1–7 (2016). (Article ID 4912520)

25. Ouannas, A., Mahmoud, E.E.: Inverse matrix projective synchronization for discrete chaotic systems with different dimensions. J. Comput. Intell. Electron. Syst. **3**(3), 188–192 (2014)

26. Ouannas, A., Odibat, Z.: Generalized synchronization of different dimensional chaotic dynamical systems in discrete time. Nonlinear Dyn. **81**(1), 765–771 (2015)

27. Ouannas, A., Azar, A.T., Abu-Saris, R.: A new type of hybrid synchronization between arbitrary hyperchaotic maps. Int. J. Mach. Learn. Cybernet. **8**(6), 1887–1894 (2017a)

28. Ouannas, A., Azar, A.T., Vaidyanathan, S.: A new fractional hybrid chaos synchronisation. Int. J. Modell. Ident. Control **27**(4), 314–322 (2017). https://doi.org/10.1504/IJMIC.2017.084719

29. Ouannas, A., Azar, A.T., Vaidyanathan, S.: New hybrid synchronization schemes based on coexistence of various types of synchronization between master-slave hyperchaotic systems. Int. J. Comput. Appl. Technol. **55**(2), 112–120 (2017c)

30. Ouannas, A., Azar, A.T., Vaidyanathan, S.: A robust method for new fractional hybrid chaos synchronization. Math. Methods Appl. Sci. **40**(5), 1804–1812 (2017d). mma.4099

31. Ouannas, A., Azar, A.T., Ziar, T., Vaidyanathan, S.: Fractional inverse generalized chaos synchronization between different dimensional systems. In: Azar, A.T., Vaidyanathan, S., Ouannas, A. (eds.) Fractional Order Control and Synchronization of Chaotic Systems, pp. 525–551. Springer, Cham (2017e)

32. Ouannas, A., Azar, A.T., Ziar, T., Vaidyanathan, S.: On new fractional inverse matrix projective synchronization schemes. In: Azar, A.T., Vaidyanathan, S., Ouannas, A. (eds.) Fractional Order Control and Synchronization of Chaotic Systems, pp. 497–524. Springer, Cham (2017f)

33. Ouannas, A., Khennaoui, A.A., Grassi, G., Bendoukha, S.: On the Q-S chaos synchronization of fractional-order discrete-time systems: general method and examples. Discrete Dyn. Nature Soc. **2018**, 1–8 (2018). (Article ID 2950357)

34. Ouannas, A., Grassi, G., Azar, A.T., Gasri, A.: A new control scheme for hybrid chaos synchronization. In: Hassanien, A.E., Tolba, M.F., Shaalan, K., Azar, A.T. (eds.) Proceedings of the International Conference on Advanced Intelligent Systems and Informatics 2018, pp 108–116. Springer, Cham (2019)

35. Ouannas, A., Grassi, G., Azar, A.T., Singh, S.: New control schemes for fractional chaos synchronization. In: Hassanien, A.E., Tolba, M.F., Shaalan, K., Azar, A.T. (eds.) Proceedings of the International Conference on Advanced Intelligent Systems and Informatics 2018, pp. 52–63. Springer, Cham (2019)

36. Ouannas, A., Khennaoui, A.A., Zehrour, O., Bendoukha, S., Grassi, G., Pham, V.T.: Synchronisation of integer-order and fractional-order discrete-time chaotic systems. Pramana **92**(4), 52 (2019c)

37. Ounnas, A., Azar, A.T., Radwan, A.G.: On inverse problem of generalized synchronization between different dimensional integer-order and fractional-order chaotic systems. In: 2016 28th International Conference on Microelectronics (ICM), pp. 193–196 (2016)

38. Pecora, L.M., Carroll, T.L.: Synchronization in chaotic systems. Phys. Rev. Lett. **64**, 821–824 (1990)

39. Prajapati, N., Khan, A., Khattar, D.: On multi switching compound synchronization of non identical chaotic systems. Chin. J. Phys. **56**(4), 1656–1666 (2018)

40. Razminia, A.: Full state hybrid projective synchronization of a novel incommensurate fractional order hyperchaotic system using adaptive mechanism. Indian J. Phys. **87**(2), 161–167 (2013)

41. Razminia, A., Dumitru, B.: Complete synchronization of commensurate fractional order chaotic systems using sliding mode control. Mechatronics **23**, 873–879 (2013)

42. Strogatz, S.H.: Nonlinear Dynamics and Chaos: With Applications to Physics, Biology, Chemistry, and Engineering. Westview Press, Boulder (2001)

43. Vaidyanathan, S., Azar, A.T., Boulkroune, A.: A novel 4-D hyperchaotic system with two quadratic nonlinearities and its adaptive synchronisation. Int. J. Autom. Control **12**(1), 5–26 (2018a)
44. Vaidyanathan, S., Azar, A.T., Sambas, A, Singh, S., Alain, K.S.T., Serrano, F.E.: A novel hyperchaotic system with adaptive control, synchronization, and circuit simulation. In: Advances in System Dynamics and Control. IGI Global, USA (2018)
45. Vaidyanathan, S., Jafari, S., Pham, V.T., Azar, A.T., Alsaadi, F.E.: A 4-D chaotic hyperjerk system with a hidden attractor, adaptive backstepping control and circuit design. Arch. Control Sci. **28**(2), 239–254 (2018c)
46. Wu, G.C., Baleanu, D.: Discrete chaos in fractional delayed logistic maps. Nonlinear Dyn. **80**(4), 1697–1703 (2015)

PID Controller for 2-DOFs Twin Rotor MIMO System Tuned with Particle Swarm Optimization

Ahmad Taher Azar[1,2(✉)], Abdelrahman Sayed Sayed[3], Abdalla Saber Shahin[3], Hassan Ashraf Elkholy[3], and Hossam Hassan Ammar[3]

[1] College of Engineering, Prince Sultan University, Riyadh, Kingdom of Saudi Arabia
aazar@psu.edu.sa
[2] Faculty of Computers and Artificial Intelligence, Benha University, Benha, Egypt
ahmad.azar@fci.bu.edu.eg
[3] School of Engineering and Applied Sciences, Nile University Campus,
Juhayna Square, Sheikh Zayed District, 6th of October City, Giza 12588, Egypt
{ab.sayed,a.saber,h.ashraf,hhassan}@nu.edu.eg

Abstract. This paper presents the modelling and control of a 2-DOFs Twin rotor multi input multi output (MIMO) system which is a laboratory setup resembling the dynamics of a helicopter. In this paper, the system modelling process is done using the common conventional mathematical model based on Euler-Lagrange method. The transfer functions of the model are used in the different tuning methods to reach the optimal PID gain values. The study uses conventional Proportional-Integral (PI) and Proportional-Integral-Derivative (PID) controllers to obtain a robust controller for the system. Particle Swarm optimization (PSO) technique was used to tune the controller parameters. A state space model is obtained considering some design assumptions and simplifications. Statistical measurement and convergence analysis is evaluated for the optimization of gain parameters of the PID controller for 2-DOF Twin rotor system using PSO by iteratively minimizing integral of squared error (ISE) and integral of time multiplied by the squared error (ITSE). The results are verified through simulations and experiments.

Keywords: Twin rotor multi input multi output system (TRMS) ·
PID controller · Particle Swarm Optimization (PSO)

1 Introduction

The Twin rotor system is an important laboratory setup as it is a base to many applications in real life as it's used to demonstrate the principles of a non-linear MIMO system, with significant cross-coupling. Its behaviour resembles a helicopter but contrary to most flying helicopters, the angle of attack of the rotors is fixed and the aerodynamic forces are controlled by varying the speeds of the

© Springer Nature Switzerland AG 2020
A. E. Hassanien et al. (Eds.): AISI 2019, AISC 1058, pp. 229–242, 2020.
https://doi.org/10.1007/978-3-030-31129-2_22

motors. The modeling of these air vehicles dynamics is difficult due to the significant coupling effect between the rotors and the unavailability of some system conditions [18]. The setup has gained much importance among the control community being a tool for different control experiments and providing real time simulation of an air vehicle [14,22,26]. However, it is a difficult procedure to design a controller for TRMS due to its high non-linear behavior between the two axes [6,12]. The inverted pendulum is an interesting point to have a good understanding of this problem [3,4]. TRMS consists of a beam with two rotors connected at its ends which are driven by separate DC Brushless motors [21]. It has two degrees of freedom, which facilitate the movements in both horizontal and vertical direction. In order to control TRMS in a desired way, the speed of rotors is varied, while in helicopter it is done by changing the angles of attack of the rotors. Also, there is no cylindrical control in TRMS while in helicopter it is used in directional control [27]. The control problem of TRMS has attracted a lot of researchers due to the high coupling effect between the two propellers, instability and non-linear dynamics. A lot of techniques like observer based control [22], adaptive Neural Control [10], sliding mode control [17], Fuzzy-Sliding mode control [25], Robust adaptive fuzzy control [13], model predictive Control [20], Adaptive feedback control [8] and adaptive backstepping Control [24] approaches are developed to solve the nonlinear MIMO system with unknown control direction and dead zones. Optimization techniques are used in order to tune the controllers of TRMS to control the stability and nonlinear dynamics of the system [23]. PID method of strategic control is the technique currently preferred by most researchers and is used in all industries for process control due to its simplicity [1,4,5]. Research works are going on to find the key methodologies to tune the gain parameters of the PID controller to have perfect feedback control [2,11]. Particle swarm optimization (PSO) is extensively used to control highly non-linear systems and achieve stability with very low steady state error [16]. In this paper, PSO is used to tune the gains of PID controller to enhance the stability of TRMS. Statistical measurement and convergence analysis is evaluated for the optimization of gain parameters of the PID controlller for 2-DOF TRMS using PSO, by minimizing ISE and ITSE. The paper is organized as follows. Section 2 presents the TRMS model. In Sect. 3, the simulation results are illustrated. In Sect. 4, discussions of the results are presented. Paper conclusion is presented. Finally, conclusions are drawn in Sect. 5.

2   Twin Rotor Multi Input Multi Output System Modeling

Two main sections are discussed considering the TRMS model. The first section is obtaining the TRMS mathematical model specified by [19] to evaluate the transfer functions. The second section discusses the hardware system of the TRMS considering sensors, drivers and actuators.

2.1   Mathematical Model Based on Euler-Lagrange Equations

The free body diagram of TRMS laboratory setup shown in Fig. 1 is used to develop the mathematical model. TRMS system is designed with two rotors including the effect of forces like gravitational, propulsion, centrifugal, frictional and disturbance torque on movement of the propellers. To overcome the effects of these forces, a control input is provided through motors [7].

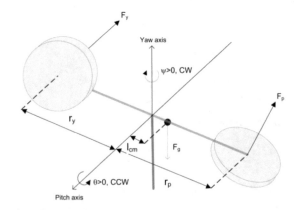

Fig. 1. The simple free-body diagram. Adopted from [14]

The dynamic equation for the vertical movement is given as:

$$I_1\ddot{\psi} = M_1 - M_{FG} - M_{B\psi} - M_G \tag{1}$$

where the main propeller thrust is given by:

$$M_1 = a_1\tau_1^2 + b_1\tau_1 \tag{2}$$

The gravity momentum id defined as:

$$M_{FG} = M_g sin\psi \tag{3}$$

The friction forces momentum is:

$$M_{B\psi} = B_{1\psi}\dot{\psi} - B_{2\psi}\sin{(2\psi)}\dot{\phi}^2 \tag{4}$$

The gyroscopic momentum is:

$$M_G = K_{gy}M_1\dot{\phi}cos\psi \tag{5}$$

The main DC motor dynamics is given by:

$$\tau_1 = (\frac{K_1}{T_{11}s + T_{10}})u_v \tag{6}$$

The dynamic equation for the horizontal movement is:

$$I_2\ddot{\phi} = M_2 - M_{B\phi} - M_R \tag{7}$$

The tail propeller thrust is given by:

$$M_2 = a_2\tau_2^2 + b_2\tau_2 \tag{8}$$

The friction forces momentum

$$M_{B\phi} = B_{1\phi}\dot{\phi} \tag{9}$$

The cross-reaction momentum

$$M_R = K_c\left(\frac{T_o s + 1}{T_p s + 1}\right)M_1 \tag{10}$$

The tail DC motor dynamics is given by:

$$\tau_2 = (\frac{K_2}{T_{21}s + T_{20}})u_h \tag{11}$$

The dynamics represented by Eq. 1 through 11 can be converted to state-space form and using Jacobian, the system can be linearized. The A, B and C matrices are defined as:

$$A = \begin{bmatrix} \frac{-T_{10}}{T_{11}} & 0 & 0 & 0 & 0 & 0 & 0 \\ 0 & 0 & 1 & 0 & 0 & 0 & 0 \\ \frac{b_1}{I_1} & \frac{-M_g}{I_1} & \frac{-B_{1\psi}}{I_1} & 0 & 0 & 0 & 0 \\ 0 & 0 & 0 & \frac{-T_{20}}{T_{21}} & 0 & 0 & 0 \\ 0 & 0 & 0 & 0 & 0 & 1 & 0 \\ 0 & 0 & \frac{-B_{1\phi}}{I_2} & \frac{b_2}{I_2} & 0 & 0 & \frac{-1}{I_2} \\ \frac{1}{T_p}(K_c - \frac{K_c T_0 T_{10}}{T_p}) & 0 & 0 & 0 & 0 & 0 & \frac{-1}{T_p} \end{bmatrix} \quad B = \begin{bmatrix} \frac{K_1}{T_{11}} & 0 \\ 0 & 0 \\ 0 & 0 \\ 0 & \frac{K_2}{T_{21}} \\ 0 & 0 \\ 0 & 0 \\ \frac{K_c T_0 K_1}{T_p T_{11}} & 0 \end{bmatrix} \tag{12}$$

$$C = \begin{bmatrix} 0 & 1 & 0 & 0 & 0 & 0 & 0 \\ 0 & 0 & 0 & 0 & 1 & 0 & 0 \end{bmatrix} \tag{13}$$

Now the system Dynamics is in the form of:

$$\dot{x} = Ax + Bu \tag{14}$$

$$y = Cx \tag{15}$$

From Eqs. 14 and 15, the linearized model is obtained as shown in Fig. 2. Pitch and yaw angles are obtained by a constructed Simulink model as shown in Fig. 3. This model is powerful in generating the transfer function for the pitch and yaw axes at 2 different step inputs. The simulink model generated four transfer functions expressing each of the two inputs with the two outputs. However, each output transfer functions were added together to provide only two transfer functions expressing each output as shown in Eqs. 16 and 17.

The pitch transfer function is expressed as:

$$\frac{(0.5015)}{(0.00051s + 50)(0.00000043s + 0.1)(0.002s^2 + 0.009022s + 0.0597)} \quad (16)$$

The yaw transfer function is expressed as:

$$\frac{60.5(s + 0.2587)}{s(s + 5)(s + 10.9091)(s + 0.5)} \quad (17)$$

These obtained transfer functions are used in the tuning process of the system to obtain the most efficient gain values in order to achieve balancing of the body even if external loads and disturbances are applied to the body.

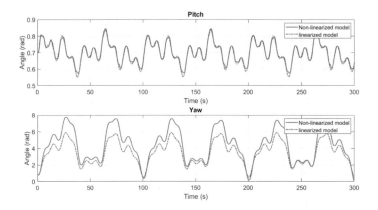

Fig. 2. Pitch and Yaw angles linearization

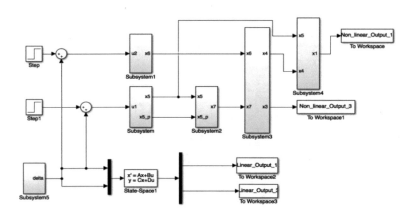

Fig. 3. Constructed simulink model

2.2   System Hardware

The actuator used in the experiment was 2122 920 KV brushless motor. Two motors have been used: one as the main rotor and one as the tail rotor. The electronic speed control (ESC) used in the experiment requires an input signal of standard frequency between 50–60 Hz which means that the Pulse Width Modulation (PWM) signal has a period of 20 ms. Because throttle speed is proportional to the duty cycle, the throttle has a duty cycle of 1 ms with a maximum of 2 ms. Figure 4 illustrates PWM signal. A Tektronix TDS 2002b oscilloscope was used to evaluate the duty cycle applied to the ESC.

3   Simulation Results

In order to determine the differences between the different optimization techniques, a group of classical tuning methods were tested like Ziegler-Nichols [29], Chien-Hrones-Reswick (CHR) [9] and Wang-Juang-Chan [28]. Also a metaheuristic optimization technique was also used which is a particle swarm optimization (PSO).

3.1   Classical Tuning Methods

Ziegler-Nichols disturbance rejection was used to force the TRMS backward from the desired set-point whenever a disturbance or load on the system causes deviation. Figures 5 and 6 show the performance criteria of the PI, PID and PID with filter controllers using Z-N method for yaw and pitch angles, respectively. It's noted that PID controller with filter has a better overshoot and settling time compared to the PI and PID controllers.

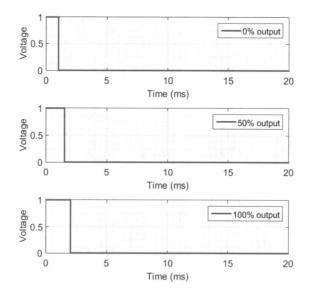

Fig. 4. Different values of duty cycles

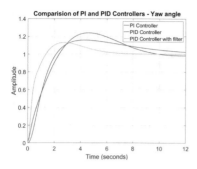

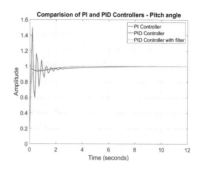

Fig. 5. Comparison of PI, PID and PID with filter controllers - yaw angle

Fig. 6. Comparison of PI, PID and PID with filter controllers - pitch angle

The Wang-Juang-Chan tuning algorithm which is based on the optimum ITAE criterion is a simple and efficient method for selecting the PID controller parameters. Figures 7 and 8 show the performance criterion of the PID and PID with filter controllers using WJC method for yaw and pitch angles, respectively. It's clear that Wang-Juang-Chan algorithm gives better performance than Ziegler-Nichols regarding the overshoot.

Chien, Hrones and Reswick (CHR) method [9] changed the step response to give better damped closed loop systems. They proposed to use either the quickest response without overshoot or the quickest response with 20% overshoot as design criteria. Figures 9 and 10 show the performance criterion of the PI, PID and PID with filter controllers using CHR method with 0Overshoot for yaw and pitch angles, respectively. Figures 11 and 12 show the performance criterion of the PI, PID and PID with filter controllers using CHR method with 20Overshoot for yaw and pitch angles, respectively. It's noted that CHR optimization technique with 20% overshoot criterion gives better performance than with 0% overshoot criterion.

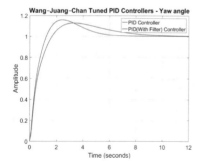

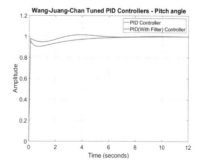

Fig. 7. System response with Wang-Juang-Chan tuned PID controllers - yaw angle

Fig. 8. System response with Wang-Juang-Chan tuned PID controllers - pitch angle

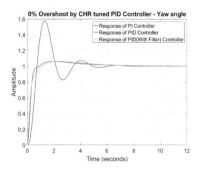

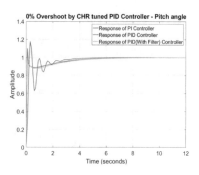

Fig. 9. System response with 0% overshoot by CHR tuned PID controller - yaw angle

Fig. 10. System response with 0% overshoot by CHR tuned PID controller - pitch angle

3.2 Metaheuristic Tuning Method

PSO is a population based stochastic search algorithm where the population dynamics are replicated with a flock of fishes or birds' activities keeping the cognitive information shared between them so that an individual can profit from the discoveries and past experience of all the other companions during the search for food [15]. As PSO is a stochastic search algorithm, it is customary to perform the search technique several times and take the best result out of that. Figure 13 illustrates the flow chart of applying the Particle Swarm Optimization (PSO) on the TRMS. The system is multi input multi output which has two inputs in the main and tail rotors, and two outputs which are pitch and yaw angles. In this paper, PSO is computed two times for each of the transfer function using 100 iterations. The performance of the controllers is analyzed based on the time domain parameters (maximum overshoot (Mp), settling time (ts)), error values like Integral of the time weighted absolute value of the error (ITAE), Integral of

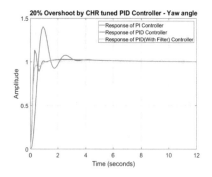

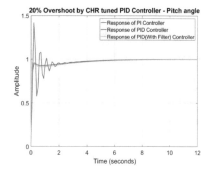

Fig. 11. System response with 20% overshoot by CHR tuned PID controller - yaw angle

Fig. 12. System response with 20% overshoot by CHR tuned PID controller - pitch angle

the square value of the error (ISE), Integral of the absolute value of the error (IAE) and Integral of the time weighted square of the error (ITSE). Figure 14 illustrates a comparison between the best cost and the number of iterations that have been applied to the transfer function obtained from the mathematical model using the tuning methods of ISE and IAE. Figure 15 shows the same comparison but using the methods of ITSE and ITAE. The results showed a great decline of the cost curves which is an indication for best convergence and best controller gains optimization.

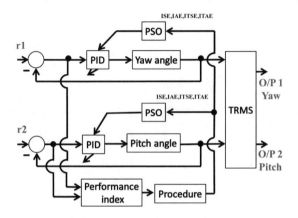

Fig. 13. PID-PSO tuning for TRMS

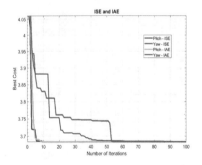 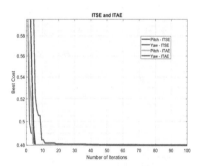

Fig. 14. Comparison of best cost vs. number of iterations - ISE & IAE

Fig. 15. Comparison of best cost vs. number of iterations - ITSE & ITAE

Figures 16 and 17 show the response of the system for yaw and pitch angles respectively after applying PSO optimization method on the tuning of PI, PID and PID with filter controllers using ISE, IAE, ITAE and ITSE performance indices. The responses of the system due to PID and PID with filter controllers are good as they nearly have minimum oscillations and nearly have no steady state error. It's noted that the best tuning technique for pitch angle that produced the most stable response is the IAE optimal method for PID controller. Moreover, the best tuning technique for yaw angle is the ITAE optimal method for PID with filter controller.

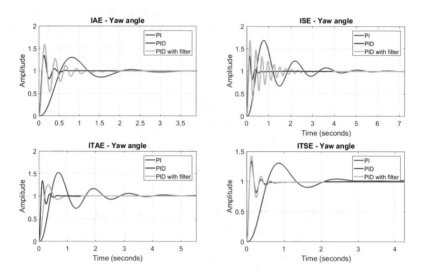

Fig. 16. Tuning of PID controller using minimum error integral criteria - yaw

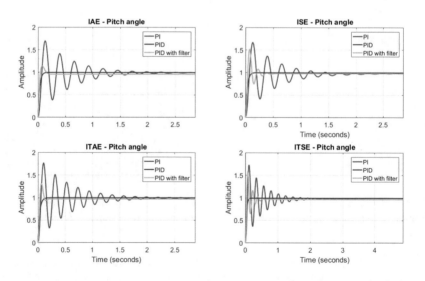

Fig. 17. Tuning of PID controller using minimum error integral criteria - pitch

After comparing the results of different tuning methods, it can be noticed that applying PSO on the minimum error integral criteria produces best responses of the system in both angles yaw and pitch according to Figs. 16 and 17. This is because PSO technique depends on selecting the best values of controller parameters as it is obvious in the time domain parameters such as maximum overshoot, rising and settling time. The performances of various controllers using different optimization techniques for TRMS for yaw and pitch angles are shown in Tables 1 and 2, respectively.

Table 1. Performance comparison of controllers - yaw angle

Table 2. Performance comparison of controllers - pitch angle

Tuning method	K_P	K_I	K_D	Tuning method	K_P	K_I	K_D
Z-N PID	4.56	14	0.5	Z-N PID	4.56	14	0.5
Z-N PID	14.8620	15.2043	11.4	Z-N PID	4.56	14	0.5
WJC PID	12.29	19.98	4.3071	WJC PID	4.3334	22.34	0.3183
CHR 0%	14.9	12.7440	13.1357	CHR 0%	4.9	22.744	0.2632
CHR 20%	18.8794	20.9600	11.3581	CHR 20%	5.8795	19.6	0.38
PSO-ISE	20	4.5010	10.1159	PSO-ISE	2.9212	19.2494	0.6393
PSO-IAE	20	20	10.1919	PSO-IAE	3.0494	20	0.6656
PSO-ITSE	20	0	10.0661	PSO-ITSE	3.0276	20	0.6666
PSO-ITAE	19.4507	20	10.1726	PSO-ITAE	2.9022	19.2045	0.6434

4   Discussion

The PID controllers performed as well as could be expected, considering the highly unstable system and the high static friction in the yaw joint. The PID controller using PSO optimization method got the pitch and yaw angles close to the set-point quickly. For evolutionary algorithms, convergence to the global solution and required execution time are considered to be important parameters through which their performance can be evaluated, analyzed and compared. The number of the individuals in the initial population was set to 100. It is clearly evident that PSO converges to a lower value. PSO showed a good degree of convergence which may be due to the inherent diversity of the population set. The effectiveness of PSO in the modelling application was analyzed through its convergence to the minimum objective value and compared with conventional tuning methods. The values of P, I, and D gains resulted from PSO optimization of pitch and yaw angles are shown in Tables 1 and 2.

5   Conclusion

In the present study, different tuning methods were applied on the TRMS to obtain most accurate PID gains. A mathematical model of the twin rotor was obtained using Euler-Lagrange method. After obtaining the TRMS model, the transfer function of each output angles; pitch and yaw were evaluated. Moreover, several classical tuning methods were applied on the transfer functions. PSO meta-heuristic technique was applied on each transfer function using 100 iterations by applying (ISE), (IAE), (ITSE) and (ITAE) criterion. The response of the system after tuning and applying PID gains on the system was robust and fast especially after applying PSO technique on the minimum error integral criteria. For the future work, the integration of fuzzy logic and neural network control methods with the TRMS model could be applied to improve the response by making the system dependent on machine learning, not on explicit programming.

References

1. Ammar, H.H., Azar, A.T., Tembi, T.D., Tony, K., Sosa, A.: Design and implementation of fuzzy pid controller into multi agent smart library system prototype. In: Hassanien, A.E., Tolba, M.F., Elhoseny, M., Mostafa, M. (eds.) The International Conference on Advanced Machine Learning Technologies and Applications (AMLTA 2018), pp. 127–137. Springer, Cham (2018)
2. Azar, A.T., Serrano, F.E.: Robust IMC-PID tuning for cascade control systems with gain and phase margin specifications. Neural Comput. Appl. **25**(5), 983–995 (2014)
3. Azar, A.T., Serrano, F.E.: Adaptive Sliding Mode Control of the Furuta Pendulum, vol. 576, pp. 1–42. Springer, Cham (2015)
4. Azar, A.T., Ammar, H.H., Barakat, M.H., Saleh, M.A., Abdelwahed, M.A.: Self-balancing robot modeling and control using two degree of freedom PID controller. In: Hassanien, A.E., Tolba, M.F., Shaalan, K., Azar, A.T. (eds.) Proceedings of the International Conference on Advanced Intelligent Systems and Informatics 2018, pp. 64–76. Springer, Cham (2019)
5. Azar, A.T., Hassan, H., Razali, M.S.A.B., de Brito Silva, G., Ali, H.R.: Two-degree of freedom proportional integral derivative (2-DOF PID) controller for robotic infusion stand. In: Hassanien, A.E., Tolba, M.F., Shaalan, K., Azar, A.T. (eds.) Proceedings of the International Conference on Advanced Intelligent Systems and Informatics 2018, pp. 13–25. Springer, Cham (2019)
6. Cajo, R., Agila, W.: Evaluation of algorithms for linear and nonlinear PID control for twin rotor mimo system. In: 2015 Asia-Pacific Conference on Computer Aided System Engineering, pp. 214–219 (2015)
7. Chalupa, P., Přikryl, J., Novák, J.: Modelling of twin rotor mimo system. Procedia Eng. **100**, 249–258 (2015). 25th DAAAM International Symposium on Intelligent Manufacturing and Automation (2014)
8. Chi, N.V.: Adaptive feedback linearization control for twin rotor multiple-input multiple-output system. Int. J. Control Autom. Syst. **15**(3), 1267–1274 (2017)
9. Chien, K., Hrones, J., Reswick, J.: On the automatic control of generalized passive systems. Trans. ASME **74**(2), 175–185 (1952)

10. Dheeraj, K., Jacob, J., Nandakumar, M.P.: Direct adaptive neural control design for a class of nonlinear multi input multi output systems. IEEE Access **7**(15), 424–15, 435 (2019)

11. Gorripotu, T.S., Samalla, H., Jagan Mohana Rao, C., Azar, A.T., Pelusi, D.: Tlbo algorithm optimized fractional-order PID controller for AGC of interconnected power system. In: Nayak, J., Abraham, A., Krishna, B.M., Chandra Sekhar, G.T., Das, A.K. (eds.) Soft Computing in Data Analytics, pp. 847–855. Springer, Singapore (2019)

12. Haruna, A., Mohamed, Z., Efe, M., Basri, M.A.M.: Dual boundary conditional integral backstepping control of a twin rotor mimo system. J. Franklin Inst. **354**(15), 6831–6854 (2017)

13. Jahed, M., Farrokhi, M.: Robust adaptive fuzzy control of twin rotor mimo system. Soft. Comput. **17**(10), 1847–1860 (2013)

14. Juang, J.G., Liu, W.K., Lin, R.W.: A hybrid intelligent controller for a twin rotor mimo system and its hardware implementation. ISA Trans. **50**(4), 609–619 (2011)

15. Kennedy, J., Eberhart, R.: Particle swarm optimization. In: Proceedings of ICNN 1995 - International Conference on Neural Networks, vol. 4, pp. 1942–1948 (1995)

16. Messaoud, R.B.: Observer for nonlinear systems using mean value theorem and particle swarm optimization algorithm. ISA Trans. **85**, 226–236 (2019)

17. Mondal, S., Mahanta, C.: Adaptive second-order sliding mode controller for a twin rotor multi-input-multi-output system. IET Control Theory Appl. **6**(14), 2157–2167 (2012)

18. Pandey, V.K., Kar, I., Mahanta, C.: Control of twin-rotor mimo system using multiple models with second level adaptation. IFAC-PapersOnLine **49**(1), 676–681 (2016). 4th IFAC Conference on Advances in Control and Optimization of Dynamical Systems ACODS 2016

19. Raghavan, R., Thomas, S.: Mimo model predictive controller design for a twin rotor aerodynamic system. In: 2016 IEEE International Conference on Industrial Technology (ICIT), pp. 96–100 (2016)

20. Raghavan, R., Thomas, S.: Practically implementable model predictive controller for a twin rotor multi-input multi-output system. J. Control Autom. Electrical Syst. **28**(3), 358–370 (2017)

21. Rahideh, A., Bajodah, A., Shaheed, M.: Real time adaptive nonlinear model inversion control of a twin rotor mimo system using neural networks. Eng. Appl. Artif. Intell. **25**(6), 1289–1297 (2012)

22. Rashad, R., El-Badawy, A., Aboudonia, A.: Sliding mode disturbance observer-based control of a twin rotor mimo system. ISA Trans. **69**, 166–174 (2017)

23. Reynoso Meza, G., Blasco Ferragud, X., Sanchis Saez, J., Herrero Durá, J.M.: Multiobjective Optimization Design Procedure for Controller Tuning of a TRMS Process, pp. 201–213. Springer, Cham (2017)

24. Sodhi, P., Kar, I.: Adaptive backstepping control for a twin rotor mimo system. IFAC Proc. **47**(1), 740–747 (2014). 3rd International Conference on Advances in Control and Optimization of Dynamical Systems (2014)

25. Tao, C., Taur, J., Chang, Y., Chang, C.: A novel fuzzy-sliding and fuzzy-integral-sliding controller for the twin-rotor multi-input-multi-output system. IEEE Trans. Fuzzy Syst. **18**(5), 893–905 (2010)

26. Tastemirov, A., Lecchini-Visintini, A., Morales-Viviescas, R.M.: Complete dynamic model of the twin rotor mimo system (TRMS) with experimental validation. Control Eng. Pract. **66**, 89–98 (2017)

27. Tiwalkar, R.G., Vanamane, S.S., Karvekar, S.S., Velhal, S.B.: Model predictive controller for position control of twin rotor mimo system. In: 2017 IEEE International Conference on Power, Control, Signals and Instrumentation Engineering (ICPCSI), pp. 952–957 (2017)
28. Wang, F.S., Juang, W.S., Chan, C.T.: Optimal tuning of PID controllers for single and cascade control loops. Chem. Eng. Commun. **132**(1), 15–34 (1995)
29. Ziegler, J., Nichols, N.B.: Optimum settings for automatic controllers. Trans. ASME **64**(8), 759–768 (1942)

Direct Torque Control of Three Phase Asynchronous Motor with Sensorless Speed Estimator

Arezki Fekik[1,2], Hakim Denoun[2], Ahmad Taher Azar[3,4(✉)],
Nashwa Ahmad Kamal[5], Mustapha Zaouia[2], Nacera Yassa[1],
and Mohamed Lamine Hamida[2]

[1] Akli Mohand Oulhadj University, Bouira, Algeria
[2] Electrical Engineering Advanced Technology Laboratory (LATAGE),
Tizi Ouzou, Algeria
arezkitdk@yahoo.fr, akim_danoun2002dz@yahoo.fr, zmust@yahoo.fr,
yassa.nacera@yahoo.fr, ml_hamida@yahoo.com
[3] College of Engineering, Prince Sultan University, Riyadh, Kingdom of Saudi Arabia
aazar@psu.edu.sa
[4] Faculty of Computers and Artificial Intelligence, Benha University, Benha, Egypt
ahmad.azar@fci.bu.edu.eg
[5] Electrical Power and Machine Department, Faculty of Engineering,
Cairo University, Giza, Egypt
nashwa.ahmad.kamal@gmail.com

Abstract. Direct torque control is undoubtedly a very promising solution to the problems of robustness and dynamics encountered in the directional flow vector control of the rotor of induction machines. Current research aims to improve the performance of this technique like the evolution of the switching frequency, the ripple on the torque, the flow and the current, and assists the cost of the sensor position. Therefore, this article presents a solution for the direct torque control without speed sensor. The simulations results showed a good dynamic performance of this control technique.

Keywords: Direct Torque Control (DTC) · Induction machine · Speed estimator · Torque control

1 Introduction

The asynchronous machine, because of its simplicity of design and maintenance, has been favored by manufacturers since its invention by Nikola Tesla when he discovered the rotating magnetic fields generated by a polyphase current system [21]. This simplicity, however, is accompanied by a great physical complexity related to the electromagnetic interactions between the stator and the rotor. On the other hand, unlike the DC motor where it is sufficient to vary the supply voltage of the armature to vary the speed, the asynchronous motor requires the

© Springer Nature Switzerland AG 2020
A. E. Hassanien et al. (Eds.): AISI 2019, AISC 1058, pp. 243–253, 2020.
https://doi.org/10.1007/978-3-030-31129-2_23

use of alternating currents of variable frequency. One of the main blockages was the inverter that had to operate in forced commutation [28]. The asynchronous machine has long been used essentially at constant speed and unable to properly control the dynamics of the engine-load [21]. The appearance of the thyristors and subsequently the transistors has allowed the development of efficient and reliable pulse modulation inverters at a non-prohibitive cost.

Recently, different control approaches have been proposed for designing nonlinear power systems for many practical applications [1,2,9,17,18,22]. The direct torque control (DTC) method of asynchronous machines appeared in the second half of the 1980s as competitive with conventional methods [8,25]. In DTC method, the stator flux is added with a differential flux linkage vector or voltage vector to accelerate or decelerate it away from the rotor flux, thereby achieving the required torque both in dynamic and steady-state conditions [5,10,20]. As opposed to Time Control Laws, these direct control strategies belong to the category of amplitude control laws and were originally designed for a two-level inverter [4,14]. They have advantages in comparison with conventional techniques such as the reduction of the torque response time and the robustness with respect to the variations of machine parameters [6,11–13,15,16,27]. By their nature, they can be functional without a speed sensor avoiding coordinate transformation, Pulse Width Modulation (PWM) generation and current regulators [23]. This paper will expose the principle of the direct torque control of an asynchronous motor without speed sensors, i.e., the reference speed is compared with an estimated speed to deduce the reference torque. The results of dynamic performance of the DTC are acceptable from the point of view of torque and speed.

The paper is organized as follows. Modeling of the asynchronous motor is discussed in Sect. 2. Then in Sect. 3, principle and modeling of the DTC is discussed. Section 4 presents the speed estimation and control approach. Section 5 includes the simulation results and discussion. Finally, Sect. 6 deals with conclusion of the research work.

2   Modeling of the Asynchronous Motor

The stator of the machine is formed of three fixed winding offset by 120° in space and traversed by three variable currents. The rotor can be modeled by three identical winding offset in the space of 120°. These winding are short-circuit and the voltage at their terminals is zero.

2.1   Electrical Equations

The six windings (a, b, c, A, B, C) is done in the following matrix equations [3,7,24]:

$$[V_s] = [R_s][i_s] + \frac{d[\psi_s]}{dt}$$

$$[V_r] = [R_r][i_r] + \frac{d[\psi_r]}{dt} \tag{1}$$

2.2 Magnetic Equations

The simplifying hypotheses quoted previously lead to linear relations between the flows and the currents of the asynchronous machine and can be written as [7, 19, 24, 26]:

For the stator:

$$[\psi_s] = [L_S][i_S] + [M_{sr}][i_r] \tag{2}$$

For the rotor:

$$[\psi_r] = [L_r][i_r] + [M_{rs}][i_s] \tag{3}$$

where $[L_S]$ and $[L_r]$ are the stator and rotor inductance matrices. $[M_{rs}]$ corresponds to the matrix of stator-rotor mutual inductances and is designated by:

$$[L_s] = \begin{bmatrix} L_s & M_s & M_s \\ M_s & L_s & M_s \\ M_s & M_s & L_s \end{bmatrix}$$

$$[L_r] = \begin{bmatrix} L_r & M_r & M_r \\ M_r & L_r & M_r \\ M_r & M_r & L_r \end{bmatrix}$$

Finally, the voltage equations become:

$$[V_{sabc}] = [R_s][i_{sabc}] + \frac{d[L_s][i_{sabc}]}{dt} + [M_{sr}][i_{rabc}]$$

$$[V_{rabc}] = [R_r][i_{rabc}] + \frac{d[L_r][i_{rabc}]}{dt} + [M_{rs}][i_{sabc}] \tag{4}$$

2.3 Mechanical Equations

The study of the asynchronous machine characteristics introduces variation not only in electrical parameters (voltage, current, flux) but also in mechanical parameters (torque, speed):

$$C_{em} = p[i_{sabc}]^T + \frac{d[M_{sr}][i_{rabc}]}{dt} \tag{5}$$

The motor mechanical equation is:

$$J\frac{d\Omega}{dt} = C_{em} - C_r - f_r\Omega \tag{6}$$

3   Principle and Modeling of the DTC

This control law has remarkable dynamic performance as well as a good robustness for engine parameter deviations. The structure of the proposed direct control is illustrated in Fig. 1.

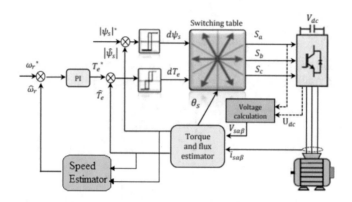

Fig. 1. Block diagram of the proposed direct control

The vector expressions of the machine are used in the stator-related repository:

$$\overline{V_s} = R_s \overline{I_s} + \frac{d}{dt}\overline{\psi_s}$$

$$\overline{V_r} = 0 = R_r \overline{I_r} + \frac{d}{dt}\overline{\psi_s} + jw\overline{\psi_r} \tag{7}$$

where $\overline{V_s}$ and $\overline{I_s}$ correspond to stator voltage and current vectors, $\overline{\psi_s}$ and $\overline{\psi_r}$ are stator and rotor flux vectors. Variable ω holds for electrical rotor speed. Parameters R_s, R_r are stator and rotor resistance, respectively. From the expressions of the flows, the current of the rotor is written as:

$$\overline{I_r} = \frac{1}{\sigma}[\frac{\psi_r}{L_r} - \frac{L_m}{L_r L_s}\overline{\psi_s}] \tag{8}$$

with

$$\sigma = 1 - \frac{L_m^2}{L_s L_r}$$

where σ is the dispersion coefficient.

The equations become:

$$\overline{V_s} = R_s \overline{I_s} + \frac{d}{dt}\overline{\psi_s}$$

$$\frac{d\overline{\psi_r}}{dt} + (\frac{1}{\sigma T_r} - Jw) = \frac{L_m}{L_s}\frac{1}{\sigma T_r}\overline{\psi_s} \tag{9}$$

so, the equation of the couple is in the form:

$$C_{em} = \frac{2P}{3}(\psi_{\alpha s}I_{\beta s} - \psi_{\beta s}I_{\alpha s}) \tag{10}$$

3.1 Choice of Voltage Vector V_s

The choice of vector $\overline{V_s}$ depends on the position of $\overline{\psi_s}$, the desired variation for the module of ψ_s, the desired variation for the torque and direction of rotation of $\overline{\psi_s}$. The fixed complex plane (α, β) of the stator is subdivided into six S_i with: $i = 1, ..., 6$ such that:

$$(2i - 3)\frac{\pi}{6} \le S_i \le (2i - 1)\frac{\pi}{6} \tag{11}$$

Each sector will contain an active voltage space vector of the inverter as shown in Fig. 2. The flow turns in the trigonometrical direction [3,7,24].

These voltage vectors are chosen from the switching table as a function of the errors of the torque and the position of the vector of stator flux. However, it is no longer necessary to position the rotor to choose the voltage vector. This particularity gives the advantage to the DTC of not using a mechanical sensor. The voltage vector at the output of the inverter is deduced from the estimated torque and flux exchanges with respect to their reference as well as the position of the vector.

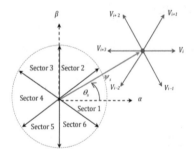

Fig. 2. Voltage vector selection when the stator flux vector is located in sector i

3.2 Development of the Control Table

The control table is built according to the state of the variables k_f and k_c and the position area of $\overline{\psi_s}$. Thus, It's in the following form (Table 1):

Table 1. Switching Table

S_i		S_1	S_2	S_3	S_4	S_5	S_6
$k_f = 1$	$k_c = 1$	V_2	V_3	V_4	V_5	V_6	V_1
$k_f = 1$	$k_c = 0$	V_0	V_7	V_0	V_7	V_0	V_7
$k_f = 1$	$k_c = -1$	V_1	V_2	V_3	V_4	V_5	V_6
$k_f = 0$	$k_c = 1$	V_4	V_5	V_6	V_1	V_2	V_3
$k_f = 0$	$k_c = 0$	V_7	V_0	V_7	V_0	V_7	V_0
$k_f = 0$	$k_c = -1$	V_5	V_6	V_1	V_2	V_3	V_4

4   Speed Estimation and Control

The produced electromagnetic torque of the induction motor can be determined using the cross product of the stator quantities (i.e., stator flux and stator currents). The torque formula is expressed as:

$$C_{em} = P(\psi_{\alpha s} I_{\beta s} - \psi_{\beta s} I_{\alpha s}) \tag{12}$$

For rotor flux estimation, the measured stator current and rotor speed are counted. The derivative of the estimated rotor flux can be computed using machine parameters as:

$$\frac{d\psi_{r\alpha\beta est}}{dt} = R_r \frac{M_{sr}}{L_r} i_{s\alpha\beta} - \omega_r \psi_{r\alpha\beta est} \tag{13}$$

The rotor speed ω_{est} can be computed from a difference between the synchronous speed and the slip speed using the following equation:

$$\omega_{est} = \omega_s - \omega_{slip} \tag{14}$$

with

$$\omega_s = \frac{\dot{\psi}_{s\beta}\psi_{s\alpha} - \dot{\psi}_{s\alpha}\psi_{s\beta}}{\sqrt{\psi_{s\alpha}^2 + \psi_{s\beta}^2}} = \frac{\dot{\psi}_{r\beta}\psi_{r\alpha} - \dot{\psi}_{r\alpha}\psi_{r\beta}}{\sqrt{\psi_{r\alpha}^2 + \psi_{r\beta}^2}} \tag{15}$$

and the ω_{slip} is given by:

$$\omega_{slip} = \frac{R_s C_{em}}{\psi_r^2 P} \tag{16}$$

By substitution of Eqs. (15) and (16) into Eq. (14), the estimated rotational speed can be obtained.

5   Numerical Simulations and Discussions

In this part, the simulation and the validation of the DTC with speed estimation for an asynchronous motor is presented where the simulation parameters are summarized in Table 2. In this case, two tests are carried out to illustrate the validity and the performance of the proposed control, namely a test with a constant reference rotation speed and the other with variation on different speed levels so that the performance of the control can be tested.

Table 2. Simulation parameters

Parameter	Value
Resistance of the stator winding R_s	1.77 Ω
Resistance of the rotor winding R_r	1.34 Ω
Inductance of the stator winding L_s	13.93e$-$3 H
Inductance of the rotor winding L_r	12.12e$-$3 H
Mutual inductance M_{sr}	369e$-$3 H
Moment of inertia J	0.025 Kg m^2
Coefficient of viscous friction f_r	1e$-$5
Rated rotation speed Ω	1490 tr/min
Number of pole pairs P	2 pole

Figure 3 shows that the flux of the stator with reference constants $alpha - beta$. From this figure, it can be seen that the form is sinusoidal which implies that the shape of the current absorbed by the induction machine is also sinusoidal.

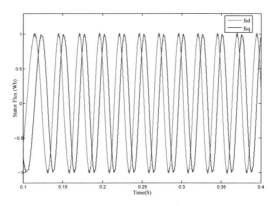

Fig. 3. Stator Flux with $\omega_{ref} = cst$

Figure 4 shows the circular path and it can be noted that that the flow ripples are considerably eliminated in the proposed DTC method without speed sensor. Figure 5 shows the estimated and actual speed of the rotor where a good dynamic at startup is achieved. It can be noticed that the estimated rotation speed follows the actual speed of the machine with a small deviation.

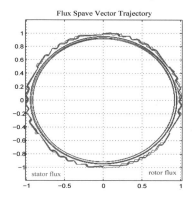

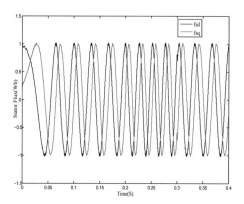

Fig. 4. Flux vector trajectory with $\omega_{ref} = cst$

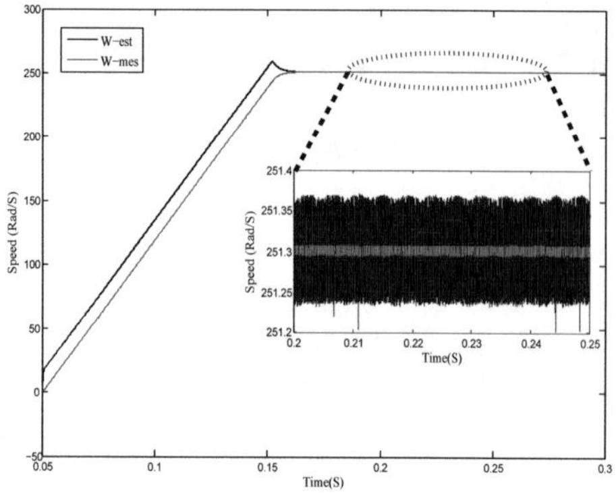

Fig. 5. Speed estimation and speed measurement with $\omega_{ref} = cst$

Figure 6 shows that the stator flux with different reference speed levels in the benchmark $alpha - beta$. From this figure, it can be seen that the shape is sinusoidal which implies that the shape of the current absorbed by the induction machine is also sinusoidal even with the abrupt variation of the speed of rotation which confirms that the estimate of the speed is well achieved.

Fig. 6. Stator Flux with variation of ω_{ref}

Figure 7 shows the circular path with different reference speed levels. The results presented in this figure demonstrated that the flux ripples are considerably reduced in the proposed control, namely DTC without a speed sensor.

Figure 8 shows the estimated speed and its measured value of the rotor. This technique shows a good dynamic at startup and it can be seen that the estimated rotational speed follows well its sensed speed with different reference levels.

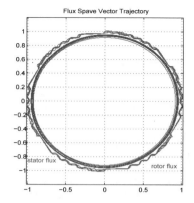

Fig. 7. Flux vector trajectory with variation of ω_{ref}

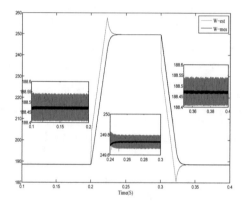

Fig. 8. Speed estimation and speed measurment with variation of ω_{ref}

6   Conclusion

In this study, the three-phase asynchronous motor direct torque control design
with sensorless speed estimator is proposed. The results of the theoretical and
numerical simulation demonstrated that the proposed controller improves the
robustness and performance of the system even when the reference speed is
changed over several levels of references. It is important to note that the velocity
estimated by the propose algorithm follows suitably its measured value. Also, the
waveform of the flux are purely sinusoidal even with changing the reference speed
which demonstrated that the current absorbed by the induction machine is pure
sinusoidal. A control sensorless speed is a hot topic for industrial applications,
especially in terms of the constraints of parameters adaptation and dynamic
stability.

References

1. Abdelmalek, S., Rezazi, S., Azar, A.T.: Sensor faults detection and estimation for a DFIG equipped wind turbine. Energy Procedia **139**, 3–9 (2017). Materials & Energy I (2015)
2. Abdelmalek, S., Azar, A.T., Dib, D.: A novel actuator fault-tolerant control strategy of dfig-based wind turbines using Takagi-Sugeno multiple models. Int. J. Control Autom. Syst. **16**(3), 1415–1424 (2018)
3. Ammar, A., Bourek, A., Benakcha, A.: Nonlinear SVM-DTC for induction motor drive using input-output feedback linearization and high order sliding mode control. ISA Trans. **67**, 428–442 (2017)
4. Arnanz, R., Miguel, L.J., Perán, J.R., Mendoza, A.: A modified direct torque control with fault tolerance. Control Eng. Pract. **19**(9), 1056–1065 (2011). Special Section: DCDS 2009 – The 2nd IFAC Workshop on Dependable Control of Discrete Systems
5. Ayrir, W., Ourahou, M., Hassouni, B.E., Haddi, A.: Direct torque control improvement of a variable speed DFIG based on a fuzzy inference system. Math. Comput. Simul. (2018)
6. Carmeli, M., Mauri, M.: Direct torque control as variable structure control: existence conditions verification and analysis. Electr. Power Syst. Res. **81**(6), 1188–1196 (2011)
7. Casadei, D., Profumo, F., Serra, G., Tani, A.: FOC and DTC: two viable schemes for induction motors torque control. IEEE Trans. Power Electron. **17**(5), 779–787 (2002). https://doi.org/10.1109/TPEL.2002.802183
8. Depenbrock, M.: Direct self-control (DSC) of inverter-fed induction machine. IEEE Trans. Power Electron. **3**(4), 420–429 (1988). https://doi.org/10.1109/63.17963
9. Ghoudelbourk, S., Dib, D., Omeiri, A., Azar, A.T.: MPPT control in wind energy conversion systems and the application of fractional control (PI^{α}) in pitch wind turbine. Int. J. Model. Ident. Control **26**(2), 140–151 (2016)
10. Sudheer, H., Kodad, S.F., Sarvesh, B.: Improvements in direct torque control of induction motor for wide range of speed operation using fuzzy logic. J. Electr. Syst. Inf. Technol. **5**(3), 813–828 (2018)
11. Hassan, A., Shehata, E.: High performance direct torque control schemes for an ipmsm drive. Electr. Power Syst. Res. **89**, 171–182 (2012)
12. Heinbokel, B.E., Lorenz. R.D.: Robustness evaluation of deadbeat, direct torque and flux control for induction machine drives. In: 2009 13th European Conference on Power Electronics and Applications, pp. 1–10 (2009)
13. Karpe, S., Deokar, S.A., Dixit, A.M.: Switching losses minimization by using direct torque control of induction motor. J. Electr. Syst. Inf. Technol. **4**(1), 225–242 (2017)
14. Liu, H., Zhang, H.: A novel direct torque control method for brushless DC motors based on duty ratio control. J. Franklin Inst. **354**(10), 4055–4072 (2017)
15. Lokriti, A., Salhi, I., Doubabi, S.: IM direct torque control with no flux distortion and no static torque error. ISA Trans. **59**, 256–267 (2015)
16. Marino, P., D'Incecco, M., Visciano, N.: A comparison of direct torque control methodologies for induction motor. In: 2001 IEEE Porto Power Tech Proceedings (Cat. No. 01EX502), vol. 2, p. 6 (2001). https://doi.org/10.1109/PTC.2001.964724
17. Meghni, B., Dib, D., Azar, A.T.: A second-order sliding mode and fuzzy logic control to optimal energy management in wind turbine with battery storage. Neural Comput. Appl. **28**(6), 1417–1434 (2017)

18. Meghni, B., Dib, D., Azar, A.T., Saadoun, A.: Effective supervisory controller to extend optimal energy management in hybrid wind turbine under energy and reliability constraints. Int. J. Dyn. Control **6**(1), 369–383 (2018)
19. Naik, V.N., Panda, A., Singh, S.P.: A three-level fuzzy-2 DTC of induction motor drive using SVPWM. IEEE Trans. Industr. Electron. **63**(3), 1467–1479 (2016). https://doi.org/10.1109/TIE.2015.2504551
20. Pimkumwong, N., Wang, M.S.: Full-order observer for direct torque control of induction motor based on constant V/F control technique. ISA Trans. **73**, 189–200 (2018)
21. Razik, H.: Handbook of Asynchronous Machines with Variable Speed. Wiley-ISTE (2013)
22. Smida, M.B., Sakly, A., Vaidyanathan, S., Azar, A.T.: Control-based maximum power point tracking for a grid-connected hybrid renewable energy system optimized by particle swarm optimization. In: Azar, A.T., Vaidyanathan, S. (eds.) Advances in System Dynamics and Control, pp. 58–89. IGI Global (2018)
23. Sutikno, T., Idris, N.R.N., Jidin, A.: A review of direct torque control of induction motors for sustainable reliability and energy efficient drives. Renew. Sustain. Energy Rev. **32**, 548–558 (2014)
24. Swierczynski, D., Wojcik, P., Kazmierkowski, M.P., Janaszek, M.: Direct torque controlled PWM inverter fed PMSM drive for public transport. In: 2008 10th IEEE International Workshop on Advanced Motion Control, pp 716–720 (2008). https://doi.org/10.1109/AMC.2008.4516155
25. Takahashi, I., Noguchi, T.: A new quick-response and high-efficiency control strategy of an induction motor. IEEE Trans. Ind. Appl. IA **22**(5), 820–827 (1986). https://doi.org/10.1109/TIA.1986.4504799
26. Tazerart, F., Mokrani, Z., Rekioua, D., Rekioua, T.: Direct torque control implementation with losses minimization of induction motor for electric vehicle applications with high operating life of the battery. Int. J. Hydrogen Energy **40**(39), 13827–13838 (2015)
27. Wang, Y., Niimura, N., Lorenz, R.D.: Real-time parameter identification and integration on deadbeat-direct torque and flux control (DB-DTFC) without inducing additional torque ripple. In: 2015 IEEE Energy Conversion Congress and Exposition (ECCE), pp. 2184–2191 (2015). https://doi.org/10.1109/ECCE.2015.7309968
28. Wildi, T.: Electrical Machines, Drives and Power Systems, 6th edn. Pearson (2005)

Robust H-Infinity Decentralized Control for Industrial Cooperative Robots

Ahmad Taher Azar[1,2(✉)], Fernando E. Serrano[3], Ibrahim A. Hameed[4], Nashwa Ahmad Kamal[5], and Sundarapandian Vaidyanathan[6]

[1] College of Engineering, Prince Sultan University, Riyadh, Kingdom of Saudi Arabia
aazar@psu.edu.sa
[2] Faculty of Computers and Artificial Intelligence, Benha University, Benha, Egypt
ahmad.azar@fci.bu.edu.eg
[3] Universidad Tecnologica Centroamericana (UNITEC), Tegucigalpa, Honduras
serranofer@eclipso.eu
[4] Faculty of Information Technology and Electrical Engineering,
Department of ICT and Natural Sciences, Norwegian University of Science and
Technology (NTNU), Larsgårdsvegen 2, 6009 Ålesund, Norway
ibib@ntnu.no
[5] Electrical Power and Machine Department, Faculty of Engineering,
Cairo University, Giza, Egypt
nashwa.ahmad.kamal@gmail.com
[6] Research and Development Centre, Vel Tech University Avadi,
Chennai 600062, Tamilnadu, India
sundarvtu@gmail.com

Abstract. In this paper, a robust H-infinity controller is proposed for industrial cooperative robots in which disturbances are taken into account for an appropriate controller design. Considering the disturbance affects the system performance is important to design an efficient control strategy to solve this problem. On the other side, robust control is necessary taking into account that disturbances, uncertainties and other unmodelled dynamics affecting the system performance for tracking control purposes. The main objective of this study is to control in a synchronized way several industrial robotic manipulators in a work-cell so the obtained results will be useful for different kinds of manufacturing processes. For this objective, the distributed dynamical model of the robot in the form of the Euler-Lagrange equation is established to derive the proposed control strategy that in this case consists of a robust H-infinity controller to deal with the disturbances in the robot manipulators and other kinds of uncertainties. For this purpose, a norm index is implemented for robust performance attributes of the system so in this way the controller is designed efficiently. One of the most important contributions of this study is that the position tracking of an object grasped by two or more industrial robotic manipulators is driven accurately. This can be done by obtaining the desired position and orientations of the end effectors of each robot and reducing the tracking error to zero as times goes to infinity while following a predefined path trajectory for the grasped object. Finally, a numerical simulation example and

© Springer Nature Switzerland AG 2020
A. E. Hassanien et al. (Eds.): AISI 2019, AISC 1058, pp. 254–265, 2020.
https://doi.org/10.1007/978-3-030-31129-2_24

conclusions will be offered to validate and analyze the theoretical results obtained in this study.

Keywords: Cooperative robotics · Distributed control · Robust control · H-infinity control

1   Introduction

Cooperative robotics is a field of robotics in which there is an interaction of two or more robots to make simultaneous tasks in a synchronized way. Cooperative robotic tasks can be done by industrial robotics manipulators, mobile robots and it can achieved by unmanned aerial vehicles also. Considering the importance of designing several cooperative robotics architectures in the industrial field, nowadays new control strategies are required due to the need of novel manufacturing processes in which the complexity of the tasks are increasing [3–6,8,14,15,20,21]. Disturbances and unmodelled dynamics are some of the obstacles that must be surpassed to design efficient distributed control techniques for synchronized trajectory tracking in a manufacturing cell.

Before beginning the explanations related to the obtained results in this study, it is important to remember some control approaches that have been developed nowadays for cooperative robotics. One example can be found in [7] in which a position/force controller that consists in an adaptive fuzzy backstepping approach is evinced, in this study. The handling of a rigid object with unknown dynamical model and external disturbances is achieved in order to reduced the position/force tracking error. One interesting example of cooperative robotics can be found in [12] in which a distributed algorithm is designed and implemented to manipulate an object by mobile robotics manipulators. In [9], a strategy to design mobile robotic manipulators to transport and manipulate payloads is proposed, where two degrees of freedom manipulators are mounted on each mobile robot to move the payload. Then, another example of mobile cooperative robots is evinced in [1]. This strategy consist in the optimal formation of mobile wheeled robots to manipulate a common object where a cost or performance function is used to obtain the grasping point of an object.

Distributed control is closely related to control strategies implemented in cooperative robotics, taking into account that distributed control provides a solution to a vast amount of control problems found in science and engineering. For the objective of this study it is important to mention some studies found in the literature that are helpful to design controller strategies for cooperative robotics tasks. So for example, in [13] a decentralized controller for cooperative robotics without velocity measurements is shown maintaining the tracking and observation as well as force errors to zero. In [10], another interesting cooperative robotics controller example is provided where a robust distributed consensus for a tethered space net robot is designed by implementing a Lyapunov approach and graph theory reducing the tracking error of the maneuverable units. Other two interesting examples are found in [19,23] where in the first case a decentralized cooperative control strategy for arbitrary set-point and collision avoidance is designed by using the null space behavioral approach to obtain the set-point tracking of multiple robots. In the second case, the distributed control of heterogeneous underactuated mechanical systems is done where a distributed passivity based law is designed for underactuated mechanical systems. Finally, another related control strategy for cooperative robotics is found in [2] in which a decentralized robust controller for needle insertion is used considering communication delays and uncertainties, the strategy used in this study consists in a robust force reflecting control of a haptic device.

Robust H-infinity control for robotic manipulators has been extensively used for industrial robotic manipulators offering an adequate theoretical background that can be extended to cooperative robotics. Two interesting examples can be found in [17,18] where nonlinear H-infinity controllers for multi-degrees of freedom robotic manipulators are shown in which in both studies the solutions of the Ricatti equations are needed to solve the robust controller conditions. In [16], the trajectory tracking of an autonomous robotic vehicle is achieved by solving an H-infinity control problem. As a last example, a robust controller design for an underwater vehicle manipulator system is shown where a force/position control is obtained successfully [citation].

In this paper, a robust H-infinity decentralized control for industrial cooperative robot is shown. First, the dynamic model of each robotic unit is derived according to its position and orientation and the joint variables. This is done by obtaining the desired trajectory of each robot based on the grasped object kinematics using the transformation matrix that relates the coordinates of the grasped object. Then, the robust H-infinity controller is designed based on [11] using the required conditions for disturbance attenuation and a suitable Lyapunov function to find an appropriate control law. The paper is organized as follows. In Sect. 2 presents problem definition. In Sect. 3, robust H-infinity distributed controller design for the cooperative robotic setup is discussed. In Sects. 4 and 5, numerical experiments and discussions are presented. Finally, conclusions is drawn in Sect. 6.

2   Problem Definition

Consider the following distributed robot dynamic models with a disturbance input:

$$M_i(q_i)\ddot{q}_i + C_i(q_i, \dot{q}_i)\dot{q}_i + G_i(q_i) = \tau_i + \tau_{di} \qquad (1)$$

where $q_i \in \mathbb{R}^n$ is the angular position vector, $M_i \in \mathbb{R}^{n \times n}$ is the inertia matrix, $C_i \in \mathbb{R}^{n \times n}$ is the coriolis matrix $G_i \in \mathbb{R}^n$ is the gravity vector, $\tau_i \in \mathbb{R}^n$ is the input torque or force vector and $\tau_{di} \in \mathbb{R}^n$ is the disturbance input vector for $i = 1, ..., k$. In order to design the proposed distributed controller strategy, the following change of variable is needed $x_{1i} = q_i$ and $x_{2i} = \dot{q}_i$ to transform (1) to a state-space model, so the following model is obtained:

$$\dot{x}_{1i} = x_{2i}$$
$$\dot{x}_{2i} = -M_i^{-1}(x_{1i})C_i(x_{1i}, x_{2i})x_{2i} - M_i^{-1}(x_{1i})G_i(x_{1i})$$
$$+ M_i^{-1}(x_{1i})\tau_i + M_i^{-1}(x_{1i})\tau_{di} \quad (2)$$

so the following tracking errors are defined to obtain the error variables for the angular position of the joints of each robot:

$$e_{i1} = x_{1id} - x_{1i}$$
$$\dot{e}_{i1} = \dot{x}_{1id} - \dot{x}_{1i}$$
$$e_{i2} = x_{2id} - x_{2i} = \dot{x}_{1id} - \dot{x}_{1i}$$
$$\dot{e}_{i2} = \dot{x}_{2id} - \dot{x}_{2i} \quad (3)$$

where $x_{1id} \in \mathbb{R}^n$ is the reference of the angular position and $x_{2id} \in \mathbb{R}^n$ is the reference of the angular velocity of each robot. The position and orientation of the object that is carried by the cooperative robot setup is related to the angular position of each robot in the following way, so the reference position of the joints of each robot can be obtained by the following relation [7]:

$$\xi_0 = T_{0,i}(x_{1i})$$
$$\dot{\xi}_0 = J_{0,i}(x_{1i})x_{2i} \quad (4)$$

where $T_{0,i}(x_{1i})$ is the transformation matrix obtained by the forward kinematics of each robot and $J_{0,i}(x_{1i})$ is the Jacobian matrix. With these problem definition, the derivation of the robust decentralized H-infinity controller for the cooperative robot setup is obtained.

3 Robust H-Infinity Distributed Controller Design for the Cooperative Robotic Setup

Before deriving the robust H-infinity controller design, it is necessary to establish the following definitions to obtain the robust H-infinity distributed control law for the cooperative robot setup [11].

Definition 1. The closed loop system is $\mathcal{L}_2$ stable if there exist constants $\delta_i(x_{mi}(0))$ that met the following conditions:

$$\int_0^T \|e_{im}\|^2 dt \leq \gamma^2 \int_0^T \|\tau_{dim}\|^2 dt + \delta_i(x_{mi}(0)) \quad (5)$$

for any initial condition $T > 0$, $\gamma > 0$ and $m = 1, ..., k$ as the index of the robot manipulator in the cooperative robotic setup.

The following definition is important for the derivation of the robust H-infinity control law by using a Lyapunov approach:

Definition 2. The following condition is met to obtain the constant $\delta_i(x_{mi}(0))$

$$\gamma^2 = \left\| \frac{e_{i2}^T M_i^{-1}(x_{1i})\tau_{di}}{\alpha} \right\| > 0 \tag{6}$$

with $\alpha = \|\tau_{di}\|^2$

In order to obtain the robust H-infinity control law the following theorem is needed based on a Lyapunov approach.

Theorem 1. *The robust H-infinity control law (9) is obtained by selecting the appropriate Lyapunov function (7) and by using Definitions 1 and 2 in order to obtain a stable closed loop system.*

Proof. Consider the following Lyapunov function:

$$V_i = \frac{1}{2}e_{i1}^T e_{i1} + \frac{1}{2}e_{i2}^T e_{i2} \tag{7}$$

Now deriving (7), the following result is obtained:

$$\dot{V}_i = e_{i1}^T(\dot{x}_{1id} - \dot{x}_{1i}) + e_{i2}^T \dot{x}_{2id} + e_{i2}^T M_i^{-1}(x_{1i})C_i(x_{1i}, x_{2i})x_{2i} \\ + e_{i2}^T M_i^{-1}(x_{1i})G_i(x_{1i}) - e_{i2}^T M_i^{-1}(x_{1i})\tau_i - e_{i2}^T M_i^{-1}(x_{1i})\tau_{di} \tag{8}$$

yielding the following robust H-infinity control law:

$$\tau_i = M_i(x_{1i})\dot{x}_{2id} + C_i(x_{1i}, x_{2i})x_{2i} + G_i(x_{1i}) + M_i(x_{1i})\frac{e_{i2}}{\|e_{i2}\|^2}e_{i1}^T \dot{x}_{1id} \\ - M_i(x_{1i})\frac{e_{i2}}{\|e_{i2}\|^2}e_{i1}^T \dot{x}_{1i} - M_i(x_{1i})e_{i2} \tag{9}$$

substituting (9) in (8) and taking the norm at both sides of (8) yields:

$$\dot{V}_i \leq \|e_{i2}^T e_{i2}\| - \left\| \frac{e_{i2}^T M_i^{-1}(x_{1i})\tau_{di}}{\alpha} \right\| \|\tau_{di}\tau_{di}^T\| \tag{10}$$

now using definition 2 in (10) yields:

$$\dot{V}_i \leq \|e_{i2}^T e_{i2}\| - \gamma^2\|\tau_{di}\tau_{di}^T\| \tag{11}$$

integrating at both sides of (11) the following result is obtained:

$$V_i(T) - V_i(0) \leq \int_0^T \|e_{i2}^T e_{i2}\|dt - \gamma^2 \int_0^T \|\tau_{di}^T \tau_{di}\|dt \tag{12}$$

Finally re-arranging and using Definition 1 yields:

$$\int_0^T \|e_{i2}^T e_{i2}\|dt \leq \gamma^2 \int_0^T \|\tau_{di}^T \tau_{di}\|dt + \delta_i(x_{mi}(0)) \tag{13}$$

assuring the robust closed loop stability of the system in the presence of disturbances.

With these results a suitable robust control law is obtained ensuring the stability of the system in the presence of disturbances. The results obtained in Theorem 1 allow to reduce the trajectory tracking error to zero in a cooperative robotic setup when a synchronized actuator trajectory reference is needed to be followed by each robot.

4   Numerical and Simulation Example

In this section, a numerical and simulation example to test and validate the theoretical results of this study is shown. The cooperative robots setup is shown in Fig. 1 in which both two links robots are aligned in the Y axis. The objective of this experiment is to move the grasped object in almost a circular trajectory and as explained before the position and orientation of the grasped object is transformed to the actuators angular positions references.

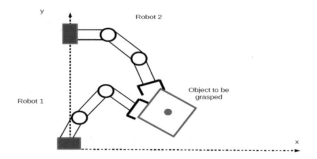

Fig. 1. Cooperative robots setup

Table 1. Cooperative robots setup

Parameter	Value
l_1 (m)	0.1
l_2 (m)	0.1
l_{c1} (m)	0.05
l_{c2} (m)	0.05
m_1 (Kg)	0.6
m_2 (Kg)	0.6
I_1 $(Kg.m^2)$	0.6
I_2 $(Kg.m^2)$	0.6
g m/s^2	9.81

In Table 1, the parameters of the two 2-links robots used in this cooperative robotic setup are shown. The kinematic and dynamic model of this manipulator can be found in [22] where the parameters for the robot 1 and robot 2 are the same. The disturbance inputs τ_{d1} and τ_{d2} are defined by:

$$\tau_{d1} = [0.001sin(\omega_{d1}t), 0.001sin(\omega_{d1}t)]^T \tag{14}$$

$$\tau_{d2} = [0.001sin(\omega_{d2}t), 0.001sin(\omega_{d2}t)]^T \tag{15}$$

where $\omega_{d1} = \omega_{d2} = (2\pi)/0.00002$ are the disturbance angular velocities constants used in this numerical simulation example for the cooperative robotic setup.

In Fig. 2 it is seen that the trajectory of the object is manipulated in the $X - Y$ plane following an almost circular trajectory. As it is shown, the desired trajectory and the path followed by the object is in a clockwise direction then it is proved that the tracking error difference is zero approximately in 0.5 seconds, proving that the tracking error is reduced to zero in finite time. It is worth noting that the trajectory reference of the object is slower than the obtained trajectory with predefined initial conditions.

In Fig. 3, the angular positions of the joints provided by the actuators for robot 1 are shown in which the desired trajectory for each joint is obtained by transforming the object desired trajectory by an appropriate transformation matrix (4) obtained by the forward kinematics of the grasped object.

As similar to the previous figure, in Fig. 4 the angular positions for the two joint of robot 2 are shown, in which it is noticed that there is a jump approximately at 0.03 s due to the synchronization of both robot desired trajectory as shown in Figs. 3 and 4. As explained before, the desired joint angular positions of robot

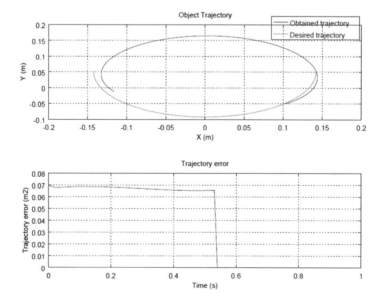

Fig. 2. Grasped object trajectory

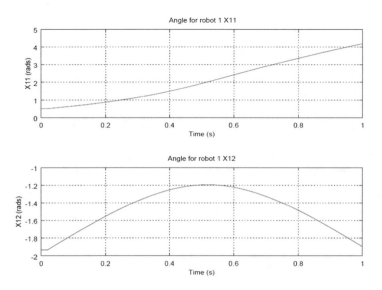

Fig. 3. Angular positions q_1 and q_2 for robot 1

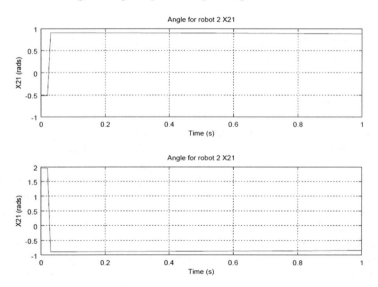

Fig. 4. Angular positions q_1 and q_2 for robot 2

1 and the desired angular positions of robot 2 are obtained by a transformation matrix as shown in (4).

Finally, in Fig. 5 the torque inputs for robot 2 are shown in which the required control input torques are applied in the actuators of robot 2 necessary to move the object to the almost circular path followed by the grasped object. It is important to remark, that the control actions of robot 2 reach zero in a smaller

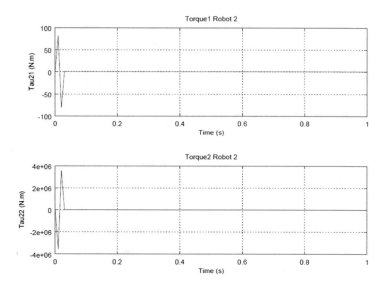

Fig. 5. Torque input τ_1 and τ_2 for robot 2

time interval that is why the control action of robot 1 is also needed to move the object in a specified path.

5 Discussion

The theoretical and simulation results shown in this paper evince the contribution of this study for the design of a distributed robust H-infinity controller for cooperative robotics tasks. As explained in the theoretical derivations of this paper, the robust closed loop stability conditions are achieved by selecting an appropriate robust control law to drive the angular positions of the robot joints in the setup according to the grasped object desired position and orientation. The simulation results show how the desired trajectory reference of the grasped object is followed efficiently reducing the tracking error to zero. For this purpose, the position and orientation of the grasped object is converted to the desired angular position of each robot in the cooperative robotic setup. This makes a difference in comparison with other studies in which the grasped object position is transformed to another dynamic model that can increase the computational effort if the controller strategy is tested in an experimental setup. The accuracy in reducing the tracking error of the grasped object is obtained by designing an appropriate control law based on the robust stability condition of the closed loop system, specifically, making the closed loop system $\mathcal{L}_2$ stable. Something that is important to remark, is that the distributed robust control law obtained by a Lyapunov approach avoids oscillations in the control action of each robot in the cooperative robot setup, even when disturbance are found in the system. To test and validate this fact, in the numerical simulation example the disturbance inputs of each robots are excited with high frequency oscillations. It is noticed that the disturbance suppression provided by the robust decentralized control law eliminates oscillations and poor performance or even instability.

6   Conclusion

In this paper a robust decentralized H-infinity controller for cooperative robotics is derived. Considering that the disturbances found in a cooperative robotic setup can yield a poor performance or instability, the proposed strategy provides an efficient solution to drive the robots joints synchronously to move a grasped object in a desired trajectory. The theoretical results ensure the closed loop system stability for the trajectory tracking of each robot even in the presence of disturbances avoiding oscillations in the controller action that can deteriorate the performance or even damage the robots structures. The numerical simulation results show that the grasped error tracking trajectory was reduced to zero in finite time. As in opposition to other approaches found in the literature, the desired angular position trajectory for each robot is obtained by a transformation of the position and orientation of the grasped object. As validated in the numerical simulation example, the trajectory tracking is accurate and with a lower computational effort. In the future this study will be extended to design other control strategies based on sliding mode control, for example, to deal with matched and mismatched uncertainties.

References

1. Abbaspour, A., Alipour, K., Zare Jafari, H., Moosavian, S.: Optimal formation and control of cooperative wheeled mobile robots. C. R. Mec. **343**(5–6), 307–321 (2015)
2. Agand, P., Motaharifar, M., Taghirad, H.: Decentralized robust control for teleoperated needle insertion with uncertainty and communication delay. Mechatronics **46**, 46–59 (2017)
3. Ammar, H.H., Azar, A.T., Tembi, T.D., Tony, K., Sosa, A.: Design and implementation of fuzzy PID controller into multi agent smart library system prototype. In: Hassanien, A.E., Tolba, M.F., Elhoseny, M., Mostafa, M. (eds.) The International Conference on Advanced Machine Learning Technologies and Applications (AMLTA2018), pp. 127–137. Springer International Publishing, Cham (2018)
4. Azar, A.T., Serrano, F.E.: Adaptive decentralised sliding mode controller and observer for asynchronous nonlinear large-scale systems with backlash. Int. J. Model. Ident. Control **30**(1), 61–71 (2018)

5. Azar, A.T., Ammar, H.H., Barakat, M.H., Saleh, M.A., Abdelwahed, M.A.: Self-balancing robot modeling and control using two degree of freedom PID controller. In: Hassanien, A.E., Tolba, M.F., Shaalan, K., Azar, A.T. (eds.) Proceedings of the International Conference on Advanced Intelligent Systems and Informatics 2018, pp. 64–76. Springer International Publishing, Cham (2019)

6. Azar, A.T., Hassan, H., Razali, M.S.A.B., de Brito Silva, G., Ali, H.R.: Two-degree of freedom proportional integral derivative (2-DOF PID) controller for robotic infusion stand. In: Hassanien, A.E., Tolba, M.F., Shaalan, K., Azar, A.T. (eds.) Proceedings of the International Conference on Advanced Intelligent Systems and Informatics 2018, pp. 13–25. Springer International Publishing, Cham (2019)

7. Baigzadehnoe, B., Rahmani, Z., Khosravi, A., Rezaie, B.: On position/force tracking control problem of cooperative robot manipulators using adaptive fuzzy backstepping approach. ISA Trans. **70**, 432–446 (2017)

8. Fekik, A., Denoun, H., Azar, A.T., Hamida, M.L., Zaouia, M., Benyahia, N.: Comparative study of two level and three level PWM-rectifier with voltage oriented control. In: Hassanien, A.E., Tolba, M.F., Shaalan, K., Azar, A.T. (eds.) Proceedings of the International Conference on Advanced Intelligent Systems and Informatics 2018, pp. 40–51. Springer International Publishing, Cham (2019)

9. Hichri, B., Fauroux, J.C., Adouane, L., Doroftei, I., Mezouar, Y.: Design of cooperative mobile robots for co-manipulation and transportation tasks. Rob. Comput. Integr. Manuf. **57**, 412–421 (2019)

10. Liu, Y., Huang, P., Zhang, F., Zhao, Y.: Robust distributed consensus for deployment of tethered space net robot. Aerosp. Sci. Technol. **77**, 524–533 (2018)

11. Liu, Y., Liu, X., Jing, Y., Zhou, S.: Adaptive backstepping H infinity tracking control with prescribed performance for internet congestion. ISA Trans. **72**, 92–99 (2018)

12. Marino, A., Pierri, F.: A two stage approach for distributed cooperative manipulation of an unknown object without explicit communication and unknown number of robots. Rob. Auton. Syst. **103**, 122–133 (2018)

13. Martinez-Rosas, J., Arteaga, M., Castillo-Sanchez, A.: Decentralized control of cooperative robots without velocity-force measurements. Automatica **42**(2), 329–336 (2006)

14. Meghni, B., Dib, D., Azar, A.T., Saadoun, A.: Effective supervisory controller to extend optimal energy management in hybrid wind turbine under energy and reliability constraints. Int. J. Dyn. Control **6**(1), 369–383 (2018)

15. Mekki, H., Boukhetala, D., Azar, A.T.: Sliding modes for fault tolerant control. In: Azar, A.T., Zhu, Q. (eds.) Advances and Applications in Sliding Mode Control systems, pp. 407–433. Springer International Publishing, Cham (2015)

16. Rigatos, G., Siano, P.: A new nonlinear H-infinity feedback control approach to the problem of autonomous robot navigation. Intell. Ind. Syst. **1**(3), 179–186 (2015)

17. Rigatos, G., Siano, P., Raffo, G.: An H-infinity nonlinear control approach for multi-dof robotic manipulators. IFAC-PapersOnLine **49**(12), 1406–1411 (2016)

18. Rigatos, G., Siano, P., Raffo, G.: A nonlinear H-infinity control method for multi-dof robotic manipulators. Nonlinear Dyn. **88**(1), 329–348 (2017)

19. Sabattini, L., Secchi, C., Levratti, A., Fantuzzi, C.: Decentralized control of cooperative robotic systems for arbitrary setpoint tracking while avoiding collisions. IFAC-PapersOnLine **48**(19), 57–62 (2015)

20. Smida, M.B., Sakly, A., Vaidyanathan, S., Azar, A.T.: Control-based maximum power point tracking for a grid-connected hybrid renewable energy system optimized by particle swarm optimization. In: Azar, A.T., Vaidyanathan, S. (ed.) Advances in System Dynamics and Control, pp. 58–89. IGI Global (2018)

21. Soliman, M., Azar, A.T., Saleh, M.A., Ammar, H.H.: Path planning control for 3-omni fighting robot using PID and fuzzy logic controller. In: Hassanien, A.E., Azar, A.T., Gaber, T., Bhatnagar, R., Tolba, M.F. (eds.) The International Conference on Advanced Machine Learning Technologies and Applications (AMLTA2019), pp. 442–452. Springer International Publishing, Cham (2020)
22. Spong, M., Hutchinson, S., Vidyasagar, M.: Robot Modeling and Control. Wiley, Hoboken (2006)
23. Valk, L., Keviczky, T.: Distributed control of heterogeneous underactuated mechanical systems. IFAC-PapersOnLine **51**(23), 325–330 (2018)

Adaptive Terminal-Integral Sliding Mode Force Control of Elastic Joint Robot Manipulators in the Presence of Hysteresis

Ahmad Taher Azar[1,2(✉)], Fernando E. Serrano[3], Anis Koubaa[4,5,6], Nashwa Ahmad Kamal[7], Sundarapandian Vaidyanathan[8], and Arezki Fekik[9]

[1] College of Engineering, Prince Sultan University, Riyadh, Kingdom of Saudi Arabia
aazar@psu.edu.sa
[2] Faculty of Computers and Artificial Intelligence, Benha University, Benha, Egypt
ahmad.azar@fci.bu.edu.eg
[3] Universidad Tecnologica Centroamericana (UNITEC), Tegucigalpa, Honduras
serranofer@eclipso.eu
[4] Prince Sultan University, Riyadh, Kingdom of Saudi Arabia
akoubaa@psu.edu.sa
[5] CISTER, INESC-TEC, ISEP, Polytechnic Institute of Porto, Porto, Portugal
[6] Gaitech Robotics, Shanghai, China
[7] Electrical Power and Machine Department, Faculty of Engineering,
Cairo University, Giza, Egypt
nashwa.ahmad.kamal@gmail.com
[8] Research and Development Centre,
Vel Tech University Avadi, Chennai 600062, Tamilnadu, India
sundarvtu@gmail.com
[9] Electrical Engineering Advanced Technology Laboratory (LATAGE),
University Mouloud Mammeri of Tizi-Ouzou, Tizi Ouzou, Algeria
arezkitdk@yahoo.fr

Abstract. In this paper, an adaptive terminal-integral sliding mode force control of elastic joint robot manipulators in the presence of hysteresis is proposed. One of the most important issues that is solved in this study is that the hysteresis phenomenon is considered something that provokes losses in the manipulator motion and controller errors. Force control is necessary because it can be implemented and very useful in the area of industrial robotics such as collaborative and cooperative robotics. Therefore, it can be implemented for precise control in which robot-operator or robot-robot interaction is needed. An adaptive terminal-integral sliding mode force control is proposed by considering the hysteresis and the effects between the end effector and a flexible environment. Force control has not been studied extensively nowadays and even less for elastic joint robot manipulators. Thus, to improve the system precision control, the adaptive sliding mode controller (ASMC) is designed by a Lyapunov approach obtaining the adaptive and controller laws, respectively. As an experimental case study, two links elastic joint robot manipulator is considered by obtaining the elastic joint model with hysteresis using a Bouc-Wen model.

© Springer Nature Switzerland AG 2020
A. E. Hassanien et al. (Eds.): AISI 2019, AISC 1058, pp. 266–276, 2020.
https://doi.org/10.1007/978-3-030-31129-2_25

Keywords: Hysteresis · Flexible structures · Robotics ·
Force control · Integral sliding mode control ·
Terminal sliding mode control

1   Introduction

Hysteresis and other non-linearities such as backlash sometimes are found in several kinds of actuators like linear and rotational, so for this reason, hysteresis in actuators has been studied since several years ago [3–5]. Hysteresis phenomenon must be avoided as it yields unwanted effects such as loss in precision and instability when classical and novel controller approaches are implemented resulting in negative and undesirable system performance. There are several hysteresis models that can be found in literature, but one of the most important hysteresis models found nowadays are the Bouc-Wen model [11,29–31]. The Bouc-Wen Model consists of a bending flexion force implementing an extra variable that is resolved by numerical integration. The Bouc-Wen model is important to be mentioned because a multi-variable Bouc-Wen model is used in this research. Other interesting studies found in literature related to hysteresis in robotics are explained in [20] where a control strategy is implemented to solve the performance deterioration of servo elastic drive systems, and in [25] where a torsion-angular velocity feedback is used to solve the vibrations of the tip of a robotic arm.

Force/position control has been extensively implemented for the control of different kinds of elastic robots [1,8,10,15,17,24,33]. For example, in [28], a hybrid force and position adaptive fuzzy sliding mode control for robotic manipulators is shown where the robot dynamics is decomposed in the force and position parts taking into account the environment and dynamic structure properties. In [27], another adaptive hybrid position control of robot arms with a rigid surface is proposed where the controller is designed in such a way constraints and uncertainties are considered to obtain a zero force and position error. In [15], a position/force controller for robotic arms in a flexible surface is implemented considering resistance and environment elasticity.

Sliding mode control (SMC) is an effective robust control strategy which has been successfully applied to a wide variety of complex systems and engineering [1,2,5,7,12,16,22,23,36,37]. The SMC system has various attractive features such as fast response, good transient performance, and robustness with respect to uncertainties and external disturbance and so on [6]. A major undesired phenomenon faced by SMC is the high frequency oscillations known as chattering which can cause instability and damage to the system. Other works have added an integral term to the sliding surface (ISMC) in order to reduce the steady-state error (SSE) compared to SMC [13,38]. Sliding mode control and its variation such as second order, integral and terminal sliding mode has been extensively used in the position and force control of robotics manipulator of any kind [9,18,21,26,32,39,40]. For example, in [19], the unknown dead-zone and disturbance phenomenon has been considered to design a finite-time sliding surface controller for robot manipulators something that is very important for this study considering that the dead-zone non-linearity has some similarities with the hysteresis treated in this work. Other examples can be found in [14,35], where a second order integral sliding mode controller for experimental applications and a guidance law for a missile intersection with impact angle constraint are shown.

In this study, an adaptive terminal-integral sliding mode controller is designed and implemented for the force control of flexible robotic manipulators in the presence of hysteresis. As explained before, a multi-variable Bouc-Wen hysteresis model is used to represent the flexion structure. The adaptive terminal-integral sliding mode force controller is designed by selecting an appropriate sliding surface with a suitable switching law and a control law and then by selecting an appropriate Lyapunov function, the adaptive laws for the controller gains can be obtained. Finally, an illustrative example is proposed where this strategy is tested in a two-links robotic manipulator.

The paper is organized as follows. In Sect. 2 presents the hysteresis modeling. In Sect. 3, the terminal-integral sliding mode force controller design is discussed. In Sect. 4, numerical experiments and discussions are presented. Finally, conclusions are drawn in Sect. 5.

2   Hysteresis Modeling

This section presents the hysteresis modeling process used in this study. In Fig. 1, a rotary actuator and joint with a longitudinal flexion are shown in which the phenomenon of hysteresis occurs yielding a difference in the rotation angle of $q - x$ difference that influences in the position and force exerted at the tip of the structure [29].

The hysteresis model used in this study is the Bouc-Wen model which is defined as [11,29–31]:

$$f(x, \dot{x}) = d\dot{x} + h(x) \tag{1}$$

where $f()$ is the restoring force, d is the linear damping and h is the hysteresis restoring force. $h(x)$ is given by [29]:

$$h(x) = \omega\eta x + (1 - \omega)\eta z \tag{2}$$

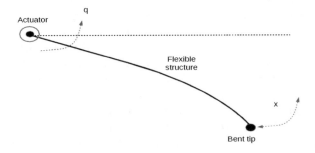

Fig. 1. Joint flexion

where η is a position dependent map and $0 \leq \omega \leq 1$ is the ratio of linear to nonlinear restoring force [29]. The hysteresis state z is defined by the following differential equation [29]:

$$\dot{z}(t) = \dot{x}(t) - \beta|\dot{x}(t)||z(t)|^{n-1}z(t) - \gamma\dot{x}|z(t)|^n \tag{3}$$

where β, γ, n determine the shape of the hysteresis. These equations are transformed in a multi-variable form considering that a robotic manipulator has several degrees of freedom. In the next section, this will be explained with more details.

3   Dynamic Model Equations of the Flexible Robotic Manipulator

The dynamic model of the flexible manipulator represented in Euler-Lagrange form with a flexible environment is presented in this section and is given by: [15, 27, 28]:

$$\mathbf{M}(q)\ddot{q} + \mathbf{C}(q, \dot{q})\dot{q} + \mathbf{G}(q) = \tau - \mathbf{J}^T(q)\bar{\mathbf{F}} \tag{4}$$

where $q \in \mathbb{R}^n$ is the actuators position and/or orientation vector, $\mathbf{M} \in \mathbb{R}^{n \times n}$ is the inertia matrix, $\mathbf{D} \in \mathbb{R}^{n \times n}$ is the coriolis matrix, $\mathbf{G} \in \mathbb{R}^n$ is the gravity vector, $\tau \in \mathbb{R}^n$ is the input control vector, $\mathbf{J} \in \mathbb{R}^{m \times n}$ is the Jacobian matrix and $\bar{\mathbf{F}} \in \mathbb{R}^m$ is the interaction force between the end effector and the environment. $\bar{\mathbf{F}}$ is given by:

$$\bar{\mathbf{F}} = \mathbf{F}(\mathbf{X}, \dot{\mathbf{X}}) - \bar{\mathbf{K}}\mathbf{X}_e \tag{5}$$

where the force $\bar{\mathbf{F}}$ is given by the difference between the forces of the flexible end effector and the flexible surface, $\mathbf{F}(\mathbf{X}, \dot{\mathbf{X}})$ is the hysteresis non-linear restoring force in vector form as shown in (1), $\mathbf{X}_e$ is the position and orientation of the deformable environment, $\mathbf{X}$ is the position and orientation of the end effector and $\bar{\mathbf{K}}$ is the environment stiffness matrix [28]. $\mathbf{F}(\mathbf{X}, \dot{\mathbf{X}})$ is given by:

$$\mathbf{F}(\mathbf{X}, \dot{\mathbf{X}}) = \mathbf{D}\dot{\mathbf{X}} + \mathbf{h}(\mathbf{X}) \tag{6}$$

So, (6) is the vector form of (1) where $\mathbf{D}$ is a matrix of appropriate dimensions and $\mathbf{h}(\mathbf{X})$ is a vector of appropriate dimensions. Substituting (6) and (5) in (4) yields:

$$\mathbf{M}(q)\ddot{q} + \mathbf{C}(q, \dot{q})\dot{q} + \mathbf{G}(q) = \tau - \mathbf{J}^T(q)\mathbf{D}\dot{\mathbf{X}} - \mathbf{J}^T(q)\mathbf{h}(\mathbf{X}) + \mathbf{J}^T(q)\bar{\mathbf{K}}\mathbf{X}_e \tag{7}$$

The difference of forces obtained by the difference in the position and orientation of the end effector when this structure is in flexion is given by:

$$\mathbf{K}_q q - \mathbf{K}_x \mathbf{X} = \mathbf{F}(\mathbf{X}, \dot{\mathbf{X}}) \tag{8}$$

where $\mathbf{K}_q$ and $\mathbf{K}_x$ are matrices of appropriate dimensions. Now substituting (6) into (8) and using the vector form of (3) yields:

$$\mathbf{D}^{-1}\mathbf{K}_q q - \mathbf{D}^{-1}\mathbf{K}_x \mathbf{X} - \mathbf{D}^{-1}\mathbf{h}(\mathbf{X}) = \dot{\mathbf{X}}$$
$$\dot{\mathbf{Z}} = \dot{\mathbf{X}} - \mathbf{B}\|\dot{\mathbf{X}}\|\|\mathbf{Z}\|^{n-1}\mathbf{Z} - \mathbf{C}\dot{\mathbf{X}}\|\mathbf{Z}\|^n \tag{9}$$

where $\mathbf{B}$ contains all the parameters β and $\mathbf{C}$ contains all the parameters γ, both matrices must be of appropriate dimensions. The whole dynamic model is given by (7) and (9). Before deriving the adaptive integral-terminal sliding mode controller it must be considered that the force error is given by:

$$\mathbf{e} = \bar{\mathbf{F}} = \mathbf{F}(\mathbf{X}, \dot{\mathbf{X}}) - \bar{\mathbf{K}}\mathbf{X}_e \tag{10}$$

4   Terminal-Integral Sliding Mode Force Controller Design

The terminal-integral sliding mode force controller for an elastic joint robot in the presence of hysteresis in a flexible environment is discussed in this section. The proposed controller is based on [21,40] by designing the variable s and σ with an appropriate switching law. Consider the following integral sliding variable:

$$s = \mathbf{e}(t) + \int_0^t \mathbf{e}(\tau)d\tau \tag{11}$$

where $\mathbf{e}$ is the error variable defined in (10). Now the following sliding variable is designed [40]:

$$\sigma = \dot{s} + k_1 s + k_2 \lambda(s) \tag{12}$$

where $k_1, k_2 \in \mathbb{R}$ are the sliding gain variables adjusted by the adaptation law and $\lambda(s)$ [40]:

$$\lambda(s) = [\lambda(s_1)...\lambda(s_m)]$$

$$\lambda(s_i) = \begin{cases} s_i^{a/p} & if \ \bar{\sigma}_i = 0 \ or \ \bar{\sigma}_i \neq 0 \ |s_i| \geq \mu \\ \gamma_1 s_i + \gamma_2 sign(s_i)s_i^2 \ if \ \bar{\sigma}_i \neq 0, & |s_i| < \mu \end{cases} \tag{13}$$

All the constants are explained in detail in [40]. The terminal-integral sliding mode control law is obtained by making $\sigma = 0$ as shown below:

$$\dot{s} + k_1 s + k_2 \lambda(s) = 0$$
$$\dot{e} + e + k_1 s + k_2 \lambda(s) = 0 \tag{14}$$

the error in (10) is equivalent to the following equation:

$$\mathbf{e} = \bar{\mathbf{F}} = \mathbf{J}^{T(-1)}(q)\tau - \mathbf{J}^{T(-1)}(q)\mathbf{M}(q)\ddot{q} - \mathbf{J}^{T(-1)}(q)\mathbf{C}(q,\dot{q})\dot{q}$$
$$- \mathbf{J}^{T(-1)}(q)\mathbf{G}(q) \tag{15}$$

Therefore, the control law is:

$$\tau = -\mathbf{J}^T(q)\dot{\mathbf{e}} + \mathbf{M}(q)\ddot{q} + \mathbf{C}(q,\dot{q})\dot{q} + \mathbf{G}(q) - \mathbf{J}^T(q)k_2\lambda(s) + \mathbf{J}^T(q)k_2 s \tag{16}$$

The adaptive gains laws for the terminal-integral sliding mode force controller for an elastic robotic manipulator are given in the following theorem:

Theorem 1. *The adaptive gains laws k_1 and k_2 given by:*

$$\dot{k}_1 = -\sigma^T \dot{s}$$
$$\dot{k}_2 = -\sigma^T \dot{s} \tag{17}$$

are obtained by selecting an appropriate Lyapunov function in order that the system must be globally asymptotically stable.

Proof. First, substitute the torque input control variable (16) in (12) obtaining the following result:

$$\sigma = k_1 s + k_2 s$$
$$\dot{\sigma} = k_1 \dot{s} + k_2 \dot{s} \tag{18}$$

and selecting the following Lyapunov function:

$$V = \frac{1}{2}\sigma^T \sigma + \frac{1}{2}k_1^2 + \frac{1}{2}k_2^2 \tag{19}$$

Taking the derivative of (19) yields:

$$\dot{V} = \sigma^T k_1 \dot{s} + \sigma^T k_2 \dot{s} + k_1 \dot{k}_1 + k_2 \dot{k}_2 \tag{20}$$

obtaining the following adaptive laws that makes the system stable:

$$\dot{k}_1 = -\sigma^T \dot{s}$$
$$\dot{k}_2 = -\sigma^T \dot{s} \tag{21}$$

concluding that the Lyapunov derivative meets the following condition:

$$\dot{V} \leq 0 \tag{22}$$

with this conclusion, the system is stable with the control law (16) and the adaptive laws (21).

In the following section a numerical example is shown to test the validity of the theoretical results.

5   Simulation Results and Discussions

In this experiment, a two links robotic manipulator is implemented with their parameter values as shown in Table 1 [34]. A comparative analysis is done using MATLAB for which the results and performances of the strategies shown in [15, 29] are compared with the proposed control strategy. In Fig. 2, it can be seen that

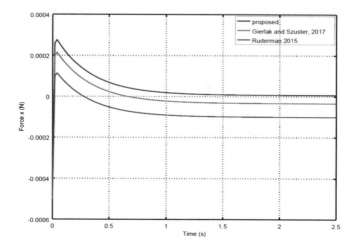

Fig. 2. Force variable $\bar{\mathbf{F}}_1$

Table 1. Two links robot parameters

Parameter	Value
m_1 Kg	0.0006
m_2 Kg	0.0006
l_1 m	0.5
l_2 m	0.5
l_{c1} m	0.25
l_{c2} m	0.25
I_1 $Kg.m^2$	0.6
I_2 $Kg.m^2$	0.6
g m/s^2	9.81

the force error $\bar{\mathbf{F}}$ approximates zeros as times goes to infinity so it is proved that the force controller provides an excellent performance. It is important to notice that the force error approaches zero as time goes to infinity in contrast with the results obtained by the other two strategies [15, 29] that do not approaches zero as time goes to infinity.

Then in Fig. 3, it can be seen how the sliding variables σ and s reach zero as times goes to infinity proving the optimal performance of this terminal-sliding mode controller. Finally in Fig. 4, the torque variable τ_1 for the actuator 1 is shown where it can be noticed that the torque provided by the proposed control strategy tends to decrease in comparison with the comparative approaches [15, 29].

The most important issue that was considered in this study is that a suitable hysteresis mathematical model is selected taking into account that the Bouc-Wen model provides an efficient mathematical representation of hysteresis.

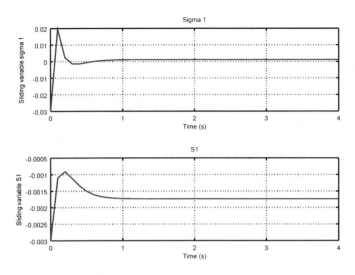

Fig. 3. Sliding variables σ and s

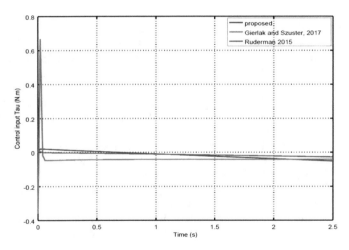

Fig. 4. Torque variable τ_1

Another important point is that this model is relatively simple and it can be implemented in a real experimental setup, so for these advantages this hysteresis model was used for the restoration forces modeling. The Bouc-wen model used in this study is extended to a multi-variable system for an efficient computation and then the derivation of the terminal-integral sliding mode control is eased by this multi-variable hysteresis model. Simulation results prove the superior performance of other force control robots in the presence of hysteresis and flexible environment.

6 Conclusion

In this paper, the design of an adaptive terminal-integral sliding mode force control for robotic manipulators in the presence of hysteresis in flexible environments is presented. The proposed controller provides a superior performance in comparison with other similar approaches found in literature. The selection of the hybrid terminal-integral sliding mode surface is done in a suitable way in order to obtain the control and adaptive gains laws that drive the force error to zero as times goes to infinity. Using of fractional order control can be considered as a future direction. Optimizing the sliding mode controller parameters with new metaheuristic techniques to obtain the best values for the sliding mode control law parameters is also another future work of this study.

Acknowledgments. This work is supported by the Robotics and Internet of Things lab of Prince Sultan University, Saudi Arabia.

References

1. Adhikary, N., Mahanta, C.: Sliding mode control of position commanded robot manipulators. Control Eng. Pract. **81**, 183–198 (2018)
2. Azar, A.T., Serrano, F.E.: Adaptive sliding mode control of the Furuta Pendulum, vol. 576, pp. 1–42. Springer, Cham (2015). https://doi.org/10.1007/978-3-319-11173-5_1
3. Azar, A.T., Serrano, F.E.: Stabilizatoin and control of mechanical systems with Backlash. In: Advances in Computational Intelligence and Robotics (ACIR), pp 1–60. IGI-Global (2015)
4. Azar, A.T., Serrano, F.E.: Stabilization of mechanical systems with backlash by PI loop shaping. Int. J. Syst. Dyn. Appl. **5**(3), 21–46 (2016)
5. Azar, A.T., Serrano, F.E.: Adaptive decentralised sliding mode controller and observer for asynchronous nonlinear large-scale systems with backlash. Int. J. Model. Ident. Control **30**(1), 61–71 (2018)
6. Azar, A.T., Zhu, Q.: Advances and applications in sliding mode control systems. In: Studies in Computational Intelligence, vol. 576. Springer (2015)
7. Azar, A.T., Kumar, J., Kumar, V., Rana, K.P.S.: Control of a two link planar electrically-driven rigid robotic manipulator using fractional order SOFC, pp. 57–68. Springer, Cham (2018). https://doi.org/10.1007/978-3-319-64861-3_6
8. Ba, K., Yu, B., Gao, Z., Zhu, Q., Ma, G., Kong, X.: An improved force-based impedance control method for the hdu of legged robots. ISA Trans. **84**, 187–205 (2019)
9. Baigzadehnoe, B., Rahmani, Z., Khosravi, A., Rezaie, B.: On position/force tracking control problem of cooperative robot manipulators using adaptive fuzzy backstepping approach. ISA Trans. **70**, 432–446 (2017)
10. Chen, F., Zhao, H., Li, D., Chen, L., Tan, C., Ding, H.: Contact force control and vibration suppression in robotic polishing with a smart end effector. Robot. Comput. Integr. Manuf. **57**, 391–403 (2019)
11. Colangelo, F.: Interaction of axial force and bending moment by using Bouc-Wen hysteresis and stochastic linearization. Struct. Saf. **67**, 39–53 (2017)

12. Deng, Y., Wang, J., Li, H., Liu, J., Tian, D.: Adaptive sliding mode current control with sliding mode disturbance observer for PMSM drives. ISA Trans. **88**, 113–126 (2019)
13. Fan, C., Hong, G.S., Zhao, J., Zhang, L., Zhao, J., Sun, L.: The integral sliding mode control of a pneumatic force servo for the polishing process. Precis. Eng. **55**, 154–170 (2019)
14. Furat, M., Eker, I.: Second-order integral sliding-mode control with experimental application. ISA Trans. **53**(5), 1661–1669 (2014)
15. Gierlak, P., Szuster, M.: Adaptive position/force control for robot manipulator in contact with a flexible environment. Robot. Auton. Syst. **95**, 80–101 (2017)
16. Gracia, L., Solanes, J.E., Muoz-Benavent, P., Esparza, A., VallsMiro, J., Tornero, J.: Cooperative transport tasks with robots using adaptive non-conventional sliding mode control. Control Eng. Pract. **78**, 35–55 (2018)
17. Gracia, L., Solanes, J.E., Muoz-Benavent, P., Miro, J.V., Perez-Vidal, C., Tornero, J.: Adaptive sliding mode control for robotic surface treatment using force feedback. Mechatronics **52**, 102–118 (2018)
18. Haghighi, D.A., Mobayen, S.: Design of an adaptive super-twisting decoupled terminal sliding mode control scheme for a class of fourth-order systems. ISA Trans. **75**, 216–225 (2018)
19. Han, S.I., Lee, J.: Finite-time sliding surface constrained control for a robot manipulator with an unknown deadzone and disturbance. ISA Trans. **65**, 307–318 (2016)
20. Helma, V., Goubej, M., Jezek, O.: Acceleration feedback in PID controlled elastic drive systems. IFAC-PapersOnLine **51**(4), 214–219 (2018)
21. Jing, C., Xu, H., Niu, X.: Adaptive sliding mode disturbance rejection control with prescribed performance for robotic manipulators. ISA Trans. (2019)
22. Ma, Z., Sun, G.: Dual terminal sliding mode control design for rigid robotic manipulator. J. Franklin Inst. **355**(18), 9127–9149 (2018). special Issue on Control and Signal Processing in Mechatronic Systems
23. Mekki, H., Boukhetala, D., Azar, A.T.: Sliding modes for fault tolerant control. In: Azar, A.T., Zhu, Q. (eds.) Advances and Applications in Sliding Mode Control systems, pp. 407–433. Springer International Publishing, Cham (2015)
24. Navvabi, H., Markazi, A.H.: Hybrid position/force control of Stewart Manipulator using extended adaptive fuzzy sliding mode controller (e-afsmc). ISA Trans. **88**, 280–295 (2019)
25. Oaki, J.: Physical parameter estimation for feedforward and feedback control of a robot arm with elastic joints. IFAC-PapersOnLine **51**(15), 425–430 (2018)
26. Peng, J., Yang, Z., Wang, Y., Zhang, F., Liu, Y.: Robust adaptive motion/force control scheme for crawler-type mobile manipulator with nonholonomic constraint based on sliding mode control approach. ISA Trans. (2019)
27. Pliego-Jimenez, J., Arteaga-Perez, M.A.: Adaptive position/force control for robot manipulators in contact with a rigid surface with uncertain parameters. Eur. J. Control **22**, 1–12 (2015)
28. Ravandi, A.K., Khanmirza, E., Daneshjou, K.: Hybrid force/position control of robotic arms manipulating in uncertain environments based on adaptive fuzzy sliding mode control. Appl. Soft Comput. **70**, 864–874 (2018)
29. Ruderman, M.: Feedback linearization control of flexible structures with hysteresis. IFAC-PapersOnLine **48**(11), 906–911 (2015)
30. Ruderman, M., Bertram, T.: Modeling and observation of hysteresis lost motion in elastic robot joints. IFAC Proc. Volumes **45**(22), 13–18 (2012)
31. Ruderman, M., Bertram, T., Iwasaki, M.: Modeling, observation, and control of hysteresis torsion in elastic robot joints. Mechatronics **24**(5), 407–415 (2014)

32. Seo, I.S., Han, S.I.: Dual closed-loop sliding mode control for a decoupled three-link wheeled mobile manipulator. ISA Trans. **80**, 322–335 (2018)

33. Solanes, J.E., Gracia, L., Muoz-Benavent, P., Miro, J.V., Carmichael, M.G., Tornero, J.: Humanrobot collaboration for safe object transportation using force feedback. Robot. Auton. Syst. **107**, 196–208 (2018)

34. Spong, M., Hutchinson, S., Vidyasagar, M.: Robot Modeling and Control. Wiley, Hoboken (2006)

35. Sun, L., Wang, W., Yi, R., Xiong, S.: A novel guidance law using fast terminal sliding mode control with impact angle constraints. ISA Trans. **64**, 12–23 (2016)

36. Vaidyanathan, S., Azar, A.T.: Hybrid synchronization of identical chaotic systems using sliding mode control and an application to vaidyanathan chaotic systems. In: Azar, A.T., Zhu, Q. (eds.) Advances and Applications in Sliding Mode Control Systems. Studies in Computational Intelligence, vol. 576, pp. 549–569. Springer, Berlin (2015)

37. Vaidyanathan, S., Sampath, S., Azar, A.T.: Global chaos synchronisation of identical chaotic systems via novel sliding mode control method and its application to zhu system. Int. J. Modell. Ident. Control **23**(1), 92–100 (2015)

38. Wang, Y., Xia, Y., Li, H., Zhou, P.: A new integral sliding mode design method for nonlinear stochastic systems. Automatica **90**, 304–309 (2018)

39. Wang, Y., Chen, J., Yan, F., Zhu, K., Chen, B.: Adaptive super-twisting fractional-order nonsingular terminal sliding mode control of cable-driven manipulators. ISA Trans. **86**, 163–180 (2019)

40. Yi, S., Zhai, J.: Adaptive second-order fast nonsingular terminal slidingmode control for robotic manipulators. ISA Trans. (2019)

Controlling Chaotic System via Optimal Control

Shikha Singh[1] and Ahmad Taher Azar[2,3(✉)]

[1] Department of Mathematics, Jesus and Mary College, University of Delhi, New Delhi, India
sshikha7014@gmail.com
[2] College of Engineering, Prince Sultan University, Riyadh, Kingdom of Saudi Arabia
aazar@psu.edu.sa
[3] Faculty of Computers and Information, Benha University, Benha, Egypt
ahmad.azar@fci.bu.edu.eg

Abstract. Chaos is a bounded unstable dynamic behavior that exhibits sensitive dependence on initial conditions and includes infinite unstable periodic motions. This article examines the controlling of a chaotic system via optimal control technique which is based on the Pontryagin minimum principle. A 3D chaotic system is considered to apply this scheme which have 5 equilibrium points. Finally, numerical simulations are presented to demonstrate the effectiveness of the proposed method. The simulation results illustrated the stabilized behaviour of states and control functions for different equilibrium points.

Keywords: Chaotic system · Equilibrium points · Chaos control · Optimal control

1 Introduction

Chaos have been detected in many natural and engineering fields [18,34,38]. Chaotic systems are being used in the field of communication [9,21], signal processing [13,39], medicine [6,14,28], biology [33], encryption [4], hydrology [32] and complex dynamical networks [12] because of their important properties, potential mechanism for signal design and generation. Aperiodic long time behavior and sensitivity to initial conditions of a chaotic system plays a vital role in synchronization and anti-synchronization [1]. It is required to synchronize or anti-synchronize between chaotic systems because of the above mentioned properties and applications.

Chaos control refers to purposefully manipulating chaotic dynamical behaviors of some complex nonlinear systems [23,25]. As a new and young discipline, chaos control has in fact come into play with many traditional scientific and technological advances today. Automatic control theory and practice, on the other hand, is a traditional and long-lasting engineering discipline. It has recently rapidly evolved and expanded, to overlap with and sometimes completely encompass many new and emerging technical areas of developments, and chaos control is one of them [2,3,8,11,22,24,26,27].

© Springer Nature Switzerland AG 2020
A. E. Hassanien et al. (Eds.): AISI 2019, AISC 1058, pp. 277–287, 2020.
https://doi.org/10.1007/978-3-030-31129-2_26

This new technology of chaos control promises to have a major impact on many novel, perhaps not-so-traditional, time- and energy-critical engineering applications. Examples include such as data traffic congestion control in the Internet, encryption and secure communication at different levels of communications, high-performance circuits and devices (e.g., delta-sigma modulators and power converters), liquid mixing, chemical reactions, power systems collapse prediction and protection, oscillators design, biological systems modelling and analysis (e.g., the brain and the heart), crisis management (e.g., jet-engine serge and stall), nonlinear computing and information processing, and critical decision-making in political, economic as well as military events. In fact, this new and challenging research and development area has become an attractive scientific inter-discipline involving control and systems engineers, theoretical and experimental physicists, applied mathematicians, and physiologists alike [7,10,21].

There are many practical reasons for controlling or ordering chaos. In a system where chaotic response is undesired or harmful, it should be reduced as much as possible, or totally suppressed. Examples of this include avoiding fatal voltage collapse in power networks, eliminating deadly cardiac arrhythmias, guiding disordered circuit arrays (e.g., multi-coupled oscillators and cellular neural networks) to reach a certain level of desirable pattern formation, regulating dynamical responses of mechanical and electronic devices (e.g., diodes, laser machines, and machine tools), and organizing a multi-agency corporation to achieve optimal performance. In recent years, the control of chaotic systems has received great attention due to its potential applications in physics, chemical reactor, biological networks, artificial neural networks, telecommunications, etc. Basically, chaos controlling is the stabilization of an unstable periodic orbit or equilibria by means of tiny perturbations of the system. For chaos control, some useful and powerful methods have been developed for sustained development of humanity. These may include active control [5,15,30,31], optimal control [20], adaptive control [17,38], state-feedback control [18], sliding mode control [29,35,36], adaptive sliding mode control [16,19], backstepping controller [37] etc. In this paper optimal control is presented for controlling chaotic system based on the Pontryagin minimum principle.

This article is organized as follows. Section 2 describes the behavior of the 3D chaotic system. The main results are presented in Sect. 3 in which controllers are computed via optimal control technique. In Sect. 4, numerical experiments are given. Finally, Sect. 5 is the brief conclusion.

2   System Description

A 3D continuous autonomous chaotic system [17] having seven terms with three quadratic terms which is given by:

$$\begin{cases} \dot{u}_1 = u_2 - au_1 + 10u_2u_3 \\ \dot{u}_2 = cu_2 + 5u_1u_3 \\ \dot{u}_3 = bu_3 - 5u_1u_2 \end{cases} \tag{1}$$

where u_1, u_2, u_3 are the variables and a, b, c are the parameters.

2.1 Chaotic Attractor of the System

When the parameters $a = 0.4$, $b = 0.175$, $c = -0.4$ and initial conditions $(0.349, 0.113, 0.2)$ are chosen, then the system displays chaotic attractor (two strange attractors) as shown in Fig. 1. The corresponding Lyapunov exponents of the novel chaotic attractor are $\gamma_1 = 0.071025$, $\gamma_2 = -0.000032932$, $\gamma_3 = -0.69599$. The Kaplyan-Yorke dimension is defined by:

$$D = j + \sum_{i=1}^{j} \frac{\gamma_i}{|\gamma_{j+1}|}$$
$$= 2.10209$$

where j is the largest integer satisfying $\sum_{i=1}^{j} \gamma_j \geq 0$ and $\sum_{i=1}^{j+1} \gamma_j < 0$. Therefore, Kaplan-Yorke dimension of the chaotic attractor is $D = 2.10209$ which means that the Lyapunov dimension of the chaotic attractor is fractional.

2.2 Behavior of Equilibria

The equilibria of system given by (1) is obtained by solving the following equations:

$$\begin{cases} u_2 - au_1 + 10u_2u_3 = 0 \\ cu_2 + 5u_1u_3 = 0 \\ bu_3 - 5u_1u_2 = 0 \end{cases} \tag{2}$$

The system has five equilibrium points:
$E_0 = (0, 0, 0)$,
$E_1 = (-0.052915, -0.016865, 0.0254983)$,
$E_2 = (-0.052915, 0.083009, -0.125498)$,
$E_3 = (0.052915, -0.083009, -0.125498)$,
$E_4 = (0.052915, 0.016656, 0.0254983)$.

Proposition 1. The equilibrium point E_0 is a saddle point and unstable.
Proof. The Jacobian matrix of system given by (1) at the equilibrium point E_0 is given by:

$$Jac = \begin{bmatrix} -a & 1 & 0 \\ 0 & c & 0 \\ 0 & 0 & b \end{bmatrix} \tag{3}$$

The eigenvalues of the Jacobian (Jac) are $\lambda_1 = -a = 0.4$, $\lambda_2 = c = -0.4$ and $\lambda_3 = b = 0.175$. Here λ_3 is a positive real number, λ_1 and λ_2 are two negative real numbers. Therefore, the equilibrium point E_0 is a saddle point and unstable. This proof is completed.

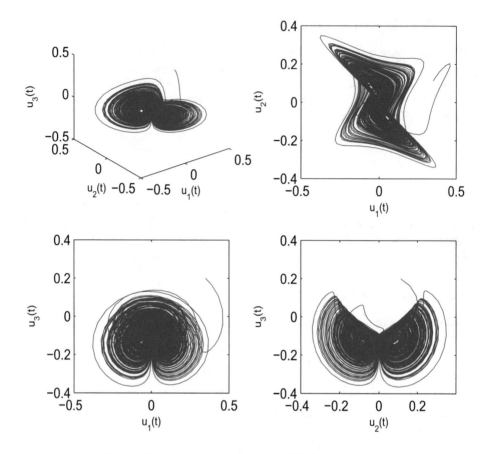

Fig. 1. Phase portrait of the novel 3D chaotic system

Proposition 2. The equilibrium points E_1 is a saddle focus point and unstable.
Proof. The Jacobian matrix of system given by Eq. 2 at the equilibrium point
E_1 is given by:

$$Jac = \begin{bmatrix} -a & 1 + 10u_3 & 10u_2 \\ 5u_3 & c & 5u_1 \\ -5u_2 & -5u_1 & b \end{bmatrix} \qquad (4)$$

$$= \begin{bmatrix} -0.4 & 0.25498 & 0.83 \\ 0.62749 & -0.4 & -0.264575 \\ 0.415 & 0.264575 & 0.175 \end{bmatrix} \qquad (5)$$

The eigenvalues of the Jacobian Jac are $\lambda_1 = -0.8$, $\lambda_2 = 0.0875 + 0.6378\iota$ and
$\lambda_3 = 0.0875 - 0.6378\iota$. Here λ_1 is a negative real number, λ_2 and λ_3 become a
pair of complex conjugate eigenvalues with positive real parts. So the equilibrium
point E_1 is a saddle focus point and unstable. This proof is completed.

Remark 1. In the same way, we can prove that E_2 , E_3 and E_4 are also saddle-
focus points and unstable.

3 Optimal Control of Chaotic System

In this section, we use Pontryagin minimum principle (PMP) to achieve optimal chaos control of the 3D chaotic system (2) at its equilibrium points. For the purpose of optimal chaos control, controllers U_1, U_2 and U_3 are applied to the 3D chaotic system (2). The controlled 3D chaotic system is defined as:

$$\begin{cases} \dot{u}_1 = u_2 - au_1 + 10u_2u_3 + U_1 \\ \dot{u}_2 = cu_2 + 5u_1u_3 + U_2 \\ \dot{u}_3 = bu_3 - 5u_1u_2 + U_3 \end{cases} \tag{6}$$

where U_1, U_2 and U_3 are the control inputs which should be satisfied by the optimal conditions at its equilibrium points obtained by PMP with respect to the cost function $\mathbf{J}$. The main strategy to control the system is to design the optimal control inputs U_1, U_2 and U_3 such that the state trajectories tend to the unstable equilibrium points in a given finite time interval $[0, t_f]$. Thus, the boundary conditions are :

$$u_1(0) = u_{1o}, \quad u_1(t_f) = \overline{u_1}$$
$$u_2(0) = u_{2o}, \quad u_2(t_f) = \overline{u_2}$$
$$u_3(0) = u_{3o}, \quad u_3(t_f) = \overline{u_3}$$

where $\overline{u_1}$, $\overline{u_2}$ and $\overline{u_3}$, denote the coordinates of the equilibrium points.

The objective function to be minimized is defined as:

$$\mathbf{J} = \frac{1}{2} \int_0^{t_f} (\alpha_1(u_1 - \overline{u_1})^2 + \alpha_2(u_2 - \overline{u_2})^2 + \alpha_3(u_3 - \overline{u_3})^2 + \beta_1 U_1{}^2 + \beta_1 U_2{}^2 + \beta_1 U_3{}^2)dt \tag{7}$$

where α_i and β_i (i = 1, 2, 3) are positive constants.

Now, the optimality conditions can be derived as a nonlinear two point boundary value problem (TPBVP) arising in the Pontryagin minimum principle(PMP). The corresponding Hamiltonian function $\mathbf{H}$ will be:

$$\mathbf{H} = \lambda_1[u_2 - au_1 + 10u_2u_3 + U_1] + \lambda_2[cu_2 + 5u_1u_3 + U_2] + \lambda_3[bu_3 - 5u_1u_2 + U_3]$$
$$- \frac{1}{2}[\alpha_1(u_1 - \overline{u_1})^2 + \alpha_2(u_2 - \overline{u_2})^2 + \alpha_3(u_3 - \overline{u_3})^2 + \beta_1 U_1{}^2 + \beta_1 U_2{}^2 + \beta_1 U_3{}^2] \tag{8}$$

where λ_1, λ_2 and λ_3 are the costate variables.

On applying the Pontryagin minimum principle (PMP), the Hamiltonian equations can be obtained:

$$\begin{cases} \dot{\lambda}_1 = -\dfrac{\partial H}{\partial u_1} \\[2mm] \dot{\lambda}_2 = -\dfrac{\partial H}{\partial u_2} \\[2mm] \dot{\lambda}_3 = -\dfrac{\partial H}{\partial u_3} \end{cases} \tag{9}$$

From (8) and (9), we have:

$$\begin{cases} \dot{\lambda}_1 = \alpha_1(u_1 - \overline{u_1}) + a\lambda_1 - 5\lambda_2 u_3 + 5\lambda_3 u_2 \\ \dot{\lambda}_2 = \alpha_2(u_2 - \overline{u_2}) - \lambda_1 - c\lambda_2 + 5\lambda_3 u_1 - 10\lambda_1 u_3 \\ \dot{\lambda}_3 = \alpha_3(u_3 - \overline{u_3}) - 10\lambda_1 u_2 - 5\lambda_2 u_1 - b\lambda_3 \end{cases} \quad (10)$$

The optimal control functions that have to be used are determined from the condition $\frac{\partial H}{\partial U_i} = 0 (i = 1, 2, 3)$. Hence,

$$U_i^* = \frac{\lambda_i}{\beta_i} \quad (i = 1, 2, 3.) \quad (11)$$

After substituting the value from (11) into boundary conditions, the controlled non linear state equations can be obtained:

$$\begin{cases} \dot{u}_1 = u_2 - au_1 + 10u_2u_3 + \dfrac{\lambda_1}{\beta_1} \\[2mm] \dot{u}_2 = cu_2 + 5u_1u_3 + \dfrac{\lambda_2}{\beta_2} \\[2mm] \dot{u}_3 = bu_3 - 5u_1u_2 + \dfrac{\lambda_3}{\beta_3} \end{cases} \quad (12)$$

The above system of nonlinear ordinary differential equations together with the (10) forms a complete system for solving the optimal control scheme. By solving the above nonlinear two point boundary value problem, the optimal control law and the optimal state trajectories can be obtained.

4   Numerical Simulation Results

In this section, the effectiveness and feasibility of the proposed optimal control scheme is presented. The system (12) is solved along with (10) and using the boundary conditions described above. For solving, we choose the finite time interval as: [0,6], initial values as: $u_1(0) = -2$, $u_2(0) = 4$, $u_3(0) = 2$, final values as: $u_1(6) = \overline{u_1}$, $u_2(6) = \overline{u_2}$, $u_3(6) = \overline{u_3}$ where, u_1, u_2 and u_3 are the coordinates of the equilibrium points, and system parameters for the chaotic system as: a = 0.4, b = 0.175 and c = −0.4.

Also,the positive constants in cost function **J** for all the five equilibrium points are chosen as: $\alpha_i = 50$ and $\beta_i = 25$ for i = 1, 2 and 3. In Fig. 2, it can be clearly seen that the state variables $(u_1, u_2, u_3) \rightarrow E_0$ and as soon as the state variables converges to the equilibrium point, the controllers U_1, U_2 and U_3 tend to zero. Similarly, Figs. 3, 4, 5 and 6 exhibit the controlled behaviour of the states variables (u_1, u_2, u_3) and controllers U_1, U_2, U_3 for its equilibrium points E_1, E_2, E_3 and E_4 of the controlled chaotic system respectively.

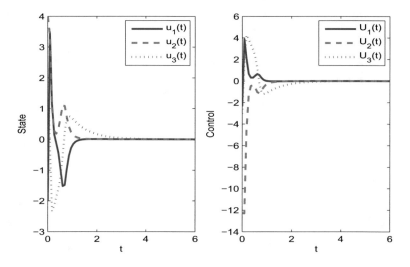

Fig. 2. The stabilized states and control functions for the equilibrium points E_0

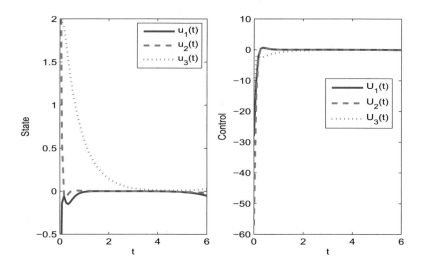

Fig. 3. The stabilized states and control functions for the equilibrium points E_1

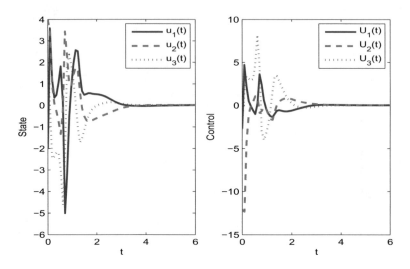

Fig. 4. The stabilized states and control functions for the equilibrium points E_2

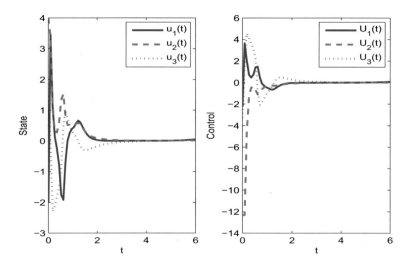

Fig. 5. The stabilized states and control functions for the equilibrium points E_3

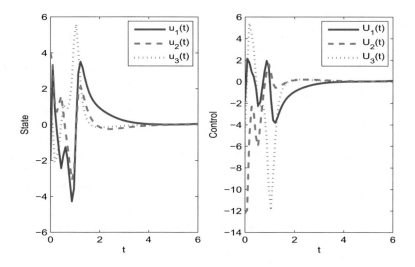

Fig. 6. The stabilized states and control functions for the equilibrium points E_4

5   Conclusion

Equilibrium points play a critical role in the discovery of chaotic systems. Recent developments in the field of chaos have led to a renewed interest in the number of equilibrium points in chaotic systems. In this article, the chaotic system is controlled to its five equilibrium points. The suitable controllers are constructed using the optimal control technique. Numerical results are given to confirm the efficiency of the proposed control scheme. The practical application of the new system will be assessed as a future works. The optimal control technique discussed in this article can be applied to discuss multi-switching synchronization, combination synchronization etc which holds important application in the field of secure communication.

References

1. Azar, A.T., Vaidyanathan, S.: Advances in Chaos Theory and Intelligent Control. Springer, Berlin (2016)
2. Azar, A.T., Ouannas, A., Singh, S.: Control of new type of fractional chaos synchronization. In: International Conference on Advanced Intelligent Systems and Informatics, pp. 47–56. Springer, Cham (2017)
3. Azar, A.T., Vaidyanathan, S., Ouannas, A.: Fractional order control and synchronization of chaotic systems. In: Studies in Computational Intelligence, vol. 688. Springer, Berlin (2017)
4. Babaei, M.: A novel text and image encryption method based on chaos theory and DNA computing. Nat. Comput. **12**(1), 101–107 (2013)
5. Bhat, M.A., Shikha: Complete synchronisation of non-identical fractionalorder hyperchaotic systems using active control. Int. J. Autom. Control **13**(2), 140–157 (2019)

6. Bozoki, Z.: Chaos theory and power spectrum analysis in computerized cardiotocography. Eur. J. Obstet. Gynecol. Reprod. Biol. **71**(2), 163–168 (1997)
7. Çavuşoğlu, Ü., Panahi, S., Akgül, A., Jafari, S., Kaçar, S.: A new chaotic system with hidden attractor and its engineering applications: analog circuit realization and image encryption. Analog Integr. Circ. Sig. Process **98**(1), 85–99 (2019)
8. Chekan, J.A., Ali Nojoumian, M., Merat, K., Salarieh, H.: Chaos control in lateral oscillations of spinning disk via linear optimal control of discrete systems. J. Vib. Control **23**(1), 103–110 (2017)
9. Chien, T.I., Liao, T.L.: Design of secure digital communication systems using chaotic modulation, cryptography and chaotic synchronization. Chaos, Solitons Fractals **24**(1), 241–255 (2005)
10. Din, Q.: Stability, bifurcation analysis and chaos control for a predator-prey system. J. Vib. Control **25**(3), 612–626 (2019)
11. Edelman, M.: On stability of fixed points and chaos in fractional systems. Chaos Interdisc. J. Nonlinear Sci. **28**(2), 023112 (2018). https://doi.org/10.1063/1.5016437
12. Čelikovský, S., Lynnyk, V., Chen, G.: Robust synchronization of a class of chaotic networks. J. Franklin Inst. **350**(10), 2936–2948 (2013)
13. Frey, D.R.: Chaotic digital encoding: an approach to secure communication. In: IEEE Transactions on Circuits and Systems II: Analog and Digital Signal Processing, vol. 40, no. 10, pp. 660–666 (1993)
14. Garfinkel, A.: Controlling cardiac chaos. Science (1992)
15. Khan, A., Singh, S.: Mixed tracking and projective synchronization of 6D hyperchaotic system using active control. Int. J. Nonlinear Sci. **22**(1), 44–53 (2016)
16. Khan, A., Singh, S.: Combination synchronization of time-delay chaotic system via robust adaptive sliding mode control. Pramana **88**(6), 91 (2017)
17. Khan, A., Singh, S.: Hybrid function projective synchronization of chaotic systems via adaptive control. Int. J. Dyn. Control **5**(4), 1114–1121 (2017)
18. Khan, A., Singh, S.: Chaotic analysis and combination-combination synchronization of a novel hyperchaotic system without any equilibria. Chin. J. Phys. **56**(1), 238–251 (2018)
19. Khan, A., Singh, S.: Generalization of combination-combination synchronization of n-dimensional time-delay chaotic system via robust adaptive sliding mode control. Math. Methods Appl. Sci. **41**(9), 3356–3369 (2018)
20. Khan, A., Tyagi, A.: Analysis and hyper-chaos control of a new 4-D hyper-chaotic system by using optimal and adaptive control design. Int. J. Dyn. Control **5**(4), 1147–1155 (2017)
21. Lau, F., Tse, C.K.: Chaos-based digital communication systems. Springer (2003)
22. Ott, E., Spano, M.: Controlling chaos. In: AIP Conference Proceedings, vol. 375, pp. 92–103. AIP (1996)
23. Ott, E., Grebogi, C., Yorke, J.A.: Controlling chaos. Phys. Rev. Lett. **64**, 1196–1199 (1990)
24. Ouannas, A., Grassi, G., Azar, A.T., Singh, S.: New control schemes for fractional chaos synchronization. In: International Conference on Advanced Intelligent Systems and Informatics, pp. 52–63. Springer (2018)
25. Ouannas, A., Grassi, G., Azar, A.T., Gasri, A.: A new control scheme for hybrid chaos synchronization. In: Hassanien, A.E., Tolba, M.F., Shaalan, K., Azar, A.T. (eds.) Proceedings of the International Conference on Advanced Intelligent Systems and Informatics 2018, pp. 108–116. Springer International Publishing, Cham (2019)

26. Ouannas, A., Grassi, G., Azar, A.T.: Fractional-order control scheme for Q-S chaos synchronization. In: Hassanien, A.E., Azar, A.T., Gaber, T., Bhatnagar, R., F Tolba, M. (eds.) The International Conference on Advanced Machine Learning Technologies and Applications (AMLTA 2019), pp 434–441. Springer International Publishing, Cham (2020)

27. Ouannas, A., Grassi, G., Azar, A.T.: A new generalized synchronization scheme to control fractional chaotic systems with non-identical dimensions and different orders. In: Hassanien, A.E., Azar, A.T., Gaber, T., Bhatnagar, R., F Tolba M. (eds.) The International Conference on Advanced Machine Learning Technologies and Applications (AMLTA2019), pp. 415–424. Springer International Publishing, Cham (2020)

28. Schiff, S.J., Jerger, K., Duong, D.H., Chang, T., Spano, M.L., Ditto, W.L., et al.: Controlling chaos in the brain. Nature **370**(6491), 615–620 (1994)

29. Singh, S., Azar, A.T., Ouannas, A., Zhu, Q., Zhang, W., Na, J.: Sliding mode control technique for multi-switching synchronization of chaotic systems. In: 2017 9th International Conference on Modelling, pp. 880–885. Identification and Control (ICMIC). IEEE (2017)

30. Singh, S., Azar, A.T., Bhat, M.A., Vaidyanathan, S., Ouannas, A.: Active control for multi-switching combination synchronization of non-identical chaotic systems. In: Advances in System Dynamics and Control, pp. 129–162. IGI Global (2018)

31. Singh, S., Azar, A.T., Vaidyanathan, S., Ouannas, A., Bhat, M.A.: Multiswitching synchronization of commensurate fractional order hyperchaotic systems via active control. In: Mathematical Techniques of Fractional Order Systems, pp. 319–345. Elsevier (2018)

32. Sivakumar, B.: Chaos theory in hydrology: important issues and interpretations. J. Hydrol. **227**(1–4), 1–20 (2000)

33. Strogatz, S.H.: Nonlinear Dynamics and Chaos: With Applications to Physics, Biology, Chemistry, and Engineering. Westview Press, Boulder (2001)

34. Vaidyanathan, S., Azar, A.T.: Analysis, control and synchronization of a nine-term 3-D novel chaotic system. In: Chaos Modeling and Control Systems Design. Springer, pp. 19–38 (2015)

35. Vaidyanathan, S., Azar, A.T.: Anti-synchronization of identical chaotic systems using sliding mode control and an application to vaidyanathan-madhavan chaotic systems. In: Azar, A.T., Zhu, Q. (eds.) Advances and Applications in Sliding Mode Control Systems, Studies in Computational Intelligence, vol. 576, pp. 527–547. Springer, Berlin (2015)

36. Vaidyanathan, S., Azar, A.T.: Hybrid synchronization of identical chaotic systems using sliding mode control and an application to Vaidyanathan chaotic systems. In: Azar, A.T., Zhu, Q. (eds.) Advances and Applications in Sliding Mode Control Systems, Studies in Computational Intelligence, vol. 576. Springer, Berlin, pp. 549–569 (2015)

37. Vaidyanathan, S., Idowu, B.A., Azar, A.T.: Backstepping controller design for the global chaos synchronization of sprott's jerk systems. In: Chaos Modeling and Control Systems Design, pp. 39–58. Springer (2015)

38. Vaidyanathan, S., Azar, A.T., Boulkroune, A.: A novel 4-D hyperchaotic system with two quadratic nonlinearities and its adaptive synchronisation. Int. J. Autom. Control **12**(1), 5–26 (2018)

39. Xing-yuan, W., Yong-feng, G.: A switch-modulated method for chaos digital secure communication based on user-defined protocol. Commun. Commun. Nonlinear Sci. Numer. Simul. **15**(1), 99–104 (2010). sI: Chaos, Complexity and Transport: Theory and Applications

Implementation of PID Controller with PSO Tuning for Autonomous Vehicle

Ahmad Taher Azar[1,2]($\boxtimes$), Hossam Hassan Ammar[3], Zahra Fathy Ibrahim[3], Habiba A. Ibrahim[3], Nada Ali Mohamed[3], and Mazen Ahmed Taha[3]

[1] College of Engineering, Prince Sultan University, Riyadh, Kingdom of Saudi Arabia
aazar@psu.edu.sa
[2] Faculty of Computers and Artificial Intelligence, Benha University, Benha, Egypt
ahmad.azar@fci.bu.edu.eg
[3] School of Engineering and Applied Sciences, Nile University Campus, Sheikh Zayed District, Juhayna Square, 6th of October City, Giza 12588, Egypt
{hhassan,z.fathy,h.ibrahim,n.ali,m.taha}@nu.edu.eg

Abstract. In the use of automatic control and its optimization methods, this research discusses how Proportional Integral Derivative (PID) controller is used to provide a smooth auto-parking for an electrical autonomous car. Different tuning methods are shown, discussed, and applied to the system looking forward to enhancing its performance. Time domain specifications are used as a criterion of comparison between tuning methods in order to select the best tuning method to the system with a proper cost function. Results show that Particle Swarm Optimization (PSO) method gives the best results according the criteria of comparison.

Keywords: PID controller · Particle Swarm Optimization (PSO) · Ackerman steering model · Lagrangian mechanics

1 Introduction

Technological advances in automotive industry is enormous, by virtue of researchers interests and efforts throughout years to automate vehicles systems and even to automate the driving itself. These efforts started in 1926 with controlling a car through radio signals, then takes a new greater step in 1980s when computer vision comes into play in the model proposed by Mercedes-Benz [9]. Scientifically, autonomous cars are vehicles derived based on sensing the surrounding environment without human interaction. Based on the definition, there are two main promising areas of research in the process of autonomous cars development: acknowledgment by surroundings and controlling the car movement. Researches focused on finding the best sensing method to inform the controller by surroundings. Main focus of sensing was for using sensors, Internet of Things (IoT), and on-board computer vision. Concerns about the use of sensors was huge because of reading disturbances and devices limitations. In 2018, a study

© Springer Nature Switzerland AG 2020
A. E. Hassanien et al. (Eds.): AISI 2019, AISC 1058, pp. 288–299, 2020.
https://doi.org/10.1007/978-3-030-31129-2_27

was done on using ultrasonic sensors in autonomous cars and have detected malfunction in case of collusion. This result motives the researches to propose dense mechanisms to provide safety to the car [23]. Seeking more reliability, IoT and cloud-based driving systems were proposed by researches multiple times [11]. Moreover, computer vision represents a new sensory method that is independent of the use of radio sensors; image processing is used instead. Apparently, interest in using vision in sensing the environment begins from 1980s till today with hopes of researches to provide the most accurate and adaptable system to guide the car in all environmental conditions [21, 22].

Regarding the controller, its capability of fast response plays an important role in determining the autonomous car functionality; accordingly, Proportional Integral Derivative (PID) controller lies in the heart of automatic control due to its preferred control dynamics represented by small to no overshoot, fast response, stability, and eliminated steady state error [1, 2, 4, 7, 17, 19]. It is even used by autonomous cars to reach their optimal path. Ensuring the best results, adjusting the values of the proportional, derivative and integral gains, tuning, is crucial. Researches had shown great interest and efforts in developing various methods for PID tuning [3, 5, 6, 8, 12].

This research provides a model for a steering car driven by the feedback of a stationary camera in a parking area. System modeling and testing are provided as well as PID tuning optimization using PSO to ensure the best accuracy of path planing and execution.

In this paper, Sect. 2 discusses the dynamic model. And, Sect. 3 shows the PID controller design. Then, Sect. 4 is for the results and PID implementation. Finally, Sect. 5 shows the conclusion.

2 System Dynamics Model

The design of the autonomous car proposed model aims to simulate an actual vehicle based on using steering mechanism for controlling the system angular motion by the use of a servo motor connected to the front wheels. Moreover, all the four wheels are connected to DC motors with the same speed. Based on the two subsystems linear and angular actuation, simulation using MATLAB Simscape and V-Rep software was made to visualize the behavior of both systems and compare them to the hardware behavior using a prototype. In simulation, integration between Simscape and V-rep software was made. Desired coordinates inputs for parking area are input using V-rep, then passed as external signals to the linear and angular actuators modeled in Simscape circuit diagram as shown in Fig. 1.

Afterwards, using the planned path (as shown in Fig. 2) and Simulink systems inputs, the vehicle moves in V-rep, and that dynamics are analyzed in Fig. 3. Arduino was programmed and manipulated using Arduino toolbox in MATLAB. It gives the advantage of encrypting the signal, then decrypting upon being received by Arduino to attenuate noise. Then, camera detection of available spots using distinguished colors and optimization for the planned path were done

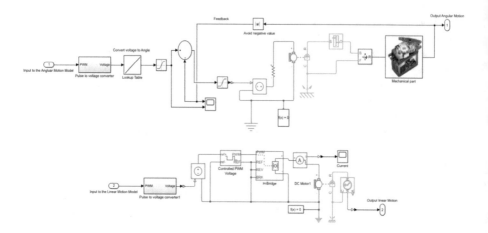

Fig. 1. Simscape model of autonomous car system

using Computer Vision Toolbox and Image Acquisition Toolbox in MATLAB. PID control is applied twice in system, one is to enhance linear motion and the other is executed on the Ackerman steering model. The control system block diagram is shown in Fig. 4.

The Ackerman steering model takes into consideration the four wheeled vehicles that have a differential mechanism of the front axle in which the turning of the directional wheels marks the instantaneous center of rotation (ICR) as in Eq. 1.

$$\cot(\delta o) - \cot(\delta i) = \frac{w}{l} \tag{1}$$

Taking 'o' as a center of the turning inner and outer wheel angles presented by δ_i and δ_o respectively. Considering 1 as the car length and w as its width in Fig. 3, Lagrangian mechanics along with geometric mechanics were used to represent the model of Ackerman steered car. Using Lagrangian in the shown simplified form:

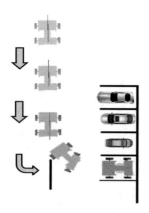

Fig. 2. Path planning for the car

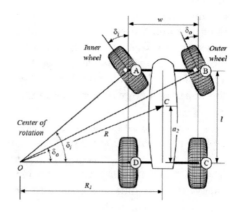

Fig. 3. Front-wheel-steering vehicle

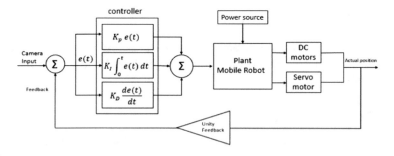

Fig. 4. Control system block diagram

$$L = \frac{1}{2}m\left(\dot{x}^2 + \dot{y}^2\right) + J\dot{\theta}^2 + J\dot{\theta}\dot{\Phi} + \frac{1}{2}J\dot{\Phi}^2 + J_k\dot{\psi}^2 \tag{2}$$

Where m is total weight, J is moment of inertia, and J_k is wheels' moment of inertia. Three nonholonomic constraints were considered for the system due to use of Ackerman model; coordinates are represented with respect to x and y:

$$\dot{x}\,sin\theta - \dot{y}cos\theta = 0$$
$$\dot{x}\,\sin(\theta + \Phi) - \dot{y}\cos(\theta + \Phi) - l\dot{\theta}cos\Phi = 0 \tag{3}$$
$$\dot{x}\,cos\theta + \dot{y}sin\theta - R\dot{\psi} = 0$$

This result setting the equation of motion for the system

$$m\ddot{x} + \lambda_1 sin\theta + \lambda_2 \sin(\theta + \Phi) + \lambda_3 cos\theta = 0$$
$$m\ddot{y} - \lambda_1 cos\theta - \lambda_2 \cos(\theta + \Phi) + \lambda_3 sin\theta = 0$$
$$2J\ddot{\theta} + J\ddot{\Phi} - \lambda_2 l \cos\theta = 0 \tag{4}$$
$$J\ddot{\Phi} + J\ddot{\theta} = \tau_1$$
$$J_k\ddot{\psi} - \lambda_3 R = \tau_2$$

The Ackerman steered car has three degree of freedom (x, y, θ), and two control variables (Φ, ψ), that classifies this model as under-actuated nonholonomic mechanical system result in having three kinematics equations of motion.

$$\begin{bmatrix} 0 \\ 0 \\ 0 \end{bmatrix} = \begin{bmatrix} \sin\theta & -\cos\theta & 0 & 0 & 0 \\ \sin(\theta + \Phi) & -\cos(\theta + \Phi) & -l\cos\theta & 0 & 0 \\ \cos\theta & \sin\theta & 0 & 0 & -R \end{bmatrix} \begin{bmatrix} \dot{x} \\ \dot{y} \\ \dot{\theta} \\ \dot{\Phi} \\ \dot{\psi} \end{bmatrix} \tag{5}$$

To obtain reconstruction equation in the form, shape variable is separated and equation is multiplied representing the equation as.

$$\begin{bmatrix} \varepsilon^x \\ \varepsilon^y \\ \varepsilon^\theta \end{bmatrix} = \begin{bmatrix} \cos\theta & \sin\theta & 0 \\ -\sin\theta & \cos\theta & 0 \\ 0 & 0 & 1 \end{bmatrix} \begin{bmatrix} \dot{g}_x \\ \dot{g}_y \\ \dot{g}_\theta \end{bmatrix}, \varepsilon = \begin{bmatrix} 0 & R \\ 0 & 0 \\ 0 & \frac{R}{l}\tan\Phi \end{bmatrix} \tag{6}$$

3   PID Controller Design

Applying PID controller to the system requires selecting the most suitable and appropriate values of Proportional, integral, and derivative controller gains, namely tuning.

3.1   PID Tuning Methods

Tuning methods are, generally, classified into two categories: Open-loop methods and Closed-loop Methods. For a system that needs feedback and work in the automatic state, closed-loop methods are preferred while open-loop methods are mainly used when system does not include response feedback [20]. In this research, Open-loop tuning methods are applied. Resulted parameters are compared based on their effect on the real system response when a step input is applied [13,18]. Classical tuning methods are applied in this research such as Ziegler and Nichols method [24] and Cohen-Coon method [14,20] and then compared with PSO tuning method.

PSO method depends mainly on population assumption for the swarm [10,15]. Each particle has an assumption in a random direction, considering that particles of the set has different directions. Also, each particle has a set of parameters forming a vector with a size of application targeted vector; in this case, vector contains the three parameters of PID controller. Once particles reach their end positions, performance of each particle is measured by plugging values of its vector parameter into the objective function, it indicates how close the particle is to the desired solution. Then, the particle with the most acceptable result or position, namely the global best particle, is given a new direction with respect to other particles. Note that all particles are updated with values of the best vector [15]. This cycle is continuously repeated until particles get an identical final position which represent the final optimal value of PID controller gains. PSO algorithm starts with initializing particles with all possible sets of PID parameters. These values are then adjusted in order to minimize objective function. The objective function relies on minimum integral error indices; additionally, this is the best way to evaluate particles performance. Minimum integral criteria is used to apply PSO as it aims to minimize the area developed by time when Process Variable (PV) deviates from its reference input. Several criterion indexes are then executed [16,18].

4   Results and Implementation

System identification is approached by imposing system data to MATLAB. Then, several commands are executed to introduce system inputs, outputs, and get system transfer functions. Actually, system is described by two transfer function modelling the steering represented by servo motor, and car motion represented by DC Motors. Matlab is, again, used to implement all previously mentioned tuning methods to PID controller of autonomous car. Results are obtained by giving a step response to the system and analyzing its output.

By implementing system identification for both linear motion subsystem and angular motion subsystem, the transfer functions for both of them are obtained. Regarding the linear motion subsystem, the system input is the voltage, while the output is the velocity. Similarly, for the angular motion subsystem, the system input is the voltage, while the output is angular velocity in radians per second. The transfer function for angular motion has two poles and no zeros, and is represented by:

$$TF_{angular} = \frac{0.121}{s^2 + 0.619s + 0.1636} \tag{7}$$

Additionally, the transfer function for linear motion has two poles and no zeros, and it is represented by:

$$TF_{linear} = \frac{0.008936}{s^2 + 0.1258s + 0.02384} \tag{8}$$

The next step after obtaining the transfer functions that describes the system dynamics in linear motion and in angular motion, different tuning techniques were applied on both subsystems by using PSO MATLAB Library. The range for the PID gains was set based on the transfer function for the system and the suitable gains were bounded by the number of iterations. For the linear movement, Table 1 shows the PID gains for each method while Table 2 shows their dynamical results.

Table 1. The gains of PID controller for vehicle linear motion

Gains/Methods	ZN	CC	ITAE	ITSE	IAE	PSO
Proportional gain	2.5	3.02	2.91	2.89	2.69	93.1
Integral gain	0.582	0.472	0.451	0.84	0.604	13
Differential gain	4.271	2.81	3.47	4.45	2.98	100

Table 2. The dynamical results for the linear motion

Results/Methods	ZN	CC	ITAE	ITSE	IAE	PSO
Rise time	8.091	7.404	6.237	6.354	7.048	0.914
Settling time	115.2	124.0	130.4	131.0	153.5	8.741
Overshoot	26.49	29.62	26.52	55.50	43.18	34.67
Undershoot	0	0	0	0	0	0
Peak	1.264	1.296	1.931	1.555	1.432	1.347
Peak time	17.48	16.44	18.26	15.06	15.91	2.653

For the angular movement, Table 3 shows the PID gains for each method while Table 4 shows their dynamical results. The motion optimization was applied using PSO technique due to its fast operation and low computational time. The technique was applied using different algorithms of minimum integral error criteria. The comparison is based on plotting the cost functions of each algorithm against the number of iterations. By analyzing the graphical results, Figs. 5 and 6 show the performance optimization for both subsystem: linear and angular. This optimization enhances the tuning process and leads the system under PID controller to give the best performance characterized by reduced overshoot, fast rise and settling times, stability and minimized steady state error, if any. Therefore, implementing them was a necessary step in order to control the autonomous car motion.

Table 3. The gains of PID controller for vehicle angular motion

Gains/Methods	ZN	CC	ITAE	ITSE	ISE	IAE	PSO
Proportional gain	1.94	2.22	2.86	3.21	3.01	2.68	4
Integral gain	1.02	1.01	0.895	0.830	1.5	0.674	0.132
Differential gain	0.922	0.745	2.29	1.85	2.64	1.35	4

Table 4. The dynamical results for the angular motion

Results/Methods	ZN	CC	ITAE	ITSE	ISE	IAE	PSO
Rise time	2.724	2.542	4.7	4.563	4.509	5.051	7.231
Settling time	27.03	25.31	0.824	0.658	0.903	0.633	0.1093
Overshoot	34.72	34.17	56.30	51.24	58.40	51.03	0.666
Undershoot	0	0	0	0	0	0	0
Peak	1.347	1.357	1.563	1.512	1.584	1.510	1.006
Peak time	7.42	7.013	0.7	0.65	0.5	0.8	0.25

Then, the tuning methods with the specified tuning parameters were applied to the linear and angular system in order to model the response of each system under these methods. Both systems were subjected to tuning by the following methods: Ziegler-Nichols (ZN), Cohen-Coon (CC), ITAE, ITSE, IAE, ISE and particle swarm optimization. By analyzing Figs. 7 and 8 which describe the behavior of angular motion subsystem, tuning by PSO results shows the best behavior which is stable and steady. This behavior is similar to that experiences when the system was tuned by IAE and ITAE; however, PSO becomes the superior over them because of its contribution in giving the system fast response compared to the other methods, while maintaining stability. On the contrary, tuning by ZN and CC resulted in dynamics with high oscillations and overshoot.

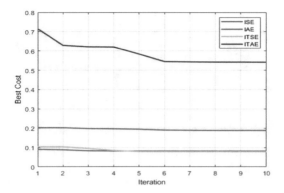

Fig. 5. Performance optimization of angular motion

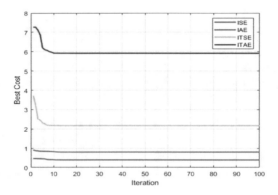

Fig. 6. Performance optimization for linear motion

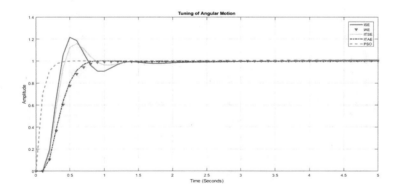

Fig. 7. PSO and different tuning methods applied on angular motion transfer function

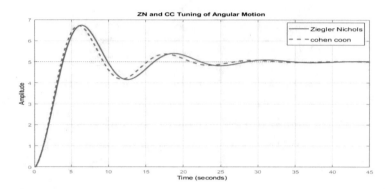

Fig. 8. Classical tuning methods applied on angular motion transfer function

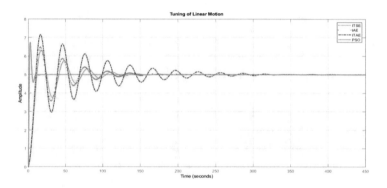

Fig. 9. PSO and different tuning methods applied on linear motion transfer function

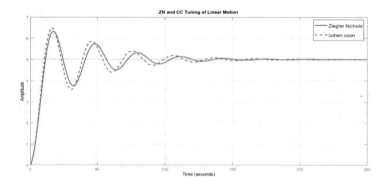

Fig. 10. Classical tuning methods applied on linear motion transfer function

Moving to the subsystem for linear motion, Figs. 9 and 10 display the behavior of the system under the selected methods. Again, as expected, PSO based tuning had resulted in the best dynamics for the system. Although overshoot is still present under the tuning parameters used in PSO, the response is very fast and the behavior stabilize quickly with nearly no steady state error. The exact same opposite is apparent in the behavior under other classical tuning methods where the frequency of oscillations is high and long time is needed for the behavior to be steady. Accordingly, in a nutshell, it is obvious that PSO based tuning was the best tuning method for controlling the motion of autonomous car using PID controlling, both in terms of linear motion and in terms of angular motion.

5   Conclusion

Tuning the PID controller for autonomous car motion, linearly or angular, based on PSO algorithm shows the superiority over the classical tuning methods in obtaining the most stable results according to the criteria of comparison. This study faces some limitations in applying the proposed optimized control on passenger cars and in making the control algorithm able to search for more than one path and choose between them in case of obstacles in the road.

References

1. Ammar, H.H., Azar, A.T.: Robust path tracking of mobile robot using fractional order PID controller. In: Hassanien, A.E., Azar, A.T., Gaber, T., Bhatnagar, R., F. Tolba, M. (eds.) The International Conference on Advanced Machine Learning Technologies and Applications (AMLTA 2019), pp. 370–381. Springer, Cham (2020)
2. Ammar, H.H., Azar, A.T., Tembi, T.D., Tony, K., Sosa, A.: Design and implementation of fuzzy pid controller into multi agent smart library system prototype. In: Hassanien, A.E., Tolba, M.F., Elhoseny, M., Mostafa, M. (eds.) The International Conference on Advanced Machine Learning Technologies and Applications (AMLTA2018), pp. 127–137. Springer, Cham (2018)
3. Azar, A.T., Serrano, F.E.: Robust IMC–PID tuning for cascade control systems with gain and phase margin specifications. Neural Comput. Appl. **25**(5), 983–995 (2014). https://doi.org/10.1007/s00521-014-1560-x
4. Azar, A.T., Serrano, F.E.: Design and Modeling of Anti Wind Up PID Controllers, pp. 1–44. Springer, Cham (2015)
5. Azar, A.T., Serrano, F.E.: Fractional order sliding mode PID controller/observer for continuous nonlinear switched systems with PSO parameter tuning. In: The International Conference on Advanced Machine Learning Technologies and Applications (AMLTA 2018), pp. 13–22. Springer (2018). https://doi.org/10.1007/978-3-319-74690-62
6. Azar, A.T., Serrano, F.E.: Fractional order sliding mode PID controller/observer for continuous nonlinear switched systems with PSO parameter tuning. In: Hassanien, A.E., Tolba, M.F., Elhoseny, M., Mostafa, M. (eds.) The International Conference on Advanced Machine Learning Technologies and Applications (AMLTA 2018), pp. 13–22. Springer, Cham (2018)

7. Azar, A.T., Vaidyanathan, S.: Handboook of Research on Advanced Intelligent Control Engineering and Automation. IGI Global, New York (2015)

8. Azar, A.T., Serrano, F.E., Vaidyanathan, S.: Proportional integral loop shaping control design with particle swarm optimization tuning. In: Advances in System Dynamics and Control, pp. 24–57. IGI Global (2018). https://doi.org/10.4018/978-1-5225-4077-9.ch002

9. Bimbraw, K.: Autonomous cars: Past, present and future a review of the developments in the last century, the present scenario and the expected future of autonomous vehicle technology. In: 2015 12th International Conference on Informatics in Control, Automation and Robotics (ICINCO), vol. 01, pp. 191–198 (2015)

10. Eberhart, R., Shi, Y., Kennedy, J.: Swarm Intelligence. Elsevier, Amsterdam (2001)

11. Gerla, M., Lee, E.K., Pau, G., Lee, U.: Internet of vehicles: from intelligent grid to autonomous cars and vehicular clouds. In: 2014 IEEE World Forum on Internet of Things (WF-IoT). IEEE (2014). https://doi.org/10.1109/wf-iot.2014.6803166

12. Gorripotu, T.S., Samalla, H., Jagan Mohana Rao, C., Azar, A.T., Pelusi, D.: TLBO algorithm optimized fractional-order pid controller for AGC of interconnected power system. In: Nayak, J., Abraham, A., Krishna, B.M., Chandra Sekhar, G.T., Das, A.K. (eds.) Soft Computing in Data Analytics, pp. 847–855. Springer, Singapore (2019)

13. Hussain, K.M., Zepherin, R.A.R., Kumar, M.S.: Comparison of tuning methods of PID controllers for FOPTD system. IJIREEICE **2**(3), 1177–1180 (2014)

14. Joseph EAAMCR: Cohen-coon PID tuning method: A better option to ziegler nichols-pid tuning method. Computer Engineering and Intelligent Systems (2018)

15. Kennedy, J., Eberhart, R.: Particle swarm optimization. In: Proceedings of ICNN 1995 - International Conference on Neural Networks, vol. 4, pp. 1942–1948 (1995)

16. Maiti, D., Acharya, A., Chakraborty, M., Konar, A., Janarthanan, R.: Tuning PID and $pi^\lambda d^\mu$ controllers using the integral time absolute error criterion. In: 2008 4th International Conference on Information and Automation for Sustainability. IEEE (2008). https://doi.org/10.1109/iciafs.2008.4783932

17. Ogata, K.: Modern Control Engineering. Prentice Hall, Upper Saddle River (2010)

18. Shahrokhi, M., Zomorrodi, A.: Comparison of PID controller tuning methods. Department of Chemical & Petroleum Engineering Sharif University of Technology, pp. 1–2 (2013)

19. Soliman, M., Azar, A.T., Saleh, M.A., Ammar, H.H.: Path planning control for 3-OMNI fighting robot using PID and fuzzy logic controller. In: Hassanien, A.E., Azar, A.T., Gaber, T., Bhatnagar, R., F Tolba, M. (eds.) The International Conference on Advanced Machine Learning Technologies and Applications (AMLTA 2019), pp. 442–452. Springer, Cham (2020)

20. Stone, R., Ball, J.K.: Automotive Engineering Fundamentals. SAE International (2004)

21. Sukthankar, R.: Raccoon: a real-time autonomous car chaser operating optimally at night. In: Proceedings of the Intelligent Vehicles '93 Symposium, pp. 37–42 (1993). https://doi.org/10.1109/IVS.1993.697294

22. Tseng, Y., Jan, S.: Combination of computer vision detection and segmentation for autonomous driving. In: 2018 IEEE/ION Position, Location and Navigation Symposium (PLANS), pp. 1047–1052 (2018). https://doi.org/10.1109/PLANS.2018.8373485

23. Xu, W., Yan, C., Jia, W., Ji, X., Liu, J.: Analyzing and enhancing the security of ultrasonic sensors for autonomous vehicles. IEEE Internet Things J. **5**(6), 5015–5029 (2018). https://doi.org/10.1109/jiot.2018.2867917
24. Ziegler, J., Nichols, N.B.: Optimum settings for automatic controllers. Trans. ASME **64**(8), 759768 (1942)

Tuning of PID Controller Using Particle Swarm Optimization for Cross Flow Heat Exchanger Based on CFD System Identification

Omar Khaled Sallam[1], Ahmad Taher Azar[2,3(✉)], Amr Guaily[1,4], and Hossam Hassan Ammar[1]

[1] School of Engineering and Applied Sciences, Nile University, Sheikh Zayed District, Giza 12588, Egypt
{okhaled, aguaily, hhassan}@nu.edu.eg
[2] College of Engineering, Prince Sultan University, Riyadh, Kingdom of Saudi Arabia
aazar@psu.edu.sa
[3] Faculty of Computers and Artificial Intelligence, Benha University, Benha, Egypt
[4] Faculty of Engineering, Cairo University, Giza 12613, Egypt

Abstract. This paper illustrates the design of proportional–integral–derivative controller (PID) controller of 10 KW air heaters for achieving the set point temperature as fast as possible with minimum response overshoot. Computational fluid dynamic (CFD) numerical simulations are utilized to predict the natural response of 10 KW input power for the air heater. CFD results are validated with experimental empirical correlations that insure the reliability of open loop results. The open loop response of CFD transient simulations is used to model the air heater transfer function and design the classical PID controllers. Particle swarm optimization (PSO) technique is used to tune the PID controller with various error fitness functions which leads to improve the closed loop response of the temperature control system compared to the classical tuning methods.

Keywords: PID controller · Particle swarm optimization · Heat exchanger · Computational fluid dynamic · Turbulence models

1 Introduction

Heat exchanger is one of the most important equipment used in modern life in both industrial and residential categories. The working principle of heat exchangers is to transfer the heat from higher temperature region (source) to lower temperature one (sink). Heat exchanger's shape and type varies dependent on the application. Most of the heat exchangers are double fluid flow which cold fluid gains the total amount of heat released from the hot fluid in the ideal case. In the single-fluid heat exchanger, the cold fluid is heated by passing past solid heating elements such as electrical resistance

© Springer Nature Switzerland AG 2020
A. E. Hassanien et al. (Eds.): AISI 2019, AISC 1058, pp. 300–312, 2020.
https://doi.org/10.1007/978-3-030-31129-2_28

coils used in heating system of air conditioning for residential, industrial and medical applications [1]. Design and control of heat exchangers are based on the requirements of the system where the equipment is installed. There are many requirements of heat exchangers, the most crucial requirements are the exit temperature of the controlled fluid (set point) and the time period required to reach this temperature (settling time). Control of heat exchangers or any input/output equipment requires a control algorithm that controls the output variables by taking specific decisions [2–5]. Many researches are working on the design of controller for heat exchangers to achieve the set point accurately in a small time without overshoot [6–10]. Padhee [11] considered the control of shell and tube heat exchanger with PID algorithm to control the exit temperature of the cold fluid. A comparative study is presented in among different approaches of controller implementation such as feedback, feedforward, and internal model control; the comparison criteria were limited between the error and overshoot of the system response. Yu and Yin [12] studied the effect of tuning the PID controller using Neural network algorithm and compared the results with traditional and classical tuning methods and observed that neural network algorithm for PID tuning gives faster rise and settling time. Sungthong [13] studied the PID parameters tuning using particle swarm optimization metaheuristic technique [20] for an air heater and compared among traditional PID tuning techniques such as Ziegler Nichols; The results show faster response with lower overshoot for exit temperature. The present work aims to develop a PID controller for single fluid cross flow heat exchanger with cylindrical solid heating elements with maximum heating power of 10 kW. The present system is a single-input/single-output so that PID controller is implemented to control the input heating power to the solid heating cylinders. To reach the target temperature with minimum overshoot and fastest response possible, PID controller is designed and tuned based on traditional techniques and PSO using different objective functions.

The rest of the paper is organized as follows. In Sect. 2, the mathematical model is presented to solve the flow in the heat exchanger with physical domain geometry definition and boundary conditions. While Sect. 3 discusses the CFD results and validate the numerical solution with empirical correlation. Section 4 covers the system transfer function modeling and PID controller different design techniques. Lastly, Sect. 5 contains the conclusion of the research work and future directions.

2 Mathematical Modeling

The process of controller design for electric heater power starts with the mathematical model and system identification of the heater as this is the milestone to predict the natural behavior of the system. Mathematical modeling maps the physical processes into a system of nonlinear partial differential equations (PDEs) that describe the fluid flow over the heating cylinders and energy transfers between the two regions. Being nonlinear system of PDEs; computational fluid dynamics (CFD) finite volume numerical technique is used to simulate the system.

2.1   Governing Equations

A finite volume numerical method using ANSYS[1] FLUENT is used to solve the nonlinear unsteady partial differential equations of the fluid flow by an implicit scheme for a total time of 10 s with consideration of air compressibility. The governing equations for non-polar Newtonian turbulent incompressible fluid are [14]:
The continuity equation is:

$$u_{i,i} = 0 \tag{1}$$

The Navier–Stokes equations are:

$$\rho \left(\frac{\partial u_i}{\partial t} + u_j u_{i,j} \right) = -p_{,i} + \left[(\mu + \mu_t) u_{i,j} \right]_{,j} \tag{2}$$

The Energy equation is:

$$\rho c_p \left(\frac{\partial T}{\partial t} + u_i T_{,i} \right) = \rho \phi - \left[(h + h_t) T_{,i} \right]_{,i} - p u_{i,i} + 2(\mu + \mu_t) d_{ij} d_{ij} \tag{3}$$

The turbulent kinetic energy equation is:

$$\frac{\partial (\rho k)}{\partial t} + (\rho k u_i)_{,i} = \left(\left(\mu + \frac{\mu_t}{\sigma_k} \right) \frac{\partial k}{\partial x_i} \right)_{,i} + P_k - \rho \varepsilon \tag{4}$$

The turbulent dissipation equation is:

$$\frac{\partial (\rho \varepsilon)}{\partial t} + (\rho \varepsilon u_i)_{,i} = \left(\left(\mu + \frac{\mu_t}{\sigma_\varepsilon} \right) \frac{\partial \varepsilon}{\partial x_i} \right)_{,i} + \frac{\varepsilon}{k} C_{1\varepsilon} P_k - C_{2\varepsilon} \frac{\rho \varepsilon^2}{k} \tag{5}$$

The turbulent viscosity is:

$$\mu_t = \rho C_\mu \frac{k^2}{\varepsilon} \tag{6}$$

The turbulent thermal conductivity is:

$$h_t = \frac{\mu_t c_p}{\mathrm{Pr}_t} \tag{7}$$

Where ρ is the density, u_i are the velocity components, p is the thermodynamics pressure, μ is the dynamic shear viscosity, c_p is the specific heat at constant pressure, T is the temperature, h is the thermal conductivity, d_{ij} are the components of the symmetric part of the velocity gradient tensor, k is the turbulence kinetic energy, P_k is

---

[1] https://www.ansys.com/.

the production of turbulence, Pr_t is the turbulent Prandtl number, and ε is the turbulence dissipation. And the typical constants are:

$$\sigma_k = 1.0, \sigma_\varepsilon = 1.3, C_{1\varepsilon} = 1.44, C_{1\varepsilon} = 1.92, C_\mu = 0.09, \Pr_t = 0.9$$

2.2   Physical Domain and Boundary Conditions

Physical domain of the crossflow heat exchanger shown in Fig. 1 contains 28 cylinders placed normal to the flow direction with length of 1 m. The staggered arrangement allows for higher streamlines' mixing which plays a vital role in convection heat transfer enhancement [15, 16]. Table 1 summarizes the numerical values of the geometric parameters.

Cylindrical tubes are the source of the electrical heating power with maximum capacity of 10 kW. The no-slip boundary condition is imposed for the coils and the walls of the heater. At the inlet, velocity, density, and the temperature of the fluid are imposed. The outlet boundary condition is set as atmospheric pressure.

The surface domain is meshed with quadrilateral shell elements. Since $k - \varepsilon$ turbulence model is used, boundary element inflation is added to the walls of the heater and cylindrical tubes to enable the solver to resolve the boundary layer development near to the walls in the first 2 turbulent layers (viscous buffer) layers. The first element height is chosen to assure that the y+ value remains near to 1 which is required by the turbulence model. Total number of elements is 101,000 with 104,000 nodes.

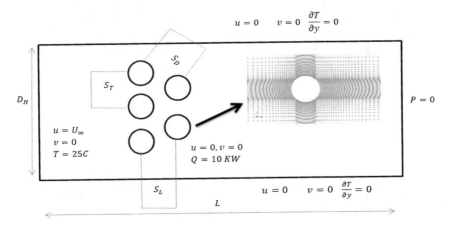

Fig. 1. Physical domain and boundary conditions and quad mesh

Table 1. Geometric parameters

Symbol	Parameter	Numerical value
S_1	Horizontal tube rows distance	50 mm
S_D	Diagonal Distance between tube	52.5 mm
D	Tube Diameter	20 mm
S_T	Vertical distance between tubes centers	32 mm
D_H	Heat Exchanger inlet vertical length	200 mm
N_t	Number of tubes	28
N_L	Number of vertical rows	5

3   Fluid Flow Results and Discussion

In this section, the flow results from the numerical simulation of the governing equations are presented. Figure 2 shows the contours of the flow temperature where the flow inters the heat exchanger with 25 °C initial temperature. By passing across the tube bundle starts to gain heat that results in a gradual temperature increase till it reaches the last row of tubes bundle. The development of the thermal boundary layer is obvious as temperature profile contours changes gradually from the centerline of the domain to the walls of the heater. The area weighted average temperature at the domain exit with respect to time is shown in Fig. 5; There are three distinct regions in the figure, namely the no-response region which takes about less than a second followed by the fast-response region in which an increase about 90 °C per second is observed. Then, the system reaches the steady state at 193 °C which is the saturation temperature that can be achieved by the input heat flux for the specified velocity inlet boundary condition 0.25 m/sec. Figure 3 shows the velocity contours of the fluid where the vortex shedding appears behind the cylinders where inlet Reynolds's number is 920 based on the cylinder diameter. Contours show that shedding disappears after the last row and streamlines start to attach again to form a uniform laminar flow at the exit of the heater.

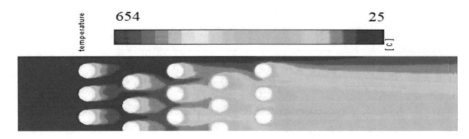

Fig. 2. Plane symmetry temperature contours

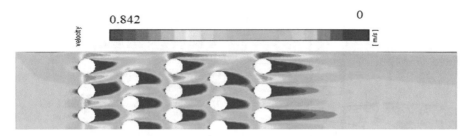

Fig. 3. Plane symmetry velocity contours

3.1   CFD Results Validation

CFD simulation is repeated for 5 scenarios of different Reynolds numbers with respect to coil diameter and maximum velocity between coil passages as described in Eqs. (7) and (8) respectively. Table 2 shows the list of Reynolds number values related to average inlet velocity conditions and the corresponding maximum stream velocity. Each scenario of simulation is validated with an empirical equation derived by Zukauskas [17] namely Eq. (9) of the pressure drop for flow over tube bundle in staggered arrangement. Friction factor f can be attained from Zukauskas chart [17] with respect to values of Reynolds number. Figure 4 shows how CFD pressure drop results meet the experimental ones from Zukauskas empirical equation for all simulated conditions of Reynolds number values.

$$Re = \frac{\rho VmaxD}{\mu} \tag{7}$$

Maximum velocity between tube passages is calculated as:

$$Vmax = \frac{ST}{ST - D} * V \tag{8}$$

Zukauskas [17] pressure drop empirical Eq. (9) is calculated as:

$$\Delta P = NLf \frac{\rho Vmax^2}{2} \tag{9}$$

Table 2. Maximum velocity and Reynolds number related to inlet average velocity

Inlet velocity m/sec	Maximum velocity m/sec	Reynolds number
0.15	0.4	551
0.25	0.667	918
0.35	0.933	1284
0.45	1.2	1650
0.5	1.33	1834

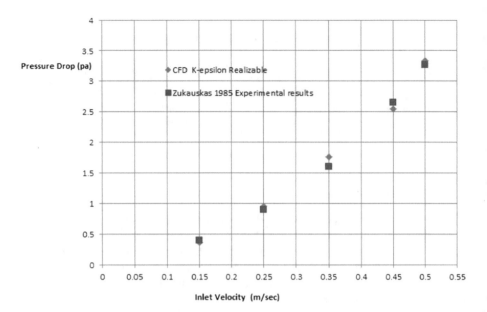

Fig. 4. Pressure drop CFD result validation

4   System Identification

Transient response for the history of temperature profile rise at the exit of air heater calculated from CFD simulation is used to build the transfer function of the system. Figure 5 shows the temperature rise through 10 s of system simulation where the profile is divided into 2 parts. Part 1 represents the delay time where exit temperature still unchanged and equals 25 °C which is the initial temperature value and inlet temperature as well. From the graph, the delay time equals. 9 s which is used as time value for time delay transfer function part in Eq. (10). Part 2 represents the temperature response to step input with value 10 kW and saturation temperature reached to 193 °C after 4.5 s. By the use of data series and MATLAB system identification toolbox; the second part transfer function is modeled as shown in Eq. (11) where the fitting accuracy reaches to 97%. The overall system transfer function is the product of part 1 and part 2 transfer functions as shown in Eq. (12).

- Time delay Transfer Function

$$D(s) = e^{-0.9s} \tag{10}$$

- Temperature Rise transfer function

$$H(s) = \frac{0.1364}{s^2 + 5.17s + 7.081} \tag{11}$$

- Overall Air heater Transfer function

$$Overall \ TF = \frac{0.1364 \ e^{-0.9s}}{s^2 + 5.17s + 7.081} \tag{12}$$

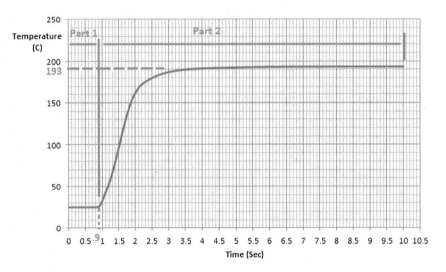

Fig. 5. Exit temperature delay and increase regions

4.1   PID Controller Design

Reaching a certain temperature with a small error in a short time period without overshoot is the most important objective of controller design for the air heater. Closed loop system with PID controller is designed to achieve system requirements as shown in Eq. (13) and Fig. 6 [18, 19]. Classical tuning methods such as Ziegler Nichols (ZN), Chien–Hrones–Reswick and Cohen Coon are used to tune PID controller gains. Particle swarm optimization (PSO) as a metaheuristic tuning technique is also used with various objective functions of response error minimization [21].

$$\text{PID Controller T.F :} \quad C(t) = K_p e + K_d \frac{de}{dt} + K_i \int_0^t e(t)dt \tag{13}$$

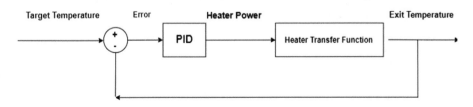

Fig. 6. Air heater closed loop block diagram

PID Classical Tuning Methods

By the aid of open loop step response done by CFD simulation for the air heater, the temperature profile is analyzed as shown in Fig. 7 to extract the response gain which is the saturation output of the system Kc equals 193 °C. By the use of Eqs. (14) and (15), dead time and time constant are calculated to be used for classical PID tuning method.

$$\text{Time Constant}: \quad T_c = \frac{2}{3}(t_2 - t_1) \tag{14}$$

$$\text{Dead Time}: \quad T_d = t_2 - T_c \tag{15}$$

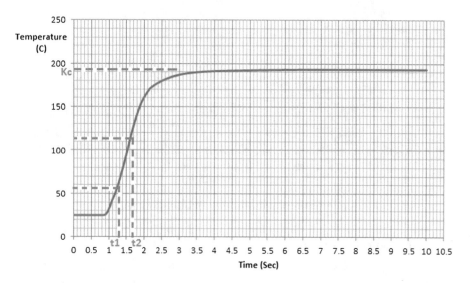

Fig. 7. Exit Temperature response profile analogy

For testing the controller of Heater model, a target temperature 90 °C is assigned as a set point for the closed loop PID controller. Figure 8 shows that Chien–Hrones–Reswick PID tuning method gives advanced results compared to ZN as the overshoot and settling time are lower compared to ZN response. For both tuning methods, the response shows an oscillation in the temperature profile that lasts for long time period around the set point. Cohen Coon Method gives an unstable temperature response that leads to temperature profile diverge.

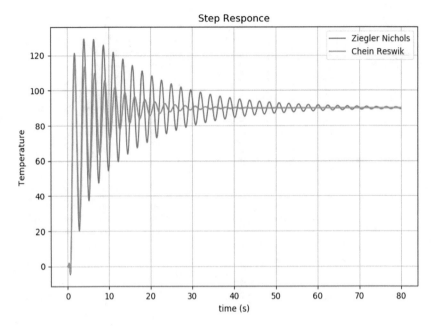

Fig. 8. Exit temperature response with Classical PID tuning methods set point 90 °C

PID-Particle Swarm Optimization Tuning

Particle swarm optimization method [21] is used as a metaheuristic tuning technique to tune the PID controller parameters with local particle coefficient 1.5 and global particle coefficient 2, number of particles is 50 and total number of iterations is 100 times. Four cost functions are tested for the PSO technique that aims to reduce the error between the closed loop temperature response and the set point temperature.

The Four fitness error functions (Integral Square Error (ISE), Integral Absolute Error (IAE), Integral Total Absolute Error (ITAE) and Integral Total Square Error (ITSE)) are used in the comparative analysis with particle swarm optimization. Figure 9 shows the variation of cost functions with respect to iteration number; from the figure it's obvious that all cost functions stabilized after only 20 iterations, and the Integral total square error fitness functions results in the lowest accumulative cost with 0.62 compared to other fitness functions.

A target temperature of 90 C is assigned as a set point for the closed loop PID controlled model and response comparative analysis is plotted for the step response as shown in Fig. 10 where all PSO Tuning methods give better response compared to classical tuning methods as over shoot decreased and settling time as well. ITAE fitness function tuned by the PSO gives the most satisfying results in overshoot reduction and faster response with low settling time compared to other error fitness functions. Also, response shows that ISE-PSO gives the highest overshoot and longest settling time compared to other PSO tuning methods. Table 3 summarizes the results of settling time, Rise time, overshoot and cost function for all classical tuning methods and PSO. Also, the PID parameters are listed for the PID parallel controller.

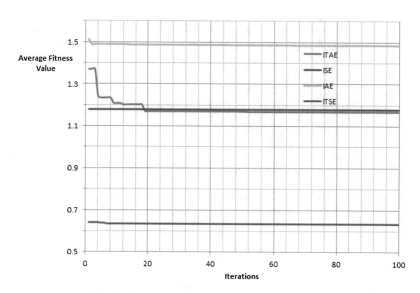

Fig. 9. Fitness function values with PSO iterations

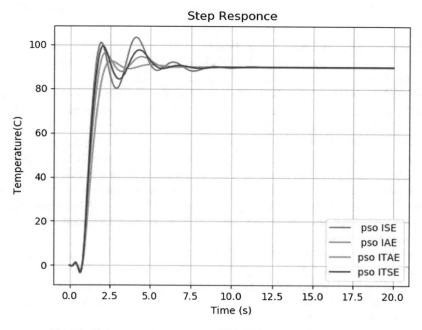

Fig. 10. Exit temperature response PSO PID tuning set point 90 °C

Table 3. Tuning methods comparative analysis

Method		Kp	Ki	Kd	Rise time	Settling time	Overshoot	Cost
Cohen Coon		81.9	45	25.4	0.32	∞	∞	
Ziegler Nichols		62	30.8	31.2	1.27	43.56	55.6	
Chien– Hrones– Reswick		62.9	31.2	26.6	1.33	26.7	23.85	
PSO	IAE	39.5	44	17	1.67	6.92	5.21	1.48
	ISE	41.1	50	23.8	1.48	14.8	6.62	1.17
	ITAE	36.4	38.2	12.5	1.9	2.88	2.83	1.17
	ITSE	41.5	46.9	20	1.56	10.4	5.1	.635

5 Conclusion

Computational fluid dynamic simulations showed a powerful tool to predict the natural operation and response of thermal dynamical system represented in the air heater equipment, modeling its transfer function and closed loop controller design. Particle swarm optimization metaheuristic technique shows a satisfactory improvement of tracking the set point temperature of the air heater by designing PID controller via iterative method based on error minimization between the response and set point temperature that leads to improve the closed system response. PSO-PID tuning technique give advance over classical tuning methods for the overshoot minimization and settling time as well. Integral total absolute error fitness function gives the minimum overshoot and settling time compared to other fitness functions of error minimization used for PSO algorithm. Other metaheuristic optimization techniques may be used instead of PSO to be compared with the results of the proposed technique.

References

1. Çengel, Y.A.: Heat Transfer Practical approach, 2nd edn. McGraw-Hill, New York (2002)
2. Cengel, Y.A., Cimbala, J.M.: Fluid Mechanics Fundamentals and Applications. McGraw-Hill, Boston (2006)
3. Ogata, K.: Modern Control Engineering, 5th edn. Prentice Hall, NJ (2010)
4. Azar, A.T., Vaidyanathan, S.: Advances in system dynamics and control. In: Advances in Computational Intelligence and Robotics (ACIR). IGI Global, USA (2018). ISBN 9781522540779
5. Azar, A.T., Vaidyanathan, S.: Handbook of research on advanced intelligent control engineering and automation. In: Advances in Computational Intelligence and Robotics (ACIR). IGI Global, USA (2015). ISBN 9781466672482
6. Vasickaninová, A., Bakošová, M.: Control of a heat exchanger using neural network predictive controller combined with auxiliary fuzzy controller. Appl. Therm. Eng. **89**(2015), 1046–1053 (2015)

7. Maidi, A., Diaf, Moussa, Corriou, Jean-Pierre: Optimal linear PI fuzzy controller design of a heat exchanger. Chem. Eng. Process. **47**(5), 938–945 (2008)
8. Vasickaninová, A., Bakošová, M., Cirka, L., Kalúz, M., Oravec, J.: Robust controller design for a laboratory heat exchanger. Appl. Therm. Eng. **128**(2018), 1297–1309 (2018)
9. Wang, Y., You, S., Zheng, W., Zhang, H., Zheng, X., Miao, O.: State space model and robust control of plate heat exchanger for dynamic performance improvement. Appl. Therm. Eng. **128**(2018), 1588–1604 (2018)
10. Jain, M., Rani, A., Pachauri, N., Singh, V., Mittal, A.P.: Design of fractional order 2-DOF PI controller for real-time control of heat flow experiment. Eng. Sci. Technol. Int. J. **22**(1), 215–228 (2019)
11. Padhee, S.: Controller design for temperature control of heat exchanger system: simulation studies. WSEAS Trans. Syst. Control **9**, 485–491 (2014)
12. Yu, Y., Yin, D.: Application of the BP neural network PID algorithm in heat transfer station control. In: Xie, A., Huang, X. (eds.) Advances in Computer Science and Education. AISC, vol. 140. Springer, Heidelberg (2012)
13. Sungthong, A., Assawinchaichote, W.: Particle swam optimization based optimal PID parameters for air heater temperature control system. Procedia Comput. Sci. **86**(2016), 108–111 (2016)
14. Hoffmann, K.A., Chiang, S.T.: Computational Fluid Dynamics, Engineering Education System, 4 edn., V4 (2000)
15. Gherasim, I., Galanis, N., Nguyen, C.T.: Heat transfer and fluid flow in a plate heat exchanger. Part II: Assessment of laminar and two-equation turbulent models. Int. J. Thermal Sci. **50**(8), 1499–1511 (2011)
16. Allegrini, J., Dorer, V., Defraeye, T., Carmeliet, J.: An adaptive temperature wall function for mixed convective flows at exterior surfaces of buildings in street canyons. Build. Environ. **49**(2012), 55–66 (2012)
17. Zukauskas, A.: Convection heat transfer in cross flow. In: Hartnett, J.P., Irvine Jr., T.F. (eds.) Advances in Heat Transfer, vol. 8, pp. 93–106. Academic Press, New York (1972)
18. Azar, A.T., Serrano, F.E.: Robust IMC-PID tuning for cascade control systems with gain and phase margin specifications. Neural Comput. Appl. **25**(5), 983–995 (2014)
19. Azar, A.T., Ammar, H.H., de Brito Silva, G., Razali, M.S.A.B.: Optimal Proportional Integral Derivative (PID) controller design for smart irrigation mobile robot with soil moisture sensor. In: The International Conference on Advanced Machine Learning Technologies and Applications (AMLTA 2019). AISC, vol 921, pp. 349–359. Springer, Cham (2020)
20. Hassanien, A.E., Emary, E.: Swarm Intelligence: Principles, Advances, and Applications. CRC Press, Boca Raton (2018)
21. Azar, A.T., Serrano, F.E.: Fractional order sliding mode PID controller/observer for continuous nonlinear switched systems with PSO parameter tuning. The International Conference on Advanced Machine Learning Technologies and Applications (AMLTA 2018). AISC, vol. 723, pp. 13–22. Springer, Cham (2018)

Design and Implementation of a Ball and Beam PID Control System Based on Metaheuristic Techniques

Ahmad Taher Azar[2,3], Nourhan Ali[1(✉)], Sarah Makarem[1(✉)],
Mohamed Khaled Diab[1(✉)], and Hossam Hassan Ammar[1]

[1] School of Engineering and Applied Science, Nile University, Giza, Egypt
{n.kamel, s.zaky, mkhalid, hhassan}@nu.edu.eg
[2] College of Engineering, Prince Sultan University,
Riyadh, Kingdom of Saudi Arabia
aazar@psu.edu.sa
[3] Faculty of Computers and Artificial Intelligence,
Benha University, Benha, Egypt
ahmad.azar@fci.bu.edu.eg

Abstract. The paper introduces a comparative analysis between three meta-heuristic techniques in the optimization of Proportional-Integral-Derivative (PID) controller for a cascaded control of a ball and beam system. The meta-heuristic techniques presented in this study are Particle Swarm Optimization (PSO), Artificial Bee Colony (ABC) and Bat Algorithm Optimization (BAO). The model uses a DC motor with encoder to move the beam and a camera as a feedback for the ball position on the beam. The control theory of the system depends on two loops; the first (inner) loop is the DC motor for position control. The three meta-heuristic techniques are applied for the tuning of the PID parameters then the efficiency of each algorithm is compared based on the time response, overshoot and steady state error. The BAT algorithm has proved to be more efficient in optimizing the controller for the motor position control. The same three algorithms are then applied for the outer loop: the Simulink model of the ball and beam system. Having the time response, overshoot and steady state error as the criteria, the PSO algorithm showed better performance in optimizing the controller for the overall system.

Keywords: PID controller · Cascaded control · Meta-heuristics · Optimization problems · Ball and beam

1 Introduction

The ball and beam system is a dynamically unstable system which acts as an equivalent to lots of real life problems that cannot be brought into the laboratory to be solved for example the horizontal stabilizer of airplane in the cases of landing or facing turbulent airflow [1]. The system has two degrees of freedom and mainly categorized into two configurations, one of them is where the beam is rotating about its central axis and it is called "ball and beam balancer" which is easier to build and simpler to be mathematically

A. E. Hassanien et al. (Eds.): AISI 2019, AISC 1058, pp. 313–325, 2020.
https://doi.org/10.1007/978-3-030-31129-2_29

modelled. The other configuration which is used in this paper is where the beam is supported from both ends, one end is attached to a lever connected to a motor and the other end acts as a pivot for the beam to rotate around which is called "ball and beam module" [2]. The main objective of the system is to control the position of the ball on the beam to a desired reference and to reject any disturbances applied to the system in the form of pushing the ball. In the system, both the position of the ball on the beam and the rotational angle of the motor is measured and controlled. When the lever attached to the motor, it changes the angle of the beam hence the ball changes its position and slide on the beam due to gravity in a nonlinear way. To regulate the ball motion, the beam angle needs to be controlled properly with time. The open loop form of the system is unstable since the ball accelerates on the beam even at a fixed beam angle. Therefore, a closed loop system with feedback is crucial to achieve stability of the system [3–7]. Using a proportional, derivative and integral (PID) control, a cascade control system is designed to bring the system to stability [8]. By feeding back the position information of the ball on the beam through a camera, the control signal is sent through Arduino to the motor in order to control and achieve the desired angle of the beam.

Tuning the PID parameters to result in the optimal solution is a challenging problem. Various classical methods have always been used for this task such as Zeigler Nichols (ZN) [9, 10]; however, it requires multiple trials and does not always lead to good results. Lots of modifications have been applied to ZN, yet the conventional methods to tune the PID parameters are sometimes limited especially in real life applications due to the systems nonlinearity and unknown disturbances [11, 12]. In addition to the conventional methods, there have been a growing research to develop more advanced and intelligent algorithms to tune the PID parameters [13, 14]. Some of these algorithms inspire their way of working and searching for optimal solutions from nature such as Genetic Algorithm (GA), Particle Swarm Optimization (PSO), Ant Colony Optimization (ACO), Fruit Fly Optimization (FOA), etc. [15].

In this paper, a comparative analysis of the performance of three meta-heuristic methods known as Particle Swarm Optimization (PSO) [16], Artificial Bee Colony (ABC) [17] and Bat Algorithm Optimization (BAO) [18] is performed to tune the PID parameters for the cascaded control system of the ball and beam. The experimental analysis is performed on a 60 × 30 cm system. The design is made using Solid works and laser cut from acrylic to ensure light weight mechanism. A DC motor interfaced with Arduino microcontroller is used to move the mechanism. The three meta-heuristic techniques are applied to optimize the PID parameters for controlling the motor shaft angle and position of the ball on the beam.

The rest of the paper is organized as follows. In Sect. 2, Ball and Beam system mechanical configuration is presented. Section 3 describes Ball and Beam system machine vision. Section 4 presents the control system design of ball and beam. Lastly, Sect. 5 contains the conclusion of the research work.

2 Ball and Beam System Mechanical Configuration

In this paper, the second configuration is used where the beam is supported from both sides. The main mechanical part of this system is the support base made of acrylic which has length of 60 cm and width of 30 cm. Acrylic beam from point A_1 to A_2 has a length of 38 cm, acrylic leaver arm from point A_2 to A_3 has a length of 9.15 cm, acrylic offset arm from point A_4 to A_3 has a length of 4.51 cm and the fixed lever arm from point A_0 to A_1 has a length of 12 cm. The points of the physical system are shown Fig. 1 while the final design of the Ball and Beam system used in the experimental analysis is shown in Fig. 2.

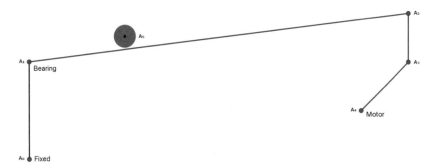

Fig. 1. Points of the physical model

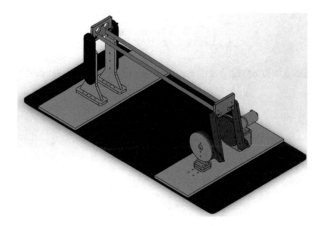

Fig. 2. Final design of the Ball and Beam system

2.1 Ball and Beam State Space Model

Similar to the DC motor, the ball and beam transfer function is a standard equation that is specified to each model by changing certain parameters (see Fig. 3) [19]. In the case of the ball and beam model, those parameters are: ball mass (m), Ball radius (R), offset of the lever arm (d), gravitational acceleration (g), beam length (L), ball's moment of inertia (J), coordinates of ball position (r), coordinates of beam angle (alpha) and servo gear angle (theta).

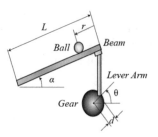

Fig. 3. Ball and Beam model

The second derivative of r is affected by the second derivative of the input angle α. The Lagrange equation of motion for the ball is:

$$\left(\frac{J}{R^2} + m\right)\ddot{r} + mg\sin\alpha - mr\dot{\alpha}^2 = 0 \tag{1}$$

Linearization of this equation about the beam angle, $\alpha = 0$, gives the following linear approximation of the system:

$$\left(\frac{J}{R^2} + m\right)\ddot{r} = -mg\alpha \tag{2}$$

The equation which relates the beam angle to the angle of the gear can be approximated linearly by:

$$\alpha = \frac{d}{L}\theta \tag{3}$$

Substituting (3) into (2):

$$\left(\frac{J}{R^2} + m\right)\ddot{r} = -mg\frac{d}{L}\theta \tag{4}$$

Then, the linearized system equations can be represented in the form of state-space. The state variables are the ball's position r and velocity $\dot{r}$ while the gear angle θ is the input to the system. The state-space representation is:

$$\begin{bmatrix} \dot{r} \\ \ddot{r} \end{bmatrix} = \begin{bmatrix} 0 & 1 \\ 0 & 0 \end{bmatrix}\begin{bmatrix} r \\ \dot{r} \end{bmatrix} + \begin{bmatrix} 0 \\ -\frac{mgd}{L\left(\frac{J}{R^2}+m\right)} \end{bmatrix}\theta \tag{5}$$

However, some slight changes are done to the state-space example used. The main change is that the torque applied to the beam is controlled directly instead of controlling the position by the gear angle, Θ, So, the system representation is:

$$\begin{bmatrix} \dot{r} \\ \ddot{r} \\ \dot{\alpha} \\ \ddot{\alpha} \end{bmatrix} = \begin{bmatrix} 0 & 1 & 0 & 0 \\ 0 & 0 & \frac{-mg}{\left(\frac{J}{R^2}+m\right)} & 0 \\ 0 & 0 & 0 & 1 \\ 0 & 0 & 0 & 0 \end{bmatrix} \begin{bmatrix} r \\ \dot{r} \\ \alpha \\ \dot{\alpha} \end{bmatrix} + \begin{bmatrix} 0 \\ 0 \\ 0 \\ 1 \end{bmatrix} u, \qquad y = [1 \quad 0 \quad 0 \quad 0] \begin{bmatrix} r \\ \dot{r} \\ \alpha \\ \dot{\alpha} \end{bmatrix} \quad (6)$$

3 Ball and Beam System Machine Vision

Essentially, the control of the whole system is the position feedback for the ball on the beam. In this paper, acquiring the exact position of the ball is done by a camera. One advantage of machine vision system is the ability to perform contact-free measurement, which is important especially when contact is difficult. One of the challenges in designing a machine vision system is how to be robust to any noise in an image to reject any disturbances caused by unwanted background and foreground objects as well as to deal with different illumination conditions [20].

Let $A_1 = [X_0, Y_0]$ be the image coordinate of the point where A_1 is the lowest pixel position of the beam along the vertical line X. Let $A_2 = [X_1, Y_1]$ be the image coordinate of the point where A_2 is the biggest pixel position of the beam along the vertical line X, and $A_5 = [X_2, Y_2]$ be the image coordinate of the red ball (see Fig. 4). A red ball is used here in order to simplify ball detection with colure filter and to improve measurement accuracy. Let $\Delta X_{Beam} = X_2 - X_0$ and $\Delta X_{Ball} = X_1 - X_2$ be the image feedback output variables. We have $s = \frac{\Delta X_{Beam}}{L}$, where L is a constant beam length in cm; therefore, the ball position can be expressed as:

$$Ball\,position = s\,\Delta X_{Ball} \qquad (7)$$

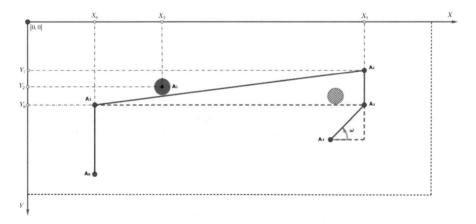

Fig. 4. Machine vision for Ball and Beam system

The values of A_1, A_2 and A_5 can be controlled in terms of distance between the camera and the Ball-Beam system, resolution of the camera, Illumination conditions and color. On the other hand, the proposed strategy for ball detection has the following stages:

1. Image acquisition.
2. Conversion from RGB image to gray scale image.
3. Subtract Color plane "Red plane".
4. Thresholding.
5. Area Filter to detect the ball.
6. Image component labeling.
7. Ball centroid estimation.
8. Calibration to convert from pixels to cm.

4   Control System Design of Ball and Beam

As shown in Fig. 5, the control of the system is a cascaded control of the two plants, where there are two loops of control; the inner loop is the motor's and it is responsible for controlling the angle of the motor shaft. The DC motor PID compensator calculates the required voltage to achieve the desired shaft angle. The outer loop PID compensator computes the desired motor shaft angle to reach the required ball position on the beam.

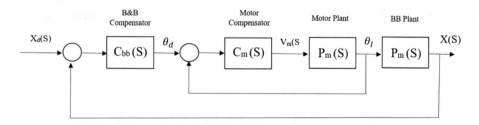

Fig. 5. Cascaded control for the ball and beam system

4.1   DC Motor Inner Loop: Position Controller

As shown in Fig. 6, the inner loop implements a PID controller in order to achieve the desired motor shaft angle. The design of this PID is applied through on-line tuning of the DC motor using three meta-heuristic techniques: PSO, ABC and BAT.

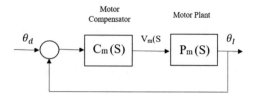

Fig. 6. DC motor closed-loop system

However, prior to apply the meta-heurist techniques, certain parameters need to be adjusted for all meta-heuristic algorithms. One of these is the search space as the tuning and search for optimal PID parameters is done on a real motor. To save time and minimize the number of required iterations, manual tuning with trial and error is performed on the motor to determine the range of parameters that will result in performance that match the required specifications for the system. For all the three algorithms the same limits where applied, where proportional gain, K_P, ranges from 0 to 0.03, the integral gain, K_I, ranges is from 0 to 0.003 and derivative gain, K_D, ranges from 0 and 0.01. The three algorithms are then tested in terms of the speed of convergence and this is done by using the same number of iteration and the same population size. The objective function used for the three algorithms which gives an indication of the performance of each solution is,

$$z = \left(1 - e^{-\beta 1}\right)\left(M_p * 0.01 + e_{ss}\right) + e^{-\beta 2}\left(t_s - t_r\right) \tag{8}$$

Where M_p, t_s, t_r and e_{ss} are maximum overshoot, settling time, rise time and steady state error, respectively. Both $\beta 1$ and $\beta 2$ are set to be equal 1. The objective is to minimize the cost of this function.

For the on-line tuning, the motor is given a setpoint of one revolution (360 degrees) and an encoder is used to measure the shaft angle. The three algorithms are run for 5 iterations and with 10 population size. For PSO, the minimum cost function was equal to 2.0787. The convergence curve is shown in Fig. 7a. The PSO algorithm is found to converge after the third iteration and the best solution values for the PID parameters were $K_p = 0.0238$, $K_i = 0.0019$ and $K_d = 0.0076$.

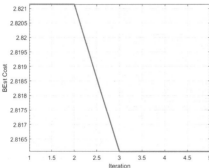

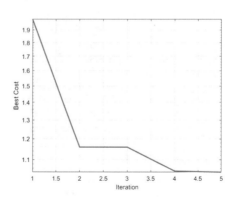

Fig. 7a. PSO convergence curve for DC motor

Fig. 7b. ABC convergence curve for DC motor

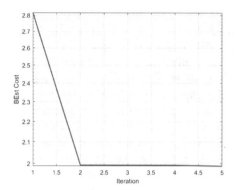

Fig. 7c. BAT convergence curve for DC motor

Applying the same steps for ABC, the minimum cost was 2.1660 and the convergence curve is shown in Fig. 7b. The best solution for the PID parameters tuned by ABC was $K_p = 0.0080$, $K_i = 0.011$ and $K_d = 0.0043$. The BAT algorithm reached a minimum cost function of 1.1645 and it's converged after the second iteration which is considered to be the fastest algorithm as shown in Fig. 7c. The best PID parameters tuned by BAT algorithms were $K_p = 0.0250$, $K_i = 0.0021$ and $K_d = 0.0039$. The step response characteristics and cost function of PSO, ABC and BAT algorithms are shown in Table 1. As shown in Table 1, the BAT algorithm in the limited iterations analysis showed an overall better performance compared to the other algorithms and it is the fastest one to converge as indicated by bold values in the table. ABC has achieved the best rising time; however, the overshoot and settling time are high in comparison with the other algorithms.

Table 1. Sep response characteristics for PSO, ABC and BAT algorithms

	Tr	Ts	M_p	e_{ss}	Cost
PSO	0.4804	0.7963	**0.1048**	4.1036	2.0787
ABC	**0.4351**	1.1741	1.2017	2.9845	2.1660
BAT	0.4646	**0.7428**	0.1562	**1.6788**	**1.1645**

In comparison between the three algorithms, the step response performance is illustrated in Fig. 8 and the overall step response is computed in Table 1.

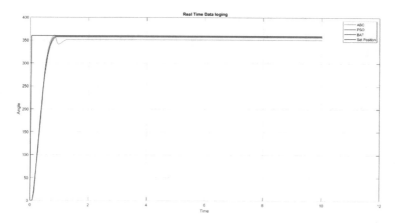

Fig. 8. Step response comparison between PSO, ABC and BAT for DC motor

4.2   Outer Loop Controller Design

To present the state space model of the outer loop accurately, Simulink was used. The state space model on Simulink is shown Fig. 9.

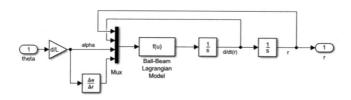

Fig. 9. Ball and beam Simulink model.

The ball and beam model on its own would be facing many stability issues. it was proven to be more difficult to be paired with the DC motor model for the whole system. Thus, a lead compensator was first used to stabilize the output of the ball and beam model before applying the PID controller and integrating the model with the DC motor model as shown in Fig. 10.

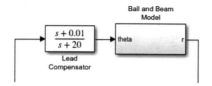

Fig. 10. Lead compensator for Ball and beam.

The used lead compensator is tuned to improve both steady-state error and transient response as shown in the comparison (see Fig. 11) between the step response with and without the lead compensator.

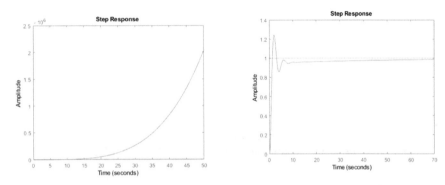

Fig. 11. Ball and beam step response before and after lead compensator.

After gaining the PID parameters for the inner loop DC motor, the same optimization techniques are applied on the entire system. The first technique was the PSO algorithm and the resulting parameters from running the PSO algorithm are K_p = 1.0023, K_i = 0.0001 and K_d = 0.0052. As shown in the convergence graph in Fig. 12a, the PSO algorithm converges very quickly with little variation in the best cost of each iteration.

ABC algorithm showed more variation in the results than PSO as shown in Fig. 12b. The ABC algorithm also converges before the 10th iteration and the tuned PID gains were K_p = 0.9892, K_i = 0.0008 and K_d = 0.0220. The best cost for PSO is evidently less than that of ABC algorithm. This shows the efficiency of PSO with the ball and beam model specifically. Figure 12c shows the convergence and the pattern of the tuning parameters for BAT optimization technique. The best results of PID gains with BAT algorithm were K_p = 0.9842, K_i = 0.0036 and K_d = −0.5502.

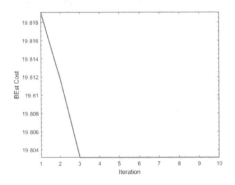

Fig. 12a. PSO cost convergence.

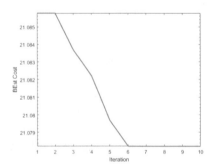

Fig. 12b. ABC cost convergence.

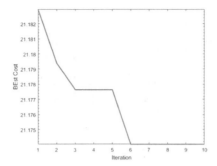

Fig. 12c. BAT cost convergence.

PSO was proved to be the best optimization technique for PID controller tuning for a ball and beam system. The cost value for PSO was 19.8041 and the results were heavily concentrated around the required region. The step response of the controlled system (see Fig. 13) is used to gauge the performance of the estimated PID parameters.

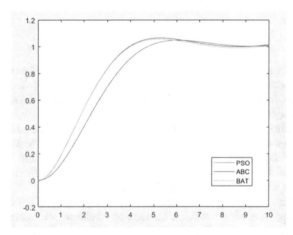

Fig. 13. Step response comparison between PSO, ABC and BAT algorithms

In Table 2, the comparison between PSO, ABC and BAT algorithms is summarized in terms of step response characteristics, PID gains and cost function. As seen, the result from BAT and ABC are very noticeably similar. PSO, on the other hand, showed a slower response but similar settling time. Yet, according to the cost calculation, PSO has the lowest cost of the three used algorithms.

Table 2. PID parameters and step response characteristics of PSO, ABC and BAT algorithms

Method	K_p	K_i	K_d	Rise time (s)	Settling time (s)	Overshoot (%)	Peak	Peak time (s)	Cost
BAT	0.9842	0.0036	−0.5502	3.0521	7.2762	6.208	1.0456	6.0384	21.1750
PSO	1.0023	0.0001	0.0052	2.6445	6.6690	5.7975	1.0624	5.2719	19.8041
ABC	0.9892	0.0008	−0.5502	2.6259	6.6528	6.20063	1.0539	5.2415	21.0791

5   Conclusion

In this paper, the control of the ball and beam system has been presented. The proposed system is a cascaded control consisting of two loops; the inner loop of DC motor and the outer loop of the ball-beam. The system was tested by three metaheuristic techniques to tune and optimize the parameters of the PID controller. The analyzed qualities and the performance criteria were the minimization of steady state error, overshoot and settling time. The simulation and experimental results showed that the BAT algorithm has proven to be of a better performance in controlling the inner loop and PSO is slightly better in controlling the outer loop in comparison with other techniques.

References

1. Negash, A., Singh, N.P.: Position control and tracking of ball and plate system using fuzzy sliding mode controller. In: Abraham, A., Krömer, P., Snasel, V. (eds.) Afro-European Conference for Industrial Advancement. Advances in Intelligent Systems and Computing, vol. 334. Springer, Cham (2015)
2. Yang, D.: Tuning of PID parameters based on particle swarm optimization. IOP Conf. Ser. Mater. Sci. Eng. **452**, 042179 (2018). https://doi.org/10.1088/1757-899X/452/4/042179
3. Azar, A.T., Vaidyanathan, S.: Handbook of Research on Advanced Intelligent Control Engineering and Automation. Advances in Computational Intelligence and Robotics (ACIR) Book Series. IGI Global, USA (2015). ISBN 9781466672482
4. Kumar, J., Azar, A.T., Kumar, V., Rana, K.P.S.: Design of fractional order fuzzy sliding mode controller for nonlinear complex systems. In: Mathematical Techniques of Fractional Order Systems, Advances in Nonlinear Dynamics and Chaos (ANDC) Series, pp. 249–282 (2018)
5. Abdelmalek, S., Azar, A.T., Dib, D.: A novel actuator fault-tolerant control strategy of DFIG-based wind turbines using Takagi-Sugeno Multiple models. Int. J. Control Autom. Syst. **16**(3), 1415–1424 (2018)
6. Ammar, H.H., Azar, A.T., Tembi, T.D., Tony, K., Sosa, A.: Design and implementation of fuzzy PID controller into multi agent smart library system prototype. In: The International Conference on Advanced Machine Learning Technologies and Applications (AMLTA2018), AMLTA 2018. Advances in Intelligent Systems and Computing, vol. 723, pp. 127–137. Springer, Cham (2018)
7. Meghni, B., Dib, D., Azar, A.T.: A Second-order sliding mode and fuzzy logic control to optimal energy management in PMSG wind turbine with battery storage. Neural Comput. Appl. **28**(6), 1417–1434 (2017)

8. Azar, A.T., Serrano, F.E.: Design and modeling of anti wind up PID controllers. In: Zhu, Q., Azar, A.T. (eds.) Complex System Modelling and Control Through Intelligent Soft Computations, Studies in Fuzziness and Soft Computing, vol. 319, pp. 1–44. Springer (2015)

9. Ziegler, J., Nichols, N.B.: Optimum settings for automatic controllers. Trans. ASME **64**(8), 759–768 (1942)

10. Meshram, P.M.R., Kanojiya, R.G.: Tuning of PID controller using Ziegler-Nichols method for speed control of DC motor. In: IEEE-International Conference on Advances in Engineering, Science and Management (ICAESM-2012), 30–31 March 2012, Nagapattinam, Tamil Nadu, India (2012)

11. Liu, G.P., Daley, S.: Optimal-tuning PID control for industrial systems. Control Eng. Pract. **9**(11), 1185–1194 (2001)

12. Azar, A.T., Serrano, F.E.: Robust IMC-PID tuning for cascade control systems with gain and phase margin specifications. Neural Comput. Appl. **25**(5), 983–995 (2014)

13. Azar, A.T., Ammar, H.H., de Brito Silva, G., Razali, M.S.A.B.: Optimal proportional integral derivative (PID) controller design for smart irrigation mobile robot with soil moisture sensor. In: The International Conference on Advanced Machine Learning Technologies and Applications (AMLTA2019), AMLTA 2019. Advances in Intelligent Systems and Computing, vol. 921, pp. 349–359. Springer, Cham (2020)

14. Gorripotu, T.S., Samalla, H., Jagan Mohana Rao, C., Azar, A.T., Pelusi, D.: TLBO algorithm optimized fractional-order PID controller for AGC of interconnected power system. In: Nayak, J., Abraham, A., Krishna, B., Chandra Sekhar, G., Das, A. (eds.) Soft Computing in Data Analytics. Advances in Intelligent Systems and Computing, vol. 758, pp. 847–855. Springer, Singapore (2019)

15. Luke, S.: Essentials of Metaheuristics. Lulu (2013)

16. Kennedy, J., Eberhart, R.: Particle swarm optimization. In: Proceedings of ICNN 1995 - International Conference on Neural Networks, vol. 4, pp. 1942–1948 (1995)

17. Karaboga, D.: An idea based on honey bee swarm for numerical optimization. Tech. Rep. J. Funct. Program. **15**(4), 615–650 (2005)

18. Yang, X.S.: A new metaheuristic bat-inspired algorithm. In: González, J.R., Pelta, D.A., Cruz, C., Terrazas, G., Krasnogor, N. (eds.) Nature Inspired Cooperative Strategies for Optimization (NICSO 2010). Studies in Computational Intelligence, vol. 284. Springer, Heidelberg (2010)

19. Ahmad, B., Hussain, I.: Design and hardware implementation of ball & beam setup. In: 2017 Fifth International Conference on Aerospace Science & Engineering (ICASE), Islamabad, Pakistan, 14–16 November 2017, pp. 1–6 (2017). https://doi.org/10.1109/icase.2017.8374271

20. Corke, P.: Robotics, Vision and Control: Fundamental Algorithms In MATLAB. Springer Tracts in Advanced Robotics Book Series, STAR, vol. 73. Springer (2011)

Sentiment Analysis, E-learning and Social media Education

Student Sentiment Analysis Using Gamification for Education Context

Lamiaa Mostafa[(✉)]

Business Information System Department, Arab Academy for Science
and Technology and Maritime Transport, Alexandria, Egypt
Lamiaa.mostafa31@aast.edu, Lamiaa.mostafa31@gmail.com

Abstract. Internet users are expressing their sentiments (opinions) online using blogs and social media. Sentiment Analysis is a new technology that is used to improve the quality of the institutions including Higher education institution (HEI). Egypt educational institutions face a difficulties based on student motivation and learning engagement. Gamification provides a great help for educational institutions to motivate student and increase their learning ability however it depends on the teacher skills to use the Gamification tools. The paper reviews work in sentiment analysis related to education field, Gamification in learning. The paper will propose a Sentiment Analysis Classifier that will analyze the sentiments of students while using Gamification tools in an educational course.

Keywords: Sentiment Analysis · Gamification · WordNet · Learning · Slang Arabic

1 Introduction

Education systems depend on four key elements which are teacher, the student, the university, and the curriculum. Student motivation affects the student performance in the educational institution. Gamification is defined as "the use of game design elements in non-game contexts [1]. Gamification can be used to solve actual problems in different fields.

Internet helped people to express their thoughts and feelings [2]. Blog post and online forums allow users to write their reviews. The people are connecting with each other with the help of the internet through the blog post, online conversation forums, and many more online user-generated reviews which is also called Sentiments also known as opinion mining [3] can be used in different ways: used in decision making process of consumers, sentiments provide efficient low cost feedback channel, improve the quality of companies products and services which will also enhance their repetition [4]. Authors in [2] proposed a tool which decides the quality of text based on scientific papers annotations.

The rest of this paper is organized as follows: Sentiment Analysis, Gamification in the second and third Section, sentiment analysis classifier will be described in Sect. 4; results will be analyzed in Sect. 5; Sect. 6 include the paper conclusion and future work.

© Springer Nature Switzerland AG 2020
A. E. Hassanien et al. (Eds.): AISI 2019, AISC 1058, pp. 329–339, 2020.
https://doi.org/10.1007/978-3-030-31129-2_30

2 Sentiment Analysis

Sentiments and opinions can be recognized in different ways. Overall sentiment which is sentiment as a whole piece of text can be analyzed using classification tools. Classifiers divide sentiments into positive or negative sentiment [2]. Different researchers classify overall sentiment analysis [5].

Another type of sentiment classification is Aspect-based sentiment analysis [6, 7]. Aspect sentiment include two major tasks, the first is detect hidden semantic aspect from given texts, the second is identify fine-grained sentiments [2].

Data has two perspectives: objective and subjective. Objective based on emotions and personal feelings and subjective is based on facts. Sentiment analysis aim is to analyze subjective perspective of data [8]. Text mining techniques used to analyze sentiments of students [9]. Sentiment analysis recognize student emotions [10]. Researchers are working on sentiment analysis in different patterns: Twitter sentiment analysis [11] and cross lingual portability [12] and others. The following subsections will discuss Sentiment processing steps and previous research in sentiment analysis used in learning.

2.1 Sentiment Processing

Authors in [13] divide the sentiment process into 4 stages: data acquisition, data preparation, review analysis, and sentiment classification. Sentiment process consists of many steps such as stop word removal in which irrelevant terms are removed such as "the" or "and") [14]. Remove html tags; remove numbering and punctuation conversion to lower case and stemming words (removing suffixes and prefixes to identify the stem [15]. Term-frequency-inverse document frequency (TF-IDF) should be used to count the frequency of keywords on the document [14]. Clustering algorithms should be used to define correlation between the terms and the topic of the document.

Classify sentiments depends on two techniques: machine learning divided into supervised and unsupervised or lexicon-based divided into dictionary based and corpus based [16]. Machine learning uses traditional mining algorithms such as Naive Bayes (NB), Support Vector Machines (SVM) and Neural Networks (NN). Naïve Bayes is the simplest and most used classifier [17], Support Vector Machine apply learning models to analyze data. Neural Networks requires a large corpus with three classes: positive, negative and neutral opinions to learn and classify new sentiment into one of the three classes [18].

Lexical based approach depends on dictionaries of terms such as WordNet. The lexicon-based approach classifies a text according to the positive, negative and neutral words and it does need training phase. Lexicon-based approach divided into dictionary-based and corpus-based approach. Dictionary-based approach uses synonyms, antonyms, and hierarchies found in lexical databases, Corpus-based approach find word patterns occurrence to determine the polarity of a text [13]. After understanding the processing of sentiment analysis, the paper will explore some of the researches that implement the sentiment analysis method in leaning field.

2.2   Sentiment in Learning

Opinions written by online users have been studied using SA [19]. Opinions text should be cleaned since it is full of spelling mistakes and implicit meaning. Online text processing is a very important process before extracting sentiments [20]. Educational Data Mining (EDM) task is to extract knowledge to enhance students' learning processes [21]. Li [22] proposed a sentiment-enhanced learning model for the online language learning, the experimental results show that sentiment learning is effective tool online language learning. Permana et al. [23] proposed a sentiment model that evaluates student satisfaction in learning using NB in classification which resulted in enhanced of 16.49% over the existing system.

3   Gamification

The previous section discussed using sentiment analysis in learning, the aim of the paper is to understand student sentiments on using Gamification in learning, and this section will discuss Gamification concept and previous researchers that uses Gamification in different sectors. Game design techniques help motivate people to finish the required duties [24]. Games can be used in different services and increase the involvement of people in non-game services [25]. The Gamification has 2 main benefits: context role in game and the user qualities [26, 27]. Gamification can be used in different contexts tourism [28] and job recommendation in which a framework proposed by [29] to understand the level of student in each working track and enhance his learning path through Gamification tools. When person chooses to play a game over responsibility, this is due to pressure and reduce stress [30]. Using Gamification in higher learning assignment enhance the academic progress while reducing stress, providing learning milestones and improving personal satisfaction [31]. It is important to understand the previous researchers that use Gamification in learning so the next subsection will discuss the previous work in Gamification.

Educational games "pretend to be games, as the fun factor is an additive, and not the goal of those creations, however it is a specific learning outcome goal" [32]. Gamification in education aims to engage and motivate students [33], Gamification develops the student's skills such as collaboration, self-regulation and creativity. There are many Gamification tools such as ClassDojo, Classcraft, Moodle plugin, Open Badges [32] Games development has 2 main types: Game based learning and serious games, game-based learning is using games to support learning however specific games are games developed for the aim of teaching [1].

Different researchers use Gamification techniques in learning [34–36]. Authors in [34] implemented a Gamification tool (Funprog) for teaching Programming fundamentals for 80 students of first-year students from University of Ecuador. The result of the survey shows an acceptance level from the student and the teacher. Authors in [1] provides a strategy to use e-learning platform and Gamification techniques in information system courses.

Five interactivity story-based games where added by Arizona State University to scientific curriculum to enhance student engagement [37]. Authors in [36] tested

Gamification on gender, age, and type of institution. They concluded that no differences in use of gamification by age, gender or type of institution. [35] applied Gamification on software teaching in a specific class and the results had shown that Gamification motivate students and enhance the learning tools used by the teacher. [38] Focuses on the social gamification of e-learning and learner performance.

Authors in [25] proposed a Gamification model to be applied in Human Resource Management course for Masters in Engineering. Authors in [39] designed a Gamification tool that include levels based on storyline to enhance the teaching productivity and learning path of the student concept for teaching in universities, they presented Gamification for EPUB using Linked Data (GEL) is a framework combines Gamification concepts and digital textbook they proposed Gamification Ontology (GO) that share knowledge of gamified books between applications.

Briers in [40] applied gamified PMBOK for project management course targeting junior project managers in Europe. PMBOK involves testing the following skills: evaluating the reactions, cognitive learning, and the behavior changes.

Sobocinski in [32] proposed properties of Gamification tool used in education as follows: narrative, points-badges, no grades, missions, small tasks, skills-badges, individual and team tasks, boss fight, window of opportunity, optionality, schedule, realoutcome. Narrative is to explain the course components, points and badges motivate the students to reach different levels, grades are associated with difficulty so it is replaced by points and badges, mission which is tasks enhance the student ability in learning, small missions represent milestones, skills- badges reflect the development of student level, team tasks create collaboration scenario between the student and his colleagues, boss fight remove the fear of the final exams associated works in the thinking of the students that is replaced by badges and points, window of opportunity in which milestones and tests can be retaken, optionality of attendance gives the students the freedom to stay home and learn, schedule flexibility enhance the student ability to strength his skills, real outcomes represented in social experiment outside class which give the student ability to present his progress to friends and family [32].

Learning motivations affect the student performance in class [41, 42]. Authors in [41] use sentiment analysis to improve the learning process in an e-learning environment. Sentiment analysis used to analyze the opinions of the students for better understanding their opinion. After exploring the literature review in Sentiment Analysis and Gamification, the Sentiment Analysis Classifier is designed and implemented.

4   Sentiment Analysis Classifier (SAR)

Sentiment Analysis classifier is used to analyze the students' opinion in using Gamification tool in learning, sentiment analysis classifier passes by different stages including text processing, feature selection and classification. The section is divided into 4 subsections including Text Processing, Features selection, Machine Learning Classifiers and Sentiment Dataset. Figure 1 represents the Sentiment Analysis Classifier steps.

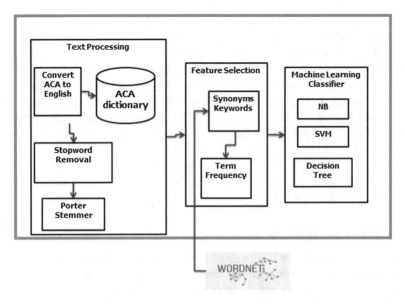

Fig. 1. Sentiment analysis classifier steps.

4.1   Text Processing

The phase of text processing passes by two steps: the first step is the preprocessing step and the second, is the feature selection or the representation of a document. The steps are executed sequentially in which the output of the preprocessing step is the input of the feature selection step [43].

Preprocessing is the step before the text is fed into a classifier. Preprocessing includes many steps. Classifier preprocessing steps include parsing, removing stop word and stemming [43]. Since the text is written in Arabic Chat Alphabet (ACA) also known as (Slang Arabic)," is a writing system for Arabic in which English letters are written instead of Arabic ones" [20], for example: "gama3aha feh haad ye3rf el gamification deh eh", the translation is "people, anyone knows about gamification?". ACA dictionary was created to be used in the Sentiment analysisClassifier. Student Sentiment is reviewed by two experts to confirm the accuracy of the ACA dictionary. Documents are preprocessed before classification. Preprocessing include stop word removal and stemming. One of the stemmers that usually used in classification is Porter Stemmer [14]. Parsing process converts the HTML file into stream of terms and remove the HTML tags, stop word removal is the process of removing words that does not affect the meaning of document sentences e.g. (and, a, or, the, on) and stemming is the process of reducing a word to its stem or root form. Thus, stems represent the key terms of a query or document rather than by the original words, e.g. "computation" might be stemmed to "compute" [43].

Word Net[1] is a large lexical database of English. George A. Miller is the main director. WordNet works as follows: it groups networks called sunsets, these sunsets include nouns, verbs, adjectives and adverbs, and each sunsets aim is to express a distinct concept. Synsets include conceptual-semantic and lexical relations. WordNet can be downloaded freely. WorldNet's structure which is the form of networks, makes it a useful tool for text mining techniques as classification and natural language processing. Different members develop WordNet as George A. Miller, Christiane Fellbaum, Randee Tengi, and Helen Langone. WordNet is maintained at the Cognitive Science Laboratory of Princeton University under the direction of psychology professor George A. Miller. WordNet's goal is to develop a system that can acquire knowledge extracted from human thinking's. One of these examples is the ability of a human being in classifying different items into groups.

4.2   Classification Features

Any Webpage or document is represented using a combination of features. The basic idea of feature selection algorithm is "Searching through all possible combination of features in the data to find which subset of features work best for prediction. The selection is done by reducing the number of features of the feature vectors, keeping the most meaningful discriminating ones, and removing the irrelevant or redundant ones" [45]. For the extraction of these features there are many known methods, including Document Frequency (DF), Information Gain (IG) [43], Mutual Information (MI) which is biased to rare terms opposite to IG [44]. Sentiment Analysis Classifier uses DF technique and features are expanded using synonyms feature extracted from WordNet.

4.3   Machine Learning Classifiers

After defining the features, sentiments can be classified. Statistical methods and Machine Learning classifiers are usually used in sentiment classification including Multivariate Regression Models, Bayes Probabilistic Approaches, Decision Trees, Neural Networks, Support Vector Machine (SVM) [46], Symbolic Rule Learning, Concept Vector Space Model (CVSM) [46], and Naïve Bayes (NB) is the most used classifier in sentiment analysis [19]. Sentiment analysis classifier will use SVM and NB.Knime[2] is used in preprocessing and keyword extraction and classification process.

4.4   Sentiment Dataset

Sentiments were collected from 1000 students in Arab Academy for Science and Technology and Maritime Transport University (AAST) University in a business course. 700 students agreed to examine Gamification in learning while 300 students were not interested. Before the process of using machine learning classifier, two datasets must be

---

[1] http://wordnet.princeton.edu.

[2] http://www.knime.com.

defined: the first dataset is agree sentiment (700) and the second dataset is disagree sentiment (300). ACA dictionary was created and mapped to every word in sentiments. Two professors in College of linguistics in the AAST revised the dictionary to ensure the accuracy level. After mapping the words of ACA to equivalent English words, the text processed by the regular text processing steps including stop word removal, stemming, WordNet is used for the enriching the keywords selected that represent agree and disagree set. Synonyms relationship was selected such as "agree = good idea = agree = happy", all there words represent an agree sentiment. Two machine learning classifiers were used: SVM and NB, both are trained and tested in the 1000 sentiments with the percent 80% for training and 20% for testing (Lewis and Catlett 1994).

Table 1. Sample for ACA dictionary

ACA	English
Ana mesh fahem	I cannot understand
3agbny awy fekrt el game	I do like the game idea
Brnamg bared	Application is not good
Mewafee gedan	I do strongly agree

After collecting the sentiments of the students, a written exam was executed on the two groups: agree and disagree, to test the performance of the student when using gamification and without using gamification and to confirm whether using gamification enhance the learning curve of the students in the learning process or not. The following section will discuss the results from using the proposed Sentiment Analysis Classifier (Table 1).

5 Sentiment Analysis Classifier Results

Sentiment analysis classifier tests the machine learning classifiers based on 1000 sentiments, sentiments are classified into agree sentiment and disagree sentiment. Naive Bayes, SVM and Decision Tree were used for the classification of the sentiments. Sentiments are converted in to English language using ACA dictionary created by the author and revised by two linguistic experts. Table 2 shows the results of the sentiment model.

Table 2. Sentiment analysis classifier results

Classifier	Accuracy	Precision	Recall
NB	83%	0.76	0.67
SVM	79%	0.61	0.83
Decision Tree	76%	0.51	0.8

The results of sentiment analysis classifier agreed with the conclusion of [1, 34] proving that using Gamification can enhance the efficiency of student learning.

Sentiment analysis techniques is a very good reflector of the student emotions towards learning, this is equivalent to the conclusion of [21, 22]. NB had shown the highest accuracy results which is agreed with [23] who uses NB to evaluate the sentiment option of students in learning.

Gamification affects the performance of student in learning. The two groups were tested to compare the performance when using Gamification and the results had shown that performance is affected by Gamification in which the highest grade reached by the Gamification group whose sentiment is agree reached 98 which is equivalent to A + in the credit hour system. Table 3 summarize the test results.

Table 3. Summary of test results.

	Traditional (Disagree) group	Gamification (Agree) group
Total number of students	300	700
Minimum grade	47	78
Maximum grade	91	98

In conclusion, sentiment analysis classifier compared to the other previous researchers focuses on ACA which is used by the Arabic internet users; sentiment analysis classifier uses an ACA dictionary that can be used on the following researchers as a base for converting ACA to English.

6 Conclusion and Future Work

In this paper we proposed a sentiment analysis classifier, the sentiment classifier data set was 1000 student sentiments that are divided into 700 agree sentiment and 300 disagree sentiments. The sentiment analysis classifier passes by text processing, feature selection and machine learning classification, three classifiers were used NB, SVM and decision tree. The results had shown that the best classifier accuracy results is the NB, also when executing a test on the 1000 students, the agree group of using Gamification in learning shown a better results comparing to the disagree group, this proves that Gamification will enhance the student performance in learning. The limitation of the Sentiment Analysis classifier is listed as follows: Sentiment analysis classifier should be tested in a larger dataset and should also be applied in different student's categories such as: different learning year, different type of courses and even different universities and the author should consider using different Machine learning classifiers can be used in classification such as Neural Networks and more semantic relationships can be extracted from the sentiment not only the synonyms WordNet relationship. The author will focus on future plans to expand the results of the sentiment analysis classifier to be tested on more students in different majors.

References

1. Elabnody, M., Fouad, M., Maghraby, F., Hegazy, A.: Framework for gamification based E-learning systems for higher education in Egypt. Int. J. Intell. Comput. Inf. Sci. IJICIS **17**(4), 59 (2017)

2. Raut, S., Mune, A.: Review paper on sentiments analyzer by using a supervised joint topic modeling approach. IJESC **8**(1) (2018)

3. Vinodhini, G., Chandrasekaran, R.: Sentiment analysis and opinion mining: a survey. Int. J. Adv. Res. Comput. Sci. Softw. Eng. **2**, 282–292 (2012)

4. Liu, B.: Sentiment analysis and opinion mining. In: Synthesis Lectures on Human Language Technologies, vol. 5, no. 1, pp. 1–167 (2012)

5. Yang, B., Cardie, C.: Context-aware learning for sentence-level sentiment analysis with posterior regularization. In: Proceedings of the 52nd Annual Meeting of the Association for Computational Linguistics, ACL 2014, 22–27 June 2014, Baltimore, MD, USA, vol. 1, Long Papers, pp. 325–335 (2014)

6. Yang, Z., Kotov, A., Mohan, A., Lu, S.: Parametric and nonparametric user-aware sentiment topic models. In: Proceedings of the 38th International ACM SIGIR Conference on Research and Development in Information Retrieval, ser. SIGIR 2015, pp. 413–422. ACM, New York (2015)

7. Rahman, M., Wang, H.: Hidden topic sentiment model. In: Proceedings of the 25th International Conference on World Wide Web, ser. WWW 16. Republic and Canton of Geneva, Switzerland: International World Wide Web Conferences Steering Committee, pp. 155–165 (2018)

8. Mandinach, E., Cline, H.: Classroom Dynamics: Implementing a Technology-Based Learning Environment. Taylor & Francis, New York (2013)

9. Schouten, K., Frasincar, F.: Survey on aspect-level sentiment analysis. IEEE Trans. Knowl. Data Eng. **28**(3), 813–830 (2016)

10. Yu, L.-C., Lee, L.-H., Hao, S., Wang, J., He, Y., Hu, J., Zhang, X.: Building Chinese affective resources in valence-arousal dimensions. Paper presented at the Proceedings of NAACL-HLT (2016)

11. Rosenthal, S., Nakov, P., Kiritchenko, S., Mohammad, S.M., Ritter, A., Stoyanov, V.: Sentiment analysis in Twitter. Paper presented at the Proceedings of the 9th International Workshop on Semantic Evaluation (2015)

12. Xu, R., Gui, L., Xu, J., Lu, Q., Wong, K.-F.: Cross lingual opinion holder extraction based on multi-kernel SVMs and transfer learning. World Wide Web **18**(2), 299–316 (2015)

13. Baidal, K., Vera, C., Avilés, E., Espinoza, A., Zambrano, J., Tapia, E.: Sentiment Analysis in Education Domain: A Systematic Literature Review, CCIS 883, pp. 158–173. Springer (2018)

14. Mostafa, L., Farouk, M., Fakhry, M.: An automated approach for webpage classification. In: ICCTA09 Proceedings of 11th International Conference on Computer Theory and Applications, Alexandria, Egypt (2009)

15. Lenco, D., Meo, R.: Towards the automatic construction of conceptual taxonomies, vol. 5182. Springer (2008)

16. Anitha, N., Anitha, B.: Sentiment classification approaches – a review. Int. J. Innov. Eng. Technol. **3**, 22–31 (2013)

17. Zhang, H.: The optimality of Naive Bayes. Am. Assoc. Artif, Intell (2004)

18. Tang, D., Qin, B., Liu, T.: Document modeling with gated recurrent neural network for sentiment classification. In: Proceedings of the 2015 Conference on Empirical Methods in Natural Language Processing, pp. 1422–1432 (2015)

19. Mostafa, L., Abd Elghany, M.: Investigating game developers' guilt emotions using sentiment analysis. Int. J. Softw. Eng. Appl. (IJSEA) (2018). ISSN: 0975–9018
20. Mostafa, L.: A Survey of Automated Tools for ARAB CHAT ALPHABET. Lab LAMBERT Academic Publishing (2016). ISBN:978-3-330-01845-7
21. Romero, C., Ventura, C., García, E.: Data mining in course management systems: moodle case study and tutorial. Comput. Educ. 5(1), 368–384 (2008)
22. Li, L.: Sentiment-enhanced learning model for online language learning system (2018)
23. Permana, F., Rosmansyah, Y., Abdullah, A.: Naive Bayes as opinion classier to evaluate students satisfaction based on student sentiment in Twitter Social Media. In: The Asian Mathematical Conference 2016 (AMC 2016) (2016)
24. Werbach, K., Hunter, D.: For the Win: How Game Thinking Can Revolutionize Your Business. Published by Wharton Digital Press, The Wharton School, University of Pennsylvania, 3620 Locust Walk - 2000 Steinberg Hall-Dietrich Hall, Philadelphia, PA 19104 (2012)
25. Martins, H., Freire, J.: Games people play – creating a framework for the gamification of a master's course in a Portuguese University. In: Proceedings of ICERI 2015 Conference 16th–18th November 2015, Seville, Spain, pp. 2821–2827 (2015). ISBN: 978-84-608-2657-6
26. Surendeleg, G., Murwa, V., Yun-Kyung, H., Kim, Y.: The role of gamification in education– a literature review. Contemp. Eng. Sci. 7(29), 1609–1616 (2014) www.m-hikari.com
27. Hamari, J., Koivisto, J., Sarsa, H.: Does gamification work?—A literature review of empirical studies on gamification. In: 47th Hawaii International Conference on System Science, pp. 3025–3034, 978-1-4799-2504-9/14 (2014)
28. Alčaković, S.: Millennials and gamification – a model proposal for gamification application in Članci/papers millennials and gamification – a model proposal for gamification application in tourism destination (2018)
29. Mostafa, L., Elbarawy, A.: Enhance job candidate learning path using gamification. In: ICCTA18 Proceedings of 28th International Conference on Computer Theory and Applications, Alexandria, Egypt (2018)
30. McGonigal, J.: Reality is Broken: Why Games Make Us Better and How They Can Change The World. Penguin Group, New York (2011)
31. Urh, M., Vukovic, G., Jereb, E., Pintar, R.: The modeleducation. Procedia Soc. Behav. Sci. 197, 388–397 (2015). https://doi.org/10.1016/j.sbspr0.2015.07.154
32. Sobocinski, M.: Necessary definitions for understanding gamification in education a short guide for teachers and educators (2018)
33. Tsai, M.-J., Huang, L.-J., Hou, H.-T., Hsu, C.-Y., Chiou, G.-L.: Visual behavior, flow and achievement in game-based learning. Comput. Educ. 98, 115–129 (2016)
34. Tejada-Castro, M., Aguirre-Munizaga, M., Yerovi-Ricaurte, E., Ortega-Ponce, L., Contreras-Gorotiza, O., Mantilla-Saltos, G.: Funprog: a gamification-based platform for higher education, CCIS 883, pp. 255–268. Springer (2018)
35. Şahin, M., Arsla, N.: Gamification and effects on students' science lesson achievement. Int. J. New Trends Educ. Their Implications 7(1), 41–47 (2016). Article: 04, ISSN 1309-6249
36. Martí-Parreñoa, J., Seguí-Masa, D., Seguí-Masb, E.: Teachers' attitude towards and actual use of gamification. Procedia Soc. Behav. Sci. 228(1877–0428), 682–688 (2016)
37. ASU Online Pilots Games for Environmental Science. http://asuonline.asu.edu/about-us/newsroom/asu-online-pilots-games-environmental-science. Accessed 21 Sept 2018
38. De-Marcos, L., García-Lopez, E., García-Cabot, A.: Social network analysis of a gamified e-learning course: small-world phenomenon and network metrics as predictors of academic performance". Comput. Hum. Behav. 60, 312–321 (2016)

39. Müllera, B., Reise, C., Seligera, G.: Gamification in factory management education – a case study with Lego Mindstorms. In: 12th Global Conference on Sustainable Manufacturing, Procedia CIRP 26, pp. 121–126 (2015)
40. Briers, B.: The gamification of project management. In: Proceedings of PMI Global Congress, New Orleans, Louisiana, New Orleans (2013)
41. Kechaou, Z., Alimi, A.: Improving e-learning with sentiment analysis of users' opinions. In: IEEE Global Engineering Education Conference – Learning Environment Ecosystem for Engineering Education, pp. 1032–1038 (2011)
42. Yu, L., Lee, C., Pan, H., Chou, C., Chao, P., Chen, Z., Tseng, S., Chan, C., Lai, K.: Improving early prediction of academic failure using sentiment analysis on self-evaluated commen. J. Assess. Learn. (2018)
43. Lin, Z., Deng, K., Hong, Y.: Research of web pages categorization. In: IEEE International Conference on Granular Computing, GRC, GRC 2007, p. 691 (2007)
44. Wang, S., Jing, F., He, J., Du, Q., Zhang, L.: IGroup: presenting web image search results in semantic clusters. In: ACM CHI, pp. 377–384 (2007)
45. Zhang, J., Qin, J., Yan, Q.: The role of URLs in objectionable web content categorization. In: IEEE/WIC/ACM International Conference on Web Intelligence, 18–22 December 2006
46. Gyongyi, Z., Molina, H., Pedersen, J.: Web content categorization using link information, Technical report, Stanford University (2006)

A Modified Fuzzy Sentiment Analysis Approach Based on User Ranking Suitable for Online Social Networks

Magda M. Madbouly$^{(\boxtimes)}$, Reem Essameldin$^{(\boxtimes)}$, and Saad Darwish$^{(\boxtimes)}$

Department of Information Technology, Alexandria University,
Alexandria, Egypt
{mmadbouly, igsr.reemessamledin,
saad.darwish}@alexu.edu.eg

Abstract. Processing Sentiment Analysis (SA) in social networks has lead decision makers to value opinion leaders who can sway people's impressions concerning certain business or commodity. Yet, a tremendous scarceness of considering perspectivism, while computing text polarity, has been spotted. Considering perspectivism in SA can help in the production of polarity scores that represent the perceptible sentiment within the content. This emphasizes the necessity for integrating social behavior (user influence factor) with SA (text polarity scores), providing a more pragmatic portrayal of how text-recipients comprehend the message. In this paper, a novel model is proposed to intensify SA process in Twitter. In the achievement of such, UCINET tool and Artificial Neural Networks (ANN) are used for social network analysis (SNA) and users ranking respectively. For sentiment classification, a hybrid approach is presented —lexicon-based technique (TextBlob) along with fuzzy classification technique —to handle language vagueness as well as for an inclusive analysis of tweets into seven classes; for the purpose of enhancing final results. The proposed model is practiced on data collected from Twitter. Results show a significant enhancement in the final polarity scores, associated with the analyzed tweets, representing more realistic sentiments.

Keywords: Sentiment Analysis · Social Network Analysis · Influential users · Fuzzy logic · Artificial Neural Networks

1 Introduction and Related Work

In social media networks, the arithmetic study of a text to define people's opinions, feelings and behavior concerning something, is known as SA which as a matter of fact is a patented process. It relies on every individual's perception; readers can disagree on what's thought-about to be positive. Also, famous writers' powerless negative tweets could be found quite powerful by some people due to their influence degree on readers [1]. SA methods today concentrate on attempting numerous ways to reinforce the analysis outcomes. However, a gap still exists once it involves reflect perspectives of readers throughout content analysis. It is favorable to ameliorate SA process through investigating influence of users on each other [2].

© Springer Nature Switzerland AG 2020
A. E. Hassanien et al. (Eds.): AISI 2019, AISC 1058, pp. 340–349, 2020.
https://doi.org/10.1007/978-3-030-31129-2_31

Accurate and computationally adequate text polarity scores that decision makers can willingly rely on is the main contribution of this work. The proposed model can enhance the process of SA with an eye on mirroring a real polarity that coincides with people's perspectives, while not neglecting the authority and impact of online network users.

In light of existing researches, Fuzzy logic techniques are not popular in the field of SA, even though Fuzzy techniques can improve the process with their weight-ages and strengths; which can end in a more accurate analysis [3, 4]. Priyanka et al. [3] tried to prove the effectiveness of fuzzy logic in classifying sentiments. They suggested a fuzzy based model that classified online reviews into seven classes (weak positive, moderate positive, strong positive, weak negative, moderate negative and strong negative) through combining fuzzy logic with sentiment based lexicon SentiWordNet to reach a close imitation of human behavior. An accuracy of 72% to 75% was the outcome of experimentation on three databases with reviews of electronic devices. The superiority of this technique is obvious in the fact of the absence of labeled data, which is costly, time consuming, and requires a lot of previous human work. More recent related works that apply Fuzzy Logic can be found in [5–7].

In different circumstances, many research works have been done based on influential users and behavioral analysis [8–10]. Jianqiang et al. [11] discussed a new technique to figure out influential users in the Chinese micro-blog network "Sina Weibo". The method took into consideration the influence of user's tweets by counting retweets and replies and the influence of users in the network by calculating influence score through follower graph representation. It has surpassed other existing methods in accuracy, recall and F1-Measure value as proved by experiments.

For analyzing social network data and getting different topology measures out of social network graphs, Li et al. [12] recommended the usage of a software package known as the UCINET tool (University of California at Irvine Network). The significance of UCINET in processing social network data was discussed by Yang et al. [13], as they explained that the complicated properties of social networking served as a grave barrier that prevented the development of SNA methods. Thus, an advanced mathematical theory was critical. UCINET helps in building social network models and designing social network diagrams based on information transition between individuals and the fullness information of individuals in the social network. This analysis tool provides a trusted guarantee for social networking data extraction and model.

Adding to recent works on SA and SNA, this work further introduces an advanced SA approach to express not only tweet's text sentiment polarity but also the real perceived sentiment associated with text. To the best of our knowledge, this is the first attempt to integrate users' behavior through SNA with the process of SA for the sake of making SA process reflects the real perceived sentiment in the text's content. Moreover, it is the first time to use ANN for ranking users according to their behavior on online social networks.

The rest of this paper is built up in the following way. In Sect. 2, illustrates the research problem, while in Sect. 3; the proposed model is pointedly defined. Experiments and results' analysis are charted and clarified in Sect. 4. Conclusion and impending breadth for development are reported in Sect. 5.

2 Problem Definition

When humans try to label a text as positive or negative, they use their own perspectives, feelings, and emotions to reflect how they perceive the text. One important factor is how much they trust the author of the text. When reader came under the sway of the writer, reader perceives the writer's text in a different way [1]. This issue highlighted the necessity to improve SA process specifically in social media networks to handle perspectivism. To achieve this, SNA could be integrated with SA to represent as close sentiment as possible to what's perceived by people on social media networks. The problems that this work is trying to solve are summarized as follows:

- Scarcity considering users' behavior and influence while computing text's polarity,
- Desire to effectively calculate users' influence. (I.e. what factors to be considered?),
- Desire to integrate SNA with SA towards polarity that represents more realistic sentiments to what others perceived,
- Besides the continuous desire to enhance the accuracy of text classification results.

3 Proposed Model

The proposed model substantially targets all previously declared problems. It integrates SNA with SA in Twitter to express the real perceived sentiment associated with text. The contribution of the proposed model is as follows:

- The proposed model considers user's influence while computing sentiment polarity scores.
- A hybridization technique is adopted for sentiment classification; hybrid sentiment classification methods proved its ability to improve accuracy. This is done by combining the standard lexicon using the scores from a popularly used sentiment based lexicon TextBlob to imitate close to human behavior with fuzzy classification technique which can handle the vagueness of language.
- A combination between topological measures, named users' network influence (Centrality, Betweenness, and Closeness) and tweet's influence (Retweet "RT" and Replies) is adopted for SNA to obtain more accurate influential scores as it was found that topological measures alone fail to accurately conclude influence of users. Moreover, these influential factors (tweet's and users' network influence) are inputted to ANN to acquire users' ranks and their corresponding weights. ANN was chosen for this mission in consequence of the complicatedness of network behavior and the fact that is unexplainable while ANN is wide utilized in similar situations by virtue of its "black box" feature because it isn't seemingly to know the explanation for the made solutions.
- For integrating results from ANN (users' weights) with the text polarity score achieved from TextBlob, fuzzy technique is adopted to deliver an inclusive classification of tweets into seven classes (e.g. weak positive, positive, strong positive, .. etc.). Figure 1 presents the proposed model and its components. The phases in the proposed model are described as the following:

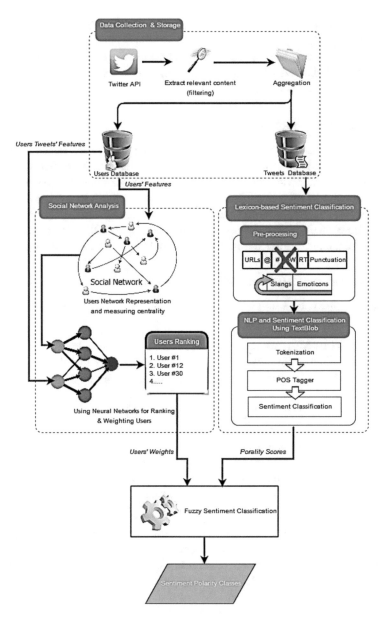

Fig. 1. The proposed fuzzy based SA model

3.1   Data Collection and Storage

This process is vital for the experimental validation. There is no established benchmark data for evaluating user influence detection on Twitter. Therefore, a major effort of this work is to build the data sets. ***Twitter API*** is used to collect data (twurl tool of twitter API in the form of 100 requests per time). Tweets are collected by querying the API for tweets of specific hashtags *(Filtering)*. Tweets for different users are then manually aggregated daily *(Aggregation)*. Finally, Data is stored in database that encompasses two portions, Portion1 has tweets *(Tweets Database)* and Portion2 has the user's features, such as followers and following lists, and the features of the user's tweet, like retweet and reply counts *(Users Database)*.

3.2   Lexicon-Based Sentiment Classification

In this phase, sentiment polarity score is calculated for each of the stored tweets using TextBlob—as a lexicon based classification technique. It was chosen as its results verified to be comparatively higher than alternative lexicons (e.g. SentiWordNet) [14]. TextBlob can also be used to perform pre-sentiment classification process that includes two main steps; text preprocessing and Natural Language Processing (NLP). See [1, 3–5, 15] for more details. After performing pre-processing then NLP on tweets, TextBlob performs sentiment classification giving tweets polarity scores ranged from −1 to 1.

3.3   Social Network Analysis and Users Ranking

Inputs to this phase are stored in "Users Database" (Portion2). Three main operations are to be performed in this phase:

Influence of Tweet Measurement: The influence of tweet is calculated by favoriting, retweeting and commenting behavior. $sp(u)$ is the probability that the tweet relocated from user u to the neighbors of user u's fans. The $sp(u)$ can be defined as [11]:

$$sp(u) = \sum\nolimits_{t \in Tweets(u)} RT(t) + Rr(t) + Fav(t) \qquad (1)$$

where, the influence of node u's tweets is $sp(u)$, t is tweet, $Tweets(u)$ is the set of user u's original tweets, $RT(t)$ represents the retweet to read ratio of t which is retweet counts per tweet stored in users database, $Rr(t)$ is the comment to read ratio of t which is replies counts per tweet also stored in users database, and $Fav(t)$ is favorite counts per tweet. If read count is zero, then $Fav(t), RT(t)$ and $Rr(t)$ is zero.

Influence of User Network Measurement: A user's tweets can go viral according to his situation on the follower relationship network and his fans influence [11]. Using data kept in Portion2, the follower relationship is extracted to form a directed graph using the NetDraw tool of UCINET for visualizing data (See Fig. 2).

$$G = (V, E) \qquad (2)$$

where, V = $\{u_1, u_2, u_3, ..., u_n\}$ is the set of nodes (users) in the micro-blog networks, and E = $\{e_1, e_2, e_3, ..., e_m\}$ is the set of edges (relationships between users) in the micro blog networks. If u_i is u_j's follower, then there is a directed edge $e_{u_i u_j} \in E$ from u_j to u_i. $sa(u)$ is defined to measure network influence of user u, the $sa(u)$ can be defined as [11]:

$$sa(u) = C_d^-(u) + c_b(u) + C_c^-(u) \qquad (3)$$

where, $sa(u)$ is influence of node u's information dissemination, $C_d^-(u)$ is degree centrality (out-degree) of node u—the number of out-going edges from a node [16], $C_b(u)$ is betweenness centrality of node u—the number of times a certain node is in the shortest paths between nodes [16], $C_c^-(u)$ is closeness centrality (out-degree) of node u—the value that is proportional to the harmonic mean of the length of the shortest paths between the i-th node and the rest of it in a network.

In the follower relationship network, if a user has a superior degree centrality, a higher read probability would be associated with his tweets, and there would be a higher opportunity for his tweets to go viral. If a user's closeness centrality is high, the user's aptitude to control information diffusion gets stronger, and the quicker a user spread information, the easier it becomes for a user to stops information from going viral. The greater a user's betweenness centrality is, the user can spread more quickly the message to the entire network through the fewer users, and the faster the user spreads information [17].

UCINET also calculates out-degree centrality, betweenness centrality, and out-degree closeness centrality measures, the three indicators for users' network influence according to Eq. (3). Influence of nodes calculated using Eq. (3) in combination with influence of node u's tweets calculated using Eq. (1) are inputted to ANN to get users' ranking value and their corresponding weights for each and every user in the network.

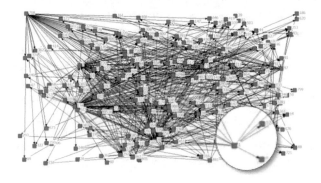

Fig. 2. Part of the communication network of some Twitter users using Netdraw of UCINET.

Authors Ranking and Weighting. Based on trial and error method, the constructed ANN is feed-forward; Single-layer ANN includes 10 elements in the hidden layer. The activation function used is sigmoid function. In the input layer of the network, we used 6 variables or 6 neurons in total; 3 variables as "input 1" concerning users' topologies measures (representing users' network influence)—called out-degree centrality, betweenness centrality, and out-degree closeness measures. And 3 other variables as "input 2" concerning tweets' features (representing tweets' influence)—called RT, Rr, and Favorite counts. Output is one neuron representing users' rank value from 1 to 100; the ranks were inspired from Klout—a website and mobile app that apply social media logical analysis on its users to assess their online social impact. The learning technique for the constructed ANN is supervised learning and it was trained using a dataset of approximately ten thousand samples.

3.4 Fuzzy Based Sentiment Classification

The antecedent variables of the rules of our fuzzy inference system are: user influence level—Low Influence (LI), Moderate Influence (MI), and High Influence (HI) with a crisp range of [0 1] and tweet's polarity level—Strong Positive (SP), Positive (P), Weak Positive (WP), Neutral (N), Weak Negative (WN), Negative (Neg), and Strong Negative (SN) with a crisp range of [−1 1]. The minimum operator "AND" is used between our antecedent variables. Table 1 shows the designed rules for obtaining real-felt polarities for the collected tweets. The accumulated fuzzy rules outputs are converted to a single crisp value which represents the final polarity of tweet using Centre of Gravity (CoG). See [3, 4] for more details.

3.5 Sentiment Polarity Classes

The proposed algorithm considered seven polarity levels [1], see Table 2 where, n represents the polarity scores of text.

Table 1. The designed fuzzy rules.

Rule #	Rule
1, 2	**IF** Input1 is 'LI' and Input2 is 'SN/SP', **THEN** Final tweet's polarity is 'NEG/P'
3, 4	**IF** Input1 is 'LI' and Input2 is 'NEG/P', **THEN** Final tweet's polarity is 'WN/WP'
5, 6	**IF** Input1 is 'LI' and Input2 is 'WN/WP', **THEN** Final tweet's polarity is 'N'
7, 8, 9	**IF** Input1 is 'LI/MI/HI' and Input2 is 'N', **THEN** Final tweets polarity is 'N'
10, 11	**IF** Input1 is 'MI' and Input2 is 'SN/SP', **THEN** Final tweets polarity is 'SN/SP'
12, 13	**IF** Input1 is 'MI' and Input2 is 'NEG/P', **THEN** Final tweets polarity is 'NEG/P'
14, 15	**IF** Input1 is 'MI' and Input2 is 'WN/WP', **THEN** Final tweets polarity is 'WN/WP'
16, 17	**IF** Input1 is 'HI' and Input2 is 'SN/SP', **THEN** Final tweets polarity is 'SN/SP'
18, 19	**IF** Input1 is 'HI' and Input2 is 'NEG/P', **THEN** Final tweets polarity is 'SN/SP'
20, 21	**IF** Input1 is 'HI' and Input2 is 'WN/WP', **THEN** Final tweets polarity is 'NEG/P'

Table 2. Seven polarity levels [1].

Score	Polarity
$n > 0.75$	Strong positive
$0.25 < n \leq 0.75$	Positive
$0 < n \leq 0.25$	Weak positive
0	Neutral
$-0.25 \leq n < 0$	Weak negative
$-0.75 \leq n < -0.25$	Negative
$n < -0.75$	Strong negative

4   Experimental Results

A desktop program is implemented using Python 3 integrated with MATLAB libraries to evaluate the performance of the proposed system. The used dataset consists of tweets with their features and authors' information collected from twitter as there was no benchmark dataset available. World Cup 2018 has been chosen as a topic of concern by collecting tweets with hashtags like: #WorldCup2018, #VAR.

The accuracy (correct instances) of the proposed lexicon based sentiment classifier (TextBlob) was compared to the suggested classifier in [14], see Table 3. The classifier used in [14] was a hybrid system of TextBlob with the Naïve Bayes classifier. Results under the same condition of 400 tweets as test data showed that, the proposed classifier has the highest accuracy for analyzing tweets' sentiment with 340 correct instances and 85% accuracy. From the illustrated results, our system outperforms the other one by 9%. One reason is the applied preprocessing steps. It assured the significance of text preprocessing in refining the accuracy of sentiment classification.

Table 3. Results from the comparative study

Sentiment classifier	Test data	Accuracy	Correct instances
The proposed classifier	400	85.00%	340
The suggested classifier in [14]	400	76.00%	304

Another experiment was conducted to view the difference between tweets' sentiment polarity before and after considering their authors' influential behavior in Twitter. This experiment showed how incomplete and different were the sentiment associated with tweets when applying any SA technique (TextBlob in our case) without considering how people perceive words. Figure 3 shows statistical summary of sentiment polarity associated with tweets of our database in the form of pie charts and Table 4 shows a sample of sentiment polarity scores before and after implementing our proposed model. This difference in polarities is consequent because of considering how people can be influenced by others thus perceived their words in a different way than a machine could.

Table 4. Sample of sentiment polarity scores before and after implementing the proposed model.

Author's ID	Sentiment polarity score (Before)	Sentiment polarity score (After)
User #1	0.8 (SP)	0.47 (P)
User #5	0.5 (P)	0.17 (WP)
User #10	0.8 (SP)	0.70 (SP)
User #117	−0.3 (Neg)	0 (N)
User #1015	−0.167 (WN)	0 (N)

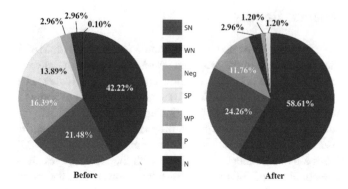

Fig. 3. Statistical representation of the sentiment polarity scores before and after implementing the proposed model.

5   Conclusion

The proposed model tried for the first time to integrate user influence measurements with polarity scores of texts for the purpose of achieving text polarities that mirrors how likely texts are perceived. The proposed model outperformed other models by implementing preprocessing steps that result in a relatively better classification accuracy. Moreover, this model assured the need for considering users' influence while computing text polarity as users were found to be different in their influence level on their followers resulting in different polarity scores than obtained by existing methods. Consequently, this model can be considered a step forward to get sentiment scores that squarely reflect reality. For future work, higher type of fuzzy logic can be used to enhance the model accuracy.

References

1. Benedetto, F., Tedeschi, A.: Big data sentiment analysis for brand monitoring in social media streams by cloud computing. In: Studies in Computational Intelligence (2016)
2. Riquelme, F., González-Cantergiani, P.: Measuring user influence on Twitter: a survey. Inf. Process. Manag. (2016)
3. Priyanka, C., Gupta, D.: Fine grained sentiment classification of customer reviews using computational intelligent technique. Int. J. Eng. Technol. (2015)
4. Darwish, S.M., Madbouly, M.M., Hassan, M.A.: From public polls to Tweets: developing an algorithm for classifying sentiment from Twitter based on computing with words. J. Comput. (2016)
5. Haque, A., Rahman, T.: Sentiment analysis by using fuzzy logic. Int. J. Comput. Sci. Eng. Inf. Technol. (2014)
6. Wang, B., Huang, Y., Wu, X., Li, X.: A fuzzy computing model for identifying polarity of Chinese sentiment words. Comput. Intell. Neurosci. (2015)
7. Jefferson, C., Liu, H., Cocea, M.: Fuzzy approach for sentiment analysis. In: 2017 IEEE International Conference on Fuzzy Systems (FUZZ-IEEE), pp. 1–6. IEEE (2017)

8. Anjaria, M., Guddeti, R.M.R.: A novel sentiment analysis of social networks using supervised learning. Soc. Netw. Anal. Min. (2014)
9. Neves-Silva, R., Gamito, M., Pina, P., Campos, A.R.: Modelling influence and reach in sentiment analysis. Procedia CIRP **47**, 48–53 (2016)
10. Bingol, K., Eravci, B., Etemoglu, C.O., Ferhatosmanoglu, H., Gedik, B.: Topic-based influence computation in social networks under resource constraints. IEEE Trans. Serv. Comput., 1 (2016)
11. Jianqiang, Z., Xiaolin, G., Feng, T.: A new method of identifying influential users in the micro-blog networks. IEEE Access (2017)
12. Hu, R.J., Li, Q., Zhang, G.Y., Ma, W.C.: Centrality measures in directed fuzzy social networks. Fuzzy Inf. Eng. (2015)
13. Yang, H., Li, Z.-F., Wei, W.: Instance analysis of social network based ucinet tool. Inf. Technol. J. **13**(8), 1532–1539 (2014)
14. Hasan, A., Moin, S., Karim, A., Shamshirband, S.: Machine learning-based sentiment analysis for Twitter accounts. Math. Comput. Appl. (2018)
15. Yoshida, S., Kitazono, J., Ozawa, S., Sugawara, T., Haga, T., Nakamura, S.: Sentiment analysis for various SNS media using Naive Bayes classifier and its application to flaming detection. In: 2014 IEEE Symposium on Computational Intelligence in Big Data (CIBD), pp. 1–6. IEEE (2014)
16. Aleskerov, F., Meshcheryakova, N., Shvydun, S., Yakuba, V.: Centrality measures in large and sparse networks. In: 2016 6th International Conference on Computers Communications and Control, ICCCC (2016)
17. Hanneman, R., Riddle, M.: Introduction to Social Network Methods. University of California (2005)

An Empirical Investigation of Students' Attitudes Towards the Use of Social Media in Omani Higher Education

Noor Al-Qaysi[1]([⊠]) , Norhisham Mohamad-Nordin[1,2],
and Mostafa Al-Emran[3]

[1] Faculty of Art, Computing and Creative Industry, Universiti Pendidikan Sultan
Idris, Tanjung Malim, Malaysia
noor.j.alqaysi@gmail.com, norhisham@fskik.upsi.edu.my
[2] College of Economics, Management and Information Systems,
University of Nizwa, Nizwa, Oman
[3] Applied Computational Civil and Structural Engineering Research Group,
Faculty of Civil Engineering, Ton Duc Thang University,
Ho Chi Minh City, Vietnam
al.emran@tdtu.edu.vn

Abstract. Although social media usage in higher education has been frequently examined, little is known regarding the differences in students' attitudes towards its use in the Omani higher education. Therefore, this study provides an opportunity to advance the understanding of these differences in attitudes by focusing on three different attributes, namely age, gender, and social media application. A total of 169 students enrolled at Sultan Qaboos University in Muscat, Oman took part in the study by the medium of an online survey. The empirical data analysis triggered out that there was a significant difference in attitudes with regard to age groups, in which students aged between 18 and 22 years old are much interested in using social media than the others. Nevertheless, the findings also showed that there was no significant difference in attitudes with respect to social media application and gender. More interesting, the outcomes pointed out that WhatsApp is the most frequent application used by students for educational purposes.

Keywords: Students · Attitudes · Social media · Higher education · Oman

1 Introduction

The traditional media use has been collapsed since the prevalent usage of social media [1]. Social media use has been extensively evidenced in the daily practices for millions of users [2]. The term "social media" refers to online social networks and social networking sites (SNSs) [3]. It also refers to the technological systems that are utilized to share messages and create personal profiles [3], sharing content [4], and sharing thoughts, comments, and opinions with other users in a virtual community [5]. Thus, defining "social media" is challenging as SNSs developers constantly create new and enhanced features to meet the users' demands [2]. Social media applications are

A. E. Hassanien et al. (Eds.): AISI 2019, AISC 1058, pp. 350–359, 2020.
https://doi.org/10.1007/978-3-030-31129-2_32

classified into five main categories regarding the users' needs, preferences, and life-styles. These five categorize include SNSs (e.g., "Facebook"), content communities (e.g., "YouTube"), blogs or microblogs (e.g., "Blogger", "Twitter"), common projects (e.g., "Wikipedia"), and virtual games or social worlds (e.g., "Second Life", "HumanSim") [6].

The excessive usage of social media applications is mostly witnessed among high school and undergraduate students all over the world [7–9]. Social media applications are considered valuable resources for research and education [10, 11]. In education, social media facilitates and enhances the learning process, since its use allows individuals to obtain different experiences such as "critical thinking", "problem-solving", "good communication", "collaboration", "lifelong learning", "information literacy", and "creative innovation" [12]. Social media also supports interaction among learners who suffer from negotiating their studies [13].

Social media has affected different sectors, like pharmacy [4], medicine [14], marketing [6], and education [10]. Given this fact, social media have proved its effectiveness in affecting students' learning experiences in the higher educational institutions [2]. An increasing amount of literature has been published to explore the usage of social media and its applications [15–20]. This includes the attitudes towards using social networking sites [21–23], social media acceptance in the higher education [24, 25], and the effects of social networks on academic performance level [26, 27].

According to the existing literature, much research was published to explore the learners' attitudes towards using social media in developed countries. Nonetheless, little is known regarding the investigation of the determinants affecting the social media use in the higher educational institutes in developing countries in general, and the Gulf region in specific [24]. The understanding of these determinants would enable the stakeholders in these institutes to signify the benefits and drawbacks of social media and set the appropriate infrastructure for its deployment [28]. In keeping with the previous assumption, the current study is an attempt to study the students' attitudes towards the use of social media in Oman. The study aims to understand the differences in age, gender, and the most common social media applications. More specifically, the study seeks to find answers to the following research questions:

RQ1. Does gender affect the students' attitudes towards social media use?
RQ2. Does age affect the students' attitudes towards social media use?
RQ3. Do social media applications affect the students' attitudes towards social media use?

2   Literature Review

Social media applications have been extensively used in higher educational institutes as these applications change the landscape in which the students learn, communicate, and collaborate [2]. The understanding of students' attitudes towards a specific technology is regarded to be one of the influential factors that facilitate the technology acceptance [29–31], in which social media is not an exception.

A wide range of researchers attempted to explore the learners' attitudes towards using social media in the higher educational institutions worldwide. In the United States, Weiler et al. [32] investigated using social media applications among 948 pharmacy students. The findings indicated that "Facebook" was the most frequent social media application used, followed by "YouTube". In addition, social media websites were reported to be highly used among the students inside and outside the classrooms. In the same context, Yoo et al. [33] studied the impact of expression and reception of smoking-related content on 366 College students' attitudes towards smoking on social media using path analysis. The findings showed that social media play an effective role in creating, sharing, receiving, and commenting on pro-smoking messages among College students. Similarly, Peluchette and Karl [34] examined a total of 433 undergraduate business students' usage of social media profiles and their attitudes towards the relevance of content they post, and how such content is used and accessed by other users. Significant gender differences concerning the type of posted information were found. The results also showed that employers were not comfortable seeing the items posted by students on their sites.

In Turkey, Avcı et al. [21] carried out a cross-sectional study to measure the medical students' usage of social media and their perceptions towards its use. It was revealed that guidelines of social media should be developed for medical students as they still did not fully perceive the potential ethical matters. In Malaysia, Mahadi et al. [35] examined the social media effect on students' attitudes. The findings revealed that the majority of students were effectively engaged in social media applications and they have successfully perceived the effect of social media in their attitudes. In Jordan, Al-Shdayfat [36] tried to examine the nurse students' perceptions towards the use of social media. The results showed that students have positive intentions towards the usage of social media academically and professionally. Additionally, Roblyer et al. [37] examined faculty members and students' perceptions of Facebook. Students were found to be more interested in using Facebook and similar technologies instructionally than faculty members for supporting classroom work. However, faculty members were more positive to the idea of using emails and other traditional technologies.

In alignment with the existing literature, it is clearly shown that little research has been undertaken to study the students' intention towards the usage of social media in the higher educational institutes in the Arab world in general, and the Gulf region in specific. In line with this limitation, the essential purpose of this study is to measure the students' attitudes towards the usage of social media in the Omani higher educational context. More specifically, the study emphasizes on the students' differences in terms of age, gender, and social media application used.

3 Method

3.1 Context and Subjects

The data were gathered from a population of undergraduate students studying at Sultan Qaboos University in Muscat, Oman. An online survey was administrated through the public relations department between September 2018 and January 2019. The students' participation was voluntary, and 174 students have successfully filled the survey. Five responses were discarded due to the large number of missing values; and hence, these responses were excluded from the final data analysis. Among the 169 valid responses, there were 105 females and 64 males. Besides, 74.6% of the students are aged between 18 and 22 years old. Concerning the usage of social media applications, it is imperative to report that about 48% of the students are frequently using WhatsApp; this is followed by Instagram (28.4%), Twitter (16%), YouTube (6.5%), and Facebook (1.2%).

3.2 Survey Structure

The survey involved in this study is developed to be consisted of three parts. The first part was related to demographic characteristics such as gender, age, and year of study. The second part consisted of six items that are dedicated to measure the social media usage information. The third part included 13 items, and these are the items used for measuring the students' attitudes towards the usage of social media in education. The items in the third part were in the form of statements in which the students' entries were interpreted using a five-point Likert-scale with "strongly disagree (1), disagree (2), neutral (3), agree (4), and strongly agree (5)". The items in part three were adopted from prior studies conducted for measuring the students' attitudes towards a specific technology with further adjustments to suit the scope of the present study [38–42].

4 Results

RQ1. Does gender affect the students' attitudes towards social media use?

An "independent samples t-test" was calculated to examine the statistically significant difference among the students' attitudes toward the usage of social media with respect to gender. As exhibited in Table 1, the findings triggered out that there is no significant difference among the students' attitudes with regard to their genders ($t = 0.167$, $p > 0.05$).

Table 1. Differences in terms of gender

Gender	N	Mean	Std. deviation	t-value	Df	Sig.
Male	64	3.7849	0.92594	0.167	167	0.867
Female	105	3.8103	0.97418			

RQ2. Does age affect the students' attitudes towards social media use?

As presented in Table 2, means and standard deviations for the students' age groups are calculated to measure the significant difference among the students' attitudes towards the usage of social media. Furthermore, a "one-way analysis of variance (ANOVA)" was performed to calculate the statistically significant differences between the mean values. Table 3 triggers out that there are statistically significant differences ($p = 0.022$, $p < 0.05$) among the students' attitudes with respect to their age and the calculated F score is (3.891). It is clearly shown that the age group (18 to 22) is the most positive among the others towards using social media.

Table 2. Mean and standard deviation for age groups

Age	N	Mean	Std. deviation
18 to 22	126	3.9176	0.81612
23 to 28	26	3.4260	1.24039
Above 28	17	3.5068	1.22928
Total	169	3.8006	0.95350

Table 3. ANOVA results for age groups

	Sum of squares	df	Mean square	F	Sig.
Between groups	6.840	2	3.420	3.891	0.022
Within groups	145.899	166	0.879		
Total	152.739	168			

RQ3. Do social media applications affect the students' attitudes towards social media use?

Figure 1 illustrates the students' usage of social media application and the activities performed under each application. It is important to report that WhatsApp is the most frequent application used by the students, followed by Instagram, Twitter, YouTube, and Facebook, respectively. In terms of social media activity, it can be seen that WhatsApp is highly used for communication purposes, followed by entertainment, knowledge sharing, education, and marketing, respectively. In addition, we have observed that Instagram, Twitter, and YouTube are the most used applications for entertainment purposes. In terms of educational purposes, WhatsApp is found to be the most effective application used in this regard. This is considered a new finding to be added to the literature of social media as most of the prior research indicated that Facebook was the most common application used by students for educational activities [32, 38].

For determining the significant difference among the students' attitudes towards the usage of social media with respect to social media applications, means and standard deviations for the social media applications (i.e., Facebook, Twitter, Blog, Instagram, WhatsApp, YouTube, and Wiki) have been calculated as Table 4 shows. Furthermore, "a one-way analysis of variance (ANOVA)" was performed to test the statistically significant difference between the mean values. Table 5 revealed that there are no significant differences ($p = 0.207$, $p > 0.05$) among the students' attitudes with respect to social media application and the resulted F value is (1.493).

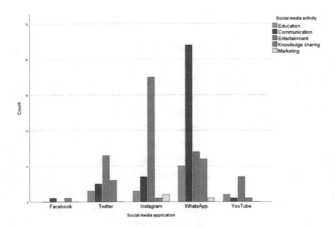

Fig. 1. Students' usage of social media application vs. social media activity

Table 4. Mean and standard deviation for social media applications

Social media application	N	Mean	Std. deviation
Facebook	2	2.5000	2.12132
Twitter	27	3.9858	0.90012
Instagram	48	3.6795	1.00149
WhatsApp	81	3.8196	0.94008
YouTube	11	3.9720	0.62990
Total	169	3.8006	0.95350

Table 5. ANOVA results for social media applications

	Sum of squares	df	Mean square	F	Sig.
Between groups	5.365	4	1.341	1.493	0.207
Within groups	147.373	164	0.899		
Total	152.739	168			

5 Discussion

Very little was known in the previous literature regarding the students' attitudes towards the use of social media in Oman. Thus, this study was designed to answer three research questions. The first research question is mainly concerned with examining the students' attitudes towards using social media in terms of gender. The results exhibited that there was no significant difference in attitudes with respect to gender. This result contradicts with the conclusion drawn by Arteaga Sánchez et al. [38], in which the predominance of social media was much favored by female users than males. The discrepancy between the outcomes observed in this study and the past literature might result from the fact that both genders in Oman are equally interested in utilizing social media for educational activities.

The second research question concerns investigating the students' attitudes towards the usage of social media with respect to age. The results indicated that students who aged between 18 and 22 are the most positive among the other age groups to use social media. This result matches those observed in a prior study conducted by Arteaga Sánchez et al. [38], in which students aged between 18 and 23 were the most effective users of Facebook. A possible explanation for this result might be that students in this age group are much experienced and motivated in using social media applications.

Regarding the third research question, which is concerned with studying the students' attitudes towards the use of social media in terms of social media application, the findings triggered out that there was no significant difference in attitudes with respect to social media application. Although the observed results indicated that WhatsApp is the most used application by students, this result does not affect the students' attitudes towards using social media in education. This could be explained by the fact that students are interested in using most of the social media applications regardless of the activity performed under these applications.

6 Conclusion and Future Work

Social media applications are being employed extensively in higher educational institutes as these applications altered the way in which students communicate, collaborate, and learn [2, 24]. The understanding of students' attitudes is one of the effective factors that affects technology adoption [43–46]. Thus, investigating the students' attitudes towards social media is still an ongoing research trend and further investigation is highly encouraged.

Based on the existing literature, little is known regarding the understanding of students' attitudes towards the use of social media in Oman. Thus, the current study mainly aims to measure the students' attitudes toward the usage of social media in Sultan Qaboos University in Muscat, Oman by considering three different attributes, namely gender, age, and social media application. A total of 169 students took part in the study. The results triggered out that there was a significant difference in attitudes with regard to age groups, in which students aged between 18 and 22 years old are much interested in social media than the others. On the other side, the findings also exhibited that there was no significant difference in attitudes with regard to gender and social media application. More interesting, the findings revealed that WhatsApp is the most commonly used application by students for educational purposes.

Similar to other studies, the current study posits some limitations that need to be considered. First, this study was mainly concerned with quantitative data analysis through the use of questionnaire surveys. Therefore, further research should consider other qualitative methods like interviews and focus groups. Second, the current study has concentrated on three different attributes (i.e., gender, age, and social media application) in studying the students' attitudes towards using social media. Thus, future attempts should consider other factors.

References

1. Al-Qaysi, N., Mohamad-Nordin, N., Al-Emran, M.: A systematic review of social media acceptance from the perspective of educational and information systems theories and models. J. Educ. Comput. Res. (2018)
2. Tess, P.A.: The role of social media in higher education classes (real and virtual) – a literature review. Comput. Human Behav. 29(5), A60–A68 (2013)
3. Boyd, D.M., Ellison, N.B.: Social network sites: definition, history, and scholarship. J. Comput. Commun. 13(1), 210–230 (2007)
4. Benetoli, A., Chen, T.F., Aslani, P.: The use of social media in pharmacy practice and education. Res. Soc. Adm. Pharm. 11(1), 1–46 (2015)
5. Weber, L.: Marketing to the Social Web: How Digital Customer Communities Build Your Business, 2nd edn. Wiley, Hoboken (2007)
6. Çiçek, M., Ozcan, S.: Examining the demographic features of Turkish social media users and their attitudes towards social media tools. In: 2013 International Conference on Control, Decision and Information Technologies, CoDIT 2013, pp. 511–515 (2013)
7. Al-Gamal, E., Alzayyat, A., Ahmad, M.M.: Prevalence of internet addiction and its association with psychological distress and coping strategies among university students in Jordan. Perspect. Psychiatr. Care 52(1), 49–61 (2016)
8. Omekwu, C.O., Eke, H.N., Odoh, N.J.: The use of social networking sites among undergraduate students of University of Nigeria, Nsukka. Libr. Philos. Pract. (2014)
9. Rice, E., Barman-Adhikari, A.: Internet and social media use as a resource among homeless youth. J. Comput. Commun. 19(2), 232–247 (2014)
10. Ambika, P., Kumar, R., Samath, A.: Comparative study of microblogging behavior on Twitter and Facebook to foster continuous learning-educational sector. In: Proceedings - 2015 IEEE International Conference on Cloud Computing in Emerging Markets, CCEM 2015 (2016)
11. Chytas, D.: Use of social media in anatomy education: a narrative review of the literature. Ann. Anatomy-Anatomischer Anzeige (2019)
12. Yussiff, A.S., Ahmad, W.F.W., Oxley, A.: An exploration of Social Media Technologies and their potential uses in higher educational institutions: a case study of Universiti Teknologi PETRONAS. In: 2013 IEEE Conference on e-Learning, e-Management and e-Services, IC3e 2013 (2013)
13. Selwyn, N.: Faceworking: exploring students' education-related use of Facebook. Learn. Media Technol. 34(2), 157–174 (2009)
14. Cartledge, P., Miller, M., Phillips, B.: The use of social-networking sites in medical education. Med. Teach. 35(10), 847–857 (2013)
15. Salloum, S.A., Al-Emran, M., Shaalan, K.: The impact of knowledge sharing on information systems: a review. In: 13th International Conference, KMO 2018 (2018)
16. Salloum, S.A., Al-Emran, M., Shaalan, K.: Mining social media text: extracting knowledge from Facebook. Int. J. Comput. Digit. Syst. 6(2), 73–81 (2017)
17. Salloum, S.A., Al-Emran, M., Shaalan, K.: Mining text in news channels: a case study from Facebook. Int. J. Inf. Technol. Lang. Stud. 1(1), 1–9 (2017)
18. Salloum, S.A., Al-Emran, M., Abdallah, S., Shaalan, K.: Analyzing the Arab Gulf Newspapers using text mining techniques. In: International Conference on Advanced Intelligent Systems and Informatics, pp. 396–405 (2017)
19. Mhamdi, C., Al-Emran, M., Salloum, S.A.: Text mining and analytics: a case study from news channels posts on Facebook, vol. 740 (2018)

20. Salloum, S.A., Mhamdi, C., Al-Emran, M., Shaalan, K.: Analysis and classification of Arabic Newspapers' Facebook pages using text mining techniques. Int. J. Inf. Technol. Lang. Stud. 1(2), 8–17 (2017)
21. Avcı, K., Çelikden, S.G., Eren, S., Aydenizöz, D.: Assessment of medical students' attitudes on social media use in medicine: a cross-sectional study. BMC Med. Educ. 15(1), 18 (2015)
22. Baföz, T.: Pre-service EFL teachers' attitudes towards language learning through social media. Procedia Soc. Behav. Sci. 232, 430–438 (2016)
23. Al-Emran, M., Malik, S.I.: The impact of Google apps at work: higher educational perspective. Int. J. Interact. Mob. Technol. 10(4), 85–88 (2016)
24. Alshurideh, M., Salloum, S.A., Al Kurdi, B., Al-Emran, M.: Factors affecting the Social Networks Acceptance: an empirical study using PLS-SEM approach. In: 8th International Conference on Software and Computer Applications (2019)
25. Dumpit, D.Z., Fernandez, C.J.: Analysis of the use of social media in Higher Education Institutions (HEIs) using the technology acceptance model. Int. J. Educ. Technol. High. Educ. 14(1) (2017)
26. Habes, M., Alghizzawi, M., Khalaf, R., Salloum, S.A., Ghani, M.A.: The relationship between social media and academic performance: Facebook perspective. Int. J. Inf. Technol. Lang. Stud. 2(1) (2018)
27. Paul, J.A., Baker, H.M., Cochran, J.D.: Effect of online social networking on student academic performance. Comput. Hum. Behav. 28(6), 2117–2127 (2012)
28. Al-Qaysi, N., Al-Emran, M.: Code-switching usage in social media: a case study from Oman. Int. J. Inf. Technol. Lang. Stud. 1(1), 25–38 (2017)
29. Al-Emran, M., Alkhoudary, Y.A., Mezhuyev, V., Al-Emran, M.: Students and Educators Attitudes towards the use of M-Learning: gender and smartphone ownership differences. Int. J. Interact. Mob. Technol. 13(1), 127–135 (2019)
30. Al-Maroof, R.A.S., Al-Emran, M.: Students acceptance of Google classroom: an exploratory study using PLS-SEM approach. Int. J. Emerg. Technol. Learn. 13(6), 112–123 (2018)
31. Malik, S.I., Al-Emran, M.: Social factors influence on career choices for female computer science students. Int. J. Emerg. Technol. Learn. 13(5), 56–70 (2018)
32. Weiler, M.I., Santanello, C.D., Isaacs, D., Rahman, A., O'Donnell, E.P., Peters, G.L.: Pharmacy students' attitudes about social media use at five schools of pharmacy. Curr. Pharm. Teach. Learn. 7, 804–810 (2015)
33. Yoo, W., Yang, J.H., Cho, E.: How social media influence college students' smoking attitudes and intentions. Comput. Human Behav. 64, 173–182 (2016)
34. Peluchette, J., Karl, K.: Social networking profiles: an examination of student attitudes regarding use and appropriateness of content. CyberPsychology Behav. 11(1), 95–97 (2008)
35. Mahadi, S.R.S., Jamaludin, N.N., Johari, R., Fuad, I.N.F.M.: The impact of social media among undergraduate students: attitude. Procedia Soc. Behav. Sci. 219, 472–479 (2016)
36. Al-Shdayfat, N.M.: Undergraduate student nurses' attitudes towards using social media websites: a study from Jordan. Nurse Educ. Today 66, 39–43 (2018)
37. Roblyer, M.D., McDaniel, M., Webb, M., Herman, J., Witty, J.V.: Findings on Facebook in higher education: a comparison of college faculty and student uses and perceptions of social networking sites. Internet High. Educ. 13(3), 134–140 (2010)
38. Arteaga Sánchez, R., Cortijo, V., Javed, U.: Students' perceptions of Facebook for academic purposes. Comput. Educ. 70, 138–149 (2014)
39. Avcı Yücel, Ü.: Perceptions of pedagogical formation students about Web 2.0 tools and educational practices. Educ. Inf. Technol. 22(4), 1571–1585 (2017)
40. Balakrishnan, V.: Key determinants for intention to use social media for learning in higher education institutions. Univers. Access Inf. Soc. 16(2), 289–301 (2017)

41. Choi, G., Chung, H.: Applying the technology acceptance model to Social Networking Sites (SNS): impact of subjective norm and social capital on the acceptance of SNS. Int. J. Hum. Comput. Interact. **29**(10), 619–628 (2013)
42. Suki, N.M., Ramayah, T., Ly, K.K.: Empirical investigation on factors influencing the behavioral intention to use Facebook. Univers. Access Inf. Soc. **11**(2), 223–231 (2012)
43. Al-Emran, M., Shaalan, K.: Attitudes towards the use of mobile learning: a case study from the Gulf region. Int. J. Interact. Mob. Technol. **9**(3), 75–78 (2015)
44. Al-Emran, M., Salloum, S.A.: Students' attitudes towards the use of mobile technologies in e-evaluation. Int. J. Interact. Mob. Technol. **11**(5), 195–202 (2017)
45. Al-Emran, M., Shaalan, K.: Academics' awareness towards mobile learning in Oman. Int. J. Comput. Digit. Syst. **6**(1), 45–50 (2017)
46. Salloum, S.A., Al-Emran, M.: Factors affecting the adoption of E-payment systems by university students: extending the TAM with trust. Int. J. Electron. Bus. **14**(4), 371–390 (2018)

Understanding the Impact of Social Media Practices on E-Learning Systems Acceptance

Said A. Salloum[1,2] ⓘ, Mostafa Al-Emran[3(✉)] ⓘ, Mohammed Habes[4],
Mahmoud Alghizzawi[5], Mazuri Abd. Ghani[5], and Khaled Shaalan[2] ⓘ

[1] Faculty of Engineering and IT, University of Fujairah,
Fujairah, United Arab Emirates
ssalloum@uof.ac.ae
[2] Faculty of Engineering and IT, The British University in Dubai,
Dubai, United Arab Emirates
khaled.shaalan@buid.ac.ae
[3] Applied Computational Civil and Structural Engineering Research Group,
Faculty of Civil Engineering, Ton Duc Thang University,
Ho Chi Minh City, Vietnam
al.emran@tdtu.edu.vn
[4] Faculty of Applied Social Sciences, Universiti Sultan Zainal Abidin,
Kuala Terengganu, Malaysia
mohammedhabes88@gmail.com
[5] Faculty of Economics and Management Sciences, Universiti Sultan Zainal
Abidin, Kuala Terengganu, Malaysia
dr.alghzawi87@gmail.com, mazuri@unisza.edu.my

Abstract. There have been several longitudinal studies concerning the learners' acceptance of e-learning systems using the higher educational institutes (HEIs) platforms. Nonetheless, little is known regarding the investigation of the determinants affecting the e-learning acceptance through social media applications in HEIs. In keeping with this, the present study attempts to understand the influence of social media practices (i.e., knowledge sharing, social media features, and motivation and uses) on students' acceptance of e-learning systems by extending the technology acceptance model (TAM) with these determinants. A total of 410 graduate and undergraduate students enrolled at the British University in Dubai, UAE took part in the study by the medium of questionnaire surveys. The partial least squares-structural equation modeling (PLS-SEM) is employed to analyze the extended model. The empirical data analysis triggered out that social media practices including knowledge sharing, social media features, and motivation and uses have significant positive impacts on both perceived usefulness (PU) and perceived ease of use (PEOU). It is also imperative to report that the acceptance of e-learning systems is significantly influenced by both PU and PEOU. In summary, social media practices play an effective positive role in influencing the acceptance of e-learning systems by students.

Keywords: E-Learning · Social media · TAM · PLS-SEM · UAE

© Springer Nature Switzerland AG 2020
A. E. Hassanien et al. (Eds.): AISI 2019, AISC 1058, pp. 360–369, 2020.
https://doi.org/10.1007/978-3-030-31129-2_33

1   Introduction

Social networking sites (SNSs) have gradually become widespread with the growth of the 2nd generation of "web-based communities" due to the augmented collaboration and sharing among individuals through different applications such as "blogs", "podcasts", and "feeds" [1–3]. For education, social networks afford an exclusive opportunity for instructors to feel the sense of community among the learners and motivate personal communications that can lead to the construction of new knowledge [4, 5]. Prior research suggested that SNSs employed for educational purposes, such as "Ning in Education" can be used as a new technological tool in distance education for delivering seminar courses [6]. These platforms provide a set of features like creating online courses, registering, and monitoring evaluation activities of learners and educators [7]. With the widespread usage of e-learning over the Internet, e-learning through various platforms has provided a new, flexible, and portable way for learners to acquire their basic knowledge from various sources [8]. Students can even interact with teachers and classmates simultaneously through these platforms [9].

According to the existing literature, much research was published to explore the learners' acceptance of e-learning systems using educational platforms in higher educational institutes (HEIs) [10]. Nonetheless, little is known regarding the investigation of the determinants affecting the e-learning acceptance through social media applications in HEIs. In keeping with the previous limitation, the current study is an attempt to study the students' acceptance of e-learning systems through social media applications in HEIs in the UAE. More specifically, the present study aims to understand the influence of social media practices (i.e., knowledge sharing, social media features, and motivation and uses) on students' acceptance of e-learning systems by extending the technology acceptance model (TAM) with these determinants.

2   Hypotheses Development and Theoretical Model

The theoretical research model examined in the present study is demonstrated in Fig. 1. It is suggested that the students' acceptance of e-learning systems is affected by perceived ease of use (PEOU) and perceived usefulness (PU). It is also anticipated that both PEOU and PU are affected by social media practices like knowledge sharing, social media features, and motivation and uses. The description of each factor along with the developed hypotheses are discussed in the following subsections.

2.1   Knowledge Sharing (KS)

Knowledge sharing (KS) is defined as "the business processes that distribute knowledge among all individuals participating in process activities" [11]. Prior research pointed out that KS has a significant influence on the adoption of various technologies [8, 12, 13]. Accordingly, the following hypotheses were proposed:

H1a: KS has a positive effect on PU.
H1b: KS has a positive effect on PEOU.

2.2   Social Media Features (SMF)

SNSs such as "Facebook", "Twitter", "Instagram", "Snapchat", "LinkedIn", and "YouTube" are among many other platforms that allow the exchange of videos and photos, sharing files and immediate talks, and interacting with others [14, 15]. These features influence the learners' attitudes and academic performance directly, either negatively or positively, and strengthen their convictions in these means [16]. Concerning learning through social media, previous studies revealed that ease of use and usefulness of social media afford adequate opportunities for users to participate in online discussions, which in turn, led to the acceptance of e-learning through the features that these platforms provide [17, 18]. Consequently, the following hypotheses were proposed:

H2a: SMF have a positive effect on PU.
H2b: SMF have a positive effect on PEOU.

2.3   Motivation and Uses (MU)

Motivation and usage (MU) of electronic education systems play an effective role in the ability of learners and teachers to adopt and accept these systems [19]. Several previous studies in the domain have confirmed that MU have a positive impact on the PEOU and PU of e-learning systems [20–22]. Further, the psychosocial needs and self-motivation of users could be useful in predicting the intention to accept e-learning systems [23]. Accordingly, the following hypotheses were proposed:

H3a: MU have a positive effect on PU.
H3b: MU have a positive effect on PEOU.

2.4   Perceived Ease of Use (PEOU) and Perceived Usefulness (PU)

The "technology acceptance model (TAM)" presented by Davis [24] has become the most famous and influential model for explaining the acceptance of a specific technology and has acquired extensive experiential props in many studies [25]. Davis suggested two main constructs for explaining the acceptance of any technology. These two constructs are "perceived ease of use (PEOU)" and "perceived usefulness (PU)" [24]. PEOU refers to "the degree to which a person believes that using a particular system would be free from effort", whereas PU refers to "the degree to which a person believes that using a particular system would enhance his or her job performance" [24]. PEOU and PU were shown to have significant positive impacts on the acceptance of numerous technologies [26–30]. Accordingly, the following assumptions were proposed:

H4: PEOU has a positive effect on PU.
H5: PU has a positive effect on e-learning system acceptance.
H6: PEOU has a positive effect on e-learning system acceptance.

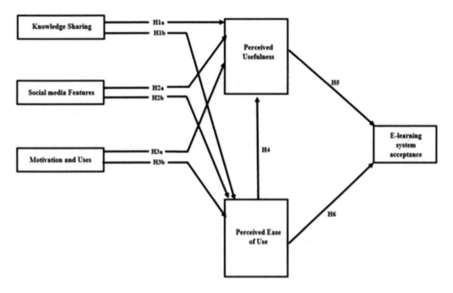

Fig. 1. Research model

3 Research Methodology

3.1 Context and Subjects

This study was undertaken at the British University in Dubai (BUiD) in the UAE. Both graduate and undergraduate students were involved in the study. The data were gathered via a questionnaire survey using the convenience sampling technique. This is the same procedure followed in prior research pertaining to technology acceptance [31–36]. Out of 480 surveys circulated, a total of 410 valid responses were obtained, with a response rate of 85.4%. Among the 410 valid responses, there were 230 females and 180 males. Besides, 75% of the participated students were aged between 18 and 29 years old. It is imperative to report that about 85.6% of the students are frequently using social media applications like Facebook, YouTube, and Twitter on daily-basis.

3.2 Data Analysis

The analysis of the proposed theoretical model was undertaken using the "partial least squares-structural equation modeling (PLS-SEM)" using SmartPLS [37]. The main purpose of employing PLS-SEM in the present study stems from the exploratory perspective of the proposed theoretical model. This study follows the general guiding procedures for employing PLS-SEM in the IS domain [38]. As suggested in prior research [39], a two-step technique (i.e., "measurement model" and "structural model") needs to be carefully followed for analyzing the theoretical research model. In the present study, these two-steps were strictly followed.

4 Findings

4.1 Measurement Model Assessment

Research pertaining to the employment of PLS-SEM reported that reliability and validity need to be confirmed while assessing the measurement model [39]. Reliability is typically evaluated by two measures, namely "Cronbach's alpha" and "composite reliability (CR)" [39]. The values of these two measures should be equal to or greater than 0.70 in order to be accepted [39]. As indicated in Table 1, both reliability measures are confirmed.

Table 1. Reliability and convergent validity results

Constructs	Items	Factor loadings	Cronbach's alpha	CR	AVE
E-learning system acceptance	ELA_1	0.919	0.851	0.930	0.869
	ELA_2	0.945			
Knowledge sharing	KS_1	0.916	0.845	0.897	0.686
	KS_2	0.867			
	KS_3	0.781			
	KS_4	0.738			
Motivation and uses	MU_1	0.708	0.806	0.873	0.634
	MU_2	0.882			
	MU_3	0.843			
	MU_4	0.740			
Perceived ease of use	PEOU_1	0.838	0.793	0.865	0.618
	PEOU_2	0.708			
	PEOU_3	0.868			
	PEOU_4	0.720			
Perceived usefulness	PU_1	0.802	0.886	0.921	0.746
	PU_2	0.875			
	PU_3	0.894			
	PU_4	0.880			
Social media features	SMF_1	0.726	0.842	0.901	0.705
	SMF_2	0.900			
	SMF_3	0.925			
	SMF_4	0.936			

For validity assessment, both convergent and discriminant validities need to be ascertained [39]. Concerning the "convergent validity", the "average variance extracted (AVE)" and "factor loadings" need to be examined. To comply with the accepted threshold values [39], the values of factor loadings should be equal to or greater than 0.70, while the values of AVE should be equal to or greater than 0.50. The findings in Table 1 reveal that the values of AVE and factor loadings are accepted; therefore, the "convergent validity" is ascertained. With respect to the "discriminant validity", the "Heterotrait-Monotrait ratio (HTMT)" of correlations need to be examined [40]. A value of less than 0.85 should be ascertained. As per the readings in Table 2, the HTMT results are regarded to be satisfactory; thus, the "discriminant validity" is confirmed.

Table 2. HTMT results

	E-learning system acceptance	KS	MU	PEOU	PU	SMF
E-learning system acceptance						
KS	0.032					
MU	0.113	0.493				
PEOU	0.072	0.469	0.615			
PU	0.074	0.309	0.713	0.548		
SMF	0.076	0.503	0.628	0.717	0.616	

4.2 Structural Model Assessment

Table 3 and Fig. 2 illustrate the results of the structural model. It can be noticed that all the proposed hypotheses are significant and positive. In that, KS is found to have significant positive influences on PU (beta = 0.601, $p < 0.05$) and PEOU (beta = 0.594, $p < 0.05$); thus, H1a and H1b are accepted. Additionally, the findings triggered out that SMF have significant positive effects on PU (beta = 0.382, $p < 0.05$) and PEOU (beta = 0.232, $p < 0.05$); hence, H2a and H2b are supported. Moreover, it can be noted that MU have significant positive impacts on PU (beta = 0.797, $p < 0.05$) and PEOU (beta = 0.505, $p < 0.05$); therefore, H3a and H3b are accepted. Further, the findings exhibited that PEOU has a significant positive influence on PU (beta = 0.412, $p < 0.05$); thus, H4 is supported. Interestingly, the findings reported that the e-learning system acceptance is significantly and positively affected by PU (beta = 0.606, $p < 0.05$) and PEOU (beta = 0.363, $p < 0.05$); hence, H5 and H6 are both supported.

Table 3. Hypotheses testing results

H	Relationship	Path coefficient	p-value	Decision
H1a	KS → PU	0.601	0.000	Supported
H1b	KS → PEOU	0.594	0.000	Supported
H2a	SMF → PU	0.382	0.011	Supported
H2b	SMF → PEOU	0.232	0.035	Supported
H3a	MU → PU	0.797	0.000	Supported
H3b	MU → PEOU	0.505	0.000	Supported
H4	PEOU → PU	0.412	0.015	Supported
H5	PU → e-learning acceptance	0.606	0.002	Supported
H6	PEOU → e-learning acceptance	0.363	0.031	Supported

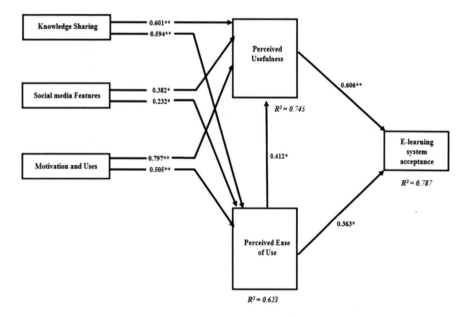

Fig. 2. Path analysis results

In addition, KS, SMF, MU, and PEOU together accounted for 74.5% of the explained variance (R^2) in PU. Besides, KS, SMF, and MU together accounted for 62.3% of the explained variance (R^2) in PEOU. The direct effects of PU and PEOU together with the indirect effects of KS, SMF, and MU accounted for 78.7% of the explained variance (R^2) in e-learning system acceptance.

5 Conclusion and Future Work

E-learning systems are being employed extensively in higher educational institutes as these systems altered the way in which students learn, collaborate, and share knowledge. The adoption of any technology is usually affected by several factors that are varied according to the context and participants. In the situation of the present study, investigating the students' acceptance of e-learning systems is still an ongoing research trend and further investigation is highly encouraged.

Based on the existing literature, much research was published to explore the learners' acceptance of e-learning systems using HEIs platforms [10]. Nonetheless, little is known regarding the investigation of the determinants affecting the e-learning acceptance through social media applications in HEIs. In keeping with the previous limitation, the current study is an attempt to study the students' acceptance of e-learning systems through social media applications in HEIs in the UAE. More specifically, the present study aims to understand the influence of social media practices (i.e., knowledge sharing, social media features, and motivation and uses) on students' acceptance of e-learning systems by extending the TAM with these determinants.

The results drawn from the structural model provide evidence that social media practices including knowledge sharing, social media features, and motivation and uses have significant positive influences on both PU and PEOU of e-learning systems. It is also imperative to report that the acceptance of e-learning systems is significantly affected by both PU and PEOU. In summary, social media practices play an effective positive role in influencing the students' acceptance of e-learning systems in the UAE.

Similar to other studies, the current study posits some shortcomings that need to be considered. First, the sample of this study was gathered from only one institution in the UAE. Thus, the generalization of the results to the entire UAE institutes should be treated with cautions. Second, this study mainly concerned with quantitative data analysis through the use of questionnaire surveys. Therefore, further research should consider other qualitative methods like interviews and focus groups.

References

1. Mhamdi, C., Al-Emran, M., Salloum, S.A.: Text mining and analytics: a case study from news channels posts on Facebook **740** (2018)
2. Salloum, S.A., Al-Emran, M., Shaalan, K.: Mining text in news channels: a case study from Facebook. Int. J. Inf. Technol. Lang. Stud. **1**(1), 1–9 (2017)
3. Salloum, S.A., Mhamdi, C., Al-Emran, M., Shaalan, K.: Analysis and classification of Arabic Newspapers' Facebook pages using text mining techniques. Int. J. Inf. Technol. Lang. Stud. **1**(2), 8–17 (2017)
4. Habes, M., Salloum, S.A., Alghizzawi, M., Alshibly, M.S.: The role of modern media technology in improving collaborative learning of students in Jordanian universities. Int. J. Inf. Technol. Lang. Stud. **2**(3), 71–82 (2018)
5. Habes, M., Alghizzawi, M., Khalaf, R., Salloum, S.A., Ghani, M.A.: The relationship between social media and academic performance: Facebook perspective. Int. J. Inf. Technol. Lang. Stud. **2**(1), 12–18 (2018)
6. Brady, K.P., Holcomb, L.B., Smith, B.V.: The use of alternative social networking sites in higher educational settings: a case study of the e-learning benefits of Ning in education. J. Interact. Online Learn. **9**(2) (2010)
7. Costa, C., Alvelos, H., Teixeira, L.: The use of Moodle e-learning platform: a study in a Portuguese University. Procedia Technol. **5**, 334–343 (2012)
8. Salloum, S.A., Al-Emran, M., Shaalan, K.: The impact of knowledge sharing on information systems: a review. In: 13th International Conference, KMO 2018 (2018)
9. Boyd, D.M., Ellison, N.B.: Social network sites: definition, history, and scholarship. J. Comput. Commun. **13**(1), 210–230 (2007)
10. Salloum, S.A., Al-Emran, M., Shaalan, K., Tarhini, A.: Factors affecting the E-learning acceptance: a case study from UAE. Educ. Inf. Technol. **24**(1), 509–530 (2019)
11. Lee, C., Lee, G., Lin, H.: The role of organizational capabilities in successful e-business implementation. Bus. Process Manag. J. **13**(5), 677–693 (2007)
12. Al-Emran, M., Mezhuyev, V., Kamaludin, A.: Students' perceptions towards the integration of knowledge management processes in M-learning systems: a preliminary study. Int. J. Eng. Educ. **34**(2), 371–380 (2018)
13. Al-Emran, M., Mezhuyev, V., Kamaludin, A., Shaalan, K.: The impact of knowledge management processes on information systems: a systematic review. Int. J. Inf. Manage. **43**, 173–187 (2018)

14. Al-Qaysi, N., Al-Emran, M.: Code-switching usage in social media: a case study from Oman. Int. J. Inf. Technol. Lang. Stud. **1**(1), 25–38 (2017)
15. Salloum, S.A., Al-Emran, M., Shaalan, K.: Mining social media text: extracting knowledge from Facebook. Int. J. Comput. Digit. Syst. **6**(2), 73–81 (2017)
16. Lüders, M., Brandtzæg, P.B.: 'My children tell me it's so simple': a mixed-methods approach to understand older non-users' perceptions of Social Networking Sites. New Media Soc. **19**(2), 181–198 (2017)
17. Chatti, M.A., Jarke, M., Frosch-Wilke, D.: The future of e-learning: a shift to knowledge networking and social software. Int. J. Knowl. Learn. **3**(4–5), 404–420 (2007)
18. Rennie, F., Morrison, T.: E-Learning and Social Networking Handbook: Resources for Higher Education. Routledge, New York (2013)
19. Keller, J., Suzuki, K.: Learner motivation and e-learning design: a multinationally validated process. J. Educ. Media **29**(3), 229–239 (2004)
20. Lee, Y.-C.: An empirical investigation into factors influencing the adoption of an e-learning system. Online Inf. Rev. **30**(5), 517–541 (2006)
21. Sun, P.-C., Tsai, R.J., Finger, G., Chen, Y.-Y., Yeh, D.: What drives a successful e-Learning? An empirical investigation of the critical factors influencing learner satisfaction. Comput. Educ. **50**(4), 1183–1202 (2008)
22. Zacharis, N.Z.: Predicting college students' acceptance of podcasting as a learning tool. Interact. Technol. Smart Educ. **9**(3), 171–183 (2012)
23. Law, K.M.Y., Lee, V.C.S., Yu, Y.-T.: Learning motivation in e-learning facilitated computer programming courses. Comput. Educ. **55**(1), 218–228 (2010)
24. Davis, F.D.: Perceived usefulness, perceived ease of use, and user acceptance of information technology. MIS Q. **13**(3), 319–340 (1989)
25. Al-Qaysi, N., Mohamad-Nordin, N., Al-Emran, M.: A systematic review of social media acceptance from the perspective of educational and information systems theories and models. J. Educ. Comput. Res. (2018)
26. Al-Maroof, R.A.S., Al-Emran, M.: Students acceptance of Google classroom: an exploratory study using PLS-SEM approach. Int. J. Emerg. Technol. Learn. **13**(6), 112–123 (2018)
27. Hsia, J.-W., Tseng, A.-H.: An enhanced technology acceptance model for e-learning systems in high-tech companies in Taiwan: analyzed by structural equation modeling. In: 2008 International Conference on Cyberworlds, pp. 39–44 (2008)
28. Mezhuyev, V., Al-Emran, M., Fatehah, M., Hong, N.C.: Factors affecting the metamodelling acceptance: a case study from software development companies in Malaysia. IEEE Access **6**, 49476–49485 (2018)
29. Salloum, S.A., Al-Emran, M.: Factors affecting the adoption of E-payment systems by university students: extending the TAM with trust. Int. J. Electron. Bus. **14**(4), 371–390 (2018)
30. Salloum, S.A., Mhamdi, C., Al Kurdi, B., Shaalan, K.: Factors affecting the adoption and meaningful use of social media: a structural equation modeling approach. Int. J. Inf. Technol. Lang. Stud. **2**(3), 96–109 (2018)
31. Al-Emran, M., Alkhoudary, Y.A., Mezhuyev, V., Al-Emran, M.: Students and educators attitudes towards the use of M-learning: gender and smartphone ownership differences. Int. J. Interact. Mob. Technol. **13**(1), 127–135 (2019)
32. Malik, S.I., Al-Emran, M.: Social factors influence on career choices for female computer science students. Int. J. Emerg. Technol. Learn. **13**(5), 56–70 (2018)
33. Al-Emran, M., Salloum, S.A.: Students' attitudes towards the use of mobile technologies in e-evaluation. Int. J. Interact. Mob. Technol. **11**(5), 195–202 (2017)
34. Al-Emran, M., Shaalan, K.: Academics' awareness towards mobile learning in Oman. Int. J. Comput. Digit. Syst. **6**(1), 45–50 (2017)

35. Salloum, S.A., Maqableh, W., Mhamdi, C., Al Kurdi, B., Shaalan, K.: Studying the social media adoption by university students in the United Arab Emirates. Int. J. Inf. Technol. Lang. Stud. **2**(3), 83–95 (2018)
36. Al-Emran, M., Shaalan, K.: Attitudes towards the use of mobile learning: a case study from the Gulf region. Int. J. Interact. Mob. Technol. **9**(3), 75–78 (2015)
37. Ringle, C.M., Wende, S., Becker, J.-M.: SmartPLS 3. Bönningstedt: SmartPLS (2015)
38. Al-Emran, M., Mezhuyev, V., Kamaludin, A.: PLS-SEM in information systems research: a comprehensive methodological reference. In: 4th International Conference on Advanced Intelligent Systems and Informatics (AISI 2018), pp. 644–653 (2018)
39. Hair Jr., J.F., Hult, G.T.M., Ringle, C., Sarstedt, M.: A Primer on Partial Least Squares Structural Equation Modeling (PLS-SEM). Sage Publications (2016)
40. Henseler, J., Ringle, C.M., Sarstedt, M.: A new criterion for assessing discriminant validity in variance-based structural equation modeling. J. Acad. Mark. Sci. **43**(1), 115–135 (2015)

A Unified Model for the Use and Acceptance of Stickers in Social Media Messaging

Rana Saeed Al-Maroof[1] (ID), Said A. Salloum[2,3]([✉]) (ID),
Ahmad Qasim Mohammad AlHamadand[3] (ID), and Khaled Shaalan[3] (ID)

[1] Department of English Language and Linguistics, Al Buraimi University
College, Al Buraimi, Oman
rana@buc.edu.om
[2] Faculty of Engineering and IT, The British University in Dubai, Dubai, UAE
ssalloum@uof.ac.ae
[3] Faculty of Engineering and IT, University of Fujairah, Fujairah, UAE
aqdl4@yahoo.com, khaled.shaalan@buid.ac.ae

Abstract. The combination of two technology model which are the Technology Acceptance Model (TAM) and Use of Gratifications Theory (U&G) to create an integrated model is the first step in predicting the importance of using emotional icons and the level of satisfaction behind this usage. The reason behind using these two theories into one integrated model is that U&G provides specific information and a complete understanding of usage, whereas TAM theory has proved its effectiveness with a variety of technological applications. A self-administered survey was conducted in University of Fujairah with college students to find out the social and cognitive factors that affect the usage of stickers in WhatsApp in the United Arab of Emirates. The hypothesized model is validated empirically using the responses received from an online survey of 372 respondents were analyzed using structural equation modeling (SEM-PLS). The results show that ease of use, perceived usefulness, cognition, hedonic and social integrative significantly affected the intention to use sticker by college students. Moreover, personal integrative had a significant influence on the intention to use sticker in UAE.

Keywords: Stickers · Technology Acceptance Model (TAM) ·
Uses and Gratification Theory (U&G) ·
Structural Equation Modeling (SEM-PLS) · United Arab of Emirates

1   Introduction

The use of "social media applications" has become a part of our daily lives [1–3]. Users nowadays enjoy to share and communicate social and psychological factors via the use of the interactive social media features [4–6]. These features have their own media contents that have changed the ways users interact with each other [7–9]. The newly implemented media features especially the stickers in WhatsApp application deserve more attention to show their real benefits and usages [10]. Emojis have been extensively used to make use of emotional expression and enrich user experience. The

© Springer Nature Switzerland AG 2020
A. E. Hassanien et al. (Eds.): AISI 2019, AISC 1058, pp. 370–381, 2020.
https://doi.org/10.1007/978-3-030-31129-2_34

"ubiquitous" usage of emojis paves the way to establish a comparison study to show the differences among users' behaviors and preferences all around the world. The differences could have stem from cultural and social basses [11]. However, the usage of stickers that are newly employed in the WhatsApp has imposed certain novelty and creativity. Novelty and creativity are two influential factors that have never been given attention to them as far as emojis are concerned. The lack of big, animated and static figures impose limitations on the part of the users among different countries. To put in other words, emojis are small, rely on Unicode, available via standardized keyboards. Users can not edit them to suit their personal needs. These facts stand in contrast with stickers because stickers are bigger, static and animated.

The main purpose of the current study is to build a model that can analyzed users' attitude and perception towards using stickers through WhatsApp by integrating TAM and Uses and Gratification approach (U&G). The importance of integrating these two theories into one unified conceptualized model has two main justifications. The first one is to use TAM theory to focus on the characteristics of the technology itself. The second justification is to use U&G approach to focus on the experiences generated by the technology. Based on the data available, it seems that WhatsApp users' attitudes and perceptions have not been investigated thoroughly due to the recent appearance and usage of stickers. Therefore, this study is entirely intended to focus on the importance of integrating two models, namely TAM and U&G theories to investigate the usage of stickers through WhatsApp text messages among students at college or university level. Accordingly, the study is supposed to bridge this gap by giving answers to the given questions:

RQ1 What are the aspects that make the stickers perceived usefulness is more conceptualized in terms of socialization, self-presentation, enjoyment, novelty, unique function via WhatsApp?

RQ2 How do stickers have a direct effect on users' intention and reaction?

RQ3 How do stickers –usage assist novelty and enjoyment?

2   Theoretical Framework and Research Model

2.1   TAM Theory

TAM examines behavioral intention when using information systems. The TAM theory is governed by two belief constructs. The first one is the perceived usefulness which is meat to refer to users' beliefs as he/she makes use of a particular system, that is, it deals with the degree of users' satisfaction of performance. However, the second is the ease of use which is meant to refer to the users' belief concerning the effectiveness and easiness of using a particular system [12]. These two core variables have both direct and indirect influences on user's intention to use particular application [13–16]. The role of TAM in explaining the effectiveness of Internet-based technology within college and university students is massive. Many researchers have emphasized this issue. For instance, [17] state that TAM is intend to give a detail background knowledge for the measuring the effectiveness behind the usage of new technology and it shows students' behavior and factors of acceptance when they are being exposed to new technology [17]. A similar attitude is given by [18] where they state that TAM can be used to measure University students' acceptance of Moodle within e-learning platform.

2.2 Model Uses and Gratification Theory (U&G)

U&G theory is a type of theory that focuses on the reasons behind using specific media. It also investigates the gratifications that are derived from media usage and access and assumes that people use media for specific and crucial reasons. U&G theory is applicable to a vast majority of mediated communication such as media, interactive media and the Internet [19]. According to [10], the main essential factors that are embedded inside the U&G theory are the inclusion of both 'novelty' and 'being there'. Novelty is considered a funny way to create innovation, whereas 'being there' refers to that social media users may click social buttons to feel immersed in a mediated reality. The two factors are crucial to the usage of stickers in Internet/Mobile App. Given that stickers are one of the newest designs in WhatsApp, users may use it to fulfill the sense of novelty. In addition, U&G theory pays attention to the 'being there' as it is one of the distinctive features behind using stickers.

2.3 The Research Model

Though TAM theory has been used intensively to measure the acceptance of technologies by a significant number of researchers all over the world [20, 21], it does not give specific and detailed reasons behind the usage of a particular technology. Due to the narrow applicability of TAM in relation to the usefulness of use, U & G theory comes as an assistant tool that gives more details in this respect. Furthermore, the U&G approach provides in-depth understanding of students' views to enhance the TAM model [22]. U & G covers three utilities that have a relation with the TAM dimensions which are social utility, hedonic utility, and functional utility. Social utility is used to reflect the interpersonal usage that has a close relationship with the perceived usefulness. Similarly, the hedonic utility is related to the perceived usefulness from the novelty and enjoyment perspectives. Finally, when the users of technology use a particular form of stickers to accomplish a particular task, then it is connected with the functional utility. The main hypotheses that can be formulated are connected to perceived usefulness that is in turn embedded within three types of utilities which are: social, hedonic and, functional. The social utility includes socialization and self-presentation, whereas hedonic includes novelty and enjoyment and finally functional utility embraces the unique functions of the stickers used. The proposed model is based on a group of hypotheses where the TAM model is extended to include related important factors that help in predicting users' attitude. See the research model (Fig. 1).

H1. Subjective norm usually show active and clear effects on perceived usefulness for (a) socialization, (b) self-presentation, (c) enjoyment, (d) novelty, (e) unique function.

H2. Perceived ease of use usually show active and clear on perceived usefulness for (a) socialization, (b) self-presentation, (c) enjoyment, (d) novelty, (e) unique function.

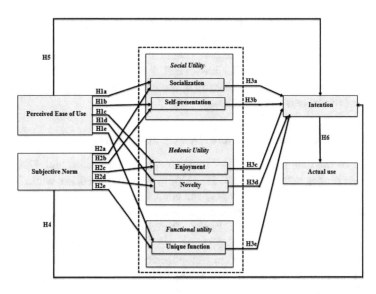

Fig. 1. Research Model

H3. Perceived usefulness for various motives including (a) socialization, (b) self-presentation, (c) enjoyment, (d) novelty, (e) unique function will implement users' intention deeply on the use of Stickers.

H4. Subjective norm usually show active and clear effect on users' intention to use Stickers.

H5. Perceived ease of use usually show active and clear effect on users' intention to use Stickers.

H6. Users' intentions to use stickers usually show active and clear effect on their actual use intensity.

3   Research Methodology

3.1   Sample and Data Collection

Data collection is implemented within the winter semester 2018/2019 between 15.02.2019 and 20.03.2019 with hard-copy surveys amongst the students in the United Arab Emirates. The total number of 400 questionnaires were randomly given out, 372 questionnaires were distributed to the respondents. Accordingly, the total response rate is 93%. It is important to note that 28 of these filled questionnaires were rejected as they had missing values. Based on the previous statistics, it seems that a total number of 372 questionnaires are filled with the required information/values. These questionnaires were assessed, with a practical response rate of 93%. This study follows the general guidelines of using PLS-SEM in IS research [23]. The analysis of the data was conducted based on the inclusion of the partial least squares-structural equation modeling (PLS-SEM) in the SmartPLS V.3.2.7 software [7, 24–27]. This study has made use of a two-step assessment approach to deal with the collected data, the approach included the measurement model and structural model [28].

3.2 Students' Personal Information/Demographic Data

The sample of the data include 53% of female students was 53% and 47% of males. Student's age ranges between 18 and 29 for 66% of the respondents while 34% of the respondents are above 29. Focusing on the educational background, 38% of the students belonged to Business Administration major while students in "Mass Communication and Public Relations", "Arts, Social Sciences & Humanities", and "Information Technology" were 25%, 17%, and 20%, respectively. Most of the students who participated in the study are well-educated and most of them own university degrees. To put in other words, the percentage of students who have a bachelor degree is 59% whereas the percentage of students who have master degree is 41. As per the opinion of [29–31], the "convenience sampling approach" was used when the respondents are willing to volunteer and that they are easily available. This study sample included students from different colleges; enrolled in different programs at different levels and have different ages. IBM SPSS Statistics ver. 23 was employed to measure the demographic data.

4 Findings and Discussion

4.1 Measurement Model Analysis

According to [28], construct reliability (consisting of Cronbach's alpha and composite reliability) and validity (consisting of convergent and discriminant validity) should be calculated to assess the measurement model. To determine the construct reliability, it is evident from the findings given in the following table that the range of values for Cronbach's alpha are between 0.759 and 0.921, which were both over the threshold value of 0.7 [32]. It is also shown in the findings of Table 1 that the range of values for composite reliability (CR) is from 0.743 to 0.891, which were also more than the recommended value of 0.7 [33]. These findings confirm the construct reliability, and consider all constructs to be free from error.

It is suggested that factor loading and average variance extracted (AVE) should be evaluated for measurement of convergent validity [28]. The findings in Table 1 has illustrated that all factor loadings have a value more than the recommended value of 0.7. In addition, the results in Table 1 has verified that the range of values of AVE is from 0.549 to 0.766, which were more than the threshold value of 0.5. These findings show that for all constructs, convergent validity has been met sufficiently. One criteria were put forward for measurement of discriminant validity, which included the Heterotrait-Monotrait ratio (HTMT) [28]. The HTMT ratio findings can be seen in

Table 1. Convergent validity results which assures acceptable values (Factor loading, Cronbach's Alpha, composite reliability $\geq$ 0.70 & AVE > 0.5).

Constructs	Items	Factor loading	Cronbach's Alpha	CR	AVE
Actual use	AU1	0.833	0.834	0.891	0.610
	AU2	0.874			
Behavioral intention	BI1	0.824	0.801	0.764	0.647
	BI2	0.758			
	BI3	0.901			
Enjoyment	ENJ1	0.854	0.794	0.773	0.549
	ENJ2	0.711			
	ENJ3	0.800			
Novelty	NOV1	0.878	0.759	0.784	0.661
	NOV2	0.819			
	NOV3	0.845			
Perceived ease of use	PEOU1	0.850	0.921	0.798	0.766
	PEOU2	0.899			
	PEOU3	0.779			
Self-presentation	SELP1	0.796	0.799	0.743	0.728
	SELP2	0.718			
	SELP3	0.880			
Socialization	SOC1	0.926	0.813	0.818	0.757
	SOC2	0.757			
	SOC3	0.846			
Subjective Norm	SUBN1	0.834	0.887	0.789	0.552
	SUBN2	0.856			
	SUBN3	0.785			
Unique function	UNIF1	0.892	0.877	0.874	0.759
	UNIF2	0.768			
	UNIF3	0.761			

Table 2, and it is evident that the threshold value of 0.85 has not exceeded by each construct's value [34]. Hence, the HTMT ratio is confirmed, implying that the discriminant validity is attained. The outcomes of the analysis show that no issues were faced during the assessment of the measurement model with respect to its reliability and validity. Accordingly, the proposed model can also be dealt with based on the usage of the collected data.

4.2 Structural Model Analysis

The coefficient of determination (R^2 value) measure is normally used to assess the structural model [35]. The purpose of developing this coefficient is to measure the model's predictive accuracy, The predicative accuracy is usually processed as the squared correlation among the particular endogenous which are the actual constructs

Table 2. Heterotrait-Monotrait Ratio (HTMT).

	AU	BI	ENJ	NOV	PEOU	SELP	SOC	SUBN	UNIF
AU									
BI	0.468								
ENJ	0.591	0.487							
NOV	0.481	0.400	0.245						
PEOU	0.510	0.634	0.336	0.159					
SELP	0.747	0.711	0.611	0.638	0.389				
SOC	0.710	0.544	0.536	0.755	0.668	0.611			
SUBN	0.677	0.738	0.705	0.614	0.778	0.674	0.715		
UNIF	0.710	0.666	0.482	0.325	0.436	0.699	0.691	0.633	

and predicted values [28, 36]. The coefficient represents the combined impact of the exogenous latent variables and an endogenous latent variable, and its effect on the second variable. The coefficient signifies the squared correlation among the actual and predicted values of the variables; therefore, it by default incorporates the meaning of variance-degree within the endogenous constructs. Every exogenous construct supports this fact. In addition, it makes easily recognized. According to [37], the value is considered as high when it is over 0.67, suggesting that the qualities ranging from 0.33 to 0.67 are direct, whereas the qualities ranging from 0.19 to 0.33 are weak. In addition, the estimation is inadmissible when it is less than 0.19. Both Table 3 and Fig. 2 illustrate this fact in a model that has a moderate predictive power. This in turn implies that the percentages of the variance in the actual use and Intention are approximately 62% and 58%, respectively.

Table 3. R^2 of the endogenous latent variables.

Constructs	R^2	Results
Actual use	0.623	Moderate
Intention	0.581	Moderate

The structural model analysis aims at analyzing the given hypothesis. To do so, the path coefficient analysis has been employed. The model was intended to deal with a group of data through a bootstrap re-sampling routine to obtain the path significances [38]. In this study, a total of 5000 re-samples were used. A one-tailed t-test was employed in this study. The usage of the one-tailed t-test can be justified based on the hypotheses are directional. The hypotheses testing results of the integrated model adopted in this study are shown in Fig. 2 and Table 4. Generally speaking, fourteen out of eighteen hypotheses were supported. This implies that the hypotheses H1a, H1b, H1c, H1d, H2a, H2c, H2d, H2e, H3a, H3c, H3d, H4, H5, and H6 were supported by the empirical data, while H1e,

H2b, H3b, and H3e were rejected. The results revealed that Subjective norm (SUBN) is significantly influencing the Socialization (SOC) (β = 0. 0.781, p < 0.001), Self-presentation (SELP) (β = 0.332, p < 0.05), Enjoyment (ENJ) (β = 0.233, p < 0.05), and Novelty (NOV) (β = 0.771, p < 0.001); thus, supporting hypothesis H1a, H1b, H1c, and H1d, respectively. The results also showed that Perceived ease of use (PEOU) significantly influenced Socialization (SOC) (β = 0.303, p < 0.001), Enjoyment (ENJ) (β = 0.911, p < 0.001), Novelty (NOV) (β = 0.530, p < 0.01) and Unique function (UNIF) (β = 0.980, p < 0.001); thus, supporting hypothesis H2a, H2c, H2d, and H2e, respectively. The results showed that Intention (BI) significantly influenced Socialization (SOC) (β = 0.658, P < 0.001), Enjoyment (ENJ) (β = 0.377, P < 0.05), Novelty (NOV) (β = 0.680, P < 0.001), Subjective norm (SUBN) (β = 0.456, P < 0.05), and Perceived Ease of Use (PEOU) (β = 0.164, P < 0.05) supporting hypothesis H3a, H3c, H3d, H4 and H5, respectively. The relationships between Intention (BI) and Self-presentation (SELP) (β = −0.055, p = 0.422), Unique function UNIF)

Table 4. Results of structural Model-Research Hypotheses Significant at p** = <0.01, p* <0.05)

H	Relationship	Path	t-value	p-value	Direction	Decision
H1a	Subjective norm -> Socialization	0.781	19.456	0.000	Positive	Supported**
H1b	Subjective norm -> Self-presentation	0.332	2.181	0.019	Positive	Supported*
H1c	Subjective norm -> Enjoyment	0.233	2.288	0.011	Positive	Supported*
H1d	Subjective norm -> Novelty	0.771	14.577	0.000	Positive	Supported**
H1e	Subjective norm -> Unique function	−0.082	1.434	0.166	Negative	not Supported
H2a	Perceived ease of use -> Socialization	0.303	8.713	0.000	Positive	Supported**
H2b	Perceived ease of use -> Self-presentation	−0.134	1.914	0.039	Negative	not Supported
H2c	Perceived ease of use -> Enjoyment	0.911	9.958	0.000	Positive	Supported**
H2d	Perceived ease of use -> Novelty	0.530	7.486	0.008	Positive	Supported**
H2e	Perceived ease of use -> Unique function	0.980	24.525	0.000	Positive	Supported**
H3a	Socialization -> Intention	0.658	15.158	0.000	Positive	Supported**
H3b	Self-presentation -> Intention	−0.055	1.802	0.422	Negative	not Supported
H3c	Enjoyment -> Intention	0.377	3.155	0.030	Positive	Supported*
H3d	Novelty -> Intention	0.680	24.343	0.000	Positive	Supported**
H3e	Unique function -> Intention	0.032	0.101	0.366	Positive	not Supported
H4	Subjective norm -> Intention	0.456	3.712	0.023	Positive	Supported*
H5	Perceived ease of use -> Intention	0.164	1.163	0.041	Positive	Supported*
H6	Intention -> Actual use	0.787	18.307	0.000	Positive	Supported**

(β = 0.032, p = 0.366) were found to be statistically not significant, thus, the hypotheses H3b, and H3e are generally not supported. It is important to note that Unique function (UNIF) have a negative impact on the Subjective norm (SUBN) (β = −0.082; p = 0.166) which means that H1e is not supported. Furthermore, the effect of Perceived ease of use (PEOU) has a negative impact on Self-presentation (SELP) (β = −0.134; p = 0.039). Accordingly, it was found to be not significant; that is, H2b is not supported. The results also revealed that Intention (BI) is significantly influencing the Actual use (AU) (β = 0.787, p < 0.001) and this gives support to hypothesis H6. The second step that follows the measurement of the model is the confirmation of the structural aspects of the model. To be able to do that properly, the coefficient of determination (R^2) and the path coefficients need to be estimated using a bootstrapping process involving 5,000 re-samples [28]. With respect to path analysis, the path coefficients, t-values, and p-values for each hypothesis can be seen in Table 4. All of the hypotheses are supported, except hypotheses 6 and 7, which were not supported.

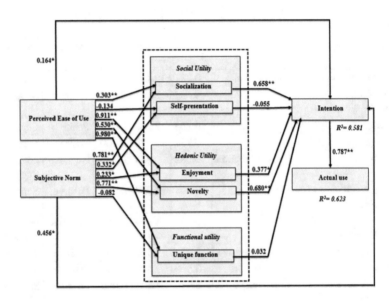

Fig. 2. Path coefficient results (significant at p** < = 0.01, p* < 0.05)

5   Discussion and Implication

The research results have shown that users' attitude towards using stickers is highly significant and stickers play an effective role in day-to-day communication. Thus, the study bridges a gap in the current field of study because it prevails users' attitude based on an integrated and conceptualized model that comprises both TAM and U&G theories. TAM functions on the acceptance level of stickers via WhatsApp whereas the U&G focuses on the experience and implication levels. Stickers can build a special conciseness in the mind of users that urge them to create unique experience and different level of implication or usage. Stickers can be used more frequently to reflect their own unique and special personal needs.

Accordingly, the newly implemented model focuses on two dominant and effective factors in the usage of stickers among college students. It shed lights on the technical aspects or experiences that are supposed to be more vital in influencing the acceptance of sticker-usage. The two basic elements that are part of the scope of the current study is the relationships between motivations to use the stickers on one hand and the college students' perceptions about using the stickers on the other hand.

Taking into consideration the finding of the study, it is concluded that all the suggested factors in the model are dominant factors in predicting users' intention to use stickers. Moreover, personal integrative had a significant influence on the intention to use sticker in UAE. Thus, students' willingness to use stickers is connected directly by the variations in forms and animations. The variations, mobility and big size urge students to use them successively in their online and everyday WhatsApp conversation. In addition, they are expected to use them frequently in their daily life to replace long sentences and expressions. In line with this, [39] state that whenever these types become more familiar and accessible, their frequency and range of functions are expected to increase. Thus, the instructional designers of WhatsApp are supposed to support the usage of these stickers in their database to ensure the ease of use. Finally, it seems that the continuous elaboration and update of the forms and functions of stickers will encourage users to create their own stickers that are either personally- based or socially -based

References

1. Salloum, S.A., Al-Emran, M., Shaalan, K.: Mining social media text: extracting knowledge from Facebook. Int. J. Comput. Digit. Syst. **6**(2), 73–81 (2017)
2. Salloum, S.A., Al-Emran, M., Abdallah, S., Shaalan, K.: "Analyzing the Arab gulf newspapers using text mining techniques. In: International Conference on Advanced Intelligent Systems and Informatics, pp. 396–405 (2017)
3. Mhamdi, C., Al-Emran, M., Salloum, S.A.: Text mining and analytics: a case study from news channels posts on Facebook. **740** (2018)
4. Alghizzawi, M., Salloum, S.A., Habes, M.: The role of social media in tourism marketing in Jordan. Int. J. Inf. Technol. Lang. Stud. **2**(3) (2018)
5. Habes, M., Alghizzawi, M., Khalaf, R., Salloum, S.A., Ghani, M.A.: The relationship between social media and academic performance: Facebook perspective. Int. J. Inf. Technol. Lang. Stud. **2**(1), 12–18 (2018)
6. Al-Qaysi, N., Al-Emran, M.: Code-switching usage in social media: a case study from Oman. Int. J. Inf. Technol. Lang. Stud. **1**(1), 25–38 (2017)
7. Alshurideh, M., Salloum, S.A., Al Kurdi, B., Al-Emran, M.: Factors affecting the social networks acceptance: an empirical study using PLS-SEM approach. In: 8th International Conference on Software and Computer Applications (2019)
8. Salloum, S.A., Al-Emran, M., Shaalan, K.: Mining text in news channels: a case study from Facebook. Int. J. Inf. Technol. Lang. Stud. **1**(1), 1–9 (2017)

9. Salloum, S.A., Mhamdi, C., Al-Emran, M., Shaalan, K.: Analysis and classification of Arabic newspapers' Facebook pages using text mining techniques. Int. J. Inf. Technol. Lang. Stud. **1**(2), 8–17 (2017)
10. Shao, C., Kwon, K.H.: Clicks intended: an integrated model for nuanced social feedback system uses on Facebook. Telemat. Informatics (2018)
11. Lu, X., et al.: Learning from the ubiquitous language: an empirical analysis of emoji usage of smartphone users. In: Proceedings of the 2016 ACM International Joint Conference on Pervasive and Ubiquitous Computing, pp. 770–780 (2016)
12. Davis, F.D.: Perceived usefulness, perceived ease of use, and user acceptance of information technology. MIS Q. **13**(3), 319–340 (1989)
13. Salloum, S.A., Al-Emran, M.: Factors affecting the adoption of E-payment systems by university students: extending the TAM with trust. Int. J. Electron. Bus. **14**(4), 371–390 (2018)
14. Mezhuyev, V., Al-Emran, M., Fatehah, M., Hong, N.C.: Factors affecting the metamodelling acceptance: a case study from software development companies in Malaysia. IEEE Access **6**, 49476–49485 (2018)
15. Salloum, S.A., Shaalan, K.: Adoption of e-book for university students. In: International Conference on Advanced Intelligent Systems and Informatics, pp. 481–494 (2018)
16. Salloum, S.A., Mhamdi, C., Al Kurdi, B., Shaalan, K.: Factors affecting the adoption and meaningful use of social media: a structural equation modeling approach. Int. J. Inf. Technol. Lang. Stud. **2**(3), 96–109 (2018)
17. Al-Maroof, R.A.S., Al-Emran, M.: Students acceptance of google classroom: an exploratory study using PLS-SEM approach. Int. J. Emerg. Technol. Learn. (2018)
18. Escobar-Rodriguez, T., Monge-Lozano, P.: The acceptance of moodle technology by business administration students. Comput. Educ. **58**(4), 1085–1093 (2012)
19. Luo, M.M., Remus, W.: Uses and gratifications and acceptance of web-based information services: an integrated model. Comput. Human Behav. **38**, 281–295 (2014)
20. Al-Qaysi, N., Mohamad-Nordin, N., Al-Emran, M.: A systematic review of social media acceptance from the perspective of educational and information systems theories and models. J. Educ. Comput. Res. (2018)
21. Al-Emran, M., Mezhuyev, V., Kamaludin, A.: Technology acceptance model in m-learning context: a systematic review. Comput. Educ. (2018)
22. Aburub, F., Alnawas, I.: A new integrated model to explore factors that influence adoption of mobile learning in higher education: an empirical investigation. Educ. Inf. Technol. 1–14 (2019)
23. Al-Emran, M., Mezhuyev, V., Kamaludin, A.: PLS-SEM in information systems research: a comprehensive methodological reference. In: 4th International Conference on Advanced Intelligent Systems and Informatics (AISI 2018), pp. 644–653 (2018)
24. Ringle, C.M., Wende, S., Becker, J.-M.: SmartPLS 3. Bönningstedt: SmartPLS (2015)
25. Habes, M., Alghizzawi, M., Khalaf, R., Salloum, S.A., Ghani, M.A.: The relationship between social media and academic performance: Facebook perspective. Int. J. Inf. Technol. Lang. Stud. **2**(1) (2018)
26. Alshurideh, M., Masa'deh, R., Alkurdi, B.: The effect of customer satisfaction upon customer retention in the Jordanian mobile market: an empirical investigation. Eur. J. Econ. Financ. Adm. Sci. **47**, 69–78 (2012)
27. Al-dweeri, R.M., Obeidat, Z.M., Al-dwiry, M.A., Alshurideh, M.T., Alhorani, A.M.: The impact of e-service quality and e-loyalty on online shopping: moderating effect of e-satisfaction and e-trust. Int. J. Mark. Stud. **9**(2), 92 (2017)
28. Hair Jr, J.F., Hult, G.T.M., Ringle, C., Sarstedt, M.: A primer on partial least squares structural equation modeling (PLS-SEM). Sage Publications (2016)

29. Al-Emran, M., Salloum, S.A.: Students' attitudes towards the use of mobile technologies in e-evaluation. Int. J. Interact. Mob. Technol. **11**(5), 195–202 (2017)
30. Al-Emran, M., Alkhoudary, Y.A., Mezhuyev, V., Al-Emran, M.: Students and educators attitudes towards the use of m-learning: gender and smartphone ownership differences. Int. J. Interact. Mob. Technol. **13**(1), 127–135 (2019)
31. Salloum, S.A., Maqableh, W., Mhamdi, C., Al Kurdi, B., Shaalan, K.: Studying the social media adoption by university students in the United Arab Emirates. Int. J. Inf. Technol. Lang. Stud. **2**(3), 83–95 (2018)
32. Nunnally, J.C., Bernstein, I.H.: Psychometric Theory (1994)
33. Kline, R.B.: Principles and Practice of Structural Equation Modeling. Guilford publications (2015)
34. Henseler, J., Ringle, C.M., Sarstedt, M.: A new criterion for assessing discriminant validity in variance-based structural equation modeling. J. Acad. Mark. Sci. **43**(1), 115–135 (2015)
35. Dreheeb, A.E., Basir, N., Fabil, N.: Impact of system quality on users' satisfaction in continuation of the use of e-learning system. Int. J. e-Educ. e-Bus. e-Manage. e-Learn. **6**(1), 13 (2016)
36. Senapathi, M., Srinivasan, A.: An empirical investigation of the factors affecting agile usage. In: Proceedings of the 18th international conference on evaluation and assessment in software engineering, p. 10 (2014)
37. Chin, W.W.: The partial least squares approach to structural equation modeling. Mod. methods Bus. Res. **295**(2), 295–336 (1998)
38. Efron, B., Tibshirani, R.J.: The jackknife. In: An introduction to the bootstrap. Springer, pp. 141–152 (1993)
39. Herring, S., Dainas, A.: Nice picture comment!' Graphicons in Facebook comment threads. In: Proceedings of the 50th Hawaii International Conference on System Sciences (2017)

The Relation Between Social Media and Students' Academic Performance in Jordan: YouTube Perspective

Mohammed Habes[1] , Said A. Salloum[2,3](✉) ,
Mahmoud Alghizzawi[4,5] , and Chaker Mhamdi[5,6]

[1] Faculty of Applied Social Sciences, University Sultan Zainal Abidin,
Kuala Terengganu, Terengganu, Malaysia
mohammedhabes88@gmail.com
[2] Faculty of Engineering & IT, The British University in Dubai, Dubai, UAE
ssalloum@uof.ac.ae
[3] Faculty of Engineering & IT, University of Fujairah, Fujairah, UAE
[4] Faculty of Economics and Management Sciences, University Sultan Zainal
Abidin, Kuala Terengganu, Malaysia
dr.alghzawi87@gmail.com
[5] University of Manouba, Manouba, Tunisia
shaker@buc.edu.om
[6] Al Buraimi University College, Al Buraymi, Oman

Abstract. This study aims mainly at analyzing the relationship between social media and students' academic performance in Jordan in the context of higher education from a YouTube perspective. It intends to explore the benefits this relationship may have in enhancing students; leaning and improving their academic performance. To successfully reach its aims, this study proposes a new model aiming at verifying the relationship of social Bookmarking, YouTube Features, Perceived Usefulness, Use of Social Media, on Jordanian students' academic performance. To verify the validity of the proposed model, data were analyzed using Smart PLS using structural equations modeling (SEM). Data were collected from Yarmouk University in Jordan covering all the levels of study at the university. An electronic questionnaire was conducted for a target of 360 students who participated in this study. The findings of the study revealed that Social Bookmarking, YouTube Features, Perceived Usefulness, Use of Social Media are important factors to predict students' academic performance in relation to using social networking media for e-learning purposes in Jordan.

Keywords: Social media · Academic performance · YouTube · Jordan

1 Introduction

Today, the improvements in communication technologies have expanded the scope of information and instant communication [1, 2]. There is no doubt that modern technology in the field of communication has been fully transferred to the "small village" bringing about a lot of changes that are said to be both positive and negative [3].

© Springer Nature Switzerland AG 2020
A. E. Hassanien et al. (Eds.): AISI 2019, AISC 1058, pp. 382–392, 2020.
https://doi.org/10.1007/978-3-030-31129-2_35

Today, most young people and students have YouTube accounts [4]. The most important reason lying behind this is the growing use of YouTube in various aspects of our life [5–7]. Recently, it has been noticed that a large number of university students are increasingly addicted to social media and the huge flow and sharing of information online to the extent that students tend to lose focus on academic tasks which negatively affect their academic results [2, 3, 8–10]. Social media is a source of information and communication among students [11]. It has become an integral part of individuals' everyday lives [8, 12–15]. Facebook and YouTube are two of the most popular social media applications for Internet users in general and in the Middle East in particular [3, 7, 9]. In 2016, 90% of social media usage was attributed to Jordan alone, which makes it the country with the largest users of the Internet relative to social media applications. Jordanians favour WhatsApp, Facebook, YouTube and use them more frequently than other social media sites, [16, 17]. According to the studies of [18] adolescents are the ones who use social media mostly. Also, the ratio varies with the range and availability of Internet services as well as perceived usefulness [19]. It was significantly found that use of social media Facebook and YouTube positively affect self-development, creativity, students' knowledge enhancement, information exchange and increases technical skills [10, 17]. Additionally, the YouTube Features enhance students' ability to excel in academic performance and deepen their collaborative learning [20]. Social media technology is used more for teenage students and equally for both genders [21, 22]. With regard to social bookmarking, it helped students join educational networks, interact with their peers, and deepen the concept of e-learning. Activating the role of the family in raising awareness of the risks and advantages of these means and employing them in the academic process are the focus of a number of studies, [23–27]. In light of what has been stated, the purpose of this study is to investigate the rapport between social media and students' academic performance in the higher education context in Jordan. The focus of this investigation is YouTube and whether academic performance is positively or negatively influenced by YouTube application utilization from the part of Jordanian university students.

2 Research Model and Hypotheses

This paper proposes a framework for the relation between social media and student academic performance in Jordan, namely YouTube perspective in Yarmouk University within the Jordanian higher education institutions based on the Constructivist Theory [28], as illustrated in Fig. 1, "the framework of the research with hypotheses". The current study revealed the integration of the academic performance of students in Jordan with the variables of social media, specifically YouTube channels, such as Social Bookmarking (SB), YouTube Features (YTF), Perceived Usefulness (PU), Use of Social Media (UOSM), in addition to Variable of academic performance (AP).

2.1 YouTube Features with Academic Performance (YTF)

YouTube Features is a likeable video sharing website where users can upload, view, and share video clips [6, 22, 29]. The site offers a variety of features besides the ability

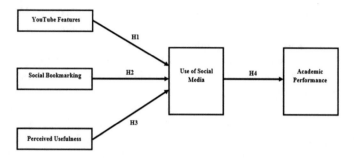

Fig. 1. Research model

to upload and deliver videos that encourage communication with other friends including commentaries on videos [30]. Currently, the demand for knowledge and science has increased due to the increasing competition in all sectors and the increasing number of researchers and students. According to [21], YouTube is a helpful and stimulating learning tool for students and can be used for educational purposes as it has a positive impact on students' learning outcomes. YouTube now appears to be a promoting learning channel. Since it began in 2005, YouTube has become the most favorite free video-sharing website for user-created content. YouTube is filled with large amounts of easily accessible learning content that directly show to users the video how to learn [19, 31]. Through the Features of the UT opus, the user mainly explores whether social features can help in a more convenient recovery videos for queries when combined with key features. According to the study of [32], the choice of YouTube depends on the availability of a rich set of educational content and features [29] and those features in this platform for the use of YouTube. The researchers believe that the findings apply to the text, search for images and/or videos on other platforms that support similar types of features which support the academic performance of students and affect their results [17]. On the basis of the discussion above, the following assumption is proposed.

H1: YouTube Features (YTF) have a positive effect on Use of Social Media (UOSM).

2.2 Social Bookmarking (SB) and Academic Performance

Social Bookmarking sites in social media have become hubs where people can express and share ideas and experiences with the world [33, 34]. They allow users to save the search and organize links to various Internet resources and websites [35]. Additionally, these services allow labeling of links to be easily shared and searched. Business teachers can use them through social media in a variety of ways to improve learning and learning processes and gain multiple benefits for collaborative professional development and academic performance [23–26].On the basis of the discussion above, the following assumption is proposed.

H2: Social Bookmarking (SB) has a positive effect on Use of Social Media (UOSM).

2.3   YouTube Uses and Perceived Usefulness (PU)

You Tube has been linked with increased student need for learning [36, 37]. The shared comments between student make an increased awareness and usage of how students describe how they acquired learning programs [10, 38] as well as how they perceived derived usefulness of electronic learning through You Tube videos by providing helpful and new ways to learn [18, 39]. Practical tasks are determined by the variables perceived usefulness. The conceptual framework of YouTube makes usefulness for both teachers and students in learning by video in addition to the easy access which make the relationship between YouTube uses and academic performance of direct effect [19]. Perceived usefulness was found to affect various technologies in the past [40, 41]. On the basis of the discussion above, the following assumption is proposed.

H3: Perceived Usefulness (PU) has a positive effect on Use of Social Media (UOSM).

2.4   Use of Social Media with Academic Performance (UOSM)

The results of previous studies showed that the use of social media platforms has a positive impact on students' academic performance [4, 10, 17, 42]. There is a clear influence of the means of social media on education for both teachers and students, which has become a motivational and main supporter of education [43, 44]. Many educational institutions have been steadily relying more on e-learning usage data [45]; seeking more interaction and effectiveness in the educational process [46]. Social media has unique and large features, easily facilitating connections to others for sharing information and knowledge [47–50]. It is, therefore, a way to exchange ideas and share pictures and videos among students [51]. Consequently, this active interaction and involvement leads to an effective participatory community to education [52]. On the basis of the discussion above, the following assumption is proposed.

H4. Use of Social Media (UOSM) has a positive effect on Academic Performance (AP).

3   Research Methodology

Data collection included undergraduate and postgraduate students from Yarmouk University in Jordan. Questionnaires, being the basis tool of data collection, were distributed to test the various research hypotheses and the effect of the factors under investigation. The questionnaire contained fourteen items under different factors including Academic Performance, Perceived Usefulness, Social Bookmarking, Use of Social Media and YouTube Features. A total of 400 respondents agreed to take part in the questionnaire where 40 of them submitted incomplete surveys. Therefore, only 360 survey questionnaires were considered complete figuring out 90% of the response rate. This study uses the "convenience sampling technique" while collecting the data. This is the same procedure followed in prior studies related to technology use and acceptance [53–55]. Having collected the questionnaires, the different respondents were classified according to various criteria including age, gender, and the use of social media. Among the total number of 360 respondents, 165 were males and 195 were females, representing 46% and 54%, respectively. 19% of the respondents were above the age of 30 years old, while 81% of them were between 18 and 29 years old. 97% of the respondents reported they were tech-savvy and 98% stated they use social media on a daily basis. Facebook, Twitter, Instagram, and YouTube were said to be the most popular social media among the respondents, respectively.

4 Findings and Discussion

4.1 Measurement Model Analysis

There has been an increase in the prevalence of using SmartPLS for the Partial Least Squares-Structural Equation Modeling (PLS-SEM) [56] are behind the development of this software. The measurement and structural models were assessed in this study with the utilization of the PLS-SEM by following the general guidelines of employing the technique in IS research [57]. The structural model is how the latent constructs are associated with each other and the measurement model, which is also called the outer model, is how the indicators are related to each other. There was a utilization of SEM-PLS alongside the probability method for the measurement of the proposed model. For the determination of reliability and convergent reliability, Composite Reliability, Average Variance Extracted and Factor Loadings were the different measurements to be conducted [58]. Factor loadings were used for demonstrating the weight and correlation value of each questionnaire variable. In addition to that, factors' dimensionality can be represented with the help of a bigger load value. To measure reliability, there has been an introduction of the Composite Reliability (CR) measure. CR provides a precise value as it uses factor loadings in the constructed formula; therefore it can also be used for the same purpose. The average quantity of variance in the given variable explaining the latent construct is known as Average Variance Extracted (AVE). The convergence of each factor can be analyzed using AVE if the discriminate validity is greater than one factor. The condition for the reliability and convergent validity has been surpassed by our experiment outcome for the questionnaire reliability and convergent validity, according to Table 1. Analysis results for all factors are presented in Table 1 in the form of the variable acquired.

The comparative amount of convergent validity was estimated using factor loadings, variance extracted and reliability, having Cronbach's Alpha and composite reliability, as indicators, in accordance with [58]. Internal consistency was found to be between a number of measurements of a construct as the reliability coefficient and composite reliability (CR) for all of constructs are more than 0.7. According to Table 1, composite reliabilities of constructs are between 0.741–0.920 and Cronbach's alpha scores surpass 0.7, which is the acceptable value [59]. Furthermore, the condition that a minimum of 50% variance extracted amongst a group of items that are fundamental to the latent construct has been fulfilled by every average variance extracted (AVE) value that was between 0.538–0.744 [60]. Consequently, it is considered that convergent validity can result under the influence of the scales that evaluate the constructs.

The squared correlation between the constructs in the measurement model has been surpassed by every AVE value [61]. According to the condition of discriminate validity, the values of HTMT have to be under 0.85. As it can be seen in Table 2, we have also satisfied the criterion. Henceforth, there has been an establishment of the discriminate validity.

Table 1. Convergent validity results which assures acceptable values (Factor loading, Cronbach's Alpha, composite reliability $\geq$ 0.70 & AVE > 0.5).

Constructs	Items	Factor loading	Cronbach's alpha	CR	AVE
Academic performance (AP)	AP_1	0.815	0.772	0.741	0.551
	AP_2	0.833			
Perceived Usefulness (PU)	PU_1	0.880	0.881	0.920	0.744
	PU_2	0.738			
	PU_3	0.902			
Social Bookmarking (SB)	SB_1	0.897	0.883	0.779	0.632
	SB_2	0.879			
	SB_3	0.899			
Use of Social Media (UOSM)	UOSM_1	0.832	0.807	0.786	0.652
	UOSM _2	0.878			
	UOSM _3	0.941			
YouTube Features (YTF)	YTF_1	0.833	0.763	0.825	0.538
	YTF_2	0.727			
	YTF _3	0.805			

Table 2. Heterotrait-Monotrait Ratio (HTMT).

	AP	PU	SB	UOSM	YTF
AP					
PU	0.033				
SB	0.245	0.270			
UOSM	0.102	0.211	0.335		
YTF	0.021	0.306	0.210	0.440	

The coefficient of determination (R^2 value) measure is normally used for analyzing the structural model. The model's predictive accuracy is determined using this coefficient, which is used as the squared correlation between a certain endogenous construct's actual and predicted values. The coefficient shows how the exogenous latent variables' collectively impact an endogenous latent variable. It can be seen how much variance is there in the endogenous constructs, which is defended by each exogenous construct identified with it, due to the coefficient being the squared correlation between the actual and predicted values of the variables. According to [62], high values include those above 0.67, direct values are the qualities between 0.33–0.67 and weak values are the qualities ranging between 0.19–0.33. Moreover, the estimation is inadmissible when it is under 0.19. In Table 3 and Fig. 2 it can be seen that the model has high predictive power, which supports almost 73% and 70% of the variance in the Academic Performance and Use of Social Media, respectively.

Table 3. R^2 of the endogenous latent variables.

Constructs	R^2	Results
Academic performance (AP)	0.726	High
Use of Social Media (UOSM)	0.699	High

4.2 Structural Model Analysis

How the theoretical constructs for the structural model are related to each other were assessed using a structural equation model with SEM-PLS having the most chances of estimation? With all this, the proposed hypotheses were tested. Table 6 and Fig. 2 show the results. According to the results, all the hypotheses are significant. Based on the data analysis hypotheses H1, H2, H3, and H4 were supported by the empirical data. The results showed that Use of Social Media (UOSM) significantly influenced You-Tube Features (YTF) ($\beta = 0.624$, $P < 0.001$), Social Bookmarking (SB) ($\beta = 0.492$, $P < 0.001$), and Perceived Usefulness (PU) ($\beta = 0.716$, $P < 0.001$) supporting hypothesis H1, H2, and H3, respectively. Furthermore, the Academic Performance (AP) was determined to be significant in affecting Use of Social Media (UOSM) ($\beta = 0.323$, $P < 0.001$) supporting hypothesis H4. FA summary of the hypotheses testing results is shown in Table 4.

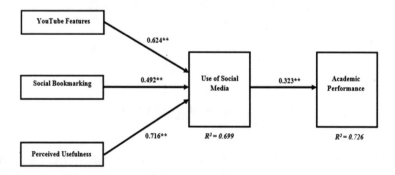

Fig. 2. Path coefficient results (significant at $p^{**} < = 0.01$, $p^* < 0.05$)

Table 4. Results of structural Model-Research Hypotheses Significant at p** =< 0.01, p* < 0.05)

H	Relationship	Path	*t*-value	*p*-value	Direction	Decision
H1	YouTube Features -> Use of Social Media	0.624	18.126	0.011	Positive	Supported**
H2	Social Bookmarking -> Use of Social Media	0.492	14.301	0.000	Positive	Supported**
H3	Perceived Usefulness -> Use of Social Media	0.716	22.434	0.000	Positive	Supported**
H4	Use of Social Media -> Academic performance	0.323	8.161	0.003	Positive	Supported**

5 Conclusion

The researchers suggested a novel model intended at verifying the connection between students' academic performance and use of social media, perceived usefulness, social bookmarking, and YouTube features. The validity of the suggested model was verified through an analysis of the data employing Smart PLS using Structural Equation Modeling (SEM). Data were collected from Yarmouk University in Jordan covering the various study levels at the university. An electronic questionnaire was conducted for a target of 360 participants. Hypotheses 1, 2, 3, and 4 were supported by a positive effect of "Use of Social Media", "YouTube Features", "Social Bookmarking", and "Perceived Usefulness" on students' academic performance. The study also revealed the significance of students' abilities and trust in utilizing social media technology. The findings of this study confirm those of previous researchers [3, 10, 17, 18, 20, 21, 23, 26, 37, 42, 63, 64]. Accordingly, this study recommends that legislators and managers of social media applications focus on the major features that are essential in motivating digital learning.

References

1. Malik, S.I., Al-Emran, M.: Social factors influence on career choices for female computer science students. Int. J. Emerg. Technol. Learn. (2018)
2. Al-Emran, M., Salloum, S.A.: Students' attitudes towards the use of mobile technologies in e-evaluation. Int. J. Interact. Mob. Technol. **11**(5), 195–202 (2017)
3. Olasina, G.: An evaluation of educational values of YouTube videos for academic writing. African J. Inf. Syst. **9**(4), 2 (2017)
4. Habes, M., Salloum, S.A., Alghizzawi, M., Alshibly, M.S.: The role of modern media technology in improving collaborative learning of students in Jordanian universities. Int. J. Inf. Technol. Lang. Stud. **2**(3), 71–82 (2018)
5. Kaplan, A.M., Haenlein, M.: Users of the world, unite! The challenges and opportunities of social media. Bus. Horiz. **53**(1), 59–68 (2010)
6. Duffy, P.: Engaging the YouTube Google-eyed generation: strategies for using web 2.0 in teaching and learning. Electron. J. E-Learn. **6**(2), 119–130 (2008)
7. Cheng, X., Dale, C., Liu, J.: Statistics and social network of YouTube videos. IEEE Int. Work. Qual. Serv. IWQoS, 229–238 (2008)
8. Salloum, S.A., Al-Emran, M., Shaalan, K.: Mining text in news channels: a case study from Facebook. Int. J. Inf. Technol. Lang. Stud. **1**(1), 1–9 (2017)

9. Salloum, S.A., Al-Emran, M., Shaalan, K.: The impact of knowledge sharing on information systems: a review. In: 13th International Conference, KMO 2018 (2018)
10. Abu-Shanab, E., Al-Tarawneh, H.: The influence of social networks on high school students' performance. Int. J. Web-Based Learn. Teach. Technol. **10**(2), 49–59 (2015)
11. Al-Qaysi, N., Al-Emran, M.: Code-switching usage in social media: a case study from Oman. Int. J. Inf. Technol. Lang. Stud. **1**(1), 25–38 (2017)
12. Al-Mohammadi, S.A., Derbel, E.: To whom do we write?: Audience in EFL composition classes. In: Methodologies for Effective Writing Instruction in EFL and ESL Classrooms, pp. 197–208. IGI Global (2015)
13. Mhamdi, C., Al-Emran, M., Salloum, S.A.: Text mining and analytics: a case study from news channels posts on Facebook, vol. 740 (2018)
14. Salloum, S.A., Mhamdi, C., Al-Emran, M., Shaalan, K.: Analysis and classification of arabic newspapers' Facebook pages using text mining techniques. Int. J. Inf. Technol. Lang. Stud. **1** (2), 8–17 (2017)
15. Alshurideh, M., Salloum, S.A., Al Kurdi, B., Al-Emran, M.: Factors affecting the social networks acceptance: an empirical study using PLS-SEM approach. In: 8th International Conference on Software and Computer Applications (2019)
16. Alghizzawi, M.: The role of digital marketing in consumer behavior: a survey. Int. J. Inf. Technol. Lang. Stud. **3**(1), 24–31 (2019)
17. Habes, M., Alghizzawi, M., Khalaf, R., Salloum, S.A., Ghani, M.A.: The relationship between social media and academic performance: facebook perspective. Int. J. Inf. Technol. Lang. Stud. **2**(1), 12–18 (2018)
18. Fralinger, B., Owens, R.: YouTube as a learning tool. J. Coll. Teach. Learn. **6**(8), 15–28 (2009)
19. Lee, D.Y., Lehto, M.R.: User acceptance of YouTube for procedural learning: an extension of the technology acceptance model. Comput. Educ. **61**(1), 193–208 (2013)
20. Raikos, A., Waidyasekara, P.: How useful is YouTube in learning heart anatomy? Anat. Sci. Educ. **7**(1), 12–18 (2014)
21. Orús, C., Barlés, M.J., Belanche, D., Casaló, L., Fraj, E., Gurrea, R.: The effects of learner-generated videos for YouTube on learning outcomes and satisfaction. Comput. Educ. **95**, 254–269 (2016)
22. Habes, M.: The influence of personal motivation on using social TV: a uses and gratifications approach. Int. J. Inf. Technol. Lang. Stud. **3**(1), 32–39 (2019)
23. Colwell, J., Gregory, K.: Exploring how secondary pre-service teachers' use online social bookmarking to envision literacy in the disciplines. Read. Horizons **55**(3), 3 (2016)
24. Gormley, K.A., McDermott, P.: How social bookmarking can help the 21st century teacher. Lang. Lit. Spectr. **20**, 5–14 (2010)
25. Lightfoot, S.: 'Delicious Politics'—the use of social bookmarking in politics teaching. J. Polit. Sci. Educ. **8**(1), 94–101 (2012)
26. Redden, C.S.: Social bookmarking in academic libraries: trends and applications. J. Acad. Librariansh. **36**(3), 219–227 (2010)
27. Derbel, E., Al-Mohammadi, S.A.: Integration of language skills and culture in english language teaching: rationale and implications for practice. In: Issues English Education. Arab world, pp. 216–232 (2015)
28. Nunnally, J.C., Bernstein, I.H.: Psychometric Theory (1978)
29. Chau, C.: YouTube as a participatory culture. New Dir. Youth Dev. **2010**(128), 65–74 (2010)
30. Paolillo, J.C.: Structure and network in the YouTube core. In: Hawaii International Conference on System Sciences, Proceedings of the 41st Annual, p. 156 (2008)

31. Alghizzawi, M., Salloum, S.A., Habes, M.: The role of social media in tourism marketing in Jordan. Int. J. Inf. Technol. Lang. Stud. 2(3) (2018)
32. Khan, M.L.: Social media engagement: what motivates user participation and consumption on YouTube? Comput. Human Behav. **66**, 236–247 (2017)
33. Shachar, M., Neumann, Y.: Twenty years of research on the academic performance differences between traditional and distance learning: summative meta-analysis and trend examination. MERLOT J. Online Learn. Teach. 6(2) (2010)
34. Alwagait, E., Shahzad, B., Alim, S.: Impact of social media usage on students academic performance in Saudi Arabia. Comput. Human Behav. **51**, 1092–1097 (2015)
35. Wongchokprasitti, C., Brusilovsky, P., Parra-Santander, D.: Conference Navigator 2.0: community-based recommendation for academic conferences (2010)
36. Burgess, J., Green, J.: YouTube: Online video and participatory culture. Wiley, Cambridge (2018)
37. Chelaru, S.V., Orellana-Rodriguez, C., Altingovde, I.S.: Can social features help learning to rank Youtube videos? In: International Conference on Web Information Systems Engineering, pp. 552–566 (2012)
38. Salloum, S.A.S., Shaalan, K.: Investigating students' acceptance of e-learning system in higher educational environments in the UAE: applying the extended technology acceptance model (TAM). The British University in Dubai (2018)
39. Alghizzawi, M., et al.: The impact of smartphone adoption on marketing therapeutic tourist sites in Jordan. Int. J. Eng. Technol. **7**(4.34), 91–96 (2018)
40. Salloum, S.A., Al-Emran, M.: Factors affecting the adoption of e-payment systems by university students: extending the TAM with trust. Int. J. Electron. Bus. **14**(4), 371–390 (2018)
41. Mezhuyev, V., Al-Emran, M., Fatehah, M., Hong, N.C.: Factors affecting the metamodelling acceptance: a case study from software development companies in Malaysia. IEEE Access **6**, 49476–49485 (2018)
42. Jeffrey Mingle, D.M.A.: Social media network participation and academic performance in senior high schools in Ghan. Libr. Philos. Pract. (e-journal). Paper 1286, 7–21 (2015)
43. Selwyn, N.: Social media in higher education. Eur. world Learn. **1**, 1–10 (2012)
44. Silius, K., Miilumaki, T., Huhtamaki, J., Tebest, T., Merilainen, J., Pohjolainen, S.: Students' motivations for social media enhanced studying and learning. Knowl. Manag. E-Learn. **2**(1), 51 (2010)
45. Saa, A.A., Al-Emran, M., Shaalan, K.: Factors affecting students' performance in higher education: a systematic review of predictive data mining techniques. Technol. Knowl. Learn. (2019)
46. Friedman, L.W., Friedman, H.: Using social media technologies to enhance online learning. J. Educ. Online **10**(1), 1–22 (2013)
47. Salloum, S.A., Al-Emran, M., Shaalan, K.: Mining social media text: extracting knowledge from Facebook. Int. J. Comput. Digit. Syst. **6**(2), 73–81 (2017)
48. Salloum, S.A., Al-Emran, M., Monem, A.A., Shaalan, K.: A survey of text mining in social media: facebook and twitter perspectives. Adv. Sci. Technol. Eng. Syst. J. **2**(1), 127–133 (2017)
49. Salloum, S.A., AlHamad, A.Q., Al-Emran, M., Shaalan, K.: A survey of Arabic text mining, **740** (2018)
50. Salloum, S.A., Al-Emran, M., Abdallah, S., Shaalan, K.: Analyzing the Arab gulf newspapers using text mining techniques. In: International Conference on Advanced Intelligent Systems and Informatics, pp. 396–405 (2017)

51. Salloum, S.A., Shaalan, K.: Factors affecting students' acceptance of e-learning system in higher education using UTAUT and structural equation modeling approaches. In: International Conference on Advanced Intelligent Systems and Informatics, pp. 469–480 (2018)
52. Greenhow, C., Lewin, C.: Social media and education: reconceptualizing the boundaries of formal and informal learning. Learn. Media Technol. **41**(1), 6–30 (2016)
53. Al-Emran, M., Alkhoudary, Y.A., Mezhuyev, V., Al-Emran, M.: Students and educators attitudes towards the use of m-learning: gender and smartphone ownership differences. Int. J. Interact. Mob. Technol. **13**(1), 127–135 (2019)
54. Al-Emran, M., Shaalan, K.: Academics' awareness towards mobile learning in Oman. Int. J. Comput. Digit. Syst. **6**(1), 45–50 (2017)
55. Al-Emran, M., Mezhuyev, V., Kamaludin, A.: Students' perceptions towards the integration of knowledge management processes in m-learning systems: a preliminary study. Int. J. Eng. Educ. **34**(2), 371–380 (2018)
56. Ringle, C.M., Wende, S., Will, A.: SmartPLS 2.0 (Beta). Hamburg (2005). http://www.smartpls.de
57. Al-Emran, M., Mezhuyev, V., Kamaludin, A.: PLS-SEM in information systems research: a comprehensive methodological reference. In: 4th International Conference on Advanced Intelligent Systems and Informatics (AISI 2018), pp. 644–653 (2018)
58. Hair Jr, J.F., Hult, G.T.M., Ringle, C., Sarstedt, M.: A primer on partial least squares structural equation modeling (PLS-SEM). Sage Publications (2016)
59. Gefen, D., Straub, D., Boudreau, M.-C.: Structural equation modeling and regression: guidelines for research practice. Commun. Assoc. Inf. Syst. **4**(1), 7 (2000)
60. Falk, R.F., Miller, N.B.: A Primer for Soft Modeling. University of Akron Press (1992)
61. Fornell, C., Larcker, D.F.: Evaluating structural equation models with unobservable variables and measurement error. J. Mark. Res. **18**(1), 39–50 (1981)
62. Chin, W.W.: The partial least squares approach to structural equation modeling. Mod. methods Bus. Res. **295**(2), 295–336 (1998)
63. Chang, Y.S., Yang, C.: Why do we blog? From the perspectives of technology acceptance and media choice factors. Behav. Inf. Technol. **32**(4), 371–386 (2013)
64. Al-Rahmi, W., Othman, M.: The impact of social media use on academic performance among university students: a pilot study (2013)

Critical Success Factors for Implementing Artificial Intelligence (AI) Projects in Dubai Government United Arab Emirates (UAE) Health Sector: Applying the Extended Technology Acceptance Model (TAM)

Shaikha F. S. Alhashmi[1] [ID], Said A. Salloum[1,2]([⊠]) [ID], and Sherief Abdallah[1] [ID]

[1] Faculty of Engineering and IT, The British University in Dubai, Dubai, UAE
`shaikha.alattar@hotmail.com, ssalloum@uof.ac.ae,`
`sherief.abdallah@buid.ac.ae`
[2] Faculty of Engineering and IT, University of Fujairah, Fujairah, UAE

Abstract. Recently, the government of United Arab of Emirates (UAE) is focusing on Artificial Intelligence (AI) strategy for future projects that will serve various sectors. Health care sector is one of the significant sectors they are focusing on and the planned (AI) projects of it is aiming to minimize chronic and early prediction of dangerous diseases affecting human beings. Nevertheless, project success depends on the adoption and acceptance by the physicians, nurses, decision makers and patients. The main purpose of this paper is to explore out the critical success factors assist in implementing artificial intelligence projects in the health sector. Besides, the founded gap for this topic was explored as there is no enough sharing of multiple success factors that assist in implementing artificial intelligence projects in the health sector precisely. A modified proposed model for this research was developed by using the extended TAM model and the most widely used factors. Data of this study was collected through survey from employees working in the health and IT sectors in UAE and total number of participants is 53 employees. The outcome of this questionnaire illustrated that managerial, organizational, operational and IT infrastructure factors have a positive impact on (AI) projects perceived ease of use and perceived usefulness.

Keywords: United Arab Emirates · Critical Success Factor (CSF) · Technology Acceptance Model (TAM) · Artificial Intelligence (AI) · Health sector project

1 Introduction

The advancement in artificial intelligence technologies has employed software and system developers to come up with new techniques in improving medical care for patients. The term of artificial intelligence (AI) is a part of advanced technology trends and some of the intelligent tools were used to assist doctors and medical practitioners in

© Springer Nature Switzerland AG 2020
A. E. Hassanien et al. (Eds.): AISI 2019, AISC 1058, pp. 393–405, 2020.
https://doi.org/10.1007/978-3-030-31129-2_36

making decisions for their patients based on their health conditions and history. Furthermore, [1] illustrated a clear initiatives toward artificial intelligence implementation strategy by 2021 that is pushing gradually in the Gulf region, precisely Dubai government in United Arab Emirates (UAE), for the public sector [2]. Great example for Dubai Electricity and Water Authority (DEWA) partners with Dubai start up on (AI) innovative solutions for future labs to implement a pilot (AI) system for distribution system. Exploring out the necessity of government entities to take the initiatives seriously and start their artificial intelligence projects within their environments suiting their business requirements. Artificial Intelligence (AI) can be stated the simulation of different processes of human intelligence by machines, more so computer-related systems [3]. However, [4] has a view that "Artificial intelligence (AI) state the intelligence of machines and the branch of computer science that aims to create it" [4]. While giving its history, [5] refers to artificial intelligence as the idea that inanimate objects are made into intelligent beings that can reason like humans, some of human intelligence processes that are simulated by computer systems include learning, reasoning, problem solving, speech recognition and planning. The purpose of this paper is to bridge the gap for Dubai government health sector to continue improving artificial intelligence practice in providing services for patient monitoring (Dubai Health Authority 2018). It's justifiable that this paper will have its main focus on the critical success factors (CSF) for implementing artificial intelligence projects within the healthcare sector. It's significant to provide the critical success factors that must be under consideration while developing artificial intelligence projects within the health care domain, by investigating theoretical existing literature studies in this paper.

2 Research model and hypotheses

Proposed framework for the modified technology acceptance model in this section to assist in describing the existing research problem in the study and implement a research model.

2.1 Managerial Factors

According to [6], managerial factors refer to the influencers within the organization that impact on the various functioning aspects of the given organization such as adoption of the technology. Moreover, management role is to build up a trust and sets out organizational norms in the work environment. Trust element is important to be among employees in the health sector and it goes with mentality of the individual towards another person. It is considered as a way to illustrate individual's gratitude or faith of having a total belief on certain thing. Scenarios of some individuals developing systems for others confirming their needs and requirements, hence dictating the ability of the subject individual to do or not do regarding the task in the question. For instance, people in the question fail to focus on their own emotional behavior as well as beliefs. Hence, there are two hypotheses that can be formulated as shown:

H1a: Managerial factor has a positive impact on perceived usefulness.
H1b: Managerial factor has a positive impact on ease of use.

2.2   Organizational Factors

Beginning with the organizational factor, it is considered as variables within a firm that allow individuals to either accept or reject any new technology in their environment [7]. Having fit in training programs in an organization, assist individual's to accept advanced technologies and enhance their skills [8]. According to [9], the availability of local expertise within the work environment increases the possibilities of accepting and adopting new technology by the target users. Moreover, the presence of global partnership with companies promotes the acceptability and adaptability of the new technology in organizations. From the findings, it can be said that the organizational factors not only have positive impacts on the perceived usefulness but also on the perceived ease of use of artificial intelligence projects within the healthcare sector. Therefore, the following hypotheses were formulated:

H2a: Organizational factor has a positive impact on perceived usefulness.
H2b: Organizational factor has a positive impact on ease of use.

2.3   Operational Factors

Operational factor reflects the variables that are used to evaluate the choices with concern of the capability to achieve the requirements of desired service [10]. Perceived enjoyment can be described as the method of using a specified system that tends to perceive the action as enjoyable. An association between enjoyment and acceptance of new technology has been analyzed [11, 12]. Also, perceived enjoyment is an important factor that uses to determine the acceptance or adoption of artificial intelligence projects in the healthcare sector. It is therefore understood that perceived enjoyment has a significant effect on the perceived usefulness of artificial intelligence projects in the health sector. Elsewhere, the systematic research conducted also presented in the literature section show that perceived enjoyment has a positive effect on the perceived ease of use of artificial intelligence projects in healthcare. From this finding, two hypotheses can be formulated as follows:

H3a: Operational factor has a positive impact on perceived usefulness.
H3b: Operational factor has a positive impact on ease of use.

2.4   Strategic Factors

According to [13], strategic factors are the way that an organization has to get right for it to succeed with different stakeholders in order to reach the success of the industry. Users' satisfaction in the health domain is the key strategy success while implementing (AI) projects. Satisfaction refers to the level of the person being pleasant with the current situation. Perceived satisfaction can be illustrated as the level of which doctors and project leaders in healthcare get contented with the artificial intelligence projects. According to the previous researches, it was observed that perceived satisfaction is accompanied with the development of artificial intelligence projects in health sector. Precisely, the systemic review conducted shows that perceived satisfaction has an impact on the perceived usefulness of artificial intelligence projects in the healthcare sector. Also, it is highlighted in the data from the systemic review that perceived satisfaction has an impact on the perceived ease of use of artificial intelligence project within the healthcare sector. Therefore, the following two hypotheses can be formulated from this explanation:

H4a: Strategic factor has a positive impact on perceived usefulness.
H4b: Strategic factor has a positive impact on ease of use.

2.5   The Information Technology (IT) infrastructure factor

Described as the variables that dictate the physical systems such as hardware of the systems aspects within the organization [14]. Three system factors are including in IT infrastructure system quality, content quality and information quality. Besides, system quality plays respectively in term of determining the way that feature of the systems include such as availability, usability, adaptability, and reliability of the given technology. It was found that information quality has a vital impact on the perceived ease of use of the artificial intelligence projects in the healthcare sector [15]. Besides, the same systematic review revealed that the information quality significantly impacted on the perceived usefulness of (AI) projects in the healthcare sector. From these assertions, two hypotheses can be formulated concerning the impact of information quality on both perceived ease of use and perceived usefulness of the AI projects in medicine domain.

H5a: Infrastructure factor has a positive impact on perceived usefulness.
H5b: Infrastructure factor has a positive impact on ease of use.

2.6   The constructs of Technology Acceptance Model

TAM model explains the general elements of acceptance that illustrates user's behavior, including two beliefs of Perceived Usefulness (PU) and Perceived Ease of Use (PEU).

2.6.1   Perceived Ease of Use (PEOU)
The perceived ease of use of any given system is defined as the level of technology used having the perception of proper use of the defined technology. Apart from the healthcare sector, there are numerous fields in which the implementation of the artificial intelligence topic has been studied [16, 17]. There are numerous published studies in the existing literature showing that perceived ease of use has a significant relationship with the specific Behavioral Intention to Use either indirectly or in a direct way [18, 19]. Based on the implementation of the artificial intelligence projects in the health sector, perceived ease of use can be said to be the extent to which the physicians view using the artificial intelligence projects will not need a lot of efforts as the task will be easy as [17] explain. From the findings, a hypothesis can be formulated stating that:

H2a1: Perceived ease of use affects positively on the behavioral intention to implement artificial intelligence projects in the healthcare sector.

H2a2: Perceived ease of use affects positively on the perceived usefulness to implement artificial intelligence projects in the healthcare sector.

H2a3: Perceived ease of use affects positively on the attitude towards the implementation of artificial intelligence projects in the healthcare sector.

2.6.2   Perceived Usefulness (PU)

According to [20], perceived usefulness refers to the level at which the users should expect the way using new technology promotes the performance of their jobs. In healthcare, artificial intelligence can be used to improve the performance of the physicians [16]. However, artificial intelligence projects within healthcare will only be accepted or adopted if the physicians perceive that those projects will improve their performance at the job. Also, previous studies confirmed that perceived usefulness has a significant relationship with attitude towards implementation of (AI) projects in the healthcare sector [21].

H2b1: Perceived usefulness affects positively on the attitude towards the implementation of artificial intelligence projects in the healthcare sector.

H2b2: Perceived use ease of use affects positively on the behavioral intention to implement artificial intelligence projects in the healthcare sector.

2.6.3   Attitude Towards Use (ATU)

Attitude refers to the degree to which people depict or rather portray their positive or negative feelings towards something. In this case, the attitude of the physicians will refer to the level in which they portray either positive or negative feelings towards the implementation of the artificial intelligence projects in the healthcare sector. [22] explain that attitude has been significantly associated with behavioral intention. In addition, physicians' attitude affects the acceptance, adoption, and use of artificial intelligence projects in health care sectors. Therefore, the following hypothesis can be formulated:

H2c: Attitude towards use affects positively on the behavioral intention to implement artificial intelligence projects in the healthcare sector.

2.6.4   Behavioral Intention (BI)

Physicians' intention to implement (AI) projects in the healthcare sector can define the proposed or preferred behavioral intention to use in a particular context. Different studies have shown that behavioral intention to use has a direct and significant influence on the actual system use of implementing (AI) projects in the healthcare [23, 24]. With this, the following hypothesis can be formulated:

H2d: the behavioral intention to use effects positively on the actual system used to implement artificial intelligence projects in the healthcare sector.

2.6.5   Actual System Use (ASU)

The actual system use refers to the precise period that the technology system is being utilized in its intended use after being accepted and adopted by the respective subject as captured by [25]. In this case, it means the time that (AI) concepts are put into practice after IT and health staff within heal sector implements the stated projects. However, this construct is dependent on other (TAM) constructs, particularly the behavioral intention to use.

H2e: the actual system use is affected positively by the behavioral intention to use to implement artificial intelligence projects in the healthcare sector.

The following research model is authenticated from the above hypotheses, which was developed in line with the TAM model (Fig. 1).

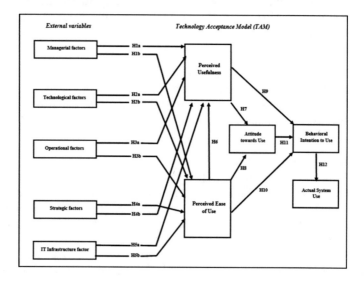

Fig. 1. Research model

3   Research Methodology

The population targeted by this study comprised of both IT and health staff. In total, 53 surveyors were considered to take part in this study. Three health centers in Dubai that physicians selected for the online survey questionnaire were picked from Al Barsha Health Center, Nad Al Hamar Health Center, Al Mamzar Health Centere. The sample population was selected based on the availability of the physicians as the tight schedules of the research participants were put into consideration. It can be said that 74% of the participants were females with only 26% of the respondents being males. Based on ages, 39 respondents were between 25 and 34 which represent 74% of the total population of the IT and health staff that took part in this study. Besides, 1 respondent was between 18 and 24 years while 13 out of the total of 53 participants were between 35 and 45 years old in Fig. 2 captures this information. The level of education was the third prompt that respondents gave their demographic data on. In this dissertation, 13% of participants had a diploma, 38% Bachelor, 45% Master and 4% Doctorate qualifications. On the other hand, the sector in which the respondents worked in based on government (60%), semi-government (25%) and private (15%).

4 Findings and Discussion

4.1 Measurement Model Analysis

The Smart PLS for Partial Least Squares Structural Equation Modeling (PLS-SEM) software was considered appropriate in this study as it has a user interface that is friendly with reporting features that are of advanced level [26]. The study follows the general guidelines of employing PLS-SEM in IS research [27]. Measurement model is responsible for describing the relationship that links indicators to the latent construct. In this particular study, two approaches were employed while evaluating the convergent reliability. The first process was done based on the corresponding constructs of the individual measures. The Partial Least Squares (Smart PLS ver. 3.2.6) was employed to determine the convergent validity. Besides, a pair of analyses was conducted in which the initial PLS operation that used boot trapping approach comprised of 300 resamples was significant in creating loadings, t-values, weights, average variance extracted (AVE) and composite reliabilities for each item of measurement that aligned with its anticipated construct [28, 29]. A relatively higher value was obtained in every measurement item whose loadings were evaluated rather than the 0.70 suggested [30, 31]. As indicated in the table below, it was depicted that loadings for each measurement item were very as compared to the 0.70 value that is recommended. According to [30], 0.70 or a higher value show that the measurement item shares a variance of over 50% with its hypothesized construct. In the table below, composite reliabilities, average variance extracted and subsequent number of each item are depicted. Moreover, the 0.70 composite reliabilities value makes the internal consistency to be satisfactory. A keen look at the table below also shows that the individual values range between 0.777 and 0.922. It is also important to note that the Cronbach's alpha score for each of the TAM's constructs was above 0.70 showing that they all met the reliable measure. According to [30], discriminate validity is used to determine the level at which an individual construct differs from other constructs within any given research model. In this particular study, discriminate validity was conducted with the help of two distinct processes. The analysis of the correlations of the measurements of the latent variable with the measurement items was done. Table 1 below shows the AVE analysis. In this table, the AVE scores' square roots are presented in the bold diagonal table constituents whereas the constructs' correlations are presented in the offload diagonal table constituents. By lying between 0.837 and 0.926, the AVE values' square roots are perceived to be higher as compared to the 0.5 which is a suggested value. According to [31], the average variance extracted is usually higher than other correlations with the model's constructs. In this case, there is a vivid depiction of the greater variance of each construct with its measures instead of other constructs in the (TAM) model, hence resulting to a discriminate validity (Table 2).

Table 1. Convergent validity results which assures acceptable values (Factor loading, Cronbach's Alpha, composite reliability $\geq$ 0.70 & AVE > 0.5).

Constructs	Items	Factor loading	Cronbach's Alpha	CR	AVE
Managerial factors	MF1	0.921	0.873	0.922	0.797
	MF2	0.894			
	MF3	0.862			
Organizational factors	TF1	0.773	0.771	0.797	0.569
	TF2	0.807			
	TF3	0.876			
Operational factors	OF1	0.898	0.800	0.885	0.723
	OF2	0.929			
	OF3	0.706			
Strategic factors	SF1	0.960	0.791	0.828	0.619
	SF2	0.861			
	SF3	0.825			
IT Infrastructure factor	IF1	0.829	0.842	0.905	0.761
	IF2	0.900			
	IF3	0.887			
Perceived Usefulness	PU1	0.874	0.787	0.875	0.701
	PU2	0.759			
	PU3	0.874			
Perceived Ease of Use	PEOU1	0.887	0.864	0.777	0.527
	PEOU2	0.852			
	PEOU3	0.862			
Attitude towards use	AT1	0.772	0.706	0.830	0.619
	AT2	0.777			
	AT3	0.811			
Behavioral intention to use	BI1	0.798	0.821	0.897	0.748
	BI2	0.929			
	BI3	0.945			
Actual System Use	AU1	0.938	0.803	0.909	0.833
	AU2	0.887			

A coefficient of determination or as commonly referred to as 'R^2' is the popular measure that is employed while analyzing the structural model [32]. It is called by the measure upon while determining the predictive accuracy of the (TAM) model. [33] state that this measures are presented as a squared correlation between the specific actual as well as predictive values of the endogenous construct. Besides, this measure shows the level of variance in such constructs as are validated be each related construct. [30] recommends that an R^2 value more than 0.67 is high, between 0.33 and 0.67 to be moderate and between 0.19 and 0.33 should be said to be weak. According to Table 3, the R^2 values for the behavioral of actual system use, attitude towards use, behavioral

Table 2. Fornell-Larcker scale

	AU	AT	BI	IF	MF	OF	PEOU	PU	SF	TF
AU	**0.913**									
AT	0.417	**0.887**								
BI	0.301	0.478	**0.865**							
IF	0.178	0.253	0.246	**0.872**						
MF	0.321	0.222	0.322	0.216	**0.893**					
OF	0.133	0.119	0.250	0.533	0.278	**0.850**				
PEOU	0.129	0.177	0.120	0.665	0.296	0.418	**0.926**			
PU	0.085	0.259	0.128	0.294	0.315	0.497	0.684	**0.837**		
SF	0.040	0.234	0.305	0.449	0.325	0.660	0.373	0.531	**0.887**	
TF	0.165	0.422	0.480	0.501	0.613	0.686	0.419	0.555	0.526	**0.854**

intention to use, perceived ease of use, and perceived usefulness were found to be between 0.33 and 0.67; and hence, the predictive power of these constructs is considered as moderate.

Table 3. R^2 of the endogenous latent variables.

Constructs	R^2	Results
Actual System Use	0.636	Moderate
Attitude towards use	0.519	Moderate
Behavioral intention to use	0.392	Moderate
Perceived Ease of Use	0.422	Moderate
Perceived Usefulness	0.464	Moderate

4.2   Structural Model Analysis

To analyze the various hypothesized associations, the structural equation modeling was used (see Table 4). [34, 35] stated that the values of fit indices that were computed showed that there was the suitable fit of the structural model to the data for the given research model. As per the opinion of [34] this study recommends the intended values of fit indices, there is fitting structural model fit to the data for the research model [28], [36] (see Fig. 2). It can be seen in the Table 4 that all the values were in the given range. In addition to it, few direct hypotheses also showed support [37]. The resulting path coefficients of the suggested research model are shown in Figure 2. Generally, the data supported sixteen out of seventeen hypotheses. All endogenous variables were verified in the model (PU, PEOU, AT, BI, and AU). Based on the data analysis hypotheses H1a, H2a, H3a, H3b, H4a, H5a, H5b, H6, H7, H8, H9, H10, H11, and H12 were supported by the empirical data, while H4b was rejected. The results showed that Perceived Usefulness significantly influenced Managerial factor ($\beta = 0.312$, $P < 0.05$), organizational factors ($\beta = 0.189$, $P < 0.05$), Operational factors supporting ($\beta = 0.147$, $P < 0.05$), Strategic factors ($\beta = 0.206$, $P < 0.05$), IT Infrastructure factor ($\beta =$

0.527, P < 0.001) and Perceived Ease of Use (β = 0.242, P < 0.001), hypothesis H1a, H2a, H3a, H4a, H5a and H6 respectively. Perceived Usefulness and Perceived Ease of Use were determined to be significant in affecting Attitude towards use (β= 0.116, P < 0.001) and (β = 0.256, P < 0.001) supporting hypotheses H7 and H8. Perceived Usefulness and Perceived Ease of Use were determined to be significant in affecting Behavioral intention to use (β = 0.502, P < 0.001) and (β = 0.105, P < 0.01) supporting hypotheses H9 and H10. Furthermore, Perceived Ease of Use was significantly influenced by four exogenous factors: Managerial factor (β = 0.163, P < P < 0.05), organizational factors (β = 0.139, P < 0.05), Operational factors (β = 0.255, P < 0.05), and IT Infrastructure factor (β = −0.605, P < 0.001) which support hypotheses H1b, H2b, H3b, H4b and H5b. The relationship between Strategic factors and Perceived Ease of Use (β = 0.033, P = 0.696) is statistically not significant, and Hypotheses H4b is generally not supported. Finally, the relationship between Attitude towards use and Behavioral intention to use (β = 0.697, P < 0.001) is statistically significant, and Hypotheses H11 is generally supported, and the relationship between Behavioral intention to use and Actual System Use (β = 0.901, P < 0.001) is statistically also significant, and Hypotheses H12 supported. A summary of the hypotheses testing results is shown in Table 4.

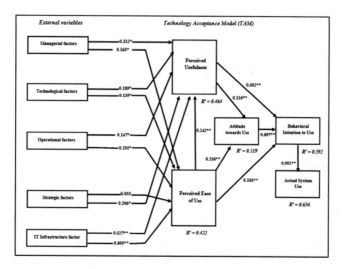

Fig. 2. Path coefficient results (significant at p** < = 0.01, p* < 0.05)

5   Conclusion and future works

Total of 17 hypotheses 16 out of it was supported to this paper and the findings illustrates positive relation between the factors and the variables of the model. Managerial factors, organizational factors, operational factors and IT infrastructure factors demonstrated a positive relationship with the Perceived Ease of Use and Perceived Usefulness. However, the strategic factors illustrated a negative relationship with the

Table 4. Results of structural Model-Research Hypotheses Significant at p** = <0.01, p* < 0.05)

Hyp.	Relationship	Path	t-value	p-value	Direction	Decision
H1a	Managerial factor -> Perceived Usefulness	0.312	3.235	0.015	Positive	Supported*
H1b	Managerial factor -> Perceived Ease of Use	0.163	2.163	0.031	Positive	Supported*
H2a	Organizational factors -> Perceived Usefulness	0.189	2.499	0.013	Positive	Supported*
H2b	Organizational -> Perceived Ease of Use	0.139	4.405	0.035	Positive	Supported*
H3a	Operational factors -> Perceived Usefulness	0.147	1.873	0.042	Positive	Supported*
H3b	Operational factors -> Perceived Ease of Use	0.255	1.604	0.046	Positive	Supported*
H4a	Strategic factors -> Perceived Usefulness	0.206	3.129	0.002	Positive	Supported*
H4b	Strategic factors -> Perceived Ease of Use	0.033	0.391	0.696	Positive	Not supported
H5a	IT Infrastructure factor -> Perceived Usefulness	0.527	7.428	0.000	Positive	Supported**
H5b	IT Infrastructure factor -> Perceived Ease of Use	0.605	7.197	0.000	Positive	Supported**
H6	Perceived Ease of Use -> Perceived Usefulness	0.242	3.268	0.001	Positive	Supported**
H7	Perceived Usefulness -> Attitude towards use	0.116	4.737	0.004	Positive	Supported**
H8	Perceived Ease of Use -> Attitude towards use	0.256	6.883	0.008	Positive	Supported**
H9	Perceived Usefulness -> Behavioral intention to use	0.502	7.036	0.005	Positive	Supported**
H10	Perceived Ease of Use -> Behavioral intention to use	0.105	7.379	0.006	Positive	Supported**
H11	Attitude towards use -> Behavioral intention to use	0.697	17.897	0.000	Positive	Supported**
H12	Behavioral intention to use -> Actual System Use	0.901	77.891	0.000	Positive	Supported**

Perceived Ease of Use. Different adaptions have been left for future research to be examined due to time shortage (interview physicians, doctors and health staff is usually time and effort consuming, requiring long process and procedures. It could be significant if future studies consider only government hospitals across UAE and propose a unique modified model with respect to the physicians practice process with patients. The modified (TAM) model in his dissertation can be constructed by focusing on one critical success factor and study it deeply from different angels.

Acknowledgment. This work is a part of a dissertation submitted in fulfilment of MSc Informatics (Knowledge & Data Management) Faculty of Engineering & Information Technology at The British University in Dubai.

References

1. ValueStrat: How can Artificial intelligence transform the healthcare sector in UAE (2018). Accessed 27 Oct 2018
2. Government.ae, "No Title," UAE Strateg. Artif. Intell. Gov. https://www.government.ae/en/about-the-uae/strategies-initiatives-and-awards/federal-governments-strategies-and-plans/uae-strategy-for-artificial-intelligence. Accessed 2018

3. Mokyr, J.: The past and the future of innovation: some lessons from economic history. Explor. Econ. Hist. **69**, 13–26 (2018)
4. Swarup, "No Title," Artif. Intell. Int. J. Comput. Corp. Res., vol. 2(4) (2012)
5. Mijwel, M.M.: History of artificial intelligence. Comput. Sci. Coll. Sci., 1–6 (2015)
6. Costantino, F., Di Gravio, G., Nonino, F.: Project selection in project portfolio management: an artificial neural network model based on critical success factors. Int. J. Proj. Manag. **33** (8), 1744–1754 (2015)
7. Mezhuyev, V., Al-Emran, M., Fatehah, M., Hong, N.C.: Factors affecting the metamodelling acceptance: a case study from software development companies in Malaysia. IEEE Access **6**, 49476–49485 (2018)
8. Navimipour, N.J., Charband, Y.: Knowledge sharing mechanisms and techniques in project teams: Literature review, classification, and current trends. Comput. Human Behav. **62**, 730–742 (2016)
9. Venkatesh, V., Thong, J.Y.L., Xu, X.: Unified theory of acceptance and use of technology: a synthesis and the road ahead. J. Assoc. Inf. Syst. **17**(5), 328–376 (2016)
10. Christensen, L.B.R., Thomas, G., Calleya, J., Nielsen, U.D.: The effect of operational factors on container ship fuel performance. In: Proceedings of Full Scale Ship Performance, The Royal Institution of Naval Architects (2018)
11. Bennani, A.-E., Oumlil, R.: The Acceptance of ICT by Geriatricians reinforces the value of care for seniors in Morocco. IBIMA Publ. J. African Res. Bus. Technol. J. African Res. Bus. Technol. **2014**, 1–10 (2014)
12. Who, X.: Extending TAM: success factors of mobile marketing'. Am. Acad. Sch. Res. J. **1** (1), 1–5 (2011)
13. Zare, S.: Identifying and Prioritizing Supply Chain Management Strategic Factors Based on Integrated BSC-AHP Approach (2017)
14. Dahiya, D., Mathew, S.K.: IT assets, IT infrastructure performance and IT capability: a framework for e-government. Transform. Gov. People, Process Policy **10**(3), 411–433 (2016)
15. Safdari, R., Saeedi, M.G., Valinejadi, A., Bouraghi, H., Shahnavazi, H.: Technology acceptance model in health care centers of Iran. Int. J. Comput. Sci. Netw. Secur. **17**(1), 42 (2017)
16. Alloghani, M., Hussain, A., Al-Jumeily, D., Abuelma'atti, O.: Technology acceptance model for the use of M-Health services among health related users in UAE. In: 2015 International Conference on Developments of E-Systems Engineering (DeSE), pp. 213–217 (2015)
17. Phatthana, W., Mat, N.K.N.: The Application of Technology Acceptance Model (TAM) on health tourism e-purchase intention predictors in Thailand. In: 2010 International Conference on Business and Economics Research, vol. 1, pp. 196–199 (2011)
18. Salloum, S.A., Al-Emran, M.: Factors affecting the adoption of E-payment systems by university students: extending the TAM with trust. Int. J. Electron. Bus. **14**(4), 371–390 (2018)
19. Alshurideh, M., Salloum, S.A., Al Kurdi, B., Al-Emran, M.: Factors affecting the social networks acceptance: an empirical study using PLS-SEM approach. In: 8th International Conference on Software and Computer Applications (2019)
20. Alharbi, S., Drew, S.: Using the technology acceptance model in understanding academics' behavioural intention to use learning management systems. Int. J. Adv. Comput. Sci. Appl. **5** (1), 143–155 (2014)
21. Emad, H., El-Bakry, H.M., Asem, A.: A modified technology acceptance model for health informatics (2016)

22. Baharom, F., Khorma, O.T., Mohd, H., Bashayreh, M.G.: Developing an extended technology acceptance model: doctors' acceptance of electronic medical records in Jordan. In: ICOCI (2011)
23. Fayad, R., Paper, D.: The technology acceptance model e-commerce extension: a conceptual framework. Procedia Econ. Financ. **26**, 1000–1006 (2015)
24. Helia, V.N., Indira Asri, V., Kusrini, E., Miranda, E.: Modified technology acceptance model for hospital information system evaluation–a case study (2018)
25. Teeroovengadum, V., Heeraman, N., Jugurnath, B.: Examining the antecedents of ICT adoption in education using an extended technology acceptance model (TAM). Int. J. Educ. Dev. using ICT **13**(3), 4–23 (2017)
26. Ringle, C.M., Wende, S., Will, A.: SmartPLS 2.0 (Beta). Hamburg (2005). http://www.smartpls.de
27. Al-Emran, M., Mezhuyev, V., Kamaludin, A.: PLS-SEM in information systems research: a comprehensive methodological reference. In: 4th International Conference on Advanced Intelligent Systems and Informatics (AISI 2018), pp. 644–653 (2018)
28. Salloum, S.A.S., Shaalan, K.: Investigating students' acceptance of E-learning system in higher educational environments in the UAE: applying the extended Technology Acceptance Model (TAM). The British University in Dubai (2018)
29. Salloum, S.A., Al-Emran, M., Shaalan, K., Tarhini, A.: Factors affecting the E-learning acceptance: a case study from UAE. Educ. Inf. Technol. **24**(1), 509–530 (2019)
30. Chin, W.W.: The partial least squares approach to structural equation modeling. Mod. Methods Bus. Res. **295**(2), 295–336 (1998)
31. Hair Jr., J.F., Hult, G.T.M., Ringle, C., Sarstedt, M.: A primer on partial least squares structural equation modeling (PLS-SEM). Sage Publications, Thousand Oaks (2016)
32. Dreheeb, A.E., Basir, N., Fabil, N.: Impact of system quality on users' satisfaction in continuation of the use of e-Learning system. Int. J. e-Education, e-Business, e-Management e-Learning **6**(1), 13 (2016)
33. Senapathi, M., Srinivasan, A.: An empirical investigation of the factors affecting agile usage. In: Proceedings of the 18th International Conference on Evaluation and Assessment in Software Engineering, p. 10 (2014)
34. Milošević, I., Živković, D., Manasijević, D., Nikolić, D.: The effects of the intended behavior of students in the use of M-learning. Comput. Hum. Behav. **51**, 207–215 (2015)
35. Al-Emran, M., Salloum, S.A.: Students' attitudes towards the use of mobile technologies in e-Evaluation. Int. J. Interact. Mob. Technol. **11**(5), 195–202 (2017)
36. Salloum, S.A., Al-Emran, M., Shaalan, K., Tarhini, A.: Factors affecting the E-learning acceptance: a case study from UAE. Educ. Inf. Technol. **24**, 1–22 (2018)
37. Ma, W., Yuen, A.: 11. E-learning system acceptance and usage pattern. Technol. Accept. Educ. Res. Issues, 201 (2011)

Examining the Main Mobile Learning System Drivers' Effects: A Mix Empirical Examination of Both the Expectation-Confirmation Model (ECM) and the Technology Acceptance Model (TAM)

Muhammad Alshurideh[1,2] [iD], Barween Al Kurdi[3] [iD],
and Said A. Salloum[2,4(✉)] [iD]

[1] Marketing Department, School of Business, The University of Jordan,
Amman, Jordan
m.shurideh@uof.ac.ae
[2] Faculty of Business, University of Sharjah, Sharjah, UAE
{m.shurideh, ssalloum}@uof.ac.ae
[3] Marketing Department, Amman Arab University, Amman, Jordan
balkurdi@aau.edu.jo
[4] Faculty of Engineering and IT, The British University in Dubai, Dubai, UAE

Abstract. This study aims to investigate the intention to use and actual use of Mobile Learning System (MLS) drivers by students within the UAE higher education setting. A set of factors were chosen to study and test the issue at hand. These factors are social influence, expectation-confirmation, perceived ease of use, perceived usefulness, satisfaction, continuous intention and finally the actual use of such MLS. This study adds more light to the MLS context because it combines between two models which are the Information Technology Acceptance Model (TAM) and Expectation-Confirmation Model (ECM). A set of hypotheses were developed based on such theoretical combination. The data collected from 448 students for the seek of primary data and analyzed using the Structural Equation Modeling (SEM) in particular (SmartPLS) to evaluate the developed study model and test the prepared hypotheses. The study found that both social influence and expectation-confirmation factors influence positively perceived ease of use, perceived usefulness and satisfaction and such three drivers influence positively students' intention to use MLS. Based on previous proposed links, the study confirms that intention to use such mobile educational means affect strongly and positively the actual use. Scholars and practitioners should take care of learners' intention to use and actual use of MLS and their determinants into more investigation especially the social influence and reference group ones within the educational setting. A set of limitation and future research venues were mentioned in details also.

Keywords: Mobile Learning System (MLS) ·
Technology Acceptance Model (TAM) · Structural Equation Modeling (SEM) ·
Expectation-Confirmation Model (ECM)

© Springer Nature Switzerland AG 2020
A. E. Hassanien et al. (Eds.): AISI 2019, AISC 1058, pp. 406–417, 2020.
https://doi.org/10.1007/978-3-030-31129-2_37

1   Introduction

Students nowadays prefer adapting and taking electronic educational courses because learners need flexible learning systems such as m-learning or e-learning that suite their needs [1, 2]. Adapting such systems also help students to choose when and where to study easily [3]. Nowadays, some of the new teaching and learning technologies have shown up lately reduce the possibility to adapt the university classical education approaches and give much opportunity for those who have full-time job to study and learn [4]. This study aimed at providing a comprehensive approach to adapt and use the m-learning systems by students. This study is essential because it is from the initial studies that combine between two models which are the Information Technology Acceptance Model (TAM) and Expectation-Confirmation Model (ECM). The TAM model has been used widely in m-learning context [5]. However, ECM has rarely use to study students' intention to use m-learning. According to [6], The ECM is a theoretical psychological framework in marketing and specifically in consumer behavior that gives a clear approach to explain how consumer intention to buy a product or/and service is shaped by comparing both the initial (pre-purchase) and later (post-purchase) expectation. Such model is important to be used to explain when studying a consumer or user's intention and continued intention to use of e-learning according to [7]. However, using such model in m-learning, which the situation in this study, will be add value to the knowledge. Such model has been used in studying a consumer or a user in different settings such as web-customer satisfaction [8], mobile internet context [9], e-Learning continuance intention [10], continuance intention to use mobile Apps [11], and mobile payment systems use [12]. Thus, using such model in m-learning is rarely used.

2   Research Model and Hypotheses

2.1   Social Influence Effect on Both Perceived Ease of Use and Perceived Usefulness

Many studies such as [13–16] denoted that flexibility in time, place and access push students to adapt and use different educational approaches such as self-direct systems, mobile-learning and e-learning and the scale of adapting such approaches have been increased and spread worldwide in a large scale especially for students who prefer taking distance learning courses. A study by [17] explored the social influence effect on perceived usefulness and ease of use. The study denoted that both mentor and instructor influence significantly the students' perception the usefulness of online course delivery systems and just mentor influence the students' perception ease of use of such e-learning systems. Such findings are also confirmed by [18] who declared that social influence clearly affect both the users' perception of usefulness and ease of use. To add more, [19] studied the influence of both perceived ease of use and perceived usefulness on information technology use. The study found that there is a valid and significant effect for both perceived both usefulness and ease of use of information technology application. The relationship of social influence and both usefulness and ease of use can be proposed as:

H1: There is a positive influence of users' social influence on users' perceived ease of mobile-learning systems use.

H2: There is a positive influence of users' social influence on users' perceived usefulness of mobile-learning systems use.

2.2 Expectation-Confirmation Effect on Both Perceived Usefulness and Satisfaction

The effect of satisfaction and e-satisfaction on behavior and repeat purchase behavior has been widely researched and practically proofed in literature [20–26]. Also, the expectation-confirmation approach proposed that users' level of satisfaction with a service and/or a product shape their repeat purchase intention [9]. By another mean, users' level of satisfaction is not just influenced by their initial expectation (before purchase) but also by the difference between their prior expectation and actual perceived performance. This highlight the idea that users and/or consumers confirm or disconfirm their initial expectation of an object used/consumed after they perceived the usefulness of such object supported by being satisfied. Many scholars such as [27–29] highlighted that users or consumers' experience moderate such relationship. For example, [6] studied how to understanding Information Systems (IS) continuance using the ECM. The goal of this paper is to examine how both cognitive beliefs and affect influence the online banking users' intention to use the continuous IS. The study found that users' continuous intention is determined by the users' satisfaction and perceived usefulness and users' satisfaction is influenced by their confirmation of expectation from prior use of IS in addition to perceived usefulness. This study also highlighted the idea that users' post-acceptance of perceived usefulness is influenced clearly by users' confirmation level. The relationship between expectation-confirmation and both perceived usefulness and satisfaction can be proposed as:

H3: There is a positive influence of users' expectation-confirmation on users' perceived usefulness of mobile-learning systems use

H4: There is a positive influence of users' expectation-confirmation on users' satisfaction of mobile-learning system use.

2.3 Perceived Ease of Use, Perceived Usefulness, Satisfaction and Continuous Intention

Developing e-learning programs has gaining the interests of educational institutions, training institutions and even governments at different levels of users' applications. Such interest makes such learning programs on high demands [30]. Lee (2010) studied how to predict users' continuous intention to use e-learning employing the ECM, the (TAM) and the Theory of Planned Behavior (TPB) from data collected from learners used a Web-based learning program designed for continuing education. The study found that users' continuance intention is influenced heavily by users' satisfaction followed by usefulness, concentration, attitude, perceived behavior and subjective norms.

[31] studied if customers prefer using Self-Service Technologies (SSTs) to offer their own service encounters via the interaction of electronic service interfaces or machines instead of interacting with a firms' services personally. The study designed a model to predict if customers prefer using (SSTs) using a set of combined models which are the TPB, TAM and the concepts of Technology Readiness (TR). A sample of 481 users were used to collect the required data which analyzed using Structural Equation Modeling method to test the developed model and proposed hypotheses. The study mainly found that consumers' continuance intention is significantly influenced by their satisfaction and such satisfaction is influenced clearly by perceived ease of use, perceived usefulness, perceived behavioral control and subjective norm concurrently. Moreover, in a study using the TAM model prepared by [32] to understand why customers reluctant to use voluntarily the mobile banking systems while millions of dollars have been spent to prepare and install such systems. This study added value to the knowledge by extends the TAM's applications by adding additional variables which are trust-based construct 'perceived credibility' and both "perceived self-efficacy' and 'perceived financial cost' as two resource-based constructs to the model. The study results denoted that the extended TAM model and its main elements (perceived usefulness, perceived ease of use and perceived usefulness) were able to predict mobile users' intention to use mobile banking systems. Thus, the relation of perceived ease of use, perceived usefulness, satisfaction and continuous intention can be proposed as:

H5: There is a positive influence of users' perceived ease of use on continuous intention of mobile-learning systems use
H6: There is a positive influence of users' perceived usefulness on continuous intention of mobile-learning systems use
H7: There is a positive influence of users' satisfaction on continuous intention of mobile-learning systems use.

2.4   Continuous Intention Effect on Actual Use

Mobile applications become part of unlimited number of banking and educational activities. Thus, the mobile technology adaption has taken both scholars and practitioners' interests. The mobile users' intention is playing a critical role on adapting, actual use and repeat use of different mobile technologies and applications [33, 34]. This idea is confirmed by [35–37] who found that there is a set of key factor that influence heavily the continuance intention of mobile application use. Part of these factors are the easiness, information seeking motivation, task performance enhancement, others' opinions, social connection motivation and entertainment motivation. Mobile payment service is also found important to be highlighted in this situation. [38] studied the key indicators of factors affecting continuance intention of using the mobile payment. The scholar found that a set of factors that determine the continuance intention of mobile payment. The study findings denoted that trust and system quality affect mobile users' satisfaction and also both information quality and service quality facilitate users' continuance usage of mobile payment technologies. The relation between continuous intention and actual use can be proposed as:

H8: There is a positive influence of users' continuous intention on their actual use of mobile-learning systems.

Based on the above explanation, the study model can be drawn as seen in Fig. 1 below.

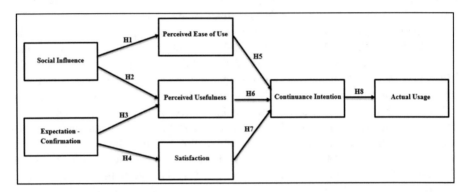

Fig. 1. Proposed research model

3   Research Methodology

Between 15 October 2018 and 20 December 2018, hard-copy surveys were distributed amongst the students in the United Arab Emirates to acquire the data during the fall semester 2018/2019. The researchers have randomly given 500 questionnaires, out of which, the respondents had returned 448 questionnaires, which constituted a response rate of 90 percent. Moreover, the research team had rejected 52 of these filled questionnaires because they found missing values in them. Consequently, they had evaluated 448 correctly filled and helpful questionnaires. Most importantly, the current theories were deriving the Hypotheses; however, it was also accustomed to the M-learning context. To evaluate measurement model, researchers employed the Structural equation modeling (SEM) (SmartPLS Version 3.2.7) by following the general guidelines of employing PLS-SEM in IS domain [39]. The model was eventually handled with final path model.

The male students made up the 49% of the population, while the females were found to be 51% of the population. For 56% of the respondents, student's age was found between 18 and 29, while 44% of the respondents are above 29 years of age. As far as the academic background is concerned, 44% of the students pertained to Business Administration major, whereas students in Engineering, Mass Communication and Public Relations, Arts, Social Sciences & Humanities, and Information Technology were 20%, 15%, 11%, and 10% respectively. Most of the respondents were well educated and they were graduates having university degrees. The bachelor degree was owned by 56% of the individuals, while master degree was acquired by 20%. Moreover, a doctoral degree was finished by 19% of respondents and the remaining ones had diploma education. According to the prior research in the field, this study employs the "convenience sampling approach" [40–44]. This approach needs to be employed when the respondents show their tendency to be volunteer and that they are obtainable without any trouble. The students of different ages from different colleges; registered in different programs at different levels have constituted this study sample. The demographic data was measured by employing the IBM SPSS Statistics ver. 23.

4   Findings and Discussion

4.1   Measurement Model Analysis

In this ponder, the data analysis was embraced utilizing the partial least squares-structural equation modeling through SmartPLS program [45]. The collected data were analyzed employing a two-step assessment approach counting the measurement model and structural model [46]. The choice of PLS-SEM in this considers is ascribed to a few reasons. To begin with, when the investigate beneath consider points to create an existing theory, PLS-SEM is the most excellent choice [47]. Moment, PLS-SEM can best serve exploratory considers that include complex models [48]. Third, PLS-SEM analyzes the total model as one unit, instead of part it into pieces [49]. Fourth, PLS-SEM gives concurrent analysis for both measurement and structural model, which in turn, leads to more exact estimations [50]. [46] recommended evaluating the construct reliability (counting Cronbach's alpha and composite reliability) and validity (counting convergent and discriminant validity) for assessing the measurement model. In arrange to degree the construct reliability, the comes about in Table 1 uncover that the values of Cronbach's alpha are extended between 0.731 and 0.891 which were all over the threshold esteem of 0.7 [51]. The comes about in Table 1 moreover show that the values of composite reliability (CR) are extended between 0.758 and 0.971 which were all over the suggested esteem of 0.7 [52]. In line with these comes about, the construct reliability is affirmed, and all the constructs were respected to be enough error-free. For the measurement of convergent validity, the factor loading, and average variance extracted (AVE) are proposed to be tired [46]. The comes about in Table 1 indicate that the values of all factor loadings are higher than the proposed esteem of 0.7. Besides, the comes about in Table 1 uncover that the values of AVE are extended between 0.850 and 0.922 which were all over the threshold esteem of 0.5. Given these comes about at hand, the convergent validity for all constructs has been adequately satisfied.

The coefficient of determination (R^2 value) measure is usually used to assess the structural model [53]. The predictive accuracy of the model is determined through the coefficient, which is used as the squared correlation between the actual and predicted values of a given endogenous construct [54]. The coefficient shows the combined impact of the exogenous latent variables on an endogenous latent variable. The coefficient represents the squared correlation between the actual and predicted values of the variables; therefore, it also shows the extent of variance in the endogenous constructs explained by every exogenous construct determined with it. According to [55], when the value exceeds 0.67, it is considered high, even though the qualities in the range of 0.33 to 0.67 are direct, whereas the qualities in the range of 0.19 to 0.33 are deemed to

Table 1. Internal consistency reliability and convergent validity of the measurement model

Constructs	Items	Factor loading	Cronbach's Alpha	CR	AVE
Actual usage	AU1	0.928	0.852	0.856	0.931
	AU2	0.938			
Continuance intention	CI1	0.889	0.849	0.877	0.906
	IC2	0.871			
	CI3	0.858			
Expectation-confirmation	EXP1	0.890	0.891	0.971	0.922
	EXP2	0.933			
	EXP3	0.952			
	EXP4	0.954			
Perceived ease of use	PEOU1	0.909	0.831	0.848	0.899
	PEOU2	0.799			
	PEOU3	0.883			
Perceived usefulness	PU1	0.696	0.731	0.758	0.850
	PU2	0.906			
	PU3	0.814			
Satisfaction	SAT1	0.857	0.787	0.789	0.877
	SAT2	0.905			
	SAT3	0.750			
Social influence	SI1	0.904	0.837	0.849	0.890
	SI2	0.826			
	SI3	0.798			
	SI4	0.742			

be weak values [56]. Furthermore, when the values are less than 0.18, they are considered to be inadmissible. In Table 2 it can be seen that the model has moderate predictive power, which supports almost 33%, 40%, 44%, 49 and 45% of the variance in the Actual Usage, Continuance Intention, Perceived Ease of Use, Perceived Usefulness, and Satisfaction respectively.

Table 2. R^2 of the endogenous latent variables.

Constructs	R^2	Results
Actual Usage	0.327	Moderate
Continuance Intention	0.394	Moderate
Perceived Ease of Use	0.437	Moderate
Perceived Usefulness	0.490	Moderate
Satisfaction	0.446	Moderate

4.2 Structural Model Analysis

The A structural equation model utilizing SEM-PLS with the greatest probability estimation was used to test the proposed hypothesis by examining the correlations between the theoretical constructs for the structural model. The outcomes are demonstrated in Table 2 and Fig. 2, and it is clear that 4 out of 5 hypotheses are found to be significant [57, 58]. It can be seen that all hypotheses are found to be significant. Based on the data analysis hypotheses H1, H2, H3, H4, H5, H6, H7 and H8 were supported by the empirical data. The results showed that Perceived Usefulness significantly influenced Social Influence ($\beta = 0.075$, P < 0.01), Expectation-Confirmation ($\beta = 0.578$, P < 0.001), supporting hypothesis H2 and H3 respectively. Continuance Intention was determined to be significant in affecting Perceived Ease of Use ($\beta = 0.363$, P < 0.01), Perceived Usefulness ($\beta = 0.477$, P < 0.001), and Satisfaction ($\beta = 0.396$, P < 0.001) supporting hypotheses H5, H6, and H7 respectively. Finally, Social Influence significantly influenced Perceived Ease of Use ($\beta = 0.460$, P < 0.001), Expectation-Confirmation significantly influenced Satisfaction ($\beta = 0.458$, P < 0.001), and, Continuance Intention significantly influenced Actual Usage ($\beta = 0.852$, P < 0.001), supporting hypothesis H1, H4 and H8 respectively. A summary of the hypotheses testing results is shown in Table 3.

Table 3. Results of hypothesis relationship path

Hyp.	Relationship	Path	t-value	p-value	Direction	Decision
H1	Social Influence -> Perceived Ease of Use	0.460	11.079	0.000	Positive	Supported**
H2	Social Influence -> Perceived Usefulness	0.075	2.274	0.013	Positive	Supported*
H3	Expectation -Confirmation -> Perceived Usefulness	0.578	14.434	0.000	Positive	Supported**
H4	Expectation -Confirmation -> Satisfaction	0.458	9.952	0.000	Positive	Supported**
H5	Perceived Ease of Use -> Continuance Intention	0.363	4.552	0.021	Positive	Supported**
H6	Perceived Usefulness -> Continuance Intention	0.477	8.598	0.000	Positive	Supported**
H7	Satisfaction -> Continuance Intention	0.396	6.663	0.000	Positive	Supported**
H8	Continuance Intention -> Actual Usage	0.852	58.592	0.000	Positive	Supported**

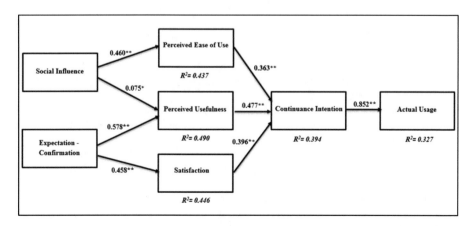

Fig. 2. Path diagram of the relationships among the dimensions

5   Conclusion and Future Works

The determining factors affecting the UAE university students' intention to use and actual use of mobile-learning systems has been considered in this paper. The study model has been proposed by constructing both the ECM and the TAM to study the mobile-learning adoption issue. The findings were indicative of the fact that the important indicators of students' intention to utilize mobile-learning system are non-other than the perceived usefulness, perceived ease of use, and satisfaction of university platform technology use. The developers must possess sound knowledge about the factors influencing such adoption. Since, due to this knowledge, they can design and develop online mobile-learning systems those consistent with the requirements of their students.

References

1. Al-Emran, M., Alkhoudary, Y.A., Mezhuyev, V., Al-Emran, M.: Students and educators attitudes towards the use of M-Learning: gender and smartphone ownership differences. Int. J. Interact. Mob. Technol. **13**(1), 127–135 (2019)
2. Salloum, S.A., Al-Emran, M., Shaalan, K., Tarhini, A.: Factors affecting the E-learning acceptance: a case study from UAE. Educ. Inf. Technol. **24**(1), 509–530 (2019)
3. Al-Emran, M., Mezhuyev, V., Kamaludin, A., Al Sinani, M.: Development of M-learning application based on knowledge management processes. In: 2018 7th International conference on Software and Computer Applications (ICSCA 2018), pp. 248–253 (2018)
4. Sife, A., Lwoga, E., Sanga, C.: New technologies for teaching and learning: challenges for higher learning institutions in developing countries. Int. J. Educ. Dev. ICT **3**(2), 57–67 (2007)
5. Al-Emran, M., Mezhuyev, V., Kamaludin, A.: Technology acceptance model in M-learning context: a systematic review. Comput. Educ. **125**, 389–412 (2018)
6. Bhattacherjee, A.: Understanding information systems continuance: an expectation-confirmation model. MIS Q. **25**, 351–370 (2001)
7. Lin, C.S., Wu, S., Tsai, R.J.: Integrating perceived playfulness into expectation-confirmation model for web portal context. Inf. Manag. **42**(5), 683–693 (2005)
8. McKinney, V., Yoon, K., "Mariam" Zahedi, F.: The measurement of web-customer satisfaction: an expectation and disconfirmation approach. Inf. Syst. Res. **13**(3), 296–315 (2002)
9. Thong, J.Y.L., Hong, S.-J., Tam, K.Y.: The effects of post-adoption beliefs on the expectation-confirmation model for information technology continuance. Int. J. Hum. Comput. Stud. **64**(9), 799–810 (2006)
10. Chou, H.K., Lin, I.C., Woung, L.C., Tsai, M.T.: Engagement in e-learning opportunities: an empirical study on patient education using expectation confirmation theory. J. Med. Syst. **36**(3), 1697–1706 (2012)
11. Tam, C., Santos, D., Oliveira, T.: Exploring the influential factors of continuance intention to use mobile Apps: extending the expectation confirmation model. Inf. Syst. Front. **21**, 1–15 (2018)
12. Almazroa, M., Gulliver, S.: Understanding the usage of mobile payment systems-The impact of personality on the continuance usage. In: 2018 4th International Conference on Information Management (ICIM), pp. 188–194 (2018)

13. Keller, C., Cernerud, L.: Students' perceptions of e-learning in university education. J. Educ. Media **27**(1–2), 55–67 (2002)
14. Koper, R.: Use of the semantic web to solve some basic problems in education: increase flexible, distributed lifelong learning; decrease teacher's workload. J. Interact. Media Educ. **1**, 2004 (2004)
15. Bates, A.W.T.: Technology, E-learning and Distance Education. Routledge, London (2005)
16. Pituch, K.A., Lee, Y.: The influence of system characteristics on e-learning use. Comput. Educ. **47**(2), 222–244 (2006)
17. Shen, D., Laffey, J., Lin, Y., Huang, X.: Social influence for perceived usefulness and ease-of-use of course delivery systems. J. Interact. Online Learn. **5**(3), 270–282 (2006)
18. Karahanna, E., Straub, D.W.: The psychological origins of perceived usefulness and ease-of-use. Inf. Manag. **35**(4), 237–250 (1999)
19. Adams, D.A., Nelson, R.R., Todd, P.A.: Perceived usefulness, ease of use, and usage of information technology: a replication. MIS Q. **16**, 227–247 (1992)
20. Alshurideh, M., Masa'deh, R., Alkurdi, B.: The effect of customer satisfaction upon customer retention in the Jordanian mobile market: an empirical investigation. Eur. J. Econ. Financ. Adm. Sci. **47**, 69–78 (2012)
21. Al-dweeri, R.M., Obeidat, Z.M., Al-dwiry, M.A., Alshurideh, M.T., Alhorani, A.M.: The impact of e-service quality and e-loyalty on online shopping: moderating effect of e-satisfaction and e-trust. Int. J. Mark. Stud. **9**(2), 92 (2017)
22. Alshurideh, M.: Customer Service Retention–A Behavioural Perspective of the UK Mobile Market. Durham University (2010)
23. Alshurideh, M.: Scope of customer retention problem in the mobile phone sector: a theoretical perspective. J. Mark. Consum. Res. **20**, 64–69 (2016)
24. Alshurideh, M.: The factors predicting students' satisfaction with universities' healthcare clinics' services: a case-study from the Jordanian Higher Education Sector. Dirasat Adm. Sci. **41**(2), 451–464 (2014)
25. Alshurideh, M.T.: Exploring the main factors affecting consumer choice of mobile phone service provider contracts. Int. J. Commun. Netw. Syst. Sci. **9**(12), 563–581 (2016)
26. Alshurideh, D.M.: Do electronic loyalty programs still drive customer choice and repeat purchase behaviour? Int. J. Electron. Cust. Relatsh. Manag. **12**(1), 40–57 (2019)
27. Alshurideh, M.: A behavior perspective of mobile customer retention: an exploratory study in the UK Market. The End of the Pier? Competing perspectives on the challenges facing business and management British Academy of Management Brighton–UK. Br. Acad. Manag., 1–19 (2010)
28. Alshurideh, M., Nicholson, M., Xiao, S.: The effect of previous experience on mobile subscribers' repeat purchase behaviour. Eur. J. Soc. Sci. **30**(3), 366–376 (2012)
29. Alshurideh, M., Bataineh, A., Alkurdi, B., Alasmr, N.: Factors affect mobile phone brand choices-studying the case of Jordan Universities students. Int. Bus. Res. **8**(3), 141 (2015)
30. Gunasekaran, A., McNeil, R.D., Shaul, D.: E-learning: research and applications. Ind. Commer. Train. **34**(2), 44–53 (2002)
31. Chen, S.-C., Chen, H.-H., Chen, M.-F.: Determinants of satisfaction and continuance intention towards self-service technologies. Ind. Manag. Data Syst. **109**(9), 1248–1263 (2009)
32. Luarn, P., Lin, H.-H.: Toward an understanding of the behavioral intention to use mobile banking. Comput. Hum. Behav. **21**(6), 873–891 (2005)
33. Chung, N., Kwon, S.J.: The effects of customers' mobile experience and technical support on the intention to use mobile banking. Cyberpsychology Behav. **12**(5), 539–543 (2009)

34. Al-Dmour, H., Al-Shraideh, M.T.: The influence of the promotional mix elements on Jordanian consumer's decisions in cell phone service usage: an analytical study. Jordan J. Bus. Adm. **4**(4), 375–392 (2008)

35. Kang, S.: Factors influencing intention of mobile application use. Int. J. Mob. Commun. **12** (4), 360–379 (2014)

36. Al Dmour, H., Alshurideh, M., Shishan, F.: The influence of mobile application quality and attributes on the continuance intention of mobile shopping. Life Sci. J. **11**(10), 172–181 (2014)

37. Alshurideh, M.T.: A theoretical perspective of contract and contractual customer-supplier relationship in the mobile phone service sector. Int. J. Bus. Manag. **12**(7), 201 (2017)

38. Zhou, T.: An empirical examination of continuance intention of mobile payment services. Decis. Support Syst. **54**(2), 1085–1091 (2013)

39. Al-Emran, M., Mezhuyev, V., Kamaludin, A.: PLS-SEM in information systems research: a comprehensive methodological reference. In: 4th International Conference on Advanced Intelligent Systems and Informatics (AISI 2018), pp. 644–653 (2018)

40. Al-Emran, M., Salloum, S.A.: Students' attitudes towards the use of mobile technologies in e-evaluation. Int. J. Interact. Mob. Technol. **11**(5), 195–202 (2017)

41. Al-Emran, M., Mezhuyev, V., Kamaludin, A.: Students' perceptions towards the Integration of knowledge management processes in M-learning systems: a preliminary study. Int. J. Eng. Educ. **34**(2), 371–380 (2018)

42. Al-Qaysi, N., Al-Emran, M.: Code-switching usage in social media: a case study from oman. Int. J. Inf. Technol. Lang. Stud. **1**(1), 25–38 (2017)

43. Salloum, S.A., Maqableh, W., Mhamdi, C., Al Kurdi, B., Shaalan, K.: Studying the social media adoption by university students in the United Arab Emirates. Int. J. Inf. Technol. Lang. Stud. **2**(3), 83–95 (2018)

44. Alshurideh, M., Salloum, S.A., Al Kurdi, B., Al-Emran, M.: Factors affecting the social networks acceptance: an empirical study using PLS-SEM approach. In: 8th International Conference on Software and Computer Applications (2019)

45. Ringle, C.M., Wende, S., Becker, J.-M.: SmartPLS 3. Bönningstedt: SmartPLS (2015)

46. Hair, J., Hollingsworth, C.L., Randolph, A.B., Chong, A.Y.L.: An updated and expanded assessment of PLS-SEM in information systems research. Ind. Manag. Data Syst. **117**(3), 442–458 (2017)

47. Urbach, N., Ahlemann, F.: Structural equation modeling in information systems research using partial least squares. J. Inf. Technol. theory Appl. **11**(2), 5–40 (2010)

48. Hair Jr., J.F., Hult, G.T.M., Ringle, C., Sarstedt, M.: A Primer on Partial Least Squares Structural Equation Modeling (PLS-SEM). Sage Publications, Thousand Oaks (2016)

49. Goodhue, D.L., Lewis, W., Thompson, R.: Does PLS have advantages for small sample size or non-normal data?. MIS Q. **36**(3) (2012)

50. Barclay, D., Higgins, C., Thompson, R.: The Partial Least Squares (PLS) approach to casual modeling: personal computer adoption and use as an illustration (1995)

51. Nunnally, J.C., Bernstein, I.H.: Psychometric theory (1994)

52. Kline, R.B.: Principles and practice of structural equation modeling (2015)

53. Lin, Y.-C., Chen, Y.-C., Yeh, R.C.: Understanding college students' continuing intentions to use multimedia e-learning systems. World Trans. Eng. Technol. Educ. **8**(4), 488–493 (2010)

54. Lin, S.-C., Persada, S.F., Nadlifatin, R.: A study of student behavior in accepting the blackboard learning system: a Technology Acceptance Model (TAM) approach. In: Proceedings of the 2014 IEEE 18th International Conference on Computer Supported Cooperative Work in Design (CSCWD), pp. 457–462 (2014)

55. Chin, W.W.: The partial least squares approach to structural equation modeling. Mod. Methods Bus. Res. **295**(2), 295–336 (1998)

56. Liu, S.-H., Liao, H.-L., Peng, C.-J.: Applying the technology acceptance model and flow theory to online e-learning users' acceptance behavior. E-Learning **4**(H6), H8 (2005)
57. Salloum, S.A., Mhamdi, C., Al Kurdi, B., Shaalan, K.: Factors affecting the adoption and meaningful use of social media: a structural equation modeling approach. Int. J. Inf. Technol. Lang. Stud. **2**(3), 96–109 (2018)
58. Salloum, S.A., Al-Emran, M.: Factors affecting the adoption of E-payment systems by university students: extending the TAM with trust. Int. J. Electron. Bus. **14**(4), 371–390 (2018)

Evaluation of Different Sarcasm Detection Models for Arabic News Headlines

Pasant Mohammed[1(✉)], Yomna Eid[1(✉)], Mahmoud Badawy[1(✉)], and Ahmed Hassan[2,3(✉)]

[1] Faculty of Engineering, Mansoura University, Mansoura, Egypt
Pasantamin@gmail.com, yomnaeid96@gmail.com,
engbadawy@mans.edu.eg
[2] Nile University, Giza, Egypt
ahassan@nu.edu.eg
[3] Ain-Shams University, Cairo, Egypt

Abstract. Being sarcastic is to say something and to mean something else. Detecting sarcasm is key for social media analysis to differentiate between the two opposite polarities that an utterance may convey. Different techniques for detecting sarcasm are varying from rule-based models to Machine Learning and Deep Learning models. However, researchers tend to leverage Deep Learning in detecting sarcasm recently. On the other hand, the Arabic language has not witnessed much improvement in this research area. Bridging the gap in sarcasm detection of the Arabic language is the target behind this work. In this paper, efficient models in short text classification are tested for detecting sarcasm in the Arabic news headlines for the first time. The dataset used to train and test these different architectures was manually collected by scrapping two different websites, sarcastic and non-sarcastic. Detailed results for each model were also represented, based on different performance metrics, such as accuracy, precision, recall and F1 score.

Keywords: Arabic language · Deep Learning · Sarcasm detection · Sentiment Analysis

1 Introduction

Sarcasm is an incorporation of both negative and positive sentiment. For example: "Oh, how I love being rejected", a positive word "love" is followed by a negative action "rejected". Sarcasm detection is a task of uncovering such vagueness between the intended meanings of a sentence against the spoken one. Due to the increasing freedom available for everyone to express opinions, social media content sometimes conveys sarcasm [1]. Additionally, the impossibility for each company to track its customer reviews is reflecting the substantial role of sarcasm detection. Sarcastic utterances appear frequently in the political field; as people tend to comment on the surrounding events in a sarcastic manner. Sarcasm detection and other Sentiment Analysis tasks can be referred to as digitalizing people's opinions, reactions and emotions for analytical purposes [2]. Additionally, sarcasm detection has a crucial role in diagnosing depression. Moreover, in the case of autistic people, sarcasm detection can provide a haven for those people who are often victims of bullying.

© Springer Nature Switzerland AG 2020
A. E. Hassanien et al. (Eds.): AISI 2019, AISC 1058, pp. 418–426, 2020.
https://doi.org/10.1007/978-3-030-31129-2_38

According to the fact that sarcasm is conveyed through many forms, it is represented in different text data types: short text datasets (tweets), long text datasets (discussion forums) and other datasets type as data from call centers [3]. Various techniques for developing sarcasm detection models can be classified as rule-based approaches and supervised Machine Learning and Deep Learning approaches. In rule-based models, the aim is to capture knowledge from a sentence by applying a set of if-else rules. While Machine Learning and Deep Learning models provide more intelligent systems, a function is implemented and generalized on the input to be mapped to the output, in lieu of if-else rules.

Over a sentence-level, sarcasm has many clues: laugh expressions (haha, LOL), heavy punctuation usage (!!!), and interjection – words for emotions expressing (Yaah) and letters capitalization (SARCASM) [3]. Sarcasm is detected relying on the understanding of many aspects such as the topic that the utterance contains [4], past tweets of a user [5], online decision forums rather than tweets [6] and also, the characteristics of a sarcastic person who posts sarcasm online [7]. Researchers in the Arabic language face many challenges, such as its complicated structure, lack of resources, rich morphology and the various dialects it includes [8].

In social media content, colloquial Arabic is used more often than standard Arabic. This arouses challenges in vocabulary as well as in sentence structure [8]. Reported results of recently proposed architectures in [9] and [10] stated that sarcasm in Arabic is still in need of enhancement. An examination was discussed in [11] for Sentiment Analysis on Arabic tweets. It proposed an ensemble architecture of CNN (Convolutional Neural Network) and LSTM (Long Short Term Memory) models to classify sentiment in Arabic tweets. Implementation of the models proposed in this paper is done based on analysis of [12–15]. This paper represents different architectures for detecting sarcasm in Arabic headlines. Additionally, different word-embedding models were utilized as input features.

The main objective of this paper is to be a stepping stone for building different sarcasm detection architectures with state-of-the-art accuracy. Training and testing were performed on manually collected Arabic headlines. Different results for the proposed architectures are documented in the experiments section. The remainder of this paper is organized as follows: Sect. 2 comprises a glimpse of related work done in this task. Section 3 demonstrates scrapped the dataset and its representation, also different architectures were illustrated. Section 4 discusses experiments performed to evaluate models with reported results. Section 5 provides a conclusion of the aforementioned work.

2   Related Work

2.1   Sarcasm Detection Models

Recent work done in detecting sarcasm is illustrated in this section. These different approaches are varying from being rule-based approaches, ones that follow a set of rules for knowledge extraction, to being more artificial approaches, ones that build rules rather than follow them. Another new rule-based approach is to identify incongruity in target text that was represented in [16]. The model predicted the rest of the sentence and computed the similarity between the actual word in the sentence and the predicted (non-sarcastic) word, if they are distinct, then the sentence tends to be sarcastic.

The role of word-embedding is illustrated in [17], which addressed the case of incongruity without the existence of sentimental words. The proposed model utilized semantic similarity and dissimilarity as a feature. It also was experimented by four-word-embedding models. Among these embedding models, Word2Vec achieved a high result. An investigation of short text data was performed in [18], which stated that the most effective representation for short text was a sentiment-aware one. While for the longer text, it was an emotion-aware one. The proposed approach was a BiLSTM (Bidirectional Long Short Term Memory) that was trained on two large corpora of product reviews. An approach for overcoming the shortage of recurrent neural networks and most neural attention mechanisms was represented in [19] that ignored word-word relationships.

A method based on two essential features that revealed sentence polarity, lexical features, and hyperbole features is illustrated in [20]. Another perspective was presented in [21] which focused on markers of sarcasm. It indicated that people tend to use emoji and emoticons more when expressing sarcasm on twitter. While tags, questions, exclamation, and metaphors are more frequent to make ridicule on Reddit. Incorporation of both attention mechanisms with LSTM was discussed in [1]. The proposed model assigned weights for words depending on their contribution in the final representation of the sentence. It was observed that this model outperforms other baseline model proposed in [14], except the one that uses CNN, LSTM, and DNN (Dense Neural Network). A study of the author's historical tweets was presented in [5] which implemented a bi-directional gated RNN (Recurrent Neural Network) that captured syntactic, semantic and contextual features over tweets.

Kolchinski and Potts [22] demonstrated that exploring the author's features can significantly improve model performance. Two methods: a Bayesian approach and an intensive embedding approach were proposed. It was deduced that the Bayesian method is sufficient in homogenous contexts, while for diverse ones, it is better to use the dense embedding approach. Hazarika et al. [6] stated that user embedding features along with discourse features were crucial for performance. It aimed at building a hybrid approach that deals with two types of sarcasm, explicit (content-based) and implicit (context-based). The main contribution of [23] was to provide an error analysis, relying on their empirical tests, with also recommendations to handle such errors. On the other hand, some other models for detecting sarcasm in the Arabic language were represented in [9] and [10]. Karoui et al. [9] developed a detection model based on some features as sentiment, contextual and other features. Work of that paper stated that the model can be generalized to detect sarcasm in other languages. In [10], a classifier that detects sarcasm in Arabic tweets was built. The model was trained by supervised Naïve Bayes algorithm for text classification. Features for that model relied on extracting some common distinctive features as an exclamation and question mark, dots, brackets, and some other words as "ههههه".

2.2 Datasets

Different types of datasets are used for training and testing sarcasm detection models. People transfer and generate data every day and hence, it is a mandatory task of data collection to make use of this data flow [24]. Datasets are differently represented and varying from the short text, long text, dialogs, and other miscellaneous types like book quotes [16]. The most familiar example of short text datasets is tweets. In 280 characters, people are free to share their opinions, some of which may convey sarcasm. Long text datasets are represented as discussion forums, books, films, product reviews and news articles [16]. Another representation of datasets is dialogues such as TV shows and scripts [16]. Some social-media platforms provide data that are already labeled. In Twitter, tweets are annotated by their hashtags. While, Reddit is a social media website that comprises variant communities. Data in Reddit is represented in tags as a category indicator.

3 Models Construction

3.1 Scrapped Dataset

Despite the availability of variant datasets online, some research tasks lack datasets. Accordingly, the task of sarcasm detection in Arabic headlines reflects this fact. The dataset used in this paper was manually collected, by scraping headlines of two different websites using the Scrapy Python library. The scrapped dataset is written in standard Arabic and is balanced, with 2999 sarcastic headlines and 2999 non-sarcastic ones. Different topics such as sports, politics and religion were covered in the dataset. Scraping the website الحدود resulted in sarcastic headlines and scraping مصرس website resulted in non-sarcastic headlines. Hence, the dataset is already labeled and required no further annotation. Some sarcastic and non-sarcastic scrapped headlines are illustrated in Table 1. The dataset is available for download https://github.com/PasantAmin/ pasantamin.github.io/blob/master/code/Arabic%20News%20Headlines%20Dataset. json in a JSON file format.

Table 1. Some sarcastic and non-sarcastic scrapped news headlines examples.

Sarcastic headlines	Non-sarcastic headlines
السرطان، الإيدز، الموت، وأمراض أخرى تعالجها حبّة البركة	أحد أبطال أكتوبر: ذكرى تحرير سيناء تشعرنا بالفخر والعزة
شاب يعتزل رفاق السوء خوفًا من جرّه إلى تدخين السجائر العادية	أول رد من "الأوقاف" علي منع مكبرات الصوت "بـ"التراويح
غادة السمان تنشر رسالة مارك زوكربيرغ لها	العربية للتصنيع» تبحث دعم مستشفى سرطان « الأطفال
إشارة الواي فاي تصل كافة أرجاء المنزل باستثناء أكثر كنبة مريحة	لاجارد: مبادرة بكين تقود الصين للشئون الدولية

3.2 Data Representation

Text classification is to label different texts according to a common feature or pattern. A substantial sarcasm detection concept is to look for sentimental words that cause incongruity in an utterance. Word-embedding as a feature is a numerical representation of words in fixed-size vectors. Additionally, it is a method for discovering semantic and syntactic information. Most common used pre-trained word-embedding is Word2Vec by Google, Glove by Stanford, and FastText by Facebook which supports the Arabic language. Researchers at Facebook AI Research leveraged Skip-Gram models to represent a language at character level. Another pre-trained model for Arabic word embedding is explained in and called "AraVec". AraVec provides various word representation models for different text domains, based on CBOW (Continuous Bag of Words) and Skip-Grams.

3.3 Architectures

In this section, the proposed models' architectures are illustrated. The dataset used in these architectures does not require cleaning or annotating. Headlines classification was performed on the following architectures: CNN, LSTM, BiLSTM, and CNN with LSTM and CNN with BiLSTM. The general process of these models is as follows: News headline are fed to a word-embedding layer to model word representation, then one of the proposed Deep Learning architectures was utilized to train these representations, finally, a softmax function is applied to classify news headline as sarcastic or non-sarcastic. All implemented models are available online.

The CNN Architecture. Recent Deep Learning researchers have revealed that CNN is remarkably powerful in classification tasks [25]. The CNN architecture proposed in this paper is based on the CNN model built in [26]. First, a representation of headlines is performed using FastText. For each input headline, each token is represented by the corresponding embedding vector to construct the embedding matrix. Consequently, an input sentence can be treated as an image. Then three of 1D convolution layers are applied on that headline using two filters for each convolutional layer.

Each filter generated a feature map. A max pooling layer is added to get a fixed-length vector from the preceding feature maps. Then, resultant outputs of the preceding stage are concatenated into a fixed-size filter map. Finally, this filter map is fed through a softmax function to determine whether the input headline is sarcastic or non-sarcastic. Figure 1 shows the CNN architecture. Kernels for this model were in size of (300×2), (300×3) and (300×4), also no padding or stride was not applied. For all layers, RELU activation function was used, except a softmax function for the last classification layer.

LSTM Architecture. LSTM as a special type of RNN that is efficient in solving the vanishing and exploding gradients problems. Simple RNN architectures suffer from such gradient problems [27]. For standard RNN models, the cell is one layer and almost tanh. However, in this model, an LSTM cell is represented in more complex structures with a higher number of gates to update memory. By keeping important information and forgetting others, LSTM is much eager to solve the problem of long-term

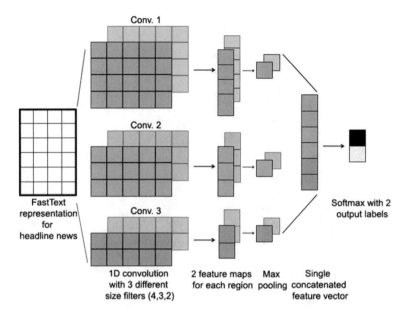

Fig. 1. A CNN model for detecting sarcasm using three of 1D convolution layers with max pooling and softmax functions [14].

dependencies. In this proposed LSTM model, the output of the embedding layer is fed to two LSTM layers, then passed to two fully connected layers (DNN) and finally, to the softmax layer.

While LSTM learns only dependencies of the preceding word's information, a crucial rule for utilizing BiLSTM was revealed. BiLSTM has the ability to learn the dependencies of both the preceding and following information of a word in a sentence. Adding a BiLSTM layer instead of the two LSTM layers in the previous architecture was experimented. Figure 2 illustrates the proposed LSTM architecture. In each LSTM layer, the number of hidden units was equal to 128.

Hybrid Architectures. Building this approach represents an avenue to leverage the advantages of both CNN and LSTM architectures. LSTM architecture is efficient in feature extraction and semantic analysis. While CNN architecture is more efficient in the accuracy of classification. Hybrid models implemented in this paper are represented in different architectures as follows: First, the structure of this proposed model is as follows: a CNN architecture that was represented in [26], on top of a multi-layer LSTM model. Second, the LSTM architecture on top of a CNN model, the resulting architecture followed the one represented in [13]. Finally, the third one has a BiLSTM architecture on top of a CNN model.

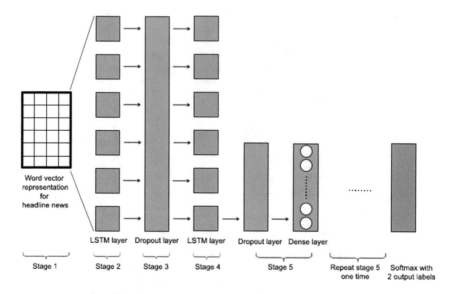

Fig. 2. LSTM model for detecting sarcasm [13].

4   Experiment and Results

In this section, the experiments performed on the different models were discussed and the results were reported. The dataset used to train and test the architecture is the scrapped sarcastic and non-sarcastic Arabic news headlines. Dataset was divided into 75% for training and 25% for testing. Moreover, 20% of the training data was used for validation. Changing word-embedding used in the CNN model from FastText to AraVec decreased the performance of the model. This was due to some missed word-representations that was not represented in the AraVec model. Hence, training more text data by the AraVec model is suggested to increase this model performance Also, Implementing additional regularization functions were ignored, as model performance is not highly affected by regularization [26].

Different optimizers were implemented to overcome the problem of over-fitting, among all of which, the Adam and RMSprop optimizers achieved the highest results. Learning rate is equal to 0.001 for all models, learning rate decay of RMSprop optimizer is equal to 0.2, learning rate decay of Adam optimizer is equal to 0.3. For the CNN model, a number of epochs were equal to 200, however, a number of 50 epochs was applied to all of the other models. For all models, the batch was in size of 32.

The following Table 2 illustrates a summary of all achieved accuracies and F1 scores. The highest performance metrics were achieved by the CNN-LSTM model with an accuracy of 0.928 and F1 score of 0.931. Work represented in this paper was performed on the Google Colab, a free cloud GPU service for AI developers. All codes were implemented in Python with the Keras open source Machine Learning libraries. All implementations of the preceding models were performed with a number of 15 epochs and a batch size of 32.

Table 2. A summary of results achieved by all architectures

	Accuracy		F1		Precision		Recall	
	Train	Test	Train	Test	Train	Test	Train	Test
CNN-RMSprop	0.840	0.825	0.849	0.811	0.850	0.812	0.851	0.815
CNN-Adam	0.843	0.823	0.853	0.832	0.849	0.824	0.851	0.830
LSTM-RMSprop	0.852	0.849	0.859	0.857	0.848	0.853	0.858	0.856
LSTM-Adam	0.918	0.890	0.921	0.893	0.921	0.893	0.928	0.893
BiLSTM-RMSprop	0.885	0.884	0.897	0.892	0.894	0.889	0.896	0.892
BiLSTM-Adam	0.918	0.902	0.924	0.904	0.921	0.902	0.922	0.904
CNN_LSTM-Adam	0.926	0.891	0.927	0.883	0.936	0.891	0.924	0.885
CNN_BiLSTM-Adam	0.925	0.904	0.919	0.895	0.93	0.895	0.923	0.892
LSTM_CNN-Adam	0.840	0.838	0.854	0.849	0.846	0.842	0.851	0.847

5 Conclusion

This paper illustrated various approaches for detecting sarcasm in Arabic news headlines. Different approaches were illustrated and for the first time tested on a manually collected dataset. Reported results were also presented, highlighting the state-of-the-art performance metrics for the most efficient models. Furthermore, the scrapped dataset and implementations are available online and ready to be used. For the AraVec model, training more text data to provide more words embedding was also suggested. As future work, an exploration of the proposed models' performance will be evaluated by addressing other different Arabic dialects, rather than just processing the standard one.

Acknowledgment. We would like to show gratitude to our faculty teaching assistant, Eng. Hossam Balaha, who provided guidance and assistance in this research work.

References

1. Martini, A.T., Farrukh, M., Ge, H.: Recognition of ironic sentences in Twitter using attention-based LSTM. Int. J. Adv. Comput. Sci. Appl. **9**(8), 7–11 (2018)
2. Medhat, W., Hassan, A., Korashy, H.: Sentiment analysis algorithms and applications: a survey. Ain Shams Eng. J. **5**(4), 1093–1113 (2014)
3. Raghav, S., Kumar, E.: Review of automatic sarcasm detection. In: 2nd International Conference Telecommunication Networks TEL-NET 2017, vol. 2018, pp. 1–6 (2018)
4. Joshi, A., Jain, P., Bhattacharyya, P., Carman, M.: 'Who would have thought of that!': a hierarchical topic model for extraction of sarcasm-prevalent topics and sarcasm detection (2016)
5. Zhang, M., Zhang, Y., Fu, G.: Tweet sarcasm detection using deep neural network. In: Proceedings of COLING 2016, 26th International Conference Computational Linguistic Technical Paper, pp. 2449–2460 (2016)
6. Hazarika, D., Poria, S., Gorantla, S., Cambria, E., Zimmermann, R., Mihalcea, R.: CASCADE: contextual sarcasm detection in online discussion forums, pp. 1837–1848 (2018)

7. Sun, Y., Song, H., Jara, A.J., Bie, R.: Internet of things and big data analytics for smart and connected communities. IEEE Access **4**, 766–773 (2016)
8. El-beltagy, S.R., Ali, A.: Open issues in the sentiment analysis of arabic, vol. 16, pp. 1–6 (2013)
9. Karoui, J., Zitoune, F.B., Moriceau, V.: SOUKHRIA: towards an irony detection system for Arabic in social media. Procedia Comput. Sci. **117**, 161–168 (2017)
10. Al-Ghadhban, D., Alnkhilan, E., Tatwany, L., Alrazgan, M.: Arabic sarcasm detection in Twitter. In: Proceedings of 2017 International Conference Engineering MIS, ICEMIS 2017, vol. 2018, pp. 1–7 (2018)
11. Heikal, M., Torki, M., El-Makky, N.: Sentiment analysis of Arabic tweets using deep learning. Procedia Comput. Sci. **142**, 114–122 (2018)
12. Ghosh, A., Veale, T.: Magnets for sarcasm: making sarcasm detection timely, contextual and very personal, no. 2002, pp. 482–491 (2018)
13. Ghosh, A., Veale, T.: Fracking sarcasm using neural network. NAACL-HLT, Assoc. Comput. Linguist., 161–169 (2016)
14. Zhang, Y., Wallace, B.C.: A sensitivity analysis of (and practitioners' guide to) convolutional neural networks for sentence classification. In: Proceedings of 8th International Joint Conference Natural Language Processing, pp. 253–263 (2017)
15. Ingole, P., Bhoir, S., Vidhate, A.V.: Hybrid model for text classification. In: Proceedings 2nd International Conference Electronics, Communication and Aerospace Technology ICECA 2018, no. Iaeac, pp. 7–15 (2018)
16. Joshi, A., Bhattacharyya, P., Carman, M.J.: Automatic sarcasm detection. ACM Comput. Surv. **50**(5), 1–22 (2017)
17. Joshi, A., Tripathi, V., Patel, K., Bhattacharyya, P., Carman, M.: Are word embedding-based features useful for sarcasm detection? In: Proceedings 2016 Conference Empirical Methods Natural Language Process, pp. 1006–1011 (2016)
18. Agrawal, A., An, A.: Affective representations for sarcasm detection, pp. 1029–1032 (2018)
19. Tay, Y., Tuan, L.A., Hui, S.C., Su, J.: Reasoning with sarcasm by reading in-between. In: Proceedings 56th Annual Meeting Association Computational Linguistics (Long Paper), pp. 1010–1020 (2018)
20. Parmar, K., Limbasiya, N., Dhamecha, M.: Feature based composite approach for sarcasm detection using MapReduce. In: Proceedings 2nd International Conference Computing Methodology Communication ICCMC 2018, no. Iccmc, pp. 587–591 (2018)
21. Ghosh, D., Muresan, S.: With 1 follower I must be AWESOME :P. Exploring the role of irony markers in irony recognition. Assoc. Adv. Artificial Intell. (2018)
22. Kolchinski, Y.A., Potts, C.: Representing social media users for sarcasm detection. In: Proceedings 2018 Conference Empirical Methods Natural Language Process, pp. 1115–1121 (2018)
23. Parde, N., Nielsen, R.D.: Detecting sarcasm is extremely easy;-). In: Proceedings Workshop Computational Semantics Beyond Events Roles, pp. 21–26 (2018)
24. Roh, Y., Heo, G., Whang, S.E.: A survey on data collection for machine learning: a big data - ai integration perspective, pp. 1–19 (2018)
25. Kim, Y.: Convolutional neural networks for sentence classification. In: Proceedings of 2014 Conference Empirical Methods Natural Language Process, pp. 1746–1751 (2014)
26. Zhang, Y., Wallace, B.: A sensitivity analysis of (and practitioners' guide to) convolutional neural networks for sentence classification (2015)
27. Hochreiter, S.: The vanishing gradient problem during learning recurrent neural nets and problem solutions. Int. J. Uncertainty Fuzziness Knowl. Based Syst. **06**(02), 107–116 (2003)

The Impact of De-marketing in Reducing Jordanian Youth Consumption of Energy Drinks

Motteh S. Al-Shibly[1] 📵, Mahmoud Alghizzawi[2] 📵,
Mohammed Habes[3] 📵, and Said A. Salloum[4,5(✉)] 📵

[1] Amman Arab University, Amman, Jordan
motteeshibly@gmail.com
[2] Faculty of Economics and Management Sciences,
University Sultan Zainal Abidin, Kuala, Terengganu, Malaysia
dr.alghzawi87@gmail.com
[3] Faculty of Applied Social Sciences, University Sultan Zainal Abidin,
Kuala Terengganu, Terengganu, Malaysia
mohammedhabes88@gmail.com
[4] Faculty of Engineering and IT, The British University in Dubai, Dubai, UAE
ssalloum@uof.ac.ae
[5] Faculty of Engineering and IT, University of Fujairah, Fujairah, UAE

Abstract. This paper explores the world of energy drinks and its negative effects on the youth and how to de-market it, the study uses a quantitative method to evaluate the impacts of consuming energy drinks among minors and the youth, in this paper we distributed surveys on the targeted sample who are the youth of Jordan and got back some very interesting results, the results of the surveys were analyzed thoroughly using the latest mathematical analyzing methods and software to give out a clear big picture of what is happening in reality with the consumers of energy drinks from the youth. the results that we arrived to showed some trends that are worrying which require immediate actions before the issues go out of hand, and for the sake of countering and solving these issues we made a list of applicable recommendations that we found best at the end of the paper to act as a starting point to apply measures to solve the issues at hand in order to save the society from a danger they are unaware of, and help preserve the health of youth.

Keywords: De-marketing · Anti-marketing energy drinks ·
Counter marketing energy drinks · Effects of energy drinks ·
Youth consumption

1 Introduction

In the recent years, a new product invaded the markets as well as the minds of the young generation which resembled a new form of addiction. It is believed to have a mystic influence on the performance of these people; it is energy drinks (ED) in its various forms and tastes. Not only teenagers and school students have this blind belief

© Springer Nature Switzerland AG 2020
A. E. Hassanien et al. (Eds.): AISI 2019, AISC 1058, pp. 427–437, 2020.
https://doi.org/10.1007/978-3-030-31129-2_39

in the effects of this product, but also their parents, who whenever they feel their children are getting week and incapable of performing well, especially during exams time, they immediately rush to the nearest supermarket and buy a pack of ED for their children hoping it will affect their energy immediately and help them perform better. This is blind ignorance of the real effect that such products may have on those young vulnerable creatures. In fact, those parents are not aware of the dangers they are exposing their children and teenagers to. ED only have the word energy in its name but no real energy, and in fact they have adverse effect, in the sense that they negatively affect the health of those who drink it and specifically It's more harmful to the youth [1–3]. In order to evaluate a certain product, we have to analyze its ingredients and study the scientific effects it has on its consumers [4]. Also we have to make a scientific survey confirming to the results of some study cases which are made in various parts of the world conducted by reliable and nonbiased parties to evaluate such product [5]. The manufacturers of ED will try to convince the consumers of the advantages such products promise to provide them with. Those who are against the product will try to point out the negative aspects of such products. All that we need is to analyze certain studies and researches conducted all around the world, to evaluate the impact of the consumption of these goods [6, 7]. We will try to include in our survey such studies and the results of such research so that we can judge such product and either advise people to use it or render them aware of the dangers of consuming it. We will try to explain the dangers of each serving and of multiple servings. The study will also tackle the effects of combining ED with other products such as ethanol or alcohol, because these days, consumers are keen on obtaining a double effect by combining ED with alcohol or ethanol.

People are used to drink soft drinks as a refreshment or a softener after a heavy meal to provide them with relief, the most famous of them are Colas and other gaseous drinks. However, in the past twenty years or so, ED have started to invade the markets with a strong force and they made their way easily in the hearts and minds of young generations [8, 9]. The new product has been described to have magical powers of boosting their energy every time they consume it. Life has changed a lot and younger generations seem to be weak and fragile, and their weakness augmented with the consumption of unhealthy food such as fast and junk foods, unlike previous generations who were used to consume more natural food ingredients. As a result, a special need for energy boosting products arose and consequently those products came to fill the new gap in their lives [10]. According to recent studies and analysis of ED, it was found that these ED contain high portions of sugar and caffeine, two product that give some sort of weak energy, the sugar and the caffeine serves to keep the consumers alert and awake. They try to include also other herbal ingredients and certain vitamins to give the impression that they contain [12] everything people need to stay energetic and awake. Moreover, Colas soft drinks also contain a good quantity of caffeine and sugar along with many other ingredients and they never dared to say that their products are ED [12, 13]. It seems that the producers of ED have built their theory on the psychological effect of calling their product energy drinks and they have succeeded in planting this illusion in the minds of youth consumers [14, 15].

2 Paper Hypotheses

H1: There is no significant impact of De-Marketing strategies on youth consumption of energy drinks.

H2: The price of product has no significant effect on youth consumption of energy drinks and psychological variables.

H3: Type of product has no significant effect on youth consumption of energy drinks.

H4: The distribution channel has no significant effect on youth consumption of energy drinks.

H5: The promotion has no significant effect on youth consumption of energy drinks and psychological variables (Fig. 1).

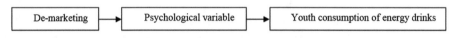

Fig. 1. Research model

3 Literature Review

Social marketing focuses greatly on fear appeals to discourage people from carrying on with a bad habit they are getting used to, It also affects their habits, mental and attitudes, and the most important of which is social media [4, 11, 15–18]. Thus, the second phase of our anti-marketing campaign is to show the consumers of ED the negative effects these brands and products can have on them. As previously mentioned in this paper it is important to tackle the adverse impacts on the health of the consumer, stating some factual information about study cases from all over the world about, not only the side effect of these products, but also about the dangers resulting from their consumption as well as the overdoses when wrongly consumed by ignorant people who think the more they take, the stronger they become [19]. ED are originally products which will hopefully provide people with energy to help them perform certain tasks or keep them awake and alert in certain situations such as exam times. Unfortunately, people are not often aware of the health dangers that come with their consumption [20].

3.1 Health Issues Caused by Energy Drinks

According to a recent medical paper, it has been discovered that the caffeine included in ED is absorbed within 30–35 min, and then it is metabolized in the liver to produce its desired effect [21]. Accordingly, the caffeine content was investigated by FDA (food and drug administration), new caffeine standards were imposed only 71 mg per 12 oz serving. The paper showed that the undesired effects of consuming such substance include: anxiety, gastrointestinal complaints, nervousness, cardiac arrhythmias, tachycardia, insomnia, restlessness, and in rare cases death. as well Low birth weight for gestational age, for women consuming 200/300 mg of caffeine daily [21]. Another important factor to mention is the relationship of ED which contain caffeine that is related to the blood pressure and hypertension. Those who consume high doses of coffee (which contains caffeine) showed a significant increase in their systolic blood pressure and peripheral vascular resistance pressure, as well diastolic blood. Not only that, but we can also consider the impact of consuming ED on fatty deposits, as well as the concentration of triglycerides (glucose). As it is always the situation, many medical studies are often carried out on animals mainly rats. There were thirty rats divided into three groups who were provided normal food and one group were administered 3 ml of ED and later tap water. Analyzing the results of the studies, the animals receiving the Energy drink was additionally characterized through declining content of Fatty tissue peril-intestinal the muscle, as well as peril-cardiac fatty tissue. Moreover, blood plasma Laboratory specimens of these animals contained higher concentration of glucose. The metabolic effects of ED are attributed to caffeine, saccharide and B-group vitamins occurring in those products [22]. The same thing would apply to human beings consuming ED at frequent intervals [23].

Another important discovery from the previous studies was the relationship between ED consumption and the social and behavioral effects of these products on these people [24]. According to a study conducted in 2008, it was discovered that there is strong relation between ED consumption and risk-taking. Frequent ED consumers were found to have smoked cigarettes, abused drug prescriptions, and were involved in violent activities. They also engaged themselves in unsafe sex, doing extreme sports, and dangerous stunts [20]. The study showed also that the consumption of ED also leads to other bad and dangerous habits of consuming other stronger stimulants and drugs which can destroy the career of people especially pilots [10, 14]. The results are usually lack of concentration, headaches, and risky behaviors.

4   Research Methodology

This research is considered to be a basic academic paper which has a multipurpose approach: exploratory, descriptive and explanatory. Exploratory in a sense that its purpose is to define the nature of the problem at hand and make it clear; descriptive because it describes the characteristics of the phenomenon we are studying and explanatory because it attempts to explain how things are going on and to recognize the cause and effects between different factors. The population represents the Jordanian youth. A diversified sample of youth that represents a wide range of the youth in Jordan was surveyed. We distributed 100 survey forms and received back 87 survey forms which are fair and square. The use of questionnaire surveys for data collection refers to fact that these tools could identify the relationship among the studied items [25].

4.1   Demographic Variables

Demographic data were evaluated for the respondents and the results were determined where the percentage of males was 63.2% for females 36.8%. The respondent's age ranged from 21–23 in the rate of 56.3%, while the educational level (Bachelor Degree) accounted for 80% of the respondents.

5 Data Analysis

In experimental studies, the dependent variable is the response that is measured by the researcher or is the presumed effect: (Jordanian youth consumption of energy drinks). An independent variable is the presumed cause, and is the variable that is varied or manipulated by the researcher: De-marketing, Intervening Variable: psychological variable.

6 Descriptive Analysis

To describe the attitude towards questions, mean and standard deviation are used as follows

Products

Table 1. Descriptive statistics (products)

Statement	N	Min	Max	Men	Std. Deviation
P1	87	1	5	3.41	1.244
P2	87	1	5	2.84	1.284
P3	87	1	5	3.44	1.309
P4	87	1	5	2.36	1.438
P5	87	1	5	2.67	1.387
Valid number	87	1.00	5.00	2.9425	.96685

The results of the previous Table 1 show negative attitudes towards q (2,4,5) Due to results less than the mean of the scale (3) but, there are positive attitudes towards the rest of the questions because the means are higher than scale (3).

Table 2. Descriptive statistics (price)

Statement	N	Minimum	Maximum	Mean	Std. Deviation
PR1	87	1	5	3.66	1.108
PR2	87	1	5	3.02	1.338
PR3	87	1	5	3.49	1.170
PR4	87	1	5	3.18	1.147
PR5	87	1	5	2.84	1.302
Price	87	1.40	5.00	3.2391	.79641
Valid number	87				

The results of the previous Table 2 show negative attitudes towards q (5) Due to results less than the mean of the scale (3) but, there are positive attitudes towards the rest of the questions because the means are higher than scale (3).

Table 3. Descriptive statistics (promotion)

Statement	N	Minimum	Maximum	Mean	Std. Deviation
PRO1	87	1	5	3.24	1.329
PRO2	87	1	5	3.17	1.287
PRO3	87	1	5	3.34	1.413
PRO4	87	1	5	3.37	1.192
PRO5	87	1	5	2.98	1.285
Promotion	87	1.20	5.00	3.2207	.96883
Valid number	87				

The results of the previous Table 3 show negative attitudes towards q (5) Due to results less than the mean of the scale (3) but, there are positive attitudes towards the rest of the questions because the means are higher than scale (3).

Table 4. Descriptive statistics (place)

Statement	N	Minimum	Maximum	Mean	Std. Deviation
PL1	87	1	5	3.11	1.409
PL2	87	1	5	3.44	1.168
PL3	87	1	5	3.43	1.178
PL4	87	1	5	2.82	1.253
PL5	87	1	5	3.31	1.323
Place	87	1.20	5.00	3.2207	.87606
Valid number	87				

The results of the previous Table 4 show negative attitudes towards q (4) Due to results less than the mean of the scale (3) but, there are positive attitudes towards the rest of the questions because the means are higher than scale (3).

Table 5. Descriptive statistics (psychological)

Statement	N	Minimum	Maximum	Mean	Std. Deviation
PY1	87	1	5	3.23	1.460
PY2	87	1	5	2.79	1.511
PY3	87	1	5	3.32	1.393
PY4	87	1	5	3.46	1.218
PY5	87	1	5	3.54	1.362
Psychological	87	1.00	5.00	3.2690	1.06829
Valid number	87				

The results of the previous Table 5 show negative attitudes towards q (2) Due to results less than the mean of the scale (3) but, there are positive attitudes towards the rest of the questions because the means are higher than scale (3) (Table 6).

7 Hypothesis Analysis

Table 6. Variables Entered/Removed[b]

Model	Variables entered	Variables removed	Method
1	Place, Products, promotion, Price		Enter

[a] All requested variables entered.
[b] Dependent variable: q2

Table 7. Model summary

Model	R	R Square	Adjusted R Square	Std. Error of the Estimate
1	.421[a]	.177	.137	1.337

[a] Predictors: (Constant), Place, Products, promotion, Price

Table 8. ANOVA[b]

Model		Sum of squares	DF	Mean square	F	Sig.
1	Regression	31.638	4	7.910	4.423	.003[a]
	Residual	146.637	82	1.788		
	Total	178.276	86			

[a] Predictors: (Constant), place, products, promotion, price
[b] Dependent variable: q2

Table 9. Coefficients[a]

Model		Unstandardized coefficients		Standardized coefficients beta	t	Sig.
		B	Std. Error			
1	(Constant)	2.490	.634		3.928	.000
	products	-.228-	.235	-.153-	-.971-	.334
	price	-.339-	.324	-.187-	-1.044-	.299
	promotion	.468	.242	.315	1.935	.056
	place	.486	.269	.296	1.806	.075

Multiple regression is used to test above hypothesis and it is found that F value = 4.423 is significant at 0.05 level so that H1 will be rejected and Ha will be accepted that mean there is significant impact of De-Marketing strategies on youth consumption of energy drinks Also it is found that the relationship is moderate because r = 0.421 and the independent variables explain 17.7% of the variance in the dependent variable. The Table 9 of Coefficients summarizes the following results:

- The price of product has no significant impact on youth consumption of energy drinks and psychological variables because t value = −1.044 is not significant at 0.05 level
- Type of product has no significant impact on energy drinks because t value = −0.971 is not significant at 0.05 level
- The distribution channel has no significant impact on youth consumption of energy drinks because t value = 1.806 is not significant at 0.05 level (Table 7)
- The promotion has no significant impact on youth consumption of energy drinks and psychological variables because t value = 1.935 is not significant at 0.05 level (Table 8).

8   Conclusion

In view of the paper and the survey related to the original topic of the investigation about the impact of de-marketing in reducing Jordanian youth consumption of energy drinks. Our original purpose is to help de-market this item because of its negative influence on people. We gave out a hundred forms of questionnaire to various types of people mainly of the youth group. We received eighty seven filled in forms which we analyzed and came up with the following results: All the respondents said they consumed energy drinks at various frequencies. The final outcome indicated that 48.3% of the sample rarely consumed energy drinks. Another important outcome was that 46% of the sample drinks more than one type of Energy Drinks. It is found that 81.6% of the sample consider themselves not to be addicted to energy drinks. This defies the idea that there is brand addiction. However, this does not rule out the possibility that there are people who are addicted to this product. As for the best location for marketing Energy Drinks, it was found that 50.6% of the sample believes that supermarkets are the best places to market these products. This conforms to what we previously stated that energy drinks are best marketed at supermarkets and grand shops. The external appearance of the E.D. This is consistent with the study of [1, 6, 26–28] products has a strong impact on the desire to consume a certain brand. The survey indicated that 63.2% of the sample finds that the external appearance of E.D. product to be appealing and inviting to customers. No wonder as the look of the product has a strong impact on the customers' decision when choosing a certain product, and there is a full science with specialty related to the design of the product, and it comes under graphic designs and promotion. These days, the content of the product is not the main decisive factor in promoting a certain product. As for the feeling achieved from consuming E.D. [8, 14] it was found that 49.4% of the sample does not immediately feel refreshed and energetic after consuming energy drinks. It is this point that requires further study and analysis, as it is very much related to our original issue, which de-marketing energy drinks. When about 50% of the sample mentioned that they did not feel refreshed and energetic immediately after consuming this product, this can be used as a starting point that the feeling developed later on could be psychological and not realistic. People tend to convince themselves that they are becoming strong and energetic and with further experimentation, the whole idea could be proven false and mere illusion. When referring to the points suggested in the hypothesis, we came out with various unexpected results worth pointing out. To start off, it has been discovered that the price of the product has no significant impact on the consumption of energy drinks and psychological variables. because the consumers value the rewards expected from consuming energy drinks higher that any cost they may have to pay for. [1, 20, 27], In fact, the value is less than the significant level. The same thing applies to the type of the product because here also because the t value is less than the significant level. The distribution channel similarly has no significant impact of youth consumption because the t value is not at the required significant level. Finally, the promotion did not exert any significant impact on the youth consumption, because it did not reach the required significant level. However, it has been discovered that the four factors combined together have a significant impact on the consumption of the youth for energy drinks and psychological variables.

9 Recommendations

In order to spread awareness in our country about a topic that is not yet seen by them as a problem, and reduce the harmful health risks and sicknesses that energy drinks might cause, we will work on introducing new ideas to help the cause and de-market a serious threat in our society. This is a list of our recommendations: Firstly Establish an awareness campaign to spread the word about the harmful side effects and health risks of drinking energy drinks, and especially spreading awareness of the dangers of youth consumptions of such drinks and how it's even more harmful to them to consume than it is to adults, in addition to how it could develop an addiction to whomever is consuming it. Second, there should be campaigns that deliver a wakeup call for the consumers exposing the lies of energy drinks companies who only care about increasing their customer base and increasing their profits regardless if their marketing base was true or not. Finally, there should be developed strict rules and regulations and severe punishments and sanctions enforced to deter energy drinks companies from such activities, where energy drinks companies always distribute free samples to attract new customers in Jordan.

References

1. Peacock, A., Droste, N., Pennay, A., Miller, P., Lubman, D.I., Bruno, R.: Awareness of energy drink intake guidelines and associated consumption practices: a cross-sectional study. BMC Public Health 16(1), 6 (2015)
2. Mhamdi, C.: Framing 'the Other' in times of conflicts: CNN's coverage of the 2003 Iraq war. Mediterr. J. Soc. Sci. 8(2), 147–153 (2017)
3. Mhamdi, C.: Interpreting games: meaning creation in the context of temporality and interactivity. Mediterr. J. Soc. Sci. 8(4), 39–46 (2017)

4. Habes, M., Alghizzawi, M., Salloum, S.A., Ahmad, M.: The use of mobile technology in the marketing of therapeutic tourist sites: a critical analysis. Int. J. Inf. Technol. Lang. Stud. **2**(2), 48–54 (2018)

5. Mhamdi, C.: The use of political cartoons during popular protests: the case of the 2011 Tunisia uprising. J. English Stud. **15**, 193–220 (2017)

6. Arria, A.M., Bugbee, B.A., Caldeira, K.M., Vincent, K.B.: Evidence and knowledge gaps for the association between energy drink use and high-risk behaviors among adolescents and young adults. Nutr. Rev. **72**(suppl_1), 87–97 (2014)

7. Al-Mohammadi, S.: Integrating reading and writing in ELT. In: Focusing on EFL Reading: Theory and Practice. Cambridge Scholars Publishing, pp. 260–274 (2014)

8. Spierer, D.K., Blanding, N., Santella, A.: Energy drink consumption and associated health behaviors among university students in an urban setting. J. Community Health **39**(1), 132–138 (2014)

9. Habes, M., Salloum, S.A., Alghizzawi, M., Alshibly, M.S.: The role of modern media technology in improving collaborative learning of students in Jordanian universities. Int. J. Inf. Technol. Lang. Stud. **2**(3), 71–82 (2018)

10. Medway, D., Warnaby, G., Dharni, S.: Demarketing places: rationales and strategies. J. Mark. Manag. **27**(1–2), 124–142 (2010)

11. Alghizzawi, M., Salloum, S.A., Habes, M.: The role of social media in tourism marketing in Jordan. Int. J. Inf. Technol. Lang. Stud. **2**(3) (2018)

12. Derbel, E.: Feminist graphic narratives: the ongoing game of eluding censorship. Mediterr. J. Soc. Sci. **10**(1), 49 (2019)

13. Mhamdi, C.: What can video add to the learning experience? challenges and opportunities. Int. J. Inf. Technol. Lang. Stud. **1**(1), 17–24 (2017)

14. Gerstner, E., Hess, J., Chu, W.: Demarketing as a differentiation strategy. Mark. Lett. **4**(1), 49–57 (1993)

15. Habes, M., Alghizzawi, M., Khalaf, R., Salloum, S.A., Ghani, M.A.: The relationship between social media and academic performance: Facebook perspective. Int. J. Inf. Technol. Lang. Stud. **2**(1), 12–18 (2018)

16. Habes, M.: The influence of personal motivation on using social TV: a uses and gratifications approach. Int. J. Inf. Technol. Lang. Stud. **3**(1) (2019)

17. Alghizzawi, M., et al.: The impact of smartphone adoption on marketing therapeutic tourist sites in Jordan. Int. J. Eng. Technol. **7**(4.34), 91–96 (2018)

18. Alghizzawi, M.: The role of digital marketing in consumer behavior: a survey. Int. J. Inf. Technol. Lang. Stud. **3**(1), 24–31 (2019)

19. Mhamdi, C.: Transgressing media boundaries: news creation and dissemination in a globalized world. Mediterr. J. Soc. Sci. **7**(5), 272 (2016)

20. Bliss, T.J., Depperschmidt, C.L.: Energy drink consumption and its effects on student pilots: perceptions of collegiate flight students. Viation Rev., 1 (2011)

21. Chrysant, S.G., Chrysant, G.S.: Cardiovascular complications from consumption of high energy drinks: recent evidence. J. Hum. Hypertens. **29**(2), 71 (2015)

22. Sadowska, J.: Evaluation of the effect of consuming an energy drink on the concentration of glucose and triacylglycerols and on fatty tissue deposition. a model study. Acta Sci. Pol. Technol. Aliment. **11**(3), 311–318 (2012)

23. Lawther, S., Hastings, G.B., Lowry, R.: De-marketing: putting Kotler and levy's ideas into practice. J. Mark. Manag. **13**(4), 315–325 (1997)

24. Moore, R.S.: The sociological impact of attitudes toward smoking: secondary effects of the demarketing of smoking. J. Soc. Psychol. **145**(6), 703–718 (2005)

25. Al-Emran, M., Mezhuyev, V., Kamaludin, A.: PLS-SEM in information systems research: a comprehensive methodological reference. In: 4th International Conference on Advanced Intelligent Systems and Informatics (AISI 2018), pp. 644–653 (2018)
26. Lin, Y.-C., Chen, Y.-C., Yeh, R.C.: Understanding college students' continuing intentions to use multimedia e-learning systems. World Trans. Eng. Technol. Educ. **8**(4), 488–493 (2010)
27. Al-dweeri, R.M., Obeidat, Z.M., Al-dwiry, M.A., Alshurideh, M.T., Alhorani, A.M.: The impact of e-service quality and e-loyalty on online shopping: moderating effect of e-satisfaction and e-trust. Int. J. Mark. Stud. **9**(2), 92 (2017)
28. Malhotra, N.K.: Marketing Research: An Applied Orientation, 4th edn. Prentice Hall, Saddle River (2004)

The Relationship Between Digital Media and Marketing Medical Tourism Destinations in Jordan: Facebook Perspective

Mahmoud Alghizzawi[1]📷, Mohammed Habes[2]📷,
and Said A. Salloum[3,4(✉)]📷

[1] Faculty of Economics and Management Sciences,
University Sultan Zainal Abidin, Kuala Terengganu, Malaysia
dr.alghzawi87@gmail.com
[2] Faculty of Applied Social Sciences, University Sultan Zainal Abidin,
Kuala Terengganu, Terengganu, Malaysia
mohammedhabes88@gmail.com
[3] Faculty of Engineering & IT, The British University in Dubai, Dubai, UAE
[4] Faculty of Engineering & IT, University of Fujairah, Fujairah, UAE
ssalloum@uof.ac.ae

Abstract. This study aimed to analyze and discover the relation of using digital media sites (Facebook) on promoting medical tourism destinations in Jordan, and its impact on the behavior of tourists through the technologies provided by these means. Away from the traditional methods in marketing, the researchers used the survey methodology for a sample of 560 tourists distributed at central of Jordan in Dead Sea area to realize the study objective, a new framework was suggested to show the impact of Facebook on the behavior of tourists through: demographic variables, Facebook features, advertising, by using the TAM model in adoption of social media technology in tourism marketing for tourist destinations in Jordan. The proposed data were analyzed using the Smart PLS system by modeling structural equations (SEM). The outcome of the study showed that the advantages of Facebook, advertising and demographic variables have a favorable effect on the (PEOU) of the tourist and the PU in the adoption of tourism behavior, in addition to the (PU) and (PEOU) (ATT), which led to the adoption of behavior around therapeutic tourism destinations in Jordan. By determining the impact of Facebook in marketing tourism in Jordan, it would be useful to conduct further research to provide better proposals for marketing tourist therapeutic destinations in Jordan.

Keywords: Facebook · TAM model · Digital media · Medical tourism ·
Jordan · Social media

1 Introduction

The gigantic development in social media channels has paved the way for many individuals to associate with each other and with companies to see all tourism marketing services [5, 6]. And thus, change the way that companies and organizations do

© Springer Nature Switzerland AG 2020
A. E. Hassanien et al. (Eds.): AISI 2019, AISC 1058, pp. 438–448, 2020.
https://doi.org/10.1007/978-3-030-31129-2_40

their work in the promotion and attraction of tourism sites and the preservation of consumers in general. Where a study showed [7]. The channels of social networking sites such as Facebook and Twitter are applied through US companies for communicate with their customers or with potential customers, especially with tourists, by 70% using Facebook, 37% using LinkedIn, and 25% using YouTube. A study showed [8] that the most important features of social networking sites in marketing for tourism purposes is to be effective in improving marketing practices in general. According to a study [9] In view of the high cost of traditional advertising 50% of Facebook users believe that any page for company on Facebook is more beneficial and effective than their website, and 95% of them use this network for travel-related activities before taking a vacation. [8] This is likely explain a tourist knows that tourism activity in general must be connected to a publication or recommended through Facebook [7]. Facebook is an active social platform with many potential customers It is also very easy to use and can be quickly created. It can also provide hotel information in an easy way and answer user requests, So Facebook is an effective tool for users looking for tourist services according to [10]. The concept of therapeutic tourism has developed along with the development of technology, and many countries have introduced their tourism services through social networking applications as in the Asian countries [4]. There are different ways to promote therapeutic tourism including Facebook, where "tourist resort marketing" occurs when tourist choose tour through international borders for the purpose of gaining some form of medical treatment [11]. According to [12] that the use of social networking platforms in the promotion of therapeutic tourism sites participate to enhance the marketing role of tourist sites and the competitive environment of tourism and archaeological sites that rely on technology and data. This increases the tourist traffic coming to countries including Jordan and promotes pattern of therapeutic tourism [3] in Jordan many of important medical and therapeutic sites that aim at achieving the new definition of therapeutic tourism according to leading international standards [13]. Social networking sites are great importance to people interested in the fields of communication, information, political science and social sciences, because of the great role of these sites and through the advantages in path of ease of use and efficiency to reach a large number of people in the whole world [2].

2   Research Model

This study suggest a framework to use of social media in marketing therapeutic tourism destinations in Jordan from the perspective of Facebook, it was suggested: Demographic variables, Facebook features, advertising and tourist behavior through their application to TAM Fig. 1 illustrates the instance of the proposed study model.

2.1   TAM

The technology acceptance model is most favorite and influential frameworks for understanding the acceptance of social media technology for users [14] it received experimental and extensive support for previous studies [15] According to the TAM model, (EOU) is the stage in which a individual feels the using of any technology

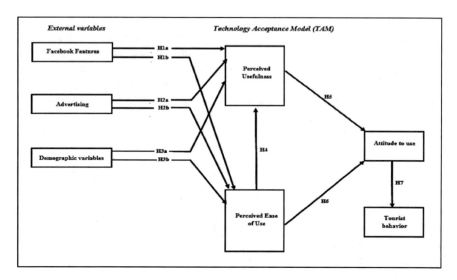

Fig. 1. Research model

system does not require hard working, It achieves usefulness through this use and reinforces the performance of the results. Which effect together on user behavior when using technology, The (PU) is outlined as "the degree to which a person believes that the use of a particular system will reinforces his or her performance" [16] Ease of use is defined as "the degree to which a person believes that the use of a particular system will be effortless" [14]; the proposed research model merges 3 categories of variables. The first category includes key determinants (independent variables) that may have an impact on behavioral intent (BI) and actual use (AU) The following hypothesis can be inferred from the above:

H4: There is a favorable and supportive relationship between (PEU) and (PU).

H5: There is a favorable and supportive relationship between (PEU) and (ATT).

H6: There is a favorable and supportive relationship between (PU) and (ATT).

H7: There is a favorable and supportive relationship between (AU) and (TOU).

2.2 TAM Technology Acceptance Model with Facebook Features

Facebook become Widely spread social network for people to participation life details and multimedia materials with their contacts, as well as dynamic user-generated media content [17] Facebook is one of the most popular platforms for applications and features [2] The advantages of Facebook is the most important element in attracting users to adopt the Facebook to use the technologies easy to use [18, 19] Sharing, video, audio, communication, information access and marketing without effort and ease. [5, 20] The technology acceptance model supposes two fixed beliefs, (PU) and (PEOU) are the opener determinants of attitudes (AT) to use of a new technicality. The following assumptions can be inferred:

H1a: There is a favorable and supportive relationship between the features of Facebook and (PU).

H1b: There is a favorable and supportive relationship between Facebook features and (PEOU).

2.3   TAM Technology Acceptance Model and Tourism Advertising via Facebook

Facebook provides a new model for advertising and interaction with consumer, the Social networking sites are now one in each five announcements users see online. Since the most important social networking platforms can realize Large reach and against target audiences at minimum expense [13, 21, 22]. Social media platforms technologies We can use them to get fresh customers, stay in contact with them, and market them with new goods, thus achieving excellent quality public relations between the tourist and the advertiser easily and reciprocally [1, 5]. According to [5]. Entertainment, informatics, interactivity, validity and secrecy fright have a direct favorable impact on the attitude of Jordanian consumers towards advertising on Facebook in the field of tourism. Thus, the attitudes of the users through the social communication technology are influenced by the perceived usefulness of these means (PU) and their perceived ease of use and availability (PEOU) According to [23–26] The following hypotheses can be inferred from the above:

H2a: There is a favorable and supportive relationship between Facebook ads and (PU).

H2b: There is a favorable and supportive relationship between Facebook ads and (PEOU).

2.4   Demographic Variables and Technology Acceptance Model TAM

About TAM approval and Implementation Previous literature, a huge number of models were used, for example, TRA, TPB, TAM, UTAUT to study and explore behavior determinants Used towards adoption and the use of information technology [27, 28]. The TAM technology acceptance model [29] is the most widespread among these models,

And effective framework for expound technology acceptance and adoption. ago its development TAM Previously vastly used demographic, social, cultural and organizational factors through the TAM acceptance model effect on the use of social media (Facebook) in marketing and more specifically in marketing tourist destinations [14], The proposed study model expands on the use of the TAM model to include (POU) and (PU). moreover, it is assumed that individual differences – age, gender, nationality – increase the effects of these variables on the intent and behavior of the tourist [27, 30]. The following hypotheses can be deduced:

H3a: There is a favorable and supportive relationship between the demographic variables and (PU).

H3b: There is a favorable and supportive relationship between demographic variables and (POU).

3 Research Methodology

3.1 Target Population and Unity of Analysis (UOA)

The tourists who's come to visit tourism sites in Jordan it's the target population in this study. to achieve the study's purpose of determining whether marketing through (Facebook) has a significant effect on promotion the tourism in Jordan, In this study, quantitative research planning used to realize this goal in order to make quantitative data more active and fit to exam the hypothesis [32] The questionnaire was designed and spread to 560 tourists in Dead Sea area of central Jordan as a sample of the survey using. In the end, 357 questionnaires were accumulating 357 valid. to measure the relationships and connection among the variables and their effect on tourist behavior in Jordan Was used (PLS-SEM). Demographic data were evaluated for the respondents and the results were determined where the percentage of males was 56.6% for females and 44.4%. The respondents ranged from 29–45% to 62%, while Jordanians accounted for 49.6% of the respondents while the rest were from other countries.

4 Findings and Discussion

4.1 Analysis of the Measurement Model

Some specific indicators are used in modeling the Smart-PLS equation to predict convergent validity: loading of measurement factors, parameter determination (alpha-Cronbach and composite reliability). Which in turn we can demonstrate the internal consistency between the various building components [32]. And when the reliability coefficient and the reliability of the compound for all compositions are more than 0.7. In Table 1, the Cronbach's Alpha score is over 0.7 according to [34] and The combined reliability range (CR) ranges from 0.861 to 0.951 and AVE between 0.613 and 0.859. Conforms to the criteria and explains at least 50% of the discrepancy between the set of elements under construction [35] Finally, metrics that measure construction are supposed to achieve convergent validity.

When the AVE values are greater than the quadratic correlation between the combinations in the Smart-Plus measurement model, therefore the requirements for the correctness of discrimination achieve [32, 36] Thus, when the value of (AVE) is more than 0.5, the structures must have at least 50% of the measurement variance. To verify the discriminating value, the Smart-PLS was used. In reference to Table 2 to load measurement factors, when the cross-sectional load and cross-load components are examined in more detail, they show that all elements of measurement are loaded only on the underlying structures of the variable The AVE analysis is shown in Table 2 so that the square roots of the AVE grades are displayed by the dark-colored diagonal elements in Table 1. On the other hand, the links between the structures are shown by the outbound elements. Between 0.832 and 0.922 is the square root of AVE results. As shown in Table 1. This value is greater than 0.5. It is also seen that AVE is larger than comparison with other links with the structure involved. In fact, there is a lot of variation between all the constructions with the feedback for each building. In addition,

Table 1. Convergent validity outcomes which assures acceptable values (Factor loading, Cronbach's Alpha, composite reliability $\geq$ 0.70 & AVE > 0.5).

Constructs	Items	Factor loading	Cronbach's alpha	CR	AVE
Advertising	AD_1	0.717	0.769	0.861	0.859
	AD_2	0.826			
	AD_3	0.768			
Attitude to use	ATT_1	0.745	0.831	0.900	0.776
	ATT_2	0.962			
	ATT_3	0.884			
Demographic variables	DEM_1	0.818	0.782	0.879	0.731
	DEM_2	0.822			
	DEM_3	0.744			
Facebook features	FAC_1	0.898	0.827	0.770	0.613
	FAC_2	0.901			
	FAC_3	0.716			
(PEOU)	PEOU_1	0.732	0.723	0.805	0.719
	PEOU _2	0.755			
	PEOU _3	0.918			
(PU)	PU_1	0.768	0.891	0.911	0.769
	PU_2	0.777			
	PU_3	0.894			
Tourist behavior	TOU_1	0.706	0.822	0.951	0.755
	TOU_2	0.923			

Table 2. Heterotrait-monotrait ratio (HTMT).

	Advertising	ATT	DEM	FAC	PEOU	PU	TOU
Advertising							
ATT	0.022						
DEM	0.121	0.391					
FAC	0.162	0.222	0.415				
PEOU	0.061	0.209	0.200	0.242			
PU	0.036	0.303	0.428	0.617	0.511		
TOU	0.371	0.412	0.522	0.625	0.319	0.013	

other structures in the model prefer discrimination. In accordance with the rules for the application of discrimination, the loading of each element must be greater than the load of the corresponding variable [34] Thus, the values in Table 2 have met the second criteria in Table 2. Also there is another condition in the criteria that indicates that HTMT values should be less than 0.85. Consequently, these criteria have been met as shown in Table 3. As a result, the validity of discrimination is now firmly established.

This parameter is utilized to verify the predictive accuracy of the framework. It is regarded as a quadratic relationship amidst the actual construction values and the predicted values and thorough examination of the structural model is performed using the coefficient of determination (R2 value) [37] and the coefficient includes the combined effect of the external variables on the internal variable. It also shows the degree of variation in the internal structures explained by the authorized external variable [33]. The values between 0.33 and 0.67 are moderate values, and the value above 0.67 is high and between 0.19 and 0.33 is the weak value and the value under 0.19 is unacceptable. As shown in Fig. 2, in Table 4 and Fig. 2, it can be observed that the framework has a wide predictive ability, so its help approximately 70% and 52% of the variance of the impact of the use of social media (Facebook) in marketing therapeutic destinations in Jordan on the behavior of TOU, PU and ATT respectively. The (PEOU) was also found between 0.33 and 0.67; hence the predictive power of these structures is moderate.

Table 3. R^2 of the endogenous latent variables.

Constructs	R^2	Results
Attitude to use	0.696	High
Perceived ease of use	0.600	Moderate
Perceived usefulness	0.525	Moderate
Tourist behavior	0.469	Moderate

4.2   Structural Model Analysis

An organizational model was used to use the SEM-PLS system to verify the proposed hypotheses [14, 21, 31, 38]. According to the model of the study, the opportunity to evaluate the relationship between the theoretical structures was evaluated in the study's structural model. Figure 2 show the results of relationships between variables. According to the results, all hypotheses have been strongly supported. Based on the hypotheses of data analysis, H1a, H1b, H2a, H2b, H3a, H3b, H4, H5, H6, and H7 were supported by experimental data (SEM-PLS). The results showed that the perceived usefulness (PU) (B = 0.505, P < 0.00). While the percentage of demographic variables (β = 0.272, P < 0.001) and ease of use (PEOU) (β = 0.110, P < 0.001) Thus supporting hypothesis H1a, H2a, H3a and H4, respectively. furthermore, ease of use (PEOU) directly supported the advantages of Facebook (β = 0.337, P < 0.00) and declarations (β = 0.113, P < 0.00) and demographic variables (β = 0.432, P < 0.00) (B) (β = 0.209, P < 0.00), and (3), respectively. Thus, the H5 and H6 hypotheses support the effect of the ATT on the visitor's behavioral intent using Facebook (β = 0.637, P < 0.00) and thus support H7 and Fig. 2 test the hypothesis results.

Table 4. Results of structural Model - Research Hypotheses Significant at p** =< 0.01, p* < 0.05)

H	Relationship	Path	t-value	p-value	Direction	Decision
H1a	Facebook Features-> Perceived Usefulness	0.505	12.489	0.001	Positive	Supported**
H1b	Facebook Features -> Perceived Ease of use	0.337	17.134	0.000	Positive	Supported**
H2a	Advertising -> Perceived Usefulness	0.476	3.132	0.010	Positive	Supported**
H2b	Advertising -> Perceived Ease of use	0.133	5.511	0.005	Positive	Supported**
H3a	Demographic variables -> Perceived Usefulness	0.272	11.434	0.000	Positive	Supported**
H3b	Demographic variables -> Perceived Ease of use	0.432	9.139	0.000	Positive	Supported**
H4	Perceived Ease of use -> Perceived Usefulness	0.110	6.266	0.011	Positive	Supported*
H5	Perceived Usefulness -> Attitude to use	0.209	7.379	0.003	favorable	Supported**
H6	Perceived Ease of use -> Attitude to use	0.164	4.142	0.021	favorable	Supported*
H7	Attitude to use -> Tourist behavior	0.637	1.112	0.041	favorable	Supported*

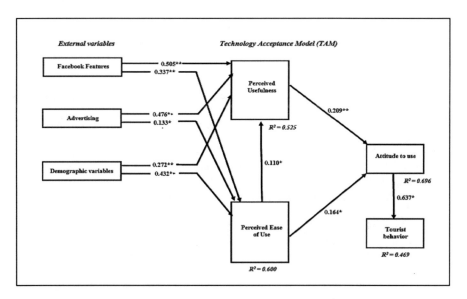

Fig. 2. Path coefficient results (significant at p** <= 0.01, p* < 0.05)

5 Conclusion

The purpose of this research was to analyze and discovery of tourist behavior the medical tourism in Jordan away from the traditional techniques of marketing. The research was proceed on tourists come to visit Dead Sea this area in central Jordan through the use of social media Facebook) and if they consider it as an effective and easy way to identify Medical tourism destinations in Jordan. The tourism authorities will be able to develop tourism destinations in Jordan. It is essential that legislators and administrators involved in the marketing of medical tourism destinations in Jordan consider the factors that have a significant impact on the marketing of tourist destinations through the means of communication Social in general and Facebook in particular and to increase the marketing and advertising methods through the design and implementation of procedures to achieve greater benefit in marketing therapeutic destinations in Jordan over the best utilize of social networking and Facebook and use in future research. The participants in this study are tourists to one region in the Dead Sea in the middle of Jordan. Therefore, they cannot be judged entirely to represent all tourists visiting the treatment sites in Jordan, which creates further restrictions on the study. Other restrictions These results can represent the view of tourists with the aim of medical tourism does not necessarily reflect the destination of all tourists visiting Jordan in different types of tourism, to identify the points of difference and similar in their view, with regard to the impact of the factors that have been searched and Recommended by the TAM model, thus further research and study is needed on the subject.

References

1. Al-sameeM, A.: Role of social media sites in marketing Egypt as an international tourist destination (2012)
2. Habes, M., Alghizzawi, M., Khalaf, R., Salloum, S.A., Ghani, M.A.: The relationship between social media and academic performance: Facebook perspective. Int. J. Inf. Technol. Lang. Stud. 2(1) (2018)
3. Magatef, S.G.: The impact of tourism marketing mix elements on the satisfaction of inbound tourists to Jordan. Int. J. Bus. Soc. Sci. 6(7), 41–58 (2015)
4. Connell, J.: Medical tourism: Sea, sun, sand and… surgery. Tour. Manag. 27(6), 1093–1100 (2006)
5. Alghizzawi, M., Salloum, S.A., Habes, M.: The role of social media in tourism marketing in Jordan. Int. J. Inf. Technol. Lang. Stud. 2(3) (2018)
6. Alghizzawi, M.: The role of digital marketing in consumer behavior: a survey. Int. J. Inf. Technol. Lang. Stud. 3(1) (2019)
7. Arsal, I.: The influence of electronic word of mouth in an online travel community on travel decisions: a case study (2008)
8. Leung, X.Y., Bai, B., Stahura, K.A.: The marketing effectiveness of social media in the hotel industry: a comparison of Facebook and Twitter. J. Hosp. Tour. Res. 39(2), 147–169 (2015)
9. MacKenzie, S.B., Lutz, R.J., Belch, G.E.: The role of attitude toward the ad as a mediator of advertising effectiveness: a test of competing explanations. J. Mark. Res. 23(2), 130–143 (1986)

10. Mariani, M., Ek Styven, M., Ayeh, J.K.: Using Facebook for travel decision-making: an international study of antecedents. Int. J. Contemp. Hosp. Manag. **31**(2), 1021–1044 (2019)
11. Lynch Jr, J.G.: Introduction to the journal of marketing research special interdisciplinary issue on consumer financial decision making. J. Mark. Res. **48**(SPL), Siv–Sviii (2011)
12. Bazazo, I.K., Alananzeh, O.A., Taani, A.A.A.: Marketing the therapeutic tourist sites in Jordan using geographic information system. Marketing **8**(30) (2016)
13. Alghizzawi, M., et al.: The impact of smartphone adoption on marketing therapeutic tourist sites in Jordan. Int. J. Eng. Technol. **7**(4.34), 91–96 (2018)
14. Venkatesh, V., et al.: Perceived usefulness, perceived ease of use, and user acceptance of information technology. Manage. Sci. **46**(2), 319–340 (2000)
15. Tarhini, A., Elyas, T., Akour, M.A., Al-Salti, Z.: Technology, demographic characteristics and e-learning acceptance: a conceptual model based on extended technology acceptance model. High. Educ. Stud. **6**(3), 72–89 (2016)
16. Davis, F.D., Bagozzi, R.P., Warshaw, P.R.: User acceptance of computer technology: a comparison of two theoretical models. Manage. Sci. **35**(8), 982–1003 (1989)
17. Habes, M., Salloum, S.A., Alghizzawi, M., Alshibly, M.S.: The role of modern media technology in improving collaborative learning of students in Jordanian universities. Int. J. Inf. Technol. Lang. Stud. **2**(3) (2018)
18. Wadie, N.: An exploration of facebook. com adoption in Tunisia using technology acceptance model (TAM) and theory of reasoned action (TRA) (2012)
19. Carter, B., Levy, J., Levy, J.R.: Facebook Marketing: Leveraging Facebook's Features for Your Marketing Campaigns. Que Publishing (2012)
20. Debatin, B., Lovejoy, J.P., Horn, A.-K., Hughes, B.N.: Facebook and online privacy: attitudes, behaviors, and unintended consequences. J. Comput. Commun. **15**(1), 83–108 (2009)
21. Curran, K., Graham, S., Temple, C.: Advertising on Facebook. Int. J. E-bus. Dev. **1**(1), 26–33 (2011)
22. Habes, M.: The influence of personal motivation on using social TV: a uses and gratifications approach. Int. J. Inf. Technol. Lang. Stud. **3**(1) (2019)
23. Jung, T., Chung, N.: The impact of interaction and ubiquity on trust, benefits, and enjoyment in social media continuance use (2015)
24. Kim, Y.B., Joo, H.C., Lee, B.G.: How to forecast behavioral effects on mobile advertising in the smart environment using the technology acceptance model and web advertising effect model. KSII Trans. Internet Inf. Syst. **10**(10), 4997–5013 (2016)
25. Kim, S., Baek, T.H., Kim, Y.-K., Yoo, K.: Factors affecting stickiness and word of mouth in mobile applications. J. Res. Interact. Mark. **10**(3), 177–192 (2016)
26. Koufaris, M.: Applying the technology acceptance model and flow theory to online consumer behavior. Inf. Syst. Res. **2**(13), 205–223 (2002)
27. Salloum, S.A., Al-Emran, M., Shaalan, K., Tarhini, A.: Factors affecting the e-learning acceptance: a case study from UAE. Educ. Inf. Technol., 1–22 (2018)
28. Yang, B., Kim, Y., Yoo, C.: The integrated mobile advertising model: the effects of technology-and emotion-based evaluations. J. Bus. Res. **66**(9), 1345–1352 (2013)
29. Davis, F.D.: Perceived usefulness, perceived ease of use, and user acceptance of information technology. MIS Q., 319–340 (1989)
30. Rauniar, R., Rawski, G., Yang, J., Johnson, B.: Technology acceptance model (TAM) and social media usage: an empirical study on Facebook. J. Enterp. Inf. Manag. **27**(1), 6–30 (2014)
31. Venkatesh, V., Bala, H.: Technology acceptance model 3 and a research agenda on interventions. Decis. Sci. **39**(2), 273–315 (2008)

32. Hair, J.F., Black, W.C., Babin, B.J., Anderson, R.E., Tatham, R.L.: Multivariate Data Analysis (1998). (5. Baski)
33. Chin, W.W.: The partial least squares approach to structural equation modeling. Mod. methods Bus. Res. **295**(2), 295–336 (1998)
34. Straub, D., Boudreau, M.-C., Gefen, D.: Validation guidelines for IS positivist research. Commun. Assoc. Inf. Syst. **13**(1), 24 (2004)
35. Falk, R.F., Miller, N.B.: A primer for soft modeling. University of Akron Press (1992)
36. Fornell, C., Larcker, D.F.: Evaluating structural equation models with unobservable variables and measurement error. J. Mark. Res., 39–50 (1981)
37. Tenenhaus, M., Vinzi, V.E., Chatelin, Y.-M., Lauro, C.: PLS path modeling. Comput. Stat. Data Anal. **48**(1), 159–205 (2005)
38. Hamouda, M.: Understanding social media advertising effect on consumers' responses: an empirical investigation of tourism advertising on Facebook. J. Enterp. Inf. Manag. **31**(3), 426–445 (2018)

Examining the Effect of Knowledge Management Factors on Mobile Learning Adoption Through the Use of Importance-Performance Map Analysis (IPMA)

Mostafa Al-Emran[1]([⊠]) [iD] and Vitaliy Mezhuyev[2] [iD]

[1] Applied Computational Civil and Structural Engineering Research Group,
Faculty of Civil Engineering, Ton Duc Thang University,
Ho Chi Minh City, Vietnam
al.emran@tdtu.edu.vn
[2] Faculty of Computer Systems & Software Engineering,
Universiti Malaysia Pahang, Gambang, Malaysia
vitaliy@ump.edu.my

Abstract. A tremendous amount of research indicated that KM factors have significant impacts on different technologies at the organizational level. What is not yet clear is the influence of these factors on technology adoption at the individual level. On the other hand, the understanding of students' behavioral intention to use m-learning systems is still an ongoing research issue. Thus, the main theoretical contribution of this study is to investigate the impact of KM factors (i.e., acquisition, sharing, application, and protection) on m-learning adoption at the individual level, and to identify the importance and performance of each factor using the importance-performance map analysis (IPMA) technique through SmartPLS. A total of 319 IT undergraduate students enrolled at Al Buraimi University College in Oman took part in the study by the medium of online survey. In terms of importance, the empirical data analysis through IPMA exhibited that knowledge protection is the most important factor in determining the students' behavioral intention to use m-learning. Concerning performance, the findings also triggered out that both knowledge sharing and knowledge protection perform well in determining the students' behavioral intention to use m-learning.

Keywords: Knowledge management factors · Mobile learning · Adoption · Importance-performance map analysis · IPMA

1 Introduction

Information and communication technologies (ICTs) have been extensively used in higher educational institutes as these technologies change the landscape through which the students learn, communicate, and collaborate [1–4]. Mobile learning (m-learning) as one of such technologies, refers to the learning that is engaged with the mobility of

© Springer Nature Switzerland AG 2020
A. E. Hassanien et al. (Eds.): AISI 2019, AISC 1058, pp. 449–458, 2020.
https://doi.org/10.1007/978-3-030-31129-2_41

the student through the use of small computerized digital devices in anytime anywhere settings [5–7]. In this study, m-learning is defined as the learning that is concerned with the mobility of the student where knowledge can be personally managed using mobile applications.

From the m-learning standpoint, research reported that knowledge management factors (i.e., acquisition, sharing, application, and protection) need to be incorporated in m-learning systems in order to improve the learners' learning capabilities [8]. With regard to KM, m-learning is one of the KM practices that supports a reliable learning atmosphere, in which knowledge has been effectively acquired and shared among individuals [9, 10].

A recent systematic review indicated that KM factors have significant effects on adopting many technologies [11]. For instance, Lin and Lee [12] investigated the effect of knowledge acquisition, knowledge sharing, and knowledge application on the adoption of e-business systems. The findings pointed out that the adoption of e-business systems is positively influenced by all KM factors. Additionally, Garrido-Moreno et al. [13] examined the effect of knowledge acquisition, sharing, and application on customer relationship management (CRM) success. The findings triggered out that CRM success is positively affected by all KM factors. Further, Cheung and Vogel [14] explored the influence of knowledge sharing on Google applications adoption. The findings revealed that knowledge sharing has a positive influence on Google applications adoption. Moreover, Arpaci [15] examined the influence of knowledge creation & discovery, knowledge sharing, knowledge application, and knowledge storage on cloud computing adoption. His results indicated that perceived usefulness of cloud computing services is significantly influenced by knowledge sharing, knowledge creation & discovery, and knowledge storage.

A tremendous amount of research indicated that KM factors have significant impacts on different technologies at the organizational level [12, 16, 17]. What is not yet clear is the influence of these factors on technology adoption at the individual level. Although studies conducted by [15] and [14] concerned the individual level, these studies did not fully explore the four KM factors highlighted in this study. Further, research suggested examining the influence of KM factors on educational technologies (e.g., e-learning, m-learning, b-learning) as educational institutes are the essential pillars for generating knowledge [11]. On the other side, there has been little agreement on the factors affecting m-learning adoption and further research is highly encouraged to examine this phenomenon [18–20]. Therefore, the core theoretical contribution of the present study is to explore the influence of KM factors on m-learning adoption at the individual level, and to identify the importance and performance of each factor using the importance-performance map analysis (IPMA).

2 Hypotheses Development and Research Model

The theoretical model examined in the current study is demonstrated in Fig. 1. It is suggested that the behavioral intention to use m-learning is influenced by four KM factors (i.e., acquisition, sharing, application, and protection). The description of each factor along with the developed hypotheses are discussed in the following subsections.

2.1 Knowledge Acquisition (KA)

Knowledge acquisition (KA) refers to "the business processes that use existing knowledge and capture new knowledge" [16]. A considerable amount of research reported that KA has a significant effect on adopting several technologies [21, 22]. Therefore, we hypothesize the following:

> **H1:** KA has a significant positive impact on the behavioral intention to use m-learning.

2.2 Knowledge Sharing (KS)

Knowledge sharing (KS) refers to "the business processes that distribute knowledge among all individuals participating in process activities" [16]. Research triggered out that KS has a significant impact on adopting various technologies [23, 24]. Hence, this leads to the following hypothesis:

> **H2:** KS has a significant positive impact on the behavioral intention to use m-learning.

2.3 Knowledge Application (KAP)

Knowledge application (KAP) refers to "the business processes through which effective storage and retrieval mechanisms enable a firm to access knowledge easily" [16]. Several studies concluded that KAP has a significant influence on adopting various technologies [12, 17, 22]. Thus, we suggested the following hypothesis:

> **H3:** KAP has a significant positive impact on the behavioral intention to use m-learning.

2.4 Knowledge Protection (KP)

Knowledge protection (KP) refers to "the ability to protect organizational knowledge from illegal or inappropriate use or theft" [25]. Prior research pointed out that KP has a significant effect on adopting e-business systems [25]. Based on that, this study suggests the following hypothesis:

> **H4:** KP has a significant positive impact on the behavioral intention to use m-learning.

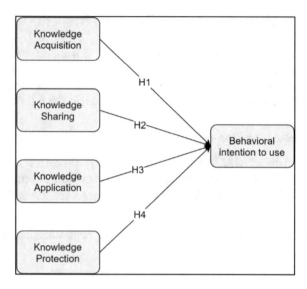

Fig. 1. Research model

3   Research Methodology

3.1   Context and Subjects

The sample of this study involves the IT undergraduate students registered at Al Buraimi University College (BUC) in Oman. The targeted students were enrolled in three different majors, including information systems, computer science, and software engineering. The selection of this sample refers to the fact that students at BUC are highly knowledgeable in using educational technologies for learning activities [26–30]. The students were requested to take part in the study through emails that include the survey link. A total of 319 students has successfully completed the online survey. Out of the 319 respondents, there were 166 females and 153 males. Moreover, about 69% of the respondents are aged between 18 and 22 years old.

3.2   Research Instrument

The survey instrument consists of two parts. The first part collects the demographic information from the students, whereas the second part gathers answers for the indicators involved under each factor. The indicators that represent the factors were adopted from the past literature and were evaluated through the five-point Likert scale ranging from "1 = Strongly Disagree" to "5 = Strongly Agree". The items used for measuring KA, KS, KAP, and KP were adapted from Al-Emran et al. [8].

3.3 Data Analysis

The proposed research model has been analyzed using the partial least squares-structural equation modeling (PLS-SEM) using SmartPLS [31]. The main purpose of employing PLS-SEM in this study stems from the exploratory essence of the theoretical model and the absence of previous related literature. This study follows the general guidelines and procedures for employing the PLS-SEM in information systems research [32]. As suggested in prior research [33], a two-step approach (i.e., measurement model and structural model) needs to be carefully followed to analyze the research model. In the current study, the authors mainly concentrate on evaluating the measurement model and examining the importance-performance map analysis (IPMA).

4   Results

4.1   Measurement Model Assessment

Research reported that both reliability and validity need to be verified during the assessment of the measurement model [34]. Reliability is usually evaluated by "Cronbach's alpha" and "composite reliability (CR)" [34]. The values of both Cronbach's alpha and CR should be equal to or greater than 0.70 in order to be accepted [34]. As indicated in Table 1, both reliability measures are confirmed.

Table 1. Reliability and convergent validity results

Constructs	Items	Factor loading	Cronbach's alpha	CR	AVE
BI	BI1	0.888	0.871	0.921	0.795
	BI2	0.894			
	BI3	0.893			
KA	KA1	0.838	0.865	0.903	0.650
	KA2	0.803			
	KA3	0.782			
	KA4	0.797			
	KA5	0.809			
KAP	KAP1	0.849	0.864	0.908	0.711
	KAP2	0.841			
	KAP3	0.854			
	KAP4	0.827			
KP	KP1	0.849	0.861	0.906	0.706
	KP2	0.855			
	KP3	0.872			
	KP4	0.783			
KS	KS1	0.821	0.860	0.899	0.642
	KS2	0.790			
	KS3	0.831			
	KS4	0.793			
	KS5	0.767			

For the purpose of confirming validity, both "convergent validity" and "discriminant validity" need to be ascertained [34]. In order to assess the "convergent validity", the indicators loadings and the Average Variance Extracted (AVE) need to be examined. To comply with the accepted threshold values [33], the values of AVE should be equal to or greater than 0.50, whereas the values of indicators loadings should be equal to or greater than 0.70. The results in Table 1 indicate that the values for both indicators loadings and AVE are accepted; therefore, the "convergent validity" is established. In order to assess the "discriminant validity", Henseler et al. [35] suggested examining the "Heterotrait-Monotrait ratio (HTMT)" of correlations. A value of less than 0.85 should be ascertained. As per the readings in Table 2, the HTMT results are regarded to be satisfactory; thus, the "discriminant validity" is confirmed.

Table 2. HTMT results

	BI	KA	KAP	KP	KS
BI					
KA	0.662				
KAP	0.734	0.747			
KP	0.782	0.670	0.810		
KS	0.657	0.789	0.836	0.840	

4.2   Importance-Performance Map Analysis

In this study, we have employed the importance-performance map analysis (IPMA) as an advanced technique in PLS-SEM using the behavioral intention to use m-learning as the target construct. Ringle and Sarstedt [36] argued that IPMA enhances the understanding of PLS-SEM analysis outcomes. As a substitute for only examining the path coefficients (i.e., importance measure), IPMA also involves the average value of the latent constructs and their indicators (i.e., performance measure) [36]. The IPMA assumes that the total effects indicate the predecessor factors' importance in framing the target factor (i.e., behavioral intention to use), whereas the average of latent constructs' values indicates their performance.

Figure 2 depicts the results of IPMA. In that, the importance and performance of the four KM factors (i.e., acquisition, sharing, application, and protection) were calculated. The findings triggered out that knowledge protection exhibits the highest values with respect to both importance and performance measures. Further, it can be seen that knowledge application exhibits the second highest values with regard to importance and performance measures. Moreover, knowledge acquisition exhibits the third highest value concerning the importance measure, but it shows the lowest value on the performance measure. Despite the fact that knowledge sharing has the lowest value on importance measure, it shows the highest similar value to the knowledge protection on the performance measure.

5 Implications for Research and Practice

The findings of the present study offer two valuable implications. For research, this study is considered the first that attempt to explore the impact of four KM factors (i.e., acquisition, sharing, application, and protection) on the behavioral intention to use m-learning at the individual level. For practice, the decision-makers of the higher educational institutes in Oman need to develop their policies and procedures by embedding KM factors into the deployment of m-learning systems due to the significant roles that these factors play in affecting students' intentions to use such systems.

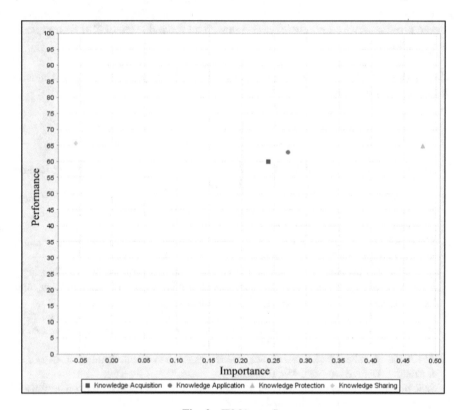

Fig. 2. IPMA results

6 Conclusion and Future Work

A tremendous amount of research indicated that KM factors have significant impacts on different technologies at the organizational level [12, 16, 17]. What is not yet clear is the impact of these factors on technology adoption at the individual level. On the other hand, the understanding of students' behavioral intention to use m-learning systems is still an ongoing research issue [37, 38]. Thus, the core theoretical contribution of this study is to investigate the impact of KM factors (i.e., acquisition, sharing, application, and protection) on m-learning adoption at the individual level, and to identify the importance and performance of each factor using the IPMA technique. In terms of importance, the findings exhibited that knowledge protection is the most important factor in determining the students' behavioral intention to use m-learning, followed by knowledge application, knowledge acquisition, and knowledge sharing, respectively. Concerning performance, the findings triggered out that both knowledge sharing and knowledge protection perform well in determining the students' behavioral intention to use m-learning, followed by knowledge application and knowledge acquisition, respectively.

The main constraint of the present study is the lack of analyzing the structural model. As a future work, we are currently working on analyzing the structural model in order to provide more comprehensive insights about the influence of KM factors on m-learning adoption.

Acknowledgment. This work was supported by Universiti Malaysia Pahang, Malaysia, under the grant RDU190311 and RDU1703107.

References

1. Ismail, M.I.B., Arshah, R.B.A.: The impacts of social networking sites in higher learning. Int. J. Softw. Eng. Comput. Syst. **2**(1), 114–119 (2016)
2. Al-Qaysi, N., Mohamad-Nordin, N., Al-Emran, M.: A systematic review of social media acceptance from the perspective of educational and information systems theories and models. J. Educ. Comput. Res. (2018)
3. Malik, S.I., Al-Emran, M.: Social factors influence on career choices for female computer science students. Int. J. Emerg. Technol. Learn. **13**(5), 56–70 (2018)
4. Al-Emran, M., Shaalan, K.: Learners and educators attitudes towards mobile learning in higher education: state of the art. In: International Conference on Advances in Computing, Communications and Informatics, pp. 907–913 (2015)
5. Al-Emran, M., Alkhoudary, Y.A., Mezhuyev, V., Al-Emran, M.: Students and educators attitudes towards the use of m-learning: gender and smartphone ownership differences. Int. J. Interact. Mob. Technol. **13**(1), 127–135 (2019)
6. Al-Emran, M., Shaalan, K.: Academics' awareness towards mobile learning in Oman. Int. J. Comput. Digit. Syst. **6**(1), 45–50 (2017)
7. Salloum, S.A., Al-Emran, M., Shaalan, K.: The impact of knowledge sharing on information systems: a review. In: 13th International Conference, KMO 2018 (2018)
8. Al-Emran, M., Mezhuyev, V., Kamaludin, A.: Students' perceptions towards the integration of knowledge management processes in m-learning systems: a preliminary study. Int. J. Eng. Educ. **34**(2), 371–380 (2018)
9. Jeong, H.Y., Hong, B.H.: A practical use of learning system using user preference in ubiquitous computing environment. Multimed. Tools Appl. **64**(2), 491–504 (2013)
10. Viberg, O., Grönlund, Å.: Cross-cultural analysis of users' attitudes toward the use of mobile devices in second and foreign language learning in higher education: a case from Sweden and China. Comput. Educ. **69**, 169–180 (2013)
11. Al-Emran, M., Mezhuyev, V., Kamaludin, A., Shaalan, K.: The impact of knowledge management processes on information systems: a systematic review. Int. J. Inf. Manage. **43**, 173–187 (2018)

12. Lin, H.F., Lee, G.G.: Impact of organizational learning and knowledge management factors on e-business adoption. Manag. Decis. **43**(2), 171–188 (2005)
13. Garrido-Moreno, A., Lockett, N., García-Morales, V.: Paving the way for CRM success: the mediating role of knowledge management and organizational commitment. Inf. Manag. **51** (8), 1031–1042 (2014)
14. Cheung, R., Vogel, D.: Predicting user acceptance of collaborative technologies: an extension of the technology acceptance model for e-learning. Comput. Educ. **63**, 160–175 (2013)
15. Arpaci, I.: Antecedents and consequences of cloud computing adoption in education to achieve knowledge management. Comput. Hum. Behav. **70**, 382–390 (2017)
16. Lee, C., Lee, G., Lin, H.: The role of organizational capabilities in successful e-business implementation. Bus. Process Manag. J. **13**(5), 677–693 (2007)
17. Migdadi, M.M., Abu Zaid, M.K.S., Al-Hujran, O.S., Aloudat, A.M.: An empirical assessment of the antecedents of electronic-business implementation and the resulting organizational performance. Internet Res. **26**(3), 661–688 (2016)
18. Almaiah, M.A.: Acceptance and usage of a mobile information system services in University of Jordan. Educ. Inf. Technol. **23**(5), 1873–1895 (2018)
19. Mohammadi, H.: Social and individual antecedents of m-learning adoption in Iran. Comput. Hum. Behav. **49**, 191–207 (2015)
20. Al-Emran, M., Mezhuyev, V., Kamaludin, A.: Technology acceptance model in m-learning context: a systematic review. Comput. Educ. **125**, 389–412 (2018)
21. He, M., Wang, T., Xiao, X.: Knowledge application process and assimilation of inter-organizational information systems: an empirical study. In: International Conference on Management Science & Engineering, pp. 916–922 (2010)
22. Yee-Loong Chong, A., Ooi, K.-B., Bao, H., Lin, B.: Can e-business adoption be influenced by knowledge management? An empirical analysis of Malaysian SMEs. J. Knowl. Manag. **18**(1), 121–136 (2014)
23. Maditinos, D., Chatzoudes, D., Sarigiannidis, L.: Factors affecting e-business successful implementation. Int. J. Commer. Manag. **24**(4), 300–320 (2014)
24. Salloum, S.A., Al-Emran, M., Shaalan, K., Tarhini, A.: Factors affecting the e-learning acceptance: a case study from UAE. Educ. Inf. Technol. **24**(1), 509–530 (2019)
25. Lin, H.F.: The effects of knowledge management capabilities and partnership attributes on the stage-based e-business diffusion. Internet Res. **23**(4), 439–464 (2013)
26. Al-Emran, M., Salloum, S.A.: Students' attitudes towards the use of mobile technologies in e-evaluation. Int. J. Interact. Mob. Technol. **11**(5), 195–202 (2017)
27. Al-Qaysi, N., Al-Emran, M.: Code-switching usage in social media: a case study from Oman. Int. J. Inf. Technol. Lang. Stud. **1**(1), 25–38 (2017)
28. Al-Emran, M., Malik, S.I.: The impact of Google apps at work: higher educational perspective. Int. J. Interact. Mob. Technol. **10**(4), 85–88 (2016)
29. Al-Maroof, R.A.S., Al-Emran, M.: Students acceptance of Google classroom: an exploratory study using PLS-SEM approach. Int. J. Emerg. Technol. Learn. **13**(6), 112–123 (2018)
30. Alfarsi, G., Alsinani, M.: Developing a mobile notification system for al Buraimi University College students. Int. J. Inf. Technol. Lang. Stud. **1**(1), 10–16 (2017)
31. Ringle, C.M., Wende, S., Becker, J.-M.: SmartPLS 3. Bönningstedt: SmartPLS (2015)
32. Al-Emran, M., Mezhuyev, V., Kamaludin, A.: PLS-SEM in information systems research: a comprehensive methodological reference. In: 4th International Conference on Advanced Intelligent Systems and Informatics (AISI 2018), pp. 644–653 (2018)
33. Hair, J., Hollingsworth, C.L., Randolph, A.B., Chong, A.Y.L.: An updated and expanded assessment of PLS-SEM in information systems research. Ind. Manag. Data Syst. **117**(3), 442–458 (2017)

34. Hair Jr., J.F., Hult, G.T.M., Ringle, C., Sarstedt, M.: A Primer on Partial Least Squares Structural Equation Modeling (PLS-SEM). Sage Publications, Thousand Oaks (2016)
35. Henseler, J., Ringle, C.M., Sarstedt, M.: A new criterion for assessing discriminant validity in variance-based structural equation modeling. J. Acad. Mark. Sci. **43**(1), 115–135 (2015)
36. Ringle, C.M., Sarstedt, M.: Gain more insight from your PLS-SEM results: the importance-performance map analysis. Ind. Manag. Data Syst. **116**(9), 1865–1886 (2016)
37. Al-Emran, M., Mezhuyev, V., Kamaludin, A., AlSinani, M.: Development of M-learning application based on knowledge management processes. In: 2018 7th International conference on Software and Computer Applications (ICSCA 2018), pp. 248–253 (2018)
38. Al-Emran, M., Shaalan, K.: Attitudes towards the use of mobile learning: a case study from the gulf region. Int. J. Interact. Mob. Technol. **9**(3), 75–78 (2015)

Toward the Automatic Correction of Short Answer Questions

Zeinab E. Attia$^{(\boxtimes)}$, W. Arafa, and M. Gheith

Institute of Statistical Studies and Research (ISSR),
Cairo University, Giza, Egypt
eng.zeinabezz@gmail.com

Abstract. The manual correction for short answer questions is a tedious and time consuming task. This issue leads to the importance of having automatic correction systems for short answer questions. Thus this paper proposes three different approaches for correcting short answer questions. One of them is NLP-based while the others are machine learning based approaches. The three approaches are tested and compared across another two short answer correction systems that are evaluated on the same dataset. Results show that the proposed approaches outperform others that are evaluated on the same dataset.

Keywords: Short answer correction system · Text similarity · Machine learning

1 Introduction

The automatic correction for short answer questions is one of the most important semantic similarity applications where the similarity degree between a model answer and a student answer is calculated. Several automatic short answer correction systems are proposed in the literature [12]. Mohler [1] proposed a similarity measure based correction system and apply it on computer science domain using texas dataset. Results show that it reaches correlation degree 0.4628 using LSA. On the other hand, Gomaa & Fahmy [3, 6] reaches correlation degree 0.504 using n-gram then Disco1 on the same dataset. However, IndusMarker [2] reaches 0.694 using a structure matching based system on an object oriented programming course. C-rater [4] proposed an NLP based approach on a mathematics course. It reaches kappa value 0.8.

The paper introduces three different approaches for correcting short answer questions one of them is NLP-based while others are machine learning based (a feature-engineering approach and a deep learning approach). All approaches are tested on texas dataset and their performance is evaluated using Pearson correlation coefficient. Then their performance is compared with other correction systems that are tested on the same dataset.

In the feature-engineering approach, features are extracted from both answers depending on their Abstract meaning Representation (AMR). AMR is a sentence semantic representation language that aims to abstract away from the syntactic representation such that sentences with the same basic meaning assign the same AMR

© Springer Nature Switzerland AG 2020
A. E. Hassanien et al. (Eds.): AISI 2019, AISC 1058, pp. 459–469, 2020.
https://doi.org/10.1007/978-3-030-31129-2_42

representation even if having different syntax. JAMR[1] is the first AMR parser that reaches accuracy 90% using F1 measure [16]. In 2015, Microsoft publishes its online AMR parser[2] [8]. Smatch[3] is an AMR evaluator that compares two monolingual AMRs [7].

The rest of this paper is organized as follows. An introduction about machine learning is presented in Sect. 2. Sections 3 and 4 discuss the proposed short answer question correction approaches. Section 5 evaluates and tests the proposed approaches. Finally in Sect. 6, the paper is concluded.

2   Machine Learning

Machine learning is a sub-field of Artificial Intelligence (AI) that is concerned with giving computers the ability to learn without being explicitly programmed. Machine learning has two main approaches namely feature engineering and feature learning (Deep learning) approach. The feature engineering approach works well on a variety of important problems. However, they have not succeeded in solving the central problems in AI, such as recognizing speech or recognizing objects [5]. Deep learning was motivated to generalize well on such AI tasks [10] using Recurrent Neural Network (RNN) [17], Long Short Term Memory (LSTM) [15]…etc. The rest of this section introduces some deep learning applications.

2.1   Text Similarity in Deep Learning

This subsection presents some deep learning based text similarity systems using Siamese network. Mueller [11] proposed LSTM Siamese model that calculates the similarity between two sentences. A sentence is entered as a sequence of word vectors to an LSTM network. Then the similarity degree between the two sentences is calculated using Manhattan distance. The model is trained and tested on SICK data. On testing, it achieves correlation degree 0.8822.

Hu [13] proposed a convolutional sentence model. The model takes as its input the embedding of words in a sentence aligned sequentially. Then it summarizes the meaning of a sentence through layers of convolution and pooling.

Also, Hu [13] proposed a Siamese convolutional based model for determining whether two sentences are similar or not. It finds the representation of each sentence, and then compares the representation for the two sentences with a multi-layer perceptron (MLP). It is evaluated using the benchmark MSRP dataset. Results shows that it reaches the accuracy degree 69.6 and F1 score 80.27.

---

[1] http://github.com/jflanigan/jamr

[2] Available at: http://research.microsoft.com/msrsplat

[3] Smatch: amr.isi.edu/evaluation.html

2.2   Text Classification in Deep Learning

Zichao [14] proposed the Hierarchical Attention Network (HAN) to classify documents. We concerned only with how Zichao represents a document as a vector. HAN is designed to capture two basic insights about document structure. First, since documents have a hierarchical structure (words form sentences, sentences form a document), Zichao constructs a document representation by first building representations of sentences and then aggregating those into a document representation. Second, it is observed that different words and sentences in a document are differentially informative. Thus, HAN has two levels of attention mechanisms, one at the word level and one at the sentence level. HAN architecture consists of four parts: a word sequence encoder, a word-level attention layer, a sentence encoder and a sentence-level attention layer.

Assume that a document has L sentences s_i and each sentence contains T_i words w_{it} with $t \in [1, T]$ represents the words in the ith sentence.

Word Encoder: Given a sentence with words wit, $t \in [0, T]$, first embed the words to vectors through an embedding matrix W_e, ($x_{it} = W_e w_{it}, t \in [1, T]$). Then use a bidirectional GRU [9] to get annotations of words by summarizing information from both directions for words,

$$\overrightarrow{h_{it}} = \overrightarrow{GRU}(x_{it}), t \in [1, T], \qquad \overleftarrow{h_{it}} = \overleftarrow{GRU}(x_{it}), t \in [1, T]$$

Then obtain the annotation for a given word wit by concatenating the forward hidden state $\overrightarrow{h_{it}}$ and backward hidden state $\overleftarrow{h_{it}}$, i.e., hit = [$\overrightarrow{h_{it}}, \overleftarrow{h_{it}}$], which summarizes the information of the whole sentence centered around wit.

Word Attention: Attention mechanism is introduced to extract important words affecting the sentence meaning and aggregate the representation of those informative words to form a sentence vector. Specifically,

$$u_{it} = \tanh(W_w h_{it} + b_w), \qquad \alpha_{it} = \frac{\exp(u_{it}^T u_w)}{\sum_t \exp(u_{it}^T u_w)}, \qquad s_i = \sum_t \alpha_{it} h_{it}$$

u_{it} is a hidden representation of h_{it}, α_{it} measures the importance of the word and is calculated as the similarity of uit with a word level context vector u_w. The word context vector u_w is randomly initialized and jointly learned during the training process. The sentence vector s_i is a weighted sum of the word annotations based on the weights.

Sentence Encoder: Given the sentence vectors s_i, a document vector can be obtained in a similar way. Use a bidirectional GRU to encode the sentences:

$$\overrightarrow{h_i} = \overrightarrow{GRU}(s_i), i \in [1, L], \qquad \overleftarrow{h_i} = \overleftarrow{GRU}(s_i), i \in [1, L]$$

Then concatenate $\overrightarrow{h_i}$ and $\overleftarrow{h_i}$ to get an annotation of sentence i, i.e., h_i = [$\overrightarrow{h_i}, \overleftarrow{h_i}$]. h_i summarizes the neighbor sentences around sentence i but still focus on sentence i.

Sentence Attention: Use attention mechanism and introduce a sentence level context vector us to measure the importance of the sentences. Thus:

$$u_i = \tanh(W_s h_i + b_s), \qquad \alpha_i = \frac{\exp(u_i^T u_s)}{\sum_i \exp(u_i^T u_s)}, \qquad v = \sum_i \alpha_i h_i$$

where v is the document vector that summarizes all the information of sentences in a document. Similarly, the sentence level context vector u_s can be randomly initialized and jointly learned during the training process.

3   The Proposed NLP-Based Short Answer Correction System

The proposed NLP-based system consists of three stages namely, the complete model answer predicate extraction stage, the student answer correction stage, and the student answer scoring stage. The complete model answer predicate extraction stage extracts the complete model answer and represents it in a predicate form via expanding the model answer using the domain ontology. Thus, this stage takes as its input the model answer, and the domain ontology (representing the course material). Then, it returns the complete model answer predicates. However in the correction stage the student answer is compared with the model answer to find the matching part(s) if exists. Finally, the scoring stage scores the student answer.

3.1   The Complete Model Answers Predicate Extraction Stage

This stage runs as follow:

1. For both the model and student answers, stem each of them. Then return each pronoun to its referring noun using a coreference tool.
2. For both the model and the student answers, divide the answer into a set of statements using ".", ";", "and", "or", "but", "while", "whereas", "thus", "so", "however", "normally", "therefore".
3. Represent the stemmed model answer text in a predicate form. All the predicate elements (e.g., relation, concept) must exist in the ontology.
4. Note that: a sentence's predicate is extracted via using the sentence AMR then AMR is converted to predicates using hand-crafted rules.
5. Expand the model answer predicate's subject and verb using the ontology.

3.2   The Student Answer Correction Stage

This stage runs as follows:

1. For each statement in the student answer, use the domain dictionary to replace each student word with its synonym respecting the complete model answer predicates.
2. Match each statement in the student answer with the complete model answer predicates. Make sure that the matched statement and predicate are not opposite to each other using negative keywords, e.g., not, except …etc.

3.3   The Student Answer Scoring Stage

In this stage, the student answer is scored with score ranged from '0' through the question full mark. It assumes either:

- Each part in the model answer is assigned a score by the instructor
- The model answer is assigned the full mark by the instructor. Then, the system will divide it on the number of the model answer predicates.

The scoring stage scores the student answer respecting the number of matched predicates detected from the correction stage.

4   The Proposed Machine Learning Short Answer Correction System

This section introduces our two proposed approaches for correcting short answer questions. The next subsection illustrates the feature-engineering approach. The deep learning approach is presented in Subsect. 4.2.

4.1   The Proposed Short Answer Correction Based on Feature-Engineering

This proposed system works through two main phases; namely concept extraction phase and similarity degree calculation phase.

Concept Extraction Phase. It is concerned with extracting domain concepts from the model answer. In this phase, we propose a machine learning based concept extraction system that takes as input an initial set of domain concepts to train on besides a set of sentences labeled with their contained concepts (training dataset). The dataset is preprocessed using stemmer, coreference resolution, removing stop words, and dividing each sentence into tokens with 1, 2, and 3-grams. Each token is an instance in the dataset where the following features are extracted from it.

- number of n-grams (1, 2, or 3-grams),
- its head word POS,
- the token shape,
- is a domain concept or not, indicated using the initial set of domain concepts.
- co-occurrence of this token with its the next token,
- $\dfrac{\text{the number of words in the token that are exist in the domain}}{\textit{total number of words in this token}}$
- $\dfrac{\sum_{i=1}^{t} \text{co–occurence between word (or its synonym)}_{w_i \text{ and } w_{i+1}}}{\textit{total number of words in this token}}$

where w_i is a word in the token, t is the number of words in the token.

Similarity Degree Calculation Phase. This phase calculates the similarity degree between the student and the model answer using our proposed similarity degree calculation system. It is a machine learning feature-engineering based system that takes as its input a set of manually graded student answers associated with their model answers (dataset) to train on it and the domain thesaurus. It works through two main stages namely dataset preparation stage and feature extraction stage.

Dataset Preparation Stage. The student answer is prepared via representing it as AMR. On the other hand, the model answer is prepared through: extract all its contained domain concepts, expand each extracted concepts with its synonyms using the domain thesaurus, expand each extracted concept with its definition using the domain thesaurus, represent the model answer and each expanded answer as AMR, and expand the model answer via replacing concepts' definition with its scientific concept.

Feature Extraction Stage. The following features are extracted from the prepared dataset:

- Smatch between both answers,
- $\frac{\text{the number of matched concepts between both answers}}{\text{number of concepts in the model answer}}$,
- Student answer's subsumption degree.

4.2   The Proposed Short Answer Correction Based on Deep Learning

This approach corrects short answer question using two deep learning models. Each model is a Siamese neural network that takes as its input; model answer, student answer, course material and a domain thesaurus. The course material is the course book. It is used to construct the embedding matrix using 300 dimensional word embedding.

The difference between the two models is in the type of the used network. Each model is illustrated in details below.

1. CNN-LSTM-Cos model: is a Siamese neural network that deals with each answer as one sentence by removing its punctuations. It is a combination of CNN and LSTM models. The CNN is used to represent each answer as a fixed vector using Hu convolutional sentence model [13] which is then encoded by LSTM. Subsequently, the result of the two LSTM networks is compared using the cos-similarity metric that calculates the similarity between both answers.
2. HAN-Cos model: is a Siamese neural network that uses HAN model [14] to represent the meaning of each answer. Then, the result of the two networks is compared using the cos-similarity metric that calculate the similarity between the two answers.

5   Results and Discussion

This section evaluates the three proposed approaches then discussing their results.

5.1   The Proposed NLP-Based Short Answer Correction System

Dataset: The proposed correction system is trained and tested on Texas dataset[4]. It is divided into: Ten assignments with between four and seven questions each, and two exams with ten questions each. This dataset was assigned to an introductory computer science class at the University of North Texas. It includes questions, model answers, and set of student answers with the average grades of two annotators. The total number of student answers in it is about 2500 answer. Moreover, the dataset includes the course material (lecture notes).

Evaluating the Proposed System: To evaluate the proposed system, Pearson's correlation coefficient is measured. The proposed system is compared with Texas [1] and Gomaa [3, 6] systems as they are tested on the same dataset. Texas reaches the correlation values 0.328, 0.395, and 0.281 by applying LSA, ESA, and tf*idf respectively. However, Gomaa reaches the correlation value 0.504 by combining n-gram and Disco1algorithms in an unsupervised way. The proposed system raises the correlation results to be about 0.65, see Fig. 1. Unfortunately, this approach is biased on the tested dataset, due to extracting predicates from AMR is rule based and such rules are biased on the tested dataset. Thus we propose the second approach, feature engineering.

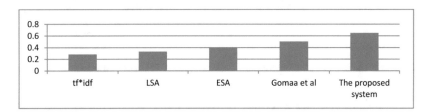

Fig. 1. Comparison between the proposed system, Mohler [1], and Gomaa et al. [3, 6] systems

5.2   The Proposed Short Answer Correction Based on Feature-Engineering

Dataset: The proposed correction system is trained and tested on Texas dataset including the course material (Book). This section evaluates the proposed concept extraction system and the proposed text similarity system.

Evaluating the Proposed System

The Proposed Concept Extraction System: The proposed concept Extraction system is trained on about 2600 text from texas dataset. To preprocess the dataset, StanfordCoreNLP is used as a coreference resolution tool. The over-stemming algorithm

---

[4] https://web.eecs.umich.edu/~mihalcea/downloads.html

Paice/Husk is used. We use 0.25% from the dataset for testing. Since the training dataset is non-linear separable, the support vector machine algorithm is used with cross validation 5 folds. Results show that the best accuracy degree reaches 95% in training and about 89% in testing using the RBF kernel and gamma value 0.1.

The Proposed Text Similarity System: The proposed feature-engineering based text similarity system is trained and tested on texas dataset which contains 2500 student answer. We use 0.2% from the dataset for testing. Linear regression and Multilayer perceptron algorithm are used. Results show that the best correlation degree achieves about 83% for training and about 65% for testing using the Multilayer perceptron algorithm with 2 hidden layers and 0.01 learning rate.

Figure 2 compares the proposed correction system with Mohler [1] and Gomaa [3, 6] systems as they are tested on the same dataset. The correlation values of applying LSA, ESA, and tf*idf is 0.328, 0.395, and 0.281 respectively. However, Gomaa [3, 6] has the correlation value 0.504 by combining String-based and Corpus-based similarity. The proposed system raises the correlation results to be about 0.65.

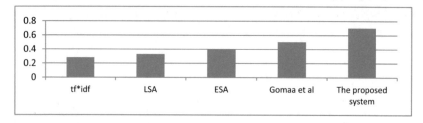

Fig. 2. Comparison between the proposed system, Mohler [1], and Gomaa et al. [3, 6] systems

5.3   The Proposed Short Answer Correction Based on Deep Learning

Dataset: The two deep learning based correction models are trained and tested on Texas dataset. The dataset is split into training dataset (0.8%) and testing dataset (0.2%).

Baseline Models: Since there is no deep learning based short answer correction model to compare our deep learning models with, we compare them with two text similarity based on deep learning siamese models which are Mueller model based on LSTM [11] and Hu model based on CNN [13] model.

The LSTM model is modified to deal with text that contains more than one sentence via removing its punctuations. On the other hand, the CNN model is modified to deal with text that contains more than one sentence via removing the text punctuations. Also it is modified via adding another layer after representing each text using Hu [13] convolutional sentence model to calculate their similarity degree using the cos-similarity function.

Evaluating the Proposed Models: To evaluate the two models, the dataset is augmented, using the domain thesaurus, with synonyms and definitions. Thus, the dataset size is increased from 2500 instance to 9735 instance, such that 7788 instance is for training and 1947 instance is for testing. A dropout layer with value 0.25 is added on the augmented dataset. Before training the two models, a preprocessing step is applied on the dataset by removing punctuations and stop words then stem the remaining words. Training and testing results of the two models are depicted in Fig. 3. HAN-cos model outperforms the rest models. It achieves 0.9 correlation degree in training and achieves 0.82 in testing. On the other hand, the CNN-LSTM-cos model reaches 0.8 correlation degree in training and 0.76 in testing.

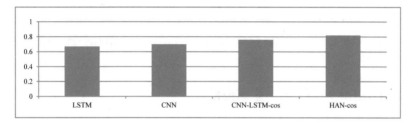

Fig. 3. The result of applying LSTM, CNN, CNN-LSTM, HAN models

5.4 The Result

From Fig. 4, it is observed that our proposed models outperform other models that are tested on the same dataset. The NLP- based approach reaches correlation degree 55%. However, this approach is biased on the tested dataset since the used statement separators and the predicate construction rules using AMR are extracted from the dataset. Thus, we propose our feature engineering approach that overcomes such limitation and reaches correlation degree 65%. On the other hand, the feature learning based approach removes the headache of extracting the best set of features from the dataset where the system will learn it by itself. The feature learning approach reaches 76% during testing by using CNN-LSTM-cos model. This value is raised to 82% by using HAN-cos model. HAN-cos model outperforms other deep learning models as:

- It works on the fact that words form a sentence and sentences from a text, unlike other deep learning models deal with a text as one sentence.
- Unlike other models, HAN-cos model works only on words and sentences that contribute to the meaning of their contained sentences and text respectively.

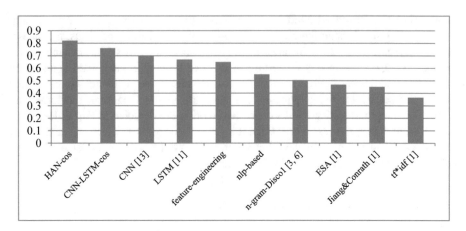

Fig. 4. Comparing the result of the proposed models with Mohler [1] and Gomaa [3, 6], LSTM [11], CNN [13]

6   Conclusion and Future Work

HAN-cos model outperforms the other models because its text representation depends on aggregating the representation of its important sentences that affect its meaning. Also the representation of its sentences depends on aggregating the representation of important words that affects its meaning. As our Future work,

- Use the paragraph2vec to represent the sentence meaning instead of the word2vec.
- Instead of having a model answer for each question, generate it via a question answer system. To have a question answering system use AMR or CBOW then compare their performance.
- Correct short answer questions using a hybrid approach (NLP-based, machine learning, deep learning).

References

1. Mohler, F., Mihalcea, R.: Text-to-text semantic similarity for automatic short answer grading. In: Proceedings of the 12th Conference of the European Chapter of the ACL, Association for Computational Linguistics, Athens, Greece, pp. 567–575 (2009)
2. Siddiqi, R.: Impact of Automated short-answer marking on students' learning: IndusMarker, a case study. In: International Conference on Information & Communication Technologies (2013)
3. Gomaa, W.H., Fahmy, A.A.: Automatic Arabic Essay Assessment. Ph.D. thesis, Faculty of Computer Science, Cairo University (2014)
4. Leacock, C., Chodorow, M.: C-rater: automated scoring of short-answer questions. Comput. Humanit. **37** (2003)
5. Goodfellow, I., Bengio, Y., Courville, A.: Deep Learning. MIT Press, New York (2016)

6. W. Gomaa, W., Fahmy, A.: Short answer grading using string similarity and corpus-based similarity. Int. J. Adv. Comput. Sci. Appl. (2012)
7. Cai, S., Knight, K.: Smatch: an evaluation metric for semantic feature structures. In: Proceedings of the 51st Annual Meeting of the Association for Computational Linguistics, pp. 748–752 (2013)
8. Vanderwende, A., Menezes, A., Quirk, V.: An AMR parser for English, French, German, Spanish and Japanese and a new AMR-annotated corpus. In: Microsoft Research, Proceedings of NAACL-HLT 2015, pp. 26–30 (2015)
9. Bahdanau, D., Cho, K., Bengio, Y.: Neural machine translation by jointly learning to align and translate. arXiv preprint arXiv:1409.0473 (2014)
10. Du, K.L., Swamy, M.N.S.: Neural Networks and Statistical Learning. Chap. 11, Canada, pp. 351–363 (2013)
11. Mueller, J., Thyagarajan, A.: Siamese Recurrent architectures for learning sentence similarity. In: Proceedings of the Thirtieth AAAI Conference on Artificial Intelligence (2016)
12. Attia, Z.E., Arafa, W., Gheith, M.: An automatic short answer correction system based on the course material. Int. J. Intell. Eng. Syst. **11**(3) (2018)
13. Hu, B., Lu, Z., Li, H., Chen, Q.: Convolutional neural network architectures for matching natural language sentences. In: NIPS (2014)
14. Yang, Z., Yang, D., Dyer, C., He, X., Smola, A., Hovy, E.: Hierarchical attention networks for document classification. In: Proceedings of the North American Chapter of the Association for Computational Linguistics: Human Language Technologies, pp. 1480–1489 (2016)
15. Hochreiter, S., Schmidhuber, J.: Long short-term memory. Neural Comput. **9**(8), 1735–1780 (1997)
16. Flanigan, J., Thomson, S., Carbonell, J., Dyer, C., Smith, N.A.: A discriminative graph-based parser for the abstract meaning representation. In Proceedings of the 52nd Annual Meeting of the Association for Computational Linguistics, pp. 1426–1436 (2014)
17. Salehinejad, H., Sankar, S., Barfett, J., Colak, E., Valaee, S.: Recent Advances in Recurrent Neural Networks. arXiv:1801.01078 (2018)

Machine and Deep Learning Algorithms

A Novel Automatic CNN Architecture Design Approach Based on Genetic Algorithm

Amr AbdelFatah Ahmed[1]($\boxtimes$), Saad M. Saad Darwish[2],
and Mohamed M. El-Sherbiny[3]

[1] Department of Computer Engineering,
Alexandria High Institute of Engineering and Technology, Alexandria, Egypt
amr.abdelfatah@aiet.edu.eg
[2] Department of Information Technology, Institute of Graduate Studies
and Research, Alexandria University, Alexandria, Egypt
saad.darwish@alex-igsr.edu.eg
[3] Department of Material Science, Institute of Graduate Studies and Research,
Alexandria University, Alexandria, Egypt
mmmsherbiny@gmail.com

Abstract. The deep "Convolutional Neural Networks (CNNs)" gained a grand success on a broad of computer vision tasks. However, CNN structures training consumes a massive computing resources amount. The researchers in this field are concerned on designing CNN structures to maximize the performance and accuracy. The main design methods are human hand-crafted fixed model structures and automatic generated models. We proposed an automatic CNN structure design approach based on genetic algorithm that concerned with generating light weight CNN structures. We also introduce a chromosome novel representation for the structure of CNN. Unlike existing approaches, the proposed methodology is designed to work on limited computing assets with achieving high accuracy. It utilizes advanced training methods to decrease the overhead on the computing resources that are involved in the process. Our experimental results denote the proposed model effectiveness over the related work methods.

Keywords: Convolutional Neural Networks · CNNs · Genetic algorithm · Automatic model design

1 Introduction

Convolutional Neural Networks (CNNs) have accomplished the state-of-the-art in solving many real-world problems of computer vision like image and video classification, object recognition, object tracking, scene labelling and detection of text [1]. It also utilized for Natural Language Processing including audio and speech. There are many various architectures that are hand crafted by researchers such as VGGnet [2], GoogLeNet [3] and ResNet [4] but these CNN structures are large models that consume massive amount resources and time in training. There are another hand-crafted CNN structures that concerned with the time and the resources consumed in process of training such as Mobilenet [5], Squeezenet [6] and Shufflenet [7]. On the other hand,

© Springer Nature Switzerland AG 2020
A. E. Hassanien et al. (Eds.): AISI 2019, AISC 1058, pp. 473–482, 2020.
https://doi.org/10.1007/978-3-030-31129-2_43

another research direction as represented in Fig. 1 focuses on automatic CNN structure design methods as in [8–18]. These methods are automated search techniques that attempt to find in the search space for the optimal arrangement of the CNN building blocks that achieve the highest accuracy. The hand-crafted CNN designs need the human experts and a set of trial and error to find efficient design of CNN model. The automated design methods solve the problem of needing human experts. The automatic CNN architecture design approaches will be discussed in the following section.

The rest of this paper is organized as follow. Section 2 discusses the Evolutionary algorithms (EA) and Reinforcement Learning (RL). Section 3 represents the proposed model. The experimental results are represented in Sect. 4. Finally, Sect. 5 concludes the paper and represents the future directions.

2　Related Work

In this section, we will discuss the related work for automatic design methods of CNN architecture. We will classify the related work as the taxonomy in Fig. 1 according the methodologies related work based on which are Evolutionary Algorithms (EA)- based, Reinforcement Learning (RL)-based and other general methods.

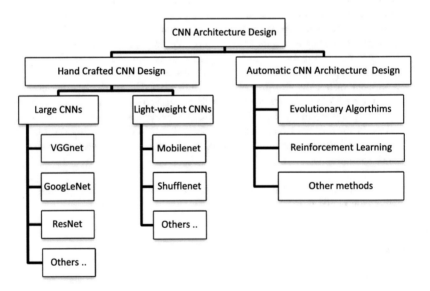

Fig. 1. CNN architecture design taxonomy

2.1 Evolutionary Algorithms-Based

The EA based methods are utilizing Genetic Algorithm (GA) as in [8, 9] or Genetic Programming" [10]. In [9], they utilize GA to find high accuracy architecture model but with time tradeoff in searching. In [10], they proposed an automatic approach based on Cartesian Genetic Programming (CGP) for construction the CNN structures but the process consumes more time than [8].

2.2 Reinforcement Learning-Based

Other related works utilize RL as in [11–16]. In [11], an approach which called Neural Architecture Search (NAS). NAS utilizes Recurrent Neural Network (RNN) that is trained in loop. The controller of RNN generates a child model utilizing the REIN-FORCE algorithm. In [12], the approach is considered an extend to the [11] work by replacing REINFORCE algorithm with Proximal Policy Optimization (PPO), but in this approach a huge computing resources are needed as it uses 450 GPUs for 3 to 4 days. In [13], NAS model search space is utilized but by replacing Q-learning instead of policy gradient as it enhances the performance as it needs 32 GPUs for 3 days which is still very time and resource consuming. While [14] utilizes Q-learning with preserving the cell structure. In [15], they trained the RNN using Policy gradient as [12] but the actions are done to make an existing layer wide or to make the network deep. In [16], the method is called "Efficient Neural Architecture Search (ENAS)". The research team used the transfer learning concept by which the generated child models share parameters among them that leads to speed the process up and much fewer GPU-hours which can be performed by a single GPU for 16 h, but the process is memory exhaustive.

2.3 Other Methods

The other methods that are based on other search techniques as in [17, 18]. In [17], the approach called "AmoebaNet" was even more computationally intensive than NAS as the method was implemented on 450 "K40" GPUs for 7 days to complete the process. In [18], a method called "Differentiable Architecture Search (DARTS)". DARTS consumed 4 GPU days to find an optimal network for benchmark dataset.

In conclusion, the challenge mainly in all automated design methods that they need high computing machines to perform their search process. This due that the search space is huge as there are a building blocks variety for CNN models. In addition, the hyper-parameters diversity for each block and there are endless stacking blocks arrangements. Our objective is to propose an effective automated design method of CNN model that minimize the search time consumed to find the CNN optimal model targeting the limited computing machines. The proposed method is based on GA which produces light-weight efficient CNNs that compete with peer CNN models.

3 Methodology and the Proposed Model

3.1 CNN Network Representation

We proposed a genotype representation chromosome as represented in Fig. 2. The chromosome is used generate and represent CNN architecture. Every chromosome consists of three building blocks which are distinguished by a number and ends with a final block. Each of these three blocks consists of layers with identity numbers (id), the input (In), the output (Out) and the Sub-sampling (Sub). Each layer (Lay) in these blocks are consisted from these components which are Convolution layer (c), activation layer (r), "Dropout" (d) [19] and "Batch Normalization" layer (b) [20].

Block32					Block16					Block8					Final
Id	Lay	In	Out	Sub	Id	Lay	In	Out	Sub	Id	Lay	In	Out	Sub	T
0	rbc	-1	1	F	0	cbr	-1	1,0,2	T	0	bcr	-1	2,1	T	
1	dcrb	0			1	rbdc	-1			1	bcr	0			
					2	rbdc	0			2	drcb	0			

Fig. 2. An example on a Genotype CNN representation chromosome.

3.2 Genetic Operations

The genetic operations which are done on our proposed CNN chromosome are selection, mutation and crossover. The fitness function controls the selection process as the validation loss in our case that must be evaluated for each generated CNN chromosome. The Crossover operation is an exchange operation used by certain rate between two chromosomes to produce generate offspring chromosomes. The Mutation operation that is used by certain rate to edit gene values in a chromosome to produce a new chromosome different from the initial chromosome.

3.3 The Proposed Model

Our proposed model process as represented in Fig. 3 depends on GA in its core with the addition of two concepts in "machine learning" which are "Transfer Learning" [21] and "Ensemble Learning" [22]. Transfer learning avoids training the CNN structures in the population from scratch by utilizing pre-trained architectures. This process saves time in training especially that we train a big number of different structures in each population. Utilizing the ensemble learning adds an advantage to our proposed model to achieve more accuracy especially using new datasets. The process of "majority voting" ensemble makes the CNN architectures found to work together to achieve the best validation accuracy when we use it in predicting a new test dataset. Algorithm 1 shows the details of proposed genetic algorithm applied to give clear view on the whole process.

Algorithm 1 The Proposed Model Algorithm

Input:
    Number of iterations (n)
    Population size (p)
    Number of Elites (ne)
    Crossover rate (X)
    Mutation rate (Y)
Output:
    Image classification prediction accuracy

Generate p of pre-trained Convolutional Neural Network (CNN) structures chromosomes $\{C_1, C_2, \ldots, C_p\}$
Save them in population (pop) $\leftarrow \{C_1, C_2, \ldots, C_p\}$
for $i = 1$ to n **do**
  Train pop with given training Dataset
  Select the best ne CNN structures and save them in pop_1
  Validate pop_1 with given validation Dataset
  Select and save the best CNN structure in the best architecture file
  Calculate the number of random parents (r) $\leftarrow p - ne$
  for $a = 1$ to r **do**
    Select two CNN structures C_j and C_k according to validation loss fitness
    from pop_1
    Generate offspring $C_j{'}$ and $C_k{'}$ by one-point crossover to C_j and C_k under the
    crossover rate X
    Save $C_j{'}$ and $C_k{'}$ to pop_2
    Select CNN structure C_m
    Mute C_m under the mutation rate Y and generate a mutated CNN structure $C_m{'}$
    While $C_m{'}$ is not unique **do**
      Update and repair $C_m{'}$ to be unique
    end While
    Save $C_m{'}$ to pop_2
  end for
  Update $pop \leftarrow pop_1 + pop_2$
end for
Select a group of CNN architectures from the best CNN architectures file
Apply majority voting ensemble using the given test Dataset to get image classification prediction accuracy

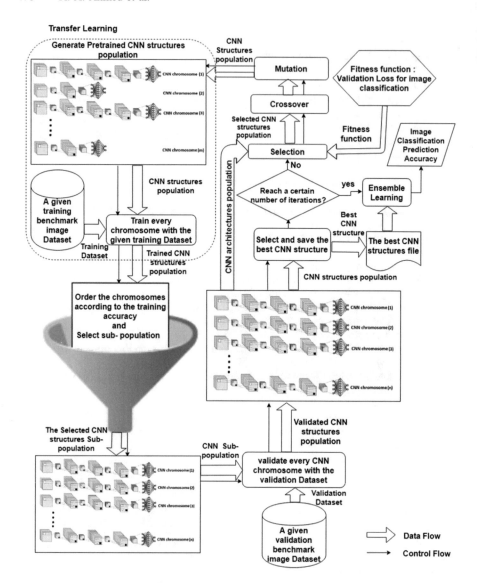

Fig. 3. The proposed model process

4   Experimental Results

In this section, we focus on evaluating our proposed model performance. We conducted a variety of the experiments on image classification tasks. There are a variety of well-known benchmark datasets as "MNIST" [23] and "CIFAR-10" [24]. Both of these

Datasets are composed from 50,000 images for training dataset and 10,000 images for the testing dataset which it is divided to 10 classes. The main difference that MNIST dataset is 28 × 28 pixels images with one channel for the color but CIFAR-10 dataset is 32 × 32 pixels images with 3 channels for the color (RGB). The researchers in image classification problem domain preferred using this dataset due to the following: They used widely to measure the performance of deep learning models, and most of the chosen compared algorithms have publicly reported their classification accuracy on them. Another reason is the objects to be classified in these benchmark datasets usually occupy various areas of the whole image, and their locations are not the same in different images. We selected these datasets to validate our proposed approach against some peer competitors. For experimental settings for our proposed model, we set the crossover operation to "single point" with rate of 0.3, the mutation rate of 0.8, population size of ten chromosomes for five generation due to computing resources constraints. The optimizer used is "Adam" stochastic optimization algorithm [25] with 128 for the batch size for 50 epochs. We repeated the experiments using a combination of hyper-parameter configurations for generated CNN chromosomes which utilizing Depth-wise Separable Convolution (DSC) [5] and Normal Convolution (NC) with Early Stopping (ES) training method and Adaptive Learning Rate (ALR) as the learning rate is decreased by a factor each time validation loss plateaus [26]. The "softmax" cross-entropy loss is used as the loss function. We run the proposed model on a computer machine with "intel core-i7-8700 K CPU of 3.7 GHz", 16 GB RAM with one "NVIDIA GeForce GTX 1080" GPU. We compare our proposed model in terms of classification Top-1 validation accuracy on benchmark datasets to other related peer competitors as stated in Tables 1 and 2.

Table 1. The comparisons between the proposed model and the peer competitors on "MNIST" dataset.

		Top-1 validation accuracy	GPU days
Hand-Crafted	MobilenetV1 [5]	99.44%	–
	MobilenetV2 [27]	99.63%	–
	Shufflenet [7]	99.65%	–
Auto	Effnet [28]	99.12%	–
	GeNet [8]	99.62%	0.21
	Ours (DSC + ES)	**99.00%**	**0.08**
	Ours (NC + ES)	**99.17%**	**0.09**
	Ours (DSC + NC + ALR)	**99.23%**	**0.15**
	Ours (DSC + ALR)	**99.14%**	**0.15**
	Ours (NC + ALR)	**99.32%**	**0.16**

Table 2. The comparisons between the proposed model and the peer competitors on "CIFAR-10" dataset.

		Top-1 validation accuracy	GPU Days
Hand-Crafted	MobilenetV1 [5]	74.81%	–
	MobilenetV2 [27]	65.07%	–
	Shufflenet [7]	77.30%	–
	Effnet [28]	80.2%	–
Auto	Enhanced Mobilenet [29]	82.1%	–
	GeNet [8]	76.24%	2
	DARTS [18]	87.67%	1
	ENAS [16]	70%	0.5
	SNAS [30]	88.54%	1.5
	Ours (DSC + ES)	**81.6%**	**0.07**
	Ours (NC + ES)	**82.3%**	**0.08**
	Ours (DSC + NC + ALR)	**84.2%**	**0.15**
	Ours (DSC + ALR)	**81.2%**	**0.14**
	Ours (NC + ALR)	**88.65%**	**0.47**

5   Conclusion

In this paper, an automatic design genetic algorithm based approach is proposed to generate light-weight CNN structures. The experimental results denoted that the proposed process competes in the time of computation compared with other related work methods. The future work direction is to enhance the genetic algorithm to generate more effective CNN structure designs with higher validation accuracy. In addition, utilizing more regularization techniques, advanced weights initialization methods and test our proposed model using more benchmark datasets.

References

1. Sermanet, P., Chintala, S., LeCun, Y.: Convolutional neural networks applied to house numbers digit classification. In: International Conference on Pattern Recognition, pp. 10–13. Tsukuba, Japan (2012)
2. Simonyan, K., Zisserman, A.: Very deep convolutional networks for large-scale image recognition. In: International Conference on Learning Representations ICLR, pp. 1–14. San Diego, USA (2015)
3. Szegedy, C., Liu, W., Jia, Y., Sermanet, P., Reed, S., Anguelov, D., Erhan, D., Vanhoucke, V., Rabinovich, A.: Going deeper with convolutions. In: The IEEE Conference on Computer Vision and Pattern Recognition, pp. 1–9. Boston, USA (2015)
4. He, K., Zhang, X., Ren, S., Sun, J.: Deep residual learning for image recognition. In: The IEEE Conference on Computer Vision and Pattern Recognition, pp. 770–778. Las Vegas, USA (2016)

5. Howard, A.G., Zhu, M., Chen, B., Kalenichenko, D., Wang, W., Weyand, T., Andreetto, M., Adam, H.: Mobilenets: Efficient Convolutional Neural Networks for Mobile Vision Applications. arXiv preprint arXiv:1704.04861 (2017)
6. Iandola, F.N., Han, S., Moskewic, M.W., Ashraf, K., Dally, W.J., Keutzer, K.: Squeezenet: Alexnet-level Accuracy with 50x Fewer Parameters and < 0.5 MB Model Size. arXiv preprint arXiv:1602.07360 (2016)
7. Zhang, X., Zhou, X., Lin, M., Sun, J.: ShuffleNet: an extremely efficient convolutional neural network for mobile devices. In: The IEEE Conference on Computer Vision and Pattern Recognition (CVPR), pp. 6848–6856. Utah, USA (2018)
8. Xie, L., Yuille, A.L.: Genetic CNN. In: The International Conference on Computer Vision ICCV, pp. 1388–1397. Venice, Italy (2017)
9. Baldominos, A., Saez, Y., Isasi, P.: Evolutionary convolutional neural networks: an application to handwriting recognition. Int. J. Neurocomput. **283**, 38–52 (2018)
10. Suganuma, M., Shirakawa, S., Nagao, T.: A genetic programming approach to designing convolutional neural network architectures. In: Genetic and Evolutionary Computation Conference, pp. 497–504. ACM, Berlin (2017)
11. Zoph, B., Le, Q.V.: Neural architecture search with reinforcement learning. In: International Conference on Learning Representations ICLR, pp. 1–16. Toulon, France (2017)
12. Zoph, B., Vasudevan, V., Shlens, J., Le, Q.V.: Learning transferable architectures for scalable image recognition. In: The IEEE Conference on Computer Vision and Pattern Recognition CVPR, pp. 8697–8710. Utah, USA (2018)
13. Zhong, Z., Yan, J., Liu, C.L.: Practical block-wise neural network architecture generation. In: The IEEE Conference on Computer Vision and Pattern Recognition CVPR, pp. 2423–2432. Utah, USA (2018)
14. Baker, B., Gupta, O., Naik, N., Raskar, R.: Designing neural network architectures using reinforcement learning. In: International Conference on Learning Representations ICLR, pp. 1–18. Toulon, France (2017)
15. Cai, H., Chen, T., Zhang, W., Yu, Y., Wang, J.: Efficient architecture search by network transformation. In: International Conference on Artificial Intelligence AAAI, pp. 2787–2794. Louisiana, USA (2018)
16. Pham, H., Guan, M., Zoph, B., Le, Q., Dean, J.: Efficient neural architecture search via parameter sharing. In: International Conference on Machine Learning, pp. 4092–4101. Stockholm, Sweden, (2018)
17. Real, E., Aggarwal, A., Huang, Y., Le, Q.V.: Regularized Evolution for Image Classifier Architecture Search. arXiv preprint arXiv:1802.01548 (2018)
18. Liu, H., Simonyan, K., Yang, Y.: DARTS: differentiable architecture search. In: International Conference on Learning Representations, New Orleans (2019). https://openreview.net/forum?id=S1eYHoC5FX
19. Srivastava, N., Hinton, G., Krizhevsky, A., Sutskever, I., Salakhutdinov, R.: Dropout: a simple way to prevent neural networks from overfitting. Int. J. Mach. Learn. Res. **15**, 1929–1958 (2014)
20. Ioffe, S., Szegedy, C.: Batch normalization: accelerating deep network training by reducing internal covariate shift. In: The 32nd International Conference on Machine Learning, PMLR, pp. 448–456. Lille, France (2015)
21. Yosinski, J., Clune, J., Bengio, Y., Lipson, H.: How transferable are features in deep neural networks? In: Advances in Neural Information Processing Systems, pp. 3320–3328. Quebec, Canada (2014)
22. Zhou, Z.H.: Ensemble Methods: Foundations and Algorithms. Chapman & Hall/CRC, Boca Raton (2012)

23. Deng, L.: The MNIST database of handwritten digit images for machine learning research. IEEE Sig. Process. Mag. **29**(6), 141–142 (2012)
24. Krizhevsky, A., Hinton, G.: Learning Multiple Layers of Features from Tiny Images. Technical Report, University of Toronto (2009)
25. Kingma, D.P., Ba, J.: Adam: a method for stochastic optimization. In: International Conference on Learning Representations ICLR, pp. 1–15. San Diego, USA (2015)
26. Goodfellow, I., Bengio, Y., Courville, A.: Deep Learning. MIT press, Cambridge (2016)
27. Sandler, M., Howard, A., Zhu, M., Zhmoginov, A., Chen, L.C.: Mobilenetv2: inverted residuals and linear bottlenecks. In: The IEEE Conference on Computer Vision and Pattern Recognition, pp. 4510–4520. Utah, USA (2018)
28. Freeman, I., Roese-Koerner, L., Kummert, A.: Effnet: an efficient structure for convolutional neural networks. In: The 25th IEEE International Conference on Image Processing (ICIP), pp. 6–10, Athens, Greece (2018)
29. Chen, H.Y., Su, C.Y.: An enhanced hybrid MobileNet. In: The 9th International Conference on Awareness Science and Technology (iCAST), pp. 308–312, Fukuoka, Japan (2018)
30. Xie, S., Zheng, H., Liu, C., Lin, L.: SNAS: Stochastic Neural Architecture Search. arXiv preprint arXiv:1812.09926 (2018)

Machine and Deep Learning Algorithms for Twitter Spam Detection

Dalia Alsaffar, Amjad Alfahhad, Bashaier Alqhtani, Lama Alamri,
Shahad Alansari, Nada Alqahtani, and Dabiah A. Alboaneen[✉]

Computer Department, College of Science and Humanities, Imam Abdulrahman Bin
Faisal University, P.O. Box 31961, Jubail, Saudi Arabia
Dabuainain@iau.edu.sa

Abstract. Twitter allows users to send short text-based messages with
up to 280 characters which is called "tweets". The reputation of Twitter attracts the spammers to spread malevolent programming through
URLs attached in tweets. Twitter spam has become a critical problem.
Spam refers to a variety of prohibited behaviours that violate the Twitter rules. In this paper, different machine and deep learning algorithms
are used to detect if the tweet is spammer or not. The performance of
six machine learning algorithms, namely Random Forest (RF), Naive
Bayes (NB), Bayesian Network (BN), Support Vector Machine (SVM),
K-Nearest Neighbour (KNN), and Multi-Layer Perceptron (MLP) and
one deep learning algorithm which is Recurrent Neural Network (RNN)
are evaluated. Different test options are used, namely cross validation
and percentage split tests. Results show that RF predicts the best result
with lowest error rate and highest classification accuracy rate with different test options comparing to all algorithms.

Keywords: Classification · Machine learning · Deep learning ·
Twitter · Spam

1 Introduction

Data mining aims to identify patterns and establish relationships to solve problems through data analysis. The main data mining parameters include classification, clustering and prediction. Classification is the process of finding a model
that describes and distinguishes data classes. The main goal of a classification
problem is to identify to which of a set of categories, a new observation belongs
to, on the basis of a training set of data containing observations.

There are many real-life classification problems such as text categorisation
(e.g., spam filtering), fraud detection, face detection, natural language processing (e.g., spoken language understanding), market segmentation (e.g. predict if
customer will respond to promotion), and bioinformatics (e.g., classify proteins
according to their function).

© Springer Nature Switzerland AG 2020
A. E. Hassanien et al. (Eds.): AISI 2019, AISC 1058, pp. 483–491, 2020.
https://doi.org/10.1007/978-3-030-31129-2_44

Machine learning algorithms are widely used to solve classification problems such as Logistic Regression, Nave Bayes (NB), Support Vector Machines (SVM), K-Nearest Neighbour (KNN), Decision Tree (DT) and Neural Network (NN).

Deep learning is a form of machine learning aims to solve perceptual problems such as speech and image recognition. A deep learning model is designed to analyse data with a logic structure similar to how a human's brain work. To achieve this, deep learning uses a layered structure of algorithms called an Artificial Neural Network (ANN). Deep ANNs contain multiple hidden layers to recognise patterns and structure in large datasets. Each layer learns a concept from the data from subsequent layers. The core difference between machine learning and deep learning lies on the feature engineering. In machine learning, feature engineering can be done manually while deep learning does not depend on prior data processing and automatically extracts features.

Twitter is one of the most popular social communication tool, where millions of users participate and discuss everything including their mood, news and events around them through a simple interface that enables the post of messages, photos and videos, you find in the trend many of the topics that may contain spam messages. Due to the popularity of Twitter, it becomes an attractive platform for spammers to spread spam. It has become a severe issue on Twitter.

This paper presents a comparative analysis of various classification algorithms in classifying tweets into spammers or non spammers. The contributions of this paper are:

1. To compare the accuracy, prediction error, precision, recall and F-measure of different machine learning algorithms on the problem of predicting the Twitter spammers.
2. To compare the accuracy, prediction error, precision, recall and F-measure between machine learning and deep learning algorithms on the problem of predicting the Twitter spammers.
3. To compare the results when using two test options, namely 10-fold cross-validation and percentage split.

The remainder of this paper is arranged as follows. Section 2 presents related work on using machine and deep learning for classification in general and in Twitter spam especially. A brief preliminary on different classification algorithms which will be compared is given in Sect. 3. Section 4 puts forward the proposed model for detecting spam tweets. Experimental evaluations are discussed in Sect. 5. Finally, Sect. 7 draws conclusion and sets future work.

2   Related Work

In [1], the authors have compared different classification techniques using WEKA for Breast Cancer patients. The algorithms that were applied are Bayes Network (BN), Radial Basis Function, DT and Pruning, and the closest neighbour algorithm. The main findings of this study that the best algorithm for breast cancer data is the BN, which has the ability to improve data classification for use in medical or in bioinformatics field. In [1], the BN algorithm had the highest accuracy rate with 89.7143%, and DT came at the third place with 85.7143% accuracy rate.

In [2], the authors have considered the problem of predicting smoking status using different machine learning algorithms including NB, MLP, Logistic, J48, and DT using WEKA. The results showed that applying these algorithms could be determine the smoking status of patients who have had blood tests and vital readings. Our study is consistent with this study, which the five algorithms are implemented for classification purpose and worked effectively with large datasets. In [2], the Logistic algorithm had the highest accuracy rate at 83.44%, followed by j48 with 83.11%, while our study shows that the RF algorithm yields the highest accuracy rate with 95.7% in cross-validation test and with 95.3% in percentage split test.

In [3], the authors have considered the problem of spam detection in Twitter. Several classic classification algorithms have been applied and evaluated, such as Bayesian, DT, NN, SVM, and KNN. The results showed that the Bayesian classifier had a better overall performance with the highest F-score. The Bayesian classifier is applied to large amount of data and achieves about 89% precisions.

In [4], the authors have considered the problem of detecting spammers on Twitter. SVM is applied to classify tweets into spam or non-spam using WEKA. Twitter dataset is used which includes more than 54 million users, 1.9 billion links, and almost 1.8 billion tweets. The main finding is that the SVM could correctly classify approximately 70% of spammers and 96% of non-spammers.

Deep learning algorithms are used in different classification problems and spam detection is one of them. In [5], NB, SVM and Recurrent Neural Network (RNN) algorithms are applied for spam detection. Recall, precision, accuracy and F-measure are used to evaluate the performance of classifiers. The results showed that RNN algorithm has the best performance compared with other algorithms with almost 18% improvement in performance.

However, this paper aims to predict whether the tweet is a spammer or a non-spammer using RF, NB, BN, SVM, KNN, and MLP and one deep learning algorithm, namely, RNN through WEKA and will determine which classification algorithm has the highest accuracy and lowest error rate.

3 Preliminaries

3.1 Machine Learning Algorithms

Random Forests: It is a popular ensemble algorithm. It is a supervised algorithm that consists of a set of decision trees to find the root node and split the features in a random way. It is mostly used in classification and regression tasks because it helps to solve the over-fitting problem and handle the missing data values well. It is used in wide applications such as banking, medicine, stock market and E-commerce. It runs efficiently on huge datasets. In addition, it considers an efficient technique to estimate the missing data and maintain accuracy [6].

Naive Bayes: It is a popular subset of Bayesian decision theory. It is a simple probabilistic classifier which calculates a set of probabilities by counting the frequency and combinations of values in a given dataset. Due to the simplicity of Naive Bayes, it can classify documents surprisingly well. It is applied to time-storage critical applications, such as automatically classifying web pages into types and spam filtering [7].

Bayesian Networks: BNs are annotated directed graphs that encode probabilistic relations among variables in uncertain problems. BNs are usually formed through cause and effect frameworks and are suitable for computing all probabilities of interest since they determine joint probability distributions for domains [8].

Support Vector Machines: It is a supervised machine learning algorithm learns by example to assign labels to objects. SVMs have been developed to solve the classification problems such as handwritten recognition. It can efficiently manage a number of data and a high dimensional problems [9].

K-Nearest Neighbours: It is a machine learning algorithm used for both classification and regression. In KNN classification, the output represents a class. Instance is classified by a plurality vote of its neighbours, to be assigned to the most common class among its k nearest neighbours [10].

Multi-layer Perceptron: It is one of the most popular and effective forms of learning systems. ANNs are composed of nodes or neurons connected by directed links. If the connections are in one direction, it's called a feed-forward network. Every neuron receives input from "upstream" neurons and delivers output to "downstream" neurons. MLP is one of the ANNs algorithms. MLP has one or more hidden layers that are connected to the outputs of the network. The standard learning algorithm of MLP is Back Propagation (BP). The learning process includes updating weights and biases of MLP network [11].

3.2   Deep Learning Algorithm

Recurrent Neural Network: In RNN, the network contains at least one feedback connection, so the activations can flow round in a loop. RNN architectures can have many different forms. One common type consists of a standard MLP plus added loops. These can exploit the powerful non-linear mapping capabilities of the MLP, also have some form of memory. Others have more uniform structures, potentially with every neuron connected to all the others, and may also have stochastic activation functions [12].

RNNs process an input sequence one element at a time, maintaining in their hidden units a "state vector" that implicitly contains information about the history of all the past elements of the sequence. When we consider the outputs of the hidden units at different discrete time steps as if they were the outputs of different neurons in a deep multilayer network, it becomes clear how we can apply BP to train RNNs [13]. In this paper, Long Short-Term Memory (LSTM) is used which is an artificial RNN architecture.

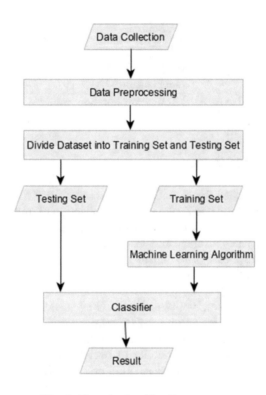

Fig. 1. Tweets classification process

4   Model

The aim of this model is to classify a tweet into spam or not spam. Identifying spam users cannot consider each spam tweet as spammer. Spammers can create another account and restart spamming activity. It is best to consider spam at the tweet level and this has been done by checking the tweet contents, the links attached or the number of duplicate tweets by more than one account [14].

To do so, different classifiers are applied on the testing dataset. After the classification model is created and tested, parameters of classification accuracy, MAE, precision, recall and F-measure are to be calculated. The tweets classification can be illustrated in Fig. 1. Basically, a classifier would consider a tweet as an input. Firstly, begin pre-process phase on the dataset to extract the important features that help in the classification process, then divide the dataset into a training set and testing set. The training set goes through the machine learning algorithm to be trained and the test data is classified directly. Then categorise it as spammer or non-spammer and present it as its output.

Table 1. Parameters setting

Algorithm	Parameter	Value
SVM	Tolerance	0.001
MLP	Learning rate	0.3
	Momentum	0.2
RNN	Activation function	Sigmoid

5   Experimental Setup

WEKA toolkit is applied for classification purpose. WEKA is developed at the University of Waikato in New Zealand. It is a collection of machine learning algorithms, which contains tools for data preparation, classification, regression, clustering, association rules mining, and visualisation [15]. Deep Learning package is developed to incorporate deep learning techniques into WEKA. It is provided by the Deeplearning4j Java library, all functionality of this package is accessible via the WEKA GUI [16].

[1]The Twitter dataset was obtained from Nuclear Science and Security Consortium labs. The dataset contains 10000 tweets with 12 attributes which are number of account age, number of followers, number of following, number of user favourites, number of lists, number of tweets, number of retweets, number of hashtag, number of user mentions, number of URLs, number of chars and number of digits [17].

The parameter settings for algorithms have been presented in Table 1.

5.1   Performance Metrics

Accuracy is the level of effectively recognised cases (spammers and non-spammers) in the total number of tweets, which can be calculated using Eq. 1.

$$Accuracy = \frac{(TP + TN)}{(TP + TN + FP + FN)} \tag{1}$$

where TP is the true positive, TN is the true negative, FP is the false positive and FN is the false negative.

---

[1] http://nsclab.org/nsclab/resource.

Mean Absolute Error (MAE) the prediction error can be defined as the difference between the predicted value and the actual value for each test instance. MAE of any model refers to the mean of the absolute values of each prediction error on all instances of the test dataset as in Eq. 2

$$MAE = \frac{\sum_{i=1}^{n} |y_i - x_i|}{n} \tag{2}$$

where y is the predicted value and x is the true value.

Precision is defined as the ratio of correctly classified tweets spammers to the total number of tweets that are classified as spammers as shown below.

$$Precision = \frac{TP}{(TP + FP)} \tag{3}$$

Recall is the ratio of correctly classified tweets spammers to the total number of tweets that are classified as non-spammers as shown by the Eq. 4

$$Recall = \frac{TP}{(TP + FN)} \tag{4}$$

F-Measure is combining both the precision and recall. F-measure can be calculated as in Eq. 5.

$$F - measure = 2 * \frac{Precision * Recall}{Precision + Recall} \tag{5}$$

6   Experimental Results

The results for the seven classification algorithms, namely, RF, NB, BN, SVM, KNN, MLP and RNN are presented in Table 2. We use 10-fold cross-validation on the dataset to split dataset into training and test data. Cross validation method selects 90% for training data and 10% for test data. The training data is used to train the classifier, while the test data is used to test the classifier. The results show that RF has the highest accuracy rate of 95.7% with lowest MAE value. In addition, RNN classifier outperforms NB and SVM only.

The results for the seven classification algorithms with different test option, which is percentage split test are presented in Table 3. The dataset is divided into two parts, 66% of the dataset is used as training data and the remaining 34% of the dataset is used as test data. The comparison has been done in terms of accuracy, MAE, precision, recall, and F-measure as well. The results show that RF has the highest accuracy rate of 95.3% with lowest MAE value as well. Moreover, the results are quite similar when using cross validation or percentage split test.

In addition, RNN classifier outperforms NB and SVM only due to the deep learning algorithms require huge number of data to efficiently work, more data means better performance. The minimum data is based on the complexity of the problem, In general, dataset of 100,000 instances is good to run.

Table 2. The performance of different classifiers with 10-fold cross-validation test

Type of learning	Classifier	Accuracy %	MAE %	Precision %	Recall	F-measure %
Machine learning	RF	**95.7%**	**0.0842**	**95.8%**	**95.7%**	**95.7%**
	NB	60%	0.4	73.1%	60%	53.4%
	BN	88.13%	0.1289	88.5%	88.1%	88.1%
	SVM	76.87%	0.2313	78%	76.9%	76.6%
	KNN	86.4%	0.1695	86.4%	86.4%	86.4%
	MLP	86.05%	0.1926	86.7%	86.1%	86%
Deep learning	RNN	80.39%	0.3178	80.5 %	80.4%	80.4%

Table 3. The performance of different classifiers with percentage split test

Type of learning	Classifier	Accuracy %	MAE %	Precision %	Recall %	F-measure %
Machine learning	RF	**95.3%**	**0.0842**	**95.2%**	**95.2%**	**95.2%**
	NB	60.5 %	0.4	73.7%	60.5%	54%
	BN	87.38%	0.1403	87.7%	87.4%	87.4%
	SVM	76.85%	0.2315	77.8%	76.9%	76.6%
	KNN	83.61%	0.1693	83.6%	83.6%	83.6%
	MLP	85%	0.2009	86.1%	85%	84.9%
Deep learning	RNN	79.3%	0.339	79.4%	79.3%	79.3%

7   Conclusion and Future Work

Spammer and non-spammer tweets have been classified using machine learning and deep learning algorithms. We used a Twitter dataset of 10000 instances to analyse the best classifier amongst RF, NB, BN, SVM, KNN, MLP and RNN algorithms. We found that RF has highest accuracy amongst machine and deep learning algorithms with 95.3% and lower error rate when using cross validation and percentage split test. For future work, more deep learning algorithms such as Conventional Neural Networks can be investigated for detecting Twitter spam. In addition, more datasets can be used to evaluate the performance of algorithms.

References

1. bin Othman, M.F., Yau, T.M.: Comparison of different classification techniques using WEKA for breast cancer. In: 3rd Kuala Lumpur International Conference on Biomedical Engineering, pp. 520–523. Springer, Heidelberg (2007)
2. Frank, C., Habach, A., Seetan, R.: Predicting smoking status using machine learning algorithms and statistical analysis. J. Comput. Sci. Coll. **33**, 66 (2018)
3. Wang, A.H.: Don't follow me: spam detection in Twitter. In: 2010 International Conference on Security and Cryptography (SECRYPT), pp. 1–10. IEEE (2010)
4. Benevenuto, F., Magno, G., Rodrigues, T., Almeida, V.: Detecting spammers on Twitter. In: Collaboration, Electronic Messaging, Anti-abuse and Spam Conference (CEAS), vol. 6, pp. 12. (2010)

5. Gao, Y., Mi, G., Tan, Y.: Variable length concentration based feature construction method for spam detection. In: 2015 International Joint Conference on Neural Networks (IJCNN), pp. 1–7. IEEE (2015)
6. Liaw, A., Wiener, M.: Classification and regression by randomForest. R News **2**(3), 18–22 (2002)
7. Zhang, H.: The optimality of naive Bayes. AA **1**(2), 3 (2004)
8. Jensen, F.V.: An Introduction to Bayesian Networks. UCL Press, London (1996)
9. Cortes, C., Vapnik, V.: Support-vector networks. Mach. Learn. **20**(3), 273–97 (1995)
10. Altman, N.S.: An introduction to kernel and nearest-neighbor nonparametric regression. Am. Stat. **46**(3), 175–85 (1992)
11. Russell, S.J., Norvig, P.: Artificial Intelligence: A Modern Approach. Pearson Education Limited, Malaysia (2016)
12. Haykin, S.S.: Neural Networks and Learning Machines. Pearson Education, Upper Saddle River (2009)
13. LeCun, Y., Bengio, Y., Hinton, G.: Deep learning. Nature **521**(7553), 436 (2015)
14. Sedhai, S., Sun, A.: Semi-supervised spam detection in Twitter stream. IEEE Trans. Comput. Soc. Syst. **5**(1), 169–175 (2017)
15. Witten, I.H., Frank, E., Trigg, L.E., Hall, M.A., Holmes, G., Cunningham, S.J.: Weka: practical machine learning tools and techniques with Java implementations (1999)
16. Team, D.: Deeplearning4j: open-source distributed deep learning for the JVM. Apache Software Foundation License 2 (2016)
17. Chen, C., Zhang, J., Chen, X., Xiang, Y., Zhou, W.: 6 million spam tweets: a large ground truth for timely Twitter spam detection. In: IEEE International Conference on Communications (ICC), pp. 7065–7070. IEEE (2015)

An Optimized Deep Convolutional Neural Network to Identify Nanoscience Scanning Electron Microscope Images Using Social Ski Driver Algorithm

Dalia Ezzat[1,2]([⊠]), Mohamed Hamed N. Taha[1,2],
and Aboul Ella Hassanien[1,2]

[1] Faculty of Computers and Artificial Intelligence, Cairo University,
Cairo, Egypt
dalia.Azzat@yahoo.com
[2] Scientific Research Group in Egypt (SRGE), Cairo, Egypt
http://www.egyptscience.net

Abstract. In this paper, transfer learning from a pretrained Convolutional Neural Network (CNN) model called VGG16 in conjunction with a new evolutionary optimization algorithm called social ski driver algorithm (SSD) were applied for optimizing some hyperparameters of the CNN model to improve the classification performance of the images which was produced by the SEM technique. The results of the proposed approach (VGG16-SSD) are compared with the manual search method. The obtained results showed that the proposed approach was able to find the best values for the CNN hyperparameters that helped to successfully classify around 89.37% of a test dataset consisting of SEM images.

Keywords: CNNs · Social ski driver · Hyperparameters optimization

1 Introduction

Due to the tremendous developments in image acquisition devices, data is very large (moving to big data) [1], which makes it challenging for image analysis tasks such as image classification task. Image classification task, has become an important task in nanoscience because most characterization techniques produce huge number of images such as scanning electron microscopy (SEM) [2]. SEM [3], is a common characterization technique used in nanotechnology to recognize the structure of the samples at higher magnifications down to 1 nm. The remarkable progress in SEM technology, especially the recent desktop SEMs, has contributed to the production of a large number of images. Due to the ease of using the desktop SEMs (less than an hour is sufficient to conduct measurements independently) and the running time of the desktop SEMs is relatively short (only a few minutes from sample loading to imaging) [4]. Automatic image classification of SEM images can be extremely useful for researchers in the field of nanoscience, because it will avoid the manual classification of the images produced. Over the past few years, CNNs have achieved state-of-the-art performances in solving problems related to computer vision especially in image classification [5]. This is due to

A. E. Hassanien et al. (Eds.): AISI 2019, AISC 1058, pp. 492–501, 2020.
https://doi.org/10.1007/978-3-030-31129-2_45

that CNN is one of the most effective ways to extract features from images for non-trivial tasks [6]. This motivation leads scientists, who work with microscopy techniques and have a special interest in recognizing the important features within microscope images [2] to use CNNs to extract the important features from microscope images. Although CNNs have proven to be highly efficient in image classification, these architectures have some drawbacks. Notably, these architectures admit a large variety of specific hyper-parameters, it is challenging and expensive to manually identify these optimal hyper-parameters [7]. CNNs are sensitive to the setting of these hyperparameters, which have a significant impact on their performance. In addition, the hyperparameters need to be adjusted for each dataset due to the hyperparameters are vary from one dataset to another. Selecting the appropriate values for the hyper-parameters of a particular dataset depends on trial and error because there is no mathematical formula which means that the hyperparameters are tuned manually. The choice of hyperparameters values requires specialized knowledge; therefore recently the selecting of the hyperparameters values has been formulated as optimization problem by the researchers [8]. Therefore, this paper proposes a CNN model in conjunction with a new evolutionary optimization algorithm called the SSD algorithm [9] to optimize some hyperparameters of the CNN model by eliminating the requirement of a manual search for optimal hyperparameters to classify the images which was produced by the SEM technique into ten different categories. Classification of imbalanced data indicating that the number of the samples is not the same for all the classes is a common problem in the machine learning (ML) techniques. Usually, ML techniques can not learn well from this kind of data, especially for minority categories because the information is covered by majority categories data. There are many researches that are trying to deal with this imbalance such as [10, 11] by using different techniques. The simplest and straightforward technique is the use of sampling methods such as random oversampling (ROS) and random undersampling (RUS). The dataset used in this paper is imbalanced data, therefore the ROS and RUS methods have been used to address the imbalance. In this paper, in order to improve the accuracy level, the transfer learning (TL) method was used. The remainder of this paper is structured as follows: the theoretical background is explained in Sect. 2. Section 3 presents an explanation of the used SEM dataset. In Sect. 4, the proposed approach is explained. The experimental results are notified in Sect. 5. Lastly, the concluding remarks of this paper is presented in Sect. 6.

2 Theory and Method

2.1 Convolutional Neural Networks

The structure of CNN, which is built from scratch, consists of two parts: convolutional and classifier.

Convolutional Part: At this part, the features of the images are extracted using a set of layers are convolutional layers, activation function layers, and pooling layers. Convolutional layers are important component in the CNN, where they contain a set of filters used to detect the important features that are useful for characterizing images. Activation function layers are responsible for applying a non-linear transformation to the output of the previous convolutional layer. There are various types of activation functions such as the Exponential Linear Unit (ELU) [12]. Pooling layers are designed to down-sample the dimensionality of the feature maps to focus only on the most important features.

Classifier Part: At this part the images are categorized based on the features that extracted from the convolutional part using a combination of the layers, namely, the fully connected layers (FC), dropout layers and batch normalization layers (BN). FC layers contain a number of neurons, where each FC layer is followed by an activation function layer. The last FC typically has the same number of neurons as the number of categories. The activation function that used for the last FC layer is different from the other FC layers. The softmax or sigmoid function is usually used for this layer. Softmax function is used in multi-class classification problems, while sigmoid function is used in binary classification problems. Dropout layers are a method of regulation, only applied in the network training to prevent the network from overfitting. BN layers aim to accelerate the training of the network by normalizing each scalar feature x_i inside a mini batch $B = \{x_{1....m}\}$ independently as shown in Eq. (1) [13].

$$\hat{x}_i = \frac{x_i - \mu_B}{\sqrt{\sigma_B^2 + \epsilon}} \tag{1}$$

Where ϵ is a small positive constant to prevent division by zero, μ_B which represents the mean of the mini batch and σ_B^2 which represents the variance of the mini batch are given by Eqs. (2) and (3) respectively.

$$\mu_B = \frac{1}{m} \sum_{i=1}^{m} x_i \tag{2}$$

$$\sigma_B^2 = \frac{1}{m} \sum_{i=1}^{m} (x_i - \mu_B)^2 \tag{3}$$

Then, the normalized value for each training mini-batch is scaled and shifted by the scale and shift parameters γ and β as shown in Eq. (4) where γ and β are parameters to be trained along with the original model parameters.

$$y_i = \gamma \hat{x}_i + \beta \tag{4}$$

Training a CNN Network: The Backpropagation algorithm is the most important method for training neural networks where the optimization function and the loss function play key tasks. The dissimilarity between the true label distribution y and the predicted label distribution ŷ from the network are measured using a loss function such as cross entropy (CE). The CE is mathematically calculated as Eq. (5) [14]. To minimize the loss function, the optimization function is used such as Adam [15].

$$L_{\text{cross entropy}}(\hat{y}.y) = - \sum_{i=0}^{N} y_i \log(\hat{y}_i) \quad \text{for } i = 1...N \tag{5}$$

Where N = Number of classes

2.2   Transfer Learning from Pre-trained CNN Models

There are various CNN models that can be used as a pretrained model such as VGG [16]. VGG released two different CNN models are VGG16 and VGG19. TL from a pretrained CNN models involves two common techniques are feature extraction (FE) and fine-tuning (FT). These techniques are usually used together in many tasks to improve the results. In the FE, the convolutional part of the pretrained model is kept in its original form and then use its outputs to train a new classifier part fits the new dataset instead of the original classifier. In the FT, the first layers of the convolutional part of the pretrained model are kept fixed but the final layers are jointly training with the newly added classifier part.

2.3   The Social Ski Driver Algorithm

SSD is a new evolutionary algorithm, proposed by Tharwat and Gabel [9]. In [9], the experimental results indicated that the SSD algorithm does not require high computational speeds or memories and it has the ability to outperform other evolutionary algorithms such as the particle swarm optimization (PSO) algorithm in finding the optimal values for the support vector machine (SVM) parameters. Therefore, in this paper the SSD algorithm is applied to optimize some hyperparameters of the CNN and this combination is summarized in Fig. 1. The key parameters of the SSD algorithm are listed below.

- $X_i \in \mathcal{R}^n$ represents the agents' positions where n is the dimensions number, the agent's position are utilized to calculate the objective function at these points. The agent' positions are randomly initialized.
- P_i: represents the previous best position. The fitness function is used to calculate the fitness values of all agents, then the fitness value for each agent is compared with its current position and the best positions is stored.
- M_i: represents the mean of the best three solutions and calculated by the following formula:

$$M_i^t = \frac{X_a + X_y + X_z}{3} \tag{6}$$

Where X_a, X_y and X_z are the best three solutions.

- V_i: represents the velocity of the agents. The velocity V_i added to the positions of the agents to update them as Eq. 7. The agents' velocities are randomly initialized and are updated as Eq. 8.

$$X_i^{t+1} = X_i^t + V_i^t \tag{7}$$

Where

$$V_i^{t+1} = \begin{cases} c\sin(r_1)\left(P_i^t - X_i^t\right) + \sin(r_1)\left(M_i^t - X_i^t\right) & \text{if } r_2 \le 0.5 \\ c\cos(r_1)\left(P_i^t - X_i^t\right) + \cos(r_1)\left(M_i^t - X_i^t\right) & \text{if } r_2 \le 0.5 \end{cases} \tag{8}$$

Where V_i represents the velocity of X_i, r_1 and r_2 are random numbers between 0 and 1. P_i represents the best solution of ith agent. c is a parameter which is used to make a balance between exploration and exploitation and it is identified as follows: $c^{t+1} = \alpha c^t$ where t is the current iteration and $0 < \alpha < 1$ is used to minimize the value of c.

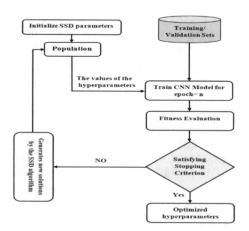

Fig. 1. Flowchart illustrate the process of optimizing the hyperparameters of CNN using SSD algorithm. N = number of epochs

3   SEM Dataset Description

The dataset consists of 21,272 SEM images with 1,024 × 728 pixels [17]. The images in the dataset were categorized into ten categories as shown in Table 1. Representative image for each category in the dataset is shown in Fig. 2.

Table 1. The dataset categories and N = the number of the images in each category

Category	N
Biological	972
Fibres	162
Films Coated Surface	326
MEMS devices and electrodes	4,590
Nanowires	3,820
Particles	3,925
Patterned surface	4,755
Porous Sponge	181
Powder	917
Tips	1,624
Total	**21,272**

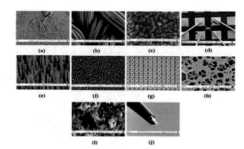

Fig. 2. Representative image for each category in the dataset. (a) Biological, (b) Fibres, (c) Films Coated Surface, (d) MEMS devices and electrodes, (e) Nanowires, (f) Particles, (g) Patterned surface, (h) Porous Sponge, (i) Powder, (j) Tips

4   Proposed Approach

In this paper the proposed approach is based on TL from a pretrained CNN model. The pretrained model that used in this paper is VGG16. To obtain the best performance of this model, its hyperparameters were optimized using the SSD algorithm. After determining the optimal values for the hyperparameters, the VGG16 was trained using the transfer learning techniques. Once the training of this model is completed, it is evaluated using a separate test set. To illustrate the procedure of the proposed approach, it has been summarized in four main phases as shown in Fig. 3.

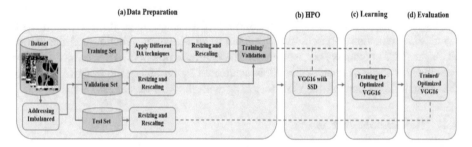

Fig. 3. The graphical representation of the proposed model

4.1   Data Preparation Phase

As shown in Table 1, each category in the used dataset has different number of images, which means it is imbalanced. In this paper, the imbalanced of the dataset is addressed by two methods are RUS and ROS which used to adjust the class distribution of a dataset. After manually applying ROS and RUS methods to the dataset, it became balanced as each category contains a similar number of 750 images. After that the dataset was divided into three sets are training, validation and test sets, where the training set contains 70% of the dataset (525 images in each category), and each of the validation and testing set contains 15% of the dataset (112 images in each category). The training and validation sets were used to determine the optimal values for the hyperparameters of the VGG16 and trained it, whereas the trained VGG16 is then evaluated using the test set. To reduce the overfitting and improve generalization, different data augmentation techniques [18] are applied for the training samples. Finally, all the images in the training, validation and test sets were resized to a resolution of 256 × 256 and rescaled by dividing the color value of each pixel by 255 to obtain values in the range [0, 1].

4.2   The Hyperparameters Optimization Phase

The pre-trained CNN model has limitations, notably, that most of its hyperparameters have already been set before and cannot be changed. There are very little hyperparameters that need to be adjusted such as the batch size of the training, the value of the learning rate, the number of neurons in the fully connected layer and the rate of the dropout layer. In this paper, the SSD algorithm optimizes the batch size (BS) of the training and the dropout rate (DR). Therefore, the search space is two-dimensional and each point in the space represents a combination between BS and DR. The searching range of BS is bounded by BS_{min} 1 and BS_{max} 128 and the search range of DR is bounded by DR_{min} 0.1 and DR_{max} 0.9.

4.3   The Learning Phase

The FE and FT techniques are used to adapt VGG16 to the new dataset that used in this paper. In the FE technique, the convolutional part is kept in its original form without any changes, but the classifier part is replaced with a new one. The new classifier consists of a flatten layer, two FC layers, a BN layer and a dropout layer. The first FC layer (FC1) contains 260 neurons with ELU as an activation function and the second FC layer (FC2) contains four neurons with a Softmax function. After trained the new classifier for some of epochs, the FT is conducted by retraining the last two blocks of the convolutional part of the VGG16 jointly with the new classifier.

4.4   The Evaluation Phase

In this paper, five measurements are utilized to evaluate predictive ability of the proposed approach on the unseen test set. These measurements are accuracy, precision, recall, F1 score, and confusion matrix. To measure the overall performance of the precision, recall and f1-score, the micro and macro averages are used [19].

5   Results and Analysis

In this paper, Google Colaboratory [20] was used for the proposed model training and testing. The methodology of the proposed approach was implemented in Python code using Keras library with Tensorflow as a backend [21]. The Keras library's ImageDataGenerator function was used to perform the data augmentation techniques that used in this paper. The data augmentation techniques used in this paper are: rotation, width shift, height shift, shear and zoom, whose values are set to 30, 0.3, 0.2, 0.3, 0.4, respectively. Keras library's ImageDataGenerator function is also used to resize and rescale all the images in the training, validation and test sets. in this experiment, the number of iterations of SSD was set to 6 and the number of agents was set to 6. The aim of the SSD is to minimize the loss rate of the validation set. After each iteration of the SSD the fitness of the proposed solutions was evaluated based on the loss value that the proposed solutions obtained in the validation set after n training epochs of CNN. After applying several experiments, it is noted that when the training epoch of CNN is too low (less than two), the tendency of the CNN converge was low or uncertain. When the training epoch of CNN is high (more than five) the SSD training process takes time exponentially. Therefore, for each SSD run, a constant number of five training epochs was used to calculate the fitness of proposed solutions based on the objective function of CNN to minimize the computation time without relapsing the convergence of network. After the SSD algorithm terminated the best value for BS was

13, and the best value for DR was 0.517. The hyperparameters of the pretrained model (VGG16) are determined by using the SSD algorithm. VGG16 is trained for n epochs on the training set and evaluated on the validation set during the training phase. After applying several experiments, it is noted that the considered model achieved its best results on the validation set around the twenty iteration in the FE technique and around the fifty iteration in the FT technique and that there was no improvement after that. Thus, the value of n has been fixed to 20 in FE and 50 in FT. To avoid overfitting, the training process was forced to stop before iterating n times if no improvement was noted for more than ten iterations. This monitoring was done using early stopping. As the dataset that used in this paper is a multi-class, the pretrained model is compiled with the CE. To optimize the cost function, the Adam optimizer was used with a constant learning rate (2e-5) in the FE technique and with a small learning rate (1e-5) in the FT technique. The reason for using a low learning rate in the FT is to ensure that the magnitude of the updates remains very small and does not break the previously learned features of the FE technique. After completing the network training, the proposed approach has been evaluated on the test set using the five indexes that mentioned in Sect. 4.4. The proposed approach achieved an accuracy of 89.37% and test loss rate of 0.397. Table 2 shows the precision (P), recall (R) and F1-score (F) of the proposed approach where C1 = Biological, C2 = Fibres, C3 = Films coated surface, C4 = MEMS devices and electrodes, C5 = Nanowires, C6 = Particles, C7 = Patterned surface, C8 = Porous sponge, C9 = Powder, C10 = Tips. The last two lines show the micro- and macro averaged scores over all classes. Figure 4 represents the confusion matrix of the proposed approach.

Table 2. Results obtained using the proposed model

Categories	P	R	F
C1	0.86	0.96	0.91
C2	0.96	1	0.98
C3	0.95	0.89	0.02
C4	0.89	0.92	0.90
C5	0.92	0.79	0.85
C6	0.81	0.93	0.87
C7	0.88	0.81	0.85
C8	0.92	0.93	0.92
C9	0.96	0.84	0.90
C10	0.86	0.91	0.89
Micro Avg.	**0.90**	**0.90**	**0.90**
Macro Avg.	**0.90**	**0.90**	**0.90**

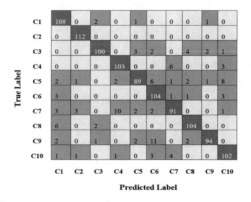

Fig. 4. Overlapping results of confusion matrix obtained with test set using the proposed approach.

In order to show the effectiveness of the proposed approach, it is compared with the performance of the manual search for the hyperparameters. In the comparison, two experiments were conducted as shown in Table 3, where the values of the BS and DR are chosen manually. In the first experiment, the BS is set to 16 and DR is set to 0.4. In the second experiment, the BS is set to 32 and DR is set to 0.3.

Table 3. Comparison of the performance of the comparative methods and proposed approach on the used test set. Macro-P = Macro averaged precision, Macro-R = Macro averaged recall, Macro-F = Macro averaged F1-score.

Method	Accuracy	Macro-P	Macro-R	Macro-F
First experiment	88.41%	0.89	0.89	0.89
Second experiment	85.34%	0.86	0.86	0.86
The proposed approach	**89.37%**	**0.90**	**0.90**	**0.90**

6   Conclusion and Future Work

This paper presented an approach for classification of nanoscience SEM images into ten classes using a pretrained CNN model in conjunction with a new evolutionary algorithm called SSD algorithm and increase the level of the accuracy. The SSD algorithm was used to optimize some hyperparameters of the CNN network. The results obtained demonstrate the high performance of the proposed approach. There are several plans for the future work. In this paper, only 7500 images were used from the original dataset, but more data should be used to achieve higher accuracy. As well as, different evolutionary algorithms can be applied to optimize the hyperparameters of the CNN network.

References

1. Vijayarani, S., Sharmila, S.: Research in big data – an overview. Inform. Eng. Int. J. (IEIJ), **4**(3) (2016)
2. Modarres, M.H., Aversa, R., Cozzini, S., Ciancio, R., Leto, A., Brandino, G.P.: Neural network for nanoscience scanning electron microscope image recognition. Sci. Rep. (2017)
3. Choudhary, O.P., Choudhary, O.P.: Scanning electron microscope: advantages and disadvantages in imaging components. Int. J. Curr. Microbiol. Appl. Sci. (IJCMAS) **6**, 1877–1882 (2017)
4. Kaplonek, W., Nadolny, K.: Advanced desktop SEM used for measurement and analysis of the abrasive tool's active surface. Acta Microscopica **22**(3) (2013)
5. Zhou, L., Li, Q., Huo, G., Zhou, Y.: Image classification using biomimetic pattern recognition with convolutional neural networks features. Comput. Intell. Neurosci. **2017** (2017)
6. Albeahdili, H.M., Alwzwazy, H.A., Islam, N.E.: Robust convolutional neural networks for image recognition. Int. J. Adv. Comput. Sci. Appl. (IJACSA) **6**(11) (2015)

7. Srivastava, N., Hinton, G.E., Krizhevsky, A., Sutskever, I., Salakhutdinov, R.: Dropout: a simple way to prevent neural networks from overfitting. J. Mach. Learn. Res. **15**, 1929–1958 (2014)

8. Albelwi, S., Mahmood, A.: A Framework for designing the architectures of deep convolutional neural networks. Entropy (2017)

9. Tharwat, A., Gabel, T.: Parameters optimization of support vector machines for imbalanced data using social ski driver algorithm. Neural Comput. Appl. (2019)

10. Kaur, P., Gosain, A.: Comparing the behavior of oversampling and undersampling approach of class imbalance learning by combining class imbalance problem with noise. In: Advances in Intelligent Systems and Computing. Springer, Singapore (2018)

11. Yıldırım, P.: Pattern classification with imbalanced and multiclass data for the prediction of albendazole adverse event outcomes. Procedia Comput. Sci. **83**, 1013–1018 (2016)

12. Pedamonti, D.: Comparison of non-linear activation functions for deep neural networks on MNIST classification task, arXiv preprint arXiv:1804.02763v1 (2018)

13. Ioffe, S., Szegedy, C.: Batch Normalization: Accelerating Deep Network Training by Reducing Internal Covariate Shift, arXiv preprint arXiv:1502.03167v3 (2015)

14. Mohamad, R., Harun, H.: Enhancement of cross-entropy based stopping criteria via turning point indicator. In: 2017 7th International Conference on Modeling, Simulation, and Applied Optimization, I (2017)

15. Indolia, S., Kumar, A., Mishra, S.P., Asopa, P.: Conceptual understanding of convolutional neural network - a deep learning approach. Procedia Comput. Sci. **132**, 679–688 (2018)

16. Simonyan, K., Zisserman, A.: Very deep convolutional networks for large-scale image recognition, arXiv preprint arXiv:1409.1556 (2014)

17. Aversa, R., Modarres, M.H., Cozzini, S., Ciancio, R.: NFFA-EUROPE Project (2018). http://doi.org/10.23728/b2share.19cc2afd23e34b92b36a1dfd0113a89f

18. Liu, B., Zhang, Y., He, D., Li, Y.: Identification of apple leaf diseases based on deep convolutional neural networks. Symmetry **10**(11) 2018

19. Maria Navin, J.R., Balaji, K.: Performance analysis of neural networks and support vector machines using confusion matrix. Int. J. Adv. Res. Sci. Eng. Technology. **3**(5), 2106–2109 (2016)

20. Google Colab. https://colab.research.google.com. Accessed 16 June 2019

21. Keras, F.C.: Deep learning library for Theano and TensorFlow (2015). https://keras.io. Accessed 16 June 2019

Heartbeat Classification Using 1D Convolutional Neural Networks

Abdelrahman M. Shaker[✉], Manal Tantawi, Howida A. Shedeed,
and Mohamed F. Tolba

Faculty of Computer and Information Sciences, Ain Shams University,
Cairo, Egypt
{Abdelrahman.shaker,manalmt,
dr_howida}@cis.asu.edu.eg, fahmytolba@gmail.com

Abstract. Electrocardiogram (ECG) is an essential source of information for heart diseases classification. Hence, it is used by the cardiologist to diagnose heart attacks and detect the abnormalities of the heart. The automatic classification of the ECG signals is playing a vital role in the clinical diagnosis of heart diseases. In this paper, an end-to-end classification method is proposed using 1D Convolution Neural Networks (CNN) to extract the important features from the input signals and classify it automatically. The main advantage of CNN compared to the related work methods is that it gets rid of the hand-crafted features by combining the feature extraction and the classification into a single learning method without any human supervision. The proposed solution consists of data filtering, dynamic heartbeat segmentation, and 1D-CNN consisting of 10 layers without the input and the output layers.

Our experimental results on 14 classes of the public MIT-BIH arrhythmia dataset achieved a promising classification accuracy of 97.8% which outperforms several ECG classification methods.

Keywords: ECG classification · Heart diseases ·
Convolution Neural Networks

1 Introduction

ECG is used to measure the electrical activity of the heart over time by putting a set of electrodes on the body surface such as chest, arms and neck. These electrodes can detect the electrical changes in the heart. ECG signals are mainly consisting of three complex waves such as P, QRS, and T waves. The P wave shows the atria contractions, while QRS complex shows the ventricular contractions and the T wave reflects the electrical activity produced while recharging the ventricles for the next contractions [1]. Hence, different cardiac activities can be represented by these complex waves and their study plays a vital role in the diagnosis of different arrythmias [2]. However, it is not an easy task for a cardiologist to analyze huge amount of ECG data due to its complexity as well as time requirements [3]. Moreover, the life threating types of arrhythmias need early and accurate detection [4].

© Springer Nature Switzerland AG 2020
A. E. Hassanien et al. (Eds.): AISI 2019, AISC 1058, pp. 502–511, 2020.
https://doi.org/10.1007/978-3-030-31129-2_46

There are two main categories for arrhythmias; the first one causes cardiac arrest and sudden death, such as ventricular fibrillation and tachycardia [5, 6]. The second category, which is our focus in this paper, needs care to avoid deterioration but it is not as dangerous as life-threatening diseases in the first category [7]. Detection and classification of arrythmias require the classification of heartbeats. We can determine the heart rhythm category by recognizing the classes of consecutive heartbeats [8]. Beat by beat human-based classification is a time-consuming task and too difficult of a process. Therefore, the automation process of ECG analysis is very important to discover cardiac disorders which need immediate medical aid in clinical situations, and it also will save a lot of time and efforts for the cardiologist.

In the recent past, different approaches have been presented for automatic classification of heartbeats such as Support Vector Machines (SVM) [9, 10], Back Propagation Neural Networks (BPNN) [11], regression neural networks [12] and Recurrent Neural Networks (RNN) [31].

Deep learning has advanced rapidly since the start of the current decade. Recently, its techniques have shown promising results and demonstrated state-of-the-art performances in tremendous tasks, due to the availability of a huge amount of data and the dramatic increase in the current computational power specially in the modern GPUs, in various fields such as Bioinformatics [13] and Medical diagnosis [14]. One of the advantages of the deep learning techniques is its structure in which both feature extraction and classification stages are performed together without requiring handcrafted features, which is called end-to-end learning [15].

In this paper, the proposed solution consists of two steps; preprocessing and classification steps. Firstly, the preprocessing step consists of data filtering and dynamic heartbeat segmentation. Data filtering is utilized to remove the noise from the ECG records as well as improving the quality of the signals and increasing the Signal-to-Noise Ratio (SNR). Heartbeats are dynamically segmented to avoid heart rate variability. Secondly, a robust end-to-end method is used to extract the most important features and classify them into 14 classes of the public MIT-BIH arrythmia dataset, using Convolution Neural Networks (CNN), with superior performance than the existing studies.

The remainder of this paper is organized as follows: in section two, the related work is provided. Section three discusses the proposed architecture and methodology in detail. Section four represents the achieved results and finally, the conclusion and the future work are provided in section five.

2   Related Work

The conventional way of ECG heartbeat classification is to develop an algorithm to extract the most important features from the signal and then choose an appropriate method to be used in the classification stage. Undoubtedly, there are many researchers in the literature who have done a lot of studies of such way for ECG classification.

Yu [16] used Independent Component Analysis (ICA) to extract the features and the classification is done between eight classes using Neural Networks and they have achieved accuracy of 98.71%, while in [17] the classification is done between four classes using Support Vector Machine (SVM) and the features are extracted using

Discrete Wavelet Transform (DWT) and they have achieved accuracy of 98.39%. In [18] the authors combined the ICA with DWT features with the use of Probabilistic Neural Networks (PNN) for classification between five classes and an accuracy of 99.28% is achieved.

Yazdanian et al. [20] have considered five different classes and they have achieved accuracy of 96.67% using a set of wavelet, morphological and time-domain features. Furthermore, The same five classes have been considered by Martis et al. [19] and an accuracy of 98.11% is achieved with comparison for different approaches for feature extraction such as Principal Component Analysis (PCA) and DWT.

In [21] the feature set is a combination of linear and non-linear features to improve the classification of ECG data, SVM is used for classification between five classes and they have achieved 98.91% recognition accuracy. The classification is done in [22] between 13 classes and the features are generated using two methods; the first method is the Higher Order Statistics (HOS) and the second is the Hermite characterization, the achieved classification accuracy is 98.18%. Khazaee [23] used morphological and time-duration features to represent each beat and the classification is done using SVM between three different classes (Normal, Premature Ventricular Contraction and Others) and an accuracy of 99.9% is achieved. In [24], non-linear transformation is utilized followed by PCA to reduce the dimensionality of the features. Thereafter, the reduced feature vector is fed into SVM for classification, five beats are considered and an accuracy of 98.70% is achieved.

Finally, in [25] they proposed a two-stage hierarchical method to classify the heartbeats of MIT-BIH arrythmia dataset into 15 classes. DWT is used to extract the morphological features and then reduced by PCA. After that, these features are concatenated with four RR features and fed into SVM classifier, an overall accuracy of 94.94% is achieved across ten trials.

To summarize, the most widely used arrythmia dataset in the literature is MIT-BIH. There are many researchers who use fixed window to segment the heartbeats. PCA, ICA and DWT are the most common methods used to extract the features and SVM is the dominant method in the classification stage. Finally, most of the existing studies consider few classes. The aim of this paper is to get rid of the hand-crafted methods of feature extraction used in the literature by using one of the most promising techniques in the deep learning (CNN). In addition, more arrythmia classes are considered.

3 Methodology

This section discusses the used approaches for preprocessing and classification. Figure 1 describes the proposed method which has two main steps; data preprocessing and classification. A detailed description of the proposed method will be discussed in the following sub-sections.

3.1 Preprocessing

This step is essential to improve the signal-to-noise ratio as well as the classification accuracy. The noise is reduced by removing both low and high frequencies out of the

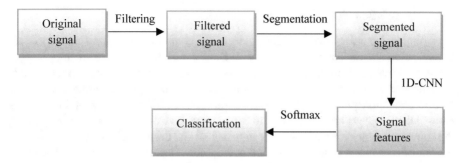

Fig. 1. Overview of the proposed method

ECG signal. Butterworth bandpass filter is applied with a range 0.5–40 because this range contains the most valuable information of the signal [28]. The effect of such filter is shown in Fig. 2. Using the R peak locations which are associated with the database in an annotation file, each ECG record is segmented into multiple heartbeats, and each heartbeat contains P wave, QRS complex, and T wave. Due to the difficulty of detecting the start and the end for each beat, a fixed segmentation method is usually applied. However, it is not always reliable because that assumption cannot consider the variations of the heart rate. Hence, a dynamic segmentation strategy is utilized to overcome the variability of the heart rate as proposed in [10]. The dynamic segmentation strategy counts the number of samples before and after each R peak according to the duration between the current and previous R peaks (RR previous) in addition to the duration between the current and the next R peaks (RR next). After that, it takes a portion from each interval that ensures to include all three complex waves. Such method is invariant to the heart rate variability. Finally, all heartbeats are resampled to have 300 samples per each heartbeat as done in [3, 25].

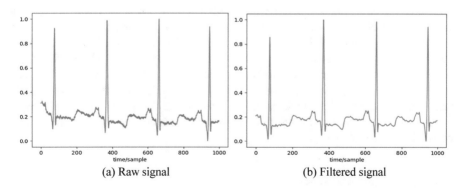

(a) Raw signal   (b) Filtered signal

Fig. 2. ECG signal before and after filtering

3.2   Convolution Neural Networks

Convolution Neural Networks (CNN) is mainly composed of feature extraction and classification stages. The feature extraction stage is responsible for extracting the most useful information from the ECG signals automatically. These features are fed into the classification part to accurately classify these features to one of the target classes. The feature extraction part consists of convolution and pooling (down sampling) layers. Convolutional layers are responsible to extract features from the input data by applying the convolution operation between the input signal and the convolution kernels. Finally, the result is calculated by passing the computed value to an activation function to add non-linearity to these features. Convolutional layer's output can be represented by Eq. 1:

$$y_k^l = \phi(\sum x_k^l * w_k^l + b_k^l) \tag{1}$$

Where y_k^l is the output vector of the lth layer with k^{th} convolution kernel and x_k^l is the input vector, while w_k^l is the weights of the convolution kernel and b_k^l is the bias coefficient and ϕ is the activation function.

Max pooling is applied after the convolutional layer to reduce the dimensionality and preserve the useful information. In the output layer, the Softmax function is used to give a probability to each class of the target classes. The loss is computed by comparing the output with the target vector according to the cross-entropy function in Eq. 2:

$$Loss(y, t) = -\sum_{i=1}^{C} t_i.\log(y_i) \tag{2}$$

Where y is the output vector and t is the target vector, and C is the total number of classes. Cross-entropy is a loss function used to measure the performance of the model whose output is a probability between 0 and 1. The values of the convolution kernels are initialized with random values from normal distribution and then adjusted according to the error.

The proposed model of 1D-CNN for heartbeat classification is shown in Fig. 3. It consists of 10 layers in addition to input and output layers, the first two layers are convolutional layers with 32 filters and kernel size of three, followed by Max pooling layer with pool size of three, followed by two convolutional layers with 64 filters and kernel size of three, followed by another Max pooling layer with pool size of five and final convolutional layer with 128 filters and kernel size of three. After that, three fully connected layers are added with number of neurons 256, 128, and 64 respectively. Finally, Softmax function is used in the output layer which contains 14 neurons corresponding to the 14 classes.

4   Experimental Results

4.1   Dataset

MIT-BIH dataset [26] is the most popular dataset for the existing studies. It consists of 48 records, each one is a 30-min-long with a sampling frequency of 360 Hz.

ANSI/AAMI EC57: 1998 standard [27] recommends only 44 records that can be utilized because there are four paced records. Each record is attached with a file containing the beats annotations and the R peak locations. These given annotations and locations have been used as the ground truth in the training and the evaluation steps. ECG data from lead1 only have been considered.

In this study, the beats of the 44 records are divided into training and testing portions. The data division in [3, 25] have been followed exactly for comparison sake. The percentage of training and testing portions is not equal for all classes because the number of samples for the classes is not equally distributed. The train set of the normal class consists of 13% of the total number of beats as it contains thousands of numbers of beats. On the other hand, training percentage of 40% is considered for other classes that have lower number of beats. Finally, for the classes that have a very limited number of beats, the percentage is 50%. The division of the beats is described in Table 1.

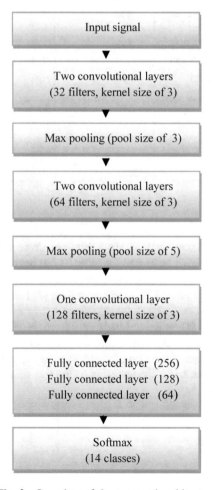

Fig. 3. Overview of the proposed architecture

Table 1. Training and testing percentages used in the experiments

Heartbeat type	Number of total beats	Training ratio	Number of training beats
Normal beat (N)	75017	13%	9753
Left Bundle Branch block (LBBB)	8072	40%	3229
Right Bundle Branch block (RBBB)	7255	40%	2902
Atrial Premature Contraction (APC)	2546	40%	1019
Premature Ventricular Contraction (PVC)	7129	40%	2852
Aberrated Atrial Premature (AP)	150	50%	75
Ventricular Flutter Wave (VF)	472	50%	236
Fusion of Ventricular and Normal (VFN)	802	50%	401
Blocked Atrial Premature (BAP)	193	50%	97
Nodal (junctional) Escape (NE)	229	50%	115
Fusion of Paced and Normal (FPN)	982	50%	491
Ventricular Escape (VE)	106	50%	53
Nodal (junctional) Premature (NP)	83	50%	42
Unclassifiable (UN)	16	50%	7
14 Classes	103052	20.6%	21272

4.2   Results

In this study, MIT-BIH records are segmented into heartbeats. Thereafter, the train set is selected randomly of 21272 beats and the other 81780 beats were used as the test set. There is no duplication between the train and test sets. After training, the CNN is utilized to classify the test data. The Adam optimizer [32] is utilized to train our deep neural network, the network weights are initialized with random distribution.

The proposed method is applied to Lead1 only from the MIT-BIH arrythmia dataset and 14 classes are considered. The evaluation is done by measuring the overall accuracy. Using the average of ten trials, the achieved overall accuracy is 0.978 after 15 epochs of training for each trial. The accuracy of the test set and the loss through each epoch are shown in Fig. 4.

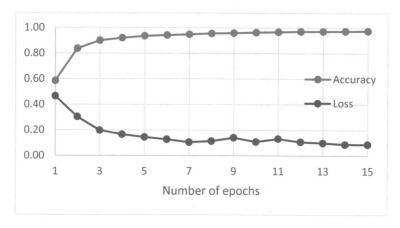

Fig. 4. Accuracy and loss for each epoch

The comparison between our work and the other existing studies is given in Table 2. It shows that the proposed method is applied to large number of classes and the overall accuracy has been improved compared to the published results.

Table 2. Records are segmented into heartbeats

Study	Year	# of classes	Feature set	Classifier	Accuracy
Yu and Chou [16]	2008	8	ICA	NN	98.71%
Sahoo et al. [17]	2017	4	DWT	SVM	98.39%
Martis et al. [18]	2013	5	DWT + ICA	PNN	99.28%
Martis et al. [19]	2012	5	PCA	SVM	98.11%
Yazdanian [20]	2013	5	Wavelet	SVM	96.67%
Elhaj et al. [21]	2016	5	PCA + ICA + DWT + HOS	SVM	98.91%
Yang et al. [29]	2018	5	PCA	SVM	97.77%
Acharya [30]	2017	5	End-to-end	1D-CNN	94.03%
Yildirim [31]	2018	5	End-to-end	LSTM	99.25%
El-Sadawy [25]	2018	15	DWT	SVM	94.94%
Proposed method	**2019**	**14**	**End-to-end**	**1D-CNN**	**97.80%**

5   Conclusion

To conclude, an end-to-end method has been applied to classify different heartbeats into 14 classes of MIT-BIH arrythmia dataset. Dynamic heartbeat segmentation method is used to be invariant to the variability of the heart rate. Overall accuracy of 97.80% is achieved using ten trials, which outperforms the existing studies as well as more classes

(14 classes) are considered. This means that CNN succeeded to learn the most important features automatically without any hand-engineering. Since this study deals with single lead signal (Lead1), further research work will be done to deal with the two channels. Furthermore, different deep learning techniques will be utilized to improve the accuracy in addition to measuring the performance of the models with more metrics such as precision and recall.

References

1. Tantawi, M., Revett, K., Salem, A.-B., Tolba, M.F.: Electrocardiogram (ECG): a new burgeoning utility for biometric recognition. In: Hassanien, A., Kim, T.H., Kacprzyk, J., Awad, A. (eds.) Bio-Inspiring Cyber Security and Cloud Services: Trends and Innovations, Intelligent Systems Reference Library, vol. 70, pp. 349–382. Springer, Heidelberg (2014)
2. Artis, S.G., Mark, R.G., Moody, G.B.: Detection of atrial fibrillation using artificial neural networks. In: Proceedings of the Computers in Cardiology, Venice, Italy, 23–26 September 1991, IEEE, Piscataway, NJ, USA, pp. 173–176 (1991)
3. Ye, C., Kumar, B.V.K.V., Coimbra, M.T.: Heartbeat classification using morphological and dynamic features of ECG signals. IEEE Trans. Biomed. Eng. **59**(10), 2930–2941 (2012)
4. Martis, R.J., Acharya, U.R., Min, L.C.: ECG beat classification using PCA, LDA, ICA and discrete wavelet transform. Biomed. Signal Process. Control **8**(5), 437–448 (2013)
5. Alfonso, V.X., Tompkins, J.: Detecting ventricular fibrillation. IEEE Trans. Biomed. Eng. **54**(1), 174–177 (2007)
6. Minami, K., Nakajima, H., Toyoshima, T.: Real-time discrimination of ventricular tachyarrhythmia with Fouriertransform neural network. IEEE Trans. Biomed. Eng. **46**(2), 179–185 (1999)
7. Ye, C., Vijaya Kumar, B.V.K., Coimbra, M.T.: Heartbeat classification using morphological and dynamic features of ECG signals. IEEE Trans. Biomed. Eng. (2012)
8. Kastor, J.A.: Arrhythmias, 2nd edn. W.B. Saunders, London (1994)
9. Moody, G.B., Mark, R.G.: The impact of the MIT-BIH arrhythmia database. IEEE Eng. Med. Biol. Mag. **20**, 45–50 (2001). [CrossRef] [PubMed]
10. El-Saadawy, H., Tantawi, M., Shedeed, H.A., Tolba, M.F.: Electrocardiogram (ECG) classification based on dynamic beats segmentation. In: Proceedings of the 10th International Conference on Informatics and Systems - INFOS 2016 (2016). https://doi.org/10.1145/2908446.2908452
11. Gautam, M.K., Giri, V.K.: A neural network approach and wavelet analysis for ECG classification. In: Proceedings of the 2016 IEEE International Conference on Engineering and Technology (ICETECH), Coimbatore, India, 17–18 March 2016, IEEE: Piscataway, NJ, USA, pp. 1136–1141 (2016)
12. Zebardast, B., Ghaffari, A., Masdari, M.: A new generalized regression artificial neural networks approach for diagnosing heart disease. Int. J. Innov. Appl. Stud. **4**, 679 (2013)
13. Min, S., Lee, B., Yoon, S.: Deep learning in bioinformatics. Briefings Bioinform. **18**, 851–869 (2017)
14. Bakator, M., Radosav, D.: Deep learning and medical diagnosis: a review of literature. Multimodal Technol. Interact. **2**, 47 (2018). https://doi.org/10.3390/mti2030047
15. Bengio, Y.: Learning deep architectures for AI. Found. Trends Mach. Learn. **2**, 1–127 (2009)
16. Yu, S.N., Chou, K.T.: Integration of independent component analysis and neural networks for ECG beat classification. Expert Syst. Appl. **34**, 2841–2846 (2008)

17. Sahoo, S., Kanungo, B., Behera, S., Sabut, S.: Multiresolution wavelet transform based feature extraction and ECG classification to detect cardiac abnormalities. Measurement **108**, 55–66 (2017)
18. Martis, R.J., Acharya, U.R., Min, L.C.: ECG beat classification using PCA, LDA, ICA and discrete wavelet transform. Biomed. Signal Process. Control **8**, 437–448 (2013)
19. Martis, R.J., Acharya, U.R., Mandana, K., Ray, A.K., Chakraborty, C.: Application of principal component analysis to ECG signals for automated diagnosis of cardiac health. Expert Syst. Appl. **39**, 11792–11800 (2012)
20. Yazdanian, H., Nomani, A., Yazdchi, M.R.: Autonomous detection of heartbeats and categorizing them by using support vector machines. IEEE (2013)
21. Elhaj, F.A., Salim, N., Harris, A.R., Swee, T.T., Ahmed, T.: Arrhythmia recognition and classification using combined linear and nonlinear features of ECG signals. Comput. Meth. Programs Biomed. **127**, 52–63 (2016)
22. Osowski, S., Hoai, L.T., Markiewicz, T.: Support vector machine-based expert system for reliable heartbeat recognition. IEEE Trans. Biomed. Eng. **51**, 582–589 (2004)
23. Khazaee, A.: Combining SVM and PSO for PVC detection. Int. J. Adv. Eng. Sci. **3**(4) (2013)
24. Khalaf, A.F., Owis, M.I., Yassine, I.A.: A novel technique for cardiac arrhythmia classification using spectral correlation and support vector machines. Expert Syst. Appl. **42**, 8361–8368 (2015)
25. El-Saadawy, H., Tantawi, M., Shedeed, H.A., Tolba, M.F.: Hybrid hierarchical method for electrocardiogram heartbeat classification. IET Signal Process. **12**(4), 506–513 (2018). https://doi.org/10.1049/iet-spr.2017.0108
26. MIT-BIH Arrhythmias Database. http://www.physionet.org/physiobank/database/mitdb/. Accessed 12 Oct 2017
27. Testing and reporting performance results of cardiac rhythm and ST segment measurement algorithms. ANSI/AAMI EC57:1998 standard, Association for the Advancement of Medical Instrumentation (1998)
28. Thakor, N.V., Webster, J.G., Tompkins, W.J.: Estimation of QRS Complex Power Spectra for Design of a QRS Filter. IEEE Trans. Biomed. Eng. BME-**31**(11), 702–706 (1984). https://doi.org/10.1109/tbme.1984.325393
29. Yang, W., Si, Y., Wang, D., Guo, B.: Automatic recognition of arrhythmia based on principal component analysis network and linear support vector machine. Comput. Biol. Med. **101**, 22–32 (2018)
30. Acharya, U.R., Oh, S.L., Hagiwara, Y., Tan, J.H., Adam, M., Gertych, A., Tan, R.S.: A deep convolutional neural network model to classify heartbeats. Comput. Biol. Med. **89**, 389–396 (2017)
31. Yildirim, Ö.: A novel wavelet sequence based on deep bidirectional LSTM network model for ECG signal classification. Comput. Biol. Med. **96**, 189–202 (2018). https://doi.org/10.1016/j.compbiomed.2018.03.016
32. Kingma, D.P., Ba, J.: Adam: a method for stochastic optimization. CoRR, abs/1412.6980 (2014)

A Transfer Learning Approach
for Emotion Intensity Prediction
in Microblog Text

Mohamed Osama[✉] and Samhaa R. El-Beltagy

Nile University, Sheikh Zayed, Giza, Egypt
mohamed.osama0342@gmail.com
http://nu.edu.eg

Abstract. Emotional expressions are an important part of daily communication between people. Emotions are commonly transferred non verbally through facial expressions, eye contact and tone of voice. With the rise in social media usage, textual communication in which emotions are expressed has also witnessed a great increase. In this paper automatic emotion intensity prediction from text is addressed. Different approaches are explored to find out the best model to predict the degree of a specific emotion in text. Experimentation was conducted using the dataset provided by SemEval-2018 Task 1: Affect in Tweets. Experiments were conducted to identify regression systems and parameter settings that perform consistently well for this problem space. The presented research highlights the importance of the Transfer Learning approach in inducing knowledge from state of the art models in sentiment analysis for use in the task of emotion intensity prediction.

Keywords: NLP · Emotion intensity prediction · Transfer learning · Sentiment analysis · Regression

1 Introduction

Humans use language to communicate and transfer information. Language can be communicated by speech, facial expressions, body language or text. Sentiment analysis is concerned with deriving the semantic polarity of text whether positive, negative or neutral rather than detecting a specific emotion. As an example, sadness and anger both have negative text polarities but distinguishing between them and predicting the intensity of these emotion in text is important for many businesses in analyzing user reviews to identify the emotional feedback about a product or service. It can also provide governments with the reaction of the public towards a specific event or news. In customer services, automatic emotion detectors from customers emails can inspire marketing departments with innovative ideas to market, advertise and generate new campaigns for new products.

Many people rely on Twitter in social interaction, and for discussing issues of life and community problems through messages which are called tweets. Since

© Springer Nature Switzerland AG 2020
A. E. Hassanien et al. (Eds.): AISI 2019, AISC 1058, pp. 512–522, 2020.
https://doi.org/10.1007/978-3-030-31129-2_47

Twitter is one of the most important tools of communication between people, so the problem of detecting the degree of emotion in tweets is addressed in this paper.

The emotion theory most widely applied in computational linguistic research is Ekman's [1] theory. Ekman classifies emotions into six classes which are called the basic emotions. Plutchik argues that there are eight basic and prototypical emotions [2]. Izard defines ten categories as basic emotions [3]. The two theories are not used as much in computational linguistics, because they are less trivial to implement computationally.

The goal of this work is to predict the intensity of a specific emotion E in a given tweet T as a score ranges from 0 to 1, where 0 indicates that the emotion E is not found in the Tweet T and 1 indicates the maximum intensity of emotion E that can be felt in the tweet. Emotion intensity prediction in this work faces many obstacles because twitter text is considered as informal text and has its own characteristics such as inclusion of abbreviations, emojis, emoticons, slang, misspelled words, new spelled words (happee) and hash-tagged words. Different experiments were held in order to find out the best model for predicting emotion intensity in text, as discussed in the following sections.

2   Background

This section reviews the NLP features and the learning algorithms that can be used in the development of emotion predication models. Those can be divided into three main categories: lexical features, learning algorithms, and manually created features.

The lexicon based features are features that are extracted based on some input lexicons. The emotions can be expressed in text by specific words that also appears in the lexicon. The performance of a particular lexicon based feature is determined based on the used lexicon and how well it covers different emotional related words. The most used emotion and affect lexicons in the field of sentiment and emotion analysis are summarized in Table 1.

Supervised learning, unsupervised learning and transfer learning algorithms are the three main categories in learning algorithms. Supervised learning algorithms use training data annotated with predefined labels. Unsupervised learning algorithms don't depend on a labeled dataset. Instead they assign a data point to a cluster based on hidden patterns and relations between different data points. Transfer Learning is a learning approach where a model's features that were developed for a specific task A are transferred as an input for another task B. Supervised learning can be divided into statistical methods and neural networks.

In terms of supervised learning, support vector machines (SVMs) [12] are popular in text classification as they can handle large number of features and can achieve better results than other algorithms [13]. SVMs were used in emotion classification as they perform well and overcome overfitting on new datasets [14].

One popular unsupervised learning method in this problem space is latent semantic analysis (LSA), which is used in evaluating the similarity between different words in a text [15].

Table 1. List of popular emotion/affect lexicons

Lexicon name	Type	Context	Size
ANEW [4]	Affect	General	1,034 words
LIWC [5]	Emotion, cognition, structural components	General	915 words (affective)
NRC Emotion Lexicon (EmoLex) [6]	Emotion, sentiment polarity (positive, negative)	General	14,182 words
WordNet-Affect [7]	Affect	General	4,787 words
AFINN [8]	Affect	Twitter	2,477 words
Affect database [9]	Affect	Instant messaging	364 emoticons 337 acronyms abbreviations 1,620 words
Fuzzy Affect Lexicon [10]	Affect	General	3,876 words
Depheche Mood [11]	Mood	News articles	37,771 words

Transfer learning was popularized in the field of computer vision, where different problem models (classification, object detection and etc.) are not trained directly on the task data-set, but get their features from popular pretrained models like ImageNet [16], MSCOCO [17], and other models.

In this work, various state-of-the-art machine learning models [18] were experimented with in order to transfer knowledge from their source task to the target task. These include: DeepMoji [19], Skip-Thought Vectors [20] and Unsupervised Sentiment Neuron [21].

In DeepMoji [19], distant supervision was employed to train a model using a very large dataset (1246 million tweets). In this data set, entries were labeled by emojis, to enable the model to learn richer representations of emotions in text, and in order to obtain better performance on benchmarks for detecting sentiment and emotions in text.

Skip-Thought vectors [20] is an encoder that can produce generic sentence vectors.

The unsupervised Sentiment Neuron [21] was inspired by the skip-thought vectors [20], and has a single unit that encodes a sentence to a vector in an unsupervised manner.

Manually created features are custom features generated in an intuitive way after observing cues during the data exploration phase. These custom features enhance the performance of the learning model, but are related to the specific dataset in which they were observed and it is hard to expect them to contribute to enhanced performance on different datasets.

3   Dataset

The dataset used in this work was provided by SemEval-2018 [22] Task 1: Affect in Tweets. The dataset was divided into four groups based on Ekman's basic emotions: fear, joy, sadness, and anger, respectively. Each emotion tweets are divided internally into three parts: training, development, and testing. The training and the development datasets are concatenated, and used as training set for the experimented models. The number of instances for each emotion is shown in Table 2.

Table 2. Number of instances

Emotion	Train	Dev	Test	Total
Anger	1701	388	1002	3091
Fear	2252	289	986	3627
Joy	1616	290	1105	3011
Sadness	1533	397	975	2905

Every tweet in a specific emotion is annotated by a score that varies form 0 to 1 based on the degree of emotion felt by the author of the tweet.

The dataset used in this task was created by searching Twitter API with some query terms for every particular emotion [23]. As an example angry, frustrated, disappointed and etc. were the query terms used to collect the angry emotion Tweets. Best-Worst Scaling (BWS) [24] was used for annotation.

4   Experiments and Results

We present the results of the machine learning experiments conducted, in three parts: Machine Learning experiments, feature related experiments and transfer learning experiments.

Since the used dataset is that of the first task in semEval 2018 [22] so the evaluation metrics used in this task were also used here. For every emotion category, performance of various models was evaluated by calculating the Pearson Correlation Coefficient between the predicted emotion intensity and the Gold ratings. As a secondary metric, the Pearson Correlation Coefficient for tweets with emotion intensity above or equal 0.5 is used.

4.1   Baseline Experiments

The goal of these experiments to find out the baseline model for this task. Tweets were pre-processed, where stop words and punctuation were eliminated. The Snow Ball stemmer [25] was used which produced slightly better results compared to the raw tweets text. A uni-gram (i.e., bag-of-words) model used in this part of experiments as the input features to the experimented set of algorithms.

Table 3. Baseline experiments

Algorithm	Pearson r (all instances)					Pearson r (gold in 0.5–1)				
	Avg.	Anger	Fear	Joy	Sadness	Avg.	Anger	Fear	Joy	Sadness
Linear regression	39.48	41.80	38.79	43.13	34.20	31.32	35.82	29.16	26.71	33.60
Bayesian ridge	**56.78**	58.23	58.37	57.36	53.19	44.12	51.12	39.58	39.99	45.82
Decision tree	37.195	39.79	46.33	26.07	36.59	27.64	32.81	33.82	16.60	27.34
SVM	**56.84**	58.37	58.47	57.66	52.89	44.18	51.32	39.57	40.31	45.52
KNN	41.54	39.59	48.44	39.71	38.42	32.017	38.08	32.47	27.81	29.71
Random baseline	−1.91	−1.59	2.14	−4.29	−3.9	−3.52	−2.34	−3.96	0.428	−8.24

Traditional machine learning algorithms like SVM, Naive Bayes, Decision Trees and etc, were experimented. Based on the micro-averaged for the Pearson correlation coefficient as shown in Table 3, the two machine learning algorithms that yielded the best performance in this problem space were SVM with linear kernel and Bayesian ridge regression.

4.2   Feature Related Experiments

The features used in this work are divided into two feature sets: Corpus based features, and Lexicon based features. Corpus based features consist of statistical features generated from the training and development corpus. We tested two different features generated from the corpus: unigrams, and bigrams with different weighting schemes, in addition to pre-trained word embeddings [26]. The machine learning model used in these experiments, was SVM because it resulted in the best performance metrics in the baseline experiments.

Bigrams with normalized sub-linear tf, i.e. where tf is replaced with 1 + log(tf) yielded the best performance over different feature groups as shown in Table 4.

Lexicon-based features were generated based on the terms in lexicons described in Table 1. In order to use all of these lexical resources and for the sake of saving effort that is done in text processing and converting words into lexical scores, AffectiveTweets [27] which is a WEKA package for analyzing emotion and sentiment of tweets was used, to extract lexicon based features from tweets.

The results of the corpus based and lexicon based features are presented in Table 4.

Based on the results, the Bigrams with normalized sublinear tf model yielded the best performance among the corpus based features models regarding the primary metric, while the Unigrams with normalized sublinear tf model is the best

Table 4. Corpus and Lexicon based features

Pearson r (all instances)					Pearson r (gold in 0.5–1)				
Feature group	Avg.	Anger	Fear	Joy	Sadness	Avg.	Anger	Fear	Joy	Sadness
Unigrams with normalized tf-idf	55.90	57.83	57.53	56.25	52.02	43.28	51.56	38.92	37.75	44.92
Unigrams with normalized tf	56.77	58.23	58.60	57.66	52.61	44.12	51.59	39.84	40.01	45.07
Unigrams with normalized sublinear tf	56.84	58.37	58.47	57.66	52.89	**44.18**	51.32	39.57	40.31	45.52
Bigrams with normalized tf-idf	56.33	59.09	58.03	55.69	52.54	43.17	52.41	38.84	36.98	44.45
Bigrams with normalized tf	56.89	58.99	58.61	56.91	53.05	43.90	51.86	39.56	39.54	44.67
Bigrams with normalized sublinear tf	**56.92**	59.12	58.39	56.96	53.21	43.93	51.68	39.29	39.85	44.93
Word embedding [26]	46.87	45.06	45.02	50.05	47.36	30.57	33.34	26.95	28.38	33.62
Lexicon based features	**60.24**	63.44	59.01	59.34	59.18	**43.35**	49.09	36.26	39.74	48.33

performing model in the secondary metric. The lexicon based model outperforms the two models in the primary metric. A combination between these models will be discussed in Sect. 5.

4.3 Transfer Learning Experiments

Inspired by the transfer learning in computer vision and the results achieved by these model in their tasks, transfer learning using DeepMoji model [19], Skip-Thought vectors [20], and Sentiment Neuron model [21] was performed. In this section implementation details are discussed per each model. For the DeepMoji model [19], two different feature sets were extracted which are the embeddings from the softmax and the attention layer from the pre-trained DeepMoji model. The vector from the softmax layer consists of 64 dimensions and the vector from attention layer is of 2304 dimensions. Skip-Thought vectors [20] is an encoder that can produce generic sentence representations. The representation is of dimension 4800. Similar to Skip-Thought vectors, a 4096 dimension was extracted from the Sentiment Neuron model [21]. The vector extracted from each model is considered as input features to a Random Forest Regressor of sklearn [28].

Based on the results presented in Table 5. The DeepMoji attention layer yielded the best performance in the primary and secondary metric across other transfer learning models.

Table 5. Transfer learning experiments

Pearson r (all instances)					Pearson r (gold in 0.5–1)				
Model	Avg.	Anger	Fear	Joy	Sadness	Avg.	Anger	Fear	Joy	Sadness
DeepMoji (attention layer)	**70.37**	76.64	64.38	71.07	69.40	**51.72**	63.03	42.88	47.17	53.82
DeepMoji (softmax layer)	68.66	75.66	63.12	68.22	67.66	48.52	60.04	42.95	40.24	50.88
Skip-Thoughts	59.25	60.63	58.71	56.15	61.52	42.95	50.13	40.97	34.46	46.27
Sentiment neuron	59.38	59.96	54.58	61.82	61.16	40.79	48.15	38.16	35.44	41.43

5   Proposed Model

Using an ensemble of more than one model is common in achieving the best performance in many tasks. Ensemble methods placed first in many prestigious machine learning competitions [29]. Ensembling of the models with the best performance in each category led to the proposed system, where every model added to the ensembled system increases the primary and secondary metric during the experiments. An ensemble stacking method was used for combining five models: the custom features model, the DeepMoji attention layer embeddings model, the DeepMoji softmax layer embeddings model, Skip-Thoughts vector model, and sentiment neuron vector model. The custom features model is a mixture between the best performing models in corpus based, lexicon based and manually created features with the use of SVM as a machine learning algorithm, where this methodology achieved the best results in combining the corpus and lexicon based features through experiments. Bigrams with normalized sub-linear tf, AffectiveTweets features in addition to some manually created features: average word length, Emoji score and hashtag score are the features of this model.

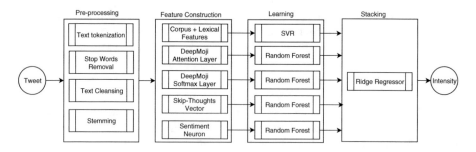

Fig. 1. Proposed system

The other four models are pre-trained models, where the output of each model is considered as a sentence representation vector, and this representation is transferred from the original model, that was designed to solve other NLP problem to help in solving the emotion prediction problem. These representations are considered as an input to a random forest regressor as in Fig. 1.

The output of the five models is stacked using a ridge regressor as a meta-learner algorithm, which was observed to achieve the best results across different regression algorithms. The proposed system achieved the best results in the experiments as shown in Table 6.

Table 6. Comparison between proposed system and other models

Pearson r (all instances)					Pearson r (gold in 0.5–1)				
Model	Avg.	Anger	Fear	Joy	Sadness	Avg.	Anger	Fear	Joy	Sadness
Bigrams with normalized sublinear tf	56.92	59.12	58.39	56.96	53.21	43.93	51.68	39.29	39.85	44.93
Lexicon based features	60.24	63.44	59.01	59.34	59.18	43.35	49.09	36.26	39.74	48.33
DeepMoji (attention layer)	70.37	76.64	64.38	71.07	69.40	51.72	63.03	42.88	47.17	53.82
Proposed system	**77.45**	**81.63**	**72.89**	**77.81**	**77.47**	**59.415**	**69.13**	**53.03**	**53.74**	**61.76**

6   Error Analysis

Our focus is to inspect the error in the tweets that have a big difference between the predicted and gold scores. The error metric that was used in this analysis is the Mean Absolute Percentage Error [31], as shown in Eq. 1.

$$MAPE = \frac{100\%}{n} \sum \left| \frac{y - \hat{y}}{y} \right| \tag{1}$$

Where y is the gold score, $\hat{y}$ is the predicted score and n is the number of data points.

Tweets with Absolute Percentage Error (APE) greater than 50% were analyzed in order to enhance the performance of the proposed model in the future. An example of these tweets for is:

"My heart is so happy I want to explode." (anger)

The proposed model considered the word "explode" as an anger word and ignored the context of the tweet.

"never had a dull moment with u guys" (anger)

Here this tweet was affected by negation word "never" where it wasnt considered to remove the effect of the anger word "dull".

> *"although I laugh and I act like a clown beneath this mask I am wearing a frown my tears are falling like rain from the sky"* (joy)

For this tweet, the underlined expressions did not affect the predicted score properly, where the gold score is very small that causes a big APE.

The result of the angry category is considered in the second place compared to the submitted systems as presented in Table 7 [30].

Table 7. Comparison between proposed system and systems that participated in SemEval

Pearson r (all instances)					Pearson r (gold in 0.5–1)				
Model	Avg.	Anger	Fear	Joy	Sadness	Avg.	Anger	Fear	Joy	Sadness
SeerNet	**79.9**	**82.7**	77.9	79.2	79.8	**63.8**	70.8	60.8	56.8	66.6
NTUA-SLP	77.6	78.2	75.8	77.1	79.2	61.0	63.6	59.5	55.4	65.4
Proposed system	77.45	81.63	72.89	77.81	77.47	59.415	69.13	53.03	53.74	61.76
PlusEmo2Vec	76.6	81.1	72.8	77.3	75.3	57.9	66.3	49.7	54.2	61.3

7   Conclusion and Future Work

This paper proposed a model for emotion intensity prediction from tweets. The proposed system results place the system in third place using the primary and secondary metric when comparing it to the submitted systems during the evaluation period of SemEval 2018. For Future work, we intend fine tune the transfer learning models on our task dataset to get a better results. So, fine tuned models will be experimented with to figure out how to enhance the performance. Transfer learning in Arabic needs to be investigated, as there are few models available in Arabic. Best results for the same task in Arabic was achieved by applying Traditional machine learning algorithms (Random Forest, SVR and Ridge regression) as mentioned in [30]. So, implementing pre-trained models in Arabic will be a promising area in research.

References

1. Ekman, P.: An argument for basic emotions. Cogn. Emot. **6**(3–4), 169–200 (1992)
2. Plutchik, R.: The Emotions: Facts, Theories, and a New Model. Random House, New York (1962)
3. Izard, C.E.: The face of emotion. Appleton-Century-Crofts, East Norwalk, CT, US (1971)
4. Bradley, M.M., Lang, P.J.: Affective norms for English words (ANEW): Instruction manual and affective ratings. Technical report, Citeseer (1999)

5. Pennebaker, J., Chung, C., Ireland, M., Gonzales, A., Booth, R.: The development and psychometric properties of liwc2007: Liwc. net. Google Scholar (2007)
6. Mohammad, S.M., Turney, P.D.: Emotions evoked by common words and phrases: Using mechanical turk to create an emotion lexicon. In: Proceedings of the NAACL HLT 2010 Workshop on Computational Approaches to Analysis and Generation of Emotion in Text, pp. 26–34. Association for Computational Linguistics (2010)
7. Strapparava, C., Valitutti, A., et al.: Wordnet affect: an affective extension of wordnet. In: Lrec, vol. 4, pp. 1083–1086. Citeseer (2004)
8. Nielsen, F.Å.: A new anew: evaluation of a word list for sentiment analysis in microblogs. arXiv preprint arXiv:1103.2903 (2011)
9. Neviarouskaya, A., Prendinger, H., Ishizuka, M.: Analysis of affect expressed through the evolving language of online communication. In: Proceedings of the 12th International Conference on Intelligent user Interfaces, pp. 278–281. ACM (2007)
10. Subasic, P., Huettner, A.: Affect analysis of text using fuzzy semantic typing. FUZZ-IEEE 2000, San Antonio (2000)
11. Staiano, J., Guerini, M.: Depechemood: a lexicon for emotion analysis from crowd-annotated news. arXiv preprint arXiv:1405.1605 (2014)
12. Cortes, C., Vapnik, V.: Support-vector networks. Mach. Learn. 20(3), 273–297 (1995)
13. Yang, Y., Liu, X.: A re-examination of text categorization methods. In: Proceedings of the 22nd Annual International ACM SIGIR Conference on Research and Development in Information Retrieval - SIGIR 1999, pp. 42–49. ACM Press, New York (1999)
14. Chaffar, S., Inkpen, D.: Using a Heterogeneous Dataset for Emotion Analysis in Text, pp. 62–67. Springer, Heidelberg (2011)
15. Strapparava, C., Mihalcea, R.: Learning to identify emotions in text. In: Proceedings of the ACM Symposium on Applied Computing - SAC 2008, p. 1556. ACM Press, USA (2008)
16. Deng, J., Dong, W., Socher, R., Li, L.-J., Li, K., Li, F.-F.: Imagenet: A large-scale hierarchical image database. In: CVPR (2009)
17. Lin, T.-Y., Maire, M., Belongie, S., Hays, J., Perona, P., Ramanan, D., Dollár, P., Zitnick, C.L.: Microsoft coco: common objects in context. In: European Conference on Computer Vision, pp. 740–755. Springer (2014)
18. Duppada, V., Jain, R., Hiray, S.: SeerNet at SemEval-2018 task 1: domain adaptation for affect in tweets. arXiv preprint arXiv:1804.06137 (2018)
19. Felbo, B., Mislove, A., Søgaard, A., Rahwan, I., Lehmann, S.: Using millions of emoji occurrences to learn any-domain representations for detecting sentiment, emotion and sarcasm. arXiv preprint arXiv:1708.00524 (2017)
20. Kiros, R., Zhu, Y., Salakhutdinov, R.R., Zemel, R., Urtasun, R., Torralba, A., Fidler, S.: Skip-thought vectors. In: Advances in Neural Information Processing Systems, pp. 3294–3302 (2015)
21. Radford, A., Jozefowicz, R., Sutskever, I.: Learning to generate reviews and discovering sentiment. arXiv preprint arXiv:1704.01444 (2017)
22. Mohammad, S.M., Bravo-Marquez, F., Salameh, M., Kiritchenko, S.: SemEval-2018 task 1: affect in tweets. In: Proceedings of International Workshop on Semantic Evaluation (SemEval-2018), New Orleans, LA, USA (2018b)
23. Mohammad, S.M., Kiritchenko, S.: Understanding emotions: a dataset of tweets to study interactions between affect categories. In: Proceedings of the 11th Edition of the Language Resources and Evaluation Conference, Miyazaki, Japan (2018)

24. Louviere, J.J., Woodworth, G.G.: Best-worst scaling: a model for the largest difference judgments. University of Alberta: Working Paper (1991)
25. Bird, S., Klein, E., Loper, E.: Nltk book (2009)
26. Godin, F., Vandersmissen, B., De Neve, W., Van de Walle, R.: Multimedia lab @ ACL WNUT NER shared task: named entity recognition for Twitter microposts using distributed word representations. In: Proceedings of the Workshop on Noisy User-Generated Text, pp. 146–153 (2015)
27. Mohammad, S.M., Bravo-Marquez, F.: WASSA-2017 Shared Task on Emotion Intensity (2017)
28. Pedregosa, F., Varoquaux, G., Gramfort, A., Michel, V., Thirion, B., Grisel, O., Blondel, M., Prettenhofer, P., Weiss, R., Dubourg, V., et al.: Scikit-learn: machine learning in python. J. Mach. Learn. Res. **12**, 2825–2830 (2011)
29. Gomes, H.M., Barddal, J.P., Enembreck, F., Bifet, A.: A survey on ensemble learning for data stream classification. ACM Comput. Surv. (CSUR) **50**(2), 23 (2017)
30. Mohammad, S., Bravo-Marquez, F., Salameh, M., Kiritchenko, S.: Semeval-2018 task 1: affect in tweets. In: Proceedings of the 12th International Workshop on Semantic Evaluation, pp. 1–17 (2018a)
31. Swamidass, P.M., (ed.): MAPE (mean absolute percentage error) Mean Absolute Percentage Error (MAPE), pp. 462–462. Springer US, Boston (2000)

Convolutional Neural Networks for Biological Sequence Taxonomic Classification: A Comparative Study

Marwah A. Helaly, Sherine Rady$^{(\boxtimes)}$, and Mostafa M. Aref

Faculty of Computer and Information Sciences, Ain Shams University,
Cairo 11566, Egypt
{marwah.ahmad.helaly,srady}@cis.asu.edu.eg, mostafa.m.aref@gmail.com

Abstract. Biological sequence classification is a key task in Bioinformatics. For research labs today, the classification of unknown biological sequences is essential for facilitating the identification, grouping and study of organisms and their evolution. This paper compares three of the most recent deep learning works on the 16S rRNA barcode dataset for taxonomic classification. Three different CNN architectures are compared together with three different feature representations, namely: k-mer spectral representation, Frequency Chaos Game Representation (FCGR) and character-level integer encoding. Experimental results and comparisons have shown that representations that hold positional information about the nucleotides in a sequence perform much better with accuracies reaching 91.6% on the most fine-grained classification task.

Keywords: DNA · RNA · Biological sequences · Deep learning ·
classification · Convolutional neural networks · Feature representation

1 Introduction

A biological sequence is a single molecule of nucleic acid or protein. Nucleic acids contain the genetic code that is considered the instruction book for creating proteins [1]. Deoxyribo-nucleic Acid (DNA) and Ribo-nucleic Acid (RNA) are composed of a series of monomer molecules called the *nucleotides*. There are four types of nucleotides *bases* in DNA: adenine (A), guanine (G), cytosine (C) and lastly thymine (T), which is replaced with uracil (U) in RNA.

Recent large biological projects such as the Human Genome project [2] have made an incredible amount of biological data available. As a result, *Bioinformatics* emerged as the science needed to analyze and manage this data, in which DNA sequence classification is a key task. *Biological systematics* is involved with the identification and study of evolutionary relationships using taxonomies that organize organisms into hierarchies (i.e. species, genera, etc.) [3].

The task of DNA sequence classification has been attempted using various Machine Learning (ML) methods. Deep Learning (DL) is a trending area of ML

© Springer Nature Switzerland AG 2020
A. E. Hassanien et al. (Eds.): AISI 2019, AISC 1058, pp. 523–533, 2020.
https://doi.org/10.1007/978-3-030-31129-2_48

found to have promising performance in many applications, including bioinformatics. DL uses computational neural networks constituting of multiple nonlinear information processing layers, that learn hierarchical representations of complex data representations with multiple increasing levels of abstraction. DL models require large memory and powerful computation resources due to its need for a large amount of training data and the possibility that the biological sequence representation method results in a representation very large in size.

This work concentrates on taxonomic classification on five different taxonomic ranks, using the commonly-used 16S rRNA gene barcode dataset. Three of the most recent works on this dataset were selected and compares the methods proposed by the authors, where each proposed method consists of a representation method and a CNN architecture, highlighting the benefits and drawbacks of every feature representation method and CNN model design.

The rest of the paper is organized as follows: Sect. 2 reviews some of the recent work in DNA and RNA sequence classification using both traditional ML methods and DL methods. The selected feature representations for the 16S rRNA dataset and CNN models are explained in detail in Sect. 3. Section 4 presents the various experiments and results on the different CNN models and feature representations. Finally, the conclusions are discussed in Sect. 5.

2 Related Works

This section summarizes recent work done for biological sequence classification. It is categorized into traditional ML methods and DL methods.

In previous traditional ML works authors commonly use small-sized datasets. In [4], was an attempt to differentiate between eukaryotic and prokaryotic cells. The authors used a Support Vector Machine (SVM) and a Multi-layer Perceptron (MLP), further optimizing their work in [5] using Particle Swarm Optimization (PSO). The sequences were represented by a normalized frequency vector of occurring constant-sized subsequences. The training and test dataset consisted of 30 and 25 sequences, respectively. Both the MLP and SVM gave a best performance of approximately 85% and increased to 99% and 90% respectively with PSO.

In [6] a SVM and a clustering-K-nearest-neighbor (C-K-NN) classifier were used for species identification. The 317-sequences dataset unevenly belonged to 4 prokaryotic species, with a 288-base average length. The sequences were partitioned, the features mathematically extracted, the networks trained then tested on *parts* of the sequences. Using 20% of a sequence for classification gave the highest accuracy and reduced the classification time by approximately 83%. The SVM generally outperformed the C-K-NN.

More recent works started using larger datasets to train their models. In [7], four DNA sequence datasets of different types were used with Wavelet Neural Networks (WNNs), ranging in sizes from 106 to 3190 sequences and 7 to 1595 bases. Three phases were proposed: (1) transformation of DNA sequences, (2) approximation and optimization of network parameters and (3) clustering of

DNA sequences. The proposed method outperformed an ANN and the run-time increased with the increasing sequence sizes.

For splice site recognition, [8] experimented with four different classifiers. The 140-nucleotide sequences used a TD-FDTF encoding in all experiments that was similar to [9]. The authors attempted three experiments, concluding that their TD-FDTF representation combined with AdaBoost.M1 outperformed similar work [9] and almost all of the state-of-the-art classifiers - with the maximum difference in accuracy being less than only 1.6% in one of the cases.

Many works compared traditional and DL methods for the various tasks of DNA sequence classification, which proved DL to be more effective.

Several recent works have explored the task of taxonomic species classification, with the common usage of the 16S rRNA barcode dataset. In [10], a LeNet-5-inspired CNN was employed on the 16S rRNA dataset, for classification on five taxonomic ranks. A k-mer spectral representation of the sequences was used. A k-mer is a subsequence of a variable size k bases from a sequence, captured by a sliding window of size k and a stride. The CNN was trained once but tested twice on both the full-sequence and fragment sequences. Compared to five traditional classifiers, it was concluded that the CNN has better generalization than other classifiers. In [11] the same previous model was used but with a frequency chaos game representation (FCGR) representation. Increasing the value of k only slightly increased the accuracy but greatly increased the computational cost. For the full-length sequences, the CNN was slightly better than a SVM, but completely outperformed than SVM in the sequence fragments test.

Also using the 16S rRNA dataset, in [12] it was used to compare a Long Short-Term Memory (LSTM) RNN and a CNN similar to that of [10,11]. Sequences were represented using a one-hot encoding, then passed to an embedding layer that resulted in a denser continuous input vector for each sequence. *Multi-task learning* improved the generalization of the LSTM model, but harmed that of the CNN. Experimental studies showed the CNN superior for the first four taxonomic ranks and the LSTM superior for the last most fine-grained rank.

A CNN was applied on 12 different datasets in [13], and compared its performance with that of other papers that used a SVM classifier. Dataset sizes ranged from approximately 100 to 37,000 samples. A combination of k-mer extraction, region grouping and a one-hot sequence representation resulted in a 2D numerical matrix input to the network [14]. It was shown that CNNs significantly improved the accuracy 1% to 6% for all datasets.

The authors in [15] used two different enhancer datasets, employing *transfer learning* from the first to the second to improve the model's generalization and performance. Using one-hot vectors and a CNN with a general enhancer dataset, the best CNN architecture produced 0.916 AUROC and 0.917 AUPRC. Their CNNs were superior in performance to the compared related works of [16]. Using *transfer learning* to fine-tune the model, the training continued on 9 specific datasets and produced better accuracy for all types than [16], with an average increase of 7% in AUROC and AUPRC.

3   Considered DL Methods

This work compares the performance of the proposed representations and CNN architectures of [10,11] and [12] on the 16S rRNA gene dataset used by all of them. Figures 1 and 2 illustrate the considered DL approach and the considered representations, respectively.

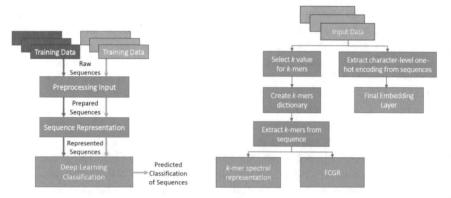

Fig. 1. Proposed DL DNA sequence classification process.

Fig. 2. Representation methods considered.

3.1   Pre-processing Input

This phase includes the parsing and shuffling of the input sequences from a FASTA file. Ambiguous characters are not considered and were removed from the sequences. Shorter sequences are zero-padded to unify the lengths of all the sequences.

3.2   Sequence Representation

ML methods receive numerical input while biological sequences commonly come in a textual form. Hence, they must be represented by a meaningful numerical form. Common representations include k-mer based representations, image representations, integer encodings, and combinations of several together. In this work we consider the following methods (Fig. 2) to represent the input biological sequences:

1. k-mer spectral representation: k-mers are similar to n-grams, where k is a variable value. Considering only the four basic nucleotide bases A, G, C and T and all their possible k-length subsequence combinations, this results in a dictionary of 4^k possible k-mers. To extract the k-mers of a sequence, a sliding window of size k is passed over it to capture regions of length k-bases. Then the window is slid with a *stride* (commonly 1), while keeping count of

the number of occurrences of each k-mer into a *frequency vector* or a *bag-of-words* representation. This work experiments with three different values of k = 4, 5 and 6. An example of the k-mer extraction and representation process for $k = 4$ is illustrated in Fig. 3.

2. Frequency Chaos Game Representation (FCGR): A CGR is also uses k-mers to produce a fractal-like image. Given the four basic nucleotide bases, the 2D-matrix representation is split into 4 quadrants, with each quadrant representing one of the four bases. The quadrants are similarly recursively split for a total of k splits. Figure 4 shows an example of the creation of a $k = 6$ CGR. This results in a 2D-matrix of size:

$$size = \sqrt{4^k} \times \sqrt{4^k}$$

To fill the FCGR matrix, the k-mer *frequencies* are counted from the sequence then placed in their respective k-mer slot in the 2D-matrix (commonly normalized). Similar to the k-mer spectral representation, the same three values of k were also experimented.

3. Character-level integer encoding: The ambiguous characters were also removed to keep the usage of the dataset similar to that of the previous two papers. Besides the four basic nucleotide bases, the alphabet includes an extra character Z to pad shorter sequences. Each character is assigned a unique integer value from $[0, 4]$, resulting in integer vectors with the length of the maximum-lengthed sequence in the dataset. This is a slight variation from the representation method used in the considered work [12].

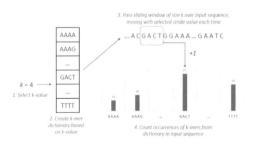

Fig. 3. Example: k-mer spectral representation steps.

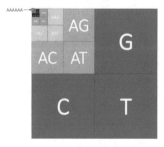

Fig. 4. Example: CGR creation for $k = 6$.

3.3   Deep Learning Classification

Table 1 shows the three considered representations in this work and their CNN architectures:

1. The *CNN1-kmer*, from [10] uses a k-mer sequence representation along with the CNN model. The CNN employs two convolutional layers with an increasing number of filters 10 and 20, respectively, and filters of size 5. After each of them are maxpooling layers with sizes of 2 and a stride of 2. Finally, there are two dense fully-connected layers, the first of size 500 neurons and the second output layer is based on the number of categories in the taxonomic rank considered.
2. In [11], a Frequency Chaos Game Representation (FCGR) sequence representation is used with a CNN model, abbreviated as *CNN2-FCGR*. The CNN architecture differs from that of *CNN1-kmer* with the usage of 2-dimensional convolutional layers with 5×5 filters and 2×2 maxpooling layers.
3. Similar to [12], the *CNN3-IntEnc* uses an integer encoding and embedding layer combination for sequence representation, followed by a CNN. This architecture is also similar to *CNN1-kmer*, with an embedding layer before the network and larger maxpooling layers of size and stride 5.

Table 1. Description of the architectures for the three CNN approaches.

Layers	Layer sizes		
	CNN1-kmer	*CNN2-FCGR*	*CNN3-IntEnc*
Input	4^k	$\sqrt{4^k} \times \sqrt{4^k}$	SeqLength
Embedding	None	None	10×1
Conv	$10 \times (5 \times 1)$	$10 \times (5 \times 5)$	$10 \times (5 \times 1)$
Maxpool	(1×2)	(2×2)	(1×5)
Conv	$20 \times (5 \times 1)$	$20 \times (5 \times 5)$	$20 \times (5 \times 1)$
Maxpool	(1×2)	(2×2)	(1×5)
Flatten	–	–	–
Dense	500	500	500
Dense	NumCategories	NumCategories	NumCategories

4   Experiments

4.1   Dataset Description

Barcode (i.e. *marker*) gene sequences are short specific regions of biological sequences that represent genetic markers in species. Providing discriminative characteristics, they are favorable for species identification. This work considers the *16S rRNA dataset*; a bacterial barcode dataset of 3000 16S rRNA gene sequences. The dataset was originally collected from the 'Hierarchy browser' of the RDP II repository [10,17] from 3 bacteria phyla: Actinobacteria, Firmicutes and Proteobacteria with 1000 sequences for each phyla. The dataset was used for the training and testing of the CNN models using a 10-fold cross validation.

It is organized into 5 different taxonomic ranks where each results in a different number of categories as shown in Table 2 [10]. The categories are imbalanced, where Actinobacteria starts with 1 *phyla* and reaches a total 79 *genera* categories, while Firmicutes reaches 110 and Proteobacteria reaches 204. For the different taxonomic ranks starting from the phylum rank and reaching the genus rank, there are a total of 3, 6, 23, 65 and 393 categories, respectively.

Table 2. Categories' distribution on different phyla for different ranks in the 16S rRNA dataset.

Phyla	Categories				
	Phylum	Class	Order	Family	Genus
Actinobacteria	1	2	5	12	79
Firmicutes	1	2	5	19	110
Proteobacteria	1	2	13	34	204
Total	**3**	**6**	**23**	**65**	**393**

4.2 Experimental Setup

The experiments were implemented on an Intel i7 7th generation processor and an AMD Radeon R7 M440 GPU, using the Keras Python library with the PlaidML backend. The selected DL models were tested using 10-fold cross validation with a fixed number of epochs per fold (200) and a fixed batch-size (64). The AdaDelta optimizer and a categorical cross-entropy loss function were used, which proved to yield the best results. Regarding the performance measures used, the *average* (1) precision, (2) recall, (3) F1-score, (4) accuracy, (5) AUPRC, (6) AUROC and (7) the general *average* 10-fold execution run time for any of the ranks within a certain representation (in minutes), are all recorded for the experiments.

4.3 Experimental Results and Discussion

Table 3 shows performance results for *CNN1-kmer* using each of the three considered k-values: 4, 5 and 6. The performance measures of the first two broadest ranks: Phylum and Class, were the same over the three k-values, hence the $k = 4$ representation was preferred for its lower execution time. The best performance measures for the last three ranks were primarily found at $k = 6$.

The performance results for *CNN2-FCGR* are shown in Table 3. The general performance of all the ranks were slightly less than that of *CNN1-kmer*. Similar to *CNN1-kmer*, the best performance of the Phylum and Class ranks was also preferred at $k = 4$, while it was concentrated at $k = 6$ for the remaining ranks.

For the *CNN3-IntEnc* representation in Table 4, the performance measures for all the ranks were less than that of the other two experiments, but heavily degraded at the Genus rank. The *CNN3-IntEnc* had a constant representation size (i.e. no k-value), but had the highest execution time out of all the three experiments.

Having obtained the best accuracies through the k-mer representation, Fig. 5a, b, c compares between the different experimented k-values approaches in the k-mer based approaches *CNN1-kmer* and *CNN2-FCGR* against the constant-sized representation *CNN3-IntEnc* approach. As shown in the figure, the performance decreases as the ranks become more fine-grained. The cause for this may be that the dataset size needed to be larger to capture more variations and features. The dataset was also imbalanced, which may have cause biasing to certain categories during training. However, increasing the value of k in *CNN1-kmer* and *CNN2-FCGR* slightly improved the accuracy but exponentially increased the execution time, as shown in Fig. 5d. CNNs typically have high computational costs, as well as needing to tune many hyperparameters. The goal of experimenting with different values of k was to test the effect of capturing longer-term interactions between different regions of a sequence. However, an average of k = 5 seems to be balanced in terms of performance and execution time. Using a CGR, an image-based representation of the sequence, was slightly less in performance, but the execution time improved. Such improvement may be accredited to the 2D convolutions in the CNN.

Table 3. Performance results for *CNN1-kmer* and *CNN2-FCGR*.

k	Rank	CNN1-kmer							CNN2-FCGR						
		Prec	Recall	F1	Acc	AUPRC	AUROC	Time	Prec	Recall	F1	Acc	AUPRC	AUROC	Time
4	Phylum	1.0	1.0	1.0	1.0	1.0	1.0	35.2	0.999	0.999	0.999	0.999	0.999	0.999	25.8
	Class	0.997	0.999	0.998	0.999	0.999	0.999		0.997	0.998	0.997	0.998	0.998	0.999	
	Order	0.986	0.987	0.986	0.987	0.987	0.993		0.976	0.974	0.974	0.974	0.975	0.986	
	Family	0.977	0.981	0.977	0.981	0.981	0.99		0.935	0.931	0.928	0.931	0.932	0.965	
	Genus	0.877	0.901	0.885	0.901	0.901	0.95		0.724	0.741	0.721	0.741	0.741	0.87	
5	Phylum	1.0	1.0	1.0	1.0	1.0	1.0	49.5	0.999	0.999	0.999	0.999	0.999	0.999	37.6
	Class	0.997	0.999	0.998	0.999	0.999	0.999		0.995	0.997	0.996	0.997	0.997	0.998	
	Order	0.99	0.991	0.99	0.991	0.991	0.995		0.978	0.978	0.977	0.978	0.978	0.988	
	Family	0.984	**0.987**	0.984	**0.987**	**0.987**	**0.993**		0.958	0.956	0.954	0.956	0.956	0.978	
	Genus	0.892	0.913	0.899	0.913	0.913	0.956		0.826	0.84	0.824	0.84	0.84	0.92	
6	Phylum	1.0	1.0	1.0	1.0	1.0	1.0	137.5	0.999	0.999	0.999	0.999	0.999	0.999	94.9
	Class	0.997	0.999	0.998	0.999	0.999	0.999		0.996	0.998	0.997	0.998	0.998	0.999	
	Order	**0.991**	**0.992**	**0.991**	**0.992**	**0.992**	**0.996**		**0.983**	**0.984**	**0.983**	**0.984**	**0.985**	**0.992**	
	Family	**0.985**	0.987	**0.985**	0.987	0.987	**0.993**		**0.974**	**0.975**	**0.973**	**0.975**	**0.976**	**0.987**	
	Genus	**0.9**	**0.916**	**0.904**	**0.916**	**0.916**	**0.958**		**0.864**	**0.88**	**0.867**	**0.88**	**0.88**	**0.94**	

Table 4. Performance results for *CNN3-IntEnc*.

	Rank	Precision	Recall	F1-score	Accuracy	AUPRC	AUROC	Execution time
Integer encoding	Phylum	0.996	0.996	0.996	0.996	0.997	0.997	
	Class	0.991	0.993	0.992	0.993	0.994	0.996	
	Order	0.958	0.958	0.955	0.958	0.959	0.978	147.95
	Family	0.917	0.921	0.912	0.921	0.921	0.96	
	Genus	0.721	0.76	0.728	0.76	0.76	0.88	

In Fig. 5, the k-mer based representations (spectral and FCGR) provided more stability over all of the ranks compared to the integer-embedding representation, which heavily decreased in performance at the Genus rank. Integer encoding does not preserve nucleotides' positional information, opposite to the k-mer-based representations. It may be deduced that position-preserving representations are preferred, especially in more fine-grained classification tasks where short-term interactions between nucleotides may be more significant.

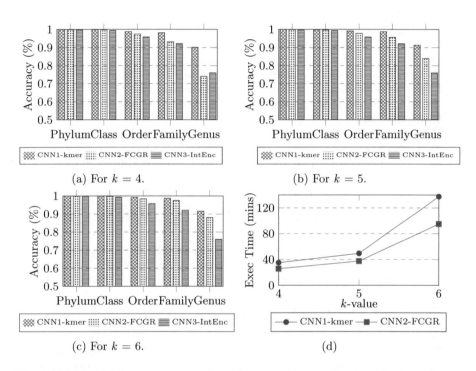

Fig. 5. (a), (b), (c) Accuracy comparisons between the considered methods on the five different taxonomic ranks for different values of k. (d) The effect of the k value on the execution time.

5   Conclusions

This paper presented a comparison of different feature representation and CNN architectures proposed by three of the most recent works on the 16S rRNA gene dataset for the task of taxonomic classification. The three different feature representation methods were a k-mer spectral representation, a FCGR and an integer-encoding with an embedding layer. The studied CNNs differed in the type of convolutions and the sizes of the maxpooling layers. Experiments have proven that representations holding positional information about the nucleotides in a sequence perform better, such as k-mer based representations. The CNN models for the k-mer spectral and FCGR representations recorded the best accuracy values of 91.6% and 88%, respectively, for the most fine-grained classification rank, with acceptable execution times. Future work will consider study of the effect of wider maxpooling layers, observing more accurate representations (possibly using different types of information together) and deeper and wider CNNs that may be able to capture more information from the sequences.

References

1. Brandenberg, O., et al.: Introduction to Molecular Biology and Genetic Engineering (2011)
2. National Human Genome Research Institute: The Human Genome Project (HGP). https://www.genome.gov/human-genome-project. Accessed 17 June 2019
3. Reece, J.B., et al.: Biology: Concepts & Connections, 7th edn. Pearson Benjamin Cummings, San Francisco (2012)
4. Kristensen, T., Guillaume, F.: Classification of DNA sequences by a MLP and SVM network. In: Proceedings of the International Conference on Bioinformatics and Computational Biology, The Steering Committee of The World Congress in Computer Science (2013)
5. Kristensen, T., Guillaume, F.: Different regimes for classification of DNA sequences. In: IEEE 7th International Conference on Cybernetics and Intelligent Systems and IEEE Conference on Robotics, Automation and Mechatronics, IEEE, pp. 114–119 (2015)
6. Alhersh, T., et al.: Species identification using part of DNA sequence: evidence from machine learning algorithms. In: Proceedings of the 9th EAI International Conference on Bio-Inspired Information and Communications Technologies, ICST, pp. 490–494 (2016)
7. Dakhli, A., Bellil, W.: Wavelet neural networks for DNA sequence classification using the genetic algorithms and the least trimmed square. Procedia Comput. Sci. **96**, 418–427 (2016)
8. Pashaei, E., Aydin, N.: Frequency difference based DNA encoding methods in human splice site recognition. In: International Conference on Computer Science and Engineering, IEEE, pp. 586–591 (2017)
9. Huang, J., et al.: An approach of encoding for prediction of splice sites using SVM. Biochimie **88**(7), 923–929 (2006)
10. Rizzo, R., et al.: A deep learning approach to DNA sequence classification. In: International Meeting on Computational Intelligence Methods for Bioinformatics and Biostatistics, pp. 129–140. Springer (2015)
11. Rizzo, R., et al.: Classification experiments of DNA sequences by using a deep neural network and chaos game representation, pp. 222–228 (2016)
12. Lo Bosco, G., Di Gangi, M.A.: Deep learning architectures for DNA sequence classification, pp. 162–171 (2017)
13. Nguyen, N.G., et al.: DNA sequence classification by convolutional neural network. J. Biomed. Sci. Eng. **9**, 280–286 (2016)
14. Yin, B., et al.: An image representation based convolutional network for DNA classification. arXiv preprint arXiv:1806.04931 (2018)

15. Min, X., et al.: DeepEnhancer: predicting enhancers by convolutional neural networks. In: IEEE International Conference on Bioinformatics and Biomedicine, IEEE, pp. 637–644 (2016)
16. Ghandi, M., et al.: Enhanced regulatory sequence prediction using gapped k-mer features. PLoS Comput. Biol. **10**(7) (2014)
17. Michigan State University Center for Microbial Ecology. Ribosomal Database Project (RDP). https://rdp.cme.msu.edu/. Accessed 18 June 2019

Recognition and Image Processing

Statistical Metric-Theoretic Approach to Activity Recognition Based on Accelerometer Data

Walid Gomaa[1,2]($\boxtimes$)

[1] Cyber-Physical Systems Lab, E-JUST, Alexandria, Egypt
walid.gomaa@ejust.edu.eg
[2] Faculty of Engineering, Alexandria University, Alexandria, Egypt

Abstract. Providing accurate information on people's actions, activities, and behaviors is one of the key tasks in ubiquitous computing and it has a wide range of applications including healthcare, well being, smart homes, gaming, sports, etc. In the domain of Human Activity Recognition, the primary goal is to determine the action a user is performing based on data collected through some sensor modality. Common modalities adopted to this end include visual and Inertial Measurement Units (IMUs), with the latter taking precedence in recent times due to their unobtrusiveness, low cost and mobility. In this work we consider the accelerometer signals streamed through a wearable IMU unit and use this data to recognize the user's activity. We develop a novel approach based on representing the coming signal as a probability distribution function and then use some distance metric to infer the dissimilarity between probability distributions corresponding to different accelerometer signals in order to infer the correct activity. Experiments are performed on 14 activities of daily living with results showing promising potential for this technique.

Keywords: Human activity recognition ·
Probability distribution function · Distance metric ·
Inertial Measurement Unit · Accelerometer

1 Introduction

Researchers are continuously thinking in making everyday environment intelligent. For this purpose, *human activity modeling and recognition* is the basis of this research trend [2, 7, 13], as providing accurate information on people's activities and behaviors is one of the key tasks in widespread ubiquitous computing and it has a wide range of applications including healthcare, well being, smart homes, gaming, sports, etc. A comprehensive survey of applications can be found in [14, 20].

Due to the wide availability of personal mobile and wearable devices, it is now easier than ever before to obtain information from a wide range of sources,

© Springer Nature Switzerland AG 2020
A. E. Hassanien et al. (Eds.): AISI 2019, AISC 1058, pp. 537–546, 2020.
https://doi.org/10.1007/978-3-030-31129-2_49

including the actions performed by the users or the environment in which a user is located. This is especially beneficial in a number of scenarios, such as health monitoring in medical and therapeutic settings, where the manner in which an action is performed or the conformance to some treatment or living regime may be relevant to treatment or health outcomes. Another possible application of this is in the domain of intelligent environments or smart spaces. Activity Recognition broadly enables the space itself to be contextually aware of the users activities, such that it can adapt itself accordingly for the maximum utility/comfort/safety of the occupant(s).

There are many methods and data acquisition systems which are based on different sensory readings for recognizing people's actions. Previously, heavy devices were used to collect accelerometer data such as data acquisition cards (DAQs) [17]. Later, smaller integrated circuits connected to PDAs were used for this aim, for instance, camera-based computer vision systems and inertial sensor-based systems. In computer vision, activities are recorded by cameras [21]. However, the drawback of the camera-based approach is that it may not work due to the absence of full camera coverage of all person's activities. In addition, cameras are meddlesome as people don't feel free being observed constantly.

In the *wearable* based systems, the measurements are taken from mobile sensors mounted to human body parts like wrists, legs, waist, and chest. Many sensors can be used to measure selected features of human body motion, such as accelerometers, gyroscopes, compasses, and GPSs. These kinds of sensors are essentially based on embedded MEMS sensors such as IMU (Inertial Measurement Unit) containing accelerometer and gyroscope. On the contrary of the fixed-sensor based systems, wearable are able to measure data from the user everywhere, while sleeping, working, or even traveling anywhere since it is not bounded by a specific place where the sensors are installed. Also, it is very easy to concentrate on directly measuring data of particular body parts efficiently without a lot of preprocessing that are needed, for example, in fixed depth cameras. Examples of wearable include smart watches, smart shoes, sensory gloves, hand straps, and clothing [3, 8, 15, 18]. Recent statistics show that the total number of smartphone subscribers reached 3.9 billion in 2016 and is expected to reach 6.8 billion by 2022 [1]. A comprehensive review on the use of wearable in the medical sector can be found in [19].

In this paper we develop an approach for activity recognition based on estimating the empirical distribution of the time series data and use metric-theoretic techniques to estimate the dissimilarity between two distributions. We focus of activities of daily living (ADLs) which are routine activities that people tend to do every day without needing assistance [4–6, 16]. We have used a public dataset of accelerometer data collected with a wrist-worn accelerometer [11,12]. The data are streamed while performing 14 activities using several human subjects and with varying number of samples for the different activities. Table 1 provides a summary of the monitored activities along with each activity sample size, the activity name is self-explanatory. Each experiment consists of three time series data corresponding to the tri-axial accelerometer directions. To the best of our knowledge, almost all techniques for activity recognition are based on methods from the machine learning literature (particularly, supervised learning). So in the current paper we explore new direction that has shown computational efficiency with reasonable predictive performance. We hypothesize that the predictive performance can be boosted by including more sensory data such as the gyroscope rotational speed.

The paper is organized as follows. Section 1 is an introduction. Section 2 gives detailed description of the dataset used for our experimentation. Section 3 describes our approach along with experimentation and results. Section 4 concludes the paper with prospective future work.

2   Data Description

We have obtained our labeled accelerometer activity data from the UCI machine learning public repository. The dataset is: 'Dataset for ADL Recognition with Wrist-worn Accelerometer Data Set'.[1] A detailed description of the data can be found in [10].

The data are collected using a wrist-worn accelerometer with sampling rate 32 Hz. The measurement range is $[-1.5, 1.5]$ (it is in g-force units). One data sample is a tri-axial acceleration: x-axis pointing toward the hand, y-axis pointing toward the left, and z-axis perpendicular to the plane of the hand. Acceleration data are collected for 14 activities with varying number of samples for each. Table 1 lists the activities along with the number of collected samples per each. The labels of the activities are self-explanatory.

Table 1. List of activities and associated number of samples.

No.	Activity	Number of samples	No.	Activity	Number of samples
1	Brush teeth	12	8	Pour water	100
2	Climb stairs	102	9	Eat meat	5
3	Comb hair	31	10	Walk	100
4	Drink glass	100	11	Liedown bed	28
5	Getup bed	101	12	Standup chair	102
6	Sitdown chair	100	13	Descend stairs	42
7	Use telephone	13	14	Eat soup	3

Figure 1 shows snapshots of the collected data signals for various activities in different accelerator directions. The results obtained in all experiments are essentially the confusion matrices and accuracies averaged over a number of experiments (15) of the same settings.

---

[1] https://archive.ics.uci.edu/ml/datasets/Dataset+for+ADL+Recognition+with+Wrist-worn+Accelerometer.

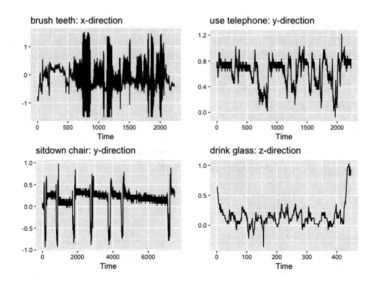

Fig. 1. Snapshots of some signals from various activities in different accelerator directions

3   Probabilistic and Measure Theoretic Approach to Activity Recognition

In this approach we have used a variety of tools from the theory of probability and measure theory. Here we deal with the statistical properties of the raw signal without taking time into consideration. In this approach we rely on probabilistic modeling of the signals and measuring the distance among these probability distributions in order to classify the given activity signal.

3.1   Histogram Based Measures

This technique can be summarized as follows:

1. For each sample s of an activity A, we build a histogram for each of the tri-axial accelerometer directions. So we have three histograms: x-histogram, y-histogram, and z-histogram. The parameters of the histogram are as follows:
   - Lower bound of interval: -1.5 (lower bound of the accelerometer measurements).
   - Upper bound of interval: 1.5 (upper bound of the accelerometer measurements).
   - Step size: 0.1 (we have tried finer resolution, but the results were the same).
2. For each activity A, some of its collected samples s are randomly selected to act as a model of the underlying activity. We call these samples 'support samples'. They act as temples for the activity they are drawn from. The remaining samples are used for testing. In experimentation we use 65% of the samples as 'support samples'.

Given a test sample s, to classify to which activity s belongs we perform the following procedure:

1. For each activity A, find an overall distance between the sample s and the activity A as follows:
   (a) For each sample s' of the 'support samples' of A find the distance between s and s'; this is done as follows:
      - Compute the distance between each corresponding pairs of histograms $h_s = (h_x, h_y, h_z)$ and $h_{s'} = (h_{x'}, h_{y'}, h_{z'})$ using the *Manhattan distance*. For example, the distance between the histograms corresponding to the x-direction is computed by:

$$d_M(h_x, h_{x'}) = \sum_i |h_x(i) - h_{x'}(i)| \tag{1}$$

   where i runs over the bins of the histogram. (the histograms are assumed to be normalized)
      - Combine the three tri-axial scores by taking their average: $d(s, s') = \frac{d_M(h_x, h_{x'}) + d_M(h_y, h_{y'}) + d_M(h_z, h_{z'})}{3}$.
   (b) Now the distance between sample s and activity A is taken to be the minimum among all computed distances with support samples:

$$d(s, A) = \min_{s' \in Supp(A)} d_M(s, s') \tag{2}$$

2. Now assign s to the activity A such that $d(s, A)$ is minimum:

$$s \in A \iff A = \underset{A'}{argmin}\; d(s, A') \tag{3}$$

Ties are broken arbitrarily.

Note that the above procedure is equivalent to k-NN with $k = 1$ using the Manhattan distance. Figure 2 shows the accuracy curve as it runs of subsets of activities ranging from 2 activities to the whole set of 14 activities. For most of the activities (about 10), the classification accuracy is above 90%. And the worst accuracy among all the 14 activities is about 80%.

This technique is very simple and computationally efficient. The training phase comprises of just computing the histograms of the template samples and the predictive phase comprises of computing the Manhattan distance between the given test sample and all the templates. The number of templates is relatively small and as well as the number of bins in the histogram, hence, computing such distance function is rather efficient.

Figure 3 shows the confusion matrix for the 14 activities. It can easily be seen that the performance loss is due mainly to the confusion between two activities: 'getup bed' and 'sitdown chair' where most of their misclassifications (false positives) are attributed to the activity 'standup chair'.

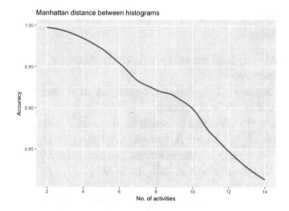

Fig. 2. Manhattan distance between histograms.

	brush_teeth	climb_stairs	comb_hair	drink_glass	getup_bed	sitdown_chair	use_telephone	pour_water	eat_meat	walk	liedown_bed	standup_chair	descend_stairs	eat_soup
brush_teeth	4	0	0	0	0	0	0	0	0	0	0	0	0	0
climb_stairs	1	34	0	0	1	0	0	0	0	2	0	0	3	0
comb_hair	0	0	10	0	0	0	0	0	0	0	0	0	0	0
drink_glass	0	0	0	32	0	0	1	3	0	0	0	0	0	0
getup_bed	0	0	0	0	23	1	0	0	0	0	3	1	0	0
sitdown_chair	0	0	0	0	2	23	0	0	0	0	1	7	0	0
use_telephone	0	0	0	0	0	0	2	0	0	0	0	0	0	0
pour_water	0	0	0	3	1	0	1	31	0	0	0	0	0	0
eat_meat	0	0	0	0	0	0	0	0	2	0	0	0	0	0
walk	0	1	0	0	0	0	0	0	0	31	0	0	0	0
liedown_bed	0	0	0	0	5	0	0	0	0	0	4	0	0	0
standup_chair	0	0	0	0	4	11	0	0	0	1	1	27	0	0
descend_stairs	0	1	0	0	0	0	0	0	0	0	0	0	12	0
eat_soup	0	0	0	0	0	0	0	0	0	0	0	0	0	2

Fig. 3. Confusion matrix using Manhattan distance between histograms.

3.2 Kernel Density Estimate

Here we use an approach that is also based on estimating a density function for the given time series. However, there are two main differences. The first is the use of *kernel density* estimates instead of building histograms to approximate the empirical density of the data samples. The second difference is the use of the KullbackLeibler KL-divergence as a dissimilarity measure instead of the Manhattan distance. The KL-divergence is not a proper distance metric since it is asymmetric with respect to its two arguments, so we compute both $KL(p \parallel q)$ and $KL(q \parallel p)$ and take the average to be the final distance between p and q.

The KL-divergence can be computed by the following formula, assuming continuous random variables as is the case for accelerometer data:

$$KL(p \parallel q) = \int_{-\infty}^{\infty} p(x) \ln \frac{p(x)}{q(x)} dx \tag{4}$$

Kernel density estimation is a *non-parametric method* for estimating the probability density function of a continuous random variable (here the acceleration). It is non-parametric since it does not presuppose any specific underlying

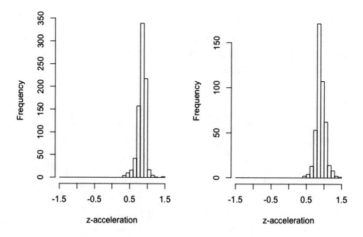

(a) Histograms for two samples of activity 'pour water' (z-direction)

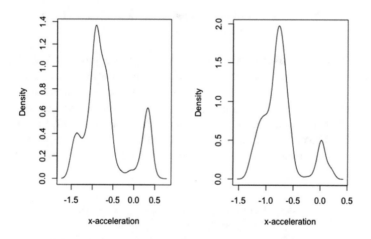

(b) Kernel density estimates for two samples of activity 'descend stairs' (x-direction)

Fig. 4. Sample histograms and kernel density estimations.

model or distribution for the given data (see section 2.5.1 in [9]). At each data point, a kernel function (we assumed Gaussian for our analysis here) is created with the data point at its center. The density function is then estimated by adding all of these kernel functions and dividing by the size of the data. Intuitively, a kernel density estimate is a sum of 'bumps'. A 'bump' is assigned to every datum, and the size of the bump represents the probability assigned at the neighborhood of values around that datum.

Figure 5a shows the performance of the KL-divergence for predicting the given ADL activities; followed by Fig. 5b for comparing the performance of the two techniques: Manhattan distance between histograms and KL-divergence between kernel density estimations.

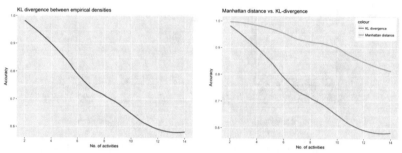

(a) KL-divergence between empirical distributions.

(b) Manhattan distance vs. KL-divergence.

Fig. 5. Performance of KL-divergence and Manhattan distance.

As is evident, the Manhattan distance achieves about 20% performance improvement over the KL-divergence. Also the computational complexity for building the histograms and computing the corresponding Manhattan distance outperform that required for building the kernel estimation and computing the corresponding KL-divergence. Figure 4 shows the histograms and kernel densities for some samples from two activities in order to give a sense of how time series data can be transformed into random variables with empirical distributions.

	brush_teeth	climb_stairs	comb_hair	drink_glass	getup_bed	sitdown_chair	use_telephone	pour_water	eat_meat	walk	liedown_bed	standup_chair	descend_stairs	eat_soup
brush_teeth	5	0	0	0	0	0	0	0	0	0	0	0	0	0
climb_stairs	0	23	0	1	1	1	0	2	0	10	1	3	2	1
comb_hair	0	1	9	0	1	1	0	1	0	0	1	1	0	0
drink_glass	0	0	0	26	2	1	2	4	0	0	1	2	0	0
getup_bed	0	0	0	1	16	5	0	2	0	1	2	4	0	0
sitdown_chair	0	0	0	2	6	18	0	1	0	1	2	3	0	0
use_telephone	0	0	0	0	0	0	1	0	0	0	0	0	0	0
pour_water	0	1	0	3	3	1	0	22	1	0	1	0	0	0
eat_meat	0	0	0	0	0	0	0	0	1	0	0	0	0	0
walk	0	8	0	1	0	1	0	1	0	18	0	1	1	0
liedown_bed	0	1	1	0	4	3	0	1	0	0	1	1	0	0
standup_chair	0	0	0	1	3	4	0	1	0	1	1	22	0	0
descend_stairs	0	2	0	0	0	0	0	0	0	3	0	0	12	0
eat_soup	0	0	0	0	0	0	0	0	0	0	0	0	0	1

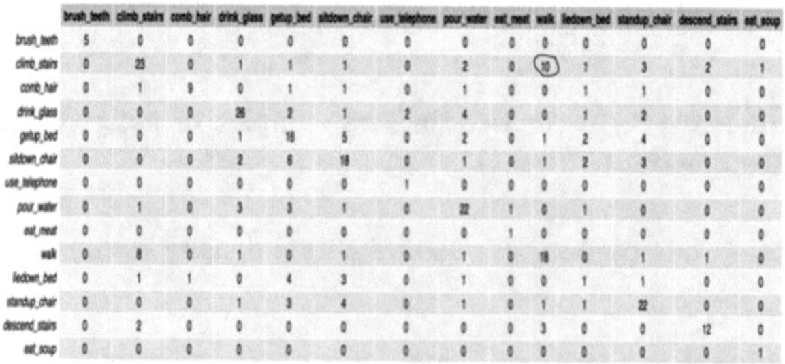

Fig. 6. Confusion matrix using KL-divergence between empirical distributions.

Figure 6 shows the confusion matrix for the 14 activities. By studying the sensitivity and specificity of each activity we have found a general remark: the specificity is very high for all activities in both the Manhattan distance and

KL-divergence cases. This means that for all activities the true negatives are high and the false positives are low which implies that the misclassifications in any activity are roughly distributed among many classes (of course, up to a factor that takes into consideration the distribution of sample data across all activities).

4 Conclusion and Future Work

In this paper we have studied a new approach to activity recognition where the sensory data are extracted from a wearable accelerometer. Our approach is based on representing the time series progression as a probability distribution. For each activity a set of samples chosen at random from the given dataset are chosen to act as templates of the activity. Then, at test/operation time a new test sample is then compared (using their representations as probability distributions) against each template set of each activity to decide its activity membership. The comparison is based on some chosen metric distance. We have tried several configurations for the probabilistic representation as well as trying several dissimilarity metrics. It turned out that using normalized histograms as probability distributions along with the Manhattan distance metric gave the best performance. In the future, we will investigate more diverse methods within the same paradigm of using probability distributions as representations of the signals along with multitude of distance metrics.

References

1. Ericsson Mobility Report on The Pulse of The Networked Society. Technical report, Ericsson, 11 (2016)
2. Abdu-Aguye, M.G., Gomaa, W.: Novel approaches to activity recognition based on vector autoregression and wavelet transforms. In: 2018 17th IEEE International Conference on Machine Learning and Applications (ICMLA), IEEE, pp. 951–954 (2018)
3. Abdu-Aguye, M.G., Gomaa, W.: Competitive feature extraction for activity recognition based on wavelet transforms and adaptive pooling. In: The 2019 International Joint Conference on Neural Networks (IJCNN) (2019)
4. Abdu-Aguye, M.G., Gomaa, W.: Robust human activity recognition based on deep metric learning. In: The 16th International Conference on Informatics in Control, Automation and Robotics (ICINCO) (2019)
5. Abdu-Aguye, M.G., Gomaa, W.: VersaTL: versatile transfer learning for IMU-based activity recognition using convolutional neural networks. In: The 16th International Conference on Informatics in Control, Automation and Robotics (ICINCO) (2019)
6. Abdu-Aguye, M.G., Gomaa, W., Makihara, Y., Yagi, Y.: On the feasibility of on-body roaming models in human activity recognition. In: The 16th International Conference on Informatics in Control, Automation and Robotics (ICINCO) (2019
7. Ashry, S., Elbasiony, R., Gomaa, W.: An LSTM-based descriptor for human activities recognition using IMU sensors. In: Proceedings of the 15th International Conference on Informatics in Control, Automation and Robotics, ICINCO, INSTICC, SciTePress, vol. 1, pp. 494–501 (2018)

8. Bamberg, S.J.M., Benbasat, A.Y., Scarborough, D.M., Krebs, D.E., Paradiso, J.A.: Gait analysis using a shoe-integrated wireless sensor system. IEEE Trans. Inf. Technol. Biomed. **12**(4), 413–423 (2008)
9. Bishop, C.: Pattern Recognition and Machine Learning. Springer, Boston (2006)
10. Bruno, B., Mastrogiovanni, F., Sgorbissa, A.: Laboratorium - laboratory for ambient intelligence and mobile robotics, DIBRIS, University of Genova, via Opera Pia 13, 16145, Genova, Italia (IT), Version 1 released on 11 2014
11. Bruno, B., Mastrogiovanni, F., Sgorbissa, A., Vernazza, T., Zaccaria, R.: Human motion modelling and recognition: a computational approach. In: CASE, IEEE, pp. 156–161 (2012)
12. Bruno, B., Mastrogiovanni, F., Sgorbissa, A., Vernazza, T., Zaccaria, R.: Analysis of human behavior recognition algorithms based on acceleration data. In: ICRA, IEEE, pp. 1602–1607 (2013)
13. Chong, N.-Y., Mastrogiovanni, F.: Handbook of Research on Ambient Intelligence and Smart Environments: Trends and Perspective. Information Science Reference (2011)
14. Elbasiony, R., Gomaa, W.: A survey on human activity recognition based on temporal signals of portable inertial sensors. In: Hassanien, A.E., Azar, A.T., Gaber, T., Bhatnagar, R., Tolba, M.F. (eds.) The International Conference on Advanced Machine Learning Technologies and Applications (AMLTA 2019), pp. 734–745. Springer, Cham (2019)
15. Gomaa, W., Elbasiony, R., Ashry, S.: ADL classification based on autocorrelation function of inertial signals. In: 2017 16th IEEE International Conference on Machine Learning and Applications (ICMLA), pp. 833–837, December 2017
16. Katz, S., Chinn, A., Cordrey, L.: Multidisciplinary studies of illness in aged personsii: a new classification of functional status in activities of daily living. J. Chronic Dis. **9**(1), 55–62 (1959)
17. Mantyjarvi, J., Himberg, J., Seppanen, T.: Recognizing human motion with multiple acceleration sensors. In: 2001 IEEE International Conference on Systems, Man, and Cybernetics, IEEE, vol. 2, pp. 747–752 (2001)
18. Maurer, U., Rowe, A., Smailagic, A., Siewiorek, D.P.: eWatch: a wearable sensor and notification platform. In: International Workshop on Wearable and Implantable Body Sensor Networks, 2006, BSN 2006, IEEE, p. 4 (2006)
19. Patel, S., Park, H., Bonato, P., Chan, L., Rodgers, M.: A review of wearable sensors and systems with application in rehabilitation. J. NeuroEng. Rehabil. **9**(1), 21 (2012)
20. Ranasinghe, S., Machot, F.A., Mayr, H.C.: A review on applications of activity recognition systems with regard to performance and evaluation. Int. J. Distrib. Sens. Netw. **12**(8), 1550147716665520 (2016)
21. Turaga, P., Chellappa, R., Subrahmanian, V.S., Udrea, O.: Machine recognition of human activities: a survey. IEEE Trans. Circuits Syst. Video Technol. **18**(11), 1473–1488 (2008)

High Efficient Haar Wavelets for Medical Image Compression

E. A. Zanaty[1](✉) and Sherif M. Ibrahim[2](✉)

[1] Mathematics Department, Faculty of Science, Sohag University, Sohag, Egypt
zanaty22@yahoo.com
[2] Computer Science and Mathematics Department, Faculty of Science,
South Valley University, Qena, Egypt
sherif_hashad@hotmail.com

Abstract. In this paper, we proposed an improved high efficient Haar wavelets (HEHW) algorithm to improve the quality of image compression ratio (CR) rate and peak signal to noise ratio for medical imaging. The proposed algorithm starts by partitioning the original image into 2 * 2 submatrices. Then the wavelets transform coefficients obtained by working on the submatrices instead of the rows and on the columns in the original image. We re-compute the resulting coefficients for sub-matrices to obtain the approximation and the sub details of the original image. Then, we calculate statistical thresholds on the details subbands to complete the compression process. The proposed algorithm is applied to five different medical image structures to prove its efficiency using the evaluation factors like CR, peak signal to noise ratio, mean square error and transform time. The comparison between the proposed and the well-known modified Haar method also compared to the results of existing wavelets techniques: like Coiflet, Daubechies, Biorthogonal, Dmeyer, and Symlets to prove the proposed algorithm efficiency.

Keywords: Haar transform (HT) · Compression ratio (CR) ·
Peak signal to noise ratio (PSNR) · High efficient haar wavelet (HEHW) ·
Statistical thresholds

1 Introduction

Medical data is one of the fundamental aspects of image processing these days. Due to the rising interest in medical care, there is a steady increase in medical patient information that needs to be saved and transmitted throw the hospital's departments. So we need to find a fast and efficient algorithm to obtain a good compression while preserving the details of the medical images needed by the physicians to detect the different diseases.

There are several techniques for medical image compression that can be categorized into two main types depending on the redundancy removal way, namely lossless and loss [1–3]. Lossless compression as Huffman coding, arithmetic coding, run length coding, and Lempel-Ziv algorithm [4] allow the original image to be reconstructed exactly from the compressed image with low compression rate.

© Springer Nature Switzerland AG 2020
A. E. Hassanien et al. (Eds.): AISI 2019, AISC 1058, pp. 547–557, 2020.
https://doi.org/10.1007/978-3-030-31129-2_50

Lossy image compression characterized by degrading image quality based upon the threshold that was used so source image can't be reconstructed exactly from the compressed image. However, the compression ratio (CR) will be higher than the other algorithm based on lossless techniques. There are several techniques can be found in [1, 5, 6] such as vector quantization, joint photographic experts group (JPEG), block truncation coding, fractal compression, transform coding, Fourier-related transform (FT), and discrete cosine transform (DCT).

Wavelet transform can be considered as Lossy or Lossless compression based on the threshold value that is calculated in the details subbands. Wavelet has several types like Haar wavelet, Daubechies wavelet, Biorthogonal wavelet, Coiflet wavelet, Symlet wavelet, the 5/3 and 9/7 lifting scheme. Haar wavelet able to compress medical images with fast calculations [7]. The major disadvantage of the Haar wavelet is its discontinuity, which makes it difficult to simulate a continuous signal [8]. Haar transform coefficients obtained by performing an average and difference on a pair of values. Then, shifting by two values and calculates again the average and difference on the next pair.

Benyahia et al. [9] applied a compression approach on medical images using wavelet packet coupled with the progressive coder SPIHT. The method allows for preserving fine structures and produces images with better quality, which is important for medical applications.

In [10], the Haar wavelet-based perceptual similarity index (Haar PSI algorithm) was improved as a similarity measure for full reference image quality assessment. The algorithm utilized the coefficients obtained from a Haar wavelet decomposition to assess local similarities between two images, as well as the relative importance of image areas. Latha [11] developed image averaging and transform coding (IATC) for the compression of image sets by calculating mean and difference images and discrete wavelet transformed up to 2nd level. These wavelet coefficients are then thresholded and Huffman coded to achieve more bit saving with allowable distortion in the reconstructed images. Heydaria [12] introduced low complexity wavelets based image quality assessment that aims to analyze the relation between image details of different granularities.

A mathematical framework for Haar wavelet transformation was stated in [13] based on lossless image compression by starting from the input image to regenerate the original image without losing any data. Vasanth [14] proposed a model that merging Haar wavelet together with Huffman encoding comparing the results with other compression techniques like run-length encoding achieving better results, especially in CR values.

Nahar in [15] proposed a system for image compression using Double Density Wavelet Transform (DDW). Its filter divides the image transform into nine levels instead of four levels in DWT. DDWT also can be used in image de-noising having a promising future in image enhancement.

In this paper, we will alleviate the problems by presenting a high efficient Haar wavelet (HEHW) through modifying the wavelet's procedures. The proposed HEHW divides the original image matrix into 2 * 2 submatrices. We state new transform equations to obtain reliable transform coefficients. Then we rearrange the resultant coefficients to obtain the approximations and three details subbands for the original image. A statistical threshold is performed on the resultant details subband in order to achieve the best compression ratio.

The performance of the proposed algorithm (HEHW) is estimated by calculating CR, peak signal to noise ratio and the mean square error results to meet the quality required for medical image compression. We compare the (HEHW) results by well modified Haar wavelet results.

The rest of the paper is organized as follows. Section 2 give an overview of the mathematical framework for Haar wavelet transform. Section 3 describes the proposed method in detail. Section 4 discusses the experimental results and compared with classical Haar wavelet. Finally, Sect. 5 offers the conclusion & future work.

2 A Mathematical Framework for Haar Wavelet Transformation

Haar wavelet is one of important transform that able to compress medical images with fast calculations [7]. The major disadvantage of the Haar wavelet is its discontinuity, which makes it difficult to simulate a continuous signal [8]. Haar transform coefficients obtained by performing an average and difference on a pair of values. Then, shifting by two values and calculates again the average and difference on the next pair.

Haar transform is widely used to decompose each signal into two components, one is called average (approximation) or trend and the other is known as difference (detail) or fluctuation [9].

Haar wavelet transform has a number of advantages [10]:

1. Efficiency since it can be calculated in place without the need for a temporary array.
2. The scaling and wavelet function are the same for both forward and inverse transform.
3. It provides high CR and good PSNR (peak signal to noise ratio).
4. It increases detail in a recursive manner.

In modified fast Haar wavelet transform (MFHWT) [6] the calculation depends on four nodes at a time instead of two nodes in the Haar transform. At first the average of the sub signal, $a_i = (a_1, a_2, ..., a_{N/2})$, at one level for signal of length N, i.e. $f = (f_1, f_2, ...$ $f_N)$ is:

$$a_m = \frac{f4m-3+f4m-2+f4m-1+f4m}{4} \quad m = 1,2,3,...N/4 \quad (1)$$

Then the first detail sub band signal, at the same level is given as:

$$d_m = \begin{cases} \frac{(f4m-3+f4m-2)-(f4m-1+f4m)}{4} & m = 1,2,3.....\frac{N}{4} \\ 0 & m = \frac{N}{2}......N \end{cases} \quad (2)$$

3   The Proposed Improved Haar Wavelets for Medical Image Compression

In this section, we disused the proposed (HEHW) to avoid these disadvantages rebuilding the limitation Haar transform processing. We work with sub images instead performing the Haar wavelet transform levels on the rows and columns. Therefore, we subdivide the original image into several 2 * 2 matrices. For each subdivision coefficients of HEHW transform obtained by proposed transform equations. Then, we reform the positions of the resulting coefficients that A's will be the approximation and B's, C's and D's will form the details. We can apply the threshold [16] on the detailed part only (B's, C's, and D's) as:

$$T(X) = \begin{cases} 0 & if \mu - \sigma < x < \mu + \sigma \\ or & if |x - \mu| < \sigma \\ x & otherwise \end{cases} \tag{3}$$

Where μ is the mean of the coefficients and σ is the standard deviation of the coefficients hence the threshold algorithm is based upon the statistics of the DWT coefficients. Figure 1 describes the main process of the proposed Haar wavelet for medical image compression. Where $A = (a + b + c + d)/4$, $B = (a - b + c - d)/4$, $C = (a + b - c - d)/4$, and $D = (a - b - c + d)/4$. The algorithm starts by convert the input image (matrix) into 2 * 2 matrices, then calculate the four coefficients A, B, C, and D. It's followed by calculating and redistribute the position of the resultant coefficients using Eq. (4).

$$A(\frac{i}{2},\frac{j}{2}), B(\frac{i}{2},\frac{n+j}{2}), C(\frac{n+i}{2},\frac{j}{2}), D(\frac{n+i}{2},\frac{n+j}{2}) \tag{4}$$

The mean of the coefficients (μ) and the standard deviation of the coefficients (σ) are computed. Then applying the threshold on the detailed parts B, C and D using Eq. (3).

We can perform multilevel transform by applying the previous steps on the approximation part (A's of previous step).

For decoding the image, we calculate the values a, b, c and d back by equations from (5–8) as follows:

$$a = A + B + C + D \tag{5}$$

$$b = A - B + C - D \tag{6}$$

$$c = A + B - C - D \tag{7}$$

$$d = A - B - C + D \tag{8}$$

Then update the resultant values to their original positions in the original matrix.

Figure 1 illustrates the main steps of the proposed Harr wavelet for medical image compression.

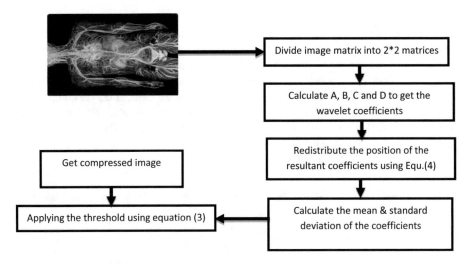

Fig. 1. The general flowchart of the proposed modified Harr wavelet medical image compression

4   Experimental Results and Discussion

The proposed algorithm is implemented on a PC processor Intel Core i7-4510U - CPU 2.00 GHz - internal memory 8.00 GB RAM. To prove the capability of the proposed algorithm, different five image structures from X-Ray, Computed tomography CT and Magnetic resonance imaging MRI types as shown in Fig. 2a, b, c and d respectively. Figure 2a shows the posterior cruciate ligament with dimensions 1280 * 1280. CT image as shown in Fig. 2b depicts the horizontal view for the human brain with

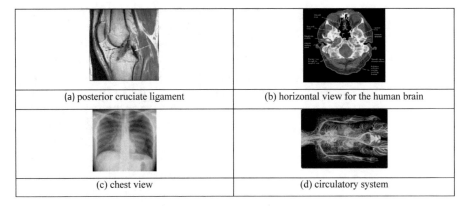

Fig. 2. Different five image structures from X-Ray, Computed tomography CT and Magnetic resonance imaging MRI types

dimensions 937 * 937. Figure 2c shows X-ray image where the chest view with dimensions 1060 * 1227 and Fig. 2d shows MRI image of the circulatory system view with dimensions 1024 * 610.

To show the capability of the proposed algorithm over other methods, we compare the proposed algorithm with modified fast Haar wavelet transform (MFHWT) and the existing wavelets techniques: like Coiflet, Daubechies, Biorthogonal, Dmeyer and Symlets. The final images (reconstructed images) are presented in last column of Figs. 3, 4, 5 and 6 respectively. From Table 1, we noted that proposed HEHW achieved better results than compared algorithms for all evaluation parameters CR, PSNR and MSE. Furthermore, the largest factor compression of HEHW is over than seven times MFHWT in case of circulatory system image. Best CR obtained for the circulatory system while the PSNR obtained in the human chest X-Ray image exceed the other results, also for MFHWT.

Table 1. Results of CR, PSNR, MSE, compression time for the proposed and compared algorithms

Image	Method	CR	PSNR	MSE	Time
The posterior cruciate ligament image	MFHWT	2.5682	26.3769	2.8233	0.0771
	HEHW	18.1963	46.5263	1.4469	1.0065
	Coiflet	1.7829	40.0881	0.6635	4.0043
	Daubechies	2.0491	30.1964	0.6803	4.0581
	Biorthogonal	2.0239	32.0253	0.1036	4.5106
	Symlets	2.9024	35.0815	0.2095	3.9491
	Dmeyer	1.1393	30.2387	0.6869	8.7209
CT brain image	MFHWT	2.6466	25.2462	2.17622	0.08116
	HEHW	16.9143	44.343	2.3920	0.9686
	Coiflet	2.1025	36.6308	0.2993	4.5709
	Daubechies	2.3615	25.4495	0.2280	4.0529
	Biorthogonal	2.406	28.2878	0.4383	3.9271
	Symlets	2.905	29.9912	0.6489	3.4985
	Dmeyer	1.359	25.9514	0.2559	7.2386
The human chest X-Ray image	MFHWT	2.5855	23.3738	1.4140	0.1009
	HEHW	15.6599	54.2994	0.2416	1.5012
	Coiflet	2.3429	37.1228	0.3352	3.9281
	Daubechies	2.6775	30.1408	0.6716	4.5241
	Biorthogonal	2.5292	30.4901	0.7279	2.5619
	Symlets	2.8059	32.1218	0.1059	3.2985
	Dmeyer	1.4911	30.1177	0.6681	6.6049
The circulatory system	MFHWT	2.4085	29.0192	5.1879	0.6807
	HEHW	30.1051	43.877	2.6630	0.9556
	Coiflet	1.9801	35.5678	0.2343	4.6809
	Daubechies	2.2243	30.6045	0.7473	4.6929
	Biorthogonal	2.3521	30.8809	0.7964	3.9497
	Symlets	2.3521	27.706	0.3834	4.6591
	Dmeyer	1.2881	30.8763	0.7956	7.6586

Results depicted in Table 1 showed that the proposed algorithms give better results than the existing wavelets like MFHWT, Coiflet, Daubechies, Biorthogonal, Dmeyer, and Symlets. We got the best PSNR and MSE values for the posterior cruciate ligament image Also, we have HEHW CR exceeded the FMHWT over six times. HEHW preserving better PSNR and MSE values than compared algorithms in case of CT brain image.

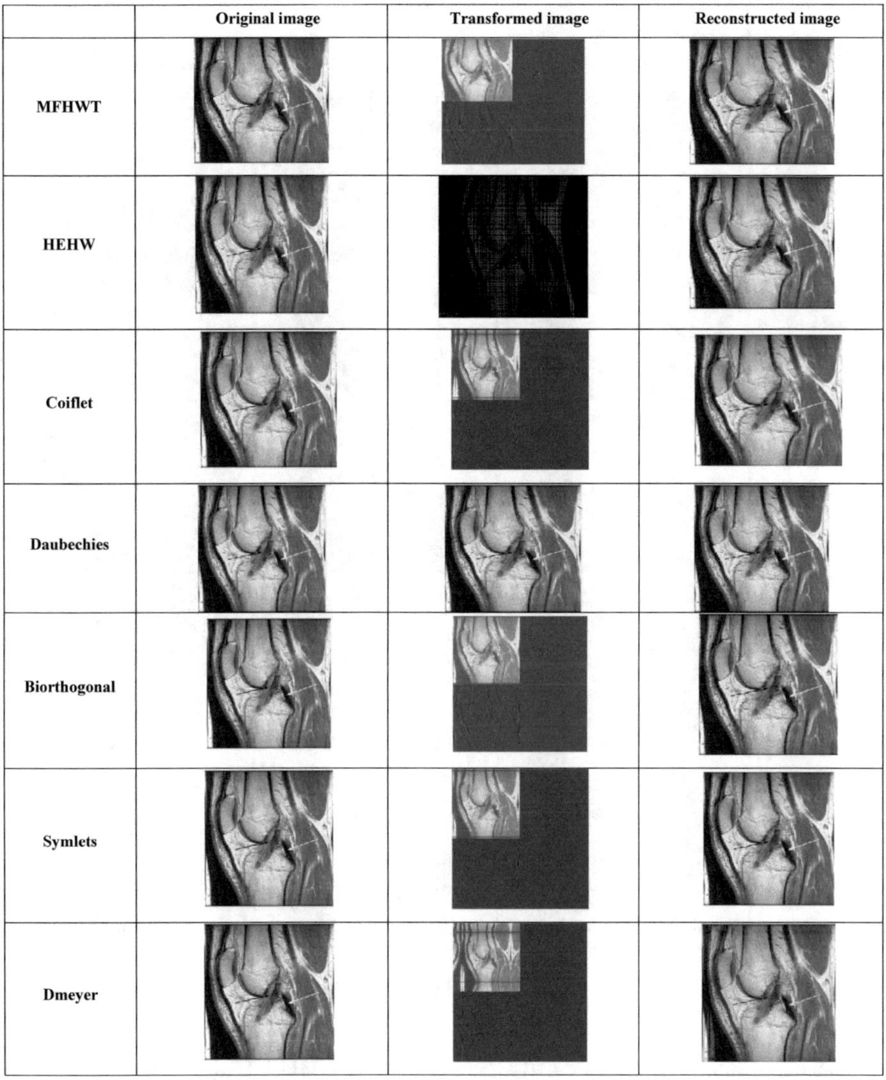

	Original image	Transformed image	Reconstructed image
MFHWT			
HEHW			
Coiflet			
Daubechies			
Biorthogonal			
Symlets			
Dmeyer			

Fig. 3. The output of the MFHWT & HEHW method when applied to the posterior cruciate ligament image.

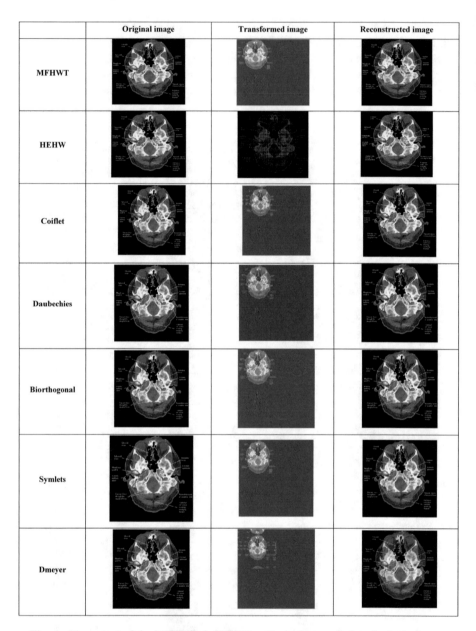

	Original image	Transformed image	Reconstructed image
MFHWT			
HEHW			
Coiflet			
Daubechies			
Biorthogonal			
Symlets			
Dmeyer			

Fig. 4. The output of the MFHWT & HEHW method when applied to the brain image.

It is clear that there is superiority of HEHW over FMHWT in X-Ray image by six and twice times for CR and PSNR respectively.

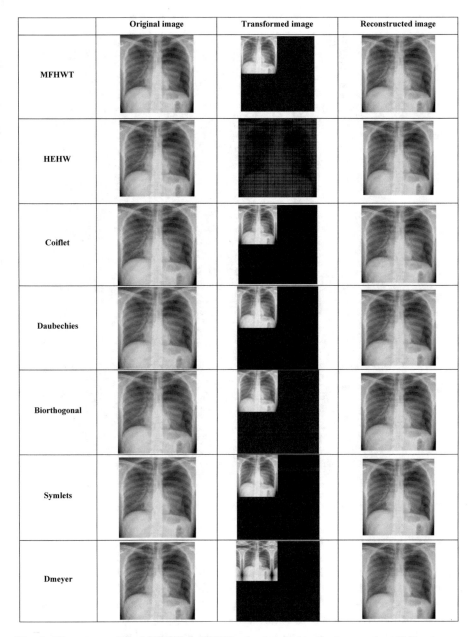

	Original image	Transformed image	Reconstructed image
MFHWT			
HEHW			
Coiflet			
Daubechies			
Biorthogonal			
Symlets			
Dmeyer			

Fig. 5. The output of the MFHWT & HEHW when applied to the human chest X-Ray image.

For the output results of HEHW and MFHWT for MRI image HEHW achieve the best for CR and PSNR. Although HEHW have provided the best results for CR, PSNR and MSE, it needs some time for re-distributions the resultant of coefficient. This is acceptable if a large factor compression is demanded.

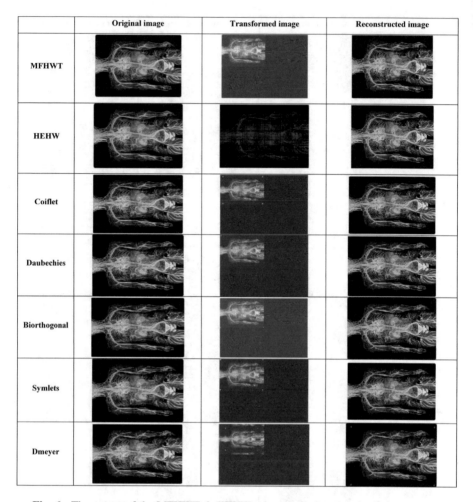

Fig. 6. The output of the MFHWT & HEHW when applied to the circulatory system.

5 Conclusion and Future Work

We have presented high efficient Haar wavelets (HEHW) for improving medical image compression. The proposed method called HEHW and worked faithfully on all test images. In the proposed HEHW, we have adopted the Haar transform to work with submatrices instead of the whole original matrix.

The HEHW starts by dividing the original image matrix into 2 * 2 submatrices. The new transform equations are induced to provide good transform coefficients. Then we perform the statistical threshold on the resultant details subband to gain a high compression ratio. Then, we have computed CR, PSNR and MSE values for each test images to judge the performance of the proposed method. In our test, more CR and PSNR obtained with lower MSE prove the efficiency of the method. Proposed HEHW algorithm has been tested on 3 varying types of test images X-ray, CT, MRI to prove its capabilities. Furthermore, the results of HEHW have compared with the output of

FMHWT when they applied to the same data set. In the future a segmentation technique can be embedded in this algorithm to decompose the background from the details of the image, applying HEHW on the image portions that have important details and apply another technique like SPIHT on the background portions.

References

1. Weinberger, M.J., Seroussi, G., Sapiro, G.: The LOCO-I lossless image compression algorithm: principles and standardization into JPEG-LS. IEEE Trans. Image Process. **2**(8), 1309–1324 (2000)
2. Chouakri, S.A., Djaafri, O., Taleb-Ahmed, A.: Wavelet transform and Huffman coding based electrocardiogram compression algorithm: application to tele cardiology. In: 24th Conference on Computational Physics, vol. 454, pp. 1–16 (2012)
3. Khalifa, O.: Wavelet coding design for image data compression. Int. Arab. J. Inf. Technol. **6** (2), 118–127 (2009)
4. Yan-li, Z., Xiao-ping, F., Shao-qiang, L., Zhe-yuan, X.: Improved LZW algorithm of lossless data compression for WSN. In: 3rd International Conference on Computer Science and Information Technology, vol. 4, pp. 523–527. IEEE Xplore Digital Library (2010)
5. Sridhar, S., Rajesh Kumar, P., Ramanaiah, K.V.: Wavelet transform techniques for image compression – an evaluation. Int. J. Image Graph. Sig. Proc. **6**(2), 54–67 (2014)
6. Ghorpade, A., Katkar, P.: Image compression using Haar transform and modified fast Haar wavelet transform. Int. J. Sci. Technol. Res. **3**(7), 302–305 (2014)
7. Porwik, P., Lisowska, A.: The Haar wavelet transform in digital image processing: its status and achievements. Mach. Graph. Vis. **13**(1), 79–98 (2004)
8. Horgan, G.: Wavelets for SAR image smoothing. Photogram. Eng. Remote Sens. **64**(12), 1171–1177 (1998)
9. Benyahia, I., Beladgham, M., Bassou, A.: Evaluation of the medical image compression using wavelet packet transform and SPIHT coding. Int. J. Electr. Comput. Eng. **8**(4), 2139–2147 (2018)
10. Reisenhofer, R., Bosse, S., Kutyniok, G., Wiegand, T.: A Haar wavelet-based perceptual similarity index for image quality assessment. Sig. Process. Image Commun. **61**, 33–43 (2018)
11. Latha, P.M., Fathima, A.A.: Collective compression of images using averaging and transform coding. Measurement **135**, 795–805 (2019)
12. Heydaria, M., Cheraaqeea, P., Mansouria, A., Mahmoudi-Aznavehb, A.: A low complexity wavelet-based blind image quality evaluator. Sig. Process. Image Commun. **74**, 280–288 (2019)
13. Sengupta, A., Roy, D.: Mathematical validation of HWT based lossless image compression. In: International Symposium on Nanoelectronic and Information Systems, vol. 1, pp. 20–22. IEEE (2017)
14. Vasanth, P., Rajan, S., Lenin Fred, A.: Efficient oppositional based optimal Harr wavelet for compound image compression using MHE. Biomed. Res. **29**(10), 2169–2178 (2018)
15. Nahar, A.K., Jaddar, A.S.: A compression original image based on the DDWT technique and enhancement SNR. Int. J. Eng. Technol. Sci. **5**(3), 73–89 (2018)
16. Nashat, A.A., Hussain Hassan, N.M.: Image compression based upon wavelet transform and a statistical threshold. In: International Conference on Optoelectronics and Image Processing, pp. 20–24. IEEE Xplore Digital Library (2016)

Modified Value-and-Criterion Filters for Speckle Noise Reduction in SAR Images

Ahmed S. Mashaly$^{(\boxtimes)}$ ⓘ and Tarek A. Mahmoud ⓘ

Military Technical College, Cairo 11766, Egypt
{mashaly, t.mahmoud}@mtc.edu.eg

Abstract. Speckle noise disturbance is the most essential factor that affects the quality and the visual appearance of the synthetic aperture radar (SAR) coherent images. For remote sensing systems, the initial step always involves a suitable method to reduce the effect of speckle noise. Several non-adaptive and adaptive filters have been proposed to enhance the noisy SAR images. In this paper, two proposed non-adaptive filters have been introduced. These proposed filters utilize traditional mean, median, root-mean square (RMS) values, and large size filter kernels to improve the SAR image appearance while maintaining image information. The performance of the proposed filters is compared with a number of non-adaptive filters to assess their ability to reduce speckle noise. For quantitative measurements, four metrics have been used to evaluate the performance of the proposed filters. From the experimental results, the proposed filters have achieved promising results for significantly suppressing speckle noise and preserving image information compared with other well-known filters.

Keywords: Synthetic Aperture Radar (SAR) · Remote sensing ·
Non-adaptive filter · Despeckling filter · Image enhancement

1 Introduction

Synthetic Aperture Radar (SAR) is the essential system for ground mapping and remote sensing applications since its performance is independent on both weather conditions and daytime. Generally, SAR senses the interaction between the incidents radiated waves from SAR and Earth surface. However, the quality of the composed images from raw SAR data suffer from speckle noise distribution. The coherent nature of radiated wave of SAR is considered as the main source of speckle noise. For more details, let consider a certain ground patch that has a various number of scatter elements with different level of backscatter coefficients. The final response of the constructed SAR image at each pixel location is the addition of amplitude and phase of each scatter element as shown in Eq. (1) [1–4].

$$Ae^{j\Phi} = \sum_{k=1}^{N} A_k e^{j\Phi_k} \tag{1}$$

A. E. Hassanien et al. (Eds.): AISI 2019, AISC 1058, pp. 558–567, 2020.
https://doi.org/10.1007/978-3-030-31129-2_51

Where A ... is the scatters amplitude response, Φ ... is the scatters phase response, and N ... is the total number of scatters for one pixel.

From Eq. (1), it is obvious that the total response of each pixel location depends on both scatter phase response and the total number of scatters. Therefore, it is accepted to get a random pixel response due to the phase differences between scatters. In fact, speckle noise can be produced due to an interference phenomenon in which the principal source of the noise is the distribution of the phase terms Φ_k. Then, the sum of Eq. (1) seems like a random walk in the complex plane, where each step of length A_k is in a completely random direction [1–4].

Finally, speckle noise is a serious problem that not only degrades the quality of the processed SAR images but also reduces the opportunity of image interpretation and analysis. Consequently, using suitable image processing method to improve SAR image quality and maintain image information is not a trivial task [1–4].

Generally, different methods and techniques are used to eliminate the effect of speckle noise. These methods could be classified as onboard methods and offline methods [1–4].

For onboard methods, the main idea of this technique is to apply multi-look processing on a certain ground patch and the final resultant image depends on the average of these multiple looks to eliminate the effect of speckle noise. This technique has promising results while its performance relies on several constraints to suppress speckle noise [3].

For offline methods, these methods utilize the great development of image processing research field to find the optimum solution to reduce speckle noise. Theoretically, two types of filters are used to eliminate the speckle noise that are adaptive and non-adaptive filters. The main difference between these filters could be noticed as follows "*adaptive filters adapt their weightings and functions across the image to the speckle level, and non-adaptive filters apply the same weightings and functions uniformly across the entire image*". Moreover, such filtering also loses actual image information as well. In particular high-frequency information and the applicability of filtering and the choice of filter type involve tradeoffs [3, 4].

Intuitively, adaptive filters have better performance compared with that of the non-adaptive filter in preserving image information in high-texture areas and resulting in non-blurred edges. On the other hand, adaptive filters require high computational power and their implementation is a complicated task. While, the non-adaptive filters seem to be simple and require less computational power but with some losses in image information and blurred edges [5, 6]. In this paper, we will concern with the non-adaptive filters in reducing speckle noise effect.

2   Non-adaptive Despeckling Filters

In this section, we are going to introduce the most popular non-adaptive despeckling filters that have a wide usage for speckle noise reduction in several remote sensing and ground mapping applications.

In the beginning, we start with the well-known Mean filter, which could be classified as a linear filter. The main operation and final response for the center pixel of the filter mask depends on the average value of the grey levels neighborhoods defined by the filter mask [7]. On the other hand, Median filter is classified as a nonlinear filter. The response of median filter for the center pixel of the filter mask depends on median value of the grey levels neighborhoods defined by the filter mask. The latter is better than the mean filter at preserving edges whilst eliminating noise spikes [7].

Another nonlinear filter is called value-and-criterion filter. This filter has a "value" function (V) and a "criterion" function (C), each operating separately on the original image, and a "selection" operator (S) acting on the output of (C). The selection operator (S) chooses a location from the output of (C), and the output of (V) at that point is the output of the overall filter. The value-and-criterion structure allows the use of different linear and nonlinear elements in a single filter [8].

Minimum Coefficient of Variant (MCV) filter is an example of the value-and-criterion filter. The MCV filter, therefore, has V = sample mean, C = sample coefficient of variation, and S = minimum of the sample coefficient of variation. At every point in an image, this filter effectively selects the ($n * n$) sub-window with an overall window ($2n - 1 * 2n - 1$) that has the smallest measured coefficient of variation, and outputs the mean of that sub-window. The Mean of Least Variance (MLV) filter is another value-and-criterion filter that uses the sample variance as the criterion for selecting the most homogeneous neighborhood and the sample mean for computing a final output value from that neighborhood. Since the sample variance requires the sample mean for its computation, the V and C functions are dependent in this case. Therefore, the general implementation for MLV filter will be as follows: V = mean, C = variance, and S = minimum. The MLV filter resembles several earlier edge-preserving smoothing filters, but performs better and is more flexible and more efficient. The MLV filter smoothes homogeneous regions and enhances edges, and is therefore useful in segmentation algorithms [9, 10].

Morphological filters are nonlinear image transformations that locally modify geometric features of images. They stem from the basic operations of a set-theoretical method for image analysis, called mathematical morphology, which was introduced by Matheron [11], Serra [12] and Maragos and Schafer [13]. The most common morphological filters that are used in speckle noise reduction are morphological opening filter. This filter can be easily expressed in terms of the value-and-criterion filter as both V, C are the minimum (erosion) operators, and S is the maximum (dilation) operator respectively. Furthermore, morphological closing filter also can be easily expressed in terms of the value-and-criterion filter as both V, C is the maximum (dilation) operator and S is the minimum (erosion) operator [8]. Finally, Tophat filter is a closing-opening filter i.e. it consists of two stages where the first stage will be morphological closing (two steps dilation followed by erosion with the same structure element (SE)). The second stage will be morphological opening (two steps dilation to the output of first stage followed by erosion with the same SE of first stage). For speckle reduction, morphological tools perform even better than classical filters in preserving edges, saving the mean value of the image and computational time according to ease of implementation [14, 15].

3   The Proposed Non-adaptive Filters

Generally, most of non-adaptive filters performances degrade rapidly by increasing the filter mask size (MS) and the final enhanced image suffers from blur appearance and great loss in image information. The purpose of increasing MS is to speed up the process of speckle noise reduction while keeping image information. However, most of non-adaptive filters are incompatible with this idea where their performances are very limited to small MS (from (3 * 3) to (9 * 9)).

As a result, our task is to propose a non-adaptive filter that has an effective response with large filter MS. Firstly, it is important to note that achieving this goal requires using the most informative pixels within the used filter mask to estimate accurate filter response. Therefore, we will be concerned only with the pixel locations that are expected to represent the image features. In other words, not all the pixels within filter mask will be processed. We will select only the pixels in the predefined directions (Vertical, Horizontal, Right, and Left diagonals) as shown in Fig. 1.

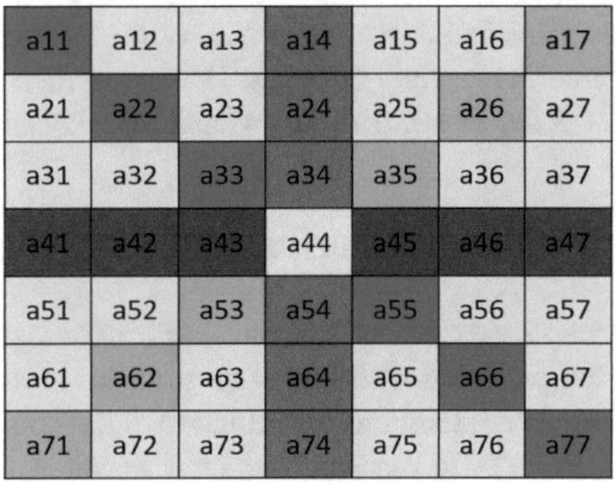

a11	a12	a13	a14	a15	a16	a17
a21	a22	a23	a24	a25	a26	a27
a31	a32	a33	a34	a35	a36	a37
a41	a42	a43	a44	a45	a46	a47
a51	a52	a53	a54	a55	a56	a57
a61	a62	a63	a64	a65	a66	a67
a71	a72	a73	a74	a75	a76	a77

Fig. 1. Example of the proposed filter mask (7 * 7) with the 4-predefined directions.

The choice of these 4-directions is related to the case where the pixel under-test (a44) may be probably a part of an image edge. In this case, we will search for the direction that has the smallest grey-scale variations. Therefore, we need to apply a suitable quantitative index to measure the variation in each direction and selecting the direction that has the smallest variation value (*the speckle existence in this direction is very low*). Sometimes, the center pixel of the filter mask may be a part of homogenous region or the whole filter mask covers a homogenous region where there is no image features. In this case, the proposed idea is still valid where the final filter response at this location will be related to the direction, which has the smallest grey-scale variations. Theoretically, as the filter mask domain decreases to a significant level, the probability to estimate both accurate response value and robust to noise increases as well. This is the key of the proposed non-adaptive filters. Finally, the detailed steps of the proposed filters could be summarized as follows:

- For each image location covered by filter mask, assign the pixels values in the four predefined directions to four one-dimensional arrays.
- For each array, estimate the root-mean square value as shown in Eq. (2) [16]

$$RMS = \sqrt{\frac{\sum_{i=1}^{n}(\widehat{a} - a_i)^2}{n}} \tag{2}$$

Where $\widehat{a}$ mean value of each array, a_i grey-scale levels, n number of grey-scale pixels in each array or direction.

- Select the direction that has the smallest RMS value.
- The filter response in this case could be the mean or median value of the selected direction.

This methodology of determining the filter response will improve the overall performance of the traditional mean and median filter. Where using edge pixel or the direction of minimum RMS will reduce the blurring effect at edge locations caused by traditional Mean and Median filters when applying large filter masks. Finally, the proposed filter could be classified as a special case of the previously mentioned value-and-criterion filter. Where the "value" function (V) is the mean or median value, the "criterion" function (C) is the RMS, and the "selection" operator (S) is the minimum.

4   Experimental Results and Performance Assessment

To facilitate performance assessment of the simulated non-adaptive filters and to validate the operation of the proposed filters, we are going to introduce four quantitative evaluation indexes. The first quantitative evaluation index that is called "Target to Clutter ratio" (TCR) index. TCR presents the ratio between target signal to clutter (background) signal for the filtered image (i.e. as this ratio increases, it indicates better speckle reduction level). TCR could be estimated as shown in Eq. (3) [15].

$$TCR = 10\log\left(\frac{\sigma_T^2}{\sigma_C^2}\right) \tag{3}$$

Where σ_T^2 target variance, and σ_C^2 clutter variance.

The second quantitative evaluation index is Normalized Mean (NM). This evaluation index indicates the filter ability to attenuate speckle noise while preserving image information. NM equals the ratio between mean value of a filtered homogenous region and the mean value of the original noisy region. The closer NM to one, the better filter ability to preserve image information. NM could be calculated as shown in Eq. (4) [15].

$$NM = \frac{\mu_{filtered}}{\mu_{original}} \tag{4}$$

Where $\mu_{filtered}$ and $\mu_{original}$ are the means of the background segments of the filtered and the original image, respectively.

Standard Deviation to Mean (STM) index assess the ability of the filter to suppress speckle noise for a predefined homogenous patch. As STM decreases as speckle noise suppression gets better. STM could be estimated as shown in Eq. (5) [15]

$$STM = \frac{\sigma_{filtered}}{\mu_{filtered}} \tag{5}$$

Where $\sigma_{filtered}$ is the standard deviation of filtered homogenous background patch.

For edge quality, Edge Index (EI) could be used to show the statement of image edges after despeckling operation. EI equals the ratio between filtered image pixels to the original image pixels. Mathematically, EI might be less than one that shows edge blurring or EI might be greater than or equals to one that indicates edge enhancement. EI could be calculated as shown in Eq. (6) [15].

$$EI = \frac{\sum P_f(i,j) - P_f(i-1,j+1)}{\sum P_O(i,j) - P_O(i-1,j+1)} \tag{6}$$

Where $P_f(i, j)$ and $P_o(i, j)$ are the filtered and the original pixel values respectively.

Finally, the Average Processing Time (APT) per frame indicates the capability of the tested algorithm for real-time applications. Intuitively, there is no specific value that could be used as a reference value. Generally, such value differs from one application to another. While, the measured value gives a good indication about the tested algorithm complexity and the validity for ease implementation.

The performance of the proposed filters and the previously mentioned non-adaptive filters will be assessed based on real noisy SAR images with resolution of (2500 * 1650). In addition, all the aforementioned non-adaptive filters are implemented using LabVIEW 2018 and NI Vision Development System 2018 running on Lenovo notebook Z50-70 with Intel processor core i7 (4510u) and 8 GB RAM.

For visual inspection index, Fig. 2 introduces a sample of the resultant enhanced images of the simulated filters, the proposed mean and median filters based on RMS error value. It is obvious that the proposed mean and median filters have promising results compared with the results of the other simulated non-adaptive filters especially, when these filters use large filter masks. We can note that MCV, MLV, and the morphological (Closing, Opening, and Tophat) filters have distorted images with large filter masks and SE. While, conventional Mean and Median filters have acceptable results with small amount of blurring at image edges. As a result, most of non-adaptive filters performance degrade rapidly by increasing filter mask. On the other hand, the enhanced SAR images of the proposed Mean and Median filters have effectively eliminated the speckle noise while retaining image integrity.

Furthermore, Table 1 presents the average processing time (APT) per frame for the aforementioned non-adaptive filters. It is clear that the traditional non-adaptive filters have high computational power for large filter MS. While the predefined directions of the proposed filter masks have low computational power, less complexity, and ease of implementation.

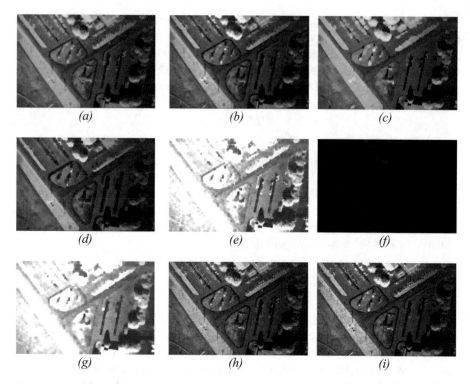

Fig. 2. Sample of enhanced SAR images using (27 * 27) for both filter MS and SE: (a) Mean (b) Median (c) MCV (d) MLV (e) Closing (f) Opening (g) Tophat (h) Proposed Mean-RMS (i) Proposed Median-RMS.

Table 1. Quantitative evaluation indexes results

Indexes	Tophat	Opening	Closing	MLV	MCV	Median	Mean	Proposed median	Proposed mean
APT (sec)	63	**30**	33	330	300	102	70	**60**	**58**
TCR	20.94	9.95	14.05	14.68	15.61	**24.42**	21.06	**24.25**	**23.15**
NM	1.237	0.087	2.397	0.953	**0.974**	0.935	1.068	**0.905**	**0.935**
STM	0.042	0.171	0.067	0.123	**0.035**	0.044	0.1	**0.16**	**0.14**
EI	0.223	0.159	0.228	0.225	0.306	0.84	**0.982**	**0.941**	**0.981**

For the other quantitative evaluation indexes like TCR, NM, STM, and EI, it seems that these evaluation indexes may not introduce an accurate assessment values. Since most of non-adaptive filters (MCV, MLC, Tophat, Closing, and Opening), have distorted resultant images with large filter masks. However, these evaluation indexes still show the enhancement amount achieved by the proposed filters compared with the results of other filters.

Table 1 presents the estimated TCR value for different non-adaptive filters results (Note that the original TCR value for noisy SAR image is 13.86). It is obvious that the proposed filters achieve the highest values with small edge blurring or distortion and with low processing time. Table 1 shows the different resultant levels of NM index for the non-adaptive filters under test. It is clear that the proposed non-adaptive filters have the ability to preserve image information of the enhanced images compared with the original noisy image. Since the estimated NM values of the proposed filters are close to the value of "1".

Table 1 introduces the different estimated STM levels for the non-adaptive filter results. It can be noticed that the proposed Mean and Median filters present moderate STM values (i.e. these proposed filters have the capability to attenuate speckle noise to an acceptable level while keeping image visual appearance and information). In addition, we can see that there are some non-adaptive filters, which show very small STM values. However, these filters have a completely distorted output images. Therefore, we cannot consider these filters values as the best STM results.

Table 1 introduces the evaluation index (EI) results of the non-adaptive filters under test. It is clear that the resultant values of EI are very near to "1" that means there is small edge blurring. Theoretically, these results seem to be normal since all non-adaptive filters suffer from edge blurring output after enhancement processing. Moreover, EI values for the proposed filter are closer to "1" compared to that of other filters. As a result, edge blurring of the proposed filters is very low compared with the other filters.

Finally, we are going to study the performance of the proposed filters (Mean-RMS and Median-RMS) with different filter MS to ensure the objective and motivation of the proposed filters. Figures 3, 4, 5 and 6 show the different evaluation indexes (TCR, NM, STM, and EI) with different sizes of used filter mask respectively. For TCR, it is clear that there is a value improvement with increasing filter mask size. In addition, NM evaluation index value still around "1" without any loss in image information. Furthermore, STM values decrease smoothly with increasing filter mask to confirm the objective of the proposed filters. Finally, EI values still less than "1" this means that there is small amount of edge blurring.

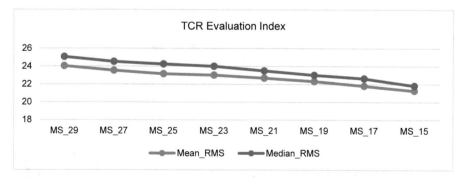

Fig. 3. TCR evaluation index with different filter MS.

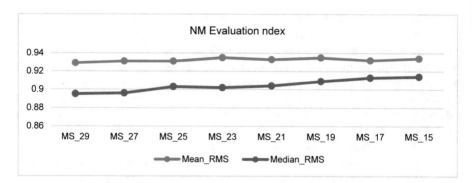

Fig. 4. NM evaluation index with different filter MS.

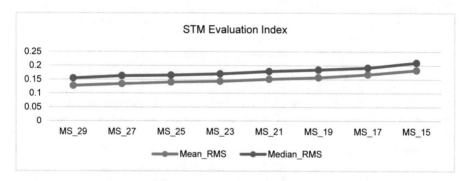

Fig. 5. STM evaluation index with different filter MS.

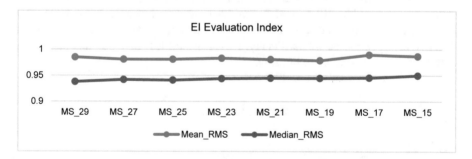

Fig. 6. EI evaluation index with different filter MS.

5   Conclusion

SAR images are useful sources of information for moisture content of the Earth surface, geometry, and roughness. As an active senor, suitable performance for different day-time, and all-weather conditions remote sensing system, SAR images can provide information for both surface and subsurface of the Earth. Speckle noise is an additive kind of random noise, which degrades the quality of SAR images appearance. Speckle noise reduces image contrast and has a negative effect on texture-based analysis. Moreover, as speckle noise changes the spatial statistics of the images, it makes the classification process a difficult task to do. Thus, it is a vital task to reduce the effect of speckle noise. In this paper, we introduce two proposed non-adaptive filters based on RMS error value that rely on large filter mask to speedup processing time of the noisy SAR images and to enhance the signal to noise ratio. Moreover, the proposed filters aim to overcome the limitations of the traditional non-adaptive filters with large mask sizes. Different types of evaluation indexes have been used to assess the performance of the proposed filters. From the experimental results, the proposed filters achieve the best visual appearance, the highest TCR value, promising values for NM, STM, and EI values especially with large filter masks.

References

1. Sivaranjani, R., Roomi, S.M.M., Senthilarasi, M.: Speckle noise removal in SAR images using multi-objective PSO (MOPSO) algorithm. Appl. Soft Comput. **76**, 671–681 (2019)
2. Singh, P., Shree, R.: A new SAR image despeckling using directional smoothing filter and method noise thresholding. Eng. Sci. Technol. Int. J. **21**(4), 589–610 (2018)
3. Oliver, C., Quegan, S.: Understanding Synthetic Aperture Radar Images, 1st edn. SciTech, Inc., Havre de Grace (2004)
4. Massonnet, D., Souyris, J.: Imaging with Synthetic Aperture Radar, 1st edn. EPFL Press, Lausanne (2008)
5. Tso, B., Mather, P.: Classification Methods for Remotely Sensed Data, 2nd edn. CRC Press, Boca Raton (2000)
6. Franceschetti, G., Lanari, R.: Synthetic Aperture Radar Processing. CRC Press, Boca Raton (1999)
7. Gonzalez, R., Woods, R.: Digital Image Processing, 3rd edn. Addison-Wesley Inc., Boston (2008)
8. Schulze, M., Pearce, J.: Value-and-criterion filters: a new filter structure based upon morphological opening and closing. In: Nonlinear Image Processing IV, Proceedings of SPIE, vol. 1902, pp. 106–115 (1993)
9. Schulze, M., Wu, Q.: Noise reduction in synthetic aperture radar imagery using a morphology based nonlinear filter. In: Digital Image Computing: Techniques and Applications, pp. 661–666 (1995)
10. Matheron, G.: Random Sets and Integral Geometry. Wiley, New York (1975)
11. Serra, J.: Image Analysis and Mathematical Morphology. Academic Press, New York (1982)
12. Maragos, P., Schafer, R.: Morphological filters-part I: their set-theoretic analysis and relations to linear shift-invariant. IEEE Trans. Acoust. Speech Signal Process. **ASSP-35(8)**, 1153–1169 (1987)
13. Shih, F.: Image Processing and Mathematical Morphology Fundamentals and Applications, 1st edn. CRC Press, Boca Raton (2009)
14. Gasull, A., Herrero, M.A.: Oil spills detection in SAR images using mathematical morphology. In: Proceedings of EUSIPCO, Toulouse, France, 3–6 September 2002
15. Huang, S., Liu, D.: Some uncertain factor analysis and improvement in space borne synthetic aperture radar imaging. Signal Process. **87**, 3202–3217 (2007)
16. GISGeography. https://gisgeography.com/root-mean-square-error-rmse-gis/. Accessed 19 Mar 2019

A Universal Formula for Side-Lobes Removal Optimum Filter in Multi-samples Phase Coded Pulse Compression Radars

Ibrahim M. Metwally$^{(\boxtimes)}$, Abd El Rahman H. Elbardawiny,
Fathy M. Ahmed, and Hazem Z. Fahim

Military Technical College, Cairo, Egypt
Ibrahimmetwally77@gmail.com,
{Bardawiny, Fkader, hkamel}@mtc.edu.eg

Abstract. Using multi samples per sub pulse with phase-coded waveforms used in Pulse Compression (PC) radars aims to enhance the resultant signal at the Matched Filter (MF) output. Recently, a generic analytical formula for a side-lobes cancellation Optimum Filter (OP-F) in PC radar with phase coded waveforms considering only one sample per sub pulse has been introduced by the author [1]. In the present work, two new universal analytical formulas for this OP-F transfer function are derived considering multi samples per sub pulse to get the benefits of multi samples consideration. The first formula, form-1, depends on the phase values of the code elements. The second formula, form-2, relays on the frequency elements contained in the autocorrelation function (ACF) of the considered waveform. The side-lobes suppression is achieved by multiplying the MF output with the OP-F coefficients in frequency domain, after that the side-lobe behavior is investigated by converting it into the time domain. Simulation study is carried out to validate the applicability of the derived formulas and to evaluate the OP-F performance in comparison with standard MF and recently reported side-lob reduction techniques. Results exhibit that the performance of OP-F is significantly superior compared to other techniques.

Keywords: Pulse Compression (PC) · Matched filter (MF) ·
Optimum filter (OP-F) · Autocorrelation function (ACF)

1 Introduction

Pulse Compression (PC) techniques are extensively used in modern radar systems as a mean to allow combining the energy of a long pulse width in transmission and simultaneously high resolution of a short pulse width in reception through phase or frequency modulation of the transmitted pulse. The received echo is processed in receiver by a matched filter (MF) to maximize the signal-to-noise ratio (SNR) of the received signal at its output. Ideally, the MF output is the autocorrelation function (ACF) of the transmitted signal. The transmitted waveforms are chosen so as to have an ACF with a narrow peak at zero time shifts (main lobe) and as low as possible values at all other time shifts (side-lobes). These side-lobes have the undesirable effect of forming spurious targets or masking small targets, which are in close vicinity to large

© Springer Nature Switzerland AG 2020
A. E. Hassanien et al. (Eds.): AISI 2019, AISC 1058, pp. 568–578, 2020.
https://doi.org/10.1007/978-3-030-31129-2_52

targets, such as clutter returns [2]. To overcome these problems, searching for good pulse compression waveforms and techniques for reduction of these unwanted side-lobes has been an active research area since the invention of PC.

To minimize the peak side-lobe levels (PSL), different techniques have been developed in literature depending on frequency modulation and phase modulation codes. Mismatched filters are used to improve the main lobe to peak side-lobe ratio (MSR) at the cost of some deterioration in the SNR [3].

A multilayer artificial neural network (MLANN) with back propagation (BP) [4] and extended Kalman filtering (EKF) [5] algorithms were employed for reduction of range side-lobes of binary phase coded waveforms. Neuro-fuzzy network based pulse compression (NFNPC) was reported in literature [6]. Radial basis function neural network (RBFNN) based method was shown to yield better side-lobes reduction compared to other competitive methods [7]. Baghel and Panda had proposed a hybrid model for the phase-coded waveforms in which MF output was modulated by the output of radial function for different Barker codes (MF-RF) [8]. In [9], a model had been developed in which a mismatched filter comprised of a matched filter was cas-caded with a parameterized multiplicative finite impulse response (FIR) filter. A hyper chaotic coding scheme and corresponding optimal selection method were proposed in [10], to obtain the phase code signal, which exhibits great performance for side-lobe reduction. Also, side-lobe cancellers based on Woo filter for polyphase codes had been developed and resulted in a significant improvement in PSL and integrated side-lobe level (ISL) had been presented in [11, 12].

Recently, an optimum filter (OP-F) with a generic analytical formula, considering one sample per sub pulse, for range side-lobes cancellation (SLC) in PC radar with phase-coded waveforms has been introduced and verified with different phase coded waveforms by the author [1]. This OP-F is based on the inverse filter concept and applicable in both frequency domain and time domain. It gave a better SLC effect than other ways. However, increasing the number of samples per sub pulse for the phase-coded waveform enhances the resultant signal at the MF output. So, application of SLC techniques considering increasing the number of samples inside the sub pulse of the phase coded waveform is an objective to get its benefits.

In this paper, analytical derivation of a general formula for the OP-F presented in [1] is derived considering any number of samples (S) per sub pulse. To validate the derived formula, the OP-F performance is compared with standard MF and other mentioned side-lobe reduction techniques considering phase coded signals such as binary phase coded (Barker code) and polyphase coded (P4) waveforms. PSL, ISL, and main-lobe width are the performance parameters that are used to evaluate different side-lobe reduction techniques.

This paper is organized as follows. After the introduction, Sect. 2 illustrates the ACF at the MF output for single and multi-samples phase coded waveforms. Mathe-matical derivation of the universal analytical formulas of the OP-F transfer function is detailed in Sect. 3. Measuring parameters used for performance evaluation and com-parison with different considered side-lobes reduction techniques are presented in Sect. 4. Finally, in Sect. 5, conclusions and future works are presented.

2   MF Output of a Single and Multi-samples Phase Coded Waveforms

Generally, the received echo signal is processed at receiver in MF to maximize SNR. The MF output is the ACF of the signal. The ACF consists of a main peak, which is used for target detection, in addition to a series of side-lobes.

The ACF of a phase coded sequence s = {s(0), s(1), ..., s(N−1)} of length N, is [2]:

$$r(n) = \sum_{k=0}^{N-1-k} s_k s_{k+n}^*, \quad n = 0, \ldots, N-1 \tag{1}$$

Unfortunately, increasing the number of samples per sub pulse by digital sampling increases both the main peak and side-lobes by the same ratio and does not enhance MSR. Table 1 lists the absolute values of the main lobe amplitude (MLA) and the maximum side-lobes amplitude (SLA) for different samples (S) within sub pulse.

Table 1. MLA and SLA for b-13 and p4-13 with different S.

	B-13		P4-13	
	MLA	SLA	MLA	SLA
S = 1	13	1	13	1.77
S = 3	39	3	26	3.54
S = 10	130	10	130	17.7

3   Mathematical Derivation of the Universal Analytical Formulas

In this section, two different formulas, form-1, and form-2, for side lobes cancellation (SLC) OP-F transfer function considering any number of samples (S) within each sub pulse of the phase-coded waveform are derived. Form-1, denoted as phase values formula, depends on the relation between the phase elements that construct the phase-coded waveform and the ACF at the MF output. While, Form-2, which referred as frequency elements formula, relies on the frequency components entire the ACF of the phase-coded waveform. In the following subsection, mathematical derivation of each formula is detailed.

3.1   Phase Values Formula (Form-1)

Considering a phase coded waveform with length N = 3, for example, and S = 2 samples per sub pulse. Mainly, the phase elements (ϕ) of this signal before sampling is given by:

$$\phi = \{a_0, a_1, a_2\} \tag{2}$$

After sampling process, the length of the phase sequence will be $(L = N * S)$, $L = 6$, and can be written as:

$$\phi = \{a_0, a_0, a_1, a_1, a_2, a_2, a_3, a_3\} \tag{3}$$

Where, a_i is the phase value, according to the type of phase coded waveform whether binary phase or polyphase coded, and $i = 0,...., N - 1$.

The corresponding phase coded complex signal (x) can be written as:

$$x(i) = e^{j\phi} = \left\{e^{ja_0}, e^{ja_0}, e^{ja_1}, e^{ja_1}, e^{ja_2}, e^{ja_2}, e^{ja_3}, e^{ja_3}\right\}, i = 0, 1...L - 1 \tag{4}$$

The ACF, r(n), of the signal x at the MF output is calculated by Eq. (1) as:

$$r(n) = \left\{ \begin{array}{l} e^{j(a_0-a_2)}, 2e^{j(a_0-a_2)}, e^{j(a_0-a_1)} + e^{j(a_0-a_2)} + e^{j(a_1-a_2)}, \\ 2e^{j(a_0-a_1)} + 2e^{j(a_1-a_2)}, e^{j(a_0-a_1)} + e^{j(a_1-a_2)} + 3, \\ 6, e^{-j(a_0-a_1)} + e^{-j(a_1-a_2)} + 3, 2e^{-j(a_0-a_1)} + 2e^{-j(a_1-a_2)}, \\ e^{-j(a_0-a_1)} + e^{-j(a_0-a_2)} + e^{-j(a_1-a_2)}, 2e^{-j(a_0-a_2)}, e^{-j(a_0-a_2)} \end{array} \right\} \tag{5}$$

Frequency domain representation of the ACF, $R(e^{j\omega})$, with length $M = 2L - 1$ is given by [1]:

$$R(e^{j\omega}) = \sum_{n=0}^{M-1} r(n) * e^{-j\omega n} \tag{6}$$

Substitution of Eq. (5) into Eq. (6) for L and M yields:

$$R(e^{j\omega}) = e^{-j(L-1)\omega}\{R_0 + ... + R_{M-1}\} \tag{7}$$

Thus,

$$R(e^{j\omega}) = e^{-j5\omega}\left\{ \begin{array}{l} R_0 + R_1 + R_2 + R_3 + R_4 + R_5 + R_6 + R_7 + \\ R_8 + R_9 + R_{10} \end{array} \right\} \tag{8}$$

Where,

$$R_0 = 6 \tag{8.1}$$

$$R_1 = \{2\cos(a_0 - a_2)\cos(5\omega) - 2\sin(a_0 - a_2)\sin(5\omega)\} \tag{8.2}$$

$$R_2 = \{4\cos(a_0 - a_2)\cos(4\omega) - 4\sin(a_0 - a_2)\sin(4\omega)\} \tag{8.3}$$

$$R_4 = \{2\cos(a_0 - a_1)\cos(3\omega) - 2\sin(a_0 - a_1)\sin(3\omega)\} \tag{8.4}$$

$$R_5 = \{4\cos(a_0 - a_1)\cos(2\omega) - 4\sin(a_0 - a_1)\sin(2\omega)\} \tag{8.5}$$

$$R_6 = \{2\cos(a_0 - a_1)\cos(\omega) - 2\sin(a_0 - a_1)\sin(\omega)\} \tag{8.6}$$

$$R_7 = \{2\cos(a_1 - a_2)\cos(3\omega) - 2\sin(a_1 - a_2)\sin(3\omega)\} \tag{8.7}$$

$$R_8 = \{4\cos(a_1 - a_2)\cos(2\omega) - 4\sin(a_1 - a_2)\sin(2\omega)\} \tag{8.8}$$

$$R_9 = \{2\cos(a_1 - a_2)\cos(\omega) - 2\sin(a_1 - a_2)\sin(\omega)\} \tag{8.9}$$

$$R_{10} = 6\cos(\omega) \tag{8.10}$$

Equation (8) represents the input to the OP-F. It consists of two parts; the samples inside the main lobe $ML(e^{j\omega})$ and those of the side-lobes $SL(e^{j\omega})$ and can be written as:

$$R(e^{j\omega}) = ML(e^{j\omega}) + SL(e^{j\omega}) \tag{9}$$

$$ML(e^{j\omega}) = e^{-j5\omega}\{R_0 + R_6 + R_9 + R_{10}\} \tag{10}$$

$$SL(e^{j\omega}) = e^{-j5\omega}[R_1 + R_2 + R_3 + R_4 + R_5 + R_7 + R_8] \tag{11}$$

So, based on the inverse filter concept, the transfer function of the OP-F that cancels the side-lobes is given by:

$$H_{opt}(e^{j\omega}) = ML(e^{j\omega}) * \left[R(e^{j\omega})\right]^{-1} \tag{12}$$

By following this methodology through considering a phase coded waveforms with different N and S then calculating their ACF $R(e^{j\omega})$ and corresponding OP-F transfer function $H_{opt}(e^{j\omega})$. After that, studying the effect of changing N and S on the elements contained in the main lobe and those of side-lobes then universal formulas for the ACF and OP-F transfer function have been deduced as follow.

The first formula which is a general formula representing the ACF at the output of the MF $R(e^{j\omega})$ can be written as:

$$R(e^{j\omega}) = L + \beta1 + \beta2 - \beta3 \tag{13}$$

Such that,

$$\beta1 = \sum_{i=1}^{S} [2N(i-1)]\cos(S+1-i)\omega \tag{13.1}$$

$$\beta2 = \sum_{q=1}^{N-1} \sum_{\substack{y=N, \\ z=q-1}}^{q+1} \sum_{i=1}^{2S-1} \alpha_i \cos(a_q - a_y)\cos((L-i-zS)\omega) \tag{13.2}$$

$$\beta3 = \sum_{q=1}^{N-1} \sum_{\substack{y=N, \\ z=q-1}}^{\substack{N-2, \\ q+1}} \sum_{i=1}^{2S-1} \alpha_i \ \sin(a_q - a_y)\sin((L - i - zS)\omega) \qquad (13.3)$$

Where,

α_i is a vector of coefficients. The length and values of its elements depends on the number of samples per sub pulse S such that:

Number of elements in $\alpha_i = 2S - 1$.

The values of α_i elements depends on the vector length such that, for the considered example S = 2, hence number of elements in $\alpha_i = 3$ thereby the element values are {2, 4, 2}. Also, if S = 3 then number of elements in $\alpha_i = 5$ and the values are {2, 4, 6, 4, 2} and so on.

The second formula which is the main contribution of this work, a general formula for the SLC OP-F transfer function has been derived and is given by:

$$H_{opt}(e^{j\omega}) = \frac{L + \beta1 + \gamma1 + \gamma2}{R(e^{j\omega})} \qquad (14)$$

Where,

$$\gamma1 = \sum_{n=1}^{N-1} \sum_{m=0}^{S-1} (2m)\cos(a_n - a_{n+1})\cos(m\,\omega) \qquad (14.1)$$

$$\gamma2 = \sum_{n=1}^{N-1} \sum_{m=0}^{S-1} (2m)\sin(a_n - a_{n+1})\sin(m\,\omega) \qquad (14.2)$$

It enables calculating the OP-F coefficients that result in cancelling the range time side-lobes of any phase-coded waveform with any number of samples per sub pulse.

3.2   Frequency Elements Formula(Form-2)

This formula depends on the frequency elements entire the ACF that are calculated at the MF output as follows:

For the considered example above where N = 3, S = 2 and L = 6, the ACF sequence, r(n), given in Eq. (5) can be written as:

$$r(n) = \left\{ r_1, r_2, r_3, r_4, r_5, r_6, r_5^*, r_4^*, r_3^*, r_2^*, r_1^* \right\} \qquad (15)$$

Where (*) denote complex conjugate.

The Z-domain representation, R(Z), of Eq. (14) is given by:

$$R(Z) = \left\{ \begin{array}{l} (r_1 + r_1^* Z^{-10}) + (r_2 Z^{-1} + r_2^* Z^{-9}) + (r_3 Z^{-2} + r_3^* Z^{-8}) \\ + (r_4 Z^{-3} + r_4^* Z^{-7}) + (r_5 Z^{-4} + r_5^* Z^{-6}) + r_6 Z^{-5} \end{array} \right\} \tag{16}$$

By resolving the elements of the ACF sequences (r) to real part (r_R) and imaginary part (r_I) and according to Eq. (6), frequency domain representation of the ACF, $R(e^{j\omega})$, represented in Eq. (16) is given by:

$$R(e^{j\omega}) = e^{-j5\omega} \left\{ \begin{array}{l} 2r_{1R} \cos(5\omega) - 2r_{1I} \sin(5\omega) + 2r_{2R} \cos(4\omega) - 2r_{2I} \sin(4\omega) \\ + 2r_{3R} \cos(3\omega) - 2r_{3I} \sin(3\omega) + 2r_{4R} \cos(2\omega) - 2r_{4I} \sin(2\omega) \\ + 2r_{5R} \cos(\omega) - 2r_{5I} \sin(\omega) + r_6 \end{array} \right\}$$

$$\tag{17}$$

Similarly, according to the concept of inverse filter represented in Eq. (8), the OP-F transfer function that results in side-lobes cancellation can be obtained.

So, a general formula for SLC OP-F transfer function with any length and any sampling rate, which is the third contribution, can be written as:

$$H_{opt}(e^{j\omega}) = \frac{r_L + \mu_1}{r_L + \mu_2} \tag{18}$$

Where,

$$\mu_1 = 2 \left\{ \sum_{i=1}^{S-1} r_{(L-i)R} \cos(i\,\omega) - r_{(L-i)I} \sin(i\,\omega) \right\} \tag{18.1}$$

$$\mu_2 = 2 \left\{ \sum_{i=1}^{L-1} r_{iR} \cos((L-i)\omega) - r_{iI} \sin((L-i)\omega) \right\} \tag{18.2}$$

Special case, when considering one sample per sub pulse S = 1; the second term of the numerator μ1 is equal to zero.

4 Performance Evaluation

4.1 Performance Measures

Performance of the OP-F is measured using the peak side-lobe ratio (PSR), integrated side-lobe level (ISL), and main lobe width to evaluate the compressed pulse at the MF output. Since there is no article in the literature has been discussed the SLC for multiple samples in sub pulse, and for the purpose of comparison, single sample per sub pulse scenario is used in comparison with previous methods. Barker code of length 13 (B-13) and P4 code of length 1000 (P4-1000) are considered for validation and evaluation of the proposed formulas in noiseless and noisy cases.

4.2   Noiseless Condition

In this case, results obtained in [9, 13] for B-13 code and those obtained in [12] for P4 code are used for comparison. The obtained PSLs from the universal OP-F and different mentioned methods are listed in Tables 2 and 3. It is clear that the performance of the universal OP-F is significantly better than that of the other mentioned methods.

Table 2. PSRS for B-13 with different SLS methods under noiseless condition

Method	B-13 PSRs (dB)
MF	−22.27
MLANN	−42.08
MF-RF	−74.39
Simple cascaded mismatched Filter	−62.02
(R-G-2)	−36.18
(R-G-2)LP	−46.35
Universal OP-F	−308.7

Table 3. PSRS for P4-1000 with different SLS methods under noiseless condition.

Method	P4-1000		
	PSRs (dB)	ISL (dB)	Main lobe width
MF	−36.37	−16.99	1
Woo filter	−57.81	−27.08	2.03
Modified Woo filter form-1	−104.1	−75.34	2.57
Modified Woo filter form-2 with Blackman window	−111.8	−81.44	4.77
Universal OP-F	−287.79	−267.97	1

4.3   Noisy Conditions

Normal Gaussian noise is used in performance study for B-13 and P4-13, Tables 4 and 5 list the PSR and ISL values of both standard MF and OP-F outputs at various SNRs considering different values of S. Results show that the OP-F offers distinct superiority over standard MF in terms of PSR and ISL.

Table 4. PSR and ISL for B-13 with different SNR and S.

	MF output		OP-F output	
	PSR	ISL	PSR	ISL
SNR = 3 dB, S = 1	−12.02	−5.5	−12.5	−6.29
S = 5	−16.43	−8.99	−18.29	−12.22
SNR = 10 dB, S = 1	−19.46	−11.03	−26.53	−20.39
S = 5	−20.89	−11.34	−32.05	−26.12

Table 5. PSR and ISL for P4-13 with different SNR and S.

	MF output		OP-F output	
	PSR	ISL	PSR	ISL
SNR = 3 dB, S = 1	−12.1377	−4.3046	−13.7352	−5.7856
S = 5	−15.3627	−7.7065	−16.6122	−8.6594
SNR = 10 dB, S = 1	−16.4618	−7.7108	−27.6876	−19.7938
S = 5	−16.9306	−9.1746	−30.6349	−22.6646

4.4   Detection Performance

Detection performance is carried out considering two closely spaced targets and normal Gaussian noise with a smallest of constant false alarm rate (SO-CFAR) processor. Two different scenarios are considered for probability of false alarm, $P_{fa} = 10^{-6}$.

First Scenario. It is considered to investigate the ability of OP-F to overcome the problem of weak target masking in the close vicinity of strong target. Let the first target has SNR = 10 dB while the second one has SNR = 5 dB. Figure 1 shows that, when the MF is used alone, the target with high SNR is detected and the target with smaller SNR is missed while the two targets are detected when the OP-F is used.

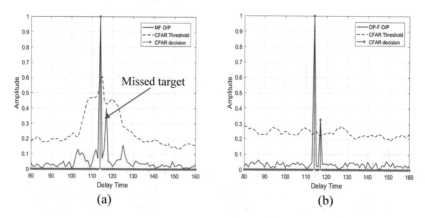

(a)   (b)

Fig. 1. CFAR detection of two close targets (SNR = 10 dB and 5 dB) for (a) MF and (b) OP-F outputs for polyphase code P4-1000.

Second Scenario. This scenario considers the problem of appearing high side-lobes as spurious targets therefore; the two close targets have high SNR (15 dB and 10 dB respectively). When using the MF alone, Fig. 2 shows that the CFAR detects only one of the two targets together with high side-lobes act as false alarms. Meanwhile, the OP-F removes all the side-lobes and detected the two targets without any false alarms.

Finally, the receiver detection curve in case of first scenario is plotted to evaluated the Probability of Detection (P_d) at $P_{fa} = 10^{-6}$ of the OP-F output to illustrate the overall enhancement in detection performance achieved by increasing the number of samples within each sub pulse considering S = 1 and 3 as shown in Fig. 3.

A significant improvement in detection is achieved by increasing the number of samples per sub and applying the corresponding OP-F. For example, for P4-1000 when S = 1 the P_d = 10% at SNR = −7.5 dB while it is 100% when S = 3.

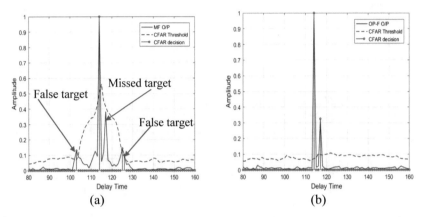

(a)   (b)

Fig. 2. CFAR detection of two close targets (SNR = 15 dB and 10 dB) for MF and OP-F output for polyphase code P4-1000.

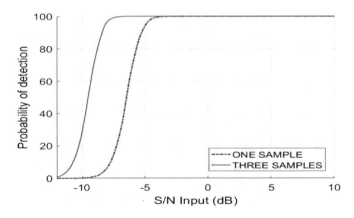

Fig. 3. Detection curves for small target with S = 1 and 3 at OP-F output (P_{fa} = 10-6) for polyphase code P4-1000.

5 Conclusion

This paper has developed a two universal formula for multi samples OP-F transfer function. This efficient OP-F is applicable to any phase coded pulse compression waveform with any number of samples per sub pulse. The PC performance of 13-bit Barker code and 1000-bit P4 code are studied using this OP-F. It gives significant improvement in PSR and ISL without disturbing the main lobe width compared to other recently reported methods. A ten times enhancement in detection performance due to applying the OP-F in case of multiple samples inside sub pulse is achieved compared to that of single sample.

References

1. Metwally, I.M., Elbardawiny, A.E.R.H., Ahmed, F.M., Fahim, H.Z.: A generic analytical formula for range side-lobes cancellation filters in pulse compression phase coded waveforms. In: 18th International Conference on Aerospace Sciences & Aviation Technology, 9–11 April 2019
2. Skolnik, M.I.: Radar Handbook, 3rd edn. McGraw-Hill, New York (2008)
3. Daniels, R.C., Gregers-Hansen, V.: Code inverse filtering for complete sidelobe removal in binary phase coded pulse compression systems. In: Proceedings of IEEE International Radar Conference, 9–12 May 2005, pp. 256–261 (2005)
4. Rihaczek, A.W., Golden, R.M.: Range sidelobe suppression for Barker code. IEEE Trans. Aerospace Electron. Syst. 7(6), 1087–1092 (1971)
5. Kwan, H.K., Lee, C.K.: A neural network approach to pulse radar detection. IEEE Trans. Aerosp. Electron. Syst. 29(1), 9–21 (1993)
6. Deergha, R., Sridhar, G.: Improving performance in pulse radar detection using neural networks. IEEE Trans. Aerosp. Electron. Syst. 31(3), 1193–1198 (1995)
7. Duh, F.B., Juang, C.F., Lin, C.T.: A neural fuzzy network approach to radar pulse compression. IEEE Geosci. Remote Sens. Lett. 1(1), 15–20 (2004)
8. Khairnar, D.G., Merchant, S.N., Desai, U.B.: Radial basis function neural network for pulse radar detection. IET Radar Sonar Navig. 1(1), 8–17 (2007)
9. Baghel, V., Panda, G.: Development of an efficient hybrid model for range sidelobe suppression in pulse compression radar. Aerosp. Sci. Technol. 27(1), 156–162 (2013)
10. Indranil, S., AdlyFam, T.: Multiplicative mismatched filters for sidelobe suppression in Barker Codes. IEEE Trans. Aerospace Electron. Syst. 44(1), 349–359 (2008)
11. Yunkai, D., Yinghui, H., Xupu, G.: Hyper chaotic logistic phase coded signal and its sidelobe suppression. IEEE Trans. Aerosp. Electron. Syst. 46(2), 672–686 (2010)
12. Thakur, A., Talluri, S.R.: A novel pulse compression technique for side-lobe reduction using woo filter concepts. In: 2017 International Conference on Communication and Signal Processing (ICCSP), pp. 1086–1090. IEEE (2017)
13. Elbardawiny, A.H., Sobhy, A., Fathy, M.A., Mamdouh, H.: A novel sidelobe cancellation method for barker code pulse compression. In: Proceedings International Conference Aerospace Sciences and Aviation Technology, Cairo, Egypt, April 2017

Estimation of Lead-Acid Battery State of Charge Based on Unscented Kalman Filtering

Yuan Yu[✉]

College of International Education, Shenyang Institute of Technology,
Shenyang 110136, China
1160660141@qq.com

Abstract. In order to realize the on-line estimation of the state of charge (SOC) of lead-acid batteries, an unscented Kalman filtering (UKF) algorithm is proposed. Thevenin's circuit is used as the equivalent circuit model, and the state space expression is established. The least squares algorithm is used to identify the model parameters. On this basis, the functional relationship between the state of charge of the battery and various parameters of the model is fitted. By analyzing the principle of unscented Kalman filtering, the equivalent circuit model verification experiment and battery SOC test experiment are designed. The experimental results show that under constant current conditions the proposed method has the advantages of online estimation, high estimation accuracy, and high environmental adaptability.

Keywords: Unscented Kalman filtering · State of charge ·
Thevenin equivalent circuit

1 Introduction

The battery is capable of storing electrical energy in other forms of energy, primarily chemical energy. The most common batteries are mainly lead-acid batteries. In 1860, the French plant Plante invented the lead-acid battery. It has been more than 150 years old. It uses lead and lead dioxide as the positive and negative electrodes and is the earliest battery. In 1890, Edison invented a reusable iron-nickel battery, which was commercialized 20 years later [1, 2]. At the same time of technological advancement, there are more and more types of batteries, and their use is becoming more widespread. Since the advent of lead-acid batteries, due to its advantages of easy access to raw materials, low manufacturing cost, good stability, large capacity, and low pollution, it has a huge market in the field of energy storage and is widely used in the production and living sectors.

In this paper, lead-acid batteries and their working characteristics are studied [3–6]. The state monitoring of the battery cannot be directly derived from the parameters. In recent years, the state of charge (SOC [7], State of Charge) is used to characterize the battery, and the state of charge can also be referred to as the remaining capacity. The SOC determines the actual discharge capacity of the battery, and the accuracy of its prediction affects the operation strategy of the battery management system, thereby

© Springer Nature Switzerland AG 2020
A. E. Hassanien et al. (Eds.): AISI 2019, AISC 1058, pp. 579–587, 2020.
https://doi.org/10.1007/978-3-030-31129-2_53

affecting whether the battery can work at an optimal state and reasonably prolong the battery life. During the use of the battery, a very complicated electrochemical reaction occurs inside the battery, which is affected by parameters such as temperature, a number of uses, discharge current, internal resistance, etc. [8], and the remaining capacity is difficult to accurately estimate. This paper will adopt prediction algorithm to propose an optimization algorithm suitable for battery capacity prediction.

2   Kalman Filtering Algorithm Principle

Kalman filtering is derived from the recursive formula of the mathematical formula to estimate the state of the target process, and the real-time correction is performed by using the principle of minimum mean square error, and the estimation of the state variable is updated by using the estimated value of the previous moment and the observed value of the current moment [9]. Finding the estimated value of the moment of occurrence is suitable for real-time processing and computer operations (Fig 1).

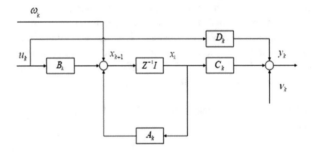

Fig. 1. Kalman filtering structure

The estimated value of the moment of occurrence is defined as follows.

$$x_{k+1} = A_k x_k + B_k u_k + \omega_k \tag{1}$$

$$y_k = C_k x_k + D_k u_k + v_k \tag{2}$$

3   Battery Dynamic Model

The simplest battery model is shown in Fig. 2. The model consists of an ideal battery, an equivalent constant internal resistance, and the terminal voltage of the battery [10]. Where, the current flowing through the battery is based on the Ohm's law of the circuit theory as defined in Eq. (2).

$$I = \frac{V_0 - E_0}{r} \tag{3}$$

This model is too simple, does not take into account the dynamic changes in the internal resistance of the battery, and cannot truly represent the internal working mode of the battery. Therefore, this model is only suitable for ideal simulation circuits that can be used without regard to other constraints.

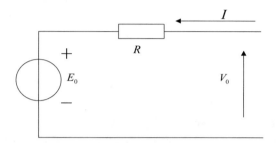

Fig. 2. The equivalent ideal model of battery

An improved model has been developed for the ideal equivalent model. The internal resistance is a variable that varies with the operating conditions of the battery, which better simulates the actual operating state of the battery. Compared with the ideal model, the proposed model has more practical reference. The internal resistance is defined as follws.

$$r = \frac{R_0}{S^k} \tag{4}$$

Based on the improved model, Thevenin battery equivalent model is used to treat the power supply as an ideal voltage source in series with the resistor, which is used to represent the battery internal resistance, capacitance and resistance to form a parallel RC loop, as shown in Fig. 3 [11]. The internal resistance and RC parallel circuit can simulate the load response of the battery under any SOC state. The voltage across the polarized capacitor is recorded as Eq. 5 according to the circuit law:

$$rC_0 \frac{dU_{C_0}}{dt} + \left(1 + \frac{r}{R_0}\right)U_{C_0} = V_0 - E_0 \tag{5}$$

The Thevenin model can more fully reflect the working principle of lead-acid batteries, and more accurately reflect the relationship between the electromotive force and the terminal voltage [12]. The main disadvantage of this model is that this model does not take into account the numerical variation of the parameters, which will be affected by other factors in actual work.

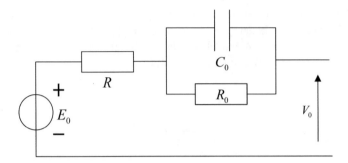

Fig. 3. The equivalent the venin model of battery

This paper uses a third-order dynamic model which can more vividly reflect the mathematical relationship between parameters such as SOC, internal resistance, temperature and battery current and voltage. The model structure is shown in Fig. 4.

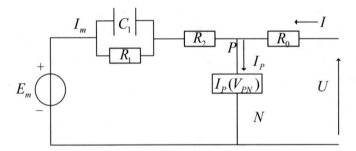

Fig. 4. Battery third-order dynamic model

According to the basic knowledge of the circuit, the discretization equation is established by inputting the equation and the state equation as follows.

$$\begin{bmatrix} U_{k+1}^{RC} \\ U_{k+1}^{PN} \end{bmatrix} = \begin{bmatrix} e^{-\frac{\Delta t}{\tau}} & 0 \\ 0 & 1 \end{bmatrix} \times \begin{bmatrix} U_{k+1}^{RC} \\ U_{k+1}^{PN} \end{bmatrix} + \begin{bmatrix} R_1\left(1 - e^{-\frac{\Delta t}{\tau}}\right) & 0 \\ 0 & 1 \end{bmatrix} \times \begin{bmatrix} i_k \\ i_k^P \end{bmatrix} \tag{6}$$

The output equation of the system:

$$y_k = U - i_k(R_0 + R_2) - U_k^{RC} \tag{7}$$

4 Unscented Transformation

Ordinary Kalman filtering is a dynamic estimation of the target using the least mean square error criterion in the case of linear Gaussian, which is suitable for systems with linear measurement process and errors conforming to Gaussian distribution. For the lead-acid battery monitoring system, the equation of state and the measurement equation are nonlinear [13]. In this case, the ordinary Kalman filtering method may have a large error in the result, so the nonlinear optimization of the Kalman filtering is performed. It is essential. Unscented Kalman filtering has a good predictive effect on nonlinear systems. It is based on the Unscented Transformation and is superior to the extended Kalman filtering in processing results.

The Unscented Transformation approximates the probability distribution of the nonlinear system by deterministic sampling, and the sampling point can accurately obtain the mean and covariance matrix of the random variable [14]. The basic idea of Unscented Transformation is to select a set of sampling points (Sigma point set) based on the mean and variance of the known state, and use this point set to describe the mean and variance of the nonlinear variation.

Select the Sigma point set based on the mean and the variance of the current state x.

$$X_i = \begin{cases} \bar{x} + \left(\sqrt{(n+\kappa)P_x}\right)_i & i = 1, 2, \ldots, n \\ \bar{x} - \left(\sqrt{(n+\kappa)P_x}\right)_i & i = n+1, n+2, \ldots, 2n \\ \bar{x} & i = 0 \end{cases} \tag{8}$$

Constructor Performs nonlinear processing on the point set to obtain a new Sigma point set.

$$Y_i = f(x_i) i = 0, 1, \ldots, 2n \tag{9}$$

This new set of points is represented by a set $\{Y_i\}$. The Sigma point set is weighted to obtain the mean and variance of the values y.

$$\bar{y} = \sum_{i=0}^{2n} W_i^{(m)} Y_i \tag{10}$$

$$P_y = \sum_{i=0}^{2n} W_i^{(c)} (Y_i - \bar{y})(Y_i - \bar{y})^T \tag{11}$$

$W_i^{(m)}, W_i^{(c)}$, respectively, the weighted value corresponding to the mean and the variance, and the weighted value can be changed to obtain the target mean and variance.

$$W_0^{(m)} = \frac{\kappa}{n + \kappa} \tag{12}$$

$$W_0^{(c)} = \frac{\kappa}{n + \kappa} + \left(1 - \alpha^2 + \beta\right) \tag{13}$$

$$W_i^{(m)} = W_i^{(c)} = \frac{\kappa}{2(n+\kappa)} i = 1, 2, \ldots, 2n \tag{14}$$

Among them, the value range of λ is $(10^{-4}, 1)$, the value of α is 0, and the value of β is 2. The specific transformation process is shown in Fig. 5.

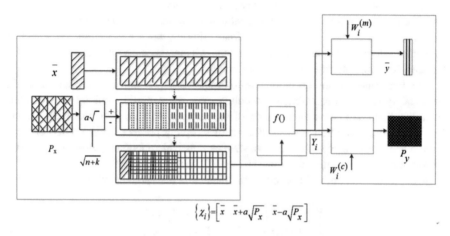

$$\{\chi_i\} = \begin{bmatrix} \bar{x} & \bar{x} + a\sqrt{P_x} & \bar{x} - a\sqrt{P_x} \end{bmatrix}$$

Fig. 5. UT transformation flowchart

5   Unscented Kalman Filtering Algorithm

The UKF algorithm mainly relies on the UT transform to convert the nonlinear problem into a linear problem. The accuracy of the UT transform largely determines the accuracy of the UKF algorithm. Different UT transform methods should be adopted for different sampling methods.

For lead-acid battery capacity estimation, the unscented Kalman filtering is suitable for variable-scale sampling strategies as shown below.

In the case of considering noise

$$x_{k+1} = f(x_k) + w_k \tag{15}$$

$$y_k = h(x_{k+1}) + v_k \tag{16}$$

Assuming a Gaussian random variable, the range of state noise w_k and measurement noise v_k is:

$$w_k \sim N(0, Q_k) \tag{17}$$

$$v_k \sim N(0, R_k) \tag{18}$$

Select the initial value of the vector.

$$\hat{x}_0 = E[X_0], P_0 = E\left[(x - \hat{x}_0)(x - \hat{x}_0)^T\right] \tag{19}$$

State equation based on Sigma point set.

$$\chi_{(k|k-1)} = f(\chi_{k-1}) \tag{20}$$

Weighted variance of Sigma point sets.

$$P_k^- = \sum_{i=0}^{2L} W_i^{(c)} \left(\chi_{(i,k|k-1)} - \hat{x}_{k-1}^-\right)\left(\chi_{(i,k|k-1)} - \hat{x}_{k-1}^-\right)^T + Q_k \tag{21}$$

Corresponding to the value of the prediction equation

$$y_{(k|k-1)} = h\left(\chi_{(k|k-1)}\right) \tag{22}$$

$$\hat{y}_k = \sum_{i=0}^{2L} W_i^{(m)} y_{(i,k|k-1)} \tag{23}$$

K-th update of covariance.

$$P_{x_k y_k} = \sum_{i=0}^{2L} W_i^{(c)} \left(\chi_{(i,k|k-1)} - \hat{x}_k^-\right)\left(\chi_{(i,k|k-1)} - \hat{x}_k^-\right)^T \tag{24}$$

The k-th update of the new covariance of the observation equation.

$$P_{y_k y_k} = \sum_{i=0}^{2L} W_i^{(c)} \left(y_{(i,k|k-1)} - \hat{y}_k^-\right)\left(\chi y_{(i,k|k-1)} - \hat{y}_k^-\right)^T + R_k \tag{25}$$

Kth Kalman Gain Prediction Update

$$K_k = P_{x_k y_k} P_{P_{y_k y_k}}^{-1} \tag{26}$$

Corresponding to the kth optimal prediction variance update.

$$P_k = P^{-1} - K_k P_{y_k y_k} K_k^T \tag{27}$$

6   Experimental Design and Results

The experimental object selected 12 V, capacity 65 Ah lead-acid battery, repeated charge and discharge experiments, SOC set to 100%, U_{oc} is 14 V, R_0 is 32.57 mΩ, R_p is 8.57 mΩ, C_P is 23.21F, load charging and discharging experiments by different methods and detecting SOC parameters.

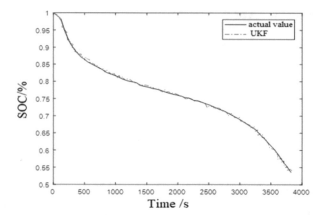

Fig. 6. Comparison of estimated battery residual capacity before and after UKF improvement

It can be seen from Fig. 6 that the curve composed of the point set is the UKF estimation curve, and the actual value of the battery SOC during the actual discharge process. From the results of the constant current discharge experiment, the estimation result of the UKF algorithm is effective. This fully validates the validity of the UKF algorithm.

7   Conclusion

In this paper, the Thevenin equivalent circuit model is established for the remaining capacity of lead-acid batteries, and the curve of the equivalent circuit model parameters and SOC function is fitted. Finally, the unscented Kalman filtering (UKF) algorithm is used to estimate the battery SOC. Experiments conducted under constant current conditions show that the proposed method is effective for estimating the SOC of the battery with less error.

Acknowledgments. This work is supported by the Science and Technology Program of Shenyang under Grant 18-013-0-18 and the National Nature Science Foundation of Liaoning Province under Grant 20180550922.

References

1. Charkhgard, M., Farrokhi, M.: State-of-charge estimation for lithium-ion batteries using neural networks and EKF. IEEE Trans. Industr. Electron. **57**(12), 4178–4187 (2011)
2. Partovibakhsh, M., Liu, G.J.: An adaptive unscented Kalman filtering approach for online estimation of model parameters and state-of-charge of lithium-ion batteries for autonomous mobile robots. IEEE Trans. Industr. Electron. **23**(1), 357–363 (2015)
3. Nejad, S., Gladwin, D.T., Stone, D.A.: A systematic review of lumped-parameter equivalent circuit models for real-time estimation of lithium-ion battery states. J. Power Sources **316**, 183–196 (2016)

4. Piller, S., Perrin, M., Jossen, A.: Methods for state-of-charge determination and their applications. J. Power Sources **96**(1), 113–120 (2001)
5. He, W., Williard, N., Chen, C., et al.: State of charge estimation for li-ion batteries using neural network modeling and unscented Kalman filter-based error cancellation. Int. J. Electr. Power Energy Syst. **62**(11), 783–791 (2014)
6. Vaishnava Dhaatri, N.: Cloud storage systems in service diversity. Int. J. Cloud Comput. Super Comput. **4**(1), 15–20 (2017). https://doi.org/10.21742/IJCS.2017.4.1.03
7. Eom, H.M., Jin, J.H., Lee, M.J.: A technique for sending email in association with the truffle ethereum development framework. Int. J Private Cloud Comput Environ. Manage. 5(1), 1–6 (2018)
8. Ying-nan, L., Shu-juan, Y.: Based on the pulse of the 51 single-chip microcomputer measuring instrument design. Int. J. Internet of Things Big Data **1**(1), 55–64 (2016). https://doi.org/10.21742/IJITBD.2016.1.1.07
9. ReddiPrasadu: Auditing for dynamic data with user revocation. Int. J. Adv. Res. Big Data Manage. Sys. 1(1), pp:25–30 (2017). https://doi.org/10.21742/IJARBMS.2017.1.1.03
10. Ragu, V., Kim, Y., Kangseok, C., Park, J., Cho, Y., Yang, S.Y., Shin, C.: A best fit model for forecasting Korea electric power energy consumption in IoT environments. Int. J. Internet of Things Appl. **2**(1), 7–12 (2018). https://doi.org/10.21742/IJIoTA.2018.2.1.02
11. Ming, C.: Implementing the K-mean using R tool for chosen the optimal K. J. Stat. Comput. Algorithm **2**(3), 13–20 (2018). https://doi.org/10.21742/JSCA.2018.2.3.02
12. Min, K.S.: Hybrid authentication for secure data. Int. J. Secur. Technol. Smart Device. **3**(2), 15–26 (2016). https://doi.org/10.21742/IJSTSD.2016.3.2.03
13. Praveen Kumar, K.: Crawler for efficiently harvesting web. Int. J. Commun. Technol. Soc. Networking Serv. 5(1), 7–14 (2017)
14. Kim, J.J., Lee, Y.S., Moon, J.Y., Park, J.M.: Clustering method based on genetic algorithm and WordNet. Int. J. Hum. Smart Device Interact. **4**(2), 1–6 (2017). https://doi.org/10.21742/IJHSDI.2017.4.2.01

Image Protection Against Forgery and Pixel Tampering Based on a Triple Hybrid Security Approach

Ahmad M. Nagm, Mohamed Torky$^{(\boxtimes)}$, Mohammed Mahmoud Abo Ghazala, and Hosam Eldin Fawzan Sayed

Scientific Research Group in Egypt (SRGE), Giza, Egypt
ahmadnagm@gmail.com, mtorky86@gmail.com
http://www.egyptscience.net/

Abstract. Due to the widespread of advanced digital imaging devices, forgery of digital images became more serious attack patterns. In this attack scenario, the attacker tries to manipulate the digital image to conceal some meaningful information of the genuine image for malicious purposes. This leads to increase security interest about protecting images against integrity tampers. This paper proposes a novel technique for protecting colored images against forgery and pixel tamper. The proposed approach is designed as a hybrid model from three security techniques, Message Digest hashing algorithm (MD5), Advanced Encryption Standard-128 bits (AES), and Stenography. The proposed approach has been evaluated using set of image quality metrics for testing the impact of embedding the protection code on image quality. The evaluation results proved that protecting image based on Least Significant Bit (LSB) is the best technique that keep image quality compared with other two bit-substitution methods. Moreover, the results proved the superiority of the proposed approach compared with other technique in the literature.

Keywords: Image forgery detection · Cyber crime · MD5 hashing · Cryptography · AES · Steganography

1 Introduction

Data forgery, the crime of falsely and fraudulently altering or tampering data content represent a classical problem since many decades. With the rapid technological advances in representing and manipulating data in different forms such as texts, images, videos, or sounds make the digital forgery is an important problem [1]. The ability to create digital photographs opened up the doors for producing forged images for malicious purposes. Due to the vast using of digital images across Internet, social media, TV channels, electronic newspapers and magazine, the attackers can execute variety of image forgery attack patterns for malicious purposes such as misusing the reputation of individuals, companies, and countries using fake profiles attacks across social media [2,3]. Digital check

© Springer Nature Switzerland AG 2020
A. E. Hassanien et al. (Eds.): AISI 2019, AISC 1058, pp. 588–597, 2020.
https://doi.org/10.1007/978-3-030-31129-2_54

forgery attack [4] is another attack scenario by which, the attacker uses some digital image processing techniques for forging the financial activities across online banking. A Copy-Move forgery attack [5] is a another attack scenario of image tampering where a piece of the image is copied and pasted on another part to hide unwanted portions of the image. The attacker's goal is covering the truth or to enhance the visual effect of the image for deceiving purposes. Hence, authenticating digital images and insuring its integrity against pixel tampering and forgery represents a major security problem [6,7].

The literature on detecting digital image forgery has highlighted several techniques and approaches. Copy-move and Splicing image forgery detection [8,9] is a common method for detecting images forgery based on dividing digital image into blocks and use block-matching algorithms for finding the similarities between blocks. The similar blocks will yield similar features. Hence, forgery detection decision is made only if similar features are detected within the same distance of features associated to connected blocks [10]. Discrete Cosine Transform (DCT) [11] is a good example of copy-move detection mechanism which has the ability to detect tampered regions of pixels accurately. Another example of copy-move forgery detection is the Scale-Invariant Feature Transform (SIFT) [12] which is a feature detection technique for detecting and describing local features in the digital images. Discreet Wavelet Transformation and un decimated dyadic wavelet transform (DyWT) [13] are another techniques that can be used to transform image pixels into wavelets, which are then used for wavelet-based compression and coding for detecting copy-move pixel tampering or alteration.

Watermarking [1,15,16,22,23] can be used as a popular means for efficient image tamper detection. The watermarking-based forgery detection is executed by marking small blocks of an image with watermarks that depend on a secret ID of that particular digital camera and later verify the presence of those watermarks for detecting image forgery.

In this paper, hybrid approach is proposed for detecting digital image forgery and pixels tampering. The proposed model is designed based on three security techniques, Message Digest hashing algorithm (MD5), Advanced Encryption Standard-128 bits (AES), and Steganography. The rest of this paper can be organized as follows: Sect. 2 presents and discusses the proposed model for detecting image forgery. Section 3 presents the experimental results. Section 3 discusses and compare obtained results. Finally, Sect. 4 is devoted to the conclusion of this study.

2   Triple Hybrid Security Approach

A number of techniques have been developed in the literature for addressing the problem of protecting images against pixel tampering and forgery [17]. One of the most well-known techniques is watermarking-based algorithms [18,19]. However, watermarking- based techniques are vulnerable against some attack models such as copy attacks, Ambiguity attacks, collision attacks, and scrambling attacks, etc [20]. Other robust forgery detection methods is based on Speed up Robust

Features (SRF) [21,24] which partly inspired by the Scale-Invariant Feature Transform (SIFT) [25]. However these methods still suffer from some advanced geometric attacks such as one pixel attack [26].

In this study, we try to introduce a novel security approach for securing images against advanced forgery attacks such as one pixel attack. The proposed approach integrated three security techniques, MD5 hashing, AES-128 bit, and steganography to produce a triple hybrid security model for protecting images against forgery and pixel tampering. For achieving this objective, the proposed approach is designed based on three security techniques, MD5 hashing [27], AES-128 bit [28] and steganography [29].

1. MD5 hashing based on SHA-160 is used as a secure hash function for protecting image integrity against altering and modification by malicious adversary [29]. It used here for converting the camera ID code into a fixed length-160 bit hash code called Secret Originality Identifier (SOI).
2. AES-128 is a symmetric block cipher based on 128- bits. AES is utilized in this work for encrypting the input colored image (only green and blue Matrices after XOR operation) using the first 16 bits of SOI as a key and produces a cipher matrix. 4*4 pixels.
3. Steganography is a hiding technique used for concealing a file, message, image, or video within another file, message, image, or video. In this work the steganography is used as follows: the cipher matrix is substituted with the red matrix for producing a modified matrix. Then, the modified matrix is demosaicated with the blue and green matrices for producing the protected RGB image. Three substitution techniques are performed, LSB (Least Significant Bit), MSB (Most Significant Bit), and Fourth Bit (#4 bit).

The methodology of the proposed approach is explained in Fig. 1. The technique can be conceived through major five stages:

1. The MD5-SHA 160 hash function is used for converting the camera ID code (16 bits) into a fixed length hash code called Secret Originality Identifier (SOI) (16 bits). The main advantage of MD5 is that it is a secure cryptographic hash technique for protecting data integrity and able to detect unintentional data tampering efficiently.
2. The captured image is filtered using the common Color Filter Array (CFA) for specifying input image into RGB architecture in the form Red, Green, and Blue matrices.
3. The Green Matrix (GM) is XORed with the Blue matrix (BM) into a new matrix which then encrypted using AES algorithm and SOI as the encryption key (16-bits) for producing a Cipher Matrix (CM) for the input image.
4. The Cipher Matrix (CM) then substituted with the Red matrix (RM) pixel by pixel in three forms of substitutions for each byte for producing a Modified Matrix (MM). Least Significant Bit (LSB) substitution, Most Significant Bit substitution, and 4-bit substitution are the three substitution techniques applied in this work.

5. The Modified Matrix (MM) then demosaicated with the Blue matrix (BM) and Green Matrix (GM) for producing the protected Watermarked RGB image again.

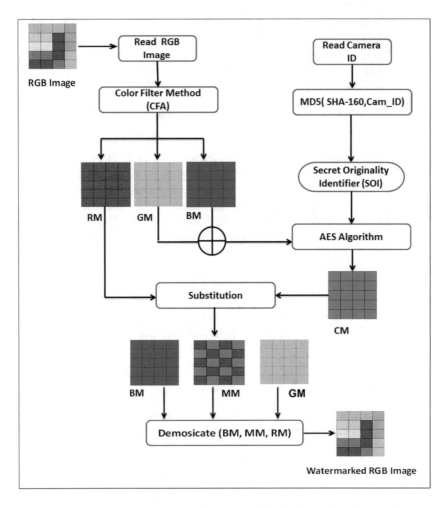

Fig. 1. Image protection based on a triple hybrid security Approach

3   Experimental Results

For evaluating the proposed image protection technique, we simulated it using MATLAB 8.5 on Five colored images from the dataset MCC-F220 [31]. Table 1 depicts the image samples and their description. Three experiments-based on

Image	Image Name	Size in Pixels	Types	Size in KB
	Mug	800*532	JPEG	31KB
	Lane	800*532	JPEG	33.7KB
	Sea	800*532	JPEG	57KB
	Library	737*492	JPEG	93KB
	Garden	800*532	JPEG	123KB

Fig. 2. Selected images for simulation experiment

three bit substitution methods (least Significant Bit (LSB), 4-Significant Bit (4-SB), and Most Significant Bit (MSB)) are performed on the five images for getting the best RGB watermarked. We used five metrics for testing the efficiency of the proposed technique: Mean Absolute Error (MAE), Mean Square Error (MSE), Peak Signal to Noise Ratio (PSNR), Structural Similarity Index (SSIM), and Universal Image Quality Index (UIQI) [32] as depicted in Eqs. 1, 2, 3, 4, 5 respectively.

$$MAE = (\frac{1}{N}) \times (\sum_{i=1}^{N} |x_i - v_i|) \tag{1}$$

where N is the number of column vectors, and v_i are variables of paired observations, where Examples of x_i versus v_i include comparisons of predicted versus observed.

$$MSE = (\frac{1}{M * N}) \times (\sum_{i=1}^{m} \sum_{j=1}^{n} [x(i,j) - v(i,j)]^2) \tag{2}$$

Where M is the number of Row vectors, N is the number of column vectors, $x(i,j)$ and $v(i,j)$ are variables of paired observations at pixel (i,j).

$$PSNR = 10 \times \log_{10}(\frac{R^2}{MSE}) \tag{3}$$

where R is the maximum possible pixel value of the image which is 255, and MSE is the Mean Square Error.

$$SSIM(x,v) = (2 \times \overline{x} \times \overline{v} + c1) \times \frac{2 \times \delta_{xv} + c2}{(\overline{x}^2 + \overline{v}^2 + c1) \times (\delta_x^2 + \delta_v^2 + c2)} \tag{4}$$

where, $\overline{x}$ is the average of x. $\overline{v}$ is the average of v, δ_x^2 is the variance of x, δ_v^2 is the variance of v, δ_{xv} is the covariance of x and v, c1 and c2 are two variables to stabilize the division with weak denominator.

$$UIQ = (\frac{\delta_{xv}}{\delta_x \times \delta_v}) \times (\frac{2 \times \overline{x} \times \overline{v}}{\overline{x}^2 + \overline{v}^2}) \times (\frac{2 \times \delta_x \times \delta_v}{\delta_x^2 + \delta_v^2}) \tag{5}$$

Table 1. Image quality measurements based on Least Significant Bit (LSB)

Images/Size/JPEG	MAE	MSE	PSNR	SSIM	UIQI
Mug/800*532, 31 KB	0.0842	0.165982587	55.93017831	0.998701529	0.999811542
Lane/800*532, 33.7 KB	0.0832	0.166969838	55.90442334	0.998094754	0.999707432
Sea/800*532, 57 KB	0.0856	0.166371269	55.92002033	0.999529746	0.999900692
Library/737*492, 93 KB	0.0835	0.16685455	55.90742308	0.999590449	0.99995085
Gardun/800*532, 123 KB	0.0824	0.166457556	55.91776847	0.999512826	0.999937976

Table 2. Image quality measurements based on # 4-Significant Bit (4-SB)

Images/Size/JPEG	MAE	MSE	PSNR	SSIM	UIQI
Mug/800*532, 31 KB	0.1664	0.666554726	49.89244549	0.994976996	0.999243778
Lane/800*532, 33.7 KB	0.1671	0.663404851	49.91301718	0.992629609	0.998838349
Sea/800*532, 57 KB	0.1648	0.66324005	49.91409618	0.998210627	0.999603839
Library/737*492, 93 KB	0.1653	0.664903457	49.9032177	0.998542376	0.999803962
Gardun/800*532, 123 KB	0.1627	0.663383085	49.91315967	0.998118126	0.999752758

Tables 1, 2 and 3 provide the simulation results based on three bit substitution methods: least Significant Bit (LSB), 4-Significant Bit (4-SB), and Most Significant Bit (MSB). The results proves that least Significant Bit (LSB) is the best bit-substitution method that can be used with the proposed approach for protecting colored images against forgery and pixel tampering.

On the issue of measuring the quality of reconstruction of lossy compression codecs, the PSNR evaluation results shows an interesting findings where the proposed approach achieved notable superiority compared with Self-Generated Verification Code (SGVC) [31] and Multiple Watermarks (MW) [33] as depicted in Table 4 and Fig. 3. Another interesting result is on the cumulative squared error between the watermarked and the original image, the Mean Square Error (MSE) result proved also the superiority of the proposed approach compared to Self-Generated Verification Code (SGVC) [30] and Multiple Watermarks (MW) [33] as depicted in Table 4 and Fig. 4. Moreover, regarding perception-based model that considers image degradation as perceived change in structural information such as luminance masking and contrast masking, the proposed approach

Table 3. Image quality measurements based on Most Significant Bit (MSB)

Images/Size/JPEG	MAE	MSE	PSNR	SSIM	UIQI
Mug/800*532, 31 KB	0.3435	2.681579602	43.84689667	0.981208471	0.996961411
Lane/800*532, 33.7 KB	0.338	2.682263682	43.84578892	0.972259217	0.995314413
Sea/800*532, 57 KB	0.3356	2.668532338	43.86807891	0.992819986	0.99840746
Library/737*492, 93 KB	0.3385	2.684631171	43.84195732	0.993361917	0.99920829
Gardun/800*532, 123 KB	0.3266	2.666044776	43.87212922	0.992400712	0.999005641

Table 4. Comparison results based on MAE, MSE, PSNR, SSIM, and UIQI metrics

Image Quality Metric (IQM)	SGVC [30]	MW [32]	Proposed approach
AVG-PSNR	52.16478	42.34178	55.91465308
AVG-MSE	0.57906	0.97638	0.166577359
AVG-SSIM	0.95148	0.92584	0.999078513
AVG-MAE	Not Evaluated	Not Evaluated	0.083816667
AVG-UIQI	Not Evaluated	Not Evaluated	0.999873788

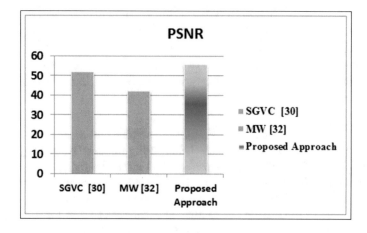

Fig. 3. PSNR comparison results

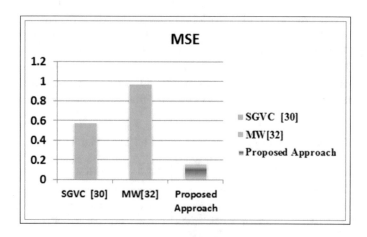

Fig. 4. MSE comparison results

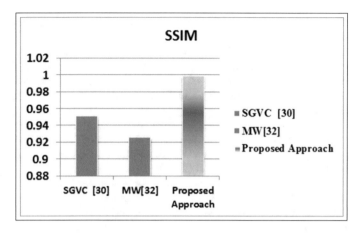

Fig. 5. SSIM comparison results

achieved superior results compared to elf-Generated Verification Code (SGVC) [31] and Multiple Watermarks (MW) [33] as depicted in Table 4 and Fig. 5.

4   Conclusions

This study introduced a novel hybrid security approach for protecting colored images against forgery and pixel tampering. The proposed approach is based on Message Digest hashing algorithm (MD5), Advanced Encryption Standard-128 bits (AES), and Stenography. The study has identified the impact of embedding the protection code on image quality measures. The main advantages of the proposed approach are, (1) adding security code to image has no significant impact on image quality. (2) The approach is testable for detecting image tamper even if the manipulated block size is one pixel. (3) the protection code is dynamic with respect to the camera's ID and the RGB structure of the handled image. The experimental results proved the efficiency of the proposed approach as a novel protection technique for colored images compared with other technique in the literature. This study lays the groundwork for future research into developing more sensitive image forgery detector system and evaluating its performance on different types of colored images and videos.

References

1. Walia, S., Kumar, K.: Digital image forgery detection: a systematic scrutiny. Aust. J. Forensic Sci. **6**, 1–39 (2018)
2. Meligy, A.M., Ibrahim, H.M., Torky, M.F.: A framework for detecting cloning attacks in OSN based on a novel social graph topology. Int. J. Intell. Syst. Appl. **7**(3), 13 (2015)

3. Torky, M., Meligy, A., Ibrahim, H.: Recognizing fake identities in online social networks based on a finite automaton approach. In: 2016 12th International Computer Engineering Conference (ICENCO), 28 December 2016, IEEE, pp. 1–7 (2016)
4. Gjomemo, R., Malik, H., Sumb, N., Venkatakrishnan, V.N., Ansari, R.: Digital check forgery attacks on client check truncation systems. In: International Conference on Financial Cryptography and Data Security, vol. 3, pp. 3–20. Springer, Heidelberg (2014)
5. Yang, B., Sun, X., Guo, H., Xia, Z., Chen, X.: A copy-move forgery detection method based on CMFD-SIFT. Multimedia Tools Appl. **77**(1), 837–855 (2018)
6. Ansari, M.D., Ghrera, S.P., Tyagi, V.: Pixel-based image forgery detection: a review. IETE J. Educ. **55**(1), 40–46 (2014)
7. Birajdar, G.K., Mankar, V.H.: Digital image forgery detection using passive techniques: a survey. Digital Invest. **10**(3), 226–245 (2013)
8. Asghar, K., Habib, Z., Hussain, M.: Copy-move and splicing image forgery detection and localization techniques: a review. Aust. J. Forensic Sci. **49**(3), 281–307 (2017)
9. Qureshi, M.A., Deriche, M.: A review on copy move image forgery detection techniques. In: 2014 IEEE 11th International Multi-Conference on Systems, Signals & Devices (SSD14), 11 February 2014, IEEE, pp. 1–5 (2014)
10. Bayram, S., Sencar, H.T., Memon, N.: A survey of copy-move forgery detection techniques. In: IEEE Western New York Image Processing Workshop, September 2008, IEEE, pp. 538–542 (2008)
11. Alkawaz, M.H., Sulong, G., Saba, T., Rehman, A.: Detection of copy-move image forgery based on discrete cosine transform. Neural Comput. Appl. **30**(1), 183–192 (2018)
12. Abdel-Basset, M., Manogaran, G., Fakhry, A.E., El-Henawy, I.: 2-levels of clustering strategy to detect and locate copy-move forgery in digital images. Multimedia Tools Appl. **1–9**, (2018)
13. Muhammad, G., Hussain, M., Bebis, G.: Passive copy move image forgery detection using undecimated dyadic wavelet transform. Digital Invest. **9**(1), 49–57 (2012)
14. Hu, W.C., Chen, W.H., Huang, D.Y., Yang, C.Y.: Effective image forgery detection of tampered foreground or background image based on image watermarking and alpha mattes. Multimedia Tools Appl. **75**(6), 3495–3516 (2016)
15. Benrhouma, O., Hermassi, H., El-Latif, A.A., Belghith, S.: Chaotic watermark for blind forgery detection in images. Multimedia Tools Appl. **75**(14), 8695–8718 (2016)
16. Qin, C., Ji, P., Zhang, X., Dong, J., Wang, J.: Fragile image watermarking with pixel-wise recovery based on overlapping embedding strategy. Signal Process. **1**(138), 280–293 (2017)
17. Kashyap, A., Parmar, R.S., Agrawal, M., Gupta, H.: An evaluation of digital image forgery detection approaches. arXiv preprint arXiv:1703.09968, 29 March 2017
18. Singh, D., Singh, S.K.: DCT based efficient fragile watermarking scheme for image authentication and restoration. Multimedia Tools Appl. **76**(1), 953–977 (2017)
19. Qin, C., Ji, P., Wang, J., Chang, C.C.: Fragile image watermarking scheme based on VQ index sharing and self-embedding. Multimedia Tools Appl. **76**(2), 2267–2287 (2017)
20. Tanha, M., Torshizi, S.D., Abdullah, M.T., Hashim, F.: An overview of attacks against digital watermarking and their respective countermeasures. In: 2012 International Conference on Cyber Security, Cyber Warfare and Digital Forensic (CyberSec), 26 June 2012, IEEE, pp. 265–270

21. Bo, X., Junwen, W., Guangjie, L., Yuewei, D.: Image copy-move forgery detection based on SURF. In: 2010 IEEE International Conference on Multimedia Information Networking and Security, 4 November 2010, pp. 889–892 (2010)
22. Dadkhah, S., Manaf, A.A., Hori, Y., Hassanien, A.E., Sadeghi, S.: An effective SVD-based image tampering detection and self-recovery using active watermarking Signal Processing. Image Commun. **29**(10), 1197–1210 (2014)
23. Hassanien, A.E.: A copyright protection using watermarking algorithm. Informatica **17**(2), 187–198 (2006)
24. Shivakumar, B.L., Baboo, S.S.: Detection of region duplication forgery in digital images using SURF. Int. J. Comput. Sci. Issues (IJCSI) **8**(4), 199 (2011)
25. Warif, N.B., Wahab, A.W., Idris, M.Y., Salleh, R., Othman, F.: SIFT-symmetry: a robust detection method for copy-move forgery with reflection attack. J. Vis. Commun. Image Represent. **1**(46), 219–232 (2017)
26. Su, J., Vargas, D.V., Sakurai, K.: One pixel attack for fooling deep neural networks. IEEE Trans. Evol, Comput (2019)
27. Gupta, S., Goyal, N., Aggarwal, K.: A review of comparative study of MD5 and SSH security algorithm. Int. J. Comput. Appl. **104**(14) (2014)
28. Devi, A., Sharma, A., Rangra, A.: A review on DES, AES and blowfish for image encryption & decryption. Int. J. Comput. Sci. Inf. Technol. **6**(3), 3034–3036 (2015)
29. Tiwari, A., Yadav, S.R., Mittal, N.K.: A review on different image steganography techniques. Int. J. Eng. Innov. Technol. (IJEIT) **3**(7), 121–124 (2014)
30. Wen, C.Y., Yang, K.T.: Image authentication for digital image evidence. Forensic Sci. J. **5**(1), 1 (2006)
31. Mani, P.R., Bhaskari, D.L.: Image tamper detection and localization based on self-generated verification code during image acquisition. Int. J. Appl. Eng. Res. **13**(5), 2110–2118 (2018)
32. Jagalingam, P., Hegde, A.V.: A review of quality metrics for fused image. Aquatic Procedia **1**(4), 133–142 (2015)
33. Pongsomboon, P., Kondo, T., Kamakura, Y.: An image tamper detection and recovery method using multiple watermarks. In: 13th International Conference on Electrical Engineering/Electronics, Computer, Telecommunications and Information Technology (ECTI-CON), Chiang Mai, 2016, pp. 1–6 (2016)

Heart Rate Measurement Using Remote Photoplethysmograph Based on Skin Segmentation

M. Somaya Abdel-Khier[1]([⊠]), Osama A. Omer[1,2]([⊠]),
and Hamada Esmaile[1]([⊠])

[1] Faculty of Engineering, Aswan University, Aswan 85142, Egypt
somaya.abdelkhier@eng.aswu.edu.eg,
{omer.osama,h.esmaiel}@aswu.edu.eg
[2] Technology and Maritime Transport,
Arab Academy for Science, Aswan, Egypt

Abstract. The remote Photoplethysmograph (rPPG) is recently used in evaluating heart rate (HR) from registered skin color differences through a camera. It is based on extracting HR from the small periodic color difference in the skin due to heartbeats. Selection of Region of Interest (ROI) is a critical process and its necessity includes as many skin pixels as likely. In this paper, we focus on improving the quality of pulsatile signals by using skin segmentation to select the ROI. So, an automatically enhanced marker built in the watershed algorithm is offered for segmentation of human skin regions from the detected face. The suggested rPPG measurements are evaluated by using root mean square error (RMSE), and complexity time. Based on the simulation result, the proposed algorithm showed that using skin segmentation can significantly improve the performance of the rPPG process.

Keywords: Heart rate (HR) · Remote photoplethysmograph (rPPG) ·
Region of Interest (ROI) · Skin segmentation

1 Introduction

The heart rate estimated by the number of beats per minute (bpm) and shows the body's physiological activity. Conventionally, observing of HR is done by utilizing sensors, electrodes, and chest straps which may cause discomposure if utilized for long times. One of the most famous and accurate methods of measuring heart rate is Electrocardiography (ECG), but it needs attaching medical electrodes to the subject the authentic criterion is the utilization ECG [1] which puts viscous gel electrodes on the participants' limbs or chest surface. This method is robust and cost-efficient; however, the drawback of it is that requires direct contact with the subject.

Recently, there have been numerous studies investigating contactless HR measuring methods. [2, 3] in these works give an overview of some non-contact methods for monitoring of physiological activity. These methods include Thermal Imaging [4], Laser Doppler [5], and imaging photoplethysmography (iPPG) [6, 7] techniques.

© Springer Nature Switzerland AG 2020
A. E. Hassanien et al. (Eds.): AISI 2019, AISC 1058, pp. 598–606, 2020.
https://doi.org/10.1007/978-3-030-31129-2_55

IPPG is similarly mentioned to remote photoplethysmography (rPPG); the rPPG measurement is built a standard such as the conventional PPG, where Photoplethysmography (PPG) is an optical technology which discovers blood size variations in the microvascular tissue bed beneath the skin [8]. There are two main techniques used for measuring heart rate using rPPG: (1) heart rate measurement dependent on the periodic variation of the subject's skin color (2) heart rate measurement dependent on periodic head motion. In that article, we are centering on the technique of periodic variation of the subject's skin color.

Different rPPG methods [9, 10] share a typical structure: firstly regions of interest (ROI) are identified and tracked through frames; RGB channels are combined then to evaluate the pulse signal. Finally to extract the physiological information the pulse signal is filtered and analyzed, for example, heart rate.

Selecting of suitable ROI for the rPPG-based HR measuring is the essential and challenging first step. In previous studies, many rPPG methods were suggested to select ROI in the face, where the ROI was selected manually in [9]. In every video frame Pixels in the ROI are regularly spatially averaged and the effect of that process is a time sequence that is used to get rPPG signal. It has been presented in [11]. The segmentation of the ROI has an effect on the value of the rPPG signal, this is because the small number of pixels causes errors, we can see that the value of rPPG signal weakens while down the ROI. Also, the value is affected by the level of non-skin pixels in the ROI. If the ROI is not correctly selected, rPPG suffer poor at performance.

This paper presents heart rate measurement approach focuses on refining the quality of the extracted cardiovascular wave by using robust skin segmentation for defining the ROI. In this article, using watershed skin segmentation algorithm for ROI definition is investigated when it is compared with steady face detection/tracking and skin detection approach. Finally, we compare our rPPG measurements using the publicly available dataset called UBFC-RPPG [12], which is dedicated to rPPG, and it is accessible on request.

The rest of this paper is organized as follows. Section 2 discuss the watershed algorithm and reviews the related work and discusses their limitation. Section 3 illustrates the proposed approach. Section 4 discusses the experimental results. The conclusion is presented in Sect. 5.

2   Materials and Related Work

2.1   Watershed Algorithm

The Skin segmentation of the ROI has an effect on the value of the rPPG; watershed algorithm is presented to formulate the skin segmentation problem. In mathematical morphological the watershed algorithm is the best way to segment the image. Watershed segmentation mainly related to the gradient magnitude, in this technique by using a Prewitt or Sobel or any other suitable filter can compute gradient magnitude. Those pixels whose have the highest gradient magnitude strengths relate to watershed lines which represent the region boundaries on that image. We have selected the contour of the face using watershed algorithm [21] to divide the human skin area on the detected face. Where the foreground objects (skin regions) and background location (non-skin regions) can be identified for segmentation.

2.2   Related Works

The existing (rPPG) systems for finding HR from face videos can be categorized to: (1) the skin color variation based-systems (2) the head motion-based systems [13–15]. Our Survey is based on the skin color variation; these systems originate from PPG signals taken by digital cameras.

Verkruysse et al. in [17], explained that from videos with a human face can measure HR utilizing PPG. In this method manually, the forehead was selected as an area of the ROI, then through the average pixels in ROI the signal can be calculated, finally to analyze the computed signals in frequency domain FFT was used. The authors have found that the strongest plethysmographic signal can be found in the green channel. The main drawback of these methods is that the data is processed manually. The influential study on rPPG [9], has motivated a growing number of publications and progressions in this area, which was indicated by Rouast et al. [16] in their technical literature review. Recent advances made it possible for rPPG-based methods to be used in realistic conditions. Analysis and comparison of systems is built on their performance.

There are two main challenges in the RFPG as following: how to reduce the dimensionality and how to select a suitable ROI. Different dimensionality reduction systems are utilized to decrease the dimensionality from signals so as to more visibly detect HR data. The first approach suggested. In [17] was used an algorithm for blind source separation (BSS) to define the best combination of raw signals and in [18] there was a comparative study for linear and nonlinear techniques for BSS.

Facial ROI choice is utilized to get blood features and get the HR signal. The bounding box obtained from Viola-Jones system was used as the ROI in [9, 19] a common drawback of these methods bounding box contains background noise and non-pixel skin. Another research used a subset of bounding box, a public technique is to include 60% of its width [10, 17, 18].

In [20] is examined the result of segmenting the face skin into three ROIs (forehead, left cheek and right cheek) and the drawback on it if the chosen ROI is small. The raw signals lose some information that may be present in other skin areas of the face. If the ROI is not adjusted correctly, the rPPG signal does not spread regularly on the face, this results in a degradation of the rPPG system performance.

From these observations, to estimate the execution of the ROI skin segmentation system of heart rate measurements. We proposed a new approach to formulate the skin segmentation issue. It based on watershed algorithm [21] to divide the human skin area on the detected face. The main theme can be extracted by the proposed method and related works. The ROI definition is essential since it contains the raw HR signals.

3   The Proposed Heart Rate Measurement Approach

The basic frame of rPPG is presented in [16]. This framework consists of determining the return on investment through detection and tracking, capturing initial signals, processing and filtering primary signals, combining RGB channels to find the heart wave, finally extracting a vital interest signal. The main differences between different rPPG methods are how to determine ROI.

Selecting a suitable ROI for the rPPG-based HR measuring is important and challenging first stage and it has a direct effect on the accuracy and reliability of the overall algorithm. Our approach focuses on refining the quality of the extracted cardiovascular wave by using skin segmentation for defining the ROI, which enables us to estimate the HR from skin pixels in all the face.

The proposed approach is divided into three main phases. They are raw signals extraction, signal estimation, and HR estimation. Figure 1 shows the main phases of the proposed heart rate measurement approach. These phases are discussed in details in the next subsections.

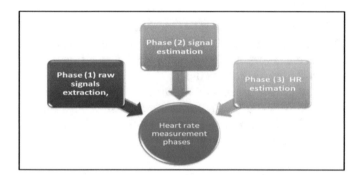

Fig. 1. The heart rate measurement phases

3.1   Raw Signals Extraction

The goal of this part is to found the RGB signals which are used to measure the rPPG signal in a next step. This part has an important impact on the quality of the rPPG method; extracting a pure signal with distinguishable pulsating rPPG wave, directly affect the system accuracy.

Face Detection and Tracking. In this step, we use Viola-Jones method [22] in our system to detect subject's face, This step delivers a bounding box coordinates defining the subject face. Applying face detection at every frame consumes a lot of computational power resources. Also, it causes undesired noises because the output bounding box of the face is not the same in the successive frames.so we track the face using KLT tracker [23]. In which only specific features of the face are tracked over the time.

ROI Definition. We present many implemented ROI segmentation methods and proposed a new method to improve the quality of pulsatile signals by using skin segmentation to select the ROI. An automatic efficient modified marker-based watershed method is offered for skin segmentation on the detected face.

Constitute Temporal RGB Signals. The raw RGB signals are composed by calculating the average pixel value of the skin-pixels within the ROI region over time.

3.2 Signal Estimation Phase

After extracting the raw signals, a number of signal processing techniques are employed including detrending, normalization, smoothing, and filtering.

Detrending: Due to the noises caused by changing of the environmental parameters, in this step, we remove the linear trends from the raw signal.

Normalization: Because the level of raw signals has no meaning when evaluating periodicity Therefore, the raw signal is normalized by dividing the raw signal by its maximum absolute value.

Smoothing: The normalized RGB signals are smoothed using a sliding average filter. This step leaves out noise, and it clarifies the signal.

Filtering: In this step, the raw signal is applied to a band-pass-filter with ideal behavior to eliminate high and low frequency noise. A public select of band is [0.7 Hz, 4 Hz], this band is commonly used in previous studies, which corresponds to an HR between 42 and 240 beats per minute (bpm).

Because the RGB signals contains information about the HR in mixed components, Therefore, In order to analyze signal, dimensional reduction methods are used to decrease the dimension. In the proposed system, FastICA method is used to analyze the RGB signals to reveal the original source signals.

3.3 HR Estimation Phase

The estimation of the heart rate is done using the following three steps.

(1) **Fast Fourier Transform (FFT)** - In this step, the spectrum of the chosen ICA component is calculated by using FFT algorithm.
(2) **Peaks Detection** - The peaks in the ICA components power are determined, and the index frequency of the highest peak corresponds to the HR frequency.
(3) **Estimate Heart Rate** - The heart rate is estimated by multiplying the estimated heart rate frequency by 60, this return a value between 42 and 240 bpm.

4 Experimental Results and Discussions

4.1 Dataset

In order to verify the validity of our rPPG method we use a new dataset composed of 45 videos, these videos show object is connected with A CMS50E pulse oximeter to get the ground truth PPG and the objects were asked to sit still, The distance between the object and the camera while recording the video from data set is 1–2 m. Each video is almost 2 min long and recorded with a low cost webcam (Logitech C920 HD pro) at 30 frames per second with a resolution of 640 × 480 in uncompressed 8-bits RGB format. All subjects are recorded using ambient light.

4.2 The Evaluation Metrics

Using MATLAB, the calculation stages from the segmentation to the metrics were achieved. The next metrics are applied for all the systems presented and utilized for comparison, Root mean square error (**RMSE**), average calculation time for one frame on the whole dataset and computational complexity.

4.3 Evaluation Methods

We compare the proposed ROI skin segmentation with three systems, cropped, face and many ROI with skin. In fact, the bounding box obtained from Viola-Jones algorithm was used as the ROI [10]. In cropped [19] the center 60% width and full height of the box is selected as the ROI. Lastly, in many ROI with skin [20] used only the face skin into three ROIs (forehead, left cheek and right cheek). We have chosen the contour of the face skin utilizing watershed skin segmentation. Figure 2 shows examples of segmentation result using 2 subjects from data set against state of the art for all the methods.

Method	face [10]	cropped [19]	many ROI with skin [20]	The Proposed
Subject 1				
Subject 2			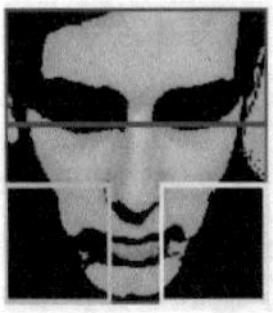	

Fig. 2. The segmentation results

4.4 Results

The overall experiments result for 2 subjects are summarized in Table 1. We can note the choice of the ROI has actually a large effect on the heart rate measurement precision. One example for this idea is that RMSE obtained with Cropped method is 18.626 while our proposed method has a RMSE of 4.463, The results demonstrate the finding that skin segmentation is an effective way to improve the accuracy of the rPPG technique by filtering out non-skin pixels and extracting the rPPG signal from all face with skin provide a better accuracy than extracting it from many ROI with skin.

Table 1. The overall experiments result.

Data set video	Method	RMSE (bpm)	Comp. time (s/frame)
1	Cropped [19]	18.626	0.152
	Face [10]	10.48	0.148
	Many ROI with skin [20]	5.812	0.135
	The proposed method	4.463	0.171
2	Cropped [19]	14.197	0.132
	Face [10]	13.612	0.183
	Many ROI with skin [20]	4.6709	0.097
	The proposed method	3.733	0.224

Computational complexity is very important metric, where complexity measured by number of operation per frame for all methods. The results are presented in Fig. 3. It is obvious the cropped method has the highest computational complexity because the ROI is selected based on many operation to determine the center 60% width and full height of the detected face per iteration. But our proposed skin segmentation is used once to determine ROI and this reduces the number of operation which affects complexity and reduces it. These results show that using skin segmentation based on watershed algorithm can significantly improve the performance and quality of the rPPG process.

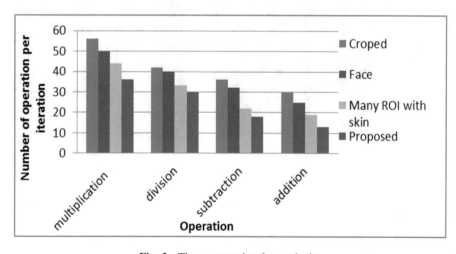

Fig. 3. The computational complexity

5   Conclusion

In this paper, a skin segmentation using watershed algorithm for ROI definition in rPPG was examined, then it was compared with many previous real-time rPPG-based methods. From the result of our investigation it is possible to conclude that skin segmentation significantly improve the quality of rPPG signals. In our future work, we plan to improve and evaluate our rPPG algorithm in more challenging conditions like intense movements, varying illumination, and varying skin color tones.

References

1. Da He, D., Winokur, E.S., Sodini, C.G.: A continuous, wearable, and wireless heart monitor using head ballistocardiogram (BCG) and head electrocardiogram (ECG). In: 2011 Annual International Conference of the IEEE Engineering in Medicine and Biology Society, pp. 4729–4732 (2011)
2. Teichmann, D., Brüser, C., Eilebrecht, B., Abbas, A., Blanik, N., Leonhardt, S.: Non-contact monitoring techniques-principles and applications. In: 2012 Annual International Conference of the IEEE Engineering in Medicine and Biology Society, pp. 1302–1305 (2012)
3. Al-Naji, A., Gibson, K., Lee, S.-H., Chahl, J.: Monitoring of cardiorespiratory signal: principles of remote measurements and review of methods. IEEE Access 5, 15776–15790 (2017)
4. Garbey, M., Sun, N., Merla, A., Pavlidis, I.: Contact-free measurement of cardiac pulse based on the analysis of thermal imagery. IEEE Trans. Biomed. Eng. 54, 1418–1426 (2007)
5. Ulyanov, S.S., Tuchin, V.V.: Pulse-wave monitoring by means of focused laser beams scattered by skin surface and membranes. In: Static and Dynamic Light Scattering in Medicine and Biology, pp. 160–168 (1993)
6. Kranjec, J., Beguš, S., Drnovšek, J., Geršak, G.: Novel methods for noncontact heart rate measurement: a feasibility study. IEEE Trans. Instrum. Meas. 63, 838–847 (2014)
7. Fallet, S., Schoenenberger, Y., Martin, L., Braun, F., Moser, V., Vesin, J.-M.: Imaging photoplethysmography: a real-time signal quality index. Comput. Cardiol. (CinC) 2017, 1–4 (2017)
8. Hertzman, A.B.: Observations on the finger volume pulse recorded photoelectrically. Am. J. Physiol. 119, 334–335 (1937)
9. Verkruysse, W., Svaasand, L.O., Nelson, J.S.: Remote plethysmographic imaging using ambient light. Opt. Express 16, 21434–21445 (2008)
10. Poh, M.-Z., McDuff, D.J., Picard, R.W.: Non-contact, automated cardiac pulse measurements using video imaging and blind source separation. Opt. Express 18, 10762–10774 (2010)
11. Bousefsaf, F., Maaoui, C., Pruski, A.: Continuous wavelet filtering on webcam photoplethysmographic signals to remotely assess the instantaneous heart rate. Biomed. Signal Process. Control 8, 568–574 (2013)
12. Bobbia, S., Macwan, R., Benezeth, Y., Mansouri, A., Dubois, J.: Unsupervised skin tissue segmentation for remote photoplethysmography. Pattern Recogn. Lett. (2017)
13. Balakrishnan, G., Durand, F., Guttag, J.: Detecting pulse from head motions in video. In: Proceedings of the IEEE Conference on Computer Vision and Pattern Recognition, pp. 3430–3437 (2013)

14. Shan, L., Yu, M.: Video-based heart rate measurement using head motion tracking and ICA. In: 2013 6th International Congress on Image and Signal Processing (CISP), pp. 160–164 (2013)

15. Irani, R., Nasrollahi, K., Moeslund, T.B.: Improved pulse detection from head motions using DCT. In: 2014 International Conference on Computer Vision Theory and Applications (VISAPP), pp. 118.-124 (2014)

16. Rouast, P.V., Adam, M.T., Chiong, R., Cornforth, D., Lux, E.: Remote heart rate measurement using low-cost RGB face video: a technical literature review. Front. Comput. Sci. **12**, 858–872 (2018)

17. Poh, M.-Z., McDuff, D.J., Picard, R.W.: Advancements in noncontact, multiparameter physiological measurements using a webcam. IEEE Trans. Biomed. Eng. **58**, 7–11 (2011)

18. Wei, L., Tian, Y., Wang, Y., Ebrahimi, T., Huang, T.: Automatic webcam-based human heart rate measurements using Laplacian eigenmap. In: Asian Conference on Computer Vision, pp. 281–292 (2012)

19. Hsu, Y., Lin, Y.-L., Hsu, W.: Learning-based heart rate detection from remote photoplethysmography features. In: 2014 IEEE International Conference on Acoustics, Speech and Signal Processing (ICASSP), pp. 4433–4437 (2014)

20. Fouad, R., Omer, O., Ali, A.-M.M., Aly, M.: Refining ROI selection for real-time remote photoplethysmography using adaptive skin detection (2019)

21. Das, A., Ghoshal, D.: Human skin region segmentation based on chrominance component using modified watershed algorithm. Procedia Comput. Sci. **89**, 856–863 (2016)

22. Viola, P., Jones, M.: Rapid object detection using a boosted cascade of simple features. In: null, p. 511 (2001)

23. Becker, B.C., Voros, S., Lobes, L.A., Handa, J.T., Hager, G.D., Riviere, C.N.: Retinal vessel cannulation with an image-guided handheld robot. In: 2010 Annual International Conference of the IEEE Engineering in Medicine and Biology, pp. 5420–5423 (2010)

Intelligent Systems and Applications

Ontology-Based Food Safety Counseling System

Yasser A. Ragab[1(✉)], Essam F. Elfakhrany[1], and Ashraf M. Sharoba[2]

[1] College of Computing, Arab Academy for Science,
Technology, & Maritime Transport, Alexandria, Egypt
{eng.yasser,essam.elfakharany}@aast.com
[2] Faculty of Agriculture, Benha University, Benha, Egypt
ashraf_sharoba@yahoo.com

Abstract. The lack of knowledge about food ingredients can be risky and lead to serious complications for those with chronic diseases and food allergies. This paper proposes ontology-based Food Safety Counseling System (FSCS) that help the users with chronic diseases and food allergies to select their food. FSCS' aim to investigate the level of appropriate food products chosen by the users. The proposed models aim is to measure semantic relations between effected elements (The Main Item) of food product and the personal health status of users to provide professional advice about risky ingredient through six level of risk factor. It contains the Egypt food safety ontology (EFSO) and set of rules created based on knowledge elicited from experts in food safety domain to discover the tacit knowledge and allow semantic integration by applying forward-chaining deduction method and inference mechanism to knowledge base. The results indicate the model works effectively with high accuracy.

Keywords: Food Safety Counseling System (FSCS) ·
Egyptian Food Safety Ontology (EFSO) ·
Semantic Web Rule Language (SWRL) ·
Rule Base Management System (RBMS) · International Article Number (EAN)

1 Introduction

Recently the food safety attracted significant interest in Egypt represented in establishing the National Food Safety Authority (NFSA) is an open and independent authority. That is to protect consumers' health and consumers' interests by ensuring that food consumed, distributed, marketed, or produced in Egypt meets the highest standards of food safety and hygiene [1]. While the food plays an important role in one's life. One of the most important points in food safety domain the right choice of food product for avoidance the risky ingredient. An important principle in food safety (knowledge equal protection) the lack of knowledge of the consumer food ingredients that may be risky and occur complications lead to death, especially those with chronic diseases and allergies. Our food safety counseling system (FSCS) based on preconstructed domain ontology provides the users professional advice about risky ingredient through six levels not only risky but also suitability or no relation. In this paper, we propose the Food Safety Counseling System (FSCS) based on ontology.

© Springer Nature Switzerland AG 2020
A. E. Hassanien et al. (Eds.): AISI 2019, AISC 1058, pp. 609–620, 2020.
https://doi.org/10.1007/978-3-030-31129-2_56

Section 2 presents background and related work. Section 3 presents the system architecture and components (Egyptian Food safety ontology, semantic rules groups, reasoning mechanism, and Rule engine). Section 4 describes the experiments and results. Finally, conclusions and future work.

2   Background and Related Work

Ontology is a conceptual model and used to represent knowledge within domain of the abstract standardized instructions and rich semantic relationship. Ontology is define as a formal explicit specification of a shared conceptualization Gruber [2]. The Web Ontology Language (OWL) is knowledge representation languages. Ontology is the heart of semantic web layer stack. Ontology can be defined $O = <C, R, I, A>$; C is a set of Classes, R is a set of Relations, I is asset of Instances, A is a set of Axioms.

Food safety refers to describing handling, preparation, and storage of food.Food safety often overlaps with food appropriateness to prevent harm to consumer. Still some consumers do not know what they should eat and should not. Especially, high chronic patients such as diabetic and heart patients. The term of food safety used in this paper, refers to the appropriateness the food for user according to medical history.

More studies concern with food safety for diabetic and few studies concern with side effect of food additives and no study concerns with food safety according the main item of food ingredients especially on Egyptian dataset. Asanee Kawtok et al. [3] proposed Knowledge portal by using ontology driven extraction with natural language processing (NLP). The study was applied on Thai crops diseases. Huan Chung et al. [4] proposed automated food ontology for diabetes diet care includes generating an ontology skeleton with hierarchical clustering algorithms (HCA). The study was applied to Taiwanese diabetics. Yuehua Yang et al. [5] proposed food safety ontology reasoning on JENA designs. Some ontology reasoning rules for deducing food safety knowledge base to provide knowledge requester with accurate problem solving and decision support. The study was applied with china dataset. Bowei He et al. [6] provide us semantic query expansion method for food safety domain by using concept similarity computational model is applied to set the expansion rang of domain concept. The results show precision of the semantic query expansion method is higher than the precision of the traditional retrieval method. Ampaphun Wijosika et al. [7] provide us Thai Food Safety document search system based on ontology was built from keywords of published documents in Thailand. The results provided the relevant documents that related to given keyword search with 71% precision and 81% recall. DuyguCelik [8] some studies for the same author applied on Turkey foods FoodWiki is a mobile application examines side effects of food additives via semantic web technology. The system uses advanced semantic knowledge base can provide recommendation of appropriate food before consumption by individual. Method applied (Jaro-Winkler distance [9], Semantic-matching algorithm) (Fig. 1).

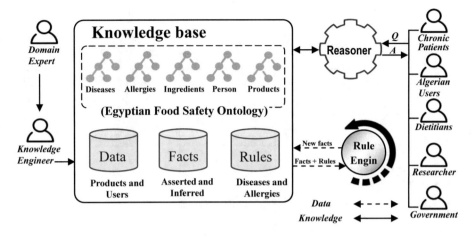

Fig. 1. Food safety counseling system

3   System Architecture and Methodology

The system structure and ontology model are briefly described in this section. This system consists of knowledge base contains the Egyptian food safety ontology and three database (data, facts and rules) addition to the reasoning mechanism include the Reasoner and Rule Based Management System (BRMS).

Developing the system consisting of six phases is firstly knowledge acquisition phase, secondly knowledge representation (building the ontology and rules), thirdly selecting the reasoning mechanism and rule engine, fourthly insert dataset for food products, fifthly execute the experiments six phases knowledge validation.

3.1   Research Scope and Parameters

The study scope of this paper is consider only packaged products and considered only some diseases and allergies show in following Table 1. We considers six levels of risk factor (High risk, Medium risk, Low risk, Suitable, Very suitable and no relation)

International Article Number (EAN): The International Article Number is a standard describing a barcode. EAN barcodes are used worldwide for looking at retail point of sale, but can also be used as numbers for other purposes such as wholesale ordering or accounting [10].

Real-time working mechanism: After insert, the data (EAN number of selected product + medical data of user) to the system. The system starts the semantic analysis. Then it checks the matched entry by searching the knowledge base. The inference engine retrieves the rules related with the user diseases and allergies according to ontology output for analysis the main item and classified the product of food category to retrieve the answer to user about the risk factor from six levels (H, L, M, S, VS, N) Addition to the reason.

Table 1. Food category, diseases, allergies, and level of risk factor

Food Category	
1	Bakery products
2	Candy, Sugar and Chocolate
3	Cereal Food
4	Chicken products
5	Egg products
6	Fat & Oil
7	Fish products
8	Fruits & Vegetables
9	Herbs & Spicy
10	Jam & Jelly
11	Juice & Beverages
12	Leguminous
13	Meat products
14	Milk & their products
15	Nuts

Allergies	
1	Celiac disease allergy
2	Egg allergy
3	Fish allergy
4	Fruits & Vegetable
5	Lactose intolerance
6	Leguminous
7	Milk protein allergy
8	Nut seed

Diseases	
1	Diabetic
2	Heart
3	Hypertension
4	Obesity

Risk factor	
H	Hi Risk
M	Medium Risk
L	Low Risk
S	Suitable
VS	Very Suitable
N	No Relation

Table 2. Decision table for diseases and allergies risk factor

Food Category		Diseases				Allergies							
		Diabetic	Heart	Obesity	Hyper-	Celiac	Egg	Fish	Fruits	Lactose	Legum-	Milk	Nut
1	Bakery products	S	VS	S	VS	H	N	N	N	N	N	N	N
2	Candy, Sugar	H	H	H	L	N	N	N	N	H	N	H	H
3	Cereal Food	L	S	L	S	H	N	N	N	N	N	N	N
4	Chicken products	S	S	S	S	N	N	N	N	N	N	N	N
5	Egg products	S	L	L	S	N	H	N	N	N	N	N	N
6	Fat & Oil	H	H	H	M	N	N	N	N	N	N	N	N
7	Fish products	S	VS	S	VS	N	N	H	N	N	N	N	N
8	Fruits & Vegetables	M	VS	M	VS	N	N	N	M	N	N	N	N
9	Herbs & Spicy	S	S	S	S	N	N	N	N	N	N	N	N
10	Jam & Jelly	H	H	H	S	N	N	N	L	N	N	N	N
11	Juice & Beverages	L	S	M	S	N	N	N	L	N	N	N	N
12	Leguminous	L	S	L	S	N	N	N	N	N	M	N	N
13	Meat products	S	L	L	L	N	N	N	N	N	N	N	N
14	Milk & their products	L	L	L	S	N	N	N	N	H	N	M	N
15	Nuts	M	L	M	S	N	N	N	N	N	N	N	L

(Packaged Food Only)

Knowledge acquisition: The aim of this phase extracts the knowledge from domain expert by using one or some knowledge acquisition techniques. In our case, we used "protocol generation technique because more suitable for study. Start with knowledge elicitation (Immediate sink) encode in natural language and (Ultimate sink) encoded with formal language protocol generation technique [11] based of various type of interviews unstructured, semi-structure and structure to produce protocol, i.e. a record of behavior to produce concepts, terminology and relation between food category with diseases and allergies. The output of this phase shows in Table 2 Decision table for Diseases and Allergies risk factor.

Egyptian food safety ontology: The current food safety ontology contains of five main classes (Thing -> Diseases, Thing -> Allergies, Thing-> Ingredients, Thing -> Person, and Thing -> Product) addition to 28 subclass, 43 object properties, 6 sub object properties, 4 data type properties, 199 individual and 86 semantic rule developed by using Small and medium-sized enterprises (SME) [12] on Protégé ontology editor version 5.2.0 [13] (Figs. 2, 3, and 4) (Table 3).

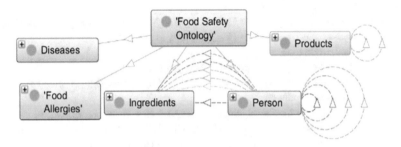

Fig. 2. (EFSO) main classes

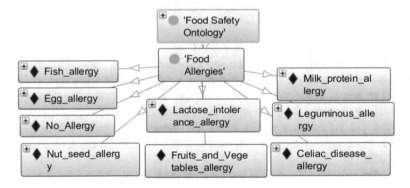

Fig. 3. (EFSO) food allergies class

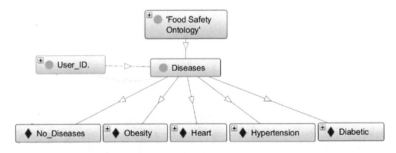

Fig. 4. (EFSO) diseases class

Table 3. Some of Semantic rules built in system

Person(?p) ∧ Has_EAN_No(?p, ?m) ∧ Has_main_item(?m, Fat_and_Oil) ∧ Has_Diseases(?p, Diabetic) -> Has_Diabetic_Hi_Risk(?p, Fat_and_Oil)
Person(?p) ∧ Has_EAN_No(?p, ?m) ∧ Has_main_item(?m, Fruits_and_Vegetables) ∧ Has_Diseases(?p, Obesity) -> Has_Obesity_Medium_Risk(?p, Fruits_and_Vegetables)
Person(?p) ∧ Has_Allergy(?p, Lactose_intolerance_allergy) ∧ Has_main_item(?m, Candy_Sugar_and_Chocolate) ∧ Has_EAN_No(?p, ?m) -> Has_Lactose_intolerance_Hi_risk (?p, Candy_Sugar_and_Chocolate)
Person(?p) ∧ Has_EAN_No(?p, ?m) ∧ Has_main_item(?m, Candy_Sugar_and_Chocolate) ∧ Has_Allergy(?p, Nut_seed_allergy) -> Has_Nut_seed_Hi_risk(?p, Candy_Sugar_and_Chocolate)
Person(?p) ∧ Has_Allergy(?p, Egg_allergy) ∧ Has_main_item(?m, Cooling_and_Freezing_products) ∧ Has_EAN_No(?p, ?m) -> Has_Egg_Hi_risk(?p, Cooling_and_Freezing_products)
Person(?p) ∧ Has_Allergy(?p, Celiac_disease_allergy) ∧ Has_EAN_No(?p, ?m) ∧ Has_main_item(?m, Bakery_products) -> Has_Celiac_disease_Hi_risk(?p, Bakery_products)
Person(?p) ∧ Has_Allergy(?p, Celiac_disease_allergy) ∧ Has_EAN_No(?p, ?m) ∧ Has_main_item(?m, Cereal_Food) -> Has_Celiac_disease_Hi_risk(?p, Cereal_Food)
Person(?p) ∧ Has_EAN_No(?p, ?m) ∧ Has_main_item(?m, Leguminous) ∧ Has_Allergy(?p, Leguminous_allergy) -> Has_Leguminous_Medium_risk(?p, Leguminous)

In EFSO modeling, ontology represents an important part of the system. The following provides a short description about the five ontology are: (1) Disease ontology- represent information about the diseases according research scope, (2) Allergy ontology: represent information about the allergy according research scope, (3) Ingredient ontology: describe the main item for food and classify of any category, (4) Product ontology: information model about the category of product and EAN, and (5) Person ontology: describe the user model. It is responsible for user information and medical data.

Semantic web rule: To encode the data extracted from domain expert we develop 86 semantic rule according Table 2. Semantic rule set created with Protégé SWRL Tab. Semantic Web Rule Language (SWRL) combining OWL and RuleML [14]. Figure 6-show small portion of semantic rules.

3.2   Reasoning Mechanism

(a) The Reasoner
"A semantic reasoner, reasoning engine, is a piece of software able to infer logical consequences from a set of asserted facts or axioms" [15]. We used two type of reasoner (Pellet and HermiT). Pellet is a free open-source Java-based reasoner for SROIQ with simple datatypes. It implements a tableau-based decision procedure for general TBoxes (subsumption, satisfiability, classification) and ABoxes (retrieval, conjunctive query answering) [16]. HermiT is a free reasoner for OWL2 datatype and supports for description graphs. It implements a hypertableau-based decision procedure [17].

(b) The Rule Engine

Business rules can be visualized as system requirement statements in the form of "IF… THEN…" sentences. These systems typically used a language that described "IF … THEN …" rules, a working memory to hold asserted facts." [18]. **Forward chaining deduction method** "is a bottom-up computational model. It starts with a set of known facts and applies rules to generate new fact." According to [19] (Fig. 5).

Table 4. Volunteers types

Number of volunteers	The type of volunteers	User have
5	0D	0 Diseases
5	1D	1 Diseases
5	2D	2 Diseases
5	0A	0 Allergy
5	1A	1 Allergy
5	2A	2 Allergy
5	0A_0D	0 Allergy & 0 Diseases
5	1A_0D	1 Allergy & 0 Diseases
5	2A_0D	2 Allergy & 0 Diseases
5	0A_1D	0 Allergy & 1 Diseases
5	0A_2D	0 Allergy & 2 Diseases
2	1A_1D	1 Allergy & 1 Diseases
1	1A_2D	1 Allergy & 2 Diseases
1	2A_1D	2 Allergy & 1 Diseases
1	2A_2D	2 Allergy & 2 Diseases
60	Total	

Table 5. volunteers main groups

Number of volunteers	Group of volunteers
15	Having no diseases and no allergies
10	Having diseases
10	Having allergies
25	Having diseases and allergies
60	Total

4 Experiments and Discussions

This study based on the dataset provided from the Egyptian Ministry of Agriculture Egypt [20] these data sets grouped into 15 major category that is composed of 45 item and consider the data quality dimension Accuracy, Completeness and Consistency.

The FSCS was implemented on machine with an Intel core i5 CPU with 4 G of Ram and MS windows 10 pro operating system. EFSO developed using on Protégé version 5.2.0 with OWL 2.0. The Pellet and HermiT reasoners, SWRL Tab for rule editor and Drool 5 as rule engine.

The experiments involved 60 volunteers having high chorionic diseases and food item allergies according to research scope show in Table 4. Volunteers divided to 4 main group and 9 types show in Tables 4, 5. The system input was user information (Name, Type, and Age) + (EAN for selected product) + (Diseases and Allergy). The system output was one or more of six level of risk factor (High risk, Medium risk, Low risk, Suitable, Very suitable and no relation) + the reason.

The user A06 as an example. He has two allergies (Lactose intolerance allergy and milk protein allergy), he selects product with EAN (6224001105005) for (كامل الدسملمار حليب). After assert this data to the system and run the reasoners (Pellet & HermiT) output was (Hi Risk for Lactose allergy with milk and their products category) and (Medium Risk for milk protein allergy with milk and their products category) shows in Fig. 6.

After execute all experiments the performance of two reasoners we used was similar for output but different for Reasoning time for HermiT was 460 ms while pellet was 1024 ms. Reasoning time t include the time of a = Pre-computing inferences, b = class hierarchy, c = object property hierarchy, d = data property hierarchy, e = class assertions, f = object property, g = assertions same individuals.

$$t = a + b + c + d + e + f + g$$

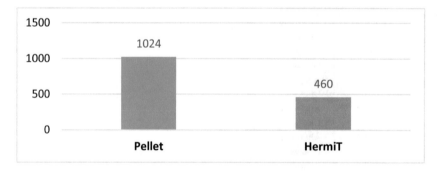

Fig. 5. Reasoning time graph

E3 show in Table 6 diseases user results one of false positive case according to Table 2 Decision table for diseases and allergies risk factor the user has obesity and heart diseases he select product EAN 6270000430112 مكسرات ابو عوف the correct result is medium risk and low risk respectively but the system result was negative result.

In the Fig. 6 show experiment for user (A06) the first 4 row represent the data inserted by the user but the rows number 6 and 7 represent the result from the system includes the level of risk factor and the reason.

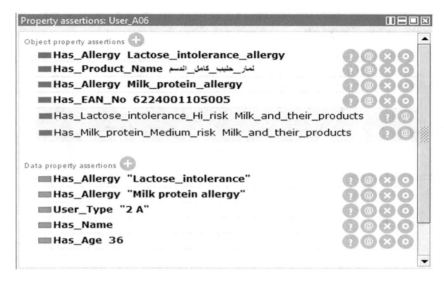

Fig. 6. Experiment for user (A06)

Table 6. Diseases user results

User	User Type	Diseases	Age	EAN no.	Product	Reasoner output		
D01	0D	No_Diseases	30	6270000430112	مكسرات ابوعوف	-	-	-
D02	0D	No_Diseases	47	6221024140457	هيرو مربى الكرز	-	-	-
D03	0D	No_Diseases	50	6223004834219	قبليو سمن طبيعى	-	-	-
D04	0D	No_Diseases	31	5038572210102	كرنشوس مكسرات	-	-	-
D05	0D	No_Diseases	32	6224001138980	ورق عنب	-	-	-
D06	1D	Diabetic	30	6270000430112	مكسرات ابوعوف	M_risk		Nuts
D07	1D	Heart	47	6221024140457	هيرو مربى الكرز	Low_risk	Hi_risk	Jam , Jelly
D08	1D	Hypertension	50	6223004834219	قبليو سمن طبيعى	M_risk		Fat & Oil
D09	1D	Obesity	31	5038572210102	كرنشوس مكسرات	M_risk		Nuts
D10	1D	Obesity	32	6224001138980	ورق عنب	M_risk	M_risk	Fruits & Vegetables
D11	2D	Heart & Hyper	29	6221134002638	جلاكسى فلوتس	Low_risk	Hi_risk	Candy ,Sugar
D12	2D	Diabetic & Hyper	45	6223000760109	البوادى حلاوه طحينيه	Hi_risk	Low_risk	Jam , Jelly
D13	2D	Diabetic & Heart	47	6270000430112	مكسرات ابوعوف	E3		Nuts
D14	2D	Obesity & Heart	29	6491044001186	بيض مجفف	Low_risk	Low_risk	Egg Products
D15	2D	Obesity & Hyper	38	6222000506014	كريستال زيت عباد	Hi_risk	M_risk	Fat & Oil

After execution, all the experiments and validated by the domain expert in food safety we consider the results of this study are good because the system can meet the users requirements show in Tables 7, 8. The results validated from experts and all system accuracy 91.6% show in Fig. 7.

Table 7. Confusion matrix with the value for measures of **Allergies, Diseases, (Allergies & Diseases), and all the system**

	Actual	Predicted					
		Positive		Negative			Total
Allergies	Positive	True positive (TP) =	8	True Negative (TN) =	5	13
	Negative	False positive (FP) =	2	False Negative (FN) =	0	2
	Total		10		5	15
Diseases	Positive	True positive (TP) =	9	True Negative (TN) =	5	14
	Negative	False positive (FP) =	1	False Negative (FN) =	0	1
	Total		10		5	15
Allergies and Diseases	Positive	True positive (TP) =	23	True Negative (TN) =	5	28
	Negative	False positive (FP) =	2	False Negative (FN) =	0	2
	Total		25		5	30
all the system	Positive	True positive (TP) =	40	True Negative (TN) =	15	55
	Negative	False positive (FP) =	5	False Negative (FN) =	0	5
	Total		45		15	60

Table 8. The Results for Allergies, Diseases and all the system

	Allergies	Diseases	Allergies and diseases	All system
Accuracy	86%	93%	93%	91.6%
Sensitivity	100%	100%	100%	100%
Specificity	74%	83%	71%	75%
PPV	80%	90%	92%	88.8%
NPV	100%	100%	100%	100%

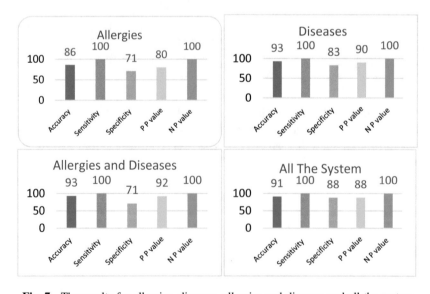

Fig. 7. The results for allergies, diseases, allergies and diseases and all the system

5 Conclusion and Future Work

The results from the 60 volunteers provide that the food safety concealing system FSCS is capable of investigate of risky ingredient about food product through six levels not only risky but also suitability or no relation. The results indicate the model works effectively with high accuracy 91.6% and Sensitivity is 100%. The FSCS can be extended to more user health status (Physical activity level, basal metabolic rate, and blood pressure level and body mass index) addition to extend the main item of ingredients to food additives.

References

1. Egyptian Food Safety Authority (2017) The EFSA register. http://www.nfsa.gov.eg. Accessed 10 Oct 2017
2. Gruber, T.R.: A translation approach to portable ontology specifications. Knowl. Acquisition **5**(2), 199–220 (1993)
3. Kawtrakul, A., Pechsiri, C.: Ontology driven k-Portal construction and K-service provision. In: Paper Presented at the Fifth International Conference on Language Resources and Evaluation, Italy, 24–26 May 2006 (2006)
4. Chuang, H., Ko, W.-M.: Automated food ontology construction mechanism for diabetes diet care. In: Paper Presented at the Sixth International Conference on Machine Learning and Cybernetics, Hong Kong, 19–22 August 2007 (2007)
5. Yang, Y., Junping, D., Hu, J.: Study on food safety ontology reasoning application based on Jena. In: Paper Presented at the International Conference on Communication and Technology Applications, China, 14–16 Oct 2011 (2011)
6. He, B., Yang, Y.: Semantic query expansion method for food safety domain. In: Paper Presented at IEEE International Conference on Computer Science and Automation Engineering, China, 22–24 June 2012 (2012)
7. Wijosika, A., Srivihok, A.: Thai food safety document searching system by ontology. In: Paper Presented at the 2nd International Conference on Research in Science, Engineering and Technology, Dubai, 21–22 March 2014 (2014)
8. Çelik, D.: FoodWiki: ontology-driven mobile safe food consumption system. Sci. World J. (2015). https://doi.org/10.1155/2015/475410
9. Jaro, M.A.: Advances in record-linkage methodology as applied to matching the 1985 census of Tampa, Florida. J. Am. Stat. Assoc. **84**(406), 414–420 (1989). https://doi.org/10.1080/01621459.1989.10478785
10. International Article Number. http://www.barcodeisland.com/ean13.phtml. Accessed 12 Feb 2018
11. Angel, M.: Knowledge acquisition Techniques (2011). http://www.epistemics.co.uk/No-te/174-0-0.htm. Accessed 15 Mar 2018
12. Öhgren, A., Sandkuhl, K.: Toward a methodology for ontology development in small and medium-sized enterprises. In: Paper presented at the International Conference on Applied Computing, Portugal, 22–25 February 2005 (2005)
13. Gon, R., et al.: Protégé platform (2017). https://protege.stanford.edu/. Accessed 15 Des 2017
14. Horrocks, I., et al.: SWRL: a semantic web rule language combining OWL and rule MI (2004). https://www.w3.org/Submission/SWRL/. Accessed 10 Oct 2017

15. Sattler, U.: Description Logic Reasoner (2015). http://www.cs.man.ac.uk/ ~ sattler/resoners.-html. Accessed 10 Oct 2017
16. Sirin, E., Parsia, B.: Pellet: a practical OWL-DL reasoner. J. Web Seman. 51–53 (2005). https://doi.org/10.1016/j.websem.2007.03.004
17. Glimm, B., Horrocks, I.: HermiT: an OWL 2 reasoner. J. Autom. Reasoning Manuscript **53**, 245–269 (2014). https://doi.org/10.1007/s10817-014-9305-1
18. Joshi, M., Meech, A., et al.: Business Rule Using OWL and SWRL. In: Advances in Semantic Computing (2010)
19. Al-Ajlan, A.: The comparison between forward and backward chaining. Int. J. Mach. Learn. Comput. https://doi.org/10.7763/ijmlc.2015.v5.492
20. Egyptian Ministry of Agriculture Egypt (2018) The EMAE register. http://far-malr.gov.eg. Accessed 20 May 2018

Orbital Petri Nets: A Petri Net Class for Studying Orbital Motion of Tokens

Mohamed Torky$^{(\boxtimes)}$ and A. E. Hassanein

Scientific Research Group in Egypt (SRGE), Cairo, Egypt
mtorky86@gmail.com, aboitcairo@gmail.com
http://www.egyptscience.net/

Abstract. Petri Nets is very interesting tool for studying and simulating different behaviors of automatic systems. It can be used in different applications based on the appropriate class of Petri Nets (e.g. colored, timed, or higher order Petri Nets). In this paper, a novel Petri Net approach called Orbital Petri Nets (OPN) is proposed for studying the dynamic behaviour of orbital auto-systems such as satellites motion. This initial study introduced a theoretical analysis of OPN with highlighting the problem of space debris collision as a case study. By this novel approach of Petri Nets, new smart algorithms can be implemented and simulated for handling several space problems, such as satellites' maneuvers or debris disposal.

Keywords: Petri Nets · Orbital Petri Nets (OPN) · Space debris

1 Introduction

Petri Nets (PN) is mathematical and modeling tools invented in August in 1939 by Car Adam. They are interesting and popular tools for simulating systems that are characterized as being parallel, distributed, concurrent, asynchronous, nondeterministic, and/or stochastic [1]. The classical Petri Nets is a directed bipartite graph with two node types called places and transitions. Places are represented by circles and transitions by bars. Places may contain zero or more tokens, which drawn as black dots, and the number of tokens may change during the execution of the net. A place p is called an input place of a transition t if there exists a directed arc from p to t, and p is called an output place of t if there exists a directed arc from t to p. Always, tokens in the input places represent a specific information in the form preconditions, input data, input signals. Tokens in the output places represent post-conditions, output data, output signals. The transition represents a specific function, event, computational step, signal processor, or clause in logic, whether, Places in Petri Nets works as a repository of input/output tokens. The change of system behaviour can be described as set of marking states $M = \{M_0, M_1, M_2, ...M_n\}$ based on the enabling/firing rule of transitions. Hence, the system's state can be simulated as a set of marking states in the handled Petri Net model.

© Springer Nature Switzerland AG 2020
A. E. Hassanien et al. (Eds.): AISI 2019, AISC 1058, pp. 621–631, 2020.
https://doi.org/10.1007/978-3-030-31129-2_57

Adding more features to tokens, places, transitions, and enabling/firing rule results in more classes of high level Petri Nets such as Colored Petri Nets [2], Timing Petri Nets [3]. For example, in Timing Petri Nets (TPN), time feature is associated with tokens, and transitions determine delays of tokens. Each token has a time-stamp which models the time the token becomes available for consumption. The time-stamp of a produced token is equal to the *firing time* plus the *firing delay* of the corresponding transition.

Although the variety classes of Petri Nets for studying different systems behaviours, there is no a Petri Net approach for studying the behavior of orbital systems (i.e. systems that move in orbits). In this paper, we propose a novel approach of Petri Nets called Orbital Petri Nets (OPN) for studding and simulating orbital systems' behaviour, such as satellites motion. The feasibility of OPN has been analyzed and verified on a case study for highlighting space collision problem. The rest of this paper can be arranged as, Sect. 2 presents the recent applications of Petri Nets in the literature. Section 3 presents the proposed Orbital Petri Net approach, Sect. 4 presents the mathematical representation of OPN. Section 6 presents a case study of applying OPN in space collision problem. Section 7 formulates the conclusion of this work.

2   Related Work

The recent literature has highlighted several Petri Net applications in different fields. For instance, Fuzzy Petri nets (FPNs) [4] has been introduced for modeling knowledge and rule- based systems. An adaptive Petri net (APN) [5] is introduced to model a self-adaptive software system with learning Petri Nets. Controllable Siphon Basis of Petri Nets [6] has been used to prevent deadlock for flexible manufacturing systems. Moreover, a novel Petri Net model is proposed [7] for analyzing the performance of concurrent system based on a polynomial algorithm and ordinary differential equations.

Ye et al. [8] introduced a decentralized supervision policy for a Petri net through collaboration between a coordinator and subnet controllers as a discrete event system (DES). Timed Colored Petri Nets [9] approach has been proposed for addressing the deadlock (DL)-free scheduling problem of flexible manufacturing systems (FMS). Zhou et al. [10] proposed a fuzzy-timing Petri nets for modeling and analyzing the performance of networked virtual environments (net-VEs). Analyzing the compatibility of Web services under temporal constraints [11] is another promising approach that utilized the timing Petri Net in a holistic manner and in modular way for modeling and analyzing the problem of message mismatches between services in a composition. Huang et al. [12] proposed a new approach to design an urban traffic network control system by using the Synchronous Timing Petri Nets (STPNs). Modeling and validating e-commerce system [13], biology systems [14], Robot control systems [15] and IoT-based smart systems [16] is another applications of Colored Petri Nets [17].

3   Orbital Petri Nets Approach

One of the most well-known tools for modeling system behaviours is Petri Nets. Although the variety of Petri Net classes, there is no Petri Net approach can be utilized for studying the dynamic behavior of orbital systems. In response to this challenge, we introduce new Petri Net approach called Orbital Petri Nets (OPN) for studying and simulating the behaviour of orbital systems such as satellites and space debris motions. The graphical representation of OPN can be described as, places are drawn as directed circle for representing the directed motion of tokens(e.g. satellites or debris) in orbits. Transitions are represented as bars or boxes for representing processing function (or computational routine) on the rotating tokens on its input places. Arcs are labeled with weight values. Weights may be orbital metric (e.g. orbital velocity, inclination, or altitude, etc). The marking in OPN assigns number of tokens in each place (i.e. number of rotating tokens in the orbit).

A transition t is said to be enabled if there is any token in any input place can be called by the input arc weight and the motion direction of tokens in input and output places is the same. Firing a transition t moves the rotated tokens from input places to the output places according to input/output arc-weight callings. A marking M_i assigns to each place p a positive integer k-tokens after transitions firing process.

It is important here to clarify that enabling/firing rule in OPNs is different from the other classes of Petri Nets. In OPN enabling/firing rule of transition t change state of tokens in any input place from state to another state or move it to another output place according to token calling in the input/output arc-weights expressions. Moreover, firing transition t is not conditioned with all input places, but only with the input place that make it enabled. An interesting advantage of OPN is that OPN may import some features of other known classes of Petri Nets such as Timing, Coloring, or inhibitor arcs with respect to the modeling case.

The Formal definition of Orbital Petri Nets (OPN) of order N can be formulated as follows:

$$\text{OPN} = (P^{+/-}, T, A, \textstyle\sum, W, G, M_0)$$

where,

- $P = \{p_1, p_2, p_3, ...\}$ is a finite set of directed places. The motion direction of its tokens may be in clockwise $\circlearrowright$ or anticlockwise $\circlearrowleft$.
- $T = \{t_1, t_2, t_3, ...\}$ is a finite set of transitions.
- $A \subseteq (P \times T) \cup (T \times P)$ is a set of input/output arcs
- $\sum$ is the set of all tokens' types ($C = \{c_1, c_2, ..c_m\}$)
- $W : F \rightarrow \{e_1, e_2, e_3, ..\}$ is a weight function, where e_j is weight expression that defines the method of tokens motion with respect to the token calling in each weight expression e_j
- G is a guard function that it maps each transition $t \in T$ to a boolean guard expression g (optional).

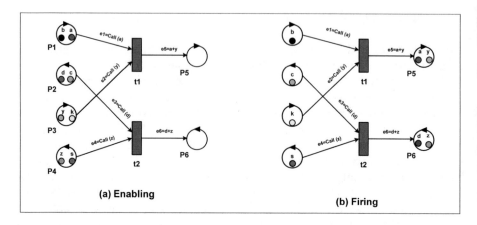

Fig. 1. Orbital Petri Net Model of order 6 (a) Enabling, (b) Firing $t1$, and $t2$

- $M_0 : P_j \rightarrow \{(nc_1 \cup nc_2, ... \cup nc_m)\}$ is the initial marking state that defines number token types in all places, $n = 1, 2, 3, ...$

Figure 1 shows an example of Orbital Petri Net Model of order 6. $t1$ and $t2$ are enabled (Fig. 1(a)) since there is a mapping between input tokens in their input places and the corresponding arc weights. Moreover, both input and output places have clockwise direction. In Fig. 1(b), Firing $t1$ moves the tokens 'a', and 'y' from $P1$ and $P3$ respectively to $P5$, whilst, Firing $t2$ moves the tokens 'd', and 'z' from $P2$ and $P4$ respectively to $P6$.

4   Behaviour Properties

Modeling orbital systems using OPN is supported with some behaviour properties, which control in token firing process such as, *Reachability*, *Liveness*, *Boundedness*, and *Reversibility*.

4.1   Reachability

Reachability is a fundamental property for studying the dynamic behaviors of Orbital Petri Net models. A firing of an enabled transition will change the token distributions (i.e. marking) according to the firing rule of OPN. Hence, a sequence of firing will result in sequence of a set of marking states $R(M_0) = \{M_1, M_2, ...M_n\}$. A marking state M_d is reachable from M_0 if there is a sequence of transition firing $\sigma = \{t1, t2, ...tm\}$ that transform M_0 to M_d. In other words, M_d is reachable of M_0 if and only if $M_d \in R(M_0)$. Reachability can be implemented using reachability graph. Figure 2 shows an OPN(3) and the corresponding reachability graph.

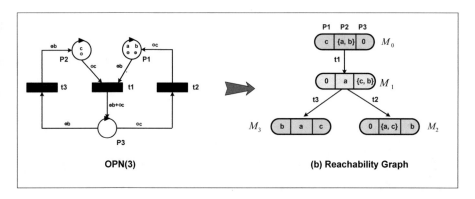

Fig. 2. Orbital Petri Net Model of order 3 and the corresponding reachability graph

4.2 Liveness

In orbital petri net, the Liveness property is closely related to tokens life cycles through transitions firing. An OPN model is said to be a completely live if all tokens are live regardless number of firings of transitions. For more verification to liveness, we can relax liveness condition into different levels of liveness as follows:

a token 'k' is said to be:

- $(l_0 - live)$ if k can never be fired or changed from state to another state in all firing sequence $\Sigma\sigma_i, i = 1, 2, ...m$
- $(l_1 - live)$ if k can be fired only one time or changed from state to another state in all firing sequence $\Sigma\sigma_i, i = 1, 2, ...m$
- $(l_n - live)$ if k can be fired $n - times$ or changed from state to another state in all firing sequence $\Sigma\sigma_i, i = 1, 2, ...m$
- $(l_\infty - live)$ if k can be $infinetly$ fired or changed from state to another state in all firing sequence $\Sigma\sigma_i, i = 1, 2, ...m$

Figure 3 illustrates four levels of liveness of OPN. Figure 3(a) shows $l_0 - live$ example, where token 'a' can't fire because of the difference in motion direction of P1 and P2. Hence token 'a' resides swivel in P1. Figure 3(b) shows $l_1 - live$, where token 'b' can fire only one time according to the boolean expression associated with $t1$. Figure 3(c) shows that the token 'c' is $l_2 - live$ since it can be fired two times by transitions $t1$ and $t2$, whilst the token 'd' is $l_3 - live$ since it can be fired three times by $t1$, $t3$, $t2$ respectively. Figure 3(d) shows that the token 'e' is infinitely fired by transitions $t1$ and $t2$, so it is $l\infty - live$.

4.3 Boundedness

An orbital petri net (OPN, M_0) is said to be $k - bounded$ if number of tokens in all places doesn't exceed a finite number k for any marking state M_d reachable from M_0, i.e., $M(p) \leq k$. For example, the net shown in Fig. 3(c) is 2-bounded.

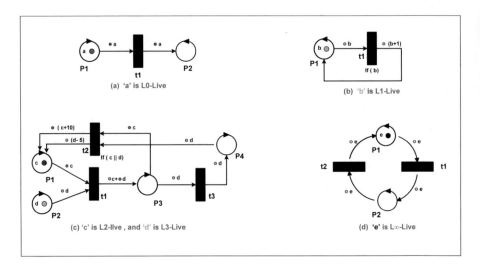

Fig. 3. Four levels of liveness in orbital Petri Nets

4.4 Reversibility

An orbital petri net (OPN, M_0) is said to be *Reversible* if the initial marking M_0 is reachable from any marking state M_d. In many cases, it is not necessary to get back to the initial marking M_0 but may get back to a home state M_z. So, we can redefine the reversibility using home state M_z. A marking M_z is said to home state if for each marking M_d, M_z is reachable from M_d. For example, the (OPN, M_0) shown in Fig. 3(d) is a reversible net.

5   Analysis Methods of Orbital Petri Nets

The dynamic behavior of OPN models can be represented by some algebraic equations. In this spirit, we can use *Incidence Matrix* and *State Equations* to govern the dynamic behavior of OPN models (Fig. 4).

5.1   Incidence Matrix

For OPN models of order N with n transition and m directed places. The Incidence Matrix $A = [a_{ij}]$ is $n \times m$ matrix of values and its typical entry is given by:

$$a_{ij} = \{a_{ij}\}^+ - \{a_{ji}\}^- \tag{1}$$

Where, $\{a_{ij}\}^+$ is the set of all tokens transferred from the transition i to its output place j and $\{a_{ji}\}^-$ is the set of all tokens transferred to transition i from its input place j, and a_{ij} is the change of tokens states in place j after firing the

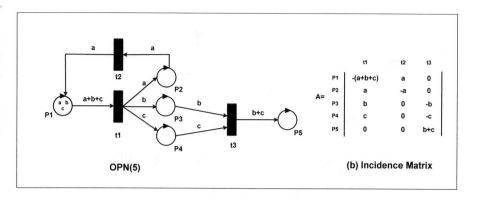

Fig. 4. Representing Orbital Petri Net in Incidence Matrix

transition i. The transition i is enabled at marking state M if there exist place j defines the following:

$$M(j) \geq 1, \quad and \quad j(\bullet) \equiv w(j \to i) \tag{2}$$

where, $M(j)$ is the marking (i.e. number of tokens) of place j, $j(\bullet)$ is the target token in place j, and $w(j \to i)$ is the weighted arc from place j to transition i.

For example, Fig. 3 depicts an Orbital Petri Net model of order 5 (Fig. 3(a)) and the corresponding Incidence matrix (Fig. 3(b)).

5.2 State Equation

In OPN, a marking state M_k is an $m \times 1$ column vector. The jth entry of M_k represents the set of tokens in place j immediately after the kth firing in some firing sequence. The control vector u_k is $n \times 1$ column vector which represent number of firing times of transition i in the kth firing. The change from marking state M_{k-1} to M_k as a result of firing transition i can be formulated as in Eq. 3:

$$M_k = M_{k-1} + A \times u_k \tag{3}$$

Necessary Reachability Condition: suppose that the destination marking state M_d is reachable from the initial marking state M_0 through the firing sequence $\{u_1, u_2, ...u_d\}$, the state Eq. 3 can be reformulated as in Eq. 4:

$$M_d = M_0 + A \times \Sigma_{k=1}^{d} u_k \tag{4}$$

Which can be rewritten as:

$$\Delta M = A \times X \tag{5}$$

Where, $\Delta M = M_d - M_0$ and $X = \Sigma_{k=1}^{d} u_k$.

6   Case Study: Space-Debris Collision Problem

Space-Debris Collision Problem [18–21] has been occurred since June 2007 when China tested anti-satellite missiles, fired a solid-fuel rocket from the Zigang base to hit a Chinese satellite. This collision resulted in 2,300 to 2,500 space particles, which considered the biggest event in the formation of space debris around the earth. This amount of resulted debris has changed the orbit of Terra environmental spacecraft to avoid potential collisions with the Chinese space debris [22]. We can model this problem using an OPN Model in Fig. 5. Figure 5(a) shows a situation in which a satellite and a space particle turn around the earth in two different orbits but they close to collide with each other. The challenge her is how to avoid the collision between the satellite and the particle?, and how to get ride of the particle to prevent occurring another collisions by this particle in the space orbits? Figure 5(b) depicts an OPN model for simulating the collision avoidance between the satellite S and Debris D. Firing transition $t1$ will automatically executes a maneuver between the satellite and Debris, where, the token S in $P1$ moves to $P3$, and token D in $P2$ moves to $P4$ according to a specified maneuver procedures. In this case $t2$ and $t3$ are enabled, thereby, firing $t3$ require only $M(P4) \geq 1$ and $P4(\bullet) \equiv w(p4 \rightarrow t3)$ which get ride of the token D and consume it. On the other hand, firing $t2$ require only $M(P3) \geq 1$ and $P3(\bullet) \equiv w(P3 \rightarrow t2)$. Hence, firing $t2$ will return back the token S to $P1$. This scenario simulates returning back the satellite to its valid orbit after avoiding the collision with the particle. Figure 6(a) depicts firing $t1$, Fig. 6(b) depicts firing $t2$, and $t3$.

Mathematically, we can represent the collision prevention between satellite and debris by solving Eq. 6, which formulate the Necessary Reachability Condition. Firing transition $t1$, will produce the marking state M_1, which can be formulated as:

$$M_1 = \begin{pmatrix} P_1 \\ P_2 \\ P_3 \\ P_4 \end{pmatrix} = \begin{pmatrix} S \\ D \\ 0 \\ 0 \end{pmatrix} + \begin{pmatrix} -S & S & 0 \\ -D & 0 & 0 \\ S & -S & 0 \\ D & 0 & -D \end{pmatrix} \times \begin{pmatrix} 1 \\ 0 \\ 0 \end{pmatrix} \tag{6}$$

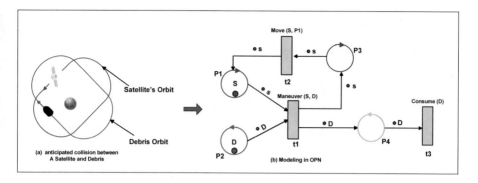

Fig. 5. Modeling spacecraft-debris collisions case in Orbital Petri Net

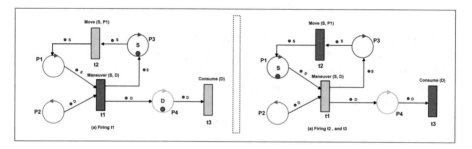

Fig. 6. Spacecraft-debris collision prevention: (a) Firing $t1$ and (b) Firing $t2$ and $t3$

So,

$$M_1 = \begin{pmatrix} P_1 \\ P_2 \\ P_3 \\ P_4 \end{pmatrix} = \begin{pmatrix} S \\ D \\ 0 \\ 0 \end{pmatrix} + \begin{pmatrix} -S \\ -D \\ S \\ D \end{pmatrix} = \begin{pmatrix} 0 \\ 0 \\ S \\ D \end{pmatrix} \tag{7}$$

Equation 7 represents moving token S from $P1$ to $P3$, and moving token D from $P2$ to $P4$.

Firing transition $t2$, and $t3$ will produce the marking state M_2, which can be formulated as:

$$M_2 = \begin{pmatrix} P_1 \\ P_2 \\ P_3 \\ P_4 \end{pmatrix} = \begin{pmatrix} 0 \\ 0 \\ S \\ D \end{pmatrix} + \begin{pmatrix} -S & S & 0 \\ -D & 0 & 0 \\ S & -S & 0 \\ D & 0 & -D \end{pmatrix} \times \begin{pmatrix} 0 \\ 1 \\ 1 \end{pmatrix} \tag{8}$$

So,

$$M_2 = \begin{pmatrix} P_1 \\ P_2 \\ P_3 \\ P_4 \end{pmatrix} = \begin{pmatrix} 0 \\ 0 \\ S \\ D \end{pmatrix} + \begin{pmatrix} S \\ 0 \\ -S \\ -D \end{pmatrix} = \begin{pmatrix} S \\ 0 \\ 0 \\ 0 \end{pmatrix} \tag{9}$$

Equation 9 represents returning the token S back to $P1$ and getting ride of the token D.

It is easy to verify and prove the necessary reachability condition in Eq. 10 as:

$$M_2 = \begin{pmatrix} S \\ D \\ 0 \\ 0 \end{pmatrix} + \begin{pmatrix} -S & S & 0 \\ -D & 0 & 0 \\ S & -S & 0 \\ D & 0 & -D \end{pmatrix} \times \begin{pmatrix} 1 \\ 1 \\ 1 \end{pmatrix} \qquad (10)$$

Hence,

$$M_2 = \begin{pmatrix} S \\ D \\ 0 \\ 0 \end{pmatrix} + \begin{pmatrix} 0 \\ -D \\ 0 \\ 0 \end{pmatrix} = \begin{pmatrix} S \\ 0 \\ 0 \\ 0 \end{pmatrix} \qquad (11)$$

Equation 11 prove that M_2 is reachable state from The token S is L2-live, whilst the token D is dead one. This scenario simulates returning the satellite back to its valid orbit and consuming the debris for preventing the collision between them. The analysis results prove that The token S is L2-live, whilst the token D is dead one such that it has been consumed by $t3$. Moreover, the net is 1-bounded and not reversible.

7   Conclusion

This paper has introduced a novel theoretical approach of Petri Nets called Orbital Petri Net for modeling orbital auto systems such as satellites and space-crafts. Behaviour properties such as Reachability, Liveness, Boundedness, and reversibly have been redefined in new manner for specifying tokens' behaviour. In this initial study, a proposed OPN model has been proposed for modeling the problem of colliding satellites and space debris. Mathematically, Reachability, Liveness, Boundedness, and reversibility of the proposed model have been verified and analysed. In the next study, we will investigate in more implementation details developing new smart algorithms which can be simulated by orbital Petri Nets for handling satellites behaviours problems in its orbits.

References

1. Murata, T.: Petri nets: properties, analysis and applications. Proc. IEEE **77**(4), 541–580 (1989)
2. Jensen, K.: Coloured Petri Nets: Basic Concepts, Analysis Methods and Practical Use. Springer, Heidelberg (2013)
3. Van der Aalst, W.M.: Petri net based scheduling. Oper. Res. Spektrum **18**(4), 219–229 (1996)
4. Liu, H.C., You, J.X., Li, Z., Tian, G.: Fuzzy Petri nets for knowledge representation and reasoning: a literature review. Eng. Appl. Artif. Intell. **1**(60), 45–56 (2017)

5. Ding, Z., Zhou, Y., Zhou, M.: Modeling self-adaptive software systems with learning Petri nets. IEEE Trans. Syst. Man Cybern. Syst. **46**(4), 483–498 (2016)
6. Liu, H., Xing, K., Wu, W., Zhou, M., Zou, H.: Deadlock prevention for flexible manufacturing systems via controllable siphon basis of Petri nets. IEEE Trans. Syst. Man Cybern. Syst. **45**(3), 519–529 (2015)
7. Ding, Z., Zhou, Y., Zhou, M.: A polynomial algorithm to performance analysis of concurrent systems via Petri nets and ordinary differential equations. IEEE Trans. Autom. Sci. Eng. **12**(1), 295–308 (2015)
8. Ye, J., Li, Z., Giua, A.: Decentralized supervision of Petri nets with a coordinator. IEEE Trans. Syst. Man Cybern. Syst. **45**(6), 955–966 (2015)
9. Baruwa, O.T., Piera, M.A., Guasch, A.: Deadlock-free scheduling method for flexible manufacturing systems based on timed colored Petri nets and anytime heuristic search. IEEE Trans. Syst. Man Cybern. Syst. **45**(5), 831–846 (2015)
10. Zhou, Y., Murata, T., DeFanti, T.A.: Modeling and performance analysis using extended fuzzy-timing Petri nets for networked virtual environments. IEEE Trans. Syst. Man Cybern. Part B (Cybern.) **30**(5), 737–756 (2000)
11. Du, Y., Tan, W., Zhou, M.: Timed compatibility analysis of web service composition: a modular approach based on Petri nets. IEEE Trans. Autom. Sci. Eng. **11**(2), 594–606 (2014)
12. Huang, Y.S., Weng, Y.S., Zhou, M.: Modular design of urban traffic-light control systems based on synchronized timed Petri nets. IEEE Trans. Intell. Transp. Syst. **15**(2), 530–539 (2014)
13. Yu, W., Yan, C., Ding, Z., Jiang, C., Zhou, M.: Modeling and validating e-commerce business process based on Petri nets. IEEE Trans. Syst. Man Cybern. Syst. **44**(3), 327–341 (2014)
14. Liu, F., Blätke, M.A., Heiner, M., Yang, M.: Modelling and simulating reaction-diffusion systems using coloured Petri nets. Comput. Biol. Med. **1**(53), 297–308 (2014)
15. Bonilla, B.L., Asada, H.H.: A robot on the shoulder: coordinated human-wearable robot control using coloured petri nets and partial least squares predictions. In: 2014 IEEE International Conference on Robotics and Automation (ICRA), pp. 119–125. IEEE, 31 May 2014
16. Teslyuk, V.M., Beregovskyi, V.V., Pukach, A.I.: Development of smart house system model based on colored Petri nets. In: 2013 XVIIIth International Seminar/Workshop on Direct and Inverse Problems of Electromagnetic and Acoustic Wave Theory (DIPED), pp. 205–208. IEEE, 23 September 2013
17. Huang, H., Kirchner, H.: Secure interoperation design in multi-domains environments based on colored Petri nets. Inf. Sci. **1**(221), 591–606 (2013)
18. Liou, J.C., Johnson, N.L., Hill, N.M.: Stabilizing the future LEO debris environment with active debris removal. Orbital Debris Quarterly News **12**(4), 5–6 (2008)
19. Nishida, S.I., Kawamoto, S., Okawa, Y., Terui, F., Kitamura, S.: Space debris removal system using a small satellite. Acta Astronaut. **65**(1–2), 95–102 (2009)
20. Than, K.: Taking out the space trash. Popular Sciences, 27 June 2008
21. Klotz, I.: Debris briefly forces astronauts from space station. Reuters, 12 March 2009
22. Burger, B.: NASA's Terra Satellite Moved to Avoid Chinese ASAT Debris. Space.com Site: www.space.com/news/070706-sn-china-terra.html

An Efficient Novel Algorithm
for Positioning a Concrete Boom Pump

Mohammad Fathy[1]([✉]), Mustafa M. Shiple[2], and Mostafa R. A. Atia[3]

[1] Higher Technological Institute Tenth of Ramadan, Tenth of Ramadan, Egypt
mohamed.sadek88@hti.edu.eg
[2] National Telecommunication Institute, Cairo, Egypt
mustafa.shiple@nti.sci.eg
[3] Arab Academy for Science Technology & Maritime Transport, Alexandria, Egypt
mrostom1@aast.edu

Abstract. Robots are used in many fields as they have remarkable benefits. Some of these benefits are robots precision, high productivity, time-saving and the ability to work in harsh working environments. The boom of the Concrete Boom Pump (CBP) is one of serial robot arm forms, and it is used in the field of construction. The boom base positioning is crucial since it is essential to accomplish the tasks efficiently. The boom base positions are determined according to the operator experience which consumes effort and time and consequently operation cost. Boom base positions could be determined using an automated method to save operation time. The proposed algorithm determines the boom base positions depending on the position of the concrete pouring points. The results show the ability of the proposed algorithm to determine the CBP base positions with high efficiency over other methods. Moreover, the proposed method shows low processing time than others.

Keywords: Robot base positioning · Mobile robots · Planning · Real-time systems · Boom pumb

1 Introduction

The boom of the Concrete Boom Pump (CBP) is considered a form of serial robot arms. No doubt those CBPs have significant effects on the field of construction. CBPs are more efficient and more economical by decreasing the labor cost, operation time and operation cost. Although CBPs have these benefits, they still have some shortcomings. It is required to determine the minimum number of base positions in which CBP could perform any task properly. The CBP base positions are determined according to the experience of the operator which differs from one on other and consumes effort and time. Thus to determine the CBP base positions in an automated method will save effort and time and reduce the operation cost. Nevertheless, many CBP automation methods focus only on how to automate the boom without addressing on how to locate

© Springer Nature Switzerland AG 2020
A. E. Hassanien et al. (Eds.): AISI 2019, AISC 1058, pp. 632–642, 2020.
https://doi.org/10.1007/978-3-030-31129-2_58

the CBP base position [1–4]. Thus, it is essential to determine the suitable CBP base positions to perform the required task.

To achieve the required task, the minimum number of CBP base positions should be calculated. These CBP base positions will be located along the allowable paths which are determined according to the instructions of the site engineer and the CBP specifications. The information about concrete blocks number, dimensions, locations and the coordinates of the allowable paths are provided by the Workspace Layout Sheet (WLS).

Many algorithms are being proposed to calculate the robot arm base positions required to accomplish various tasks. All these algorithms have marvelous benefits to the field of robotics. However, they have some limitations.

Kapusta et al. [5] proposed an algorithm "Task-centric initial Configuration Selection (TCS)" evaluates base positions. The algorithm calculates the base positions which have the highest score. The highest score means the best base position. The algorithm limitations are as follows. All tasks were performed with six degrees of freedom (6-dof) goal poses (base positions and joints configuration). However, some tasks require different position and orientation requirements. All the addressed tasks are under the assumption of collision-free paths, in addition to, they do not require complex motions or high power such as lifting. The algorithm is based on discretization which may cause errors and the possibility of missing solutions. High computation cost as a result of discretization (discretization increases space and resolution). The results show that no single initial configuration can reach all poses. Moreover, some initial configurations put the robot arm close to its joint limits.

Dong et al. [6] proposed an algorithm based on calculating the reachable map of the end-effector. Orientation-based reachability map or "OBR map" algorithm is developed to determine robot base location to perform tasks using several tools. The algorithm limitations are as follows; The validity of every base position is checked by computing the inverse kinematics (IK) solutions for all trajectory points from the base position. The target reachability is affected as a result of the number of orientation constraints.

Vahrenkamp et al. [7] proposed an Oriented Reachability Map (ORM) algorithm. Information about the robot parameters help in describing the capabilities of the robot regarding reaching and manipulation. A spatial grid covers the workspace is built to represent the robot's workspace. The algorithm limitations are as follows. The algorithm is based on sampling the workspace offline to generate voxel grid. The efficiency of object reachability is affected by the efficiency of the IK solver. If the resolution of the voxel grid is increased, the computation cost would be increased. The algorithm determines the base position based on target position rather than target region.

Chen et al. [8] proposed object-oriented hand posture (OOHP) algorithm to find the optimal base position for arm-base coordinated grasping. Given the information about object position, the operating environment, and the robot pose, the algorithm generates a coordinated arm-base trajectory for the grasping task. The algorithm limitations are the uncertainty of object perception, and the workspace environment does not change with time.

Yang et al. [9] proposed an inverse Dynamic Reachability Map (iDRM) algorithm which can find valid and sufficient robot base end-poses for floating base system in complex and changing environments in real-time. The input to such algorithm is the end-pose information. The algorithm uses this information to generate a trajectory to move the robot to a suitable pre-grasp stance. This stance is used to generate whole-body motion to reach the target efficiently. The algorithm limitations are as follows. The algorithm should include parallelization of the selected solution as it introduces viable solutions for human operators and high-level decision agents.

Leidner et al. [10] proposed an algorithm for a service robot house tasks. The task requires suitable base position, manipulate tools over vast areas, suitable whole body motion and the right amount of force to complete the task successfully. The algorithm is calculating the region of interest (ROI) based on the capability maps and the inverse reachability map. ROI indicates the reachability index in colors. Dark red regions represent the unreachable areas while dark blue regions represent the entirely reachable areas. The algorithm limitations are as follows. The complexity of the task affects the parameterization process. The approach needs to be tested by using mobile robots with higher dofs.

Stulp et al. [11, 12] proposed an algorithm for a mobile manipulator to be able to learn the concept of PLACE. The mobile manipulator will obtain this concept through performing a specific task many times and computes the probability of successful manipulation. Therefore, the result is an area of positions, not a specific point. This approach considers the mobile manipulator hardware and control schemes, task context and the environment.

Zacharias et al. [13] proposed an algorithm for a mobile robot to help a planner to decide if the robot base is needed to perform the required task, in addition, to determine a proper base position of the robot. The algorithm takes into account the task requirements, environment, and robot capabilities to generate reachability spheres to describe the workspace.

Nektarios et al. [14] proposed an algorithm to determine the optimal manipulator base position during workcell design to maximize manipulator's velocity performance. Manipulator velocity ration is computed from finite values sampled along the end-effector path. The algorithm optimizes cycle time even in tasks requires smooth end-effector curved paths with high velocity.

Gadaleta et al. [15] proposed an algorithm to determine the optimal base position of the industrial robot (IR) to save energy consumption. A grid is generated to describe the workspace. The energy consumption is computed for every grid point to determine the areas in which the robot base could be placed have the minimum energy consumption.

Spensieri et al. [16] proposed an algorithm of a robot used in assembly tasks, e.g., spot welding tasks. The algorithm is required to optimize the robot base positions to generate reachable paths to perform the assembly tasks correctly with minimum cycle time. The algorithm inputs are required tasks, e.g., welding points, environment and the pose of the welding gun. The algorithm is consists of two parts; the first part is to determine the base position suitable to perform a task according to the minimum cycle time. The second part is to determine the base position required to perform multiple tasks. The algorithm limitations are as follows. Processing time is increased in complex CAD models and when the number of tasks being performed is increased.

Abolghasemi et al. [17] and [18] proposed an algorithm for a robot arm mounted on a wheelchair to help people with disabilities to perform their activities of daily living (ADL). Ease-of-Reach Score (ERS) algorithm is required to find a suitable base position of the robotic arm. The required task is to grasp an object with identifier c. The algorithm performs the task in two steps. The first step is to find the base position while the other is to grasp the object. The algorithm searches the area for the best base positions have the highest ERS. The ERS is based on the Count of Distinct Grasp Trajectories CDGT. CDGT is calculated by counting the number of distinct robot arm poses can grasp the object from position $p = p(x, y)$ among obstacles $O = o1, o2, \ldots, on$. The algorithm limitations are as follows. In case of having many positions have the same ERS, more factors could be considered for decision making. Errors generated from computing ERS for all workspace. For complex working spaces, the discretization of the state space would be less efficient.

The previous diverse algorithms calculate the base positions according to 1. The distribution form of concrete blocks, 2. The density of concrete blocks per unit area. Therefore, the results of these two research lines show low efficiency of base position numbers. It is required to develop an algorithm satisfies the requirements, and at the same time is not greatly affected by the distribution form or density per unit area of concrete blocks.

Therefore, the work aims to determine the minimum number of base positions to achieve the required tasks. These base positions could be determined without any restrictions on the robot base placement (base positions are determined statically) [11]. In this paper, a new approach is developed to determine the minimum number of base positions dynamically (the base positions will be placed within the boundary limits of the path) based on the position of the pouring points. The paper is organized in the following way. Section 2 describes the method of the proposed algorithm showing the inputs of the algorithm and the results of the proposed algorithm compared with old approaches. Section 2.1 describes the improvements of the proposed algorithm showing the reason of the improvements and the results of the improved proposed algorithm compared with the older version of the proposed algorithm and the older approaches. Section 3 shows the conclusion of the work.

2   The Proposed Algorithm

The purpose of the proposed algorithm is to attempt to reduce the number of base positions of the Concrete Boom Pump (CBP) required to place concrete into concrete blocks. Meanwhile, the minimum number of base positions is determined, the operation time and cost will be saved.

Blocks distribution and number over the workspace are essential information to determine the suitable base positions to accomplish the tasks.

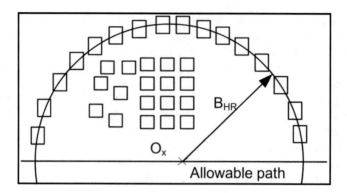

Fig. 1. Boom horizontal reach and the outer shell blocks

2.1 The Input of the Algorithm

The information required for the algorithm are boom horizontal reach (BHR), workspace layout sheet (WLS) and the allowable paths. BHR is extracted from the boom specifications. WLS shows concrete blocks positions, dimensions, and distribution. The allowable paths are where the boom base could be located.

Every corner of the concrete block is represented by (x, y) pair. To represent the (x, y) pairs for the four corners, eight vectors will be developed. A transformation matrix TM is developed from these vectors represents the orientation of every concrete block and the position of the center point of the concrete block.

2.2 The Proposed Idea

The algorithm should calculate the base positions according to the distribution form of the outer shell of concrete blocks. This method will not be affected by the distribution form and density per unit area of concrete blocks inside the outer shell (as all the blocks inside the shell will be within the reachable area of the boom) as shown in Fig. 1. Therefore, this method is considered a dynamic approach to calculating the required base positions.

1. Considers (x_i, y_i) to be the center points of all blocks.
2. Considers the block has minimum x coordinate. Consider (x_j, y_{jm}). Where: $x_j = min\{x_i\}$ and y_{jm} is the y coordinate of all blocks have $x_i = x_j$. Take (x_j, y_j*), Where: $y_j* = max\{y_{jm}\}$.
3. Calculates the base position (O_x) related to the block has maximum y coordinate. $O_x = f(x_j, y_j^*)$.
4. Checks if there are any blocks outside the reachable range of the calculated base positions (O_x) but have x coordinate lies between x_j and O_x as shown in Fig. 2.
   Take (x_i, y_i), Where: $O_x > x_i > x_j$ and y_i: y coordinate of the blocks have x_i.
   Define: $D_{mn} = \sqrt{((x_i - O_x)^2 + (y_i - O_y)^2)}$

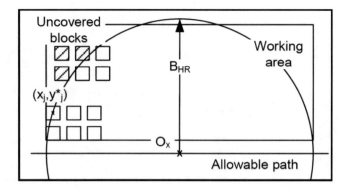

Fig. 2. Uncovered blocks but have x coordinate lies between x_j and O_x

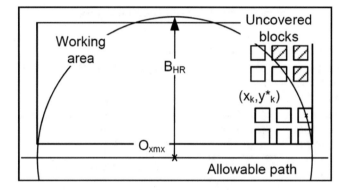

Fig. 3. Uncovered blocks but have x coordinate lies between x_k and O_{xmx}

Define: (x_o, y_o) to be the points satisfy $D_{mn} > B_{HR}$ B_{HR} is the boom horizontal reach.

5. Calculates the new base position (O_{xf})
   Consider (x_c, y_{ce}), Where: $x_c = min\{x_o\}$ and y_{ce} is the y coordinate of all blocks have $x_c = x_o$.
   Take (x_c, y_c^*), where $y_c^* = max\{y_{ce}\}$ and $O_{xf} = f(x_c, y_c^*)$

6. Considers the block has maximum x coordinate. Consider $(x_k, y_k n)$, Where: $x_k = max\{x_i\}$ and y_{kn} is the y coordinate of all blocks have $x_i = x_k$.
   Take (x_k, y_k^*), where: $y_k^* = max\{y_{kn}\}$.

7. Calculates the base position (O_{xmx}) related to the block has maximum y coordinate. $O_{xmx} = f(x_k, y_k^*)$

8. Checks if there are any blocks outside the reachable range of the calculated base positions (O_{xmx}) but have x coordinate lies between x_k and O_{xmx} as shown in Figure 3.
   Take (x_r, y_r), Where: $x_k > x_r > O_{xmx}$ and y_r: y coordinate of the blocks have x_r.
   Define: $D_{mx} = \sqrt{((x_r - O_{xmx})^2 + (y_r - O_y)^2)}$.
   Define: (x_t, y_t) to be the points satisfy $D_{mx} > B_{HR}$.

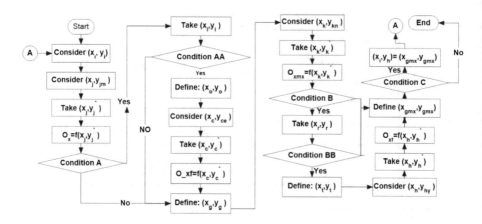

Fig. 4. The proposed algorithm flowchart

9. Calculates the new base position (O_{xmxf})
   Consider (x_h, y_{hy}), Where: $x_h = max\{x_r\}$ and y_{hy} is the y coordinate of all blocks have $x_h = x_r$.
   Take (x_h, y_h^*), where $y_h^* = max\{y_{hy}\}$, and $O_{xmxf} = f(x_h, y_h^*)$
10. Considers the blocks outside the reachable range of the calculated base positions to calculate the proper base positions to cover them.
11. Blocks outside the reachable range of O_x or O_{xf}.
    $$\text{Define: } D_1 = \begin{cases} \sqrt{((x_j - O_x)^2 + (y_j^* - O_y)^2)} & \text{in case of } O_x \\ \sqrt{((x_c - O_x)^2 + (y_c^* - O_y)^2)} & \text{in case of } O_{xf} \end{cases}$$
    Define: (x_g, y_g) to be the points satisfy $D_1 > B_H R$
12. Blocks outside the reachable range of $O_x mx$ or O_{xmxf}.
    $$\text{Define: } D_2 = \begin{cases} \sqrt{((x_k - O_x)^2 + (y_k^* - O_y)^2)} & \text{in case of } O_{xmx} \\ \sqrt{((x_h - O_x)^2 + (y_h^* - O_y)^2)} & \text{in case of } O_{xmxf} \end{cases}$$
    Define: (x_{gmx}, y_{gmx}) to be the points satisfy $D_2 > B_H R$
    General form equation being used by the algorithm to determine the base positions:
    $$B_{HR}{}^2 = (O_x \pm x_j)^2 + (y_{jm} - O_y)^2$$

Where: B_{HR} is the boom horizontal reach, O_x and O_y are the base position coordinates, x_j and y_{jm} are the block coordinates. Figure 4 shows the proposed algorithm flowchart.

2.3   Results

All algorithms are simulated with two sets of examples. The first set of examples are proposed by ref [1] and [2], while the other set of examples are proposed by the author to find the boundary limits of the algorithms. The first set of examples show horizontal, vertical lines pouring and circular line pouring distribution forms of concrete blocks as shown in Fig. 5(a). These distribution forms could be found in crossbeams of the foundations and the platforms of buildings in the construction field. On the other hand, the second set of examples show a uniform and non-uniform distribution forms of concrete blocks as shown in Fig. 5(a).

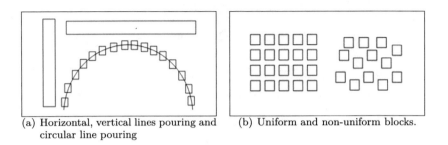

(a) Horizontal, vertical lines pouring and circular line pouring

(b) Uniform and non-uniform blocks.

Fig. 5. Different distribution forms of concrete

Figure 6 shows the number of base positions calculated by every algorithm. Examples are horizontal line, vertical line [1], circular [2], uniform and non-uniform distribution form of concrete blocks respectively.

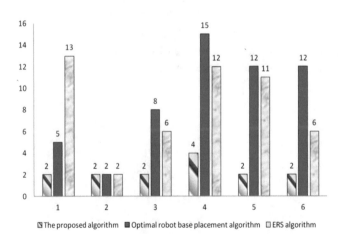

🖾 The proposed algorithm   ■ Optimal robot base placement algorithm   ☐ ERS algorithm

Fig. 6. Number of base positions calculated by every algorithm for six examples

The results show that the proposed algorithm achieves remarkable results with high efficiency over other algorithms. The results show that the proposed algorithm is not affected by the distribution form or density per unit area of concrete blocks. However, the proposed algorithm is affected only by the outer shell of concrete blocks.

Figure 7 shows the concrete blocks distribution form of some examples. For non-uniform distribution forms (as provided in examples 5 and 6), the density of

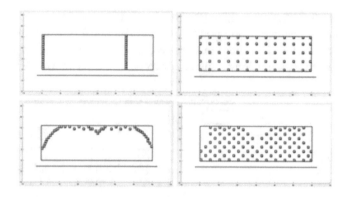

Fig. 7. Concrete blocks distribution form of some examples

blocks per unit area affects the results of ERS algorithm. The density of blocks per unit area indicates the number of blocks to be covered. The higher density tends to a higher number of blocks. The densities are 0.056 and 0.1 respectively. The equation of density of blocks per unit area is as follows:

$$Density = \frac{number\ of\ blocks}{W_{ws} \times H_{ws}} \tag{1}$$

Where: W_{ws} and H_{ws} are the width and height of the workspace.

Figure 8 shows the optimization ratio in the number of base positions achieved by the proposed algorithm over other algorithms.

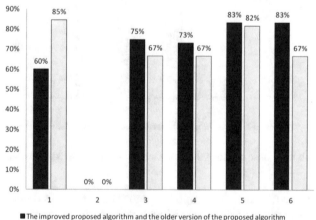

■ The improved proposed algorithm and the older version of the proposed algorithm

□ Ratio between the improved proposed algorithm and Optimal robot placement algorithm

Fig. 8. Optimization ratio in number of base positions achieved by the proposed algorithm over other algorithms

From the vertical line pouring distribution form to uniform distribution form, it is cleared that the improvement ratios starting from 0% to 85%, by neglecting the results of the vertical line pouring distribution form the ratios will start from 60% up to 85%.

It is clear that the superiority of the proposed algorithm is increased when the distribution form of concrete blocks expanded horizontally. This notice is highlighted in the example of vertical lines pouring, since Fig. 6 shows the results are identical. From the previous results of the vertical line distribution form show no enhancement among experimented algorithms.

3   Conclusion

The proposed algorithm shows high efficiency over other algorithms, because, the proposed algorithm calculates proper base positions dynamically. The distribution form of the outer shell of concrete blocks that could be reached by the boom affects the number of base positions. This proposed idea enhances the base position calculations in execution time and provides the minimal number of base positions. Different case studies are analyzed; vertical lines, horizontal lines, circular forms and random forms either are uniform and non-uniform distributions. From 60% to 83% enhancement ratios in the number of base positions is achieved by proposed algorithm over Optimal robot placement algorithm. The proposed algorithm achieves enhancement ratios in the number of base positions from 67% to 85% over ERS algorithm. From these two points of views (minimum number of base positions and execution time), the proposed algorithm shows high efficiency over other algorithms. The results show that the number of base positions of all algorithms converges as the distribution form of concrete blocks approaches to the vertical distribution form of concrete blocks.

References

1. Zhou, S., Zhang, S.: Co-simulation on automatic pouring of truck-mounted concrete boom pump. In: 2007 IEEE International Conference on Automation and Logistics, pp. 928–932 (2007)
2. Zhang, S., Zhou, S., Sun, J.: Study on automatic planning of truck-mounted concrete boom pump system. Appl. Mech. Mater. **236–237**, 1265–1269 (2012)
3. Wang, T., Wang, G., Liu, K.: Simulation control of concrete pump truck boom based on PSO and adaptive robust PD. In: 2009 Chinese Control and Decision Conference, pp. 960–963 (2009)
4. Ren, W., Li, Z., Bi, Y., Zhao, S., Peng, B., Zhou, L.: Modeling and analysis of truck mounted concrete pump boom by virtual prototyping. J. Robot. **2017**, 1–10 (2017)
5. Kapusta, A., Park, D., Kemp, C.: Task-centric selection of robot and environment initial configurations for assistive tasks. In: 2015 IEEE/RSJ International Conference on Intelligent Robots and Systems (IROS), pp. 1480–1487 (2015)
6. Dong, J., Trinkle, J.: Orientation-based reachability map for robot base placement. In: 2015 IEEE/RSJ International Conference on Intelligent Robots and Systems (IROS), pp. 1488–1493 (2015)

7. Vahrenkamp, N., Asfour, T., Dillmann, R.: Robot placement based on reachability inversion. In: 2013 IEEE International Conference on Robotics and Automation, pp. 1970–1975 (2013)

8. Chen, D., Liu, Z., von Wichert, G.: Grasping on the move: a generic arm-base coordinated grasping pipeline for mobile manipulation. In: 2013 European Conference on Mobile Robots, pp. 349–354 (2013)

9. Yang, Y., Ivan, V., Li, Z., Fallon, M., Vijayakumar, S.: iDRM: humanoid motion planning with real-time end-pose selection in complex environments. In: 2016 IEEE-RAS 16th International Conference on Humanoid Robots (Humanoids), pp. 271–278 (2016)

10. Leidner, D., Dietrich, A., Schmidt, F., Borst, C., Albu-Schaffer, A.: Object-centered hybrid reasoning for whole-body mobile manipulation. In: 2014 IEEE International Conference on Robotics and Automation (ICRA), pp. 1828–1835 (2014)

11. Stulp, F., Fedrizzi, A., Beetz, M.: Learning and performing place-based mobile manipulation. In: 2009 IEEE 8th International Conference on Development and Learning, ICDL, pp. 1–7 (2009)

12. Stulp, F., Fedrizzi, A., Beetz, M.: Action-related place-based mobile manipulation. In: 2009 IEEE/RSJ International Conference on Intelligent Robots and Systems, pp. 3115–3120 (2009)

13. Zacharias, F., Sepp, W., Borst, C., Hirzinger, G.: Using a model of the reachable workspace to position mobile manipulators for 3-D trajectories. In: Proceedings of IEEE-RAS International Conference on Humanoid Robots (Humanoids), pp. 55–61 (2009)

14. Nektarios, A., Aspragathos, N.: Optimal location of a general position and orientation end-effector's path relative to manipulator's base, considering velocity performance. Robot. Comput. Integr. Manuf. **26**(2), 162–173 (2010)

15. Gadaleta, M., Genovesi, A., Balugani, F.: Determining the energy-optimal base position of industrial robots by means of the modelica multi-physics language. Appl. Mech. Mater. **680**, 320–325 (2014)

16. Spensieri, D., Carlson, J., Bohlin, R., Kressin, J., Shi, J.: Optimal robot placement for tasks execution. Procedia CIRP **44**, 395–400 (2016)

17. Abolghasemi, P., Rahmatizadeh, R., Behal, A., Boloni, L.: Real-time placement of a wheelchair-mounted robotic arm. In: 2016 25th IEEE International Symposium on Robot and Human Interactive Communication (RO-MAN), pp. 1032–1037 (2016)

18. Abolghasemi, P., Rahmatizadeh, R., Behal, A., Boloni, L.: A real-time technique for positioning a wheelchair-mounted robotic arm for household manipulation tasks. In: Workshops at the Thirtieth AAAI Conference on Artificial Intelligence, pp. 2–7 (2016)

Experimental Procedure for Evaluation of Visuospatial Cognitive Functions Training in Virtual Reality

Štefan Korečko[1]([✉]), Branislav Sobota[1], Marián Hudák[1], Igor Farkaš[2],
Barbora Cimrová[2,3], Peter Vasiľ[1], and Dominik Trojčák[1]

[1] Department of Computers and Informatics, Faculty of Electrical Engineering
and Informatics, Technical University of Košice, Košice, Slovakia
{stefan.korecko,branislav.sobota,marian.hudak.2}@tuke.sk
[2] Department of Applied Informatics, Comenius University in Bratislava,
Bratislava, Slovakia
farkas@fmph.uniba.sk
[3] Institute of Normal and Pathological Physiology, Centre of Experimental Medicine,
Slovak Academy of Sciences, Bratislava, Slovakia

Abstract. In recent years, visuospatial cognitive functions, which play
a crucial role in human cognition, have sparked interest among psychologists and neuroscientists, focusing on assessment, training and restoration of these functions. Virtual reality, recognized as a modern technology, addressing the real-life aspects of visuospatial processing, provides
an immersive environment that can be used for stimulation of cognitive
functions and its effects that can be measured afterwards. In this paper,
we describe an experimental design that involves cognitive testing and
targeted, cognitively-oriented, stimulation in an immersive 3D virtual
environment, rendered by a unique CAVE system. We focus primarily on
a game, designed and developed to serve as the virtual environment. We
also describe the experimental procedure that includes the measurement
of an electrophysiological neural correlate of spatial working memory
capacity – contralateral delay activity.

Keywords: Virtual reality · Training · Spatial working memory ·
Change detection task

1 Introduction

Visuospatial cognitive functions allow us to detect, represent, manipulate and
store visual and spatial information [5]. This entails the ability to perceive visual
objects, locate their position in space, orient our attention, infer various spatial
relations and remember the scene. In addition, visuospatial cognitive abilities
enable to perform judgments related to direction and distance among external
objects and thus allow individuals to navigate in the environment [1].

© Springer Nature Switzerland AG 2020
A. E. Hassanien et al. (Eds.): AISI 2019, AISC 1058, pp. 643–652, 2020.
https://doi.org/10.1007/978-3-030-31129-2_59

Due to their importance in everyday life, visuospatial functions attracted attention of both psychologists and neuroscientists who have developed various means how to measure them in humans (see [4] and references therein). Most-widely applied visuospatial tests target specific functions ranging from relatively automatic perceptual and attentional abilities to more complex and deliberative cognitive faculties, such as visuospatial short-term or working memory, mental rotations and executive visual attention [2,11]. At the same time, visuospatial training and restoration programs, exploiting the known principles of brain plasticity, have recently manifested a great interest in cognitive neuroscience [9]. The primary goal of such assessment methods and interventions is to diagnose, maintain, improve, or at least delay cognitive and brain decline of the visuospatial cognitive functions, to improve the quality of human life [1].

In order to advance the current state of research methodology, by addressing the real-life aspect of visuospatial processing, we will utilize virtual reality (VR) technologies for training the selected cognitive functions. VR has been already considered suitable for these purposes before, when the corresponding equipment was considerably less developed and affordable [10]. The idea that virtual environments may modulate neuropsychological measures is supported by several studies, such as [7], where a virtual office environment, experienced via a VR headset, was used for assessing the learning and memory in individuals with traumatic brain injury. A recent survey [8] also advocates for VR-based function-led assessments that are closer to the real-world functioning.

In this paper, we report about an experiment that aims at evaluating how an experience in an immersive VR environment can stimulate selected cognitive functions, namely working visual memory. As a matter of fact, although our visuospatial capacities allow us to understand and infer relationships of three-dimensional (3D) objects in space, these 3D aspects of visuospatial processing are profoundly neglected in laboratories. The experiment will use VR as an experimental condition that aims at maximally exploiting the immersive 3D environment. The computerized cognitive tests will be in 2D, following the current standards, focusing on measuring targeted visuospatial functions.

The VR experience will be represented by a game, developed solely for the purpose of the experiment. The preliminary outline of the experiment was introduced in [4], where two game prototypes were initially considered: a logical, Tetris-like construction game in 3D and a first-person shooter of the tower defense genre. After considering pros and cons of both games, with respect to their suitability, the latter game was eventually chosen and the experiment procedure has been refined to its final form. Both the procedure and the game are described in the rest of the paper, which is organized as follows. Section 2 provides the basic information about the LIRKIS CAVE, a VR facility to be used in the experiment. Section 3 specifies the final form of the experiment procedure and Sect. 4 deals with the design and implementation of the tower defense game. Section 5 concludes the paper.

2 LIRKIS CAVE

The virtual environment for the experiment will be provided by LIRKIS CAVE [3], a compact and transportable cave automatic virtual environment, built at the Technical University of Košice. It features a $2.5 \times 2.5 \times 3$ m display area, made of twenty LCD panels. The panels are 55 in. LCD TV sets with passive stereoscopy, made by LG. Fourteen of these panels are positioned vertically, forming seven sides of a decagon. Thank to this arrangement, the CAVE offers a 250° panoramic space. Six panels are positioned horizontally. They form the ceiling (3 panels) and the floor (3 panels). The whole display area is mounted inside a self-supporting steel frame. This means that the CAVE doesn't need to be fixed to the walls, ceiling or the floor of the room where it is installed. The interaction between the CAVE and its human users is provided by a variety of input devices. These include the usual ones, such as mouse and keyboard, the gaming devices (joystick, gamepad) and VR-specific peripherals such as the Myo armband[1] and OptiTrack[2]. The OptiTrack system of 7 cameras captures the user movement. Rendering of virtual scenes, user interaction and control over the whole system are the responsibility of a cluster of 7 computers, equipped with the NVIDIA Quadro graphics cards. The CAVE is shown in Fig. 1(b).

3 Experimental Procedure

For each participant, the experiment procedure starts with filling a form with basic information (name, age, gender, education, handedness), exclusion criteria and a questionnaire about mood, mental energy and fatigue. Exclusion criteria are psychological or neurological diagnosis, traumatic brain injury, learning disability and psychoactive medicaments taking. Due to the lateralized nature of the main cognitive test used in this study, only right-handed subjects can participate. The entire procedure, applied to the experimental group, can be split into two phases – preparation and training.

3.1 Preparation

First two visits consist of performing a change detection task (CDT) in which the capacity of spatial working memory is assessed for each participant. The task consists of a specific number of red (targets), blue and green (distractors) rectangles in various orientations. At the beginning of each trial, an arrow (cue) indicates visual hemi-field to which the participant is supposed to pay attention. The rectangles (sample array) are presented only for a limited time (200 ms) followed by a retention period of 1 s. The task of the participant is to keep in his or her memory the orientation of all target (red) rectangles from the hemi-field indicated by the cue. Afterwards, a probe (test array) containing the same sample array, with or without a change in orientation of one target rectangle, is

---

[1] https://support.getmyo.com.
[2] https://optitrack.com/.

displayed and the subject should indicate whether he or she noticed any change in orientation in comparison to the sample array. During this task, participants' eye movements are recorded using electrooculography (EOG) to assure they are moving only their attention and not their gaze (as the neural correlate connected to the spatial working memory capacity is lateralized, see below, contralateral delay activity, CDA).

For CDT evaluation and for measuring the performance (accuracy in detecting the change in one of the targets), we need to identify invalid trials, i.e. those during which undesirable eye movements have been detected (because those would violate the experimental design). As a preset threshold, the number of invalid trials must not exceed 20% for including the participant into the group. The semi-automatic procedure for detecting eye movements is based on evaluating pre-collected EOG signals, when the subject was instructed to move his or her eyes, and blink, in a controlled manner, such that appropriate thresholds could be set based on visual inspection of the preprocessed eye signals (detecting horizontal and vertical eye movements).

3.2   Training

The first day of training in LIRKIS CAVE starts with the same questionnaire mentioned above and is followed by CDA pre-test assessment. The task is the same CDT as described above, and in addition, event-related electroencephalographic (EEG) potentials are being recorded. Approximately 250 ms after the sample array onset, a large negative electrical deflection can be seen over the posterior parietal sides opposite (contralateral) to the targeted side (target hemifield indicated by a cue). This deflection, named Contralateral Delay Activity or CDA is considered a neural correlate of working memory capacity as its amplitude changes with the number of items held in memory [6]. The difference wave is calculated as a difference of an average signal from all contralateral and ipsilateral (same side) electrodes. EEG is measured from four left and four right posterior electrode sites plus three midline spots (P3, P7, PO7, O1, P4, P8, PO8, O2, Fz, Cz, Pz) using a high-performance and high-accuracy biosignal amplifier g.USBamp with ground electrode on AFz and reference electrode attached to left earlobe and digitally re-referenced to linked earlobes. If the person performing the task cannot filter out the distracting stimuli, the number of items in their memory increases and their CDA changes correspondingly. Therefore, we can use this as a sensitive method for estimating changes induced by training in VR.

After the CDA pre-test, the participants undergo first training in LIRKIS CAVE. Each training session takes 30 min and consists of three atomic sessions. One *atomic session* refers to the 8-min block of continuous playing and there are 3-min breaks between them. In subsequent two weeks the participants continue with training resulting in 5 sessions (days) before the next CDA measurement (mid-test). In the next two weeks, the training continues with five more sessions and it finishes the day after the last training with the final CDA (post-test) assessment. During each training day, participants fill in the questionnaire. The whole protocol is performed by an experimental group. The same protocol, except the training in the CAVE, is followed by a control group. The procedure was tested in a small pilot version on five subjects to adjust the protocol and its parameters. As a result we decided to use a combination of two, three and four target stimuli (red rectangles) and zero or two distractors (blue or green rectangles). One CDA session consists of 240 trials in total and takes 20 min plus four brakes for approximately 3 min.

4 Tower Defense Game

After an evaluation of the two prototypes, described in [4], the tower defense has been selected for the experiment. It has been chosen primarily because its prototype was further developed than the construction game, it is more adjustable and the fact that no performance-related issues were observed during its testing.

4.1 Design

To ensure a high level of user immersion, the game is designed in such a way that the physical constraints of the CAVE are a natural part of it. The most significant constraints are

1. *Limited user movement.* The user can move freely inside the CAVE, but the LCD panels present an impassable barrier.
2. *Fixed position of the CAVE.* The CAVE itself cannot move in the real world. Therefore, at least the horizontal position should be maintained during a VR experience to minimize the risk of the simulation sickness. It is also best to avoid simulated movements where the sensations related to the acceleration or deceleration should be felt.
3. *Visible LCD panel bezels.* While the distance between LCD panels is kept at minimum, their bezels are still visible in the CAVE.

To incorporate the constrains, the CAVE itself represents an interior of the operator cabin of a fictional defense tower, armed with laser cannons. The LCD panels serve as the glazing of the cabin and their bezels form the frame of the glazed part. As can be seen in Fig. 1(a), the only movement allowed for the turret and the cabin itself is a rotation around its vertical axis. Just the cannons can move up or down. The final appearance of the game is shown in Fig. 1(b). The dark background with stars has been chosen to make important objects well recognizable. The most noticeable difference between the concept and the actual game is the absence of the cannons. We decided not to show the cannons as they may block the view significantly and because of potential performance issues. In the actual game, only the cannon beam is visible when fired and it originates from underneath the operator cabin.

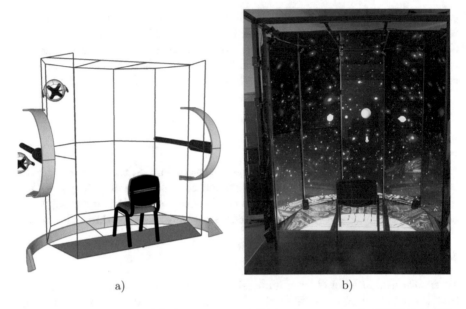

a)   b)

Fig. 1. Game appearance: (a) the concept and (b) implementation in LIRKIS CAVE.

4.2 Gameplay

From the player's perspective, the goal of the game is to defend the turret from invaders for a given period of time. The invaders are represented by drones flying towards the turret in groups (of various sizes). Six different 3D models are used for the drones, shown in Fig. 2.

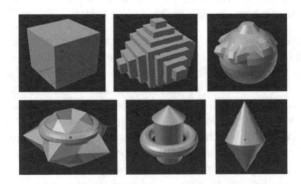

Fig. 2. Drone models used in the game.

The drones may seem very basic and strange-looking, however this is intentional: As the shapes in the CDT, the drones are divided into two groups - *targets* and *distractors*. The targets are enemies to be shot down by the user, while the distractors are friendly drones to be ignored. All the drones of the same model are either targets or distractors. And because of the nature of the tests, used in the experiment, it was required that all drones are the same color. Therefore, the only way to easily tell one from each other was to use very different models for them.

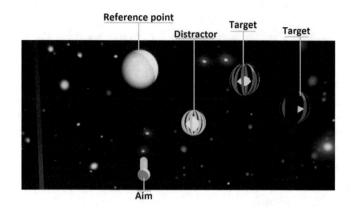

Fig. 3. The middle section of the view from the user's position in the CAVE with important objects labeled.

Each 8-min-long atomic session is a sequence of separate episodes. Each episode involves an attack, performed by a group of drones, and proceeds as follows:

1. The number of targets (n_T) and distractors (n_D) in the attack is computed as
$$n_T = n_T^l + \delta_T$$
$$n_D = n_D^l + \delta_D \tag{1}$$
where n_T^l and n_D^l is the number of targets and the number of distractors for the difficulty level of the session and δ_T and δ_D are random values, drawn from the discrete uniform distribution on the interval $[-r, r]$. The value r is set to 1.
2. A group of n_T target and n_D distractor drones is generated. 3D models and trajectories are randomly assigned to the drones.
3. The group is displayed in the game.
4. For one second, the drones are marked (Fig. 3) to inform the player which ones are distractors and which are targets.
5. The attack begins. All the drones in the group move towards the turret at a constant speed.
6. Once per episode, a blackout event occurs. The time of the event is chosen randomly from the interval $[0.2T_e, 0.6T_e]$, where T_e is the supposed duration of the episode. The duration of the blackout is random, from the interval $[600\,\text{ms}, 900\,\text{ms}]$. The drones disappear from the screens during the blackout.

7. When all the drones pass the turret or are shot down, the episode ends.

All random values mentioned in the episode description are uniformly distributed. The supposed episode duration (T_e) is the time when the last drone in the group passes the turret. The next episode in the sequence starts only after the T_e of the previous one passes, even in the case that the player shoots down all the drones sooner. This is because all atomic sessions must have the same duration and all episodes in a session have to be completed.

The visual output of the game during the first second of an episode can be seen in Fig. 3. There are three drones: one distractor with green marking and two targets with red marking. The rightmost target is just passing the border between two LCD panels. The aim is used for targeting the drones and can be moved using joystick, gamepad or keyboard. If a drone is aimed, it is marked in the same way as all the drones during the first second, but in yellow. The reference point is positioned in the center of the virtual environment and helps the player to navigate. It has been added when the tests of the game prototype revealed difficulties with returning to the central position after an episode ended.

4.3 Levels of Difficulty

Each participant will spend 240 min in total playing the game and the game should train his or her visuospatial cognitive abilities, so it is only natural that the difficulty cannot remain the same all the time. The game has several configurable properties and four of them have been selected to define the levels of difficulty:

- *Drone speed* (v_d in Table 1). The constant speed by which the drones move towards the player can be set to *low* (L in Table 1), *medium* (M) or *high* (H). The speed also defines the number of episodes in a single atomic session.
- *Drone placement* (p_d). It determines how the drones are positioned in the virtual world. Again, there are three options: The easiest setting is *centered* (c), where the drones are concentrated around the reference point. In the *normal* (n) setting the drones appear inside the three middle columns of the LCD panels. Such situation is captured in Fig. 1(b). The most difficult is the *widespread* (w) setting, where the drones may also appear on the next two columns of the LCD panels.
- *Number of drones*. It is defined separately for targets (n_T^l) and distractors (n_D^l). The actual number for given episode is then computed by (1).

Based on these properties, 30 levels have been defined (Table 1); one for each atomic episode, played by a participant during the experiment. The difficulty doesn't increase automatically, but only if the condition (2) holds for the previous atomic session, played by the participant.

$$(n_T^d - n_D^d)/n_T > 0.5 \tag{2}$$

In (2), n_T^d is the number of destroyed targets, n_D^d is the number of destroyed distractors and n_T is as in (1).

Table 1. Difficulty settings for the levels (Lv.) of the game.

Lv.	1	2	3	4	5	6	7	8	9	10	11	12	13	14	15	16	17	18	19	20	21	22	23	24	25	26	27	28	29	30
v_d	L	L	L	L	L	L	L	L	L	L	M	M	M	M	M	M	M	M	M	M	H	H	H	H	H	H	H	H	H	H
p_d	c	c	c	n	n	n	n	w	w	w	c	c	c	n	n	n	w	w	w	w	c	c	n	n	n	w	w	w	w	w
n_T^l	2	2	3	3	3	4	4	4	4	5	3	3	4	4	5	5	4	5	5	5	3	4	4	5	5	4	5	5	5	6
n_D^l	2	3	3	3	4	3	4	4	5	4	3	4	4	5	4	5	5	5	6	7	4	4	5	4	5	5	5	6	7	7

5 Conclusion

The experimental procedure, described in this paper, should bring us closer to understanding how VR experience may improve our visuospatial cognitive functions. The procedure and the tower defense game are results of several testing trials, performed with the corresponding prototypes. The game has both the typical features of the tower defence genre and features specific for the procedure. The most prominent of the specific features are an inclusion of the targets and distractors in a way similar to the CDT and the blackout event, which resembles the retention period from the task.

While the main results of the experiment will be derived from the CDT and CDA outcomes, collected before, in the middle and after the training, additional statistics are also computed and stored by the game itself. These include identification of the player, episode, difficulty level and information about every shot fired. The information consists of the time-stamp and success of the shot. The availability of these statistics promises to find correlations between the test results and performance in the game, beyond the main goal of the experiment.

Acknowledgements. This work has been supported by the APVV grant no. APVV-16-0202: "Enhancing cognition and motor rehabilitation using mixed reality".

References

1. de Bruin, N., Bryant, D., MacLean, J., Gonzalez, C.: Assessing visuospatial abilities in healthy aging: A novel visuomotor task. Front. Aging Neurosci. **8**, 1–9 (2016)
2. Dijkstra, N., Zeidman, P., Ondobaka, S., Gerven, M.V., Friston, K.: Distinct top-down and bottom-up brain connectivity during visual perception and imagery. Sci. Rep. **7**(1), 1–9 (2017)
3. Korečko, Š., Hudák, M., Sobota, B.: LIRKIS CAVE: architecture, performance and applications. Acta Polytech. Hung. **16**(2), 199–218 (2019)
4. Korečko, Š., Hudák, M., Sobota, B., Marko, M., Cimrová, B., Farkaš, I., Rosipal, R.: Assessment and training of visuospatial cognitive functions in virtual reality: proposal and perspective. In: Proceedings of 9th IEEE International Conference on Cognitive Infocommunications (CogInfoCom), Budapest, pp. 39–43 (2018)
5. Kravitz, D., Saleem, K., Baker, C., Mishkin, M.: A new neural framework for visuospatial processing. Nat. Rev. Neurosci. **12**(4), 217–230 (2011)

6. Luria, R., Balaban, H., Awh, E., Vogel, E.: The contralateral delay activity as a neural measure of visual working memory. Neurosci. Biobehav. Rev. **62**, 100–108 (2016)
7. Matheis, R.J., Schultheis, M.T., Tiersky, L.A., DeLuca, J., Millis, S.R., Rizzo, A.: Is learning and memory different in a virtual environment? Clin. Neuropsychol. **21**(1), 146–161 (2007)
8. Parsons, T.D., Carlew, A.R., Magtoto, J., Stonecipher, K.: The potential of function-led virtual environments for ecologically valid measures of executive function in experimental and clinical neuropsychology. Neuropsychol. Rehabil. **27**(5), 777–807 (2017)
9. Paulus, W.: Transcranial electrical stimulation (tES - tDCS; tRNS, tACS) methods. Neuropsychol. Rehabil. **21**(5), 602–617 (2011)
10. Schultheis, M.T., Himelstein, J., Rizzo, A.A.: Virtual reality and neuropsychology: upgrading the current tools. J. Head Trauma Rehabil. **17**(5), 378–394 (2002)
11. Shipstead, Z., Harrison, T., Engle, R.: Working memory capacity and visual attention: top-down and bottom-up guidance. Q. J. Exp. Psychol. **65**(3), 401–407 (2012)

A Literature Review of Quality Evaluation of Large-Scale Recommendation Systems Techniques

Hagar ElFiky[(⊠)] [iD], Wedad Hussein [iD], and Rania El Gohary [iD]

Information Systems, Faculty of Computer and Information Sciences,
Ain Shams University, Cairo, Egypt
hagar_elfiky@hotmail.com,
wedad.hussein@fcis.asu.edu.eg,
rania.elgohary@cis.asu.edu.eg

Abstract. The sudden increase of online or internet based cooperates have led to the migration to RS. These systems shall provide accurate predictions and recommendations of services and products to the users of the same interest. The recommendation and prediction performance should strictly audited and evaluated to maintain the optimum quality of service that is being served and to ensure the continuity of such technologies through various considered factors. However, due to the exponential growth of number of the services available online, new challenges have erupted leading to many defects that affect drastically the quality of accurate prediction and recommendation of these systems such as data sparsity, the problem of scalability and cold start. These challenges have attracted many researchers and data scientists to investigate and further exploration of the main source of these raising issues especially in large scale and distributed systems.

Keywords: System quality · System accuracy · Rating prediction ·
Large-scale recommendation systems

1 Introduction

The intensive expansion of the available and accessible digital data and the number of active users to the internet have a direct impact on the performance and efficiency of predicting the users' interests and preferences in an accurate manner. Many of internet-based applications, require users' interactions to express their satisfaction, opinion, and ratings towards a certain item or product of interest through recommendation systems (RS).

This paper is focused on illustrating the major techniques of recommender systems followed by the various challenges of RS. On the other hand, we discussed how the recommendation system got evaluated through common quality measurers. The purpose of this paper is to represent the latest and most recent strategies of evaluating larger-scale internet-based RS in order to recommend accurately the right item for the right user especially if its newly registered user. Besides, the exploration of data sparsity issues that are caused by such systems. Readers will be exposed to distinct

© Springer Nature Switzerland AG 2020
A. E. Hassanien et al. (Eds.): AISI 2019, AISC 1058, pp. 653–662, 2020.
https://doi.org/10.1007/978-3-030-31129-2_60

factors and strategies in evaluating the quality as well as the challenges that affect the performance of such RS.

The rest of the paper is organized as follows: Sect. 2 presents an overview of RS. In Sect. 3 a comparative analysis between the advantages and disadvantages of RS. Section 4, we discuss the prediction of large-scale systems and their main issues. Section 5 sheds the light on the quality issues and challenges of RS. In Sect. 6, quality evaluation measures are discussed. Section 7, we have provided a brief explanation of the literature review and some related work on RS and their applications. Finally, our conclusion and future work is summarized in Sect. 8.

2   Overview of Recommendation Systems

RS are systems or applications that attempt to suggest items to significant potential customers, based on the available data and information or based on predefined interests and preferences [5, 6] besides they are not field specific [8].

In [11], the authors showed the following flowchart in Fig. 1 that illustrates the variety of techniques that could be implemented in any recommendation system.

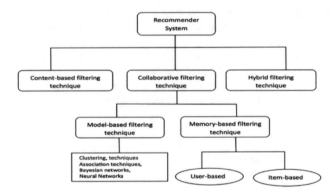

Fig. 1. Diagram of RS techniques[1]

Collaborative filtering (CF), constructs a model from the user's history such as items viewed or purchased by the user, in addition to equivalent decisions taken by other users. Accordingly, the constructed model is utilized to predict items like the users' interests. Content-based filtering (CB), uses a sequence of distinguished features of an item so that further items could be recommended with the same characteristics.

Hybrid filtering technique is a mixture of two or more recommendation approaches in order to dominate the drawbacks of using any approach individually and to improve the overall prediction quality [18, 19, 40] and computational performance [41].

---

[1] https://www.sciencedirect.com/science/article/pii/S1110866515000341.

Social filtering (SF) or community-based RS is based on the interests of the user's circle of friends. These RS are built upon the saying of "Tell me who your friends are and I will tell you who you are" [17].

Knowledge-based systems (KB), are usually depending on the certain domains to get their knowledge from. Knowledge-based systems uses the similarity equation computes the degree of matching the user's interests with the final recommendations [24].

3 Advantages and Dis-advantages of Recommendation Techniques

After reviewing multiple scientific studies and researches, it was discovered that the disadvantages of one algorithm can be overcome by another algorithm [20]. While other techniques like CB cannot afford data sparsity, scalability and accuracy [55]. Due to the nature of hybrid filtering techniques, they can overcome each other's weakness in order to gain better efficiency [18, 19, 40]. Unlikely the KB techniques made their recommendations depending on their knowledge base so any preference change in it may reflect in the recommendations as they map the user interests to the products features as well as considering the product's value factors like delivery schedule and warranty conditions not only the product's specific features [16].

Tables 1 and 2 are produced to conclude these strengths and weaknesses respectively of various recommendation techniques from different quality perspectives.

3.1 Advantages

Table 1. Advantages of various recommender systems

Quality perspective	Recommender system				
	CF	CB	Hybrid	SF	KB
Accuracy and efficiency	X		X		
Ignoring Domain Knowledge	X	X			
Quality improvement over time	X	X			
Preference change sensibility					X
Considering non-product properties					X
Matching users' needs to the recommended item					X
Recommendations sensitivity				X	
Scalability adaptation	X			X	
Overcome Latency		X			

3.2   Dis-advantages

Table 2. Dis-advantages of various recommender systems

Quality perspective	Recommender system				
	CF	CB	Hybrid	SF	KB
Latency	X				
Privacy				X	
Plenty of past data affects the quality	X				
Requires domain knowledge					X
Low performance when limited resource	X				
Recently added user	X	X			
Recently added item	X				

4   Large Scale Recommender Systems

The main challenge for large scale RS is to enhance the recommendation process for large volume, diversity of data, and their ability to scale large graphs [33, 36, 45]. The major large-scale shortcomings are abated in the following challenges:

4.1   Data Diversity

Social networks have helped a lot in solving the poor performance of cold start recommendation of new users [29] despite the variety and multiplicity. Thus, amendments and extensions of traditional techniques have been introduced like matrix factorization, social based matrix factorization [37, 38].

4.2   Data Size

The data size issue started to appear when the number of users and items increase exponentially exceeding the performance of CF. At this point, the computational processors became highly demanding to be practically utilized [34]. However, this problem got solved in [35].

To sum up, our point here is that most of deep learning researches were focusing on reaching more precise and relevant recommendations rather than attacking the large-scale problems. In the light of this, this area needs to be the upcoming challenge for researchers to pay attention for and solve due to its significance.

5   Quality Challenges of Large-Scale Recommenders

5.1   Cold Start Issue

This problem arises when a new item or a new user added in the database [28]. Since any newly added item has no ratings. In the CB, a new user problem becomes more difficult to find similar users without knowing the prior user's interests [16]. Authors in [43] have shown great effectiveness and efficiency in new spatial items recommendation.

5.2 Sparse

Not all users rate most of the items. Consequently, the accuracy of the resultant rating matrix is not of that good quality and it became difficult to find users with a similar rating [28]. In [42], authors have proposed an algorithm that merges both the user and item information to generate higher recommendation quality.

5.3 Large Scale and Scalability

With the tremendous increase of data, the systems become the large scale and having millions of items. Authors in [21] have succeeded to use the CF to search in few clusters instead of searching in the whole database either by reducing dimensionality [22] or by pre-processing approaches [23].

5.4 Latency Problem

CF systems struggle from latency problems [26]. Latency problems mean taking recommendations from already existing rated items instead of the newly added items as they are not rated yet. Being biased to existing items, would eventually cause degradation in the efficiency and quality of the recommended results [27].

5.5 Privacy Problem

Cryptographic algorithms were introduced to handle selling personal data to third parties [26] and other systems [25], allow users to distribute their personal data anonymously without revealing their identity.

6 Quality Evaluation Measures

The type of metric is highly dependable on the type of filtering technique used [12]. Moreover, the objective of quality examination may differ depending on the scenario presented in the system [14]. The main categories of these metrics are presented well below [13, 15] as the following:

6.1 Statistical Metrics

Statistical or probabilistic metrics are extremely effective when they are assessing the validity of predictions through comparing the predicted ratings with the user's actual ratings. The key examples of these metrics include Mean Absolute Error (MAE) and Root Mean Squared Error (RMSE).

6.2   Qualitative Metrics

Qualitative metrics are popular when the main objective is to have a model that cuts down the number of errors. In [16], these measures are heavily used in various RS. The key instances include accuracy, recall and coverage.

6.3   Ranking Metrics

These metrics are built based on the concept of how well the recommender ranks the recommended items. Accordingly improves the accuracy of the ranked output. The main examples of these measures include precision and recall [16].

7   Related Work

7.1   Background

The exponential increase of web services has led the recommender systems to become the backbone of the internet-based systems such as Netflix, YouTube and Google. The newly proposed method in [46] was to improve the performance of RS and overcome data sparsity, cold start and scalability issues. The target was met using singular value decomposition (SVD) [47, 48].

Similar researches such as [49, 50], were conducted to address the most accurate clustering technique for RS. In other researches [51–53] were focused on problems issued by big data. Data mining applications have solved a lot of accuracy and performance problems [1]. Their data mining application model was developed on the weka [2] tool. Their proposed model has succeeded in reaching an accuracy of 80%.

7.2   Applications

Social Networks
In [29], the authors have discussed different context aware RS. The user context can be fixed or dynamically changes over time according to [30]. A post-filtering step that re-ranks and filters the result of traditional recommender in [31]. CARS has solved the cold start, and scalability problems [32].

Movie Ratings
In [3], the authors have managed to reach a RMSE score of 0.8985 in Netflix challenge. They succeeded to achieve performance improvement of 5.91% over Netflix's recommendation system.

Deep learning and neural networks also have shown an intervention in the movie rating prediction as well [39]. However, none of them paid much attention to the large scale and scalability challenge. Our research among the studies has shed light on this gap in order to be inspected later. We have collected many of movie ratings predictions researches in Table 3.

Table 3. Movie ratings prediction techniques

Authors	Objectives	Technique	Dataset
Felfernig and Jeran in [4]	– Presented a comparative analysis between el CF, CB and KB – Suggested a hybrid approach	CF, CB and KB	MovieLens dataset [9]
Zhou and Wilkinson in [3]	– Defined movie rating predictive model with performance improvement 5.91% over Netflix's Cinematch	Alternating-least-squares with weighted-λ-regularization based on CF	Netflix dataset holding 100 million movie rating [8]
Kabinsingha and Chindasorn in [1]	– New movie rating application based on data mining with accuracy 80%	Weka tool and data mining application using decision tree	IMDB dataset [10]
Ponnam in [55]	– Improves prediction accuracy versus CB – Work well with data sparsity – Reduces the number of big error predictions	Item based collaborative filtering technique	Netflix dataset
Basu [39]	– Proposed new movie rating prediction system based on artificial neural networks with accuracy 97.33% and a sensitivity of 98.63%	Levenberg-Marquardt back propagation algo. which has steep degradation in prediction errors	IMDB dataset [10]
Viard and Fournier-S'niehotta in [54]	Extend the graph-based CF for correctly modelling the graph dynamics without losing info	Graph and link stream features intercorporate with CB	MovieLens 20M [9]

8 Conclusion and Future Work

This paper presents a systematic review of many quality challenges and quality evaluation factors for large scale and distributed RS. We have investigated the quality of RS from different perspectives. Also, we have shed light on the gap that can be covered in the future to enhance the quality of cold start and data sparsity problems. This study has highlighted also on the relation between the movie ratings prediction and the movie RS. Since the more accurate the movie rating predictions are, the better quality can be obtained for the RS. In future work, a model will be developed to evaluate & enhance a large-scale recommendation system's quality.

References

1. Kabinsingha, S., Chindasorn, S., Chantrapornchai, C.: Movie rating approach and application based on data mining. IJEIT **2**(1), 77–83 (2012)
2. Weka 3: Data Mining Software in Java. University of Waikato. http://www.cs.waikato.ac.nz/ml/weka/. Accessed 4 Feb 2019
3. Zhou, Y., Wilkinson, D., Schreiber, R., Pan, R.: Large-scale parallel collaborative filtering for the netflix prize. In: Proceedings of 4th International Conference on Algorithmic Aspects in Information and Management 2008, LNCS 5034, pp. 337–348. Springer (2008)
4. Felfernig, A., Jeran, M., Ninaus, G., Reinfrank, F., Reiterer, S., Stettinger, M.: Basic approaches in RS. In: RS in Software Engineering, pp. 15–37. Springer (2014)
5. Burke, R., Felfernig, A., Goeker, M.: Recommender systems: an overview. AI Mag. **32**(3), 13–18 (2016)
6. Jannach, D., Zanker, M., Felfernig, A., Friedrich, G.: Recommender Systems – An Introduction. Cambridge University Press, Cambridge (2010)
7. Linden, G., Smith, B., York, J.: Recommendations – item-to-item collaborative filtering. IEEE Internet Comput. **7**(1), 76–80 (2003)
8. https://www.kaggle.com/netflix-inc/netflix-prizedata#movie_titles.csv
9. https://grouplens.org/datasets/movielens/
10. Internet movie database, IMDb.com, Inc. https://www.imdb.com/interfaces/. Accessed 11 Feb 2019
11. Isinkaye, F.O., Folajimi, Y.O., Ojokoh, B.A.: RS: principles, methods and evaluation. Egypt. Inf. J. **16**(3), 261–273 (2015)
12. Friedman, N., Geiger, D., Goldszmidt, M.: Bayesian network classifiers. Mach. Learn. **29**(2–3), 131–163 (1997)
13. Véras, D., Prota, T., Bispo, A., Prudêncio, R., Ferraz, C.: A literature review of recommender systems in the television domain. Expert Syst. Appl. **42**(22), 9046–9076 (2015)
14. Herlocker, J.L., Konstan, J.A., Terveen, L.G., Riedl, J.T.: Evaluating collaborative filtering recommender systems. ACM Trans. Inf. Syst. (TOIS) **22**(1), 5–53 (2004)
15. Ferri, C., Hernández-Orallo, J., Modroiu, R.: An experimental comparison of performance measures for classification. Pattern Recogn. Lett. **30**(1), 27–38 (2009)
16. Kumar, B., Sharma, N.: Approaches, issues and challenges in recommender systems: a systematic review. Indian J. Sci. Technol. **9**(47) (2016). https://doi.org/10.17485/ijst/2016/v9i47/94892
17. Arazy, O., Kumar, N., Shapira, B.: Improving social recommender systems. IT Prof. Mag. **11**(4), 31–37 (2009)
18. Adomavicius, G., Zhang, J.: Impact of data characteristics on recommender systems performance. ACM Trans. Manage. Inform. Syst. **3**(1), 3:1–3:17 (2012)
19. Stern, D.H., Herbrich, R., Graepel, T.: Matchbox: large scale online bayesian recommendations. In: Proceedings of the 18th International Conference on World Wide Web, pp. 111–120. ACM, New York, USA (2009)
20. Schafer, J.B., Frankowski, D., Herlocker, J., Sen, S.: Collaborative filtering recommender systems. In: Brusilovsky, P., Kobsa, A., Nejdl, W. (eds.) The Adaptive Web, LNCS, vol. 4321, pp. 291–324. Springer, Heidelberg (2007)
21. Su, X., Khoshgoftaar, T.M.: A survey of collaborative filtering techniques. Adv. Artif. Intell. **2009**, 4 (2009)
22. Billsus, D., Pazzani, M.J.: Learning collaborative information filters. In: ICML, pp. 46–54 (1998)

23. Shahabi, C., Banaei-Kashani, F., Chen, YS., McLeod, D.: Yoda: an accurate and scalable web-based recommendation system. In: Cooperative Information Systems, pp. 418–432 (2001)
24. Bridge, D., Göker, M.H., McGinty, L., Smyth, B.: Case-based recommender systems. Knowl. Eng. Rev. **20**(03), 315–320 (2005)
25. Polat, H., Du, W.: Privacy-preserving collaborative filtering using randomized perturbation techniques (2003)
26. Khusro, S., Ali, Z., Ullah, I.: Recommender systems: issues, challenges, and research opportunities. In: Information Science and Applications (ICISA) 2016, LNEE, vol. 376, pp. 1179–1189. Springer, Singapore (2016)
27. Sollenborn, M., Funk, P.: Category-based filtering and user stereotype cases to reduce the latency problem in recommender systems. In: Advances in Case-Based Reasoning, pp. 395–405. Springer (2002)
28. Ma, H., Yang, H., Lyu, MR., King. I.: Sorec: social recommendation using probabilistic matrix factorization. In: Proceedings of the 17th ACM Conference on Information and Knowledge Management, October 2008, pp. 931–940 (2008)
29. Eirinaki, M., Gao, J., Varlamis, I., Tserpes, K.: Recommender systems for large-scale social networks: a review of challenges and solutions. Future Gener. Comput. Syst. **78**, 412–417 (2017)
30. Ricci, F.: Context-aware music recommender systems: workshop keynote abstract. In: Proceedings of the 21st International Conference Companion on World Wide Web, pp. 865–866. ACM, Lyon (2012)
31. Adomavicius, G., Tuzhilin, A.: Context-aware recommender systems. In: Ricci, F., Rokach, L., Shapira, B. (eds.) Recommender Systems Handbook, pp. 191–226. Springer (2015)
32. Song, L., Tekin, C., Schaar, M.V.D.: Online learning in large-scale contextual recommender systems. IEEE Trans. Serv. Comput. **9**(3), 433–445 (2016)
33. Sardianos, C., Varlamis, I., Eirinaki, M.: Scaling collaborative filtering to large-scale bipartite rating graphs using lenskit and spark. In: IEEE Second International Conference on BigDataService, pp. 70–79 (2017)
34. Su, X., Khoshgoftaar, T.: A survey of collaborative filtering techniques. Adv. Artif. Intell. **4** (2009)
35. Sardianos, C., Tsirakis, N., Varlamis, I.: A survey on the scalability of recommender systems for social networks. In: Social Networks Science: Design, Implementation, Security, and Challenges, pp. 89–110. Springer, Heidelberg (2018)
36. Shanahan, J.G., Dai, L.: Large scale distributed data science using apache spark. In: Conference Proceedings of the 21th ACM SIGKDD International Conference on Knowledge Discovery and Data Mining, pp. 2323–2324 (2015)
37. Ma, H., King, I., Lyu, M.R.: Learning to recommend with explicit and implicit social relations. ACM Trans. Intell. Syst. Technol. **2**(3) (2011). Article No. 29
38. Gurini, D.F., Gasparetti, F., Micarelli, A., Sansonetti, G.: Temporal people-to-people recommendation on social networks with sentiment-based matrix factorization. Future Gener. Comput. Syst. **78**(P1), 430–439 (2018)
39. Basu S.: Movie rating prediction system based on opinion mining and artificial neural networks. In: Kamal, R., Henshaw, M., Nair, P. (eds.) International Conference on Advanced Computing Networking and Informatics. Advances in Intelligent Systems and Computing, vol. 870. Springer, Singapore (2019)
40. Unger, M., Bar, A., Shapira, B., Rokach, L.: Towards latent context aware RS. Knowl. Based Syst. **104**, 165–178 (2016)

41. Verma, J.P., Patel, B., Patel, A.: Big data analysis: recommendation system with hadoop framework. In: Computational Intelligence & Communication Technology, IEEE, pp. 92–97 (2015)

42. Niu, J., Wang, L., Liu, X., Yu, S.: FUIR: fusing user and item information to deal with data sparsity by using side information in RS. J. Network Comput. Appl. **70**, 41–50 (2016)

43. Yin, H., Sun, Y., Cui, B., Hu, Z., Chen, L.: LCARS: a location-content-aware recommender system. In: Proceedings of the 19th ACM SIGKDD International Conference on Knowledge Discovery and Data Mining, ACM, pp. 221–229 (2013)

44. Tao, Y.: Design of large-scale mobile advertising recommendation system. In: 2015 4th International Conference on Computer Science and Network Technology (ICCSNT), IEEE, pp. 763–767 (2015)

45. Smith, B., Linden, G.: Two decades of recommender systems at amazon.com. IEEE Internet Comput. **21**(3), 12–18 (2017)

46. Zarzour, H., Al-Sharif, Z., Al-Ayyoub, M., Jararweh, Y.: A new collaborative filtering recommendation algorithm based on dimensionality reduction and clustering techniques. In: 9th International Conference on Information and Communication Systems (ICICS), Irbid, pp. 102–106 (2018)

47. Arora, S., Goel, S.: Improving the accuracy of recommender systems through annealing. Lecture Notes in Networks and Systems, pp. 295–304 (2017)

48. Koren, Y., Bell, R., Volinsky, C.: Matrix factorization techniques for recommender systems. Computer **42**(8), 30–37 (2009)

49. Koohi, H., Kiani, K.: User based collaborative filtering using fuzzy C-means. Measurement **91**, 134–139 (2016)

50. Kim, K., Ahn, H.: Recommender systems using cluster-indexing collaborative filtering and social data analytics. Int. J. Prod. Res. **55**(17), 5037–5049 (2017)

51. Meng, S., Dou, W., Zhang, X., Chen, J.: KASR: a keyword-aware service recommendation method on MapReduce for big data applications. IEEE Trans. Parallel Distrib. Syst. **25**(12), 3221–3231 (2014)

52. Zarzour, H., Maazouzi, F., Soltani, M., Chemam, C.: An improved collaborative filtering recommendation algorithm for big data. In: Proceedings of the 6th IFIP International Conference on Computational Intelligence and Its Applications 2018, CIIA, pp. 102–106 (2018)

53. Tian, H., Liang, P.: Personalized service recommendation based on trust relationship. Sci. Program. **2017**, 1–8 (2017)

54. Viard, T., Fournier-S'niehotta, R.: Movie rating prediction using content-based and link stream features. Arxiv (2018). https://arxiv.org/abs/1805.02893

55. Ponnam, L.T., et al.: Movie recommender system using item based collaborative filtering technique. In: International Conference on Emerging Trends in Engineering, Technology and Science (ICETETS), vol. 1, IEEE, pp. 56–60 (2016)

Intelligent Transport Systems and Its Challenges

Elezabeth Mathew$^{(\boxtimes)}$ (ID)

British University in Dubai (BUiD), Dubai, UAE
2016152110@student.buid.ac.ae

Abstract. The paper is an extension of 'Vision 2050 of the UAE in Intelligent Mobility' (Mathew and Al Mansoori 2018). Information and communication technology has renovated several industries, from education to health care to government, and at present it is stage of makeover of transportation systems. While people consider enlightening a country's transportation system merely means construction of new roads or repairing aging infrastructures, the future of transportation lies not only in concrete and steel, but also progressively more in using IT. IT empowers components within the transportation system—vehicles, roads, traffic lights, message signs, etc.—to become intelligent by implanting them with microchips and sensors and empowering them to interconnect with each other through wireless technologies. This paper evaluates intelligent transport system as a system of systems in the cyber-physical world. Moreover, this paper tries to identify the challenges and opportunities. The emphasis is on long-term challenges in SOSE research, and ideas for attacking these challenges.

Keywords: AI · Big data · CPS · CPSS · IoT · ITS · IT · Intelligent · SoS · SoSE · Systems · Transport · Vehicle

1 Introduction

Many years ago, 3C (computing, communication, and control) technology has been effectively used to various corporal systems like defense, energy, critical infrastructure, healthcare, manufacturing and transportation, vividly improved their controllability, adaptability, autonomy, efficiency, functionality, reliability, safety, and usability [1]. These corporal systems are now controlled by IoT, social network, cloud computing, big data and intelligent systems due to the technical and social development. A Cyber-Physical System (CPS) integrate the physical, cyber, and human factors in a single framework. Cyber systems are gradually implanted in all types of physical systems and are making the systems more intelligent, energy-efficient and comfortable, e.g., intelligent transportation systems (ITS), factories, and cities [1]. The CPSs are suitably ever-present in daily life and their complexity is growing endlessly. CPS by itself cannot approximate the influence of humans, organization, and societies which are unpredictable, dissimilar and complex. This implies that CPS may become useless in some situations where human interference is necessary. Fortunately, the social behavior can be attained, evaluated and exploited to elevate to control and manage the performance of complex systems based on the worldwide internet and big data technologies [1].

© Springer Nature Switzerland AG 2020
A. E. Hassanien et al. (Eds.): AISI 2019, AISC 1058, pp. 663–672, 2020.
https://doi.org/10.1007/978-3-030-31129-2_61

Recently, the Cyber-physical social system (CPSS) has been proposed by integrating the social components into CPS. This paper presents compelling arguments in favor of new research directions in Intelligent Transport Systems (ITS) area that is based on a cyber-physical systems (CPS) perspective [2].

2 Intelligent Transport System (ITS)

The territory of transport which include aircraft, roads, waterways, sea and potentially, undersea are all in fast changes in the size, shape, their destinations, and especially how these mobility vehicles deal with their customers [2]. The entire business models, processes, and procedures that use conveyance as means to distribute goods and people have also been changes drastically with the emergence of artificial intelligence (AI). AI integrated into software in these transport modules enhances a faster, better and cheaper delivery, at the same time cutbacks in emissions and other waste from the perception of sustainability [2].

3 Is ITS a SoS?

A software-intensive system is, by definition, any system where software influences to a large extent the design, construction, deployment, and evolution of the system as a whole [2]. The system of Systems Engineering (SoSE) is an evolving interdisciplinary approach that emphasis on ultra-large scale interoperation of many independent self-contained constituent systems in order to satisfy a global need [1].

Table 1 explains each property of SoS and equates it with ITS. Hence, as in concept map of SoS below ITS depicts all the characteristics of SoS, so definitely intelligent transport system is coming under the umbrella of the acknowledged system of systems [1].

Table 1. Properties of SoS

	Property	Proof
1	Operational independence	Each constituent system in ITS like rail, road, air etc. can operate independently and is capable of achieving its own goals in the absence of the other systems
2	Managerial independence	The constituent systems of ITS can managed independently and can be added or removed from the system of systems
3	Evolutionary in nature	The SOS adapts to fulfill its, possibly evolving, objectives as its underlying technologies and needs evolve with time
4	Emergent behaviour	The functionality and behavior of the SOS develops in ways not achieved by the individual systems
5	Geographical distribution	ITS has a geographical distribution that limits the interaction of the constituent systems to information exchange

The ITS has operational independence, managerial independence, is evolutionary in nature, has got emergent behavior and it is distributed geographically. Moreover, ITS belongs to acknowledges category of SoS. The concept map of SoS is given below for clarifications. As depicted in the concept of SoS contains systems that are independently managed and operated. It does not have clear boundaries and exhibits complex properties. SoS also have other properties like heterogeneity, geographically distributed, emergence and evolutionary in nature (Table 2).

Table 2. ITS compared with SoS characteristics

Categories	Directed	The interoperable SOS is built to fulfill a specific purpose
	Collaborative	The constituent system 'voluntarily' collaborate in an agnostic way to achieve an agreed-upon central purpose
	Acknowledged	There are recognised objectives, a designated manager, and resources for the SOS. The constituent systems return their independent ownership, objectives, funding, development, and sustainment approaches
	Virtual	There is no centrally agreed purpose for the SOS

Hence, as displayed in concept map below ITS depicts all the characteristics of SoS, so definitely intelligent transport system is coming under the umbrella of the acknowledged system of systems [1].

Each constituent system in ITS has a designated manager and resources for each SoS. However, the constituent systems return their independent ownership, objectives, funding, development, and sustainment approaches and hence ITS is categorized as acknowledged Systems.

4   Challenges in System of Systems

1. Emergence - Unpredictable emergent behaviour (and particularly undesirable emergent behaviour) is the fundamental challenge of SoS. That is to say the development of theories, methods, and tools to predict the emergent behaviours of combinations of SoS forms a critical underlying challenge for SoS Engineering [3].
2. Effective Systems Architecting - The challenges are to develop a better understanding of the relationship of architectural features to behaviours, the assessment and evaluation of architectures, and the use of architectures by a range of SoS stakeholders. The improved understanding must be realised through new methods and approaches to architecting [1].
3. Situational Awareness - The root cause of many SoS failures can be attributed to a lack of situational awareness of either human or non-human decision making agents. In particular, the inability of an agent to foresee and analyse the behaviour of future SoS states. This is different from "bad" emergence where adverse behaviour arises due to all systems within the SoS behaving normally (sometimes called a normal error), instead, a lack of situational awareness leads an actor in the SoS to

behave inappropriately. This could be because of partial information, too much information, or mis-analysis of information [1].

4. Governance and relationships - There are two main challenges associated with this theme: firstly, clarity of responsibility in SoS and, secondly, the approaches that are needed to encourage the owners/developers of individual systems to build and manage their systems in such a way that the overall goals of any SoS within which such systems are incorporated, are achieved and/or the SoS benefits are maximized [1].

5. This requires research at quite a fundamental level into models of SA in SoS, information management, and presentation and, also, training and education for both systems developers and systems users [1].

6. The evolutionary nature of SoS causes issues with the interoperation of component systems that have been designed and built at different times, maybe to different standards, and with different performance characteristics [1].

5   Cyber-Physical Systems and Limitations

These concern the increasing use of embedded software and, in particular, the direct and real-time interface between the virtual and physical worlds, including Internet of Things! There are many similarities with SoS, but the main distinction concerns emphasis. CPS is mainly concerned with control, whereas SoSE is mainly concerned with configurations, authorities and responsibilities, interoperability, and emergent behavior. CPS could be considered to be a special case of a SoS [3].

The main restrictions of cyber-physical systems, from a systems software perspective, are as follows:

– There is an incompatibility between application needs and system service supplies due to insufficient APIs making it difficult for applications to precisely identify the service they desire [3].
– System services tend to be application-agnostic, typically focusing on fairness and efficiency rather than timing, safety, security and reliability constraints, and
– Current systems are too inflexible to be easily extended with specific services for target applications [3].

The challenges faced by future systems support for CPS include:

– The design of an underlying software architecture that organizes itself with the most appropriate methods of communication and isolation between services [3].
– The automatic composition of services to satisfy application constraints, given underlying hardware limitations, and
– Careful design of APIs and consideration of hardware heterogeneity in the generation and verification of a software system for a given application [3].

In summary, the unresolved questions are:

1. How to guarantee the functional correctness of complex interactions between related vehicle features.

2. How to guarantee compliance with parafunctional requirements in distributed implementations of vehicle features.
3. What types of affordable run-time architectures to deploy to maintain prescribed levels of functionality and performance (ranging from full to degrade to failsafe) in the presence of hardware and software faults?

6 Challenges in ITS

The challenges are expected in technical and non-technical areas are listed below:

1. AI - empowering self-governing administration, better controlling of procedures and declines in cost [4].
2. IoT - progressions in what it can process and the speed with superior networks. In transportation, we can solve challenges with rail safety, revenue collection, congestion management, and customer experience with intelligent IoT architectures and analytics software.
3. Energy - The concept of alternate energy resource instead of fossil fuels are in high priority and experiments are still on-going.
4. 3D printing - manufacturing towards 'weightless distribution'
5. Big data- Big Data, IoT and AI together are considered the superhero team that is currently emerging.

Although these carters are strong, emerging and prevailing necessity there are still some obstacles in both technical and non-technical side. Amongst the technical barriers are the present inadequate understanding of AI in the real world, the absence of standards and protocols for proficient operation of forthcoming transport networks, and the slow development in streamlining energy networks for transport and other needs [4].

At the same time, in non-technical challenge, there should be a new guiding environment that is well-matched with autonomous vehicles and networks, safety and security etc. Moreover, the communities should adjust themselves to the new disruptive changes to their daily life and accepts the impacts on jobs, skills, and employment. Most important co-ordination plans will be the expansion of all-encompassing information networks and big data applications that are safe and secure to enable the whole transport collaboration to operate successfully and proficiently and to establish flexibility and fast recovery when situations mandate this [4].

The need to develop autonomous vehicles in all transport modes as CPS devices that are harmless, protected and competent to fulfil customer needs is a mandate from the perception of technology. Moreover, a vital requirement to provide for human authority and responsibility for their operations. It is also equally important to have standards and protocols for the interoperation of these vehicles. Simultaneously, the information networks will make available them with the on the spot information essential for smooth, safe operation. Everything necessitate support; the links, the vehicles, the information and knowledge both over the lifecycles of the individual components and of the whole Systems of Systems represented beneath as modification frequently happens [5]. This is a demand on systems for education.

7   Opportunities in ITS

Fully simulated autonomous vehicles will still need a long way to go the opportunities to implement the same are plentiful.

- Access to all:- Even though the figures show there is a deterioration in the road accidents in the UAE, that drop is not a big one. Still the futurists of the country are trying to investigate ways to decrease road accidents most of which is caused by inattentiveness, sloppiness, alcoholism or wrong decision making. The system which is having a better safety and warning system that also has better security features with latest sensing and control systems could make a big difference in the statistics. Additionally, tryouts are carried on in the area of lane changing, automatic braking systems, automated parallel parking, and autonomous override control [12].
- Changing markets:- Since the sensors are becoming cheaper day by day, the vendors are utilizing it to the maximum, which will make the active control system more reliable and verifiable [12].
- Demand for new infrastructure:- the recent revolution in transportation like metro and transit has been overlooked on the arrival of accompanying services such as car sharing or bike sharing [14]. Freedom in transportation is an on-going challenge in the region. It is still a dream of both people in city and village area to have a convenient pods which travel across places, communities, and sectors [11].
- Behavioral change:- The competitors in the field could make a system that can give more customer satisfaction by enabling the vehicle facility with more flexible and faster service. The latest development in big data analytics has helped the new entrants to understand the change in behavior of customer demands to have more flexible journey planning on demand and better monitoring services to vehicle [12].
- Personalized service:- The demand of personalized mobility is increasing day by day giving more opportunity to enhance the standardization of transportation by allowing the customers to choose the seating and standing capacity on public transport [19]. This will allow flexibility to serve all users, including socially challenged and aging populations. The custom-made service which emphases on customer satisfaction will help in breaching the obstacle of privacy. Moreover, the personalized services will prompt the customers to have more trust in users and the agents get a chance to have better service to the customer [11].
- 5G Network:- It launches a wide-ranging wireless network with no limitation. Wireless World Wide Web, www is highly supported by 5G network. The TRA (Telecommunications Regulatory Authority) has already announced the launch of 5G network in Dubai, which will revolutionize the internet usage of UAE [7] [24]. This technology falls in line with the directives of prudent leaders and UAE Vision 2021.
- Industrial revolution:- the intelligent mobility has changed a lot due to the revolution in the industrial sector. Firstly MaaS (Mobility as a service) has concentrated on customer-centric transportation to deliver an integrated transportation system [17].

Then big data has also traveled a long way where now BDA has great influence on decision making and prediction. As a result the smart roads are expected to take a maximum benefit and provide better driving experience [16]. Recently IoT has become the trend revolution where interconnected communication of vehicles especially with emergency services [18].

8   Sustainability in ITS

1. Traffic Management:- ITS deployment is expected to increase the efficiency of traffic management. The biggest impact would in reducing the congestion and cut down on travel time. By giving the real-time traffic information to whoever is driving can enhance traffic efficiency. By reducing congestion and better traffic efficiency mobility becomes more punctual and reliable and thus increasing the trust in customers. It also includes a quick response from transport managers to traffic incidents [5].
2. Carbon emission:- Effective and efficient ITS will result in public using more public transport which in turn result in a reduction of carbon dioxide(CO_2) and other airborne toxins. Moreover, the vehicles integrated with ITS services can yield an approximate 15% reduction in carbon emissions. People's use of public transportation will likely increase to that of personally owned vehicles [5].
3. Economic values:- The enhanced efficiency in ITS will yield economic dividends as well. This is achieved by optimized use of existing infrastructure and transportation systems. Simultaneously, moving both people and freight will benefit overall economic activity and traffic management. Moreover, it will assist in the development of industrial sectors such as automobiles, electrical equipment, communication networks, software, and engineering. Eventually, all of these will result in a reduction of time, costs and the stress associated with travel time [5].

9   ITS Applications

The major applications used in ITS can be grouped into 5 (Table 3).

ITS provide five key modules of benefits by (1) enhanced safety, (2) refined operational performance, especially by decreasing bottleneck, (3) improving freedom of movement and accessibility, (4) supplying eco-friendly benefits, and (5) furthering throughput and growing economic and occupational growth [6, 15].

10   ITS Products

1. Technology and Infrastructure Support:- Advancement and installation of the DSRC (Dedicated Short Range Communication), the GPS (Global Positioning System) and CALM (Communication Air Interface and Long and Medium Range)

Table 3. ITS applications

Categories	Used for:
Advanced traveler information systems	Providing drivers with real-time information, such as transit routers and schedules; navigation directions; and information about delays due to congestion, accidents, weather conditions or road repair work
Advanced transportation management systems	Include traffic control devices, such as traffic signals, ramp meters, variable message signs, and traffic operations centers
ITS-enabled transportation pricing systems	Include systems such as electronic toll collection (ETC), congestion pricing, fee-based express (HOT) lanes, and vehicle miles traveled (VMT) usage-based fee systems.
Advanced public transportation systems	Allow trains and buses to report their position so passengers can be informed of their real time status (arrival and departure information)
Fully integrated intelligent transportation systems	Vehicle-to-infrastructure (V2I) and vehicle-to-vehicle (V2V) integration, enable communication among assets in the transportation system, for example, from vehicles to roadside sensors, traffic lights, and other vehicles

technologies improved the direction-finding and monitoring systems. Being a short-range communication protocol, DSRC offers a consistent and economical means for vehicle to vehicle (V2V) and vehicle to infrastructure (V2I) interaction [9, 13].

2. "Vehicle-to-Everything" communication:- The emerging core technology, V2X, transmits valuable traffic statistics with each other using DSRC, GPS, IoT, big data analytics, etc. ITS countries are directing their research and business development into V2X that's expected to be a doorway towards the future of this communication technology [13].

3. The Co-operative ITS model:- The advanced "ubiquitous" communication technology enhances the communication between systems and roads, vehicles and drivers by using C-ITS. The International Organization of Standardization (ISO) defines C-ITS stations facilitate actions that result in improved safety, sustainability, efficiency and comfort beyond the range of unconnected systems. These modern systems use a wireless connection to communicate with other vehicles or roadside infrastructure [9].

4. Advanced Transport Management systems:- Advanced Transport Management Systems (ATMS) operate with infrastructure fitted with vehicle detection systems, automatic vehicle identification, and CCTV to allow for real-time traffic data to be both sent and delivered through various service devices or facilities including Variable Message Sign (VMS), the World Wide Web or mobile devices. The data collected are transformed into various ITS services including real-time traffic information, Bus Information Services (BIS) and Electronic Toll Collection Services (ETCS) [9].

5. SoS interference:- Systems of Systems field is also promising more advancement in prediction and trust systems by installing 18–20 different sensors in the vehicle.

Few examples are Light Detection and Ranging (LADAR), Radar, cameras, and GPS Mapping. The artificial intelligence will provide advanced solutions for validation and verification, which includes how a smart car reacts towards dump cars in the road. Vigorous deviations will also reach the network system as it is probable to have the 6G network via air instead of regular fiber optic and cables and broadband calculations [9].

11   The Way Forward

ITS needs more in-depth and systematic research in policymaking. Autonomous vehicles will revolutionize the transport sector with numerous significant policy suggestions. Further research is required to measure how much ITS can lessen bottleneck and contamination and thus to make a better sustainable ecology. Also, similar in-depth research is required in enhancing safer and more effectual rail systems in making better the expedition efficacy and cross-border enablement [9]. The secretariat of the Economic and Social Commission for Asia and the Pacific (ESCAP) could also bring out additional exploration on mixed mode commuting and interconnected transport systems, with a vision to promote smooth trade in possessions and the development of tourism in the province. A similar study should also be done in domestic waterways, navigation, air traffic control, and maritime transport [9].

References

1. Xiong, G., Zhu, F., Liu, X., Dong, X., Huang, W., Chen, S., Zhao, K.: Cyber-physical-social system in intelligent transportation. IEEE/CAA J. Autom. Sin. **2**(3), 320–333 (2015)
2. Gokhale, A., McDonald, M., Drager, S., McKeever, W.: A cyber-physical systems perspective on the real-time and reliable dissemination of information in intelligent transportation systems. Netw. Protoc. Algorithms **2**(3) (2010)
3. Sadek, A., "Brian" Park, B., Cetin, M.: Special issue on cyber transportation systems and connected vehicle research. J. Intell. Transp. Syst. **20**(1), 1–3 (2014)
4. Elloumi, M., Kamoun, S.: Adaptive control scheme for large-scale interconnected systems described by hammerstein models. Asian J. Control **19**(3), 1075–1088 (2017)
5. Kant, V.: Cyber-physical systems as sociotechnical systems: a view towards human-technology interaction. Cyber-Phys. Syst. **2**(1–4), 75–109 (2016)
6. Reimann, M., Rückriegel, C.: Road2CPS priorities and recommendations for research and innovation in cyber-physical systems (2017)
7. R. Team: A smart information system for public transportation using IoT. Int. J. Recent Trends Eng. Res. **3**(4), 222–230 (2017)
8. Misauer, L.: IoT, Big Data and AI - the new 'Superpowers' in the digital universe | Striata, Striata (2018). https://striata.com/posts/iot-big-data-and-ai-the-new-superpowers-in-the-digital-universe/. Accessed 13 June 2018
9. The Information and Communications Technology and Disaster Risk Reduction Division: Intelligent Transportation Systems for Sustainable Development in Asia and the Pacific, ESCAP, UN (2018)

10. Campbell, M., Egerstedt, M., How, J., Murray, R.: Autonomous driving in urban environments: approaches, lessons and challenges. Philos. Trans. Roy. Soc. A Math. Phys. Eng. Sci. **368**(1928), 4649–4672 (2010)
11. Uhlemann, E.: Transport ministers around the world support connected vehicles [Connected Vehicles]. IEEE Veh. Technol. Mag. **11**(2), 19–23 (2016)
12. Smart Mobility UAE Forum: Smart Mobility UAE Forum (2018). https://smartmobilityuae.iqpc.ae/. Accessed 21 Mar 2018
13. Mervis, J.: Are we going too fast on driverless cars? Science (2017)
14. Google Drive - Cloud Storage & File Backup for Photos, Docs & More (2018). Drive.google.com
15. Waldrop, M.: Autonomous vehicles: no drivers required. Nature **518**(7537), 20–23 (2015)
16. Pavlova, L.: Wi-Fi and IoT in focus. LastMile **69**(8), 56–60 (2017)
17. Acknowledgement to Reviewers of Robotics in 2016. Robotics **6**(4), 1 (2017)
18. Tomory, L.: Technology in the british industrial revolution. Hist. Compass **14**(4), 15 (2016)
19. Mathew, E., Al Mansoori, S.: Vision 2050 of the UAE in Intelligent Mobility. In: 2018 Fifth HCT Information Technology Trends (ITT), Dubai, United Arab Emirates, pp. 213–218 (2018)

A Conceptual Framework for the Generation of Adaptive Training Plans in Sports Coaching

Laila Zahran[1,2]([⊠]), Mohammed El-Beltagy[1]([⊠]),
and Mohamed Saleh[2]([⊠])

[1] Optomatica, Giza, Egypt
{laila,mohammed}@optomatica.com
[2] Faculty of Computers and Information, Cairo University, Giza, Egypt
m.saleh@fci-cu.edu.eg

Abstract. Planning training sessions is one of the coaches' main responsibilities in Sports Coaching. Coaches watch their athletes during training, identify key aspects of their performance that can be improved and plan training sessions to address the problems that they have observed. Limited work has been proposed and applied to the generation of training plans using technology. There is great potential for improving the generation of personalized training plans by using Machine Learning techniques. Recently, many methods and techniques were proposed in theory and practice in order to help athletes in sports training generally. Integrating some of these methods and techniques would result in the generation of automated, adaptive and personalized training plans. In this paper, we propose a conceptual framework for training plan generation in an adaptive and personalized way for athletes. This framework integrates performance indicators such as training load measures, physiological constraints, and behavior-change features like goal setting and self-monitoring. It provides a training plan, being adopted by the athlete, and its goal adapts to the athlete's behavior.

Keywords: Sports training plan · Training load · Behavior-change features · Physiological constraints

1 Introduction

In recent years, the number of sports participants has remarkably increased. Casual sports events with a small number of participants suddenly became mass sports events that assemble from hundreds or thousands of participants. In order to participate in such events, people have to be prepared, and preparation demands proper sports training.

Sports coaching is a complex and sophisticated process where athletes try to prepare themselves optimally for a sports event or competition. The European Coaching Council defines coaching as the process of guided improvement and development of sports participants in a single and at identifiable stages of sportsperson pathway [2]. It is a very challenging task and it is viewed as an episodic process, with the coach working on a week-to-week basis to improve the athlete's performance [1]. One of the

© Springer Nature Switzerland AG 2020
A. E. Hassanien et al. (Eds.): AISI 2019, AISC 1058, pp. 673–684, 2020.
https://doi.org/10.1007/978-3-030-31129-2_62

coach's main responsibilities is to prescribe training plans to the athletes to increase their performances. This would be a very long process if coaches do not know the characteristics of the athletes' bodies in details.

The presence of a sports coach was inevitable until recently. Unfortunately, coaches are generally very expensive. Using computer technology to help athletes in training was a vibrant topic for research in the past 10–15 years. Many useful coaching applications emerged in theory with practical applicability. It is necessary to develop new cost-effective, scalable approaches to increase physical activity and participation in sports. One promising direction is the use of Smartphone in the delivery and personalization of programs that motivate athletes to have proper sports coaching [8]. To help athletes with a proper sports training plan, artificial sports applications were designed that is an inexpensive variant of the real coach. These applications support a wide spectrum of tracking mobile devices and are based on sports training theories. Smartphone have powerful computation and communication capabilities that enable the use of machine learning and other data-driven analytics algorithms for personalizing the physical activity programs to each individual. Furthermore, the past several generations of Smartphone integrate reliable activity tracking features [3–6], which makes possible the real-time collection of fine-grained physical activity data from each individual.

Personalizing generated training plans has caught attention in the research area. Some models were proposed but they lack features of a real sports coaching process for monitoring, motivating and adapting to athletes' performance [7, 9]. Some other models were proposed to encourage increase general physical activity by setting personalized, adaptive goals, in addition to, behavior-change features such as self-monitoring and goal setting [8]. Currently developed sports coaching applications that help generate training plans for athletes have only caught fixed or rule-based generated training plans. For example, Racefox application offers rule-based generated training plans. While both Garmin and Fitbit offer fixed training plans [23–25]. Applications are not exposed yet to the latest research regarding the optimization of the personalized generation of training plans using Machine Learning techniques.

Since, limited work has been done on the generation of personalized training plans. There is a great potential for improving the generation of training plans, taking into account: proper and accurate choice of training load measure when using it as a performance indicator, taking advantage of personal characteristics and behavior-change features, and constructing training plans in an adaptive fashion. In this paper, we propose a conceptual framework for generating adaptive sports' training plans using training load and physiological constraints as performance indicators, in addition to, applying behavior-change features like dynamic goal setting and self-monitoring. The framework deals with psychological measures of training load, and include exercises variation constraints. It provides a training plan with adaptive goals to the athlete's behavior.

In the next section, we will discuss and review related work. Then, we will discuss the training plans, their types and their limitations. Next, we will present our proposed framework. Finally, we will discuss the framework strength points and future work.

2 Related Work

In this section, we review the latest work of training plan generation using swarm intelligence algorithms. Then, we review work done on the generation of training plans in Cycling while including some physiological constraints. Next, we describe the work done on setting adaptive personalized step goals.

A swarm intelligence algorithm solution-based was proposed to optimally select sports sessions from predefined course clusters that are tracked using mobile sports tracking [7]. In this research training plans were generated using the training load as a performance indicator. The proposed algorithm consists of preprocessing (assembling and identifying activities, parsing characteristics, determining training load indicators and clustering), optimization, and visualization stages. Reviewing this proposed work we find that clustering was based on historical data that lacked heart rate variation, they used a week measure for training load and only captured heart rate and duration taken by the athlete for each exercise. As for the optimization, the model proposed would yield very little variation of exercises within any given cluster, and it doesn't contain any constraints regarding exercises or their choice according to the athlete's personal features.

Physiological constraints are addressed in research when generating training plans [9]. In this research they presented a model for cycling 8-week training plan generation while satisfying some physiological constraints such as monotony, chronic training load (CTL) ramp rate, and daily training impulse. It used Adaptive Particle Swarm Optimization using ε-constraint methods to formulate such a plan and simulate the likely performance outcomes. The model produced a safe, and a high-performance training plan. Limitations of this model are that it lacks adaptation to both the athlete's physical activity on a particular sport, as well as, interruptions that intervene in the generated plan. Training plan reorganization would be needed to cover any unpredictable event to maintain the athlete's performance as much as possible in the remaining training plan number of days.

CalFit is a novel fitness application that promotes physical activity. It is designed to automatically set personalized, adaptive daily step goals and adopt behavior-change features such as self-monitoring [8]. The daily step goals are computed using a reinforcement learning algorithm [10, 11] adapted to the context of physical activity interventions. The application uses inverse reinforcement learning to construct a predictive quantitative model for each user, and then uses this estimated model in conjunction with reinforcement learning to generate challenging but realistic step goals in an adaptive fashion. The study was done on a small sample size of students of a small age range. The model proposed was not used for a particular sport, it was only addressed to optimize steps goals.

3 Training Plans

Training plans are developed to meet the individual needs of the athlete and take into consideration many factors: gender, age, strengths, weaknesses, goals, training facilities, etc. We categorized the existing training plans into two main types that are currently implemented in sports coaching applications. These two types are: traditional/fixed training plans and customized/rule-based training plans. Real sports coaching applications like Racefox offers rule-based generated training plans. While as, both Garmin and Fitbit offer fixed training plans. In this section, we define each of these types describing their main features, and mention their limitations.

3.1 Traditional/Fixed Training Plans

Fixed training plans are one in which training exercises are predefined for certain goals over a particular training period. It has many limitations concludes that it is never considered as a good option for automating the generation of training plans. It does not provide any interaction between the athlete and the coach, ignores the health and psychological state of the athlete, and it is not optimal for the athlete's level in a specific sport.

3.2 Customized/Rule-Based Training Plans

Training exercises in rule-based training plans are set for the athlete based on prescribed rules or procedures. These prescribed rules may include some personal parameters that are defined based on the athlete's movement data. Its limitations include the probability of inaccuracy of these rules, and it also ignores the health and psychological state of the athlete. Rules are limited to domain knowledge of experts, and adaptable but its flexibility remains limited. Moreover, its modification gets complicated by time.

4   Proposed Framework

On the basis of related work and currently existing types of training plans offered in real sports coaching applications discussed in previous sections, the need of having an adaptive generated complete cycle of training plan for athletes has emerged. If we want to imitate the real process of sports coaching using Machine Learning and technological devices when planning for athlete's training, then we need to do what the real coach does.

The coach identifies key aspects of the athlete's performance that can be improved and plan training sessions to address the problems that they have observed. Coaching aims at improving the performance qualitative level and the learning capacity. It involves providing feedback, but also the use of some techniques, such as motivation, the efficient question asking and the management style conscious adaptation to the athlete's training level, in relation to the objective to be fulfilled [12]. The coach should be capable of assisting athletes to prepare training programs, communicate effectively with athletes, assist athletes to develop new skills, use evaluation tests to monitor training progress and predict performance [26]. Accordingly, you can see that coaching is a very challenging task and it requires an awareness of how things should be.

The contribution of the proposed framework lies in the integration of ideas proposed in the area of generating personalized training plans [7–9]. The framework explains steps to be followed for the generation of adaptive sports training plan by interrelating some key components. In this section we first describe the three main components (training load, physiological constraints, and behavior-change features), Table 1 defines some of the measures used in them (proposed framework's model parameters). Then, we present the framework's flow.

4.1 Training Load

Tracking training load is one way of applying training monitoring which is about keeping track of what athletes accomplish in training, for the purposes of improving the interaction between coach and athlete [13]. It could be defined as the stress placed upon the body as a result of a training session. The main principle of training can be reduced to a simple 'dose-response' relationship. The 'dose' is the physical work that the athlete must do during the entire training [14], it is the physiological stress associated with the load of exercise training. While the 'response' is the adaptations that are derived from the dose and could be measured as a change in performance [15–17].

Manipulating training load to maximize adaptation and performance is the main purpose of systematically designed training programs [18]. They are adjusted at various times during the training cycle to either increase or decrease fatigue depending on the phase of training (e.g. baseline or competition phase) [19, 20]. Appropriate load monitoring can aid in determining whether an athlete is adapting to a training program and in minimizing the risk of developing non-functional overreaching, illness, or injury [20]. It can provide a scientific explanation for changes in performance. Therefore, this can aid in enhancing the clarity and confidence regarding possible reasons for changes in performance and minimizing the degree of uncertainty associated with the changes.

There are two ways to measure training load, external and internal. External load is defined as the work completed by the athlete, measured independently of his or her internal characteristics [22]. It is important in understanding the work completed and the capabilities and capacities of the athlete (e.g., distance, speed, and power). Internal load or the relative physiological and psychological stress imposed is also critical in determining the training load and subsequent adaptation (e.g., training impulse (TRIMP), rate of perceived effort (RPE) refer to Table 1 As both external and internal loads have merit for understanding the athlete's training load, a combination of both may be important for training monitoring [20].

4.2 Physiological Constraints

In order to minimize the risk of injuries and overtraining, sports training plan needs to be satisfied by related physiological constraints. Some of these constraints were addressed in previous work, which are training monotony, strain, chronic training load (CTL) ramp rate, and training impulse (TRIMP) which is also an internal training load measure.

Training monotony is a factor of training with a monotonous pattern may consequence becoming overtrained. It represents the variation in training done by an athlete, with higher values indicating more monotonous training [9, 21]. While training strain is an extension of cumulative training volume that incorporates a weighting factor based on the amount of daily variation. The CTL was used as the progressive increase restriction of training load so that athletes can avoid being overtrained. The last introduced constraint is the daily training load limitation. This constraint aims to eliminate excessive training sessions [9].

4.3 Behavioral-Change Features

These are the features that can effectively initiate and maintain the behavioral changes necessary to increase physical activity. Examples of key behavior change features include objective outcome measurements, self-monitoring, personalized feedback, behavioral goal-setting, individualized program, and social support. In particular, researchers recommend that self-monitoring should be conducted regularly and in real-time, so as to target activity with precise tracking information and emphasize performance successes. Moreover, personalized feedback is most effective when it is specific; this could happen by comparing current performance to past accomplishments and previous goals. Personalized goal setting is also a critical factor for facilitating behavior change and it has the potential to increase the effectiveness of physical activity interventions. Setting "sweet spot" goals should be based on the athlete's past behavior in a personalized and adaptive fashion [8].

Table 1. Definitions of proposed framework's model parameters

Internal training load	
Training impulse (TRIMP)	Is a unit of physical effort that is calculated using training duration and average heart rate during the training session
Rate of perceived exertion (RPE)	Involves multiplying the athlete's RPE (on a 1–10 scale) by the duration of the training session
Physiological constraints	
Monotony	Is the average training load in the previous week divided by the standard deviation of daily loads over the same time
Strain	Is the sum of load in the previous week multiplied by the training monotony
Behavioral-change features	
Rate of increase in happiness	Happiness increases, as the athlete gets closer to the goal. It is calculated using athlete's learned helplessness, and self-efficacy
Self-efficacy	Is defined as the athlete's beliefs in their capabilities to successfully execute a course of action (plays an essential role in the theory of goal setting). [8]

4.4 Framework Flow

Constructing a training plan including all the mentioned components would yield to a well designed real personalized training plan customized for the athlete and adapts to the athlete's behavior. The flow (see Fig. 1) proposed for the adaptive training plan generation uses training load measures and physiological constraints as performance indicators, optimizes athletes parameters and future goals based on their behavior, and include exercises variation constraints.

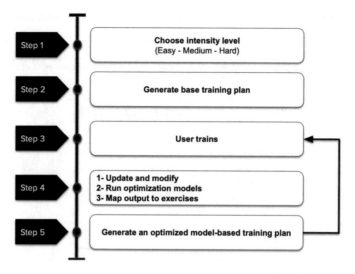

Fig. 1. Training plan generation framework

Here are the proposed steps for generating the training plan (See an example of running training plan in Table 2):

- Step 1:
  - Athlete picks a goal in a specified sport. Then, chooses a desired intensity level (e.g. easy, medium, low).
- Step 2:
  - According to this choice, the athlete will be offered a one-week base training plan. This plan includes exercises that are chosen from a predefined list designed for the chosen goal.
  - Each athlete has a personalized two-column table. The first one includes all set of exercises could be played by the athlete. While the second column has corresponding measured training load values calculated when this exercise was performed by the athlete previously (history data).
  - Training load value is calculated using an internal measure, external measure, or a combination of measures.
- Step 3: Athlete trains.
- Step 4: Actions performed on the athlete's training data are
  - Update and modify:
    - Athlete's personalized two-column table is updated with new measured values of training load.
    - Key performance indicators (KPIs) are analyzed. Pain and injury data is collected.
    - Some exercises might be excluded for the next training week (e.g. If the athlete is injured in the right leg, we exclude exercises that mainly focus on it).
    - In this stage, we have collected daily training load values for the previous week.

Table 2. Example of running training plan generation for an athlete

Step 1 Intensity level: Medium

Step 2

Offered base plan - with estimated training load values (measured by TRIMP) for each exercise based on athlete's history:

Day	Exercises	Training load measure(s) value
1	Interval running	12
2	Frequency running	27
3	Distance running	90
4	Running drills	36
5	Free run	20

Training load values array: (12, 27, 90, 36, 20).

Step 3 Athlete trains

Step 4

Update and modify:

- Updated training load values: (12, 30, 88, 37, 24)
- Check injuries, analyze KPIs and accordingly exclude some exercises for next week (e.g.: exclude 'Distance training 5km')

Run optimization models:

- Calculate personal parameters (behavior-change features) u_b, μ, p_0 (using BAA's first stage [8]):
  - Input: Athlete's goal (estimated numerical value calculated based on athlete's history data) and training load values.
  - Output: Estimated u_b, μ, p_0
- Maximize smallest training load on any given day in the future u_{min} (using BAA's second stage [8])
  - Input: Estimated u_b, μ, p_0
  - Output: Estimated u_{min}
- Estimate target/ total training load (TL) value for the next week
- Now we have all the values ready for the optimization model: minimize the summation of the difference between the defined exercises (x) training load values (r) and the target/ total training load (TL) over the number of weekdays (D) using eq. (1)
  - Add constraints of monotony, strain, exercises variation, exercises distribution over the week (see Table 1)
  - Output: Next week's daily adapted training load values (23, 30, 54, 96, 10)

Map output to exercises

Step 5 Go to step 3

- Run optimization models:
  - The next week training plan given to the athlete is required to enhance performance level, avoid overtraining and injuries, be achievable.
  - Accordingly, we first calculate personal parameters using the athlete's goal and training data. By using a reverse reinforcement learning to construct a predictive quantitative model from the Behavioral Analytics Algorithm (BAA) first stage [8].
  - Then, we maximize smallest training load on any given day in the future to generate challenging, realistic goal in an adaptive fashion using a reinforcement model BAA's second stage [8].
  - Next, we use both personalized parameters, maximized the smallest training load value to estimate a target/total training load (TL) for the next week.
  - After that, we use a designed optimization model to minimize the summation of the difference between the defined exercises (x) training load values (r) and the target/total training load (TL) over the number of weekdays (D). This model is dynamic to include exercises variation, monotony, strain, and any physiological constraints.

$$Min \sum_{d=0}^{D} |\hat{r}_{xd} - TL| \tag{1}$$

  - The output of this model will be a set of daily training load values for next week's training plan.
- Map output to exercises:
  - The set of daily training load values will be mapped back to exercises from the athlete's two-column table
- Step 5:
  - Now, the athlete has an adapted training plan that includes exercises to be performed next week. We repeat the cycle and go to Step 3 again.

5 Discussions

The proposed framework has linked many concepts and techniques based on a theoretical background. The integration done based on related work previously proposed [7–9] and discussed in the flow has provided a powerful tool for generating the training plan. There are both dynamicity and flexibility in the framework that appear in the ability to add, or enhance any of its steps. In addition to the ease of adjusting rules and constraints for different goals to different athletes in different sports. Here are the strength points of the framework proposed.

5.1 Training Load Measures

Training load is an accurate and representative tool for performance measuring. Quantifying training load gives the capability to measure stress and all variables related to the training process to understand the athlete's improvement. As mentioned before, training load could be measured for the athlete externally (e.g.: distance, speed, power) or internally (e.g., training impulse TRIMP, rate of perceived effort RPE). Both provide means of understanding athlete's training load, and a combination of them gives a great value towards training monitoring [20].

5.2 Athlete's History Data and Behavior Adaptation

Making use of the athlete's history data is a key element, and adaptation is one of the main components of this framework. Usage of both has appeared in the 4th step (update of athlete's table, calculating athlete's personal parameters, and maximizing the smallest training load value). The personal parameters represent the rate of increase in happiness when the athletes get closer to the goal, self-efficacy which is the athlete's beliefs in their capabilities to successfully execute courses of action and it plays an essential role in the theory of goal setting, and athlete's natural or baseline training load value per day [8]. Adaptation has also occurred when exercises were updated and some were excluded.

5.3 Exercises Variation Control and Injury Prevention

There is a lot of dynamicity that would exist in the designed optimization model constraints (see Eq. 1). We can design constraints so that it would allow us to control exercises variation, monotony, strain CTL, overtraining and prevent injury. Accordingly, we can add multiple of different rules that guide the model towards a great level of personalization and adaptation to athlete's behavior. We can distinguish between different ages, genders, and different levels of athletes performance levels. Manipulating measures of training load would also allow us to control overtraining.

6 Conclusion and Future Work

This paper proposes a conceptual framework for generating an adaptive and personalized sports training plans. The framework interrelates performance indicators such as training load measures, physiological constraints, and behavior-change features like goal setting and self-monitoring. It deals with psychological measures of training load, and include exercises variation constraints. Training plan's goal is adaptive to the athlete's behavior. Construction of the framework flow relies on the knowledge of existing proposed research in the area of sports training plans generation as well as currently implemented types of training plans in sports coaching AI applications. The proposed framework shows dynamicity and flexibility that appear in the ability to easily add, enhance, or adjust any of its steps, rules, and constraints when designing a training plan with different goals to different athletes in different sports. The proposed framework is subject to many modifications, additions, and enhancements. There are many directions for framework's improvement that include:

1. Adding rules and constraints that differentiate between genders and their adaptability to the exercises assigned to them in the training plan, as well as the training load they can handle in different days throughout the month
2. Support athlete's goals for marathons and deadlines handling when assigning the athlete a training plan, taking into consideration interruptions that could happen (e.g. illness, travel, and injuries).
3. Test the framework applicability on multisports (e.g. triathlon, duathlon, etc.).

In addition to the framework improvement, this framework can be used to conduct experiments and construct comparisons between fixed, rule-based, and adaptive training plans (generated using the framework) on athlete's performance improvement and goal achievement. These experiments will show to what extent the main components of training load measures, physiological constraints, and behavior-change features are leading to much greater and guaranteed results of goal achievement. This will lead to a leap in the area of personalized training plans generation. Furthermore, the logic of generating the adaptive training can be integrated into mobile applications, hence athletes training and history data can be easily tracked, used and analyzed to guarantee a suitable environment for a real sports coaching process.

References

1. Knowles, Z., Borrie, A., Telfer, H.: Towards the reflective sports coach: issues of context, education and application. Ergonomics **48**, 1711–1720 (2005)
2. Farrow, D., Baker, J., MacMahon, C.: Developing Sport Expertise: Researchers and Coaches Put Theory into Practice, 1st edn. Routledge (2013)
3. Althoff, T., Sosič, R., Hicks, J.L., King, A.C., Delp, S.L., Leskovec, J.: Large-scale physical activity data reveal worldwide activity inequality. Nature **547**, 336–339 (2017)
4. Case, M.A., Burwick, H.A., Volpp, K.G., Patel, M.S.: Accuracy of smartphone applications and wearable devices for tracking physical activity. JAMA **313**, 625–626 (2015)
5. Fujiki, Y.: iPhone as a physical activity measurement platform. In: CHI 2010 Extended Abstracts on Human Factors in Computing Systems, pp. 4315–4320. ACM, Atlanta (2013)
6. Hekler, E.B., Buman, M.P., Grieco, L., Rosenberger, M., Winter, S.J., Haskell, W., King, A.C.: Validation of physical activity tracking via Android Smartphones compared to ActiGraph accelerometer: laboratory-based and free-living validation studies. JMIR mHealth and uHealth **3**, e36 (2015)
7. Fister Jr., I., Fister, I.: Generating the training plans based on existing sports activities using swarm intelligence. In: Patnaik Srikanta, N.K., Xin-She, Y. (eds.) Nature-Inspired Computing and Optimization: Theory and Applications, pp. 79–94. Springer International Publishing (2017)
8. Zhou, M., Mintz, Y.D., Fukuoka, Y., Goldberg, K.Y., Flowers, E., Kaminsky, P., Castillejo, A., Aswani, A.: Personalizing mobile fitness apps using reinforcement learning. In: Companion Proceedings of the 23rd International on Intelligent User Interfaces: 2nd Workshop on Theory-Informed User Modeling for Tailoring and Personalizing Interfaces (HUMANIZE) (2018)

9. Kumyaito, N., Yupapin, P., Tamee, K.: Planning a sports training program using Adaptive Particle Swarm Optimization with emphasis on physiological constraints. BMC Res. Notes **11**, 9 (2018)

10. Aswani, A., Kaminsky, P., Mintz, Y., Flowers, E., Fukuoka, Y.: Behavioral modeling in weight loss interventions. Eur. J. Oper. Res. **272**, 1058–1072 (2016)

11. Mintz, Y., Aswani, A., Kaminsky, P., Flowers, E., Fukuoka, Y.: Behavioral Analytics for Myopic Agents (2017)

12. Teodorescu, S., Urzeală, C.: Management tools in sports performance. Procedia Soc. Behav. Sci. **81**, 84–88 (2013)

13. Foster, C., Rodriguez-Marroyo, J.A., de Koning, J.J.: Monitoring training loads: the past, the present, and the future. Int. J. Sports Physiol. Perform. **12**, 2–8 (2017)

14. Wallace, L.K., Slattery, K., Coutts, A.J.: A comparison of methods for quantifying training load: Relationships between modelled and actual training responses. Eur. J. Appl. Physiol. **114**, 11–20 (2013)

15. Amatori, S.: Training load: study and comparison of the main training load quantification methods used | A new proposal for triathlon (2015)

16. Lambert, M., Borresen, J.: Measuring training load in sports. Int. J. Sports Physiol. Perform. **5**, 406–411 (2010)

17. Hernández-Cruz, G., López-Walle, J., Chacón, J.T.Q., Sanchez, J.C.J., Rangel, B.R., Reynoso-Sánchez, L.F.: Impact of the internal training load over recovery-stress balance in endurance runners. Revista de Psicologia del Deporte **26**, 57–62 (2017)

18. Leo, P.: A contemporary approach of training load quantification in endurance athletes. BSc thesis, University of Innsbruck, Austria (2016)

19. Pyne, D., Martin, D.: Fatigue - insights from individual and team sports. Regulation of Fatigue in Exercise. In: Marino, F.E. (ed.) Regulation of Fatigue in Exercise, pp. 177–185. Nova Science, New York (2011)

20. Halson, S.: Monitoring training load to understand fatigue in athletes. Sports Med. (Auckland, N.Z.) **44**, 139–147 (2014)

21. Foster, C., Florhaug, J.A., Franklin, J., Gottschall, L., Hrovatin, L.A., Parker, S., Doleshal, P., Dodge, C.: A new approach to monitoring exercise training. J. Strength Conditioning Res. **15**, 109–115 (2001)

22. Wallace, L.K., Slattery, K.M., Coutts, A.J.: A comparison of methods for quantifying training load: Relationships between modelled and actual training responses. Eur. J. Appl. Physiol. **114**, 11–20 (2014)

23. Racefox Homepage. https://racefox.se/. Accessed 13 May 2019

24. Garmin Homepage. https://explore.gar. Accessed 13 May 2019

25. Fitbit Homepage. https://www.fitbit.com. Accessed 13 May 2019

26. Coach's role. https://www.topendsports.com/fitness/coach.htm. Accessed 13 May 2019

Synonym Multi-keyword Search over Encrypted Data Using Hierarchical Bloom Filters Index

Azza A. Ali[1,2(✉)] and Shereen Saleh[2]

[1] Department of Computer Science, College of Science and Humanities, Jubail,
Imam Abdulrahman Bin Faisal University, Dammam, Saudi Arabia
aaaali@iau.edu.sa
[2] Math and Computer Science, College of Science, Minufiya University,
Shebeen El-Kom, Egypt
shereensaleh357@gmail.com

Abstract. Search on encrypted data refers to the capability to identify and retrieve a set of objects from an encrypted collection that suit the query without decrypting the data. Users probably search not only exact or fuzzy keyword due to their lack of data content. Therefore, they might be search the same meaning of stored word but different in structure, so, this paper presents synonym multi-keyword search over encrypted data with secure index represented by hierarchical bloom filters structure. The hierarchical index structure improves the search process, and it can be efficiently maintained and constructed. Extensive analysis acquired through controlled experiments and observations on selected data, show that the proposed scheme is efficient, accurate and secure.

Keywords: Searchable encryption · Synonym keyword · Bloom filter · Hierarchical bloom filter · Locality Sensitive Hashing

1 Introduction

Cloud storage service has become more popular as providing a lot of benefits over traditional available storage solutions. Companies use storage to store and search data used in various aspects of their businesses. The data may include several million records; at least some of these companies wish to keep some records private such as the customer private information. This information may be of value to others who may have a malicious intent. If the adversary be able to obtain such private information, he will create problems for the company, its customers, or both.

One common method used to protect sensitive information and to comply with privacy regulations or policies is encryption. However, use of encrypted data raises other issues, such as searching for particular data without decrypting all of data and performing search on it. Most papers support search over encrypted data [1–8], and they are based on storing additional information (i.e. metadata) which by itself does not seep out any information about the data, but enables the predicates to be evaluated over the encrypted representation.

© Springer Nature Switzerland AG 2020
A. E. Hassanien et al. (Eds.): AISI 2019, AISC 1058, pp. 685–699, 2020.
https://doi.org/10.1007/978-3-030-31129-2_63

With huge increasing in data size, especially with Cloud data, searching mechanisms becomes easier by using bloom filters structure. Recently most papers used bloom filter data structure in encrypted data search [9–13], and it leads to solve most problems of searching over Cloud data.

The proposed scheme (Synonym Multi-Key-SHBlooFi) uses Hierarchical index structure for bloom filters HBlooFi in searching over encrypted data over cloud. The proposed Synonym Multi-Key-SHBlooFi scheme enhances the search process of Wang scheme [9]. Our results show that Hierarchical Bloom Filter Index introduces efficient and a scalable solution for searching through a huge number of bloom filters compared with Wang scheme [9].

The rest of the paper is organized as follows. Section 2 represents the related work such as Wang scheme [9], and hierarchical index structure for bloom filters [14]. Section 3 represents the proposed Synonym Multi keyword search scheme over encrypted data (Synonym Multi-Key-SHBlooFi). Evaluation of the proposed Synonym Multi-Key-SHBlooFi scheme is represented in Sect. 4. Finally, we conclude the paper in Sect. 5.

2   Related Work

2.1   Hierarchical Index Structure for Bloom Filters

Bloom filter is a data structure used to check whether a specific element exists in a specific set or not. Every bloom filter is represented by an array of bit values (0's and 1's) with length m. Values of bloom filter array are created by utilizing a set of k hash functions. The blank bloom filter has value zero in all bits. A new element can be inserted into bloom filter by applying k-hash functions on that new element, and then every result of k-hash functions denotes to a specific position in the bloom filter. The bit value of each position in bloom filter becomes *one* as illustrated in Fig. 1. To check whether that element exists in the set of bloom filters or not, k of hash functions of the specific element is computed and then searching for the positions in the bloom filters with the same values resulted from k hash functions. If any used hash function result index position in the bloom filter has value *zero*, the test element does not exist within the set, with probability 1. But in case of all used hash functions result index position in the bloom filter have values one, it means that the test element exist within the bloom filter. It should be taking into consideration that, having one in the specified places in bloom filter does not necessarily mean existence of the element in the set which represented by the bloom filter; it may be a false positive or a true positive.

Dataset index file contains the bloom filter of each file. Bloom filters structure can be utilized to examine existence of specific object in the set (match) or whether the object without doubt not exists within the set (no match). There is a possibility of existence false positives, but false negatives unlikely to occur.

Wang et al. [9] proposed a privacy-preserving multi-keyword fuzzy search over encrypted data in the Cloud. Wang built an index for each file; this index file contains all words in file and represents it as a bloom filter. The index file supports multiple keyword searches without needing for increasing the index. The search operation in

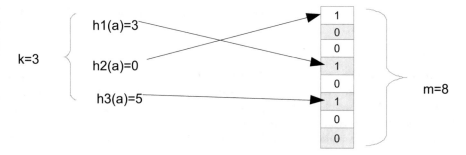

Fig. 1. Using k hash functions with bloom filter

Wang scheme passes all files within the dataset. Thus, the search time of Wang scheme grows linearly with growing of size of the documents set.

Hierarchical Bloom filter Index (HBlooFi) was proposed by Adina [14] as a searching data structure, which can accommodate to tens of thousands of bloom filters with low maintenance cost. This paper suggests a scheme to check existence of specific object in data. Searching over HBlooFi is based on the following idea, first it must be noted that the leaves of the tree are represented the original bloom filters index, and the bloom filters of parent nodes are acquired by applying a bitwise-OR on child nodes. This process keeps going until the root is reached. Query starts from the root bloom filter traverses the tree down the path until reach to the leaf level. To examine that specific object exist in specific node in leaf level, it should satisfy the condition that all the position's value in bloom filter that match result from k hash functions must be equal one.

Construction of HBlooFi tree is based on the following steps: *firstly*, generating the outer nodes (leaves) of the tree which represent the original Bloom filters index of data. *Secondly*, by using a bitwise-OR on the child nodes, we can obtain the parent nodes within the bloom filters tree. This process keeps going till the root node is reached. The HBlooFi has a feature that every inner nodes bloom filter among the tree represents the union of the sets that formed by the bloom filters among the sub-tree rooted at that node. Figure 2 illustrates an example of HBlooFi of order two. Suppose the threshold $d = 2$ (then every inner node has between d and $2d$ child nodes). The leaf nodes of the tree include the actual bloom filters, with identifiers 1, 2, 3, 4, 5, and 6.

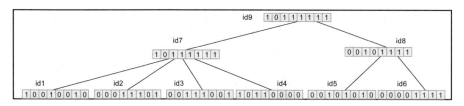

Fig. 2. HBlooFi tree of order 2

Now we will introduce algorithms of inserting, deleting, and updating Bloom filters by HBlooFi structure.

- Insert

In the insertion operation, HBlooFi scheme starts searching for a leaf node within the tree that is "close" to the input bloom filter wanted to be inserted into the tree. Searching for the closest leaf node depends on the distance measured between leaf node bloom filter and the desired inserted bloom filter.

Once the closest node found, if the parent node for the closest node still has no more than $2d$ children's (where d specific threshold), add the new one bloom filter with its child leave nodes, and then the insertion process is completed. On the other hand, if the number of parent child exceeds the specified threshold $2d$; therefore the parent node must be split. Splitting is done and new parent node has been created contains the closest child leaf node and the desired inserted node. A new splitting node (new parent node) has new bloom filter value computed by performing bitwise-OR operation between the desired inserted bloom filter node and the closest bloom filter child node. The desired inserted bloom filter node and the closest bloom filter child node will be modified as child nodes of the new parent node. For example, to insert the bloom filter with value "00101100" into the tree of bloom filters in Fig. 2. First search for the nearest node to the new node by using hamming distance as the metric distance. As a result, it is found that distances, measured between the new node and nodes id_7 and id_8, are equal to *four*. Let's assume that we chose id_7 as the nearest node. Therefore the new node is inserted as a child node of id_7 which has more than *four* child nodes after inserting new one, so id_7 should be split and create new inner node as shown in Fig. 3.

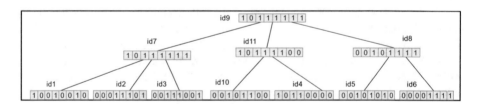

Fig. 3. Bloom filter tree after insert and split

- Delete

The deletion algorithm operates on remove specific node from the bloom filter index tree. If the node, needed to be deleted, is a leaf node, then delete the pointer to this node from its parent node, and check if the parent node is not under flowing (at least 2 children are exist). The bloom filter values for the remaining nodes must be recomputed by using bitwise-OR of their children in the path from the parent node to the root node. If there is an underflow happened (mean there are node contains children under specific threshold after the delete process), then parent node tries to redistribute its child nodes with a sibling and update the sub tree contain the moving node to the root. For Example, assume that id_5 node, with value "00101010" from Fig. 2, is needed to be removed. The final tree result after deletion process of node id_5 and redistribution between id_7 and id_8 is illustrated in Fig. 4.

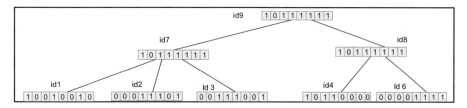

Fig. 4. Bloom filter tree after delete and redistribute

- Update

The input parameters of the update algorithm are the new value of bloom filter and the desired updated leaf node. Once update specific bloom filter value within the tree, all bloom filters values will be updated, using bitwise-*OR*, within the path from the leaf to the root with the new value. For Example: to update node id_6 with new value "11001111", values of all the pass nodes, from node id_6 to the root, will be recomputed using bitwise-*OR* with "11001111" and the final structure of tree is shown in Fig. 5.

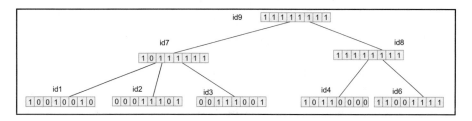

Fig. 5. Bloom filter tree after update

2.2   Multi Keyword Search Scheme over Encrypted Data (Wang Scheme)

In 2014, Wang et al. presented multi-keyword fuzzy search scheme over encrypted data in the Cloud [9]. Wang scheme builds an index I_D for each file D. The index I_D contains all the keywords in file D. Convert the file to m-bit and then treated as bloom filter [10]. In Wang scheme, read file word by word and convert each word into its bigram vector representation using this bigram as input to specific Locality Sensitive Hashing *LSH* functions family using the result from this function in insert process within the Bloom filter. Generating query which containing multiple keywords to be searched can be done by following the same steps used to create the index. The search process includes a simple inner product operation done between the index bloom filter values and the query bloom filter values. If the keyword(s) in the query exist in specific file, therefore the bloom filter represent this file contains bit 1 corresponding to the same bit position in query bloom filter which result that the inner product between them can came a high value. The steps of Wang scheme are explained as following steps:

1. *keyGen*: Given a security parameter m, output the secret key $sk(M_1, M_2, S)$, where $M_1, M_2 \in R_{m \times m}$ are invertible matrices and $S \in \{0, 1\}m$ is a vector.
2. *BuildIndex*(D, SK, L): choose L independent LSH functions, build a m-bit bloom filter I_D to represent the index for every file D:

   - Obtain the keywords set from D.
   - For each keyword, using specific *LSH* functions family to insert it into bloom filter index.
   - Encrypt the index I_D using **Index_Enc**(SK, I_D) and output $Enc_{sk}(I_D)$.

3. *Index_Enc*(SK, I): The bloom filter index I is divided into two vectors $\{I', I''\}$ by using the rule: for each element $i_j \in I$, set $i'_j = i''_j = i_j$ if $s_j \in S$ is one; else $i'_j = \frac{1}{2}i_j + r$, $i''_j = 1/2i_j - r$, as r is a random number. Then encrypt $\{I', I''\}$ with (M_1, M_2) into $\{M_1^T.I', M_2^T.I''\}$. Output $Enc_{sk}(I) = \{M_1^T.I', M_2^T.I''\}$ as the secure index.
4. *Trapdoor*: Create bloom filter of length m-bit for the query Q. For every keyword wanted to be searched, insert it into bloom filter using the same *LSH* functions family used in creating index. Then using $Query - Enc(SK, Q)$ to encrypt query.
5. $Query - Enc(SK, Q)$: The query Q is divided into two vectors $\{Q', Q''\}$ using the rule: $q'_j = q''_j = q_j$ if $s_j \in S$ is zero; else $q'_j = 1/2q_j + r'$, $q''_j = 1/2q_j - r'$ where r' is another random number. Then encrypt two vectors $\{Q', Q''\}$ as $\{M_1^{-1}.Q', M_2^{-1}.Q''\}$ which result the trapdoor represented by $Enc_{sk}(Q) = \{M_1^{-1}.Q', M_2^{-1}.Q''\}$.
6. *Search* $(Enc_{sk}(Q), Enc_{sk}(I_D))$:   Perform   the   inner   product   operation $\langle Enc_{sk}(Q), Enc_{sk}(I_D) \rangle$ as the output of the search process between the query Q and the document D.

Wang scheme is distinguished from the previous search schemes [6, 15] as it enables the update process efficiently, due to the fact that each document is indexed individually. So when trying adding, deleting and modifying file can be done efficiently, compared to the other schemes, including only indexes of the files that will be modified, not affecting other files.

3   The Proposed Synonym Multi Keyword Search Scheme Over Encrypted Data

The proposed Synonym Multi keyword search scheme over encrypted data can be illustrated in the following subsections.

3.1 Constructing the Bloom Filter for Each File in Dataset Collection (Algorithm 1)

First read word by word from file then extract all synonyms of each word and converting each word to its bigram vector and using LSH Euclidean hash function family to insert each word in bloom filter as illustrated in Fig. 6.

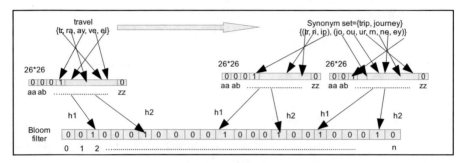

Fig. 6. Construct bloom filter

Algorithm1: Creating bloom filter

---

1. **Input:** File (F). **Output:** bloom filter (bf).
2. **For each** word W_i in File (F), find list of all synonym words for W_i.
3. Words$\{W_1, W_2, W_3, \ldots . W_n\}$. Where n is the number of synonym words for W_i.
4.   **For** $i = 1\ to\ n$
5.   Compute $LSH(W_i)$. Where LSH locality sensitive hash function.
6.   Find all positions of bloom filter bf that equal to all the values of $LSH(W_i)$ result.
7.   Change position's values in bloom filter bf to be equal one.
8.   **End for**
9. **End for**

3.2 Constructing the Bloom Filter Tree (Algorithm 2)

For each document generate a leaf node where store bloom filter, then build the tree from down to up traversal with all leaf nodes generated where internal nodes created from union of its child's values.

Algorithm 2: Constructing the bloom filter tree
---
1.   **Data:** bloom filter list $bfList$ represent all files in dataset
2.   **Sort** bloom filters in $bfList$ according to specific distance (hamming distance).
3.   **For** each bloom filter bf of $bfList$ **do**
4.   Create leaf node in the tree.
5.   Set name and value (bloom filter) for each created leaf node
6.   Insert right each bloom filter to leaf node
7.   **End** for
8.   Perform bitwise OR bloom filters for all interior nodes, recursively computing them traverse all the tree from down to up.

3.3   Encrypt Bloom Filter Index

Split each bloom filter in leaf node into two vectors, and then encrypt two vectors with transpose of two matrices. Output the secure index.

Algorithm 3: encrypt tree T
---
1.   **Input:** tree of bloom filters
2.   generate secret key as two matrix M_1, M_2 and split vector S
3.   **for** each leaf node in tree **do**
4.   split bloom filter I into two vectors I', I''
5.   **for** each element $i_j \in I$
6.   **If** $s_j = 1$ **(where** $s_j \in S$**)**
7.   set $i'_j = i''_j = ij$ **(where** $i' \in I'$, $i'' \in I''$**)**
8.   **Else**
9. $i'_j = 1/2 \, i_j + m$, $i''_j = 1/2 \, i_j - m'$ where m is a random number
10.   **End for**
11. $Enc_{sk}(I) = \{M_1^T.I', M_2^T.I''\}$
12.   **End for**

3.4   Search

The search algorithm starts from root node traverse down to leaf nodes in tree. Following steps described in Algorithm 4. When algorithm find node is inner node, first check the condition that the hash result of query has values *one* in that inner node value. If the condition met then the search function is called recursively for all child nodes of this inner node until reach to leaf nodes. When reaches to leaf-level which rooted to specific node, algorithm computes inner product between query bloom filter value and leaf nodes values and return the identifier of the leaf node that produce the biggest inner product between its value and the value of the query. Else if the condition is not met, exit from this sub-tree and search in other inner node value.

Algorithm 4 : Search over encrypted tree

1. **Input**: Encrypted tree T, query hash result H, two vectors of encrypted query $v1, v2$)
2. **For** each node in the tree T **do.**
3.   **If** node is interior node N and contain all H.
4.   **for** (i = 0 ; i < $N.nbchildren$; $i + +$) **do**
5.   Multiply $v1, v2$ into leaf node values.
6.   **end for**
7.   set all result of multiplication in hash table HT < result, file name >
8.   Return file name correspond to max result.
9.   **end if**
10.   **Else if** node is interior node N and not contain all H.
11.   Exit from this inner node
12.   Start search in other inner node in the tree calling *search* algorithm from step3.
13.   **End else if**
14.   **Else**
15.   this node is a leaf
16.   just call step 5, 7
17.   **End else**
18.   **End for**

4 Performance Analysis for the Proposed Scheme (Synonym Multi-Key-SHBlooFi)

In this section, we will clarify overall analysis and experiments of the proposed algorithm on specific dataset: the Enron Email Dataset [16]. We choose randomly variety number of emails to create dataset. The entire experiment system is implemented by java language on windows 7 with Intel Core i5-2450M Processor 2.50 GHZ. Using JWNL (Java WordNet Library) which is API used for accessing WordNet dictionary which consider a lexical database for the English language connecting with it and find set of words that have the same meaning for that word. WordNet is containing set of words that are synonyms of each other. Each of these sets is called a synset. Since a word can have more than one meaning, so more than one synset can be contain the same word.

Tables 1 and 2 represent the storage size, construction time and search time regard to the proposed encrypted data searching over BF-tree scheme compared with Wang scheme. Also, the results of Tables 1 and 2 show the memory space, that the index used, for each of the proposed scheme and Wang scheme. The proposed Synonym Multi-Key-SHBlooFi scheme uses more index storage space compared to Wang scheme. Also, two Tables 1 and 2 demonstrate the searching time with each scheme which is affected by the increasing number of documents uploaded. Two tables illustrate the changes in construction index time once the numbers of files increased. In the proposed Synonym Multi-Key-SHBlooFi scheme, the index construction time is increased compared to the index construction time of Wang scheme.

Table 1. Experimental results obtained from the multi keyword fuzzy search (bloom filter list index) of Wang scheme.

Text files size	Size of index in memory			Primitives						Construction Time/s	Search Time/s
	Objects	Non null references	Null references	Int	Long	Double	Char	Float	Boolean		
1 KB	166	165	325	84	–	32800	–	–	–	0.832	0.004
40 KB	642	641	1280	322	–	128000	–	–	–	2.239	0.05
300 KB	2662	2661	5320	1332	–	532000	–	–	–	11.572	0.09
512 KB	5350	5349	10696	2676	–	1069600	–	–	–	15.592	0.379
1 M	12674	12673	25344	6338	–	2554400	–	–	–	48.377	0.602
4 M	36918	36917	73832	18460	–	7383200	–	–	–	347.646	4.712

Table 2. Results of the proposed encrypted data searching over the proposed Synonym Multi-Key-SHBlooFi scheme.

Text files size	Size of index in memory		Null references	Primitives						Construction Time/s	Search Time/s
	Objects	Non null references		Int	Long	Double	Char	Float	Boolean		
1 KB	1791	2222	8142	1974	16400	146310	589	298	59	1.402	0.001
40 KB	7223	8982	32808	7948	64400	591642	2215	1204	241	2.568	0.003
300 KB	29956	37272	136161	32954	266400	2456200	9527	4993	999	13.238	0.007
512 KB	60228	74944	273941	66252	548584	4959040	19484	10039	2009	16.589	0.011
1 M	142551	177390	647033	156810	1299276	11739272	64375	23758	4751	56.086	0.025
4 M	415304	516816	1888967	456838	3784288	34201944	141125	69217	13843	379.429	0.105

Tables 1 and 2 show the results of searching time for the proposed algorithm compared to Wang scheme algorithm. Encrypted data searching over proposed Synonym Multi-Key-SHBlooFi scheme is better than encrypted data searching over Wang scheme. When evaluating search time over the proposed scheme, search over Synonym Multi-Key-SHBlooFi, for different number of bloom filters, it seems better than search time in Wang scheme, as we don't need to traverse all bloom filters in tree, we only checked bloom filters that have high probability contained the keyword search. On the other hand, the search operation in Wang scheme includes passes all the files within the dataset perform inner product function. Thus, the search time of Wang scheme grows linearly with growing of size of the documents set, this can be a result of the search process must re-examine all the files within the dataset before the ultimate result get.

The performance analysis of the proposed scheme is illustrated the remaining of this section. The search cost, the search efficiency, efficiency of index construction, and the storage requirements are computed compared with Wang scheme.

4.1   Search Cost

The time consumed for search is given by the invocations number of search process. In the proposed scheme there are three cases:

- **The best case:** in case of the tree does not contain leaf-level, contains only the root node, then only one node will be checked for matching, which can be considered as the best case.
- **Average case:** in case of existing the leaf-level, number of search invocations is equal to $O(d * log_d N)$, where N is the number of nodes, and d is number of checked nodes to find bloom filter that matches a query. In this case, one path will be followed and the nodes of this path will only be checked to find the one that match.
- **Worst case:** the worst case is $O(N)$, it will exist when examining all nodes in the tree to find the matched one.

4.2   Search Efficiency

The search time grows linearly with growing of variety number of the documents with the same size. Figure 7(a) and (b) Show results of searching time of the proposed scheme compared with Wang scheme, when using different number of documents with the same size. The results from Fig. 7 show that the search time of the proposed scheme is smaller than in Wang scheme for various varieties of documents. This is often as a result of the search process in Wang scheme has to reconsider all the files within the dataset before the ultimate result get compared to our proposed scheme, as don't need to traverse all bloom filters of files within the same tree.

4.3   Efficiency of Index Construction

Also in Fig. 8, we are able to see the changes in construction index time with the proposed scheme and Wang scheme, once variety number of documents uploaded increases as described in the following figures (Fig. 8(a), (b)). In the proposed Synonym Multi-Key-SHBlooFi scheme, time is increased compared with time Wang scheme. As the index construction of our proposed scheme includes first creating bloom filters and furthermore building the search tree. Expected it requires more time than the Wang scheme.

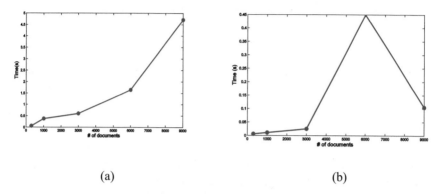

(a)   (b)

Fig. 7. (a) Searching time of encrypted data using Wang scheme, (b) Searching time of encrypted data using the proposed scheme.

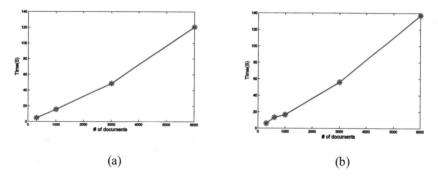

(a)   (b)

Fig. 8. (a) Construction time of encrypted data using Wang scheme, and (b) Construction time of encrypted data using the proposed scheme.

4.4   Storage Requirements

In our experiments, different numbers of files are uploaded as a whole and monitoring the changes within the size of the index file. It is noticed that the index file created by Wang scheme smaller than the index file created by our proposed scheme.

This reduction in index file size proved that Wang scheme is efficient than our proposed scheme in terms of storage requirements.

From the above we can summarize the comparison between the proposed scheme and Wang's scheme as the following:

Wang's scheme	Proposed scheme
Support multi-keyword fuzzy search	Support synonym multi-keyword search
Need less storage than proposed scheme	Need more storage than Wang's scheme
Less efficiency in search time than proposed scheme	Fast and more efficient in search time
Need less construction time than proposed scheme	Need more construction time than Wang's scheme
Need complexity search time $O(n)$	Need complexity search time $O(d * log_d N)$

5 Conclusion

This paper introduced a proposed, synonym multi-keyword search over encrypted data with secure index, scheme for enhancing the searching time over encrypted data.

The proposed scheme was based on combining Wang scheme with secure index represented by hierarchical bloom filters structure. The experiments showed that the proposed scheme improved the search time compared with Wang scheme. Although, the proposed scheme increased the storage needed for the index and the time of the index construction, the proposed scheme achieving the desired goal of fasting search time and also support the synonym multi-keyword search property. In future work, we suggest implement the proposed algorithm in parallel using map-reduce mechanism to reduce construction time.

References

1. Song, D., Wagner, D., Perrig, A.: Practical techniques for searches on encrypted data. In: Proceedings of IEEE Symposium on Security and Privacy, pp. 44–55 (2000)
2. Goh, E.-J.: Secure indexes. In Cryptology ePrint Archive on October 7th, pp. 1–18 (2003)
3. Boneh, D., DiCrescenzo, G., Ostrovsky, R., Persiano, G.: Public key encryption with keyword search. In: Proceedings of Euro Crypto 2004, pp. 506–522 (2004)
4. Li, J., Wang, Q., Cao, N., Ren, K., Lou, W.: Fuzzy keyword search over encrypted data in cloud computing. In: Proceedings of IEEE INFOCOM, pp. 1–5 (2010)
5. Jin Shi, X.., Ping Hu, Sh.: Fuzzy multi-keyword query on encrypted data in the cloud. In: 2016 4th International Conference on Applied Computing and Information Technology/3rd International Conference on Computational Science/Intelligence and Applied Informatics/1st International Conference on Big Data, Cloud Computing, Data Science & Engineering (ACIT-CSII-BCD), pp. 419–425 (2016)
6. Xia, Z., Wang, X., Sun, X., Wang, Q.: A secure and dynamic multi-keyword ranked search scheme over encrypted cloud data. IEEE Trans. Parallel Distrib. Syst. 27, 340–352 (2015)
7. Pandiaraja, P., Vijayakumar, P.: Efficient multi-keyword search over encrypted data in untrusted cloud environment. In: 2017 Second International Conference on Recent Trends and Challenges in Computational Models (ICRTCCM), pp. 251–256 (2017)

8. Orencik, C., Kantarcioglu, M., Savas, E.: A practical and secure multi-keyword search method over encrypted cloud data. In: IEEE Sixth International Conference on Cloud Computing (CLOUD), pp. 390–397 (2013)
9. Wang, B., Yu, S., Lou, W., Hou, Y.: Privacy preserving multi-keyword fuzzy search over encrypted data in the cloud. In: IEEE INFOCOM, IEEE Conference on Computer Communications, pp. 2112–2120 (2014)
10. Ali, F., Lu, S.: searchable encryption with conjunctive field free keyword search scheme. In: 2016 International Conference on Network and Information Systems for Computers (ICNISC), pp. 260–264 (2016)
11. Fu, Z., Wu, X., Guan, Ch., Sun, X., Ren, K.: Toward efficient multi-keyword fuzzy search over encrypted outsourced data with accuracy improvement. IEEE Trans. Inf. Forens. Secur. **11**, 2706–2716 (2016)
12. Umer, M., Azim, T., Pervez, Z.: Reducing communication cost of encrypted data search with compressed bloom filters. In: IEEE 16th International Symposium on Network Computing and Applications (NCA), pp. 1–4 (2017)
13. Poon, H., Miri, A.: A low storage phase search scheme based on bloom filters for encrypted cloud services. In: 2015 IEEE 2nd International Conference on Cyber Security and Cloud Computing, pp. 253–259 (2015)
14. Crainiceanu, A.: Bloofi: a hierarchical bloom filter index with applications to distributed data provenance. In: Proceedings of the 2nd International Workshop on Cloud Intelligence Article No. 4 (2013)
15. Jivane, A.: Time efficient privacy-preserving multi-keyword ranked search over encrypted cloud data. In: 2017 IEEE International Conference on Power, Control, Signals and Instrumentation Engineering (ICPCSI), pp. 497–503 (2017)
16. Cohen, W.W.: Enron email dataset. http://www.cs.cmu.edu/~enron/

GloSOPHIA: An Enhanced Textual Based Clustering Approach by Word Embeddings

Ehab Terra[✉], Ammar Mohammed[✉], and Hesham A. Hefny[✉]

Department of Computer Science, Institute of Statistical Studies and Research (ISSR), 5 Ahmed Zweil Street, Orman, Giza 12613, Egypt
ehabterra@hotmail.com, ammar@cu.edu.eg,
hehefny@ieee.org

Abstract. Textual case based reasoning (TCBR) is a challenging problem because a single case may consist of different topics and complex linguistic terms. Many efforts have been made to enhance retrieval process in TCBR using clustering methods. This paper proposes an enhanced clustering approach called GloSOPHIA (GloVe SOPHIA). It is based on extending SOPHIA by integrating word embeddings technique to enhance knowledge discovery in TCBR. To evaluate the quality of the proposed method, we will apply the GloSOPHIA to an Arabic newspaper corpus called watan-2004 and will compare the results with SOPHIA (SOPHisticated Information Analysis), K-means, and Self-Organizing Map (SOM) with different types of evaluation criteria. The results show that GloSOPHIA outperforms the 3 other clustering methods in most of the evaluation criteria.

Keywords: Text clustering · Case based reasoning ·
Textual case based reasoning · Word embeddings · Knowledge discovery

1 Introduction

Case-Based Reasoning (CBR) [1] is a method used to understand and solve problems 'reasoning' based on previous experiences 'cases'. Generally, to design an effective CBR system, the first step is to select appropriate features. The second is to retrieve cases correctly [2]. CBR is used in several domains, including, machine learning [3] and information retrieval (IR) [4]. Applying CBR on unstructured text is called Textual CBR (TCBR) [5]. Unlike CBR where knowledge embedded in cases are easily acquired and represented by objects. TCBR the case knowledge is more complicated, since a single case may be consists of different topics and complex linguistic terms. Therefore, it is challenging problem in TCBR to search for similar cases that contains different words of similar topics. Moreover, Arabic text manipulation is a challenging task because it contains many different interpretations and meanings of the same word (e.g.: the word "شعر" could mean hair, poetry, or feel). Usually, the retrieval process in TCBR using clustering to organize cases into subsets can enhance the retrieval process from TCBR systems by searching only into similar group of cases. Clustering [6] is searching for potential patterns and partitioning the dataset according to these patterns. Text clustering is crucial for several applications such as, organizing web content and

A. E. Hassanien et al. (Eds.): AISI 2019, AISC 1058, pp. 700–710, 2020.
https://doi.org/10.1007/978-3-030-31129-2_64

text mining [7, 8]. The problem of TCBR clustering methods is that it must be efficient and incremental. This means that it enables the addition of new cases without the vital change in its structure [9]. Moreover, today Natural Language Processing (NLP) techniques [10] (for example: N-gram) can be achieved on much larger datasets and they distinctively surpass the simple models [11] Fast and high-quality clustering of documents can greatly facilitate the summarization, successful navigation, organization of huge information. It is also used to determine the content and structure of unknown text sets.

Several efforts have been made to improve the clustering of TCBR. Cunninghamet et al. [12] proposed an algorithm based on graph; however the proposed algorithm is not able to distinguish between problem and solution cases. Proctor et al. [13] proposed a method based on Information Gain (IG). Fornells et al. [14] used Self-Organizing Map (SOM) [15] to manage case retrieval. Among those, Recio-Garcıa et al. [2] were they proposed TCBR clustering technique based on a Lingo method [16] that is based on Singular value decomposition (SVD). However, their previous works are not expanding well to larger dataset. Another effort which is called "SOPHIA-TCBR" (SOPHisticated Information Analysis) is introduced by Patterson et al. [9]. The main idea of their proposed approach is to build clusters based on the entropy of conditional probabilities to intelligently extract narrow themes from the case-base where a theme is a single word, thus group semantically related cases based on these themes. However, their experiment's was set to a high number of clusters (for example: initial number of clustering N = 1000) the quality of clustering is decreasing due to forming empty clusters or clusters with only few cases. Additionally, using just a single word to form a cluster theme is not sufficient (for example: "Poetry" can be a theme and "Poet" would be another theme).

Word embeddings aims at measuring words similarity to obtain meaningful representation of text. In order to perform that, the text is represented in a D-dimensional embedding space. In 2014, Pennington et al. [17] have proposed a word embedding method called GloVe (Global Vector). It is unsupervised way to find relationships between words that make it a good alternative to the external ontologies or the predefined classes to find these relationships.

To enhance the textual case based clustering, this paper proposes a method to integrate the power of GloVe word embeddings technique into SOPHIA clustering method to form the so-called GloSOPHIA (GloVe SOPHIA). In the proposed method an algorithm is introduced to construct themes based on a threshold (for example, "Poetry", "Poem", "Poetic", and "Poet") instead of a single word theme, this could attracts as many documents as possible with a vector of closest alternative or complement words. GloSOPHIA can enhance and increase the quality of clustering compared to SOPHIA. The proposed method is evaluated on the watan-2004 dataset [18], some evaluation criteria are used to evaluate the performance of the proposed method in comparison to other clustering methods, namely, SOPHIA, K-means [19], and SOM. It is worth mentioning here that few clustering methods have been implemented on the watan-2004 dataset. For example, Kelaiaia et al. [20] have compared the Latent Dirichlet Allocation (LDA) to K-means technique. The best scores they achieved are F measure = 0.66 and Rand Index = 0.87. They have used all documents in the dataset in training process and that may cause problems such as overfitting. In addition, they mentioned that the text is stemmed and not lemmatized.

The rest of this paper is organized as the following. Section 2 describes the proposed method. Section 3 shows the evaluation criteria that are used to test the performance. Section 4, discussed the experimental results. Section 5 concludes the paper.

2   Proposed Method

The main idea of the GloSOPHIA is to feed word embeddings in SOPHIA. The proposed method, as shown in Fig. 1, is described in the following steps: Preprocessing, Normalization, Build word embeddings vector space, Construct word group vector space, Narrow theme discovery, Similarity knowledge discovery, and Case assignment discovery. The preprocessing is the initial step as demonstrated in Fig. 1. The following steps are described here:

- Normalize the characters to a single form ﺍ, ﺍ, ﺍ => ﺍ
- Remove Arabic diacritics using "arabicStemR" package.
- Lemmatize words and removes "pronoun, punctuation, coordinating conjunction, determiner, numeral, adposition, particle" using the excellent package called "ud-pipe" (it used as Part of speech tagging tool (POS)) [21, 22].
- Remove Arabic stop words collected more than 1500 words and noises.
- Stripe whitespaces, remove numbers and punctuations.
- Documents tokenization and convert to Document-Term matrix with term frequencies tf.
- Remove sparse terms that sparsity ratio is more than 99%.

In Step 1, the normalization of the case-base by the conditional probability is demonstrated. Step 2, GloVe model is used to get the matrix of word-word co-occurrence. Step 3, the vectors is constructed of the closest words Ω that have a minimum threshold of cosine similarity. Step 4, subset of themes is selected to construct the clusters. Steps 5 and 6 respectively, similarity are calculated between the clusters and all cases and the most similar clusters are assigned. It worth here to emphasize that, steps from 2 to 4 do not depend on step 1. Figure 1 describes steps in details:

Step 1. Normalization:
In this step, case knowledge is discovered for feature (word) space [23] as step 1 in SOPHIA. X is denoted by the set of all documents in the corpus and Y is the set of all words in X. The normalization can be then represented by the following formula [9]:

$$P(y|x) = \frac{tf(x, y)}{\sum_{t \in Y} tf(x, t)} \tag{1}$$

Where, x is a document in X, y is a term in Y and tf(x, y) is term frequency for y term in x document.

Step 2. Build word embeddings vector space:

The word embeddings vector space is generated using GloVe model from the case-base. The main formula for GloVe is as following [17]:

$$J = \sum_{i,j=1}^{v} f(X_{ij}) \left(w_i^T \tilde{w}_j + b_i + \tilde{b}_j - \log X_{ij} \right)^2 \tag{2}$$

Where, $w \in R_d$ are word vectors, $\tilde{w} \in R_d$ are independent context word vectors, V is the size of the vocabulary, X_{ij} the frequency of term j in the context of term i. $f(X_{ij})$ is the weighting function,

$$w_i^T \tilde{w}_j = \log\left(P_{ij}\right) = \log\left(X_{ij}\right) - \log(X_i) \tag{3}$$

Where, P_{ij} is the conditional probability of the term j in the context of the term i. Therefore, P_{ij} can be represented by the following equation:

$$P_{ij} = P(j|i) = \frac{X_{ij}}{X_i} \tag{4}$$

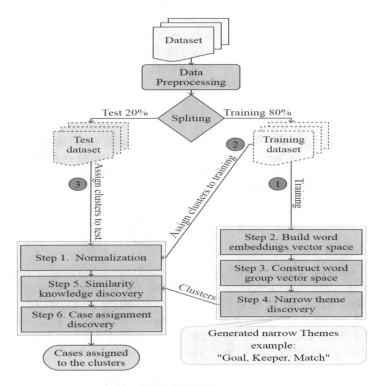

Fig. 1. GloSOPHIA processes.

Where, X_i is all the terms that appeared in context of the term i and can be calculated as follows:

$$X_i = \sum_k X_{ik} \tag{5}$$

Step 3. Construct word group vector space:
In this step we propose an algorithm to construct the themes from the generated word embeddings vector space in the previous step. The following algorithm takes word-word matrix that is generated from GloVe method, with a threshold that ranges from 0 to 1 (where 0 means no similarity and 1 means typical word) and returns a vector space with the closest terms (we could not add more details due to the limited space). Table 1 demonstrates an example of themes after narrowed by step 4.

Input: Terms vector space from GloVe algorithm → X, threshold → t
Output: Closest terms Vector space which is based on a threshold t → Ω
1- Normalizing X using L2 norm
2- Calculates pairwise cosine similarities for X.
3- Sort all related terms descending by similarity value.
4- Ω = all terms with similarity values above or equal the threshold t.
5- Eliminates all vectors in Ω that have terms ≤ 1.
6- For each vector Z in Ω vector space
   6.1- $\Omega(Z)$ = all vectors that have mutual terms with Z vector.
   6.2- uniqueTerms = all unique terms in $\Omega(Z)$.
   6.3- For each Z in $\Omega(Z)$: Z = uniqueTerms
7- Remove duplicated vectors from Ω.

Step 4. Narrow theme discovery
In this step we updated SOPHIA formula's to allow the theme to be represented as a vector (for example: Elected, presidential, election). The theme in GloSOPHIA is a conditional probability over a set of words that co-occur with any of the theme terms $z \in Z$. The updated equation here can be represented by:

$$p(y|Z) = \frac{\sum_{x \in X(Z)} tf(x,y)}{\sum_{x \in X(Z), t \in Y} tf(x,t)} \tag{6}$$

Where, y is the word that co-occurs with any from Z terms in the same document, X(Z) is all documents that contains any word from Z terms, tf(x, y) is the term frequency for the term y in document x, and Y: all terms that co-occur with any term from Z.
By applying Eq. 6 a vector of normalized values for every theme in Ω is ready to be compared with documents, selection of the subset of Ω that will be most significant theme as a cluster attractor is performed. This has been done by calculating the entropy:

$$H(Y|Z) = -\sum_{y \in Y} p(y|Z) \log(p(y|Z)) \tag{7}$$

Where, $Z \in \Omega$ and Y is all words that co-occur with any term in Z vector. $Z_1 \in \Psi_i$, $Z_2 \in \Omega - \Psi_i \rightarrow H(Y|Z_1) \leq H(Y|Z_2)$, then $\Psi = \bigcup_i^N Z_i$ where Ψ is a set of selected narrow theme. Based on the rule that states that only themes with in-between case frequency are informative, only the terms with case frequencies range between *min_cf* and *max_cf* is included. Also only themes Ω from narrow themes that have case frequencies range between *theme_min_cf* and *theme_max_cf* is included. Note that $|\Omega|$ is the length of vector space which must be greater or equal to N.

Table 1. Narrow themes Ψ from watan-2004 dataset when N = 7 and threshold = 0.72

Arabic	English
هريرة ,رضي ,ابي	May be pleased with, Abi, Huraira
اولمبياد ,اثينا ,اولمبي	Olympic, Olympiad, Olympiad
مبار ات ,خسر ,منتخبنا	Match, lost, our team
خام ,برميل ,نفط ,نفطي	Crude, oil, barrels
جورج ,بوش ,واشنطن	George, Bush, Washington
بمحافظة ,ظفار ,مديرية	Governorate of, Dhofar, Directorate
انتخاب ,رئاسي ,انتخابي	Elected, presidential, election

Step 5. Similarity knowledge discovery:
As in SOPHIA, Jensen-Shannon divergence (JS) [24] is used to measure the semantic similarity between document and cluster theme because it is found to get better results than Kullback–Leibler divergence (KL) and cross-entropy.

$$JS_{\{0.5,0.5\}}[p_1, p_2] = H[\bar{p}] - 0.5H[p_1] - 0.5H[p_2] \qquad (8)$$

Where, $H[\bar{p}]$ is the entropy of probability distribution p and $\bar{p}$ average probability distribution $= 0.5p_1 + 0.5p_2$, where the lower value represents the higher similarity between two probabilities.

Step 6. Case assignment discovery:
Z attractor theme will be assigned to case when:

$$Z = \arg\min_{t \in \Omega} JS_{\{0.5,0.5\}}[p(Y|x), p(Y|t)] \qquad (9)$$

3   Evaluation Criteria

In order to evaluate the performance of the proposed method, F measure on beta = 0.4 [9], Purity [4], Normalized Mutual Information (NMI) [4], Rand Index (RI) [4], Silhouette [25], Dunn index [26], and Connectivity [27] are used. These methods can be categorized into two main groups, either external evaluation or internal evaluation. External evaluation is based on labels. On the contrary, internal evaluation is based on the quality of the clusters and the degree of its separation, although good results in internal evaluations does not means it is effective in application [4].

4   Experimental Results

In this section, the results of the 7 types of evaluation criteria are presented on watan-2004 [20] dataset; Subsect. 4.1 is describing the dataset structure. The experiment environment is a laptop with Intel Core i5 2.5 GHz processor, 16 GB Ram. The language used for processing is R version 3.5.3 that is running on Ubuntu 18.04 operating system. The execution time of all experimental test data (4057 documents) is ranging from 0.45 to 1.77 min depending on the number of clusters 6–24.

4.1   Dataset Description

In the following experiments, an Arabic dataset called Watan-2004 news which contains 2866 distinct terms after preprocessing and 20,291 documents, organized in 6 main topics, is used. The dataset is divided into 6 directories named with the class names. Each directory contains file for each document. Figure 2 shows the distribution of the dataset. Caret package is used in R to partition data randomly by a ratio into training and test datasets. The selected ratios are 80%, 20% for training and test datasets respectively. The case frequency used here for terms ranges from min_cf = 100 to max_cf = 1000, and theme_min_cf = 500 to theme_max_cf = 5000 as a case frequencies for the themes has to be narrowed (see Step 4 in Sect. 2). In addition, all themes which have only one word are eliminated.

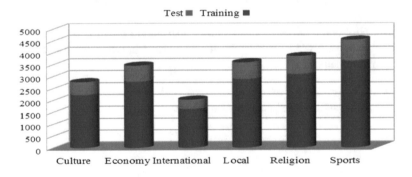

Fig. 2. Watan-2004 dataset structure after splitting to training and test sets

4.2   Evaluation

Figure 3 depicts a comparison between GloSOPHIA (threshold = 0.72), SOPHIA, K-means, and SOM using 7 different evaluation criteria: F measure on beta = 0.4, Purity, NMI, RI, Silhouette, Dunn index, and Connectivity. The criteria are described in details in Sect. 3. The number of clusters is set in the range from 6 to 24 to allow testing intensively the behavior of each clustering technique. It shows that the contributed algorithm outperforms SOPHIA in all criteria and also has a higher performance than K-means and SOM in some criteria, which makes our technique superior to the other techniques in most of the tests. Table 2 summaries the best score for each method. As an example for the case-base distribution over clusters using GloSOPHIA when $N = 7$, the major classes for cluster translated labels measured by number of cases are: ["May be pleased with, Abi, Huraira": (Religion: 732/881), "Olympic, Olympiad, Olympiad": (Sports: 363/403), "Match, lost, our team": (Sports: 458/488), "Crude, oil, barrels": (Economy: 417/527), "George, Bush, Washington": (International: 181/339; Culture: 119/339), "Governorate of, Dhofar, Directorate": (Local: 622/1098), "Elected, presidential, election": (International: 214/321)].

Table 2. Best values comparison between clustering methods.

	GloSOPHIA	SOPHIA	K-means	SOM
F measure	**0.8097**	0.6147	0.7275	0.7472
Purity	**0.8097**	0.6170	0.7259	0.7375
NMI	**0.5547**	0.4249	0.5153	0.4641
RI	**0.8707**	0.8288	0.8121	0.8189
Silhouette	**0.03812**	0.027366	0.009602	0.03099
Dunn index	**0.3376**	0.2245	0.12771	0.27864
Connectivity	2752	3333	1528	**1091**

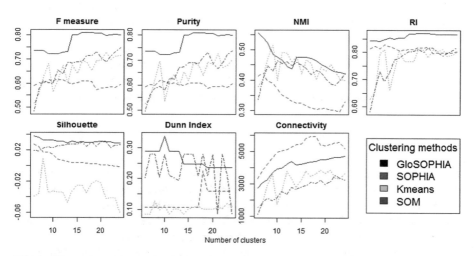

Fig. 3. Comparison between GloSOPHIA (on threshold = 0.72), SOPHIA, K-means, and SOM using evaluation criteria. Note that in Connectivity the lower is better

4.3 Investigating the Influence of Threshold on Clustering

The main goal is to investigating the effect of changing threshold on clustering performance, two examples of evaluations (Purity, Silhouette) are taken. The threshold range is changed from 0.65–0.85, we have chosen this range because, if the threshold is decreased to low scores (e.g., 0.5), it would place a burden on the algorithm. This is caused by the construction of group with weaker relationships, which reduces the performance and quality of the algorithm. On the other hand, increasing the threshold with very high scores (e.g.: 0.9) will make it harder to construct a number of groups that fit the number of clusters. The number of clusters is changed from 6–24. Figures 4 and 5 show that, the threshold affects the purity and silhouette when the number of clusters is low. Moreover, when the number of clusters is increased the effect of the threshold on the clustering quality is decreased gradually. Also it's obvious that there is an inverse relationship between silhouette and the number of clusters.

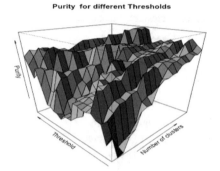 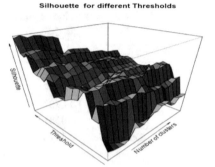

Fig. 4. The relationship between threshold, Purity, and number of clusters.   **Fig. 5.** The relationship between threshold, Silhouette, and number of clusters.

Therefore, it is better to choose high threshold when the number of clusters is low. However, when the clusters number is increased to reduce the threshold, more groups of similar words can be generated by step 3 and hence N is always $\leq |\Omega|$.

5   Conclusion

In this paper, we have introduced an enhanced clustering model called GloVe SOPHIA (GloSOPHIA). It based on extending SOPHIA by integrating word embeddings technique to enhance knowledge discovery in textual case based reasoning (TCBR). To evaluate the quality of the proposed method, we applied GloSOPHIA to an Arabic newspaper corpus called watan-2004 and compared the results with (SOPHIA, K-means, and SOM) using evaluation criteria. The results showed that GloSOPHIA has outperformed in 6 from 7 evaluations (F measure 0.80, Purity reached 0.80, NMI 0.55, RI 0.87, Silhouette 0.038, and Dunn Index 0.33).

References

1. Aamodt, A., Plaza, E.: Case-based reasoning: foundational issues, methodological variations, and system approaches. AI Commun. **7**(1), 39–59 (1994)
2. Recio-García, J.A., Díaz-Agudo, B., González-Calero, P.A.: Textual CBR in jCOLIBRI: from retrieval to reuse. In: Proceedings of the ICCBR 2007 Workshop on Textual Case-Based Reasoning: Beyond Retrieval (2007)
3. Witten, I.H., et al.: Data Mining: Practical Machine Learning Tools and Techniques. Morgan Kaufmann, Boston (2016)
4. Manning, C., Raghavan, P., Schütze, H.: Introduction to information retrieval. Natural Lang. Eng. **16**(1), 100–103 (2010)
5. Weber, R.O., Ashley, K.D., Brüninghaus, S.: Textual case-based reasoning. Knowl. Eng. Rev. **20**(3), 255–260 (2005)
6. Aggarwal, C.C., Zhai, C.: A survey of text clustering algorithms. In: Mining Text Data, pp. 77–128. Springer, Boston (2012)
7. Allahyari, M., et al.: A brief survey of text mining: Classification, clustering and extraction techniques. arXiv preprint arXiv:1707.02919 (2017)
8. Silge, J., Robinson, D.: Text Mining with R: A Tidy Approach. O'Reilly Media, Sebastopol (2017)
9. Patterson, D., et al.: SOPHIA-TCBR: a knowledge discovery framework for textual case-based reasoning. Knowl. Based Syst. **21**(5), 404–414 (2008)
10. Hirschberg, J., Manning, C.D.: Advances in natural language processing. Science **349** (6245), 261–266 (2015)
11. Mikolov, T., et al: Efficient estimation of word representations in vector space. arXiv preprint arXiv:1301.3781 (2013)
12. Cunningham, C., et al.: Investigating graphs in textual case-based reasoning. In: European Conference on Case-Based Reasoning. Springer, Heidelberg (2004)
13. Proctor, J.M., Waldstein, I., Weber, R.: Identifying facts for TCBR. In: ICCBR Workshops (2005)
14. Fornells, A., et al.: Integration of a methodology for cluster-based retrieval in jColibri. In: International Conference on Case-Based Reasoning. Springer, Heidelberg (2009)
15. Kohonen, T.: The self-organizing map. Proc. IEEE **78**(9), 1464–1480 (1990)
16. Osiński, S., Stefanowski, J., Weiss, D.: Lingo: search results clustering algorithm based on singular value decomposition. In: Intelligent Information Processing and Web Mining, pp. 359–368. Springer, Heidelberg (2004)
17. Pennington, J., Socher, R., Manning, C.: Glove: global vectors for word representation. In: Proceedings of the 2014 Conference on Empirical Methods in Natural Language Processing (EMNLP) (2014)
18. Abbas, M., Smaili, K., Berkani, D.: Evaluation of topic identification methods on arabic Corpora. J. Digit. Inf. Manage. **9**(5), 185–192 (2011)
19. Hartigan, J.A., Wong, M.A.: Algorithm AS 136: a k-means clustering algorithm. J. Roy. Stat. Soc. (Appl. Stat.) **28**(1), 100–108 (1979)
20. Kelaiaia, A., Merouani, H.F.: Clustering with probabilistic topic models on arabic texts: a comparative study of LDA and K-means. Int. Arab J. Inf. Technol. **13**(2), 332–338 (2016)
21. Hajič, J., et al.: Prague Arabic Dependency Treebank 1.0. (2009)
22. Smrz, O., Bielicky, V., Hajic, J.: Prague Arabic dependency treebank: a word on the million words (2008)

23. Joachims, T.: A Probabilistic Analysis of the Rocchio Algorithm with TFIDF for Text Categorization. No. CMU-CS-96–118. Carnegie-mellon Univ. Pittsburgh dept. of computer science (1996)
24. Lin, J.: Divergence measures based on the Shannon entropy. IEEE Trans. Inf. Theory **37**(1), 145–151 (1991)
25. Rousseeuw, P.J.: Silhouettes: a graphical aid to the interpretation and validation of cluster analysis. Comput. Appl. Math. **20**, 53–65 (1987)
26. Dunn, J.C.: Well-separated clusters and optimal fuzzy partitions. J. Cybern. **4**(1), 95–104 (1974)
27. Handl, J., Knowles, J.: Exploiting the trade-off—the benefits of multiple objectives in data clustering. In: International Conference on Evolutionary Multi-Criterion Optimization. Springer, Heidelberg (2005)

Assessing the Performance of E-government Services Through Multi-criteria Analysis: The Case of Egypt

Abeer Mosaad Ghareeb[1(✉)], Nagy Ramadan Darwish[2],
and Hesham A. Hefney[1]

[1] Computer and Information Sciences Department, Institute of Statistical Studies
and Research, Cairo University, Giza, Egypt
abeer_mosad@yahoo.com, hehefney@ieee.org
[2] Department of Information Systems and Technology, Institute of Statistical
Studies and Research, Cairo University, Giza, Egypt
drnagyd@yahoo.com

Abstract. E-government projects have been mostly supply driven with relatively less information about the performance of e-government services as well as the perception of the citizens regarding them. Less demanding of the available e-government services may be an indication of some difficulties in using or accessing these services. Therefore, the performance of the available e-government services needs to be assessed. This paper aims to propose a methodology to assess the performance of e-government services. The proposed methodology combines two of the most well-known and extensively used MCDM methods. These methods are PROMETHEE and AHP. The proposed methodology has been applied to assess the performance of e-government services available on Egyptian national portal. The research outcomes inform the policy makers about how e-government services have performed from citizen's perspective and help them to take suitable corrective actions to better the ranks of the underperforming services and fulfill the citizens' needs.

Keywords: E-government · PROMETHEE · AHP · Adoption

1 Introduction

E-government refers to utilization of ICTs to transform and enhance the relationship of the public sector and its clients through an improved range and quality of service [1]. The countries around the globe have invested heavily in e-government projects and have experienced substantial progress in provision of online services [2]. However, E-government projects have been mostly supply driven with relatively less information about the performance of the e-government services. However, this paradigm has begun to change. Citizens are placing numerous stresses on governments by less demanding e-government services. Having high demand and use of the available e-government services is crucial to the success of e-government projects. The failure of e-government projects comes with a high cost [3].

© Springer Nature Switzerland AG 2020
A. E. Hassanien et al. (Eds.): AISI 2019, AISC 1058, pp. 711–721, 2020.
https://doi.org/10.1007/978-3-030-31129-2_65

Citizen's usage of e-government services mainly depends on multiple criteria. Multi-Criteria Decision Making (MCDM) has been widely used for evaluation problems containing multiple criteria. However, the quantitative approach supported by statistical analysis is the most dominant used approach in the area of e-government adoption [4]. Preference Ranking Organization Method for Enrichment Evaluations (PROMETHEE) and Analytical Hierarchy Process (AHP) are widely used MCDM methods. This paper aims to assess the performance of e-government services using PROMETHEE and AHP. The remainder of this paper is organized as follows. Section 2 provides an overview of MCDM. Section 3 presents the mathematical background of PROMETHEE. Section 4 provides the proposed methodology. Section 5 presents a case study. Finally, Sect. 6 gives conclusion.

2 Multi-criteria Decision Making

MCDM has grown as a part of operations research concerned with designing computational and mathematical tools for supporting the subjective evaluation of performance criteria by decision-makers [5]. MCDM methods have been broadly classified into two categories: Multiple Attribute Decision Making (MADM) and Multiple Objective Decision Making (MODM) [6]. MADM is the most well-known branch of decision making [6]. The two main families in MADM methods are those based on the Multi-Attribute Utility Theory (MAUT) and Outranking methods. MAUT methods allow complete compensation between criteria. AHP is one of the more widely applied MAUT methods. It was proposed by Saaty [7]. It builds on complete aggregation of additive type characteristic of the American school [8]. AHP is based on three principles as follows: construction of a hierarchy; priority setting; and synthesis of the priorities [8]. AHP has been used to propose evaluation methodology for government websites [9].

PROMETHEE is one of the main outranking methods typical of the European school [8]. It is quite simple in conception and application compared to other MCDM methods [10]. PROMETHEE has been widely used for practical MCDM problems in various domains such as finance, transportation, and information technology strategy selection [10]. PROMETHEE has been applied to rank a sample of 13 water development projects in Jordan [11]. An integrated approach combining PROMETHEE and Geometrical Analysis for Interactive Aid methods has been applied for evaluating the performance of 20 national institutions in India [12]. PROMETHEE has some strength in comparison with other MCDM methods [13].

3 PROMETHEE

The mathematical background of PROMETHEE has been described in several references [14, 15]. Let us define a multi-criteria problem as follows.

$$Optimizing\{f_1(a), f_2(a), \ldots\ldots f_k(a) | a \in A\} \tag{1}$$

Where A is a set of alternatives and $f_j, j = 1, 2, 3, \ldots\ldots, k$, is criterion to be maximized or minimized.

PROMETHEE procedure starts by filling the evaluation table. The alternatives are assessed on each criterion where $f_j(a)$ is the performance measure of alternative (a) with respect to j^{th} criterion. Then, a specific preference function $P_j(a, b)$ is associated to each criterion to translate the deviation between the assessments of two alternatives on that criterion into a preference degree as given in Eq. 2.

$$P_j(a, b) = P_j\big(f_j(a) - f_j(b)\big) \tag{2}$$

The preference degree represents the preference of alternative (a) over alternative (b) for criterion f_j. This degree is normalized so that

$$0 \le P_j(a, b) \le 1 \tag{3}$$

There are six known preference functions as follows: Usual; U-shape; V-shape; level; V-shape with indifference area; and Gaussian preference functions [15, 16]. The preference degrees are used to calculate a global preference index $\pi(a, b)$ for each pair of alternatives as given in Eq. 4. It represents the preference of alternative (a) over (b) considering all criteria and taking into account the weights of criteria.

$$\pi(a, b) = \sum_{j=1}^{k} w_j P_j(a, b) \tag{4}$$

Where, w_j is the weight of jth criterion and it is a choice of the decision maker and

$$\sum_{j=1}^{k} w_j = 1 \tag{5}$$

Global indices are used to calculate three preference flows for each alternative. Leaving Flow $\varphi^+(a)$ denotes how much alternative (a) dominates the other as given by Eq. 6. Entering flow $\varphi^-(a)$ denotes how much alternative (a) is dominated by other alternatives as given by Eq. 7. Net flow is calculated by subtracting entering flow from leaving flow as given by Eq. 8.

$$\varphi^+(a) = \frac{1}{n-1} \sum_{y \in A} \pi(a, y) \tag{6}$$

$$\varphi^-(a) = \frac{1}{n-1} \sum_{y \in A} \pi(y, a) \tag{7}$$

$$\varphi(a) = \varphi^+(a) - \varphi^-(a) \tag{8}$$

Where n is the number of alternatives.

PROMETHEE ranks the alternatives based on the values of the preference flows. Two main PROMETHEE tools can be used. They are PROMETHEE I partial ranking, and PROMETHEE II complete ranking. PROMETHEE I uses the values of leaving and entering flows. Alternative (a) is preferable to alternative (b) if alternative (a) has a greater leaving flow than the leaving flow of alternative (b) and a smaller entering flow than the entering flow of alternative (b). PROMETHEE II uses the net flow to rank the alternatives. In addition, the uni-criterion net flow can be calculated, as given by Eq. 9, to show the contribution of one criterion to the net flow score of the alternative.

$$\varphi_j(a) = \frac{1}{n-1} \sum_{y \in A} \left[P_j(a, y) - P_j(y, a) \right] \tag{9}$$

4   The Proposed Methodology

The proposed methodology for assessment the performance of e-government services contains five phases (see Fig. 1). In Phase 1, a set of e-government services are selected for evaluation purpose. In Phase 2, the citizen requirements are identified and translated into a set of acceptance and adoption criteria. In Phase 3, AHP is used to set up the decision hierarchy. In Phase 4, the PROMETHEEE procedure described in Sect. 3 is implemented. The measurement scale is defined. The criterion can be assessed quantitative or qualitative. Each criterion has to be decided whether it has to minimized or maximized. The preference functions are chosen by decision makers. The evaluation outcomes depend on both choice of preference function and its parameters [16]. PROMETHEE II is used to rank the e-government services. In Phase 5, the results are analyzed using PROMETHEE rainbow. The basic idea of the PROMETHEE rainbow is to calculate the uni-criterion net flow assigned to each alternative. For each alternative, a bar is drawn. The different slices of each bar are colored according to the

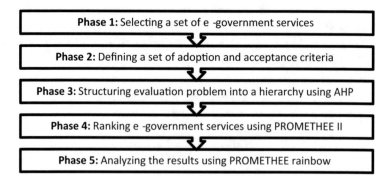

Fig. 1. Proposed methodology to assess the performance of e-government services

criteria. Each slice is proportional to the contribution of one criterion to the net flow score of the alternative.

5 Case Study

A case study is presented to demonstrate effectiveness of the proposed methodology. Egypt provides a perfect context for this study. Egypt started its e-government initiative early with clear vision and top management support. Several successful projects have been undertaken such as Family Card system [17] and University Enrolment project [18]. The country achieved a great deal at the beginning and won some awards [19]. However, Egypt declined in its e-government development rank over the last years [2, 20]. It lags behind other Arab countries [21]. Challenges of e-government in Egypt include lack of coordination among governmental ministries [17], poor marketing efforts [18], and lack of e-government services quality measurement [21]. Therefore, there is a need to study the e-government development for the Egyptian case to better the position of Egypt and enhance citizen satisfaction with public services.

5.1 Selecting a Set of E-government Services

The e-government services available on the Egyptian national portal [19] have been examined to select a set of services for assessment. Citizen-oriented services are the

Table 1. E-government services

No	Name	Label	Provider
1	Birth certificate extract	BCE	Ministry of Interior- Civil Status Organization
2	Death certificate extract	DCE	Ministry of Interior- Civil Status Organization
3	Marriage document extract	MDE	Ministry of Interior- Civil Status Organization
4	Divorce document extract	DDE	Ministry of Interior- Civil Status Organization
5	Family record extract	FRE	Ministry of Interior- Civil Status Organization
6	Replacement of national ID card	RID	Ministry of Interior- Civil Status Organization
7	Traffic Fines Payment Certificate	TFP	Ministry of Justice
8	Bus reservation	BUS	West middle delta, east delta, and upper Egypt companies
9	Trains tickets reservation	TTR	Egyptian National Railways
10	Equivalence of scientific degree	ESD	Supreme Council of Universities
11	Searching in laws and legalization	SLL	Ministry of Justice
12	Payment of phone bill	PPB	Telecom Egypt
13	Egypt-Air e-ticketing	EAT	Egypt-Air

main focus of this study. 44 services were found. 20 of these 44 are either under construction or not found. Running e-government services are further sub-divided into two categories. They are paid and free services. The current study focuses on evaluating paid e-government services (see Table 1).

5.2 Defining a Set of Adoption and Acceptance Criteria

Based on extensive literature review, nine important acceptance and adoption criteria are defined (see Table 2). These criteria are cost (CO), payment method (PM), time (TI), responsiveness (RE), usability (US), accessibility (AC), information quality (IQ), marketing campaign (MC) and mandate (MA).

Table 2. Adoption and acceptance criteria.

Criterion	Description
CO	Cost of e-government services should not exceeds the cost of traditional ones
PM	Payment systems should be convenient with citizens' economic and social conditions, educational level, and their lifestyle
TI	Ability of e-government services in time saving. Displaying citizen charter online which provides the minimum number of days that a particular public organization takes to deliver the service
RE	Ability to make inquiries online. Replying to citizens' inquiries or complaints through various e-government delivery channels like emails. Providing on-going technical support to citizens
US	Supporting a complete set of navigational aids such as main menu, site map, home link, and click-ability identification. Adherence of uniformity in terminology, color, style, labeling, abstraction, and positioning of elements. Easy of reading by taking into account some features like font type and font size. Providing service description, hints, guidelines and examples. Explaining the steps of using the service
AC	Available through the national portal anytime from anywhere. Concise and easy to remember URL. Easing of locating intended service after accessing intended application. Download speed. Compatibility with a variety of browsers
IQ	Accurate, consistent, and up to date
MC	Promotion of services through various channels such as TV talk shows, newspaper ads, and word of mouth. Informing the citizens about the web addresses, the laws and legislations of electronic dealing and about relative advantages of e-services such as cost and time saving, avoiding long queues, synchronizing and updating records among different governmental organizations, and tracking the status of the conducted service
MA	Services, that don't involve personal interaction for identity authentication and authorization or that don't require physical inspection, should be totally complete online

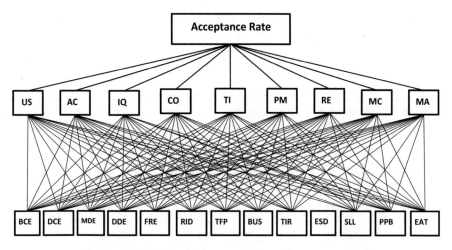

Fig. 2. A 3-level hierarchy structure for assessment problem.

5.3 Structuring the Evaluation Problem into a Hierarchy Using AHP

The decision problem is structured into a 3-level hierarchy as depicted in Fig. 2. Level 1 represents the goal of improving acceptance rate. Level 2 contains the nine identified criteria. Level 3 contains 13 paid e-government services.

5.4 Ranking E-government Services Using PROMETHEE II

All the nine criteria are evaluated using a qualitative scale. The cost and time criteria should be minimized. The other criteria should be maximized. For mandate criterion, a simple yes/no scale is used. A 7-point scale, ranging from strongly agree (SA) to strongly disagree (SDA), is selected for cost and time criteria. A 7-point scale, ranging from strongly disagree to strongly agree, is used for the rest of the criteria (see Table 3). PROMETHEE method provides a technique to handle the missing or not

Table 3. Measurement scale

Value	Level
1	Strongly Disagree (SDA)
2	Disagree (DA)
3	Little Disagree (LDA)
4	No comment (NC)
5	Little Agree (LA)
6	Agree (A)
7	Strongly Agree (SA)

available values (N/A) at the pairwise comparison level [22]. It set both preference degrees equal to zero as given in Eq. 10.

$$P_j(a, b) = P_j(b, a) = 0 \tag{10}$$

By the time of data collection (which began on 1 January and finished on 31 May, 2018) there has been no change in the design and content of the examined e-government services. Table 4 gives the evaluation data of e-government services.

Table 4. Evaluation table

	US	AC	IQ	CO	TI	PM	RE	MC	MA
BCE	LA	A	NC	SDA	NC	SA	A	DS	NO
DCE	LA	A	NC	DS	NC	SA	A	DS	NO
MDE	LA	A	NC	DS	NC	SA	A	DS	NO
DDE	LA	A	NC	DS	NC	SA	A	DS	NO
FRE	A	A	NC	LA	A	SA	A	DS	NO
RID	A	A	NC	A	A	SA	A	DS	NO
TFPC	NC	DA	A	A	NC	SA	NC	DS	NO
BUS	LDA	LDA	LDA	A	A	DS	NC	DS	NO
TTR	A	A	A	A	A	DS	A	DS	NO
ESD	A	A	A	A	A	A	NC	NC	YES
SLL	LDA	A	NC	N/A	A	NC	NC	DS	NO
PPB	A	A	SA	A	A	DS	SA	NC	NO
EAT	LA	A	A	A	A	DS	LDA	D	NO

For all criteria, except mandate, the V-shape preference function is employed. V-shape function depends on a preference parameter. This parameter is the lowest value of the difference between two evaluations, above which there is a strict preference of one of the corresponding alternatives. As long as the difference is lower than this threshold, the preference increases linearly with the difference [16]. The values of measurement scale range from 1 to 7. So, when the preference threshold is set to 6.00, a small difference, as much as larger one, is accounted. Usual preference function is a good choice for mandate criterion, which includes a small number of evaluation levels. All weights are set to be equal. The multi-criteria decision aid software Visual PROMETHEE 1.4 is used to compute preference flows and rank the e-government services. Ranks and preference flows are shown in Table 5. ESD has the higher net flow score. It dominates the others. PPB, RID, and FRE are very close to every other. They have good positive scores. TTR is the closest to zero. It is more average positive item. DCE, MDE, and DDE have the same negative score. EAT, BCE, TFP, and SLL have also very close but negative scores. BUS service is dominated by all the others.

Table 5. Ranks and preference flows.

Rank	Service	$\varphi(a)$	$\varphi^+(a)$	$\varphi^-(a)$
1	ESD	0.2361	0.2747	0.0386
2	PPB	0.1157	0.1883	0.0725
3	RID	0.0957	0.1296	0.0340
4	FRE	0.0772	0.1219	0.0448
5	TTR	0.0355	0.1173	0.0818
6	DCE	−0.0386	0.0756	0.1142
6	MDE	−0.0386	0.0756	0.1142
6	DDE	−0.0386	0.0756	0.1142
7	EAT	−0.0448	0.0880	0.1327
8	BCE	−0.0571	0.0741	0.1312
9	TFP	−0.0648	0.0957	0.1605
10	SLL	−0.0926	0.0417	0.1343
11	BUS	−0.1852	0.0463	0.2315

5.5   Analyzing Results Using PROMETHEE Rainbow

PROMETHEE rainbow is constructed based on the values of the uni-criterion net flow (see Fig. 3). It demonstrates that ESD reveals very little weakness compared to other e-government services. All criteria, except responsiveness, contribute positively to its net flow score. Mandate is the most important feature of this service. Payment method is the most criterion of the positive effect on the net flow score of RID, FRE, DCE, MDE, DDE, BCE, and TFP services. On the other side, payment method is the most criterion

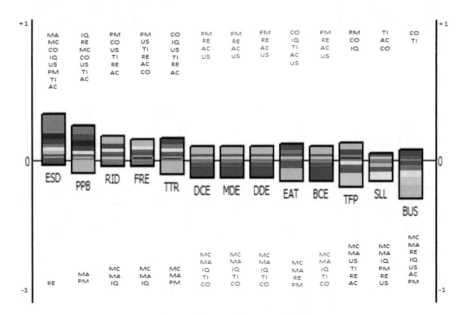

Fig. 3. PROMETHEE Rainbow

of the negative effect on the net flow score of PPB, TTR, EAT, and BUS services. Cost may be the most criterion of the negative effect on the citizen's acceptance and adoption to Civil Status Office online services. Accessibility is the most influential weakness of TFP service. BUS service shows little strengths represented in a limited positive affect of cost and time criteria. However, the rest of the criteria are drawn in the form of large downward slices which means that large negative contribution to its net flow score.

6 Conclusion

A methodology to assess the performance of e-government services through multi-criteria analysis has been proposed. The proposed methodology has been applied to assess the performance of citizen-oriented e-government services available on Egyptian National Portal. Each e-government service has been analyzed against a pre-defined set of acceptance criteria covering application and service aspects. Most of the examined e-government services have inadequacy regarding some criteria. Although some e-government services tend to be performing quite acceptably, there is a room for improvement. Mandate seems to be a very influential criterion. The vast majority of the examined services employ traditional and electronic ways and this is in favor of the traditional one. Payment method can be considered the most criterion of the positive or negative impact on e-government services. If the cost of an online service exceeds the cost of the traditional one, the extent to which the citizen accepts this increase is affected by the amount of increase and the frequency of the demand of the service. Live chat is an effective way to communicate with citizens while using the services. However, this feature doesn't take enough attention. Given the scarcity in literature in performance measurement of e-government services, it is hoped that the proposed methodology shared in this paper fills the gap left in the literature.

References

1. Yonazi, J.J.: Enhancing adoption of e-government initiatives in Tanzania. Ph.D. thesis, University of Groningen, SOM Research School (2010)
2. United Nations: Gearing e-government to support transformation towards sustainable and resilient societies. Department of Economic and Social Affairs. New York: United Nations (2018)
3. Heeks, R.: Most e-government for development projects fail: How can risks be reduced? Institute of Development Policy and Management (IDPM), University of Manchester (2003)
4. Irani, Z., Weerakkody, V., Kamal, M., Hindi, N.M., Osaman, I.H., Anouze, A.-L., et al.: An analysis of methodologies utilized in e-government research. J. Interprise Inf. Manage. **25**, 298–313 (2012)
5. Zavadskas, E., Turskis, Z., Kildiene, S.: State of art surveys of overviews on MCDM/MADM methods. Technol. Econ. Dev. Econ. **20**, 165–179 (2014)
6. Kahraman, C.: Fuzzy Multi-criteria Decision Making- Theory and Applications with Recent Developments, 1st edn. Springer, US (2008)
7. Saaty, T.L.: The Analytic Hierarchy Process. McGraw-Hill, New York (1980)

8. Macharis, C., Springael, J., Brucker, K.D., Verbeke, A.: PROMETHEE and AHP: the design of operational synergies in multicriteria analysis. Strengthening PROMETHEE with the ideas of AHP. Eur. J. Oper. Res. **153**, 307–317 (2004)

9. Buyukozkan, G., Ruan, D.: Evaluation government websites based on a fuzzy multiple criteria decision approach. Int. J. Uncertainty Fuzziness Knowl. Based Syst. **15**(3), 321–343 (2007)

10. Bilsel, R.U., Buyukozkan, G., Ruan, D.: A fuzzy preference ranking model for a quality evaluation of hospital websites. Int. J. Intell. Syst. **21**, 1181–1197 (2006)

11. Al-Kloub, B., Al-Shemmeri, T., Pearman, A.D., Brans, J.P., Mareschal, B.: Application of multicriteria analysis to rank and evaluation of water development projects (the case of Jordan). In: Multi-objective Programming and Goal Programming. Springer, Heidelberg (1996)

12. Ranjan, R., Chakraborty, S.: Performance evaluation of India technical institutions using PROMETHEE-GAIA approach. Inform. Educ. **14**(1), 103–125 (2015)

13. Behzadian, M., Kazemzadeh, R., Albadvi, A., Aghdasi, M.: PROMETHEE: a comprehensive literature review methodologies and applications. Eur. J. Oper. Res. **200**(1), 198–215 (2010)

14. Brans, J.P., Vincke, P.H.: A preference ranking organization method the PROMETHEE method for multiple criteria decision making). Manage. Sci. **31**(6), 647–656 (1985)

15. Brans, J., Vincke, P., Mareschal, B.: How to select and how to rank projects: the PROMETHEE method. Eur. J. Oper. Res. **24**, 228–238 (1986)

16. Podvezko, V., Podviezko, A.: Dependence of multi-criteria evaluation result on choice of preference functions and their parameters. Technol. Econ. Dev. Econ. **16**(1), 143–158 (2010)

17. Abdelkader, A.A.: A manifest of barriers to successful E-government: cases from the Egyptian programme. Int. J. Bus. Soc. Sci. **6**(1), 169–186 (2015)

18. El-Baradei, L., Shamma, H.M., Saada, N.: Examining the marketing of e-government services in Egypt. Int. J. Bus. Public Manage. **2**(2), 12–22 (2012)

19. Egyptian Government Portal. www.egypt.gov.eg. Accessed 31 May 2018

20. United Nations: E-government in support of sustainable development. United Nations Department of Economic and Social Affairs. New York: United Nations (2016)

21. Gebba, T.R., Zakaria, M.R.: E-government in Egypt: an analysis of practices and challenges. Int. J. Bus. Res. Dev. **4**(2), 11–25 (2015)

22. Fernandez, N.G.: The management of missing values in PROMETHEE Method. Master thesis, Universitat Politecnic De Catalunya, Barcelona, Spain (2013)

SQL Injection Attacks
Detection and Prevention
Based on Neuro-Fuzzy Technique

Doaa E. Nofal[1(✉)] and Abeer A. Amer[2]

[1] Institute of Graduate Studies and Research, Alexandria University,
Alexandria, Egypt
Doaa_ean@yahoo.com
[2] Sadat Academy for Management and Sciences, Alexandria, Egypt
abamer_2000@yahoo.com

Abstract. A Structured Query Language (SQL) injection attack (SQLIA) is one of most famous code injection techniques that threaten web applications, as it could compromise the confidentiality, integrity and availability of the database system of an online application. Whereas other known attacks follow specific patterns, SQLIAs are often unpredictable and demonstrate no specific pattern, which has been greatly problematic to both researchers and developers. Therefore, the detection and prevention of SQLIAs has been a hot topic. This paper proposes a system to provide better results for SQLIA prevention than previous methodologies, taking in consideration the accuracy of the system and its learning capability and flexibility to deal with the issue of uncertainty. The proposed system for SQLIA detection and prevention has been realized on an Adaptive Neuro-Fuzzy Inference System (ANFIS). In addition, the developed system has been enhanced through the use of Fuzzy C-Means (FCM) to deal with the uncertainty problem associated with SQL features. Moreover, Scaled Conjugate Gradient algorithm (SCG) has been utilized to increase the speed of the proposed system drastically. The proposed system has been evaluated using a well-known dataset, and the results show a significant enhancement in the detection and prevention of SQLIAs.

Keywords: SQL injection attacks · Neuro-fuzzy · ANFIS · FCM · SCG · Web security

1 Introduction

In just a few decades, the Internet has become the biggest network of connections known to users. However, with the rising dominance of the Internet in our daily lives, and the broadened use of its web-based applications, threats have become more and more prominent. SQLIA is a widely common threat, and it has been rated as the number one attack in the open web application security project (OWASP) list of top ten web application threats [1]. SQLIA is a class of code injection attacks that take advantage of a lack of validation of user input [2, 3]. There are several SQLIA techniques performed by hackers to insert, retrieve, update, and delete data from databases;

© Springer Nature Switzerland AG 2020
A. E. Hassanien et al. (Eds.): AISI 2019, AISC 1058, pp. 722–738, 2020.
https://doi.org/10.1007/978-3-030-31129-2_66

shut down an SQL server; retrieve database information from the returned error message; or execute stored procedures [4, 5]. Generally, there are several classifications of SQLIA, such as tautologies, illegal/logically incorrect queries, union query, piggy-backed queries, stored queries, inference, and alternate encodings [6–8].

The main problem of SQLIAs and other security threats is that developers did not previously consider structured security approaches and dynamic and practical policy framework for addressing threats. Moreover, when such approaches are taken into consideration, attackers aim to develop new ways that can bypass the defenses designed by developers; they began to use different techniques to perform the SQLIA [9]. The rising issue is that SQLIA techniques have become more and more complex. Thus, most of the current defense tools cannot address all types of attacks. Further-more, there is a large gap between theory and practice in the field nowadays; some existing techniques are inapplicable in real, operating applications. Some of the used techniques also need additional infrastructures or require the modification of the web application's code [4]. Lack of flexibility and scalability is another challenge. Some existing techniques solve only a subset of vulnerabilities that lead to SQLIAs [10]. The lack of learning capabilities is another significant hurdle. In the last few years, machine learning techniques were adapted to overcome the aforementioned problems [9–11]. However, most existing machine learning techniques suffer from high computational overhead. Furthermore, a number of existing solutions do not have the capability to detect new attacks. Uncertainty is a common phenomenon in machine learning, which can be found in every stage of learning [12, 13].

One of the most important machine learning techniques is the neural network (NN) model. The main characteristic of NN is the fact that these structures have the ability to learn through input and output samples of the system. The advantages of the fuzzy systems are the capacity to represent inherent uncertainties of the human knowledge with linguistic variables; simple interaction of the expert of the domain with the engineer designer of the system; easy interpretation of results, which is achieved due to the natural rules representation; and easy extension of the base of knowledge through the addition of new rules [14]. Nevertheless, an interpretation of the fuzzy information in the internal representation is required to reach a more thorough insight into the network's behavior. As a solution to such problem, neuro-fuzzy systems are employed to find the parameters of a fuzzy system (i.e., fuzzy sets and fuzzy rules) by utilizing approximation techniques from NN. By using a supervised learning algorithm, the neuro-fuzzy systems can construct an input-output mapping based on either human knowledge or stipulated input-output data pairs. Therefore, it is rendered a powerful method that addresses uncertainty, imprecision, and non-linearity [15].

This paper presents a modified approach for the detection and prevention of SQLIA. This modified approach is proposed to address the problems of uncertainty, adaptation, and fuzziness that are associated with existing machine learning techniques in the field of SQLIA. An ANFIS has the ability to construct models solely based on the target system's sample data. The FCM has been utilized in this work as a clustering method to enhance system performance by solving the fuzziness and uncertainty problems of the input data. Moreover, the SCG algorithm has been used to speed up and, thus, improve the training process. Finally, a malicious SQL statement has been prevented from occurring in the database. To the best of our knowledge, there is no previous work that uses ANFIS-FCM for detecting and preventing SQLIAs.

The subsequent parts of this paper are organized as follows: Sect. 2 provides a background and a literature survey of works related to SQLIA detection and prevention systems, an overview of the proposed is explained in Sect. 3, the experimental result and evaluation of the proposed system are discussed in Sect. 4, and Sect. 5 presents the conclusion of the work and the directions that could be taken in future works.

2   Literature Review and Related Work

There are four main categories of SQL injection attacks against databases, namely SQL manipulation, code injection, function call injection, and buffer overflow [16]. SQLI detection and prevention techniques are classified into the static, dynamic, combined analysis, and machine learning approaches, as well as the hash technique or function and black box testing [3, 17]. Static analysis checks whether every flow from a source to a sink is subject to an input validation and/or input sanitizing routine [18]; whereas dynamic analysis is based on dynamically mining the programmer's intended query structure on any input and detects attacks by comparing it against the structure of the actual query issued. The existing machine learning models deal with uncertain information, such as input, output, or internal representation. Hence, the proposed system can be classified as a machine learning technique.

The analysis for monitoring and NEutralizing SQLIAs, known as the AMNESIA technique, suggested in [3] is a runtime monitoring technique. This approach has two stages: a static stage, which automatically builds the patterns of the SQL queries that an application creates at each point of access to the database, and a dynamic stage, in which AMNESIA intercepts all the SQL queries before they are sent to the database and checks each query against the statically built patterns. If the queries happen to violate the approach, then this technique prevents executing the queries on the database server. This tool limits the SQLIAs when successfully building the query models in the static analysis, but it also has some limitations, particularly in preventing attacks related to stored procedures and in supporting segmented queries. Moosa in [5] investigates Artificial Neural Networks (ANN) in SQL injection classification—a technique that is an application layer firewall, based on ANN, that protects web applications against SQL injection attacks. This approach is based on the ability of the ANN concept to carry out pattern recognition when it is suitably trained. A set of malicious and normal data is used to teach the ANN during the training phase. The trained ANN is then integrated into a web application firewall to protect the application during the operational phase. The key drawback of this approach is that the quality of a trained ANN often depends on its architecture and the way the ANN is trained. More importantly, the quality of the trained ANN also depends on the quality of the training data used and the features that are extracted from the data. With relatively limited sets of training data, the resulting ANN seems to be sensitive to content that has an SQL keyword. Another work related to ANN-based SQLI detection is introduced in [19, 20]. It depends on limited SQL patterns for training, which renders it susceptible to generate false positives.

Shahriar and Haddad in [21] introduced a fuzzy logic-based system (FLS) to assess the risk caused by different types of code injection vulnerabilities. Their work identified several code-level metrics that capture the level of weakness originating from the source code. These metrics, along with essential MF, can be used to define the subjective terms in the FLS. There are three FLS systems, each of which has three general steps: defining linguistic terms (fuzzifying crisp inputs), defining rule sets and evaluating the rules based on inputs, and, finally, aggregating the rule outputs and defuzzifying the result. This approach can effectively assess high risks present in vulnerable applications, which is considered as an advantageous aspect. Nevertheless, fuzzy logic-based computation is flexible and tolerates inaccuracy while specifying rule and membership functions. The main problem of this technique is the overhead on the system caused by web code scanning and training. In [22], Joshi and Geetha designed an SQL injection attack detection method based on Naïve-Bayes machine learning algorithm combined with role-based access control mechanism. The Naïve-Bayes classifier is a probabilistic model, which assumes that the value of a particular feature is unrelated to the presence or absence of any other feature. The Naïve Bayes algorithm detects an attack with the help of two probabilities, the prior probability and the posterior probability. The drawback of this system is that small datasets are used in testing and evaluating the system's efficiency. Said datasets, based on and used in the test cases, are mostly derived from only three SQLIA attacks: comments, union, and tautology. Therefore, this technique cannot detect the other types of SQLI.

The main objective of this paper is to propose an adaptive approach that solves and bypasses the limitations of ANN and machine learning approaches. It also aims to reduce the occurrence of errors in the stage of producing the output of the adaptive network and improve the accuracy of the system in detecting and preventing SQLIs. This shows the ability of the system to resolve uncertainty and fuzziness problems founded in SQLI attack statement. Moreover, the proposed system can avoid the overhead problem that is encountered in neuro-fuzzy approaches by applying the SCG learning algorithm.

3   Proposed System

This paper introduces an ANFIS with FCM cluster method for detecting and preventing SQLIs, which is presented in Fig. 1. As previously mentioned in the introduction, the reason behind choosing machine learning and ANFIS lies in ANFIS's ability to avoid the issues of uncertainty and lack of flexibility, and machine learning's ability to handle and overcome the problems that arose in previous techniques. In addition, to overcome the overhead problem that initially faced the proposed system, the SCG algorithm was used instead of the BP algorithm to enhance the proposed system's training speed. The following subsections describe, in detail, the steps of the proposed system as mentioned in the following:

- Parsing and Extracting SQLI Features from Dataset
- SQL Signatures Fuzzification
- Clustering

- Building an initial and learned ANFIS
- Prevention of SQLI A

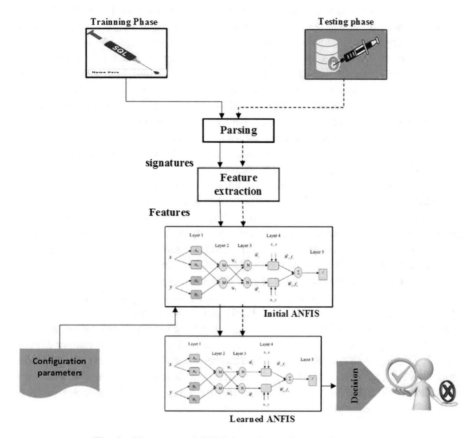

Fig. 1. The proposed SQLI detection and prevention system.

3.1   Parsing and Extracting SQLI Features from Dataset

In the proposed system, the SQL statement is treated as features, which characterize the SQLI attack keywords. Keywords that include create, drop, alter, where, table, etc., are the most well-known keywords in the SQL language, and they are used to perform operations on the tables in underlying database. The proposed system has investigated the most common features that have been used in numerous approaches related to machine learning and NN dealing with the problem of SQLIAs [3, 10, 11, 23]. These features are categorized into four signatures, namely punctuation signatures (*PU*), dangerous words signatures (*DW*), combination word signatures (*CW*), and SQL keyword signatures (*K*). Each query statement, whether it is a normal or an attack query, is converted into a vector of numerical values; each number inside the vector represents a signature. The value of a signature is calculated by the addition of the frequency of each feature in the signature. A query statement is thus parsed into a list of signatures, as illustrated in Table 1. There are 71 features that have been used in the feature extraction stage. Each feature belonged to one of the following signatures: the SQL keyword signature, a punctuations signature, a dangerous words signature, or a combination of words signature that appear in the content of the query statement.

Table 1. SQLI signatures and keywords.

Signature	Keywords
Punctuations (*PU*)	;, and, ,, or, ', +, -, !, ,;, Existence of statements that always result in true value like "1=1" or "@=@"or "A=A'
Dangerous words (*DW*)	delete, drop, create, wait, rename, exec, shutdown, sleep, load_file, userinfo, information_schema, if, else, convert, xp_cmdshell, sp_, ascii, hex, execchar(\w*\w*),xp_,sp_,
Combination words (*CW*)	\\,//,>>,<<,&#, &#x,*/,/*, *,*\,%,@@,(,),{,},[,],*,–
SQL keyword (*K*)	Union, union all, select, from, insert, where, table, into, update, set, alter, like, revoke, truncate, having, union select, join, group

The features are chosen because of their ability to identify most of the SQLIA types, such as tautologies, union, piggybacked, illegal/logically incorrect, inference, alternate encodings, and stored procedures. They also work on increasing the ability of the system to detect a new malicious code, and other features can be added. Undoubtedly, the appropriate selection of the system features is the most critical step in establishing strong and practical SQLI detection and prevention systems. Basically, the feature extraction stage can be described as follows: if a keyword is discovered in the sentence, its corresponding signature will increase the value by 1. Therefore, if a feature appears more than once, its corresponding signature's value will increase by the number of occurrences.

3.2   SQL Signatures Fuzzification

Due to its low computational requirement and capability of modelling human perception, FL is probably the most efficient and flexible method available for managing degrees of uncertainty in the detection of dangerous attacks. FL is a theory that allows the natural descriptions, in linguistic terms, of problems to be solved rather than having to use numerical values. Therefore, after the stage of feature extraction, the numerical signatures that represent SQL statement features are converted into linguistic terms. A linguistic variable is defined as a variable whose values are words or sentences in a natural or synthetic language. For example, 'frequency of dangerous words' can be a linguistic variable that takes the fuzzy sets "low", medium" and "high" as its linguistic term.

3.3   Clustering Methods

Before a final optimal model can be derived, the initial fuzzy model in the proposed system can be determined based on the fuzzy rules formed by either using the Subtractive Clustering (SC), Grid Partitioning (GP) or FCM clustering method, as SC and GP are two main clustering methods that are commonly used with ANFIS. GP is the most frequently used input partitioning method for ANFIS. In this paper, we will explore the suitability of FCM as a powerful data clustering method; in which each data point has a membership degree between 0 and 1 to each fuzzy subset [24]. As aforementioned, the data point resembles a vector of five numbers; FCM partitions a collection of n vectors x_i, $i = 1, 2..., n$ into fuzzy groups, and determine a cluster center for each group $i = 1, 2,..., c$ that are arbitrarily selected from the n points. Where n is the total number of training data set and c is the cluster center. The steps of the FCM method are explained briefly [24]:

1. The centers of each cluster c_i, $i = 1, 2,..., c$ are randomly selected from the n data patterns (training data set) $\{x_1, x_2, x_3,..., x_n\}$.
2. The membership matrix (μ) is computed with the following equation.

$$\mu_{ij} = \frac{1}{\sum_{k=1}^{c} \left(\frac{d_{ij}}{d_{kj}}\right)^{2/m-1}},$$ (1)

where μ_{ij} is the degree of membership of object j (new signature value; punctuation signature or dangerous words or combination words) in cluster I, m is the degree of fuzziness determined by the user; $m = 2$ is initially chosen, m is important because it significantly influences the fuzziness of the resulting partition and $d_{ij} = \|c_i - x_j\|$ is The Euclidean distance between c_i and x_j

3. The objective function is calculated with the following equation:

$$J(U, c_1, c_2, \ldots, c_c) = \sum_{i=1}^{c} \sum_{j=1}^{n} \mu_{ij}^m d_{ij}^2 \cdot 1 \leq m < \infty$$ (2)

where U is the partition matrix that contains the all data points and computed cluster center in each cluster.

4. The new c fuzzy cluster c_i, $i = 1, 2,..., c$ is calculated using the following equation [24]

$$c_i = \frac{\sum_{j=1}^{n} \mu_{ij}^m x_j}{\sum_{j=1}^{n} \mu_{ij}^m}.$$ (3)

By applying the FCM clustering to each class of data individually, a set of rules for identifying each class of data has been obtained. Due to this, the system must be pre-configured with the numbers of clusters that is determined by the user. The individual sets of rules are then combined to form the rule base of the classifier. Furthermore, according to the Sugeno model, each rule has a number of consequence parameters in FIS output part; where each MF is determined by two parameters c and σ (σ is used to determine the width of the MF and c is the center point of the Gaussian MF).

3.4 Building an Initial and Learned ANFIS

ANFIS is basically a graphical network representation of Sugeno-type fuzzy systems endowed with the neural learning capabilities [24]. Inputs for the proposed ANFIS for SQLI detection and prevention are the numerical values (the four signatures). The output of the system has been configured in such manner that it is equal to 2 if there is an attack, and 1 otherwise. Gaussian MF has been used for fuzzy set due to its nonlinear, smooth and continuous derivatives [23]. An ANFIS presented in Fig. 2 it is functionally equivalent to the fuzzy inference system (as seen in Fig. 3). Figure 2 illustrates the reasoning mechanism for the Sugeno model where it is not only the basis of ANFIS model; but also, it is simple in computation and easy to be combined with optimizing and self-adapting methods. Subsequently, the corresponding equivalent ANFIS architecture is as shown in Fig. 3, where nodes of the same layer have similar function [22, 23].

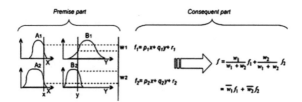

Fig. 2. First-order Sugeno fuzzy model.

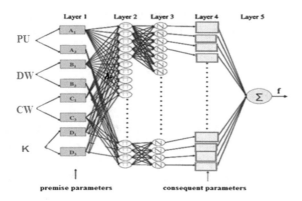

Fig. 3. ANFIS structure.

i. ANFIS Layers

The input data and output data were fed into the ANFIS model to extract the rules. The 'fuzzification' layer is set and adapted according to the parameters for the chosen membership. After that, the strength firing layer represents the IF conditions to set the rules. The output of the firing strength is normalized in the normalization layer. Before the final layer, there is another adaptation layer that works as a 'defuzzification' layer of the rules, where the consequent model parameters are tuned to derive the best matching between input and output [24]. Here, a four-input and single-output fuzzy system has been used. Three fuzzy variables including 'low', 'medium' and 'high' have been used to describe the features. Their respective MFs (μ_A) are Gaussian function that introduces the fuzzification operation of input parameters.

Layer 1: The output of this layer represents the membership grade of the inputs; for example, the MF for *PU* signature can be parameterized as given in the following equation [22, 23]

$$\mu_{Ai}(PU) = \exp\left[-\left(\frac{PU - c_i}{2\sigma_i}\right)^2\right] \tag{4}$$

$$O_{1,i} = \mu_{A_i}(PU), \text{ for } i = 1, \tag{5}$$

where $\{\sigma_1, c_i\}$ is the premise parameter set that changes the shape of the membership function; *and* $O_{1,i}$ is the output of the *i*-th node of layer 1.

Layer 2: Every node in this layer represents the firing strength of a rule:

$$O_{2,i} = w_i = \mu_{A_i}(PU)\mu_{B_i}(DW)\mu_{c_i}(CW)\mu_{D_i}(K), \ i = 1, 2 \tag{6}$$

where A_i, B_i, C_i, D_i are the fuzzy sets for each signatures. Each rule is assigned a firing strength that measures the degree to which the rule matches the inputs.

Layer 3: Each node in this layer calculates the ratio of the ith rule's firing strength to the sum of all rules' firing strengths. Outputs of this layer are called normalized firing strengths

$$O_{3,i} = \bar{w}_i = \frac{w_i}{\sum_i w_i}, i = 1, 2 \tag{7}$$

Layer 4: Every node in this layer is an adaptive node with a node function, where $\bar{w}_i$ is a normalized firing strength from layer 3 and $\{a_i, q_i, t_i, e_i r_i\}$ is the parameter set of this node. Parameters in this layer are referred to as consequent parameters. Consider a first order Sugeno type of fuzzy system the proposed system having 8 significant rules as following

$$O_{4,i} = \bar{w}_i f_i = \bar{w}_i(a_i PU + q_i DW + t_i CW + e_i K + r_i)$$

Rule 1:   **If** *PU* is *low* and *DW* is *low* and *CW* is *low* and *K* is *low* **then** *output f_1 = 0.0229* PU + 0.0101* DW +0.266* CW +0.0392*K+ 1.5112*

Rule 2:   **If** *PU* is *high* and *DW* is *high* and *CW* is high and *K* is *high* **then** *output f_2= 0.2212* PU + 0.1280* DW +0.0574* CW +0.0301*K+0.9644*

Rule 3:   **If** *PU* is *low* and *DW* is *high* and *CW* is *high* and *K* is *high* **then** *output f_1 = -0.0372* PU + 0.0583* DW +0.0647*CW +0.0886*K+0.8160*

Rule 4:   **If** *PU* is *low* and *DW* is *high* and *CW* is *low* and *K* is *high* **then** *output f_1 = 0.3932* PU + 0.3844* DW - 0.0456* CW +0.0688*K+ 1.1169*   (8)

⋮

Rule 8

Layer 5: The output of the fuzzy system in the SQLI detection system is linear and the single node in this layer computes the overall output as the summation of all incoming signals.

$$O_{5,1} = \sum_i \bar{w}_i f_i = \frac{\sum_i w_i f_i}{\sum_i w_i} \qquad (9)$$

ii. Learning Algorithm

ANFIS approach, using the Takagi-Sugeno rule format, combines optimizing the premise membership functions by gradient descent Backpropagation (BP) with optimizing the consequent equations by linear least squares estimation. There is a number of training algorithms that can be used in training the premise parameters of the ANFIS systems. In this system, two training algorithms have been evaluated, the BP and SCG algorithms. SCG algorithm has been presented as one of the algorithms that enhance the processing time. Generally, the SCG algorithm shows great performance over a wide variety of problems [15]. Therefore, in this paper, ANFIS has been improved with the SCG learning algorithm to speed up its training process [25, 26].

3.5   Prevention of SQLI A

After ANFIS model has been trained with the input and output training data, the system that has been tested using the testing dataset and SQLIA is detected; this system prevents the malicious SQL statement from accessing the database by converting it into a comment. As it is well-known, the comment statements do not execute in the database engine.

4   Experimental Results and Discussion

All these experiments were carried out on windows 10 (64-bit) operating system with i7 processor and 8 GB RAM. To evaluate the performance of the proposed system, all methods and training functions that are used are coded in MATLAB. The dataset is downloaded from Testbed [27], which is then used to evaluate the Amnesia approach in [3]. The testbed has two sets of inputs: "legit" set, which consists of legitimate inputs for the application, and "attack" set, which consists of attempted SQLIAs. All types of attacks were represented in this set with the exception of the multi-phase attacks. The multi-phase attacks include inference attacks and illegal/logically incorrect queries, such attacks require human intervention and interpretation. The testbed includes seven folders; four of which are used for the training and the remainder of the three sets are used for testing. At first, to compare the efficiency of the different clustering methods GP, SC and FCM, the popular measure, RMSE (root mean squared error), was employed for performance evaluation according to the next formula:

$$MSE = \frac{\sum_{i=1}^{N} (y_i - o_i)^2}{N} \tag{10}$$

$$RMSE = (MSE)^{1/2} \tag{11}$$

where y_i is the target value, o_i the observed output, and n is the number of data set [25].

Table 2. The RMSE for the clustering methods in different epochs based on BP learning algorithm.

Epochs no.	Clustering method	RMSE of the test data set
40	GP	0.1632
	SC	0.1342
	FCM	0.0468
60	GP	0.15838
	SC	0.1205
	FCM	**0.0402**
100	GP	0.1307
	SC	0.0984
	FCM	0.0402

The results are shown in Table 2; in which the RMSE of the testing dataset displays the error between the target output and the observed output in the testing dataset. According to the results, it is verified that with the use of 60 epochs, the FCM cluster has achieved the minimum value of the RMSE among the others. The GP and the SC clustering behaviours did not achieve the minimal error, such as the case with FCM. The reason for such results is that the interpretation of m (fuzziness degree) is different than the case of FCM, where values of m increase the sharing of points among all the clusters which will lead to better performance; they also lead to the reduction of the objective function of the dissimilarity measure. Therefore, this improvement in FCM-ANFIS is related to its ability to manage uncertainty and the fuzziness degree in dangerous attacks' statements. Moreover, any value with more than 60 epochs resulted in the overtraining of the model.

Figure 4 shows the waveforms of RMSE for the ANFIS-FCM system based on BP learning algorithm in details. It is obvious that RMSE waveforms start descending before 50 epochs. After 60 epochs, the RMSE curve tends to be stabilized with very small variation (overfitting). In this case, the network parameters are saturated as the network output matches the target; any additional epochs will decrease the accuracy performance inside NN as the reason of overtraining [28].

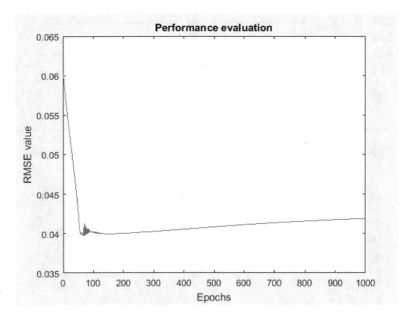

Fig. 4. RMSE curve for the ANFIS-FCM system based on BP learning algorithm.

Figure 5 indicates that the suggested system is improved by the SCG Algorithm and reaches the same RMSE in less epochs. With 60 epochs, the utilized BP learning algorithm archives 0.0402 RMSE and this value is obtained through the utilization of the SCG learning algorithm in only 5 epochs. As a result, the SCG learning algorithm needs lower computational time as SCG involves twice as much calculation work per iteration when compared with BP. As stated in [26, 29], one iteration in SCG needs the calculation of two gradients, and in addition to this it requires one call to the error function, while one iteration in standard BP needs the computation of one gradient and one call to the error function.

In the third experiment, the performance of the suggested system under SCG learning algorithm with different number of epochs is discussed. As shown in Fig. 6, the RMSE curve starts descending before 20 epochs. After 20 epochs, the RMSE curve tends to be stabilized; this is due to the error decreasing in monotonic towards zero, which is characteristic for SCG. In this case, an error increase is not allowed and second order information (second derivatives) of global error function hasn't been positively definite, only in the beginning of the minimization. This is not surprising because the closer the current point is to the desired minimum the bigger is the possibility that global error function is positive definite [26, 29].

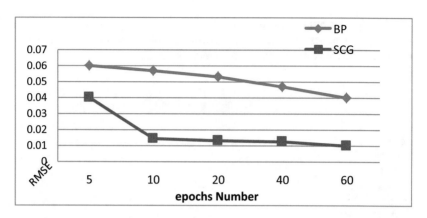

Fig. 5. RMSE curves for ANFIS based on BP and SCG learning algorithm.

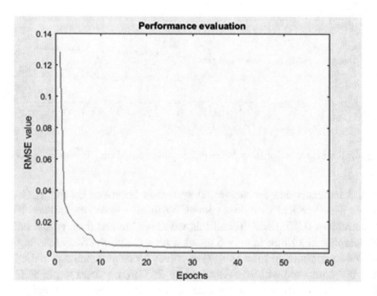

Fig. 6. RMSE curve for the ANFIS-FCM system based on SCG learning algorithm.

Table 3. Time per seconds for testing phase based on different learning algorithms.

Training folder subject	Testing folder subject	No of attack URLs	No of legit URLs	Testing time per seconds	
				BP algorithm	SCG algorithm
Bookstore	Employee	6530	1268	45.821000	39.00562
Checkers	Employee	6939	2019	58.950023	47.99056
Classifieds	Office Talk	6958	1000	43.85002	38.70112
Events	Portal	5970	1980	55.560023	45.98002

The fourth experiment compares between the two learning algorithms in terms of the computational cost (processing time) under different attack folders for each a testing phase and all phases respectively. Table 3 reveals that using SCG as a learning algorithm inside the proposed system requires less time as compared with BP in all attack folders (about 20% reduction in time) in the testing phase. For the whole system (training and testing phases), as it was expected, the SCG learning algorithm achieves reduction in the running time with 30% as shown in Table 4. This improvement is due to that SCG does not require a line search at each iteration step, unlike the other training algorithms. In another words, the used step size scaling mechanism avoids the time consuming line search per learning iteration. This mechanism makes the SCG learning algorithm faster than the BP learning algorithm.

Table 4. Comparison of the processing time between BP and SCG for training and testing phases.

Learning algorithm	Time per minutes
BP	11.3481206
SCG	7.99533351

In the fifth experiment, the accuracy of the proposed system that employs the SCG algorithm and the FCM clustering method for the detection and prevention of SQLI, this is compared with the algorithm suggested by Sheykhkanloo in [19] that utilized traditional neural network. The traditional NN model has 10 hidden layers and 32 input features. Generally, the correct response of the NN system depends on the number of hidden layers that are commonly determined by the user. From the illustrated results in Fig. 7, the proposed system outperforms the other one by 3.08%. Based on the research findings, the proposed system for SQLIA detection and prevention trained by the SCG training algorithm with four inputs and one output achieved high accuracy, less time, and avoided the computational complexity of the network.

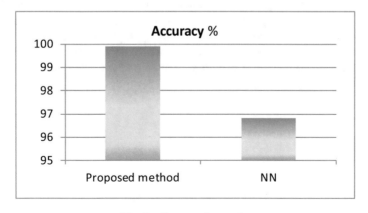

Fig. 7. Comparative study

In the last experiment, the accuracy of the proposed system compared with the algorithm suggested by Basta in [10] shows 98.4% accuracy with the duration being 217 s for 13079 attacks, i.e. 16.6 ms for one attack; while the proposed system shows 99.8% accuracy with the duration being 180.920011 s for 12241 attacks, i.e. 14.7 ms for one attack (Table 5).

Table 5. Comparative study of the accuracy and processing time.

Technique	Accuracy	Time per sec for one attack
Proposed system	99.8	14.7
Genetic fuzzy classifier	98.4	16.6

The solution has some limitations such as the dependability on the quality of the training data and the features extracted from it. With a relatively limited set of training data, the resulting classifier can be sensitive to the features of benign queries. Despite the success of the technique overall, an acceptable amount of overhead still takes place due to the learning phase.

5   Conclusions and Future Work

In this paper, a new approach based on the ANFIS system for SQLIA detection and prevention has been proposed. The proposed model includes two main elements: URL parser and feature extraction and the training ANFIS classifier. The proposed system has been implemented with the benign and malicious URLs depending on the extracted features. Furthermore, the FCM is employed as a clustering method in the first layer of ANFIS to enhance the system's performance by dealing with fuzzy and uncertain inputs. Additionally, SCG learning algorithm is employed instead of BP learning algorithm for enhancing the system's processing time. The approach that is based on the ANFIS classifier has the advantages of learning through patterns (input and output system data), and the easy interpretation of its functionality. Future work includes upgrading the system to detect XSS attacks with the ANFIS classifier. Furthermore, the learning algorithm can be replaced with another appropriate one to fine tune parameters, and then the results should be evaluated to deduce the better optimization approach. Finally, employing a new algorithm to encrypt data query for preventing SQLIA could be considered in future work.

References

1. O.W.A.S.P.Top10Vulnerabilities. https://www.owasp.org/index.php/Top_10_2013-Top_10. Accessed 1 Dec 2016
2. Shegokar, A., Manjaramkar, A.: A survey on SQL injection attack, detection and prevention techniques. Int. J. Comput. Sci. Inf. Technol. 5(2), 2553–2555 (2014)
3. Halfond, W., Orso, A.: AMNESIA: Analysis and Monitoring for Neutralizing SQL-Injection Attacks. In: 20th IEEE/ACM International Conference on Automated Software Engineering, USA, pp. 174–183 (2005)
4. Bhagat, M., Mane, V.: Protection of web application against SQL injection attack. Int. J. Sci. Res. Publ. 3(10), 1–5 (2013)
5. Moosa, A.: Artificial neural network based web application firewall for SQL injection. Int. J. Comput. Electr. Autom. Control Inf. Eng. 4(4), 12–21 (2010)
6. Nithya, V., Regan, R.: A survey on SQL injection attacks, their detection and prevention techniques. Int. J. Eng. Comput. Sci. 2(4), 886–905 (2013)
7. Gomaa, Y., El Aziz Ahmed, A., Mahmood, M., Hefny, H.: Survey on securing a querying process by blocking SQL injection. In: 3rd World Conference on Complex Systems, Morocco, pp. 1–7 (2015)
8. Som, S., Sinha, S., Kataria, R.: Study on SQL injection attacks: mode, detection and prevention. Int. J. Eng. Appl. Sci. Technol. 1(8), 23–29 (2016)
9. Valeur, F., Mutz, D., Vigna, G.: A learning-based approach to the detection of SQL attacks. In: International Conference on Detection of Intrusions and Malware, and Vulnerability Assessment, Austria, pp. 123–140 (2005)
10. Basta, C., Elfatatry, A., Darwish, S.: Detection of SQL injection using a genetic fuzzy classifier system. Int. J. Adv. Comput. Sci. Appl. 7(6), 129–137 (2016)
11. Komiya, R., Paik, I., Hisada, M.: Classification of malicious web code by machine learning. In: 3rd International Conference on Awareness Science and Technology, China, pp. 406–411 (2011)
12. Wang, X., Zhai, J.: Learning with uncertainty. CRC Press, ISBN 9781498724128 - CAT# K25713, pp. 1–227 (2016)
13. Hammer, B., Villmann, T.: How to process uncertainty in machine learning? In: European Symposium on Artificial Neural Networks, Belgium, pp. 79–90 (2007)
14. Toosi, A., Kahani, M.: A novel soft computing model using adaptive neuro-fuzzy inference system for intrusion detection. In: The 2007 IEEE International Conference on Networking, Sensing and Control, UK, pp. 15–17 (2007)
15. Ghaffari, A., Abdollahi, H., Khoshayand, M., Bozchalooi, I., Dadgar, A., Rafiee-Tehrani, M.: Performance comparison of neural network training algorithms in modeling of bimodal drug delivery. Int. J. Pharm. 327(1), 126–138 (2006)
16. Tajpour, A., Ibrahim, S., Masrom, M.: SQL injection detection and prevention techniques. Int. J. Adv. Comput. Technol. 3(7), 82–91 (2011)
17. Singh, S., Tripathi, U., Mishra, M.: Detection and prevention of SQL injection attack using hashing technique. Int. J. Modern Commun. Technol. Res. 2(9), 27–31 (2014)
18. Shar, L., Tan, H.: Defeating SQL injection. Comput. Softw. Eng. 46(3), 69–77 (2013)
19. Sheykhkanloo, N.: SQL-IDS: evaluation of SQLI attack detection and classification based on machine learning techniques. In: The 8th International Conference on Security of Information and Networks, USA, pp. 258–266 (2015)
20. Sheykhkanloo, N.: Employing neural networks for the detection of SQL injection attack. In: The 7th International Conference on Security of Information and Networks, UK, pp. 318–323 (2014)

21. Shahriar, H., Haddad, H.: Risk assessment of code injection vulnerabilities using fuzzy logic-based system. In: The 29th Annual ACM Symposium on Applied Computing, Korea, pp. 1164–1170 (2014)
22. Joshi, A., Geetha, V.: SQL injection detection using machine learning. In: The International Conference on Control, Instrumentation, Communication and Computational Technologies, India, pp. 1111–1115 (2014)
23. Batista, L., Adriano De Silva, G., Araujo, V., Rezende, T., Guimarães, A., Souza, P., Araujo, V.: Fuzzy neural networks to create an expert system for detecting attacks by SQL Injection. Int. J. Forensic Comput. Sci. **13**(1), 8–21 (2018)
24. Abdulshahed, A., Longstaff, A., Fletcher, S., Myers, A.: Thermal error modelling of machine tools based on ANFIS with fuzzy c-means clustering using a thermal imaging camera. Appl. Math. Model. **39**(7), 1837–1852 (2015)
25. Hager, W., Zhang, H.: A survey of nonlinear conjugate gradient methods. Pac. J. Optim. **2**(1), 35–58 (2006)
26. Cetişli, B., Barkana, A.: Speeding up the scaled conjugate gradient algorithm and its application in neuro-fuzzy classifier training. Soft. Comput. **14**(4), 365–378 (2010)
27. Halfond, W.: Testbed. http://wwwbcf.usc.edu/∼halfond/testbed.html
28. Nasr, M., Mahmoud, A., Fawzy, M., Radwan, A.: Artificial intelligence modeling of cadmium biosorption using rice straw. Appl. Water Sci. **7**(2), 823–831 (2017)
29. Prerana, P.S.: Comparative study of GD, LM and SCG method of neural network for thyroid disease diagnosis. Int. J. Appl. Res. **1**(10), 34–39 (2015)

An Adaptive Plagiarism Detection System Based on Semantic Concept and Hierarchical Genetic Algorithm

Saad M. Darwish and Mayar M. Moawad[✉]

Institute of Graduate Studies and Research,
Alexandria University, Alexandria, Egypt
{saad.darwish,igsr.mayar.mostafa}@alexu.edu.eg

Abstract. Plagiarism became a considerable issue; the reason is easy access to articles on the Internet. However, many issues arise as the majority of available Plagiarism Detection (PD) tools could not identify plagiarism by structural variations and paraphrasing. For applied systems, with regards to more complicated levels, those systems fail. Genetic Algorithm (GA) is now broadly utilized in accomplishing best solution in multidimensional nonlinear problems, unfortunately, system structure must be pre-defined to be optimized. This paper introduced an improved plagiarism detection system aiming to detect cases of plagiarism by semantic similarity with Hierarchal Genetic Algorithm (HGA). HGA operates without pre-defining system structure, moreover, system structure and parameters might be optimized. For discovering plagiarism, semantic similarity depending on intelligent procedures must be applied for extracting the idea. In addition, HGA is employed in finding interrelated cohesive sentences that convey the concept. Results reveal the capability of the system to present a significant improvement over compared systems.

Keywords: Plagiarism · Semantic concept · Hierarchal Genetic Algorithm · Text segmentation

1 Introduction

Plagiarism represents a terrifying problem because current technologies are growing fast [1, 2]. There exist two plagiarism types; syntactic plagiarism where plagiarizer takes the entire or part of someone's work to his own, also in semantic plagiarism plagiarizer changes various terms [3]. Moreover, different plagiarism classifications exist like replacing words by another, idea, and paraphrasing plagiarism [4]. Unfortunately, plagiarism instances increased nowadays, because challenges as pressure for publication, lacking writing skills, and lacking awareness among authors [1]. PD is a significant concept for data processing and it utilized to protect the author's innovation [5]. It comprises of looking for relative and most indistinguishable text between documents. PD is a complex task because many plagiarists will reuse the text from other documents aiming to avoid plagiarism [6]. Two basic types exist for PD; monolingual PD which manages one language and cross-lingual PD which manages two languages [3]. Many systems can distinguish same words cases, while some can follow out

© Springer Nature Switzerland AG 2020
A. E. Hassanien et al. (Eds.): AISI 2019, AISC 1058, pp. 739–749, 2020.
https://doi.org/10.1007/978-3-030-31129-2_67

arbitrary manipulations. When complexity increases, these Plagiarism Detection System (PDS) may fail.

Optimization is an interesting field, where many optimization ways are developed for solving complex problems of different fields [7, 8]. GA is stochastic optimization. GA utilizes an evolutionary process as selection, crossover, and mutation [9], to get the ideal solution. HGA is an improved GA version. It's utilized when inspired by topology optimization instead of parameter optimization [10]. HGA differ from GA in chromosomes structure, as having a hierarchical structure. The rest of this paper is organized as follow. Section 2 describes related work. Section 3 discussed the proposed model. Section 4 represents results and discussion. Finally, conclusion and future plan discussed in Sect. 5.

2   Literature Review

Recently, PD became the most alarming challenge for its nature in knowing reused texts which are commonly modified to conceal plagiarized documents. In 2012, PD technique is introduced which relays on Semantic Role Labeling (SRL) aiming to discover semantic similarity among two sentences [11]. It detects copy paste, synonym modification, and word structure replacement. However, it's shown that some of the arguments don't affect the PD and have to investigate the term role in SRL.

The new technique uses sentence ranking and SRL. It compares suspicious and source documents depending on terms allocation using SRL. Sentence ranking application in PD decreases checking time to increase performance. It's unclear, what appropriate syntax level is expected to support robust testing for semantic roles [6]. Another work discussing the semantic relations combination among words and their syntactic arrangement to boost PD performance. It is carried out by calculating the syntactic and semantic similarity of the sentence regarding another. Also, this method cannot recognize the difference between active and passive voice [12].

A new fuzzy semantic-based similarity model is presented for revealing obfuscated plagiarism [2]. Semantic relatedness of words was tested on Part of Speech (POS) tagging and WordNet measures. Fuzzy rules are used in calculating the semantic distance between suspicious and original text. However, it can't generalize written rule base. Also, it's not robust in topology modifications that would need modifications in rule base and depend on specialist presence to determine the inference rules [13]. Although PD has been studied for nearly many decades, a room still exists to make it more efficient. In web-based applications, factors like space requirements, running time, and computational difficulty must be taken into consideration. Our system considers the previous points to accomplish better performance.

3   The Proposed Plagiarism Detection System

The paper shows an enhanced PD algorithm for detecting plagiarism utilizing semantic similarity and HGA. Using HGA may optimize parameters and structure of system [10]. Also, HGA is utilized to get out interrelated cohesive sentences of source document concept. Moreover, concepts of semantic level are caught through utilizing measures of semantic similarity by WordNet. Each module involved is described in upcoming subsections. The main system components are described in Fig. 1.

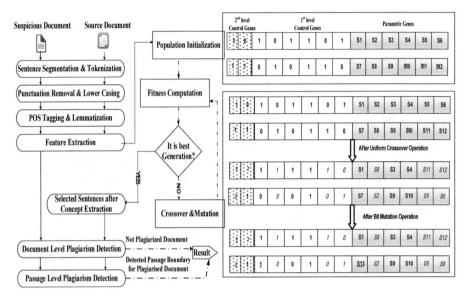

Fig. 1. Proposed system diagram

- **System and Dataset Information:** System is carried out using MATLAB 2017. The system's inputs are benchmark dataset where Summary Obfuscation (SO) training and testing sets supplied by PAN13-14 text alignment task. Also, WorldNet is employed.

- **Pre-processing and Document Representation Module:** the pre-processing of given source-suspicious documents. Where

   (a) **Sentence Segmentation and Tokenization:** Given suspicious document X_{susp} and comparing source document X_{src} undergo sentence segmentation. Then tokenizing these sentences for source-suspicious sentences. Further, basic pre-processing like punctuation removal and lowercasing are done [4, 6, 12].

   (b) **Part of Speech Tagging and Lemmatization:** Once essential pre-processing is completed and ready to get POS tagged words by Stanford Log-Linear POS Tagger for suspicious-source tokenized sentences. Current method keeps only words under verb, adverb, adjective and noun tagged for having semantic meaning in it. Words tagged functional words are trimmed out which enhances time requirements and efficiency. After that followed by lemmatization by WordNet Lemmatizer, that ease in meaningful comparisons. Pre-processed suspicious sentence in X_{susp} is known, as S_{susp} and pre-processed source sentence in X_{src} is known, as S_{src} [4, 6, 12, 13].

(c) **Feature Extraction.** Pre-processed tokenized sentences contain group of words which are POS tagged and lemmatized are given in vector space model with term frequency-inverse sentence frequency ($tf - isf$) weighting [4]. In information retrieval $tf - isf$ is a numerical statistic that reflects how important a word to a sentence [14]. The $tf - isf$ weight is computed using Eq. (3).

$$tf(t,S) = f(t,S) \tag{1}$$

$$isf(t,X) = log \frac{|X|}{|\{S \in X; t \in S\}|} \tag{2}$$

$$w(t,S) = tf(t,S) * isf(t,X) \tag{3}$$

Term frequency $tf(t,S)$ is how many the word t appeared in any generic sentence S. Frequency $f(t,S)$ is how many t appeared in S. Inverse sentence frequency (isf) affirms computation is done for S as units, not whole documents, where X applies the set of all S in given documents. S_{src} and S_{susp} vector is presented as $\overrightarrow{s_{src}}$ and $\overrightarrow{s_{susp}}$.

– **Sentence-Level Concept Extraction (SLCE) using Hierarchical Genetic Algorithm:** After pre-processing documents and representing in $tf - isf$ form is ready for concept extraction. The presented SLCE using HGA algorithm is explained in coming subsections [10, 15].

(a) **Population Initialization.** The input for HGA is S_{src} which will have $Stat()$. For each S_{src} in X_{src}, certain features are taken away for assigning sentence weights based on $w(t,S)$. This step will be performed by Relevance Score which presents document S_{src} score in terms of $if - isf$. Where $w(t,Ssrc)$ presents the weights of each t in S_{src} and $|Ssrc|$ is S_{src} length. Also, Thematic Score where initially t from given pre-processed X_{src} are extracted and arranged depends on tf. Then top words are chosen and saved in the set presented as $kw(X_{src})$, that is X_{src} keyword set. Similar words between $kw(X_{src})$, and a S_{src} in X_{src} are picked then saved in set $kw(S_{src})$.

$$Rel(S_{src}) = \frac{\sum_{i=1}^{|S_{src}|} w(t_i, S_{src})}{|S_{src}|}; Rel(S_{src})[0, 1] \tag{4}$$

$$Thm(S_{src}) = \frac{|kw(S_{src})|}{L};$$

$$Kw(S_{src}) = \{t | t \in S_{src} \wedge t \in kw(X_{src})\}; \quad Thm(S_{src}) \in [0, 1] \tag{5}$$

Where $|kw(S_{src})|$ shows number of familiar words between $kw(X_{src})$ and S_{src} in X_{src}. $kw(X_{src})$ is the number of words is known, as L. Then $Stat(S_{src})$ is determined by Eq. (6). Population is chosen randomly where each chromosome C contains control gene is 0's and 1's generated randomly [16] and parametric gene are sentences chosen randomly from X_{src} [4]. Integer encoding utilize the sentence position in X_{src} [17].

$$Stat(S_{src}) = Rel(S_{src}) + Thm(S_{src}); \; Stat(S_{src}) \in [0,2] \quad (6)$$

(b) **Fitness Function Computation:** Finding fitness function $Fit(C)$, for each sentence calculate cohesion factor $Coh()$ and add with $Stat(S_{src})$ related with it. To compute cosine similarity in each pair of $\overrightarrow{s_{src}}$ using Eq. (7). $Cos\left(\overrightarrow{S_{srci}}, \overrightarrow{S_{Srcj}}\right)$ presents cosine similarity between vector pair $\left(\overrightarrow{S_{srci}}, \overrightarrow{S_{Srcj}}\right)$ so each sentence is an element of C. Cosine similarities between sentences are calculated and saved in symmetric matrix. Then sentence $Coh()$ for i^{th} sentence S_{srci} is determined by utilizing Eq. (8). $Tot(S_{src})$ presents total score for S_{src} is determined by Eq. (9).

$$Cos\left(\overrightarrow{S_{srci}}, \overrightarrow{S_{Srcj}}\right) = \frac{\overrightarrow{S_{srci}} \cdot \overrightarrow{S_{Srcj}}}{\left\|\overrightarrow{S_{srci}}\right\|\left\|\overrightarrow{S_{Srcj}}\right\|}; \; \forall i, j \overrightarrow{S_{srci}}, \overrightarrow{S_{Srcj}} \in C \quad (7)$$

$$Coh(S_{srci}) = \frac{\sum_{j=1, j \neq i}^{|C|} \cos\left(S_{srci}, S_{srcj}\right)}{\max\left\{\left(S_{srci}, S_{srcj}\right)\right\}}, \; \forall_{ii} = \{1, 2, \ldots, |C|\}, \; i \neq j \quad (8)$$

$$Tot(S_{src}) = Stat(S_{src}) + Coh(S_{src}) \quad (9)$$

Finally, depending on calculated fitness, C with highest fitness are determined to make next generation using next step.

$$Fit(C) = \sum_{i=1}^{|C|} Tot(S_{src}) \quad (10)$$

(c) **Selection, Crossover and Mutation:** After $Fit(C)$ is determined. For selection process, parent C with high fitness are selected. Best parents are chosen and passed to next generation. The space of new population is completed by determining offspring generated from good parents. These offspring then replace the weakest C of old population. Offspring are created by chosen parents by employing crossover [16]. Uniform crossover is applied, where randomly a gene position in parents are chosen and swapped [18]. Bit mutation is employed for parametric gene and control genes where mutate novel position offspring randomly within C and replacing it with a random gene (see Fig. 1). On completion of required generations Max_Gen, which presents the superior selected source sentences known, as S_{src_sel} [4].

- **Document Level Plagiarism Detection (DLD) phase** Once important SLCE are chosen from source, word level concept is taken from S_{src_sel}. Words tagged as noun and verb are selected from S_{src_sel}[4]. Similarly, done for S_{susp} in X_{susp}. These source-suspicious words are utilized to determine plagiarism in DLD by calculating number of same word level. If calculated value is above threshold θ, then document is detected as plagiarized or not.

- **Passage Level Plagiarism Detection (PLD) phase** Semantic concept extractions are applied for calculating semantic similarity metric then passage boundary to detect plagiarized passages [4].

  (a) **Semantic-Level Concept Extraction.** Sentence wise comparisons are done between S_{susp} and all S_{src_sel}. To compute semantic similarity, synset list W_{q_syn} of w_q and W_{k_syn} of w_k are selected by WordNet. They are chosen relaying on POS. The similarity, among each pair $\left(W_{q\_syn}, W_{k\_syn}\right)$ such that $w_{q\_syn} \in W_{q\_syn}$ and $w_{k\_syn} \in W_{k\_syn}$ is calculated with wup_similarity. Output s is passed to balanced semantic value function known by $adjSVq, k()$.

$$adjSVq, k\left(wq_{syn}, wk_{syn}\right) = \begin{cases} 0.0 \text{ if } s = 0.0 \\ 0.2 \text{ if } s \in (0.0, 0.3) \\ 0.3 \text{ if } s \in [0.3, 0.5) \\ 0.5 \text{ if } s \in [0.5, 0.7) \\ 0.7 \text{ if } s \in [0.7, 1.0) \\ 1.0 \text{ if } s = 1.0 \end{cases} \tag{11}$$

  Heuristic conditions in Eq. (11) are made by enhancing conditions defined by [12]. Approach limitation is when wup_similarity is very large or very small, that was overcome by different ranges as in Eq. (11). Once semantic value is found for all pairs of (w_q, w_k), the highest value is chosen. If it's greater than 0.5, then value saved in an array known, as *mxsim*. Once process is finished for all (w_q, w_k). A combined semantic metric is employed that considers similarity found by *sim1* and *sim2*, where it eases in getting the potentials of various measures. If similarity value is greater than thresholds $\alpha 1$ and $\alpha 2$, then it's plagiarized, otherwise not.

$$sim1 = \frac{\sum(mxsim)}{|Ssusp| + |Ssrc\_sel|}$$
$$sim2 = \frac{|Count|}{\max\left(|S_{susp}|, |S_{src\_sel}|\right)}; \tag{12}$$

$$Count = \{t | (tS_{susp} \wedge tS_{src_{sel}}) \vee (t \in Syns(w_q) \wedge tS_{src_{sel}})\} \tag{13}$$

  (b) **Passage Boundary Detection.** Given sentence offset list (*offlist*), that contain m plagiarized sentence pairs. Let suspicious sentence offset as off_{susp} and source sentence offset as off_{src}. Permissible gap is introduced for suspicious and source sentence offsets as $susp_{gap}$ and src_{gap}. Each (off_{susp}, off_{src}) in *offlist* is saved ascending based on off_{susp} values. If off_{susp} satisfy this $abs(offsuspi - offsuspj + 1) - susp_{gap} - 1 < = 0; of fsuspi$ $(offsusp, offsrc)i$ and $offsuspj \in (offsusp, offsrc)\_j$ add equivalent offset to the same sub list. If not, split to another one.

If compared off_{src} satisfy this $\mathrm{abs}\left(offsrc_i - offsrc_j + 1\right) - src_{\mathrm{gap}} - 1 < \; = 0$:
$offsrc_i \in \left(offsusp, offsrc\right)_i$
and $offsrc_j \in \left(offsusp, offsrc\right)_j$ the corresponding offset pairs will preserve $\left(off_{\mathrm{susp}}, off_{src}\right)_i$ and $\left(off_{\mathrm{susp}}, off_{src}\right)_j$ in same sub list. Otherwise, split in another one. Finally, main list will contain the sub lists depending on boundary conditions. The least and most value of off_{susp} and off_{src} are taken to make start and end positions for each passage. Thus, filtering small passages is done when the length is less than threshold β. The following Adaptive Plagiarism Detection Algorithm clarifies previous steps. Figure 2 shows the case study for adaptive plagiarism detection system.

Input: Dataset X; Selected Sentences S; Pre-processed Suspicious Sentence S_{susp}; Initial Populations Pop_s; Passage Level Plagiarism Detection PLD; Fitness Value F_Val; First Semantic Metric $sim1$; Second Semantic Metric $sim2$	$Pop_s \leftarrow$ Random Selection (S)
	$F\_Val \leftarrow$ Fitness_Measure (Pop_s)
	$Pop \leftarrow$ Select Best Population (Pop_s, F_Val)
	If Termination Condition is False, then
	$New\_Pop \leftarrow$Crossover (Pop)
while n< size of documents do	$New\_Pop \leftarrow$ Mutation (New_Pop)
$S \leftarrow$ Sentence Segmentation $((X)_x)$	end
$y \leftarrow 0$	Else
while y< S ! = NULL do	Return *Best Pop* $\leftarrow New\_Pop$
$T \leftarrow$ Tokenization (S)	end
$z \leftarrow 0$	end
while z < size of T do	$sim1 \leftarrow$Semantic_Metric_1 $(Best\ Pop, S_{susp})$
$M \leftarrow$ POS Tagging (T)	$sim2 \leftarrow$Semantic_Metric_2 $(Best\ Pop, S_{susp})$
$N \leftarrow$ Lemmatization (M)	If sim1>α1 andand sim2>α2
z++	Sentence is plagiarized
end	end
TF-ISF (N)	Else
y++	Sentence is not plagiarized
end	end
n++	If Doc. Status = =Plagiarized
end	Selected Passages $\leftarrow$ PLD (Suspicious Document)
$t \leftarrow 0$	end
while $t <$ MAX_GENS do	Else
$t \leftarrow t+1$	end
	Output = Selected Passages

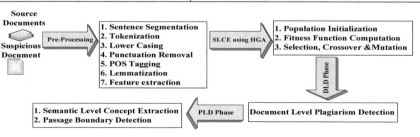

Fig. 2. Case study for adaptive plagiarism detection system

4   Experimental Results and Discussion

DLD approach and its variations are rated by utilizing IR measures; recall (Rd), precision (Pd) and F-score (Fd). That defines approach accuracy as correct and false detections. Also, Fd presents harmonic mean of Rd and Pd.

$$R_d = \frac{|\{correct\ documents\} \cap \{retrieved\ documents\}|}{|\{correct\ documents\}|} \tag{14}$$

$$P_d = \frac{|\{correct\ documents\} \cap \{retrieved\ documents\}|}{|\{retrieved\ documents\}|} \tag{15}$$

$$F_d = 2\frac{(R_d * P_d)}{(R_d + P_d)} \tag{16}$$

PLD performance is analyzed by PAN measures; Recall (Rec), Precision $(Prec)$, Granularity $(Gran)$ and Plagdet score $(Plagd)$. Let Y denote the set of plagiarism cases in the given corpus and Z denote the set of detections. Rec and $Prec$ measures the accuracy of PDS. $Gran$ represents detection of duplicate source portions for same suspicious passage. Rec, $Prec$ and $Gran$ are combined to rank a PDS by Eq. (20).

$$Rec(Y, Z) = \frac{1}{|Y|} \sum_{y \in Y} \frac{|\cup_{z \in Z} (y \cap z)|}{|y|} \tag{17}$$

$$Prec(Y, Z) = \frac{1}{|Z|} \sum_{z \in Z} \frac{|\cup_{z \in Y} (y \cap z)|}{|z|} \tag{18}$$

$$Gran(Y, Z) - \frac{1}{|YZ|} \sum_{y \in YZ} |Zy| \tag{19}$$

$$plagd = \frac{2 * Rec * Prec}{(Rec + Prec) * (\log_2(1 + Gran))} \tag{20}$$

SLCE and PLD parameters are chosen depending on various experiments on PAN13-14 as shown in Table 1. Tested approaches and their variations are examined as follow, (1) Proposed DLD; DLD-Without SLCE and DLD-With SLCE, and (2) Proposed PLD; PLD-Without SLCE, PLD-With SLCE and Combined DLD-PLD.

DLD is evaluated separately to understand its performance. Results are compared with DLD Without SLCE to understand and analyze the role of SLCE with HGA in DLD. Parameter selection, evaluation with SLCE is done at different thresholds and performance is analyzed using IR measures. The best θ value is $\theta = 8$, obtained for

Table 1. SLCE and PLD parameters

SLCE parameters		PLD parameters	
Parameter	Value	Parameter	Value
Population size (N)	50	sim1 threshold (α1)	0.4
Chromosome length (|C|)	0.5	sim2 threshold (α2)	0.5
Max. No. of generations (Max_Gen)	10	susp_gap	24
Selection	Highest fitness	src_gap	24
Likelihood of crossover	0.7	Minimum passage length (β)	150
Likelihood of mutation	0.3	sim1 threshold (α1)	0.4

both plagiarized document and non. To evaluate, proposed DLD Without SLCE, the best θ is 18. The comparison of DLD results is done against DLD Without SLCE shown in Table 2. Thus, the approach with SLCE is better than without.

Table 3 shows that amongst other approaches, highest *Rec* is demonstrated by PLD-With SLCE. Performance of *Rec*, *Prec* and *plagd* is improved compared to PLD-Without SLCE. In Combined DLD-PLD; *Prec* increases considerably compared to other approaches. It is clearly observed that in terms of *plagd*, combined DLD-PLD presents an improvement than other variation.

Table 2. DLD variations comparison.

Method	Sets	Type of document	R_d	P_d	F_d
DLD-With SLCE	Test set 1	Plagiarized	1.08	0.99	1.03
		Non-plagiarized	1.00	1.07	1.05
DLD-Without SLCE		Plagiarized	0.97	0.89	0.91
		Non-plagiarized	0.96	1.01	0.98
DLD-With SLCE	Test set 2	Plagiarized	0.99	1.00	1.00
		Non-plagiarized	1.01	1.01	1.01
DLD-Without SLCE		Plagiarized	0.79	0.73	0.73
		Non-plagiarized	0.97	0.99	0.97

Table 3. PLD Variations comparison.

PLD Approaches Variation	Sets	Rec	Prec	Gran	Plagd
PLD-Without SLCE	Train set	0.7733	0.9578	1.1402	0.8123
	Test set 1	0.7559	0.9753	1.1402	0.8028
	Test set 2	0.7427	0.9577	1.1313	0.7973
PLD-With SLCE	Train set	0.8256	1.0391	1.0823	0.8859
	Test set 1	0.8159	1.0180	1.1357	0.8494
	Test set 2	0.7921	1.0134	1.1021	0.8589
Combined DLD-PLD	Train set	0.7957	1.0323	1.0738	0.8899
	Test set 1	0.8134	1.0458	1.1093	0.8678
	Test set 2	0.7725	1.0433	1.1035	0.8599

5 Conclusion

Designing a suitable approach for PD involves multiple specifications. Finding an optimal system is not simple. Hence, by applying a modified form of GA as HGA to optimize parameters. So, adopting a new approach that applies semantic concept and HGA to reveal plagiarized documents. System performance evaluated utilized by PAN 13-14 and promising research results are obtained as described. For future work, the system may be upgraded to apply with additional control levels, also may work with variable length chromosome. WordNet has restricted word coverage so; it affects negatively algorithm performance. Another optimized technique could be utilized.

Reference

1. Sánchez-Vega, F., Villatoro-Tello, E., Montes-y-Gomez, M., Villaseñor Pineda, L., Rosso, P.: Determining and characterizing the reused text for plagiarism detection. Expert Syst. Appl. **40**(5), 1804–1813 (2013)
2. Abdi, A., Idris, N., Alguliyev, R., Aliguliyevb, R.: PDLK: plagiarism detection using linguistic knowledge. Expert Syst. Appl. **42**(22), 8936–8946 (2015)
3. Shahabeddin, G., Mahmood, A.: An efficient and scalable plagiarism checking system using bloom filters. Comput. Electr. Eng. **40**(6), 1789–1800 (2014)
4. Vani, K., Gupta, D.: Detection of idea plagiarism using syntax-semantic concept extractions with genetic algorithm. Expert Syst. Appl. **73**, 11–26 (2017)
5. Ehsan, N., Shakery, A.: Candidate document retrieval for cross-lingual plagiarism detection using two-level proximity information. Inf. Process. Manage. **52**(6), 1004–1017 (2016)
6. Paula, M., Jamalb, S.: An improved SRL based plagiarism detection technique using sentence ranking. Proc. Comput. Sci. **46**, 223–230 (2015)
7. Ng, C., Li, D.: Test problem generator for unconstrained global optimization. Comput. Oper. Res. **51**, 338–349 (2014)
8. Lin, Y., Sun, Z., Dadalau, A., Verl, A.: Efficient combination of topology and parameter optimization. Open J. Optim. **3**, 19–25 (2014)
9. Cheong, D., Kima, Y., Byun, H., Oh, K., Kim, T.: Using genetic algorithm to support clustering-based portfolio optimization by investor information. Appl. Soft Comput. **61**, 593–602 (2017)
10. Guenounou, O., Belmehdi, A., Dahhou, B.: Optimization of fuzzy controllers by neural networks and hierarchical genetic algorithms. In: Proceedings of the European Control Conference, Greece, pp. 196–203 (2007)
11. Osman, A., Salim, N., Binwahlan, M., Alteeb, R., Abuobieda, A.: An improved plagiarism detection scheme based on semantic role labelling. Appl. Soft Comput. **12**(5), 1493–1502 (2012)
12. Alzahrani, S., Salim, N., Palade, V.: Uncovering highly obfuscated plagiarism cases using fuzzy semantic-based similarity model. Comput. Inform. Sci. **27**(3), 248–268 (2015)
13. Mirrashid, M.: Earthquake magnitude prediction by adaptive neuro-fuzzy inference system (ANFIS) based on fuzzy C-means algorithm. Nat. Hazards **74**(3), 1577–1593 (2014)
14. Rajaraman, A., Ullman, J.: Data Mining, Mining of Massive Datasets. Cambridge University Press, Cambridge (2011)

15. Joeran, B., Gipp, B., Langer, S., Breitinger, C.: Research-paper recommender systems: a literature survey. Int. J. Digit. Libr. **17**(4), 305–338 (2016)
16. Sanchez, D., Melin, P.: Hierarchical Modular Granular Neural Networks with Fuzzy Aggregation, 1st edn. Springer, Heidelberg (2017)
17. Garcia-Capulin, C., Cuevas, F., Trejo-Caballero, G., Rostro-Gonzalez, H.: A hierarchical genetic algorithm approach for curve fitting with B-splines. Genet. Program. Evol. **16**, 151–166 (2015)
18. Umbarkar, A., Sheth, P.: Crossover operators in genetic algorithms: a review. ICTACT J. Soft Comput. **6**(1), 1083–1092 (2015)

Anti-jamming Cooperative Technique for Cognitive Radio Networks: A Stackelberg Game Approach

Reham M. Al-Hashmy, Mohamed AbdelRaheem[✉],
and Usama S. Mohmed

Faculty of Engineering, Assiut University, Assiut, Egypt
reham.alhashmy@gmail.com, {m.abdelraheem,
usama}@aun.edu.eg

Abstract. Cognitive Radio Networks have been proposed as a promising solution for the spectrum deficiency problem. Even though, because of the opportunistic access and heterogeneity nature of these networks, they are exposed to threats such as the presence of jammer which aims to jeopardize the network transmissions. In this article, we introduce an anti-jamming cooperative framework to mitigate or minimize the impact of the jammer on the cognitive radio network. In the proposed solution, the secondary coordinator distributes the Secondary Users (SU) over the selected channel and pairs the SUs to form cooperative coalitions which utilize cooperative transmission to establish more robust transmission links to overcome the effect of the jammer power. Stackelberg game is used to model the interaction between the jammer and the network coordinator. The outcomes show the superiority of using cooperative transmission over the traditional direct transmission. Also, the performance of the secondary network enhances as the number of SUs increases due to the higher probability of forming cooperative groups.

Keywords: Cognitive Radio Network · Anti-jamming ·
Cooperative communication · Stackelberg game

1 Introduction

The rapidly growing wireless technologies created an unsatisfied demand on the wireless spectrum which resulted in the spectrum scarcity and shortage problem. Cognitive radio networks (CRNs) [1] have attracted attention as a mean to solve this problem. CRN based on the concept of spectrum sharing where the unlicensed user (a.k.a Secondary User (SU)) can use the spectrum in the absence of the licensed user (a.k.a Primary User (PU)) and when PU returns, SU, depending on its sensing capability, has to vacate the channel immediately for PU.

Because of the opportunistic access and heterogeneous nature of CRNs they are exposed to many security threats like [2]: (1) Jamming attack, where the attacker transmits harmful packets continuously to obstruct the users from sending or receiving data. (2) Primary User Emulation attack (PUE): When a selfish PUE attacker detects a free spectrum band, it prevents other SUs from accessing this band by emulating the

© Springer Nature Switzerland AG 2020
A. E. Hassanien et al. (Eds.): AISI 2019, AISC 1058, pp. 750–759, 2020.
https://doi.org/10.1007/978-3-030-31129-2_68

transmission characteristics of the PU [3]. (3) Sinkhole attack: the attacker advertises itself as having the best route to a certain destination [4]. (4) Spectrum sensing data falsification attack (SSDF): In this type of attack, the malicious user transmits false spectrum sensing data to inform other SUs with wrong information about the PU channel status. (5) Hello flood attack: The attacker transmits a message for all users of the network with enough power to deceive them by acting like one of them. (6) Control channel saturation DOS attack (CCSD): This attack makes the CRN throughputs close to nil.

The jamming attack is one of the most important threats in CRNs where the Jammer tries to reduce the Signal to Interference plus Noise Ratio (SINR) and so increases a Bit Error Rate (BER) at the receiver side higher than an acceptable threshold. In this article, we present a secondary network anti-jamming defence strategy that depends on utilizing cooperative communication between SUs to overcome the attack of the jammer or to reduce its effect. SU with a poor channel characteristic between itself and the SAP (for example, due to the long distance between them) may not be able to communicate directly with the SAP due to the effect of the jammer. Alternatively, the SU uses an intermediate relay which has a good channel characteristic between itself and the SU and between itself and the SAP, which may not be severely degraded by the jammer, to forward the SU packets to the final destination over a two-hop transmission.

Figure 1 shows a simple example to explain our idea of the proposed cooperative anti-jamming technique. SU1 attempts to transmit to its destination (the SAP) under the effect of jammer which lies near to the SAP and its jamming power has a damaging effect on the SINR of SU1. As a result, the SAP will not be able to correctly decode the received packet due to its low SINR value. SU1 may use another SU (SU2) which acts as a Secondary Relay (SR) to help it. SU2 receives data from SU1 without being hardly affected by the power of the jammer and transmits it to the SAP with better transmission characteristic (as it has a higher level of SINR) because of the shorter transmission distance that overcomes the jamming effect at the SAP.

In CRN, the scenario is more complex due to its nature, In CRNs, the SU operates over many channels and with different PU's activities. In such a case, the jammer distributes its power budget among the available channels to harm to the secondary network in the most devastating way. The SAP protects its SUs by choosing a suitable defence strategy. In our approach, the SAP strategy is to distribute the SUs among the available channels to form cooperative coalitions. These coalition members utilize cooperative communication to enhance their transmission to the SAP and overcome the jammer effect. Here, the SU with a better channel quality acts as a relay to forward the other node to the SAP. We model the interaction among the SAP, as the network coordinator, and the jammer using Stackelberg game. In this game, the SAP acts as a leader and the jammer as a follower. The SAP chooses its best strategy, in which it distributes the SUs among channels and pair them, to maximize its total achieved throughput given the jammer strategies in which the jammer allocates its power over the PUs' channels to minimize the total throughput achieved by the network.

The rest of the article is organized as follows: An overview of the related work is listed in Sect. 2. Network model is presented in Sect. 3. In Sect. 4, we introduce the

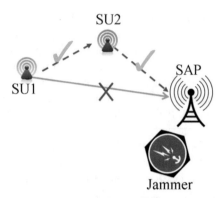

Fig. 1. The effect of AWGN jammer on direct and cooperative transmissions

game formulation. The performance evaluation section is presented in Sect. 5. Finally, we conclude the paper in Sect. 6.

2 Related Work

Recently, cognitive radio network security threats attracted the attention of the research community. In this Section, we will present some of the related work in this area. Cadeau et al. [5] focused on the performance of anti-jamming and jamming techniques in CRN. They analysed their performances using a Markov model. Chen et al. [6] used Markov decision process to model the interaction between the jamming and anti-jamming, where SUs avert the jamming power by hopping among different vacant channel. In [7], Xiao et al. assumed that the SU takes its decision to avoid the jammer under some level of uncertainty, the interactions between the SU and the jammer were modeled using prospect theory. Yang et al. [8] modeled the problem as a Stackelberg game where the user plays as a leader and the jammer plays as the follower, they study the anti-jamming problem in the presence of a jammer who can adjust its power quickly when learning the transmission power of the user to maximize the damaging effect. Chang et al. [9], suggested a robust channel hopping mechanism to resist jamming attacks for cognitive radio networks. They formulated this system by assuming that time is divided into many timeslots and several time slots make a channel hopping. Li et al. [10], concentrate on the anti-jamming communications for cooperative cognitive network where a source user and a relay user transmit the same message on one channel in the presence of a jammer. They model the relationship between transmitter and jammer as a Stackelberg game and they analyze the equilibrium point of the game. In [11] authors evaluated the effect of the observation error of a jammer on the performance of a Stackelberg game where the jammer chooses its power level depending on the SU's observed transmission. Jia et al. [12] discussed the anti-jamming defence strategies in wireless communications. They formulated a framework for an anti-jamming decision-making which uses Stackelberg game for anti-jamming modeling. Jia et al. [13] discussed the anti-jamming issue with discrete power and then modelled the interaction between the user and the jammer as Stackelberg game as the user is the leader and the jammer is the follower. They proposed a hierarchical power control algorithm (HPCA) to get the Stackelberg equilibrium (SE). Jia et al. [14] proposed the anti-jamming Bayesian Stackelberg game. They calculated the Stackelberg equilibrium point of the game under the incomplete information case.

3 Network Model

Our model comprises of a secondary CRN managed by a SAP and works over more than one PUs' channel each has a various PU activity level ρ. Here, the activity level equals the time percentage where the PU is active utilizing the channel. The channel free time equals $(1 - \rho)$ and represents the time in which the SUs are authorized to use this channel. The SAP adopts interweave spectrum sharing technique in which the SUs are only authorized to access the PU channel in its absence and depend on their sensing ability to detect the PU presence. The SUs are equipped with adaptive transceiver which is able to change the modulation technique and so the transmission data rate according to the channel condition. Also, all SUs are assumed to have equal access probability to the shared channels.

The CRN is subjected to the effect of an Additive White Gaussian Noise (AWGN) jammer that attempts to lower the SINR level and so increases the BER to a level beyond the minimum accepted one for correct decoding. The jammer should distributes its jamming power among the vacant PUs channels such that it causes the most damaging effect and reduce the secondary network throughput to its minimum. Also, we assume that the jammer power level and location are such that they do not affect the PU transmissions. The SUs has the ability to transmit to the SAP using Direct Transmission (DT) or by using Cooperative Transmission (CT) by utilizing intermediate relays. The form of cooperative communication used by the secondary network is Decode and Forward (DF). In this type, the source node (SU) transmits its data packets to the final destination (SAP) over two hops utilizing an intermediate relay (SR). If the data rate of the first hop equals R_{H1} and the second hop equals R_{H2} then the net rate of the two-hop transmission R_{CT} equals to [15]:

$$R_{CT} = \frac{R_{H1} \cdot R_{H2}}{(R_{H1} + R_{H2})} \qquad (1)$$

The SAP coordinates the SUs on the available channels where SUs can access to these channels in the absence of licensed user. The jammer distributes its jamming power among the available channels to reduce the SINR at SAP below the accepted threshold. Accordingly, the SUs has to lower their modulation order, and so the data rate. However, SUs with poor channel quality (for example, as they are located far away from the SAP) will not be able to communicate with the SAP using any of its modulation techniques. Alternatively, the SAP adopts a defence strategy against the jammer that depends on two steps. First, it distributes the SU over the PUs' channels. Second, it pairs them in a way that enhances the slower node transmission channel to decrease the effect of the jamming power and achieve the network throughput objective.

Figure 2 shows an illustrative example of SAP and jammer strategies. The secondary network consists of 10 SUs distributed around the SAP and lie in three transmission ranges areas (BPSK, QPSK, and 16QAM areas). The SUs operate over two PUs channels with different activity level ρ. In the beginning, we supposed that the jammer is idle (the jammer's power is zero). The SAP distributes the SUs over the two channels to maximize its throughput and all SUs send with its highest data rates using DT or CT. The jammer distributes its jamming power on the two channels to result in the highest degradation to the CRN throughput. Under the jammer effect, some SUs has to lower their modulation order and data rate to be able to communicate with the SAP while others are not able to communicate with the SAP using any modulation technique as they were using their lowest data rate in the case of no jammer.

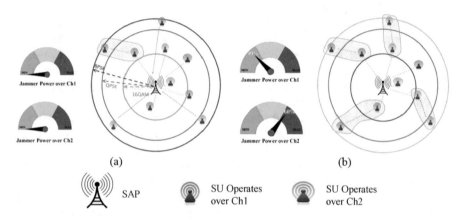

(a)   (b)

Fig. 2. SAP vs Jammer strategies

As a counteraction, the CRN SAP defends its SUs by reallocating the SUs over the channels and group them in pairs to form cooperative collations in which SUs with better channels condition act as relays and help those totally lost their communication with the SAP. By using the intermediate relay, the affected SU has to deliver its packet to the relays instead of the SAP where the link between the SU and the relay is still up in the presence of the jammer. Our goal is to calculate the secondary network performance at the equilibrium point where both the SAP and the jammer choose their best strategies. We adopt the Stackelberg game to find the targeted equilibrium point.

4   Problem Formulation

In this section, we present the mathematical formulation of our problem. First by defining the Stackelberg game. Then the proposed game model and the players' utilities.

4.1   Stackelberg Game

Stackelberg game is a strategic non-cooperative game in which there is players hierarchy as the game is played as a set of sequential actions between two players [16]. In this game, the leader starts the game by taking an action ($a_L \in A_L$). Given that the leader played strategy (action) a_L the follower will respond with the reaction $R_f(a_l)$ such that:

$$R_F(a_l) = \{a_F \in A_F : U_F(a_L, a_F) \geq U_F(a_L, a_o), \forall a_o \in A_F\} \tag{2}$$

The leader knows about the follower strategies and due to its playing order, it is in a position to impose its strategy, which achieves its objective, over the follower who tries to react by its best response to the leader. From that, we can determine the Stackelberg equilibrium utility for the leader $U_L(a_L^*, a_F)$ from:

$$U_L^* = \min_{a_F \in R_2(a_L^*)} U_L(a_L^*, a_F) = \max_{a_L \in A_L, a_F \in R_2(a_L)} . \min U_L(a_L, a_F) \tag{3}$$

It is also worth mention that, due to the playing order, the Stackelberg equilibrium utility of the leader is higher than or equal to its Nash Equilibrium one,

$$U_L^* \geq U_{NE}^*. \tag{4}$$

4.2   The Players' Utilities and the Game Formulation

In this paper, we assume that the SAP objective is to maximize its total rate. By other words, maximize the summation of all SUs throughput over the various channel. To formulate the problem, we introduce two binary variables s_{nc}, p_{nr} to accommodate for the allocation of channels and SUs pairing functions, respectively.

Where

$$s_{nc} = \begin{cases} 1 & \text{if SU n operates over channel c} \\ 0 & \text{otherwise} \end{cases} \tag{5}$$

and

$$p_{nrc} = \begin{cases} 1 & \text{if SU n receive help from SU r over channel c} \\ 0 & \text{otherwise} \end{cases} \tag{6}$$

Where $p_{nnc} = 1$ means that SU (n) adopts direct transmission instead of cooperative transmission when operating over the channel (c). The SAP utility at jammer strategy (a_f) equals to

$$U_{SAP}(a_F) = \max_{(s_{nc}, p_{ncr})} \sum_{n=1}^{N_s} \sum_{r=1}^{N_s} \sum_{c=1}^{N_C} s_{nc} \cdot p_{ncr} \cdot T_{nrc}^{(a_f)} \cdot (1 - \rho_c). \tag{7}$$

Where $T_{rnc}^{(a_f)}$ is the SU (n) rate if it received help by SU (r) over channel (c) at jammer strategy (a_F).

Here the interaction among the SAP and the jammer is a zero-sum game such that the utility of the jammer equals the wastage in the throughput of the SAP. The jammer has a total power budget of JP which can be divided into the number of power levels among the available channel such that:

$$a_F \in \left\{ a_F^{(1)}, a_F^{(2)}, a_F^{(3)} \ldots a_F^{(N_J)} \right\}. \tag{8}$$

Where (N_j) is the number of power level distribution over the channels permutations and represents the total number of jammer strategies[1]. So, the jammer utility at SAP strategy (a_L) equals to:

$$U_{Jam}(a_L) = \max_{(a_F)} . (U_{SAP}^{max} - U_{SAP}^{(a_F)}(a_L)). \tag{9}$$

Where U_{SAP}^{max} is the SAP maximum utility when the jammer is idle.

5   Performance Evaluation

In this section, we show the technique performance. We evaluate the system at various jammer power levels and SUs densities. The simulation parameters are listed in Table 1 except if not mentioned otherwise explicitly.

5.1   The Impact of Jammer Power

Here, we investigate the performance enhancement of the secondary network throughput when using Cooperative Transmission (CT) over the traditional Direct Transmission (DT) due to the jammer power effect.

Figure 3 depicts the effect of jammer power on the SAP utility in two scenarios first when only DT is allowed and the other when the CT is utilized. As can be inferred from the figure, CT outperforms DT in terms of the total achieved throughput as a result of forming a robust transmission link between the SUs and the SAP. To clearly understand the impact of the cooperation on the network performance, we study its effect on the slow SUs only. Where slow SUs are those with low data rate (BPSK modulation and rate equal to 6 Mbps) in the case when the jammer is idle. We choose this category

---

[1] As an example, in Fig. 2, it is supposed that the jammer is able to divide its total power to 10 levels each equals to 10% of the total power and the jammer inject 4 levels in channel 1 (40% of its power) and 6 levels in channel 2 (the remaining 60%).

Table 1. Simulation parameters

Parameter	Value
Modulation	BPSK, QPSK, 16QAM
Transmission rate	6, 12, 24 Mbps
Path loss exponents	3.5
SU transmission power	0.5 mW
Jammer power	2 μW
Jammer power levels	5
BER threshold	10^{-3}
Channel model	Rayleigh flat fading
Number of PU channels	2
Primary user activity level	30%, 70%

as it is the most affect one by the jammer. Figure 4 clarifies the impact of the jammer power on the slow SUs performance. It is clearly obvious that slow SUs has a much better performance when they are able to use intermediate relays to communicate with the SAP and overcome the jammer effect.

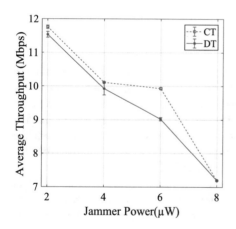

Fig. 3. The effect of jammer power on the SAP utility (total network throughput)

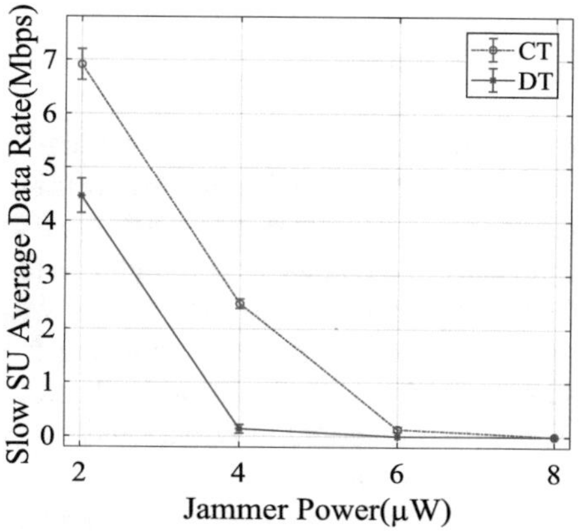

Fig. 4. The effect of jammer power on the slow SU average data rate

5.2   The Effect of Node Density

In this subsection, we investigate the impact of SUs density (number of SUs) on the performance of the secondary CRN. As can be noticed in Fig. 5, when the number of SUs increases the SAP utility (total throughput) increases. This can be justified as follows: as the number of SUs increases, their density increases and so the probability that slow SUs find useful relays in its nearby. This can be better observed by evaluating the performance of SUs only. As shown in Fig. 6, the average data rate of the slow SUs increases as the SUs density increase as a result of their better transmission performance when utilizing CT.

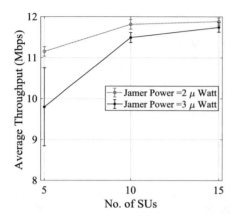

Fig. 5. The effect of SUs density on the SAP utility (total throughput).

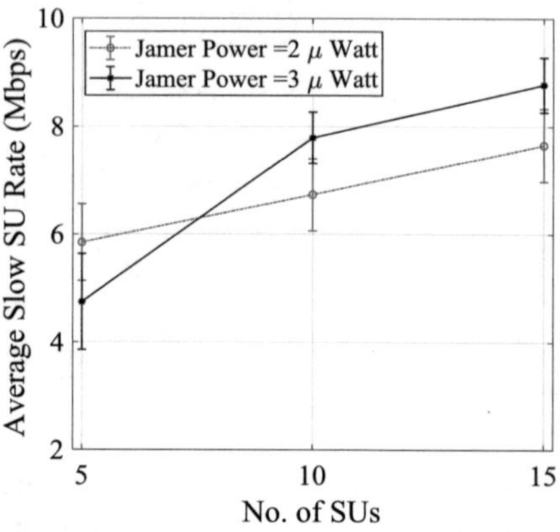

Fig. 6. The effect of SUs density on the slow SUs average data rate.

6   Conclusion

In this article, we introduced a technique to overcome the effect of AWGN jamming in the cooperative cognitive radio networks. In this technique, the network coordinator groups the SUs in cooperative coalitions where SUs are able to use decode and forward cooperative communication to mitigate the effect of the jammer power. The interaction among the network and the jammer was modelled as a Stackelberg game and the second network performance utilizing the proposed technique is compared with that when only direct transmission is allowed. The numerical evaluation shows the feasibility of the proposed technique and its ability to reduce the jammer effect at different power levels compared to the direct transmission case. Also, results clarify that the increase in SUs density raises the probability of forming cooperation groups and so further enhance the network performance.

References

1. Haykin, S.: Cognitive radio: brain-empowered wireless communications. IEEE J. Sel. Areas Commun. **23**(2), 201–220 (2005)
2. Bhattacharjee, S., Rajkumari, R., Marchang, N.: Cognitive radio networks security threats and attacks: a review. J. Comput. Appl. ICICT **975**, 8887 (2014)
3. Yu, Y., Hu, L., Li, H., Zhang, Y., Wu, F., Chu, J.: The security of physical layer in cognitive radio. J. Commun. Netw. **9**(12), 28–33 (2014)

4. Ganesh, D., Pavan Kumar, T.: A survey on advances in security threats and its countermeasures in cognitive radio networks. Int. J. Eng. Technol. **7**, 372–378 (2018)
5. Cadeau, W., Li, X.: Anti-jamming performance of cognitive radio networks under multiple uncoordinated jammers in fading environment. In: Proceedings of the 46th Annual Conference on Information Sciences and Systems (CISS), March 2012
6. Chen, C., Song, M., Xin, C., Backens, J.: A game-theoretical anti-jamming scheme for cognitive radio networks. IEEE Network **27**(3), 22–27 (2013)
7. Xiao, L., Liu, J., Li, Y., Mandayam, N., Poor, H.V.: Prospect theoretic analysis of anti-jamming communications in cognitive radio networks. In: Proceedings of the IEEE Global Communication Conference (Globecom), pp. 1–6 (2014)
8. Yang, D., Xue, G., Zhang, J., Richa, A., Fang, X.: Coping with a smart jammer in wireless networks: a Stackelberg game approach. IEEE Trans. Wireless Commun. **12**(8), 4038–4047 (2013)
9. Chang, G., Wang, S., Liu, Y.: A jamming-resistant channel hopping scheme for cognitive radio networks. IEEE Trans. Wireless Commun. **16**(10), 6712–6725 (2017)
10. Li, Y., Xiao, L., Liu, J., Tang, Y.: Power control Stackelberg game in cooperative anti-jamming communications. In: Proceedings of the 5th International Conference on Game Theory Network, Beijing, China, November 2014
11. Xiao, L., Chen, T., Liu, J., Dai, H.: Anti-jamming transmission Stackelberg game with observation errors. IEEE Commun. Lett. **19**(6), 949–952 (2015)
12. Jia, L., Xu, Y., Feng, S., Anpalagan, A.: Stackelberg game approaches for anti-jamming defence in wireless networks. IEEE Wireless Commun. **25**(6), 120–128 (2018)
13. Jia, L., Yao, F., Sun, Y., et al.: A hierarchical learning solution for anti-jamming Stackelberg game with discrete power strategies. IEEE Wireless Commun. Lett. **6**(6), 818–821 (2017)
14. Jia, L., Yao, F., Sun, Y., et al.: Bayesian Stackelberg game for anti-jamming with incomplete information. IEEE Commun. Lett. **20**(10), 1991–1994 (2016)
15. Liu, P., Tao, Z., Narayanan, S., Korakis, T., Panwar, S.S.: CoopMAC: a cooperative MAC for wireless LANs. IEEE J. Sel. Areas Commun. **25**(2), 340–354 (2007)
16. Zhu, H., Dusit, N., Saad, W., Tamer, B., Are, H.: Game Theory in Wireless and Communication Networks: Theory, Models, and Applications. Cambridge University Press, Cambridge (2012)

Geometry Aware Scheme for Initial Access and Control of MmWave Communications in Dynamic Environments

Ahmed S. Mubarak[1,2(✉)], Osama A. Omer[2,3], Hamada Esmaiel[2], and Usama S. Mohamed[4]

[1] Department of Electrical Engineering, Faculty of Engineering,
Sohag University, Sohag, Egypt
[2] Department of Electrical Engineering, Faculty of Engineering,
Aswan University, Aswan, Egypt
{ahmed.soliman, omer.osama, h.esmaiel}@aswu.edu.eg
[3] Arab Academy for Science, Technology and Maritime Transport,
Aswan, Egypt
[4] Department of Electrical Engineering, Faculty of Engineering,
Assiut University, Assiut, Egypt
usama@aun.edu.eg

Abstract. Immigration to the millimeter wave (mmWave) bands has got the attention of both industry and academia for playing a crucial role in the future 5G wireless networks. Unlike the traditional sub 6 GHz bands, large segments of contiguous spectrum are available at the mmWave bands allowing multigigabit data rates transmission. However, signals at mmWave bands are susceptible to harsh propagation and penetration losses which in turn limit its transmission range. Highly directional antennas are employed to compensate the propagation losses by concentrating the signal power towards a specific direction and providing adequate link margin. However, initial access and control of mmWave communications in dynamic environments turns to be a big challenging issue. This is due the fact that, a highly complicated conventional beamforming training (BT) process based on exhaustively searching all beam settings in all possible directions is performed whenever a new initial access or serving beam changing is required. Inspired by the fact that the mmWave communications are location and environmental driven, this paper proposes a geometry-based scheme to alleviate the high complexity of the initial access and control of mmWave communications. Simulations show that the proposed scheme provides a great complexity and energy consumption reduction over the conventional exhaustive-based scheme with comparable link quality performance.

Keywords: 5G · MmWave · Beamforming training (BT) · IEEE 802.11ad · Initial access · Geometry

© Springer Nature Switzerland AG 2020
A. E. Hassanien et al. (Eds.): AISI 2019, AISC 1058, pp. 760–769, 2020.
https://doi.org/10.1007/978-3-030-31129-2_69

1 Introduction

The worldwide rapid adoption of high-end devices such as mobile tablet, wearable devices and smartphones exacerbates the demand for high data rate services [1]. Fifth generation (5G) mobile cellular networks have been introduced to face the bottleneck of the conventional mobile networks in addressing the explosive growing demand for high data rate mobile applications [2]. Unlike legacy sub-6 GHz bands, large segments of contiguous spectrum are available at mmWave bands allowing huge bandwidths, hence, multi Gbps data rates can be supported. However, signals at mmWave bands are susceptible to harsh free space losses which in turn limit its transmission range. These sever propagation losses can be compensated by employing highly directional antennas to concentrate the signal power towards a specific direction and provides adequate link margin. However, due to the exploitation of directional transmission strategy, initial access and control of mmWave communications in dynamic environments with mobile user devices (UDs) turns to be a big challenging issue IEEE 802.11ad [3], mmWave WLAN standard was developed for communications in the 60 GHz band, where an exhaustive search medium access control (MAC) based beamforming training (BT) protocol is used for mmWave initial access. Nevertheless, this exhaustive search BT complicates the mmWave initial access, where the UD and the deployed mmWave access points (APs) should jointly sweep the large angular space to determine the best transmission and reception communication beam directions. This time consuming and huge overhead BT will dramatically degrade the system performance in terms of latency and energy consumption. In addition, transmission in narrow directional beams results in discontinuous availability of mmWave access in the space, which makes the mmWave links fragile with continuous and fast variation in quality. The consequences become more serious in the case of dynamic environments with mobile UDs, where a frequent and intensive BT should be performed to track the direction between the UD and the deployed mmWave APs to control the communication link during the connection time in order to maintain its quality. As a result, the conventional time-consuming exhaustive search BT will complicate the mmWave initial access as well as controlling the existing links to maintain its quality a above a certain threshold. Consequently, it will be difficult to meet the ultra-low latency and energy consumption requirement needed for the future 5G mobile networks.

The challenging of mmWave communication realization have been addressed in literature. To this end, many works have been introduced to overcome the complexity of the conventional exhaustive-based scheme for mmWave link initial access and control. The dominant seem in the literature is supporting the mmWave system with external legacy service to provide out-of-band information such as UD location, which in turns can significantly reduce the exhaustive BT complexity and enabling efficient mmWave communications. In this direction, authors in [4, 5] have proposed to use the LTE and Wi-Fi legacy bands to assist the mmWave AP and the UD initial synchronization, respectively. However, Despite the effectiveness of these schemes, the interoperation between the legacy and mmWave bands in hybrid architecture results in integration challenging. For example, such multi-band architectures may need additional costs and protocol stacks modifications.

On the other hand, other research direction has adopted to mitigate the complexity of the exhaustive-based scheme through using in-band schemes instead of the assistance with legacy bands. Authors in [6] have proposed to reduce the complexity of the exhaustive search BT using iterative search procedures, since a configurable antenna pattern is employed to successively narrow the search space using adaptive beam width. However, this method suffers from bad coverage due to the initial search with wide width, short range beams. Authors in [7] have exploited the sparsity of the mmWave and proposed a solution based on the estimation of angle-of-arrival/angle-of-departure (AoAs/AoDs) to reduce the mmWave communication complexity. However, the compressive sensing channel estimation schemes mainly based on the hard assumption of mmWave channel sparsity and the angle of AoAs/AoDs quantization, which in turns put uncertainty about its practical feasibility.

In this paper, inspired by the fact that directionality makes the mmWave communications to be geometry and location driven, then the exploiting the target area geometry including the UDs locations, APs coordinates, and the target area maps can extremely facilitate the mmWave communications. Accordingly, in this paper, a geometry-based scheme is proposed to alleviate the complexity of the exhaustive-based BT. In the proposed scheme, the geometric information of the target area geometric information are centrally collected, saved and processed in an AP local controller (ALC). By leveraging theses geometric information, each deployed AP can identify its dead beams and eliminate it from the BT search process. For a certain mmWave AP, any beam with coverage exists in a dead area i.e. wall is classified as a dead beam. Accordingly, the complexity of the mmWave initial access for a UD can be extremely reduced due to the reduction of the number of BT beams. Moreover, the ALC is used to manage the UD mobility between the different APs. By exploiting, the global coordinate's information of the deployed APs according to a reference point, the ALC can efficiently reduce the number of used APs for searching the UD in its next location. To this end, only the neighbours of the serving AP are informed to perform BT and search for the UD at its next location. Simulation results show that a significant BT complexity and energy consumption reduction can be achieved with the proposed scheme compared to the conventional exhaustive-based one.

The reminder of this paper is organized as follows. Section 2 presents the system model. Section 3 provides the details of the conventional and the proposed scheme. The performance of the proposed scheme is evaluated in Sect. 4 using numerical simulations. Finally, conclusion is presented in Sect. 5.

2 Proposed System

2.1 System Architecture

Figure 1 shows the mmWave system architecture. In this architecture, multiple 60 GHz mmWave APs are deployed in an outdoor target area e.g., campus, open markets, etc. The deployed APs are connected to an ALC via high speed links i.e. Fiber cables to form a WLAN and X2 interfaces. The ALC provides the processing and controlling of the connected group of APs. Also, ALC acts as an external gateway connects the

WLAN to cellular system and Internet. Also, all the geometric information of the target area including the coordinates of the APs locations are collected and processed in the ALC. The details of the geometric-based scheme and the function of the ALC and its interoperation with the deployed APs will be presented in the next sections.

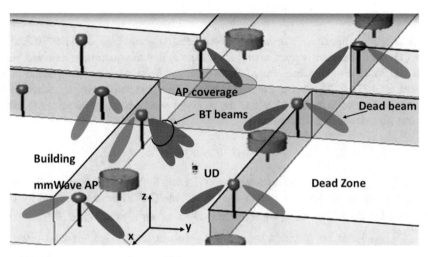

ALL APs are connected to an ALC

Fig. 1. The mmWave system architecture.

2.2 MmWave Propagation Link Models

The propagation model for outdoor urban environment defined in [8] is considered for mmWave communications, which expressed as follows:

$$PL(d)_{dB} = PL(d_0)_{dB} + 10nlog_{10}\left(\frac{d}{d_0}\right) + \chi\sigma, \tag{1}$$

Where, $PL(d)$ means the path loss and the shadowing value at distance d from a mmWave AP. d_0 represents the reference distance, which is set as 5 m. $PL(d_0)$ is the path loss at d_0 and is equal to 82.02 dB for the 60 GHz band. $\chi\sigma$ is the shadowing term in dB, which is a zero mean random variable with standard deviation, σ. n is the exponent path loss. The antenna gain $G_{dB}(\Phi, \Theta)$ in a certain azimuth and elevation angles Φ and Θ, respectively, can mathematically formulated as follows [9]:

$$G_{dB}(\Phi, \Theta) = G_m(dB) - 12\left(\frac{\Phi - \Phi_0}{\Phi_{-3dB}}\right)^2 - 12\left(\frac{\Theta - \Theta_0}{\Theta_{-3dB}}\right)^2,$$

$$G_m(dB) = \left(\frac{16\Pi}{6.76\Phi_{-3dB}\Theta_{-3dB}}\right), \tag{2}$$

Where G_m denotes the maximum beam gain, and Φ_{-3dB} and Θ_{-3dB} are the azimuth and elevation half power beamwidths, respectively, and Φ_0 and Θ_0 represent the azimuth and tilt angles of the beam center, respectively. Accordingly, the received power by a UD at a distance d from an mmWave AP, $P_r(d)$, is expressed as:

$$P_r(d)_{dB} = P_t(dBm) + G_{dB}(\Phi, \Theta) - PL(d)_{dB}, \tag{3}$$

Where, P_t is the mmWave AP TX power. $PL(d)_{dB}$ and $G_{dB}(\Phi, \Theta)$ are calculated from (1) and (3), respectively. Clearly, from Eq. (3) that the maximum received power by a UD located at distance, $P_{rmax}(d)$, is achieved with the maximum value of the antenna gain G_{mdB}. Therefore, the mmWave AP transmission range, $\Re_t$, which is the maximum distance from the AP where the received power exceeds the power threshold P_{th}, can be calculated as follows:

$$\Re_t = max\{d|P_{rmax}(d) \geq P_{th}\} \tag{4}$$

Generally, P_{th} depends on the receiver sensitivity and in this paper, is selected to be the power level essential for modulation coding scheme (MCS) zero as defined by IEEE 802.11ad standard. So, the coverage area, $\Omega_{cov}^{(i)}$, of an ith mmWave AP located at $\left(x_{AP}^{(i)}, y_{AP}^{(i)}, z_{AP}^{(i)}\right)$ can be specified by a circular disc in the UD's horizontal plane and centred at $x_{AP}^{(i)}$ and $y_{AP}^{(i)}$ in the x and y axes, respectively.

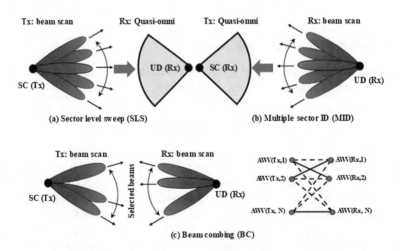

Fig. 2. The conventional exhaustive IEEE 802.11ad-based BT process

3   Initial Access and Control for MmWave Links

3.1   Conventional Exhaustive-Based Scheme

According to exhaustive IEEE 802.11ad-based schemes, as shown in Fig. 2, both the UD and the mmWave AP exhaustively scan the whole angular space to find the best transmission/reception (TX/RX) beam pairs between each other, which in turns results in large overhead for link establishment. For example, in [3], about 2 ms is required to determine the best TX/RX beam directions between an mmWave AP and a UD with 32 different beam settings. In a typical environment with several deployed mmWave APs, this not only complicates the process of initial access, but also consumes a great amount of energy especially in dense network with a large number of deployed mmWave APs. The situation will be worse in case of dynamic environment under UDs mobility, since this complicated BT process needed to repeat frequently to control the communication links for connected UDs among different mmWave APs.

3.2 Proposed Geometry-Based Scheme

In the proposed scheme, the geometry of the target area including the UDs locations, APs coordinates and the target area maps are exploited to facilitate the mmWave communications by reducing the exhaustive-based BT complexity. By leveraging the geometric information, the ALC can identify the dead beams for each deployed AP and to eliminate from the BT search process. All beams have a coverage area overlapped with a dead zone are classified as dead beams. The dead beam for a certain AP cannot provide the sufficient power for the UD initial access. Thus, over the all target area a large number of dead beams from different APs will be eliminated from the BT process. This in turns wills extremely the UD initial access compared to the conventional exhaustive-based BT.

Moreover, the global coordinate's information of the deployed APs according to a reference point is exploited by the ALC for reducing the BT complexity of controlling the established links for mobile UD. Specifically, based on the target is geometry information, the ALC can determine the APs that expected to serve a mobile UD at its next location? Thus, only the neighbours of the serving AP are informed to perform BT and search for the UD at its next location. Particularly, the AP with a coverage area intersects with the coverage area of the serving AP will be classified as an expected AP to serve the UD and informed by the ALC to perform a BT. The coverage of the ith mmWave AP, $\Omega_{cov}^{(i)}$, is determined according to the calculation of the potential coverage which have presented in Sect. 2.

Table 1. Main simulation parameters

Parameter	Value
Num. of APs MmWave	20
TX Power of mmWave AP	10 dBm
Height of mmWave AP/UD	3 m/0.75 m
Num. of total beams mmWave AP	varied
$\Phi_{-3dB}/\Theta_{-3dB}$ of mmWave beam	20°/20°
UD speed, V	2 m/s
Simulation period	60 s
Threshold received power, P_{th}	−78 dBm
Shadowing standard deviation, σ	10.3 dB

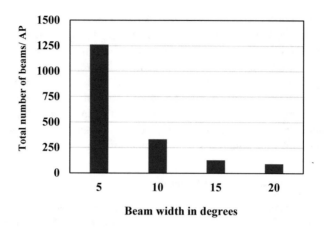

Fig. 3. The total available beam foe mmWave AP at different beam widths

4   Simulation Analysis

In this section, the performance of the proposed and the conventional schemes are evaluated and compared in terms of BT complexity.

4.1   Simulation Parameters and Scenario

To evaluate the performance of the proposed scheme, simulations are performed in an outdoor local environment having area of 200×150 m^2, as shown in Fig. 1. Multiple 60 GHz mmWave APs are uniformly distributed in the target area. In this simulation, we consider a scenario of single UD that is initially and randomly dropped in the target area and then moves with walking speed ($V = 2m/s$) in a random direction for simulation period T s. The details of the simulation parameters and its related values are listed in Table 1. At first, the simulation of the initial access process is performed and repeated for many Monte Carlo simulations. In each iteration, the UD is randomly dropped in the target area and the average total number of the used BT for the initial access process is counted for both the proposed and conventional schemes. Moreover, the BT energy consumption of the mobile UD is calculated during the simulation period and compared for both the proposed and conventional schemes.

4.2   Simulation Results

Figure 3 indicates the relationship between the total number of available beams for a mmWave AP and the beam width. In this simulation, the azimuth and elevation half power beam widths of a beam is assigned with equal values i.e. $\Phi_{-3dB} = \Theta_{-3dB}$ and defined as the beam width. As shown in Fig. 3, the total available number of a mmWave AP is decreased with increasing the beam width due to the increasing of the beam footprint.

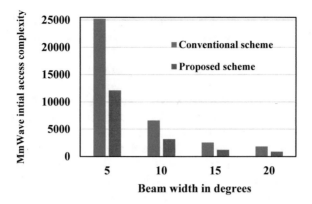

Fig. 4. Initial access complexity at different beam widths

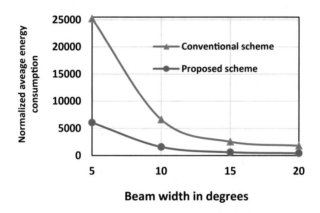

Fig. 5. The normalized energy consumption at different beam widths

Figure 4 shows the initial access complexity comparison between the proposed and the conventional scheme for different beam widths. Here, the complexity of the initial access process is expressed as the average number of used BT beams for enabling the initial connection of the UD with an mmWave AP. As shown in Fig. 4, the complexity of the initial of the proposed is smaller than those of the conventional scheme for the all beam widths. Specifically, the proposed scheme achieves about 52% complexity reduction over the conventional exhaustive-based scheme. In fact, this huge reduction in the BT complexity due to the ability of the proposed scheme to exploit the target area geometric information in eliminating a large number of dead beams form the BT process.

As a result of the large BT complexity reduction, the energy consumption due BT is greatly enhanced using the prosed scheme, especially for mobile UDs. In this simulation, the normalized averaged energy consumption (NAEC) for a mobile UD during the simulation period, T, is defined as:

$$NAEC = \frac{Total\ energy\ consumption\ over\ T\ using\ the\ used\ BT\ beams}{Energy\ consumption\ using\ one\ beam\ transmission} \tag{5}$$

Figure 5 shows the normalized energy consumption for the proposed and the conventional scheme for different beamwidths. As shown in Fig. 5, the energy consumption for control the communication with a mobile UD is greatly enhanced with the proposed scheme for all beam widths. Specifically, the proposed scheme achieves about 97% energy consumption reduction over the conventional exhaustive-based scheme. Particularly, in addition to eliminate the dead beams for each AP, this great reduction in the energy consumption is due to the reduction of the number of APs used in BT process to be only the expected ones. However, according to (3) and (4) the mmWave AP coverage is varied with the value of the beam width. Hence, in the proposed scheme simulation, Ω_{cov} for an mmWave AP is calculated according to 20-degree beamwidth. According to this simulation, only seven APs after eliminating their bad beams are used for BT to maintain the link with the UD in each step in the proposed scheme. It is worth to mention that this huge reduction in BT complexity and energy consumption is achieved with a comparable link quality performance with the conventional scheme. According to simulations only about 2 dB reduction in the average received power is resulted using the proposed scheme over the conventional one due to the shadowing effect.

5   Conclusion

In this paper, the challenges of initial access and control of mmWave communications in dynamic environments are addressed. To this end, inspired by the directionality feature of the mmWave communications, this paper proposes a geometry-based scheme facilitate the mmWave communications. In the proposed, geometric information of the target area including the UDs locations, APs coordinates and the target area maps are exploited to alleviate the complexity of the exhaustive-based BT. Also, by leveraging this information only a few numbers of expected APs are used for BT to control and maintain the link with a connected mobile UD. The simulation results show that the proposed scheme can efficiently reduce the complexity of mmWave initial access and control over the conventional exhaustive-based scheme. Specifically, the proposed scheme achieves about 52% complexity reduction over the conventional exhaustive-based scheme. Moreover, the energy consumption for control the communication with a mobile UD is greatly enhanced with the proposed scheme for all beam widths. Specifically, the proposed scheme achieves about 97% energy consumption reduction over the conventional exhaustive-based scheme.

References

1. Benisha, M., Thandaiah Prabu, R., Thulasi Bai, V.: Requirements and challenges of 5G cellular systems. In: 2nd International Conference Advances in Electrical, Electronics, Information, Communication and Bio-Informatics (AEEICB) (2016)

2. Al-Falahy, N., Alani, O.Y.: Technologies for 5G networks: challenges and opportunities. IT Prof. **19**(1), 12–20 (2017)
3. Hosoya, K.I., et al.: Multiple sector ID capture (MIDC): a novel beamforming technique for 60-GHz band multi-Gbps WLAN/PAN systems. IEEE Trans. Antennas Propag. **63**, 81–96 (2015)
4. Capone, A., Filippini, I., Sciancalepore, V.: Context information for fast cell discovery in mm-wave 5 G networks. In: Proceedings of European Wireless; 21th European Wireless Conference. VDE (2015)
5. Chandra, K., et al.: CogCell: cognitive interplay between 60 GHz picocells and 2.4/5 GHz hotspots in the 5G era. IEEE Communications Magazine **53**(7), 118–125 (2015)
6. Desai, V., Krzymien, L., Sartori, P., Xiao, W., Soong, A., Alkhateeb, A.: Initial beamforming for mmWave communications. In: 2014 48th Asilomar Conference on Signals, Systems and Computers, pp. 1926–1930 (2014)
7. Duan, Q., Kim, T., Huang, H., Liu, K., Wang, G.: AoD and AoA tracking with directional sounding beam design for millimeter wave MIMO systems. In: 2015 IEEE 26th Annual International Symposium, Personal, Indoor, and Mobile Radio Communications (PIMRC), pp. 2271–2276 (2015)
8. Rappaport, T.S., Sun, S., Mayzus, R., Zhao, H., Azar, Y., Wang, K., et al.: Millimeter wave mobile communications for 5G cellular: it will work! IEEE Access **1**, 335–349 (2013)
9. FP7-ICT-608637 MiWEBA Project Deliverable D5.1 Channel Modeling and Characterization, June 2014. http://www.miweba.eu/wp-content/uploads/2014/07/MiWEBA"D5.1v1.01.pdf

IoTIwC: IoT Industrial Wireless Controller

Tassnim Awad, Walaa Mohamed, and Mohammad M. Abdellatif[✉]

Electrical Engineering Department, The British University in Egypt, Cairo, Egypt
{Tassnim24751, Walaa25600,
Mohammad.Abdellatif}@bue.edu.eg

Abstract. Industrial controller systems are crucially essential to the cutting edge power systems industries. Electric services are expected to determine the best media tool to be used for any distribution. The communication systems contribute towards a physical intermediary layer for transferring, controlling and acquirement of data within the system from distant locations. This paper discusses the Supervisory Control and Data Acquisition (SCADA) systems and proposes a similar system that is an IoT based industrial wireless controller. The proposed system can control multiple devices through the internet without the need to be physically near the devices. Because it uses simple and cheap devices, the system is low cost and easy to install. Additionally, the system is modular because extra microcontrollers can be added to the system to control more devices should the need arise.

Keywords: IoT · Industrial controller · IIoT · SCADA

1 Introduction

A SCADA (Supervisory Control And Data Acquisition) is an automation control system that is used in industries such as energy, oil, gas, water, power, and many more. Normally, the system has a centralized location that monitors and controls entire sites, ranging from an industrial plant to a complex of plants across a whole country. SCADA could be defined as the process of controlling and supervising data collection and processing. This word is used to describe real time systems where data is gathered, processed, and maintained in real time. Some examples may include monitoring over a power plant or an irrigation system. To further illustrate the example above and to show how the system works, a home station could be fitted with that system to monitor different substations or remote stations. If an error occurs in one of the stations, the home station could analyze the data and make sure that the error may not be critical and fixable. The system uses client server architecture and has several elements, most notably, input-output sever, SCADA server, Human machine interfaces, and a control server.

The first generation of SCADA systems was designed with no networking infrastructure as it wasn't quite developed by that time. Then came the Next generation of SCADA system where the concept of networking was introduced. SCADA systems utilized networking for load balancing for efficient resource use and for increasing the dependability of the system. The third generation of SCADA system followed the same

© Springer Nature Switzerland AG 2020
A. E. Hassanien et al. (Eds.): AISI 2019, AISC 1058, pp. 770–778, 2020.
https://doi.org/10.1007/978-3-030-31129-2_70

design principles as the previous generation but it became a more open system architecture compared to the self-tailored versions done by the companies. There are a-lot of transmission technologies used in SCADA systems, for example, coaxial cables which are used by TVs providers to send data, twisted metal pair as in telephone lines, fiber optic cables which is a relatively expensive option but allows real time communication and disaster recovery because of the high speeds it could transmit data through, it could also utilize satellite communications. All these communication methods require a solid infrastructure and to deploy them in a rural area would cost a fortune. So, we need to relay on wireless communication to reduce the Installation cost.

As the current SCADA systems use the same frequency as those of the TV channels. There was a conflict with major TV companies as it may negatively affect their income from advertisements, It uses the channels between 54 and 862 MHz for maximum coverage which also happens to be the TV channels, The bandwidth is about 6 to 8 MHz per channel and It identifies the free channels and send on those only to prevent any conflict with any used TV brands. Another method that is used is utilizing the already existing infrastructure of the cellular networks as one of the building blocks for wireless SCADA systems, with base stations connecting to the main network and sending data through it. The base stations can utilize components in the SCADA system to identify the free bands to send data upon so it won't interfere with any Calls or bands which are quite busy at the time. This would allow the SCADA system to cover about 100 km of area and it could provide the same level of service relative to DSL. However, this will add extra cost to the system because of the subscription charges of the cellular service.

This paper proposes an IoT based SCADA system which is made with low cost devices and that can be controlled wirelessly through Wi-Fi from any device in the network. The proposed system is modular which simplifies expanding the system to control more devices without the need of reconfiguring the system as a whole.

The rest of the paper is organized as follows. Section 2 gives a background and literature review on SCADA systems. Section 3 describes the system architecture. And finally, Sect. 4 concludes the paper.

2 Background and Literature Review

In this section, we describe some of the earlier work done on web-based remote monitoring as well as some of the state of the art available on the same topic.

2.1 Web-Based Remote Monitoring

One of the early work done on the subject was that of Bertocco et al. [1], where the authors described client–server architecture for remote monitoring of instrumentation over the Internet. The proposed solution allowed multi-user and multi-instrument sessions by means of a queuing and instrument locking capability. A queue mechanism has been added to the remote environment along with the possibility for each client to query the actual server load. The communication between the server and clients can be obtained either at instrument level or by means of encoded requests in order to reduce the network-imposed overhead.

Tso et al. [2] presented a study that indicates that a while a number of frameworks related to global systems have been described in contemporary publications, the detailed structure and formulation of the central-monitoring mechanism of such a partnership system has not received as much attention as it deserves. The proposed framework of a service network is characterized by its coordinating as well as monitoring capabilities. The main feature of the presented system is its rule-based reasoning capability to convert a job request from clients into basic tasks which are to be carried out by a group of virtual agents equipped with various defined capabilities.

Tommila et al. [3] discussed new ways of implementing existing functions and defined that new functionality, e.g. management of hierarchical structures and exception handling should be included in the basic control platform and engineering tools. The current 'flat' collection of application modules like loops and sequences had to be organized in a more hierarchical fashion based on process structure. Each process system is seen as an intelligent resource capable of performing different processing tasks. The interaction mechanisms between different automation activities are defined on the basis of object-oriented analysis and design and emerging international standards. A standardized distribution middleware takes care of the needs specific to the control domain. Above that, a higher level working environment for the other system components of the control platform is needed.

Yang et al. [4] reported a study on Networked Control System (NCS) historical review, recent revolution and research issues on NCS. Fast development and major use of the Internet, a global information platform has been created for control engineers allowing them to do the following:

- Monitoring the condition of machinery via the Internet.
- Remotely control machine.

They also addressed many new challenges to control system designers. These challenges are summarized as follows:

- Overcoming Web-related traffic delay, i.e. dealing with Internet latency and data loss.
- Web-related safety and security, i.e. ensuring the safety and security of remote control and stopping any malicious attacks and misoperation.
- Collaborate with skilled operators situated in geographically diverse location.

Kalaitzakis et al. [5] developed a SCADA based remote monitoring system for renewable energy systems. It is based on client/server architecture and it does not require a physical connection, e.g. through network, serial communication port or 14 standard interfaces such as the IEEE-488 of the monitored system with data collection server.

Kimura et al. [6] reported remote monitoring system as one component of manufacturing support system. The proposed remote monitoring system can support single-night unmanned night time operations for diversified manufacturing from the operator's home as the remote site. According to the results, the remote monitoring system performed quite well for providing backup of manufacturing systems during unmanned nighttime operation.

Crowley et al. [7] experimentally explored the implementation of a wireless sensor network with Global system for Mobile (GSM) based communication for real-time temperature logging of seafood production. Subsequently, the network was developed and applied to the monitoring of whelk catches from harvest to delivery at the processing plant. The GSM communication was shown to have performed very well, especially in circumstances where problems with poor network coverage were expected to be encountered.

De la Rosa et al. [8] addressed the challenges and trends in the development of web-based distributed Power Quality (PQ) measurement and analysis using smart sensors. Registered users can configure the sensors, adjust sensitivity levels, and specify deployment location and email notification addresses. The developed website also provides a number of ways to view data from single or aggregated monitors. The authors addressed low cost Internet power monitor, which is cost-effective at the single user level. In addition, the reliance on standard web browsers eliminated the need for significant investment in software and hardware infrastructure that is typically required for other measurement systems.

Ong et al. [9] demonstrated existing SCADA with Java-based application in power systems. The authors also addressed the design issue in Graphical User interface (GUI). The proposed Web-based access tool can not only be used for SCADA Systems via intranet, or internet, but also can be readily used for information exchange among market operators via Internet.

Sung et al. [10] designed a test bed for an Internet-based Computer Aided Design (CAD) and Computer Aided Manufacturing (CAM) system. It was specifically designed to be a networked, automated system with a seamless communication flow from a client-side designer to a server side machining service. This includes a Web-based design tool in which Design-for-Manufacturing information and machining rules constrain the designer to manufacturable parts, a geometric representation called SIF-DSG for unambiguous communication between client-side designer and server-side process planner in an automated process planning system with several sub modules that convert an incoming design to a set of tool-paths for execution on a 3-axis Computer Numeric Control (CNC) milling machine.

Altun et al. [11] presented a study on Internet based process control via Internet. The study is to show that any process can be managed remotely with ease. Need for remote managing could appear in health-critical or dangerous conditions, being far away from job, etc. It could be extremely useful for managers to check, administer, or just for taking information as if using visual phone.

The scope of Internet based process control has been clearly specified by Yang et al. [12]. Internet-based control is only an extra control level added to the existing process control hierarchy. The objective is to enhance rather than to replace computer-based process control systems. Six essential design issues have been fully investigated which form the method for design of such Internet-based process control systems. The design issues include requirement specification, architecture selection, web-based user interface design and control over the Internet with time delay, concurrent user access, and safety checking.

Su et al. [13] presented a two WAN model on distribution management system. An integrated DMS consists of networked hardware and software capable of monitoring and controlling the operations of substations and feeders. Building a communication model allows one to determine if leased line capacity or system hardware speeds will cause a bottleneck in the system. The model contains sufficient details about the traffic load and their performance characteristics. WAN modeling is aimed to verify whether hardware design could accommodate the communications load and to avoid overpaying for network equipment. Simulation results indicate that, to cover feeder automation functions, a WAN with distributed processing capability would provide better SCADA performance than an extension of the old centralized system.

2.2   State of the Art

There are many implementation of the classic SCADA system. For example, the SCADA framework that the Lexington-Fayette Urban County Government (LFUCG) [14] depended on to run a system of 80 pump stations and two wastewater treatment plants has worked dependably for almost 20 years. In any case, time was incurring significant damage as new parts were progressively hard to discover and a great part of the framework was out of date. A large number of the current SCADA PCs and HMI programming running the plant's checking framework were old and in the need of substitution. Following a necessary appraisal, LFUCG chose to overhaul the electrical and SCADA frameworks at the Town Branch and West Hickman Creek wastewater treatment plants and supplant every one of the 80 remote terminal units (RTU) at each pump station.

The system mentioned above is shown in Fig. 1. It is bulky and requires special installation expertise to function as intended.

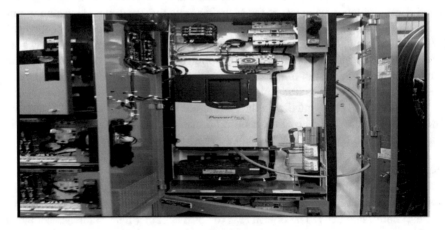

Fig. 1. Smart motor control centers (MCC) [13].

2.3 Industrial Internet of Things (IIoT)

With the increasing popularity of the Internet of Things, the idea came to incorporate SCADA systems with the internet of things. This gave birth to the so called Industrial Internet of Things (IIoT) [14].

IIoT has all of SCADA capacities and the connection between the whole frameworks through the system, in which all gadgets of the framework can gather/trade information with each other. Obviously, this information can be broken down and prepared when SCADA is working.

IIoT is a much cheaper replacement due to the cost reduction in both the price of the equipment and the installation. Additionally, since most of the communication will be done wirelessly, there is another cost reduction from the lack of wired connections.

Moreover, the system can easily incorporate sensors as well as actuator to monitor and control a large spectrum of devices. And there are always new sensors and actuators that are being built to handle the need of the user.

However, as the connection is wireless, security may be an issue as these wireless devices can be hacked remotely if they are not secured well enough.

3 System Architecture

A block diagram of the proposed IoT Industrial wireless controller is shown in Fig. 2.

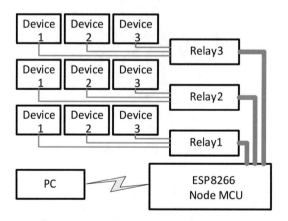

Fig. 2. Block diagram of the SCADA system.

The PC can connect to multiple microcontrollers wirelessly and each of the microcontrollers can connect to multiple devices through relays.

The microcontroller chosen is the ESP8266 Node MCU [15] which is shown in Fig. 3.

NodeMCU is an open source IoT platform. It includes firmware which runs on the ESP8266 Wi-Fi SoC from Espressif Systems, and hardware which is based on the ESP-12 module. The term "NodeMCU" by default refers to the firmware rather than the

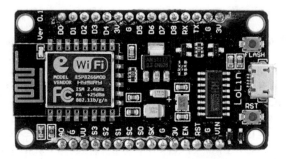

Fig. 3. ESP8266 microcontroller

development kits. The firmware uses the Lua scripting language. It is based on the eLua project, and built on the Espressif Non-OS SDK for ESP8266. It uses many open source projects, such as lua-cjson and SPIFFS. The ESP8266 receives commands from the user through its Wi-Fi module. Based on these commands, it controls the on/off operation of multiple devices. These devices are connected to the microcontroller through a relay which is shown in Fig. 4.

Fig. 4. Relay Module.

Each relay controls 4 devices, and each microcontroller can control up to 4 relays. The aim is to place one microcontroller in a room and have it controlling up to 16 devices per room. Should there be more devices to be controlled; additional microcontroller can be easily added. Additionally, the user can control the devices from a web page that is hosted on the microcontroller. This page can be accessed from any PC with in the same Wi-Fi network. The user will then have the option to switch on or off any device connected to that microcontroller.

An example of a basic webpage is shown in Fig. 5. The figure shows how you can turn on or off four devices that are connected to the microcontroller through a relay.

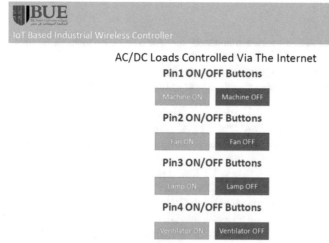

Fig. 5. Example webpage

Figure 6 shows the complete setup in real life. Ideally the microcontroller should be powered by a battery or an independent power source. In the figure, for the sake of simplicity, it is powered by the USB cable from the PC.

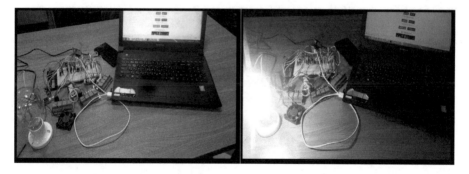

Fig. 6. Complete setup

4 Conclusion

This paper proposed an IoT based industrial wireless controller system. The proposed system was shown to be able to control multiple devices with the very low overhead and infrastructure. It is easy to install with very low cost as it uses basic devices and microcontrollers. Additionally, it can be controlled through the internet which increases the range of control with no boundaries. Moreover, the system is modular, which means that whenever the need arises to control more devices, another module can be added to the existing system without the need to change the whole system.

References

1. Bertocco, M., et al.: A client-server architecture for distributed measurement systems. IEEE Trans. Instrum. Meas. **47**(5), 1143–1148 (1998)
2. Tso, S.K., et al.: A framework for developing an agent-based collaborative service-support system in a manufacturing information network. Eng. Appl. Artif. Intell. **12**(1), 43–57 (1999)
3. Tommila, T., Ventä, O., Koskinen, K.: Next generation industrial automation–needs and opportunities. Autom. Technol. Rev. **2001**, 34–41 (2001)
4. Yang, T.C.: Networked control system: a brief survey. IEE Proc. Control Theory Appl. **153** (4), 403–412 (2006)
5. Kalaitzakis, K., Koutroulis, E., Vlachos, V.: Development of a data acquisition system for remote monitoring of renewable energy systems. Measurement **34**(2), 75–83 (2003)
6. Kimura, T., Kanda, Y.: Development of a remote monitoring system for a manufacturing support system for small and medium-sized enterprises. Comput. Ind. **56**(1), 3–12 (2005)
7. Crowley, K., et al.: Web-based real-time temperature monitoring of shellfish catches using a wireless sensor network. Sens. Actuators A Phys. **122**(2), 222–230 (2005)
8. De La Rosa, J.J.G., et al.: A web-based distributed measurement system for electrical power quality assessment. Measurement **43**(6), 771–780 (2010)
9. Ong, Y.S., Gooi, H.B., Lee, S.F.: Java-based applications for accessing power system data via intranet, extranet and internet. Int. J. Electr. Power Energy Syst. **23**(4), 273–284 (2001)
10. Sung, H.A., et al.: CyberCut: an Internet-based CAD/CAD system. Trans. ASME J. Comput. Inf. Sci. Eng. **1**, 52–59 (2001)
11. Altun, Z.G., et al.: Process control via internet. J. Integr. Des. Process Sci. **5**(2), 111–122 (2001)
12. Yang, S.H., Chen, X., Alty, J.L.: Design issues and implementation of internet-based process control systems. Control Eng. Pract. **11**(6), 709–720 (2003)
13. Su, C.-L., Lu, C.-N., Lin, M.-C.: Wide area network performance study of a distribution management system. Int. J. Electr. Power Energy Syst. **22**(1), 9–14 (2000)
14. https://www.cdmsmith.com/en/Client-Solutions/Projects/Lexington-Fayette-SCADA
15. http://www.nodemcu.com/index_en.html

Wireless Sensor Networks-Based Solutions for Cattle Health Monitoring: A Survey

Mohamed Gameil[1] and Tarek Gaber[2,3,4(✉)]

[1] Electrical Engineering and Information Technology, RUR University Bochum, Bochum, Germany
[2] Faculty of Computers and Informatics, Suez Canal University, Ismailia, Egypt
[3] School of Computing, Science and Engineering, University of Salford, Salford, UK
t.m.a.gaber@salford.ac.uk
[4] Scientific Research Group in Egypt (SRGE), Cairo, Egypt

Abstract. Wireless Sensor Networks (WSN) are nowadays becoming an active research in different fields. Precise irrigation, agriculture, earthquake, fire monitoring in forests and animal health monitoring are few applications of WSN. Animal health monitoring systems (AHMS) are usually used to monitor physiological parameters such as rumination, heart rate, and body temperature. Traditional methods to monitor animal health such as (traditional surveillance, single observation, and simple tabular and graphic techniques) are not efficient to achieve high performance in the large herds' management systems. These methods can only provide partial information and introduce a large cost in staffing and physical hardware. Thus, it is of important need to overcome a foresaid draw-back by using alternative low cost, low power consumption sensor nodes, and providing real-time communications at a sensible hardware cost. The objectives of this paper are: reviewing existing WSN solutions for cattle health monitoring models and determining the requirements needed for building an effective WSN model suitable for cattle health monitoring and detect animal diseases. From this review, requirements of the effective WSN-based solution for cattle health monitoring were suggested.

Keywords: Wireless Sensor Networks · Zigbee · GPS · UART · Cattle monitoring · Animal health monitoring · ZebraNet

1 Introduction

Humans rely on animals for food, fiber, labor and companionship. Thus, it is very crucial to keep these animals healthy and productive. According to the FAO, the world cattle population is estimated to be about 1.5 billion head. Hence cattle management becomes difficult by increasing the number of cow. In large herds, infection with enteric pathogens such as (E. coli) or Salmonella, and

© Springer Nature Switzerland AG 2020
A. E. Hassanien et al. (Eds.): AISI 2019, AISC 1058, pp. 779–788, 2020.
https://doi.org/10.1007/978-3-030-31129-2_71

foot-and-mouth disease is common and associated with poor performance and animal welfare, as well as expensive treatment costs [1,18].

These diseases can spread and infect other animals as well as humans. For these reasons, a system is needed to be in place for continuously monitoring the animal health. Technology is already part of modern farming and is playing an increasing role as more advanced systems and tools become available [1,15,19]. The new concepts and advancement in the technologies nowadays are Internet of Things (IoT). The main idea of IoT is getting real world objects connected with each other forming Wireless Sensor Network (WSN) which would help to control and prevent the eruption of diseases at large scale of cattle management. With the use of sensor, application on mobile phones and the transfer of useful data generated by the system will make it easy to use it [2]. Usually, there are two methods of monitoring animal health: indirect and direct contact (noninvasive) method. In the indirect method:

- Traditional methods (traditional surveillance/single observation) is the practices of disease reporting. Traditionally, this have been based on observation of activity with the naked eye.
- Simple tabular and Graphic techniques that analyze surveillance data, compare current data with some "expected" value and identify how these differ.

In the direct contact method (invasive/Information Technology Methods) of animal health monitoring, there are three methods.

- Video Magnification, is the act of making something look larger than it is, the act of magnifying something or the larger appearance of an object when it is seen through a microscope, telescope, etc. Video Magnification Disease revealing invisible changes in the world allowed you to see subtle changes that cannot be seen with the naked eye like respiratory motion, human pulses to extract heart rate, see invisible (tiny) motion and hear silent sounds [13].
- Location tracking using Image Processing Based on video footage from multiple cameras located in and around a pen, which houses the animals, to extract their location and determine their activity [14].
- Wireless sensor networks (WSN), as depicted in Fig. 1 is spatially distributed, collection of sensor nodes for the purpose of monitoring physical or environmental conditions, such as temperature, sound, pressure, Earthquake and Fire prediction etc, and to cooperatively pass their data through the network to a main location [20,21]. WSN consists of distributed wireless enabled devices that have the ability to handle a variety of electronic sensors. Each node of the WSN called a mote and is accompanied with one or more sensors in addition to a microcontroller, wireless transceiver, and energy source [3,16].

Wireless Sensor Networks are found to be more advantageous over traditional systems as the WSNs-based systems are founded on embedded construction and distributed nature and they are low cost, low power consumption, mesh networking scheme and inherit nature of RF communication transmission of data from one point to another among nodes in a mesh based topology network takes less

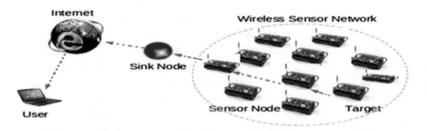

Fig. 1. Collection of sensor nodes for the purpose of monitoring physical or environmental conditions, such as temperature, sound, pressure [3]

energy. WSN has a better coverage than centralized traditional sensing technology [3]. WSNs-based solutions could keep quality of indoor environment that is very important for animal health and welfare which ultimately impacts productivity and quality [17].

Two main standard technologies are usually used in WSN: ZigBee and Bluetooth. Both of these technologies operate within the Industrial Scientific and Medical (ISM) band of 2.4 GHz. This band leads to license-free operations and huge spectrum allocation for compatibility. It is also possible to create a WSN using Wi-Fi (IEEE 802.11) which has high power consumption [3]. The objectives of this paper are reviewing existing WSN solutions for cattle health monitoring models and determining the requirements needed for building an effective WSN model suitable for cattle health monitoring and detect animal diseases.

The rest of this paper is organized as the follows: Sect. 2 presents the related work; Sect. 3 conducts a comparison among some of the existing WSN models. Finally, Sect. 4 concludes the work done on this paper.

2 Related Work

RFID technology is among different electronic means for monitoring animal/cattle health. In RFID, tags or collars were placed on the neck and microphone is incorporated and sounds are analyzed through a complex algorithm inside the tag. However, in many cases monitoring body temperature becomes important. Many new technologies have been introduced to measure body temperature of cattle at various locations including ear, rectum, reticulum-rumen, skin and milk. Most of the existing AHMS models make their decision based on the use of Wi-Fi modules such as Zigbee, Bluetooth or UART and the use of GPS modules such as in ZebraNet or LynxNet modules. Therefore we classified the literature into the following two classes: Zigbee and GPS based animal health solutions.

2.1 Zigbee-Based Animal Health Solutions

Zigbee communication has been used of the development of Animal health monitoring system (AHMS) as it is an energy efficient, high accuracy, self-configuring, and low cost communication technology. Zigbee communication has well-known applications such as environment monitoring, viginet (military), smart farms, smart building, telemedicine services, and other industrial applications. Zigbee module working on the 2.4 GHz band, but data transmits and receives serially through UART. Zigbee module has configured through X-CTU software [1]. Kumar et al. [1] identified and addressed the problem of the continuous rise in air temperature in the troposphere and the variations in temperature that has harmful effect on animal's health leading to diseases such as foot-and-mouth disease and swine fever- these diseases can spread and infect other animals as well as humans. To address these problems, Kumar et al. [1] proposed a prototype tele monitoring system consists of sensing unit and receiving unit with PC which reported the animal health monitoring system with a capability to monitor heart rate, body temperature, and rumination with surrounding temperature and humidity. The sensing unit is consisting of sensor, processor, and ZigBee module. Sensors is used to measure parameters that have been used for different animal species health monitoring. The sensed data of the developed sensor are sent to a host computer through ZigBee module. The values of a foresaid parameters can be displayed on the GUI PC. The design of AHM system is a scalable device. Nadimi et al. [2] proposed a system integrating the control of all the deployed systems in a single system. The central system (Base Station) is the heart of this system as it is responsible for communications between nodes and central server and database management as well as communication with the outer world, Experiment was carried out at Island in Denmark over 5 days with eleven sheep for 9 h per day. The node on the collar and the collar itself were fixed to prevent them from sliding to the right or left. Each sheep in the flock had a wireless sensor measuring and transmitting the head movement acceleration measurements with a sampling rate of 1 Hz. The performance of the handshaking communication protocol and the successful use of acknowledgment messages used to enhance communication reliability. A 2.4-GHz ZigBee-based mobile ad hoc wireless sensor network (MANET) aiming to monitor animal behavior parameters (head movements of each individual sheep in a herd) was successfully designed and established. The deployment of two relay nodes enhanced the network connectivity, and the multi-hop communication and handshaking protocol among the wireless nodes resulted in high communication reliability and low energy consumption. Leena et al. [4] proposed a solution consists of four sections: Raspberry Pi, Accelerometer module, Temperature and humidity sensor modules. This AHMS detects the animal parameters such as rumination, body temperature along with surrounding temperature and humidity. Zigbee protocol is used for data transmission and reception. Raspberry Pi is used as web server and only authorized persons can access the collected data. Raspberry Pi is a basic low cost computer on a single-board. It uses Linux-kernel-based operating systems. Model B+ was upgraded version of Model B which includes an improved power circuitry for attaching high powered USB devices, and switching regulators that can be used to reduce power consumption. Sonia et al. [6] proposed and implemented a disease forecasting system for pigs using a received signal through

ZigBee-based wireless network using a 3-axis acceleration sensor to detect illness at an early stage by monitoring movement of experimentally infected weaned piglet. The movement of infected piglets was altered, and the acceleration sensor could be successfully employed for monitoring pig activity. Accelerometers are sensors that can be used as motion detectors as well as for body position and posture sensing. The overall objective was to investigate to what extent physical activity/movement changes in response to oral infection with S. enteritidis and E. coli, and to establish whether or not monitored behavior altered due to infection can be used as an early sign of pending disease induced by inoculated bacteria. Myeong et al. [7] proposed an effective livestock monitoring system (LMS) using biosensors for cattle health monitoring systems. The monitoring system aims to collect biometric data directly associated with the diseases from an individual entity and prevent them from occurring or spreading. Zigbee module is employed for transmitting the collected biometric data to the forecasting system on WSN. The validity of the system was verified by comparing the results measured by a commercial ECG equipment for cattle to those of LMS in terms of the heartbeat and the breath rate.

2.2 GPS-Based Animal Health Solutions

In this section, we will review two main solution based on GPS technology for animal health problems. There are two main solutions: ZebraNet and LynxNet.

ZebraNet Module. In the ZebraNet-based solutions, GPS devices are mounted on the Zebra to routinely exchange position data with all other devices that fall within their transmission range. If sufficient memory space is available, a user could then download historical position data of multiple animals by approaching a single zebra. A number of solutions based on the ZebraNet model is discussed as follow. Kae et al. [8] have noticed a number of major animal disease outbreaks in the UK which the farming industry is an important sector of its economy. The two most significant incidents were the BSE and FMD outbreak where 4.5 million cows were burnt and over 4 million were killed to stop the spread of these diseases. ZebraNet model was used to track and monitor the health condition of individual cattle activity. However, this model is based on store and forward approach which is not efficient to achieve high performance. So, there is a need for an alternative low cost and low power consumption sensor nodes to support a real-time health monitoring application. To achieve this aim, a particular routing protocol is presented to assist multi hop connectivity that skips the time spent in creating and maintaining plain routing path that led to shorter packet delay. To address the above problem, Tsung et al. [9] suggested a new routing protocol to address the connectivity problem between collars which would lead to an unstable routing path and resulting in increased packet delay. They proposed an Implicit Routing Protocol (IRP) consisting of two phases: configuration and data forwarding. In the configuration phase, the BS periodically send a TIER message throughout entire network contains a BS's ID field, and a hop count field. In the data forwarding phase, if the collar wants to report its measured data back to the base station, it will create a packet containing its current TIER ID and measurement data. This packet is then broadcasted to its vicinity which has a smaller TIER ID. This collar, after acknowledging to the source collar, will broadcast the received packet. This forwarding rule will then be repeated until the data reaches at the BS. Thus, the IRP protocol can reduce the impact of mobility under varying "Off" probability, and number of sensor node. Dukki et al. [10] proposed a cow monitoring system consisting of real-time monitoring device, environment information device, activity parameter device and GPS device. Monitoring control middle ware is video control module, environment control module, monitoring setting module, location awareness control module and activity calculation module. Also, there are monitoring server system that can creates event for cow activity.

LynxNet Module. LynxNet system is based on tracking collars, built around T Mote Mini sensor nodes, sensors, GPS and 433 MHz radio, and stationary base stations, placed at the locations that are visited frequently by the animals. This system is quite similar to ZebraNet but Lynx animal is smaller than a zebra so the latter requires more compact and lightweight solution. Reinholds et al. [11] proposed LynxNet system with extended sensing modality and multi hop delay tolerant communication approach to track Eurasian lynx migration in Latvian forests. The challenge is to achieve long-term operation with a single set of batteries. LynxNet nodes are producing two types of packets. The first type contains GPS location and fix quality information, temperature, relative humidity and amount of ambient light. In this type, one packet is formed once every hour. The second type of packets contains data from 3D accelerometer and 2D gyroscope that can be used to calculate motion vector. Every 5 min, 5 samples of data are gathered, stored in 5 packets to help with the lynx activity classification. These collected data are then analyzed to monitor the lynx health. Similar to LynxNet-based solution, González et al. [12] proposed a method to identify the value of monitoring technologies of cattle grazing tropical pastures. To capture LW of three groups of 20 animals in an experiment of 341 days length, three remote weighing systems were set up at the water troughs. These collected LW data and data from monitored collars, sufficient detail in real-time making can give a valuable insight for early management interventions and right decisions. This consequently would result in an increased production, animal welfare and environmental stewardship. Observations of the data recorded by all three weighing stations throughout the 341 days of the experiment were 41824 observations. This means that there is 35.8% unsuccessful rate of observations between missing EID number and outliers with LW records, i.e. 64.2% success rate.

3   Comparison of WSN Solutions for Cattle Monitoring

From the reviewed solutions above, we can determine the requirements of the effective WSN-based solution for cattle health monitoring:

- Wireless communication module: the device that collect data from different sensors, transmit the biometric data through sink module to the forecasting system.
- Mobile Sensors: measure bio-signals of cattle such as the heartbeat, the breath rate, body temperature, rumination, surrounding temperature, humidity and the momentum.
- Immobile Sensors: environmental fixed sensors such as (thermal camera, video camera, thermistor for surrounding temperature and humidity).
- Energy Consumption: the amount of energy consumed during the system lifetime.
- Cost: the amount of money that has to be paid or given up in order to get the devices.
- Real Experiment/Simulation: is the researchers implemented, simulate their project or just a proposal and how is the accuracy of the result.
- Addressing Security: the prevention of unauthorized access or damage to the forecasting system by applying special polices or protocols.
- Web/mobile system: the ability of controlling and monitoring system remotely.
- Future Enhancement: the future improvements that make the system agreeable.

3.1 ZigBee-Based Solutions

In the following tables, these annotation is used:

- Symbol Y: Means that model did achieve this probability.
- Symbol X: Means that model did not achieve this probability.

Table 1. Comparison among ZigBee-based solutions

	Module (ZigBee)	Mobile sensors	Immobile sensors	Energy consumption	Low coast	Real experiment	Security	Web page	Future enhancement
Kumar et al. [1]	Y	Y	X	Y	Y	X	X	X	Y
Nadimi et al. [2]	Y	Y	X	Y	X	Y	Y	Y	Y
Leena et al. [4]	Y	Y	X	X	Y	X(sim)	X	Y	Y
Sonia et al. [6]	Y	Y	X	X	Y	Y	X	X	X
Myeong et al. [7]	Y	Y	X	Y	Y	Y	X	Y	X

Table 2. Comparison among GPS-based Solutions

	Model		Wireless sensor	Immobile sensors	Energy cons.	Low coast	Real exp.	Security	Web page	Future enh.
	ZebraNet	LynxNet								
Kae et al. [8]	Y	X	Y	X	Y	Y	X	X	X	X
Tsung et al. [9]	Y	X	Y	X	Y	Y	X	X	X	X
Dukki et al. [10]	Y	X	Y	Y	X	X	X	X	X	Y
Reinholds et al. [11]	X	Y	Y	X	Y	X	Y	X	X	X
González et al. [12]	X	Y	Y	Y	X	X	Y	X	Y	X

From Table 1, it can be noticed that the ZigBee module is very helpful device of inexpensive health care of cattle management. It is an energy efficient, high accuracy, self-configuring, and low cost communication technology and no one uses fixed sensors.

3.2   GPS-Based Solutions

Various researchers use ZebraNet to track and monitor the health condition of individual animal activity. GPS position and Zebra Net, both schemes are based on store and forward approach. LynxNet system with extended sensing modality and multi hop delay tolerant communication approach able to achieve long-term operation. Table 2 shows that ZebraNet module is not efficient to achieve high performance.

4   Conclusion

In this paper, we discussed the importance of animals in our life and why we need to keep the animals healthy and productive. Also, we highlighted the traditional methods used for monitoring cattle and how these methods are not efficient to achieve high performance in the large herds management systems. Therefore, we surveyed the solutions that are proposed to address the limitation of traditional methods through Wireless Sensor Networks (WSN) that found to be more advantageous over traditional sensing technology, GPS since the WSNs-based systems are low cost, low power consumption sensor nodes, and providing real-time communications at a sensible hardware cost. Also, a comparison among the proposed solutions were conducted based identified requirements for effective cattle health monitoring system based on WSN. In the future, the security problems would be identified and addressed. This is important as the collected and analyzed should be reliable (not tampered with during its transmission from the sensor nodes to the sink node). Otherwise, the decisions based on these data would not be effective. So security service such as integrity, availability and authentication should be addressed in cattle health monitoring systems.

References

1. Kumar, A., Hancke, G.P.: A zigbee-based animal health monitoring system. IEEE Sens. J. **15**(1), 610–617 (2014)
2. Nadimi, E.S., Jørgensen, R.N., Blanes-Vidal, V., Christensen, S.: Monitoring and classifying animal behavior using ZigBee-based mobile ad hoc wireless sensor networks and artificial neural networks. Comput. Electron. Agric. **82**, 44–54 (2012)
3. Neethirajan, S.: Recent advances in wearable sensors for animal health management. Sens. Bio-Sens. Res. **12**, 15–29 (2017)
4. Narayan, L., Muthumanickam, D.T., Nagappan, D.A.: Animal health monitoring system using raspberry PI and wireless sensor. Int. J. Sci. Res. Educ. (IJSRE) **3**(5), 3386–3392 (2015)
5. Taylor, S.M., Andrews, A.H.: Endoparasites and ectoparasites. In: Andrews, et al. (eds.) Bovine Medicine Diseases and Husbandry of Cattle, 1st edn. Blackwell Scientific Publications, Hoboken (1992)
6. Ahmed, S.T., Mun, H.S., Islam, M.M., Yoe, H., Yang, C.J.: Monitoring activity for recognition of illness in experimentally infected weaned piglets using received signal strength indication ZigBee-based wireless acceleration sensor. Asian-Australas. J. Anim. Sci. **29**(1), 149–156 (2016)
7. Park, M.C., Ha, O.K.: Development of effective cattle health monitoring system based on biosensors. Adv. Sci. Tech. **117**, 180–185 (2015)
8. Kwong, K.H., Wu, T.T., Goh, H.G., Stephen, B., Gilroy, M., Michie, C., Andonovic, I.: Wireless sensor networks in agriculture: cattle monitoring for farming industries. Piers Online **5**(1), 31–35 (2009)
9. Wu, T., Goo, S.K., Kwong, K.H., Michie, C., Andonovic, I.: Wireless sensor network for cattle monitoring system, pp. 173–176 (2009). https://strathprints.strath.ac.uk/14614/. Accessed 25 Jul 2019
10. Kim, D., Yoe, H.: Cow monitoring system based on event using wireless sensor network. Adv. Sci. Technol. Lett. **95**, 173–176 (2015). (CIA 2015)
11. Zviedris, R., Elsts, A., Strazdins, G., Mednis, A.: LynxNet: wild animal monitoring using sensor networks. In: REALWSN 2010. LNCS, vol. 6511, pp. 170–173. Springer-Verlag, Berlin, Heidelberg (2010)
12. González, L.A., Bishop-Hurley, G., Henry, D., Charmley, E.: Wireless sensor networks to study, monitor and manage cattle in grazing systems. Anim. Prod. Sci. **54**(10), 1687–1693 (2014)
13. Takeda, S., Akagi, Y., Okami, K., Isogai, M., Kimata, H.: Video magnification in the wild using fractional anisotropy in temporal distribution. In: Proceedings of the IEEE Conference on Computer Vision and Pattern Recognition, pp. 1614–1622 (2019)
14. Dao, T.K., Le, T.L., Harle, D., Murray, P., Tachtatzis, C., Marshall, S., Michie, C., Andonovic, I.: Automatic cattle location tracking using image processing. In: 2015 23rd European Signal Processing Conference (EUSIPCO), pp. 2636–2640. IEEE, August 2015
15. Molapo, N.A., Malekian, R., Nair, L.: Real-time livestock tracking system with integration of sensors and beacon navigation. Wireless Pers. Commun. **104**(2), 853–879 (2019)
16. Behera, T.M., Mohapatra, S.K., Samal, U.C., Khan, M.S.: Hybrid heterogeneous routing scheme for improved network performance in WSNs for animal tracking. Internet Things **6**, 100047 (2019)

17. Priya, M.K., Jayaram, B.G.: WSN-based electronic livestock of dairy cattle and physical parameters monitoring. In: Sridhar, V., Padma, M., Rao, K. (eds.) Emerging Research in Electronics, Computer Science and Technology, pp. 37–45. Springer, Singapore (2019)
18. Kiani, F.: Animal behavior management by energy-efficient wireless sensor networks. Comput. Electron. Agric. **151**, 478–484 (2018)
19. Moreno-Moreno, C.D., Brox-Jiménez, M., Gersnoviez-Milla, A.A., Márquez-Moyano, M., Ortiz-López, M.A., Quiles-Latorre, F.J.: Wireless sensor network for sustainable agriculture. Multidiscip. Digit. Publ. Inst. Proc. **2**(20), 1302 (2018)
20. Gaber, T., Hassanien, A.E.: An overview of self-protection and self-healing in wireless sensor networks. In: Hassanien, A., Kim, T.H., Kacprzyk, J., Awad, A. (eds.) Bio-Inspiring Cyber Security and Cloud Services: Trends and Innovations, pp. 185–202. Springer, Heidelberg (2014)
21. Gaber, T., Abdelwahab, S., Elhoseny, M., Hassanien, A.E.: Trust-based secure clustering in WSN-based intelligent transportation systems. Comput. Netw. **146**, 151–158 (2018)

Mobile Computing and Networking

Mobile Intelligent Interruption Management: A New Context-Aware Fuzzy Mining Approach

Saad M. Darwish[1], Ahmed E. El-Toukhy[2(✉)], and Yasser M. Omar[2]

[1] Institute of Graduate Studies and Research,
Alexandria University, Alexandria, Egypt
Saad.Darwish@alexu.edu.eg
[2] College of Computing and Information Technology,
AASTMT, Alexandria, Egypt
A.El-Toukhy@outlook.com, Yasser.Omar@aast.edu.eg

Abstract. In recent days, phones recognized as more significant personal communication device for daily life. Usually, ringing notifications are utilized in notifying users on incoming calls. Notifications of inappropriate incoming calls occasionally cause interruptions for users and surrounding people. These unwanted interruptions have a disruptive effect on productivity, employee concentration, and error rate for tasks. A diversity of recommendation approaches for context-aware (e.g., data mining, decision tree, statistics, besides the soft computing) for limiting mobile phone interruptions was presented. However, a mutual problem for current techniques to minimize the interruptions of the mobile phone isn't sufficiently coping with noisy or inconsistency instances that may minimize prediction accuracy. Hence, we are motivated to implement an integrated approach depends upon Bays classifier that classifies noisy cases from training the dataset, and fuzzy logic to manage the nebulizer in mobile phone context situations. The integration methodology implemented through feature-in-decision-out level fusion. In these regards, current work thong to extend a commonly utilized context-based data mining approaches that take out individuals unwavering temporal patterns to fuzzy data mining which might recognize social practices patterns, that might change after some time by supporting reinforcement learning. Simulation and evaluation results on real-life datasets cell phone reveal the efficiency of the suggested model. It achieves an improvement of 5%, 7% and 9% for precision, recall, and f-measure respectively contrasted with traditional systems which include decision tree model and Apriori model.

Keywords: Mobile interruptions · Context-aware systems ·
Mobile user behavior · Prediction model · Fuzzy data mining

1 Introduction

The mobile phone represents a substantially unique tool of individual's day-to-day lifestyle that causes considerably mode exchanged that users cooperate and interconnect with others. These mobiles are enumerated for being connected devices. Typically,

© Springer Nature Switzerland AG 2020
A. E. Hassanien et al. (Eds.): AISI 2019, AISC 1058, pp. 791–800, 2020.
https://doi.org/10.1007/978-3-030-31129-2_72

notifications ring out of cell phones is employed to notify users regarding the received calls. The inform of unsuitable inbound calls from time to time trigger disruptions for both user, and surrounding people [1]. Sometimes these interruptions cause a confusing situation in the official environment (e.g., meetings) but have hard-negative leverage on other activities also as driving a vehicle. These interruptions have negative leverage on worker productivity [2–4].

Recently, several interruption management mobile approaches were dedicated to diminishing the interruptions that mainly vary from all other by modeling kinds of background information taken from user's behavior [1, 5–7]. All those methods are commonly separated into four broad groups: machine learning algorithm (e.g., KNN), optimization techniques (e.g., particle swarm), decision tree algorithm, a data mining algorithm (e.g., Apriori). Still, the critical obstacle of these approximates is mobile ringer mode found rules engaged by applications are not regularly exposed; users must determine and boost rules manually by themselves. Also, these studies did not gaze the behavior's changing that might redirect the ringer mode configuring rules; besides their inability to control nebulizer in phone context situations.

The notion of context, which might be characterized as any information that could be hard-done-to illustrates entity state [8], was vital in perceptualizing intelligent computer models for several years. Models that force utilizes this information are termed, context-aware models. Most of the up-to-date context-aware versions are settled for smartphone, that allows for manipulation of sundry independent contexts afforded by different sources as phone logs. However, contextual data might be executed for the smartphone context-aware system in many ways, in either of that system possibly will suffer difficulties initiated by suspicion of information contextual. The mobile surrounding is extremely, which entails the uncertainty treatment mechanism to regulate itself to quickly altering situations. Although it's not probable in dealing professionally with aleatoric uncertainty (statistical variability), as numerous it's not likely to gain a confident deduction from uncertain data, there is a means to counterbalance this difficulty by suspicion diminishing epistemic (lack of knowledge).

The current smartphones interruption handling systems regularly analyze the context of user's and proactively adjusts the active phone profile according to the learned rules when it recognizes context change. User does not demand to concentrate on style changing because phone adapts to changes it observes. Recently, fuzzy data mining methods utilized to alleviate that problem and handle aleatoric uncertainty. Fuzzy mining methods can get smoother mining results as the utilization of fuzzy membership characteristics. Extracted rules are specified in linguistic expression, that is more natural, and reasonable for human. Another weal of the fuzzy logic method provides classification results, which comprise the degree of possibility.

In order to diminish interruptions brought out by mobile phone, this manuscript suggests an intelligent mobile interruption management model which integrates Bayes classifier to identify and remove additional prediction rules caused by noise instances and fuzzy association mining rule that generates corresponding mode configuring rules. Those rules may handle nebulizer of user's phone log data (user behavior). The system is an intelligent cause of its adaptableness, realization for both contexts, and automatic decision-making capability. In particular, this paper contributions might be summarized as below:

- We regulate a noise threshold with dynamism by investigating an individual's single behavioral patterns, for intention to classify irregularity instances in the dataset. The system adds new rules, removes unnecessary and obsolete rules, and possibly strengthens or weakens existing rules.
- We deduce appropriate instants to limit call interruptions rely upon user inaccessibility by acquiring an individual's primary behavior at several times-of-the-day and days-of-the-week. Herein, utilizing fuzzy linguistic variables and membership degrees to define context situations.

The remaining of this work organized as follow. In Sect. 2, review of some related works. Section 3 presents this proposed model. In Sect. 4, we report some empirical results. Finally, Sect. 5 concludes this paper and shadows the future work.

2 Related Work

Due to the popularity of smartphones with innovative features, and context-aware technology, analyzing real-life mobile phone data in modeling users' behavioral actions was a hot research area currently. Many researchers utilized data extracted from mobiles in modeling and predict user pattern for various uses. However, these versions are not robust as they do not care for issues of noisy or inconsistence instances in mobile phone data while developing a prediction model. As an example, Authors in [9] reviewed a study on user's search behavior to enhance searching relevance rely upon log data. In [10] pressed a scalable model employing multiple sensor data for mining the patterns of user's daily behavioral. Mukherji et al. [11] suggested a smart modeling approach rely upon multi-dimensional context utilizing mobile phone data. Bayir et al. [12] showed a technique for smartphone apps to offer a web-based personalized service. In [13] have suggested a method to build personalization mobile game recommendation system. These approaches utilize mobile data to establish their systems rely upon multi-dimensional contexts. However, they do not consider robustness in building their methods, that we are interested in.

In addition, these approaches, numerous models were suggested for various purposes. For instance, phone call interruptions, and their Management systems. To diminish such disruptions, in [14], designed an interruption management library relies upon several sensors of Android phones. In [9], the author introduced a new technique for smartphone interruptions that maintains mitigation quality under drift concept with long-term usability. The method uses online machine learning and gathers labels for interrupt causing events (e.g., incoming calls) using implicit experience sampling without requesting extra cognitive load on the user's behalf. Besides these, several authors studied the rule-based systems that can manage these interruptions. However, researches do not take in consideration configuring rules produced by automatically analyzing mobile phone log data. For a model, Khalil and others [15] used the information of calendar to set mobile phones accordingly for managing the interruptions automatically. Authors in [16] offered a context-aware mobiles configuration manager to stop phone call without bothering users—authors [17] sophisticated mobile application for minimizing phone disruptions. The smart context-aware interruption management system was modeled by Zulkernain [12]. Valtonen et al. press a context-aware, proactive, and adaptive phone-profile control system, which was given for hands-on testing for conference attendees. The system relies upon fuzzy control and allows automatic control of active phone profile relying on time, place, and weekday.

The previous systems are not data-driven, i.e., the configuring rules utilized by phone apps are not dynamic; besides they are not individualized; it is the main drawback of those approaches. A group of automatized note rules according to the individual's phone call activity patterns can do the interruption management system more efficient and personalized; this interests us. This paper extends traditional rule-based mobile interruption systems with non-fuzzy inputs and fuzzy output by building optimal time segmentation depends upon the individual's current location. This optimal time is utilized to create fuzzy Apriori rules to diminish mobile phone interruption. The main difference between the suggested model and current fuzzy control-based mobile interruption systems lies on its ability to build fuzzy rules automatically; not relying on the user to record the rules.

3   The Proposed Model

In this section, we showed our approach of mobile phone's real-life data of individuals to diminish mobile phone interruption (Fig. 1).

3.1   Noise Reduction

Given the phone log dataset that contains several records; each record contains attributes that include (User ID, Day, Time Stamp, User Location, Phone Call Behavior). In our system, firstly, we identify and eliminate noisy instances from training dataset by determining a dynamic noise threshold that may vary from one-to-another consistent with its unique pattern [8]. So, for getting a noise-free quality dataset, Naive Bayes classifier utilized for noise identification because it determines class conditional probability for the given contextual information to identify dataset inconsistency. As we seek to develop a model founded on multi-dimensional contexts, the assumption utilized in Naive Bayes classifier for zero probabilities is always unreasonable to predict an individual's phone call behavior. Therefore, conditional probability is estimated by Laplace-estimator for contexts. Then, noisy instances are identified by calculating the "noise-threshold" utilizing values of generated probability. As an individual's phone call behavioral patterns are not familiar in the real world, this noise threshold might differ according to the individual's unique behavioral patterns. See [8] for additional details regarding Naive Bayes classifier and Laplace-estimator.

3.2   Optimal Time Segmentation for Locations

In this phase, given the classified mobile log data based on day and location for individual's, we generate the initial time slices per location by dividing each daytime into relatively slight time slices using a small base period. Additionally, we generate behavior-oriented segments. For this, we identify the dominant behavior of each slice

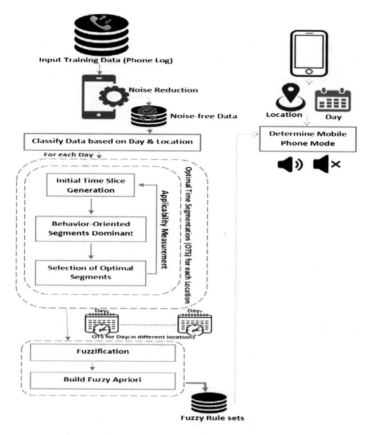

Fig. 1. Proposed interruption management model

that includes both silent and general phone mode proportion instead of the current methods in which the dominant behavior defined as the 'maximum number of occurrences' in a specific action between a list of actions at time slice. The adjacent time slices are then aggregated dynamically with the same dominant characteristics to get more significant segments of similar behavior. These aggregated segments will have more support and temporal coverage and can be utilized as the starting point for mining rules pertinent to the users. Third, we select finest segmentation. For this, we measure the *applicability* for that, which is a descriptive statistic that considers two parameters for a particular confidence threshold. See [18] for extra details. By way of having no prior knowledge around the individual's behavioral patterns, at that point, we iteratively increment the base period (BP * iteration ++) and compare the applicability of the matching segmentation through each iteration to detect the optimal base period. Time segmentation yields the maximum applicability establishes the finest time segmentation, and the corresponding base period is the optimal base period to capture the unrivaled individual's behavioral patterns. Toward the finish of this progression, the optimal segments list $OSeg_{list, location}$ is generated for the optimal base period $BP_{optimal}$ in each day.

3.3   Building Fuzzy Association Rules

Given the $OSeg_{list,location}$ in different days, we build MFs for each time segment (see Fig. 2 to produce the rules of an individual user by employing the well-known association rule mining algorithm.

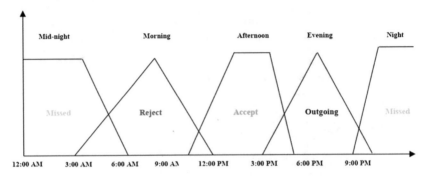

Fig. 2. MFs for each time segment

In general, fuzzy association rules are of the form: $(X$ is $A) \rightarrow (Y$ is $B)$, where X and Y are attributes, and A and B are fuzzy sets that identify X and Y respectively. An example fuzzy association rule is the following:

$$(\text{Time}, \text{night}) \text{ and } (\text{Location}, \text{home}) \rightarrow (\text{Phone\_mode}, \text{silent})$$

Fuzzy logic assigns the degree of membership between 0 and 1 (e.g., 0.3) to each element of a set, permitting for a smooth transition among non-membership and membership of a set. The measures of confidence, support, and lift have been fuzzified for fuzzy association rules. The fuzzy support termed as [21]:

$$\text{Fuzzy Support} \, (<X, A>) = \frac{\textbf{\textit{Sum of votes satisfying}} <X,A>}{\textbf{\textit{number of records in D}}} \tag{1}$$

Let $D = \{t_1, t_2, \ldots, t_n\}$ be the transaction database, and t_i represents the ith transaction in D. Let us define the itemset-fuzzy set pair $<X, A>$ where X is the set of attributes x_j and A is the set of fuzzy sets a_j. A transaction satisfies $<X, A>$ means that the vote of a transaction is calculated by the membership grade of each x_j in that transaction. The membership_grade for attribute a_j in transaction t_i is defined as the transaction is greater than zero. The vote of a transaction is calculated by the membership grade of each x_j in that transaction. The membership_grade for attribute a_j in transaction t_i is defined as:

$$\text{Membership\_grade} \, a_j(t_i[x_i]) = \begin{cases} \text{membership\_function } a_j(t_i[x_i]), \text{if} \geq \text{threshold} \\ 0, otherwise \end{cases} \tag{2}$$

The fuzzy support of a 1-itemset is defined as:

$$\text{Fuzzy Support}\,(<X,A>) = \frac{\sum_{t_i \epsilon T}\left(membership\_grade\ a_i\left(t_i\left[x_j\right]\right)\right)}{number\ of\ records\ in\ D} \tag{3}$$

The fuzzy support of k-itemsets is defined as:

$$= \frac{\sum_{t_i \epsilon T} T\ Normal\left(membership\_grade\ a_j\left(t_i\left[x_j\right]\right)\right)}{number\ of\ records\ in\ D} \tag{4}$$

Where *TNorm* can be any of T-Norm operators [21]: product, minimum, etc. When a frequent itemset <X, A> obtained, fuzzy association rules of the form "If X is A then Y is B" are generated, where $X \subset Z$, $Z = X \cup Y$, $A \subset C$ and $C = A \cup B$. The fuzzy confidence value can is computed as follows:

$$\text{Fuzzy Confidence}\,(<X,A>,<Y,B>) = \frac{Fuzzy\ Support\ of\ <X\cup Y, A\cup B>}{Fuzzy\ Support\ of\ <X,A>} \tag{5}$$

$$\text{Fuzzy Confidence}\,(<X,A>,<Y,B>) = \frac{\sum_{t_i \epsilon T} T\ Norm_{z_i\ \epsilon\ z}\left(membership\_grade\ a_j\left(t_i\left[z_j\right]\right)\right)}{\sum_{t_i \epsilon T} T\ Norm_{x_i\ \epsilon\ x}\left(membership\_grade\ a_j\left(t_i\left[x_j\right]\right)\right)} \tag{6}$$

where $Z = X \cup Y$ and $C = A \cup B$.

For mining fuzzy association rules, Apriori extended to Fuzzy Apriori. The difference between the two algorithms is that Fuzzy Apriori uses definitions of fuzzy support and fuzzy confidence instead of their crisp counterparts used in Apriori. Finally, the extracted fuzzy rules stored in a database. The quality of the proposed model depends mainly on these rules. The above steps have done offline. In the online phase, given the (location, day and time), the model links these data with its fuzzy association rules and taking action based on the related rule.

4 Experimental Results

This section illustrates results for real-life datasets on individual users mobile phone to ensure the robustness of the proposed model. These experiments are conducted on a laptop with following specifications Intel Core i7-6500U Processor, 12 GB system memory, 1 TB hard drive. Benchmark MIT dataset utilized to estimate the model. Herein, prediction accuracy in the form of precision, recall, and f-measure utilized as an evaluation metric. Also, we introduced an experimental valuation comparing our approach with the current approaches in the purpose of predicting call behavior for the phone user.

For showing full effectiveness in the form of model prediction accuracy, Fig. 3 display the relative compare of precision, recall, and f-measure by calculating average results. If observing Fig. 3, we would discover that the proposed model consistently outperforms decision tree and Apriori approaches that predict behavior phone call for individuals in terms of precision, recall, and *f*-measure. The core reason is the current decision tree and Apriori models careless about robustness while user behavior predicting, and resulting accuracy is consequently low (both are unable to hold user context nebulizer, i.e., deal with the crisp situation). In contrast, the proposed model effectively handles noisy instances by analyzing the individual phone users behavioral patterns, so improves accuracy prediction in relevant contexts. Overall, the model is entirely behavior oriented, and individualized. Contrasted with current applicable methods, user behavior accuracy prediction in the form of precision, recall, and *f*-measure improved when our model is applied, as shown in Fig. 3. Although the model is a two-level process, it's useful while demonstrating user behavior using real-life cell phone information. Also, our model does not rely upon any particular contexts number. However, we consider time, and location contexts, as these are relevant to our problem domain. For another problem domain in mobile phones, the pertinent corresponding relevant contexts can use in this model. In general, the average rules number that extracted per user is between 13 to 20, relying on users contextual information.

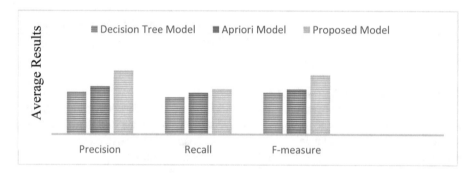

Fig. 3. Effectiveness comparison results.

5   Conclusions

In this work, we proposed a new approach for real-life mobile phone data to diminish cell phone interruption. In this model, we handled some issues effectively as noisy instances from training dataset by determining the threshold of dynamic noise utilizing Naive Bayes classifier, and Laplace estimator, that can vary from user-to-user depending on their unrivaled behavioral patterns. To build a complete model, we also classified dataset noise-free quality rely upon day and location and getting optimal time segmentation to each location. Finally given the $OSeg_{list,location}$ in different days, we build MFs to each time segment to make rules of the individual user by employing the well-known association rule mining algorithm. Extracted fuzzy rules stored in the database; these steps have done offline. Online phase, model links these data with its fuzzy association rules and taking actions rely upon related rule. By utilizing a fuzzy concept, generated rules are more comprehensible for human.

Additionally, fuzzy sets handle numerical values better than current methods for that fuzzy sets soften the impact of sharp boundaries. Experiments on real-life mobile phone data show proposed model effectiveness. In future work, we plan in identifying redundancy of association rules, and to extract a brief group of behavioral association rules which are non-redundant.

References

1. Sarker, I.H.: Silentphone: inferring user unavailability based opportune moments to minimize call interruptions. arXiv:1810.10958 (2018)
2. Spira, J.B., Feintuch, J.B.: The cost of not paying attention: how interruptions impact knowledge worker productivity. Technical report, Basex, USA (2005)
3. Zulkernain, S., Madiraju, P., Ahamed, S.I.: A context-aware cost of interruption model for mobile devices. In: International IEEE Conference on Pervasive Computing and Communications Workshops, USA, pp. 456–460 (2011)
4. Pielot, M., Church, K., Oliveira, R.: An in-situ study of mobile phone notifications. In: 16th International Conference on Human-Computer Interaction with Mobile Devices and Services, Canada, pp. 233–242 (2014)
5. Zulkernain, S., Madiraju, P., Ahamed, S.I., Stamm, K.: A mobile intelligent interruption management system. J. Univ. Comput. Sci. **16**(15), 2060–2080 (2010)
6. Pejovic, V., Musolesi, M.: InterruptMe: designing intelligent prompting mechanisms for pervasive applications. In: ACM International Joint Conference on Pervasive and Ubiquitous Computing, USA, pp. 897–908 (2014)
7. Dekel, A., Nacht, D., Kirkpatrick, S.: Minimizing mobile phone disruption via smart profile management. In: 11th International Conference on Human-Computer Interaction with Mobile Devices and Services, USA, pp. 43–48 (2009)
8. Sarker, I.H.: A machine learning based robust prediction model for real-life mobile phone data. Internet Things **5**, 180–193 (2019)
9. Song, Y., Ma, H., Wang, H., Wang, K.: Exploring and exploiting user search behavior on mobile and tablet devices to improve search relevance. In: International Conference on World Wide Web, Brazil, pp. 1201–1212 (2013)
10. Rawassizadeh, R., Momeni, E., Dobbins, C., Gharibshah, J., Pazzani, M.: Scalable daily human behavioral pattern mining from multivariate temporal data. IEEE Trans. Knowl. Data Eng. **28**(11), 3098–3112 (2016)
11. Mukherji, A., Srinivasan, V.: Adding intelligence to your mobile device via on-device sequential pattern mining. In: ACM International Joint Conference on Pervasive and Ubiquitous Computing, USA, pp. 1005–1014 (2014)
12. Bayir, A.M., Demirbas, M., Cosar, A.: A web-based personalized mobility service for smartphone applications. Comput. J. **54**(5), 800–814 (2010)
13. Paireekreng, W., Rapeepisarn, K., Wong, K.W.: Time-based personalised mobile game downloading. In: Transactions on Edutainment II, pp. 59–69 (2009)
14. Agrawal, R., Srikant, R.: Fast algorithms for mining association rules. In: International Joint Conference on Very Large Data Bases, Chile, pp. 487–499 (1994)
15. Sarker H. I.: Research issues in mining user behavioral rules for context-aware intelligent mobile applications. J. Comput. Sci., 1–11 (2018). Iran
16. Sarker, H.I., Kabir, A.M., Colman, A., Han, J.: An improved naive Bayes classifier-based noise detection technique for classifying user phone call behavior. In: The 2017 Australian Data Mining Conference, Australia (2017)

17. Sarker, H.I., Kabir, A.M., Colman, A., Han, J.: Evidence-based behavioral model for calendar schedules of individual mobile phone users. In: IEEE International Conference on Data Science and Advanced Analytics, Canada, pp. 584–593 (2016)
18. Sarker, H.I., Colman, A., Kabir, A.M.: Individualized time-series segmentation for mining mobile phone user behavior. Comput. J. **61**(3), 349–368 (2018)
19. Sarker, H., Sharmin, M., Ali, A.A., Rahman, M.M., Bari, R., Hossain, M.S., Kumar, S.: Assessing the availability of users to engage in just-in-time intervention in the natural environment. In: The 2014 ACM International Joint Conference on Pervasive and Ubiquitous Computing, USA, pp. 909–920 (2014)
20. Eagle, N., Pentland, S.A.: Reality mining: sensing complex social systems. Ubiquitous Comput. **10**(4), 255–268 (2006)
21. Anna, B.L., Christopher, G.M.: Fuzzy association rule mining for community crime pattern discovery. In: ACM SIGKDD Workshop on Intelligence and Security Informatics, USA (2010)

Hybrid Parallel Computation for Sparse Network Component Analysis

Dina Elsayad[(⊠)], Safwat Hamad, Howida A. Shedeed,
and M. F. Tolba

Faculty of Computer and Information Sciences,
Ain Shams University, Cairo, Egypt
{dina.elsayad, shamad, Dr_Howida}@cis.asu.edu.eg,
fahmytolba@gmail.com

Abstract. The gene regulatory network analysis primary goal is understanding the gene interactions topological order and how the genes influence each other. Network component analysis is a vital technique for build gene regulatory network. However, the network component analysis technique is time consuming and computational intensive. Therefore, parallel techniques are required. PSparseNCA is a parallel network component algorithm. This work present an improved version of PSparseNCA, referred as hPSparseNCA (Hybrid Parallel Computation for Sparse Network Component Analysis). hPSparseNCA uses the hybrid computational model to enhance the performance of PSparseNCA. hPSparseNCA is outperformed PSparseNCA achieving speedup reached 192.77 instead of 36.03 for PSparseNCA on 40 processing nodes. Furthermore, the speedup of the proposed algorithm reached 728.48 when running on 256 processing nodes.

Keywords: Gene data · Bioinformatics · Component analysis ·
High performance · Parallel · Regulatory network

1 Introduction

One of the Bioinformatics research areas is gene data analysis [1]. The main goal of gene data analysis is studying the genes behavior, functions and interaction. Bioinformatics main goal is processing, management, interpretation and analysis of biological data. Microarrays [2] is one of the significant research areas of bioinformatics. The others research areas of Bioinformatics are Protein-protein docking [3], Predictions of protein structure [4], Protein expression analysis [5], High throughput image analysis [6], Comparative Genomics [7], Computational evolutionary biology [8], Modeling biological systems [9], Genome annotation [10], Sequence analysis [11], Analysis of regulation [12], Analysis of mutations in cancer [13].

Microarrays [2] is laboratory tool that use high throughput screening techniques for assessment of biological material. One type of microarrays is DNA microarrays. DNA microarrays used to measure expression levels of hundreds of thousands of genes simultaneously. The DNA microarrays is effective techniques for identification of disease genes, development of new diagnostic tool and gene expression analysis. The

© Springer Nature Switzerland AG 2020
A. E. Hassanien et al. (Eds.): AISI 2019, AISC 1058, pp. 801–808, 2020.
https://doi.org/10.1007/978-3-030-31129-2_73

primary challenges of gene data analysis are huge data size and multi computational tasks that involved in the analysis process [14].

One of the gene data analysis computational tasks is gene regulatory network analysis [15]. The gene regulatory network aims to analyzing the topological organization of genes interactions. The common drawback of the recent gene regulatory network techniques is intensive computation and time consuming. To overcome these limitations parallel techniques are required.

The rest of that paper is organized as follows; the next section provides the required background and related work. The third section provides the proposed algorithm. Furthermore, the implementation and results are discussed in section four. The last section provides the conclusion.

2   Background and Related Work

One of gene data analysis computational tasks is gene regulatory network (GRN) [15]. The goal of gene regulatory network is defining the topological order of genes interaction. In addition to, understanding how the different genes effect each other. The gene regulatory network consists of set of nodes and set of edges. Where, the nodes represent genes or transcription factors while the edges represent regulatory interactions between genes or transcription factors.

As demonstrated in Fig. 1 the gene regulatory network techniques is categorized into four categories: the regression techniques [16–19], mutual information based techniques [20–23], reverse engineering techniques [24–28] and component analysis techniques. Moreover, the component analysis technique is further classified into four sub-categories. These sub-categories are Singular value decomposition (SVD) [29], Principal Component Analysis (PCA) [30], Independent component analysis (ICA) [31–33] and Network component analysis (NCA) [34–40]. The focus of this paper is NCA techniques.

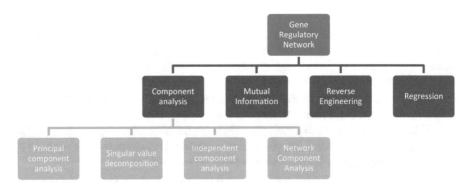

Fig. 1. Different techniques of gene regulatory network

Input	[Y] : Gene Data matrix
Step 1:	Initialize A(0) = I
Step 2:	For n = 1,2,....,N do
Step 2.1:	For j=1,2,..... do
Step 2.1.1:	Update $a_n(j)$ as clarified in eq. (2)
Step 2.1.2:	Update weight vector W as clarified in eq. (5)
Step 3:	Form updated matrix A = [a^T_1 a^T_2 ... a^T_N]^T
Step 4:	Update S = $(A^T A)^{-1} A^T Y$

Fig. 2. SparseNCA algorithm

To overcome the drawbacks and limitations of the others component analysis techniques Chang et al. presented the NCA technique. As expressed in Eq. (1) the NCA mathematical model takes into consideration the data noise (denoted by Γ matrix). The (M × K) [Y] matrix represent the gene expression data. While. The (M × N) [A] is the connectivity matrix and (N × K) [S] is transcription factor (TF) activity matrix. Where, M is number of genes, K is number of samples and N is number of transcription factor. One of the recent algorithms to solve this NCA model is Sparse Network Component Analysis technique (SparseNCA) [40]. As illustrated in Fig. 2, the connectivity matrix [A] is estimated using Eqs. (2), (3) and (4). After that, the transcription factor activity [S] matrix is estimated using Eq. (5). Where, $\in$ is regularizing that has large initial value and decremented every iteration. While, $U_0^\wedge$ is the M-N columns of $U^\wedge$ of singular value decomposition of [Y] matrix.

$$Y = AS + Γ \tag{1}$$

$$a_m(j+1) = \frac{(Q_{11} + \lambda W_m(j))^{-1} 1}{1^T (Q_{11} + \lambda W_m(j))^{-1} 1} \tag{2}$$

$$Q_{11} = H_m^T Q H_m \text{ Where } Q = U_0^\wedge U_0^{\wedge^T} \tag{3}$$

$$H_m = \begin{bmatrix} I_{(N-L_m)} \\ 0_{L_m X(N-L_m)} \end{bmatrix} \tag{4}$$

$$w_{i(j+1)} = \left[\left(a_{mi}^2(j+1) + \in (j+1) \right) \right]^{-1} \tag{5}$$

As well as the other gene regulatory network techniques, the SparseNCA is computational intensive and time-consuming algorithm. Therefore, a parallel version of SparseNCA is introduced that is referred as PSparseNCA (Parallel computation for Sparse Network Component Analysis) [41] (illustrated in Figs. 3 and 4). To enhance speedup of SparseNCA, PSparseNCA uses data parallelism technique.

Input	[Y] : Gene Data matrix
Step 1:	Initialize A(0) = I
Step 2:	Estimate [A] and weight vector W in parallel
Step 2.1:	Send chunk of the data for each processing node
Step 2.2:	Collect data from the processing nodes
Step 3:	Form updated matrix $A = [\, a^T_1 \quad a^T_2 \ldots a^T_N]^T$
Step 4:	Update $S = (A^T A)^{-1} A^T Y$

Fig. 3. PSparseNCA master node

Input	[A] : Connectivity sub matrix
	W : weight sub vector
Step 1:	For n = 1,2,....,N do
Step 1.1:	For j=1,2,..... do
Step 1.1.1:	Update $a_n(j)$ as clarified in eq. (2)
Step 1.1.2:	Update weight vector W as clarified in eq. (5)
Step 2:	Send connectivity sub matrix [X] and weight sub vector W to the master node

Fig. 4. PSparseNCA worker node

From hardware architecture aspect, there are three parallelism categories: shared memory model, distributed model and hybrid model. In shared memory model, the different processing node has a common shared memory. On the other hand, in distributed model each processing node has its own memory. The different processing nodes are communicated using send and receiving messages. Furthermore, the hybrid model is combining the advantage of the others two models. Where, the processing nodes has its own memory and each node uses its private memory as shared one. The next section discusses the hPSparseNCA (Hybrid Parallel Computation for Sparse Network Component Analysis) algorithm that implement hybrid model to enhance the performance of PSparseNCA.

3   HPSparseNCA

One of NCA based techniques for gene regulatory network analysis is SparseNCA. Unfortunately, SparseNCA is time consuming and computational intensive technique. Therefore, PSparseNCA is introduced that is a parallel version of SparseNCA. Moreover, this paper present more enhanced version refereed as hPSparseNCA (Hybrid Parallel Computation for Sparse Network Component Analysis) (shown in Fig. 5). hPSparseNCA implement the hybrid parallelism model instead of the distributed model used in SparseNCA. Where, each processing node receives data block of size N/P. Moreover, each processing node process its own data block in parallel manner using shared memory model. The next section illustrate the performance of the proposed technique.

Input	[A] : Connectivity sub matrix
	W : weight sub vector
Step 1:	Parallel For n = 1,2,....,N do
Step 1.1:	For j=1,2,..... do
Step 1.1.1:	Update $a_n(j)$ as clarified in eq. (2)
Step 1.1.2:	Update weight vector W as clarified in eq. (5)
Step 2:	Send connectivity sub matrix [X] and weight sub vector W to the master node

Fig. 5. hPSparseNCA worker node

4   Implementation and Results

Some experiments are performed to evaluate the performance of the proposed algorithm hPSparseNCA against SparseNCA and PSparseNCA. The hPSparseNCA is implemented using C++ with MPI and OpenMP. While, PSparseNCA is implemented using C++ and MPI. Moreover, SparseNCA is implemented using C++. The experiments were performed on the Bibliotheca Alexandrina High Performance Computing cluster, where each processing node is Intel ® Xeon® CPU E5-2680 v3 @ 2.50 GHz. The performance of the algorithms is compared using three gene datasets. The datasets are available from the GEO database (http://www.ncbi.nlm.nih.gov/) through their accession numbers. Table 1 shows the size (number of genes x number of samples) and accession number of each dataset.

Table 1. Datasets accession number and size

Dataset No.	Accession Number	Size
1	GSE30053	10928×49
2	GSE9195	54675×78
3	GSE6532	54675×88

Table 2 shows the speedup of the PSparseNCA and hPSparseNCA algorithms when running on 40 processing nodes using the three datasets. The table indicate that hPSparseNCA is outperformed PSparseNCA where its speedup reached 193.89 instead of 37.26 for PSparseNCA when running on second dataset. Figure 6 shows the speedup of hPPSarseNCA algorithm when running on different number of processing nodes using the third dataset. As illustrated in Fig. 6 the achieved speedup is super linear speedup reached 728.48 on 256 processing node. Super linear speedup means that the speedup is greater than its limit value, which is the number of used processing nodes (linear speedup). One of super linear speedup reasons is cache usage [42]. Where, the dataset size is greater than the cache memory size. As a result, the data cannot be fully loaded in the cache. Therefore, it is partially loaded which consume time to process all the data. On the other hand, when that dataset is divided among the N processing nodes, the data can be fully loaded in the cache.

Table 2. Speedup of PSparseNCA and hPSparseNCA on 40 processing nodes

Dataset no.	1	2	3
PSparseNCA	24.71	37.26	36.03
hPSparseNCA	287.34	193.89	192.77

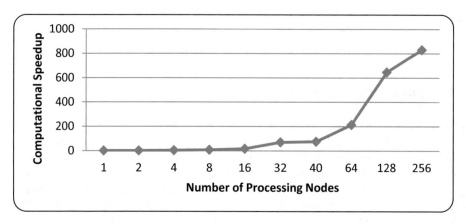

Fig. 6. Speedup of hPSparseNCA using third dataset

5   Conclusion

The inferring of gene regulatory network is one of gene data analysis tasks. The gene regulatory network aims to understanding the topological order of gene interaction. In addition to, determine how the gene effect each other. SparseNCA is a network component based technique for inferring gene regulatory network. Unfortunately, SparseNCA is time consuming and computational intensive technique. Therefore, parallel techniques are needed. PSparseNCA is a parallel version of SparseNCA that use distributed model to enhance the performance of SparseNCA. hPSparseNCA is an improved version of PSparseNCA. hPSparseNCA uses hybrid model which is combing distributed and shared memory model. Where the gene data is divided into blocks each one of size N/P where P is number of processing nodes. Moreover, each processing node handle its block using shared memory model. Three datasets are used to evaluate the performance of hPSparseNCA. The experimental result shows that hPSparseNCA is outperformed PSparseNCA providing speedup 192.77 instead of 36.03 for PSparseNCA using 40 processing nodes. Furthermore, the speedup of the hPSparseNCA algorithm reached 728.48 when running on 256 processing nodes.

References

1. Velculescu, V.E., Zhang, L., Vogelstein, B., Kinzler, K.W.: Serial analysis of gene expression. Science **270**(5235), 484–487 (1995)

2. Muller, U.R., Nicolau, D.V.: Microarray Technology and its Applications. Springer, Heidelberg (2005)
3. Dominguez, C., Boelens, R., Bonvin, A.M.: HADDOCK: a protein-protein docking approach based on biochemical or biophysical information. J. Am. Chem. Soc. **125**(7), 1731–1737 (2003)
4. Shortle, D.: Prediction of protein structure. Curr. Biol. **10**(2), 49–51 (2000)
5. Ghaemmaghami, S., Huh, W.-K., Bower, K., Howson, R.W., Belle, A., Dephoure, N., O'Shea, E.K., Weissman, J.S.: Global analysis of protein expression in yeast. Nature **425** (6959), 737–741 (2003)
6. Dowsey, A.W.: High-throughput image analysis for proteomics. Citeseer (2005)
7. Rubin, G.M., Yandell, M.D., Wortman, J.R., Gabor, G.L., Nelson, C.R., Hariharan, I.K., Fortini, M.E., Li, P.W., Apweiler, R., Fleischmann, W.: Comparative genomics of the eukaryotes. Science **287**(5461), 2204–2215 (2000)
8. Cosmides, L., Tooby, J.: From function to structure: the role of evolutionary biology and computational theories in cognitive neuroscience. The MIT Press (1995)
9. Haefner, J.W.: Modeling Biological Systems: Principles and Applications. Springer, Heidelberg (2005)
10. Kelley, L.A., MacCallum, R.M., Sternberg, M.J.: Enhanced genome annotation using structural profiles in the program 3D-PSSM. J. Mol. Biol. **299**(2), 501–522 (2000)
11. Durbin, R.: Biological Sequence Analysis: Probabilistic Models of Proteins and Nucleic Acids. Cambridge University Press, Cambridge (1998)
12. Janssen, P.J., Jones, W.A., Jones, D.T., Woods, D.R.: Molecular analysis and regulation of the glnA gene of the gram-positive anaerobe Clostridium acetobutylicum. J. Bacteriol. **170** (1), 400–408 (1988)
13. Berrozpe, G., Schaeffer, J., Peinado, M.A., Real, F.X., Perucho, M.: Comparative analysis of mutations in the p53 and K-ras genes in pancreatic cancer. Int. J. Cancer **58**(2), 185–191 (1994)
14. Yang, Y., Choi, J., Choi, K., Pierce, M., Gannon, D., Kim, S.: BioVLAB-microarray: microarray data analysis in virtual environment. In: IEEE Fourth International Conference on eScience (2008)
15. Aluru, S.: Handbook of Computational Molecular Biology. CRC Press, Boca Raton (2006)
16. Pirgazi, J., Khanteymoori, A.R.: A robust gene regulatory network inference method base on Kalman filter and linear regression. PLoS One **13**(7), e0200094 (2018)
17. Lam, K.Y., Westrick, Z.M., Muller, C.L., Christiaen, L., Bonneau, R.: Fused regression for multi-source gene regulatory network inference. PLoS Comput. Biol. **12**(12), e1005157 (2016)
18. Omranian, N., Eloundou-Mbebi, J.M.O., Mueller-Roeber, B., Nikoloski, Z.: Gene regulatory network inference using fused LASSO on multiple data sets. Sci. Rep. **6**, 20533 (2016)
19. Guerrier, S., Mili, N., Molinari, R., Orso, S., Avella-Medina, M., Ma, Y.: A predictive based regression algorithm for gene network selection. Front. Genet. **7**, 97 (2016)
20. Sales, G., Romualdi, C.: parmigene—a parallel R package for mutual information estimation and gene network reconstruction. Bioinformatics **27**(13), 1876–1877 (2011)
21. Zhang, X., Zhao, X.-M., He, K., Lu, L., Cao, Y., Liu, J., Hao, J.-K., Liu, Z.-P., Chen, L.: Inferring gene regulatory networks from gene expression data by path consistency algorithm based on conditional mutual information. Bioinformatics **28**(1), 98–104 (2011)
22. Lachmann, A., Giorgi, F.M., Lopez, G., Califano, A.: ARACNe-AP: gene network reverse engineering through adaptive partitioning inference of mutual information. Bioinformatics **32**(14), 2233–2235 (2016)
23. Barman, S., Kwon, Y.-K.: A novel mutual information-based Boolean network inference method from time-series gene expression data. PloS One **12**(2), e0171097 (2017)

24. Klinger, B., Bluthgen, N.: Reverse engineering gene regulatory networks by modular response analysis-a benchmark. Essays Biochem. **62**(4), 535–547 (2018)
25. Perkins, M., Daniels, K.: Visualizing dynamic gene interactions to reverse engineer gene regulatory networks using topological data analysis. In: 2017 21st International Conference Information Visualisation (IV) (2017)
26. Liu, Z.-P.: Reverse engineering of genome-wide gene regulatory networks from gene expression data. Curr. Genom. **16**(1), 3–22 (2015)
27. de Souza, M.C., Higa, C.H.A.: Reverse engineering of gene regulatory networks combining dynamic bayesian networks and prior biological knowledge. In: International Conference on Computational Science and Its Applications (2018)
28. Reverse engineering and identification in systems biology: strategies, perspectives and challenges. Villaverde, Alejandro F and Banga, Julio R, vol. 11, no. 91 (2014)
29. Holter, N.S., Mitra, M., Maritan, A., Cieplak, M., Banavar, J.R., Fedoroff, N.V.: Fundamental patterns underlying gene expression profiles: simplicity from complexity. Proc. Natl. Acad. Sci. **97**(15), 8409–8414 (2000)
30. Raychaudhuri, S., Stuart, J.M., Altman, R. B.: Principal components analysis to summarize microarray experiments: application to sporulation time series. In: Pacific Symposium on Biocomputing. Pacific Symposium on Biocomputing, NIH Public Access, pp. 455–466 (2000)
31. Hyvarinen, A., Karhunen, J., Oja, E.: Independent component analysis. Wiley, Hoboken (2001)
32. Aapo, H.: Fast and robust fixed-point algorithms for independent component analysis. IEEE Trans. Neural Netw. **10**(3), 626–634 (1999)
33. Liebermeister, W.: Linear modes of gene expression determined by independent component analysis. Bioinformatics **18**(1), 51–60 (2002)
34. Liao, J.C., Boscolo, R., Yang, Y.-L., Tran, L.M., Sabatti, C., Roychowdhury, V.P.: Network component analysis: reconstruction of regulatory signals in biological systems. In: Proceedings of the National Academy of Sciences (2003)
35. Chang, C., Ding, Z., Hung, Y.S., Fung, P.C.W.: Fast network component analysis (FastNCA) for gene regulatory network reconstruction from microarray data. Bioinformatics **24**(11), 1349–1358 (2008)
36. Jayavelu, N.D., Aasgaard, L.S., Bar, N.: Iterative sub-network component analysis enables reconstruction of large scale genetic networks. BMC Bioinform. **16**(1), 366 (2015)
37. Elsayad, D., Ali, A., Shedeed, H.A., Tolba, M.F.: PFastNCA: parallel fast network component analysis for gene regulatory network. In: International Conference on Advanced Machine Learning Technologies and Applications (2018)
38. Shi, Q., Zhang, C., Guo, W., Zeng, T., Lu, L., Jiang, Z., Wang, Z., Liu, J., Chen, L.: Local network component analysis for quantifying transcription factor activities. Methods **124**, 25–35 (2017)
39. Noor, A., Ahmad, A., Serpedin, E., Nounou, M., Nounou, H.: ROBNCA: robust network component analysis for recovering transcription factor activities. Bioinformatics **29**(19), 2410 (2013)
40. Noor, A., Ahmad, A., Serpedin, E.: SparseNCA: sparse network component analysis for recovering transcription factor activities with incomplete prior information. IEEE/ACM Trans. Comput. Biol. Bioinf. **15**(2), 387–395 (2018)
41. Elsayad, D., Hamad, S., Shedeed, H.A., Tolba, M.F.: Parallel computation for sparse network component analysis. In: International Conference on Advanced Machine Learning Technologies and Applications (2019)
42. Ristov, S., Prodan, R., Gusev, M., Skala, K.: Superlinear speedup in HPC systems: why and when? In: 2016 Federated Conference on Computer Science and Information Systems (FedCSIS) (2016)

An Efficient Hybrid Approach of YEAH with Multipath Routing (MP-YEAH) for Congestion Control in High-Speed Networks

Nermeen Adel[1](✉), Ghada Khoriba[2](✉), and Rowayda Sadek[1](✉)

[1] Department of Information Technology, Helwan University, Cairo, Egypt
Nonan626@yahoo.com, Rowayda_sadek@yahoo.com
[2] Department of Computer Science, Helwan University, Cairo, Egypt
Ghada@fcih.net

Abstract. Congestion occurs in any network whenever the load on the network is greater than the capacity of the network. Congestion control concerns in controlling the network traffic in the network, to prevent the congestion to happen. Congestion control trying to keep away from the unfair allocation that happens in the networks and making the suitable resource dropping the rate of packets sent. The paper represents a solution for the high dropping packets problem of YEAH algorithm by using multipath routing. The proposed algorithm is a hybrid approach of YEAH algorithm with multipath routing called Multipath-YEAH (MP-YEAH) which helps the researcher to get better performance than YEAH algorithm. Simulation results were presented for YEAH algorithm and Multipath-YEAH for controlling the congested network with high throughput, reducing delay with 7.3%, reducing energy consumption with 4.1%, and increasing packet delivery ratio with 24%.

Keywords: Congestion · Congestion control · Collision detection · Long Fat Network (LFN) · Multipath routing (MPR)

1 Introduction

Congestion refers to a network state where a node carries so many data that quality of service crumbles resulting in delay or packet loss and the blocking of new connections in the network. Congestion may occur in any network whenever the load on the network is greater than the capacity of the network which proposed by Xu and Zhao [1]. Congestion control focus on controlling the traffic in the network, to prevent the congestion by trying to avoid the oversubscription of the processing and making the proper resource reducing the rate of packets sent which proposed by Ashwini and Sahana [2]. Congestion control field has a lot of tracks to focus on such as mobility, scalability, long distance, and high-speed network to get high QOS [26]. Some researchers work on congestion control and choose to concentrate on mobility and how to control the network in an efficient manner with mobile nodes which proposed by [3, 4]. Other researchers focus on scalability and how to make a network more scalable with high

© Springer Nature Switzerland AG 2020
A. E. Hassanien et al. (Eds.): AISI 2019, AISC 1058, pp. 809–820, 2020.
https://doi.org/10.1007/978-3-030-31129-2_74

congestion control level which proposed by Sun et al. [5]. long distance part in congestion control is also an important track that some researchers work on it to have the network with good performance and controlled even if the nodes are too far from each other which proposed by [6, 7]. Some others concentrate on how to make congestion control in a high-speed network which proposed by [8–10]. The paper deliberates congestion control on a high-speed network. The high-speed network is the network that has a large bandwidth-delay product "BDP". The high-speed network is also known as a long fat network "LFN" or elephant. The network is considered a Long Fat Network if it's BDP is larger than 10^5 bits (12500 bytes). Bandwidth-delay product "BDP" refers to the product of a data link's capacity (bits/second) and its end-2-end delay (seconds) who defined by [11, 28]. The result will be an amount of data measured in bits (or bytes), is equivalent to the maximum amount of data on the network.

Congestion control is an algorithm that dynamically adjusts the rate at which data are sent to reduce the amount of congestion in the network, packet dropping, and delay which defined by Ashwini and Sahana [2]. The congestion control in LFN either basic or hybrid congestion control algorithms. Basic congestion control algorithms are loss based or delay based algorithms such as HTCP, HSTCP, BIC, CUBIC, and STCP. The hybrid congestion control algorithms are such as Africa, Compound, TCP fusion, YEAH, HCCTCP. YEAH algorithm has the highest performance for congestion control but it still suffers from packet loss or packet dropping and this is a big problem. The purpose of the research in this paper is to solve the problem of YEAH in packet losses especially it has the highest performance between congestion control algorithms and this is its only problem that researchers hoping to solve in the future. The paper proposes an efficient Multipath hybrid approach to reduce packet dropping, losses, and delays. The proposed Multipath-YEAH algorithm (MP-YEAH) is based on the original YEAH algorithm to get better performance in throughput, packet dropping and delays. There are applications based on high-speed networks such as interactive education/training, multimedia applications, medical information systems, VOIP applications, and video conferences which defined by Wang et al. [26, 29]. The remainder of this paper is organized as follows: Sect. 2, provides a full description of basic and hybrid congestion control algorithms in LFN. Section 3, explains YEAH algorithm and how it works. Section 4, provides a hybrid approach of YEAH algorithm called Multipath-YEAH (MP-YEAH) and shows the difference between it and the original YEAH algorithm. Section 5, provides a simulation result and discussion. Section 6, concludes the paper.

2   Congestion Control Algorithms in LFN

Congestion control is the algorithm that dynamically adjusts the rate at which data are sent to reduce the amount of congestion in the network, packet dropping, and delay which defined by Ashwini and Sahana [2]. There are two categories of algorithms working on congestion control in LFN which are basic and hybrid congestion control algorithms. The researchers made many experiments to test the performance of basic algorithms for congestion control in LFN and found that cubic has the best performance between basic algorithms which proposed by [9]. The hybrid congestion control algorithms have been tested in many experiments and got that YEAH has the highest performance between all the basic and hybrid algorithms which proposed by [9, 10, 22]. Figure 1 shows congestion control algorithms in LFN.

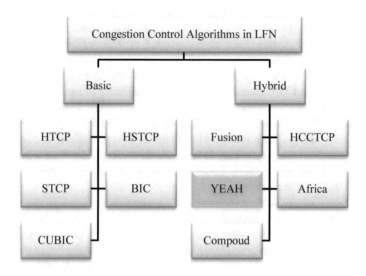

Fig. 1. Congestion control algorithms in LFN

2.1 Basic Algorithms

There are basic algorithms that deal with congestion in high-speed networks and making a good congestion control such as HTCP, STCP [14], BIC [16], CUBIC [18, 27], and HSTCP [12]. Abed et al. [12] discussed that the congestion control algorithms of the Standard TCP limit the congestion windows "Cwnd" in the networks which result in low utilization of network bandwidth. Therefore, they proposed High-speed- TCP (HSTCP) which utilizes a wide range of bandwidths in an effective manner and competes fairly with Standard TCP in congested networks. Nguyen et al. [13] Considered two major challenges while implementing TCP congestion control for the high-speed network. These two major challenges are high bandwidth utilization in the network and fairness with TCP so they proposed (Scalable TCP) to totally focus on fairness and get high utilization in the network.

Alrshah et al. [9] discussed the compatibility ratio of high-speed TCP with Standard TCP while the implementation of high-speed TCP. They Proposed (H-TCP) with focus on friendliness, fairness, and throughput. Hua et al. [15] discussed that previous congestion control algorithms in the high-speed network only solved the scalability of bandwidth and friendliness problems in the network. He indicates another important problem which is RTT unfairness and proposed (BIC TCP) as a solution to this problem in the high-speed networks. Gilkar et al. [17] proposed (CUBIC TCP) which working by using the cubic growth function which has good scalability and stability.

The cubic function is making window RTT rate growth independently when he made an improvement in "TCP-friendliness" and "RTT-fairness".

2.2   Hybrid Algorithms

Alrshah et al. [9] have proposed (TCP- Africa) which is a hybrid algorithm for high-speed networks. While deploying TCP for high BDP "bandwidth-delay product" networks, TCP Africa concentrated on maintaining the balance between the aggressive increase and fairness. Oyeyinka et al. [19] confirmed that delay-based algorithms are not working well when they compete with loss-based algorithms and proposed a hybrid algorithm (Compound TCP) which provides high scalability of bandwidth and provides fairness in the network. Kaneko et al. [20] proposed (TCP fusion) which take the helpful characteristics of TCP-Reno, TCP-Vegas, and TCP-Westwood by using its congestion avoidance mechanism and can get the highest throughput between the other TCP variants. TCP fusion is working in delay mode when the network is not congested and working in losses mode when the network is congested for making a high level of congestion control in the network.

Sheu et al. [21] discussed on high-speed TCP algorithms, although these algorithms made a successful improvement to the utilization of the bandwidth, they still have some weaknesses on their performance such as RTT fairness, TCP-friendliness. They declared that none of the existing algorithms are overwhelmingly better than the other algorithms so the development of new high-speed algorithms is still needed. They proposed (HCC TCP) which suitable for an ideal TCP variant and get high performance on throughput, TCP-friendliness, fairness in high-speed networks. Kushwaha et al. [10] declared while implementing high-speed TCP, it should be considered not only the full link utilization but also a congestion avoidance as it makes the network unstable and non-negligible degradations. They proposed (YEAH TCP) which works to achieve the balance among different opposite requirements.

To the best of our knowledge, most of the recent research papers that made a comparison between congestion control algorithms in high-speed networks in their performance with variant experiment, they proved that YEAH algorithm is the best algorithm for congestion control in his performance with the highest throughput, a low packet dropping, a low delay which proposed by [9, 10, 22, 23].

3   YEAH Algorithm

The main characteristic of YEAH-TCP algorithm is its capability to avoid the congestion in the network not only to achieve the full link utilization and this is the main point in YEAH algorithm. YEAH-TCP algorithm has two different modes: "**Fast mode**" and "**Slow mode**", like Africa TCP which proposed by Baiocchi et al. [22]. During the "**Fast**" mode, YEAH-TCP increments the congestion window "Cwnd" according to STCP algorithm rule. In the "**Slow**" mode, it works as Reno TCP. The decongestion process happens only during the slow mode. The state of the algorithm is decided according to the number of packets in the bottleneck queue. RTT_{base} is the minimum RTT measured by the sender or the source. RTT_{min} is the minimum RTT estimated in the current window of Cwnd packets. $RTT_{queue} = (RTT_{min} - RTT_{base})$ which is the current estimated queuing delay. The value of RTT_{queue} can be used to estimate the number of packets enqueued (Q) which is calculated by Eq. (1), Where G is the goodput "Gput" of the network.

$$Q = RTT_{queue} \cdot G = RTT_{queue} \cdot (RTT/cwnd_{min}) \qquad (1)$$

The ratio L has been evaluated between the queuing RTT "RTT_{queue}" and the propagation delay "RTT_{base}" which is calculated by Eq. (2).

$$L = RTT_{queue} / RTT_{base} \qquad (2)$$

Where L presents the congestion level in the network. RTT_{min} is updated once per window of data. If $Q < Q_{max}$ and $L < 1/\phi$, YEAH algorithm will work in the Fast mode "as STCP" otherwise, it will work in the Slow mode "as RENO". Q_{max} and ϕ are two parameters; Q_{max} is the maximum number of packets that allowed to keep into the buffers and $1/\phi$ is the maximum level of buffer congestion with respect to the bandwidth delay product "BDP". During the "Slow mode", a decongestion algorithm takes place: whenever $Q > Q_{max}$, Cwnd is decreased by Q. Since RTT_{min} is calculated once per RTT "Round Trip Time", the decongestion granularity is one RTT.

Research papers that made a comparison between congestion control algorithms in high-speed networks in their performance with variant experiment, they reach that YEAH algorithm is the best algorithm for congestion control in his performance with the highest throughput, low packet dropping, a low delay which proposed by [9, 10, 22, 23]. YEAH still suffers from packets dropping so in the proposed algorithm Multipath-YEAH works to solve packets dropping problem using multipath routing.

4 The Proposed Multipath-YEAH Algorithm (MP-YEAH)

The proposed algorithm Multipath-YEAH (MP-YEAH) is a hybrid approach of YEAH algorithm with multipath routing. Although YEAH has the highest performance for congestion control in high-speed networks, it still suffers from packet loss or packet dropping and this is a big problem which proposed by Alrshah [9]. This hybrid approach was to solve YEAH problem of packet dropping. In this paper studies were made to know the reason of packet dropping in YEAH algorithm to try to solve it and discover that packet dropping is happening from link failure between a source and destination that might be happening because of a lot of reasons such as that destination is so far from the source or many other reasons which proposed by [24, 25]. Link failure problem effect on the performance, it will effect on throughput and aid to the increasing number of losses or dropping so this problem has been attended to solve by using multipath routing. MP-YEAH algorithm is working such as YEAH in all steps but multipath was added to it for reducing the number of packet dropping.

Multipath routing (MPR) is an effective routing method to achieve load balancing in the network, high congestion reduction, reducing power consumption, and reducing packet losses. Multipath means if there is link failure automatically, it will select another path to send packets. It solves our main problem in YEAH algorithm which is packet dropping in the network. In MP-YEAH algorithm, when the source sends a packet and link failure was detected, the source calculates available paths between him and destination, among that available paths source calculate the shortest path and choose it for data transmission. Forward the packet to the closest neighbor's node and check if ACK received or not. If ACK received, it will start to send the next packet in the flow. If ACK not received, it will back to check link failure again. In another word, source calculates the distance between him and it's neighbor's nodes and chooses the nearest one to him which proposed by [24, 25].

Multipath-YEAH algorithm (MP-YEAH) helps the researchers to reach a better performance than the original YEAH. MP-YEAH algorithm controls the congestion with high throughput, high utilization in the network, and reducing packets losses which is the main problem in the original YEAH. The limitation of the proposed approach is the topology of the network if the nodes were so far from each other; it will take time to check all it's neighbor's nodes and calculates the nearest one and this cause delay in the network. See Fig. 2 flowchart of MP-YEAH algorithm and how it works step by step.

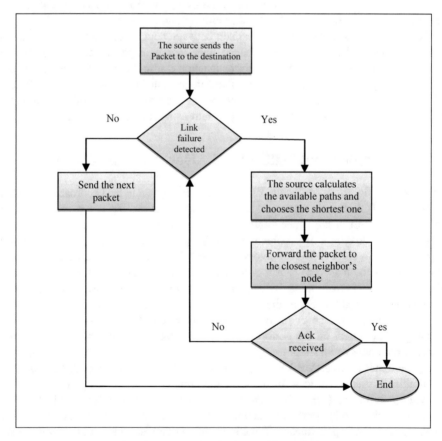

Fig. 2. Flowchart of MP-YEAH algorithm

The number of paths that will be available in the network is depending on two parameters which are the numbers of nodes and the topology of the network. For example, let us see a scenario (1) and scenario (2) know how there is a relation between the number of nodes and the number of paths available in the network. In the first scenario, there are only two nodes in our network so they have only one path to send the packet from the source "sender" to the destination "receiver". In the second scenario, there are four nodes in our network if two nodes were subtracted (source and destination) two nodes will be remained (2 paths) as an option to send the packet through it but the shortest one will be chosen. See Fig. 3 different network scenarios.

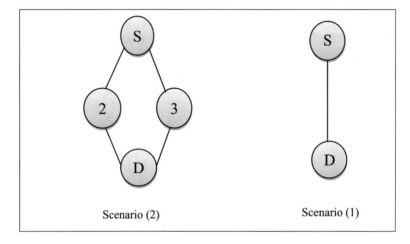

Fig. 3. Different network scenarios

5 Simulation and Result Comparison

In this paper, the simulation was done on ns2.34 simulator using RedHat operating system. The system information is Processor Intel(R) Core(TM) i3 CPU M350 @2.27 GHzm and Installed Memory 3.00 GB. The simulation was done with 50 nodes using a wireless channel, drop tail queue, and Omni antenna. Simulation makes the comparison between YEAH algorithm and our hybrid approach MP-YEAH to test if MP-YEAH has better performance than YEAH or not. Table 1 shows YEAH and MP-YEAH algorithm Comparison in their performance with different measurement criteria such as throughput, delay, energy consumption, average packet delivery ratio, and the number of dropping packets. The result shows that MP-YEAH has better performance with high throughput, decrease delay, reduce power consumption, and solved the main problem that YEAH algorithm faced by reducing the number of packet dropping.

When MP-YEAH was compared with YEAH, the throughput of MP-YEAH is greater than YEAH algorithm and proves that MP-YEAH is better in performance. MP-YEAH at first 2 s every node start to search about their neighbors to calculate their available paths to send their packets to the destination and choose the shortest one that it will send their packets through it if she faced link failure so for this reason no traffic

Table 1. YEAH and MP-YEAH comparison

T	YEAH			MP-YEAH		
	Throughput	Delay	Energy	Throughput	Delay	Energy
4	51	7.4	14.3	208	4.56	9.75
6	66	11.1	14.8	211.6	4.58	9.89
8	71	15.7	15.4	213.1	4.60	10.00
10	70	20.7	15.9	214.2	4.62	10.16
Avg packet delivery ratio	0.475			0.712		
# of packet dropping	494			49		

send at first 2 s so the throughput in 2 s is zero. Figure 4 shows MP-YEAH and YEAH algorithm throughput comparison.

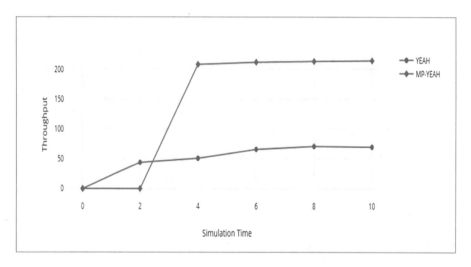

Fig. 4. MP-YEAH and YEAH algorithm throughput

When MP-YEAH was compared with YEAH, the delay of MP-YEAH is lower than YEAH algorithm and prove that MP-YEAH algorithm is better I performance. Figure 5 shows MP-YEAH and YEAH algorithm delay comparison.

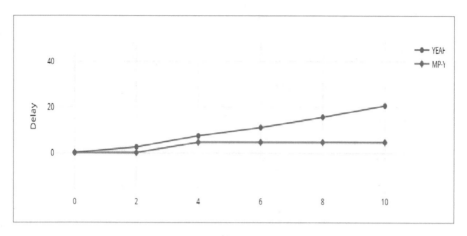

Fig. 5. MP-YEAH and YEAH algorithm delay

When MP-YEAH was compared with YEAH algorithm, the energy of MP-YEAH is lower than YEAH algorithm and Prove that MP-YEAH is better. Figure 6 shows MP-YEAH and YEAH algorithm energy comparison.

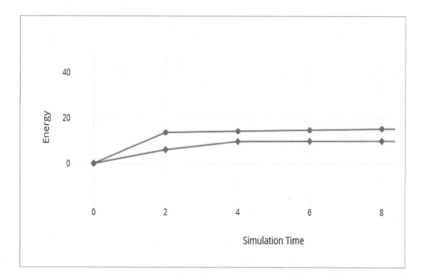

Fig. 6. MP-YEAH and YEAH algorithm energy

If MP-YEAH was compared with YEAH algorithm, the jitter of MP-YEAH is lower than YEAH algorithm and Prove that MP-YEAH is better. Figure 7 shows MP-YEAH and YEAH algorithm jitter delay comparisons which prove the effectiveness of the efficient hybrid approach of YEAH algorithm "MP-YEAH".

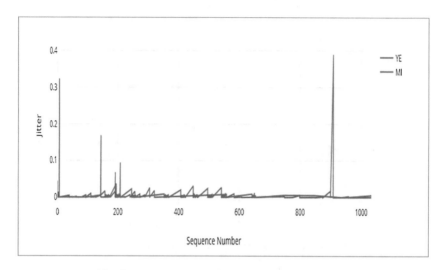

Fig. 7. MP-YEAH and YEAH algorithm jitter delay

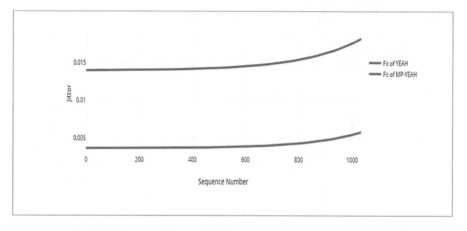

Fig. 8. Exponential trend of MP-YEAH and YEAH jitter delay

For seeing varieties of values between YEAH and MP-YEAH more clearly and see which algorithm achieves small jitter, the exponential fitting trend was made to clear this. Figure 8 shows exponential trend YEAH and MP-YEAH jitter delay.

6   Conclusion and Future Works

In this paper, it has been discussed congestion control algorithms in high-speed networks and knows that the best algorithm in performance is YEAH algorithm but it suffers from packet losses or packet dropping so hybrid approach was made of YEAH called Multipath-YEAH (MP-YEAH). The hybrid algorithm MP-YEAH works like

YEAH in all steps but the multipath technique was merged to it for solving link failure problem that was faced in YEAH and causing a burst of packet dropping. Comparison between MP-YEAH and YEAH prove that MP-YEAH has better performance than YEAH by increasing throughput, reducing delay with 7.3%, reducing power consumption with 4.1%, increasing packet delivery ratio with 24%, and reducing packet loss in the high-speed network and the problems are solved. The limitation of the proposed approach is the topology of the network if there are long distances between nodes; it will take time to check its neighbor's nodes and calculates the nearest one and this cause delay in the network. There are applications based on high-speed networks such as VOIP applications, multimedia applications, medical information systems, and video conferences. The research is still opened in this issue with mobility and scalability and long distance. In future work, the researcher should concern more about mobility and scalability field to get a higher level of congestion control. Deep packet inspection "DPI" gives high quality of service "QOS" in the network because it makes analysis for the whole packet hopping the researchers to merge it with congestion control algorithms in a high-speed network in the future.

References

1. Xu, C., Zhao, J.: Congestion control design for multipath transport protocols: a survey. IEEE Commun. Surv. Tutor. **18**, 1–22 (2016)
2. Ashwini, G.S., Sahana, B.R.: A survey on different congestion control and congestion avoidance mechanisms. Int. J. Eng. Appl. Sci. Technol. **1**(5), 103–105 (2016)
3. Elmannai, W., Razaque, A., Elleithy, K.: TCP-UB: a new congestion aware transmission control protocol variant. Int. J. Comput. Netw. Commun. (IJCNC) **4**(4), 129–141 (2012)
4. Elmannai, W., Razaque, A.: Deployment of TCP university of bridgeport (UB) to control law enforcement department over wireless mesh network. J. Commun. Comput. Eng. **3**(1), 40–47 (2013)
5. Sun, W., Xu, L., Elbaum, S.: Scalably testing congestion control algorithms of real-world TCP implementations, pp. 1–7. IEEE (2018)
6. Baiocchi, A., Mascolo, S., Vacirca, F.: TCP internal buffers optimization for fast long-distance links. Italian Ministry for University and Research (MIUR), pp. 1–5 (2015)
7. Wang, G., Ren, Y., Li, J.: Experimental study of congestion control algorithms in fast long distance network. ICIC Int. **6**(10), 2617–2624 (2012)
8. Qureshi, B., Othman, M., Sabraminiam, S., Wati, N.A.: QTCP: an optimized and improved congestion control algorithm of high-speed tcp networks. In: CCIS, vol. 179, pp. 56–67. Springer, Heidelberg (2011)
9. Alrshah, M.A., Othman, M., Alib, B., Hanapia, Z.M.: Comparative study of high-speed Linux TCP variants over high-BDP networks. J. Netw. Comput. Appl. 1–12 (2016)
10. Kushwaha, V.: Research issues in TCP based congestion control for high-speed network. Int. J. Comput. Sci. Commun. Netw. **4**(1), 18–21 (2016)
11. Kushwaha, V., Ratneshwer: A review of end-to-end congestion control algorithms for high-speed wired network. Int. J. Eng. Res. Technol. **2**(9), 1240–1244 (2013)
12. Abed, J.B., Sinda, L.: Comparison of high-speed congestion control protocols. Int. J. Netw. Secur. Appl. (IJNSA) **4**(5), 15–24 (2012)

13. Nguyen, T.A.N., Gangadhar, S., Sterbenz, J.P.G.: Performance evaluation of TCP congestion control algorithms in data center networks. In: Information and Telecommunication Technology Center, pp. 1–8, June 2016
14. Kharat, P., Kulkarni, M.: Congestion controlling schemes for high-speed data networks: a survey. J. High Speed Netw. **25**, 41–60 (2019)
15. Hua, W., Jian, G.: Analysis of TCP BIC Congestion Control Implementation. In: IEEE International Conference on Computer Science and Service System, pp. 109–112 (2012)
16. Dangi, R., Shukla, N.: A new congestion control algorithm for high speed networks. Int. J. Comput. Technol. Electron. Eng. (IJCTEE) **2**(1), 218–221 (2015)
17. Gilkar, G.A., Sahdad, S.Y.: TCP CUBIC - congestion control transport protocol. Int. J. Multidisc. Acad. Res. (SSIJMAR) **3**(5), 116–120 (2014)
18. Gaya, A.D., Danbatta, H.B., Musa, A.G.: Survey on congestion control in high speed network. Int. J. Adv. Res. Comput. Sci. Softw. Eng. **5**(3), 1237–1244 (2015)
19. Oyeyinka, K.I., Oluwatope, A.O., Akinwale, A.T.: TCP window based congestion control slow-start approach. Commun. Netw. **3**, 85–98 (2011)
20. Kaneko, K., Fujikawa, T., Su, Z., Katto, J.: TCP-Fusion: a hybrid congestion control algorithm for high-speed networks. White Paper, pp. 31–36 (2015)
21. Sheu, J.-.P., Chang, L.-J., Hu, W.-K.: Hybrid congestion control protocol in wireless sensor networks. J. Inform. Sci. Eng. 1103–1119 (2009)
22. Baiocchi, A., Castellani, A.P., Vacirca, F.: YEAH-TCP: yet another high-speed TCP. University of Roma, White Paper, pp. 1–6 (2016)
23. Kushwaha, V.: An analysis of performance parameters for congestion control in high-speed wired network. In: International Conference on Computer and Communications Technology (ICCCT), pp. 133–138, September 2013
24. Pendke, K., Nimbhorkar, S.U.: Study of various schemes for link recovery in wireless mesh network. Int. J. AdHoc Netw. Syst. (IJANS) **2**(4), 51–56 (2012)
25. Mohanapriya, M., Kavitha, P.: Multipath routing and dual link failure recovery in IP fast rerouting. Int. J. Adv. Trends Eng. Technol. **2**(2), 164–170 (2017)
26. Wang, Y., Xu, F., Chen, Z., Sun, Y.: An application-level QoS control method based on local bandwidth scheduling. J. Electr. Comput. Eng. **2018**, 1–10 (2018). Article ID 4576245
27. Al-Saadi, R., Armitage, G., But, J., Branch, P.: A survey of delay-based and hybrid TCP congestion control algorithms. IEEE Commun. Surv. Tutor. 1–30 (2019)
28. Zhang, Y., Zheng, T., Dong, P., Luo, H.: Comprehensive analysis on heterogeneous wireless network in high-speed scenarios. Wirel. Commun. Mob. Comput. **2018**, 1–12 (2018). ID 4259510
29. Ng, J.K.Y., Liu, J.W.S.: Performance of high-speed networks for multimedia applications, pp. 513–522. IEEE (1993)

NCtorrent: Peer to Peer File Sharing Over Named Data Networking Using Network Coding

Aya A. Gebriel, Taha M. Mohamed[(✉)], and Rowayda A. Sadek

Faculty of Computers and Information, Helwan University, Cairo, Egypt
ayaabdallah03@hotmail.com, tahamahdy3000@yahoo.com,
rowayda_sadek@yahoo.com

Abstract. Network coding is a technique which allows nodes to encode and decode the transmitted data which leads to robust network, reduce delay, and increase network throughput. Named Data Networking (NDN) is an entirely a new architecture in which communication is based on the name of the desired content. This paper proposes a new peer to peer system, called NCtorrent, in which network coding is used over named date network. The proposed system allows peers to share large files between each other. The simulation results show that, NCtorrent reduces downloading time, increases cache hit ratio and reduces NACK compared to other previous work. For example, NCtorrent achieves an average time reduction of 36% compared to other peer-to-peer enabled NDN systems. Moreover, the cache hit ratio is increased by 11%. The number of NACK is reduced by 70%. All experimental results are compared to other previous work.

Keywords: Peer-to-peer networks · File sharing · Network coding · Named Data Network

1 Introduction

Content distribution has made remarkable progress over peer to peer (P2P) networks. In P2P networks, each node has the same capabilities and responsibilities. It can be either a client or a server at the same time. There are two types of P2P network: pure peer to peer and hybrid peer to peer network [1]. P2P network enables participating peers to distribute digitized content such as e-books, audio, documents and video. For example, the BitTorrent [2] becomes a very popular peer-to-peer file sharing application. In BitTorrent, a seeder which has the entire shared file splits the file into blocks. Each block has a unique name identified by a sequence number.

Network coding (NC) [3] is a technique in which nodes can encode and decode the transmitted data. This process leads to reduction of delay over the whole network, increase network robustness, and increase network throughput [4, 5]. Network coding

© Springer Nature Switzerland AG 2020
A. E. Hassanien et al. (Eds.): AISI 2019, AISC 1058, pp. 821–830, 2020.
https://doi.org/10.1007/978-3-030-31129-2_75

allows intermediate nodes to encode the received messages generating a decoded output messages. Therefore, the performance of the distributed system is enhanced [6, 7].

Named Data Networking (NDN) [8] is Internet architecture. In NDN, the communication is based on the desired content not the owner of the content. The desired content is requested from content producers by sending an interest packet. These packets contain the name of the requested content. When these interest packets are received by the producer node, it replies by sending a data packet contains the requested content back to the client.

In this paper, we introduce a new P2P system based on the hybridization of NDN and NC. The new proposed system is efficient from the time perspective. Moreover, the cache hit ratio is increased, and the number of NACK is reduced. The rest of the paper is organized as follows: Sect. 2 presents the necessary background. Section 3 shows the related work. Section 4 describes the proposed system. Section 5 shows the simulation results. Finally, Sect. 6 concludes the paper.

2 Background

This section introduces the necessary background of both network coding and NDN. Also, it shows the methods of using network coding over NDN.

2.1 Network Coding

Network coding is a networking technique in which messages are encoded on the sender side and decoded on the receiver side. One of the network coding techniques is the Random Linear Network Coding (RLNC) technique. RLNC enables receivers to decode and recover the original data when enough linearly independent coded symbols are received [8]. Using RLNC enables intermediate network nodes to recode received messages without decoding them first. This increases the probability of sending new linear independent messages to the receiver. So, this recoding ability is considered as an advantage of RLNC [8].

2.2 Named Data Network

In NDN, the consumer nodes express interest (request) packets containing a name that uniquely identify the desired data. Every node in NDN has three tables. The first is the CS content store which caches data packets that pass through the node for possible later requests. The second is the PIT pending interest table. It keeps track of the interests that have not been satisfied yet. The third is the Forwarding Information Base (FIB) table. It stores information about the faces of a node. In NDN, when a router node receives an interest, it replies if it finds a matching in its cs. If not found, the node adds this data packet to the PIT. Then, the node forwards the interest to its neighbor nodes according to the FIB. When the router receives a data packet, it first checks its PIT for an entry that matches the name of the data packet. If no matching, it discards the packet. However, if matching found, the packet is forwarded over all faces specified in the corresponding PIT entry. Then, the router may cache this data packet according to the caching policy used [9].

2.3 Network Coding in NDN

In NDN, the network coding could be applied as follows [10]. Assume that, if a given file has a number of segments. Each segment consists of n data blocks called C1, C2, ..., Cn. The producer encodes the data blocks into encoded packets E1, E2, E3, . . .,En. While V1 . . . Vn represent the coefficient vectors chosen randomly from a finite field. So, the encoding formula is shown in Eq. (1) [10].

$$
\begin{pmatrix} E_1 \\ E_2 \\ \vdots \\ E_n \end{pmatrix} = \begin{pmatrix} V_1^1 & \cdots & V_n^1 \\ \vdots & \ddots & \vdots \\ V_1^n & \cdots & V_n^n \end{pmatrix} \begin{pmatrix} C_1 \\ C_2 \\ \vdots \\ C_n \end{pmatrix} \tag{1}
$$

After encoding, the user starts to request chunks one by one. The consumer appends the coefficient vectors, which already received, in the interest packet. A producer will respond with data packets. These packets contain linearly independent coefficient vectors that are different from those received in the interest packet. So, the consumer will be able to perform the decoding operation. The decoding operation is shown by Eq. (2) [10].

$$
\begin{pmatrix} C_1 \\ C_2 \\ \vdots \\ C_n \end{pmatrix} = \begin{pmatrix} V_1^1 & \cdots & V_n^1 \\ \vdots & \ddots & \vdots \\ V_1^n & \cdots & V_n^n \end{pmatrix}^{-1} \begin{pmatrix} E_1 \\ E_2 \\ \vdots \\ E_n \end{pmatrix} \tag{2}
$$

3 Related Work

Many benefits of network coding are shown in the literature. Linblom et al. [11] present an application for peer-to-peer file synchronization in NDN. Mastorakis et al. [9] present nTorrent application which is a Peer-to-Peer file sharing in Named Data Networking. Chu et al. [12], state that, RLNC has a lot of benefits in P2P content distribution. Chou et al. [8], apply the random network coding by dividing the original data into generations. Gkantsidis et al. [7], apply a random network coding over P2P networks. Lei et al. [10], used network coding technique to achieve better performance for IoT distribution. Saltarin et al. [13], propose a dynamic adaptive streaming over named data network using network coding. Franz et al. [14], discuss the relationship between network coding and security. From the previous survey, according to the best of our knowledge, network coding has not been used with P2P systems over NDN architecture. So, most of the file sharing algorithms are not efficient in the literature.

4   The Proposed System

The proposed system depends on nTorrent system introduced by Mastorakis et al. [9]. In nTorrent, when a peer wants to download a torrent, it should first download a torrent file. The torrent file contains the full names of file manifests. Each file manifest contains the full names of all data packets in this file. So, when a consumer receives the torrent file, it starts to download the file manifest for each file. Then, the consumer sends interests for all data packets contained in this file manifest. However, in the proposed NCtorrent, we use network coding to share files between peers. Moreover, NCtorrent does not use file manifests as opposite to [9]. So, the download time is reduced in our proposed system.

The proposed system share files in NDN using network coding. In NCtorrent, when a peer is interested in a torrent, it should first download the torrent file. The name of a torrent file consists of four components. The first one is the application name. The second is the name of the torrent. The third is the torrent file. The forth one is the implicit digest of the torrent file. The torrent file helps peers to discover file names and the total number of generations for each file as shown in Fig. 1. Each file in the torrent is divided into equal-sized blocks, denoted as generations. Each generation constitutes g symbols of size m where m is the data packet size, and g is called the generation size. Encoding and decoding operations are performed over each generation separately.

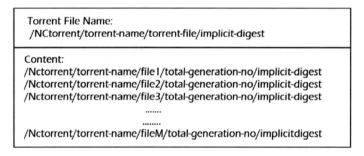

Fig. 1. Simple description for torrent file

For each generation to be decoded successfully, the consumer has to receive g encoded linearly independent data packets. A producer which has the full torrent file divides each file into a set of generations. Then, it creates g network coded data packets for each generation. *Where g is the number of encoded data packets that will be needed to decode the generation.* Each data packet name contains the file number followed by the generation number followed by the data packet number (e.g., /NCtorrent/torrent-name/file-no/gen-no/packet-no). The operations performed at the producer are showed in algorithm 1 listings. As shown in Algorithm 1, the producer starts first to generate torrent file. Each file in the torrent is divided into generations. Then the producer creates encoded data packets for each generation. The time complexity of the algorithm is O(nm) such that n is the number of files and m is the total number of generation.

Algorithm 1 Producer Operations
1: **Generate Torrent File**
2: **foreach** (File in Torrent)
3:    Divide file into generations
4:    **foreach** (generation)
5:      **for** (data packet i < no. of symbols)
6:        create encoded data packet i
7:      **End for**
8:    **End foreach**
9: **End foreach**
10: **output**:
11:    Torrent File, Encoded data packets

A consumer interested in this content must first acquire the torrent file. After receiving the torrent file, the consumer sends g interest packets for each generation for all files in the torrent. The consumer operations are showed in Algorithm 2. As shown in Algorithm 2, the consumer requests the torrent file. For each file, it sends interest packets that are sufficiently to decode each generation in the file. The consumer seeds every received data packet. We choose to apply network coding in NCtorrent this way to avoid modifying the NDN content store or pending interest table as in [13]. So, NCtorrent can be applied over original NDN architecture. A simple description for NCtorrent is shown in Fig. 2.

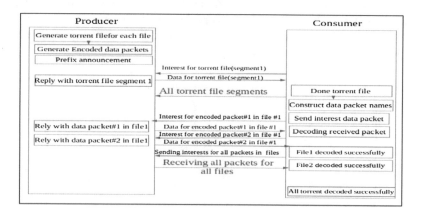

Fig. 2. Simple description of NCtorrent

Algorithm 2 Consumer Operations
1: **Retrieve Torrent File**
2: **Create Interest names**
3: **foreach** (File in Torrent)
4:      **for** (genno i < Total gen numbers)
5:          **for** (interest packet j < no. of symbols)
6:              Send interest packet j
7:              Seed received data packet
8:          **End for**
9:      **End for**
10: **End foreach**
11: **output:**
12:     Sending interest for all data packets

5 The Experimental Results

This section shows the simulation experiments to examine the performance of NCtorrent. Kodo C++ library [15] is used to enable network coding at the producers and the consumers. ndnSIM 2.5 [16, 17] is used to run the simulation scenarios. We examine NCtorrent on different topologies. We run the simulation over two files each with 10 KB size over the topology shown in Fig. 3. The main metrics used are the download time and the cache hit ratio and the number of negative acknowledgements (NACK). The topologies shown in Figs. 3 and 4 are used. The first Topology shown in Fig. 3 which has router node connected to four links. The second topology shown in Fig. 4 has bottleneck links to most of the peers.

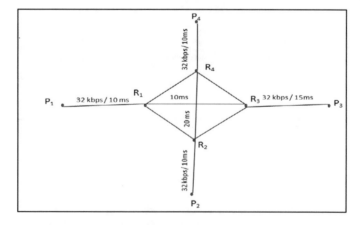

Fig. 3. Topology with router connected to four links

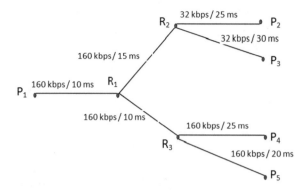

Fig. 4. Topology with router connected to three links.

NCtorrent assumes that consumer is interested in downloading a torrent file consists of two files each of 1 KB size. This torrent file can be satisfied by the producer. After running NCtorrent simulation, we conclude that, NCtorrent reduces the downloading time especially over multicast network. Our torrent file takes 1.9 s-2 s to be downloaded by all consumers, which is less than the downloading time achieved in [18] that ranges between 2.7 s–3.6 s. Moreover, running our simulation on topology shown in Fig. 3, with different number of files, reduces the downloading time by 43% compared to [18]. This comparison is shown in Fig. 5.

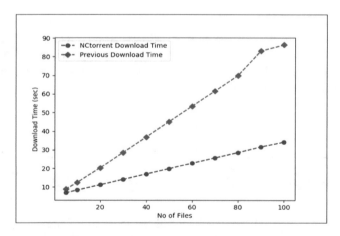

Fig. 5. Number of files vs. download time.

Additionally, we apply NCtorrent over the same topology showed in Fig. 3 to calculate the cache hit ratio using the least frequently used (LFU) caching strategy. In this case, NCtorrent achieves an average increase in the cache hit ratio by 26% compared to [18]. This is illustrated in Fig. 6.

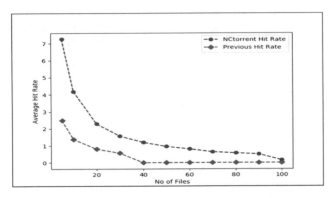

Fig. 6. Number of files vs. cache hit ratio.

Table 1 shows the comparison between NCtorrent and previous work [18] in terms of download time and cache hit ratio when the topology shown in Fig. 4 is used. Again, it is clear from Table 1 that, NCtorrent achieves less download time, higher cache hit ratio, and reduced number of NACK.

Table 1. Simulation result over topology 2

	Download time	Cache hit rate	No. of NACK
NCtorrent	6.4 s	14.3%	7
Previous work [18]	18 s	1.6%	99

Table 2 shows the comparison between NCtorrent and previous work in terms of download time and cache hit ratio when being run on topology shown in Fig. 7 with 25 nodes. Again, it is clear from Table 2 that, NCtorrent achieves less download time, higher cache hit ratio and reduce number of NACK.

Table 2. Simulation Result over Topology 3

	Avg. download time	Cache hit rate	No. of NACK
NCtorrent	1.5%	13.2%	10
Previous work [18]	2.2%	2.5%	120

NCtorrent is also tested over topology with 50 nodes with 2 files each of 1 kB size. Results show that, NCtorrent has the least average download time; the higher cache hit ratio and the least number of NACK. This is illustrated in Table 3.

Table 3. Simulation Result over Topology with 50 nodes

	Avg. download time	Cache hit rate	No. of NACK
NCtorrent	6.7%	6.4%	34
Previous work [18]	16.9%	1.8%	94

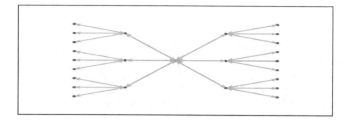

Fig. 7. Topology with 25 nodes [19]

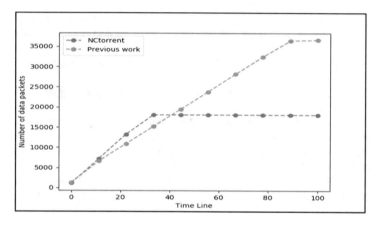

Fig. 8. Congestion

Figure 8 shows the number of successfully transmitted encoded packets per second. We run the simulation over topology shown in Fig. 3 with 100 files. In the first 30 s, the number of transmitted packets are close to each other, however, after that NCtorrent stabilizes. NCtorrent finishes download the entire torrent in the very beginning which leads to reduce network congestion.

Additionally, the goodput, the number of useful data packets delivered through the network, is increased using NCtorrent. This is due to that, NCtorrent uses every received encoded data packet as they are linear independent, which increase the network goodput.

6 Conclusion

In this paper, we introduce a new efficient P2P system called NCtorrent. The proposed system uses network coding over named data network to efficiently disseminate files. The experimental results show that, using the proposed system with multicast strategy reduces the downloading time, increases the cache hit ratio and reduces NACK. For example the proposed system reduces the downloading time by 36% on the average compared to previous work. Moreover, the proposed system increases the cache hit ratio by 11% and decreases the number of NACK by 70%.

References

1. Yang, B., Molina, H.G.: Comparing hybrid peer-to-peer systems. In: Proceedings of the 27th International Conference on Very Large Data-Bases, pp. 561–570, September 2001
2. Cohen, B.: Incentives Build roubustness in bittorrent. In: Workshop on Economics of Peer-to-Peer Systems, vol. 6, pp. 68–72 (2003)
3. Ahlswede, R., Cai, N., Li, S.Y.-R., Yeung, R.W.: Network information flow. IEEE Trans. Inform. Theory **46**(4), 1204–1216 (2000)
4. Fragouli, C., LeBoudec, J.Y., Widmer, J.: On the benefits of network coding for wireless applications. In: 4th International Symposium on Modeling and Optimization in Mobile, Ad Hoc and Wireless Networks. IEEE, Febuary 2006
5. Zhang, L., Afanasyev, A., Burke, J., Jacobson, V., Claffy, K., Crowley, P., Papadopoulos, C., Wang, L., Zhang, B.: Named data networking. SIGCOMM Comput. Commun. Rev. **44** (3), 66–73 (2014)
6. Gkantsidis, C., Miller, J., Rodriguez, P.: Anatomy of a P2P content distribution system with network coding. In: IPTPS (2006)
7. Gkantsidis, C., Rodriguez, P.: Network coding for large scale content distribution. In: IEEE Infocom (2005)
8. Chou, P., Wu, Y., Jain, K.: Practical network coding. In: Proceedings of Allerton Conference on Communication Control and Computing, October 2003
9. Mastorakis, S., Afanasyev, A., Yu, Y., Zhang, L.: nTorrent: peer-to-peer file sharing in named data networking. In: Proceedings of IEEE ICCCN (2017)
10. Lei, K., Zhu, F., Zhong, S., Xu, K., Zhang, H.: An NDN IoT content distribution model with network coding enhanced forwarding strategy for 5G. IEEE Trans. Ind. Inform. **14**, 2725–2735 (2018)
11. Lindblom, J., Huang, M., Burke, J., Zhang, L.: FileSync/NDN: peer-to-peer file sync over named data networking. In: NDN Technical Report NDN-0012, March 2013
12. Chu, X., Kong, H., Jiang, Y.: Random linear network coding for peer-to-peer applications. IEEE Netw. **24**(4), 35–39 (2010)
13. Saltarin, J., Bourtsoulatze, E., Thomos, N., Braun, T.: Adaptive video- streaming with network coding enabled named data networking. IEEE Trans. Multimed. **19**(10), 2182–2196 (2017)
14. Franz, E., Pfennig, S., Fischer, A.: Efficiency of secure network coding schemes. In: Proceedings of Communications and Multimedia Security, pp. 145–159 (2012)
15. Pedersen, M., Heide, J., Fitzek, F.H.P.: Kodo: an open and research oriented network coding library. In: Proceedings of IFIP 10th International Conference on Networking (NETWORK-ING11), Valencia, Spain, pp. 145–152, May 2011
16. Mastorakis, S., Afanasyev, A., Moiseenko, I., Zhang, L.: ndnSIM2: an updated NDN simulator for NS-3. Technical report NDN-0028, Revision 2, NDN, Technical report (2016)
17. NS-3 Simulation Framework. http://www.nsnam.org
18. https://github.com/AkshayRaman/scenario-ntorrent. Accessed 17 Mar 2019
19. http://ndnsim.net/2.2/metric.html. Accessed 17 Mar 2019

Using Resampling Techniques with Heterogeneous Stacking Ensemble for Mobile App Stores Reviews Analytics

Ahmed Gomaa, Sara El-Shorbagy$^{(\boxtimes)}$, Wael El-Gammal,
Mohamed Magdy, and Walid Abdelmoez

College of Computing and Information Technology, Arab Academy for Science,
Technology and Maritime Transport, Alexandria, Egypt
{agomaa_mis,mmagdy,walid.abdelmoez}@aast.edu,
{sara.elshorbagy86,waelgammal}@student.aast.edu

Abstract. Over the past few years, a boom in the popularity of mobile devices and mobile apps has appeared. More than 205 billion apps were downloaded in 2018. Developers directly distribute mobile apps to end users via a centralized platform like the "App Store" for iOS or the "Play Store" for Android. The Mobile app developers get continuous feedback from users' reviews added to these stores. Tools like CLAP or AR-MINER were used to crawl reviews from the stores and try to classify them into lots of classifications like (Bug, required feature, usability issue, performance issue,) to facilitate the categorization of issues or features addition to the developer. Some machine learning techniques is used to get the most accurate data classification to help the developer to classify the reported reviews on the stores. This paper presents a machine learning model that uses the Resampling techniques for handling imbalanced classes in addition to ensemble learning and stacking. The model outperforms those tools and enhances the results applied on Mobile App Stores Reviews Analytics. In addition to that the paper provides experiments applied on different kinds of datasets and showed improvements in accuracy from 85% (previous model) to 90% (our model) and ROC from 96% to 98% especially on the Reviews dataset.

Keywords: Mobile app reviews · Machine learning · Ensemble · SMOTE · Software engineering · Classification · Stacking · Heterogeneous

1 Introduction

During the last few decades, mobile devices have spread widely all over the Globe. In 2018 more than 205 billion apps were downloaded which shows an increase in the number of downloads by 27.3 billion apps than what was downloaded in 2017. It is speculated that the amount of downloads could reach 258 billion downloads by 2022 [1]. Almost all the apps are distributed directly to end users via a centralized platform like the "App Store" for iOS or the "Play Store" for Android. These stores provide a feedback module that helps the mobile applications developers to get continuous feedback from users' reviews. Low star rating reviews have a superior effect on sales

© Springer Nature Switzerland AG 2020
A. E. Hassanien et al. (Eds.): AISI 2019, AISC 1058, pp. 831–841, 2020.
https://doi.org/10.1007/978-3-030-31129-2_76

than high star rating reviews. Subsequently consumers are more expected to respond to complaints and low ratings as well as user reviews can have an extraordinary impact on number of downloads and are an important measure for app success. A number of studies make use of app review analytics to ensure high app ratings and to help developers in understanding user's reviews [2].

Thus, developers have to constantly improve their apps by fixing critical bugs and implementing the most desired features requested from the users in order to compete in the continuously increasing and competitive market of mobile apps [3]. To make an app successful, you need to use the right metrics, best analytic tools, optimize and iterate your apps to your target goals. Unfortunately, not all apps are successfully if we look to the reviews of those apps. There are different kinds of issues that faces these apps like performance, usability, security, etc.

Here comes the need of the developers to look for apps feedback to resolve the reported issues and requested features. But there are tons of issues that are declared every day which lead the developers to either delay or neglect solving them. Analyzing mobile app store reviews is important for app release decision making, as the app store receives hundreds of reviews for each app daily. The app store reviews need to be classified. The need to build a classification model that gives an accurate result is crucial. Thus, there is a need to enhance analysis techniques and choosing the appropriate technique that gives better accuracy.

According to [4] resampling techniques can enhance the imbalanced data problem. Adding to that the inspiration of the idea of minority cancer cells, majority normal class and the imbalanced data class.

In this paper, the same technique was used in app reviews after categorization. It was noticed that the minority group (under-sampled class) in the model did not have enough data to train (validate) and test on. Here came the over- sampling technique SMOTE (Synthetic Minority Oversampling Technique) that will be explained later on. Then came the majority class, which had to be under-sampled to make the numbers of items in each class close to each other so the model can train and validate without bias [5, 6]. After applying those two filtering methods, we apply staking with three heterogeneous classifiers to get the best classification accuracy on the reviews [7]. This paper will provide an experiment, showing that using resampling and ensemble learning with the use of stacking achieve better than the state-of-art model CLAP (Crowd Listener for releAse Planning).

2   Background

In this section, some of the required background in mobile applications is presented. The following mobile metrics helps to ensure mobile app success because either sides whether you are the developer or the user that uses the app you need to know what to expect according to the metrics [8].

Mobile app metrics are very important to companies as it measures their apps success or failure and makes them see the issues to start fixing them. Also, they are the indicators about everything in the app like user engagement, shares, or churn. In addition to that from user reviews (feedback) the app designers and teams can improve the product [10]. Based on a study on mobile app-pairs, mixed-methods research approach to analyze both quantitative and qualitative data is the best way for mobile app success [9].

2.1 Metrics to Ensure Mobile App Success

App Metrics can be classified into four broad kinds: (1) **Performance Metrics:** These IT measures focus on how the user is experiencing the app. Some of the performance metrics are: App crashes, API latency, End-to-end application latency, App Attention span, App load per period, Network errors, (2) **User and usage metrics:** These data points provide visibility into the user and their demographics. Some of these metrics are: Monthly/Daily Active User (MAU/DAU), Devices and OS metrics, Geo metrics, (3) **Engagement metrics:** These metrics highlight how users are engaging with the app. Some of these metrics are Session length, Session interval, Retention rate, and (4) **- Business metrics:** Focus on business (e.g. revenue etc.) flowing through the app. Some of these metrics are Acquisition cost, Transaction revenue, Abandonment rate, LTV (Lifetime value), the mother of all metrics - The app star rating [10].

2.2 Best App Analytics Tools

App analytics are a way to know how your mobile applications works, how to improve it, attracts users and adhere their loyalty. Whether it is Google or Apple tracking their apps performance is a key to a successful business [2]. App analytics can be divided into 3 macro categories. They are (1) **App marketing analytics** which can show comparisons between downloads and in-app purchases adding to that Key Performance Indicators (KPIs), (2) **In app analytics** is everything a user does within the application like how do they behave within your app?, and (3) **- App Performance** Analytics They are in-app, concerned with the "machine" itself and identifying pages that make an app crash and not the user [2].

3   The Proposed Model

Most of the datasets needs preprocessing before starting to use it in some cases it is called cleaning. In this paper the Reviews dataset needed resampling. The proposed model shown in Fig. 1 which was made due to the inspiration of the idea of over or under sampling made in cancer cells detection. Then by using the idea of stacking in [7] by combining different classifiers (Heterogeneous Ensemble) and merging them by Bagging have shown to improve the results over all. Comparing the presented model with Random forest results in accuracy: 0.858, Precision: 0.860, ROC: 0.964 our model has shown to improve the results to reach accuracy: 0.907, Precision: 0.908, ROC: 0.98 which shows that there is a great potential in the model that is explained in the experiment in details. In addition to that it was applied on other 3 datasets from the same source of CLAP, and achieved better results in all the tried datasets.

By applying the model in Fig. 1 on 4 different kinds of datasets, great potentials were shown. Choosing different resampling techniques and heterogeneous ensemble made it possible to reach better results. At first, the accuracy of the data set of reviews improved from 85% to 90%. The second was on functional bug dataset and accuracy was enhanced from 91% to 93%. The third, was on non-functional dataset and accuracy improved from 81% to 86%. The final one, was a non-functional dataset and accuracy increased from 79% to 89%.

4   Experimental Setup

Inspired by the minority cancer cells (the minority class) that needs to be over sampled and the majority class that needs under sampling, came up the idea for the experiment [11]. Using the inspired idea, we applied to the dataset to under sampling and over sampling to have a much better-balanced dataset. Then, we applied Stacking ensemble on three base different classifiers and taking the Bagging as a meta classifier to refer to the Model applied in [7] which results somehow a close progress compared to the technique in [12].

4.1   Dataset Description

CLAP (Crowd Listener for releAse Planning), is used to categorize user reviews based on the information they carry out (e.g., bug, feature, etc.,) collect related reviews together and divide them to groups that are automatically prioritized to be implemented depending on the next app release. CLAP dataset1 (RQ1) contains 3000 preprocessed reviews. After automatically categorizing the dataset using the CLAP tools in [2]. It was divided into 7 categories (Bug reporting, requesting new Feature, Security issue, Performance issue, Usability issue, Energy issue, Other) as shown in Fig. 2. Also, the dataset includes different Applications categories (game, education books, communication and social, entertainment, health and sports, travels and weather, work, system).

4.2   Experiment Evaluation (How to Evaluate It)

After building a classification model, an estimate of how accurately the classifier can predict is needed. Different methods were used to build more than one classification model and compare their accuracy. Several evaluation metrics were used to predict the classifier accuracy such as cross validation, sensitivity (or Recall), Cost-benefit and Receiver Operating Characteristic (ROC) curves [13]. Some of these estimating methods are the true positives, true negatives, false positives, and false negatives are convenient in calculating the costs and benefits (or risks and gains) associated with a classification model. They are the building blocks forming the confusion matrix (a table used to designate the performance of a classification model) used in figuring many evaluation measures and understanding them will make it easy to understanding the significance of the various measures [13].

5   Experimental Results and Discussion

In this experiment WEKA was used to implement the results for these machine learning techniques. WEKA is one of the free tools used to apply any of the built-in huge number of techniques available on demand to be applied on any dataset [11].

In the First phase, RQ1 dataset were used. The applied techniques were targeting to look for another technique than the Random Forest (RF) used in [12] to achieve better results as shown in Table 1A. The use of heterogeneous ensemble with stacking and bagging technique mentioned in [7] but with the use of K-Star, NB and J48 (a kind of tree) new results have been achieved as shown in Table 1B.

In the Second phase, the use of Resampling techniques was needed. By applying oversampling on the initial dataset by the use of Synthetic Minority Oversampling Technique known as SMOTE filtering was used [7]. The implementation of SMOTE in WEKA is only applied on the lowest category (minimum column). To enhance the minimum categories, SMOTE 250% NN5 (nearest neighbor) was used on the proposed ensemble and better results were achieved as shown in Table 1C.

In the Third phase, when analyzing the results in Table 1, we found that the dataset needs to be balanced with another way to avoid biased for the majority classes. So, it was found that the Class Spread Subsample (under sampling filter) in WEKA was used as it produces a random subsample of a dataset. It does not threshold the values that it cuts rather than that it takes random samples, so it does not ruin the variations in the dataset [11]. Then, spread subsample was used at threshold 764 to prevent the deduction of the classified instances as bug reporting while using it with over-sampling using SMOTE 250% which created a more balanced data set. But, unfortunately the results decreased as shown in Table 1D.

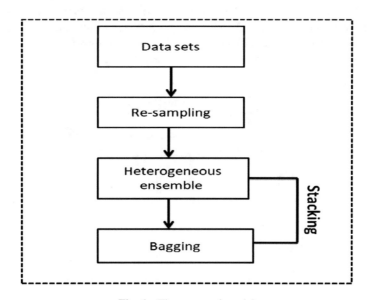

Fig. 1. The proposed model

In the Fourth phase, Checking the new balanced data set with only RF which shows great enhancement as shown in Table 1E than RF without resampling in Table 1A.

Finally, a new mythology was applied by using resampling then a heterogeneous ensemble of k-star, NB and RF as a 3 base classifiers then using a Meta classifier which is bagging all stacked together. That model has outperformed all the previous results as proposed in Table 1E and Fig. 3.

Stacking is a method for selecting multiple sub-models to create the used heterogeneous ensemble. In addition, stacking allows specifying another model (Meta Classifier) to learn how to best combine the predictions from sub-models (Base Classifiers). The base classifiers are trained based on a complete training set, and then the meta-classifier is trained on the outputs of the base level model as features. The base classifiers often consist of different learning algorithms and therefore stacking ensembles are often heterogeneous [7].

Bagging is an ensemble learning used as meta-classifier in the proposed ensemble model. In bagging each model predicts/classifies the target variable independently of all other base classifiers, and then the outputs of the entire base classifiers are combined using weighted average, majority voting, or simple averaging of output values. Bagging combines many simple learners in an ensemble and achieve good results, as well as simplicity of implementation, very quick training, and high interpretability [7]. According to [14] the use of trees (J48 or RF) lowers the bias. And with bagging especially (rep-tree interior by default in WEKA) the problem of overfitting the training data is minimized. In a study by [20] it was shown that the performance of ensembles for automatically classifying app reviews had better performance than individual classifiers.

Because of the imbalanced data problem, the model needed more training on the minimum category for example "security" in our case. That's why we had to do oversampling (SMOTE) on it so the model can have enough data to train on and not to neglect it. While on the other side undersampling (Spread subsample) was needed to minimize the gap between the highest categories and the lowest ones even though it does not always robust the model but balances the dataset.

Table 1. Experiment results

ID	Applied filter, dataset, and ensemble	Machine learning technique	Results								Time
			TP rate	FP rate	Precision	Recall	F-Measure	MCC	ROC area	PRC area	
A	10 Fold Cross-validation, Initial dataset	Random Forest	0.858	0.071	0.860	0.858	0.851	0.795	0.964	0.906	3.63
B	10 Fold Cross-validation, Initial dataset + Stacking	**Base Classifiers** J48, NB & K-Star **Meta Classifier** Bagging	0.857	0.073	0.858	0.857	0.853	0.794	0.955	0.885	527.12

(*continued*)

Table 1. (*continued*)

ID	Applied filter, dataset, and ensemble	Machine learning technique	Results								Time
			TP rate	FP rate	Precision	Recall	F-Measure	MCC	ROC area	PRC area	
C	Oversampling 250% + Stacking	**Base Classifiers** J48, NB & K-Star **Meta Classifier** Bagging	0.892	0.032	0.892	0.892	0.892	0.861	0.979	0.939	558.22
D	Oversampling 250 + under sampling 764 + Stacking	**Base Classifiers** J48, NB & K-Star **Meta Classifier** Bagging	0.855	0.023	0.886	0.885	0.886	0.862	0.980	0.934	385.24
E	Oversampling 250 + under sampling 764	**Random Forest**	0.893	0.023	0.897	0.893	0.894	0.872	0.983	0.941	4.5
F	**Oversampling 250 + under sampling 764 + Stacking**	**Base Classifiers** RF, NB & K-Star **Meta Classifier** Bagging	**0.906**	**0.020**	**0.907**	**0.906**	**0.906**	**0.886**	**0.981**	**0.942**	**397.62**

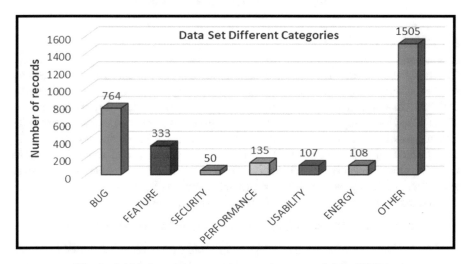

Fig. 2. Initial dataset counts and categories exported from WEKA.

To validate our model, we tried it on other 3 datasets from the same source of CLAP mentioned above and we found that the Stacking section in our model always achieves better results in all the tried datasets. The only drawback in the experiment is that it took more time but by using a GPU (graphical processing unit) in the future we can minimize the time taken dramatically.

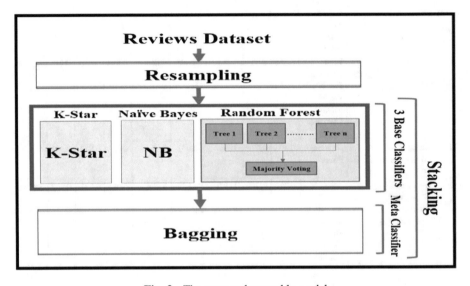

Fig. 3. The proposed ensemble model

6   Related Work

CLAP (Crowd Listener for releAse Planning) an approach to categorize user reviews into seven categories, related reviews clustered in a single request, and prioritize review clusters to be implemented in next app release [12, 15]. While App Review Miner (AR-Miner) (the closest approach to CLAP) is an approach to categorize mobile app reviews into informative and non-informative reviews. Also, AR-Miner was established for prioritization of user reviews [16]. It spontaneously cleans and ranks informative reviews. Once categorized informative from non-informative reviews, AR-Miner groups them into topics and ranks the groups of reviews by priority.

The main differences between AR-Miner and CLAP are:

1. **Bug/new feature reviews vs. informative/non- informative reviews**

CLAP shows developers the category to which each review belongs (e.g. bug reporting vs suggesting new feature), while AR-Miner only differentiates between informative and non-informative reviews. In AR-Miner a specific topic (e.g., a topic referred to a specific app's feature) could specify suggestions to improve features and bugs reports. While in CLAP clustering of reviews is implemented independently between the different review categories

2. Recommending next release features/fixes vs. ranking reviews

To prioritize the clusters to be implemented in the next app release, CLAP uses a machine learner (to learn from the actual decisions made by developers over the change history of their app). On the other hand, the importance of reviews ranked by AR-Miner is based on the prioritization score. In predicting the items that will be deployed by developers in the next release CLAP outperforms AR-Miner. Finally, the bug/feature taxonomy certificates the use of different prioritization models for different kinds of change requests.

Guzman, Maalej, and Harman et al. [17, 18] concentrated on extracting features from user reviews and to find out how users like a certain feature by applying a sentiment analysis. A large data input to train a model was not needed by Guzman et al. approach. Harman et al. [18] introduced an approach to extract features from app store descriptions targeting more clustering for comparison purposes. Elements such as near Wi-Fi, hotspot was considered as features; however, this approach is looking for fine-grained features such as upload pdf file or edits photos.

While Johann et al. [19] developed SAFE, a simple, uniform approach to extract and match the app features from the app descriptions written by developers and the app reviews commented by app users. According to [21] AUREA makes analysis easier by using fine grained mobile specific topics to classify reviews so developers reach the direct and important feedbacks to enhance maintenance.

7 Conclusion and Future Work

This paper gave a brief overview on the metrics and analytics tools used for mobile app store analytics to ensure mobile success. The paper focused on comparing the new method results with CLAP results that used RF and other approaches. Imbalanced data is a problem in many datasets which leads to biased results or even neglecting a categorization. Using Resampling has showed a great potential in training the model. Adding to that the great potentials of combining the heterogeneous ensemble model with stacking and not voting. Using Bagging, decreases the overfitting problem in addition to that the use of trees decreases bias. The results exceeded the other models according to our knowledge. In the future the use of GPU will save more time in the experiments. The existence of a lot of different over and under sampling techniques, that needs more time to investigate and see the different results [3, 6]. The five different boosting based techniques: SMOTEBoost, DATABoost–IM, EUSBoost, DATABoost–IM with SVM, and boosting support vector machines will be proposed with ensemble techniques to improve the minority class classification (imbalanced data problem).

References

1. Number of mobile app downloads worldwide in 2017, 2018 and 2022 (in billions). https://www.statista.com/statistics/271644/worldwide-free-and-paid-mobile-app-store-downloads/. Accessed 14 Mar 2019

2. Furlan, A.: Best App Analytics Tools. Business of Apps (2017). Accessed 14 Oct 2019
3. Menzies, T., Williams, L., Zimmermann, T.: Perspectives on Data Science for Software Engineering. Morgan Kaufmann (2016). [ref 10] book 2 citations
4. Punlumjeak, W., Rugtanom, S., Jantarat, S., Rachburee, N.: Improving classification of imbalanced student dataset using ensemble method of voting, bagging, and adaboost with under-sampling technique. In: IT Convergence and Security 2017, pp. 27–34. Springer, Singapore (2017). https://doi.org/10.1007/978-981-10-6451-7_4
5. Bilogur, A.: Undersampling and oversampling imbalanced data. https://www.kaggle.com/residentmario/undersampling-and-oversampling-imbalanced-data. Accessed 14 Oct 2018
6. Blagus, Rok, Lusa, L.: Joint use of over- and under-sampling techniques and cross-validation for the development and assessment of prediction models. BMC Bioinform. (2015). https://doi.org/10.1186/s12859-015-0784-9
7. El-Shorbagy, S.A., El-Gammal, W.M., Abdelmoez, W.M.: Using SMOTE and heterogeneous stacking in ensemble learning for software defect prediction. In: ICSIE 2018 Proceedings of the 7th International Conference on Software and Information Engineering, pp. 44–47 (2018)
8. Kearl, M.: 10 Essential Mobile App KPIs And Engagement Metrics (And How To Use Them), 9 March 2016. https://www.braze.com/blog/essential-mobile-app-metrics-formulas/. Accessed 23 Mar 2019
9. Ali, M., Joorabchi, M.E., Mesbah, A.: Same app, different app stores: a comparative study. In: 2017 IEEE/ACM 4th International Conference on Mobile Software Engineering and Systems (MOBILESoft), Buenos Aires, pp. 79–90 (2017). https://doi.org/10.1109/mobilesoft.2017.3
10. APPDYNAMICS: 16 metric to ensure app success (2016). https://www.appdynamics.com/media/uploaded-files/1432066155/white-paper-16-metrics-every-mobile-team-should-monitor.pdf
11. Frank, E., Hall, M.A., Witten, I.H.: The WEKA Workbench. Online Appendix for "Data Mining: Practical Machine Learning Tools and Techniques", 4th edn. Morgan Kaufmann (2016)
12. Scalabrino, S., Bavota, G., Russo, B., Di Penta, M., Oliveto, R.: Listening to the crowd for the release planning of mobile apps. IEEE Trans. Softw. Eng. (2017). https://doi.org/10.1109/tse.2017.2759112
13. Han, J., Kamber, M., Pei, J.: Data Mining: Concepts and Techniques, 3rd edn. Morgan Kufmann (2012)
14. Brownlee, J.: Bagging and Random Forest Ensemble Algorithms For Machine Learning, 22 April 2016. https://machinelearningmastery.com/bagging-and-random-forest-ensemble-algorithms-for-machine-learning/. Accessed 14 Mar 2019
15. Villarroel, L., Bavota, G., Russo, B., Oliveto, R., Di Penta, M.: release planning of mobile apps based on user reviews. In: IEEE/ACM 38th International Conference on Software Engineering (ICSE), Austin, TX, pp. 14–24 (2016). https://doi.org/10.1145/2884781.2884818
16. Chen, N., Lin, J., Hoi, S.C.H., Xiao, X., Zhang, B.: AR-miner: mining informative reviews for developers from mobile app marketplace. In Proceedings of the 36th Inter-national Conference on Software Engineering, ICSE 2014, pp. 767–778 (2014)
17. Guzman, E., Maalej, W.: How do users like this feature? A fine grained sentiment analysis of app reviews. In: 2014 IEEE 22nd International Requirements Engineering Conference (RE), Karlskrona, pp. 153–162 (2014). https://doi.org/10.1109/re.2014.6912257

18. Harman, M., Jia, Y., Zhang, Y.: App store mining and analysis: MSR for app stores. In: 2012 9th IEEE Working Conference on Mining Software Repositories (MSR), Zurich, pp. 108–111 (2012). https://doi.org/10.1109/msr.2012.6224306
19. Johann, T., Stanik, C., Maalej, W.: SAFE: A simple approach for feature extraction from app descriptions and app reviews. In: 2017 IEEE 25th International Requirements Engineering Conference (RE), Lisbon, pp. 21–30 (2017). https://doi.org/10.1109/re.2017.71
20. Guzman, E., El-Haliby, M., Bruegge, B.: Ensemble methods for app review classification: an approach for software evolution (N). In: 2015 30th IEEE/ACM International Conference on Automated Software Engineering (ASE), Lincoln, NE, pp. 771–776 (2015). https://doi.org/10.1109/ase.2015.88
21. Ciurumelea, A., Panichella, S., Gall, H.C.: Poster: automated user reviews analyser. In: 2018 IEEE/ACM 40th International Conference on Software Engineering: Companion (ICSE-Companion), Gothenburg, pp. 317–318 (2018)

TCAIOSC: Trans-Compiler Based Android to iOS Converter

David I. Salama[1]([⊠]), Rameez B. Hamza[1], Martina I. Kamel[1],
Ahmad A. Muhammad[2], and Ahmed H. Yousef[1,2]

[1] Department of Computer and Systems, Faculty of Engineering,
Ain Shams University, Cairo, Egypt
davidibrahim8888@gmail.com, rameez.barakat@gmail.com,
martina.ihab96@gmail.com, ahassan@eng.asu.edu.eg
[2] School of Information Technology and Computer Science, Nile University,
Giza, Egypt
{ah.mohammed,ahassan}@nu.edu.eg

Abstract. Cross-platform development is the practice of developing software products or services for multiple platforms or software environments. The idea of cross-platform development is that a software application or product should work well in more than one specific digital habitat. This capability is typically pursued in order to sell software for more than one proprietary operating system. In general, cross-platform development can make a program less efficient. However, in many cases, the makers of software figured out that the limitations of cross-platform development are worth dealing with in order to offer an application or product to a wider set of users. This paper discusses the development of a code converter from Android to iOS and vice versa through trans-compiler approach, the challenges encountered in the process and the outcomes resulted by this attempt.

Keywords: Cross-platform mobile development · Code conversion ·
Code reuse · Generated apps · Intelligent code conversion ·
Smart code generation

1 Introduction

With the increasing number of users of smart mobile devices, mobile app development is now highly demanded to produce apps in different fields and serve the needs of the already huge and increasing number of mobile users. Due to the variety of smartphones and different vendors, smartphones can operate on different platforms. There are many platforms such as Android, iOS, Windows Phone, etc. Given the differences between platforms, with different tools, programming languages and libraries provided by each platform vendor, the developer has to develop the exact same application for every platform to cover as many users as possible, hence arises the challenge of developing a reliable mobile app for the different platforms. Therefore, a cross platform development tool is a good possible solution.

© Springer Nature Switzerland AG 2020
A. E. Hassanien et al. (Eds.): AISI 2019, AISC 1058, pp. 842–851, 2020.
https://doi.org/10.1007/978-3-030-31129-2_77

In [1], El-Kassas et al. presented many approaches and solutions to solve the problem of building mobile applications on several platforms. It also included some new approaches such as Cloud Based Approach (where you run the code on the cloud rather than running it on your local machine), Merged Approach and Component Based Approach. In [2], Latif et al. also proposed many solutions to cross-platform applications. In this paper they emphasize on the model driven Approach (MDA) which is used to design a user interface once for multiple platforms and then generate the corresponding interface of the target platform.

The same authors of [1] created a tool to convert from Windows Phone to Android using regular expressions [3], and then enhanced the code conversion of this tool and evaluated its results compared to other commercial tools [4]. These tools included PhoneGap [5] and Xamarin [6]. However, Windows Phone becomes now an obsolete platform and most of the mobile apps use Android and IOS.

In [7], Luis Atencio et al. performed a measurement-driven experiment to test the performance of different cross-platform technology based on the different communication channels that these approaches used. They tested this technology under different scenarios so that mobile developers can choose the approach that suits them best.

In [8], Perchat proposed another cross-platform solution using component based approach. In their solution they used combination of native and universal declarative language which consist of several layers. These layers are: component layer which defines a set of component to be integrated in the application, universal layer that consists of declarative languages for describing the components, and integration layer as the name suggests, it integrate the native application.

Although these attempts tried to solve the cross platform mobile development problem, there is no agreed upon solution or approach.

2 Methodology

The trans-compiler approach was chosen to implement the solution for the following reasons:

- The proposed system's goals conform to the definition of a trans-compiler; high level language to high level language.
- Developers do not have to learn new languages in this approach.
- It can provide the flexibility to convert any code not only the most commonly used features.
- It can be able to support both source code and UI code transformation.
- It enables existing code reuse.
- It provides maximum flexibility for code generation (supporting new UI libraries or even new features in the programming language itself requires minimum effort).

2.1 Proposed Solution

The proposed solution provides an efficient way to convert the already developed applications, so it does not only target the to-be-developed applications therefore it provides less effort and less cost.

The proposed solution is to develop a system that converts a native Android application to a native iOS application, as shown in Fig. 1.

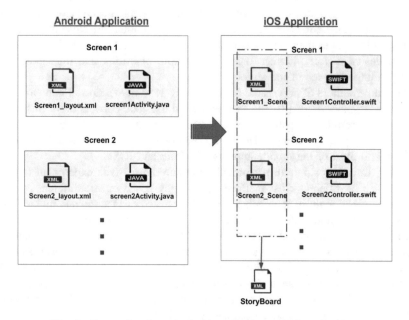

Fig. 1. Converting from Android application to iOS application.

An android application consists of multiple screens and each screen is defined by two files: **an activity java file**, and **a layout xml file**, while screen in an iOS application is defined by: **a swift controller file**, and multiple scene XML code of all the screens of the app, which are embedded in **one storyboard file**.

2.2   High Level System Architecture and Data Flows

The source project is passed through multiple stages to produce the desired destination project. In this section of the paper, the stages are demonstrated briefly, and in Sects. 2.3 and 2.4, they will be explained in detail.

Controller: The Controller is the interface unit in charge of all communication between the other components. It is responsible for receiving the source project and delivering the target project.

Code Conversion Unit: The Code Conversion Unit is responsible for processing and the backend code files and producing the equivalent ones.

UI Conversion Unit: The UI Conversion Unit receives the corresponding UI file for the backend code and the additional data collected by the Code Conversion Unit in order to parse the source UI code and produce the destination user interface file.

Database: In the Database stored the data necessary to map data types, built-in functions and their signatures (parameters, their types and return value types) and UI components from java to swift or vice versa (Fig. 2).

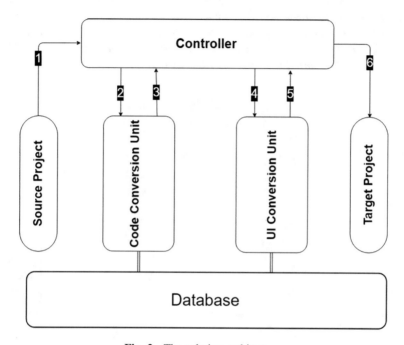

Fig. 2. The solution architecture

2.3 Implementation and Internal Design

In this section, the structure of TCAIOSC and its implementation will be illustrated, regarding the path from Android to iOS, but before that, a more detailed explanation of components that will be used in explaining the conversion process in Sect. 2.4.

Controller: The Controller is designed to take the mobile application and the desired target platform as an input. It categorizes and divides the files in the application. Then, it passes each category of files separately to the Code Converter and UI Converter, one by one. In the end, the Controller packages the output files and produces the target project.

Lexer and Parser: The basis of the converters' implementation was the parser which was based on a parser generator tool called ANTLR (ANother Tool for Language Recognition) [9]. Initially, TCAIOSC used this tool and supplied it with the grammar files of the source files (Java and XML grammar files); to generate the lexers and parsers used in both conversion units in runtime. TCAIOSC has two sets of a lexer and a parser, one for code conversion and UI conversion. They are used to build the parse tree of the input code.

Code Converter: The Converter contains classes that implement the interfaces provided by ANTLR to traverse the parse tree. The Converter propagates the parse tree, and converts the code from the source language to the target language using the stored data types, built-in functions, and operators and libraries in the database. In the end, the converter maps every single java file to a single swift file, and it produces some data if the java file was of an Activity, such as the Layout file name of the Activity, and the names of UI objects with their corresponding Layout IDs. This data is to be used by the UI Conversion Unit during conversion of the UI file.

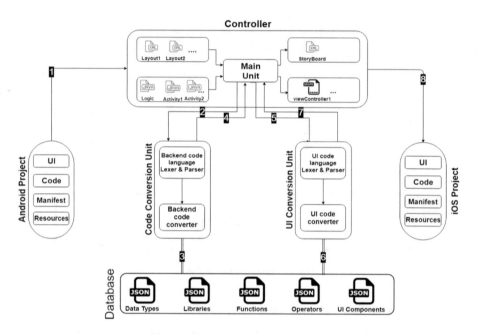

Fig. 3. The solution internal design.

UI Converter: The application user interface in Android is implemented in one or multiple XML files; each file can be the layout of an activity. An Android activity is one screen of the Android app's, which maps for a scene in iOS. Multiple scenes are implemented in one Storyboard file. Therefore, the goal is to transform multiple XML files to a single Storyboard file and vice versa, using the UI Converter. To produce the Storyboard hierarchy, the design of the proposed solution was inspired by the actual Apple UIKit in building the converter, by studying UIKit Framework [10] and UIKit Namespace Documentation [11]. The UI Converter also requires UI related code collected by the code converter, which can only be mapped in the UI destination code. Like the code converter, it uses the interfaces produced by ANTLR to traverse the nodes of the parse tree, and outputs the equivalent UI code in target platform.

2.4   Conversion Process

To begin with, the Android project consists of the following types of files: UI files (XML), Code files (Java Classes and Activities), Manifest file and Resources files. Similarly, the iOS project consists of the following types of files: UI files (Storyboard), Code files (Swift Classes and viewControllers), Manifest file and Resources files. The Android project undergoes multiple phases, starting and ending at the Controller, before the iOS project comes out. The process of conversion can be explained as follows (see Fig. 3):

1. The Controller is designed to categorize the project files, process the manifest and the resources, and pass the Java and XML files separately to the Main Unit.
2. The Code Conversion Unit receives only the Java files from the Main Unit. The lexer and parser form the parse tree of the code and forward it to the converter.
3. The Code Converter traverses the parse tree node by node, and uses the resources in the database. The product of this process is a viewController to every Activity - a Swift file to every Java file.
4. The Code Conversion Unit forwards, to the Main Unit, the Swift files, and some data concerning the Activity files, such as the Layout file name corresponding to that Activity file, and the names of Activity objects with their corresponding Layout IDs. This is all to be used by the UI Conversion Unit in connecting the viewController files to the Storyboard.
5. The UI Conversion Unit receives the Swift files as well as the Activity-concerned data (mentioned in phase 4) from the Main Unit. The lexer and parser form the parse tree of the XML file and forward it to the converter.
6. The UI Converter traverses the parse tree node by node, and uses the resources in the database. The product of this process is a Scene to every Layout.
7. All Scenes are put together into a single Storyboard file and forwarded to the Main Unit.
8. Finally, the Swift and Storyboard files are assembled in the iOS project along with the new manifest and resources files.

3   Results and Discussion

3.1   Testing

Testing is divided to three sections: Java to Swift testing, UI Conversion testing, Integrated Application. The test cases can be found in the TCAIOSC repository on Github [12].

Java to Swift Testing. In order to check the ability of TCAIOSC in converting from Java to Swift, two test cases are presented here. They are adding elements to an Arraylist of Strings with different methods and defining classes with inheritance. The results of these test cases are shown in Tables 1 and 2. The first test case creates a String ArrayList object and adds two elements to it, then gets its size. After that the code defines a new String object and assigns the second element in the ArrayList to this object. The second test case tests defining classes and inheritance. It defines a class named Calculation, with one method, and another class named MyCalculation, that has another method and inherits from Calculation. After that, it instantiates an object from the child class MyCalculation and tries calling all the two methods.

It can be seen how TCAIOSC converted it successfully despite the differences between Java and Swift. For example, in Table 1, it can be observed that defining an ArrayList, and adding an element to it in Java, is different from swift, not only in the names of the Data type and the methods, but also in the order of the method arguments. In Table 2, it can be observed that in calling a method for a class in swift, it is needed to write the parameters' names of the method definition before the argument, as in demo.addition(ab:a, y:b), and this is not needed in Java. So, to be able to convert despite that, TCAIOSC keeps track of the methods of every class with its parameters' names and data types, and the inheritance between classes.

Table 1. ArrayList test case.

Input Java Code	Generated Swift Code

ArrayList<String> x = new ArrayList<String>();
x.add("string1");
x.add(0,"string2");
x.size();
String temp = x.get(2);
	var x:Array<String> = Array()
	x.append("string1")
	x.insert("string2", at: 0)
	x.count
	var temp:String = x[2]

Table 2. Classes declaration and inheritance test case.

Input Java Code	Generated Swift Code

class Calculation {
  int z = 0;
  public void addition(int ab, int y){
  z = ab + y;
  System.out.println(z);
  }
}
public class MyCalculation extends Calculation {
  public void multiplication(int x, int y){
  z = x * y;
  System.out.println(z);
  }
  public static void main(String args[]){
  int a = 20, b = 10;
  MyCalculation demo = new
  MyCalculation();
  demo.addition(a, b);
  demo.multiplication(a, b);
  }}
	class Calculation{
	  var z:Int = 0
	  public func addition(ab:Int, y:Int) {
	  z = ab + y
	  print(z)
	  }
	}
	class MyCalculation: Calculation{
	
	  public func multiplication(x:Int,y:Int) {
	  z = x * y
	  print(z)
	  }
	}
	var a:Int = 20, b:Int = 10
	var demo:MyCalculation = MyCalculation()
	demo.addition(ab:a ,y:b)
	demo.multiplication(x:a ,david:b)

UI-Conversion Testing: The purpose of the tests in this section is to ensure that the attributes of a UI component are correctly mapped, from Android XML code to iOS XML, in a way that outputs the same appearance in the destination app as the source app. The tests were performed as follows: The first test was converting individual components, such as Button, TextView, EditText and SearchView, and the second test was converting a Constraint Layout containing multiple components together.

Table 3. UI test case 1: conversion of a button.

XML Input Code	iOS Converted XML Code

`<Button`
`    android:id="@+id/button1"`
`    android:layout_width="110dp"`
`    android:layout_height="40dp"`
`    android:background="#4056FF"`
`    android:text="save"`
`    android:textColor="#fff"`
`/>`
	`<button opaque="NO" contentMode="scaleToFill"`
	`fixedFrame="YES" contentHorizontalAlignment="center"`
	`contentVerticalAlignment="center" translatesAutoresizing-`
	`MaskIntoConstraints="NO" buttonType="roundedRect" line-`
	`BreakMode="middleTruncation" id="button1">`
	`    <rect key="frame" x="183.0" y="251.0" width="110.0"`
	`height="40.0" />`
	`    <autoresizingMask key="autoresizingMask" flexible-`
	`MaxX="YES" flexibleMaxY="YES" />`
	`    <color key="backgroundColor"`
	`red="0.25098039215686274" green="0.33725490196078434"`
	`blue="1.0" alpha="1.0" colorSpace="custom" customColor-`
	`Space="sRGB" />`
	`    <state key="normal" title="save" />`
	`</button>`

As Shown in Table 3, converting a component was a successful operation, as be the same appearance of the original component is delivered. However, some of the limitations are translating the `layout_width` and `layout_height` in terms of `"match_parent"`, `"fill_parent"` or `"wrap_conent"`. Specific measurements in `"dp"` only can be mapped. This test succeeded in transforming an Android Activity containing a Constraint Layout to iOS Storyboard. The limitation faced was positioning the components correctly in the screen, so the components are arranged next to each other in the screen. Although, this limitation can be easily overcome by a developer through manually arranging the components, until this issue is treated.

Integrated Application Testing: After testing the backend code and UI code conversion separately, it was important to test the whole application after integrating both phases. The tests focused on the connections between the backend code and UI components and checking they are converted correctly. For example, the following test case, an application to solve second degree equation, the system takes input data through UI components, uses it in the backend code to calculate results and shows it to the user through UI. Both Tables 4 and 5 show a small part of the android files and the output iOS project files. The test focuses on how the connection between the UI views and the code is implemented in android and how it is converted in iOS. The complete files can be found in equation solver in the TCAIOSC repository [12].

Table 4. Part of the converted code for the Equation Solver application of the XML file.

Android Input File	iOS output File

`<EditText`
    `android:id="@+id/a_ValueEditText"`
    `android:layout_width="40dp"`
    `android:layout_height="40dp"`
`/>`
	`<textField opaque="NO" id="a-`
	`ValueEditText">`
	    `<rect key="frame" x="80.0"`
	       `y="50.0" width="40.0"`
	       `height="40.0"/>`
	`</textField>`
	`<connections>`
	    `<outlet property="editTextA"`
	       `destination="a-ValueEditText"`
	       `id="id29"/>`
	`</connections>`

Table 5. Part of the converted code for the Equation Solver application of the Java file.

Android Input File	iOS output File

`public class MainActivity extends AppCompatActivity {`
  `private double a_value;`
  `EditText editTextA;`
 	
  `protected void onCrate(Bundle savedInstanceState) {`
  `super.onCreate(savedInstanceState);`
  `setContentView(R.layout.activity_main);`
  `editTextA = findViewById(R.id.a_ValueEditText);`
  `editTextA.addTextChangedListener(new`
   `TextWatcher() {`
  `@Override`
  `public void afterTextChanged (Editable  editable) {`
`if (editable.length() != 0)`
    `A_value = Double.parseDouble( editable.toString());`
    `else {a_value = 0;}`
  `setNewValues();`
  `}`
  `});`
  `}`
`}`
	`class MainActivity: UIViewController{`
	  `private var a_value:Double = 0.0`
	  `@IBOutlet weak var editTextA:UITextField!`
	
	  `override internal func viewDidLoad() {`
	  `super.viewDidLoad()`
	  `editTextA.addTarget(self, action:`
	   `#selector(targetMethod0(_:)), for:`
	       `.editingChanged)`
	  `}`
	
	  `@objc func targetMethod0(_  edita-`
	`ble:UITextField){`
	  `if editable.text!.count != 0 {`
	   `a_value = Double(editable.text!)!`
	  `}`
	  `else { a_value = 0}`
	  `setNewValues()`
	  `}`
	`}`

4   Conclusion and Future Work

Compiler based approach showed its superiority to convert simple mobile applications from Android to iOS. In the future, larger applications with more than one screen will be supported and more libraries that support integration with databases will be supported. The native features of mobile phone (making calls, saving contacts, using sensors and others) will be tested thoroughly.

Acknowledgement. We wish to acknowledge the help provided by Eng. El-Shimaa S. Al-Kaliouby and Eng Amany M. Shobak for their guidance throughout this project. We would also like to thank our colleagues Aly M. Aly and Mahmoud Hamdy for their assistance and advice. We cannot forget the support of Dr. Maged Ghonaima and ITAC in achieving our goal by providing us for the hardware needed for this work.

References

1. El-Kassas, W.S., Abdullah, B.A., Yousef, A.H., Wahba, A.M.: Taxonomy of cross-platform mobile applications development approaches. Ain Shams Eng. J. **8**(2), 163–190 (2017)
2. Latif, M., Lakhrissi, Y., Nfaoui, E.H., Es-Sbai, N.: Cross platform approach for mobile application development: a survey. In: 2016 International Conference on Information Technology for Organizations Development (IT4OD) (2016)
3. El-Kassas, W.S., Abdullah, B.A., Yousef, A.H., Wahba, A.: ICPMD: integrated cross-platform mobile development solution. In: 2014 9th International Conference on Computer Engineering & Systems (ICCES) (2014)
4. El-Kassas, W.S., Abdullah, B.A., Yousef, A.H., Wahba, A.M.: Enhanced code conversion approach for the integrated cross-platform mobile development (ICPMD). IEEE Trans. Softw. Eng. **42**(11), 1036–1053 (2016)
5. PhoneGap. https://phonegap.com/. Accessed 05 May 2019
6. Xamarin App Development with Visual Studio, Visual Studio. https://visualstudio. microsoft.com/xamarin/. Accessed 12 May 2019
7. Atencio, L., Aybar, B., Padilla, A.B.: Comparative analysis of cross-platform communication mechanisms. In: Proceedings of the 2nd International Conference on Compute and Data Analysis - ICCDA 2018 (2018)
8. Perchat, J., Desertot, M., Lecomte, S.: Component based framework to create mobile cross-platform applications. Procedia Comput. Sci. **19**, 1004–1011 (2013)
9. Parr, T.: The Definitive ANTLR 4 Reference, Pragmatic Bookshelf (2013)
10. UIKit. Apple Developer Documentation. https://developer.apple.com/documentation/uikit. Accessed 30 Apr 2019
11. dotnet-bot: UIKit Namespace. https://docs.microsoft.com/en-us/dotnet/api/uikit. Accessed 30 Apr 2019
12. TCAIOSC: TCAIOSC/Andriod-Vs-iOS, GitHub. https://github.com/TCAIOSC/Andriod-Vs-iOS. Accessed 13 May 2019

Cyber-Physical Systems and Security

Cyber-Physical Systems in Smart City: Challenges and Future Trends for Strategic Research

Mazen Juma[✉] and Khaled Shaalan

Faculty of Engineering and Information Technology, British University in Dubai,
Dubai International Academic City, Block 11, 1st and 2nd Floor, Dubai, UAE
Mazen.Juma@gmail.com, Khaled.Shaalan@buid.ac.ae

Abstract. Modern cities today compete with each other to be smarter, maintain a more sustainable with high-quality living, acquire talents, and provide jobs. This digital transformation through agile drivers will help address the increasing challenges of urbanization in a couple of decades. The Cyber-Physical System (CPS) is becoming pervasive in every aspect of smart city daily life and considered as one of the four fundamental conceptual approaches of the fourth generation industrial revolution (Industry 4.0). CPS used to describe the next generation of a diverse spectrum of complicated, multidisciplinary, physically comprehending engineered systems that integrates embedded cyber aspects into the physical world. It implants computation technologies, communication control, the convergence of information, and physical processes together with strategic importance internationally. CPS is still a vast research area. As a result, it opens venues for applications across multiple scales. This paper presents an in-depth survey of the related works, focusing on the design and how it relates to different research fields, current concepts, and real-life applications to understand CPS more precisely. Further, it enumerates an extensive set of CPS challenges and opportunities, introducing visionary ideas, research strategies, and future trends expected for future-oriented technological solutions, like cloud computing, Internet of Things, and Big Data. These technological solutions are to play a critical role in CPS research and have significant impacts on the smart city.

Keywords: Cyber-physical systems · Smart city · Industry 4.0 · Big Data · Cloud computing · Internet of Things · Research strategies · Future trends

1 Introduction

The World is becoming a digitally enabled network of individuals, objects, and products and services. Technology will be part of everything in the digital business of the future; the average person will live in a digitally enhanced world [25]. Cyber-physical Systems (CPS) are intelligently networked systems with sensors, instilled processors, and actuators installed within them that identify and interact with real-world aspects and human end-users; CPS support real-time, ensure quality and extent of performance within safety-oriented applications, especially in the smart city model.

© Springer Nature Switzerland AG 2020
A. E. Hassanien et al. (Eds.): AISI 2019, AISC 1058, pp. 855–865, 2020.
https://doi.org/10.1007/978-3-030-31129-2_78

In CPS systems, the cumulative behavior of the mixed physical and cyber aspects of the system consists of control, critical computing, identification, and networking embedded within all components [14]. The five major dominant trends will drive the digital life in the smart city. Firstly, Artificial Intelligence and advanced machine learning that is now at a significant changing process and continues to embolden hypothetically all technology-enabled services, things, and applications, which creates smart systems capable of learning, adapting, and potentially acting at their own accord instead of adhering to preset instructions protocols [22]. Secondly, smart apps, including virtual assistants as a second trend, possess the potential to overhaul the workplace by rendering daily tasks simpler, consequently, making users more efficient. However, such apps extend beyond digital assistants; all software types, like security tooling and applications within enterprises [2]. The third trend is the new intelligent things, spanning three categories: autonomous vehicles, robots, and drones which experience continuous evolution thus impacting a more considerable portion of the market, inviting a new era of digital business [28]. The fourth trend is the virtual & augmented reality, transforming the way people interact together and how software systems construct an immersive setting [24]. Finally, digital twins, replicas of physical assets, will be capable of illustrating billions of things; a digital twin acts as an adaptive software replica of a system or physical object. This model emulates how physics interact with the object and how the object interacts with the environment, in addition to data supplied by planted sensors in the real world [12].

2   The State-of-the-Art

In less than five decades, cities around the world tripled from 548 back in 1970 to 1,692 in 2017. Currently, over 54% of global populations are located in cities; this percentage may reach 66% by 2050. With roughly four billion people residing in cities nowadays, a ripple of global digital-era urbanization is at hand [28]. Smart cities' definition relies on capacity, political agenda, stakeholders, and the city's vision [3]. However, common definitions share a common key aspect of "smartness" in an urban context: it is the use of communication and information technology solutions as core enablers of "smart" transformation of the city [13]. Cities with resources, technological infrastructure, political leadership, and vision utilize digital technology as a facilitator and a vital infrastructural enabler that solves the challenges of population growth, urbanization, and financial and environmental issues [8]. The United Nation considers smart cities, universally, as innovative cities that combine information and communication technologies to improve the quality of life [14]. CPS, as a term, first appeared at the National Science Foundation in the US around 2006 [24]. The scientific community perceived it in different lights. Arasteh et al. [1] describe CPS as physical and engineered systems employing tracked, planned, controlled, and integrated operations by a computing and communicating core. Han and Hawken [7] describe CPSs as the integration of computation with physical processes. Lu et al. [11] describe them as systems embedded together and with their surrounding physical environment. Sivarajah et al. [20] describe them as biological, physical, and engineered systems with integrated, monitored, and controlled operations, done so via computational core. These components interconnected at every scale and the computing profoundly implemented in every physical component to find its way into materials [10]. Some studies focused on the underlying problem formulations, system level requirements, and challenges in CPS. Zafar et al. [27] introduced the CPS concept and offered research pointers for CPS design. Tokody and Schuster [22] correctly pointed out the failure of standard abstraction layers, the necessity of positive timing behavior, and deficit in temporal semantics of current programming language models for CPS. Liao et al. [10] addressed system-level aspects of CPS by tackling scientific and social impact standpoints. Preuveneers and Ilie-Zudor [15] proposed two approaches: one of them is cyberizing the physical while the other is physicalizing the cyber, both acting as means for blending the cyber aspect with the physical aspect in systems. Several existing surveys illustrate the comprehensive view of CPS. Bessis and Dobre [3] outlined CPS features, applications, and challenges generally without in-depth details. Seshia et al. [17] described CPS research directions, specifications, and design with a shallow explanation of CPS applications and system level requirements. Varghese and Buyya [24] touched on CPS characteristics, design, synergic technologies, and implementation principles. Smart cities should satisfy minimal norms related to concepts such as smart governance, mobility, living, people, environment, and economy [7].

3 Future Trends in CPS Within Smart City

The main CPS Futuristic trends in different domains within the smart city through various studies and research efforts that have addressed as the following: (1) Smart manufacturing, it is the application of infused hardware and software technologies to enhance the productivity of manufacturing products and delivery of services. It considered one of the dominant CPS direction domains because of tweaks in domestic and international marketing, mass production, and economic boom. The characterizing of smart manufacturing undertaken in Industry 4.0 revolution that aimed to pioneer in the manufacturing of the future [6]. (2) Emergency response, it is responding to threats to public safety, health and welfare, and fending for the safety and integrity of natural assets, valuable infrastructures, and properties. CPS can devise and exercise a rapid emergency response through various sensor nodes in multiple areas, ready to respond to natural or human-made disasters [6]. (3) Critical infrastructure, it is the set of valuable properties and public infrastructures necessary for the welfare and survival of the society, usually a standard among nations. The smart grid is a lucrative application in the vital infrastructure domain. It employs industrial and central power plants, renewable energy resources, energy storage and transmission facilities, and energy allocation, and management facilities in smart homes and buildings [10]. (4) Healthcare and medicine, these refer to the multiple issues concerning the physiological state of the patient. Specific attention is drawn to medical implementations in CPS research, opening venues for opportunities of research for the CPS community. These opportunities can be among the assisted living, technologies related to home care, smart medical devices, operating room technologies, and smart prescription [21]. (5) Intelligent Transportation: It is the confluence of advanced technologies of communication, sensing, computation, and control mechanisms in transportation systems to improve coordination, safety, and facilities in traffic management with real-time information sharing. These technologies expedite transportation, air, ground and maritime, applying information sharing via satellites and master planning a communication environment among infrastructure, vehicles, and passengers' portable devices [23]. (6) Robotic for service, it is the deployment of intelligent robots performing services for the wellbeing of humans in a fully autonomous, semi-autonomous, or remotely controlled manner, excluding manufacturing operations. Robots can deploy, for example, for defence, environmental survey and control, logistics, assisted the living, etc. Since next-generation robots will interact closely with humans physically, interpretation and learning of human activities by robots became an important factor [16]. (7) Building automation, it is the deployment of actuators, sensors, and control systems providing automation and optimum control over ventilation, heating, air conditioning, fire prevention, lighting, and security in smart buildings [10].

4   Research Challenges and Opportunities

4.1   CPS in Big Data: Challenges and Opportunities

Big Data is a data set; both at static and dynamic states expand over the boundaries of regular processing methodologies. It is highly diverse with multiple sources (variety), not one dimension and trusting and devoted to verification (veracity), enormous and sophisticated (volume), and being delivered rapidly and sporadically (velocity) [8]. Big Data extracts valuable information derived from data to use it intelligently. CPS acts as a master planner inceptor, collector, and allocator of data in a quick, verified, diverse, and immense manner. The meeting ground of big data and CPS expands, posing greater importance and impact to the emphasis on data as irreplaceable business assets and crucial for businesses to remain competitive [22]. The challenges involve handling massive data production, new technologies will transform how huge input of data handled and keep showing, in the case of CPS. Besides this thriving volume, the mixing diverse data from different sources will create the need for applications focusing on query, integration, analysis, high-performance computing devices, and methodologies for data reduction [5]. Distributed data storage and processing, remote storage of big data ranks as one of the recurring concern in terms of content and the technical field. Cloud-based models assisted developers to reduce the costs on storing and processing big data compared to previous models, delivering data accessibility, leading to decentralized data storage and process some problems arising such as replication, parallelism, and requirements inherent to the natural attributes of CPS. The necessity for real-time evaluation and accurate response will become increasingly important despite the increasing size of data and faster velocity it occurs at [13]. Monetizing big data stemming from CPS, the mix between big data and CPS is a bold move, from a management perspective. New approaches necessary to optimize an enterprise's standing to be able to face these challenges. Healthcare or scientific research, domains that use massive datasets, combining big data with CPS creates lucrative new areas, like manufacturing or food production [20]. Data visualization, new means for data visualization needed to deliver information for human decision-makers, due to decentralizes volatile and mobile nature of immense volumes of rapidly changing data sources [16]. While on the other hand, the opportunities include leveraging big data analytics for CPS adaptation, CPS shall deliver vast volumes of data. Real-time data allows for unique opportunities for real-time planning and thus real-time decision-making; therefore, driving the dynamic adaptation of CPS. This opens up room for techniques such as tweaking user profiles and tracking and applying environmental context data through IoT [21]. Greater customization and certification in products and services are highly demanding. Big data generates an opportunity to fill in with CPS and supposedly to create a beneficial economic effect. The embedded computing technology in different industries, such as automotive, aerospace, healthcare, and transportation, gives a high starting point [26]. Assuring CPS assets stay online, the need for new impressive, competitive, effective methods to utilize assets and to remain them running an account for unpredictable failures. More intelligent and anticipative algorithms can enhance the capacity of tracking unscheduled maintenance, ageing assets, losses in processes and boosted operational efficiency [4].

4.2 CPS in Cloud Computing: Challenges and Opportunities

Cloud computing designed to enable universal, comfortable, transparent, and on-demand admission to a joint resources pool of modifiable computing resources, quickly delivered with scant managerial effort or service provider's interaction. To reduce costs, Cloud computing helped alleviate capital expenditures and replace them with operational ones [20]. By the time, cloud services now drive agility, productivity, and optimal performance, a source of influence on the processes and many organizations, and a digitalization driver for mass users. Cloud computing delivers a convenient computing model for effortless integration of physical and computing components [21]. The challenges take in real-time data collection, analysis, and actuation are a necessary reality for CPS to excel in their purpose; subsequent analysis for and by CPS expedites appropriate decisions. Low-latency computation and actuation is a necessity within CPS due to their time-critical nature. CPS avert the need for analysis, evaluation, and decision making to be in a physical central environment and do so through a well-allocated, robust cloud structure [25]. Multi-tenancy in CPS infrastructures, compatibility of systems to place in various places permits different clients to take part in the open-world setting of CPS. Integration of cloud resources as a component of high performing and efficient CPS requires special attention for servers and resources isolation reservation. It even surpasses server isolation and requires the proper adaptation of progressive cloud paradigms to solve the isolation problem as a container or cloud-enabled and application–server isolation [27]. Dependable and predictable cloud SLAs for CPS; it requires a high level of reliability, anticipation, planning, and tolerance to deploying critical CPS functionality within a cloud. Usually, cloud SLAs are provided based on best-effort only, making them barely provide a reliable mechanism for mitigation and risk prediction. Understanding comprehends impacts of CPS workloads on infrastructure, cloud planning and risk management, audibility, and verifiable conformance to defined behavior regarding functional and nonfunctional requirements [18]. Cloud services and platforms for CPS construction and deployment related to cloud infrastructures and executive services that are critical. Therefore, this type of cloud services is crucial for CPS construction and deployment, an essential feature of a cloud development framework for CPS is to back up the right development technique and methodology [12]. The opportunities cover scalability, elasticity, and availability of clouds can efficiently fine-tune the resources needed, varying with the loading volumes; CPS dictate resources needed and use devices that can be interconnected with the cloud entirely in a quick manner and hence the creation of a CPS ecosystem is relatively fast [5]. Additionally, the derivative quality of aggregated data, usually gauged by various CPS observers, could consider achieving more traceable and cost-effective cloud solutions. CPS acting as the basis for cloud platforms, exercising replication and allocation of data on several sites, can exponentially spread globally. This is to enhance the availability and tolerance of CPS ecosystems, especially in emergencies [7]. Infrastructure costs reduction, maintaining cloud computing applications are cost-effective a convenient due to lack of needing hem to be available on each computer system; these applications are accessible from different locations. Multi-tenancy allows outsourcing of costs and resources across multiple users [23].

4.3   CPS in IoT: Challenges and Opportunities

Internet of Things (IoT) efficiently manage sourcing, development, production, sales, and logistics through software-based services and improving upon novel business models for hybrid serviced products. Beyond the physical flow of materials in fresh supply or end product form, the information accumulated by implanted systems, smart items, sensors, and end-users orchestrate opportunities for constructing unique, crafty software-based services in IoT [9]. The challenges in this context close to instance-based architecture for a real business network of things, CPS dramatically allow simple handling of physical world objects in software systems and services within IoT environments. Such an instance-based architecture needs corresponding data standards that can be processed through an independent domain and specific services representing smart things, software-based services, business-wise, architect effortless access to the smart things, and deliver standardized services based on, triggering or mirroring the movements between the participating organizations [14]. Service architecture for software-based services on top of CPS will enable business users instead of guiding them, enhancing their approaches. Software-based services must consist of four attributes that differentiate them from current business applications: to be self-explanatory so that business users are capable of designing them to tailor to their needs without IT experts' support [24]. Creating a CPS architecture delivering services based on software and services in a simple manner, which enables business users to design collaborations easily and quickly is the challenge to conquer nowadays. Hypothetically, the unified data model is required; it should allow users to focus on generic smart objects and services as per their individual needs within IoT space [26]. The opportunities encompass software-defined industries are appealing to just-in-time processing, production, and delivery of innovative products, goods, and services, often for a lot-size of only one. Programmable facilities create tailored products, contrary to large quantities of products in the manufacturing process [27]. Highly customized and optimized manufacturing plants and supply systems will be capable of rectifying fluctuations and meeting customer demands effectively. In this model, the production relies on real-life demand gauging and reconfiguration of the methodologies of production in software via CPS instead of traditional long-term prediction and anticipation modules [9]. Static lean manufacturing is going to be no more. Manufacturing enterprises should be more "Leagile" (lean + agile) or face extinction instead. "Leagile" evokes perpetual monitoring and analysis of volumes of data related to production systems, inventories, and supply chains while efficiently eliminating or minimizing waste [15]. Moreover, "Leagile" implies elastically shifting manufacturing potency on demand from the manufacturers at the edges, those who wish to pool in their manufacturing prowess on the cloud to attract more orders from the ecosystem, increasing efficiency of their machines and resources material thus coping with the volatility of demand within the IoT world [25].

5 Visionary Ideas for Research Trends

5.1 CPS in Big Data: Visionary Ideas

Big data for CPS design, evolution and maintenance that many data sources generate a plethora of raw unorganized data that can evaluate, providing evidence on usage trends of CPS frameworks. In addition, data collected from CPS to track preferences, user trends, and features and performance improvement opportunities for advanced analytics, giving informed decision support for developing and evolving CPS [6]. Big data for CPS quality assurance and diagnosis and usage of automated analyses of CPS artefacts have pondered on for some time. Today, owing to the expanse of data volumes and analytics capacitances for enormous volumes of data structured and unstructured. Deviations and failure patterns can analyze through big data analytics catering to such massive amounts of "meta-data" being collected [19]. Big data for CPS run-time monitoring and adaptation employs the available data during the functioning of CPS monitoring of services, cloud infrastructure, things, and end users shall provide information aplenty and in real-time in a timely manner [25]. Optimizing CPS resource allocation worldwide, installing shared data centers. Novel mechanisms needed to adapt to handle such complex global resource sharing. Efficient recycling of CPS units should be considered as a mean tweaking resource usage and as a result, productivity [24]. Fostering CPS skill building during research activities wherein industry and academia meet is to deem as an optimal, complementary method of delivering and educating functional people through various opportunities for designing work and joint research. Hence, CPS research should embolden their activities to promote these chances for skill building in CPS [2].

5.2 CPS in Cloud Computing: Visionary Ideas

Dynamic adaptation of CPS cloud computing facilitates the run-time adaptation of cloud infrastructures and applications to respond to context changes and system failures dynamically. Measurements of the execution parameters of infrastructure utilization and implementation, when collected in real-time and fused with the Internet of Things, allows data execution imbued with insightful data concerning the system context delivered by numerous sensors [1]. The cloud as CPS means that it will be able to develop CPS running on a vast number of allocated resources it as if it was running on a single resource while being able to tap into the nearly infinite power and being resilient to failures [4]. Leveraging CPS as key enabler crucially to understand, practice, and authenticate assumptions and prototypes while providing analysis for related intellectual property. Therefore, the CPS design and the interplay are required and needed to combine with a robust, constant, and feedback-driven approach [26]. Globally scaling CPS to initial success on new approaches driven from new mathematical models that benefit crowd designing needed to make sure adoption fostered. Complementarily, new concepts for catering to future needs for reactive design and cloud-based systems should manage perpetual delivery and updates of content as well [9]. Pursue a dynamic approach to CPS delivery in an ideal scenario, technology resulting from CPS are preferred in packaged form, i.e., in the form of application of frameworks or integrated models. CPS should bring their solution into a reusable frame for CPS perspective to foster adoption in practice [19].

5.3   CPS in IoT: Visionary Ideas

Perform CPS-IoT research exploiting real-world cases and addressing particular CPS-IoT problems allows the exploitation of outcomes. Real world use complex realistic scaling invoking problem understanding and appraisal. Rigorous and relevant empirical studies in the industry are necessities for the future of CPS. Addressing CPS-IoT challenges opens research difficulties to address and conquer the trench between theory and practice [26]. Deliver CPS pilots in IoT makes the results accessible as much as possible; CPS should construct and sustain preindustrial systems for its outcomes to attract industry premises. To sponsor the understanding of these systems, existing technology can act as templates; innovative features should be presented once available in the CPS. These systems have to interlink to vital areas of CPS to open venues for exploitation [8]. Larger-scale, integrated CPS is important for IoT research, the current work of CPS is extensive, integrated, and concentrated efforts of research and industry domains, considering various scopes, angles, and aspects simultaneously to devise relevant and significant solutions in a practical manner [8]. Pursuing an open IoT strategy for CPS research should carry out early by tapping in current ecosystems, sustaining and adopting open CPS outcomes; this develops an open-source community in the IoT. Open source inspired development open models and interfaces as a new, engaging direction. Open models serve as basic templates for standards allowing easy access for them later on [3]. Modernizing CPS curricula to cater to demands for basic and advanced CPS skills and prowess. Based on a sturdy foundation of CPS principles, Curricula need to address technology trends [26].

6   Roadmap of CPS Strategic Research

Research roadmap is essential to helping the research community achieve its mission to create robust, sustainable, and quality-oriented for cyber-physical systems. CPS provides the solutions and services necessary to implement such scenarios of smart cities.

To that end, the roadmap of strategic research for CPS of the smart city will inter-operate to integrate the smart services and help provide solutions for sustainable and efficient management of urban areas [1]. Intensify enabling sciences, CPS built on scientific conclusions and technologies from different fields of sophisticated and large-scale technical and organizational systems. Therefore, innovations in those fields are incentives for technology push factors [15]. Address human-machine interaction, CPS enable an acceptable interaction strictly with a wide range of human users. Therefore, these systems have to be fathoming of the limits and capacities, assessing the full reach, as well as the expectations and intentions of their human users [24]. Sponsor cross-disciplinary research to provide system level services in CPS, features along the value-chain for these services should be masterfully integrated, whether concerning microsystem technology via software and systems engineering up to include macro-economics [27]. Support maturation initiatives, the complexity of CPS necessitates the development of technologies that are mature in the context of forming large-scale CPS and demonstrate requirements of that relevant technology [24]. Promote available CPS infrastructure, information and communication technology backbone large-scale coordination of technical and organizational processes. Therefore, performance infor-mation and communication infrastructure in the urban areas are core factors for their implementation [7]. Coordinate installation of key systems while CPS can address critical societal issues, their inherent attributes of large-scale systems surpassing national boundaries require a huge investment in private and public infrastructure [5]. Provide reference platforms defined levels in the stack of disciplines should create the means to influence innovation by building upon the facilitate maturation and bench-marking and to support services integration in CPS [14]. Homogenize interoperability standards, CPS provide system-level functionality via the collaborative coordination of processes, the crucial regulatory and technical prerequisites for this association should be delivered without exhausting and limiting potential applications [18]. Define system-level design methodologies added-value, innovative services provided by CPS usually stem from the cross subsystem orchestration of more critical services imple-menting sequences of technical, physical, and organizational processes. Therefore, broad-spectrum aspects should be considered when designing and constructing these services [27]. Provide open standards, CPS has to promote the definition, provision, and evolution of pre-normative and normative standards for coordination and support actions [26]. Promote open source and open license as CPS sponsors the delivery of value-added services via the confluence of more basic services. Access to interoper-ability platforms support is a fundamental prerequisite to rapid innovation and is quite of high importance for enterprises [3]. Increase open data to optimize technical and social processes, CPS make detailed utilization of data for coordination and collabo-ration purposes to facilitate quickly the implementation of innovative governing schemes; the availability of open data should be increased [11]. Harden infrastructure in CPS makes use of open information and communication technology to coordinate the control of vital organizational and technical processes, including the electric grid, systems and their roadside installation, marketplaces for energy trading, or traffic control centers [22]. Protect data ownership using their deep embedding in socio-technical environments, CPS sustain a substantial amount of delicate data, from pro-filing of traffic participants or patients to closely tracking sensitive production processes

[16]. Adapt dependability regulations, CPS, owing to their characteristics, large-scale, and multiuser support, follow maintenance and implement cycle common in the domain of dependable systems [25]. Stimulate collaboration; the innovative nature of CPS does not describe the technologies involved and the engineering challenge of systems of such a scale and diversity that demands advances in both theoretical foundation and practical engineering [8]. Provide awareness platforms due to their infrastructure and set of vital services, futuristic CPS can create innovative added-value service-engineering, cross-discipline technologies to naturalize with practical and technical, rather than theoretical, capabilities [17]. Enable decision makers to seize upon the impact of CPS on society and business in order to make educated decisions, and fosters the transfer of knowledge [9].

References

1. Arasteh, H., Loia, V., Tomaseti, A., Troisi, O., Shafie, M., Siano, P.: IoT-based smart cities: a survey. In: 16th International Conference on Environment and Electrical Engineering, pp. 1–6. IEEE (2016)
2. Assunção, D., Calheiros, N., Bianchi, S., Netto, A., Buyya, R.: Big data computing and clouds: Trends and future directions. J. Parallel Distrib. Comput. **79**, 3–15 (2015)
3. Bessis, N., Dobre, C.: Big Data and Internet of Things: A Roadmap for Smart Environments, vol. 546. Springer, Basel (2014)
4. Chaâri, R., Ellouze, F., Koubâa, A., Qureshi, B., Pereira, N., Youssef, H., Tovar, E.: Cyber-physical systems clouds: a survey. Comput. Netw. **108**, 260–278 (2016)
5. Gepp, A., Linenlucke, K., Neill, J., Smith, T.: Big data techniques in auditing research and practice: current trends & future opportunities. J. Acc. Lit. **40**, 102–115 (2018)
6. Gunes, V., Peter, S., Givargis, T., Vahid, F.: A survey on concepts, applications, and challenges in cyber-physical systems. KSII Trans. Internet Inform. Syst. **8**(12), 4242–4268 (2014)
7. Han, H., Hawken, S.: Introduction: innovation and identity in next-generation smart cities. City Cult. Soc. **12**, 1–4 (2018)
8. Jin, X., Wah, B.W., Cheng, X., Wang, Y.: Significance and challenges of big data research. Big Data Res. **2**(2), 59–64 (2015)
9. Kraijak, S., Tuwanut, P.: A survey on internet of things architecture, protocols, possible applications, security, privacy, and real-world implementation and future trends. In: 16th International Conference on Communication Technology (ICCT), pp. 26–31. IEEE (2015)
10. Liao, Y., Deschamps, F., Loures, R., Ramos, P.: Past, present and future of Industry 4.0-a systematic literature review and research agenda proposal. Int. J. Prod. Res. **55**(12), 3609–3629 (2017)
11. Lu, Y.: Industry 4.0: a survey on technologies, applications and open research issues. J. Ind. Inform. Integr. **6**, 1–10 (2017)
12. Mishra, D., Gunasekaran, A., Childe, J., Papadopoulos, T., Dubey, R., Wamba, S.: Vision, applications and future challenges of the internet of things: a bibliometric study of the recent literature. Ind. Manage. Data Syst. **116**(7), 1331–1355 (2016)
13. Moura, J., Hutchison, D.: Review and analysis of networking challenges in cloud computing. J. Netw. Comput. Appl. **60**, 113–129 (2016)
14. Ojo, A., Dzhusupova, Z., Curry, E.: Exploring the nature of the smart cities research landscape. In: Smarter as the New Urban Agenda, pp. 23–47. Springer, Cham (2016)

15. Preuveneers, D., Ilie-Zudor, E.: The intelligent industry of the future: a survey on emerging trends, research challenges and opportunities in Industry 4.0. J. Ambient Intell. Smart Environ. **9**(3), 287–298 (2017)
16. Rodríguez, L., Sánchez, L., García, L., Alor, G.: A general perspective of big data: applications, tools, challenges and trends. J. Supercomput. **72**(8), 3073–3113 (2016)
17. Seshia, A., Hu, S., Li, W., Zhu, Q.: Design automation of cyber-physical systems: challenges, advances, and opportunities. IEEE Trans. Comput. Aided Des. Integr. Circ. Syst. **36**(9), 1421–1434 (2017)
18. Silva, N., Khan, M., Han, K.: Towards sustainable smart cities: a review of trends, architectures, components, open challenges in smart cities. Sustain. Cities Soc. **38**, 697–713 (2018)
19. Singh, D., Tripathi, G., Jara, J.: A survey of internet-of-things: future vision, architecture, challenges and services. In: IEEE World Forum on Internet of Things (WF-IoT), pp. 287–292. IEEE (2014)
20. Sivarajah, U., Kamal, M., Irani, Z., Weerakkody, V.: Critical analysis of big data challenges and analytical methods. J. Bus. Res. **70**, 263–286 (2017)
21. Stankovic, A.: Research directions for the internet of things. IEEE Internet Things J. **1**(1), 3–9 (2014)
22. Tokody, D., Schuster, G.: Driving forces behind Smart city implementations-the next smart revolution. J. Emerg. Res. solut. ICT **1**(2), 1–16 (2016)
23. Tsai, W., Lai, F., Vasilakos, V.: Future internet of things: open issues and challenges. Wirel. Netw. **20**(8), 2201–2217 (2014)
24. Varghese, B., Buyya, R.: Next generation cloud computing: new trends and research directions. Future Gener. Comput. Syst. **79**, 849–861 (2018)
25. Xu, D., Xu, L., Li, L.: Industry 4.0: state of the art and future trends. Int. J. Prod. Res. **56**(8), 2941–2962 (2018)
26. Yang, C., Huang, Q., Li, Z., Liu, K., Hu, F.: Big data and cloud computing: innovation opportunities and challenges. Int. J. Digit. Earth **10**(1), 13–53 (2017)
27. Zafar, F., Khan, A., Malik, R., Jamil, F.: A survey of cloud computing data integrity schemes: design challenges, taxonomy and future trends. Comput. Secur. **65**, 29–49 (2017)
28. Zhou, K., Liu, T., Zhou, L.: Industry 4.0: towards future industrial opportunities and challenges. In: 12th International Conference on Fuzzy Systems Knowledge Discovery, pp. 2147–2152. IEEE (2015)

Analysis of Substitutive Fragile Watermarking Techniques for 3D Model Authentication

Kariman M. Mabrouk[1(✉)], Noura A. Semary[1,2], and Hatem Abdul-Kader[1]

[1] Faculty of Computers and Information, Menoufia University, Menoufia, Egypt
Kariman.mamdouh@ci.menofia.edu.eg
[2] Faculty of Computers and Information Technology, Jeddah University, Khulais, Kingdom of Saudi Arabia

Abstract. Due to the importance of multimedia data and the urgent need to use it in many fields such as industry, medicine and entertainment, protecting them becomes an important issue. Digital watermarking is considered an efficient solution for multimedia security as it preserves the original media content as it is. For 3D graphical models, 3D fragile watermarking aims to detect any attacks to the models and protect the copyright and the ownership of the models. In this paper, we present a comparative analysis between two substitutive 3D fragile watermarking algorithms. The first algorithm is based on adaptive watermark generation technique using Hamming code, while the other algorithm uses Chaos sequence for 3D models fragile watermarking in the spatial domain. The study uses different assessment measures to show the points of strength and weaknesses of both methods.

Keywords: Adaptive watermarking · Hamming code · Tampering detection · Authentication

1 Introduction

Information security refers to the protection of information from unauthorized access, use, modification, or destruction to achieve confidentiality, integrity, and availability of information. There are two types of information security; *information hiding* (Steganography) or *information encryption* (Cryptography). Encryption is the science of protecting information from unauthorized people by converting it into a form that is non-recognizable by its attackers. Information hiding embeds a message (watermark) over a cover signal such that its presence cannot be detected during transmission. There are two categories of information hiding: *steganography* and *watermarking*. The main goal of steganography is to protect the message itself and hide as much data as possible in the cover signal, while the goal of the watermarking is to protect the cover signal by hiding data (watermark) in it, therefore the digital watermarking is the best solution for providing protection for multimedia copyrights and owner shipment data.

Due to the growth of 3D graphical model generation and the spread of using it in data representations of other applications like fuel or water transferring pip models, 3D cartoon models etc., and recently many researchers have a great interest in

A. E. Hassanien et al. (Eds.): AISI 2019, AISC 1058, pp. 866–875, 2020.
https://doi.org/10.1007/978-3-030-31129-2_79

watermarking of 3D models. There are different representations of the 3D models [1]. In this research we concern with polygonal representation of the 3D model [1], which consists of vertices, faces, and edges connecting the vertices. The vertices list is the 3D coordinates of each vertex in the model, the faces list describes the connectivity of the vertices to each other, and the edges list could be derived by traversing the face and vertices lists.

Watermarking may be used for a wide range of applications, such as Copyright protection and content authentication. There are three types of watermarking according to the goal to be achieved; *robust watermarking, fragile watermarking* and *semi-fragile watermarking*. The aim of the robust watermark is to protect the ownership of the digital media and keep the embedded watermark detectable after being attacked. On the other hand, the fragile watermark aims to be sensitive to any attack on the model and locate the modified regions and possibly predict how the model was before modification. Therefore, fragile watermarking is used for content authentication and verification. The semi-fragile watermark combines both the advantages of the robust watermark and the fragile watermark so that it is more robust than fragile watermark and less sensitive to classical user modifications that aims to discriminate between malicious and non-malicious attacks.

In this paper, we present a comparative analysis between two substitutive 3D fragile watermarking algorithms proposed by Wang et al. [2, 3]. Strengths and weaknesses are analyzed. The rest of this paper are organized as follows, Sect. 2 explains the fragile watermarking algorithms from the state of art. Section 3 discusses the two substitutive fragile watermarking techniques. Section 4 shows the experimental results and discussion. Finally, conclusion is provided in Sect. 5.

2 Related Work

Watermark embedding strategies primarily are divided into two classes; *additive* and *substitutive* [1]. In the case of additive strategy, a watermark is considered as a random noise pattern which is added to the mesh surface as in [4–7]. But in the case of substitutive, the watermark is embedded by a selective bit substitution in the numerical values of the mesh elements as in [2, 3, 8–10]. Based on this embedding style, the watermark may be embedded in different embedding primitives as follows:

2.1 Data File Organization

This category utilizes the redundancy of polygon models to carry information. Ichikawa et al. [11] modified the order of the triangles (the order of the triple vertices forming a given triangle). They only used the redundancy of description. Wu et al. [12] used the mesh partitioning to divide the mesh into patches with a fixed number of vertices. While the geometrical and topological information of each patch, as well as other properties (color, texture, and material), are used to produce the hash value which represents the signature embedded in the model. The goal of Bennour et al. [13] was to protect the visual presentations of the 3D object in images or videos after it has been marked. They also proposed an extension of 2D contour watermarking algorithm to a

3D silhouette. Sales et al. [14] presented a method based on the protection of the intellectual rights of 3D objects through their 2D projections.

2.2   Topological Data

These algorithms use the topology of the 3D object to embed the watermark which leads to change the triangulation of the mesh. Ohbuchi et al. [15] presented two visible algorithms where the local triangulation density is changed to insert a visible watermark depending on the triangle similarity quadruple (TSQ) algorithm. Whereas the second algorithm embedded a blind watermark by topological ordering TVR (Tetrahedral Volume Ratio) method. Mao et al. method [16] triangulated a part of a triangle mesh to embed the watermark into the new positions of the vertices, this algorithm is considered a reversible as it allows a full extraction of the embedded information and a complete restoration of the cover signal.

2.3   Geometrical Data

Most of the 3D fragile watermarking algorithms embed the watermark by modifying the geometry of the 3D object either in the spatial domain or in the frequency domain. Yeo and Yeung [4] proposed the first 3D fragile watermarking algorithm where each of the vertex information is modified by slightly perturbing the vertex based on a predefined hash function to make all vertices valid for authentication. Lin et al. [5], and Chou et al. [17] solved the causality problem raised in Yeo's method by setting both hash functions depending only on the coordinates of the current vertex. After that, they proposed a multi-function vertex embedding method and an adjusting-vertex method [17]. With considering high-capacity watermarks, Cayre and Macq [18] considered a triangle as a two-state geometrical object and classify the triangle edges based on the traversal into entry edge and exit edge, where the entry edge is modulated using Quantization index modulation (QIM) to embed watermark bits.

To immune similarity transformation attacks, Chou et al. [19] embedded watermarks in a subset of the model faces so that any changes will ruin the relationship between the mark faces and neighboring vertices. Huang et al. [20] translated the 3D model into the spherical coordinate system, then used the QIM technique to embed the watermark into the r coordinate for authentication and verification. Xu and Cai [21] used the Principal Component Analysis PCA to generate a parameterized spherical coordinates mapping square-matrix to embed a binary image (watermark). Wang et al. [2] used the hamming code to calculate the parity bits that embedded in each vertex coordinate with the LSB substitution to achieve verification during the extraction stage. According to the problem of high collision characteristic of hash function used for generating the watermark from the mesh model, Wang et al. [3] employed a chaotic sequence generator to generate the embedded watermark to achieve both the authentication and verification of the model.

3 Substitutive Fragile Watermarking Techniques

As mentioned before, watermarking techniques can be classified according to the embedding method to: *additive* and *substitutive*. In the case of substitutive embedding, the watermark is embedded by a selective bit substitution in the numerical values of the mesh elements. The watermarking techniques can be classified also according to the watermark generation pattern which relays on the application type. The watermark may be a stream of bits that is related to the model or may be an information that is not related to the model at all. Generally, there are two ways of watermark generation pattern (WGP):

1. Self-embedding: which means that the watermark is a compacted version of the same model by some embedding strategy (e.g. the hash of the cover model or error correction code).
2. External embedding: which means that the watermark is an external information related or not related to the cover model. This external information could be text data, image data or pseudo-random bit sequence, and it is needed to transform the embedded data into binary bit sequence before embedding.

According to this classification, Wang et al. [2, 3] proposed two fragile watermarking techniques based on substitutive embedding method. Where they used the Least Significant Bit (LSB) substitution embedding method. At first technique [2] used the self-embedding WGP strategy, where an adaptive watermark is generated from the cover model by using Hamming code technique for 3D objects verification. The hamming code is used to generate three parity bits from each vertex, they are used for verification during the extraction stage. These three parity check bits P_1, P_2, and P_3 are regarded as the watermark, which embedded in each vertex coordinate by the least significant bit (LSB) substitution. Leading to increasing the data hiding capacity but on the other hand, the embedding distortion to the model is uncontrollable. Authors claimed that the method to be immune to the causality, convergence and embedding holes problems.

The second technique [3] is based on the external embedding WGP strategy, where a novel Chaos sequence based fragile watermarking scheme for 3D models in the spatial domain was proposed. The authors used the chaotic sequence generated from the Chen-Lee system [22], which is considered as the embedded watermark. Then they embedded this watermark in each vertex coordinate according to a random sequence of integers generated by using a secret key K, to achieve both the authentication and verification. Instead of the hash function, the tampering region can be verified and located by the Chaos sequence-based watermark check.

Both techniques don't need the original model or the watermark for the 3D model verification and tampering detection localization, as they don't depend on using the hash function for authentication and verification. Also, they achieve high embedding capacity, since they used all the vertices of the model for embedding. For the second technique [3], from the security point of view, finding the Chaos sequence is a challenge for the attackers. Security was also achieved by using secret keys to embed the watermarks.

4 Experimental Result and Dissection

The two techniques of Wang et al. [2, 3] were implemented using a multi-paradigm numerical computing environment and a propriety programming language developed by MathWorks (MATLAB R2018a).

4.1 Assessment Methods

The main requirements to provide an effective watermark are the imperceptibility, the robustness against intended or non-intended attacks and the capacity. Based on these requirements a series of experiments were conducted to measure the imperceptibility robustness. Table 1 illustrates the assessment measures needed to evaluate watermarking systems [23].

Table 1. Performance assessment measures used in mesh watermarking

Assessment type	Assessment measure	Formula
Imperceptibility measures	Hausdorff distance (HD)	$d(v, M') = min_{v' \in M'} \|v - v'\|$
	Modified Hausdorff distance (MHD)	$D(M, M') = \max(d_H(M, M'), D_H(M', M))$
	Root mean square error (RMSE)	$d_{rms}(M, M') = \sqrt{d_{v \in M}(v, M')^2 dM}$
Robustness measures	Correlation coefficient	$corr(w^d, w) = \dfrac{\sum_{i=0}^{N-1} (w_i^d - \overline{w^d})(w_i - \overline{w})}{\sqrt{\sum_{i=0}^{N-1} (w_i^d - \overline{w^d})(w_i - \overline{w})}}$

The *RMSE* measures the differences between the values predicted by the model and the values observed. When the *RMSE* values are small this indicates insignificant positional changes during the watermark embedding. Therefore, lower values of *RMSE* indicates better fit. The Hausdorff distance *HD* measures the similarity of two sets in the metric sense. If the Hausdorff distance is small between two sets these means that they are looks almost the same. The Modified Hausdorff distance *MHD* computes the forward and reverse distances, and outputs the minimum of both. The Correlation Coefficient (*CC*) measures the degree of the relationship between two variables. The values of Correlation coefficients are ranged from −1 to 1, whereby the value of −1 refers to a perfect negative correlation and the value of 1 refers to a perfect positive correlation. However, if the value is zero that's meaning there is no relationship between the two sets. Generally, the correlation coefficient used to measure the change in the bit values of the original watermark and the extracted watermark, meanly it measures how the watermark robust to the attacks. Since in the fragile watermark, the aim is to be sensitive to any attacks, and to detect any tempering to the model, we measured the *CC* metric between the original watermark W and the extracted watermark W′, and between the original model M and watermarked model M′ to measure the sensitivity to the attacks.

We have applied the measures on both technique of [2] and [3] using 7 models. In the hamming code-based technique, the author normalized the 3D model into the range [0, 1] to embed the watermark but they didn't perform the denormalization after embedding. The normalization of 3D models is considered as a preprocessing step before watermark embedding which shifts the center of mass of the 3D model to the origin, so the model is scaled to fit in a unit cube, which makes the watermark resilient to any modification like affine and scaling transformations. In our experiment, the algorithm has been implemented as mentioned in [2] and implemented after the denormalization step as we suggested. The results of applying the technique on the models without the denormalization step are presented in Table 2. Table 3 shows the measurement metrics after applying the denormalization step to the 3D model. The results obviously show that the values of the *RMS* are less than the first values which indicate minimal positional changes during the watermark embedding.

Table 2. Hamming code based fragile technique measurements without denormalization

Model	No. vertices/faces	Imperceptibility measures			Robustness measure	
		HD	MHD	RMSE	CC (M, M')	CC (W, W')
Cow	2904/5804	0.9043	0.4848	0.4965	0.8995	1.0000
Casting	5096/10224	1.1005	0.4039	0.4912	0.9008	1.0000
Bunny	1355/2641	1.2419	0.6803	0.5303	0.9765	1.0000
Bunny_bent	1355/2641	1.4291	0.6760	0.5450	0.9549	1.0000
hemi_bumpy	1441/2816	1.5671	0.7331	0.5592	0.7941	1.0000
Bunny	34835/69666	0.9822	0.5176	0.4866	0.9544	1.0000
hand	36619/72958	1.0941	0.4169	0.4811	0.8988	1.0000

Table 3. Hamming code based fragile technique measurements - after denormalization

Model	No. vertices/faces	Imperceptibility measures			Robustness measure	
		HD	MHD	RMSE	CC (M, M')	CC (W, W')
Cow	2904/5804	1.5685e-15	5.0240e-16	3.4358e-16	1.0000	−0.0050
Casting	5096/10224	1.8388e-15	5.3842e-16	3.6531e-16	1.0000	0.0056
Bunny	1355/2641	1.9375e-15	7.2551e-16	4.7291e-16	1.0000	−0.0253
Bunny_bent	1355/2641	2.4139e-15	8.4952e-16	5.5116e-16	1.0000	−0.0057
hemi_bumpy	1441/2816	3.1563e-15	9.8454e-16	6.5371e-16	1.0000	0.0048
Bunny	34835/69666	1.7844e-15	5.2505e-16	3.4298e-16	1.0000	0.0011
hand	36619/72958	1.8113e-15	5.1365e-16	3.5427e-16	1.0000	0.0117

However the values of the Correlation Coefficient between the embedded and extracted watermark (*W*, *W'*) is equal to 1, and between the original model and the watermarked model (*M*, *M'*) is less than 1 according to the method without denormalization step. On the other side, by applying the denormalization step we found the *CC* (*M*, *M'*) equal to 1 which means zero distortion after embedding the watermark.

Figure 1 shows the model before and after embedding the watermark before and after denormalization. Moreover, Figs. 2 and 3 show the differences between the coordinates X, Y, and Z values of the vertices of the original and the watermarked model before and after denormalization respectively.

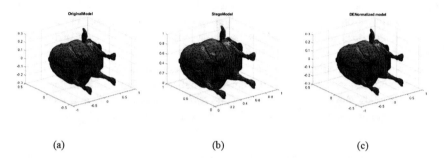

(a)   (b)   (c)

Fig. 1. (a) Original Caw model, (b) Stego Caw model after implementing technique [2] without denormalization, (c) Stego Caw model after implementing technique [2] with denormalization

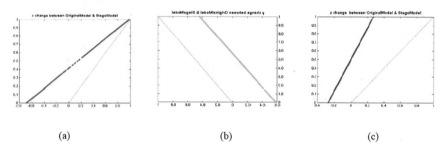

(a)   (b)   (c)

Fig. 2. Change in x, y and z coordinates after applying Hamming code technique [2] without denormalization

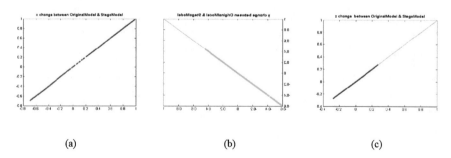

(a)   (b)   (c)

Fig. 3. Change in x, y and z coordinates after applying Hamming code technique [2] then applying the denormalization

Table 4 shows the measurement metrics of the Chaos sequence based fragile technique [3]. Which illustrates that the imperceptibility measures are less than the previous technique that means this technique doesn't distort the model after watermark embedding. Figure 4 shows the model before and after watermarking and the difference between the vertices in XYZ coordinate system.

Table 4. Chaos sequence based fragile technique measurements

Model	No. vertices/faces	Imperceptibility measures			Robustness measure	
		HD	MHD	RMSE	CC (M, M')	CC (W, W')
Cow	2904/5804	8.9034e-16	2.8523e-16	1.9411e-16	1	1
Casting	5096/10224	1.0270e-15	2.7498e-16	1.8997e-16	1	1
Bunny	1355/2641	3.9374e-16	1.5555e-15	2.6232e-16	1	1
Bunny_bent	1355/2641	4.0792e-16	1.6542e-15	2.8374e-16	1	1
hemi_bumpy	1441/2816	1.6514e-15	5.3170e-16	3.5182e-16	1	1
Bunny	34835/69666	9.0876e-16	2.3628e-16	1.6133e-16	1	1
hand	36619/72958	8.7595e-16	2.4214e-16	1.6964e-16	1	1

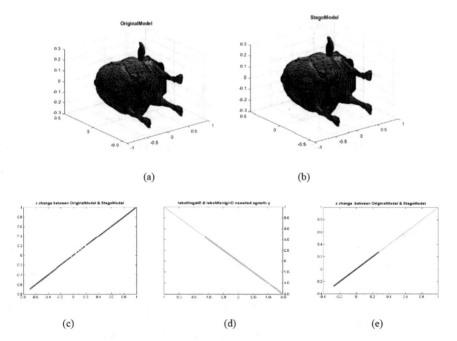

Fig. 4. (a) Original Caw model, (b) Stego Caw model after implementing Chaos based technique [3] (c), (d) and (e) the change in X, Y and Z coordinates.

5 Conclusion

In this paper, we presented a comparative analysis between two substitutive fragile watermarking algorithms of Wang et al. [2, 3] by clarifying the points of strength and weaknesses of both. The main requirements to design an effective watermark are imperceptibility, robustness against intended or non-intended attacks and capacity. The main reason of selecting both techniques for this research is that both are substitutive fragile algorithms that use all the vertices in the model (full capacity) while differing in the watermark generation pattern strategies. We have used the *RMSE, HD,* and *MHD* to measure the imperceptibility. The Correlation Coefficient *CC* was often used to measure the robustness of the watermark, but we use it to measure the sensitivity of the watermark as shown in the experiment results. By analyzing these techniques, we found that they achieved high embedding capacity as they used all of vertices for embedding that also leads to high distortion. To avoid this distortion, we suggested selecting the best vertices for embedding by using one of computational intelligent *CI* techniques like neural network.

References

1. Mabrouk, K.M., Semary, N.A., Abdul-Kader, H.: Fragile watermarking techniques for 3D model authentication: review. In: Hassanien, A., Azar, A., Gaber, T., Bhatnagar, R., Tolba, M.F. (eds.) The International Conference on Advanced Machine Learning Technologies and Applications (AMLTA2019). AMLTA 2019. Advances in Intelligent Systems and Computing, vol. 921, pp. 669–679. Springer, Cham (2020)
2. Wang, J.T., et al.: Hamming code based watermarking scheme for 3D model verification. In: 2014 International Symposium on Computer, Consumer and Control, Taichung, pp. 1095–1098 (2014)
3. Wang, J.T., et al.: A novel chaos sequence based 3d fragile watermarking scheme. In: 2014 International Symposium on Computer, Consumer and Control, Taichung, pp. 745–748 (2014)
4. Yeo, B.L., Yeung, M.M.: Watermarking 3D objects for verification. IEEE Comput. Graph. Appl. **19**(1), 36–45 (1999)
5. Lin, H.S., et al.: Fragile watermarking for authenticating 3-D polygonal meshes. IEEE Trans. Multimed. **7**(6), 997–1006 (2005)
6. Dugelay, J.L., Baskurt, A., Daoudi, M. (eds.): 3D Object Processing: Compression: Indexing and Watermarking. Wiley, Hoboken (2008)
7. Wu, H.T., Cheung, Y.M.: A fragile watermarking scheme for 3D meshes. In: Proceedings of the 7th Workshop on Multimedia and Security, pp. 117–124. ACM (2005)
8. Wang, W.B., et al.: A numerically stable fragile watermarking scheme for authenticating 3D models. Comput. Aided Des. **40**(5), 634–645 (2008)
9. Wang, Y.P., Hu, S.M.: A new watermarking method for 3D models based on integral invariants. IEEE Trans. Vis. Comput. Graph. **15**(2), 285–294 (2009)
10. Wang, J.T., et al.: Error detecting code based fragile watermarking scheme for 3D models. In: 2014 international Symposium on Computer, Consumer and Control (IS3C), pp. 1099–1102. IEEE (2014)
11. Ichikawa, S., Chiyama, H., Akabane, K.: Redundancy in 3D polygon models and its application to digital signature. J. SCG **10**(1), 225–232 (2002)

12. Wu, H.T., Cheung, Y.M.: Public authentication of 3D mesh models. In: 2006 IEEE/WIC/ACM International Conference on Web Intelligence (WI 2006 Main Conference Proceedings) (WI 2006), Hong Kong, pp. 940–948, December 2006
13. Bennour, J., Dugelay, J.L.: Protection of 3D object visual representations. In: 2006 IEEE International Conference on Multimedia and Expo, Toronto, Ontario, pp. 1113–1116, July 2006
14. Sales, M.M., RondaoAlface, P., Macq, B.: 3D objects watermarking and tracking of their visual representations. In: The Third International Conferences on Advances in Multimedia (2011)
15. Ohbuchi, R., Masuda, H., Aono, M.: Watermarking three-dimensional polygonal models through geometric and topological modifications. IEEE J. Sel. Areas Commun. 16(4), 551–560 (1998)
16. Mao, X., Shiba, M., Imamiya, A.: Watermarking 3D geometric models through triangle subdivision. In: Security and Watermarking of Multimedia Contents III, vol. 4314, pp. 253–260, August 2001
17. Chou, C.M., Tseng, D.C.: A public fragile watermarking scheme for 3D model authentication. Comput. Aided Des. 38(11), 1154–1165 (2006)
18. Cayre, F., Macq, B.: Data hiding on 3-D triangle meshes. IEEE Trans. Signal Process. 51(4), 939–949 (2003)
19. Chou, C.M., Tseng, D.C.: Affine-transformation-invariant public fragile watermarking for 3D model authentication. IEEE Comput. Graph. Appl. 29(2), 72–79 (2009)
20. Huang C.C., et al.: Spherical coordinate based fragile watermarking scheme for 3D models. In: International Conference on Industrial, Engineering and Other Applications of Applied Intelligent Systems, pp. 566–571. Springer, Heidelberg, June 2013
21. Xu, T., Cai, Z.: A novel semi-fragile watermarking algorithm for 3D mesh models. In: 2012 International Conference on Control Engineering and Communication Technology, Liaoning, pp. 782–785 (2012)
22. Chen, H.-K., Lee, C.-I.: Anti-control of chaos in rigid body motion. Chaos Solit. Fractals 21(4), 957–965 (2004)
23. Borah, S., Borah, B.: Watermarking techniques for three-dimensional (3D) mesh authentication in spatial domain. 3D Res. 9 (2018). https://doi.org/10.1007/s13319-018-0194-7

Attribute-Based Data Retrieval with Keyword Search over Encrypted Data in Cloud

Azza A. Ali[1,2] and Shereen Saleh[2(✉)]

[1] Department of Computer Science, College of Science and Humanities, Jubail,
Imam Abdulrahman Bin Faisal University, Dammam, Saudi Arabia
aaaali@iau.edu.sa
[2] Math and Computer Science, College of Science,
Minufiya University, Shebeen El-Kom, Egypt
shereensaleh357@gmail.com

Abstract. This paper presents searchable encryption with access control under multi-user setting with secure index represented by Hash Table structure encrypted by $RC2$ encryption algorithm and combine with Cipher-text Policy Attribute Based Encryption $(CP - ABE)$. It can be efficiently constructed and implemented. Extensive analysis on selected data shows that proposed scheme is efficient and accurate.

Keywords: Text Information Retrieval (TIR) ·
Ron Rivest encryption algorithm $(RC2)$ · Attribute Based Encryption (ABE) ·
Ciphertext Policy- Attribute Based Encryption $(CP - ABE)$ ·
Hash Table $RC2$ Cipher-text Policy Attribute Based Encryption
$(HTRC2 - CP - ABE)$

1 Introduction

Information is the most vital resource for all organizations either small or large and considered as valuable assets. The amount of information that should be deal with it has been raised dramatically with the appearance of Internet and alternative communication means. The issue of information overload in organizations may be solved by technologies providing intelligent information retrieval with easy accessibility to information from anyplace to anyone. The first step in information retrieval is constructing index which organize, summarize documents and facilitate the search operation. After creating index and thinking about upload data to remote server to make it available to multiple users and remove the burden of management and maintaining this huge amount of data, sensitive data will be exposed to critical issue, so data should be encrypted before outsourcing to un-trusted server. Therefore the encrypted data will be available to search and retrieve from it from anyplace. Although searchable encryption schemes present search over encrypted data, they cannot identify access control on search result. So in this paper we will introduce searchable encryption technique in addition to applying the access control mechanism by using Attribute Based Encryption ABE scheme, where Attribute Based Encryption (ABE) is a type of public-key cryptography. In Attribute Based Encryption (ABE) cipher text and private keys are

A. E. Hassanien et al. (Eds.): AISI 2019, AISC 1058, pp. 876–890, 2020.
https://doi.org/10.1007/978-3-030-31129-2_80

based on the notion of attributes or a policy over attributes to achieve the access control on search result [1, 2], where different access control attached to encrypted files to determine which users authorized to access the data based on the attributes owned by each user [3].

There are two kinds of Attribute Based Encryption:

1-Ciphertext-Policy Attribute-Based Encryption ($\mathbf{CP-ABE}$): An entity encrypting data using specific access policy generated from the users' attributes which used to specify the access rights of the users. Access policy represented as attributes combined with each other by gates (AND, OR) where AND gate means that to decrypt cipher text should contain all attributes but OR gate means that user can decrypt cipher-text if has one of two attributes. In $\mathbf{CP-ABE}$ technique the access policy is associated with cipher-text and attributes concatenated in user's private key. Only users can decrypt cipher-text if they have private key's attributes satisfy specific access policy in cipher-text. For example, if all attributes is defined as $\{G, H, I, L\}$ and user1 has private key with attributes $\{G, I\}$ and user2 has attribute $\{L\}$. If a cipher-text is encrypted with reference to the policy $(G \wedge H) \vee L$, then user2 will be able to decrypt the data [4].

2-Key-Policy Attribute Based Encryption ($\mathbf{KP-ABE}$): $\mathbf{KP-ABE}$ considered the other side of $CP-ABE$ where plaintext encrypted with set of attributes and user private key hold specific access policy which identify the appropriate access to documents, e.g. $(M \wedge L) \vee O$, and a cipher-text is related to a set of attributes, e.g. $\{M, N\}$. In this example the user does not have the identified attributes with cipher text, so cannot decrypt it and can only decrypt cipher text has attributes $(M \wedge L)$ or cipher text with O attributes [5].

Cloud computing considered as un-trusted server so sensitive data should be encrypted before outsourced to cloud to protect the privacy of user data. Recently, many papers studied searching over encrypted data and retrieve from data with no need to decrypt it. But most of these schemes focus only on single user setting although the most practical scenario in Cloud is multiple users with different access rights trying to access the data. So In this work, firstly create index as Hash Table data structure, the Hash Table contains keys and values, where key contains all dataset language alphabetic character and values contain array for each character, where each record in array contain words start with this character, name of files contain this word, frequency of this word in each file, and frequency of word in all dataset. Then encrypt the index using $RC2$ encryption algorithm word by word and put the encrypted word in array and remove the unencrypted word, then using $CP-ABE$ to encrypt file identifiers Fid with access policy and only users have attributes attached with his private key satisfy this access policy in Fid can retrieve file and decrypted it. In this paper we propose information retrieval Hash Table index combined with $RC2$ encryption algorithm and Cipher-text Policy Attribute Based Encryption technique $CP-ABE$.

The contributions of this paper are:

1. Firstly creating index as Hash Table which is implemented and maintained easily containing all words and clarify the importance of each word in file and in all dataset.

2. Using *RC*2 encryption algorithm which considered secure and return accurate result over encrypted data.
3. Using Decryption Authority Server to reduce the computational cost workload on user, where the computational cost implement Attribute Based Encryption *ABE* has workload affect on users with bounded computing resources, so using Decryption Authority Server to execute heavy computations bilinear pairing operations on *ABE* cipher-text will help user when decrypting process of Attribute Based Encryption *ABE*.

The paper organized as follows: Sect. 2 describes the related work, Sect. 3 defines the proposed algorithm, Sect. 4 contains the experimental results, and the last section is the summary and conclusion.

2   Related Work

Information retrieval based on the concept of using index which convert document to simplest view summarized the document content and enable users search and retrieve from it [6], this index represented by various data structures. Whatever the structure of the index is, but all these structures share that the index contains all words that appear in documents. Text Information Retrieval used different techniques of data structure. The first phase of Text Retrieval (TR) process is indexing process. There are many effected factors that affect the performance of indexing data structure. Construction time of the indexing and searching time considered critical factors in text retrieval process. Retrieval process is represented as component sub-processes. These sub-processes include set of files as dataset, index data structure created from that dataset and some operations performed against the index to retrieve files. When user apply specific query that represent its need to search over the index it first convert the query using the same transformation steps used in text file. Then implement query to retrieve desired documents. There is various index structure used in index and retrieval process [7]. To ensure privacy of the data, encrypting it using one of the encryption techniques. And implement search operation over this encrypted data.

Searchable encryption is a new concept used to search on data encrypted in an un-trusted third party without information leaking. Many schemes studied in this field to enhance the search and efficiency based on secret-key cryptography [8–10] or public key cryptography [11]. Song et al. [14] proposed the first searchable encryption scheme which encrypt each word in file and traverse all files to find specified keyword search. Curtmola et al. [8] constructed one index for the entire file collection. Cao et al. [9] proposed the first multi-keyword ranked query scheme over encrypted cloud data using inner product between query and index then compute highest score result and return to user. Later on, Sun et al. [10] also used "cosine similarity measure" between query and index to obtain accurate search result. In [11], Boneh et al. proposed single keyword search over encrypted data using public key cryptography. The scheme from [12, 13] enable the user to ask conjunctive keyword query to the server.

Recently many papers focus on developing attribute-based encryption to force access control on users and making this technique more flexible [17, 18]. Sahai and Waters [15] proposed fuzzy Identity-Based Encryption (*IBE*) in 2005. Yu et al. [16] also designed a selectively secure *CP − ABE* scheme with the ability of attribute revocation, classified into two forms of *ABE* which are practical when using *ABE*. Then Goyal et al. [17] designed the first Key Policy Attribute-Based Encryption *KP − ABE* scheme, where plaintext encrypted by set of attributes and only users have access structure on private key satisfy this attributes can be decrypt the data. On the other side *CP − ABE* considered plaintext encrypted using access structure and users have private key with attributes satisfy this access structure can decrypt this data [18].

In 2014, Abdellah and et al. [19] presented access control over searchable encryption. This scheme builds an index consists of two data structures: an array A and lookup table T. The array contains documents' identifiers encrypted with a set of attributes using *KP − ABE* which contain all words in the dataset and storing it in random order. Documents' identifiers which contain specific word represent in array as a chained list, that list consist of set of nodes where each node of the list is consisted of: the document' identifier which includes the word, the decryption key of the subsequent node in the array, and the address of the subsequent node in the array (subsequent node is another node in the same list contains another document's identifier which contains the same word) as illustrated in Fig. 1. The lookup table contains the position of the first element of the list in the array and its decryption key as illustrated in Fig. 2. Every element of the lookup table T is represented by a pair $<y_i, x_i>$. The field y_i used to locate specific entry in lookup table T. The second field x_i contains the address $Adr(N_{i1})$ in the array A of the first element of the list with its decrypted key K_{i1}.

Fig. 1. Structure of the array A **Fig. 2.** Structure of the lookup table T

3 Proposed Algorithm

In this paper the proposed index represented as Hash Table <key, value> is created for storing mapping between words and its location which documents containing that keyword which improve the search process. The proposed index Hash Table contains in each entry <key, value> where key represents by one of the English alphabetic character and value contains array with each record of the array consists of a word start with the character contained in the key corresponding to that value, file identifiers that contain that word concatenating with id-frequency of each file the word appeared in it, and total frequency which means the number of times the word appeared in the dataset. Then putting all words start by the same alphabetic character in specific array in one entry in Hash Table index. The strategy of the search process is getting the first character of the keyword search, and then traversing the Hash Table index to search for this character. Once it found start searching for the word in the array corresponding to this character, as soon as the word is get by the search, the whole record which is related to this word is also get, which contains all information related to that word. The whole proposed model illustrated in Fig. 4.

The main advantages of the proposed index are straightforward to build and enable Keyword-based search. After creating Hash Table index, using $RC2$ encryption algorithm to encrypt keywords in index: where plaintext is encrypted through two kind of round denoted mixing and mashing round. Mixing operation implemented by sixteen round and implements mashing operation in two rounds [20]. To achieve access control on the result of searchable encryption technique, combining ABE with searchable encryption. Where encrypting each file' identifier Fid in the index using $CP - ABE$, where most searchable encryption schemes focus on implement the system as single user setting. The proposed scheme implemented as multiple user with different access control.

Finally outsources the encrypted documents in the dataset to the server with the secure index. In this work, consider the searchable encryption scheme with Attribute Based Encryption illustrated in the Fig. 5.

3.1 Combine Encrypted Keyword Search Scheme Algorithm with Cipher-text Policy Attribute Based Encryption $HTRC2 - CP - ABE$

Which consist of the following steps:

1 - $ABE - setup$: Using the security parameters as input. The Trusted Authority generating public key PK and the master key MK using in encryption $CP - ABE -$ Encryption and $CP - ABE -$ Decryption processes.

2 - $BuildIndex$: Building index include initialize Hash Table that consist of keys and values. The keys contain all the alphabetic character of the language which is used in the dataset. And the values contain array for each key, each record within the array contains word start with the character in the key correspond to that value, name of files

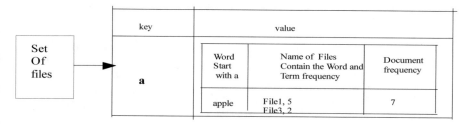

	key	value		
a		Word Start with a	Name of Files Contain the Word and Term frequency	Document frequency
		apple	File1, 5 File3, 2	7

Fig. 3. Simple part of proposed index

that contain this word, term frequency of that word in the file and document frequency for that word within all the dataset.

1. First get files from dataset (file by file) and read each file (word by word).
2. For each word, determine the first character of that word.
3. Put the word on the specified record in index which related to its first character, computing its term frequency and document frequency for this word. These steps repeat until the end of the file. As illustrated in Fig. 3.
4. At the end of this file, a new file is read from the dataset and processed as in the second step, till the last file in the dataset.

Algorithm 1: Create Hash Table index

1. **Input data**: set of files F
2. **Initialize** Hash Table $H < key, value >$ where key is $a, b, ..., z$ characters and $values$ is array, each record in array contains words start with specific character and file identifier f_{id} contains the word, term frequency tf for that word and document frequency df for word in all dataset.
3. **For each** $(f_i \in F)$, $i = 1,2, ..., n$ where n number of files **do**
4.   **For each** $(w_j \in f_i)$, $j = 1,2, ..., n$ where n number of words in each file **do**
5.   Char $C =$ Get first character (w_j)
6.   Compute term frequency tf for w_j in f_i
7.   Compute document frequency df for w_j in F
8.   Check C against each entry in HashTable H to find array X corresponding to that char.
9.   **Add** in the array X the record $(w_j, f_{id} contain\ w_j + tf, df)$
10.   **End for**
11. **End for**

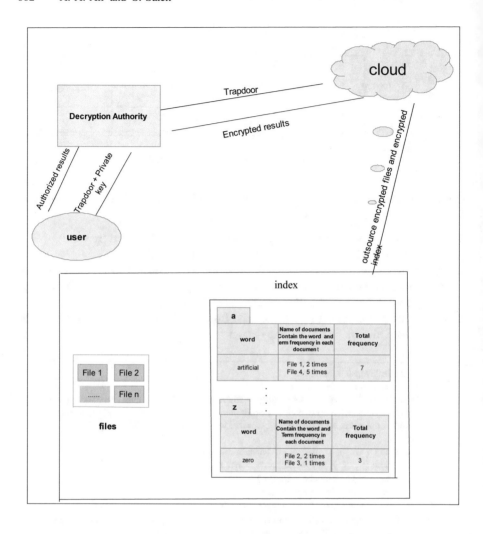

Fig. 4. Modeling of the proposed scheme $HTRC2 - CP - ABE$

3 - Index_Enc:

3.1 Using $RC2$ algorithm for encrypt alphabetic characters and each keyword in index: Firstly, the input key used for encryption must be expanded as following:

- *Key expansion*: The input parameter for key expansion is key whose length can vary. The user choose key of T bytes in the range $1 <= T <= 128$. The algorithm starts by inserting $T -$ byte key into a key buffer of 128 bytes $S[0], \ldots, S[T - 1]$. Then determine 64 subkeys, $Key[c]$ for $c = 0, \ldots, 63$ as follows: $Key[c] = S[2c] + 256 * S[2c + 1]$. The output is an extended key consisting of $64 \, dimention$ Key $[0], \ldots, Key[63]$.

- *Encryption*: This takes each word from index 64 − *bit* input placed. And encrypts it using *RC2* algorithm. The result is stored in *data*[0], ..., *data*[3] as *data* represent array of each dimension has sixteen bits.

Now the whole encryption operation can be identified as follows:

RC2 Can be implemented by using two kind of operations *MIX* and *MASH*. The word before encryption stored in array as illustrated *data*[0], ..., *data*[3]. And input key is expanded to *key*[0], ..., *key*[4] using expansion algorithm. Then execute 16 mixing rounds operation and 2 mashing rounds operation [20, 21].

3.2 CP − ABE − Encryption (File Identifier *Fid*, Security Level *s*, Access Policy *P*): Encrypt each Fid with access policy that represented as a Boolean expression of valid attributes *S* to implement the access control on each file and achieving that only authorized users can access specified files, only users have attributes that satisfy with access policy can decrypt Fid and therefore can obtain only the encrypted files of that Fid.

Main tasks in this module include:

1. Implicitly, using security level s which specified by the user to produce $AES − key(k)$.
2. With k, encrypt Fid by using *AES* generating the cipher-text CT_{AES}.
3. Data owner connected to Trusted Authority to obtain public key.
3. $CP − ABE$ Used public key *PK* and the access structure *A*, which generated from access policy *P* to encrypt k, which result *CT*.
4. The result of CP − ABE − Encryption is $\{CT_{AES}, CT, A, s\}$. This can be decrypted only by users having a set of attributes that satisfy the access structure attached to the cipher-text CT.

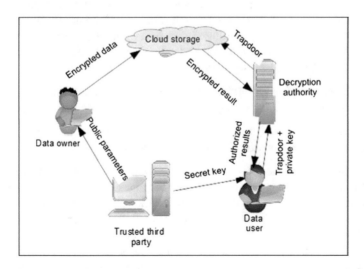

Fig. 5. Combine searchable encryption with Attribute Based Encryption

4 - Trapdoor: Sending keyword search to trusted third party Trapdoor Generation Authority. Which encrypt first character of keyword and encrypt the word by follow the same encryption algorithm $RC2$ which used to encrypt words in index. Then re-sending the encrypted keyword to the user.

5 - Search(Enc_{RC2}(first haracter of keyword search), **Enc_{RC2}**(keyword search), **Index _Enc**(Hash Table index): By knowing Enc_{RC2}(*first character of keyword search*) over $Enc(index)$, specific record is specified from the *twenty six* contains the keyword search, search is performed in this array. If the word is found, the record related to this word will be returned. Return this record to user.

So, in this work, utilizing the symmetric encryption the Advanced Encryption Standard (AES) to encrypt files and protecting the AES key used in encryption process and included in the cipher text securely by using Cipher text Policy Attribute Based Encryption $CP - ABE$ technique.

6 - CP – ABEKeyGen(PK, MK): The algorithm used to produce a private key SK which including set of attributes describing the user by using master key MK and public key PK as inputs.

7 - *CP – ABE Decrypt* (Public key PK, a cipher $-$ textCT, private keySK): $CP - ABE$ decryption algorithm used to obtain the original message by executing the following steps:

1. User connects to Trusted Authority to obtain private key SK which compatible with his attributes.
2. Trusted Authority returns the private key SK to user.

User starts connection with Decryption Authority server which has capability for robust computing, send to it private key and trapdoor then the Decryption Authority send the trapdoor to cloud, cloud start search using Search algorithm and return $CP - ABE(Fid)$ result to Decryption Authority to make partial decryption removing the computational overhead from the user and return result to user only the Fid which authorized to the user and satisfy with his attributes in his private key.

3. Only Fid that satisfy user attribute decrypted and return to user.
4. User sends this Fid to the cloud to retrieve files of this fid and decrypting it.

Thus, the access control undertaken ensures that users can be access only authorized files and decrypted it without seep any additional information.

Add or Update Process: In the updating process, the files which are added or modified in the dataset are required to update the index. The first step in the updating process is reading one file of the new or modified files of the dataset. The second step is reading word by word of the file, putting each word in the index in position which is specified for it by the first character of this word and updating its record data. This step is repeated until the end of the file. At the end of this file, another file from the new or modified files is read from the dataset and processed as in the second step.

This process is repeated until the end of the new or modified files in the dataset. Then using $RC2$ and $CP - ABE$ algorithm for encrypt each keyword and each file' identifier in the index.

Algorithm 2: Adding or updating new file to the index

1. **Input:** new file to add or update
2. **WHILE** (not end of the new and modified files in the dataset)
3.   **GET** new or modified file from the dataset (file by file)
4.   **WHILE** (not end of file)
5.   **READ** file (word by word)
6.   **DETERMINE** the first character of that word
7.   **PUT** the word on array which related to its first character and **UPDATE** its record data.
8.   Using $RC2$ to encrypt each keyword added to index
9.   Using $CP - ABE$ to encrypt file' identifier Fid in the index
10.   **END WHILE**
11. **END WHILE**

Delete Process: In the deleted process, the file which deleted from dataset is required to delete from index. Loop each entry in Hash Table index and delete file' identifier for specific file and update the record.

Algorithm 3: delete file from index

1. **Input Data:** file identifier Fid
2. **WHILE** (not end of index)
3.   **FOR each** (array in Hash Table index)
4.   **IF** Fid exist then
5.   **Remove** Fid
6.   **Update** record
7.   **END IF**
8.   **END FOR**
9. **END WHILE**

---

4 Performance Analysis for the Proposed Scheme $(HTRC2 - CP - ABE)$

In this section, we demonstrate through wide analysis and experiments of the proposed algorithm encrypted keyword search with attribute based encryption HTRC2 − CP − ABE on real-world dataset: the Enron Email Dataset [23]. We choose randomly variety number of emails to create dataset. The following charts obvious the construction time and searching time for the proposed scheme and Abdellah scheme [19]. Also explain the time needed for encrypting file' identifiers using $CP - ABE$, key generation time for proposed scheme and decryption time. The overall experiment system is implemented by java language on windows 7 with Intel Core i5-2450 M Processor 2.50 GHZ. Using Pairing-based library $DET - ABE$ library which encapsulates all the pairing map mathematics operations and make the implement for $CP - ABE$ more easier [22].

Table 1. Experimental results obtained from the proposed scheme (HTRC2 – CP – ABE)

Data set size	Construction encrypted index time/s	Searching time/s	Encrypted files with ABE time/s
20 kb	1040.131	0.11	28.96
300 kb	124434.136	0.5	231.666
512 kb	501849.728	0.59	463.332
1 m	2365670.422	1.4	643.332
4 m	3052927.46	3.612	1929.996

Table 1 represents the dataset size, construction time, search time and encrypted file' identifiers with $CP - ABE$ regard to the proposed scheme ($HTRC2 - CP - ABE$). The performance analysis of the proposed scheme is illustrated the remaining of this section. The efficiency of index construction and search efficiency for proposed scheme and Abdellah scheme [19], and the encryption time using $CP - ABE$, key generation time and decryption time for the proposed scheme.

4.1 Efficiency of Index Construction

Table 1 demonstrates the construction index time for the proposed scheme which is affected by the increasing number of documents uploaded as the create index time interacts with all documents in dataset. Also in Fig. 6, we are able to see the changes in construction index time once variety number of documents uploaded increases which need additional indexation time. On the other hand, in Abdellah scheme [19] index creation needs less time than the proposed scheme as it creates only one array and one lookup table for all the dataset. While in the proposed scheme needs to create array for each alphabetic character for the language used in the dataset.

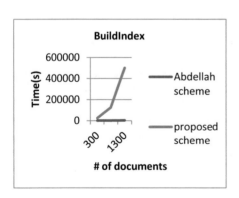

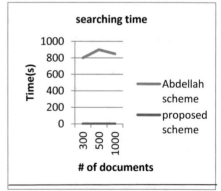

Fig. 6. Construction time of proposed scheme ($HTRC2 - CP - ABE$) and Abdellah scheme

Fig. 7. Searching time of proposed scheme ($HTRC2 - CP - ABE$) and Abdellah scheme

4.2   Search Efficiency

Figure 7 shows the results of search time for the proposed scheme does not increase significantly although the increment of the files in the dataset, as whatever the number of files in the dataset the search algorithm loop only one entry in Hash Table in the proposed algorithm $(HTRC2 - CP - ABE)$ index to find the files contain specific keyword search. It takes complexity time O(n) as the algorithm once found the first character of the keyword search in Hash Table, need to loop the array corresponding to that character to find the files contain this word. But in Abdellah scheme [19], it needs to loop more than one node in list to find the files identifiers containing keyword search. So the search time in proposed scheme is considered less than Abdellah scheme [19].

4.3   Encryption Time Using $CP - ABE$

The $CP - ABE$ encryption algorithm in Fig. 8 requires much time to achieve more computations, with the increase of the number of documents in dataset. This because it need time to compute the pairing operations and adding number of access policy to files encryption. Table 1 shows increases in time for encrypted file' identifiers Fid with access policy which using pairing map equations which need computational time with the increment of files; also $CP - ABE$ encryption algorithm need more computations with the increasing of access policy.

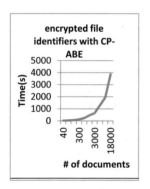

Fig. 8. Time consumed in encrypted files with ABE

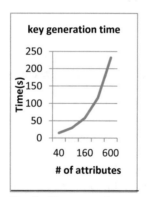

Fig. 9. Key generation time For private key

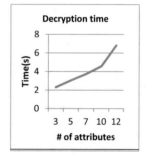

Fig. 10. Decryption time using $CP - ABE$

4.4   Private Key Generation Time

In proposed scheme using $CP - ABE$ which attach number of attributes with each private key for every user. It can obvious from implementation that the key generation time increases linearly with the increasing number of attributes. Figure 9 shows that key generation time for private key for each user increases with the number of attributes.

4.5   Decryption Time

Decryption process for file' identifiers regard to $CP - ABE$ needs more computations which become easier by using the Decryption Authority but it also increases with number of attributes contained in the policy as illustrated in Fig. 10. At the last step in decryption to obtain the plaintext data which executed by user needs a constant amount of computation.

At last we should mention that our proposed scheme enables attribute based data retrieval with searchable encryption using $CP - ABE$ which satisfy the access control on search result achieving efficiency in search time compared with [19] which combining searchable encryption with $KP - ABE$ which needs implement parallelization process to enhance search time. Also our scheme can add, update and delete files from dataset efficiently which different from [24] that present attribute based keyword search scheme with efficient user revocation that enables scalable fine-grained search authorization. Where the server contains a user list has all the legitimate users' identity information for each dataset. Also comparing with existing public key authorized keyword search scheme [25], our proposed scheme could achieve accurate search and access control at the same time. In [26] proposed search scheme with predicate encryption which differ from our scheme which enables a flexible authorized keyword search over encrypted structured data. So the proposed scheme is consider better suitable outsourcing to cloud, However in case of big data, it is possible to lead to large index which occupies a large storage space.

5   Conclusion

In this paper, we introduce and combine two concepts searching over encrypted data with Attribute Based Encryption to apply access control on search result. We also should mention that using RC2 encryption algorithm combining with $CP - ABE$ achieve the access control and enhance the performance. So the two algorithms in our proposed scheme are considered complementary to each other to achieve the required security in the proposed scheme. The experiments showed that the proposed scheme improved the search time compared with Abdellah scheme and achieving the desired goal of fasting search time. In the future, there are several modifications that can be applied in the proposed approach such as parallelizing the construction process using multi-threads techniques may be applied to improve the index construction time.

References

1. Yin, H., Zhang, J., Xiong, Y., Ou, L., Li, F., Liao, Sh., Li, K.: CP-ABSE: a ciphertext-policy attribute-based searchable encryption scheme. IEEE Access. 5682–5694 (2019)
2. Kaci, A., Bouabana-Tebibel, T., Challal, Z.: Access control aware search on the cloud computing. In: The third International Conference on Advances in Computing, Communication and Informatics-ICACCI 2014, New Delhi, India (2014)
3. Bouabana-Tebibel, T., Kaci, A.: Parallel search over encrypted data under attribute based encryption on the cloud computing. J. Comput. Secur. 77–91 (2015)

4. Fugkeaw, S., Sato, H.: An extended CP-ABE based access control model for data outsourced in the cloud. In: 2015 IEEE 39th Annual Computer Software and Applications Conference, pp. 73–78 (2015)
5. Zhu, H., Wang, L., Ahmad, H., Niu, X.: Key-policy attribute-based encryption with equality test in cloud computing. IEEE Access 20428–20439 (2017)
6. Baeza-Yates, R., Ribeiro-Neto, B.: Modern Information Retrieval, Chapter 8, Modeling. Addison Wesley, New York (1999)
7. Bat, Q., Ma, Ch., Chen, X.: Anew index model based on inverted index. In: IEEE International Conference on Computer Science and Automation Engineering, pp. 157–160 (2012)
8. Curtmola, R., Garay, J., Kamara, S., Ostrovsky, R.: Searchable symmetric encryption: improved definitions and efficient constructions. In: Proceedings of CCS, pp. 79–88. ACM (2006)
9. Cao, N., Wang, C., Li, M., Ren, K., Lou, W.: Privacy preserving multi-keyword ranked search over encrypted cloud data. In: Proceedings of INFOCOM, pp. 829–837. IEEE (2011)
10. Sun, W., Wang, B., Cao, N., Li, M., Lou, W., Hou, Y.T., Li, H.: Privacy- preserving multi-keyword text search in the cloud supporting similarity-based ranking. In: Proceedings of ASIACCS, pp. 71–82. ACM (2013)
11. Boneh, D., Di Crescenzo, G., Ostrovsky, R., Persiano, G.: Public key encryption with keyword search. In: Proceedings of Eurocrypt, pp. 506–522. Springer (2004)
12. Golle, P., Staddon, J., Waters, B.: Secure conjunctive keyword search over encrypted data. In: Proceedings of ACNS, pp. 31–45. Springer (2004)
13. Boneh, D., Waters, B.: Conjunctive, subset, and range queries on encrypted data. In: Theory of cryptography, pp. 535–554. Springer (2007)
14. Song, D.X., Wagner, D., Perrig, A.: Practical techniques for searches on encrypted data. In: Proceedings of S&P, pp. 44–55. IEEE (2000)
15. Sahai, A., Waters, B.: Fuzzy identity-based encryption. In: Advances in Cryptology–EUROCRYPT, pp. 457–473, May 2005
16. Yu, S., Wang, C., Ren, K., Lou, W.: Attribute based data sharing with attribute revocation. In: Proceedings of ASIACCS, pp. 261–270. ACM (2010)
17. Goyal, V., Pandey, O., Sahai, A., Waters, B.: Attribute-based encryption for fine-grained access control of encrypted data. In: Proceedings of CCS, pp. 89–98. ACM (2006)
18. Rathod, Ms., Ubale, S., Apte, S.: Attribute-based encryption along with data performance and security on cloud storage. In: International Conference on Information, Communication, Engineering and Technology (ICICET), pp. 1–3 (2018)
19. Kaci, A., Bouabana-Tebibel, T.: Access control reinforcement over searchable encryption. In: The 15th IEEE International Conference on Information Reuse and Integration-IEEE IRI 2014, San Francisco, USA (2014)
20. Knudsen, L.R., Rijmen, V., Rivest, R.L., Robshaw, M.J.B.: On the design and security of RC2. In: International Workshop on Fast Software Encryption, pp. 206–221 (1998)
21. Rivest, R.: A description of the RC2 (r) encryption algorithm. RFC 2268, March 1998
22. Morales-Sandoval, M., Diaz-Perez, A., Cohen, W.W.: DET-ABE: A Java API for data confidentiality and fine-grained access control from attribute based encryption. In: Proceedings of the 9th IFIP WG 11.2 International Conference on Information Security Theory and Practice – vol. 9311, pp. 104–119 (2015)
23. Cohen, W.W.: Enron email dataset. http://www.cs.cmu.edu/ ~ enron/
24. Sun, W., Yu, S., Lou, W., Thomas Hou, Y., Li, H.: Protecting your right: verifiable attribute-based keyword search with fine-grained owner-enforced search authorization in the cloud. IEEE Trans. Parallel Distrib. Syst. 1187–1198 (2014)

25. Hwang, Y.H., Lee, P.J.: Public key encryption with conjunctive keyword search and its extension to a multi-user system. In: Proceedings of Pairing, pp. 2–22. Springer (2007)
26. Li, M., Yu, S., Cao, N., Lou, W.: Authorized private keyword search over encrypted data in cloud computing. In: Proceedings of ICDCS, pp. 383–392. IEEE (2011)

Performance Evaluation of mm-Wave RoF Systems Using APSK Modulation

Ahmed A. Mohamed[1(✉)], Abdou Ahmed[1,3(✉)], and Osama A. Omer[1,2(✉)]

[1] Faculty of Engineering, Department of Electrical Engineering, Aswan University, Aswan, Egypt
a7med.abdel.ra7em@gmail.com,
{abdou.ramadan, omer.osama}@aswu.edu.eg
[2] Arab Academy for Science, Technology and Maritime Transport, Aswan, Egypt
[3] Mathematics Department, Taif University, Turabah Branch, Turabah University College, Turabah, Kingdom of Saudi Arabia

Abstract. Radio over fiber (RoF) technology is considered a promising broadband technology especially when it merged with the large unused bandwidth of millimeter-wave (mm-wave) frequencies. Selecting an appropriate modulation format for mm-wave RoF system will specify the overall performance of the system. In this paper, we proposed using amplitude phase shift keying (APSK) digital modulation for mm-wave RoF system. APSK is considered an attractive modulation scheme for digital transmission over fiber due to the power and spectral efficiencies as well as its robustness against nonlinear distortions. In comparison to the other digital modulations, simulation results show that APSK achieved a low peak-to-average power ratio (PAPR) and very limited power loss compared to quadrature amplitude modulation (QAM). But the bit error rate (BER) shows that QAM modulation has a slightly lower BER than APSK for high signal-to-noise ratio (SNR).

Keywords: RoF · mm-wave · APSK · QAM · PAPR · BER

1 Introduction

By way of introduction, broadband technologies enable end users to access the Internet at high speeds to meet the increasing demands for higher bandwidth for services and applications. Broadband has various wired and wireless access technologies, but each technology has its pros and cons. The Organization for Economic Cooperation and Development (OECD) presents detailed statistics about fixed broadband subscriptions. It shows that DSL and cable modem are the most common broadband technologies, however reports show that the number of fiber subscriptions are growing from 9% in 2010 to reach 23.4% in 2017 which illustrates that optical fiber become more popular in the last years [1].

In the wireless access domain, the fully occupied microwave bands motivated the researchers to take the advantages of the large unused bandwidth of millimeter-wave

© Springer Nature Switzerland AG 2020
A. E. Hassanien et al. (Eds.): AISI 2019, AISC 1058, pp. 891–900, 2020.
https://doi.org/10.1007/978-3-030-31129-2_81

(mm-wave) bands, between 30 and 300 GHz. Mm-waves offer large potential band-widths in their frequency ranges and the expected throughput can reach to 20 Gbit/s [4]. According to a report by SNS Research, mm-wave technology market size in 2015 was valued at 70 million USD as shown in Fig. 1 and reached to 280 million USD in 2018 with 4x increase. The report also expected that mm-wave market will keep on growing in the next decade and will reach to 1500 million USD by 2030 [2].

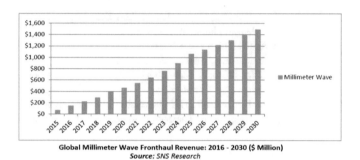

Global Millimeter Wave Fronthaul Revenue: 2016 - 2030 ($ Million)
Source: SNS Research

Fig. 1. Market size of millimeter wave technology

In [9], a summary of 60 GHz characteristics that should be considered to fully take the advantages of mm-wave communication. Mm-wave propagation characteristics differ from the other microwave frequencies. Microwave frequencies can diffract around obstacles compared to mm-waves that suffer from high path losses and dis-continuous connectivity in the non-line-of-sight situations. Also, mm-waves are not immune to rain attenuation and atmospheric oxygen absorption and reflections [3].

Actually, there is no single broadband technology can effectively deliver all the required services to all end-users anywhere anytime. Mm-wave radio over fiber (mm-wave RoF) is considered a promising solution and efficient way to exploit the advantages of optical fiber and mm-waves. The valuable idea behind fiber-wireless integration is transporting radio signals over optical fiber from central station (CS) to multiple base stations (BSs) before being radiated in the air [17] as shown in Fig. 2.

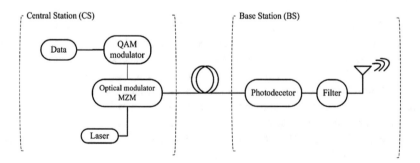

Fig. 2. Simple RoF architecture

Mm-wave access networks require many BSs to extend the coverage area; hence the cost and complexity will be increased. To minimize the BS cost, complexity should be reduced by shifting most of signal processing, routing, channel assignment, handover and mm-wave generation to CS. BS will only perform uncomplicated functions and will contain optical/electrical converter, antennas and amplifiers.

Modulation is one of system considerations that are linked to system requirements for transmitter and receiver. Consequently, the proper selection of digital modulation will specify the overall performance of the system. Higher-order QAM modulation has been proposed enormously in mm-wave systems and RoF systems that require higher data rates with a higher spectral efficiency [15]. However, QAM modulation can't satisfy the requirements of high-data rate systems due to the non-linear effects of optical fiber channel and transmission distance limitation of mm-waves.

Here we propose using amplitude phase shift keying (APSK) modulation scheme to satisfy mm-wave RoF system requirements since it has good power and spectral efficiencies as well as its robustness against nonlinear distortion compared to QAM [5, 11]. The idea behind using APSK in this paper is inspired by the recent using of APSK in commercial satellite standard such as DVB-S2 [10] that operates over nonlinear environment and require a highly efficient modulation scheme.

In this paper, we provide a short overview of mm-waves characteristics, including a discussion of design requirements for BSs in mm-wave networks. Moreover, it will be showed that APSK performs well for the both investigated metrics (peak-to-average power (PAPR) and bit error rate (BER)). The low PAPR of APSK is also statistically investigated using the MATLAB simulation and the analytical method. The paper is organized as follows. Section 2 discusses how we can get an energy efficient access network. Section 3 investigates the relationship between PA efficiency and PAPR of digital modulation while simulation results are provided in Sect. 4. Finally, Sect. 5 concludes this paper.

2 Challenges of Using mm-Wave in RoF Networks

Short coverage distance of mm-waves is considered one of the most important challenges, as the high propagation losses limits the transmission range of mm-wave to few meters. Therefore, the need for a large number of BSs in mm-wave RoF networks encouraged to design new BSs that are simple and more energy efficient. For mm-wave bands, the high transmission losses will need high effective isotropic radiation power (EIRP) at each remote BS antenna to cover the expected transmission distance. High EIRP can be obtained by using a high directive antenna as well as high output power of amplifiers.

- *High directivity antenna.* Mm-wave systems generally require high directive antenna to extend the transmission distance; however, the high directivity of the antenna limits the coverage area. In this scenario, we can achieve a highly directive signals by using beam-steering technique basing on phased-array antenna technology where the signals from multiple antennas are combined and adjusted at all antennas to directly steer the transmitted beams and provide the required EIRP [19].

- *High output power of amplifiers.* In mm-wave communication systems, we need a greater power level to achieve higher data rates. PA is the most crucial part in the radio transmitter that is mainly used to increase the power level of the radiated signal in order to be retrieved at the receiver. Hence, the transmitters demand high-power efficiency and stringent linearity from their PA in their sides.

On the other hand, PAPR determines the ratio of the highest expected transmitted power by the amplifier to the average power. PAPR is usually used to specify the needed back-off to amplify the signal with minimal distortion. A low PAPR means less back-off is needed in a PA, and hence better power efficiency can be achieved. Therefore, there is a trade-off between the PA efficiency and linearity, as the PA efficiency decreases when the PA linearity increases and vice versa. Consequently, from power efficiency and unit cost perspective low PAPR is certainly beneficial, but from a spectral efficiency perspective, it typically isn't.

3   Digital Modulation and PA Power Efficiency

It was found in [7] that BSs dominate the total energy consumption in the wireless access networks since they consume about 57% of the total energy used. The items with the highest impact on a BS's power consumption are: PAs that consume most of energy, digital signal processors (DSPs), A/D converters, cooling systems, and feeder cables. With the large number of BSs that need to be deployed for mm-wave RoF systems, the complexity of BS must be reduced, and high efficient PAs should be used for OPEX cost reduction. The power amplification stage found in mm-wave transmitters plays a key role in the energy efficiency and power requirements of the whole transmitter as PA dominates the most power consumption in BSs [18]. For any successful communication system, it has been noted that the linearity of PA is the most essential design parameter which is depending on the modulation scheme to satisfy higher-order modulation and to achieve the critical reduction in energy usage by decreasing BS energy consumption.

Traditional quadrature modulation schemes such as QAM and QPSK are considered complex modulation schemes and need linear amplifiers to accurately follow changes in phase and amplitude of the transmitted signals. In addition, when the non-constant envelop signals are passed through a saturated PA, it results in signal distortion and spectral spreading which lead to some problems for PA design. Although these modulation schemes are spectrally efficient, they have high PAPR and require a highly linear PA to process high data rate non-constant envelope signals. Therefore, using a constellation shape has equal constant amplitudes points and only change in phase will further reduce the PAPR. Generally, the higher the data rates the higher the PAPR [13], and PAPR depends strongly on the type of modulation.

In this regard, APSK is considered an attractive modulation scheme to enhance the performance of mm-wave RoF systems due to its power and spectral efficiency and its robustness against nonlinear distortion. The concept of circular APSK modulation was already proposed in 1970's [21], but it was not efficient. However, APSK modulation is mainly considered for many commercial standards related to satellite communications and DVB-S2 satellite broadcasting standard [8, 10].

APSK is a combination of amplitude and phase shift keying but in an efficient manner; it was created to obtain the best of them. PSK has an advantage of low amplitude distortion due to its constant envelop signal. But, higher order PSK results in higher BER as signal states become closer. On the other hand, the higher order QAM modulation has many different amplitude levels that will lead to higher PAPR as well as more susceptible to noise and require a linear PAs. APSK takes the QAM modulation advantage of having more bits per symbol, leading to better spectral efficiency.

APSK constellation points are configured into two or more concentric rings and each ring has constant amplitude and represents a PSK format. For instance, 16-APSK uses two rings of PSK format, the first ring contains 4 points and the second ring contains 12 points. Fewer amplitude levels of APSK enable the PA to operate closer to the nonlinear region which boosts power level and improves the efficiency. Since the PAPR of APSK is depending on the radius of the constellation circle, the proper choice of APSK parameters; number of rings, points per rings and the radii is crucial and can produce a robust constellation format with low PAPR value [16, 20]. On the other hand, in order to minimize BER, it is required to map each signal point of APSK constellation by selecting an appropriate combination of $log_2 M$ bits [12]. Higher-order APSK modulation suffers from computational complexity of de-mapping method. In recent years, many experimental studies have been devoted to produce low-complexity de-mapping methods as discussed in [6, 14].

4   Simulation Results and Evaluations

This paper concerned with power and error performance metrics to analyze M-ary digital modulation schemes. Error performance of RoF system can be measured either in wireless domain between the BS and user terminal, or in optical domain between the BS and CS. In the wireless domain, mm-wave wireless link suffers from propagation losses and sensitivity to blocking and shadowing by obstacles, which accordingly requires a greater power to achieve high data rates. On the other hand, in the optical domain, non-linear effects are caused by fiber link and arise from the optical modulation whether it was directly modulate the amplitude of the laser beam (OOK) or external using Mach-Zehnder modulator (MZM) by modulating the phase of the optical carrier. In addition to the fiber chromatic dispersion that produces phase shifts for signals and reduces their amplitude levels depending on fiber length and mm-wave frequency. In conclusion, the non-linearities and fiber distortion in the optical domain will add more power penalty to maintain the signal-to-noise ratio (SNR) and hence get better BER for the mm-wave RoF system. In this paper, we calculate the system performance, measured by the BER and power issues, in the wireless domain with different digital modulation formats. Simulations to evaluate the BER and PAPR performance for M-PSK, M-QAM and M-APSK system models were carried out by using MATLAB program. Besides, all PAPR results also were verified analytically.

4.1   BER for M-PSK, M-QAM and M-APSK

The main objective of this simulation is to analyze and measure the BER over AWGN channel for PSK, QAM and APSK modulation techniques. Figure 3 shows the BER curves for M-QAM and M-APSK modulation schemes respectively. As shown in the figure, using higher order of modulation schemes increases the bandwidth efficiency but requires higher transmission power to keep the same BER.

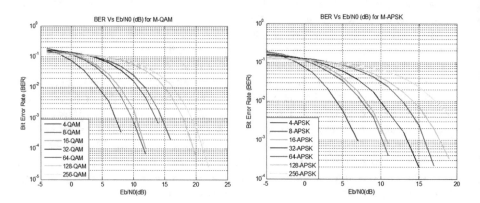

Fig. 3. BER curves for different M-ary levels modulation schemes. (a) QAM and (b) APSK.

Figure 4(a) illustrates the BER comparison between 16-ary PSK, QAM and APSK modulations. It is seen from the figure that, the achievable BER performance with QAM is slightly lower than APSK for high SNR, whereas PSK shows higher BER. The obtained result can be explained by the effect of different constellation shapes used for digital modulation. In APSK, the neighboring symbols of the outer ring have lower Euclidean distance compared to QAM constellation points. Conversely, PSK provides higher BER than QAM and APSK because the constellation points are closer and more

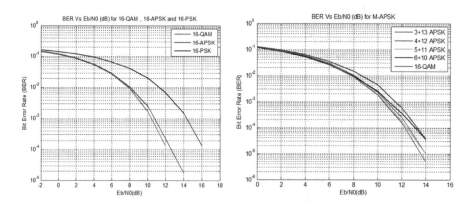

Fig. 4. BER comparison for different 16-ary levels modulation schemes (a) 16-ary PSK, QAM and APSK modulations, and (b) 16-QAM and different constellation patterns of 16-APSK.

susceptible to noise. In Fig. 4(b), BER performance of different constellation patterns for 16-APSK is plotted. As it can be seen from this figure, assigning a different number of points to each ring in APSK will result in small changes in BER.

4.2 PAPR for M-QAM, M-PSK and M-APSK

QAM. Here, we illustrate PAPR calculation for 16-QAM as an example, where there are 16 signal constellation points that are divided equally into four quadrants. The QAM signal waveforms are illustrated in Fig. 5 and each of them can be represented as a vector of length (Am) and the average of signal powers $E_{s'av}$ is represented by: $E_{s'av} = \frac{1}{M} \sum_{m=1}^{M-1} A_m^2$.

Referring to Fig. 5, we have three amplitudes levels $(\sqrt{2}A, \sqrt{10}A, \sqrt{18}A)$; the average and peak power will be $10A^2$, $18A^2$ respectively. So, PAPR can be calculated as follow: $\text{PAPR} = \frac{\text{Peak power}}{\text{Average power}} = \frac{18A^2}{10A^2} = 1.8 \rightarrow 10\log_{10}(1.8) \approx 2.56\,\text{dB}.$

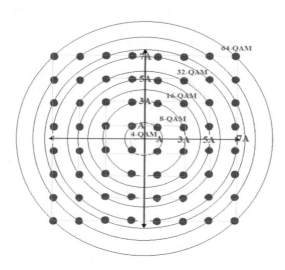

Fig. 5. Multilevel QAM signal waveform

The presented results in Table 1 show that, the higher orders of QAM have many different amplitude levels and have high average and peak powers which will results in high PAPR.

PSK. The ideal PAPR is 1:1 or 0 *dB*, and in fact, this is the ratio for M-PSK modulation schemes as all constellation points are placed in circle having the same distance from the origin. Consequently, they all have the same average and peak power. Consider the 16-PSK scheme whose radius is B and suppose that the average power of 16-PSK and 16-QAM are the same (i.e. same *Eb/N0*), $B^2 = 10A^2$, i.e. $B \approx 3.16A$. Then the peak voltage of 16-PSK will be $B \approx 3.16A$ compared to $(\sqrt{18}\,A \approx 4.24A)$ for 16-QAM. Consequently, 16-PSK is better than 16-QAM in respect of the peak voltage.

Table 1. PAPR calculation for M-QAM

QAM order	Peak power	Average power	PAPR (dB)
4	$2A^2$	$2A^2$	0
8	$10A^2$	$6A^2$	2.2185
16	$18A^2$	$10A^2$	2.5527
32	$34A^2$	$20A^2$	2.3045
64	$98A^2$	$42A^2$	3.6798
128	$170A^2$	$82A^2$	3.1664
256	$450A^2$	$170A^2$	4.2276

All calculations for any M-PSK scheme are the same as 16-PSK, since all of them are on the same ring with radius B.

APSK. For 16-APSK modulation, there are different constellation patterns, such as 3+13 APSK, 4+12 APSK, 5+11 APSK, and 6+10 APSK. Each one has 2 concentric rings and the points are arranged on the two rings with total number of 16 constellation points. The effect of the number of constellation points assigned to each ring on PAPR value is discussed here.

Constellation pattern 3+13 APSK will be used in the investigation for performance assessment. In this pattern there are two concentric rings with 3 PSK constellation points in the inner ring and 13 PSK constellation points in the outer ring. Assume that the radius of the inner and the outer ring are C, 3C respectively. Then, the average power $= 7.5C^2$, peak power $= 9C^2$ and PAPR $= \frac{9C^2}{7.5C^2} = 1.2 \approx 0.792\,\mathrm{dB}$.

Suppose that the average power of 3+13 APSK and 16-QAM are the same, then $7.5C^2 = 10A^2$, i.e. C = 1.1547A. The peak voltage for 3+13 APSK is 3C = 3.4641A compared to 4.24A for 16-QAM, then 16-APSK is better than QAM in respect of the peak voltage. Let's compare different combinations of points per ring for 16, 32 and 64-APSK in Table 2.

Relying on the above results, the proper choice of APSK parameters; number of rings, points per rings and the radii can produce a robust constellation format with low PAPR. Moreover, standard M-QAM has higher PAPR than M-APSK because higher order of QAM has more amplitude levels than APSK. By adjusting the spacing between rings and other factors, we can get lower PAPR and better resistance to distortion. In general, non-square M-QAM/APSK constellations have lower PAPR than standard square constellations because their corner points have been cutoff. Based on the overall measurement results obtained by analytical calculation and MATLAB simulation, APSK modulation has a very limited power loss compared to QAM but the BER performance shows that QAM modulation has a slightly lower BER than APSK for high SNR. In fact, digital modulation format selection is application dependent because there is always a trade-off between power and spectral efficiencies. We may need to waste more signal power to get a better data rate or vice versa.

Table 2. PAPR calculation for different M-APSK

	Peak/Average power	PAPR (dB)	Peak for same avg pwr
16-QAM	$18A^2/10A^2$	2.552	4.24A
3+13 APSK	$9C^2/7.5C^2$	0.7918	3.46A
4+12 APSK	$9C^2/7C^2$	1.09	3.58A
5+11 APSK	$9C^2/6.5C^2$	1.413	3.72A
6+10 APSK	$9C^2/6C^2$	1.76	3.87A
32-QAM	$34A^2/20A^2$	2.3	5.83A
4+12+16 APSK	$9C^2/6.12C^2$	1.67	5.42A
4+8+20 APSK	$9C^2/6.75C^2$	1.25	5.16A
5+11+16 APSK	$9C^2/6.03C^2$	1.73	5.46A
6+10+16 APSK	$9C^2/5.93AC^2$	1.80	5.50A
64-QAM	$98C^2/42C^2$	3.68	9.89A
4+12+16+32 APSK	$16C^2/11.06C^2$	1.60	7.79A
4+12+20+28 APSK	$16C^2/10.62C^2$	1.77	7.95A

5   Conclusion

This paper proposes using APSK modulation in mm-wave RoF systems and provides a performance assessment of this proposal. By assuming a 16-ary modulation order as a test case, we have compared both PAPR and BER for PSK, QAM and the different types of APSK. It has been shown that the APSK solution has a better performance and very limited power loss compared to the standardized QAM. Regarding to the indicated PAPR results, the number of constellation points per ring is considered as an important factor affecting the PAPR value of the APSK. Also, simulation curves over AWGN channels show that the BER performance of QAM modulation is very close to, and may even slightly outperform the standard APSK. From the obtained results, we can conclude that APSK is more preferable than QAM and PSK in mm-wave RoF networks to process high data rate non-constant envelope signals. Added to that, APSK provides high robustness against nonlinear amplification and it also sustains the high flexibility in terms of nonlinear detection. Further study will concern in using APSK modulation technique in OFDM systems which suffer from the high PAPR caused by the large subcarrier components.

References

1. http://www.oecd.org/internet/broadband/broadband-statistics/
2. http://www.snsintel.com/the-wireless-network-infrastructureecosystem-2017-ndash-2030.html
3. Akyildiz, I.F., Han, C., Nie, S.: Combating the distance problem in the millimeter wave and terahertz frequency bands. IEEE Commun. Mag. **56**(6), 102–108 (2018)
4. Andrews, J.G., Buzzi, S., Choi, W., Hanly, S.V., Lozano, A., Soong, A.C.K., Zhang, J.C.: What will 5 g be? IEEE J. Sel. Areas Commun. **32**(6), 1065–1082 (2014)

5. Baldi, M., Chiaraluce, F., De Angelis, A., Marchesani, R., Schillaci, S.: A comparison between APSK and QAM in wireless tactical scenarios for land mobile systems. EURASIP J. Wireless Commun. Netw. **2012**(1), 317 (2012)

6. Bao, J., Xu, D., Zhang, X., Luo, H.: Low-complexity de-mapping algorithms for 64-APSK signals. ETRI J. (2019)

7. Chen, T., Yang, Y., Zhang, H., Kim, H., Horneman, K.: Network energy saving technologies for green wireless access networks. IEEE Wirel. Commun. **18**(5), 30–38 (2011)

8. De Gaudenzi, R., Fabregas, A.G.I., Martinez, A.: Performance analysis of turbo-coded APSK modulations over nonlinear satellite channels. IEEE Trans. Wireless Commun. **5**(9), 2396–2407 (2006)

9. Emami, S.: UWB Communication Systems: Conventional and 60 GHz. Springer, New York (2013)

10. EN ETSI: 302 307 v1. 2.1 (2009–2008) digital video broadcasting (DVB). Second generation framing structure, channel coding and modulation systems for Broadcasting, Interactive Services, News Gathering and other broadband satellite applications (DVB-S2) (2009)

11. Ferrand, P., Maso, M., Bioglio, V.: High-rate regular apsk constellations. IEEE Trans. Commun. **67**(3), 2015–2023 (2019)

12. Kapoor, S., Bera, S., Sur, S.N.: Design and analysis of optimum APSK modulation technique. In: Advances in Communication, Cloud, and Big Data, pp. 11–22. Springer (2019)

13. Larson, L., Kimball, D., Asbeck, P.: Advanced digital linearization approaches for wireless RF power amplfiers. In: 2008 IEEE Dallas Circuits and Systems Workshop: System-on-Chip-Design, Applications, Integration, and Software, pp. 1–7. IEEE (2008)

14. Lee, J., Yoon, D.: Soft-decision demapping algorithm with low computational complexity for coded 4+12 APSK. Int. J. Satell. Commun. Network. **31**(3), 103–109 (2013)

15. Li, X., Yu, J., Xu, Y., Pan, X., Wang, F., Li, Z., Liu, B., Zhang, L., Xin, X., Chang, G.-K.: 60-Gbps W-band 64QAM RoF system with T-spaced DD-LMS equalization. In: 2017 Optical Fiber Communications Conference and Exhibition (OFC), pp. 1–3. IEEE (2017)

16. Liu, Z., Xie, Q., Peng, K., Yang, Z.: Apsk constellation with gray mapping. IEEE Commun. Lett. **15**(12), 1271–1273 (2011)

17. Ogawa, H., Polifko, D., Banba, S.: Millimeter-wave fiber optics systems for personal radio communication. IEEE Trans. Microw. Theory Tech. **40**(12), 2285–2293 (1992)

18. Rappaport, T.S., Murdock, J.N., Gutierrez, F.: State of the art in 60-GHz integrated circuits and systems for wireless communications. Proc. IEEE **99**(8), 1390–1436 (2011)

19. Rappaport, T.S., Sun, S., Mayzus, R., Zhao, H., Azar, Y., Wang, K., Wong, G.N., Schulz, J. K., Samimi, M., Gutierrez, F.: Millimeter wave mobile communications for 5G cellular: it will work! IEEE Access **1**, 335–349 (2013)

20. Sanchez, M.J.C., Segneri, A., Kosmopoulos, S.A., Zhu, Q., Tsiftsis, T.A., Georgiadis, A., Goussetis, G.: Novel data pre-distorter for APSK signals in solid-state power amplifiers. IEEE Trans. Circuits Syst. I Regul. Pap. (2019)

21. Thomas, C., Weidner, M., Durrani, S.: Digital amplitude-phase keying with M-Ary aphbets. IEEE Trans. Commun. **22**(2), 168–180 (1974)

Key Point Detection Techniques

Abdelhameed S. Eltanany[1]([✉]), M. SAfy Elwan[2], and A. S. Amein[1]

[1] Military Technical College (MTC), Cairo, Egypt
`abdotanany@hotmail.com`, `asamein@gmail.com`
[2] Egyptian Academy for Engineering & Advanced Technology (EAE&AT),
Cairo, Egypt
`msafy@eaeat.edu.eg`

Abstract. Image registration is a process to find the offset or misalignment between two or more images for a certain area to determine the required geometrical transformation that aligns points in one image with its corresponding in the other one. Generally, the operational goal of the registration process is a geometrical transformation for the input leading to geometrically agreement for input images, so that the matched pixels in the input images refer to the same region of the captured area. So, image registration can be applied in many applications such as change detection, mosaicking, creating super-resolution images etc. Registration process is divided into two categories: (1) Traditional methods and (2) Automated methods. For the traditional methods, the anchor, control, points are selected manually and applying the transformation model leading to time consuming and low accuracy. So, automatically detection of these points helps to recover the performance of manual selection. Registration process deals with many problems such as illumination changes, intensity variations, Different sensors, noise etc. So, its applications are mainly dependent on errors (multi temporal, multi view, or multi modal) occurred during capturing process. Feature detection, as a step of image registration process, aims to find a set of stable (invariant) distinctive key points or regions under varying conditions. Also, it is critical for the detector to be robust to changes in viewpoint, brightness, and other distortions. The goal of current paper is discussion and exhibition of the common corner detectors helping to be familiar with the various applied techniques for feature detection.

Keywords: Feature matching · Image registration · Key point detection

1 Introduction

The process of image features detection and extraction is very important for representing the image. There are two main approaches for automatic registration. Firstly, Area Based Matching methods (ABM), may be called as coarse co-registration, is pixel-to-pixel registration process. ABM methods may be Correlation methods, Fourier methods, Mutual Information (MI) methods, and Optimization methods. ABM methods emphasize the existence of features rather than their detection, i.e. no features are detected. Secondly, Feature Based Matching methods (FBM) or fine co-registration as may be called, is subpixel-to-subpixel process. Each one of these two approaches has

© Springer Nature Switzerland AG 2020
A. E. Hassanien et al. (Eds.): AISI 2019, AISC 1058, pp. 901–911, 2020.
https://doi.org/10.1007/978-3-030-31129-2_82

its own advantages and drawbacks [1–5]. Features may be points (corners, interest), edges, blobs, texture, or color [6–9]. In general, there are two types of features that can be detected, global and local features. Generally, the content of image is depicted totally using global features, whereas detecting the key points in the image requires the usage of local ones [6–10].

Global features have the drawback of image clutter and occlusions which limit the usefulness of the description. To handle the limitations of global features, its desired to do one of the following: (1) image segmentation in a limited number of regions for example the blob world system or (2) image sampling for different image subparts and this bordered the extent of potential applications [8, 10]. The expectation of local features being more useful than global ones is based on their specifications which are mainly dependent on ascendant performance, structure stability, and informativeness property [6, 7, 10]. In order to get correspondences among images, it is required to distinguish a set of noteworthy points in each one [8, 11]. Various techniques of feature detection have been proposed and were designed for a certain application such as shape matching. The best execution is mainly dependent on the parameters of the detector [8–11]. The detectors must have specific characteristics depending on the application as robustness, repeatability, accuracy, generality, efficiency, and quantity. Good feature should be repeatable, informativeness, localized [2, 8–10]. It must be noticed that distinctiveness and locality are competing properties and cannot be fulfilled simultaneously [10, 11].

The corner can be illustrated as the intersection point of two or more different directions of rims [8, 10, 11]. Accuracy of features detection, specially corners, is difficult under imperfections of data such as radiometric (illumination changes and/or different sensor characteristics) and geometric (scaling, translation, rotation, affine, projective, or nonlinear differences) differences. So, it is desired to detect the same features in the images in spite of these imperfections. Scaling, translation, and rotation problems can be easily recovered since there is no change in the shape of information. But; with respect to affine, projective, and nonlinear problems, the process is more difficult to be handled since the shape of information changed [8, 9]. Therefore, features detection methodology should be insensitive to common geometric and radiometric changes. Commonly used techniques to acquire corners are contour curvature-based methods, intensities based on derivatives methods, exploitation of color information, human visual system, model-based methods, segmentation-based methods, and machine learning techniques [1–3, 8, 11]. Examination of the performance of a corner detector, basically; started with artificially created images with various types of junctions associated with different lengths, contrasts, angles etc. This approach can make simplification for the evaluation process but cannot model all imperfections which affect the performance of a detector in a real case, so; the results of performance are often over-optimistic [1, 8–11]. Generally; the evaluations of both performance and quality of different key point detection methods were discussed where merging several key point detection techniques is useful if it is handled well [1, 8, 11].

2 Common Corner Detectors

Feature detection methods can be classified into three groups: (a) template-based corner detection, (b) contour-based corner detection, and (c) direct corner detection [8–11]. Also, it can be categorized according to the operating scale into three categories: (a) single scale, (b) multi-scale, and (c) affine invariant [1, 8–10]. As mentioned before, the aim of this paper is helping readers who is looking forward to be familiar with the different applied techniques used for feature detection.

2.1 Moravec Corner Detection

Moravec's detector [12] is a single-scale Detector. It is one of the earliest approaches that specifies a key point to be that of low essence similarity. Each image's pixel is investigated by correlating overlapping patches surrounding each neighboring pixel such that, the strength of correlation in any direction disclosures information about this point. Strength of corner is acquainted as the smallest sum of squared differences (SSD) between the patch and its neighbors for the three main directions (horizontal, vertical, and diagonals), where the moment at which SSD reaches a local maximum; a corner (interest point) is detected depending on the non-maximum suppression operation [8–11]. Existence of noise can be recovered by smoothing operations. Both existence of noise and increasing the size of window reduce the number of detected feature points and subsequently number of symmetric points. When calculating differences in eight directions, the output befits invariant to rotations that are a multiple of $45°$ [9]. Advantages of moravec's detector are as following: (a) detecting the lump of the corners, and (b) requires less computation than finding the central moment at a window of the same size. While, the disadvantages of moravec's detector are: (a) it is not isotopic, and (b) its computing is shrill [8–10].

2.2 Harris Corner Detector

Harris corner detector [13] is a single scale derivatives-based detector that had been advanced to handle the constraints of moravec's detector by getting the variation of the auto-correlation (i.e. intensity variation) over all different orientations. It can be considered as a combination of both edge and corner detectors, mainly for gray scale images [8–11, 13]. The gist idea is calculation of the eigenvalues for a small region. Then, using the largest two eigenvalues to estimate a specific function which is used together with a threshold to detect the corners which have an ill-defined gradient as shown in Fig. 1 [8, 9, 13]. The target of harris method is to find the direction of fastest and lowest change for feature orientation by using local directional derivatives covariance matrix [11, 13]. Also, an provides improvements over the Moravec method was done by developing a detector based on harris to be utilized for the color information [2, 6, 7].

Harris corner detector has many advantages as following: (1) it is invariant to translation, rotation and illumination change, (2) it is the most repetitive, most informative or discriminative, and precise localization, (3) it is the most stable one in many independent evaluations, (4) less computations requirements, (5) it provides high rotational invariance repeatability. The Disadvantages of harris corner detectors are: (1) it might be sensitive to noise, (2) its computations are more expensive, and (3) it is it is not invariant to large scale change (i.e. only repeatable up to relatively small-scale changes) [8–11]. Shi, Tomasi [14] was developed as an optimization based on harris method. Shi, Tomasi allow the usage of the minimum eigenvalues for differentiation and considerably thus regulates and facilitates the computation of harris [8, 11, 14].

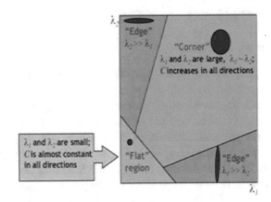

Fig. 1. Illustration of gist idea of harris detector

Since developing a detector robust to perspective transformations requires invariance to affine transformations, so; Harris Laplace/Harris Affine were developed, based on harris method, to act as a multi-scale affine invariant corner detector. The main idea, a combination of harris method and a gaussian scale space representation, allows detection of features over a range of scales. Their advantages are: (1) providing a representative set of points in the scale dimension, (2) invariant to scale changes, rotation, illumination, and addition of noise, and (3) interest points are highly repeatable. Their disadvantages are: (1) reduction the number of redundant interest points compared to multi-scale approaches as Laplacian of Gaussian (LoG) or Difference of Gaussian (DoG) detectors, and (2) failing in the case of affine transformations [8–11].

Also, a development, based on harris method, depending on local invariants was achieved allowing distinguish neighborhoods of interest points. The reason behind this trend is that edges are considered stable image features that can be detected over a multiple viewpoints, multiple scales and illumination changes. Moreover, by exploiting the edge geometry, the dimensionality of the problem can be significantly reduced [8–10].

2.3 SUSAN Corner Detectors

Susan corner detector [15] is a single scale detector which does not smooth the image or use locative derivatives. It uses a morphological approach as a different technique rather than computing the local gradients which might be noise sensitive and computationally more expensive. Smallest Univalue Segment Assimilating Nucleus (SUSAN) has been used as a corner detection and edge detection besides noise suppression where a circular mask is applied around every pixel, and checking the gray values for all pixels within the mask then compared to that of the centre pixel (nucleus) [8, 10, 11]. Corners can be distinguished as the locations where the number of pixels with similar intensity records in a local neighborhood catches a local minimum and below a specified threshold as shown in Fig. 2 where A is an edge, B is a corner, and C is neither an edge nor a corner. Increasing robustness can be achieved using weighting coefficient where Univalue Segment Assimilating Nucleus (USAN) represents the pixels having almost the same value as the nucleus are grouped together [8–11]. It is important to notice that it can work well when the noise is present since it does not require derivative [11, 15].

The advantages of SUSAN corner detector are as following: (1) its computations are fast since no derivatives are used, (2) it can be used as corner and edge detector in addition as image noise reduction, (3) it has good repeatability rate, (4) it is more efficient. (5) It is invariant to rotation, translation and illumination change. The disadvantages are: (1) it is not invariant to scaling and affine, (2) specifying a fixed threshold is not suitable for general applications; (3) it has low noise sensitivity, low robustness, and low localization, (4) it is less discriminative [2, 8–11].

2.4   Förstner Corner Detector

Förstner corner detector [16] is a single scale detector, developed in a manner to compute the location of a corner up to sub-pixel accuracy. The main goal is finding the best solution for the point closest to all tangential lines of the corner in a given window as shown in Fig. 3. For the ideal corner, all the tangential lines cut across at a unique point and this is its main gist. Förstner corner detector had been automatically adapted by Lindeberg [17] to have the ability to be multi-scale levels for computing the image gradients in the presence of the noise [8, 10].

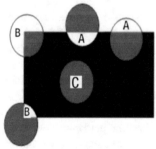
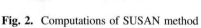
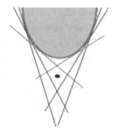

Fig. 2. Computations of SUSAN method   **Fig. 3.** Illustration of Förstner corner detector

2.5   Robust Fuzzy Rule Corner Detector

No doubt that corner detection for general purpose with high performance and efficiency is a challenge. Starting from this point, a few fuzzy approaches have specifically discussed this problem [8, 10, 11]. An algorithm has been proposed by Banerjee & Kundu [18] to extract significant corner points in gray level where the measurement of cornerness for each point is computed by means of the vague edge strength. Changing the threshold value causing different corner sets, which are fuzzy as shown in Fig. 4 [8, 10]. The main advantage of Fuzzy Rule corner detector is that it can be easily extended to detect different types of features, while using different types of detectors which operate at different stages, causing a high computational load, can be considered its main drawback [8, 10].

2.6   Hessian Corner Detector

Hessian detector [19] is a single scale blob detector, also it referred to as Determinant of Hessian (DoH) method which investigates positions in image that offer strong derivatives in two orthogonal directions. Its main idea stands behind searching for a point, center of searching window, whose determinant of hessian, based on the second derivative matrix, is maximum as shown in Fig. 5. Then, non-maximum suppression is applied retaining only pixels whose value is greater than that of all eight immediate neighbors inside the window. Finally, all lasting points, whose value is above a specific predefined threshold, are returned by the detector [2, 4, 7, 11]. Determinant of Hessian (DoH) method is used in the popular SURF algorithm [20].

The advantages of hessian detector are: (1) regions with strong texture variation could be returned with many responses, (2) it can provide further interest points that result in a densely objects, (3) it has good performance and repeatability. The drawbacks of hessian detector are: (1) its returns are only repeatable up to relatively small-scale changes, (2) it is sensitive to noise leading to low localization accuracy causing the local maxima to be less stable [2, 4, 8–11].

Hessian detector was developed by Mikolajczyk and Schmid [21] to operate as a scale invariant blob detector called Hessian-Laplace which is a combination of hessian detector and a gaussian scale space representation. The core is that it operates on local extrema, using DoH at multiple scales for spatial localization, and the Laplacian at multiple scales for scale localization. Also, it was developed to be affine invariant utilizing harris corner detector by integrating key points from several scales in a pyramid with some iterative selection gauge, and a Hessian matrix [8, 9]. Hessian-Laplace concluded that the response of hessian detector exists if significant variations over any two perpendicular ways occurred. Also, for a single image, the repeatability, discernment more reliable regions, of hessian affine approach better than that of harris affine one. In addition, the hessian affine detector responds well to regions which have a lot of corner like parts (for example, hessian affine detector performs very well for buildings scenes). The advantages of hessian affine are: (1) extraction large number of features, (2) using spatial and scale localization is more suitable for scale estimation, and (3) Hessian detectors are more stable than the Harris-based counterparts [8–11]. Although these advantages, hessian affine has a drawback that its repeatability is low compared to Harris-Laplace [8, 10, 20].

2.7   FAST Corner Detector

Rosten, E., Drummond [22] proposed Features from Accelerated Segment Test (FAST) which is a uni-scale corner detector. Interest points are identified by applying a part test to the intensity of every pixel in the image. This is achieved by considering a circle (bresenham circle) of sixteen pixels, as a base of computation, around the interest point [8, 10, 11]. The key point can be detected if its intensity value (I) is greater or smaller than the value of a collection of (n) neighboring pixels in bresenham circle as shown in Fig. 6, taking into account a predefined threshold [8].

Rosten, E., Drummond developed a high-speed test to eliminate a very large number of non-corner points by investigating only the pixels numbers 1, 5, 9 and 13. If three of these four test pixels are brighter or darker than the intensity value (I), then a corner can only exist as a first step. For final result, the remaining pixels are examined. FAST detector is similar to SUSAN, but uses a smaller window size in addition to the fact that only some of pixels are examined instead of all of them. It can be considered a relative of the local binary pattern LBP [2, 7, 10, 11].

Fast detector has many advantages as it has high performance, high speed, efficient to compute and fast to match, a good accuracy, high repeatability, its ability to average out noise is reduced and it is faster than other corner detectors. But it suffers from being not robust to high levels noise, depending on a threshold, probability to respond to one-pixel wide lines at certain angles, and it is not a scale-space detector [8, 10, 22].

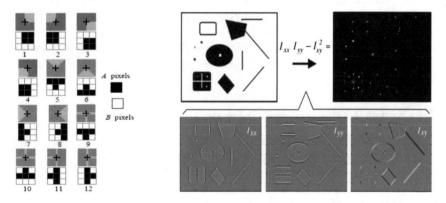

Fig. 4. Different Fuzzy Corner set   **Fig. 5.** Both Hessian matrix and determinant

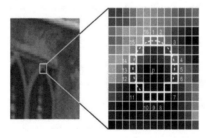

Fig. 6. Illustration of operation of FAST corner detector

2.8   Laplacian/LoG/DoG Detectors

Laplacian processing, is a method of computing the maximum rate of change (derivative) in a pixel. The Laplacian operator is described for a gray-scale image [8, 10, 11]. Laplacian of Gaussian (LoG) [23] is a multi-scale detector which is simply combines Laplacian operation over an area that had been processed using a Gaussian smoothing kernel to converge the edge [8, 9]. This means that LoG is a linear combination of second derivatives and a blob detector. The operator's output is strongly dependent on the relationship between blob structures size and smoothing Gaussian kernel size where the standard deviation of the Gaussian is used to control the scale by changing the amount of blurring [8, 11]. LoG at a pixel in the image becomes locally maximum or minimum when the pixel represents the center of a dark or bright spot respectively [8–10]. Lindeberg [17] developed an automatically method, considered to be a local curvature detection, for capturing blobs of different size where a multi-scale approach with automatic scale selection is proposed where the gradient magnitude and the local surface curvature are both high. It is important to mention that the local curvature methods are the first known to be accurate and reliable in focusing corners over scale changes [8–11]. Advantage of LoG are: (1) it is invariant to rotation, (2) it provides somewhat a good rating for other features such as corners, ridges, etc., (3) the detected features can represent extended and splitting structures. Disadvantage of LoG are: (1) its computation is time consuming, (2) less stable, (3) more sensitive to noise [8, 9].

Difference of Gaussian (DoG) [24] is a multi-scale detector which is used to speed up the computation of LoG, i.e. DoG gives a close approximation to LoG. This can be accomplished by subtracting adjacent scale levels of a Gaussian pyramid separated by a certain factor without needing for convolution. DoG is used to efficiently detection of stable features from scale-space, while it suffers from the same drawbacks of LoG [2, 6–9, 11].

2.9   Salient Regions (Entropy) Detection

Kadir and Brady [25, 26] proposed a new algorithm which is based on thought that key points over a multi-scale should present a local characteristics (entropy) that are unexpected compared to the neighborhood region. The entropy method is mainly based on locally maximizing the information content, therefore, the most distinctiveness points can be identified [8, 9, 11]. The most important note for the entropy method is being a rotationally invariant feature since location of pixel are not used in the calculation, therefore the entropy window will not alternate by rearrangement its pixels with respect to the centered window [11, 26].

2.10   Gabor-Wavelet Corner Detector

Gabor-Wavelet detector [27] is a multi-scale interest points detector which is based on the base of Gabor wavelets which are stimulated convolution kernels in the pattern of plane waves bounded by a Gaussian envelope function [8, 9, 11]. Gabor-Wavelet has many advantages as: (1) providing simultaneous optimal resolution in both spatial and space frequency domains, (2) capability of enhancing low level features such as peaks, ridges, etc., (3) capability of extraction of points from different scales, and (4) the detected points have high accuracy and adaptability to various geometric transformations [8, 9, 27].

2.11 Morphological Interest Regions Detection

Morphological operations [28, 29] can be rotation and scale invariant. The main usage of this method is creating an interest region on gray scale, binary, or color images, although preparing both gray scale and color images for morphology operations requires some form of pre-processing operations are used [2, 7–10, 26]. The binary thresholding necessitates a part of work to regulate the parameters correctly for a given application to get the highest performance. Generally, the morphological operations alone are not sufficient [2, 6–9, 26]. Since Maximally Stable Extremal Regions (MSER) [30], feature descriptor method, is based on the principle of connecting groups of pixels at maxima or minima locations, so the morphological interest regions method is similar to MSER. But it is important to know that MSER does not use morphology operators [8–11].

3   Conclusion

This paper offers a survey of different key points (corner) detection methods for remote sensing images. The aim of this paper is helping readers who is looking forward to be familiar with the different applied techniques used for feature detection. Generally, the process (method) of features selection determine the performance. Also, FBM methods is more efficient than ABM methods for reduc-ing errors and consuming time for computations.

References

1. Hassaballah, M., Hosny, K.M.: Recent Advances in Computer Vision Theories and Applications, vol. 804. Springer Nature, Switzerland (2019)
2. Bayraktar, E., Boyraz, P.: Analysis of feature detector and descriptor combinations with a localization experiment for various performance metrics. Turk. J. Electr. Eng. Comput. Sci. **25**, 2444–2454 (2017)
3. Hassanien, A.E., Tolba, M.F., Shaalan, K., Azar, A.T.: Advances in intelligent systems and computing. In: Proceedings of the International Conference on Advanced Intelligent Systems and Informatics, vol. 845, ©Springer Nature, Switzerland (2019)
4. Krishnan, R., Anil, A.R.: A survey on image matching methods. Int. J. Latest Res. Eng. Tech. **2**(1), 58–61 (2016)

5. Jiao, N., Kang, W., Xianga, Y., You, H.: A novel and fast corner detection method for sar imagery. International Archives of the Photogrammetry, Remote Sensing and Spatial Information Sciences XLII–2/W7 (2017)
6. Karim, A.A., Nasser, E.F.: Improvement of corner detection algorithms (Harris, FAST and SUSAN) based on reduction of features space and complexity time. Eng. Technol. J. 35, Part B(2) (2017)
7. Salahat, E., Qasaimeh, M.: Recent advances in features extraction and description algorithms: a comprehensive survey, vol. 1. arXiv:1703.06376 (2017)
8. Hassaballah, M., Awad, A.I.: Image Feature Detectors and Descriptors: Foundations and Applications Studies in Computational Intelligence, vol. 630. © Springer, Switzerland (2016)
9. Goshtasby, A.A.: Theory and Applications of Image Registration, 1st edn. JohnWiley & Sons Inc., New York (2017)
10. Tuytelaars, T., Mikolajczyk, K.: Local invariant feature detector: a survey. Found. Trends Comput. Graphics Vision 3(3) (2008)
11. Davies, E.R.: Computer Vision Principles, Algorithms Applications, Learning, 5th edn, ©Elsevier Inc. (2018)
12. Moravec, H.P.: Towards automatic visual obstacle avoidance. In: 5th International Joint Conference on Artificial Intelligence, pp. 584–594 (1977)
13. Harris, C., Stephens, M.: A combined corner and edge detector. In: Alvey Vision Conference, pp. 147–151 (1988)
14. Shi, J., Tomasi, C.: Good features to track. In: Conference on Computer Vision and Pattern Recognition (1994)
15. Smith, S., Brady, J.: SUSAN-A new approach to low level image processing. Int. J. Comput. Vision 23(1), 45–78 (1997)
16. Förstner, W.G.: A fast operator for detection and precise location of distinct points, corners and centers of circular features. In: International Society for Photogrammetry and Remote Sensing ISPRS, Intercommision Workshop, Interlaken (1987)
17. Lindeberg, T.: Feature detection with automatic scale selection. Int. J. Comput. Vision 30(2), 77–116 (1998)
18. Banerjee, M., Kundu, M.K.: Handling of impreciseness in gray level corner detection using fuzzy set theoretic approach. J. Appl. Soft Comput. 8(4), 1680–1691 (2008)
19. Beaudet, P.R.: Rotationally invariant image operators. In: International Joint Conference on Pattern Recognition, pp. 579–583 (1978)
20. Bay, H., Ess, A., Tuytelaars, T., Gool, L.V.: Speeded-up robust features (SURF). Comput. Vis. Image Underst. 110(3), 346–359 (2008)
21. Mikolajczyk, K., Tuytelaars, T., Schmid, C., Zisserman, A., Matas, J., Schaffalitzky, F., Kadir, T., Gool, L.: A comparison of affine region detectors. Int. J. Comput. Vision 65(1/2), 43–72 (2005)
22. Rosten, E., Drummond, T.: FAST machine learning for high speed corner detection. In: 9th European Conference on Computer Vision (ECCV 2006), pp. 430–443 (2006)
23. Gunn, S.R.: Edge detection error in the discrete Laplacian of Gaussian. In: International Conference on Image Processing ICIP Proceedings, vol. 2 (1998)
24. Lowe, D.G.: Distinctive image features from scale-invariant keypoints. Int. J. Comput. Vision 60(2), 91–110 (2004)
25. Kadir, T., Brady, J.M.: Saliency, scale and image description. Int. J. Comput. Vision 45(2), 83–105 (2001)

26. Timor, K., Zisserman, A., Brady, M.: An affine invariant salient region detector. In: European Conference on Computer Vision (2004)
27. Yussof, W., Hitam, M.: Invariant gabor-based interest points detector under geometric transformation. Digital Signal Process. **25**, 190–197 (2014)
28. Jackway, P.T., Deriche, M.: Scale-space properties of the multiscale morphological dilation-erosion. IEEE Trans. Pattern Anal. Mach. Intell. **18**(1), 38–51 (1996)
29. Soille, P., Vogt, P.: Morphological segmentation of binary patterns. Pattern Recogn. Lett. **30**(4), 456–459 (2009)
30. Donoser, M., Bischof, H.: Efficient maximally stable extremal region (MSER) tracking. In: IEEE Computer Society Conference on Computer Vision and Pattern Recognition (CVPR 2006), no. 1, pp. 553–560 (2006)

Cyber Security Risks in MENA Region: Threats, Challenges and Countermeasures

Ahmed A. Mawgoud[1] , Mohamed Hamed N. Taha[1(✉)] ,
Nour Eldeen M. Khalifa[1] , and Mohamed Loey[2]

[1] Information Technology Department, Faculty of Computers and Artificial
Intelligence, Cairo University, Giza, Egypt
aabdelmawgoud@pg.cu.edu.eg,
{mnasrtaha, nourmahmoud}@cu.edu.eg
[2] Computer Science Department, Faculty of Computers and Informatics,
Benha University, Benha, Egypt
mohamed.loey@fci.bu.edu.eg

Abstract. Over the last few years, MENA region became an attractive target
for cyber-attacks perpetrators. Hackers focus on governmental high valued
sectors (i.e. oil and gas) alongside with other critical industries. MENA nations
are increasingly investing in Information and Communication Technologies
(ICTs) sector, social infrastructure, economic sector, schools and hospitals in the
area are now completely based on the Internet. Currently, the position of ICTs
became an essential phase of the domestic future and global security structure in
the MENA Region, emphasizing the real need for a tremendous development in
cybersecurity at a regional level. This environment raises questions about the
developments in cybersecurity and offensive cyber tactics; this paper examines
and investigates (1) the essential cybersecurity threats in MENA region, (2) the
major challenges that faces both governments and organizations (3) the main
countermeasures that governments follow to achieve the protection and business
continuity in the region. It stresses the need for the importance of cybercrime
legislation and higher defenses techniques towards cyberterrorism for MENA
nations. It argues for the promotion of a cybersecurity awareness for the individuals as an effective mechanism for facing the current risks of cybersecurity in
MENA region.

Keywords: Cyber security · Malicious attack · Cybercrime · MENA region

1 Introduction

Cyber threats are an international phenomenon that keeps increasing proportionally
with the ICT sector rapid growth; they are continuously growing to advanced levels
that needs from the governments and individuals to cooperate to face this kind of cyber
risks. Cyber-crime has developed into well-organized networks with superior attack
techniques, cyberattacks grow to represent the type of risks a real war can cause; as the
rapid-growth digitization in MENA nations and the dependence on information systems increases [1]. Based on Symantec [2] survey that questioned about 30,000 person

© Springer Nature Switzerland AG 2020
A. E. Hassanien et al. (Eds.): AISI 2019, AISC 1058, pp. 912–921, 2020.
https://doi.org/10.1007/978-3-030-31129-2_83

from 24 countries, victims of cyber-attacks are growing drastically with one million attacks each day as 69% approved they were a cyber-attack victim at least once in their lifetime. Every second, 14 adults turn out to be the victim of a cyber-attack. As IstiZada [3] defined that the abbreviation MENA stands for Middle East and North Africa. MENA nations have common properties as geographically, economically and politically [2]. IstiZada started defining the acronym MENA to refer to the region spanning horizontally from Morocco to Iran [3] (Fig. 1).

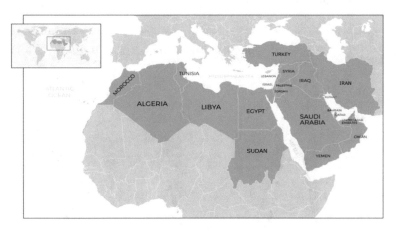

Fig. 1. MENA region map

Those countries are listed as following (Syria – Algeria – Sudan – Bahrain – Egypt – Tunisia - Iran – Iraq – Jordan - United Arab Emirates – Kuwait – Lebanon – Libya – Morocco – Oman – Qatar - Saudi Arabia - State of Palestine – Turkey - Israel - Yemen). MENA nations are pursuing digitization in all sectors, the mass adoption of linked digital technologies and software through consumers, enterprises, and governments with a rapid rate. It additionally brings with it tremendous threats perpetrated through cyber criminals and cyber hacktivists. These hackers are motivated and intent to make the best use of the existing vulnerabilities in a MENA government's digital infrastructure to gain [4]. This sustained barrage of cyber-attacks and exploitation should undermine the self-assurance the government, the enterprise sector and civil society have in derailing its development and thereby threatening the attainment of its promised benefits. The oil & gas resources of MENA nations and their fast digitization adoption have made this sector specially a desirable goal for an extensive range of cyber threats [5]. As a result, governments and worldwide organizations have sustained huge loss politically, economically and socially from cyber-attacks. With the existing of such an environment, many questions about the prevention methodologies, cyber defense tactics and legislation law both governments and individuals take to reduce those offensive cyber risks. This paper investigates the risks of cyber security in MENA region from cyber-attacks threats, impact challenges and the taken countermeasures tactics that were taken by governments and organizations in cyber surveillance, awareness and legislation [6].

2   Literature Review

MENA region is an ideal place to pose an escalated threat in the sector of practicable cyber-attacks from geopolitical uncertainty and the inflow of internet service provider overseas nationals in the region. [7] stated in his research that the high escalation rate of cybercrime acts concentrated on the MENA region. Such acts are motivated economically, financially, politically and ideologically [7]. Figure 2 shows the penetration rate and population rate in MENA.

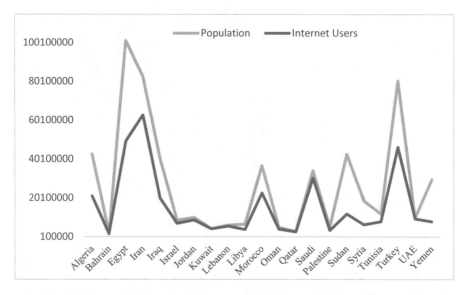

Fig. 2. The penetration rate for MENA nations' internet users in (March 2019)

In March 2019, the highest country in its population is Egypt with estimated 100 million citizens while the lowest one is Bahrain with estimated 1.6 million citizens. On the contrary, Iran has the highest number of internet users with 62 million users while Bahrain has the lowest number of internet users with about 1.5 million users. Every country's population census does not reflect the estimated ratio of internet users' for each one; Bahrain is considered as the lowest country in both population census and internet user numbers; it is considered the second highest penetration ratio in MENA nations with about 95.9% coming after Kuwait with 98%. However, Egypt has the highest population census in MENA region; it came the fourth lowest rank in penetration percentage with only 48.7% while Yemen came as a lowest penetration ratio rank of 26.7%. In this part of the world, cyber-attacks have a devastating impact now not solely on businesses, but also on national economies as well, because these are poorly diversified. In Saudi Arabia, the largest economy of the peninsula, the oil area bills for roughly 60% of GDP [8]. Strikwerda [9] stated in his research that governments are in need for more policies to face the information security challenges; however, efforts are being slowed down by using the lack of cohesive and regular countrywide and cybersecurity policies and legislation. On the other hand, the need for setting up a common framework with overarching security standards and concepts is now greater than ever before [9]. In order to study the possible cyber risks in this region, there is a need to study the development of internet usage rate in each of MENA country according to the population census of each country as shown in total in Table 1. (Internet World State) made a study on 21 countries in MENA region, according to the relation between their estimated population and the internet users in April 2019; the reason behind this comparison is to clarify the penetration ratio in each country [10].

Table 1. Populations and penetration rate in Mena countries and the rest of the world

Region	Population (2019 Est.)	World Pop. %	Internet users (April 2019)	Population (% penetration)
MENA countries	617,333,445	8%	342,590,664	67.2%
Rest of the world	7,098,889,764	92%	4043894877	56.5%
World total	7,716,223,209	100%	4,386,485,541	56.8%

In April 2019, from population number, Egypt was the highest country with estimated 100 million citizens while the lowest one was Bahrain with estimated 1.6 million citizens. On the contrary, Iran's internet users had the highest number with about 62 million users while Bahrain had the lowest number of internet users with about 1.5 million users.

3 Threats of the Cyber Attack Techniques

Cybercrime activity in the whole world in general increased drastically last years due to the rise of new attacks related to the new technologies. [11] published a statistics in Fig. 3 below compares the internet penetration rate in MENA nations to the world average in the last 10 years between April, 2009 till April, 2019 [11].

As it is shown from the Fig. 3 above, MENA countries greatly suffer from cyber attacks comparing to the rest of the world. In 2009, the internet penetration of MENA countries were 29% while the rest of the world was 27%, the difference was only 2%. However, in 2019 the ratio difference increased as MENA countries reached 78% (with about 50% increase since 2000 while the rest of the world reached 67% with about 10% diffrences. Figure 4 is a statistic that was published by (Symantic, 2018) [12].

It shows (1) Iran and Egypt are the highest MENA countries in penetration ratio while UAE is the lowest one. (2) the penetration rate in 2018 is noticeably lower than 2017 with average 20% while Fig. 5 illustrate the techniques that is commonly used in the attacks.

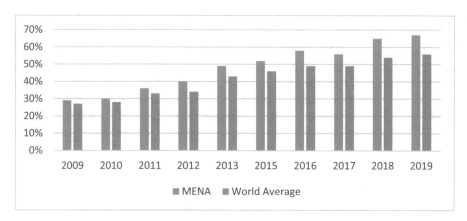

Fig. 3. Cyber-attack rate in the MENA region compared to rest of the world from 2009 to 2019

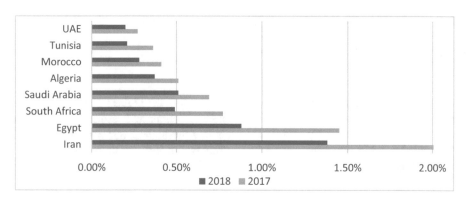

Fig. 4. Highest penetration ratio in MENA Nations (Symantic, 2018)

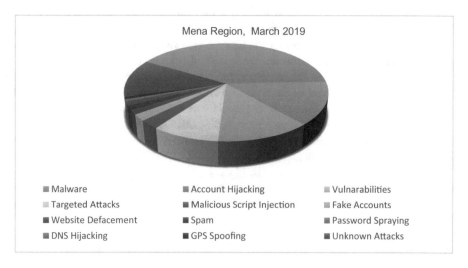

Fig. 5. Cyber-attack techniques used in MENA regions (March 2019)

As shown from Internet World Stats the statistics above in Fig. 5 [13]. Malware are presenting the highest rate threat while the GPS spoofing is the lowest one, Table 2 below illustrate each technique's rate by the attacker with its description.

Table 2. Different attack techniques rates in MENA region in March 2019

Attack technique	Ratio	Description
Malware	41.1%	An application designed to damage operating systems. "Malware" is the regular definition covering all the distinctive sorts of threats to PC protection such as anti-viruses and spyware [14]
Account hijacking	14.2%	A technique in which a person or company's account is hijacked by a hacker. It a common strategy in identity theft schemes that the hacker uses the stolen data to perform malicious or unauthorized activity [15]
Vulnerabilities	12.8%	A defect in a device that can be vulnerable for it open to attack. This vulnerability refers to any kind of weakness in the infrastructure system or any other factor that permits data security to be exposed to a threat [16]
Targeted attacks	9.2%	A risk on users pursue a target of entity's infrastructure whilst keeping anonymity. These attackers have a specific level of knowledge and have adequate sources to conduct their schemes over a long-term period [17]
Malicious script injection	2.8%	A serious security risk that allows an attacker to inject malware in the user interface factors of website to steal information or money [18]
Fake accounts	2.1%	An account where a scammer is pretending to be celebrity or a character that does not exist. Fake social profiles can consist of money owed for fake or made up individuals, pets, or organizations [19]
Website defacement	1.4%	An attack on an internet site that modifies the user interface of a website or internet page. Attackers break into the server and exchange the hosted internet web page with one of their own [20]
Spam	1.4%	A misusage messaging platforms to defraud the user with an unsolicited message, specifically advertising, and sending messages iteratively on the site [21]
Password spraying	1.4%	An attack that tries getting illegal access to a massive range of accounts (usernames) with regularly used passwords. Traditional brute-force obtains unauthorized login to accounts via guessing the password [22]
DNS hijacking	0.7%	A kind of DNS attack in which DNS queries are resolved with an error; in order to redirect internet users to malicious sites [23]
GPS spoofing	0.7%	An attack occurs when a malicious side impersonates some other de-vice or person on a network in order to launch attacks on the whole network system GPS [24]
Unknown attacks	12.8%	Recycled attacks are seen as the most inexpensive hacking method, which is why attackers regularly recycle present threats using former techniques [25]

4   The Main Cyber Security Challenges for MENA Nations

4.1   Data Exchanging Electronically in Critical Sectors

Information has not been more effortlessly accessible and transmittable. Businesses, particularly banking and financial corporations are increasingly processing, exchanging individual records electronically and throughout borders [26].

4.2   Lack of IT Security Awareness

Stated in their study about 18% of employed respondents in the MENA region are aware of the IT security regulations and policies set in their workplace. While the other 82% are not having this kind of awareness that represents a huge threat for all fields in MENA nations, this scary rate has made it an essential duty for the governments to raise the cyber security awareness in all fields [27].

5   Cyber Attacks Impacts on MENA Organizations

Statista [11] published a statistic shown Fig. 6 below that compares between MENA nations penetration rate to the rest of the world in the last 10 years [28].

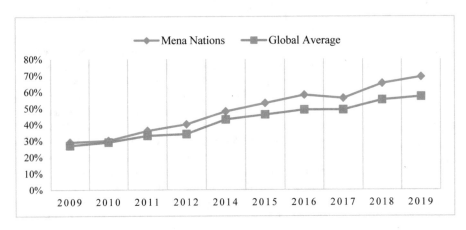

Fig. 6. Penetration rate in MENA nations to rest of the world from 2009 to 2019

MENA nations greatly suffer politically, economically and socially from the consequences of the high penetration rate comparing to the rest of the world. In 2009, the penetration rate for MENA nations were 29% while the rest of the world was 27%, the difference was only 2%. [However, in 2019 the ratio difference increased as MENA nations reached 78% (with about 50% increase since 2000) while the rest of the world reached 67% with about 10% diffrences]. Many incidents represented a huge economical threat for some of MENA countries. In Kaspersky summit in April, 2019 showed that about 23.4 million incident risks were reported in MENA countries during the first quarter of 2019, those illegal operations have targeted oil corporations, telecommunications, governments, and critical infrastructure [29].

6   Countermeasures to Tackle Cyber Risks Socially and Economically

1. *Cyber Surveillance:* Many cyber surveillances are still in the early development stages by MENA nations' governments. For example, there are reports of a cyber-group in Egypt that claims it has attacked the online propaganda of terrorists and that it aims to defend critical Egyptian infrastructure from all terrorist attacks [30].
2. *Awareness Strategy:* Attackers are constantly figuring out new methodologies of stealing information. Regrettably, the lack of cybersecurity awareness and high rate of uneducated citizens in MENA region makes them an attractive target for attackers. Raising users' awareness is essential to prevent IT security threats. It is the expert roles to clarify the risks and the techniques in their daily life [31].
3. *Legalization:* Legislation is an integral prevention part against the cyber threat. Cyber legislation in MENA nations is at an early stage. It is an essential part of anti-cybercrime strategy. Legislations were set in some of MENA nations' governments to provide a solid framework for data protection [32] (Table 3).

Table 3. Examples of cybercrime released laws in MENA nations [33]

Country	Year	Law address
Turkey	October 2004	Criminal code law no. 5237
Saudi Arabia	March 2007	Anti-Cyber Crime law decision no. 79
Algeria	August 2009	Cybercrime law no. 09-04
Oman	April 2011	Royal Decree no. 12
UAE	August 2012	The Federal Decree Law no. 5
Tunisia	May 2018	Data Protection Law, N°2004-63
Egypt	November 2018	Law 175 "Anti-Cyber Security Crimes"

7   Conclusion

MENA is an acronym refers to both Middle East and North Africa. Due to the rapid technology and network infrastructure growth, the cyber security risks rate increasingly raised to represent a huge threat to all fields in MENA nations. Internet users in MENA region highly increased in the last 10 years with a low cyber security awareness; as a result, the rate of cyber-attacks in this region affected all the MENA organisations and the impact was high in both economically and politically. The need for enhancing the network infrastructure was mandatory to face all the cyber risks challenges as well as setting a cyber-security legalisation regulate the cyber-attack operation in the region.

This paper presents a descriptive analysis of cyber-attack techniques for the cyber threats facing MENA nations with their negative impact on the countries. Finally, raising awareness and setting a cyber-security law will protect many different sectors in MENA nations as well as preventing the cyber operations in the region; as a necessary move for to expand the cyber security culture to raise the citizens awareness about safe methods of using the internet and protecting their own data from being hacked.

References

1. Barnett, D., Sell, T., Lord, R., Jenkins, C., Terbush, J., Burke, T.: Cyber security threats to public health. World Med. Health Policy 5(1), 37–46 (2013)
2. Symantec.com: Internet Security Threat Report 2019 | Symantec (2019). https://www.symantec.com/security-center/threat-report. Accessed 11 June 2019
3. IstiZada: MENA Region Countries List 2018 | IstiZada (2019). http://istizada.com/mena-region/. Accessed 11 July 2019
4. Meri, J.: Software and technology review: multilingual computing in middle east studies. Middle East Stud. Assoc. Bull. 34(1), 14–19 (2000)
5. Zeng, Q., Pu, S., Zhang, X.: Statistical tests for integrity attacks on cyber-physical systems. Asian J. Control (2018). https://doi.org/10.1002/asjc.1945
6. Efg, A.: The impact of cybercrimes on global trade and commerce. Int. J. Inform. Secur. Cybercr. 5(2), 31–50 (2016)
7. Parodi, F.: The concept of cybercrime and online threats analysis. Int. J. Inform. Secur. Cybercr. 2(1), 59–66 (2013)
8. Alshathry, S.: Cyber attack on saudi aramco. Int. J. Manage. Inform. Technol. 11(5), 3037–3039 (2016)
9. Strikwerda, L.: Should virtual cybercrime be regulated by means of criminal law? A philosophical, legal-economic, pragmatic and constitutional dimension. Inform. Commun. Technol. Law 23(1), 31–60 (2014)
10. Internetworldstats.com: Middle East Internet Statistics and Telecommunications Reports (2019). https://internetworldstats.com/stats5.htm. Accessed 09 June 2019
11. Statista: Middle East: internet penetration 2019 | Statista (2019). https://www.statista.com/statistics/265171/comparison-of-global-and-middle-eastern-internet-penetration-rate/. Accessed 03 June 2019
12. 2018.semantics.cc: SEMANTiCS 2018 (2019). https://2018.semantics.cc/. Accessed 08 June 2019
13. World Internet Users Statistics and World Population Stats. In: Internetworldstats.com. https://internetworldstats.com/stats.htm. Accessed 07 Jun. 2019
14. Kabakus, A.: What static analysis can utmost offer for android malware detection. Inform. Technol. Control 48, 235–240 (2019). https://doi.org/10.5755/j01.itc.48.2.21457
15. Thomsen, D.: IP spoofing and session hijacking. Netw. Secur. 1995, 6–11 (1995). https://doi.org/10.1016/s1353-4858(00)80045-8
16. Singh, A., Singh, B., Joseph, H.: Vulnerability Analysis and Defense for the Internet. Springer, New York (2008)
17. Bahtiyar, Ş.: Anatomy of targeted attacks with smart malware. Secur. Commun. Netw. 9, 6215–6226 (2016). https://doi.org/10.1002/sec.1767
18. Nagothu, D., Chen, Y., Blasch, E., et al.: Detecting malicious false frame injection attacks on surveillance systems at the edge using electrical network frequency signals. Sensors 19, 2424 (2019). https://doi.org/10.3390/s19112424

19. Kaveyeva, A., Gurin, K.: "VKontakte" fake accounts and their influence on the users' social network. Zhurnal Sotsiologii i Sotsialnoy Antropologii (J. Sociol. Soc. Anthropol.) **21**, 214–231 (2018). https://doi.org/10.31119/jssa.2018.21.2.8
20. Mondragón, O., Mera Arcos, A., Urcuqui, C., Navarro Cadavid, A.: Security control for website defacement. Sistemas Telemática **15**, 45–55 (2017). https://doi.org/10.18046/syt. v15i41.2442
21. Patel, K.: Recognizing spam domains by extracting features from spam emails. Int. J. Comput. Appl. **90**, 25–30 (2014). https://doi.org/10.5120/15595-4341
22. Guo, Y., Zhang, Z.: LPSE: lightweight password-strength estimation for password meters. Comput. Secur. **73**, 507–518 (2018). https://doi.org/10.1016/j.cose.2017.07.012
23. Liu, Y., Peng, W., Su, J.: A study of IP prefix hijacking in cloud computing networks. Secur. Commun. Netw. **7**, 2201–2210 (2013). https://doi.org/10.1002/sec.738
24. Jee, S., Kim, S., Lee, J.: A study to efficiently overcome GPS jamming and GPS spoofing by using data link system. J. Korea Inst. Milit. Sci. Technol. **18**, 37–45 (2015). https://doi.org/10.9766/kimst.2015.18.1.037
25. Zhao, J., Shetty, S., Pan, J., et al.: Transfer learning for detecting unknown network attacks. EURASIP J. Inform. Secur. (2019). https://doi.org/10.1186/s13635-019-0084-4
26. Rajamohan, D.S., Subha, K.K.: Information technology in financial inclusion. Paripex - Indian J. Res. **3**, 1–2 (2012). https://doi.org/10.15373/22501991/july2014/37
27. Torten, R., Reaiche, C., Boyle, S.: The impact of security awarness on information technology professionals' behavior. Comput. Secur. **79**, 68–79 (2018). https://doi.org/10.1016/j.cose.2018.08.007
28. Middle East: internet penetration | Statista. In: Statista. https://www.statista.com/statistics/265171/comparison-of-global-and-middle-eastern-internet-penetration-rate/. Accessed 08 July 2019
29. Kaspersky Security Analyst Summit (SAS) – Singapore, 8–11 April 2019. In: Kaspersky Security Summit (SAS) – Singapore, 8–11 April 2019. https://sas.kaspersky.com/. Accessed 11 July 2019
30. Bennett, C., Clement, A., Milberry, K.: Introduction to cyber-surveillance. Surveill. Soc. **9**, 339–347 (2012). https://doi.org/10.24908/ss.v9i4.4339
31. Tasevski, P.: IT and cyber security awareness – raising campaigns. Inform. Secur.: Int. J. **34**, 7–22 (2016). https://doi.org/10.11610/isij.3401
32. Restricting cybersecurity, violating human rights: cybercrime laws in MENA region. In: OpenGlobalRights (2019). https://www.openglobalrights.org/restricting-cybersecurity-violating-human-rights/. Accessed 22 Jan 2019
33. Hakmeh, J.: Cybercrime Legislation in the GCC Countries, International Security Department, Chatham House (The Royal Institute of International Affairs), July 2018

Smart Grid and Renewable Energy

Analysis of the Construction and Operation Cost of the Charging Station Based on Profit and Loss

Ce Xiu[1(✉)], Jin Pan[1], Xuefeng Wu[1], and Jinyuan Liu[2]

[1] Economic Research Institute, State Grid Liaoning Electric Power Supply Co., Ltd., Shenyang, China
xiuce@163.com
[2] Economic and Technical Research Institute, State Grid Liaoning Electric Power Supply Co., Ltd., Shenyang, China

Abstract. In recent years, electric cars become Chinese implementation of national energy strategy. The construction of resource-saving and environmentally friendly countries important strategic initiatives, in the state of charging station construction attaches great importance to, give full consideration to the charging station of economy, do inputs and outputs a reasonable current charging stations in the construction of the key. However, the current operation of the charging stations is basically in a state of loss. Electric vehicle charging infrastructure is necessary for the development of the electric car infrastructure, the construction of the type, size, and mode of operation directly restricts the development of the electric vehicle industry. Therefore, it is necessary for us to select suitable charging facilities. In view of this situation, this paper analyzes the impact of construction cost and operating cost on the profit and loss point of charging stations, and discusses the construction and operation cost control.

Keywords: Charging station · Multilevel · Profit and loss

1 Introduction

With the rapid development of economy, due to air pollution caused by the traditional fuel automobile consumption of energy and pollutant emissions accounted for the share is more and higher, the development of electric vehicle industry is to solve the scarcity of oil resources and environmental pollution are the most effective way [1–3]. And in the development of the electric car industry chain, charging station is undoubtedly the important link of the industry chain, is the basis for the development of electric vehicle industry, to accelerate the development of electric vehicle industry, infrastructure construction, the infrastructure construction of charging stations than the development of the automobile to moderate advance, has become the consensus of the development of the electric car industry at home and abroad. Electric vehicle charging station construction, operation, and maintenance of electric vehicles is the first step in the commercialization and industrialization process, is to dispel market concerns, to facilitate customers to take electricity, reduce the cost of the owner of the important

A. E. Hassanien et al. (Eds.): AISI 2019, AISC 1058, pp. 925–932, 2020.
https://doi.org/10.1007/978-3-030-31129-2_84

groundwork. Currently our charging station construction is only in its infancy, have been completed and the charging stations mostly belong to the nature of the demonstration, impossible to conduct application in accordance with the corresponding model, therefore must be combined with different electric vehicle charging demand characteristics of good charging station construction and planning, improve the convenience of electric vehicle charging, so as to attract more people to buy electric cars, so as to realize the electric car and charging the implementation of construction of positive feedback, and promote the healthy development of the electric car industry, promote the realization of the strategic objectives of the overall energy of the country.

The measure of the electric vehicle charging station construction is feasible and the most fundamental elements are after the completion of the operation of economic analysis, for the economic analysis mainly needs to consider charging station construction costs and put into operation after the profit [4–6]. Charging facilities construction investment cost contains many factors, including the cost of land, supporting the cost of power grid construction, the supporting road network construction cost, construction cost, equipment costs, and other charging station operation profits mainly from the purchase price and sale of the difference between the electricity tariff. The of electric vehicle charging station construction and operation cost of the influencing factors analysis, and according to the development needs of the EV industry in our country, income from charging station, construction cost, operation cost analysis the charging station construction of comprehensive benefits. Figure 1 illustrates the flow of the energy storage charging station.

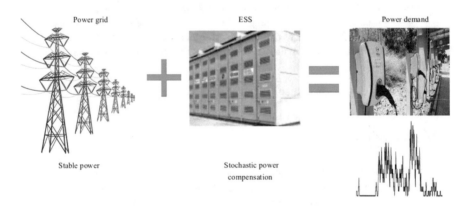

Fig. 1. Flow of energy storage fast charging station

2   Charging Station Construction and Operation Cost Analysis

Charging station construction investment cost should be the whole process of construction cost, according to the different nature of the cost can be roughly divided into the cost of land, supporting the cost of power grid construction, the supporting road network construction cost, construction cost, equipment costs, operating costs, operating costs several [7]. For land costs, the charging stations have been built, for the construction of power plants or charging piles, mainly by the government departments or charging stations to provide a trial unit. Such as bus charging stations provided by the bus company in the bus station, sanitation charging stations by sanitation company in a station or a garbage disposal plant, taxi charging stations and charging pile group construction land mainly by traffic department provides, most of which is located in overpass bridges, pilot charging pile of land mainly by pilot application units in their parking lot provides. Charging station construction increased the distribution network of electric demand capacity, where the regional power grid can meet the new requirement for charging, whether there is a need for new distribution networks or to reform the old distribution network is the need to consider the problem. At the same time, the charging station also needs to carry out the construction of the supporting grid, such as the installation of distribution transformers, the construction cost of these power investment in the estimation of the need for comprehensive consideration. The ideal situation is that the cost of supporting power grid construction or renovation costs and the cost of charging stations by the Power Grid Corp to carry out investment and construction, charging station construction will not consider the cost of external power grid construction. At the same time, the charging station also needs to carry out the construction of the supporting grid, such as the installation of distribution transformers, the construction cost of these power investment in the estimation of the need for comprehensive consideration. Charging station stationing in the best position should be close to the main road and need to consider by the main road by the introduction of the charging station ramp construction cost. In the same site construction process, as far as possible the use of compact design, the implementation of office space and production integration, reduce non-production space, reduce the area and project cost, effectively control the cost of construction [12, 13]. Cost of equipment including charger, charging pile, power distribution equipment, monitoring systems, such as charging for electric facilities and types of equipment cost. To change the station, still need to consider battery holder, split type charger, change electric robot, forklift acquisition cost, and battery charging the purchase or lease of them. Operating costs is used to maintain the normal operation of the charging station, the human cost of the normal business activities and equipment repair, and replacement costs. In charging stations in the process of operation need to fully consider the actual situation of the charging station, the allocation of personnel to implement dynamic management, through the strengthening of personnel training, improve staff quality; reduce personnel quantity to reduce labor costs [8].

Different charging stations due to the construction site and the different charging demand, the construction scale is not the same. In order to facilitate the calculation, the cost of the land cost, the matching network, and the road network construction can be neglected in the calculation process, mainly considering the cost of construction and equipment purchase and installation costs. Assuming charging stations have 10 fast charging pile, charging pile, battery maintenance equipment, charging station monitoring and safety monitoring equipment costs were 2 million, 200 thousand, 200 thousand, infrastructure costs 2 million 400 thousand Yuan. Charging station distribution cost is relatively fixed, charging stations and distribution facilities in general, including 2 sets transformer, a power distribution cabinet, 1 km 0.4 kV cable, 2 km

10 kV cable and above capacity 700 KVA active filtering device charging stations and distribution costs in 190 million Yuan. Charging stations operating costs include staff costs, station equipment consumption costs, a total of 20 million Yuan/year, distribution facilities maintenance costs are generally about 3% of distribution costs, about $60000 a year, overall operating cost 26 million Yuan/year [14, 15]. Figure 2 illustrates the state transition relation.

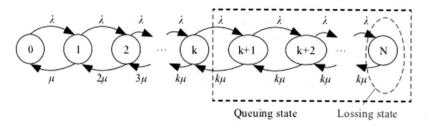

Queuing state   Lossing state

Fig. 2. State transition relation

3   Charging Station Revenue Analysis

Because electric vehicles battery is not fully discharged after charging again. At the same time charging may not be entirely the battery is full, in contrast to the conventional gas car of each gas volume and take a certain margin of, per battery charge can be according to 70% of the battery capacity of approximate computation, the ordinary electric car every time charging amount is about 14 kWh. Assuming the charging stations in each DC charging pile can serve 20 or so electric vehicles per day, and so on to calculate the typical charging station daily charge of about 2800 kWh [16, 17].

Taking Beijing as an example, a comprehensive tariff is about 0.4 yuan/kWh, charging the average price in accordance with the current 0.8 yuan/kWh calculation, electric vehicle charging station it does not consider the time value of money and charging station construction of static investment payback period, considering the time value of money and dynamic investment recovery period will be longer. Assuming the investment amount of each charging station is C, daily charge/for W, the daily operation cost is P, the price per kilowatt hour is P0, the sale price is Ps, then the static payback period of the charging station is T:

$$T = \frac{C/365}{(P_s - P_0) \times W - P} \tag{1}$$

$$C_{zy} = \sum_{i=1}^{n} \varepsilon \rho Q_i^{car} \tag{2}$$

Overall, a single charging station infrastructure and distribution facilities investment are not large, a single charging station infrastructure and distribution facilities investment in 4 million 300 thousand Yuan or so. From the point of view of the charging station cost recovery period, the basic implementation of the charging station and the power distribution facilities recovery period is 28.7 years. Through the calculation shows that, under the existing charging price, the payback period is very long, the economic value is low. But along with the strengthening of the battery life, under the premise of the operating costs unchanged, in site construction process, as far as possible the use compact design, implementation of office space and production space integration, reduce non productive space, reducing the occupied area and construction cost, effective control of the construction cost; in consideration of the cost of equipment, it is necessary to optimize the equipment selection scheme, under the premise of quality assurance by domestic equipment to replace imported equipment or joint venture brands, cancel or simplified auxiliary system, retaining only the production to meet the basic needs of some, charging station construction cost recovery period will be shortened [18, 19].

4 Operating Costs on the Charging Station Break Even

The factors that cause the change in operating cost are the change in electricity price, the salary of the staff and the management cost [9]. The application of electric vehicles can be brought about an increase in the overall social welfare greatly, and to facilitate the implementation of load scheduling, will play a better cut a peak to fill valley effect, can effectively improve the utilization rate of power transmission equipment assets and to reduce the fixed cost of investment. On the current view, the electric vehicle using the comprehensive cost to higher fuel vehicles, but from the perspective of pricing competition of the comprehensive cost should not be higher than the vehicle fuel; otherwise, the user will refuse to use of electric vehicles and the choice of ordinary fuel vehicles. This work will be difficult to sustain. Under the premise of not involving the construction costs of charging stations, electric car prices should be closely related to the price of fuel, and with the change in the price of fuel. From 2009 to date, 92/95# gasoline prices continue to go up and change dramatically, a number of changes in a year, the change is too frequent. But the price of the electricity price should not change with the same period of oil prices, the proposed annual change in accordance with the magnitude of the price of oil, the price of the same proportion of electricity to meet the needs of operating costs and charging station construction costs [20, 21]. Charging station and operation cost because there are no electric vehicles running in a batch of charging stations charging service demand to offset, will inevitably lead to the charging station "do not have enough to eat" phenomenon occurred, so that the charging stations in the market "sentinel" stage have to out of pocket to bear electric cars this technological innovation of products in the introduction period of uncertainty in the cost. With the gradual transition to the electric car fuel gas vehicles to bring the scale of the use of electric vehicles, electric vehicle charging station project will gradually reduce losses, and then embarked on the road to profitability.

About personnel wages are used in Beijing city standards, taking into account the change in staff wages and management costs, so the short term operating costs will not be a big change. In charging stations in the process of operation need to fully consider the actual situation of the charging station, the allocation of personnel to implement dynamic management, through the strengthening of personnel training, improve staff quality; reduce personnel quantity to reduce labor costs [10, 11]. When people pay a larger adjustment, it is recommended to carry out cost accounting, timely adjustment of the price.

Compared to the above comprehensive analysis showed that oil prices factors, price factors and cost factors in the relatively stable, in the short term of charging service smaller price impact. The whole charging standards of service charges is not higher than the cost of oil. Therefore, it is recommended to take charge of electricity service prices and oil prices linkage mode, the annual cost accounting, according to the same proportion of the oil price changes to adjust.

5   Analysis of Profit Model Under V2G Mode

Electric car battery can of charging and discharging two different processes. In the process of charging, the electric vehicle can be considered for network load, in electric vehicle normal running process is discharge process of battery, storage batteries for electric cars, electric can be changed into mechanical energy, and do as the power to drive the electric vehicle driving. When the electric car in the power grid in the case of the discharge, the electric car battery can be seen as a power grid power supply facilities to participate in power grid, which is the V2G model of electric vehicles access to power grid. After implement V2G, electric vehicles can also do for power grid load and power supply, and constantly changes between the source and the load, making the grid in the power distribution line flow may be bidirectional tidal flow, thus changing the original distribution network operation mode and network structure. In the V2G mode, the electric vehicle users in the lower load for the electric vehicle charging, in the grid load peak as the power supply to the power grid.

Different pricing policy may produce different charge for electricity demand, power grid and the electric vehicle charging mode of the transaction will also from simple to complicate, these need to more advanced construction of power market to adjust. Affect the charging station profit of the two factors is the key factor of charge of electricity sales and the charging price difference, a charging station, once completed, to the charging station for charging electric vehicle load has been basically fixed, the sale of electricity fluctuation range is not large also so that the charge price of the charging station put into operation after the profitability. In the absence of peak-valley electricity price, the owner at the end of the road will be immediately carried out on the electric car charging, when the user's power consumption is the largest. After the implementation of peak-valley electricity price, the owner of the start charging time to reduce the electricity, so as to form a new load curve, which means that the owners of their own part of the use of electricity in exchange for economic benefits. Therefore in order to achieve the best effect of V2G mode, power grid company improve the existing tariff policy, and formulate different charge price and reverse purchase electricity price, thus affecting the charging behavior can play a significant role in "cut peak and fill valley".

V2G mode needs to be as far as possible to realize the intelligent electric vehicle charge, discharge; the ideal situation is the electric car in the nonpeak period of automatic charging, automatic discharge in the peak hours. At the same time grid, systems need to fully consider electric vehicle battery capacity, to do contain complementary mobile distributed energy storage unit of electric vehicles and the grid system, to achieve stabilize power grid load and frequency fluctuation, improve the reliability of the power grid, power grid reserve capacity of generation capacity is reduced. In the current V2G technology, most commercial operation scheme only considers the electric cars need to accept the regulation and control scheme of grid or operators, without considering the electric car users are willing to participate in V2G that V2G theory of optimal control strategy in practice is difficult to be accepted by users. Therefore, V2G technology scheme to realize if you want to have a better, better approach is through the development of reasonable charge and discharge the price to guide the user, users change the passive acceptance of grid scheduling for price to guide users to act according to the optimal scheme of electric vehicle charge and discharge, and at this time the electricity price strategy requires according to an analysis of the electric network and electric cars in total, establish V2G guide customers to charge the price model, calculate the charge price sensitivity to a variety of factors determined optimal pricing scheme.

6 Conclusion

Electric vehicle charging facilities construction, can promote the development of electric vehicle industry, to reduce energy consumption and environmental pollution and other aspects to promote the role, has a high social benefits. Electric vehicle charging station as an important infrastructure for electric vehicles, will inevitably precede the electric car and the sudden and advance, and then drive the commercialization and industrialization of electric vehicles. To avoid capital chasing and blindness of rushing headlong into mass action and quitting behavior of social resources caused serious waste of consequences, charging station is introduced and planning, can not be separated from the state and local governments on investment, operation, the guide to unify the main players in the market the pace of action. Charging station life cycle cost of construction and running costs of scientific control and management from system point of view of optimizing the, focusing on the construction period of the project cost, charging facilities and equipment purchase channel construction and price negotiation and operation of the purchase of electricity price and other key cost elements of the project, so as to achieve charging station earnings.

References

1. Kempton, W.F., Tomic, J.S.: Vehicle-to-grid power implementation from stabilizing the grid to supporting large-scale renewable energy. Power Sour. **144**, 280–294 (2005)
2. Tomic, J.F., Kempton, W.S.: Using fleets of electric-drive vehicles for grid support. Power Sour. **168**, 459–468 (2007)
3. Clement-Nyns, K.F., Haesen, E.S., Driesen, J.T.: The impact of charging plug-in hybrid electric vehicles on a residential distribution grid. IEEE Trans. Power Syst. **25**, 371–380 (2010)

4. Kiviluoma, J.F., Meibom, P.S.: Methodology for modelling plug-in electric vehicles in the power system and cost estimates for a system with either smart or dumb electric vehicles. Energy **36**, 1758–1767 (2011)
5. Pantos, M.F.: Stochastic optimal charging of electric-drive vehicles with renewable energy. Energy **36**, 6567–6576 (2011)
6. Valentine, K.F., Temple, W.G.S., Zhang, K.M.: Intelligent electric vehicle charging rethinking the valley-fill. J. Power Sour. **196**, 10717–10726 (2011)
7. Suh, I.F.: Wireless power transfer technology and EV application. J. Int. Design Process Sci. **15**, 1–2 (2011)
8. Williamson, S.S.F., Emadi, A.S.: Comparative assessment of hybrid electric and fuel cell vehicles based on comprehensive well-to-wheels efficiency analysis. IEEE Trans. Veh. Technol. **54**, 856–862 (2005)
9. Kang, B., Kang, S., Kim, K.: A study on the enhancement of mechanical properties of metakaolin-based geopolymers using by precuring process. Int. J. Bio-Sci. Bio-Technol. **96**, 13–18 (2017)
10. Choi, S., Bok, J., Lee, H., Bae, J., Kim, Gim, G.Y.: An empirical study on crawler-based security control systems. Int. J. Reliab. Inform. Assur. **52**, 7-12 (2017)
11. Kim, D.G., Seo, D.G.: Construction robot technology for construction in outer walls of high-rise building. Int. J. Comput.-Aided Mech. Design Implement. **32**, 7–12 (2017)
12. Chandrika Sai Priya, A.: Integrated framework for multi-user encrypted query operations on cloud database services. Int. J. Cloud-Comput. Super-Comput. **3**, 1–6 (2016)
13. AlQattan, K.I., Mirzal, A.: A framework for cloud-based E-prescription in healthcare information system. Int. J. Cloud-Comput. Super-Comput. **5**, 9–22 (2018)
14. Satish Babu, J.: Identifying the complications in software project. Int. J. Private Cloud Comput. Environ. Manage. **3**, 7–12 (2016)
15. Yoo, E., Jung, G., Cha, J.: A consortium blockchain-based certificate and verification framework for apostille E-register service. Int. J. Private Cloud Comput. Environ. Manage. **5**, 1–6 (2018)
16. Kim, K.: Trustful mechanism for sharing the cost of multicast transmissions. Int. J. Urban Design Ubiquit. Comput. **5**, 17–20 (2017)
17. Kuppala, D.R., Sim, H.M., Liu, X.J.: Accurate methodology for large-scale storage systems. Int. J. Internet Things Big Data **2**, 1–8 (2017)
18. Kumar, T.R.: Secure data deduplication systems with reliability. Int. J. Adv. Res. Big Data Manage. Syst. **1**, 39–44 (2017)
19. Chang, X.: Increase the performance of K-means clustering algorithm using apache spark. Int. J. Internet Things Appl. **1**, 13–28 (2017)
20. Harika, N., Vamsilatha, M., Rao, N.T., Bhattacharyya, D., Kim, T.: Performance analysis and implementation of traffic monitoring system using wireless sensor network for reducing blockage in traffic on indian city roads. J. Stat. Comput. Algorithm **1**, 15–26 (2017)
21. Chai, S.S., Suh, D.J.: Design of risk analysis database system based on open data for South Korea. J. Stat. Comput. Algorithm **2**, 1–6 (2018)

Anti-error Technology of Secondary System of Intelligent Substation Based on Device State Recognition and Analysis

Tongwei Yu[1(✉)], Ziliang Li[1], Wuyang Zhang[1], Baotan Li[2], Baowei Li[2], and Zhe Chen[2]

[1] Electric Power Research Institute, State Grid Liaoning Electric Power Supply Co., Ltd., Shenyang 110106, China
Tongweiyul1@163.com
[2] XJ Electric Co., Ltd., Xuchang, Henan 461000, China

Abstract. An error-proof technology for power secondary system based on state identification and analysis of relay protection devices is proposed by using digital information exchange means in intelligent substation in view of the lack of secondary system error-proof links in the existing substation error-proof operation system. Firstly, the operation status is identified and abnormally diagnosed according to the SV and GOOSE message transmission and protection status and function configuration of the protection device. Three types of error prevention methods are proposed for the substation operation on this basis. Locking logic is generated by matching device status with error-proof rules to realize permissive error-proof for operation execution In the process of remote/local operation. When the system is abnormal caused by unexpected factors, according to the overall operation state and the active error prevention of the abnormal state is realized. The isolation measures are automatically generated according to the safety regulations in the sequential control operation ticket to realize the one-button isolation of the device operation. Finally, the function of the error-proof system is applied and verified by the embedded function module of the monitoring system. Error-proof technology of intelligent substation secondary system can effectively reduce the possibility of substation device misoperation and misconfiguration, and avoid the occurrence of system malignant accidents.

Keywords: Error prevention of secondary system n · State recognition · Intelligent substation · Relay protection · Generic object oriented system-wide events (GOOSE)

1 Introduction

The relay protection device uses the network, intelligent technology [1], the information transmission mode from the conventional cable to the transmission of electrical signals into the digital transmission of optical fiber [2], the function of the hard plate and the export of hard plate are converted to a soft platen with the development of the intelligent substation. These changes reduce the complexity and complexity of the

© Springer Nature Switzerland AG 2020
A. E. Hassanien et al. (Eds.): AISI 2019, AISC 1058, pp. 933–940, 2020.
https://doi.org/10.1007/978-3-030-31129-2_85

engineering configuration of the relay protection device, but it also brings some problems. The information is hidden in the optical fiber [3] because the relay protection device has no physical terminal and connection. Maintenance mode cannot meet the needs of the power grid operation; relay protection equipment maintenance should be able to operate according to the operating state Efficient and intelligent equipment overhaul work [5].

The application and popularization of intelligent substation provide the technical conditions for the relay protection device to realize on-line operation and maintenance, and the relative research on the operation and maintenance technology of relay protection device is more [1–3, 5–8]. In [1], the authors studied the condition monitoring of the relay protection device and the state monitoring technology of the secondary circuit and provided the isolation diagram of the device for the operation and maintenance personnel. In [2], the authors proposed a secondary loop fault diagnosis method by analyzing the secondary loop topology relationship and fault diagnosis principle of the device. In [3, 5], the authors studied the two loop state monitoring technology of relay protection device.

2 Relay Protection Device Status Monitoring

The relay protection device can be divided into Operation State, signal State, exit state and Maintenance State according to the state of the relay protection device and the Operation State of the equipment. The alarm is generated when the diagnostic rules are not satisfied, and the cause of the anomaly and the influence range are analyzed according to the condition of the device, the condition of the pressure plate of each device is diagnosed in accordance with the rules of the expert library.

2.1 Relay Protection Device Status

The soft platen of relay protection device mainly has protection function soft plate, GOOSE receive and transmit soft platen, SV receives soft platen, remote cast back soft platen, remote switching fixed value area and remote modification fixed value soft plate. The hard platen has the overhaul platen and the remote operation The pressure plate that affects the protection function or logic calculation has GOOSE receiving and transmitting soft pressure plate, SV receiving soft pressure plate, protection function soft pressure plate and maintenance pressure plate.

The device can be divided into operation state, signal state, exit state and maintenance state according to the input state of the pressure plate and the operation state of the primary equipment as shown in Fig. 1.

Operation State: The relay protection device must satisfy the following conditions in operation: (1) the primary equipment runs normally, the SV receiving software board of the device is put into operation, or the primary equipment is withdrawn, the maintenance pressure board of the combined unit is put into operation, and the SV receiving software board of the device is withdrawn. (2) The associated device repairs the pressure plate input, the device GOOSE receives the soft pressure plate to exit, or the

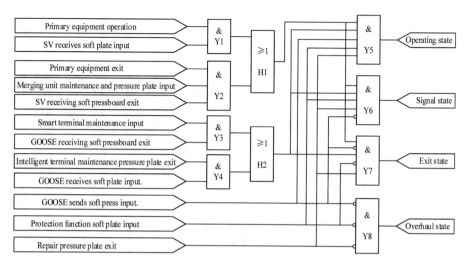

Fig. 1. Device status monitoring

associated device repairs the pressure plate to exit, and the device GOOSE receives the soft pressure plate input.

Signal State: When the relay is in a signal state, the following conditions must be met:

(1) The primary equipment runs normally, the SV receiving software board of the device is put into operation, or the primary equipment is withdrawn, the maintenance pressure board of the combined unit is put into operation, and the SV receiving software board of the device is withdrawn.
(2) The related device overhaul pressure plate input, the device GOOSE receives the soft pressure plate withdrawal, or the related device overhaul pressure plate withdrawal, the device GOOSE receives the soft pressure plate input.

Exit State: The relay protection device is in an exit state and must satisfy the following conditions: (1) the primary equipment is in normal operation, the device SV receives the soft pressure plate input; or the primary equipment exits, the combined unit repairs the pressure plate input, and the device SV receives the software board to exit [9]. (2) The associated device repairs the pressure plate input, the device GOOSE receives the soft pressure plate to exit, or the associated device repairs the pressure plate to exit, and the device GOOSE receives the soft pressure plate input. (3) The device GOOSE sends a soft pressure plate to exit [10].

2.2 State Diagnosis of Relay Protection Device

The platen rules corresponding to each device state are described in Fig. 1, and the device state can be diagnosed according to the rules [11].

The device status diagnosis is configured by two parts: status diagnosis and influence range analysis as shown in Fig. 2 [12].

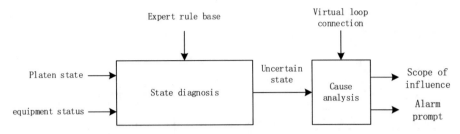

Fig. 2. Device status diagnosis

3   Device Allowable Anti-error Technology

The operation personnel may easily cause mis-projection and false retreat during the operation due to the large number of soft pressure plates of the intelligent substation relay electric protection device [13].

As shown in Fig. 3, the allowable error-proof relay protection device is composed of error-proof rule analysis and operation logic judgment [14]. The device has the characteristics of self-locking, active locking and local soft-plate operation [15].

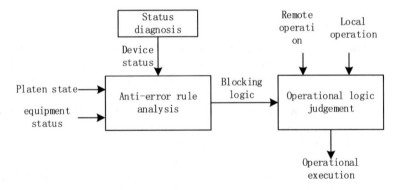

Fig. 3. Allowable error proofing process

When the device is in a certain state, only the pressure plate that can be switched to other states can be allowed to operate, and the other platen operations need to be blocked [15].

The device status parameter is added in the relay protection device model [16], and the locking state is added for each pressure plate (SV receiving soft pressure plate, GOOSE receiving and transmitting soft pressure plate [17], protection function soft pressure plate), and the error prevention rule analysis process is as follows [18]:

– The status diagnosis module performs real-time diagnosis on the device status according to the collected primary device measurement information and the pressure plate status. The diagnosis logic is shown in Fig. 1, and the diagnosis result is written into the device status parameter [19].

– The anti-error rule analysis module monitors the status of the device in real time. When the status changes, the operation logic is generated according to the device platen status and the primary device measurement information, combined with the device status condition rules, as shown in Table 1.

Table 1. Logic generation rule

Device status	Status value	Allow operation of soft platen
Operating state	1	Exit GOOSE to send soft pressure plate Cast/return GOOSE receiving soft pressure plate
Signal state	2	Put GOOSE to send soft pressure plate Exit protection function soft plate Cast/return SV receiving soft pressure plate Cast/return GOOSE receiving soft pressure plate
Exit state	3	Input protection function soft plate Cast/return SV receiving soft pressure plate Cast/return GOOSE receiving soft pressure plate
Inspection state	4	nothing
Uncertain state	0	Unlimited

– The locking state of the pressure plate which is not allowed to operate is set to true.

All logic is cleared if the device is in an indeterminate state.

Operating logic judgment is shown in Fig. 4. When pressing plate operation is carried out distantly or locally, the operation state of the pressing plate is judged to be blocked first, if not, the operation command is executed, otherwise the operation is prohibited and the cause is prompted [20].

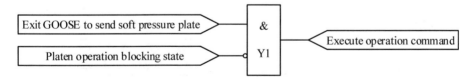

Fig. 4. Operation logic error proofing

4 Device Isolation Error Prevention Technology

Intelligent substation maintenance during the accident is usually due to the operation and maintenance personnel did not strictly follow the security process operation. The process of isolation and error prevention is as follows:

(1) The pressure plate state corresponding to each state of the relay protection device is described. The isolation process of the device is executed in the order of "operation

state - signal state - exit state - overhaul state", so the rules of switching from operation state to overhaul state can be analyzed.

(2)   The analysis device model can get the loop information and corresponding pressure plate information of the device.

5   Practical Applications

The on-line fault-proof system of relay protection is developed with the above key technologies and applied to a 220 kV substation. The rule base is built by the background monitoring system. The monitoring system presets the rule template for each state. The template is instantiated, and the state rule of the device is generated and delivered to the device when the device model is imported. The real-time calculation of the state diagnosis module is carried out according to the state rules. The current state of the device is displayed in the contact diagram of the device in real time.

(3)   It starts to calculate according to the status rule after the enabled error prevention module receives the device status change. In the device model, the locking state sign is added for the press plate (mainly GOOSE sending/receiving press plate, functional press plate, SV receiving press plate) [21]. The active error-proof module sets the locking state of the press plate according to the rules of Table 1. The pressure plate shown in Fig. 5 is a schematic diagram.

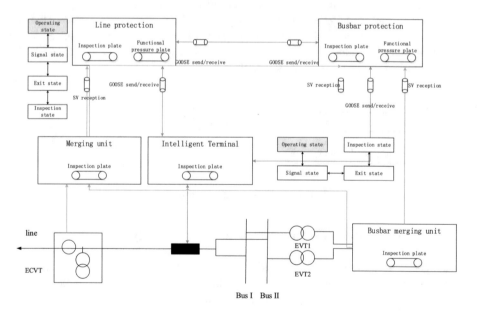

Fig. 5. Device liaison diagram

(4) Click the target state to toggle in the graph to start the one button isolation function. You can select any state to switch.

6   Conclusion

The key technology of on-line anti-misoperation system of Relay protection device is studied by analyzing the cause of accident caused by man-made operation during the maintenance of intelligent substation in this paper. This paper puts forward the equipment condition diagnosis technology, the allowable anti-misoperation technology, the active anti-misoperation technology and the device isolation and anti-misoperation technology to prevent the malfunction and rejection caused by human error operation during substation maintenance. It provides effective anti-misoperation measures for substation maintenance, and enriches the anti-misoperation mechanism of two equipment operation.

Acknowledgements. This paper is supported by Science & Technology Fund of State Grid Corporation of China under Grant NO. 521304170028.

References

1. Du, J., Ye, X., Ge, L., et al.: Key technologies of online maintenance system for relay protections in smart substation and its implementation. Electr. Power Autom. Equip. **36**(7), 163–168 (2016)
2. Ge, L., Zhao, G., Yang, F., et al.: Research on the secondary circuit fault diagnosis methods. Smart Grid **2**(6), 28–31 (2014). Author, F., Author, S., Author, T.: Book title. 2nd edn. Publisher, Location (1999)
3. Author, F.: Contribution title. In: 9th International Proceedings on Proceedings, pp. 1–2. Publisher, Location (2010)
4. Li, B., Qiu, J., Zhang, H., et al.: Style based graph auxiliary generation system in substation. Autom. Electr. Power Syst. **39**(14), 120–125 (2015). Sd
5. Ye, Y., Sun, Y., Huang, T., et al.: On-line state detection technology of relay protection relevant secondary circtuits. Autom. Electr. Power Syst. **38**(23), 108–113 (2014)
6. Xue, A., Wang, R., Wang, B., et al.: A relay replacement strategy based on the least unit life-cycle cost. Autom. Electr. Power Syst. **37**(5), 44–60 (2013)
7. Xue, A., Luo, L., Jing, Q., et al.: Research on the maintenance strategies of protective relay based on time-varying Markov model including multi-factors. Autom. Electr. Power Syst. **39**(7), 124–129 (2015)
8. Chao, H., Oren, S., Smith, S., Wilson, R.: Priority service: unbundling the quality attributes of electric power. Electric Power Research Institute, Report EA-4851(1986)
9. Ji, S., Kim, W.: Whispered speech recognition for mobile-based voice authentication system. Int. J. Image Signal Syst. Eng. **1**(1), 15–20 (2017)
10. Shulka, M., Tripathi, B.K.: On the efficient machine learning of the fundamental complex-valued neurons. Int. J. Neural Syst. Eng. **1**(1), 21–26 (2017)
11. Li, H., Tian, F.: A summary of multivariable control methods for greenhouse environment. World J. Wirel. Devices Eng. **2**(2), 13–21 (2018)

12. Vijaya, L.P.: Cloud computing for smart grids. Int. J. Cloud-Comput. Super-Comput. **3**(2), 19–26 (2016)
13. Anil, B., Asit, D.: Effective management of security of risk in cloud computing environment. Int. J. Private Cloud Comput. Environ. Manage. **3**(1), 1–10 (2016)
14. Kumar, K.P.: Performing initiative data prefetching in file systems for cloud computing. Int. J. Private Cloud Comput. Environ. Manage. **4**(1), 15–20 (2017)
15. Sarita, S.B., Deepika, J.: Performance analysis of traffic type and routing protocols in VANET for city scenario. Int. J. Urban Design Ubiquit. Comput. **4**(1), 1–12 (2016)
16. Laibao, Y.U., Zhang, T.A.O.: Research on intensity quick report and the key technology of seismic intensity monitoring. Int. J. Internet Things Big Data **1**(1), 1–10 (2016)
17. Bhargavi, N.S.P.: Service scheme for data cloud services. Int. J. Adv. Res. Big Data Manage. Syst. **1**(1), 45–50 (2017)
18. Zhu, X., Yang, M., Zhu, C., Tiejun, Z.H.A.O.: A comparison of pruning methods for pivot-based statistical machine translation. J. Stat. Comput. Algorithm **2**(3), 1–12 (2018)
19. Yu, Y., June, W.H., Kim, S.-J., Song, C.G.: Educational effectiveness of virtual museum. Int. J. Multimed. Ubiquit. Eng. **13**(2), 21–26 (2018)
20. Lee Ijsda, G., Lee, D.: Int. J. Smart Device Appl. **4**(2), 1–6 (2016)
21. Choi, Y., Woo, J., Hong, T., Yoon, J., Lee, H.: Summary on Chinese supercomputing strategies and status. Int. J. Digit. Cont. Appl. Smart Devices **3**(2), 21–26 (2016)

Homemade Electric Wire Electric Cable Breakpoint Nondestructive Testing Tools Based on X-Rays

Han Bai[✉], Dianyu Chi, Xiawen Wang, Hao Wu, and Yueyue Li

State Grid Shenyang Electric Power Supply Company, State Grid Liaoning
Electric Power Supply Co., Ltd., Shenyang, China
xiaobai1120@163.com

Abstract. Internal conductors open wire and cable is a common fault in electrical maintenance work. It is a thorny problem for maintenance. Maintenance generally has only multi-meters simple testing equipment, with little damage to the cable insulation condition often cannot judge open-circuit conductor points of a basin. Destroy the insulation not only lose wire and cable applications may also give your seat belt. As the city's rapid economic development and urbanization advancement speeding up unceasingly, the use of the power cable has become a trend. And to ensure the safe and stable operation of the cable, reduce unplanned shutdown, naturally become one of the key research topics. We carried on the thorough research of X-ray nondestructive testing technology, can ensure that the staff under the premise of safety guarantee, in the field of cable force damage, etc. all kinds of defects of internal damage to cable is convenient, efficient, judgment, provide a reliable and effective basis for further diagnostic work.

Keywords: Nondestructive testing · X ray · Power cable

1 Introduction

Cable industry in China in the present or in the outside, almost don't have to worry about market space problem, the industry has ushered in the great development of China, wire and cable industry as an important supporting industry of national economy, where the construction phase is a scat, so they need cables and therefore there is inflation in the market cables. But we also see that in the future the demand of the special cable will be bigger and bigger, special cable has been to require high technical content, high margin, high threshold, is famous for its great features such as market space, but in the special cable product technology research and development, but it is soft rib of domestic cable industry in China, thus causing a high-end market by foreign giants his situation, therefore, the domestic cable production enterprise to pry open high-end market gap, thus playing a piece of heaven and earth, is the domestic cable manufacturing enterprises to consider the important issues.

According to incomplete statistics, according to the Liaoning province power company power cable year-end summary statistics in the last three years, the results of

© Springer Nature Switzerland AG 2020
A. E. Hassanien et al. (Eds.): AISI 2019, AISC 1058, pp. 941–948, 2020.
https://doi.org/10.1007/978-3-030-31129-2_86

power cable fault and the main causes of unscheduled shutdown (Refer to Fig. 1) include external causes, installation, debugging, operation maintenance does not reach the designated position, improper design and old equipment. The most reason is the external causes and cable equipment Installation and debugging, has become the main reason for a threat to the safe and stable operation of power cable [1]. Therefore, for power cable fault And cable installation is not in place such potential defects. Looking for more effective means of detection for accurate assessment of electric cable defects and the dangers of cable operation, reducing the change caused by the additional cable equipment economic loses.

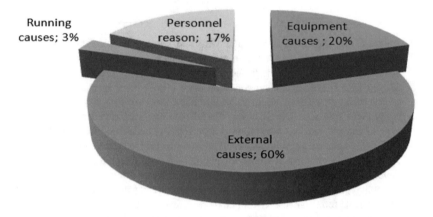

Fig. 1. Main reasons cable fault scale map.

2 Analysis on Development of Power Energy Saving Service

2.1 Detection Principle

When a conductor near another conductor capacitance was formed between them, the size of the capacity and the area of the conductor, the distance between the conductors and populates the dielectric permittivity, dielectric constant, the greater the distance is smaller capacity. As is known to all, capacitance is by alternating current. For our daily life and production used in 50 Hz ac power. Jean in three phases four wire ac power line is grounded, if we use an equal potential of the electrode to close to the surface of the power cord, the line will produce capacitive current on the surface, and the zero lines and the earth the same there is no current is generated. So by testing the conductor surface had ac signal can know this point of the potential of conductor [2]. According to this principle, we are open at both ends of the wire are respectively connected to the zero lines and the line. When the detection electrode moves from one end to the other end of the wire must be after a signal for a turning point, which is the point of the open circuit of the wire. The wire insulation layer is equivalent to medium, can make the capacitive current increases greatly, and facilitate testing of judgment. Strength is relative, of course, by detecting electrode sensing signal is still very weak, an ordinary multimeter is unable to show, but the signal is amplified or directly using a digital multimeter can be detected. Here we introduce the method of making the two detection tools, to meet the needs of the readers to different conditions.

2.2 Detect Circuit

The transistor amplifier circuit is mainly designed for readers without a digital multimeter. Use only about three little power triode amplifying circuit compound for high p amplifying circuit, working power supply from pointer multimeter resistance, for signs that again at the same time, make the whole detection circuit is very concise, with electronic knowledge reader can make a success. In the transistor do selects high p tube. Experiments show that with two high p triodes can achieve 5 to measuring sensitivity. Detection electrode is more than a bare wire, the length of the 50 mm to 80 mm advisable, detection electrodes around too long easily sensed signal, not easy to judge "dot" the change of the signal, will probe into "L" shape can increase with the measured capacitance between conductor and improve the detection sensitivity. File chooses Rx100 or resistance when measuring t Rx1 K, two file pointer deflection amplitude difference is not big, the main reason is that p affected by the working current of the transistor. Rx1 K file while the sensitivity is higher, but the working current is smaller and smaller. Amplifying circuit structure is simple, generally, do not need to be welded, no quality problem, as long as the components the pin directly fixed on the pens and twisted together. Amplifying circuit connection is complete, can be validated, armed with universal and pens and, do not have to contact amplifying circuit "to" human body.

Digital multimeter, flow the resolution of the voltage profile is 0.1 mV, and the input impedance is as high as 10 M, so can directly display detection electrode sensor signal. Comparing digital multimeter test signal is simple, just need to add a detection electrode. The use of the electrode and the producing method and the amplifying circuit are exactly the same. Detecting electrodes are "V/Ω" in the hole. Choose AC200 mV voltage gear. In the process of detection, the signal shows a general in more than 50 mV voltage value, even more than 200 mV, no signal voltage value, when under 10 mV.

Two measurement methods are not shaking to also don't need to direct contact with the human body measure t public side of the circuit. As long as a handheld multimeter, so even if the detection electrode one thousand access to the electric conductor body also won't have to get an electric shock risk. It will not cause damage to the device [3].

2.3 Detection Method

During the operation, the single strand wire is simpler and then can handheld testing equipment. Since the phase line end, detection electrodes on the conductor surface under test, to move, the zero line ends soon can find cut-off point signal strength, namely open points; longer wires can be according to a certain spacing selection of testing point, once the signal is changed back to narrowing the scope of testing again. For multi-core cable detection principle with single strand wire is the same, but some differences to the specific operation, because in addition to the measured that cable wire and other wire is tightly wrapped in a protective layer, the distribution of the signal changed. Connection wire is detected when still answer the zero line and the line and series current limit load, other wires connecting with all zero. Thus, when the detection electrodes on the surface of the cable, close to the phase of the signal are stronger, on one side of the line near the zero lines on the side of the relatively weak signal, the strength of the signal on the surface of internal wires and cables is consistent. Since as described in this paper, the sensitivity of the detection method is very high, the cable insulation layer surface, and detection electrode mobile several m signals occurs obvious change. Then, the detection cable on each is testing point to detect the circumferential surface signal. As long as the cable on the surface can also find a strong signal It hasn't reached open point location [4].

After mastering detection method can be found in the actual work of the measuring methods and applications tend to be flexible. For example, by measuring the electric wire electric cable is on the ground, because of the distributed capacitance between conductor and ground, generally, don't have to meet the zero line, only pick up line. Also, did not need concatenated current limit load. Household appliances and some industrial equipment are 220 V power supply, meet the power cord. When open closed the power switch is equivalent to give open circuit point connected to the zero lines and the line on both sides, whether the power cord is twin wire or cable, point signal is always open to change, find open circuit point just lift a finger.

3   The Related Technologies of X-Ray Nondestructive Testing

3.1   The Principle of X-Ray Nondestructive Testing

The current commonly used low energy X-ray (below 1 mev) is in X-Ray tube. X-ray tube is a cathode and anode Vacuum tube, the cathode is tungsten wire, the anode is made of metal target [3]. X-ray transmission of the law of decay of matter, such as

$$I = I_0 e^{-\mu \cdot x} \tag{1}$$

Where I_0 is the initial energy value of X-ray; I is the value of the ray energy received by the flat panel detector after penetrating the work piece to be tested; X indicates the thickness of the work piece; and u is a constant.

If there is no defect in the work piece to be tested, the wall thickness of the work piece does not change, and the initial energy value of the X-ray is I, and the X-ray energy value after penetrating the work piece to be tested is attenuated to

$$I = I_1 e^{-\mu \cdot x} \tag{2}$$

In between cathode and anode has a high dc voltage (tube voltage), when the state of the cathode heated to incandescence release a large number of hot electron, the atom electron is accelerated in the high voltage electric field, flew from cathode to anode Tube (current) that, in the end, with great speed hit on a metal target, lost most of all with kinetic energy, the kinetic energy is converted to heat energy, only a tiny part of the conversion for X-ray radiation [5, 6]. Common X ray tube structure is shown in Fig. 2.

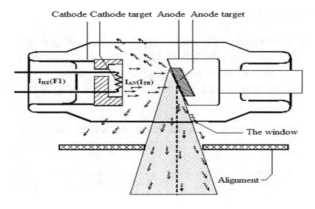

Fig. 2. X-ray tube glass structure

3.2   The Study of Power Cable Detection of Ontology Technology

The power cable is a kind of multilayer loop structure, relatively different rules, and the layers of material. Theoretically, ray radial by electricity cable, because there are differences between each layer of X-ray attenuation ability and trans-illumination thickness, cause its performance to have a different gray shadow on the negatives like [7].

From the field application experience, cable forces destroy major source ontology include: construction mechanical damage, man-made destruction and in cable trench cover plate and so on. Aluminum sleeve can reduce a certain extent less external force damage to the cable. On the other hand, only cable semi-conducting layer and insulating layer intact is certainly no need to replace or make cable connector, just an easy repair process. So, only when the X-ray detection combined with cable type and structure observe the external force damage is complete, we can determine the electric parts of each layer image cable damage degree [8–10]. Figure 3 shows the imaginary X-ray detection images of 110 kV cable.

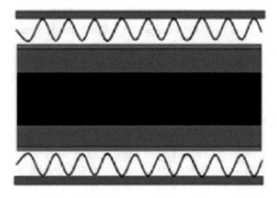

Fig. 3. Imaginary X-ray detection images of 110 kV cable.

3.3   Typical Defects Cable X-Ray Spectrum Library Research

For power cable accessories of all kinds of typical defects using X-ray machine filmed and finally gets the following conclusion. The following stress cone displacement and copper screen processing of them bad defects described in detail [11]. Table 1 shows the power cable accessories of all kinds of typical defects using X-ray machine filmed.

Table 1. Power cable accessories of all kinds of typical defects

Number	Defects	Positions	X-ray inspection situation
1	The main insulation metal powder	The middle joint orterminal	Can be seen clearly
2	Stress cone displacement	The middle joint or terminal	Can be seen clearly
3	The main insulation cut	The middle joint or terminal	Can't see
4	Semiconducting strip cut bad	The middle joint or terminal	Can be seen clearly
5	Joint adverse pressure	The middle joint or terminal	Can be seen clearly
6	Copper shield bad treatment	The middle joint or terminal	Can be seen clearly
7	Connector and cable ontology don't match	The middle joint or terminal	Can be seen clearly

Stress Cone Displacement: According to the requirements of making the cable accessories, the semi-conductive layer should be located in the stress cone [12–15]. According to its structure, as a result of the stress cone material and accessories ontology has a certain attenuation coefficient is poor, so the stress cone is shown in the image for the light arc shape. We can find that the two kinds of material, radiation attenuation coefficient difference is not too big, with lighter image, can by adjusting the contrast and blackness of software in order to achieve the best effect [16–18].

For semi-conductive layer cutting position, because this position has certain thickness difference before and after, although do not make both ends of the location and form due to the thickness difference of blackness is poor [19–21].

4   Conclusions

This paper built high voltage cross linked cable nondestructive testing and diagnostic tools for cable structure characteristics on the field application practice. Thus effectively identify the concentration of defects, cables and accessories to fill the domestic blank in the research and application of this technology, to achieve early damaged cable equipment health status, to ascertain the effective excavation affect the normal operation of cable latent defects and hidden trouble. So as to reduce the cross linking cable sudden blackouts in absolute number and relative proportion, enhance the level of power cable operation maintenance and management in order to improve the power supply company cross linking cable transmission network reliability to provide a strong guarantee.

References

1. Shi, C.F.: Ray testing technology is briefly reviewed in the development. J. Nondestr. Test. **23**(11), 15–17 (2012). (in Chinese)
2. Zhao, L.F., Ling, Q.S., Guo, L.T.: Development of X-ray cable insulated core measuring instrument. Wire Cable **03**, 36–38 (2003). (in Chinese)
3. Ren, H.F., You, Z.S.: Improvement of image quality in X-ray inspection. Instr. Technol. Sens. **01**, 36–37 (2001). (in Chinese)
4. Lina, Y.F., Jing, G.S.: An analysis of large-scale wireless sensor networks with energy-saving problem. In: Proceedings of 2014 5th International Conference, vol. 17, no.14, pp. 15–20 (2014). (in Chinese)
5. Ren, L.F., Chen, F.S., Lu, Y.T.: Non-destructive testing of power distribution equipment based on X-ray digital imaging technology. Rural Electrif. **05**, 30–32 (2018). (in Chinese)
6. Wang, Y.F., Tian, J.S., Song, L.T.: Computer-assisted X-ray tomography for texture nondestructive quantification of semi-sweet tough biscuits. Food Sci. **39**(05), 93–98 (2018). (in Chinese)
7. Shi, W.F., Xie, B.S.: Experimental study on non-destructive testing of SOFC metal seals by X-ray. J. Univ. Sci. Technol. China **47**(12), 1002–1005 (2017). (in Chinese)
8. Yang, F.F., Zeng, L.S., Ma, Y.T.: X-ray detection and analysis of tensile clamps for overhead transmission lines. J. Shaanxi Univ. Technol. (Nat. Sci. Edn.) **34**(01), 9–33 (2018)
9. Dong, J.K., Sujatha, V.: Auditing of storage security on encryption. Int. J. Reliab. Inform. Assur. (IJRIA) **5**(01), 9–14 (2017)
10. Sela, L., Young, C.: The effects of Korean genital image on marital satisfaction in Korean married women: mediating effect of sexual satisfaction using path analysis model. Int. J. Comput. Graph. **9**(01), 7–12 (2018)
11. Vardhan, K.A., Pasala, S.: Sandhya pasalastudies on the various issues relating to data storage space and safety in Cloud Computing. Int. J. Adv. Res. Comput. Inform. Secur. **2** (01), 7–14 (2018)
12. Pei, S.J., Kong, D.K., Hu, D.M., Chen, D.Y.: The application of fast pruning algorithm under map/reduce task scheduling. Int. J. Cloud-Comput. Super-Comput. **4**(1), 1–8 (2017)
13. Kuma, K.P.: Efficient approach for query based search in data mining. Int. J. Private Cloud Comput. Environ. Manage. **3**(2), 1–6 (2016)
14. Babu, J.S.: Scheme for resource matchmaking across different clouds. Int. J. Urban Design Ubiquit. Comput. **4**(2), 1–6 (2016)
15. Hai-yun, Peng: Energy consumption of wireless sensor network research problem. Int. J. Internet Things Big Data **1**(1), 29–36 (2016)
16. Han, H., Lee, J.: Effects of CGI-based number and operation activities on young children's understanding of numbers and arithmetic problem-solving abilities. Int. J. Internet Things Big Data **2**(2), 7–12 (2017)
17. Lee, Y.H.: Development of problem solving path algorithm for individualized programming education. Int. J. Adv. Res. Big Data Manage. Syst. **2**(1), 7–12 (2018)
18. Sivamani, S., Cho, K., Shin, C., Park, J.W., Cho, Y.Y.: Model fitting comparison for the risk assessment of weather conditions. Int. J. Internet Things Appl. **2**(1), 13–18 (2018)

19. John, S., Hwang, B.B.: Application of big data for development of korean christian research. J. Stat. Comput. Algorithm **2**(1), 7–12 (2018)
20. Choi, J., Kwon, J.H.: Motion synthesis for controlling a virtual ball in real-time. Int. J. Multimed. Ubiquit. Eng. **13**(4), 1–6 (2018)
21. Kim, H.S., Cho, Y.W.: A design of hovering system for quadrotor UAV using multi-sensor fusion. Int. J. Smart Device Appl. **4**(2), 13–20 (2016)

Research on Application of Power Supply Supervisory System Based on Telecommunication Network

Shunli Qiu[1(✉)] and Yi Yang[2]

[1] Information and Communication Branch, State Grid Liaoning Electric Power Supply Co. Ltd, Shenyang 110004, China
syshunli@163.com
[2] College of Information, Shenyang Institute of Engineering, Shenyang 110136, China

Abstract. As the development of the smart grid in the society, the telecommunication network's size expands unceasingly and the proliferation of devices becomes also reality, with the result that the workload of daily administration and maintenance increases. Based on the above situation, the condition of real-time operation in the power system must be monitored intensively to run the telecommunication system safely and efficiently. The telecommunication power supply supervisory system can help the operators know the operating condition of power system so that they can manage, monitor and maintain the telecommunication system available by finding some actual problems of work in it. This article aims to introduce the structure and function of telecommunication power supply supervisory system and propose some measure schemes of the system in the telecommunication networks, including the improvement of the telecommunication system's safety and security and the power device's maintenance.

1 Introduction

The progress of the smart grid promotes the growing scale of the communication network and increases the number of devices, also increasing the workload of daily management and maintenance [1]. Telecommunication system which is running safely and efficiently requires that the real-time running status of the power supply system can be monitored intensively [2]. Efficient telecommunication power supply supervisory system can make the operations staff grasps the operation condition of the power system timely, finding the actual problems of work in a telecommunication system and having the centralized management, centralized monitoring, and centralized maintenance. In the actual practice, the function of it can effectively decrease the number of attendance for maintenance's; meanwhile, the targeted maintenance of fault zone can raise the standard of the telecommunication management [3, 4].

Since the construction of telecommunication power supply supervisory system started, the monitoring mode with centralized monitoring and unified management has formed [5]. It can not only analyze the collected data by monitoring unit with warming but also make the telecommunication power system operate safely, stably and

© Springer Nature Switzerland AG 2020
A. E. Hassanien et al. (Eds.): AISI 2019, AISC 1058, pp. 949–955, 2020.
https://doi.org/10.1007/978-3-030-31129-2_87

efficiently, eliminating the potential failure and making passive repairing way turn into active maintenance way. So it ensures that the telecommunication system can maximize the risk-averse and reduce the events of cutting off the primary equipment, providing the stronger telecommunication guarantee for the production and management of the power system [6, 7].

2 The Structure and Function of Telecommunication Power Supply Supervisory System

(a) The structure of telecommunication power supply supervisory system can make the monitoring of telecommunication system become a kind of monitoring mode with the centralized monitoring and the unified management. The actual construction of telecommunication power supply supervisory system is a multi-stage distributed network monitored by computer, including three parts which are the monitoring center (control center), the monitoring station (network management center), monitoring unit (substation) generally. The monitoring unit sets up all kinds of sensors and its data are uploaded to the monitoring station, and then the monitoring station connects to the monitoring center through computer networks, refer to Fig. (1) [8, 9].

(b) Monitoring unit during the work gathers periodically the real-time data of the communication power supply, which collects mainly the equipment working log. These data collected by the monitoring unit are sent initiative into the monitoring station and the monitoring unit also receives the monitoring commands from the monitoring station. Monitoring unit can effectively receive the configuration information that is given by the upper-level host and refreshes the configuration file. When a failure happens in the power supply system, the relevant alarm data would be saved. After the power system can recover the function of transporting the information normally, the data during the communication interrupt will be uploaded again [10, 11].

(c) The computer system of the monitoring station is responsible to passing information between the upper-unit and the lower-unit, playing the role of a transit hub for information. It can take the information gathered by all the monitoring units for data analysis and disposal, and then transmitting these data to the monitoring center at the higher level. Meanwhile, it can also carry out real-time monitoring to the monitoring unit of the area and ensure the smooth exchange of information between the monitoring unit and monitoring station, monitoring station and the monitoring center. At the same time, the fault information of monitor unit can be timely received or it can also transmit the alarm information to the monitoring center. Monitoring stations need to set the parameters of the monitoring unit, which aims to collect the monitoring information of the monitoring unit and communication power supply for reporting to the monitoring center [12, 13].

(d) The monitoring center has the highest computer equipment of telecommunication power supply supervisory system, not only possessing all necessary functions of all the monitoring stations but also monitoring the work status of monitoring stations and analyzing the data and the warning information uploaded by monitoring

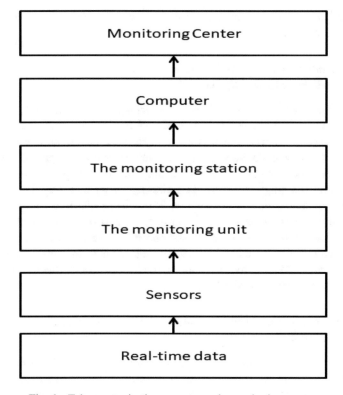

Fig. 1. Telecommunication power supply monitoring system

stations, with the result that the maintenance can be take timely [14]. It can also reduce the failure rate and avoid the happening of equipment power events [15].

3 The Actual Application of Telecommunication Power Supply Supervisory System in the Telecommunication System

By analyzing the structure and function of telecommunication power supply supervisory system, the work process of monitoring system includes the real-time monitoring of communication power supply from monitoring unit. The function means that the data is uploaded to the monitoring station, the information classified by monitoring station is uploaded to the monitoring center, at the same time, the monitoring station gives orders to the monitoring unit. The monitoring center monitors the working status of all equipments, analyzing the data and information; it also provides the service of maintenance and makes decision. In the process of actual operation, monitoring system generally settings three parts according to the actual situation, including communication scheduling monitoring center, network management center, substation monitoring substation unit [16, 17].

3.1 The Establishment of the Substation Monitoring in Telecommunication System

The safe and efficient operation of telecommunication system requires that communication power supply must run properly and reliably. Communication power supply at the present stage transforms the traditional centralized power supply mode to distributed power supply mode. So it is very necessary to detect each part of power supply system and control it according to the actual situation, integrating the monitoring information and adopting a centralized monitoring for guaranteeing the reliable operation of the communication power supply. Through the effective collection of information and the analysis of data and the warning, the maximum risk of communication power supply would be found. Data integration can also be used in the later movement of communication circuit, maintainer and the replacement of communication power supply equipment, reducing the risk and improving the stability and security of communication power supply system [18].

3.2 The Construction of the Monitoring Center Master Station with Powerful Functions

The monitoring center of communication scheduling is the core part of telecommunication power supply supervisory system which can be able to timely handle the message from each of monitoring units, regional monitoring stations and substation monitoring unit, meanwhile, it can also establish the intersection between data in the monitoring host database of communication power supply and handle these data, which effectively overhauls and grades the communication power supply in the region, Therefore, it provides the decision-making basis for the maintenance plan [19].

3.3 The Establishments of the High-Speed Information Networks Between Telecommunication Monitoring Units

The monitoring substation units of the substation are connected to the communication power supply equipment through the sensors. The monitoring unit has the ability of monitoring the communication power supply every moment and sends periodically the information to host of the monitoring center. In the process of the actual operation, the substation device can access to the monitoring substation unit after the device adopts the RS232 protocol conversion. Then the monitoring substation would store data and display data after dealing with data, at the same time, the data would be transmitted to the monitoring center in the region. The monitoring substation units of the substation adopts 1000M/1M Ethernet network interface, using optical fiber to transmit the information and using the TCP/IP protocol network. Based on the above situation, power monitoring network can become reality. Therefore, we can accomplish the transmission of information through electric power communication network, realizing the monitoring mode of the centralized monitoring and unified management.

3.4   The Process of Data Transmission in Telecommunication Power Supply Supervisory System

Monitoring data collected by monitoring units would be transmitted outside through the substation and the optical fiber communication system, eventually reaching the monitoring unit of the central station. The protocol processor of the central station can monitor the monitoring data received by the monitoring unit which is working through the terminal server. Then these data would be interpreted on the protocol. Finally, the data disposed by the protocol processor would be sent to the server. Monitoring personnel check the data on the server by the client, understanding how the communication power supply from the far side works timely and then taking corresponding measures.

3.5   Comprehensive Evaluation of Power Communication Network

As the comprehensive evaluation of power communication network is carried out, there are relatively many indexes for evaluation, and then the influencing factors are also diverse. Therefore, this paper adopts the hierarchical structure to divide the indexes, which is mainly divided into three levels: target layer, criterion layer and indicator layer. The purpose of the hierarchy is to make the structure more clear and the complex problems simpler. The analysis of data is more convenient and beneficial to data processing. This paper adopts the hierarchical structure to construct the comprehensive evaluation index system. Through the analysis of various aspects and the opinions of experts, the comprehensive evaluation index of power communication network is mainly divided into three aspects: network reliability index, network business support ability index and network developable ability index, refer to Fig. (2).

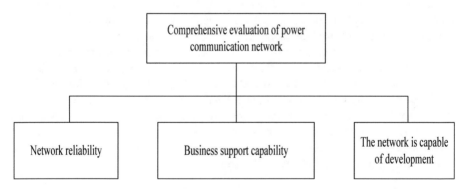

Fig. 2. Comprehensive evaluation index of power communication network

In the comprehensive evaluation of power communication network, it is assumed that A is the weight of B, and the allocation of weights can be defined using Eq. (1).

$$w_j \geq 0, \sum_{j=1}^{m} w_j = 1 \tag{1}$$

According to the comprehensive evaluation theory of attributes, the comprehensive measurement of multiple indicators A satisfies the following formula:

$$\mu_{xk} = \sum_{j=1}^{m} w_j \mu_{xjk} \tag{2}$$

According to the above Equations, the allocation of weights can directly affect the final result.

3.6 The Problem and Outlook of Telecommunication Power Supply Supervisory System

Since the centralized monitoring system of communication power supply to start construction, it has formed a centralized monitoring and unified management of the monitoring mode, the ability to monitor unit to early warning analysis of the collected data, the system of communication power supply to high efficiency and stable running safely. Through data analysis, the staff in a timely manner to the warning of communication power supply monitoring unit for maintenance, eliminate the potential failure, makes passively repair way change into active maintenance way. This ensures the electric power communication system which can maximize the risk-averse, reduce equipment power outage events, for the power system of production management to provide stronger communication support.

To some extent, it is difficult to gather the data because of a variety of communication power supply in practice. The reason is that we need connect them with the monitoring substation based on a large number of persons and money in order to solve the problem of communication protocol. Therefore, it is very necessary to establish a standardized communication interface protocol immediately. Telecommunication power supply supervisory system in the future will develop towards informational, intelligent-direction, unification and standardization. Telecommunication power supply supervisory system can meet the monitoring requirement at present, but the system has also the large development prospect in the aspects of the analyzable, statistic and managed dates' the same time, Cloud computing must be applied in telecommunication power supply supervisory system in the future.

4 Conclusion

With the development and the innovation of the technology with the integration of communication power supply and power environmental monitoring, etc., telecommunication power supply supervisory system in the future will expand its coverage constantly, and it can also be applied better and better in the telecommunication system, enhancing the stability and security of the telecommunication system. So, it also improves the level of scientific management and communication equipment maintenance.

References

1. Liang, Z.X.: The maintenance and improvement of the communication power supply centralized monitoring system. Northwest Electr. Power Technol. **3**, 39–41 (2012). (in Chinese)
2. Paw, K., Jeong, H.D.J.: On credibility of simulation studies of telecommunication networks. IEEE Commun. Mag. **40**, 132–139 (2002)
3. Sarkar, N., Gutie, J.: Revisiting the issue of the credibility of simulation studies in telecommunication networks. IEEE Commun. Mag. **52**, 218–224 (2014)
4. Lang, C., Kos, D.: Energy consumption of telecommunication networks and related improvement options. Sel. Top. Quantum Electron. **17**, 285–295 (2011)
5. Liu, X.R., Zhao, Y.L.: Application of profibus in coal mine power supply supervisory system. Taiyuan Sci. Technol. **3**, 28–32 (2008). (in Chinese)
6. Liu, D., Chen, L.: Research of EPS emergency power supply supervisory system based on Modbus. Mod. Electron. Tech. **6**, 16–19 (2013). (in Chinese)
7. Jung, J., Kang, Y.: An empirical study on the adoption intention of IoT based smart gas safety shutoff device service. Int. J. Internet Things Appl. **2**(2), 7–12 (2018)
8. Krishna, M., Vardhan, K.: Studies on the applications and performance associated topics of fog computing. J. Stat. Comput. Algorithm **1**(1), 33–38 (2017)
9. Oreste, K., Christopher, C.: Skills collaboration during infrastructure-service migration to cloud. Int. J. Cloud-Comput. Super-Comput. **5**(2), 1–8 (2018)
10. Jang, M., Byun, Y., Ryu, R., Kim, Y.: A basic study on the application of prism sheets to double-skin facades to improve indoor light environments. Int. J. Comput.-Aided Mech. Des. Implementation **4**(1), 7–12 (2018)
11. Choi, D., Jeon, H., Lim, P.: A study on the distribution ratio of optimum hot water heating load depending on the solar-powered hot water supply system. Asia-Pac. J. Adv. Res. Electr. Electron. Eng. **2**(2), 1–8 (2018)
12. Mehrotra, P., Kumar, V., Gupta, N.: A new mechanism of perpetual motion machine-3. Int. J. Energy Technol. Manage. **2**(1), 13–24 (2018)
13. Kim, Y., Choi, Y., Choi, J.: Study on noise characteristics of airport. Int. J. Energy Inf. Commun. **9**(3), 7–12 (2018)
14. Kim, D., Seok, J.: Active sonar target classification using multi-aspect based sensing and deep belief network. Asia-Pac. J. Neural Netw. Appl. **2**(2), 1–6 (2018)
15. Kweon, S., Kim, S., BinSong, M., Li, X.: Analysis of news frame semantic networks on virtual reality and augmented reality of the media. Int. J. Image Signal Syst. Eng. **2**(1), 1–6 (2018)
16. Kim, G., Rim, M., Jeong, J.: The optimum number of channels in a channel hopping environment. World J. Wirel. Devices Eng. **1**(1), 39–46 (2017)
17. Sekhar, C., Satish, S.: Effectively utilization of road divider for organized vehicular traffic using IoT. Int. J. Sci. Eng. Smart Veh. **2**(1), 1–10 (2018)
18. Yang, M., Song, S., Kim, Y.: A basic study on the sustainable urban development case studies. J. Creative Sustain. Archit. Built Environ. **7**(1), 35–42 (2017)
19. Lee, H., Lim, S.: A study on considerations of design changes through technical proposal tendering case of apartment housing. Int. J. ICT-Aided Archit. Civ. Eng. **3**(2), 39–46 (2016)

Theory and Application of Post Evaluation in Power Grid Construction Projects

Ruyu Zhang[1(✉)], Shuai Zheng[1], Jinyuan Liu[2], and Jianfeng Lin[2]

[1] Economic Research Institute, State Grid Liaoning Electric Power Supply Co. Ltd., Shenyang, China
ryzhang@163.com, zhengshuai@163.com
[2] Jinzhou Power Supply Branch, State Grid Liaoning Electric Power Supply Co. Ltd., Jinzhou, China
liujinyuan@163.com, 1179677489@qq.com

Abstract. The post-grid evaluation theory is closely related to the post-evaluation of the national economic environment. The power industry as a public service field has previously been in a state monopoly. It seems imperative that the future development of the power market after the establishment of the grid assessment mechanism. To this end, the relevant departments need to establish the implementation rules of the evaluation and management system of the grid project as soon as possible to form a real post-evaluation system. This paper is in this context, after the evaluation of the power grid construction project, the analysis, and evaluation of post-assessment, financial evaluation and the process of a comprehensive evaluation of the indicator system, and put forward some reasonable suggestions for the sustained and rapid development of the power industry.

Keywords: Electric power development · Evaluation of power grid · Economic environment assessment

1 Introduction

With the continuous development of social economy, society as a whole depends on the degree of power more and more, the power investment projects have been expanding. The grid construction project is the State Grid Corporation of important infrastructural projects, but also an important part of our infrastructure projects, power grid construction projects on China's economic development has a major role in promoting. After the project evaluation project to improve investment efficiency and decision-making is significant, currently the work has become an essential project cycle link. Therefore, the evaluation grid construction projects in an increasingly important role in showing how to do the work of the project investment have become an important topic of study power companies through scientific decision-making methods.

Currently evaluation after China's major economic benefit is still in the evaluation stage, post evaluation done better financial institutions, they are more concerned about the quality of the economic benefits of the project in order to determine whether the loan principal and interest can be recovered, and the social and environmental impacts can be sustainability aspects of attention is not enough, but also for the implementation of a number of public projects post evaluation of the lack of oversight mechanisms. But with China's further economic development, government affairs public and the importance of the ecological environment, more and more attention to its various departments will post evaluation of the work will be more perfect. This paper is in this context, the post-project evaluation grid construction issues were discussed, and then established research evaluation index system, evaluation methods, existing problems systematic research, and put forward some reasonable proposals for sustained and rapid development of the power industry it has important significance.

2 Post Evaluation Grid Construction

The post project is to evaluate the implementation process has been completed construction projects, benefit, purpose, role, and so influence the whole process of a comprehensive and objective analysis and summary, it is the least important part of the project cycle. Post evaluation of the project includes process evaluation (ie the implementation process evaluation), benefits evaluation (including economic and environmental benefits and social benefits), impact assessment (including the economic impact assessment, environmental impact assessment, social impact assessment) and continuous evaluation are content [1]. Post evaluation of the project is to run after the completion of the project or the investment was a time when its economic benefits fully revealed [2].

Re-analysis and evaluation of the project prior to the assessment carried out, it is the feedback phase of the project management. Through the project after the completion of the evaluation, it can detect whether the project to achieve the desired investment results, which summed up both positive and negative lessons provide macro guidance for the development and implementation of new projects, refine and improve investment decisions, adjust the relevant policy, management policies and procedures, improve the scientific management services; investment and financing parties in the future construction of similar projects experience; perfect for the project legal project has been built to guide projects to be built to provide advice, and then to increase and improve investment efficiency. After the project evaluation is generally made by the major investors and investors in construction projects, independent advisory body or related senior experts cooperate in complying with a series of principles, such as independence, objectivity and impartiality principles down to complete [3]. Of course, the project owner and project legal person may also be based on their own needs after the project organization and project evaluation, but evaluation should follow certain principles and procedures. The post-project evaluation in the project cycle is shown in Fig. 1.

Due to the different time points evaluated which, post evaluation of the construction project feasibility study is different from the pre-project investment decisions, has its own characteristics. After a complete and objective evaluation of the project study report is based on a comprehensive consideration of impacts from the project, the effect of the investment on the basis of various factors, so that the project should have a comprehensive evaluation of the content, it should include not only the implementation

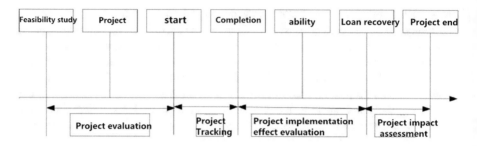

Fig. 1. Post-project evaluation in the project cycle

of the project preparation evaluation of the implementation process of work and projects but also to analyze the effect of the implementation process of its operations; it is necessary to evaluate the economic benefits of investment projects, but also to analyze environmental impacts and social benefits it generates, etc., and even need to analyze project management, if necessary capacity for sustainable development and the level of the project [4–6]. Purpose of the project is to evaluate the completed projects were retrospective to treat the construction project guidance and reference, and for the government to feedback information to improve decision-making project management level, and to discover and solve problems for future projects and macro and micro basis for decision-making, and to explore the development trend of future construction projects. Results of project evaluation often require feedback to decision-making, which summed up the experience and lessons learned and provides guidance on project basis for evaluating new projects, as well as adjust the direction of investment, which is the ultimate goal of the project evaluation. Therefore, the evaluation findings of feedback mechanisms and methods become one of the key factors in the success of the project.

After the project evaluation process, the project is pre-project condition is correct, decision-making procedures for compliance, whether the project progress according to plan and other control. Details include: survey design evaluation, geographic position, geological structure, the degree of job satisfaction survey design and construction of soil; construction design program evaluation, namely Optimization of construction scheme, construction technical feasibility and advancement of expenses rationality, etc; the project process control evaluation, namely construction preparation, project progress, project supervision, project cost and the execution of the contract and so; production and management assessment that the main material, fuel supply, the integration of resources, the production capacity of the developments, financial revenue and expenditure and balances and so on. Through the evaluation process, it can analyze project success and failure [7]. Including project financial benefit evaluation, the main principles and evaluation methods to assess the pre-project the same, but different evaluation purposes and data values. Analysis of the effectiveness of the project were evaluated before and after comparison method was used to re-index estimates derived FIRR and EIRR and project feasibility assessment were compared to identify differences and reasons and with the industry benchmark rate of return or project lending rate (the composite interest rate) compared to the financial benefits of the project evaluation.

The main contents include the preparation of financial evaluation and analysis of the project's basic financial statements, profitability analysis, liquidity analysis, and sensitivity analysis.

3   Method of Post Evaluation Grid Construction

3.1   Post Evaluation Process Engineering

For the implementation of the various stages of the project implementation process of laws and regulations, the implementation of scientific and regulatory level operability, objective evaluation of managers and executives from the various stages of the qualifications and credit management to review and evaluate whether it Science can effectively manage the work of the project, whether government agencies and other organizations to establish the necessary contacts, whether rational use of human resources and so on. Through the quality and ability of managers to set and evaluation functions, forms of organization and function, reflects the management efficiency and effectiveness, and efficiency of the decision-making level of management and other aspects of the organization, m looking for deviations from the expected target of causes and propose countermeasures and recommendations, so as to continuously improve the level of construction power construction projects [8].

Project Decision project evaluation is a necessity of the project, feasibility, design, and documentation and related decision-making throughout the project, a comprehensive assessment and reporting procedures. The establishment of the project, the implementation of the project to assess whether the project decision science, rationality validation of the project content, time-axis analysis, summarized the project decision, design experience. The main evaluation include: pre-relevant preparations, the decision to declare the project, environmental impact of the construction and building control processes. By comparing before and after construction of the project, the project construction process to find the problems, thereby rectification. Project preparation work mainly includes: project preparatory work related summary, experience analysis; the project site investigation, standardization of design, material procurement, project bidding, financing aspects of investment and building construction and other comprehensive evaluation. Project during construction of the project from the overall design to the whole process of project implementation, is an important part of project development. During construction of the project evaluation, including whether the main construction contract related to the construction unit in accordance with established program, the construction process meets the relevant requirements of construction drawings, construction quality, construction schedule and cost effective construction and cost rationalization [9]. Focus is on quality and system cost, construction cost and system analysis and evaluation, but also taking into account the completion and acceptance of, the evaluation of production preparation work.

Table 1 gives an important summary table of project process evaluation indicators.

Table 1. Table project process evaluation indicators

	x_1	x_2	x_3	x_4	x_5	x_6	x_7	x_8	x_9	x_{10}
E_1	10	9	8	9	7	6	7	9	4	10
E_2	9	10	9	10	5	7	9	10	5	9
E_3	9	9	8	9	6	7	7	8	6	8
E_4	10	7	10	7	6	8	8	9	5	9
E_5	8	10	9	10	6	7	6	10	6	7
E_6	10	9	8	9	7	6	7	9	5	7
E_7	9	8	10	8	6	6	6	10	6	8
E_8	7	9	8	9	4	7	8	7	7	6
E_9	9	6	7	10	5	8	7	8	6	9

3.2   Post Evaluation of the Financial

In the grid post evaluation of the project, the most important one is, post evaluation of the financial situation was evaluated. As for, the financial evaluation phase, mainly based on inputs and outputs of the project, from a business point of view to achieve profit maximization to consider. For the actual financial status of the project, such as the overall cost of production enterprises, sales prices, income and other aspects of profit, anticipated financial indicators and the actual financial operation will be some differences. Financial evaluation contained therein, this comparative analysis to identify the reasons for the deviation of, and reasonable evaluation and concludes in accordance with reason. Finally, post evaluation of relevant, improve operational program, lessons learned to improve the related investment income. Basic data grid construction projects included in the input data and output data are divided into two categories. The role of financial statistics data is the basis for accounting, processing a large number of basic data, the need for computer-aided processing, data accuracy and reliability of demanding. The evaluation of the financial aspects of the costs and benefits of the project construction project construction costs is needed to be consistent in the calculation caliber [10]. Evaluation and analysis of financial data mainly focus on the overall consistency. Investment scope and cost of benefits is consistent with one of the basic principles of financial evaluation.

Project construction financial indicators are mainly classified two aspects: the actual financial indicators of project construction and the difference between the pre-rating and the actual financial indicators. Actual financial business indicators project is reflected in the two dynamic static posture indicators, static indexes in fixed investment is mainly reflected in the actual profit rate, the interest rate on the investment projects reflect the specific financial operations. Also include investment recovery period, pay off the loan period, the ratio of assets and liabilities, cash flow ratios, financial tax net present value amount, the actual internal static earnings ratio, net present value of the actual financial indicators dynamic, dynamic internal gains and dynamic investment recovery and so on. After evaluating reflects actual income and project evaluation phase difference amount is on the other hand, including the rate of change in static investment recovery period, changes in the actual flow rate, changes in actual financial terms of financial indicators to assess the ratio of the rate of change in financial income inside asset indebted rate, the actual rate of change of dynamic recovery expectations. Such as: In terms of investment profit ratio is equal to the real rate of return divided by the expected rate of investment.

3.3 Post Evaluation Project

Through various sub-items of the project evaluation, can learn from the different perspectives of the project implementation effect, but it also should be noted that the various sub-items has a certain independence, and has a strong professional, that is not yet formed a complete concluding observations. Thus, after the project needs to be evaluated on the basis of a comprehensive evaluation of the various sub-item evaluation, concluding observations, to the relevant personnel to provide a simple and intuitive judgment basis. Evaluation is unstructured decision-making process of a complex object system, because the relevant factors to evaluate the effectiveness of a lot, and comprehensive evaluation of the target system are often social, economic, science and technology, education, environment, and management for a class of complex systems, Therefore, the evaluation is a very complex thing. The traditional comprehensive evaluation method in practical applications also cannot get rid of the evaluation process randomness, uncertainty and ambiguity recognized expert subjective evaluation on. So, how can we take full account of both the experience and intuition of experts evaluate the mode of thinking but also reduce the uncertainty evaluation process artificially, with both normative comprehensive evaluation method can reflect the high efficiency of problem solving, this is the key to making the right decisions [11]. At present, the commonly used method for comprehensive evaluation of domestic and foreign stars have expert evaluation, economic evaluation, multi-objective decision-making method, AHP analysis method, fuzzy comprehensive evaluation method, evaluation methods and the extension decision logical framework approach.

Here we use a logical framework approach, combined with the previous quantitative evaluation, qualitative analysis through a comprehensive system of logical thinking, to break the three questions fully answered. For example, to improve the performance of the Japanese standard of the grid through line loss rate, voltage qualification rate, capacity-load ratio and other indicators of the actual value of the original target contrast, you can verify whether the original target to achieve what extent, it did not meet, is what causes it; to achieve social goals, by improving people's living standards and promote economic development in the region, energy conservation, and other computing and integrated indicators, access to social benefits to achieving the original objective measure of the degree of realization effect of a good or poor indicators, analyze its causes, such as whether there are other factors that promote or restrict the realization of these goals. On the basis of the foregoing sub-section evaluation, the initial establishment of the overall evaluation of the structure of a power transmission project (hierarchical structure). The hierarchical structure of three levels of a comprehensive evaluation, the total evaluation of goal into the construction process evaluation, economic evaluation, performance evaluation grid technology, and social impact assessment, and chose the single factor index and sub-targets linked. When using the fuzzy comprehensive evaluation method to evaluate the practical problems and decisions need to determine the weight of each factor field factor evaluation object weight. Methods commonly used by respondents given directly based on their knowledge, experience, and preferences and other factors it deems appropriate value, and then by forming a statistical summary weight set.

4 Conclusions

After the project evaluation is a means of effective supervision of investment activities. After carry out the project evaluation, attention is to look a little further, with bit deeper, a little meticulous; so as to better play the role of evaluation in the project after project construction. Post evaluation of the project itself started late, in theory, is not perfect and combined with industry-specific theory, methods, and application of research and even less. The power industry as a public service in the field had previously been in a state monopoly, in the future with the development of the electricity market, the evaluation is bound to become important after the grid. Therefore, after the establishment of a mechanism for the evaluation of the grid, it seems imperative. For this reason, the relevant departments need to establish a grid projects as soon as possible the evaluation and management system implementation details of the formation of a true post-evaluation system. It can provide useful experience and lessons for future investment. Post evaluation of how to carry out good grid projects, the most important prerequisite is to follow the "independent, scientific and impartial" principle.

References

1. Casley, Kumar, K.: Project monitoring and evaluation in agriculture. Henan Sci. Technol. **15**(3), 28–29 (1992)
2. Cook, D.C.: Benefit cost analysis of an import access request. Food Policy **33**(3), 277–285 (2007)
3. Harmelink, M., Nilsson, L., Harmsen, R.: Theory-based policy evaluation of 20 energy efficiency instruments. Energ. Effi. **1**(2), 131–148 (2008)
4. Killc, M., Kaya, J.: Investment project evaluation by a decision making methodology based on type-2 fuzzy sets. Appl. Soft Comput. **27**, 399–410 (2015)
5. Rocha, L., Tereso, A., Couto, J.P.: Project management, evaluation of the problems in the Portuguese construction industry. Adv. Intell. Syst. Comput. **353**, 69–78 (2015)
6. Vicente: Association of project management. Syllabus for the APMA Examination (Second Edition) (1996)
7. Abdeen, M.O.: Energy, environment and sustainable development. Renew. Sustain. Energy Rev. **12**(9), 32–36 (2008)
8. Clifford, S., Goodman, Roy, A.: Methodological approaches of health technology assessment. Int. J. Med. Inform. **56**(1–3), 97–105 (1999)
9. Kim, J., Lee, Y., Moon, J., Park, J.: Bigdata based network traffic feature extraction. Int. J. Commun. Technol. Soc. Netw. Serv. **6**(1), 1–6 (2018)

10. Choi, Y.: Physical awareness symptoms and mental health according to the characteristics of smart phone usage of university students. Int. J. Hum. Smart Device Interact. **5**(1), 7–12 (2018)
11. Minh, N., Kim, M.: CMP-MAC: a cross-layer, multiple pipeline-forwarding MAC protocol for wireless sensor network. Int. J. Power Devices Compon. Smart Device **3**(1), 15–20 (2016)
12. Bai, Y., Chang, B.Y.: Optimal green-advertising incentive model in a closed-loop supply chain. Int. J. Adv. Sci. Technol. **119**, 153–170 (2018). SERSC Australia
13. Budiharjo, Soemartono, T., Windarto, A.P., Herawan, T.: Predicting tuition fee payment problem using backpropagation neural network model. Int. J. Adv. Sci. Technol. **120**, 85–96 (2018)
14. Rao, N.T., Bhattacharyya, D., Raj, S.N.M.: Queuing model based data centers: a review. Int. J. Adv. Sci. Technol. **123**, 11–20 (2019)
15. Nkenyereye, L., Jang, J.W.: Adaptive in-car external applications using nomadic smartphones and cloudlets. Int. J. Control Autom. **8**, 75–96 (2015)
16. Bai, J., Liu, H.: Improved artificial bee colony algorithm and application in path planning of crowd animation. Int. J. Control Autom. **8**, 53–66 (2015)
17. Huang, W.H., Ma, Z., Dai, X.F., Gao, Y., Xu, M.D.: Cluster load balancing algorithm based on intelligent genetic algorithm of vector. Int. J. Grid Distrib. Comput. **10**, 73–84 (2017)
18. Wan, B., Wan, I., Maryati, Y.: Mitigation strategies for unintentional insider threats on information leaks. Int. J. Secur. Appl. **12**, 37–46 (2018)
19. Shan, L.H., Fang, W.D., Qiu, Y.Z., He, W., Sun, Y.Z.: Smart mobile gateway: technical challenges for converged wireless sensor networks and mobile cellular networks. Int. J. Future Gener. Commun. Netw. **9**, 87–98 (2016)
20. Kaur, T., Sidhu, R.K.: Optimized adaptive fuzzy based image enhancement techniques. Int. J. Signal Process. Image Process. Pattern Recogn. **9**, 11–20 (2016)
21. Boyinbode, O.: Implementing RFID ubiquitous learning environment. Int. J. u - e - Serv. Sci. Technol. **10**, 47–58 (2017)

Development Route of Power Grid Enterprises Based on New Electricity Reform

Jing Gao[1] and Yong Wang[2(✉)]

[1] Economic Technology Research Institute State Grid Liaoning Electric Power
Co. Ltd., Shenyang, China
[2] State Grid Liaoning Electric Power Supply Co. Ltd., Shenyang, China
yongwang112@163.com

Abstract. The paper believes that the introduction of the new electricity reform program has an important impact on the development of power grid enterprises, and the power grid enterprises need to re-comb and locate in the development path. First, it sorts out for Chinese the electricity reform program and the reform of the core ideas over the years, focuses on the interpretation of the core elements of this reform, sorts out the important issues associated with the development of power grid enterprises, combining the development path problem of power grid enterprises. We should optimize the investment, assets, innovation and cost control in order to improve the profitability of grid enterprises in the new situation.

Keywords: New electricity reform · Power grid enterprise · Development path

1 Introduction

The development of China's power system is complex and has undergone many major reforms. Before 1985, due to the serious shortage of power supply in our country, the state implemented a high monopoly mode of "combining government with enterprise" and "state-owned state" to the power industry, and the only producer and operator in the power industry was the state. Because of the market distortion and the contradiction between supply and demand, since 1985, the country began to reform the electric power industry [1]. The main objective is to solve the problem of the integration of government and enterprises. In May 1985, a notice of the State Council approved the National Economic Commission and other departments to raise funds "Provisional Regulations on encouraging the collection of funds for the construction of electric power and the implementation of a variety of electricity prices" [2].

On July 1987, "national power system reform forum" proposed government enterprise separately, the province as the entity, the joint grid, unified scheduling, and raising power to do the twenty word guideline. On January 1993 [3]; the Department of energy, electric power company reorganized into electric power group company; On January 16, 1997, the national power company was formally established in March 1998, revoked Ministry of electric power industry and so on [4]. Through a series of initiatives, the initial realization of the separation of government and enterprises. But the power system of this stage is still the vertical integration management mode, the

A. E. Hassanien et al. (Eds.): AISI 2019, AISC 1058, pp. 964–971, 2020.
https://doi.org/10.1007/978-3-030-31129-2_89

efficiency is still low [5]. On February 10, 2002, the State Council issued document No. 5, promulgated the "reform plan of power system", and proposed the overall reform direction of "separation of factories and networks, separation of main and auxiliary services, separation of transmission and distribution, and bidding online" [6]. After more than ten years of development, the separation of factories and networks and the separation of main and auxiliary services have been basically realized. However, the separation of transmission and distribution and bidding for the Internet has not been realized, and the competitive electricity market has not been fully established. In 2013, the State Council issued a document on the plan of energy 12th Five-Year [7]. Planning to do some new deployment for electricity reform: the steady development of the separation of transmission and distribution of the pilot, the establishment of market trading platform in the regional and provincial power grid within the scope of batch release users, placing an independent power generating companies and trading enterprises; improve the power generation scheduling approach, gradually increase the economic dispatch factors; promote the overall development of rural power system reform [8]. Encourage private capital into the laws and regulations are not explicitly prohibited by the energy sector; promote the diversification of the power grid and other infrastructure investment [9].

On March 15, 2015, the high-profile "on the further deepening of power system reform opinions" issued, the focus of the electric power system reform to further improve the government enterprise separation, separate the plant and network, the main and auxiliary separation on the basis of the control system in accordance with control center, open ends, orderly liberalization of the competitive aspects of electricity transmission and distribution outside [10]. In order to social capital liberalization placing an orderly liberalization of electricity business, plans to use public welfare and regulatory outside, promote trading institutions are relatively independent, standardized operation, further to the regional power grid construction and suitable for China's national conditions of transmission distribution system, to further strengthen the government supervision, to further strengthen the power of overall planning, to further strengthen the safety and high efficiency the operation of power supply and reliable [11].

The introduction of the new electricity reform plan will have an important impact on the development of power grid enterprises, and what kind of measures should be taken to power grid enterprises are of vital importance for the development of enterprises [12].

2 Comb the Core Elements of New Electric Power Reform

Aiming at the lack of power of reform in the field of trading mechanism, resource utilization efficiency is not high, the price relationship and the market pricing mechanism has not yet fully formed, the transformation of government functions is not in place, the problem of planning coordination mechanism is not perfect, the new electricity reform program has an important breakthrough (see Fig. 1).

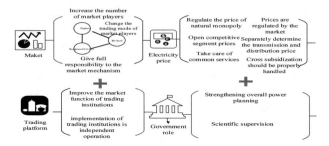

Fig. 1. Interpretation of the new electricity map

2.1 Orderly Reform of Electricity Prices, Straighten Out the Formation Mechanism of Electricity Prices

On the formation of electricity price, the new electricity reform program will account for 85% of electricity consumption of commercial and industrial electricity prices, the supply and demand sides (power generation and sales) to determine the relationship between market supply and demand [13], the price difference between the Internet and the sales of power grid enterprises no longer as the main source of income, income from government approved distribution electricity charges. Involving the residents, agriculture, and utilities, electricity consumption, although the proportion of electricity consumption is low, but the users of many people, with a wide range of social impact, so this part of the electricity prices will continue to be controlled by the government [14].

2.2 Promote the Reform of Power Generation Plans and Orderly Reduce Power Generation Plans

In terms of electricity, the number 6 article adopts the same "double track" system as the same as the electricity price management. The volume and capacity of direct trade in industry and Commerce will no longer be included in the generation plan and will be decided by both sides of the market [15]. But the government will retain some public welfare, electricity regulatory plan, to ensure that residents of electricity, agriculture, public utilities and public services, in order to maintain peak FM and the safe operation of power grid, to ensure that renewable energy power generation planning in accordance with the protection of the acquisition". In short, the government will continue to control part of the planned electricity generation to ensure that electricity is cheap, safe, and clean, while releasing the electricity [16].

2.3 Push Ahead with the Reform of the Sale of Electricity and Orderly Release and Distribution of Electric Energy to Social Capital

Encourage social capital investment in the sale of the electricity business, and gradually to the qualified market players liberalized incremental sales of the investment business, to encourage the development of mixed ownership of the sale of electricity business [17]. The market-oriented reform of the power industry should follow the two basic market characteristics: transaction market and investment diversification. In the new situation of the electricity system reform, electricity transmission and distribution pricing mode and change to effective asset-based companies approved to permit new income [18], in the sale of electricity and power distribution side of the incremental part such as the introduction of social capital to the development of the company investment scope, focus and goals have a profound impact, and put forward new requirements for the development of the company investment plan management connotation, etc.

3 Opportunities and Challenges Faced by Power Grid Enterprises

The power grid is the hardest hit party in the reform of electricity industry. In a sense, the "independent three release" is essentially stripped of the Power Grid Corp's pricing power, quantitative right and trading right. The loss of these powers has transformed the grid from electricity traders to logistics companies, and the profit model has changed radically, thus directly affecting revenues.

3.1 Weakened Monopoly and the Reduced Profitability

The diversification of market price, the sale of electricity in the main, the monopoly of the Power Grid Corp has been weakened, thus weakening the Power Grid Corp market forces, reduce the Power Grid Corp use market forces to obtain excess profit may. This also means that price competition will intensify, industry profits will be redistributed in the industrial chain. The release of competitive electricity price under the new electricity reform plan will break the enterprise electricity cost under the government electricity price control, and make the electricity price drop, thus enhancing the profit level and the market competition ability of the power consumption enterprise [19]. The rise and fall of electricity prices will be determined by the market supply and demand and the scarcity of resources. With the promotion of the electricity market reform, market competition has reduced the price difference between hydropower and thermal power. After the reform, the profit of Power Grid Corp is no longer related to the purchase and sale of electricity, and it has nothing to do with the transmission capacity of the power grid. It is only related to the effective assets of the grid. Power Grid Corp will be strictly regulated to become a high-speed channel for power transmission [20]. After the change of Power Grid Corp profit model, the power generation side and the user side will form effective competition, promote the development of the power market, and play the leading role of the market in the allocation of resources. According to the results of the previous pilot reform, the average transmission and distribution price of Shenzhen in 2015 was 0.1435 yuan/kWh, down 0.01 yuan/kWh compared with 2014, and it will continue to decline in the next two years. Therefore, we can expect that the profitability of the power grid enterprises will be greatly impacted by the context of the new electricity reform.

3.2   New Cost Plus Pricing Requires Major Changes in Grid Companies

An important task of the electricity system reform is to rationalize the price formation mechanism and the transmission mechanism. The former is closely related to electricity production, while the latter is related to the transmission, distribution, and sale of electricity. From the supporting documents, the government adopted a relatively simple "cost-plus pricing", allowing the so-called cost plus a reasonable profit model, the formation of transmission and distribution price is based on cost, and set the yield ceiling". The core of this method lies in cost accounting, including the cost of electricity production, the negative externalities of electricity production, the cost of power transmission and so on. The power cost accounting is a major challenge because of the asymmetric information between the power production enterprises or the power transmission enterprises and the electricity price supervision departments. From the international experience, the cost plus method, power companies will have excessive investment motives.

3.3   The Original Investment Model

In view of deepening the reform of the power system, the traditional mode of investment and construction and analysis methods of the power grid is no longer suitable for the development of the power grid. Combined with the introduction of competition at the offering side, the overall design idea of strengthening the supervision of the transmission and distribution terminal, the original investment system of the grid enterprise needs to be reconstructed, the investment model needs to be changed, and the scale of investment needs to be adjusted. At the same time, the reasonable distribution method and engineering project budget by analysis of the cost and the capital investment of science, to ensure the company's investment can be fully integrated into the effective assets in terms of technology and economy, maintain reasonable transmission and distribution price level protect the company's power grid investment ability.

The risk impact value model is:

$$
\begin{cases}
v_i = \dfrac{1}{n} \displaystyle\sum_{j=1}^{n} \left(w_p p_{ji} + w_t t_{ji} + w_u u_{ji} + w_r r_i + w_c c_i \right) \\
w_p + w_t + w_u + w_c + w_r = 1
\end{cases}
\tag{1}
$$

Where, v_i is the impact value of risk i; j is the number of survey object; p is the risk probability value; t is the length of delayed construction period; u is the risk uncontrollability value; r represents the risk responsibility party; c represents the risk category.

4 Analysis of Development Path of Power Grid Enterprise

4.1 Change the Investment Model to Form an Effective Asset

Because it represents the general trend of China electric power system reform and strengthens the cost constraint and income change regulation, so the emphasis of power grid enterprises to "reduce costs, improve efficiency" internal operation mode, enterprise asset management will become the focus of the field of Power Grid Corp.

In the legal and regulatory framework, reasonable use of legal means, the Power Grid Corp should be approved before the first effective asset supervision period, and lay the foundation for making the relatively favorable price level, to set aside a certain space to gradually reduce the transmission and distribution price level. After the official start of the regulatory cycle, we need to be cautious in expanding effective investment and increasing effective assets.

4.2 Whole Process Monitoring of Power Grid Project and Optimizing Cost Management

In the introduction of competition to break the monopoly regulation under the background of grid enterprise profitability will be weakened can determine the expected, the power grid enterprises in addition to the reasonable arrangement of investment, scale, investment, but also strict cost management of power grid investment projects, comprehensive combing and Optimization in the construction of power grid project the budgetary estimate and budget, estimation, accounts and other aspects of the project, for the completion of the project post evaluation, reduce project costs, improve operational efficiency, enhance profitability.

4.3 Enhance Business Innovation Capability and Enhance Core Competitiveness

The strengthening of the supervision and control of power transmission and distribution links and the introduction of competition for the sale of electric links all require grid enterprises to continue their business innovation and enhance their core competitiveness. In the sense of innovation, innovation and management continue to deepen. In innovative technology, advanced technology focused on the development of UHV transmission, smart grid, new energy consumptive access and other fields, to build global energy internet.

In the aspect of business model innovation, we should focus on the development of alternative energy projects, continue to promote the perfection of the electric car industry system, improve the scientific and technological achievements, such as incentive mechanism, promote the overall development of science and technology, management, brand, business model innovation, relying on the "Internet plus", big data and other new technologies, improve the management level and the serviceability. To speed up the construction of innovative talents, improve the scientific research and industry talents bidirectional flow mechanism, researchers mobilize entrepreneurial enthusiasm, formed to attract talented people, retain top talent, a good environment for accelerating the growth of innovative talent.

5 Conclusion

This paper discusses the new electricity reform program and sorting out the important issues associated with the development of power grid enterprises. Also, discusses the way to optimize the investment, assets, innovation and cost control in order to improve the profitability of grid enterprises in the new situation. Also, provide an analysis of development path of power grid enterprise

References

1. Desheng, C., Guoliang, Z., Hongxia, L.: Interpretation of new electricity modification scheme (2019, in Chinese)
2. Can, L., Ming, G., Zhiheng, C., Kai, Q.: Review of China's electricity market reform. Innov. Appl. Sci. Technol. **2**, 147–148 (2016)
3. Mei, B.: Logical starting point and policy influence of new electricity reform plan. Price Theory Pract. **6**, 8–13 (2015)
4. Ying, D.: Study on investment scale, risk and decision of power grid under new electricity reform. Financ. Ind. **23**, 47 (2015)
5. Yi, C.: Strategic thinking of electric power enterprises (Power Grid Corp) under the background of new electricity reform. Technol. Enterp. **11** (2015)
6. Lihua, Q.: Research on the impact of new electric power on the grid and its existing marketing model. Mod. Econ. Inform. **20**, 96 (2016)
7. Yue, L., Fenge, L.: Opportunity, challenge and countermeasure of power grid enterprise development under new electric power reform. J. China Univ. Netw. Technol. **18**(06), 69–71 (2015)
8. Kim, K.S., Lee, H.S., Park, C.Y., Jang, H.I.: Proposal and performance evaluation of automatic blind algorithm based on illuminance sensors. Asia-Pac. J. Adv. Res. Electr. Electron. Eng. **21**, 7–12 (2018)
9. Kim, K., Kim, D., Lee, H., Kim, Y.: A basic study on suggestion of sloped type light-shelf with height adjustment. Int. J. ICT-Aided Archit. Civ. Eng. **32**, 7–12 (2016)
10. Park, J.J.: The characteristic function of corenet (multi-level single-layer artificial neural networks). Asia-Pac. J. Neural Netw. Appl. **11**, 7–14 (2017)
11. Alrefai, M., Faris, H., Aljarah, I.: Sentiment analysis for arabic language: a brief survey of approaches and techniques. Int. J. Adv. Sci. Technol. **119**(24), 13–24 (2018)
12. Farozin, M., Yudha, C., Herzamzam, D.: The educational games application using smartphone in learning mathematics for elementary school students. Int. J. Adv. Sci. Technol. **122**, 1–14 (2019)
13. Bai, D., Wang, C., Zou, J.: Design and simulation of fractional order control systems based on bode's ideal transfer function. Int. J. Control Autom. **8**(3), 1–8 (2015)
14. Xue, T., You, X., Yan, M.: Research on Hadoop job scheduling based on an improved genetic algorithm. Int. J. Grid Distrib. Comput. SERSC Aust. **10**(2), 1–12 (2017)
15. Zhang, F.: Robust analysis of network based recommendation algorithms against shilling attacks. Int. J. Secur. Appl. **9**(3), 13–24 (2015)
16. Peng, Z., Yin, H., Dong, H.: A harmony search based low-delay and low-energy wireless sensor network. Int. J. Future Gener. Commun. Netw. **8**(2), 21–32 (2015)
17. Liu, X., Duan, Z., Yang, X.: Vector quantization method based on satellite cloud image. Int. J. Signal Process. Image Process. Pattern Recogn. **8**(11), 27–44 (2015)

18. Wang, C., Lu, T., Chen, X.: Evaluation model of enterprise microblog marketing effectiveness on the basis of micromatrix. Int. J. U - E Serv. Sci. Technol. **8**(2), 1–10 (2015)
19. Ren, Q., Gao, J., Gao, H.: Research on improved hadoop distributed file system in cloud rendering. Int. J. Database Theory Appl. **9**(11), 1–12 (2016)
20. Shrivastava, S., Shrivastava, L., Bhadauria, S.: Performance analysis of wireless mobile ad hoc network with varying transmission power. Int. J. Sens. Appl. Control Syst. **3**(1), 1–6 (2015)

Power Line Loss Analysis and Loss Reduction Measures

Fengqiao Li[(⊠)], Lianjun Song, and Bo Cong

Department of Marketing, State Grid Liaoning Electric Power Supply Co. Ltd.,
Shenyang, China
418484168@qq.com

Abstract. Line damage refers to the power consumption of power grid lines and equipment, an indicator of the power of enterprise self-digestion. The line loss rate is an important technical and economic indicator of the national assessment of power sector energy consumption level; also, the planning and design of power system comprehensively reflect the level of production operation, management, and economic benefits. Power grid power loss rate is an important comprehensive technical and economic indicator of power companies; it reflects a power network planning and design, production technology and operation and management level. This paper explains the definition of line loss, analyzes the cause of line loss and classifies it, and expounds how to reduce line loss from the aspects of distribution network layout, three-transformer loss, and reactive power compensation.

Keywords: Line loss · Power system planning · Economic indicators

1 Introduction

Grid power line loss is a loss for short. Line loss rate is electricity line losses accounted for a percentage of electricity supply. The national assessment of power supply enterprise important technical and economic indicators. It is one of the main power companies plan to complete the national and enterprise assessment; at the same time, line loss rate is reflected in the grid power supply enterprise planning and design, technical equipment and the level of economic operation of a comprehensive technical and economic indicators [1–4]. In the power system, transmission, transformation, distribution and marketing in all sectors will have a power loss and line loss management. It is an important part of the power supply enterprise cost management. Power requirements of the market economy to the economic efficiency of enterprises, while the purpose of energy conservation, but also requires the use of resources for maximum efficiency, reduce energy consumption, therefore, accurate and reasonable distribution network line loss theoretical calculation of line loss in the power sector analysis constituted a powerful tool for the development of loss reduction measures, to promote the power companies to reduce energy consumption, internal potential to improve economic efficiency, optimize network planning and design programs to strengthen the operation and management is important [5, 6]. The line lose is defined using Eq. (1).

© Springer Nature Switzerland AG 2020
A. E. Hassanien et al. (Eds.): AISI 2019, AISC 1058, pp. 972–980, 2020.
https://doi.org/10.1007/978-3-030-31129-2_90

$$\Delta A = 3I_{jf}^2 Rt \times 10^{-3} \tag{1}$$

At the end of the distribution network as the power grid has a low voltage level, wide distribution lines, multi-line equipment, the resulting line loss has a relatively large impact [7–9]. Distribution network loss calculation is economic operation, reactive power optimization and infrastructure such as power grids transformation, through theoretical calculation of line loss reduction and loss analysis calculations can be performed line loss analysis and management decisions for the operation and management departments to provide accurate reliable data reference information to help supply enterprise managers to develop and implement sound economic line loss rate index, found weaknesses grid structure and scheduling, production technology, power, performance measurement, and management equipment operating conditions to grasp Composition and direction of loss, targeted to reducing losses to take reasonable measures to guide the power supply enterprise energy loss, so as to achieve maximum economic benefits.

Accurate and reasonable line loss theoretical calculation of the power sector constitutes line loss analysis, developed powerful tools to loss reduction measures, to promote the power companies to reduce energy consumption, the internal potential to improve economic efficiency, optimize network planning and design programs to strengthen the operation and management of an important significance. The study found that the main reason for the error of calculation of line losses is generated: a mathematical model for the actual network usage is not accurate; the original data is not accurate. Where the original data is not accurate is major sources of error. For line loss calculation, the original data includes two aspects: First, the wiring and component parameters of the power grid, and second, real-time data grid operation, namely current, voltage, active power, reactive power, and power factor. Component parameters more accurate and easier to obtain, and the voltage involved in the calculation of network nodes, the load curve current, active power, reactive power is changing with time. In addition, the reactive power compensation device within the grid also with manual or automatic change of operating mode switching, there is greater uncertainty, making the accurate calculation of theoretical line loss and loss reduction measures for the development of science has been a problem.

2 Defining Line Loss

Power grid power loss (referred to as the line loss) is the power transmission from the power plant to the client process, power loss and loss in transmission, transformation, distribution and marketing of various links generated. Electricity line losses ranging from power plant main transformer primary power loss to all users of energy meter, power supply, and electricity sales are calculated by subtraction. The percentage of electricity line losses account called electricity supply line loss rate, referred to as line loss rate. The line loss rate is a comprehensive reflection of the power network planning and design, production and operation and management level of the main economic indicators [10, 11]. Line loss rate based on the jurisdiction and the voltage level can be divided into a net loss rate and regional line loss rate, line loss rate region can be divided into regional networks and distribution loss rate of line loss rate. In practice, the electricity line losses have two values, namely the actual electricity line losses and theoretical electricity line losses. Thus, the line loss rate also has two corresponding values, the actual loss rate with the theoretical line loss rate. Under normal circumstances, the actual power grid line loss rate was slightly higher than the theoretical line loss rate. Today, the actual power grid line loss rate of 7% to 8.5%, the line loss rate covers urban network and rural network is defined as in Eq. (2).

$$K^2 = \frac{\alpha + \frac{1}{3}(1 - \alpha^2)}{\left(\frac{1 + \alpha^2}{2}\right)^2} \tag{2}$$

Traditionally, the line loss can be divided into statistical line losses, theoretical line loss and line loss management and so are several. Statistical line loss is calculated based on meter readings that the power supply and electricity sales difference between the two. Theoretical line loss in statistical electricity line losses, there are some in the transmission and distribution of electrical energy during the inevitable, by the time parameter power grid load conditions and power supply equipment decisions, this part of the loss of electricity called technical losses of electricity, it can be calculated by theory, it is also known as the theory of electricity line losses, the corresponding line loss rate theory called line loss rate. Management line loss is unknown loss, equal to line losses and statistical difference between the theoretical line losses, requiring smaller the better.

In the electric power business line loss management, distribution network line loss management has always been important and difficult line loss management. Related statistics show that in the power supply session, 10 kV power distribution lines and loss of electricity, accounting for 60%–70% loss of power of the region, how to effectively reduce the power consumption of the power distribution network, is the pursuit of long-term line loss management goal and medium-sized urban power supply enterprise line loss management focus [12, 13]. Distribution services to customers directly reducing the damage to be effective. We must first understand the natural condition of the grid line loss: natural line losses as a yardstick to distinguish between statistical line losses constituted and understand the different nature of the electricity distribution network in the "Line Loss Management" the causes and magnitude. Quantization line loss management indicators are used to take targeted technical and management measures to reduce losses. The Line loss indicator is defined by Eq. (3).

$$\Delta A = \frac{A_P^2 + A_Q^2}{3U_{pj}^2} K^2 Rt \times 10^{-3} \tag{3}$$

Line loss theoretical calculation based on power running mode, grid computing actual loss of each active element, reactive power loss and power loss within a certain time. The main purpose of distribution network line loss is calculated by the proportion of power consumption of various types of loss calculation and transmission and distribution of electrical energy in the process of the components produced to determine the variation of the grid with line loss; thereby analyzed for the presence issues to be taken to improve the power grid structure and technical management measures to reduce energy consumption, to reduce power consumption to a more reasonable range. Accurate and reasonable distribution network line loss theoretical calculation of line loss in the power sector analysis constituted Reduction Measures to develop powerful tools for promoting supply enterprises to reduce energy consumption, internal potential to improve economic efficiency, optimize network planning and design programs to strengthen the operation and management has Significance. Currently, due to the complexity of the structure of the distribution network, the parameters and the lack of diversity and imperfect data in real-time monitoring equipment, to accurately calculate the distribution grid more difficult theoretical line loss. But its power consumption is almost half of the entire grid power loss. With distribution automation level of continuous improvement, a variety of automatic data collection systems continue to emerge, it can collect real-time distance to each load node voltage, current, active, reactive and other data, which is the theoretical line distribution network loss calculation provides a more adequate data so that accurate calculation of the wiring grid loss possible. Because data collection automation and high collection efficiency can do this by increasing the number of calculations to minimize line losses reflects the impact of load curves for line loss calculation. But, also it can achieve theoretical line loss of distribution network automatic online calculation. In order to better monitor the operation of the distribution network, reflecting the specific line loss and the loss reduction carried out in order to guide the work. Figure 1 shows the model map of distribution network with distributed generation (DG) units.

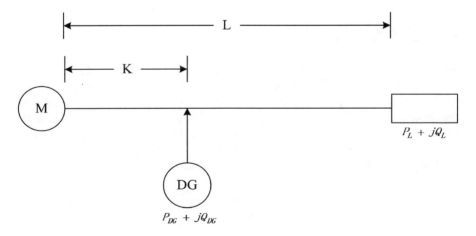

Fig. 1. Model map of distribution network with DG

There are two types of existing line loss theoretical calculation: one is the traditional method, and the other is the trend calculation. The traditional method is to load the selected representative of Japan made on behalf of the measured data with 0:00 to 24 points, including the substation lines, etc. The whole point of running power supply, output, input current, power, voltage, power factor and power throughout the day recorded the whole load point reproduces recorded data mode operation, the head-end copy see power supply and distribution of total secondary side table see electricity or load test records, large user-specific high-voltage distribution lines, distribution transformers special high or low-pressure side of the load voltage, current, power, power factor meter records. Calculation of distribution network structure parameters including the line wire gauge, length, resistance, reactance, transformers, condensers, voltage regulators, capacitors, reactors, and other nameplate data or measured power loss values, transformer, electric meter, and other required device parameters and then family line length. We mastered the grid structure parameters and operating data using line loss theoretical calculation method to find the line loss value, by using "rms" current method, the average current method, and the equivalent resistance method [14–16]. Distribution power flow algorithm is the basis for distribution grid network analysis, network distribution network reconfiguration; fault handling, reactive power optimization, and state estimation will need distribution network flow data. Distribution Network structure has a significant difference compared with the transmission grid. Distribution Network radically, in normal operation, is open loop. It is only possible for a short time in the loop while running the switch load or failure. Another feature of the distribution network is the total length of transmission lines and distribution lines to be longer than the branch and more. Distribution line diameter than the fine grid leads to the distribution network of R/X is large, and the line of charging a capacitor can be ignored. Various calculation methods can be used to calculate the trend such as forward and backward substitution method, Newton-Raphson method, and P-Q decomposition method.

3 Loss of Produce

Because of the resistance in the circuit, so the power transmission grid, the current must flow against the action of resistance and, consequent temperature rise and the heat conductor, converted to heat energy, and in the form of heat dissipation around the conductor medium, which produce power loss. Because such losses are due to the current conductor of the impediment caused, so called resistive losses; and because of this loss with the conductor by the size of the current changes, it is also known as variable loss. In this case, the lose reduction measure is defined using Eq. (4).

$$\Delta P_k = \sum_{i=1}^{m} \left(\frac{k_i S_i}{S_i} \right)^2 P_{ki} = \sum_{i=1}^{m} k_i^2 P_{ki} \tag{4}$$

In AC circuits, electrical current through the device to establish and maintain a magnetic field to the normal operation of electrical equipment, belt load acting. The motor needs to establish and maintain a rotating magnetic field to normal operation to drive mechanical load acting. Another example is the need to establish and maintain a transformer alternating magnetic field in order to play the role of step-down or step-up, the power transmission to the remote, and then be able to facilitate the transforming power users. It is known that in an *AC* circuit system, the current through electrical equipment, electrical equipment suction system reactive power, and continue to exchange, so as to establish and maintain a magnetic field, a process known as electromagnetic conversion process [17–19].

In this process and since the magnetic field is generated in the iron core of electrical equipment in hysteresis and eddy current phenomenon, then, the core temperature of the electrical equipment and heat rise resulting in a power loss. This loss is due to the establishment and maintenance of a magnetic field by an alternating current in the electrical equipment in the core generated. It is also known excitation loss; and because of this loss has nothing to do with the current size of electrical devices, and device access grid voltage level related to that grid voltage is fixed, this loss also fixed, it is also known fixed losses.

Electrical industry management department lags behind, due to sound system, causing some problems at work. For example, a user stealing and illegal use of electricity; grid insulation level is poor, resulting in leakage; meter equipped unreasonable, not timely repair school exchange, resulting in the loss of error; business management lax, resulting in core income working copy error losses. Because of this loss is no certain rules, we cannot use the meter and calculation methods to obtain estimates. It can only be determined by the final statistics, but its value is not very accurate, so it called unknown loss; and because of this loss is by an electric management factors caused business administration, it is also known management loss.

4 Reduction Measures

Distribution grid layout, especially the urban distribution, structure, and layout should optimize the rational allocation, to meet the people's demand for electricity, combined with the overall urban planning, to shorten the distance power lines. Promote the use of load tap changer, an appropriate increase in the operating voltage according to the load situation. According to transmission line capacity, transport distance and the necessary grid developments boost the transformation of large users should try to use a high voltage power supply, not only conducive to reducing losses but also can improve the transmission capacity and expanding power supply radius. In opacity, ensure that the voltage quality of the premise should select the economic current density cross-section wire.

In practice, regularly carry out theoretical line loss calculations, regular line loss calculation options towards the end, so it is beneficial for line loss to work the whole year were summarized and analyzed. The results of these calculations for line loss analysis can provide a theoretical basis. According to the changes in the actual production power company marketing the generated loads, and analysis of the theoretical results, the timely discovery of line loss changes [20, 21]. Timely development of corrective measures for weak links; meets monthly line loss analysis in a fixed time period, line loss statistics, analysis; quarterly line loss season also held meetings quarterly line loss summing up the work.

Perfection meter reading system, strengthen personnel Collecting meter reading management and assessment. In the cases specified days to complete the meter reading, the meter reading is mandatory for each person should be strictly followed. Develop an appropriate reward system, thus resolutely put out of sync meter, missing copy, copy and estimate not copy phenomenon. Full replacement of smart meters, vigorously popularize automated meter reading, meter reading to reduce due to human factors are not synchronized, not timely. Regularly carry out professional ethics education Collecting staff to improve professionalism in collecting staff to improve meter reading personnel responsibility. Improve site response capabilities and professional level meter reading staff through comprehensive training. Figure 2 shows the topology diagram of a 10 kV distribution network.

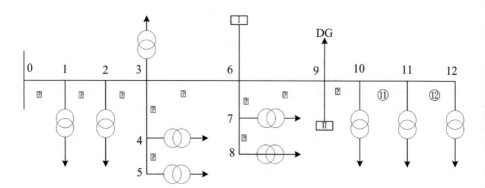

Fig. 2. Topology diagram of a 10 kV distribution network

10 kV distribution network in urban and rural areas, the transformer loss (copper loss and iron loss) accounted for a considerable proportion of the total loss of the distribution network, especially in the low power consumption, because the load is low, high operating voltage, the greater the load loss, which the proposed transformer economic operation and other issues. Improve power factor, and this is not only about reducing losses, but also can improve the transformer output; replace the transformer. The high energy consumption of new energy-saving transformer replacement transformer; voltage grid should be appropriate for the load during peak hours should be increased by the electrical voltage, low hours should be reduced by the voltage, the rational allocation of distribution transformer capacity, transformers are generally rated load 25%–75% range of the highest operating efficiency, and capacity increase for new clients should be a reasonable choice of distribution transformer capacity, the use of the distribution transformer capacity should arrange the load as much as possible by economic load operation; for parallel operation of multiple distribution transformers, according to the load, determine the transformer cast back.

Under load the power P remains unchanged, raising the load power factor, can reduce the load required reactive power Q, it is possible to reduce the reactive power through the line or transformer, thus reducing by lines or transformers current, it will also reduce the power consumption of the line or transformer. It must have a certain amount of reactive power when distribution lines, distribution transformers, motors and other devices to work, because the distribution lines to be charged, to establish a magnetic flux distribution transformers, electric motor to create a rotating magnetic field. Reactive power compensation is to the nearest (to) these reactive power supply, while minimizing the power delivery from reactive. In the distribution network, to provide reactive power equipment, mainly parallel capacitor. Distribution network mainly considered optimal allocation problems fixed capacitors in parallel, and voltage constraints to consider in order to avoid the phenomenon of high voltage compensated, but also consider the impact of changes along the line voltage transformer load loss and no-load loss. The objective function should include the net present value of the capacitor to invest savings, consists of three parts: power loss reduces the present value of the savings obtained, the peak power reduction integrated investment and operating costs and the present value of the savings obtained capacitor (this is divided into negative). It is suitable for multi-stage, multi-branch, any conductor cross-section and arbitrary load distribution of radial distribution network.

Power boost transformation is to take some technical measures to simplify the voltage level to address the imbalance in electricity load, power loss is too large. The voltage level reduces to 220/380 V, 10 kV, 66 kV, 220 kV, 500 kV voltage grade five, all other voltage through the transformer into several voltage level then more power delivery. Under the same premise to ensure the voltage quality, moderately generator terminal voltage, transformer tap and switching on the bus capacitors and condensers the regulator and other means. Reasonable load distribution lines, replace the wire diameter to reduce resistance in the circuit, thereby reducing line losses. For larger load line, use a large wire size for a smaller load line, selecting the smaller diameter. Reasonable selection of wire diameter can reduce costs, to achieve the best economic benefits.

5   Conclusion

This paper explains the definition of line loss, analyzes the cause and classification of line loss, explains how to reduce line loss from distribution network planning, loss of three transformers, and compensation for interactive capacity.

References

1. Conrad, R., Pierre, S.: Loss-of-excitation protection for synchronous generators on isolated systems. IEEE Trans. Ind. Appl. **21**(1), 81–98 (1985)
2. Merlin, A., Back, H.: Search for a minimal loss operating spanning tree configuration for an urban power distribution system. In: Proceedings of the 5th Power System Computation Conference (PSOC), Cambridge, pp. 1–18 (1975)
3. Bialek, J.W.: Tracing the flow of electricity. IEE Gener. Trans. Distrib. **143**, 313–320 (1996)
4. Hossain, M.J., Saha, T.K., Mithulananthan, N., Pota, H.R.: Robust control strategy for PV system integration in distribution systems. Appl. Energy **99**, 355–362 (2012)
5. Tsengenes, G., Adamids, G.: Investigation of the behavior of a three phase grid connected photovoltaic system to control active and reactive power. Electr. Power Syst. Res. **81**, 77–184 (2011)

6. Widyan, M.S.: Large- and Small-signal stability performance of a power system incorporated with PV generator. Int. J. Power Energy Syst. **33**(4), 1–9 (2013)
7. Conejo, A., Galiana, J.F.D.: Z-bus loss allocation. IEEE Trans. Power Syst. **16**(1), 105–110 (2001)
8. Exposito, A.G., Santos, J.M.R., Garcia, T.G., Velasco, E.A.R.: Fair allocation of transmission power losses. IEEE Trans. Power Syst. **15**(1), 184–188 (2000)
9. Choi, Y., Lee, H., Park, H.: A preliminary study on the color environment characteristics of seniors centers in Korea. J. Creat. Sustain. Archit. Built Environ. **7**(1), 15–22 (2017)
10. Pandey, K., Negi, S., Khulbe, A., Singh, P., Pandey, P.S.: Gasoline purity check prior to Tank Up. Int. J. Energy Inform. Commun. **8**(4), 11–20 (2017)
11. Ko, D., Kim, T., Kim, J., Park, J.: CPS based data collection system for smart factory. Asia-Pac. J. Model. Simul. Mech. Syst. Design Anal. **2**(1), 47–52 (2017)
12. Manhasa, P., Soni, M.: Performance evaluation of various channel equalization techniques in terms of BER for OFDM system. Int. J. Wirel. Mob. Commun. Ind. Syst. **3**(2), 55–62 (2016)
13. Kim, H.: Tracking and counting vehicles in nighttime. Int. J. Web Sci. Eng. Smart Devices **3** (2), 13–18 (2016)
14. Kang, M.W., Chung, Y.W.: A novel power saving scheme for base stations in 5G network. Int. J. Mob. Device Eng. **1**(1), 27–34 (2017)
15. Song, E., Kim, G.: A Unified gesture platform for multiple-device environments. Int. J. Artif. Intell. Appl. Smart Devices **5**(2), 13–18 (2017)
16. Kim, J., Kim, H., Kim, S.: A study on understanding internet ethics and improving practice by making UCC. Int. J. Hybrid Inform. Technol. **11**(2), 1–6 (2018)
17. Jin, J., Koo, H., Lee, M.: User-defined Beacon services for pattern-specifiable beacons. Int. J. Smart Home **11**(11), 19–24 (2017)
18. Lee, S., Cho, B., Park, H.: Design of scalable sensor and actuator interface module for smart farm. Int. J. Smart Home **12**(4), 1–6 (2018)
19. Jung, C.Y., Konagala, V.L., Bushanam, V.: Hybrid intrusion detection for ACMC physical systems. Int. J. Reliab. Informa. Assur. **4**(1), 9–14 (2016)
20. PilKwak, Y., Choi, Y., Choi, J.: A study on the human errors on the aircraft automation system. Int. J. Adv. Res. Comput. Inform. Secur. **1**(2), 1–6 (2017)
21. Kumar, T.R., Kumar, Y.P.: Dynamic authentication for privacy preserving data. Int. J. Big Data Secur. Intell. **3**(1), 9–16 (2016)

Power Demand Side Management Strategy Based on Power Demand Response

LianJun Song[⊠], FengQiang Li, and Bo Cong

Department of Marketing, State Grid Liaoning Electric Power Supply Co., Ltd.,
Shenyang 110004, China
sygcl113@163.com

Abstract. Power demand side management is an advanced resource planning methods and management techniques, to promote the coordinated development of economy, resources, and environment, and to ease the power shortage, improve energy efficiency and improve efficiency and so has a very important role. This paper describes the theory of demand side management (DSM), DSM points out the lack of process to run in present, Boot electricity customers optimize the power consumption, it helps to rational consumption, and detailed description of the specific implementation of demand-side management strategies for power companies to carry out DSM has an important role in guiding and practical significance.

Keywords: Power demand · Boot electricity · Demand side management

1 Introduction

Electricity is an important support for economic and social development and the normal operation of the city's lifeline. In recent years, as China's economic development and the continuous expansion of the city, making the surge in electricity consumption of the whole society, have occurred when the domestic electricity shortage, demand time growing imbalances, resulting in increased cost of power generation, power peaking difficulties, decreased reliability of power supply and power cuts during peak hours and other issues. In this context, in order to solve the power supply and demand, an increase in the power supply at the same time, demand-side management as to adapt to an economy operating mechanism control means and power requirements of the market economy system, power management technology, by the whole of society great attention to the application of technology can ease their demand for electricity in the whole society to some extent, for the promotion of industrial and environmental, economic, and social development of the power, and will continue to play an important role. Demand side management is in support of government regulations and policies, through effective incentive and guidance measures, with a suitable mode of operation, prompting power companies, energy service companies, intermediaries, suppliers supply energy-saving products, power users and other joint efforts, electricity to meet the same function at the same time, improve end-use efficiency and improved power mode, reducing power consumption and electricity needs and achieve the lowest cost of

A. E. Hassanien et al. (Eds.): AISI 2019, AISC 1058, pp. 981–988, 2020.
https://doi.org/10.1007/978-3-030-31129-2_91

energy services, social optimum, resource conservation, environmental protection, carried out by all parties benefit management activities.

2   Content of Demand Side Management

DSM is through improving end-use efficiency and optimizes the power consumption, electricity consumption in the completion of the same functionality while reducing power consumption and power demand, to save energy and protect the environment and achieve low-cost electric power services performed management activities. Its main contents are load management and energy efficiency management [1–3].

2.1   Load Management Measures

Load management is based on the load characteristics of the power system in some way the user demand for electricity from the grid peak load reduction, or transfer them to the grid load trough to reduce power or seasonal peak load day, prompting demand for electricity in Different timing rational distribution, increase trough equipment utilization, improve system reliability and economy. There are clipping load shaping, filling the valley, peak load shifting three [4].

Clipping occurs is to reduce demand for electricity users in the period of peak load power, stable system load Tripping avoid power outages and other circumstances, not only increasing customer satisfaction, but also to avoid the additional marginal costs are higher than the average cost of installed capacity, reduce the cost of power grid enterprises. But on the other hand, it will lead to peak clipping reduction in electricity sales and lower part of the revenue Grid Company (Fig. 1).

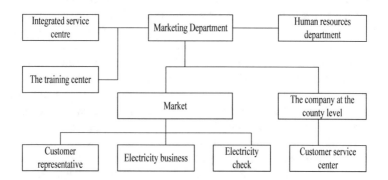

Fig. 1. Customer response organizational system

Valley, also known as strategic load growth, is through preferential tariffs to stimulate users to increase power consumption in the power grid trough period. This method is particularly suitable for large difference between peak power load, low-cost system is idle generating capacity, poor load regulation and low piezoelectric difficult power system is conducive to start idle generating capacity, reduce the system average fuel costs, and improve system operation stability of the economy and grid load [5, 6]. Because of its increased sales of electricity, while reducing fixed costs per unit of electricity, increase power Grid Company's sales revenue, more able to mobilize the enthusiasm of enterprises.

The peak load shifting by the scheduling of Time-of-use (ToU), alternating operation and other power-hungry equipment for the electricity side, and peak loading times are part of the arrangements for the trough periods and to play a dual role in cutting and filling the valley. It can make full use of idle capacity, reduce new capacity, but also stable system load, and reduce power consumption. In severe power shortages, peak disparities, limited capacity load regulation power system, has been the peak load shifting as a major task to improve the management of the power grid. Peak load shifting valley on the grid's impact is two-way, on the one hand, increases the valley of electricity, improve the power grid company's sales revenue; the other hand, reduces the peak power consumption, lower power company's sales revenue, Therefore, the power system depends on changes in income increase of valley Electric revenues and reduce operating costs to compensate for the reduced level of peak electricity revenue [7].

2.2 Energy Management Measures

Energy management means taking effective incentives, energy technology and efficient equipment to improve end-use efficiency and energy management measures include direct and indirect energy-saving, change the user's consumer behavior, the use of advanced multi-section and its fundamental purpose is to save electricity, reduce power consumption. Direct-saving is to adopt scientific management methods and advanced technology to energy-saving, energy-saving is to rely on indirect economic restructuring, improve production efficiency, productivity rational distribution, to reduce high energy product exports to achieve [8].

3 Situation of DSM

3.1 Lack of Adequate Legal Support

DSM implementation requires the government to play an active leadership role in the support and cooperation of all social workers, laws, and regulations is also essential. DSM and residents, the power company, and even the whole country concerned, this feature makes the implementation of relevant laws and regulations inseparable support of DSM, but China's current laws and regulations associated with little, now promulgated The law, only the "Energy Conservation Law" and "conserve electricity management approach" and a few other related laws and demand-side management, support was not enough, how to use legal means to ensure the implementation of DSM, will in the future for a long time within troubled relevant staff. In addition, demand-side management projects require a huge investment of money, but this lack of construction funds in our country, so that demand-side management cannot be expanded, causing many users and related power company participation in the initiative hit.

In order to give full play to the initiative of the user, the user can make decisions independently, including cutting time and reducing power. Load aggregators have certain limitations on response time and power reduction. To simplify the model, it is assumed that DRE's issue time, response time, and time limit parameters are all multiples of Δk, and the load aggregator will not issue new DRE before the end of one DRE response. Consistent with the maximum power constraint, the reduced power is also replaced by the average power.

As shown in Fig. 2, it is denoted that the initial response time is s_{DR}, and under the response time, the upper limit is a_{DR}, b_{DR} so that $m = (s_{DR} + a_{DR})/\Delta k$, $n = (s_{DR} + b_{DR})/\Delta k$, then the reduction amount and the user get the compensation C_{DR}. The relationship can be expressed as:

$$C_{DR} = c_{DR}P_{re}N \cdot \Delta k \tag{1}$$

$$N = \frac{a_{DR}}{\Delta k} + \sum_{k=m+1}^{n} r_k \tag{2}$$

Where $\Delta k \cdot N$ denotes the cutting time; c_{DR} represents the economic compensation; P_{re} for unit power reduction. Is the average power cut in this period; r_k is a variable of 0–1, indicating whether DRE, 1 is responsive during k period, while 0 means no response. Reduction meets range constraints:

$$P_{remax} \geq P_{re} \geq P_{remin} \tag{3}$$

Where P_{remin}, P_{remax} Max is the minimum and maximum power reduction allowed; if P_{remin} is 0, users are allowed not to cut.

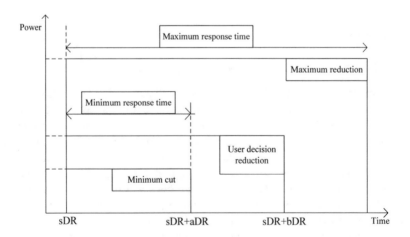

Fig. 2. Electricity reduction

3.2   Lack of Effective Economic Stimulus

Relevant government departments are developing systems and policies to promote the stable development of demand-side management so that community members and energy companies are the relevant beneficiaries. Where the members of society is the biggest beneficiary, only the majority of the members of society feel that they really are the interests of the recipient, will actively participate in demand-side management; the implementation of demand-side management, greatly improving the load factor grid, reducing energy consumption while improving equipment utilization, so that the power company has also become associated beneficiaries. Grid Company must first solve the contradiction between sales efficiency and power users, the average price of electricity sales will be reduced by the grid company benefit, which largely reduces the actual perpetrators of the DSM initiative.

3.3   Unreasonable Tariff Structure

Core DSM is implemented by the user peak load shifting way with the power, ease user peak demand, improve energy efficiency. When the power supply is in peak condition, raise the price of electricity to stimulate the user as little as possible, when the power supply at the bottom, lower electricity prices to promote electricity users. Electricity uses today are mostly two-part tariff, the basic tariff is too low to power electricity price is too high, this price mechanism does not have much incentive to users. The use of more mature international tariff pricing includes two real-time pricing and peak sea-sonal tariff, the tariff peak when raised to 8-10 times the price trough periods, can effectively alleviate network load and seasonal use electric contradictory, but both electricity pricing has not been widely used in our country, to ease China's huge demand for electricity did not play much of a role. Figure 2 shows the user the summer to load distribution [9] (Fig. 3).

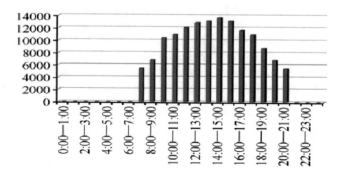

Fig. 3. User summer time-to-time load distribution map

4   Demand Side Management Suggestion

4.1   Develop and Improve Relevant Laws and Regulations

Improve laws and regulations on the implementation of demand-side management can provide support, such as to ensure that the relevant measures are in place to provide a legal basis for administrative measures related to government departments. DSM research in most countries of the world, there is one thing in common, that is, with strict and detailed regulations to support, such as Britain's "Electricity Law", while developing a large number of mandatory standards associated with that user explicitly their obligation to protect the smooth conduct of DSM from the legal level.

4.2   Establish a Variety of Economic Incentives

Relevant government departments related to the formulation of economic policy, the use of a variety of suitable means to meet the needs of the interests of all parties. DSM is recommended to collect part of the money from the electricity in the management, and the use of funds, strict supervision, and audits to ensure the effective use of funds, while the power supply on the part of enterprises appropriate compensation [10].

4.3   Establish a Rational Pricing System

Change China's current two-part electrical mechanism, giving more flexibility of the price mechanism, the implementation of seasonal price and the spot price peak within the wider manner, effectively improve power efficiency, peak power supply to ensure stability and security.

4.4   Demand Side Management Nurture Intermediary Organizations

Contract energy management has been proven in the international arena is an advanced management mode and effective mechanism, making energy services company American and European countries play an important role in the management of electricity demand side management. Therefore, China should vigorously develop to meet China's electricity consumption of commercial energy company, to provide users with a variety of reliable information and services, and to share relevant interests [11].

5   Conclusion

Construction DSM system is a large and complex systems engineering, not only to the power company and power users economic benefits, but also related to economic and social development. By establishing a rational management system, scientific pricing system, as well as a high level of automation integrated monitoring system can improve our existing demand-side management system to reach peak load shifting load, energy-saving purposes, promote intelligent DSM sustainable development system construction and power resources. At the same time, DSM as the current international advanced resource planning methods and management techniques, the implementation of China's power industry is an important way to the scientific concept of development, the need for the implementation of all subjects were widely publicized, guide electricity customers to optimize the power consumption, it has contributed to reasonable consumption and adjust the load, to allow electricity customers to science of electricity, improve energy efficiency, energy conservation.

References

1. Gellings, C.W.: The concept of demand-side management for electric utilities. Proc. IEEE **73** (10), 1468–1470 (1985)
2. Mohsenian-Rad, A.H., Wong, V.W.S., Jatskevich, J., et al.: Autonomous demand-side management based on game-theoretic energyconsumption scheduling for the future smart grid. IEEE Trans. Smart Grid **1**(3), 320–331 (2010)
3. Ruiz, N., Cobelo, I., Oyarzabal, J.: A direct load control model for virtual power plant management. IEEE Trans. Power Syst. **24**(2), 959–966 (2009)
4. Paatero Jukka, V., Lund Peter, D.: A model for generating household electricity load profiles. Int. J. Energy Res. **30**(5), 273–290 (2010)
5. Ipakchi, A., Albuyeh, F.: Grid of the future. IEEE Power Energy Mag. **7**(2), 52–62 (2009)
6. Philpott, A.B., Pettersen, E.: Optimizing demand-side bids in day-ahead electricity markets. IEEE Trans. Power Syst. **21**(2), 488–498 (2006)
7. Conejo, A.J., Morales, J.M., Baringo, L.: Real-time demand response model. IEEE Trans. Smart Grid **1**(3), 236–242 (2010)
8. Coll-Mayor, D., Paget, M., Lightner, E.: Future intelligent power grids: analysis of the vision in the European Union and the United States. Energy Policy **35**(4), 2453–2465 (2007)
9. Jang, S.K., Ko, J., Lee, Hun, J., Kim, Y.S.: A study on human body tracking and game information visualization for mobile AR. Int. J. Mob. Device Eng. **1**(1), 13–18 (2017)
10. Ren, M., Kang, S.: Korean 5W1H extraction using rule-based and machine learning method. Int. J. Artif. Intell. Appl. Smart Devices **5**(2), 1–10 (2017)
11. Shin, D., Kim, C.: Short-term photovoltaic power generation forecasting by input-output structure of weather forecast using deep learning. Int. J. Softw. Eng. Appl. **12**(11), 19–24 (2018)
12. Vyshnavi, G., Prasad, A.: Detection and location of high impedance faults in distribution systems: a review. Int. J. Adv. Sci. Technol. **119**, 53–66 (2018)
13. El Qutaany, A.Z., Hegazi, O.M., El Bastawissy, A.H.: A mapping approach for fully virtual data integration system processes. Int. J. Adv. Sci. Technol. **120**, 29–48 (2018)
14. Park, J.-K., Choi, S.Y.: A integrity checking mechanism using physical independent storage for mobile device. Int. J. Control Autom. **8**(3), 109–114 (2015)
15. Liu, Y., Xiong, R.H.: Parallel gauss-jordan elimination on two-dimensional constant bandwidth storage. Int. J. Grid Distrib. Comput. **10**(I), 31–42 (2017)
16. Biswas, P.K., Bahar, A.N., Habib, M.A., Nahid, N.M., Bhuiyan, M.M.R.: An efficient design of reversible subtractor in quantum-dot cellular automata. Int. J. Grid Distrib. Comput. **10**(5), 13–24 (2017)
17. Virgile, K., Yu, H.Q.: Securing cloud emails using two factor authentication based on password/apps in cloud computing. Int. J. Secur. Appl. **9**(3), 121–130 (2015)
18. Mohammad, K.H., Ahmad, F.I., Wahidah, H., Shayla, I.: Investigation of enhanced particle swarm optimization algorithm for the OFDMA interference management in heterogeneous network. Int. J. Future Gener. Commun. Netw. **9**(9), 15–24 (2016)

19. Qiu, L.R.: Syntactic parsing tree in tibetan language based on context free grammars. Int. J. Signal Process. Image Process. Pattern Recogn. **8**(11), 1–6 (2015)
20. Fang, X.W., Liu, L., Liu, X.W.: Analyzing the consistency of business process based on behavior petri net. Int. J. U - E Serv. Sci. Technol. **8**(2), 25–34 (2015)
21. Mandeep, K., Prince, V.: Fusion of PACE regression and decision tree for comment volume prediction. Int. J. Database Theory Appl. **9**(11), 71–82 (2016)

Micro-Grid and Power Systems

Relay Protection Hidden Fault Monitoring and Risk Analysis Based on the Power System

LiaoYi Ning[(⊠)]

State Grid AnShan Electric Power Supply Company, Anshan 114000, China
sygcllll@163.com

Abstract. With the continuous expansion of the grid, the safe operation of the power grid is also becoming more and more important than before. Relay protection is one of the important protection devices to ensure the safe operation of the power system; its failure can cause great influence on the power system. Relay protection hidden fault is a kind of the relay protection fault, however, the phenomenon of power outages caused by power system fault is the result of relay protection hidden fault in many situations, which is one of the most important hidden areas to threaten the safety of power grid. Therefore, the hidden failure of relay protection has a great influence on the electric power system, so monitoring the hidden fault of the relay protection will become more and more important. This paper introduces the concept of relay protection of hidden faults, its characteristics, and then analyzes the detection, risk and the calculation method of the relay protection of hidden fault.

Keywords: Fault monitoring · Power system · Relay protection hidden fault

1 Introduction

Electricity permeates every corner of the development of the society, scientific progress and development, and indispensable part of people life is closely related to the time progress including the penetration of electric appliances in various fields, and the position of power in the social development. The popularity of television, refrigerators, computers, and other electronics, power consumption has also begun to soar. For power supply enterprise, strong electricity load is a huge challenge to the power supply work. Take protective measures, to ensure the normal power of residents and commercial are series of major issues. Relay daily maintenance, and repair, it is one of the important factors to ensure the power supply. Figure 1 shows the main working principle of relay.

Electricity use safety is very important, while the relay is to ensure that the power supply to safety in the process. Power is able to satisfy the main function of power supply, and the measure of the strength of an electric power enterprise is to see if we can continue to the power supply. Relay protection is to maintain the normal operation in the process of power supply, thus reducing accidents occurred in the process of the power supply. If there is a power accident, also can timely analysis, processing, in a timely manner to find the cause of the accident, the maintenance staff to provide

A. E. Hassanien et al. (Eds.): AISI 2019, AISC 1058, pp. 991–997, 2020.
https://doi.org/10.1007/978-3-030-31129-2_92

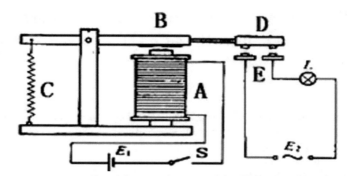

Fig. 1. Working principle of relay

information for electric safety. In the process of the electric power enterprise power, without the presence of relay protection, it is to everyone.

Economic boom needs based on electric power construction, people want to have a high quality of life, the power protection system is absolutely essential. Relay protection is in the process of power supply convoy, in view of the relay protection technology level speed, the construction scale, must also be improved. Human resource is always the core of the enterprise, must, therefore, be to ensure that the core of maintenance. Each technician requires not only rich in theoretical knowledge, on the practical ability, must also be to ensure that, to face all kinds of fault well acquainted, improve work efficiency, put an end to power accident exist, but these are also all in the electric power development process, essential conditions. Under the situation of informatization, is bound to the relay protection rise again a ladder.

2   Relay Protection Hidden Faults

2.1   Overview of Relay Protection Hidden Faults

Relay protection mainly refers to the power system failure or abnormal situation to its for testing, and according to the corresponding detection signal alarm or direct processing of a kind of power protection. Is the main function of relay protection in power system appear natural, artificial or equipment failure, failure to timely, accurate cutting the failure, so as to protect the safe operation of power system, the maximum to reduce losses [1]. Relay protection system mainly consists of relay protection devices, communication channel, the voltage current transformer and circuit breaker, any one part of the failure cause of relay protection system failure [2]. Relay protection system hidden failures is mainly refers to relay protection system of internal components exist a permanent defect, under the condition of normal operation of the defects in the system will not affect the system, and the normal running of the system is not in a state of such defects will appear, and leads to a series of failure, the most direct consequence is led to the incorrect disconnect protected components appear. Although the probability of relay protection hidden fault causing blackouts is small, but the harm is great, so once accident will cause a chain of reactions, may even make the grid collapse.

2.2   Relay Protection Hidden Fault Characteristics

Hidden fault refers to a system failure has no effect on the system during normal operation, and when to change some parts of the system, the fault will be triggered and thus lead to the occurrence of cascading failure. Hidden fault in the system during normal operation cannot be found, but once has the fault occurs, the relay correct resection, the fault trend to redistribute power system, the running status are likely to make with a hidden fault protection misoperation [3].

Relay protection system components are likely to exist in the hidden trouble, such as pt and ct, lug, connectors, relays, communication channel, and so on. The existence of hidden failure doesn't mean the relay itself is a problem with the design or the application of relay is wrong, also have not said relay calibration error [4]. The difference between it and general fault lies in the defects will not make the relay action immediately, but only when some event occurs when the other systems that can be detected. The biggest characteristic of hidden trouble, and also it is the most dangerous place their effect to system only in the power system is in a state of stress, such as in the moments after the occurrence of failure or malfunction, low voltage, overload and other switch after the incident.

2.3   The Calculation Method of Relay Protection Hidden Failure Probability

Relay protection maloperation under the hidden fault is a kind of probability event, will cause the system initial fault continues to grow, or even worse for a cascading failure, cause a blackout accidents. Is to identify the specific protection hidden faults probability there are two main ways: probability statistical analysis and probability model [5].

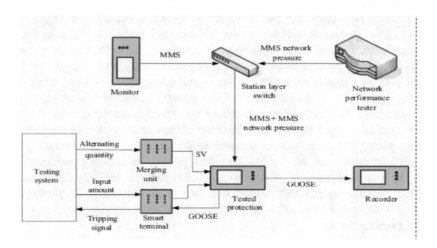

Fig. 2. Station layer network pressure test system for relay protection

Probability statistical analysis is based on the data of protection operation were analyzed, and the statistics by the adjacent components fault protection misoperation caused by the total number, and then divided by the total number of protection operation, the value that is considered to be all protect the hidden failure probability value in the system. Existing probability model used to describe the line protection mainly include distance protection hidden failure probability model and over current hidden failure probability model.

Hidden faults in the power system relay protection problem is one of the key factors in the production of large fault, so there is must need to analyze and evaluate the hidden fault. Figure 2 illustrates the main station layer network pressers test for relay protection.

2.4   Relay Protection Hidden Fault Monitoring

Relay protection with hidden failure under the condition of the normal operation of power system is not performed, and only when the system is running state will also appear abnormal, that is to say, the relay protection hidden fault only if the system is running until he has run out, so the detection of relay protection of hidden fault in traditional offline detection methods are not suitable [6]. It shows that the relay protection of hidden fault monitoring needs an online monitoring system, but there is no special monitoring system of relay protection hidden failures in protection, and it can only rely on the microcomputer protection of self-check function to ensure the safe running of the system [7].

As early as in 1996, some experts have pointed out that the protection of the famous international authority have led to a fault occurs interlocking is one of the main reason of the protection device and the hidden trouble in the system, and has carried on the detailed research, in view of the relay protection hidden trouble of the technical scheme of monitoring and control are put forward. The system is mainly to the high vulnerability index of protection device in power grid monitoring and control, the system will first analyze the internal signal input relay diagnosis, in fact is to copy the protection algorithm and function, at last the system output and relay protection device in the operation of the output of the corresponding logical relations of comparative analysis of both the output if it is the same, so will be allowed to execute protection tripping command; and if the output of the two different, so tripping commands will be banned, then, the system is equivalent to have the effect of the closure. But since the 1990s, the microcomputer protection device its own hardware and software technology and substation integrated automation has become a priority in the relay protection technology in the development of substation automation, and the monitoring and control of the relay protection hidden fault research is at a standstill. Thus formed at present is still adopts microcomputer protection self-checking function to ensure the safety of the system running status.

2.5   The Risk of Relay Protection Hidden Faults

From the analysis of the relay protection hidden faults we can see that in the relay protection hidden faults and fault lies in the difference between conventional hidden fault is not immediately trigger a system fault [8], but in the case of the abnormal operation of the system will appear, this is the relay protection hidden fault one of the most dangerous point [9]. Relay protection mechanism of hidden faults mainly in the power system failure or malfunction after moment under abnormal condition, but at the first time, from any one of the power system components are hidden faults may occur [10]. Relevant data show that in large-scale power grid disturbance events three quarters are related to hidden failures of relay protection [11], and they also there is a significant feature is: the existing defects and hidden danger can't be detected, only in the adjacent after the accident, and worsen the accident further [12]. Figure 3 describe the hidden fault of relay protection components.

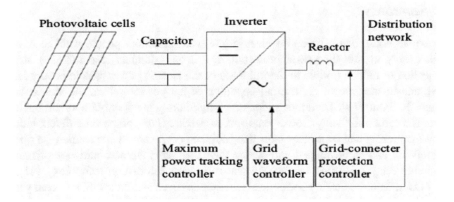

Fig. 3. Hidden fault of relay protection

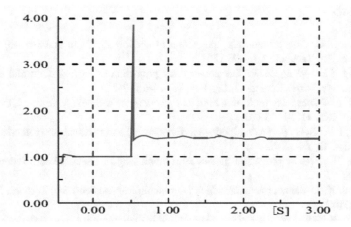

Fig. 4. Risk assessment of hidden failures

Relay protection hidden fault location is different, its harm to power system caused by degree is not the same, and it mainly depends on the occurrence of hidden fault location [13]. In order to evaluate the risk of hidden faults and, some scholars have put forward the application of the theory of risk based hidden failure risk assessment scheme [14]. Relay protection hidden failure risk assessment is the basic idea of using hidden failure probability, according to the topology of the system to establish the simulation calculation of cascading failure model [15], the main issue is all hidden failures in relay protection all for risk assessment, and then according to the evaluation results to find out the weak links of power system, and take corresponding preventive measures [16]. We can observe that all situation of risk assessment of hidden failures from Fig. 4.

3   Summary

Economic boom needs based on electric power construction, people want to have a high quality of life, the power protection system is absolutely essential [17]. Relay protection of hidden trouble to the influence of the electric power system is very large [18], and in the case of the normal operation of the power system, the hidden fault cannot be found [19]. Therefore, hidden fault monitoring is not only important but also there is a certain difficulty. Some years ago, researchers have proposed different hidden fault monitoring and control systems like [20], and others have been studied the power automation and its developed and design their own software and hardware of micro-computer relay protection technology and substation automation technology [21], in the hidden fault monitoring and control research has been at a standstill. In recent years, the relay protection of hidden fault monitoring and control of research also gradually developed.

References

1. Zhang, H.C.: The powers relay protection of shallow fault and maintenance technology. Inform. Sci. Technol. **7**, 69–75 (2011)
2. Huang, J.H., Wang, Z.H.: The power relay protection of shallow fault and maintenance technology. Urban Constr. Theory Res. **16**, 14–18 (2012)
3. Shi, C.J.: Power relay protection fault and maintenance technology analysed. Urban Constr. Theory Res. **21**, 21–25 (2011)
4. Ceng, L.L.: Relay protection hidden fault monitoring and risk analysis method study. J. Sci. Technol. **13**, 64–68 (2012)
5. Zhou, H.S.: Relay protection hidden fault monitoring methods. Silicon Valley **24**, 14–17 (2011)
6. Huang, X.H.: Relay protection hidden fault monitoring methods. Sci. Technol. Life **9**, 160–161 (2012)
7. Xiong, X.F., Cai, W.X.: Relay protection hidden failure probability model of transmission line trip chain research. Electr. Power Syst. Autom. **20**, 26–31 (2010)
8. Zhou, Y.L., Zan, R.R.: In 2003 the national grid relay protection and safety automatic device operation and analysis. Grid Technol. **20**, 48–53 (2012)
9. Niharika, V.: Effective authentication data security. Int. J. Adv. Res. Comput. Inform. Secur. **1**(1), 7–12 (2017)

10. Zhang, Q.: Assured deletion and version control in cloud computing. Int. J. Big Data Secur. Intell. **3**(2), 21–34 (2016)
11. Thota, D., Donapaneedi, S.N., Reddy, L., Kotra, S.R., Ganapathy, R.: Current prospects of change management in biopharmaceutical quality systems – a holistic cumulative review. Int. J. Bio-Sci. Bio-Technol. **9**(5), 1–12 (2017)
12. Debasis, K., Singh, M.P.: Bit-map-assisted energy-efficient MAC protocol for wireless sensor networks. Int. J. Adv. Sci. Technol. **119**, 111–122 (2018)
13. Mishra, S., Dastidar, A., Samal, U.C., Mohapatra, S.K.: Review of different image data hiding techniques. Int. J. Adv. Sci. Technol. **121**, 43–54 (2018)
14. Min, P.K., Fujinami, T., Kim, H., Ko, J.B., Min, C.K.: Masking effect of peppermint fragrance stimulus on car horn sound stimulus while driving in a graphic driving simulator. Int. J. Adv. Sci. Technol. **125**, 1–12 (2019)
15. Gao, X., Guo, S.: LMI-based H2 control for T-S fuzzy system with hard constraints. Int. J. Control Autom. **8**, 21–30 (2015)
16. Sarddar, D., Biswas, M., Sen, P., Pandit, R.: Authentication using unique identification number in cloud network using RSA algorithm. Int. J. Grid Distrib. Comput. **10**(4), 1–8 (2017)
17. Sujatha, V., Bhattacharyya, D., Chaitanya, P.S., Kim, T.: Data outsourcing based on secure association rule mining processes. Int. J. Secur. Appl. **9**(3), 41–48 (2015)
18. Kang, B., Zheng, G., Nie, F., Ma, H., Li, J., Xue, Y., Li, P.: A MAC protocol based on broadcast messages for wireless sensor network. Int. J. Future Gener. Commun. Netw. **9**(11), 1–14 (2016)
19. Rao, P.C., Patnaik, M.R., Babu, A.S.: An improved delay-limited source and channel coding of quasy-stationary sources over block fading channels: design and scaling laws. Int. J. Signal Process. Image Process. Pattern Recogn. **8**(11), 67–74 (2015)
20. Li, B., Chen, N., Wen, J., Jin, X., Shi, Y.: Text categorization system for stock prediction. Int. J. U - E Serv. Sci. Technol. **8**(2), 35–44 (2015)
21. Liu, S., Ma, X., Wang, X.: An improved model of general data publish/subscribe based on data distribution service. Int. J. Database Theory Appl. SERSC Aust. **9**(11), 83–94 (2016)

Waveform Improvement Analysis of Bridge Rectifier Circuit in Energy Internet Based on Full Feedback Regulation

Qingshen Gong[1(✉)], Xiangluan Dong[2], Dawei Li[1], Xiaoming Zhang[1], Wenrui Li[2], and Hongyu Jia[1]

[1] State Grid Chaoyang Electric Power Supply Company, State Grid Liaoning Electric Power Supply Co., Ltd., Chaoyang, China
820370889@qq.com
[2] Graduate Department, Shenyang Institute of Engineering, Shenyang, China

Abstract. With the power electronics technology improvement in the deployment of energy internet, the demand for rectifier circuit is rapidly growing. The Energy Internet (EI) also known as the integrated Internet-based smart grid and energy resources inherits all the functionality of the existing smart grid. The rectifier circuit structure of the smart grid has become inadequate in needs of energy domains in the 21st century. Then, the UPS system is also an important position, which not only ensures the uninterrupted power supply of the system but also improves the more reliable power quality of energy storage devices and the utilization rate of power greatly. In this paper, the improved bridge rectifier circuit topology is adopted to study the output waveform of the rectifier circuit, and through the logic circuit to control the thyristor on and off, the full feedback regulation of the output circuit can be realized to further control the output waveform. This study overcomes the disadvantages of high harmonic content and low power factor of traditional diode rectifier and phase controlled rectifier and has the advantages of high power factor and low harmonic content of rectifier. The problems to be studied are illustrated from the collection of test points and the realization of full feedback control. Finally, the comparative analysis of output waveform of several different rectifier circuits is made to further prove the necessity of the study of full feedback regulation of bridge rectifier circuits to improve the power factor. The research results are of great significance for developing new rectifier, filtering harmonic in power grid, improving power factor, maintaining power supply reliability and improving the power quality of energy Internet.

Keywords: Bridge rectifier · Full feedback regulation · Waveform improvement analysis

1 Introduction

Over the past 30 years, semiconductor switching devices have been developing rapidly, with high efficiency, energy saving, and high quality and reasonable power electronic devices becoming the main components of large circuits. In the traditional

A. E. Hassanien et al. (Eds.): AISI 2019, AISC 1058, pp. 998–1008, 2020.
https://doi.org/10.1007/978-3-030-31129-2_93

uncontrollable rectification circuit, the most commonly used method to obtain relatively flat output waveform is to connect a large capacitor in parallel on the load side. Although this method is reliable, it can make the grid inject more harmonics, absorb the reactive power of the grid and damage the insulation, etc. [1].

In China on March 1, 1994, formally implement the power quality and the utility grid harmonic standards, in the vast majority of electronic devices, requires the input side of the alternating current (ac) through the rectifier and filter out the high harmonics and square wave after obtaining relatively stable dc voltage, have various forms and control method of the rectifier, and even some inverter unit also needs to be installed before the rectification device, to guarantee the reliability of the voltage and current [2].

At present, UPS output fluctuates in the range of +/− 5% or so. The most important way to reduce fluctuation and improve waveform is to start from the topology of the circuit and improve the wiring of the rectifier circuit to get a better output waveform. In this paper, the traditional single phase bridge rectifier circuit connection mode to make improvements, through experiments to control editable chip, and the specific point voltage measurement circuits, detect the shape of the output waveform, by adjusting the controllable thyristor open and shut off, so as to be opened and shut off when the voltage waveform, observe the waveform of volatility and scope, get after rectifying waveform. The compensation of power is effective and the power factor is improved [16, 17].

2 Preliminary Knowledge

2.1 Basic Knowledge of Rectifier Circuits

In the three-phase controllable rectification circuit, the most basic is the three-phase semi-wave controllable rectification circuit, and the most widely used is the three-phase bridge full-control rectification circuit, the dual-reverse star controllable rectification circuit, the twelve-pulse controllable rectification circuit, etc., which can be analyzed on the basis of the three-phase half-wave. Several terms related to phase rectification [3].

(1) Control Angle: the corresponding electrical Angle from the moment when the thyristor receives positive voltage to the moment when the trigger pulse is added to make it conductive.
(2) Conducting Angle: the electric Angle corresponding to the time of conducting the thyristor in a period.
(3) The moment when the trigger pulse appears when the phase shift changes, that is, the size of the control Angle is changed, which is called phase shift. Change the size of the control Angle, so that the output rectifier average voltage changes, namely phase shift control, referred to as phase control.
(4) The change of phase shift range reduces the average value of output rectifier voltage from the maximum to the minimum, and the change range of control Angle is the phase shift range of trigger pulse.
(5) Synchronization makes the coordinated relationship of frequency and phase between the trigger pulse and the ac power supply voltage of the controllable

rectifier circuit synchronous. Keeping the trigger pulse in sync with the power supply voltage is essential to the normal operation of the rectifier circuit.

(6) In the controllable rectification circuit, the process of changing the current from one thyristor to another thyristor is called phase change or commutation [4].

Figure 1 shows the basic bridge rectifier circuit and waveform diagram.

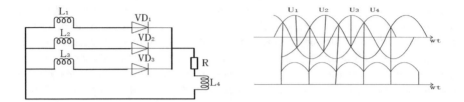

Fig. 1. Bridge rectifier circuit and waveform diagram

2.2 Some Research Status and Methods About the Rectification Problem

In [3], authors propose a method of series saturation inductance on the original side of the transformer to reduce the duty cycle loss of the secondary side. The saturation inductance has such a characteristic that when the current flowing through it is large, its inductance is low, which is conducive to the transmission of electric energy, but as the current decreases, it begins to exit the saturation state, and its inductance value starts to increase again. The full bridge topology with saturated inductance has a wider ZVS range and a smaller duty cycle loss. However, this topology also has disadvantages, that is, in the case of no load, the equivalent inductance value of the saturated inductance is large, and the loss of the saturated inductance itself will be large [5].

In [4], author presents a new topology structure based on the transformer in series and auxiliary network, ShangPin were split into two specifications of the same transformer, separate condenser is divided into two specifications of the same capacitor, so can reduce the secondary current stress of the rectifier side, voltage type auxiliary network to add also makes the topology can achieve soft switch to full load range, however, this topology structure is complex, not suitable for high-power, circuit diagram as shown in Fig. 2 [6].

In [5], the authors proposed to add two diodes on the original side of the transformer to reduce the voltage spike on the rectifier side diode through the clamping effect, as shown in the figure. In this scheme, the high-frequency transformer is connected to the lagging arm, and the diode can only be switched on once within a switching cycle, which can reduce the conduction loss [18, 19]. At the same time, the current stress on the clamping diode is better than that of the transformer connected to the leading arm. When the circuit is in zero states, the leakage current and the original side current will be smaller than before, thus reducing the conduction loss of the original side, reducing the duty ratio loss of the secondary side, and improving the efficiency of the whole circuit [7]. Figure 3 shows the diode pinion full bridge topology with clamp position as discussed in [7].

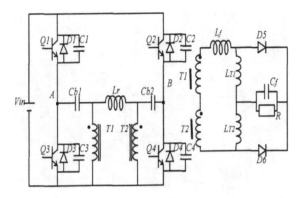

Fig. 2. A new type of phase-shifted full bridge topology

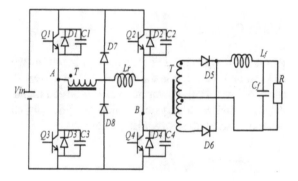

Fig. 3. Diode pinion full bridge topology with clamp position

In [6], the authors discuss the full-bridge topology, which is also commonly used. Due to the topology design of the filter inductance is small, so under the condition of light load, will reverse current through the filter inductance, in the full bridge converter zero state, reflect the current will be the primary side, prompted the original edge current increases, it is because in the filter inductor current feedback [8], so can the topology in a wider scope of zero voltage open shut off. In addition, as the leakage inductance value of the transformer is very small, the voltage oscillation on the secondary rectifier tube is very small under the condition of an ordinary diode rectifier and the output voltage is not high. Compared with the full wave rectifier, the transformer of the double current rectifier is simpler in design and smaller in volume. At the same time, since the output current is the superposition of two inductance currents, the ripple will be smaller [20, 21]. Because of the symmetry of the two filter inductors, it is beneficial to the application of magnetic integration technology. However, since the secondary rectification side requires two filtering inductors, the consumption on filtering inductors will be larger in the case of large current [9] and Fig. 4 shows the phase-shifted full bridge multiple rectifier topologies as discussed in [7–9].

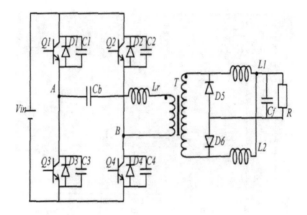

Fig. 4. The phase-shifted full bridge multiple rectifier topology

3 Design of Main System

The bridge rectifier circuit designed in this paper is composed of two parts, one is the main power circuit part, and the other is the digital control system part. The structure diagram is shown in the Fig. 5 [10]. Power main circuit design mainly includes bridge rectifier circuit design, digital control system mainly includes output voltage, current sampling circuit diagram and feedback regulation [24, 25].

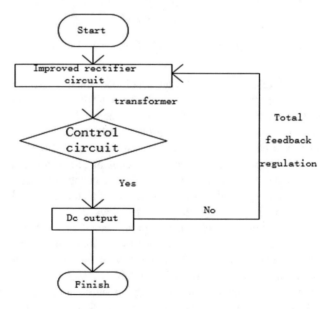

Fig. 5. System structure flowchart

The input is three-phase alternating current, the line voltage is 380 V, the grid fluctuation is 5%, and the grid frequency is 50 Hz [11].

Single-Phase Bridge Rectifier Output

$$U_d = 0.9U_s = 198v \tag{1}$$

The topology of the bridge rectifier circuit adopted in this paper is explained as follows.

Figure 6 shows the improved single-phase rectification circuit. Compared with Fig. 1, a few controllable thyristors and capacitors are added. The charging and discharging capacity of capacitors can be used to fill the valley, and the conduction Angle of thyristor can be controlled [22, 23].

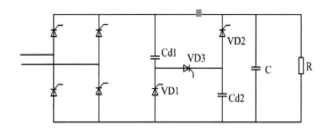

Fig. 6. Single-phase bridge controlled rectifier circuit improved circuit

The topology of the three-phase bridge rectifier circuit can be obtained from the single-phase bridge rectifier circuit, as shown in Fig. 7.

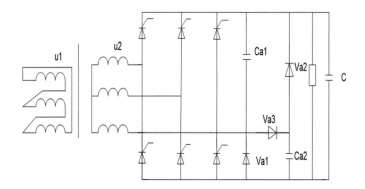

Fig. 7. Improved Triphase Bridge controlled rectifier circuit

Three-Phase Bridge Rectifier Output

$$U_d = 1.35U_S = 513v \qquad (2)$$

With the acquisition system, the voltage waveform at both ends of the load was detected every 0.25 ms, and the output of the waveform was observed [12]. The maximum value obtained was Umax and the minimum value was Umin, and the waveform was obtained, as shown in Fig. 8.

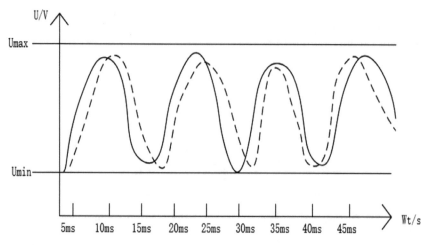

Fig. 8. Single-phase bridge controlled rectifier circuit output voltage waveform

The logic gate circuit used in this paper is diode-transistor and non-gate circuit diagram [13] as shown in Fig. 9.

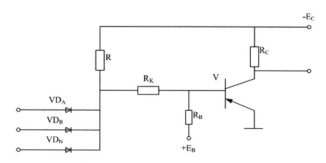

Fig. 9. Output waveform with proper capacitance in parallel

In order to obtain a relatively flat output waveform, the rectifier circuit is subject to full feedback regulation as shows in Fig. 10.

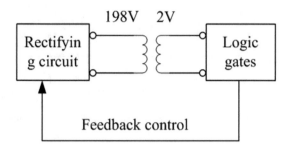

Fig. 10. Full feedback control flow chart

Logic gate detected at the ends of the load voltage waveform figure range, if the volatility still do not conform to the requirements of the smooth [14], by full feedback adjusting control conduction Angle of thyristor rectifier circuit, range from 0° to 180°, in turn, change, the output waveform was observed in the control process, found that when the thyristor conduction Angle of 30°, the output load voltage approximately 1/4 Umin − Umax, waveform diagram as shown in the Fig. 11.

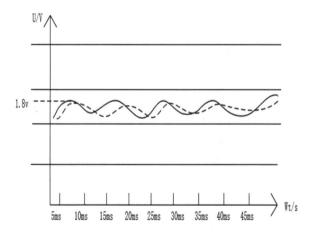

Fig. 11. The conduction Angle of 30° output waveform figure

If the filtering effect of capacitance is also considered, the capacitance is as follows when the parallel resistance of the load end is R [15]:

$$C = \frac{3(T/2)}{R} = \frac{0.03}{R} F \qquad (3)$$

Contrast diagram of the unimproved circuit and the improved circuit is shown in Fig. 12.

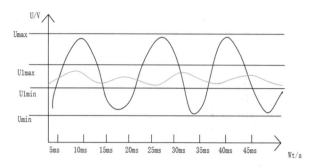

Fig. 12. Contrast diagram of the unimproved circuit and the improved circuit

The waveform diagram of the output end is shown as follows:

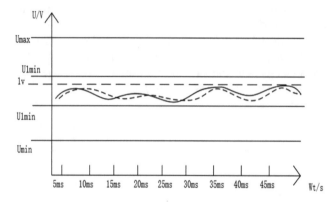

Fig. 13. Output waveform with proper capacitance in parallel

As can be seen from Fig. 13, the filtering effect of large capacitance is very obvious, and compared with Fig. 11, it has a significant effect on peak subtraction.

4   Conclusion and Discussion

During the experiment, whether the control strategy of synchronous rectification side is realized, the effect of voltage waveform peak absorption at synchronous rectification side, and the opening and closing process of synchronous rectification thyristor are mainly concerned.

Synchronous Rectifier Control Strategy Waveform
The waveform of the rectifier control strategy is shown in Fig. 12. In the figure, the black one is the original output waveform, and the light blue one is the output waveform adjusted through the full feedback. As can be seen from the waveform diagram, as a result of the thyristor conduction Angle is not consistent, namely, conduction at 30°, and compared with the original circuit, also need to use additional thyristor and

capacitor, change the storage capacitor charging and discharging each stage to improve the dc, again through the feedback to adjust and control the range of output voltage, basic implementation ultimately control strategy.

Contrast of Synchronous Rectification Peak Absorption Effect
The voltage waveform comparison between the rectifier circuit and the capacitance resistance absorption is shown in Fig. 13. It can be seen from the figure that when the capacitance resistance is added, the voltage peak reaches nearly 1.8v, and when the capacitance resistance is increased, the voltage peak is about 1v, which has a significant effect on the absorption of the voltage peak, and the voltage peak is reduced by 44.4%. There are still some deficiencies in this paper such as.

- In the aspect of data collection, there will be time intervals, and the errors will affect the waveform range;
- Logic gates have certain limitations, lack of accuracy, and error is easy to occur, which requires further improvement of follow-up work;
- Although the output waveform is close to dc, there will be harmonic components, and more accurate filtering device is needed to connect with the battery and inverter.

Acknowledge. This paper is supported by "State grid science and technology project (2019YF-29)".

References

1. Piset, C.F., Wohandech, P.S.: Study of Converter Model for Superconducting Magnetic Energy Storage. Ph.D. thesis, Rajamangara University (2015)
2. Meejit, N.F., Jaruensri, A.S.: Design and Implementation of 1 kW Diode - Clamped Three-level Inverter. PhD thesis, Rajamangara University (2014)
3. Mingkhuan, P.F.: Minimization of switching losses using SVPWM for Three-Level Inverters. Ph.D. thesis King Mongkut's Institute of Technology North Bangkok (2015)
4. Petrella, R.F., Revelant, A.S., Stocco, P.T.: A novel proposal to increase the power factor of photovoltaic grid-connected converters at light loads. In: Proceedings of International Universities Power Engineering Conference, vol. 1, no. 2, pp. 1–5 (2010)
5. Lesieutre, G.A.F., Ottman, G.K.S., Hofmann, H.T.: Damping as a result of piezoelectric energy harvesting. J. Sound Vibr. **269**(5), 991–1001 (2013)
6. Le, T.T.F., Ji, H.S., Jouanne, K.T.: Piezoelectric micro-power generation interface circuits. IEEE J. Solid State Circuits **41**(6), 1411–1420 (2011)
7. Marinkovic, D.F., Frey, A.S., Kuehne, I.T.: A new rectifier and trigger circuit for a piezoelectric microgenerator. Procedia Chem. **1**(1), 1447–1450 (2009)
8. Lee, B.F., Kim, C., Park, K., Moon, G.-W.: A new single-stage PFC AC/DC converter with low link-capacitor voltage. J. Power Electron. **7**(2), 328 (2007)
9. Ahmad, S.F., Tan, N.M.L.S.: A non-electrolytic-capacitor low-power AC-DC single-stage SEPIC-Flyback LED converter. Telecommun. Electron. **8**(31), 105–111 (2016)
10. Kutkut, N.H.F., Divan, D.M.S.: An improved full-bridge zero-voltage switching PWM convert using a two-inductor rectifier. IEEE Trans. Ind. Appl. **31**(1), 119–126 (2015)
11. Park, J.F., Ahn, J.Y.S., Cho, B.H.S.: Dual-module-based maximum power point tracking control of photo-voltaic systems. IEEE Trans. Ind. Electron. **53**(4), 1036–1047 (2016)

12. Subhadeep, B.F., Diego, M.S.: A dual three-level T-NPC inverter for high-power tracyion application. IEEE J. Emerg. Sel. Top. Power Electron. **4**(2), 668–677 (2016)
13. Mou, C.F.: A soft switched DC/DC converter with current-doubler synchronous rectification. IEEE **5**(2), 527–531 (2013)
14. Ye, Z.F.: Dual half-bridge DC-DC converter with wide-range and ZVS and zero circulating current. IEEE Power Electron. **6**(9), 3276–3286 (2012)
15. Martins, M.F., Perdigão, M.S., Mendes, A.M.T.: Alonso analysis, design, and experimentation of a dimmable resonant-switched-capacitor LED driver with variable inductor control. IEEE Trans. Power Electron. **32**(3), 3051–3062 (2017)
16. Khmou, Y.: A parametric study of binary sequence. Int. J. Adv. Sci. Technol. **119**, 89–102 (2018)
17. Rekha, Rai, M.K., Kumar, G., Lim, S.: A structured review on security and energy efficient protocols for wireless sensor networks. Int. J. Adv. Sci. Technol. **122**, 49–74 (2019)
18. Jung, H.: Analysis of subthreshold transmission characteristics for gate voltage and doping profiles of asymmetric double gate MOSFET. Int. J. Control Autom. **8**(3), 31–36 (2015)
19. Sarddar, D., Sen, P., Pandit, R., Chakra, S.: Reducing of frequency reuse distance using sectoring in mobile communication. Int. J. Grid Distrib. Comput. **10**(2), 13–20 (2017)
20. Liu, Q.H.: Accurate and diverse recommendations based on communities of interest and trustable neighbors. Int. J. Secur. Appl. **9**(3), 63–76 (2015)
21. Bertrand, M.D., Yu, C.Y.: Base station coordination towards an effective inter-cell interference mitigation. Int. J. Future Gener. Commun. Networking **8**(2), 45–58 (2015)
22. Agrawal, A.S.: Geneti algorithm optimization tool for channel estimation and symbol detection in MIMO-OFDM systems. Int. J. Signal Process. Image Process. Pattern Recogn. **8**(11), 45–54 (2015)
23. Lee, T.H., Hong, S.G.: A study on the regional innovation in an industrial cluster: focusing on a traditional textile industry of Nishijin area. Int. J. u - e Serv. **9**(4), 9–20 (2016)
24. Bachhav, A., Kharat, V., Shelar, M.: Query optimization for databases in cloud environment: a survey. Int. J. Database Theory Appl. **10**(6), 1–12 (2017)
25. Thakur, S.: Modeling and performance analysis of MEMS based capacitive pressure sensor using different thickness of diaphragm. Int. J. Sensor Appl. Control Syst. **4**(2), 21–30 (2016)

Analysis Winding Deformation of Power Transformer Detection Using Sweep Frequency Impedance Technology

Tao Wang, Yaqing Hu, Xianfeng Li, Hua Zhang[(✉)], Zhenwei E,
Lei Zhang, Zhongbin Bai, and Chunmei Guan

State Grid Fushun Electric Power Supply Company, Fushun 113008, China
ambitious9966@126.com, sky_0735@126.com

Abstract. In this paper, a new proposed transformer winding fault detection method is adopted to test the common faults of the transformer, short circuit, and deformation. Taking the 10 kV three-phase double-winding transformer as the research object, the normal and faulty transformers are detected by the sweep impedance method. And the sweep impedance curve changes before and after the fault. The test results show that the difference of the sweep impedance curve of the transformer winding short circuit and the concave deformation fault is obvious. The fault can be judged by the sweep impedance spectrum. The feasibility and effectiveness of the sweep impedance method for transformer winding deformation detection are verified.

Keywords: Sweep frequency impedance ·
Winding deformation of power transformer · On-site testing technology

1 Introduction

With the rapid development of the economy and technology, the power system is growing bigger and the demand for various power equipment is increasing. In the power system, the traditional power transformer as a Power conversion device is an important and expensive part and its failure rate is very high. If a large power transformer fails during operation, it must lead to a large-scale, long-term power outage in an area. It will cause major economic losses and will also bring inconvenience to people's lives. According to relevant statistics in recent years, many reasons such as human error can make the transformer out of service, but the transformer windings are the most common reason [1, 2]. The national 110 kV and above power transformers have been damaged by external short-circuit faults, which accounted for 50% of the total number of accidents. From the disintegration inspection, most of them are caused by winding deformation [3]. Therefore, in order to discover the hidden dangers of transformers in a timely manner to avoid sudden accidents and economic losses, it is of great significance to carry out research on transformer winding diagnostic instruments.

Therefore, the detection of transformer winding deformation has been a research focus. In the early days, the inspection mainly used overhaul maintenance. This method was time-consuming and not effective. So it was gradually replaced by non-destructive

© Springer Nature Switzerland AG 2020
A. E. Hassanien et al. (Eds.): AISI 2019, AISC 1058, pp. 1009–1016, 2020.
https://doi.org/10.1007/978-3-030-31129-2_94

testing methods. Nowadays, the common non-destructive testing methods mainly include the following four types [4, 5]: capacitance change method, which judge the transformer winding state according to the change of the capacitance; vibration signal analysis method, which use the vibration signal of the transformer to analyze the winding state; Short-circuit impedance method [6], which diagnose the winding deformation by comparing the leakage reactance of the transformer; frequency response analysis, which compare the response changes when injecting a stable sweep signal into the transformer [7]. Among them, the short-circuit impedance method and the frequency response analysis have obtained a lot of research and application due to the advantages of simple, portable test equipment and strong anti-interference ability. However, the short-circuit impedance method cannot locate the winding fault, and the sensitivity is low [8]. The result of the frequency response is not intuitive enough, and there is no quantitative judgment standard. So it requires the tester to be familiar with the historical data for horizontal and vertical comparison. Therefore, in order to judge the transformer winding deformation more effectively, a new method called sweep frequency impedance was proposed, which has quantitative judgment and high sensitivity [9].

Based on the basic principle of the sweep frequency impedance technology, this paper completes the test in the laboratory. We focus on the short circuit and fault condition of the transformer winding study [10]. By analyzing the sweep impedance curve, the actual deformation of the transformer winding is judged, and the technology is verified. The feasibility of detecting the transformer winding deformation is proved by the power field test [11].

2   Experiment Method: Sweep Frequency Impedance

Sweep frequency impedance technology is a new type of non-destructive testing technology for wide-band transformer winding deformation [12]. This technology combines the advantages of frequency response analysis and short-circuits impedance method. It can obtain transformer sweep frequency impedance curve and 50 Hz short-circuits impedance through one test, and accurately determine the type and location of the winding fault. At the meantime, it effectively reduces the error caused by multiple wiring. The test system wiring used in the experiment is shown in Fig. (1) [13].

During the test, the transformer is short-circuited to one side winding, and a sweep signal is applied to the excitation end of the non-short-circuit side winding. After the signal passes through the winding, the output signal of the winding response end is obtained by using the sampling resistor R. The transformer sweep impedance can be obtained by the following equation [14].

$$Z(j\omega) = \left(\frac{U_{in}(j\omega) - U_{out}(j\omega)}{I_1(j\omega)}\right) = R(\omega) + jX(\omega)$$

$$|Z_k(j\omega)| = \sqrt{R(\omega)^2 + X(\omega)^2}$$

Where

- $Z_k(j\omega)$ is the sweep impedance, the unit is ohm;
- $|Z_k(j\omega)|$ is the value of sweep impedance, the unit is ohm;
- $U_{in}(j\omega)$ is the voltage signal at excitation terminal, the unit is volt;
- $U_{out}(j\omega)$ is the voltage signal at response terminal, the unit is volt;
- $I_1(j\omega)$ is the current signal in the non−short side transformer winding, the unit is ampere;
- $R(\omega)$ is the real part of sweep impedance, the unit is ohm;
- $X(\omega)$ is the imaginary part of sweep impedance, the unit is ohm

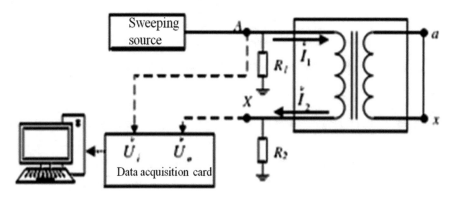

Fig. 1. Measurement system of sweep frequency impedance method

According to the frequency of the applied excitation signal, the equivalent circuit model of the test system can be divided into a low frequency equivalent circuit model and a high frequency equivalent circuit model.

(1) Low frequency equivalent circuit model

In the low-frequency band, the transformer winding can be regarded as a lumped circuit composed of components such as a resistor and an inductor. The equivalent circuit is shown in Fig. 2. Among them, R1 and R2 are sampling resistors, and both of their values are 50 Ω; Ri1, Xi1, and Z1 are the resistance, leakage reactance and impedance of the primary side coil respectively; Ri2, Xi2, and Z2 are the resistance and leakage reactance of the secondary side coil respectively. And Rim, Xim, and Zm are the excitation resistance, reactance, and impedance of the iron core; Z is the equivalent impedance of the measured winding. According to Fig. 2, the test circuit at this time is equivalent to the test circuit of the short-circuit impedance [15], so the value of the sweep frequency impedance curve at 50 Hz is the short-circuit impedance value of the transformer. The study of the impedance value can determine the operation of the winding status. At the same time, compared to the frequency response analysis method, the swept frequency impedance technology has better resistance to interference since the short-circuited wire is open, and Xim ≫ Xi1 + Xi2, Rim ≫ R i1 + R i2R. Therefore, under the same test conditions, the response signal of the frequency response analysis method is much smaller than the sweep frequency impedance technology. Since the higher response signal is the signal-to-noise ratio. Thus, the swept frequency impedance technology has better resistance to field interference [16].

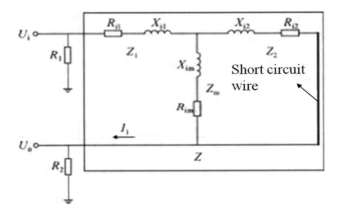

Fig. 2. Equivalent testing circuit of sweep frequency impedance method at low frequencies

(2) **High frequency equivalent circuit**

When the frequency of the high-frequency signal f > 1 kHz, the excitation of the core disappears. So the winding can be equivalent to a series of linear circuits composed of distributed parameters such as resistance, capacitance and inductance. The equivalent circuits as shown in Fig. 3. In Fig. 3, Rw1 - Rwn is the resistance of the winding under test; Ck1 - Ckn is the inter-cake capacitance of the winding under test; Ce1 - Cen is the capacitance to the ground of the winding under test; Lw1 - Lwn is the inductance of the winding under test. It can be seen from Fig. 3 that when the winding is in normal working state, the positions of various equivalent electronic components in the system are fixed, and so does the operating parameters. But when the transformer winding undergoes geometrical changes such as distortion or bulging, the parameters of the electronic device must change, causing the circuit sweep impedance change.

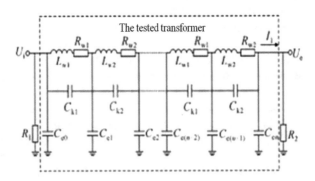

Fig. 3. Equivalent testing circuit of sweep frequency impedance method at high frequency

Therefore, the detection of transformer winding deformation can be performed using a higher frequency sweep impedance curve. If the power of the sweep signal is increased during the test to enhance the test signal, the anti-interference ability can be improved. So, the influence caused by adverse environmental factors can be greatly reduced.

In summary, the swept frequency impedance technology is a method for detecting the winding state by applying a swept sine wave signal to the other side winding of the transformer in the case where the winding on one side of the transformer is short-circuited. When using this method for detection, the transformer winding can be regarded as a circuit system composed of a series of components such as resistance, inductance, and capacitance.

When, the winding is deformed, the low-frequency leakage, and distribution parameters change the high frequency, thus changing the value of the overall sweep impedance resistance accordingly. Therefore, the winding state of the transformer can be determined by comparing the sweep resistance values before and after the fault.

3   Result and Discussion

The experiment uses a 10 kV three-phase double-winding transformer as the research object. The two sides of the high/low voltage winding, which located at the top of the transformer tank, is taken out from the casing. According to the literature [21], the transformer winding deformation can be judged by the change of the winding inductance, the capacitance to the ground and the longitudinal capacitance. Therefore, it is possible to study the winding deformation such as short-circuit and deformation by the series, parallel inductance or capacitance at the high-voltage winding tap of the 10 kV transformers.

When the transformer low-voltage winding is short-circuited in three phases and the connection mode is Yy connection group, the A-phase high-voltage winding is tested by the sweep impedance method. The sweep impedance spectrum is shown in the Fig. (4).

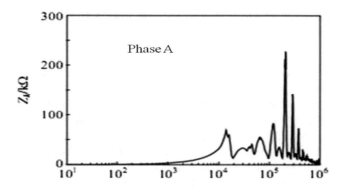

Fig. 4. Original SFI curve of normal transformer

It can be seen from the impedance spectrum in the Fig. 4 that the amplitude of the resonance point rises first and then falls over the entire frequency domain, and reaches a maximum at a frequency of 200 kHz. According to this waveform trend, the amplitude of the impedance changes greatly in the entire frequency domain. It may cause the small waveform change in the curve not to perform well, thus affect the determination of the transformer winding deformation. Therefore, the impedance $Z_k(jw)$ is subjected to secondary transformation using decibels to obtain a dBΩ impedance. At this time, the sweep impedance amplitude span is reduced, which is between 30 and 110 dBΩ. That can better reflect the slight change of the curve, and make it more conducive to high-sensitivity winding deformation detection.

$$Z_{k'} = 20\lg Z_k$$

For the winding short-circuit fault, the wire is short-circuited to the A-phase high-voltage winding 1–2 cake 53 turns coil. For the deformation and depression fault, the bottom end of the A-phase high-voltage winding is set for the axial recess 2 cm. At this time, the sweep impedance test is performed on the normal transformer winding and the faulty transformer winding, respectively.

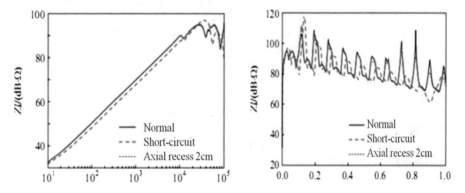

Fig. 5. Comparison of SFI curves between fault and normal: (a) The frequency is 10 Hz–10 kHz, (b) The frequency is 10 Hz–1 MHz

It can be seen from the Fig. 5 that the impedance curve of the fault winding is significantly different in the entire frequency domain. When the high voltage winding has a local sag fault, the impedance curve of the fault winding is basically coincident with the impedance curve of the normal winding in the frequency range of 10 Hz – 150 kHz. This is because the axial translation of 2 cm has less influence on the leakage inductance of the winding. In the case of low frequency, the inductance in the impedance is the main factor. So the impedance curve of the depressed fault winding at the low frequency band basically coincides with the normal curve.

When the frequency reaches 150 kHz–1 MHz, the normal curve and the concave fault curve are basically coincident. But the peaks are significantly different, especially at 170 kHz. The peak amplitude of the concave fault curve is larger than normal.

When short circuit fault occurs in the high voltage winding, then the amplitude of the impedance decreases in the entire frequency. The entire resonance point shifts toward the high frequency. This is because when the wire is short-circuited, the wire is connected in parallel with the coil. The inductance and resistance of the wire are much smaller than the coil so that the total inductance of the winding is reduced and the capacitance is increased. So the resonance point is shifted to the high frequency. At the same time, since the winding inductance is reduced, the equivalent circuit impedance is reduced. So the impedance curve amplitude of the fault winding is smaller than the impedance curve of the normal winding.

4   Conclusions

In this paper, for the common short-circuit and deformation faults of transformer windings, a 10 kV three-phase double-winding transformer is taken as an example. The sweep impedance method is used to detect and compare the impedance curves of the normal winding and the fault winding. For the faulty winding fault, the winding fault cannot be judged by the deviation of the impedance. Since the impedance curve before and after the fault changes significantly, it can be determined according to the impedance curve. For a short-circuit fault, the amplitude of the impedance curve of the fault winding decreases in the entire frequency domain. And the entire resonance point shifts toward the high frequency. Therefore, the fault of the transformer short circuit and the dent deformation can be accurately determined by the sweep impedance method

References

1. Guo, C., Wang, Y., Mei, W., et al.: Time-varying outage model for transformers representing internal latent fault. Proc. CSEE **33**(01), 63–69 (2013)
2. Liu, R., Li, J., Zhang, S., et al.: Study on the on-site heating method for large-scale power transformers. Proc. CSEE **32**(01), 193–198 (2012)
3. Wang, M.: Statistic analysis of transformer's faults and defects at voltage 110 kV and above. Distrib. Utilization **24**(1), 1–5 (2007)
4. Liu, Y., Liang, X., Ji, S., et al.: Influence of transformer bushing capacitive fault on frequency response curves. J. Xi'an Jiaotong Univ. **48**(2), 62–68 (2014)
5. Ji, S., Li, Y., Fu, C.: Application of on-load current method in monitoring the condition of transformer's core based on the vibration analysis method. Proc. CSEE **23**(6), 154–158 (2003)
6. Wang, J., Li, R.: Influence of ground lead on experiment of transformer winding deformation. Yunnan Electr. Power **35**(1), 1–2 (2007)
7. Shah, N., Fiaidhi, J., Mohammed, S.: Evaluating anti-hypertension medications usage on patient care online blogs. Int. J. Adv. Sci. Technol. **125**, 43–52 (2019)
8. Hur, K., Sohn, W.: Wearable computer network design using dual-role WUSB clustering. Int. J. Control Autom. SERSC Australia **8**(3), 127–134 (2015)
9. Jamal, H., Nasir, A., Ku-Mahamud, K.R., Kamioka, E.: Enhanced ant colony system for reducing packet loss in wireless sensor network. Int. J. Grid Distrib. Comput. **11**(1), 81–98 (2018)

10. Hordri, N.F., Ahmad, N.A., Yuhaniz, S., Sahibuddin, S., Ariffin, A., Zamani, N., Jeffry, Y.: Classification of malware analytics techniques: a systematic literature review. Int. J. Secur. Appl. **12**(2), 9–18 (2018)
11. Eddy, M., Ahmad, A., Tang, A.: Agents of things (AOT): utilizing JADE agent technology as communication middleware for vehicle monitoring system. Int. J. Future Gener. Commun. Networking **11**(1), 47–54 (2018)
12. Zhang, J., Song, W.: Recognition method based on size in depth image. Int. J. Signal Process. Image Process. Pattern Recogn. **10**(1), 81–88 (2017)
13. Espinosa, P.: Knowledge about and health practices on sexually transmitted infections among commercial sex workers in an urban community. Int. J. u - and e - Serv. Sci. Technol. **11**(1), 45–54 (2018)
14. Cheng, Q., Liu, Z., Selmoune, A.: Nonlinear distance-based dynamic pricing considering congestion-level correction. Int. J. Transp. **6**(2), 1–8 (2018)
15. Pansotra, A., Singh, S.: Additive hough transform and fuzzy c-means based lane detection system. Int. J. Disaster Recovery Bus. Continuity **8**, 11–28 (2017)
16. Pradhan, S., Sharma, K.: Cluster head rotation in wireless sensor network: a simplified approach. Int. J. Sensor Appl. Control Syst. **4**(1), 1–10 (2016)

Application of Electric Power Automation System Based on Power Distribution Network

Bo Zhao[⊠]

Department of Marketing, State Grid Liaoning Electric Power Supply Co. Ltd.,
Shenyang, China
418484168@qq.com

Abstract. As the development of the smart grid in the society, in order to meet the needs of various users, will be the smallest waste electricity to meet the interests of all users and realize make full use of electricity, power grid has gradually mixed with the application of automation elements, this to our country on the performance of the power grid, monitoring, control, maintenance has brought great convenience, at present our country on the power grid automation are research and development, has made the corresponding improvement.

Keywords: Smart grid · Automation elements · Corresponding

1 Introduction

Dispatching automation of electric power systems has there aspects area real-time monitoring, substation automation, and load control. Regional scheduling real-time monitoring system usually consists of small or microcomputer, functions as a center scheduling monitoring system, but a bit simple. Is the unattended substation automation development direction, its remote device USES a single-chip programmable? The load control of the power supply system or audio power frequency control method is often used.

Automation technology is a comprehensive technology, it cybernetics and information theory, systems engineering, computer technology, electronics, hydraulic pressure technology, automatic control and so on all has a very close relationship, and the "control theory" and "computer technology" the biggest impact on automation technology. Popular speaking, automation is to facilitate people to accomplish a goal, people on the machine or system by programming and related mechanical design or circuit design, make the change a machine or system can complete the task according to the will of the people themselves, and achieve the result of people want to. Automation technology has greatly liberated human productivity, to industrial and power system has brought revolutionary progress [1].

The feedback loop is the core of the automatic control, is to realize the data monitoring and improvement, make the output data of the key to achieving stability or dynamic balance. Than single automation function is a higher level of automation control system, automation control system not only can realize various automation relate to each other and adjust to each other, to achieve the stability of the system, such

A. E. Hassanien et al. (Eds.): AISI 2019, AISC 1058, pp. 1017–1025, 2020.
https://doi.org/10.1007/978-3-030-31129-2_95

as the electric power supply system, we should combine the electricity consumption with supply customer, when a certain stage the user of this kind of new type of electric power reached the peak, the system will think of some way to increase the user's power supply, increase the user's power supply has many aspects, one is to increase power output, the second is to ask the other storage unit transferred any extra points, and then the peak in return, such a complex thinking, and the process is automatic control system of tasks, power automatic control system can serve the whole country grid can deal with all kinds of complicated situations, to be able to suffer all kinds of foreign attack and restore balance.

Until the 1950s, in millions of kw power system capacity, capacity of less than 100000 kw, limited to the single automatic device, power system automation, and safety protection and automatic adjustment process. For example, all kinds of relay protection of power grid and generator, steam turbine of crisis protector, boiler safety valve, steam turbine speed and generator voltage automatic adjustment, interconnection of automatic device, etc. 50–60s, the scale of power system development to the tens of millions of mw, stand-alone capacity of more than 200000 kw, and form a regional network, in the aspect of system stability, economic operation, and integrated automation puts forward new requirements. Used in factory automation machine, furnace, electric unit centralized control. The system begins to furnish the analog FM device and on the basis of offline calculation of economic power distribution device, and widely used in remote communication technologies. Various kinds of new type automatic device protection devices such as transistor, thyristor excitation regulator, such as electric hydraulic governor is widely used. 70s–80s, with computer as the main body is equipped with a fully functional software and hardware of a complete set of grid real-time monitoring and control system (SCADA) began to emerge. 200000 kw or more large-scale thermal power units started using real-time security monitoring and closed-loop automatic start-stop process control. Hydropower station reservoir scheduling, dam monitoring and integrated automation of power plant computer monitoring and began to get a promotion. All kinds of automatic adjusting device and widely used microcomputer relay protection device.

2 Power System Automation Technology

2.1 The Concept of Power System Automation Technology

Power system automation technology refers to the use of various has the function of automatic detection and control device, through the data transmission system and signal system will each element of the power system, local system, or the whole system for automatic monitoring, coordination control technology, to ensure the safe and stable operation of power system can, to ensure the reliability of power supply.

2.2 Electric Power Automation System Components

Power system automation technology is the outstanding performance of the electric power industry innovation technology, the current situation, mainly exist the following components in this system including Substation automation technology, Distribution network automation technology, and Power system dispatching automation technology. Here we will discuss these components in a brief.

(a) Substation automation technology

Substation is assembling a some equipment, cut off or on the system voltage device, the electric power system, the sub-station is the concentration of power distribution and transmission, it will be better able to monitor the power supply of transportation, ensure the safety of the whole process of economic, efficient and reliable so that the substation automation technology is particularly important.

Substation automation technology is the use of advanced computer technology, modern communication technology, electronic technology and information processing technology, the implementation of substation secondary equipment of recombination and optimization design, the operation of the all equipment of substation can realize real-time monitoring, the comprehensive automatic monitoring system can improve the stability of the substation operation, reduce the cost of operation maintenance, achieve high quality transmission process, ensure the economic benefits.

(b) Distribution network automation technology

For a long time, can only adopt manual operation of distribution network control method, with the progress of technology, can use independent island automation technology gradually, but still exist deficiencies of electricity distribution, therefore, distribution network automation technology in power distribution and monitoring is very important.

Mainly includes the feeder automation and power distribution automation system in automatic drawing, equipment management, information analysis and the analysis of distribution network automation, it relies on a lot of intelligent terminal, rich background software and database information support, through the driving of information technology, realize the power distribution network automation, ensures that the full use of electricity.

(c) Power system dispatching automation technology

Power system dispatching automation technology is currently one of the fastest growing technology, its function performance of powerful enough to ensure that the power system in the accuracy, reliability and economy in the process of operation. Power system data acquisition and monitoring function is the basis of dispatching automation, in addition, the market of power system operation and decision-making is also the link cannot be ignored.

3   The Application of Automation Technology in Power Distribution Network

The automatic control technologies in power distribution in some applications mainly include the following aspects:

- *In the distribution network operation personnel training, the use of simulation technology:* Using simulation technology, can let staff in the distribution network from actual grid dangerous circumstances, to experience the actual operation process, can be quick and easy to master skills, improve the level of operation, at the same time also can ensure the staff's life safety and the safety of the running environment of the whole distribution network, improve the work efficiency. On assessment of employees' operation ability, also can use this kind of simulation technology, the need to review the abilities and skills are basically consistent with the actual operation, and the way to avoid some not correct evaluation of their employee's safety [2].

– *Information technology application in power distribution network operation management:* In order to ensure the safety of each distribution network equipment can normal run, must be monitored, and the faults in the operation and management, and timely isolation and repair, guarantee high quality of power supply service, the request is the best choice of distribution network automation, information technology is applied to distribution network equipment and the monitor management of electricity. Monitoring is the application of automation technology is an important job, it can detect the grid through various channels of each part in health, for example by sensors, automation instrument can detect transformer, transformer room of coil voltage, current, temperature, magnetic field, the situation which they can determine their health. According to the topological structure of power grid on the other hand, through the intelligent analysis can find out the fault point, fast to make corresponding measures, can greatly save the employee's workload.

– *It should be able to be found in operation of the power distribution network security hidden danger in time:* at other faults, ensure the safety of power distribution network operation with high-risk conditions such as being struck by lightning, short circuit, automatic control system can reliably to cut off the power supply in an instant, and the lightning in an instant to protect the safety of electrical equipment [3]. Than the traditional distribution box, both in terms of response speed and on the protection performance has improved significantly. Distribution network automation system must be able to do it, of course, for each line of data collection and monitoring, timely eliminate unsafe factors at the same time, good fast processing barriers, improve the reliability of power supply, reduce the running loss, good for the operation of the power distribution network to provide a safer environment. Is closely related with the monitoring, point of failure in the power grid automation test according to test data and intelligent analysis, can quickly find out the fault point, and analyzes the cause of the problem. And digital circuits, analog power grid can also be precise data analysis. Like computers for cable testing, when offline, we can enable the computer fault detection function, in a short period of time, the computer can generate fault analysis diary, is overdue bills, there was an error or a modem or port disconnect will have a clear hint, this give the work to guide the direction of maintenance personnel, provides a great convenience. We also can be transplanted in distribution power grid technology.

– *To numerical analysis, to supply power:* Automation system can according to the test automation instrument get data analysis, and through the feedback system of power allocation can be realized. When the voltage is low, the voltmeter data back to the command center, command center will appropriately increase the value of the offset voltage, automatic control system can meet the needs of more users, it can allocate more different nature of the electricity used by the user. Power consumption calculation for the user the automatic statistics to replace the original manual meter reading trouble, when people or small electricity user power failure occurs, due to the fault events more frequently, if as usual trouble employees ran a long journey just to repair the individual private users, waste the time of users has cost the electric power energy. In order to simplify the problem in this regard, can build a fault processing platform, the user can fill in fault classification according to the cause of the problem in private to submit to the processing platform, such processing personnel according to the division of geography, give the user a batch of a batch process faults, this is I just summarize information society automation management experience.

- *The application of feeder automation technology:* Feeder automation technology in the application of power distribution network, it is direct and effective technological means to enhance the reliability of power supply, the first is often in the process of power distribution automation system in distribution network of feeder automation, also is the basis of power distribution network automation, the feeder automation technology refers to the user between substation and feeder line automation, electrical equipment usually includes two aspects: one is accident situation of fault detection, isolation and fail over and restore power supply control; The other is a normal data of measurement, user testing and operation optimization. Feeder automation technology in power distribution network automation system technology can find fault in time, and according to the testing information to judge, quick protection action, as far as possible to reduce the influence of fault [4].

4 The Application of the Electric Power Automation System in Distribution Network Operation Management Principles

4.1 Safety Principles

Along with the development of the society, people growing demand for electricity, the power is higher; the dependence of electricity can help people, may also harmful to people, so to ensure the power system security, prevent power unsafe cause casualties or bring economic losses. The automatic control system must be able to protect the circuit. One is to ensure that peak electricity and ensure the quality of electric power, bring benefits to users. The second is to ensure that in time of danger immediately without electricity, to ensure staff security and safety of electrical equipment. For the special user, electricity will bring heavy casualties accident, or bring huge economic losses, or pose serious social security problem of the user, should give double power supply device, when one of the power failures occurs, automation equipment can immediately disconnect the power supply is connected to a power source.

4.2 The Principle of Reliability

Power system for reliability, can't often fail, main is to make sure the user power supply reliability, in order to avoid because of the circuit itself affect users work, bring inconvenience to the user or economic loss. The principle of the reliability of the power system mainly embodies in four aspects: one is to have reliable equipment, such as switch, FTU, network equipment; it is to have a reliable network communication system, such as the master station, station system, and communications medium; it is to have a reliable design layout, such as power lines, the arrangement of the space truss layout, etc.; Four is to have a reliable power supply.

4.3   The Principle of Seeking Truth from Facts

In determining the choice of mode, power distribution automation system should be based on national conditions, cannot blindly copying, want to combine the local actual situation, according to the actual living standards and requirements of users, from services, for the user to provide the power supply reliability, considering the factors that make the maximum effectiveness of power distribution automation system.

4.4   The Principle of Keeping Pace with the Times

As the electric power system automation, the biggest drawback is always maintenance updates, automation system has been gradually connected to the network computer, in order to prevent hackers and virus into, to increase the protective wall, password Settings such as protective measures. And is updated, and constantly enhance the technology content of automation, intelligent and integrated automation [5].

5   The Application Prospects of Electric Power Automation System in Distribution Network Operation Management

5.1   The Application of Digital and Internet Is More Widely

Intelligent era will be a highly developed stage of Internet technology, the power grid, all kinds of physical phenomenon and the related parameters will develop into the scale of the standard, through the representation of a digital method has many advantages: one is conducive to the standardization of said, is advantageous to the computer for data processing. Network access to the Internet can realize the power of globalization, and China's electricity can sell abroad, occupy the foreign markets, can also be a merger with foreign power grid, and achieve greater range of power grid. Strengthen the cooperation between the countries in space [6].

5.2   The Flexibility of Automatic Technology

With the development of science and technology, and the transplantation of technology, fault diagnosis and treatment in the grid will be more automation [7]. Especially in the case of the widespread expansion of power grid, the grids will even expansion abroad, thus for grid monitoring strength and accuracy is higher, prevent power failures is the danger and loss, prevent to steal and destruction of electric power, prevent the network hackers. Some danger and in complex maintenance work also can have an intelligent machine to replace manual maintenance, should be set up in peak period in the accident, the high-risk period of intelligent robot real-time monitoring, specific operation by the distance of the remote operation and maintenance engineers.

5.3   The Development Trend of Technology

By the open loop, monitoring is given priority to transfer to the closed-loop control is given priority to, for example, from the system power sum to automatic generation control. By the technology of single function to multifunction comprehensive direction, for example, the application of substation automation, distribution network automation technology [8, 9].

To develop in the direction of the whole system consists of single component development, such as testing control and data acquisition technology development is a very good sign [10–12]. From high to low voltage level voltage grade development, such as from the energy management system to develop in the direction of the power distribution management system [13].

The performance of the device in the direction of flexibility, technical, creative and digital shift, ensure the efficiency of the power supply system, intelligent and economy, for example, the relay protection technology innovation, the improvement of excitation control technology, etc. [14–16].

Will be the efficient, economic and safe operation of the power system and the management of automation and efficiency realize the electric transport process smooth, such as the application of the management information system in the whole process.

5.4   The Development Trend of Technology

The application of new technology should be implemented adaptability, coordination, innovation and the optimality of the perfect fit. On the technical design to conform to the machine system, as much as possible to deal with problems that may occur; In monitoring link to improve the technology content, to achieve real-time monitoring, monitoring combined with a focus on monitoring the monitoring mode [17–19]; Train technical personnel's technical and technological literacy, implement a variety of technical personnel joint operation, in practice the theory of modern technology theory, automation control technology, electronic technology application in practice.

6   Conclusions

The development of industrial society, the construction of urban civilization growing demand for electricity, electricity has become the main power to promote the development of society, but the power as a kind of indispensable energy, often appear fault and dangerous situation, loss and even is the life of the people to harm. In order to reduce power's shortcomings in this aspect, the national power grid has joined the automation elements on the electric power increase power reliability, intelligence, and security. China has made great progress in electric power automation, also has a very big development space in the future, China's intelligent power system would be to the next level. In the golden time of development, our related people, it is necessary to strengthen the research to make a certain contribution for the country's electric power industry development, the above this article is the application of electric power automation system in distribution network operation management research.

References

1. Lu, J.: Electric power automation system in distribution network operation management application. Technol. Market **13**, 21–26 (2011). (in Chinese)
2. He, M.K., Chen, B.S.: Introduction to electric power automation system and 10 kv distribution network operation and management. Sci. Technol. Innov. Appl. **4**, 13–19 (2013). (in Chinese)
3. Zhang, H.L.: The electric power automation technology in the application of power distribution network operation management system. Innov. Sci. Technol. **2**, 21–24 (2014). (in Chinese)
4. Lang, C., Kos, D.: Energy consumption of telecommunication networks and related improvement options. Sel. Topics Quantum Electron. **17**, 285–295 (2011)
5. Zeng, B.T.: Analysis of electric power automation system technology in power distribution network operation management using the point. China's New Technol. New Products **13**, 28–32 (2014). (in Chinese)
6. Li, J.: Introduction to electric power automation technology in the application of power distribution network operation management system. Henan Sci. Technol. **6**, 16–19 (2014). (in Chinese)
7. Kim, W.J., Kim, T.H., Lee, S.S.: Properties according to alkali activator mixing ratio of non-cement composite. Asia-Pac. J. Model. Simul. Mech. Syst. Des. Anal. **21**, 21–26 (2017)
8. Hong, S.W., Kim, K.S., Park, C.Y., Jang, H.-I.: Comparison analysis of interstitial condensation occurrence rate by regional weather conditions - focused on 9 Regions in Korea. Int. J. Eng. Technol. Automob. Secur. **21**, 1–6 (2018)
9. Kim, Y.A., Kim, K.S., Park, C.Y., Choi, C.H.: Discussion of current support policies of performance improvements in existing buildings by nation and comparison of characteristics of performance evaluation tools. Int. J. Energy Technol. Manag. **21**, 1–6 (2018)
10. Jayashree, J., Vijayashree, J., Iyengar, N.Ch.S.N.: Risk level prediction of diabetic retinopathy using adaptive neural fuzzy inference system. Int. J. Adv. Sci. Technol. **120**, 59–72 (2018)
11. Jin, B., Hong, L.: Improved artificial bee colony algorithm and application in path planning of crowd animation. Int. J. Control Autom. **8**, 53–66 (2015)
12. Aslam, U., Batool, E., Ahsan, S.N., Sultan, A.: Hybrid network intrusion detection system using machine learning classification and rule based learning system. Int. J. Grid Distrib. Comput. **10**, 51–62 (2017)
13. Esfahani, R., Norozi, Z., Jandaghi, G.: Cover selection for more secure steganography. Int. J. Secur. Appl. **12**, 21–36 (2018)
14. Sabor, N., Sasaki, S.B., Abo-Zahhad, M., Ahmed, S.M.: An immune-based energy-efficient hierarchical routing protocol for wireless sensor networks. Int. J. Future Gener. Commun. Network. **9**, 47–66 (2016)
15. Zhang, M.H., Cheng, X., Zong, X.J., Zhang, Y.: Switching-based SVPWM control for three-phase active power filter. Int. J. Sign. Process. **8**, 85–98 (2015)
16. Gao, Q., Ao, C.L., Yang, J., Tong, R.: Contingent valuation model based on contoured willingness to pay and empirical analysis. Int. J. u- e-Service Sci. Technol. **9**, 47–60 (2016)

17. Stamatiadis, N., Psarianos, B.: Impacts of design exceptions on highway safety. Int. J. Transp. **5**, 1–14 (2017)
18. Tomas, U., Ganiron, J.: The human impact of floods towards mega dike effectiveness. Int. J. Disaster Recovery Bus. Continuity **7**, 1–12 (2016)
19. Kumar, M.: Various factors affecting performance of web services. Int. J. Sensor Appl. Control Syst. **3**, 11–20 (2015)

Adjusting Control Method of AC Output Synchronization Based on Phase Lock Technology

Rongyu Du[1]([⊠]), Wenrui Li[2], Hongtu Wang[1], Jing Xu[1], Xiangluan Dong[2], and Bo Sun[1]

[1] State Grid Chaoyang Electric Power Supply Company, State Grid Liaoning Electric Power Supply Co. Ltd., Chaoyang, China
18842333082@163.com
[2] Graduate Department, Shenyang Institute of Engineering, Shenyang, China

Abstract. Uninterruptible Power Supply (UPS) can provide a stable, continuous and uninterrupted power supply and it has higher requirements for the power supply quality of UPS. In order to improve the reliability and engineering feasibility of uninterruptible power supply (UPS) redundant parallel operation system, a phase locking control method is proposed. This method adjusts the AC output by measuring the system frequency dynamics in real time to realize the seamless switching of the bypass input. There is no discontinuous process in this process, so as to provide stable, continuous and uninterrupted power.

Keywords: Uninterruptible Power Supply · Phase locking control · Cycle synchronization

1 Introduction

In order to ensure the continuity and reliability of power supply in important departments, eliminate the influence of grid interference on electrical equipment, and at the same time, in order to avoid the interference of load on the grid, Uninterruptible Power Supply (UPS) emerged [1–3]. Along with the market to the Uninterruptible Power Supply high performance, high efficiency and so on request Uninterruptible Power Supply has brought the development opportunity, but at the same time to its redundant parallel operation system reliability and the engineering may realize, proposed one realization instantaneous electric quantity inspection method. This method adopts fixed sampling points and adjusts the sampling frequency by measuring the system frequency dynamically in real time. This method can make the collected electric quantity information more accurate and realize the seamless switch of bypass input [4–6]. There is no discontinuous process in this process, so as to provide stable, continuous and uninterrupted power [7].

Parallel operation of UPS power supply means that multiple UPS power supply to the same load at the same time, that is, parallel machine. In parallel control, the main problem is how to reduce the inhibition of circulation between parallel machines and load sharing. If the UPS parallel system is to reach an ideal operating state, the external output characteristics of each UPS module in the system must be completely consistent, while the output of UPS is ac, so the amplitude, frequency, phase and output impedance of the output voltage must be completely consistent, so as to achieve load power sharing. But in practice, due to the dispersion of component parameters, it is difficult to achieve the same output characteristics [8–11].

Therefore, this paper proposes an algorithm to detect and track the output voltage of the bypass and the main circuit. That is, bypass input voltage, current, amplitude and phase Angle are collected, and the intelligent processor analyzes the collected voltage, current, amplitude and phase angle. According to the results of amplitude and phase Angle analysis by the intelligent processor, the voltage and current of ac output and bypass input can be judged to be synchronous or not. Through real-time monitoring of the bypass communication, the ac output inside the UPS of the device is adjusted to improve the synchronization between the bypass and the output, so as to realize the seamless switch of the bypass input. There is no discontinuous process in this process, so as to provide stable, continuous and uninterrupted power [12–14].

2 Preliminary Knowledge

Parallel operation of UPS power supply refers to that multiple UPS supply power to the same load at the same time, that is, the parallel operation of UPS is generally referred to. The main problem in parallel control is how to reduce the inhibition of circulation between parallel machines and load sharing. To realize parallel operation of UPS, the following two conditions must be met:

1. The phase, amplitude and frequency of output voltage of each UPS module shall be consistent [15–17].
2. Only when the output impedance of each UPS module is consistent can the load distribution be realized, that is, the total load current should be shared equally among all UPS units in parallel. Each UPS in the parallel system must share the load current equally; otherwise it will generate circulation between each UPS module in the system, which directly affects the operation reliability and system operation efficiency of the whole UPS parallel system [18–20].

At present, there are several control methods to suppress the circulation and load power sharing of parallel system such as centralized control, master-slave control, decentralized logic parallel control, and no interconnection line control.

(1) **Centralized control:** when the mains power is normally supplied, the control module of the system samples the phase and period of the mains power and generates the corresponding synchronization signal, which is used as the tracking reference for the output voltage of UPS in the system: when the mains power is abnormal, the control unit of the parallel system will generate a synchronous signal of power frequency. The PLL tracking of all UPS modules in the system is synchronized with the same signal, so that the phase and frequency of the output voltage of all UPS modules in the system are consistent. And the control link will be the current value of the difference as a voltage regulation of the compensation

quantity feedback to the UPS itself, in order to reduce the system circulation. The average current control method must be based on the premise that the output voltage frequency and phase of each UPS are basically the same. At this time, the circulation is mainly caused by the voltage amplitude difference. This control method has its inherent defects, that is, once the parallel control unit failure, may lead to the collapse of the parallel system [21–23].

(2) **Master-slave control:** a certain UPS module in the system is set as the main control unit of parallel control, that is, the host computer, which is responsible for the control function of the parallel system. Other UPS modules are taken as the controlled object, that is, the slave machine. When the host module fails, as long as other UPS modules are still in normal operation, it can be switched to other UPS modules as the main control unit of the parallel system. Compared with the centralized control strategy, the master-slave control strategy has higher reliability. When any UPS in the system fails, it can still keep running. But this kind of control mode still exists defects, when the host failed, may have a period of time in the process of switching the UPS lost synchronous reference signal and result in the whole system collapse, at the same time control logic circuits of the UPS modules are more complex, such not only improves the production cost, and to a certain extent, its reduce the reliability of the system.

(3) **Decentralized logic parallel control:** this control method is also known as active and reactive power control. In this scheme, the status of any UPS module in the parallel system is independent and equal, and there is no main control module. The current sharing control of UPS parallel system is realized by the communication bus of each UPS module in the system. Through the transmission of the bus, the UPS modules in the system exchange data information with each other during operation, and the control strategy can be adjusted accordingly according to the obtained real-time information, so as to realize the phase-locking of the output voltage of each UPS module, output load sharing and relevant logic switching.

(4) **No interconnection line control:** According to the principle of dropping external characteristic, there is no signal line connecting to each UPS module. Only the power line output connected to the load. Each UPS module detects its own output power. After obtaining the active power and reactive power, the phase and amplitude of its output voltage are fed back and adjusted respectively to achieve the load power distribution. In this way, the UPS modules in the system only rely on their own control strategy. There is no electrical connection between the control systems of the UPS modules, that is, electrical isolation, which makes the installation system or system maintenance more convenient and flexible. In theory, the parallel power supply is more reliable, and the expansion of the system is more flexible and convenient. The problem is that all UPS in the system do not exchange information with each other, which reduces the accuracy of current sharing control. In fact, in the operation of parallel system, there are some problems that UPS needs to switch to the bypass state, internal and external synchronous switching, etc., and the control strategy without interconnection line is still facing many difficulties in practical application.

This paper mainly discusses the first condition that must be met to realize parallel operation of UPS, that is, the phase, amplitude and frequency of output voltage of each UPS module should be consistent. When the output of each UPS meets the first condition, the consistency of phase, amplitude and frequency of ac output and bypass input on each UPS is also guaranteed. The schematic diagram of the parallel UPS is shown in Fig. 1.

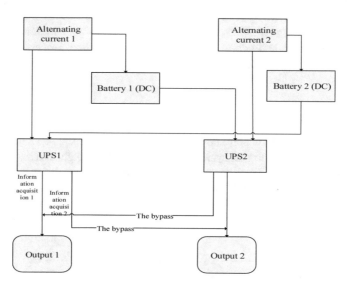

Fig. 1. Structure diagram of parallel UPS

When the voltage information of the ac output terminals of UPS1 and UPS2 collected are the same in frequency, amplitude and phase, that is, the voltage information of the ac output and the bypass input of UPS1 is consistent. At this time, no gap switching can be realized between the bypass and the main circuit to ensure the safety and reliability of the load.

3   UPS Phase-Locked Synchronization Technology

Phase-locked performance is an important index to evaluate the performance of UPS. When normal mains frequency fluctuations (allow less than 5%), in order to safely achieve the bypass switch inverter, UPS bypass the input voltage, output must be real-time tracking system make the UPS system and bypass the input voltage synchronous operation, ensure the UPS to provide continuous and stable electricity load, and can't have too much impact on the load. Bypass inverter switching should be smooth and continuous, so as not to cause UPS frequency jitter or phase mutation in the conversion process [24].

Under normal circumstances, the load of power grid is not static, but as the system running produces tiny difference, plus many other uncertain factors, causes the output voltage in the grid frequency cannot be maintained in a fixed value, so you need to take some measures to solve this problem, is usually a detection module UPS output voltage phase with the mains voltage and frequency, further the size of the phase difference and the frequency difference, again through the closed-loop feedback control to adjust the phase and frequency of UPS module output voltage, thus make UPS module output voltage and output voltage synchronized, electric, this is called phase lock.

Digital phase-locked loop (Digital Phase-locked Loop, Referred to as "DPLL") is not only has the advantages of digital circuit, but also solves some difficulties encountered in analog phase-locked loop, such as component saturation, dc zero drift, must carry out initial calibration, etc., and has the real-time processing ability for discrete sampling values. Therefore, it has been widely used in the field of power electronics.

3.1   Synchronous Principle of AC Output and Inverter

Often the UPS AC output is periodic function, as shown in Fig. 2 X T1 to represent its cycle, when the power is normal cycle fluctuations in a safe range, and the output of the UPS inverter is periodic function, using T0 said its cycle, can be controlled by the adjustment of the controller to change its cycles. Figure 2 shows the UPS inverter and bypass the input voltage phase diagram.

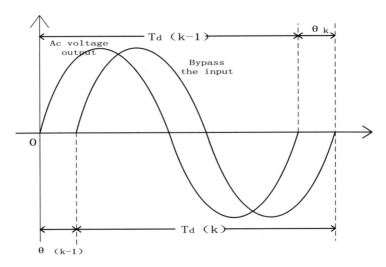

Fig. 2. Voltage phase diagram.

As shown in Fig. 3, at the beginning of the UPS inverter output AC voltage ahead of bypass entering a theta $(k - 1)$ value is the size of the phase Angle, in order to make the phase difference decreases to zero finally, you will need to increase the inverter output T0 the next cycle of cycle value (k), until meet the Td $(k + 1) -$ T0(k) < 0, the phase difference can beginning gradually began to decrease; When the inverter outputs phase lag bypass input, it is necessary to reduce the period bit of the next period output by the inverter. This is the regulation principle of phase lock.

If we want to change the output of the UPS inverter control cycle, cycle can be set by software in different carrier value, the phase in the function of simulating the voltage-controlled oscillator phase-locked loop, in the analog phase-locked loop, phase discriminator's main function is to detect the phase difference, but is mainly composed of DSP digital phase-lock link to realize the phase difference detection.

In order to monitor the phase difference, the AC output voltage needs to be shaped to form a square wave. As shown in the figure, the square wave signal after zero crossing shaping is connected to the capture pin of DSP for detection. If the preset jump edge is detected, the DSP will automatically obtain a capture interrupt. Through the software of the DSP internal counter value, the value size is equivalent to the value of the AC output and an AC output cycle before phase-lock link output cycle refers to the T0(k − 1) is representative of this counter cycle, if the counter to zero moment just after the sine signal, and for the standard sine wave, then to counter T0 (k)/2, the output signal is right can correspond to another for zero, this time through the communication software output and T0(k)/2 is the difference between the output and bypass the input phase difference theta (k).

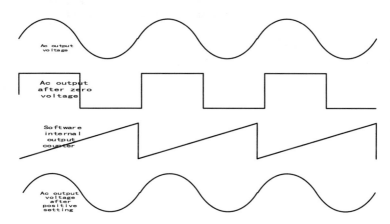

Fig. 3. Digital phase-locked schematic diagram based on DSP.

For the parallel connection of UPS, the significance of phase locking is to make the AC output of UPS and the voltage frequency, waveform and phase of the bypass input consistent, so as to make the output switch between the bypass and the pin of UPS reach no gap switch and ensure the stability of UPS power supply.

3.2 Digital Phase-Locked Loop Model

The relationship between phase difference and UPS inverter and AC output can be obtained Eq. (1).

$$\theta_{(k)} = \theta_{(k-1)} + \left(T_{d(k)} - T_{0(k+1)}\right) \tag{1}$$

In practical application, it is quite difficult to adjust directly by changing the phase Angle, and the feasibility is very low. Therefore, the purpose of tracking AC output is usually achieved by adjusting the output frequency of the inverter. Therefore, the UPS inverter is usually equivalent to an integral link. The control of the period is also the control of the frequency. Since there is a period difference between the output of the UPS module and the bypass input, it can also be equivalent to the integral link. If there is no phase-locking link, even if there is a small phase difference between the two, eventually the periodic difference will be amplified, eventually leading to a large error and the collapse of the system.

4 UPS AC Output and Bypass Input Voltage Test

When the UPS module runs in parallel, it wants to reach the ideal state, that is, the AC voltage output of each module and the no-gap switching between the bypasses. This requires that the output external characteristics of all modules in parallel operation should be exactly the same. For the AC power supply, the frequency, phase and amplitude of the output should be taken into account. Only when the three are completely the same, can the seamless switching between them be realized.

The parallel structure diagram of the system is established under the simulation of the simulation software. The output voltage waveform of the model is shown in the Fig. 4.

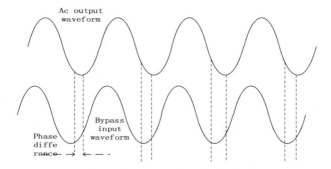

Fig. 4. Phase difference of output voltage.

After the phase-locked control technology, the AC output waveform and the bypass input voltage waveform is shown in Fig. 5.

Through the simulation experiment of the output voltage waveform, drag after phase locking technique from the waveform of the output circuit and output can bypass the waveform of the input AC voltage is the Sine wave and the waveform is relatively close, so that when the switch can be achieved without switching the gap, to ensure the reliability of UPS power supply and stability [25].

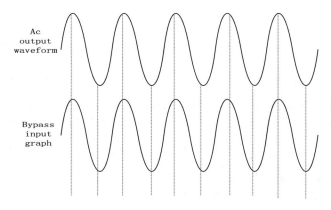

Fig. 5. Output diagram after phase-locked control.

5   Conclusion

With the wide application of UPS equipment, people put forward higher requirements on its operation characteristics, and they want UPS to achieve more accurate requirements on the switching between the main circuit and bypass. The phase-locked control method used in this paper can achieve a seamless switch between the two circuits. Through the real-time monitoring of the bypass input and AC output voltage, and then to adjust the device UPS internal inverter AC output, to improve the synchronicity between the bypass and output, so as to achieve the bypass input seamless switching, there is no uninterrupted process, thus providing a stable, continuous, uninterrupted power.

For UPS, the error between the output of the UPS module and the bypass input is compared, and then the output is made consistent by fine-tuning or eliminating the difference. In the actual operation process, the bypass input and AC output will often switch, which requires higher accuracy to ensure the stable and reliable power supply of UPS, so the phase lock control is of great significance to it. The significance of phase lock control is to make the AC output and the bypass input voltage reach the same state of frequency, phase, and amplitude by adjusting the output voltage of the system inverter. To improve the synchronicity between the bypass and the output, so as to realize the seamless switch of the bypass input and ensure the stable and reliable power supply.

Acknowledge. This paper is supported by "State grid science and technology project (2019YF-29)".

References

1. Xuejun, M., Zhigang, Z., Wei, L.: Parallel control technology of master-slave dual-module single-phase inverter based on DSP. Electric Power Autom. Equip. **29**(02), 110–112 (2009)
2. Xiaoqian, L., Wenhua, L., Zhengxi, X., Zhaohui, S., Chao, H.: Research on parallel control technology of ship power station based on multi-machine static redundancy system. Ship Sci. Technol. **31**(12), 72–75+98 (2009)

3. Weijian, W., Yunfei, L., Xiaomin, W.: Discussion on parallel control technology of UPS inverter based on no interconnection line. Sci-tech Wind (04), 138–139 (2009)
4. Yuzhen, R., Fuwen, Y.: UPS parallel operation control technology. Fujian Comput. (09), 37–39 (2003)
5. Sheng, C., Ying, S.: Research on parallel control technology of inverter. Electric Technol. (01), 24–27 (2008)
6. Hongtao, S., Yong, K., Shanxu, D., Yu, Z., Mei, Y., Yongqiao, L., Guoying, C.: New digital current sharing control technology based on modular UPS parallel connection. High Voltage Technol. (02), 289–292 (2008)
7. Peng, L., Hongjie, L.: Discussion on UPS and its parallel connection technology. Northeast Electric Power Technol. (02), 36–39 (2008)
8. Hongtao, S., Yong, K., Shanxu, D., Mi, Y., Yongqiao, L., Guoying, C.: Research on new modular digital current sharing control technology for parallel UPS. Electric Drive (05), 39–42 (2008)
9. Shucheng, D.: Application case of UPS parallel redundancy technology. Autom. Appl. (03), 3–6 (2011)
10. Taiqiang, C., Jianping, X., Shungang, X.: Parallel control technology of single-phase inverter power supply based on instantaneous reactive power theory. Electric Power Autom. Equip. **31**(05), 80–83 (2011)
11. Yingping, M.: Research on key problems of UPS parallel system. Ju she **15**, 189–190 (2018)
12. Jiangtao, L., Jiaxin, L., Xisen, Q., Yingchao, Z., Zhengming, Z.: Review on parallel control technology of UPS inverter. J. Power Supply (05), 21–27 (2013)
13. Shanxu, D., Yong, K., Jian, C.: Analysis of parallel control strategy of UPS modular power supply system. J. Electr. Technol. (01), 43–46+50 (2004)
14. Shanxu, D., Bangyin, L., Guyu, M., Jian, C.: Research on a parallel control technology for UPS without interconnection. New Technol. Electr. Energy (02), 1–4+59 (2004)
15. Lan, X., Lei, H.: Research on a simplified parallel control technology of inverter. J. Electr. Technol. (08), 19–24 (2006)
16. Bhanu Prakash, D., Divya, M., Debnath, B.: Early stage detection of cardiomegaly: an extensive review. Int. J. Adv. Sci. Technol. **125**, 13–24 (2019)
17. Woosik, L., Sangil, C., Teuk Seob, S., Namgi, K., Jong-Hoon, Y.: OR-based block combination for asynchronous asymmetric neighbor discovery protocol. Int. J. Control Autom. (03), 45–52 (2015)
18. Jinho, P., Doyoung, Y., Woo-Hyoung, L.: Architectural factors detection from plan by deep learning framework. Int. J. Grid Distrib. Comput. (01), 57–64 (2018)
19. Fiza Abdul, R., Gopinath, M., Zuraini, I., Ganthan, N.S.: A qualitative analysis of information privacy concerns among healthcare employees. Int. J. Secur. Appl. (01), 69–78 (2018)
20. Fahad, S., Asad, A., Zulfiqar Ali, M., Abdul, A., Abdul, R.: The future of internet: IPv6 fulfilling the routing needs in Internet of Things. Int. J. Future Gener. Commun. Network. (01), 13–22 (2018)
21. Gaiyun, Z., Guoping, Z., Cunhong, C., Li, M.: Research on moving target tracking based on Kalman filter in volleyball video. Int. J. Sign. Process. Image Process. Pattern Recogn. (01), 99–108 (2017)
22. Akabogu, J.U., Ibe, K.C., Omeje, J.C., Ede, M.O., Ezurike, A.C., Ezeh, E.N., Nnamani, A. P., Ulo-bethels, A.C.: Social networking and academic adjustment of English language students. Int. J. u- e-Service Sci. Technol. (01), 55–64 (2018)
23. George, B., Nikolaos, P., Vassilios, P., Athanasios, G.: Effects on road safety and functionality of installing countdown timers to traffic lights. Int. J. Transp. (01), 59–72 (2017)

24. Tomas, U., Ganiron, J.: Flood control and drainage system of Espana Boulevard in Metro Manila. Int. J. Disaster Recovery Bus. Continuity (06), 17–28 (2015)
25. Lovepreet, K., Jyoteesh, M.: Comparison of wise route and flooding network type of converge cast routing in wireless sensor network. Int. J. Sens. Appl. Control Syst. (02), 1–10 (2015)

The Influence of Smart Grid on Electric Power Automation

Yilin Liu[✉]

State Grid Liaoyang Electric Power Supply Company, Liaoyang, China
13224240163@163.com

Abstract. With the rapid development of economic construction, we need more power than ever before. In order to improve the quality and stability of the power supply, the smart grid has broad application prospects. In the construction of the smart grid, we improve the efficiency of energy utilization and management system operation and effectively promote the development of power grid system construction. The emergence of smart grid makes our lives safer, more convenient, more economical and more practical than before. Many advantages of the smart grid make its future development prospects become extremely broad. This paper studies the development of smart grid system and its development mechanism analysis, communication platform needs to cooperate with the smart grid business for unified planning; power communication platform is an open network architecture, is a general communication standard. The information between equipment and equipment is interoperable. Electric power communication network can not only extend to the terminal of relevant power generation, transmission, and transformation, terminal equipment, but also provide an effective communication network to support smart grid data acquisition, protection and control services.

Keywords: Smart grid · Electric power automation

1   Introduction

The smart grid is the development direction of the power grid in the 21st century, its significance is the ability to inspire and promote the user to take an active part in power system operation and management, meet the customers' requirements to provide efficient operation of the optimization of the power and safeguard assets. The most striking aspect of the smart grid is to support the huge amount of clean energy access. The access of distributed energy brings substantial challenges to the traditional network planning; power grid planning must fully consider the influence of distributed generation on the power grid. At present, the study of load resources optimization coordination and control in the initial stage, and the proposed algorithms are mostly based on off-line optimization or static optimization (such as linear programming, etc.), so looking for a new kind of optimal control method to solving the smart grid dynamic optimal control is the key to the development of the smart grid.

In order to adapt to the development needs of economic construction, the smart grid has been rapid development. In the smart grid construction, it combined with a variety

A. E. Hassanien et al. (Eds.): AISI 2019, AISC 1058, pp. 1036–1043, 2020.
https://doi.org/10.1007/978-3-030-31129-2_97

of modern factors, set computer network technology, modern communication technology, and sensor technology as a whole. In our country's power grid construction, the main direction of development is the efficient use of energy and large capacity, long-distance transmission. So this needs to the development of smart grid construction to improve the quality and efficiency of power supply and create a favorable environment for the development of the power grid in China. The smart grid construction is facing a series of challenges, new problems appear constantly. China is actively promoting the development and construction of a smart grid [1].

The emergence and development of a smart grid have a certain social necessity, but there is also some development chance. In addition, only to realize the optimization of the power system on the basic model can effectively promote the external driving power for the development of vision effect, achieve the comprehensive development of electric power system, realize the ideal operating condition. In addition, to focus on the market competitiveness of the power system, better accord with the requirement of the planned economy and market [2].

2   The Development of Smart Grid System

2.1   The Expansion of the Current Social Demand for Electricity

The growing demand for electricity is the main reason for the development of the smart grid in the current society. Along with the advance of social economy and culture and the growing of the national economy, in the production and living for the use of traditional energy and demand is increasing, the traditional energy consumption is increasing with the passage of time. Under the new situation, the development of the society in various fields has a strong demand for new energy. In the era of information network under the new situation, for the secondary energy, especially electrical energy demand continues to increase, in the field of social development is inseparable from the electricity. According to statistics, in 2030 the power consumption of the whole society will reach 8.9 trillion KW • h, generator capacity will reach 1.7 billion KW, and in terms of current status of the development of power industry, power industry to develop and expand the scale of the corresponding alone, these still cannot meet the social demand for electricity, so smart grid may develop effectively [3].

2.2   The Ecological Environment Pressure

The ecological environment pressure and social environmental protection is the important reason for the development of smart grid mechanism effectively. Because of being restricted by economic level and science and technology, people in production and living are generally used to traditional energy sources. With the progress and development of social economy and culture, people continuously strengthen the use of traditional energy sources, so the traditional energy is consumed at the same time, bring some pressure to the social environment. Traditional energy sources such as oil, coal in use process, will be different levels of carbon dioxide emissions, thus social-ecological environment is threatened, the global warming phenomenon is widespread and severe, the deterioration of the environment causing social environment carrying capacity is declining, not only has serious influence on social ecology, also has a potential threat to people's health. Under the new situation, it is very important to strengthen the study of electrical energy; electric power industry is one of the main ways to emit greenhouse gases, so in the ecological environment pressure the smart grid gets common development.

$$\eta = \frac{P_0(1 - P_f)}{P_0(1 - P_f) + P_1(1 - P_d)} \tag{1}$$

$$M_{stat} = R(T_p - \tau)P_0(1 - P_f(\tau)) \tag{2}$$

The smart grid will have enough flexibility and sensitivity that can effectively cope with the increasingly aging power grid infrastructure and increasing demand for electricity and negative constraint factor in the development of the power grid. For example, fossil fuel supply ability gradually reduces the rising, costs of fuel gradually rise and climate change and other environmental factors. At the same time, the smart grid may adjust measures to local conditions especially in the distribution network. The key of the implementation in the United States is to make the grid more reliable and prevent blackouts. In Malta in the islands and self-sufficient cities and Singapore, they will invest more used to improve the intelligent distribution network, to better cope with renewable energy sources and the development of electric vehicles. In high technology concentrated areas, such as Silicon Valley in the United States, the priority will be given the power supply quality, because the voltage and short supply interruption may cause serious damage of the product and the user's big losses. In China, the vast territory has complicated environment and demand. Its adjust measures to local conditions is more important.

$$P_{n,k}^* = \left[\frac{\lambda_n}{\ln 2(\eta + 1)} - \frac{1}{g_{n,k}^{D,L}}\right]^+ \tag{3}$$

$$\tilde{r}_s = \sum_{k=1}^{k} x_{s,k} \log_2\left[1 + u_{s,k}p_{s,k}/(v_{s,k} + w_{s,k}p_{s,k})\right] \tag{4}$$

3   The Smart Grid Development Mechanism Analysis

3.1   The Composition Model of Based Power System

For basic power system, the part mainly includes the electricity generation and transmission system, substation, power distribution system, basic electric power market, and the overall participation main body and scheduling structure, it realizes the organic energy of electrical energy transformation and will deliver power to the corresponding system. In the process of electric power system transport, the basic information and communication control can form the effective state of perception and data transmission, promote the overall data information effectively in accordance with the direction and timing to produce the necessary linkage. When the electric power development gradually strengthened, and the people will improve power system operation mode of the basic improvement. Based on the same power system model, through the base of power flow and the overall contact the people ensure the establishment of the organizational form and form the effective operation of the power system. According to the basic application requirements and conditions, the people establish an effective energy distribution and market organization. In addition, the electric power system has dynamic characteristics, the relevant management personnel according to the corresponding changes in the change of the actual model [4].

3.2   The Smart Grid External Drive Structure

The external drive structure of China's smart grid mainly includes four aspects. Firstly, it provides effective and high-quality power for the economic society, meanwhile, also providing the corresponding service items so as to promote the development of smart grid mechanism and the harmonization of society. Secondly, we can optimize and control the cost of the power system by the use of the basic external drive and saving energy.

Based on the development of electric power system, the external drive of smart grid is the most critical part, providing the society's main demand. The different countries also have different demands for the external drive system, for example, the reliable power supply and the high-quality power supply, new energy and cost constraints and an open interface, etc. In addition, due to the influence of different environment, the basis factors of economic and political will become the basis factors of the external driving [5]. Figure 1 shows the proposed model diagram.

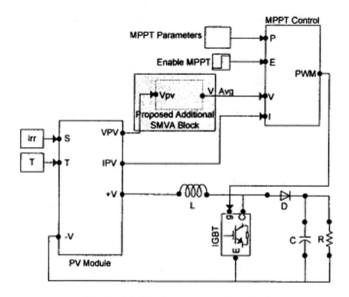

Fig. 1. Model of the proposed method

3.3   The Development Mechanism of the Smart Grid as a Whole

In the development of the smart grid, the basic contradiction of the development of power system lies in the contradictions between the basal technology and economy, but the constraints of the environment and energy more and more concentrated in the operation of the basic power system with the continuous development of environmental behavior. The whole development of mechanism related to the basic vision produced by the external drive, under the impetus of the vision, the enterprise need to make the necessary economic investment and policy support, making the central plan on the corresponding vision, and in accordance with the relevant plans, the enterprise collocate the corresponding technology and equipment, forming the optimization of management structure and standards. In addition, because the soft factors such as basic electricity substation and power transformation and distribution will affect the basic model of the power system, the related design personnel will optimize the power system management and the operation.

4   The Influence of Smart Grids Development on the Automation Technology of Power Grid

4.1   To Provide a Stable Environment for the Development of Automation Technology

In the development process of the smart grid, it is able to work for the development of automation technology to provide a relatively stable environment; it is the main influence for power grid automation technology development. In the process of smart grid mechanism development, through the corresponding intelligent electrical equipment, involved in the management of electricity market, thus effectively reduce the electric power market demand for electric power grid automation technology unit, and through price signals, strong power grid automation technology was applied to the power non-blocking area, effectively reduce the power grid in the process of the output produced by the electricity congestion cost. Under the condition of the electricity market, the smart grid with its powerful range of circle, through the corresponding signal to guide the effective operation of power grid corresponding equipment, thus the power grid automation technology in the electric power market development, to provide a relatively stable working environment.

4.2   To Promote the Development and the Operation of Automation Technology in Power Grid

A smart grid can through its unique advantages, and constantly promotes the development of power grid automation technology, and to a certain extent, promoting its running efficiency is the smart grid development mechanism of important influence on power grid automation technology. Smart grid development environment of electricity market, in the process of electric power development and major energy management as the main basis, due to the electric power market in the process of power transmission, are relatively complex, so the management of the smart grid for electricity market and the energy management system, to ask for a better and more accurate. At the same time, the energy management system in a certain extent, also can affect the operation of power grid security will smart grid based on the high requirements of the electric power market and power management system, can effectively promote the development of power grid automation technology in the electric power market, and by the high standards for improving the efficiency of its operation [6–8].

4.3 To Promote the Development and the Operation of Automation Technology in Power Grid

The smart grid development mechanism in the process of constantly updated, can strengthen the power grid automation technology integration of various functions, is the smart grid development mechanism is the key influence on automation technology. Constantly update and development of the smart grid system, can effectively promote the integration of various functions power grid automation technology, is mainly due to the smart grid in the power market development, through the use of electricity power management system, to the power distribution management system, etc., to apply for the corresponding grid system has the ability of overall monitoring, through electric power in the process of transmitting data, etc., to understand the current situation of the power transmission, various management system in the application of the smart grid, can effectively promote the power grid automation technology, the effective integration between the various functions of power grid automation system, through to the electrical energy transport in the power market situation, carries on the full analysis and discussion, which can effectively improve the efficiency of the power grid operation, for the smart grid, automation technologies provide a powerful basis for the stable operation of the various functions [9–11].

5 Electric Power Communication in the Application of the Smart Grid

Through high-speed communication networks, the smart grid realizes on-line condition monitoring for running equipment. This way may get the equipment running status, give need maintenance equipment signal in the most appropriate time, realize the equipment maintenance, and promise the equipment running in the best state at the same time. This information will provide better tools for designers, to create the best design, to provide the required data for planning personnel, so as to improve the ability and level of the power grid planning. In this way, operation and maintenance costs and power grid construction investment will get more effective management [12–14].

5.1 The New Energy Field

Because traditional energy sources such as oil and coal are non-renewable resources and can cause serious damage to the environment, they have been gradually replaced by new energy sources such as wind power and solar energy. Especially after the access to success in new energy, the smart grid may automatically regulate according to the smart grid to the quality of the access of electricity power, automatic regulation. Second, about the research of new energy power generation process, strengthen the electric power communication technology of new energy power generation and control of the process, implement start-up, and shut down, intelligent control of voltage regulation. Today's rapid development of new energy [15–17] smart grid will try new energy field, it must strengthen the electric power communication technology application in the only power grid, smart grid, the management of new energy.

5.2   The Transmission Field

With the sustainable development of the power industry, the people put a forward higher request to power distance, low loss transmission, the electric power communication technology applied in the smart grid solve these problems. For example, in the transmission process of detection, it must take reasonable power communication technology to achieve comprehensive and intelligent monitoring period, strengthen the basic information transmission process monitoring, fault warning information, power communication technology is widely used in the field of transmission in the smart grid [18–20].

5.3   Internal Construction of the Power Grid

The electricity load of power grid gradually increased because of the increase of the electricity demand, many regions in our country appeared the short circuit in the large area, causing the great influence on the stability and security of smart power grids 'operation. This kind of phenomenon needs to have a comprehensive promotion and can't be done by part adjustment [21, 22].

6   Conclusions

Smart grid construction is the inevitable trend of economic development in our country, under the background of rapid development in science and technology, computer network technology, communication technology, etc. so there is certain difficulty in management, is not conducive to the stability of power grid operation and safety. In the smart grid construction, through the application of modern equipment and technology, can improve the efficiency and quality of power supply. Although at present there exist some problems, restricting the development of smart grid, but with constant practice and the development of science and technology, the construction of the smart grid will continue to improve, in terms of quality of power supply efficiency and improve, create favorable conditions for the development of Chinese electric power industry.

References

1. He, Y.F.: The key problem of networked control in power systems. Grid Technol. **8**, 74–78 (2013)
2. Cao, L.: The application practices of east China power grid WAMAP system. Autom. Electr. Power Syst. **12**, 20–28 (2012)
3. Tong, X.Y., Ye, S.Y.: Data mining application in power system transient stability assessment were reviewed. Grid Technol. **18**, 27–34 (2015)
4. Zheng, Z.: Consider the distributed environment factors multi-objective optimal allocation of power. Chin. CSEE **26**, 29–33 (2014)
5. Yu, F.R.: Communication system for power grid integration of renewable energy. IEEE Netw. **24**, 13–20 (2012)
6. Zhou, L.S.: The list control scheme in the application of power system stability control. Grid Technol. **12**, 17–27 (2009)
7. Yu, W., Hu, Z.J.: Research review of distributed generation and its application in power system. Grid Technol. **6**, 27–31 (2015)
8. Xin, Y.Y., Luan, W.P.: Smart grid. Power Syst. Clean. Energy **14**, 25–27 (2015)
9. Barton, J.P., Infield, D.G.: Energy storage and use of the in - termite renewable energy. IEEE Energy convers. **44**, 12–19 (2009)
10. Khati, M., Bisht, V.S., Kumar, A.: Thermal & exergetic enhancement of a roughened solar air heater using matlab. Int. J. Energy Technol. Manage. **1**, 1–10 (2017)
11. Lee, S.Y., Ahn, E.Y.: Redundant channel allocation using propagation interference analysis. World J. Wirel. Devices Eng. **1**, 19–26 (2017)
12. Huh, Y.J.: Increase the performance of k-means clustering algorithm using talk aloud protocol with geneplore model on concept generation. World J. Game Sci. Eng. **1**, 7–12 (2017)
13. Kumar, K., Kiran, S.R., Prasad, A.: Analysis of solar photovoltaic source fed BLDC motor drive with double boost converter for water pumping application in irrigation system. Int. J. Adv. Sci. Technol. **120**, 73–84 (2018)
14. Essam, H., Mostafa, A., Hassanien, A.E.: A hybrid lion optimization algorithm with support vector machine for liver classification. Int. J. Adv. Sci. Technol. **123**, 1–10 (2019)
15. Bai, D.Y., Wang, C.Y., Zou, J.: Design and simulation of fractional order control systems based on bode's ideal transfer function. Int. J. Control Autom. **8**, 1–8 (2015)
16. Rao, N.T., Bhaskar, P.U., Srinivas, P., Bhattacharyya, D., Kim, H.: An efficient technique to design elasticity and reliable content based publish/subscribe system in cloud. Int. J. Grid Distrib. Comput. **10**, 21–32 (2017)
17. Wang, C.K., Zhang, Z.B.: Research on Iris localization algorithm based on the active contour model. Int. J. Secur. Appl. **9**, 131–138 (2015)
18. Sun, W., Huo, C.H., Yu, Y.: Management of basic scientific research achievements based on knowledge supernetwork. Int. J. Future Gener. Commun. Netw. **9**, 27–38 (2016)
19. Singh, M.: Analyzing the effect of channel spacing and chromatic dispersion coefficient on FWM in optical WDM system. Int. J. Signal Process. Image Process. Pattern Recogn. **8**, 99–110 (2015)
20. Xie, Y.F., Wang, R.X., Li, C.: Design and implementation of the auxiliary teaching platform in colleges and universities under the hierarchical teaching based on web. Int. J. U - E Serv. Sci. Technol. **10**, 37–46 (2017)
21. Kelebeng, K., Hlomani, H.: Capital markets prediction: multi-faceted sentiment analysis using supervised machine learning. Int. J. Database Theory Appl. **10**, 87–102 (2017)
22. Abbas, M., Machiani, S.G.: Agent-based modeling and simulation of connected corridors—merits evaluation and future steps. Int. J. Transp. **4**, 71–84 (2016)

Micro Power Grid Operation Control and Its Shortage

ChunXu Ding[✉]

State Grid Shenyang Electric Power Supply Company, Shenyang 110811, China
sygcl112@163.com

Abstract. With the wide application of distributed generation system and the request of users to power supply quality and reliability of power supply, a new grid structure appeared in the society. Micro grid has solved the problem of large-scale access of distributed power supply, giving full play to all kinds of advantages of distributed power supply, meanwhile it also has brought a variety of other benefits to the user. Micro network will radically change the traditional way to deal with load growth and has great potential to reduce the energy consumption, improve the power system reliability and flexibility. At the same time, the grid is an important network model of the intelligent power distribution system in the future because its operation control system is the key to ensure the safe, economic and reliable operation of power grid. This paper mainly studies the present research situation of the technique of micro power grid operation control and its shortage, laying the foundation for other researchers.

Keywords: Micro grid · Micro grid operation control · Intelligent power distribution system

1 Introduction

Distributed generation is the important way to improve energy efficiency and change the energy structure. But the access of distributed generation on power grid operation, control, protection has brought the profound influence to solve the problem of distributed generation access, make full use of the advantages of distributed generation, the century beginners micro power grid operation and management mode are put forward. It is using the advanced power electronic technology, in a local area directly to the micro power supply, load, energy storage device, control unit and the end user together, form a "single control unit", to optimize and improve the efficiency of energy utilization, promote the society to develop in the direction of the green, environmental protection, energy saving. Is a micro grid can realize self-control, protection and management of autonomous system, it can not only with power grid parallel operation, can also be used in large power grid with large power grid failure or need to disconnect the isolated net running, fully meet user requirements for power quality, power supply reliability and security [1] (Fig. 1).

Through the research results at home and abroad were reviewed and summarized, found in a variety of types of micro power grid power exist at the same time, is rich of control object and the operation mode and varied and complex control strategy, and the

A. E. Hassanien et al. (Eds.): AISI 2019, AISC 1058, pp. 1044–1050, 2020.
https://doi.org/10.1007/978-3-030-31129-2_98

Fig. 1. Distributed generation

user demand for power supply reliability and quality of power supply is also diversified. Micro power grid operation control system is particularly important, therefore, only if the operation control system under the management of micro power grid to safe, reliable, economic and efficient operation. But as a result of micro grid control target has bigger difference with large power grid, power grid operation control experience is very difficult to copy directly to the micro power grid. Existing micro power grid operation control systems, components and system coupling, many function modules together, when the micro grid topology and power configuration changes, it is difficult to transplant operation control system, which limits the development of micro power grid.

2 Micro Power Grid Operation Control Related Technical Research Present Situation and the Deficiencies

2.1 Micro Power Control Strategy

When the power supply without used to control the voltage or frequency of micro power grid, mainly uses the following control method for network tracking control is typically used for power control of voltage and frequency of qualified micro power supply [2]. At this point, if the power supply power output control and other micro power source or load control strategies are independent of each other, between the micro power control strategy is called a non-interactive control, such as using maximum power tracking control of photovoltaic or wind power; When need instructions for scheduling, micro power control strategy is called interactive control network with control is generally used in micro grid parallel operation situation, micro power source adopts the active, reactive power or active - voltage control strategy (Fig. 2).

When the need for micro power grid voltage or frequency control, need micro power control was formed by micro power grid. If the voltage and frequency of micro power source controlled by a single complete, then the micro power run interactive control mode, its function is equal to the ideal voltage source, at this point, the

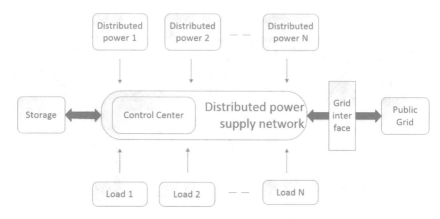

Fig. 2. Distributed power supply network

requirements for voltage, frequency stability control task of micro power supply that is active, reactive enough capacity in order to realize power balance. If borne by two or more micro power system running deviation after the change of state power, you need to through interactive control strategy of the micro power supply the size of the target to adjust power, such as through the vertical control to coordinate various micro power output, and then adjust the voltage and frequency of the whole system. Micro power grid in the operating point of micro power supply can be dynamic vertical slope, set up under the change of the voltage and frequency values. Micro grid formation control is generally used in micro grid isolated net running situation, micro power source controlled by voltage – frequency or VF droop control.

At present, the power electronic converter interface of VF control strategy, especially comparing its control characteristic and the difference between synchronous generator research is also imperfect.

2.2   Micro Power Grid Control System

In view of the characteristics of micro power grid, the grid control shall meet the following requirements: Ensure the quality of micro grid power supply, especially for the requirements of voltage and frequency; Micro power to "plug and play", and access does not affect the quality of power supply; Using local information of micro power supply to control rather than global information; Correctly flexible micro power grid interconnection and the isolated net, and can run normally under two kinds of operation modes; Has the ability of correction voltage drop and unbalanced system; Can the decoupling control of active and reactive power (Fig. 3).

For micro grid internal coordination control of various micro power supply, at present the proposed micro grid control method can be divided into the following three categories:

Decentralized control. Decentralized control without communication between micro power supply, each with a local information control plug and play and peer-to-peer control belong to a decentralized control [3]. Using distributed control, the

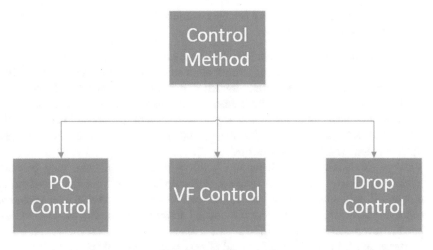

Fig. 3. Control method

commonly used method is analogy to the synchronous generator, the power electronics mouth micro power according to the micro grid under the control of target selection of vertical control, use of the system under the vertical imbalance of dynamically allocated to each power supply, ensure the power balance between supply and demand in the solitary off-line micro grid and frequency is unified, has the characteristics of simple, reliable. But at present the study of decentralized control is based on pure micro grid power electronic interface, without considering interface with traditional synchronous generator and induction generator interface coordination control.

Centralized control. Centralized control general micro grid can be divided into components, micro power grid control layer and layer to control the distribution network. Distribution network layer for coordination between micro power grid operation, let the micro grid to participate in the distribution network voltage and frequency regulation, and is responsible for the distribution network and the exchange of information between micro grid controller. The control system will be micro power grid operation and control of all sorts of problems are concentrated in the micro grid layer, not from the perspective of micro grid subdivided from the function, reduces the economy and efficiency of the micro grid.

2.3 Power Quality Analysis

Micro power grid of small geographic range, tiny disturbance of power quality quickly spread in the micro grid, and even turned into stability problems. Therefore, improving power quality is auxiliary micro power grid operation stability, important technical method to adjust the frequency, compensation of load fluctuation [4]. According to the European energy regulatory commission, the definition of power supply quality is mainly composed of three parts, respectively is the commercial quality, continuity and the voltage quality of power supply. The voltage quality and power supply continuity is called the power quality. Power supply continuity usually use occurred within a specific time without power, power outage duration, the cumulative number of indicators to measure the blackout time [5].

- **Harmonic problems:** In the micro grid harmonic originate from various power electronic interface equipment [6], the current study mostly focused on the analysis of the harmonic power electronic switch and inhibition, and through the power electronic controller on the suppression of harmonic produced by nonlinear load. Less research analysis of micro power grid operation of the isolated net, controlled by inverter as a micro power grid voltage compensation device when the source and nonlinear load impact on power quality [7].
- **Voltage fluctuation and flicker:** In a micro grid system, there may be some intermittent power supply, its frequent start-stop operation, the change of power output [8], will give users access to the system by the voltage flicker problems, especially in the micro grid under the isolated net running state [9], the problem will become even more prominent [10].
- **Voltage sag.** Grid voltage sag is a kind of very common problem; the power quality problem will be serious influence on the operation of the micro grid [11].

2.4   Small Signal Stability Analysis

Due to the location of the micro power supply with micro grid dispersion, different micro power to the rated power of the ordinary, micro power grid under different power management strategy for the analysis of small signal stability [12], generally need to build the whole micro power grid of small signal stability analysis to touch type: to evaluate the stability of the micro grid; use the power of small signal stability analysis and design of the controller control parameters and optimizing; different power management strategy for micro power grid [12], especially the influence of the isolated net running micro power grid dynamic performance; assessment of new energy and traditional energy with power electronic interface properties and dynamic performance of micro grid [13], as well as the micro power source and the interaction of the network [14].

Small signal stability analysis of micro grid model not only need to consider the traditional synchronous machine model, also need to take into consideration of the containing the power of the power electronic interface, and the impact of the effect of the micro grid control line impedance [15]. At home and abroad about the micro grid of small signal stability analysis of the research object is generally of synchronous generator and VSC network or a network composed of VSC alone [16], without considering interaction with asynchronous generator interface, and adopt the whole order model in the research, makes the system order is higher, analysis is difficult [17].

3   Results and Analysis

Through the research results at home and abroad were reviewed and summarized, found in a variety of types of micro power grid power exist at the same time [18] is rich of control object and the operation mode and varied and complex control strategy, and the user demand for power supply reliability and quality of power supply is also diversified [19]. Micro power grid operation control system is particularly important, therefore, only if the operation control system under the management of micro power grid to safe, reliable, economic and efficient operation [20]. But as a result of micro grid control target has bigger difference with large power grid, power grid operation control experience is very difficult to copy directly to the micro power grid [21]. Existing micro power grid operation control systems, components and system coupling, many function modules together, when the micro grid topology and power configuration changes, it is difficult to transplant operation control system, which limits the development of micro power grid.

References

1. Yang, Z.S.: May realize the flexible operation mode switch small micro network experiment system. Autom. Electr. Power Syst. **14**, 89–98 (2009)
2. Pei, W., Li, P.S.: The key technology of micro network operation control and its test platform. Autom. Electr. Power Syst. **1**, 94–98 (2010)
3. Liang, Y.W., Hu, Z.J.: Distributed generation and its application in power system research were reviewed. The Grid Technol. **12**, 71–75 (2007)
4. Lu, Z.X., Wang, C.X.: The micro grid research were reviewed. Autom. Electr. Power Syst. **19**, 25–31 (2007)
5. Zhang, S.L., Zhang, J.H.: Micro network parallel operation mode. J. Taiyuan Univ. Technol. **2**, 184–187 (2011)
6. Ding, M., Zhang, Y.Y.: The key technology in the study of the piconets. Grid Technol. **11**, 6–11 (2010)
7. Zhao, H.W.: Micro network technology based on distributed power supply. Electr. Power Syst. Autom. **20**, 26–31 (2011)
8. Fang, S., Chen, S., Center, G.C.: Analysis and suggestions of the present situation of the technical resources sharing laws. Sci. Technol. Manage. Res. (2015)
9. Yang, Y., Li, R., Feng, L.: Wheat varieties identification research based on sparse representation method of dictionary learning. Int. J. Hybrid Inform. Technol. **11**(2), 13–24 (2018)
10. Park, K.J.: smart city risk management model of sustainability perspective. Int. J. Smart Home **11**(11), 1–6 (2017)
11. Lee, I.F., Chaeho, C., Won, Y.: A text-learning based method of detecting personal information. Int. J. Reliab. Inform. Assur. **4**(2), 7–12 (2016)
12. Eom, J.: Reliability logic model for incomplete information in sensor monitoring databases. Int. J. Grid Distrib. Comput. **11**(1), 27–36 (2018)
13. Bassel, A., Nordin, M.J., Abdulkareem, M.: An invisible image watermarking based on modified particle swarm optimization (PSO) algorithm. Int. J. Secur. Appl. **12**(2), 1–8 (2018)
14. Lakshmibai, T., Parthasarathy, C.: Self B - adaptive key generation for primary users in cognitive radio networks for less prone primary user emulation attacks. Int. J. Future Gener. Commun. Netw. **11**(1), 1–12 (2018)
15. Ji, X., Qin, N., Wang, Y.: The target classification of optical remote sensing image based on hierarchical features and adaboost algorithm. Int. J. Signal Process. Image Process. Pattern Recogn. **10**(1), 17–26 (2017)

16. Mirzadeh, M., Ahrami, G.H., Haghighi, M., Darveshi, A., Khezri, S.: Intelligent model-reference method to control of industrial robot arm. Int. J. U - E Serv. Sci. Technol. **8**(2), 71–80 (2015)

17. Wu, D.: An empirical study of BB network teaching platform in college english teaching. Int. J. U – E Serv. Sci. Technol. **10**(1), 27–36 (2017)

18. Na, X.: Open network knowledge and independent inquiry teaching mode of college chinese course based on information retrieval. Int. Database Theory Appl. **9**(11), 95–106 (2016)

19. Alexandros, K., Panagiotis, P., Sokratis, B., Ioannis, P.: An intertemporal evaluation of highway intersections red-light running. Int. J. Transp. **5**(1), 47–58 (2017)

20. Rashid, E.: Disease detection on the basis of multiple symptoms by expert system. Int. J. Disaster Recov. Bus. Contin. **8**, 1–10 (2017)

21. Gehlot, A., Singh, R., Mishra, R.G., Kumar, A., Choudhury, S.: IoT and Zigbee based street light monitoring system with LabVIEW. Int. J. Sens. Appl. Control Syst. **4**(2), 1–8 (2016)

Design and Implementation of Micro-grid System for Station with Hybrid Photovoltaic and Wind

Peng Jin[✉]

Department of Technology and Internet,
State Grid Liaoning Electric Power Supply Co., Ltd., Shenyang 110006, China
ml8265537369@163.com

Abstract. Microgrid control is a key technology for microgrid access to the conventional grid. This paper analyzes and summarizes the control strategy of the power station. This paper analyzes and summarizes the control strategies of photovoltaic power generation and wind power generation micro-grid. Firstly, the structure and function of the power station Photovoltaic and wind power micro-grid system are introduced and demonstrated. Second, the functions and effects of the battery are accurately applied to the system. Finally, the self-healing operation of a microgrid is proposed to ensure the normal and stable operation of the system. A microgrid is the backup power source of the power station by self-repairing and using energy storage technology. This scheme can adopt different control strategies according to the operation mode of the microgrid. The program can extend the battery life of substation. Based on the principle of the most efficient utilization of renewable energy, this paper studies the development and utilization of renewable energy optimization under different working modes. On the basis of ensuring the load power supply-demand and meeting the battery power consumption state, the optimal energy scheduling strategy under different operating modes is proposed, and the corresponding optimization model with the minimum operating cost as the objective is established.

Keywords: Micro-grid control · Self healing operation mode of micro-grid · Maximization of energy utilization

1 Introduction

In grid connected station micro-grid, the photovoltaic power generation system use the inverter output to achieve powering substation with load, through substations of transformer low voltage terminal 380 V and distribution cabinets in parallel [1]. The excess energy will supply other loads by station grid. Energy storage systems smooth PV power fluctuations by automatically adjusting the charge and discharge mode and output power control, and other functions suppress voltage fluctuation and flicker, and compensate load current harmonics. Station micro-grid control power systems is used by closed-loop control strategy to ensure the normal and stable operation of the system [2].

© Springer Nature Switzerland AG 2020
A. E. Hassanien et al. (Eds.): AISI 2019, AISC 1058, pp. 1051–1056, 2020.
https://doi.org/10.1007/978-3-030-31129-2_99

2   Operating Status of the Station with Micro-grid

There are two modes in which the power system of station microgrid operates. The first one is station micro-gird run with a conventional power grid under normal circumstances. When the grid faults detected or the power quality does not meet the requirements, station micro-gird will promptly disconnect and independently run. The second type of operation is bidirectional inverter which can achieve automatic conversion mode. When the DC voltage under normal circumstances, the power began to judge whether they have the condition of grid-connected if the conditions are not met, then the power is in the standby mode; if the conditions are met, the grid will start and the current fed into the public grid after 2 min.

During operation, the power control systems automatically determine the grid conditions without human intervention. Adopting different control strategies depend on the operating mode of the station micro-grid, such as distributed power using PQ control at networked mode or using V/f control at islanding mode [3].

Next, the design applies photovoltaic systems to traditional station grid. Using the photovoltaic effect of photovoltaic panels is to convert light energy into electric energy, and then charge the battery group to supply power to the load by converting DC to AC through the inverter.

This program of station micro-grid design has a self-healing function. When the system power is restored, it automatically adjust the inverter voltage phase of micro-grid systems for micro-grid bus voltage frequency, phase and AC bus voltage frequency making the same phase or the same permission within range, and then come into bus compiler cubicle by detecting the voltage and the frequency phase difference across the bus compiler cubicle. Micro-grid system change island operation state to grid-connected operation state [4].

Again, the reasonable micro-grid control strategy is the key to ensuring that the micro-network smoothly switches between a mode to another mode of operation. Judging from the current domestic and foreign practical micro-grid technology, the master-slave control strategy based on the micro-grid system has been gradually commercialized but based on peer-control strategy has been still in wide research [5]. In this project, the micro-grid project design will use the master-slave control strategy. The main power grid is used to the main power source in grid-connected operation, and the battery energy storage device is used to the primary power source in isolated operation. Mode control achieves bidirectional inverter and the other powers control mode under the grid-connected mode and island mode [6].

3   Automation and Control Systems of the Power System

Hybrid photovoltaic wind storage automation and control system structure is described as follows [7]. System Master level, achieve HMI, graphics display and monitoring, operation, management and use of historical data, remote and other embedded communications, the control logic configuration, and modification.

Considering the DG/MG control and distribution automation analysis are a process of gradual study and providing tools that are easy to maintain and quickly edit as well as supporting the self-logic of users and self-editing algorithm.

Distributed real-time database and efficient memory management such as indexing mechanism, object-oriented, distributed, high-capacity, high-performance, developed real-time relational database management system are main part of the development to ensure real-time, consistent, predictable lines and large throughput requirements, keeping timers, memory pool, shared memory, HASH as well as to ensure timely response and quick query [8]. Real-time database system supports networking and distributed deployment by mode and tabular. Also, it supported by a multi-node database image deployment to ensure data redundancy hot standby and load balancing access request.

Based on CIM's map module and libraries integrated modeling [9] to complete once and quadratic model's establishment and maintenance of the power system by graphically model. The graphics platform is based on Auto Desk Map and QT platform. It supports the SVG standard graphics formats export and import, high-performance, high-capacity; distributed real-time database, support one million/sec read and write access. Use acquisition mode on duty supports multiple capture modes, support for multi-source collection, calculation, and integration process. It supports a variety of communication methods and provides powerful data forwarding function. Enhanced ability to maintain uninterrupted operation of systems and on-line modified mechanism. Moreover, powerful intelligent terminal access, support record filter, power quality analyzers, two-way meter, reactive switching devices, and other data are very important when designing automation and control systems of the power system [10]. Figure 1 illustrates the schematic diagram of operating mode.

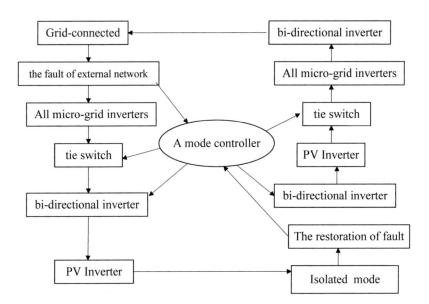

Fig. 1. Schematic diagram of operating mode

4 Control Strategy

Control strategies should also ensure that the station power systems preferentially using power distributed by generation device, and to meet the requirements of the smart battery charging and discharging. To achieve this function, it should use the following method.

4.1 The Station of Power System

The system consists of power generation unit, the inverter rectification control unit, the detection unit [11], and grid-connected unit. Power generation unit including photovoltaic panels matrix, battery, battery controller and local energy storage unit controller and system peripheral control circuitry. Inverter control unit includes an converter that be integrated by inverter and rectifier. Detection unit is one load simulation box that includes DC load simulation and AC load simulation. Figure 2 shows the scheme automation and control systems of power system.

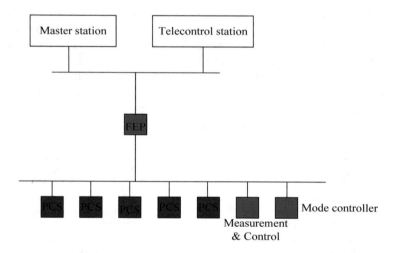

Fig. 2. Automation and control systems of power system

Overall system functions have automatic tracking function of photovoltaic panels, the control of storage battery charging and discharging, the inverter/rectifier integrated converter control, and islanding detection function [12]. Direct current of photovoltaic panels supply local DC load by the combiner box and DC load switch [13]. If the power is surplus, then the storage battery will be charged by photovoltaic panels constituted by the controller of DSP [14].

$$P_m = V_m I_m \tag{1}$$

$$\eta = \frac{p_m}{A \times P_m} \tag{2}$$

Conversely, the storage battery will be discharged by photovoltaic panels constituted by the controller of DSP [15]. The system achieve the AC load control switches controlling in order that achieve local AC load switching, through current and voltage of alternating bus sampling collected by transformers, and comparison received from the data of DSP and the data of AC load terminal voltage and current [16].

$$p = \frac{1}{2}\rho C_p \pi R^2 V^3 \tag{3}$$

$$p_{bsoc}^t = p_{bsoc}^{t-1} + P_b^t \cdot \Delta t \tag{4}$$

4.2 Energy Storage System

Energy storage systems smooth PV power fluctuations by automatically adjusting the charge and discharge mode and output power control, and other functions suppress voltage fluctuation and flicker, and compensate load current harmonics [17]. Station micro-grid control power systems by closed-loop control strategy to ensure the normal and stable operation of the system [18].

5 Conclusion

To sum up, the main innovations of this paper, described as follows. Station micro-grid with dual operating mode and the micro-grid load can be supplied power in the normal and fault operation. Automatic conversion mode is achieved. Photovoltaic power generation is a conventional power generation form with the ideal strategy for the sustainable development of green energy sources. Micro-grid is self-healing. Standby power station can be achieved by energy storage technologies. The implemented system can take different control strategies depending on the micro-grid operation mode of station. The program can extend the usage time of the substation storage battery.

References

1. Xu, S., Li, J.: Grid-connected/island operation control strategy for photovoltaic/battery micro-grid. Proc. CSEE **33**, 25–33 (2013)
2. Vidyanandan, K.V., Senroy, N.: Primary frequency regulation by deloaded wind turbines using variable droop. IEEE Trans. Power Syst. **28**, 837–846 (2013)
3. Attya, A., Hartkopf, T.: Wind turbine contribution in frequency drop mitigation: modified operation and estimating released supportive energy. IET Renew. Power Gener. **8**, 862–872 (2014)

4. Xu, X.: Research on model and operation characteristic of grid-connected photovoltaic generation. China Electric Power Research Institute (2009)
5. Ouddalov, A., Chartouni, D., Ohler, C.: Optimizing a battery energy storage system for primary frequency control. IEEE Trans. Power Syst. **22**, 1259–1266 (2007)
6. Peas, A., Moreira, C., Madureira, A.: Defining control strategies for microgrids islanded operation. IEEE Trans. Power Syst. **21**, 916–924 (2016)
7. Zhao, J., Xue, L., Yang, F.: Frequency regulation of wind/photovoltaic/diesel microgrid based on DFIG cooperative strategy with variable coefficients between virtual inertia and over speed control. Trans. China Electr. Soc. **30**, 59–68 (2015)
8. Zhang, M., Du, Z., Li, N.: Control strategies of frequency stability for islanding high-voltage microgrids. Proc. CSEE **32**, 20–26 (2012)
9. Wang, C., Xiao, Z., Wang, S.: Synthetical control and analysis of microgrid. Autom. Electr. Power Syst. **32**, 98–103 (2008)
10. Mondol, J., Yohanis, Y., Norton, B.: Comparison of measured and predicted long term connected photovoltaic system. Energy Convers. Manag. **48**, 1065–1080 (2007)
11. Zhang, W., Qiu, Q., Lai, X.: Application of energy storage technologies in power system. Power Syst. Technol. **32**, 1–9 (2008)
12. Zhang, Y., Yu, G., Shi, M.: Review of energy storage system. East China Electr. Power **4**, 91–93 (2008)
13. Liu, S., Huang, K., Liu, Y.: The development and research progress in a energy storage unit: the vanadium redox flow battery. Battery Bimon. **5**, 356–359 (2005)
14. Li, B.: Study on characteristic analysis, modeling and simulation of vanadium-redox flow battery applied in energy storage of power system. China Electric Power Research Institute, Beijing (2010)
15. Wang, Y., Xu, H.: Research of capacity limit of grid-connected photovoltaic power station on the basis of chance-constrained programming. Proc. CSEE **30**, 22–28 (2010)
16. Babu, S.S., Chimili, A.: New proof methods for improving privacy and security by using abe. Int. Cloud-Comput. Super-Comput. **3**, 1–6 (2016)
17. Lee, W., Jin, J., Lee, M.: A blockchain-based identity management service supporting robust identity recovery. Int. J. Secur. Technol. Smart Device **4**, 17–22 (2017)
18. Min, J.: Ease of comparability affects regret. The differential processing of product category and noncomparable choice alternatives affects regret. Int. J. Wearable Device **5**, 5–10 (2018)

Effect of Photovoltaic Generation on Relay Protection of Distribution Network

Bo Zhao[✉]

Department of Marketing, State Grid Liaoning Electric Power Supply Co. Ltd.,
Shenyang, China
1179677489@qq.com

Abstract. The current situation of energy shortage crisis and environmental pollution is aggravating, Renewable energy development and utilization and various green energy to achieve sustainable development has become a necessary measure. Solar energy is the cleanest, most realistic and most promising renewable energy in the world. Solar photovoltaic grid-connected generation is the main development trend of solar photovoltaic utilization. However, as the capacity of pv grid-connected system becomes larger and larger, This will have a profound impact on the power system, and a series of technical problems will change the operation of the system. This paper discusses the principle of relay protection based on traditional distribution network and the influence of photovoltaic on relay protection of distribution network. Then, the positioning method of photovoltaic power grid is expounded. The protection scheme adopted in this paper is to allow isolated island operation, which needs to consider the impact of photovoltaic power generation system on relay protection and put forward relevant protection measures.

Keywords: Photovoltaic power generation · Power distribution network · Relay protection

1 Introduction

Photovoltaic power technology is developing rapidly all over the world. The photovoltaic power system in distributed photovoltaic power grid development trend, challenges of relay protection induced the grid distribution network is more and more impact on the protection of distribution network becomes more and more serious, the problems and challenges worth re-examine photovoltaic power workers problems [1]. Photovoltaic power supply with high capacity of large-scale networks involved will affect the trend after the distribution, Change the distribution network configuration, and the current distribution network problems of relay protection are by a single power supply of the radial structure of protection based on tuning the visible. Photovoltaic grid related protection problem is a big problem in power system planning and operation and maintenance personnel needs major consider the worth of research workers in science and technology [2].

This paper discusses the principle of relay protection based on the traditional distribution network and the influence of photovoltaic on relay protection of

© Springer Nature Switzerland AG 2020
A. E. Hassanien et al. (Eds.): AISI 2019, AISC 1058, pp. 1057–1064, 2020.
https://doi.org/10.1007/978-3-030-31129-2_100

distribution network. In addition, studies the effect of photovoltaic generation on relay protection of distribution network. It is the basis of the analyzing of the photovoltaic power generation and power grid location problem [3]. For photovoltaic, the effect on distribution network protection is expounded, and analysis the influence of current protection, especially the fuse reclosers and sectionalized [4]. Finally, on the grid-connected photovoltaic power distribution network protection, the use of allowing the island to run with considering the protection of photovoltaic power generation system involved in the distribution network after the relay protection [5].

2   Traditional Distribution Network Relay Protection

2.1   Structure Characteristics of Distribution Network

In the power system, the connection mode has a very important function for the user. The system can ensure that it is safe, reliable and economical to save electricity. The connection of the power station, the main wiring of the substation, and the power grid wiring and according to the power network functions. The system can be divided into a transmission network and distribution network, the transmission network is the power station safely and economy to the loading area [6] Network transport stability and sufficient reliability change the operating system mode and operation of adaptability.

The function of economic distribution of electrical energy is to meet the transmission requirements of the distribution network where its connection mode depends on the charge character [7]. The main difference between the distribution network and the transmission network is that the distribution network is the power supply system of the power supply unilaterally, the direction of the power flow, the current is in one direction, so the relay system configuration corresponds to the power supply design, 8]. Normally, the traditional feeder protection of the distribution network is configured for the three stages of current protection including the current speed off, time limit current speed off and the timing limit over current protection [9].

2.2   Basic Requirements of Relay Protection

The relay protection setting account is calculated with some parameters including maximum operating mode, operating mode, minimum reliability indicator, branch factor, return factor, non-component factor. When requesting weeks of graduation, relay protection should meet selectivity, speed, sensitivity and reliability requirements [10].

When the relay protection device action and only the failure of the components to cut, and in order to ensure that the scope of power cuts as much as possible. Then, the system does not continue to run the fault and the selection is mainly determined by the whole calculation to consider. On the pother hand, quick troubleshooting can improve the stability of the system in parallel operation, reduce the user voltage drop of working time, minimize the extent of the damage to the fault component, and prevent the spread of the accident [11].

As a result, it should be possible to remove the fault as soon as possible. The stability of the system and equipment safety has an important influence, and important users of the action time of the protection should ensure that its speed, the necessary time can sacrifice the selectivity. In addition, the reaction ability to produce or not work normally is within the protection of the relay [12].

Regardless of the location and nature of the fault can be sensitive to the reaction, with sufficient sensitivity, generally used to describe the sensitivity coefficient. Sensitivity factor is the ratio of the fault or the setting value to the whole set value and the ratio of the whole set value and the ratio of the whole set value and the value of the whole set. Reliability is the protection of the relay protection device, the fault should be removed when the action, it should not refuse to move, and other protection should not move, it should not be mistaken. In order to ensure the reliability, the hardware and software are used in the device selection, the circuit design is as simple as possible and to reduce the auxiliary components, the installation and debugging is reliable, and the operation and maintenance management is strengthened [13]. These functions and requirements are used to protect the device through different fixed value setting to achieve, to ensure that the system operation safety and protection equipment is not affected by the injury to play a great role [14].

2.3 The Relay Protection Principle of Distribution Network

The distribution network is basically a single power supply. The circuit is usually installed with three-level current protection, which can play the role of primary protection and backup protection [15]. For distribution network, zero current protection is installed as grounding protection when the neutral point is grounded. Neutral grounding install single-phase grounding protection. Single - terminal power supply circuit mostly adopts time - limit over current protection. Normally, this has little impact on the stability of the power grid and can cut off line faults. To sum up, the distribution circuit is usually based on simple current protection or the principle of distance protection as the main protection of circuit [16]. For non-terminal lines, the feeder protection is mostly used in the three sections of the Tuen Mun flow protection and other protection. The cable and the overhead phase mixed line, but also should be configured to ensure that the circuit to ensure that the instantaneous failure of the line to quickly restore power supply. For all cable lines, the fault is permanent, so the automatic coincidence gate is not applicable to the cable line [17].

Current quick break protection is increasing in the current response to the current action of protection; in accordance with the exit of the next line to avoid the maximum short circuit current to the whole set, can quickly remove the fault. In order to ensure the selectivity of action, the whole length of the current circuit cannot be protected by quick-break protection, which is more commonly used in distribution net-work main protection [18]. The time-limited current quick-break protection is added to the circuit to protect the length of the entire circuit and to backup protection along with the current speed [19]. Different operating mode lower limit current quick break protection can protect the line length, but the disadvantage is that the action time limit is longer than the current quick break protection. Normally, the initial current of the current protection is used to escape the maximum load current. This is normal operation at startup when a fault occurs and the current flows through the protection slot causing protection action. Under normal circumstances, it can protect the circuit and the length of adjacent lines and can be used as backup protection of adjacent lines. Taking into account the protection device or circuit breaker may misoperation. The power line is also equipped with time over current protection as backup protection to protect both the line length and adjacent line length [20].

2.4   Photovoltaic Power Generation Related Protection Location

According to the high, medium and low voltage, three levels can be classified distribution network voltage level, low voltage distribution network, medium voltage distribution network and high voltage distribution network. Currently, the main developing systems focus on the feeder protection for the distribution network. Some systems operating mode change rate of the high rate of the feeder and the most important cable lines are selected using distance protection [21]. It is not suitable for the automatic switching device to ensure that the power supply should be equipped with a three-phase primary circuit when the transient fault occurs in the non-full cable line. The current into the photovoltaic power grid protection related to network location there are two situations: one is the PV system and grid-connected low-voltage bus directly, two possible situations for access:

– With a large load of the low voltage bus, local load from normal operation, and power supply system with PV.
– With the small load, low voltage bus, the normal operation of PV to the low voltage bus load and power system at the same time.

These two situations used to reduce the low voltage bus capacity and the load from the system to the photovoltaic. While the photovoltaic system and grid-connected system adapted for load power supply from high voltage bus side access to the transformer.

3   Influence on Relay Protection of Distribution Network

3.1   Photovoltaic Power Generation Related Protection Location

Single power radial is the structure of the traditional distribution network, which is simple in structure, small in investment and convenient in maintenance. The distribution network will change the original structure of the grid due to photovoltaic power generation access. The short-circuit current distribution of the grid will interact with the coordination between the relays when the photovoltaic power distribution network access capacity of large or multiple small capacities, because of the shunting effect of photovoltaic power generation branch. After the fault current flowing through the protection device may be reduced to reduce the scope of protection.

Faults of adjacent lines may occur in the distribution network where photovoltaic power generation is connected. It is due to the effect of the reverse current of photovoltaic power generation. The photovoltaic power generation line has no fault trip. Photovoltaic power generation access improves the short-circuit current level of the whole system but may reduce the short-circuit current of a single branch. The distribution and size of photovoltaic power generation capacity are obviously affected, such as the size of short-circuit current, the location in the distribution network and the access to photo-voltaic power generation may cause fusible load feeder fault-free fuse.

The three-phase short circuit fault is the biggest influence on the distribution network, so the analysis system involved in the network situation. When the safety limiting protection system is determined, then the three-phase short-circuit occurs in the system operation mode.

Photovoltaic grid related protection except for current protection. Equipped with electric distribution reclosing before and post-acceleration and acceleration of protection current, which lead to automatic reclosing, when the power supply and power system between the PV grid connected system power supply wire failure causes the protection action. Before automatic reclosing, the grid-connected power grid still in contact with the photovoltaic power. As for photovoltaic power without splitting, the photovoltaic power will continue to increase the fault current of the fault point. While, the fault point to continue because of power supply, and will cause the arc cannot be extinguished, and reclosing the fault arc step again, even cannot be extinguished. It makes a temporary fault into the permanent fault which causing huge losses.

3.2 The Influence of the Fuse, the Fuse and the Segment

Fuses are widely used in distribution systems, control system, and electrical equipment. It is a type of current protection device with the advantages of simple structure, low cost, convenient operation and so on. The fuse current of the fuse current with anti-time limit characteristic is short, and the current is small, and the fuse time is long. Often it's adopted in the scheme of feeder automation and subunit matching. It can detect fault current trip in the fixed time. One of the intelligent switching devices in distribution network automation is used for the control and protection function.

The first and second time is set to fast switching to prevent the expansion of the accident and the action program of the whole set can be quickly divided into a break action to eliminate the instantaneous fault. A time limit is provided for the back of the coincidence device, so as to cooperate with the segmented device. In the case of pressure loss or no current, the breaker switch device can be automatically divided. There are two types of voltage-time type, over current pulse count type.

4 Protection Strategy of Grid Connected Photovoltaic Power Generation System

4.1 An Overview of the Protection Scheme for Isolated Island Operation

PV system involved in the network may destroy the coordination between the original protections, resulting in normal action. For distribution network relay protection and security automatic equipment, network protection schemes involving existing. The photovoltaic power generation system has distribution network fault, take photovoltaic power grid related processing of treatment scheme, there are two aspects: when the island operation and planned islanding operation of distribution network fault of the grid. Will be from the grid, or protect the power supply from the island protection, which will protect the power grid before the photovoltaic power supply. Figure 1 illustrates the maximization of load voltage and load current, B for the combination diagram of different load mechanisms.

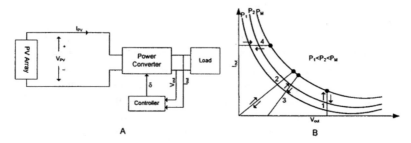

Fig. 1. Maximization of load voltage and load current, B for the combination diagram of different load mechanisms

4.2   Research on Protection Countermeasures

Using the protection scheme will allow isolated island operation. The purpose of this is to better realize the realization of the protection program of the photovoltaic power generation system. According to the previous overview and analysis of the protection scheme for the isolated island operation, consider the implications of the distribution network and photovoltaic protection schemes. In order to realize effective protection measures, some protection schemes can be applied. Including relying on the protection scheme of the communication system; protection scheme under the condition of micro grid; and protection scheme based on wide area measurement system. It can be studied from the following aspects.

Improved protection scheme of reclosing acceleration before and with the current protection based on improved; post reclosing acceleration with protection scheme based on current protection; improvement of the configuration of the fuse protection scheme; improvement of voltage time type re-closer and sectionalizer principle; improvement of distance protection; improvement of the inverse time over current protection; On this basis, the pilot protection scheme is studied (Fig. 2).

At present, the mainstream protection schemes mainly include those based on wide-area measurement system, which rely on the communication system and the protection of the micro grid. After the improvement of existing distribution network protection configuration. The classification overview method is one that will allow the inclusion of power system power to run island-wide based on protected distribution networks, the formation of the island model and the choice of the prior solution of the choice of ways to describe.

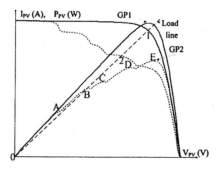

Fig. 2. IV-PV curves under different solar irradiance conditions.

5 Conclusions

This paper discusses the application analysis involving network relay protection technology in the field of the distribution network of the photovoltaic power generation system. The large capacity and high ratio described in the PV grid will affect the relay protection and automatic device safety of the distribution network. Grid-connected photovoltaic (PV) power the adverse impact on the current protection and current protection, fuse, reclosing device, and period of time, distance protection and over-current protection. In the paper, the photovoltaic grid related protection scheme is discussed to allow the use of islanding scheme, grid generation, and Qian Guangfu proposal to require access to the distribution network capacity ratio, further puts forward the protective countermeasures

References

1. Mason, C.: A new loss of excitation relay for synchronous generators. AIEE Trans. **68**, 1240–1245 (1949)
2. Brundlinger, R., Strasser, T., Lauss, G.: Lab tests: verifying that smart grid power converters are truly smart. IEEE Power Energy **13**(2), 30–42 (2015)
3. Wu, T., Chang, C., Chen, Y.: A fuzzy-logic-controlled single-stage converter for PV-powered lighting system applications. IEEE Trans. Ind. Electron. **47**(2), 287–296 (2000)
4. Cell, G., Pilo, F.: Optimal distributed generation allocation in MV distribution networks. In: Proceedings of 2001 IEEE Power Engineering Society Meeting Sydney, Australia, pp. 81–86 (2001)
5. Wang, C., Nehrir, M.: Analytical approaches for optimal placement of distributed generation sources in power systems. IEEE Trans. Power Syst. **19**(4), 2068–2076 (2004)
6. Pipattanasompom, M., Willinghan, M., Rahman, S.: Implications of on-site distributed generation for commercial industrial facilities. IEEE Trans. Power Syst. **20**(1), 206–212 (2005)
7. Brahnla, S., Girgis, A.: Development of adaptive protection scheme for distribution systems with high penetration of distributed generation. IEEE Trans. Power Deliv. **19**(1), 56–63 (2004)

8. Tan, Y., Kirschen, D.S., Jenkins, N.: A model of PV generation suitable for stability analysis. IEEE Trans. Energy Conv. **19**(4), 748–755 (2004)
9. Srikanth, H., Appalanaidu, V., Rao, V.: Determination of network intrusion detection system based on big data. Int. J. Big Data Secur. Intell. **4**(2), 7–14 (2017)
10. Han, M.: Optimal routing path calculation for SDN using genetic algorithm. Int. J. Hybrid Inform. Technol. **11**(3), 7–12 (2018)
11. Jung, W., Yim, H.: Relationship between user interface design attributes in smartphone applications and the intentions of users. Int. J. Smart Home **12**(3), 7–12 (2018)
12. Baik, N., Kang, N.: Multi-phase detection of spoofed SYN flooding attacks. Int. J. Grid Distrib. Comput. **11**(3), 23–32 (2018)
13. Thirupathi, N., Srinivas, P., Bhattacharyya, D.: A detailed review on mobile ad hoc networks: protocols, security issues and challenges. Int. J. Secur. Appl. **12**(2), 67–78 (2018)
14. Gaber, A., Ghazali, M., Iahad, N.: A conceptual model for mobile interaction using brain computer interface. Int. J. Future Gener. Commun. Netw. **11**(2), 71–78 (2018)
15. Cao, W., Li, H.: A wavelet de-noising algorithm based on adaptive threshold function. Int. J. Signal Process. Image Process. Pattern Recogn. **10**(1), 131–138 (2017)
16. Oluchukwu, J., Ezenwaji, I., Ololo, K.: Ethical and policy issues in current trends of media technological and mass communication development in nigerian society: perspective from ICT administrative management and sociology. Int. J. U E Serv. Sci. Technol. **11**(1), 65–74 (2018)
17. Nwobi, U., Mbagwu, F., Olelewe, C.: Application of data mining in adult education students' record keeping: implications for higher education administrators. Int. J. Database Theory Appl. **11**(3), 1–12 (2018)
18. Priyadarshini, R., Narayanappa, C., Chander, V.: Optics based biosensor for medical diagnosis. Int. J. Sens. Appl. Control Syst. **5**(2), 1–4 (2017)
19. Kim, S., Won, J., Kim, J.: The review on MaaS application methods for smart cities. Int. J. Transp. **6**(2), 33–46 (2018)
20. Tajuddin, N., Abdulllah, R., Jabar, M.: Effecting factors of knowledge integration through social media in small medium enterprises environment. Int. J. Disaster Recov. Bus. Contin. **8**, 31–42 (2018)
21. Liu, Y., Yi, S., Tao, G.: Animals Chew Monitoring System Design based on MSP430. Int. J. Sens. Appl. Control Syst. **5**(1), 11–24 (2017)

Micro-grid System in Auxiliary Power System of Substation

Peng Jin[(⊠)]

Department of Technology and Internet, State Grid Liaoning Electric Power
Supply Co., Ltd., Shenyang 110006, China
ml8265537369@163.com

Abstract. This paper analyzes and summarizes the operation of conventional power and proposed the station of micro-grid. Firstly, the paper introduces the structure of each part of the station with a micro-grid system and describes the role and function of each part of the power system. Then, the paper states the influence which the station of micro-grid affects to a network of distribution and summed up its own role. In the end, the paper discusses the research orientation on the station of micro-grid and the advantage of the station with micro-grid. Experiments show that when the traditional grid is in normal operation, the micro-grid connects the operation station and the substation system, and will disconnect from the grid when the power quality is not met. This paper implements intelligent and flexible control on the micro-electronic website and has formed a "plug and play" function on the micro-electronic website. This paper can also realize unattended substation, troubleshooting, automatic mode conversion and so on.

Keywords: Micro-grid · Substation system · Intelligent substation

1 Introduction

Currently, substations of 500 kV and above are occupying an important position in the power system, especially in the regional power supply system, which makes a very important role. But most of these substation power system export power by the secondary substation transformer [1]. This lead to the backup power cannot be sustained supply power and self-healing capabilities are still not perfect and so on. Once the station power system failure, a power system will directly affect the operation of primary equipment and secondary equipment, severely, it can cause blackouts in the large area [2]. With the reform and development of power systems, more ore and more unmanned substation emerge in power system [3]. When the station transformer with a load of loss of power, the staff on duty need to take action to recover the substation electrical system. Thus, the conventional substation power system also has many security risks [4].

In normal operation, the station of micro-grid connected operation of the conventional power grid. When the grid failure or power quality does not meet the requirements, the micro-electronic website will be disconnected from the grid and run independently [5]. This article is based on the method of controlling power electronics.

© Springer Nature Switzerland AG 2020
A. E. Hassanien et al. (Eds.): AISI 2019, AISC 1058, pp. 1065–1071, 2020.
https://doi.org/10.1007/978-3-030-31129-2_101

it achieves intelligent and flexible control in the station of micro-grid and form a "plug and play" function in the station of micro-grid. Meanwhile, the work in this paper achieved unmanned substation, fault removal, and the mode is automatically converted.

2   The Station of Micro-grid System

Micro-grid electricity substation system consists of photovoltaic systems, batteries, chargers, dual power switch, bi-directional inverter, and inverter components. The role and function of the station of micro-grid of each part are different on power system [6–10].

2.1   Photovoltaic Power Generation System

Solar energy can be converted into electricity by using photovoltaic modules' photovoltaic effect, so as to charge the battery group. It can convert DC power to AC power supply load by the inverter. With the application of a large number of renewable energy sources such as photovoltaic power generation and wind power generation, distributed power sources have been widely recognized [11]. Considering the principle of distributed power station with access to the power grid, development of large-scale, high-pressure access are the main principles; decentralized development, low voltage situ access [12–15].

At the moment, there are two main forms of solar photovoltaic: independent photovoltaic system and grid-connected photovoltaic system. Solar power has become one of the main forms of new energy generation in China. The solar power system has the advantages of convenient use, safety, reliability, and easy maintenance. The photovoltaic system can provide local power without long-distance transmission, save investment and reduce power loss. Grid-connected photovoltaic power generation has become the mainstream mode, which can provide active and reactive power for the power system. [16, 17]. Solar photovoltaic power generation as a new type of energy has broad prospects for development and a bright future. Solar panels are shown in Fig. 1.

Fig. 1. Solar panels

2.2   Battery

The battery is a key component in Photovoltaic system. It can store electrical energy which converted to from photovoltaic cells. At present, China has no special batteries for photovoltaic systems [18], but the use of conventional lead-acid batteries. Although lead-acid battery possesses high power density, long discharge time, and technological maturity, it informs cost-effective. Therefore it becomes the most widely used battery. When the battery sees as a regulatory power, the remaining capacity of the battery should be checked before dispatch [19].

The battery can run in power adjustable operating mode or constant power operating mode. It should follow P-f droop and Q-V droop in power adjustable operating mode. In this operating mode, the battery pack has an adaptively adjustable function that quickly tracks load changes in active and reactive power. It is output in constant power mode according to the set values of active power and reactive power.

2.3   Bidirectional Inverter

The bidirectional inverter has a series of special features, functionality, and inverter, which is suitable for a class of smart grid construction, used in storage areas, the basic characteristics of a bidirectional inverter. Energy storage sectors can effectively regulate power resources, which can be a good balance with the power differences between day and night and different seasons to ensure grid security in smart grid [20]. It will become an effective means of achieving interactive management of the grid. The smart grid realization is impossible without energy storage. The inverter of storage is suitable for a variety of needs of dynamic energy storage applications, which means energy storage in surplus power and output of power to the grid power by the inverter in insufficient power. It can play a role in micro-grid [21–24]. Two-way high-frequency inverter improved circuit shown in Fig. 2.

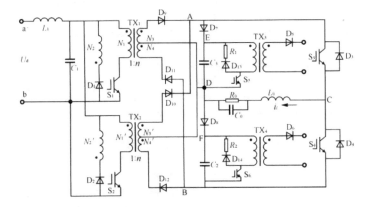

Fig. 2. Two-way high frequency inverter improved circuit.

2.4   Grid-Connected Inverter

The grid-connected inverter will be used as an output device of a photovoltaic grid-connected device. It will also be a core component of photovoltaic grid-connected systems, and one of the key technologies. At the same time, the direct current in the photovoltaic array is converted into alternating current [25]. Grid-connected inverter convert AC has strict requirements in frequency, voltage, current, phase and also Controlling active and reactive power in output AC to ensure power quality.

Grid-connected Inverter comprises a controller, converter, and inverter. In addition to this, DC power can be converted to AC power and the following main functions can be fulfilled. According to the sunlight conditions of sunrise and sunset, the different ambient temperature and solar illumination conditions, so the PV array tries to keep the maximum power output of working conditions. It also has the following functions: suppressing the inflow of higher harmonic currents of the grid, reducing and resolving grid anomalies, and achieving functions such as guarantee the normal operation of the grid [26].

The inverter has only AC conversion function and must maximize the solar energy performance and system fault protection capability. In short, it will include the following major functions: automatic operation and shutdown, maximum power point tracking and control, anti-single-point operation, automatic voltage regulation, dc detection, dc grounding detection, and other functions.

3   Micro-grid Electricity Substation System and Distribution Network

3.1   Voltage Adjustment

Under normal circumstances, the power system sets OLTC and the voltage regulator and other devices in the low-voltage distribution network. This makes the voltage offset of the load node controlled in a controllable range. For the distribution network of Voltage adjustment, it should set reasonable the operating mode of power including wind and solar energy. When the wind sufficient or plenty of sunshine, the output of power is typically large. The micro-grid substation system significantly improves the voltage of the network access point when the grid is under a light load. In this case, if the point is at the end of the line, the voltage at the access point may exceed the limit. Therefore, the operation mode of the photovoltaic power supply should be reasonably planned.

Photovoltaic power generation usually does not generate active power at night, but it can provide reactive power to the grid, thus improving the voltage quality. The system voltage which influences by the photovoltaic power cause voltage fluctuations and flicker. If the heap is changed as the load changes, it will cause greater voltage fluctuations and flicker than before. Although photovoltaic power supply will not cause large voltage fluctuation in actual operation, it is very important to plan access capacity and location reasonably when a large number of the photovoltaic power supply is connected to the power grid. The curve of the node voltage with time is shown in Fig. 3.

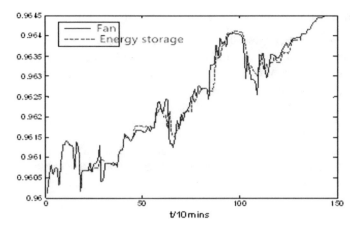

Fig. 3. Node voltage versus time curve.

3.2   Short Circuit Current Level

As the direct grid generator increases the short circuit current level of distribution network, it will improve the interrupting capacity requirements of distribution network breaker.

3.3   The Power Supply Quality of Distribution Network

As photovoltaic power generation and wind power generation are intermittent, voltage fluctuation will be caused. Grid harmonic current inevitably injected into the grid by the inverter, it led to the waveform distortion voltage.

4   The Function of Micro-grid Electricity Substation System

With the development of science and technology, power electronics and control technology have been developed rapidly. Electronic devices are widely used in electronic systems [27]. The power generation device of Micro-grid electricity substation system will bring a series of a positive impact.

4.1   Improve Power Supply Reliability

When the outage of substation system, it can maintain the power supply for All or part of critical equipment. It maintain the operation of control device in transformer substation and maintain communications system.

4.2   Reduce the Transmission Loss

After the micro-grid electricity substation system entre the grid, energy can be obtained from the power grid in the bottom of the load and send power to the grid during peak load. While electricity sent to the grid in peak load, it can play the role of load shifting load and improve grid efficiency [28].

4.3   Utilization Renewable Energy

The micro-grid electricity substation system can used in conjunction with Wind, solar and other power generation device. It may compensate for renewable energy power generation device intermittent power output.

5   Conclusions

This paper analyzed and elaborated the role of the stand with the micro-grid system in various parts. We can yield the following conclusions:

- The grid connection mode of a conventional power grid during normal operation is called micro-grid operation station.
- When the grid fails or the grid voltage fails to meet the requirements, the micro-grid substation can monitor this phenomenon and disengage from the grid timely and independently.
- It accomplished intelligent and flexible control in the station of micro-grid and formed a "plug and play" function in the station of micro-grid.
- It can also achieve unmanned substation, fault removal, the mode is automatically converted and so on. All these show that the micro-grid electricity substation system will wide range of applications and practicality.

References

1. Zhang, D., Yuan, Y., Mao, M.: Key technologies for micro-grids being researched. Power Syst. Technol. **33**, 6–11 (2009)
2. Hatziargyriou, N., Asano, H., Iravani, R.: Microgrids. IEEE Power Energ. Mag. **5**, 78–94 (2007)
3. Zhang, X.: Emergency power supply configuration analysis and suggestion of wind-PV-ES power station. Electr. Power Energy **34**(4), 412–417 (2013)
4. Yang, Q., Ma, S., Song, Y.: Comprehensive valuation of distributed generation planning scheme. Power Syst. Technol. **36**, 212–216 (2013)
5. Ding, M., Wang, W., Wang, X.: A review on effect of large-scale PV generation on power. Proc. CSEE **34**, 1–14 (2014)
6. Chen, W., Ai, X., Wu, T.: Influence of grid-connected photovoltaic system on power network. Electr. Power Equipment **32**, 26–32 (2013)
7. Liu, W.: Discussion on application prospect of new energy as station-use electric source in converter station. East China Electr. Power **38**, 1712–1713 (2010)

8. Bie, Z., Li, G., Wang, X.: Review on reliability evaluation of new distribution system with micogrid. Electr. Power Autom. Equipment **31**, 1–6 (2011)
9. Ding, M., Zhang, Y., Mao, M.: Economic operation optimization for microgrids including Na/S battery storage. Proc. CSEE **31**, 7–14 (2014)
10. Tsikalakis, A., Hatziargyriou, A.: Centralized control for optimizing micro-grids operation. IEEE Trans. Energy Convers. **23**, 241–248 (2008)
11. Frangioni, A., Gentile, C., Lacalandraf, F.: Tighter approximated MILD formulations for unit commitment problems. Trans. Power Syst. **24**, 105–113 (2004)
12. Wang, S.: Design and operation of micro-grid based on distributed generations. Electr. Power Autom. Equipment **31**, 120–123 (2011)
13. Anderson, R., Boulanger, A., Powell, W.: Adaptive stochastic control for the smart. Proc. IEEE **99**, 1098–1115 (2011)
14. Lu, H., He, B.: Application of the super-capacitor a micro-grid. Autom. Electr. Power Syst. **32**, 87–91 (2009)
15. Wang, Q., Xie, G., Zhang, L.: An integrated generation consumption dispatch model with wind power. Autom. Electr. Power Syst. **35**, 15–18 (2011)
16. MohamedRafi, A., Manikandan: Improving data reliability and integration in cloud using PRCR. Int. J. Softw. Eng. Smart Device, **3**(2), 1–20 (2016)
17. Pramodkumar, K., Munot, H., Malathi, P.: Survey on computational intelligence based routing protocols in wireless sensor network. Int. J. Wireless Mob. Commun. Ind. Syst. **3**(2), 23–32 (2016)
18. Jin, J.H., Koo, H.S., Lee, M.J.: Extending BSmart to support Beacon receivers sending context information. Int. J. Web Sci. Eng. Smart Devices **3**(2), 1–6 (2016)
19. Yun, J.S., Kim, J.H.: A study on training data selection method for EEG emotion analysis using machine learning algorithm. Int. J. Adv. Sci. Technol. **119**, 79–88 (2018)
20. Cherkaoui, H., Omar, D.A., Baslam, M., Fakir, M.: Cache replacement algorithm based on popularity prediction in content delivery networks. Int. J. Adv. Sci. Technol. **122**, 35–48 (2019)
21. Liu, Y.T., Hu, J.J.: Obstacle avoidance with formation kept for multi-agent. Int. J. Control Autom. **8**(3), 67–74 (2015)
22. Zhang, F., Liu, C., Wei, Y.S., Liu, C.: Data service hyperlink and its application in automatic data service composition. Int. J. Grid Distrib. Comput. **10**(2), 37–50 (2017)
23. Yao, J.R., Wei, M.X.: A new bionic architecture of information system security based on data envelopment analysis. Int. J. Secur. Appl. **9**, 85–98 (2015)
24. Li, T.S., Guo, C.X., Ge, Z.H.: Study of topology optimization algorithm of wireless mesh backbone networks based on directional antenna. Int. J. Future Gener. Commun. Networking **9**(9), 25–34 (2016)
25. Yang, H.Y., Tu, Y.Q., Zhang, H.T., Li, M.: Phase difference measurement based on recursive DFT with spectrum leakage considered. Int. J. Signal Process. **8**(11), 55–66 (2015)
26. Lin, Y., Yu, X.W., Ma, C.G., Zheng, D., Wu, Z.Q., Zhang, Z.P.: Genetic analysis of generalized S-transform. Int. J. u- e- Serv. Sci. Technol. **9**(4), 27–38 (2016)
27. Wu, G.H., Li, S., Han, L., Zhao, M.M.: CHI statistical text feature selection method based on information entropy optimization. Int. J. Database Theory Appl. **9**(11), 61–70 (2016)
28. Pandey, S., Mahapatra, R.P.: An adaptive gravitational search routing algorithm for channel assignment based on WSN. Int. J. Sens. Appl. Control Syst. **5**(1), 1–14 (2017)

Winding Deformation Detection of Transformer Based on Sweep Frequency Impedance

Hui Zhang[1](✉), Tao Wang[1], Yunshan Zhang[2], Yujie Pei[1],
Zhongbin Bai[1], Ling Guan[1], Yaoding Gu[1], and Jianguo Xu[1]

[1] State Grid Fushun Electric Power Supply Co., Ltd., Fushun, China
sky_0735@126.com
[2] State Grid Liaoning Electric Power Supply Co., Ltd., Shenyang, China

Abstract. To improve the timeliness and accuracy of transformer winding deformation detection, the deformation of the transformer winding is studied by sweeping frequency impedance method. Based on the principle of scanning impedance method, an experimental test circuit is constructed to perform on-site detection on a 10 kV transformer. The results of transformer sweep impedance curve show that the simulated deformation fault has little effect on the low-frequency band of the sweep impedance curve, but the impedance amplitude shifts upward in the high-frequency band. At 50 Hz, the phase relation value of impedance changes significantly, which can be used as the basis for determining the winding deformation fault. It is proved that the sweep impedance method can be used well for the detection of transformer displacement faults with high sensitivity.

Keywords: Sweeping frequency impedance method · Winding deformation · Displacement fault

1 Introduction

Socio-economic electricity consumption is growing rapidly; the reliability of the power system is becoming more and more important. Ensuring the stability and reliability of power systems and power equipment has become a research hotspot in the power industry [1]. As the most important primary equipment in the substation, the power transformer is used to realize high-voltage transmission and low-voltage power distribution. It's running status and power system security and stability has a direct link [2]. For the past few years, the power load has increased rapidly, so the transformer load has become heavier and heavier. The internal and external factors such as insulation aging and inrush current have made the winding of the power transformer vulnerable to deformation and damage. The frequency of various faults is getting higher and higher, which has seriously threatened the grid stable operation [3]. Therefore, when the transformer winding is deformed, it is very practical to quickly and effectively judge and process the winding state to reduce the fault [4].

© Springer Nature Switzerland AG 2020
A. E. Hassanien et al. (Eds.): AISI 2019, AISC 1058, pp. 1072–1079, 2020.
https://doi.org/10.1007/978-3-030-31129-2_102

At present, there are two ways to diagnose transformer winding deformation, the first is the main frequency response analysis method, the second is the low-voltage short-circuits impedance method. Both methods have been promulgated by relevant standards and played a certain role in diagnosing winding deformation. However, these two methods also have certain limitations. The frequency response analysis method is less susceptible to field interference due to the use of small excitation signals. And it is insensitive to changes in inductance [5]. Low-voltage short-circuit impedance method has no obvious response to slight deformation of winding. Therefore, the two test methods are prone to missed judgments and misjudgments in practical applications. In order to make a more comprehensive judgment of the state of the winding, an alternative new method "the sweeping frequency impedance method" has become the focus of current research [6]. This method combines the advantages of the above two test methods and can be more comprehensive. The deformation state of the winding can be accurately judged.

Based on the principle of sweeping short-circuit impedance test, this paper conducts on-the-spot detection of 10 kV transformer. The changes of sweep impedance curve and short circuit impedance of normal and fault transformers are studied by using sweep impedance method [7].

2 Deformation Detection Method of Transformer Winding

A transformer is a type of electric energy converter that uses the principle of electromagnetic induction to change the alternating voltage. It is made up of two coils wound around a metal ring. The transformer is the most important part of the winding, which constitutes the transformer's internal circuit [8]. During operation, the winding is subjected to mechanical or electric power, such as short-circuit surge current or collision during transportation, which may be deformed in a direction perpendicular or horizontal to the shaft. It mainly includes winding displacement expansion or local deformation [9]. Transformer winding deformation can be simply divided into:

(1) Radial stretching: Generally in the transformer center of the high - voltage side, amplitude is not large.
(2) Radial compression: Generally occurs in the transformer low voltage side, and the center is up and down movement.
(3) Axial extension: Usually at the top of the low pressure side.
(4) Axial compression: Located in the transformer high voltage side winding in the middle.
(5) Axial nesting: Both high and low pressure sides may appear.
(6) Short circuit between turns.

Nowadays, there are many detection methods for transformer windings. Usually, the corresponding standard requirements and technical specifications are used. And it also combined with the long-term accumulated practical experience for testing. At present, there are mainly the following detection methods [10]. Figure 1 shows the short-circuit impedance wiring diagram.

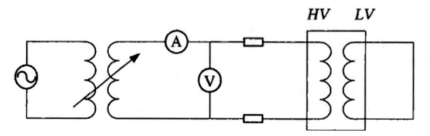

Fig. 1. Short-circuit impedance wiring diagram

Short-circuit reactance (SCR) is one of the earliest methods to detect transformer winding deformation. The detection principle is to measure the variation of leakage inductance and impedance of transformer winding. When the load impedance of the transformer is zero, the short-circuit impedance of the transformer is equal to the equivalent impedance inside the transformer. It is composed of winding resistance and leakage reactance, namely the resistance component and reactive components [11]. For large transformers such as 110 kV and above, it mainly depends on the magnitude of the reactance component because the ratio of the resistance component is very small. The reactance component is the leakage reactance of the transformer winding. When the frequency is constant, It is decided by the geometric shape and distribution of the transformer location. When the transformer winding is deformed, its geometrical dimension changes, so the short-circuit impedance also changes. The method is to judge the deformation of transformer winding by measuring the change of leakage inductance [12]. The test principle diagram is shown in Fig. 1.

The method has simple test procedure, convenient measurement, good repeatability, and high reliability of detecting winding deformation. Usually, it was used to test small currents and low voltages [13]. The different countries have their own judgment criteria. However, the current test sensitivity is low, the detection time is long and misjudgment may occur.

(1) Low Voltage Impulse

The Low Voltage Impulse Method (LVI) is the IEC and IEEE transformer short-circuits test guidelines and test standards [14]. When the transformer winding is operated at a voltage higher than 1 kHz, Each winding can be regarded as composed of a linear resistance linear distributed parameter of the passive network. Under the influence of high-frequency voltage, transformer winding mainly considers the role of capacitance and inductance because the resistance value of winding is small. Transformer winding deformation, the parameters of the components will change, and the unit impulse response will also change. By comparing the excitation and response in the time domain, the state of the winding can be judged (Fig. 2).

The method has high detection speed and high test sensitivity. However, it will be affected by some electromagnetic interference in the field test, which makes the first-end fault response of the winding insensitive [15]. It will make the position of the deformation is difficult to determine, and the repeatability is poor. So it is often necessary to rely on the experience of the technician to judge.

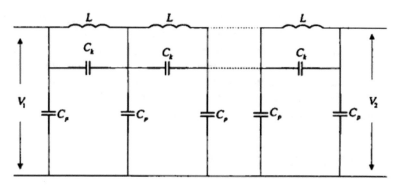

Fig. 2. Equivalent circuit for high frequency transformer winding

(2) Frequency Response Analysis method

Frequency response analysis method is to use the frequency characteristic of the equivalent two levels of network a method to detect transformer winding deformation [16]. The variation of the parameters inside the winding causes the change of the frequency response characteristic. The excitation voltage frequency changes continuously when the excitation voltage is applied, the amplitude ratio and the phase difference of the response terminal voltage and the excitation terminal voltage are measured at different frequencies. That is the winding frequency response curve obtained [17] (Fig. 3).

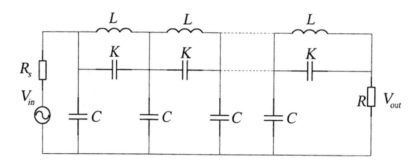

Fig. 3. Frequency response analysis method of circuit

The frequency response analysis method can be used to judge the deformation of transformer windings without ring. It has better repeatability than the LVI method, high detection sensitivity, reduced electromagnetic interference. The test tool is small in size and easy to carry. Therefore, it has been widely deployed in the power industry. However, due to the use of the sub-band method in the extraction of feature quantities, this method lacks a unified and quantifiable basis for the time being. And it requires high technical experience [18].

Since the detection methods of the three types of winding deformations need to be detected after the transformer is powered off and stopped, it belongs to off-line detection. That may have problems such as long detection period, long analysis time, poor accuracy, and difficulty in detecting sudden failures. The rapid development of the electric power industry has become more and more demanding for the detection of timeliness, real-time and effectiveness. Therefore, online detection technology has been highly valued by people and has become a new development trend [19].

3 Sweep Frequency Impedance Method

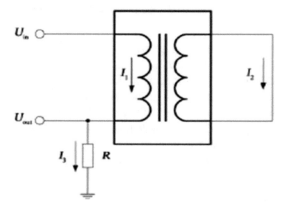

Fig. 4. Frequency sweep impedance measurement principle

Sweep impedance method is a wide-band transformer winding deformation test method. Its biggest a bit is the frequency response analysis method and the advantages of short circuit impedance method combined together. It achieves a breakthrough in test principle and analysis method. In the process of the test can get frequency response characteristics of the transformer winding. Except for the curve, At the same time also can be calculated and the fitting of the test winding short circuit impedance frequency characteristic curve. Using this test method, Transformer short-circuits impedance values of 50 Hz can be obtained on the sweep impedance curve. The winding state result will be obtained by comparing with the nameplate value and the current common short-circuit reactance test national standard. Used in conjunction with the sweep impedance test standard, the diagnostic technology for transformer winding deformation is more complete and mature. The test principle is shown in Fig. 4.

As can be seen from Fig. 4, in the test, Short circuit on high voltage side or low voltage side in transformer, The sweep signal U_{in} is loaded at the first end of the non-short-circuit side winding. After the signal passes through the winding, Through the use of grounding resistance R for winding at the end of the output signal Uout. According to the above data, the transformer sweep impedance value can be obtained as follow.

$$Z_k(j\omega) = \left(\frac{U_{in}(j\omega) - U_{out}(j\omega)}{I_1(j\omega)}\right) = R + jX(\omega)$$

In order to better judge the transformer fault, the impedance modulus should be decibelized as follows.

$$Z_I = 20 \times lg(|Z_k|)$$

4   Experiments

The experiment uses a 10 kV single-phase double-winding transformer as the research object. The three ends of the transformer's three tap-changers and the low-voltage winding are led out by the bushing. The capacitor or resistor string can be connected to the bushing of the tap-changer to simulate different types of winding faults. At the same time, the transformer winding is fixed on the iron core, and the sling can be processed by the gantry crane. The size of the hoisted winding can be changed to simulate various faults such as bulging, warping and displacement. During the test, a certain pressure was applied to the outermost layer of the right side winding of the transformer to deform the coil and form an axial displacement with a displacement distance of 4 mm. The set frequency is increased by 10 Hz. The scan of the test transformer impedance curve, and the results and compares the ordinary transformer scanning impedance curves. The sweep impedance curve is shown in Fig. 5.

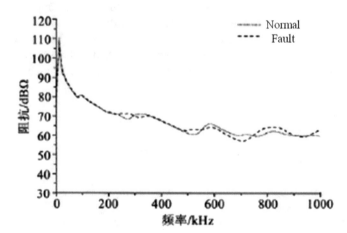

Fig. 5. Sweeping frequency impedance curves of normal and fault transformers

When the frequency is less than 200 kHz, the transformer displacement fault has almost no influence on the sweeping impedance curve. Fault scanning frequency and normal transformer impedance curve are basically the same. This is because the inductance plays a major role in the sweep curve at low frequencies, so the displacement fault caused by the winding deformation does not cause the inductance value to change. When the frequency above 200 kHz, the fault curve begins to differ from the normal curve, fluctuating above and below the normal curve. This is due to the displacement fault causing the transformer geometry to change, resulting in changes in Winding of capacitance.

At 50 Hz, the short-circuit impedance can be obtained from the figure. In the short circuit impedance test of 50 Hz, the normal transformer is 85.8 dBΩ, fault of transformer is 86.2 dBΩ. The impedance value of the transformer at 50 Hz has almost no change. However, the correlation coefficient calculated by the relevant industry standards has obvious changes. Especially at the low frequency, the correlation coefficient is close to the diagnostic critical value of the winding fault, and the high-frequency correlation coefficient is less than 0.6. Therefore, it can be determined that the winding has an abnormality and should be deepened Checked.

5   Conclusions

This paper adopts a new online technique scanning impedance method to test the transformer winding deformation failure. The basic principle and realization method of frequency sweep impedance method are described. The transformer is tested and the result is found from the sweep frequency impedance curves. It found that the scan impedance curve has no fluctuation in the low-frequency range, although the displacement fault caused the transformer geometry to change. At high frequencies, the influence of the displacement fault on the impedance curve gradually appears, causing the curve to fluctuate on the upper line of the normal transformer curve. In addition, the impedance value at 50 Hz can be also extracted from the curve. The results show that the displacement fault has no significant effect on the impedance value, but has a significant influence on the correlation coefficient. Therefore, the transformer winding fault can be detected using the sweeping frequency impedance method.

References

1. Bengtsson, C.: Status and trends in transformer monitoring. IEEE Trans. Power Delivery **11**(3), 1379–1384 (1996)
2. Wang, M., Vandermaar, A., Srivastava, K.: Review of condition assessment of power transformers in service. IEEE Electr. Insul. Mag. **18**(6), 12–25 (2002)
3. Abu-Siada, A., Hashemnia, N., Islam, S.: Understanding power transformer frequency response analysis signatures. IEEE Electr. Insul. Mag. **29**(3), 48–56 (2013)
4. Ryder, S.: Diagnosing transformer faults using frequency response analysis. IEEE Electr. Insul. Mag. **19**(2), 16–22 (2003)
5. Rahimpour, E., Christian, J., Feser, K.: Transfer function method to diagnose axial displacement and radial deformation of transformer windings. IEEE Trans. Power Delivery **18**(2), 493–505 (2003)

6. Christian, J., Feser, K.: Procedures for detecting winding displacements in power transformers by the transfer function method. IEEE Trans. Power Delivery **19**(1), 214–220 (2004)
7. Bagheri, M., Phung, B., Blackburn, T.: Transformer frequency response analysis: mathematical and practical approach to interpret mid-frequency oscillations. IEEE Trans. Dielectr. Electr. Insul. **20**(6), 62–70 (2013)
8. Mohammad, H., Stefan, T.: Effect of different connection schemes, terminating resistors and measurement impedances on the sensitivity of the FRA method. IEEE Dielectr. Electr. Insul. **26**, 1713–1720 (2017)
9. Ludwikowski, K., Siodla, K., Ziomek, W.: Investigation of transformer model winding deformation using sweep frequency response analysis. IEEE Trans. Dielectr. Electr. Insul. **19** (6), 1957–1961 (2012)
10. Thirupathi, N., Srinivas, P., Sudha, K.: Performance of M/M/1 and M/D/1 queuing models on data centers with cloud computing technology using MATLAB. Int. J. Grid Distrib. Comput. **11**(3), 11–22 (2018)
11. Al-Sharrah, M., Alkandari, M.: Towards a more secured solution in RDP: on-demand desktop local admin rights. Int. J. Secur. Appl. **12**(2), 45–58 (2018)
12. Ahuja, M., Singh, J.: Finding communities in social networks with node attribute and graph structure using jaya optimization algorithm. Int. J. Future Gener. Commun. Netw. **11**(2), 33–48 (2018)
13. Bachu, S., Manjunathachari, K.: New approach for image segmentation based on graph cuts. Int. J. Signal Process. Image Process. Pattern Recogn. **10**(1), 119–130 (2017)
14. Han, W., Li, S., Jia, H.: Research on software trustworthiness evaluation for web application based on software product. Int. J. u - e - Serv. Sci. Technol. **10**(1), 89–104 (2017)
15. Karchi, R., Munusamy, N.: Hyperspectral image classification and unmixing by using ART and SUnSPI techniques. Int. J. Database Theor. Appl. **11**(3), 13–28 (2018)
16. Kumar, M.: Various factors affecting performance of web services. Int. J. Sens. Appl. Control Syst. **3**(2), 11–20 (2015)
17. Alp, S., Özkan, T.: Modelling of multi-objective transshipment problem with fuzzy goal programming. Int. J. Transp. **6**(2), 9–20 (2018)
18. Thirupathi, N.: A review on industrial applications of machine learning. Int. J. Disaster Recovery Bus. Continuity **8**, 1–10 (2018)
19. Patgar, T., Shankaraiah, : The impact of hybrid data fusion based on probabilistic detection identification model for intelligent rail communication highway. Int. J. Sens. Appl. Control Syst. **4**(2), 9–20 (2016)

State of Charge Estimation Method of Lead-Acid Battery Based on Multi-parameter Fusion

Yuan Yu[✉]

College of International Education, Shenyang Institute of Technology,
Shenyang 110136, China
1160660141@qq.com

Abstract. Under the background of energy crisis, new energy power generation technology develops continuously along with the gradual increase of power demand. Energy storage technology plays a very important role in the new energy power generation and grid operation safety. Currently, as the most widely used and developed equipment in the field of energy storage, battery has a significant impact on the safety and stability of people's life and the steady development of economy. Therefore, it is helpful to predict the health condition of the battery and replace the used battery in time to avoid power accidents and improve the reliability of the power grid. In this paper, an improved SOC estimation method is proposed based on the combination of open-circuit voltage method and ampere hour integral method. SOC initialization algorithm combining multiple parameters and introducing weight w is studied. In the SOC estimation process, the battery capacity is calibrated according to real-time temperature, which can improve the calculation accuracy and meet the operating conditions.

Keywords: Energy crisis · Energy storage · SOC · Battery · Battery capacity

1 Introduction

As an important energy storage component, lead-acid battery plays an important role in the safety and economy of the whole vehicle power system. State of charge (SOC) of lead-acid battery is an important parameter to evaluate its internal state and guide users to use vehicles, and also an important basis for automotive power management system [1–3]. Accurate estimation of lead-acid battery SOC is one of the key technologies to realize vehicle energy recovery, power balance and extend battery life.

Existing estimation methods for battery SOC can be classified into three categories:

(1) Estimation methods based on measurement values of specific characterization parameters of the battery, including residual capacity method, impedance spectrum method, open circuit voltage method, etc. The residual capacity method is obtained by constant discharge of electricity for a long time. Impedance spectrum method requires the measurement of the impedance value with an impedance meter. Open circuit voltage method requires the battery to stand for a long time, which is inconsistent with the working environment of the battery.

© Springer Nature Switzerland AG 2020
A. E. Hassanien et al. (Eds.): AISI 2019, AISC 1058, pp. 1080–1086, 2020.
https://doi.org/10.1007/978-3-030-31129-2_103

(2) Estimation method based on ampere hour integral. This method, as the calculation core of battery SOC, has been most widely used. However, it is difficult to obtain the exact value of the initial SOC of the battery, requires high accuracy of current detection [4–6], and aging of the battery will cause changes in the reference capacity, thus reducing the calculation accuracy of this method.

(3) Fusion estimation methods based on battery model and observer, including neural network method, fuzzy controller method and support vector machine. The accuracy of this method depends on the accuracy of the model. To improve the accuracy of the model, a large number of data samples need to be trained.

2 The Working Principle of Lead Acid Storage Battery

The principle of lead-acid battery is shown in Fig. 1.

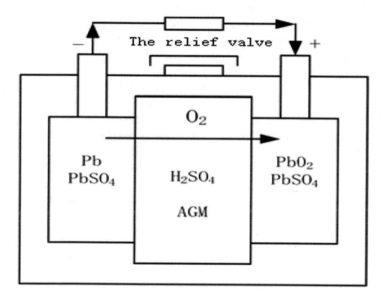

Fig. 1. Lead acid battery schematic

Positive electrode of lead-acid battery is PbO_2, which are typically brown and granular, have better access to the electrolyte, increasing the reaction area and reducing the battery's internal resistance. Battery negative pole is Pb, dark gray spongy; Electrolyte is a dilute sulfuric acid solution mixed by concentrated sulfuric acid and distilled water in a certain proportion. The battery separator is composed of insulating materials, and there are holes distributed on the separator [7–9], which can reduce the probability of short circuit generated by the positive and negative electrodes, and make the oxygen separated from the positive electrode compound with the negative electrode. According to the difference of using the separator, the battery can be divided into two types: AGM and GEL. The specific working process is shown in Figs. 2 and 3.

When the battery is connected to the external load and the circuit is connected, the battery starts to discharge externally. The positive electrode absorbs electrons and produces reduction reaction. At the same time, the negative electrode releases electrons and produces oxidation reaction. When the charging device charges the battery, contrary to the above process, the positive electrode releases electrons for oxidation reaction, while the negative electrode absorbs electrons for reduction reaction.

Due to the low voltage and capacity of single battery, it cannot meet the needs of practical application. In power network, electric vehicle, photovoltaic power generation and other systems, batteries are often connected in series or parallel to form battery groups. When connected in parallel, the storage battery capacity increases and the terminal voltage remains unchanged [10–14]. In series, the terminal voltage is the sum of all the terminal voltage of series batteries, and the capacity remains unchanged. The combination of different ways can improve the voltage level of the battery and expand the capacity of the battery, so as to apply in practice.

The internal electrochemical reaction of lead-acid battery is essentially a process of mutual conversion of chemical energy and electrical energy [15, 16]. When discharging, chemical energy is converted to electrical energy, while when charging, electrical energy is converted to chemical energy.

In 1882, J. H. Gladstone and A. Tribe proposed the "bipolar sulfuric acid salinization theory", which could perfectly explain the internal electrochemical reactions of lead-acid batteries, replacing the earlier "oxidation-reduction' theory, and was widely recognized. According to this theory, the chemical equation of battery charging is defined as follows.

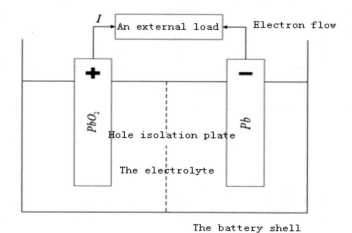

Fig. 2. Lead acid battery discharge schematic

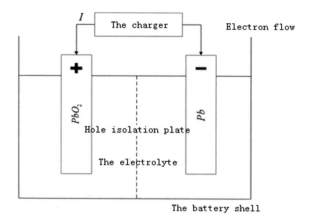

Fig. 3. Lead acid battery charging diagram

$$2PbSO_4 + 2H_2O \rightarrow PbO_2 + Pb + 2H_2SO_4 \tag{1}$$

The chemical equation of discharge is defined as follows.

$$PbO_2 + Pb + 2H_2SO_4 \rightarrow 2PbSO_4 + 2H_2O \tag{2}$$

From the above two equations, it can be concluded that sulfuric acid, as the electrolyte of lead-acid battery, not only ACTS as a medium, but also participates in the chemical reaction of the battery as a reactant. When the battery is discharged, sulfate ions and lead ions in the anode and cathode participate in the reaction to form sulfate, which reduces the number of sulfate ions and the concentration of electrolyte. When the battery is charged, the sulfate breaks down the sulfate ions, increasing the electrolyte concentration. From the above analysis, it can be seen that battery working process is closely related to electrolyte concentration change.

3   SOC Estimation Method

3.1   Initialize SOC Algorithm

When the performance of lead-acid battery reaches a completely stable state, the open circuit voltage OCV value has a good correspondence with the SOC. The 12 V/200 Ah lead-acid battery was tested, and the OCV and SOC curves were shown in Fig. 4.

Figure 4 shows that the OCV-SOC curve of lead-acid battery has a good linearity in each section of SOC, which can be used to calculate SOC. After stopping (t = 0 time), the open circuit voltage of the battery changes with time as shown in Fig. 5. The pressure tends to stabilize. However, in driving environment, it is difficult to stop for a long time and leave the battery to reach 100% stable open circuit voltage. Therefore, in the process of SOC initialization, only open circuit voltage is used to calculate SOC value, which will introduce a large error.

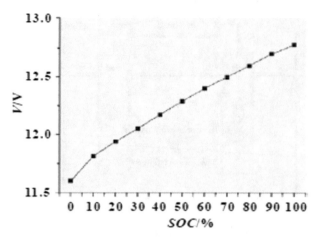

Fig. 4. Relationship between open circuit voltage OCV and SOC of lead-acid battery

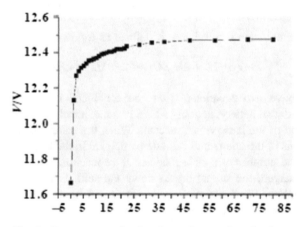

Fig. 5. Battery open circuit voltage change after shutdown

3.2   The Influence of Temperature on the Capacity of Lead-Acid Battery

The working process of lead-acid battery is the internal electrolytic liquefaction reaction process, so it's working characteristics must be affected by the temperature [17, 18]. When the temperature of the electrolyte increases, the molecular movement accelerates and the kinetic energy increases, so the permeability is also strengthened, the resistance value of the electrolyte decreases, the diffusion degree of the solution deepens, and the electrochemical reaction is enhanced, which increases the capacity of the lead-acid battery. Conversely, the capacity decreases. Figure 6 shows the relationship between the 20-h multiple discharge capacity and temperature of lead-acid battery.

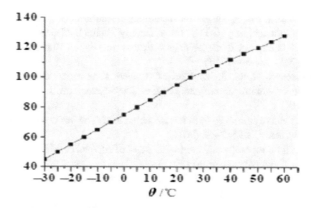

Fig. 6. The relationship between temperature and rate discharge capacity at 20 h

The output capacity of battery increases with the increase of battery temperature. When the temperature is −15 °C, the total capacity of the battery is only 25 °C rated capacity of 60%. Therefore, it is necessary to revise the capacity value CN according to the temperature during SOC calculation.

4 Prospects and Conclusion

For lead-acid battery SOC estimation, an improved SOC estimation algorithm based on open circuit voltage and ampere hour integration is proposed. By studying the SOC initialization method based on multiple parameters, the problem of long standing time required by open-circuit voltage method to determine the initial value of SOC is solved. The real-time correction of battery capacity according to temperature improves the accuracy of SOC prediction. The experimental results show that the SOC estimation algorithm of lead-acid battery has high accuracy, and the SOC estimation error can be controlled within 3%, which meets the practical application requirements.

Acknowledgments. This work is supported by the Science and Technology Program of Shenyang under Grant 18-013-0-18 and the National Nature Science Foundation of Liaoning Province under Grant 20180550922.

References

1. Choi, K., Cho, J.: A study on durability through structural analysis by configuration of torque sensor disk. Int. J. Comput. Aided Mech. Des. Implementation **4**(2), 15–20 (2018)
2. Zhang, L., Qu, M.: Research on topology structure of wind power DC microgrid. China New Commun. **20**(3), 207 (2016). (in chinese)
3. Dev, N.: Analysis of thermal system. Asia Pac. J. Adv. Res. Electr. Electron. Eng. **1**(2), 9–14 (2017)
4. Tian, L.: Topology analysis and control strategy of optical storage AC-DC hybrid microgrid. Hefei University of Technology (2016)

5. Kumar, A., Guleria, M.: A review on machining of aluminium metal matrix composite in electro discharge machining (EDM). Int. J. Energy Technol. Manag. **1**(1), 23–28 (2017)
6. Prasadu, R.: Auditing for dynamic data with user revocation. Int. J. Adv. Res. Big Data Manag. Syst. **1**(1), 25–30 (2017)
7. Park, J., Cheon, S., Eom, J.-H.: Plan of protection measures construction for military information and communication system against EMP threats. Int. J. Energy Inf. Commun. **9**(4), 7–12 (2018)
8. Chandra, P.: A Survey on deep learning its architecture and various applications. Asia Pac. J. Neural Netw. Appl. **1**(2), 7–12 (2017)
9. Jia, L.: Research on topology and control strategy of AC/DC hybrid microgrid. North China Electric Power University (Beijing) (2017)
10. Negi, S., Singh, S., Panwar, N.: CBIR using simplified wavelet-based color histogram. Int. J. Image Signal Syst. Eng. **11**(2), 1–8 (2017)
11. Ren, C., Han, X., Wang, P.: Key technology of AC/DC hybrid microgrid and its demonstration project. J. Taiyuan Univ. Technol. **48**(3), 486–491 (2017). (in chinese)
12. Oh, J., Cho, M., Lee, H.: IoT environment and personal information protection law in Korea. World J. Wireless Devices Eng. **2**(1), 1–6 (2018)
13. Kim, S.: Building a virtual reality driving simulator to study distracted driving. Int. J. Sci. Eng. Smart Veh. **2**(2), 19–26 (2018)
14. Wang, C., Zhou, Y.: Summary of Microgrid Demonstration Project. power supply 1, 16–21 (2015)
15. Zhu, Y., Tang, Z., Jia, L.: New AC/DC hybrid microgrid topology and its reliability analysis. Power Constr. **38**(09), 81–87 (2017)
16. Lim, S.: A theater in architecture and the city. J. Creative Sustain. Architect. Built Environ. **7**(1), 51–56 (2017)
17. Ma, Y., Yang, P., Wang, Y., Zhao, Z.: Typical characteristics and key technologies of microgrid. Power Syst. Autom. **39**(8), 168–175 (2015). (in chinese)
18. Kim, S.: Evaluation of public design guidelines for bus stop Jeju. Int. J. ICT Aided Architect. Civ. Eng. **4**(2), 39–44 (2017)

Author Index

Printed in the United States
By Bookmasters